Lecture Notes in Artificial Intelligence 12712

Subseries of Lecture Notes in Computer Science

More information about this subseries at http://www.springer.com/series/1244

Kamal Karlapalem · Hong Cheng ·
Naren Ramakrishnan · R. K. Agrawal ·
P. Krishna Reddy · Jaideep Srivastava ·
Tanmoy Chakraborty (Eds.)

Advances in Knowledge Discovery and Data Mining

25th Pacific-Asia Conference, PAKDD 2021
Virtual Event, May 11–14, 2021
Proceedings, Part I

Springer

Editors
Kamal Karlapalem
IIIT, Hyderabad
Hyderabad, India

Hong Cheng
Chinese University of Hong Kong
Shatin, Hong Kong

Naren Ramakrishnan
Virginia Tech
Arlington, VA, USA

R. K. Agrawal
Jawaharlal Nehru University
New Delhi, India

P. Krishna Reddy
IIIT Hyderabad
Hyderabad, India

Jaideep Srivastava
University of Minnesota
Minneapolis, MN, USA

Tanmoy Chakraborty
IIIT Delhi
New Delhi, India

ISSN 0302-9743 ISSN 1611-3349 (electronic)
Lecture Notes in Artificial Intelligence
ISBN 978-3-030-75761-8 ISBN 978-3-030-75762-5 (eBook)
https://doi.org/10.1007/978-3-030-75762-5

LNCS Sublibrary: SL7 – Artificial Intelligence

This Springer imprint is published by the registered company Springer Nature Switzerland AG
The registered company address is: Gewerbestrasse 11, 6330 Cham, Switzerland

General Chairs' Preface

On behalf of the Organizing Committee, it is our great pleasure to welcome you to the 25th Pacific-Asia Conference on Knowledge Discovery and Data Mining (PAKDD 2021). Starting in 1997, PAKDD has long established itself as one of the leading international conferences in data mining and knowledge discovery. Held during May 11–14, 2021, PAKDD returned to India for the second time, after a gap of 11 years, moving from Hyderabad in 2010 to New Delhi in 2021. Due to the unexpected COVID-19 epidemic, the conference was held fully online, and we made all the conference sessions accessible online to participants around the world.

Our gratitude goes first and foremost to the researchers, who submitted their work to the PAKDD 2021 main conference, workshops, and data mining contest. We thank them for the efforts in research, as well as in preparing high-quality online presentations videos. It is our distinct honor that five eminent keynote speakers graced the conference: Professor Anil Jain of the Michigan State University, USA, Professor Masaru Kitsuregawa of the Tokyo University, and also the National Institute of Informatics, Japan, Dr. Lada Adamic of Facebook, Prof. Fabrizio Sebastiani of ISTI-CNR, Italy, and Professor Sunita Sarawagi of IIT-Mumbai, India. Each of them is a leader of international renown in their respective areas, and we look forward to their participation.

Given the importance of data science, not just to academia but also to industry, we are pleased to have two distinguished industry speakers. The conference program was further enriched with three high-quality tutorials, eight workshops on cutting-edge topics, and one data mining contest on the prediction of memory failures.

We would like to express our sincere gratitude to the contributions of the Senior Program Committee (SPC) members, Program Committee (PC) members, and anonymous reviewers, led by the PC co-chairs, Kamal Karlapalem (IIIT, Hyderabad), Hong Cheng (CUHK), Naren Ramakrishnan (Virginia Tech). It is through their untiring efforts that the conference have an excellent technical program. We are also thankful to the other Organizing Committee members: industry co-chairs, Gautam Shroff (TCS) and Srikanta Bedathur (IIT Delhi); workshop co-chairs, Ganesh Ramakrishnan (IIT Mumbai) and Manish Gupta (Microsoft); tutorial co-chairs, B. Ravindran (IIT Chennai) and Naresh Manwani (IIIT Hyderabad); Publicity Co-Chairs, Sonali Agrawal (IIIT Allahabad), R. Uday Kiran (University of Aizu), and Jerry C-W Lin (WNU of Applied Sciences); competitions chair, Mengling Feng (NUS); Proceedings Chair, Tanmoy Chakraborthy (IIIT Delhi); and registration/local arrangement co-chairs, Vasudha Bhatnagar (University of Delhi), Vikram Goel (IIIT Delhi), Naveen Kumar (University of Delhi), Rajiv Ratn Shah (IIIT Delhi), Arvind Agarwal (IBM), Aditi Sharan (JNU), Mukesh Giluka (JNU) and Dhirendra Kumar (DTU).

We appreciate the hosting organizations IIIT Hyderabad and the JNU, Delhi, and all our sponsors for their institutional and financial support of PAKDD 2021. We also appreciate Alibaba for sponsoring the data mining contest. We feel indebted to the

PAKDD Steering Committee for its continuing guidance and sponsorship of the paper and student travel awards.

Finally, our sincere thanks go to all the participants and volunteers. There would be no conference without you. We hope all of you enjoy PAKDD 2021.

May 2021

R. K. Agrawal
P. Krishna Reddy
Jaideep Srivastava

PC Chairs' Preface

It is our great pleasure to present the 25th Pacific-Asia Conference on Knowledge Discovery and Data Mining (PAKDD 2021). PAKDD is a premier international forum for exchanging original research results and practical developments in the space of KDD-related areas, including data science, machine learning, and emerging applications.

We received 768 submissions from across the world. We performed an initial screening of all submissions, leading to the desk rejection of 89 submissions due to violations of double-blind and page limit guidelines. Six papers were also withdrawn by authors during the review period. For submissions entering the double-blind review process, each paper received at least three reviews from PC members. Further, an assigned SPC member also led a discussion of the paper and reviews with the PC members. The PC co-chairs then considered the recommendations and meta-reviews from SPC members in making the final decision. As a result, 157 papers were accepted, yielding an acceptance rate of 20.4%. The COVID-19 pandemic caused several challenges to the reviewing process, and we appreciate the diligence of all reviewers, PC members, and SPC members to ensure a quality PAKDD 2021 program.

The conference was conducted in an online environment, with accepted papers presented via a pre-recorded video presentation with a live Q/A session. The conference program also featured five keynotes from distinguished researchers in the community, one most influential paper talk, two invited industrial talks, eight cutting-edge workshops, three comprehensive tutorials, and one dedicated data mining competition session.

We wish to sincerely thank all SPC members, PC members, and external reviewers for their invaluable efforts in ensuring a timely, fair, and highly effective PAKDD 2021 program.

May 2021

Hong Cheng
Kamal Karlapalem
Naren Ramakrishnan

Organization

Organization Committee

General Co-chairs

R. K. Agrawal	Jawaharlal Nehru University, India
P. Krishna Reddy	IIIT Hyderabad, India
Jaideep Srivastava	University of Minnesota, USA

Program Co-chairs

Kamal Karlapalem	IIIT Hyderabad, India
Hong Cheng	The Chinese University of Hong Kong, China
Naren Ramakrishnan	Virginia Tech, USA

Industry Co-chairs

Gautam Shroff	TCS Research, India
Srikanta Bedathur	IIT Delhi, India

Workshop Co-chairs

Ganesh Ramakrishnan	IIT Bombay, India
Manish Gupta	Microsoft Research, India

Tutorial Co-chairs

B. Ravindran	IIT Madras, India
Naresh Manwani	IIIT Hyderabad, India

Publicity Co-chairs

Sonali Agarwal	IIIT Allahabad, India
R. Uday Kiran	The University of Aizu, Japan
Jerry Chau-Wei Lin	Western Norway University of Applied Sciences, Norway

Sponsorship Chair

P. Krishna Reddy	IIIT Hyderabad, India

Competitions Chair

Mengling Feng	National University of Singapore, Singapore

Proceedings Chair

Tanmoy Chakraborty	IIIT Delhi, India

Registration/Local Arrangement Co-chairs

Vasudha Bhatnagar	University of Delhi, India
Vikram Goyal	IIIT Delhi, India
Naveen Kumar	University of Delhi, India
Arvind Agarwal	IBM Research, India
Rajiv Ratn Shah	IIIT Delhi, India
Aditi Sharan	Jawaharlal Nehru University, India
Mukesh Kumar Giluka	Jawaharlal Nehru University, India
Dhirendra Kumar	Delhi Technological University, India

Steering Committee

Longbing Cao	University of Technology Sydney, Australia
Ming-Syan Chen	National Taiwan University, Taiwan, ROC
David Cheung	University of Hong Kong, China
Gill Dobbie	The University of Auckland, New Zealand
Joao Gama	University of Porto, Portugal
Zhiguo Gong	University of Macau, Macau
Tu Bao Ho	Japan Advanced Institute of Science and Technology, Japan
Joshua Z. Huang	Shenzhen Institutes of Advanced Technology, Chinese Academy of Sciences, China
Masaru Kitsuregawa	Tokyo University, Japan
Rao Kotagiri	University of Melbourne, Australia
Jae-Gil Lee	Korea Advanced Institute of Science and Technology, South Korea
Ee-Peng Lim	Singapore Management University, Singapore
Huan Liu	Arizona State University, USA
Hiroshi Motoda	AFOSR/AOARD and Osaka University, Japan
Jian Pei	Simon Fraser University, Canada
Dinh Phung	Monash University, Australia
P. Krishna Reddy	International Institute of Information Technology, Hyderabad (IIIT-H), India
Kyuseok Shim	Seoul National University, South Korea
Jaideep Srivastava	University of Minnesota, USA
Thanaruk Theeramunkong	Thammasat University, Thailand
Vincent S. Tseng	National Chiao Tung University, Taiwan, ROC
Takashi Washio	Osaka University, Japan
Geoff Webb	Monash University, Australia
Kyu-Young Whang	Korea Advanced Institute of Science and Technology, South Korea
Graham Williams	Australian National University, Australia
Min-Ling Zhang	Southeast University, China
Chengqi Zhang	University of Technology Sydney, Australia

Ning Zhong	Maebashi Institute of Technology, Japan
Zhi-Hua Zhou	Nanjing University, China

Senior Program Committee

Fei Wang	Cornell University, USA
Albert Bifet	Universite Paris-Saclay, France
Alexandros Ntoulas	University of Athens, Greece
Anirban Dasgupta	IIT Gandhinagar, India
Arnab Bhattacharya	IIT Kanpur, India
B. Aditya Prakash	Georgia Institute of Technology, USA
Bart Goethals	Universiteit Antwerpen, Belgium
Benjamin C. M. Fung	McGill University, Canada
Bin Cui	Peking University, China
Byung Suk Lee	University of Vermont, USA
Chandan K. Reddy	Virginia Tech, USA
Chang-Tien Lu	Virginia Tech, USA
Fuzhen Zhuang	Institute of Computing Technology, Chinese Academy of Sciences, China
Gang Li	Deakin University, Australia
Gao Cong	Nanyang Technological University, Singapore
Guozhu Dong	Wright State University, USA
Hady Lauw	Singapore Management University, Singapore
Hanghang Tong	University of Illinois at Urbana-Champaign, USA
Hongyan Liu	Tsinghua University, China
Hui Xiong	Rutgers University, USA
Huzefa Rangwala	George Mason University, USA
Jae-Gil Lee	KAIST, South Korea
Jaideep Srivastava	University of Minnesota, USA
Jia Wu	Macquarie University, Australia
Jian Pei	Simon Fraser University, Canada
Jianyong Wang	Tsinghua University, China
Jiuyong Li	University of South Australia, Australia
Kai Ming Ting	Federation University, Australia
Kamalakar Karlapalem	IIIT Hyderabad, India
Krishna Reddy P.	International Institute of Information Technology, Hyderabad, India
Lei Chen	Hong Kong University of Science and Technology, China
Longbing Cao	University of Technology Sydney, Australia
Manish Marwah	Micro Focus, USA
Masashi Sugiyama	RIKEN, The University of Tokyo, Japan
Ming Li	Nanjing University, China
Nikos Mamoulis	University of Ioannina, Greece
Peter Christen	The Australian National University, Australia
Qinghua Hu	Tianjin University, China

Rajeev Raman	University of Leicester, UK
Raymond Chi-Wing Wong	Hong Kong University of Science and Technology, China
Sang-Wook Kim	Hanyang University, South Korea
Sheng-Jun Huang	Nanjing University of Aeronautics and Astronautics, China
Shou-De Lin	Nanyang Technological University, Singapore
Shuigeng Zhou	Fudan University, China
Shuiwang Ji	Texas A&M University, USA
Takashi Washio	The Institute of Scientific and Industrial Research, Osaka University, Japan
Tru Hoang Cao	UTHealth, USA
Victor S. Sheng	Texas Tech University, USA
Vincent Tseng	National Chiao Tung University, Taiwan, ROC
Wee Keong Ng	Nanyang Technological University, Singapore
Weiwei Liu	Wuhan University, China
Wu Xindong	Mininglamp Academy of Sciences, China
Xia Hu	Texas A&M University, USA
Xiaofang Zhou	University of Queensland, Australia
Xing Xie	Microsoft Research Asia, China
Xintao Wu	University of Arkansas, USA
Yanchun Zhang	Victoria University, Australia
Ying Li	ACM SIGKDD Seattle, USA
Yue Xu	Queensland University of Technology, Australia
Yu-Feng Li	Nanjing University, China
Zhao Zhang	Hefei University of Technology, China

Program Committee

Akihiro Inokuchi	Kwansei Gakuin University, Japan
Alex Memory	Leidos, USA
Andreas Züfle	George Mason University, USA
Andrzej Skowron	University of Warsaw, Poland
Animesh Mukherjee	IIT Kharagpur, India
Anirban Mondal	Ashoka University, India
Arnaud Soulet	University of Tours, France
Arun Reddy	Arizona State University, USA
Biao Qin	Renmin University of China, China
Bing Xue	Victoria University of Wellington, New Zealand
Bo Jin	Dalian University of Technology, China
Bo Tang	Southern University of Science and Technology, China
Bolin Ding	Data Analytics and Intelligence Lab, Alibaba Group, USA
Brendon J. Woodford	University of Otago, New Zealand
Bruno Cremilleux	Université de Caen Normandie, France
Byron Choi	Hong Kong Baptist University, Hong Kong, China

Cam-Tu Nguyen	Nanjing University, China
Canh Hao Nguyen	Kyoto University, Japan
Carson K. Leung	University of Manitoba, Canada
Chao Huang	University of Notre Dame, USA
Chao Lan	University of Wyoming, USA
Chedy Raissi	Inria, France
Cheng Long	Nanyang Technological University, Singapore
Chengzhang Zhu	University of Technology Sydney, Australia
Chi-Yin Chow	City University of Hong Kong, China
Chuan Shi	Beijing University of Posts and Telecommunications, China
Chunbin Lin	Amazon AWS, USA
Da Yan	University of Alabama at Birmingham, USA
David C Anastasiu	Santa Clara University, USA
David Taniar	Monash University, Australia
David Tse Jung Huang	The University of Auckland, New Zealand
Deepak P.	Queen's University Belfast, UK
De-Nian Yang	Academia Sinica, Taiwan, ROC
Dhaval Patel	IBM TJ Watson Research Center, USA
Dik Lee	HKUST, China
Dinesh Garg	IIT Gandhinagar, India
Dinusha Vatsalan	Data61, CSIRO, Australia
Divyesh Jadav	IBM Research, USA
Dong-Wan Choi	Inha University, South Korea
Dongxiang Zhang	University of Electronic Science and Technology of China, China
Duc-Trong Le	University of Engineering and Technology, Vietnam National University, Hanoi, Vietnam
Dung D. Le	Singapore Management University, Singapore
Durga Toshniwal	IIT Roorkee, India
Ernestina Menasalvas	Universidad Politécnica de Madrid, Spain
Fangzhao Wu	Microsoft Research Asia, China
Fanhua Shang	Xidian University, China
Feng Chen	UT Dallas, USA
Florent Masseglia	Inria, France
Fusheng Wang	Stony Brook University, USA
Gillian Dobbie	The University of Auckland, New Zealand
Girish Palshikar	Tata Research Development and Design Centre, India
Giuseppe Manco	ICAR-CNR, Italy
Guandong Xu	University of Technology Sydney, Australia
Guangyan Huang	Deakin University, Australia
Guangzhong Sun	School of Computer Science and Technology, University of Science and Technology of China, China
Guansong Pang	University of Adelaide, Australia
Guolei Yang	Facebook, USA

Kaiyu Feng	Nanyang Technological University, Singapore
Kangfei Zhao	The Chinese University of Hong Kong, China
Karan Aggarwal	University of Minnesota, USA
Ken-ichi Fukui	Osaka University, Japan
Khoat Than	Hanoi University of Science and Technology, Vietnam
Ki Yong Lee	Sookmyung Women's University, South Korea
Ki-Hoon Lee	Kwangwoon University, South Korea
Kok-Leong Ong	La Trobe University, Australia
Kouzou Ohara	Aoyama Gakuin University, Japan
Krisztian Buza	Budapest University of Technology and Economics, Hungary
Kui Yu	School of Computer and Information, Hefei University of Technology, China
Kun-Ta Chuang	National Cheng Kung University, China
Kyoung-Sook Kim	Artificial Intelligence Research Center, Japan
L Venkata Subramaniam	IBM Research, India
Lan Du	Monash University, Canada
Lazhar Labiod	LIPADE, France
Leandro Minku	University of Birmingham, UK
Lei Chen	Nanjing University of Posts and Telecommunications, China
Lei Duan	Sichuan University, China
Lei Gu	Nanjing University of Posts and Telecommunications, China
Leong Hou U	University of Macau, Macau
Leopoldo Bertossi	Universidad Adolfo Ibañez, Chile
Liang Hu	University of Technology Sydney, Australia
Liang Wu	Airbnb, USA
Lin Liu	University of South Australia, Australia
Lina Yao	University of New South Wales, Australia
Lini Thomas	IIIT Hyderabad, India
Liu Yang	Beijing Jiaotong University, China
Long Lan	National University of Defense Technology, China
Long Yuan	Nanjing University of Science and Technology, China
Lu Chen	Aalborg University, Denmark
Maciej Grzenda	Warsaw University of Technology, Poland
Maguelonne Teisseire	Irstea, France
Maksim Tkachenko	Singapore Management University, Singapore
Marco Maggini	University of Siena, Italy
Marzena Kryszkiewicz	Warsaw University of Technology, Poland
Maya Ramanath	IIT Delhi, India
Mengjie Zhang	Victoria University of Wellington, New Zealand
Miao Xu	RIKEN, Japan
Minghao Yin	Northeast Normal University, China
Mirco Nanni	ISTI-CNR Pisa, Italy
Motoki Shiga	Gifu University, Japan

Shoujin Wang	Macquarie University, Australia
Shu Wu	NLPR, China
Shuhan Yuan	Utah State University, USA
Sibo Wang	The Chinese University of Hong Kong, China
Silvia Chiusano	Politecnico di Torino, Italy
Songcan Chen	Nanjing University of Aeronautics and Astronautics, China
Steven H. H. Ding	Queen's University, Canada
Suhang Wang	Pennsylvania State University, USA
Sungsu Lim	Chungnam National University, South Korea
Sunil Aryal	Deakin University, Australia
Tadashi Nomoto	National Institute of Japanese Literature, Japan
Tanmoy Chakraborty	IIIT Delhi, India
Tetsuya Yoshida	Nara Women's University, Japan
Thanh-Son Nguyen	Agency for Science, Technology and Research, Singapore
Thilina N. Ranbaduge	The Australian National University, Australia
Tho Quan	John Von Neumann Institute, Germany
Tianlin Zhang	University of Chinese Academy of Sciences, China
Tianqing Zhu	University of Technology Sydney, Australia
Toshihiro Kamishima	National Institute of Advanced Industrial Science and Technology, Japan
Trong Dinh Thac Do	University of Technology Sydney, Australia
Tuan Le	Oakland University, USA
Tuan-Anh Hoang	L3S Research Center, Leibniz University of Hanover, Germany
Turki Turki	King Abdulaziz University, Saudi Arabia
Tzung-Pei Hong	National University of Kaohsiung, Taiwan, ROC
Uday Kiran Rage	University of Tokyo, Japan
Vahid Taslimitehrani	PhysioSigns Inc., USA
Victor Junqiu Wei	Huawei Technologies, China
Vladimir Estivill-Castro	Griffith University, Australia
Wang Lizhen	Yunnan University, China
Wang-Chien Lee	Pennsylvania State University, USA
Wang-Zhou Dai	Imperial College London, UK
Wei Liu	University of Western Australia, Australia
Wei Luo	Deakin University, Australia
Wei Shen	Nankai University, China
Wei Wang	University of New South Wales, Australia
Wei Zhang	East China Normal University, China
Wei Emma Zhang	The University of Adelaide, Australia
Weiguo Zheng	Fudan University, China
Wendy Hui Wang	Stevens Institute of Technology, USA
Wenjie Zhang	University of New South Wales, Australia
Wenpeng Lu	Qilu University of Technology (Shandong Academy of Sciences), China

Wenyuan Li	University of California, Los Angeles, USA
Wilfred Ng	HKUST, China
Xiang Ao	Institute of Computing Technology, CAS, China
Xiangliang Zhang	King Abdullah University of Science and Technology, Saudi Arabia
Xiangmin Zhou	RMIT University, Australia
Xiangyu Ke	Nanyang Technological University, Singapore
Xiao Wang	Beijing University of Posts and Telecommunications, China
Xiaodong Yue	Shanghai University, China
Xiaohui (Daniel) Tao	The University of Southern Queensland, Australia
Xiaojie Jin	National University of Singapore, Singapore
Xiaoyang Wang	Zhejiang Gongshang University, China
Xiaoying Gao	Victoria University of Wellington, New Zealand
Xin Huang	Hong Kong Baptist University, China
Xin Wang	University of Calgary, Canada
Xingquan Zhu	Florida Atlantic University, USA
Xiucheng Li	Nanyang Technological University, Singapore
Xiuzhen Zhang	RMIT University, Australia
Xuan-Hong Dang	IBM T.J Watson Research Center, USA
Yanchang Zhao	CSIRO, Australia
Yang Wang	Dalian University of Technology, China
Yang Yu	Nanjing University, China
Yang-Sae Moon	Kangwon National University, South Korea
Yanhao Wang	University of Helsinki, Finland
Yanjie Fu	Missouri University of Science and Technology, USA
Yao Zhou	UIUC, USA
Yashaswi Verma	IIT Jodhpur, India
Ye Zhu	Deakin University, Australia
Yiding Liu	Nanyang Technological University, Singapore
Yidong Li	Beijing Jiaotong University, China
Yifeng Zeng	Northumbria University, UK
Yingfan Liu	Xidian University, China
Yingyi Bu	Google, USA
Yi-Shin Chen	National Tsing Hua University, Taiwan, ROC
Yiyang Yang	Guangdong University of Technology, China
Yong Guan	Iowa State University, USA
Yu Rong	Tencent AI Lab, China
Yu Yang	City University of Hong Kong, China
Yuan Yao	Nanjing University, China
Yuanyuan Zhu	Wuhan University, China
Yudong Zhang	University of Leicester, UK
Yue Ning	Stevens Institute of Technology, USA
Yue Ning	Stevens Institute of Technology, USA
Yue-Shi Lee	Ming Chuan University, China
Yun Sing Koh	The University of Auckland, New Zealand

Yunjun Gao	Zhejiang University, China
Yuqing Sun	Shandong University, China
Yurong Cheng	Beijing Institute of Technology, China
Yuxiang Wang	Hangzhou Dianzi University, China
Zemin Liu	Singapore Management University, Singapore
Zhang Lei	Anhui University, China
Zhaohong Deng	Jiangnan University, China
Zheng Liu	Nanjing University of Posts and Telecommunications, China
Zheng Zhang	Harbin Institute of Technology, China
Zhengyang Wang	Texas A&M University, USA
Zhewei Wei	Renmin University of China, China
Zhiwei Zhang	Beijing Institute of Technology, China
Zhiyuan Chen	University of Maryland Baltimore County, USA
Zhongying Zhao	Shandong University of Science and Technology, China
Zhou Zhao	Zhejiang University, China
Zili Zhang	Southwest University, China

Competition Sponsor

Alibaba Cloud

Host Institutes

Jawaharlal Nehru University

INTERNATIONAL INSTITUTE OF INFORMATION TECHNOLOGY HYDERABAD

Contents – Part I

Data Mining of Specialized Data

Contents – Part II

Contents – Part III

Learning from Data

Applications of Knowledge Discovery

Fuzzy World: A Tool Training Agent from Concept Cognitive to Logic Inference

Minzhong Luo[1,2(✉)]

[1] Institute of Information Engineering, Chinese Academy of Sciences, Beijing, China
luominzhong@iie.ac.cn

[2] School of Cyber Security, University of Chinese Academy of Sciences, Beijing, China

Abstract. Not like many visual systems or NLP frameworks, human generally use both visual and semantic information for reasoning tasks. In this paper, we present a 3D virtual simulation learning environment Fuzzy World based on gradual learning paradigm to train visual-semantic reasoning agent for complex logic reasoning tasks. Furthermore our baseline approach employed semantic graphs and deep reinforcement learning architecture shows the significant performance over the tasks.

Keywords: Visual-semantic reasoning · Gradual learning paradigm · Virtual learning tool.

1 Introduction

Although artificial intelligence based on deep learning has achieved impressive results in vision and language respectively, agents need to combine language and visual signals to complete complex tasks in real environments [30], so completing quests in natural language by agent in unstructured environments has been a research goal caught increasing attention for recent years [11,16]. This paper examines the idea of training and testing agent to combine language signals with visual signals to achieve concept cognitive and logic reasoning capabilities based on a virtual training environment with interactive functions.

1.1 Basic Idea

Even if the agent has enough information, logical reasoning tasks can not be completed efficiently through end-to-end models. Some advancing approaches that address the challenge to logic reasoning have been developed [4,5], however, these models often rely on hard-ruled inference or pure fuzzy computation, which imposes limits on reasoning abilities of agents. Agent with logical reasoning abilities is not an inseparable end-to-end system, on the contrary, logical reasoning requires three levels of ability from bottom to top(the three-level learning paradigm):

– **Basic recognition ability:** the classification system $\mathcal{P}(L_o|V)$, only requires the cognitive ability for object's label L_o based on the inputed image V;

K. Karlapalem et al. (Eds.): PAKDD 2021, LNAI 12712, pp. 3–14, 2021.
https://doi.org/10.1007/978-3-030-75762-5_1

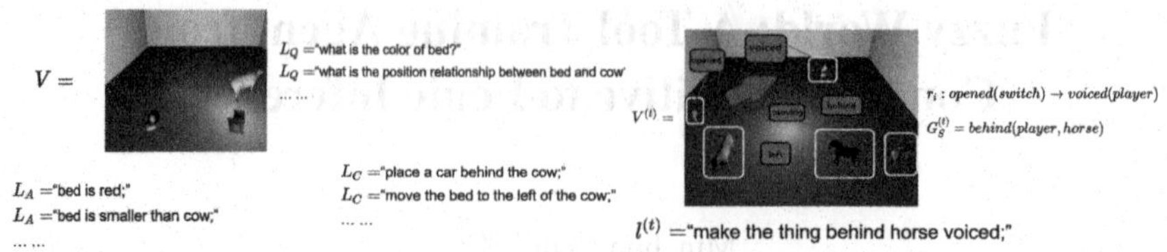

Fig. 1. (Left): A task instance of Fuzzy World static dataset: the agent receives a question L_Q and a visual signal V, gives out an answer L_A. **(Right): A task instance of Fuzzy World dynamic enviroment**: the agent receives a command $L_C^{(t)}$ and a visual signal $V^{(t)}$ at time step t, identifies the current state of the scene $G_S^{(t)}$ and give the next action $A^{(t)}$ based on the logic rule r_i from knowledge graph G_L.

- **Concept cognitive ability:** abstract concepts without clear forms such as "on the left", "red and big", the agent with both the ability to give answer L_A to question $L_Q(\mathcal{P}(L_A|V, L_Q))$ and the ability to give action $A^{(t)}$ by command $L_C(\mathcal{P}(A^{(t)}|V, L_C))$ based on inputed image V has learned the concept [20].
- **Logic reasoning ability:** the ability to use knowledge to infer the next state based on the current state of environment, let $G_S^{(t)}$ denote the state semantic graph [21] of state at time step t, then the reasoning ability is $\mathcal{P}(G_S^{(t+1)}|G_S^{(t)}, G_L)$ (where the G_L is the semantic graph summarizing the logic rules which represents the knowledge learned).

1.2 The Proposed Tool

In order to examine the proposed three-level learning paradigm, a 3D virtual simulation learning environment Fuzzy World[1] is developed, which is a simple and easy-to-use Python program can load custom 3D models and custom logic rules, and is easy to implement deep reinforcement learning with Keras [9].

Figure 1 shows examples training and testing agents based on the three-level learning paradigm in Fuzzy World. According to the different characteristics of tasks, Fuzzy World static dataset (FWS) is employed to test the concept cognitive ability of the agent (Left of Fig. 1), Fuzzy World dynamic interaction environment (FWD) is employed to test the logic reasoning ability of the agent (Right of Fig. 1). Furthermore, Fuzzy World is an AI training tool can be used for richer custom high-level intelligent training and testing tasks development (see Sect. 5 for more details).

1.3 Main Contributions

Our main contribution is proposing a 3D virtual simulation learning environment **Fuzzy World** based on the three-level learning paradigm, furthermore, baseline methods for agent-based concept cognitive and logic reasoning with semantic and visual information are also implemented.

[1] Implemented at https://github.com/Luomin1993/fuzzy-world-tool.

Table 1. For language, **Fuzzy World** constructs commands enable agent interact with environment with gradual increasing difficulty, while the questions/commands of other datasets are either too difficult or too simple. For visual navigation, **Fuzzy World** enables grid-free and discretized movement, while other datasets use topological graph navigation. For logic reasoning, The interactive environment of **Fuzzy World** contains long-range logical rules, but it will teach agent to get reasoning ability from simplest object recognition to concept cognitive gradually. By contrast, other datasets only require end-to-end learning paradigm.

	Language		Visual		Logic reasoning		
	Questions	Commands	Visual form	Objects changeable	Interactive	Navigation mode	Dynamic multi-step reasoning
Touchdown [8]	9.3k+	✗	Photos	✗	✗	By photo	✗
IQA [15]	✗	✗	3D virtual	✗	✗	Discrete	✗
EQA [11]	✗	✗	3D virtual	✗	✗	Discrete	✗
TACoS [22]	17k+	✗	Photos	✗	✗	✗	✗
VHome [28]	2.7k+	1k+	3D Virtual	Discrete	Discrete	✗	✗
ALFRED [24]	25k+	10k+	3D Virtual	Discrete	Discrete	Discrete	✗
Fuzzy World	4k+	4k+	3D Virtual	Discrete	Discrete	Grid-free	✓

2 Related Work

Table 1 summarizes the benefits of Fuzzy World relative to other datasets/tools with both language annotations and visual informations.

Concept Cognitive. Concept cognitive is a branch of Cognitive Science [19]. [2] developed a neural network model based on energy function, in which agent can learn and extract concepts from tasks and has few-shot learning and generalization ability. [27] constructed a Bayesian inference model to transform concept cognitive into numerical computable problems. [17] learned the visual concepts (color, shape, etc.) based on VAE (Variational Autoencoder).

Intelligent System for Logic Reasoning. Building an intelligent system with logic reasoning ability is a long-standing challenge [19]. Rule-based expert systems with reasoning ability are applied to various professional fields [12] [13,23]. Fuzzy reasoning which combines rule-based inference with numerical computations[18] makes intelligent system more flexible. [10] combined the perception and fitting ability of neural networks with the reasoning ability of symbolic logic AI, which can process numerical data and knowledge in symbolic forms in sequence or simultaneously.

Language Grounding for Vision. Robotics navigation based on language commands [7,26,31] requires robots to move according to language goals strategically. A virtual 3D environment is usually required for such training tasks. In the Fuzzy World, agent not only needs to navigate correctly based on language commands but more complex operations on objects in the environment(such as opening switches, moving objects, etc.) are required to complete the tasks.

Other Related Works. Our work is also related to graph-based knowledge representation [6], virtual environment based reinforcement learning [31], and visual semantic problems [14].

3 Prelimilaries

In this section we briefly show the formulized description of the problem and the proposed tasks in tool with baseline approaches.

3.1 Deep Q-Learning Based on POMDP

We formally introduce our problem as reinforcement learning based on partially observable Markov decision process(POMDP) [1] as follows. The Reinforcement Learning (RL) algorithm simulates the process of an agent learning to complete tasks in an environment, consisting of the environment state set $\{s_t\}$, the agent's action decision set $\{a_t\}$. Agent needs to evaluate the value of the decision $Q(a_{t+1}, s_t)$ at time step t, then gives the action decision $a_{t+1} = \pi(a_{t+1}|s_t)$ for the next time step according to the current state of environment, then the environment will give a reward signal R_t to the agent. The goal of reinforcement learning is to let the agent learn the optimal strategy $\pi(a_{t+1}|s_t)$ and maximize the reward expectation $v(s) = \mathbb{E}[G_t|S_t = s]$, which can be solved by the Bellman equation $v(s) = \mathbb{E}[R_{t+1} + \lambda v(S_{t+1})|S_t = s]$; We introduce a new value function(Action-value function) $Q^\pi(s, a)$ starting from the state s, executing the action a and then using the cumulative reward of the strategy π:

$$Q^*(s,a) = \max_{\pi} Q^\pi(s,a) = \mathbb{E}_{s'}[r + \lambda \max_{a'} Q^*(s',a')|s,a] \tag{1}$$

In actual engineering, only a limited number of samples can be used for solving operations. Q-Learning [29] proposes an iterative method for updating Q-values:

$$Q(S_t, A_t) \leftarrow Q(S_t, A_t) + \alpha(R_{t+1} + \lambda \max_{a} Q(S_{t+1}, a) - Q(S_t, A_t)) \tag{2}$$

3.2 Concept Cognitive Task Formulization

Examples of concepts include visual ("green" or "triangle"), spatial ("on left of", "behind"), temporal ("slow", "after"), size ("bigger", "smaller") among many others. Concept cognitive [2] is equivalent to an ability with recognition and generalization for abstract concepts [20]. Agent with recognition ability can identify specific concepts (e.g. left or right, upper or lower relations, etc.) from the scene. Agent with generalization ability can generate a conforming pattern according to the connotation of concept (e.g. placing multiple objects as a square, or circle, etc.).

The first-order predicate logic is used to describe the connection between concepts and entities. The training samples can be organized into queries as $L_A(o_1, o_2)$ which represents predicate $\text{Relation}_A(\text{object}_1, \text{object}_2)$, and then the dataset can be denoted by $\mathcal{D} = \{L_A^{(i)}, V_i, L_Q^{(i)}, L_C^{(i)}\}$. The system needs to be trained separately as the concept identification map $\mathcal{P}(L_A|V, L_Q)$ (agent predicts L_A based on the image V and question L_Q) and the concept generalization map $\mathcal{P}(A^{(t)}|V, L_C)$ (agent predicts representation vector $A^{(t)}$ of action based on the image V and command L_C) to obtain concept cognitive ability.

3.3 Logic Reasoning Task Formulization

The logic reasoning task [3] requires the agent to speculate on the state $G_S^{(t+1))}$ of the environment at next step based on the observed state $G_S^{(t)}$ now and the existing knowledge G_L, where $G_S^{(t)}$ is a semantic graph which represents the state of the objects in environment at the current time step t (e.g. the stool is red and the stool is on the left of the table), G_L is called logic-rule graph represents the correlations between events (e.g. if switch is opened, light will be on), $G_S^{(t+1)}$ is the state of the environment at the next time step $t + 1$. Graph convolutional neural network [25] is utilized to construct map $\mathcal{P}(G_S^{(t+1)}|G_S^{(t)}, G_L)$ to learn the logic reasoning ability.

In a virtual closed environment with the entity set $\mathcal{O} = \{o_k\}$, each entity o_k has state $s^{(t)}(o_k)$ at time t (e.g. the switch is open now). Now the closed environment contains a finite constant rules set $R = \{r_i\}$, each rule $r_i \in R$ can be described as an association between object states (e.g. the light turns on when the switch is turned on):

$$r_i : s^{(t)}(o_k) \rightarrow s^{(t+1)}(o_j) \tag{3}$$

Now suppose the command is "makes the bulb light at the next moment $t + 1$", denoted by $s^{(t+1)}(o_j)$, then the agent needs to find and use the rule $r_i : s^{(t)}(o_k) \rightarrow s^{(t+1)}(o_j)$ to complete this task. In this paper, we generated 5000+ rules and more 2000+ logic-based tasks in Fuzzy World to train and test the logic reasoning ability of the agent.

4 Fuzzy World and Baselines

Recall that our goal is to train an intelligent system with concept cognitive ability and logic reasoning ability. We first generated a static dataset containing visual and linguistic interactions based on Fuzzy World, concept recognition network $\mathcal{P}(L_A|V, L_Q)$ and the concept generalization network $\mathcal{P}(A^{(t)}|V, L_C)$ are developed for the concept cognitive ability. Then the agent is placed in a virtual dynamic 3D environment, logic-based tasks are used to train logic reasoning ability $\mathcal{P}(G_S^{(t+1)}|G_S^{(t)}, G_L)$.

4.1 The Fuzzy World environment

Hierarchical Learning Form of Fuzzy World: in the three-level learning paradigm of Fuzzy World, agent will gradually complete the tasks combine language and vision from object recognition, concept cognitive to logic reasoning; The Table 2 shows the examples and compares details of tasks in different learning stages.

In the three-level learning paradigm, the first level with "classification and recognition for object" is very simple, while the second level for concept cognitive and the third level for logic reasoning require the agent to use semantic graphs to assist in cognition of the states and rules in environment, which is critical and necessary to accomplish tasks, the details of tasks are shown in Fig. 3.

Table 2. An overview of tasks in different learning stages, the ability of the agent to complete the task in certain stage is based on the ability from the previous stage.

Visual	Language	Target	Learning Form	Baseline
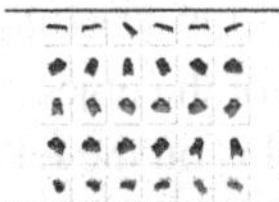	"What is this?" "What's that?"	Classification Recognition	$\mathcal{P}(y_i \mid X_i)$ X_i is the image, y_i is the label.	CNN
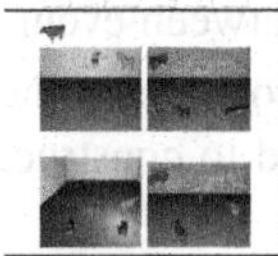	"What is the color of jeep?" "What is the size relationship between chair and horse"	Concept Cognitive	$\mathcal{P}(L_A \mid V, L_Q)$ $\mathcal{P}(A^{(t)} \mid V, L_C)$	Concept Cognitive Network
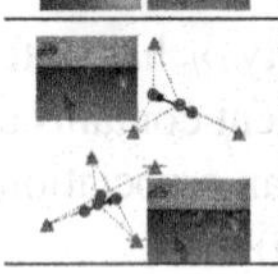	"Makes the bulb light." "Turn off that switch." "Make the temperature of the room lower."	Logic Reasoning	$\mathcal{P}(G_S^{(t+1)} \mid G_S^{(t)}, G_L)$	Semantic Graph+ Reasoning Network

4.2 The Concept Cognitive Task and Baseline

The concept recognition network $\mathcal{P}(L_A|V, L_Q)$ and the concept generalization network $\mathcal{P}(A^{(t)}|V, L_C)$ are utilized to develop the concept cognitive ability of the agent:

- **concept recognition network:** the image $V \in \mathbb{R}^{l\times 96\times 96}$ and the question L_Q (text data, which is question about the content in the image) are inputed, the output should be answer L_Q (text data, the answer to the question L_Q). The network can be denoted by $L_A = f_r(\text{CNN}(V), \text{embd}(L_Q), A_V)$, where the visual feature A_V of the concept is a trainable vector.
- **concept generalization network:** the image $V \in \mathbb{R}^{l\times 96\times 96}$ and the command L_C (text data, command to change certain state of objects in the image) are inputed, and the output should be the action representation vector $A^{(t)}$ (action choosed to change certain state of objects). The network can be denoted by $A^{(t)} = f_g(\text{CNN}(V), \text{embd}(L_C), A_S)$, where the semantic feature A_S of the concept is a trainable vector.

The model	Component details of the model
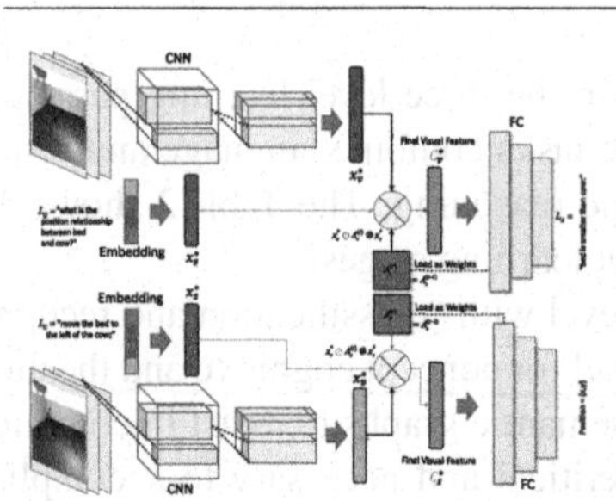	**The proposed model for concept cognitive:** **the concept recognition network:** $\mathcal{P}(L_A \mid V, L_Q)$ **and the concept generalization network:** $\mathcal{P}(A^{(t)} \mid V, L_C)$. **Notice that the weights are shared between the two networks,for example, the weights matrix A_V in the network $\mathcal{P}(L_A\|V, L_Q)$ are also weights in the FC layers from $\mathcal{P}(A^{(t)}\|V, L_C)$ and updated in two networks asynchronously.**

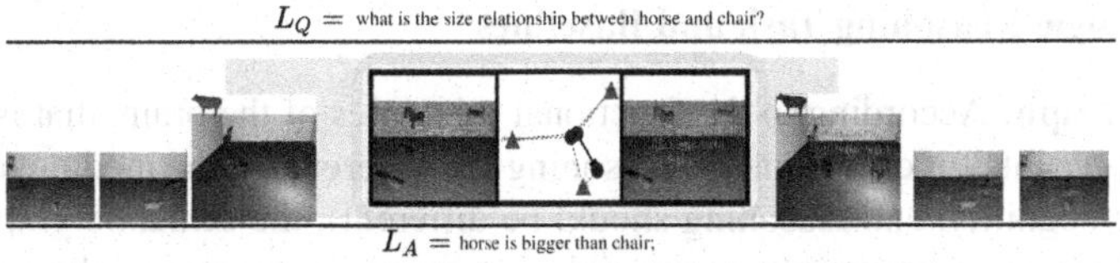

(a) Concept Cognitive Task

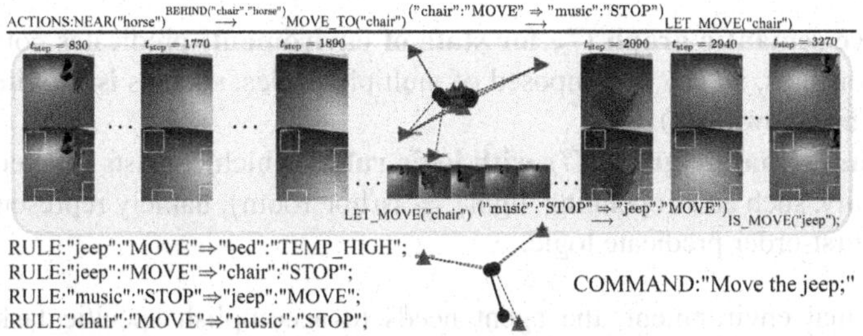

(b) Logic Reasoning Task

Fig. 2. (a): **Concept Cognitive Task**. This task is designed to be the mode with asking and answering interactively based on the picture. **(b)**: **Logic Reasoning Task**. The agent not only needs to use the state semantic graph [21] of the environment, but also needs to derive the reasoning semantic graph through repeated trial and error interactions with the environment, to accomplish tasks finally.

The specific effect of matrix $A_V^{(t)}$ as network weights is:

$$x_S^* \odot A_V^{(t)} \otimes x_V^* = \{\begin{matrix} x_S^* \odot A_V^{(t)1}, \\ x_S^* \odot A_V^{(t)2}, \\ \dots, \\ x_S^* \odot A_V^{(t)n} \end{matrix}\}^{\mathrm{T}} \otimes x_V^* \tag{4}$$

Where $x_S^* \odot A_V^{(t)i}$ represents the inner product of semantic feature x_S^* and the visual feature $A_V^{(t)i}$ for concept i, this process is equivalent to let the semantic feature x_S^* **go one by one to select which visual concept vector matches the semantic concept vector** (inspired by [31]). Notice that the above matrix $A_V^{(t)}$ is extracted from the weights of the generated model, which is equivalent to the latest visual concept features learned.

Algorithm 1. The Iterative Training Steps for Concept Cognitive

$A_V^{(0)}, A_S^{(0)} \sim \mathcal{N}(u, \sigma)$ is sampled from a Gaussian distribution at the initial $t = 0$;
for $t = 1$ to T do **do**
    train the recognition network $\mathcal{P}(L_A|V, L_Q)$ and performing a gradient descent;
    train the generation network $\mathcal{P}(A^{(t)}|V, L_C)$ and performing a gradient descent;
    Update $A_S^{(t)}$, $A_V^{(t)}$;
end for

4.3 The Logic Reasoning Task and Baselines

Semantic Graph: According to the functional principles of the brain, that is, modules in brain responsible for cognition and reasoning are different (so the module responsible for learning cognition and reasoning should be different), the semantic graph denoted by G_S and G_I are respectively responsible for cognition (corresponding to concept cognitive mentioned above) and reasoning:

- **Cognitive semantic graph G_S for state of enviroment:** predicates connect entities or concepts, so G_S is composed of multiple triples, such as **is(moving,cow)** or **biggerThan(cow,desk)**.
- **Reasoning semantic graph G_I with logic rules:** which consists of events related to causality, such as is(openning, light) $\Rightarrow$ is(hot, room), namely represented in the form of first-order predicate logic.

In a virtual environment, the agent needs to accomplish specific tasks require logic reasoning ability based on visual and language informations. The ability can be abstracted into the policy network $\pi(a^{(t)}|G_S, G_I, l^{(t)}, V^{(t)})$ responsible for decision making and the value network $Q(A, G_S, G_I, l^{(t)}, V^{(t)})$ responsible for evaluating the strategy, where $V^{(t)}$ is the inputed image, $l^{(t)}$ is the inputed language command, and $a^{(t)} \in A$ is the action decision taken, where A is the finite actions set. The loss function is defined as follows (where $S^{[t]}$ is the state of environment at current time step, $G_S^{[t]}$ is the semantic graph representation of $S^{[t]}$):

$$\begin{aligned}\mathbb{E}_{V^{[t]},S^{[t]}\sim Env}[&\underbrace{\log \pi_\theta(a^{(t)}|G_S, G_I, l^{(t)}, V^{(t)})}_{\text{Policy}} + \underbrace{\lambda v_\theta(G_S^{[t]}, G_I, S^{[t]})}_{\text{Value Estimation}} \\ &+ \underbrace{\gamma(R_\theta(V^{(t)}, S^{(t)}) + I_\theta(G_S^{[t-1]}, G_S^{[t]}))}_{\text{Value Estimation}} + \underbrace{\kappa||\theta||}_{\text{Norm}}]\end{aligned} \tag{5}$$

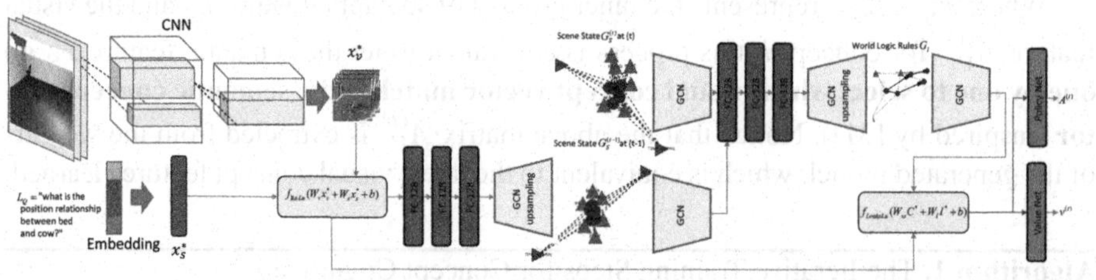

Fig. 3. The proposed logic reasoning network: the agent receives the visual signal $V^{(t)}$ and a language command $l^{(t)}$ about the current environment, and predicts the state semantic graph $G_S^{(t)}$ of the current environment, then predicts the new logic rule $r_i \in G_I$ according to the state semantic graphs from multiple time steps by interactions, finally gives action $A^{(t)}$ of next time step and evaluation $v^{(t)}$.

5 Experiments

In this section we introduce Fuzzy World and its dataset for concept cognitive and logic reasoning firstly, then the comparison of performance between our method and baseline method on the dataset is given by detailed analysis.

5.1 Experiment Tasks Based on Fuzzy World

Fuzzy World is a 3D virtual simulation environment for reinforcement learning written in Python. This program imports multiple 3D files in ".obj" format to establish a closed environment with specific tasks(see Fig. 2 for details).

- **Static dataset for concept cognitive (FWS):** We generate a large number (2k+) of images $\{V_i\}$ containing 3D objects of different scenes by the rule-based algorithm, also including corresponding language questions L_Q, language answers L_A, the command text L_C and the action representation $A^{(t)}$ (see the examples in Fig. 5); this dataset is suitable for training by end-to-end neural network.
- **Dynamic 3D environment for logic reasoning (FWD):** We construct a large number (1.5k+) of simulation environments containing 3D objects, logic rules, and specific tasks by the rule-based algorithm (average 30 rules in the form of first-order predicate logic, more than 50 tasks per environment), and corresponding commands $\{l^{(t)}\}$ describing the tasks; This environment is suitable for training agent equipped logical reasoning ability employing deep reinforcement learning architecture;

5.2 Experiment Details

Baseline method: We use the method of [31] as the baseline for the comparison with our logic inference learning method. In this method, the neural network of agent has similar inputs and outputs to our method (the reinforcement learning architecture is used, inputs include language signal $l^{(t)}$ and visual signal $V^{(t)}$, outputs are action decision $a^{(t)}$ and evaluation $v^{(t)}$ of decision). Objective function is as follows:

$$-\mathbb{E}_{s^{[t]},a^{[t]},l} = [\underbrace{(\nabla_\theta \log \pi_\theta(a^{[t]}|o^{[t]}, h^{[t-1]}, l)}_{Policy} + \underbrace{\eta \nabla_\theta v_\theta(o^{[t]}, h^{[t-1]}, l))}_{Value\ \ Estimation} \underbrace{A^{[t]}}_{Award} + \underbrace{\kappa \nabla_\theta \mathcal{E}(\pi_\theta)]}_{Norm} \quad (6)$$

The learning rate of the network is 10^{-5}, the update moment is 0.9, and the optimizer of the concept cognitive network uses the Adam optimizer with parameters $\beta_1 = 0.90, \beta_2 = 0.99$, the optimizer of the logic reasoning network adopts SGD. Program is implemented using Python with keras [9].

5.3 Experiment Results

Figure 4 visualizes the detailed results of the control experiment, where the success rate refers to the percentage of the number of successful tasks completed by the agent to the total number of tasks. Based on analysis on the experimental results, agents utilizing three-level learning paradigm obtain much better tasks accomplishing results than the comparison baseline, which is clear evidence that the gradual learning paradigm in Fuzzy World is more conducive for the agent to complete logic reasoning tasks compared with the end-to-end methods.

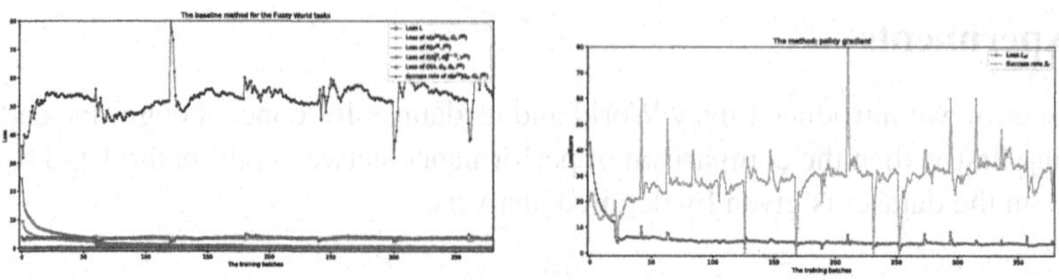

Fig. 4. The evaluation results. **(a) The baseline model** [31]. The success rate of the agent is significantly lower than the proposed model with semantic graphs. **(b) The proposed model.** The convergence of the method is better and faster, indicating that the introduced advanced logic-rule information represented by semantic graphs is more efficient for agent to accomplish complex tasks in environment.

6 Conclusion

In this paper, we presented the Fuzzy World, which provides a 3D virtual enviroment simulated with interactions based on the three-level learning paradigm. We expect the tool to be beneficial for a variety of research activities in agent integrates language and visual information to complete complex tasks. The evaluation results on challenging tasks based on the tool Fuzzy World suggest that proposed method employed gradual learning paradigm and semantic graphs can accomplish tasks required logic reasoning ability better. For future work, we will further investigate the functions of Fuzzy World. We also want to leverage the environments provided by the tool for the improvement of high-level intelligent decision system for efficient use of environmental information.

Appendix

6.1 Using Second Order Derivative Gradient for Cross Training Parameter

Notice that the prediction of the model $\mathcal{P}(L_A|V, L_Q)$ is in one-hot form of space concept like:[up and down, left and right, top left and bottom right...], then the loss of last layer employed softmax cross entropy loss is $\mathcal{L}(A_S) = -\hat{y} \odot log(f_{softmax}(A_S \odot C^{*T}))$. The next is the provement of an upper bound of $\hat{\mathcal{L}}(A + \alpha \Delta A)$.

Note that the updating of parameters A takes the simple SGD: $A^{t+1} \leftarrow A^t + \alpha \nabla_A \hat{\mathcal{L}}$.

Theorem 1. *When* $\nabla_A^2 \hat{\mathcal{L}} \leq MI$, *we have* $\hat{\mathcal{L}}(A + \alpha \Delta A) \leq \hat{\mathcal{L}}(A) + \gamma ||\nabla_A \hat{\mathcal{L}}||^2$.

Proof. Easy to know $-\nabla_A \hat{\mathcal{L}}(A) = \Delta A$, do Taylor expansion to $\hat{\mathcal{L}}(A + \alpha \Delta A)$:

$$
\begin{aligned}
\hat{\mathcal{L}}(A + \alpha \Delta A) &= \hat{\mathcal{L}}(A) + \alpha \nabla_A \hat{\mathcal{L}}(A) \odot \Delta A + \nabla_A^2 \hat{\mathcal{L}} ||\Delta A||^2 \alpha^2 / 2 \\
&\leq \hat{\mathcal{L}}(A) + \alpha \nabla_A \hat{\mathcal{L}} \odot (-\nabla_A \hat{\mathcal{L}}) + M ||\Delta A||^2 \alpha^2 / 2 \\
&= \hat{\mathcal{L}}(A) + (\alpha^2 M/2 - \alpha) ||\nabla_A \hat{\mathcal{L}}||^2
\end{aligned}
$$

Now let $\gamma = \alpha^2 M/2 - \alpha \leq 0$ then the below is satisfied:

$$
\hat{\mathcal{L}}(A + \alpha \Delta A) \leq \hat{\mathcal{L}}(A + \alpha \Delta A) + (\alpha - \alpha^2 M/2) ||\nabla_A \hat{\mathcal{L}}||^2 \leq \hat{\mathcal{L}}(A)
$$

.

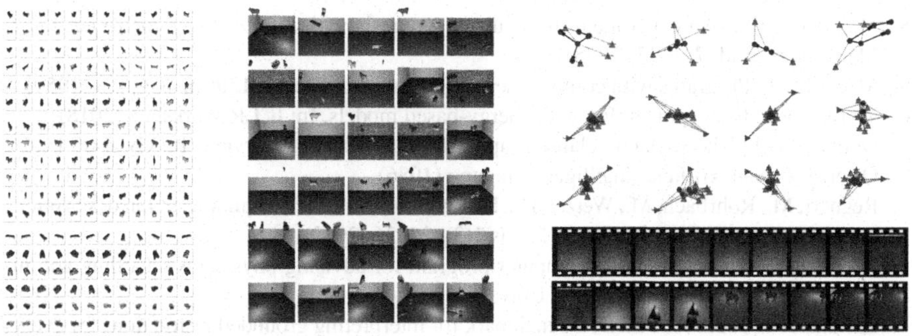

Fig. 5. The Three-level Learning Paradigm: task examples from object classification to concept cognitive and logic reasoning, semantic graphs are generated and used in logic reasoning tasks.

References

1. Kaelbling, L.P., Littman, M.L., Cassandra, A.R.: Planning and acting in partially observable stochastic domains. Artif. Intell. **101**(1–2), 99–134 (1998)
2. Angluin, D.: Queries and concept learning. Mach. Learn. **2**(4), 319–342 (1988)
3. Apt, K.R., Bol, R.N.: Logic programming and negation: a survey. J. Logic Program. **19**(94), 9–71 (1994)
4. Apt, K.R., Emden, M.H.V.: Contributions to the theory of logic programming. J. ACM **29**(3), 841–862 (1982)
5. Besold, T., et al.: Neural-symbolic learning and reasoning: a survey and interpretation
6. Chein, M., Mugnier, M.-L.: Graph-based knowledge representation: computational foundations of conceptual graphs. Univ. Aberdeen **13**(3), 329–347 (2009)
7. Chen, D.L., Mooney, R.J.: Learning to interpret natural language navigation instructions from observations. In: AAAI Conference on Artificial Intelligence, AAAI 2011, San Francisco, California, USA (2011)
8. Chen, H., Suhr, A., Misra, D., Snavely, N., Artzi, Y.: Touchdown: natural language navigation and spatial reasoning in visual street environments
9. Chollet, F., et al.: Keras (2015). https://keras.io
10. Dai, W.Z., Xu, Q.L., Yu, Y., Zhou, Z.H.: Tunneling neural perception and logic reasoning through abductive learning
11. Das, A., Datta, S., Gkioxari, G., Lee, S., Parikh, D., Batra, D.: Embodied question answering. In: 2018 IEEE/CVF Conference on Computer Vision and Pattern Recognition (2018)
12. Duda, R., Gaschnig, J., Hart, P.: Model design in the prospector consultant system for mineral exploration. Read. Artif. Intell. 334–348 (1981)
13. Feigenbaum, E.A., Buchanan, B.G. Lederberg, J.: Generality and problem solving: a case study using the DENDRAL program. Stanford University (1970)
14. Frome, A., Corrado, G.S., Shlens, J., Bengio, S., Dean, J., Ranzato, M., Mikolov, T.: DeViSE: a deep visual-semantic embedding model. In: International Conference on Neural Information Processing Systems, pp. 2121–2129 (2013)
15. Gordon, D., Kembhavi, A., Rastegari, M., Redmon, J., Fox, D., Farhadi, A.: IQA: visual question answering in interactive environments
16. Hermann, K.M., Felix Hill, S.G., Fumin Wang, P.B.: Grounded language learning in a simulated 3D world. In: NIPS Workshop (2017)
17. Higgins, I., et al.: SCAN: learning abstract hierarchical compositional visual concepts

18. Mamdani, A.S.: An experiment in linguistic synthesis with a fuzzy logic controller. Int. J. Man-Mach. Stud. **7**, 1–13 (1975)
19. Mccarthy, J.: Programs with common sense. Semant. Inf. Proces. **130**(5), 403–418 (1959)
20. Mordatch, I.: Concept learning with energy-based models. In: ICLR Workshop (2018)
21. Ohlbach, H.J.: The semantic clause graph procedure - a first overview. In: Gwai-86 Und 2 Österreichische Artificial-intelligence-tagung (1986)
22. Regneri, M., Rohrbach, M., Wetzel, D., Thater, S., Pinkal, M.: Grounding action descriptions in videos. Trans. Assoc. Comput. Lingus **1**(3), 25–36 (2013)
23. Shortliffe, E.H.: A rule-based computer program for advising physicians regarding antimicrobial therapy selection. Stanford University (1974)
24. Shridhar, M., et al.: ALFRED: a benchmark for interpreting grounded instructions for everyday tasks
25. Tang, J., Qu, M., Wang, M., Zhang, M., Yan, J., Mei, Q.: Line: large-scale information network embedding (2015)
26. Tellex, S., et al.: Understanding natural language commands for robotic navigation and mobile manipulation. In: AAAI Conference on Artificial Intelligence, pp. 1507–1514 (2011)
27. Tenenbaum, J.B.: Bayesian modeling of human concept learning. In: Conference on Advances in Neural Information Processing Systems II, pp. 59–65 (1998)
28. Torralba, X., et al.: VirtualHome: simulating household activities via programs. In: 2018 IEEE/CVF Conference on Computer Vision and Pattern Recognition (2018)
29. Watkins, C.J., Dayan, P.: Technical note: Q-learning. Mach. Learn. **8**(3–4), 279–292 (1992)
30. Winograd, T.: Procedures as a representation for data in a computer program for understanding natural language. Technical report, Massachusetts Institute of Technology (1971)
31. Yu, H., Lian, X., Zhang, H., Xu, W.: Guided feature transformation (GFT): a neural language grounding module for embodied agents

Collaborative Reinforcement Learning Framework to Model Evolution of Cooperation in Sequential Social Dilemmas

Ritwik Chaudhuri[1(✉)], Kushal Mukherjee[2], Ramasuri Narayanam[1], and Rohith D. Vallam[1]

[1] IBM Research, Bangalore, India
{charitwi,ramasurn,rovallam}@in.ibm.com
[2] IBM Research, New Delhi, India
kushmukh@in.ibm.com

Abstract. Multi-agent reinforcement learning (MARL) has very high sample complexity leading to slow learning. For repeated social dilemma games e.g. Public Goods Game(PGG), Fruit Gathering Game(FGG), MARL exhibits low sustainability of cooperation due to non-stationarity of the agents and the environment, and the large sample complexity. Motivated by the fact that humans learn not only through their own actions (organic learning) but also by following the actions of other humans (social learning) who also continuously learn about the environment, we address this challenge by augmenting RL based models with a notion of collaboration among agents. In particular, we propose Collaborative-Reinforcement-Learning (CRL), where agents collaborate by observing and following other agent's actions/decisions. The CRL model significantly influences the speed of individual learning, which effects the collective behavior as compared to RL only models and thereby effectively explaining the sustainability of cooperation in repeated PGG settings. We also extend the CRL model for PGGs over different generations where agents die, and new agents are born following a birth-death process. Also, extending the proposed CRL model, we propose Collaborative Deep RL Network(CDQN) for a team based game (FGG) and the experimental results confirm that agents following CDQN learns faster and collects more fruits.

Keywords: MARL · PGG · Collaborative

1 Introduction

Understanding the emergence of cooperation in conflicting scenarios such as social dilemmas has been an important topic of interest in the research community [4,15]. Several approaches such as MARL [20,22], simple RL [12], influences of social networks [11,19], enforcement of laws and rewards for altruism and punishment [2,10], emotions giving rise to direct reciprocity [21] or even indirect reciprocity [18] have been proposed to achieve stable cooperation among the agents in repeated multi-agent social dilemma settings. It is well-known that MARL based models have very slow learning capabilities due to large state space size [9] and non-stationary environment (composed

K. Karlapalem et al. (Eds.): PAKDD 2021, LNAI 12712, pp. 15–26, 2021.
https://doi.org/10.1007/978-3-030-75762-5_2

of agents that are themselves learning) [14]. Further MARL, has been ineffective for explaining the rapid emergence of cooperation in repeated social dilemma setting such as repeated PGGs. In this paper, we address the above research gap by augmenting the RL based models with a notion of *collaboration* among the agents, motivated by the fact that humans often learn not only through rewards of their own actions, but also by following the actions of other agents who also continuously learn about the environment. In particular, we propose a novel model, which we refer to as *Collaborative RL (CRL)*, wherein we define collaboration among the agents as the ability of agents to fully keep a track of other agent's actions(decisions) but autonomously decide whether to take an action based on the actions taken by the peers in the past. The efficacy of this approach is demonstrated on two examples; (1) A repeated public goods game (2) a 2D fruit gathering game. The repeated public goods game model consists of n agents and it runs though multiple rounds [20]. In each round an agent decides to make a contribution to a common pot. The total amount in this pot is multiplied by a factor β and this "public good" payoff is evenly divided among agents. The socially optimal strategy (that maximizes the sum of payoff for the agents) is to contribute towards the pot. However, the Nash equilibrium strategy is to not contribute. When agents are modelled using RL the free riding sort of behaviour among agents can be seen in the initial stages of the learning process and thereby making it very slow for each agent to learn that higher contributions lead to better long-term returns. Our proposed CRL framework enables the entire group of agents to learn quickly that it is their best interest to make very high contributions. In the second example, we extend our proposed notion of CRL to the setting of deep Q learning model referred to as *collaborative deep Q learning (CDQN)*. We consider matrix games form of social dilemmas (MGSD) (e.g. prisoner's dilemma and fruit gathering game). There exists work in the literature on learning in the Prisoners' Dilemma setting based on RL [7,20] and deep Q networks (DQN) have been investigated in fruit gathering games [16]. In the context of fruit gathering games, we consider multiple agents divided into two teams of equal size where each agent in one team learns about the environment through CDQN by observing actions of their peers in that team. The agents in the other team learn about the environment through the standard DQN model. We compare the number of fruits gathered by each of the team. We observe that our proposed CDQN model enables the team to gather more fruits than that of the other team based on DQN model.

The following are the key contributions of this paper: **(A)** A novel collaborative RL (CRL) model and Collaborative deep Q learning (CDQN) model (based on social learning paradigm) to enable faster learning by the agents in repeated social dilemma settings and thereby making them effective to explain the sustainability of cooperation in such settings. **(B)** Analysis of emergent behavior of such systems on two social dilemma games (repeated public goods game and fruit gathering game).

2 Mathematical Setup of CRL Model

In this section, we first present our proposed CRL model. As shown in Fig. 1, each agent follows a hierarchical RL model to take an action in each round of the game. The hierarchical RL model is divided into 2 phases. In Phase 1, agent i first finds out the best possible action he could play based on the Q-table learned from self taken actions. Then agent i also finds out the best possible action he could choose based on the observation matrix he obtained watching the actions of the peers. Finally in the Phase 2 he uses a stateless RL model with two actions, to decide whether to use the action based on Q-table or the action based on observing peers (matrix O). We describe the mathematical set up in detail for both Phase 1 and Phase 2.

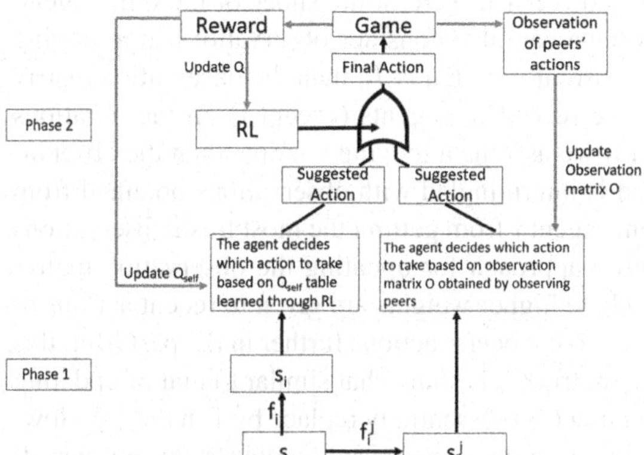

Fig. 1. CRL methodology followed by each agent

Phase 1: The MARL with n agents is typically defined by the tuple $(S, f_1, f_2, \ldots, f_n, A_1, A_2, \ldots, A_n, R_1, R_2, \ldots, R_n, P)$, where S denotes the state of the environment. Each agent i only gets a partial information about the full configurations of the environment S. Such a partial information is the perceived state of the environment for agent i. Let $f_i : S \to S_i$ be a surjective function that maps the state of the environment to a set of perceived states, call it S_i, for agent i. A_i is the set of actions for agent i, $R_i : S \times \prod_{i=1}^{n} A_i \times S \to R$ is the reward function for agent i, and P is probability distribution function. The notion of collaboration among agents come from the fact that each agent, i can track each of the other agent j's past actions a_j and states s_i^j in a observation matrix. This can be mathematically formulated by the function $f_i^j : S \to S_i$ where $f_i^j = f_i \circ SW$ and $SW : S \to S$ is defined as the state obtained after swapping the jth agent with i^{th} agent. The function f_i^j can be thought of as how the ith agent perceives the state if he were the jth agent. In Phase 1 of the Fig. 1, agent passes the perceived state $f_i(s)$ through a RL model which is being learned from the actions taken by agent i and the reward received for taking that action. The policy update rule for agent i based on the transition $\{(s_i, a, r_i, s_i')\}$ in Phase 1 using the standard Bellman equation which is given by $Q_i(s_i, a) = (1-\alpha)Q_i(s_i, a) + \alpha[r_i + \gamma \max_{a_i \in A_i} Q_i(s_i', a_i)]$. To choose an action from A_i, agent i employs a policy $\pi_i : S_i \to \Delta(A_i)$ using the well known Boltzman distribution as given in equation (2) of [20]. Here, $\Delta(\bullet)$ is a distribution on $\bullet$. Similarly, in Phase 1, agent i also observes his peers' actions, For agent i, the observation matrix that he maintains is denoted by O_i. Since the agents observe contribution levels of all other agents, accordingly each agent i increases the corresponding cells in O_i based on the actions taken by each of its peers. The update equation of O_i is $O_i(f_i^j(s), a_j) = O_i(f_i^j(s), a_j) + 1$ when agent i is swapped with agent j and a_j is the action taken by agent j. For coming up with the policy based on O_i again a Boltzman distribution is computed using the elements of O_i while being in a certain state. Finally the best action is chosen based on the Boltzman probability distribution. In other words $P(f_i(s), a_k) = \frac{e^{O_i(f_i(s), a_k)/t}}{\sum_{a_k \in A_i} e^{O_i(f_i(s), a_k)/t}}$. Also, each entry of O_i is initialized to 0 at the beginning and then it is updated based on the observations obtained. It is recursively set to 0 after some I iterations. This is done so that agents can start tracking the recent actions of the peers. For agent i, $O_i(s,a)$ (s = state, a = action) is a supervised model (frequency-based) of peer agents that predicts their actions given their states. $O_i(s,a)$ is the number of times peers have taken action a given they were in state s. Actions of peers are directly observed while the state of the peers are computed by SW function, that is, the state that agent i would see if it were in the shoes of the other agent. Hence, O_i depends on peer actions. One can also consider observations in a moving window of I iterations. That is if the current iteration is T_0 then the observation matrix O_i consists of the actions taken by the rest of the agents (except i), in the iterations $T_0 - I - 1, T_0 - I - 2, \ldots, T_0$. But if we assume a moving window then the observations obtained from the current round is intermingled with observations obtained from the past $I-1$ iterations which prevents agent i from getting the most fresh observations at the current round. Note that another approach for updating the observation matrix could be using a discounting factor where higher weights are given to recent actions of the peers while lower weights are given to the peers' actions further in the past. But, this approach of updating the observation matrix O_i is somewhat similar to that of updating the Q_i matrix where the reward amount in Q_i is essentially replace by 1 in the O_i. However, for this work we have considered O_i to be a supervised model (frequency-based) of peer agents that predicts their actions given their states.

Phase 2: In Phase 2, agent i makes a decision between two actions using a RL model whether to use the action coming out from self-learned RL model or action based on socially learnt model i.e. the action based on observation matrix. Note that the RL model defined in Phase 2 (i.e. Fig. 1) is stateless and there are 2 possible actions in $\tilde{A}_i = \{\tilde{a}_{i,1}, \tilde{a}_{i,2}\}$ of agent i, where, $\tilde{a}_{i,1}$ = use the action based on self-learned Q table and $\tilde{a}_{i,2}$ = use the action based on observation matrix O. Based on the action from $\tilde{A}_i$, the final action is played and once the reward r_i is realized the updates on Q values are made based on Bellman equation to both the RL models as shown in Fig. 1.

2.1 Extension of CRL to CDQN

We extend our proposed notion of CRL to the setting of deep Q learning model leading to *collaborative deep Q learning (CDQN)* model and present the formal model for the same. In CDQN, the steps followed by agent i are essentially same as that given in Fig. 1. However, the significant difference is that in the CDQN approach, the agents use a DQN and instead of a Q-table and a CNN instead of the peer based observation matrix. Consider the state of the entire environment to be $s \in S$. For agent i, the perceived state of the environment is given by $f_i(s)$. In phase 1, the agent passes the perceived state through a DQN. The outer-layer of the DQN has those many nodes corresponding to the final action set A_i for agent i, and each node produces Q-value for each respective action $a \in A_i$. The loss function used for training DQN is given by equation (2) in [17]. Then, agent i chooses a policy $\pi_i^1 : S_i \rightarrow \Delta(A_i)$ using an ϵ-greedy method where

$$\pi_i^1(s) = \begin{cases} argmax_{a_i \in A_i} Q_i^1(s_i, a_i), & \text{with probability } 1 - \epsilon_1 \\ \text{unformly an action from } A_i, & \text{with probability} \epsilon_1 \end{cases}$$

Agent i also passes the perceived state $f_i^j(S) = S_i^j$ through a CNN. The CNN is trained on the observations from peers. Analogous to before, once the Q-values are obtained in the outer layer for each respective action $a_i \in A_i$, a policy $\pi_i^2 : S_i \to \Delta(A_i)$ chooses an action using softmax methodology. Hence, agent i, gets an action based on the DQN. He also gets another action based on the CNN trained from the observations of the other agents' actions. Now, the algorithm moves to Phase 2 where the agent decides whether to use the action obtained from the DQN or to use the action obtained from the CNN. That means, In phase 2 each agent has an action set denoted by $\tilde{A}_i = \{\tilde{a}_{i,1}, \tilde{a}_{i,2}\}$ where, $\tilde{a}_{i,1}$ = use the action based on DQN, $\tilde{a}_{i,2}$ = use the action based on CNN. In this case we consider a stateless RL model with 2 possible actions as shown above. Using the traditional RL based learning method and to encourage exploration, agent i chooses the best policy $\tilde{\pi}_i : S_i \to \Delta(\tilde{A}_i)$ using an ϵ-greedy method where

$$\tilde{\pi}_i(.) = \begin{cases} argmax_{\tilde{a}_i \in \tilde{A}_i} \tilde{Q}_i(., \tilde{a}_i), & \text{with probability } 1 - \tilde{\epsilon} \\ \text{unformly an action from } \tilde{A}_i, & \text{with probability} \tilde{\epsilon} \end{cases}$$

If, $\tilde{a}_{i,1}$ is the action chosen, then agent i finally uses the action obtained through DQN. If, $\tilde{a}_{i,2}$ is the action chosen then agent i finally uses the action obtained through CNN. Once the action has been played and the reward r_i has been realized for the action played, agent i updates its policy exactly in the same way it is done in Phase 2 for CRL. Each agent updates the weights of its Q- network and weights of its trained CNN given a stored "batch" of experienced transitions based on the methodology suggested in [16] and [17]. Each agent follows the complete flow as shown in Fig. 1 to come up with the final action to be played in the game. In the long run of the game, it is expected that final actions being taken in the game from the set A_i are the favourable actions which can only benefit agent i and in turn, enhance the cooperation among agents. To ensure such favourable actions are being taken, each agent is allowed to make a choice whether to use the action based on DQN or the action based on CNN in Phase 2 of the process. Each agent chooses the appropriate action among the two possible actions based on whichever yielded them larger rewards.

Discussion: The proposed CRL and CDQN method hinges upon observing the peers' actions and then deciding whether to take an actions based on the observations gathered from the peers (using O_i) or take an action based on self-learned (using Q_i) actions. One of the main reason for storing O_i is to increase the number of samples that can help an agent i to decide on an action. So, the action set A_i is needed to be the same for every agent i. Also, for every agent i, when she is in a certain state s if the action taken is $a_i \in A_i$, the reward obtained r_i is needed to be the same for every agent. This leads to the main conclusion that CRL and CDQN model works only for symmetric pay-off games. We consider two settings: (i) one with repeated PGG games where we apply CRL model to validate if the agents participating in repeated PGG learn to cooperate faster as compared to only RL model. This setup was chosen because in this set up the Nash Equilibrium contribution of the agents are always zero while the social optimum is to contribute the full amount which also in turn yields a higher reward in the log run. (ii) We also consider a set up of team based fruit gathering game (FGG) where agents in teams collect fruits. As the number of possible states S for this environment is huge, for this particular problem we used CDQN to validate if a team using CDQN model against another team using DQN model ends up gathering more fruit or not.

3 Related Work

Our work is in the domain of *social learning* where every agent being in the same environment tries to follow or observe the actions of his neighbours who are also trying to understand the stochastic environment and the behaviour of the other agents. The entire paradigm of social learning is a well-researched topic. [5,6] proposed the paradigm of *observational learning* or social learning. In MARL, it is generally assumed that agents have incomplete but perfect information of other agents' immediate actions taken previously. Based on this information an agent can be modeled as a joint action learner (i.e. takes actions based on the combination of probable actions taken by all other agents) or it can also be an independent learner (i.e. takes actions based on his past actions and rewards received for such actions). In this paper, we model each agent as a joint action learner. But the learning style of each agent has two parts (as shown in Fig. 1). Each agent takes a probable action based on the reward received earlier for the actions taken. We term this learning as a self-learning method which is exactly the same as how an agent learn in MARL. But at the same time each agent takes a probable action based on his social learning that he obtained by keeping a track of the past actions of the other agents for different states in the past. This idea is different from *fictitious play* discussed in [8] where an agent keeps a track of only the past actions of the peers. In our modelling paradigm, for each agent there are two probable actions: one is based on self-learning and the other one is based on social learning. The agent chooses one of the two actions by running a table based RL technique. There are multiple papers [23,24] which are closely related to our proposed model, yet very different from our model. [8,13] consider general sum games and show that in a MARL set-up with a decaying learning rate, under certain conditions, the strategies of the independent learner agents converge to a Nash Equilibrium only if the Nash Equilibrium is unique for the game. On the contrary, [20], show that in an iterated prisoners' dilemma set up, with learning rate as constant, agents form a cycle of strategies and often take optimal strategies and sometimes sub-optimal strategies based on the wherever the strategies are in the cycle. They model each agent as joint action learners. In our paper, we have considered all the agents to be joint action learners and the learning rate of the agents to be constant. The main purpose of following such configurations is that the converged strategies of agents might be an optimal strategy or close to being an optimal strategy as compared to the Nash Equilibrium which is certainly the worst strategy for each agent in a game of PGG. In [1,3], the authors state a learning paradigm such as apprenticeship learning which is a task of learning from an expert (learning by watching, imitation learning). For this, Inverse RL algorithm is used that finds a policy that performs as good as the expert, which is learnt by the follower. Hence it is clear that in imitation learning one needs an expert from whom other agents can learn. The proposed model in this paper is based on social learning where each agent is following the same social learning paradigm while trying to figure out the best strategies to be taken in the stochastic environment. This learning paradigm may seem to be very similar to imitation learning but the basic difference in our model is there is no teacher or expert from which the other agents are socially learning.

4 CRL Model on Repeated PGGs

In this section, we first present our proposed CRL model for repeated PGGs. Consider n agents in a repeated PGG setting. In each round of the PGG, the reward of agent i is given by $r_i = 1 - e_i + \beta \sum_{j=1}^{n} e_j$ where contribution level for agent i is $e_i \in [0, 1]$. The range $[0, 1]$ of e_i is discretized into M possible different contribution levels, which are represented as $C = \{c_0, c_1, ..., c_{M-1}\}$. Hence, in the n player PGG game, the total possible configurations of the contributions from all agents is M^n. This number increases exponentially as n becomes large. For convenience of notations, we define the set of all possible configurations of the contribution levels from all agents as S; that is $S = \{(e_1, e_2, ..., e_n) \mid \forall e_i \in C, \ \forall i\}$ where $|S| = M^n$. Each agent i only gets a partial information about the full configurations of contribution. Such a partial information is the perceived state of the environment for agent i. As described in Sect. 2, $f_i : S \rightarrow S_i$ is a surjective function that maps the state of the environment to a set of perceived states, call it S_i, for agent i. For our analysis, we consider a loss less f_i that maps $s \in S$ to $s_i \in S_i = \{0, 1, \ldots, M^n - 1\}$. That is $s_i \in S_i$ can be thought of as an index of the state.

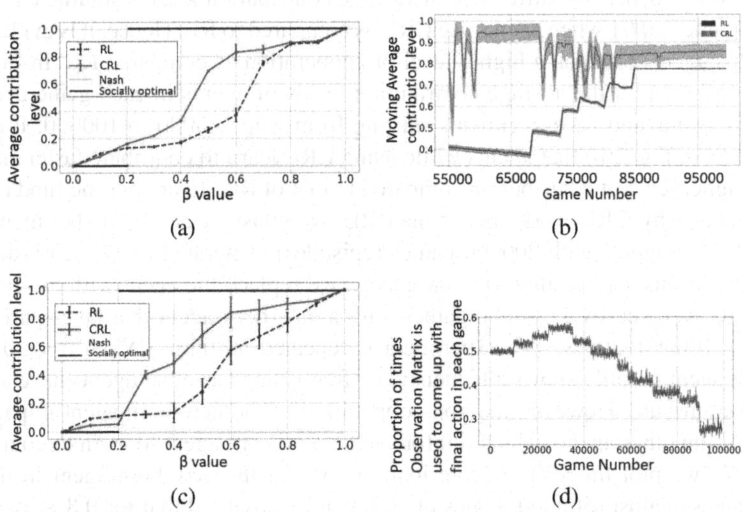

Fig. 2. (a) Average contribution level of agents for different values of β. (b) Average contribution level and standard deviation of contribution of agents for $\beta = 0.7$ from game number 55000. (c) Average contribution level of the new-born agent for different values of β (d) Proportion of times the new-born agent took the final action based on observation matrix in every game.

4.1 Experimental Results

For empirical analysis, we consider an iterated PGG with $n = 5$ agents. We conduct an experiment with 100000 iterated PGG games (episodes) where each game lasts for 30 rounds. Hence, after each 30 rounds, we restart the game but we retain the learned Q-tables. In other words, we make this an ongoing learning process for agents in every game, starting from game 1. The contribution levels of each agent is either 0 or 1. We also recursively initialize the observation matrix to 0 after every 50 games. Each experiment is repeated 50 times. We measure the sustainability of cooperation by calculating average contribution level in the last 30000 games averaged over all 5 agents and also over all 50 experiments. Mathematical representation of the measure is given by $\frac{1}{30000}\sum_{T=70001}^{100000}\frac{1}{5}\sum_{i=1}^{5}\frac{1}{30}\sum_{r=1}^{30}e_i^{r,T}$. where $e_i^{r,T}$ is the contribution level of the ith agent in the rth round of game T averaged over 50 games. In Fig. 2(a), we plot average contribution level for different values of β obtained following CRL and traditional RL. The Nash Equilibrium is 0 as expected. It is evident that, in the region of social dilemma where $\frac{1}{n} = 0.2 \leq \beta < 1$, the collective reward of the entire group of agents that is $\sum_{j=1}^{5} r_j$ is maximized when the contribution is 1 from each agent. Hence, the socially optimal contribution level is 1 when $\beta \geq 0.2$. On the contrary, when $\beta < 0.2$, $\sum_{j=1}^{5} r_j$ is maximized when the contribution is 0 from each agent. Hence, the socially optimal solution is 0 when $\beta < 0.2$. When $\beta < 0.2$, we observe that average contribution level of agents obtained using CRL is smaller than that of RL and average contribution levels are close to 0 as expected. But, when $0.2 \leq \beta \leq 0.8$ agents while using CRL method contributes more than that obtained using RL method. For, each of $\beta = [0.4, 0.5, 0.6, 0.7]$ the difference in average contribution level is significantly more with a P-value < 0.01 when agents use CRL as compared to RL. Hence, it is evident that CRL helps agents to reach a higher level of cooperation as compared RL. In Fig. 2(b) for $\beta = 0.7$ we plot the average contribution levels of agents in each game averaged over all 5 agents and 50 experiments starting from game 55000 to 100000. It can be observed from Fig. 2(b) that agents while using CRL learn to contribute faster and also attain a higher level of contribution compared to that of RL. To answer the fundamental question that why CRL works better than RL, we consider another experimental set up with $n = 5$ agents with 200000 games (episodes) of iterated PGG each lasting for 30 rounds. In this set up, after 100000 games, we replace the agent with lowest average reward averaged over 100000 games with a new-born agent that takes part in the remaining 100000 games. The experiment is repeated 50 times. When $\beta \geq 0.2$, the new-born agent initially shows a tendency of "free-riding" as other agents have already learnt to contribute. However, the non-cooperation of the new-born agent impacts the policies of the other agents which reinforces the new-born agent to learn to contribute. In Fig. 2(c) we plot the average contribution level of the new born agent in the last 30000 games against different values of β. It can be observed that for $0.3 \leq \beta \leq 0.8$, the new-born agent's average contribution level is significantly more (P-value < 0.01) when CRL method is used as compared to RL. Figure 2(d) shows the fraction of times the new born agent uses the observation matrix (O) based action for latter 100000 games (with $\beta = 0.7$). We observed that, initially the new born agent learns to rely on social/observation based learning as the RL model take longer time to train. However, as the game progresses, the agent's RL model improves, inducing it to slowly switch to the RL model for a greater fraction of time. The advantage of following what other agents (who are experts in the system after learning for first 100000 games) are doing helps in reduction of the exploration phase for the new-born agent and thereby helping it to accelerate its learning rate.

4.2 CDQN Model on Fruit Gathering Game (FGG)

Some of the recent research work has focused on the topic of successfully learning control policies directly from high-dimensional sensory input using reinforcement learning, such as playing Atari games [17], in fruit gathering game [16]. In our paper, we consider fruit gathering game with a different variation as compared to [16] by formulating it in a team game setup, wherein the agents are grouped into two teams and play the game to gather fruits from an environment. In our game set up, agents in each team can throw beams towards the agents in the opponent team similar to that in [16]. In a time step, if an agent is hit by more than 50% of the beams from agents of the other team, that agent remains static for a pre-specified number of time steps T_{tagged} and hence cannot collect fruits. For collecting a fruit an agent gets 10 units of reward.

Learning and Simulation Method: To train the agents we use CDQN model as described above. At the beginning of each round in the game, each agent chooses an action $\tilde{a}_i \in \tilde{A}_i$ to decide whether his final action will be based on the self-learned-action-based DQN or his final action will be based on the CNN learned from his peers actions (agents from his own team only). The game environment is a 31×18 grid of which a fraction is filled with fruits initially. The observation field ("patch") of agent i is a 17×17 grid at the centre of which the agent is present. In every step agent i has 8 possible actions from the set A_i where $A_i = \{$move right, move up, move left, move down, turn left, turn right, throw a beam, do nothing$\}$. DQN has two hidden layers of 32 units interleaved with Rectified Linear Layer and 8 units on the outer layer corresponding to each action. CNN has the same architecture as DQN. Each agent in its memory stores self state s_i (image), peer states s_i^j (image), Reward r_i, self action a_i, peer action a_j. For training DQN agent i uses samples s_i, a_i, r_i and for CNN, samples s_i^j, a_j (1-hot/softmax) are used. There are 10 episodes in each game and each episode lasts for 200 time-steps. Hence after 2000 time-steps the game is reset to the beginning. During training, players use ϵ-greedy policies, with ϵ decaying linearly over time from 0.8 to 0.08. Per time-step, the discount factor is kept at 0.99. A picture of the game environment is shown in Fig. 3. We denote the agents from one team with the colour Green while the agents from the other team are represented by the colour Blue. Note that the colour of the fruits is Red. On the left of the Fig. 3, a snapshot of episode 1 of the game environment is shown where agents from Team 1 (coloured green) and agents from Team 2 (coloured blue) collect fruits. In the initial phase of the game, since the environment is still filled with a lot of fruits, agents learn to collect fruits. On the contrary, a similar snapshot is taken at a later time point when the episode is 9 and the number of fruits in the environment is scarce. In such a scenario, as it can be seen, agents from Team 1 are hitting an agent from Team 2 with beams while the blue agent is one step away from capturing a fruit from North while agents from Team 1 are far away from any fruits.

Fig. 3. Left: A snapshot of episode 1 of game environment Right: A snapshot of episode 9 of game environment (Color figure online)

4.3 Experimental Results

We consider a 2-team game with 2 agents in each team. In one team the two agents collaborate by observing each others action by following the methodology as described in CDQN model. We call this team as *Collaborating team*. In the other team, the 2 agents, do not collaborate with each other and they simply learn through their own actions (i.e. traditional DQN model). We call this team as *Non-collaborating team*. Agents play 500 games with this setup and each game lasts for 2000 time steps. After 2000 time steps, we start a new game but allow the agents to retain their learned network from the past game. This entire experiment is repeated 50 times. Each game begins with an environment filled with randomly placed fruits covering 90% of the environment. Note that any fruit that is gathered does not re-spawn again in the game. At the beginning of each game, agents are placed randomly in the available spaces. For each of the games, we set $T_{tagged} = 20$. In Fig. 4(a) we show the proportion of fruits gathered by the collaborating and non collaborating teams in each of the 500 games averaged over 50 experiments. As it can be seen from this figure, both the teams collected similar proportion of fruits in a few initial games. However, the collaborating team soon starts gathering more fruit. Agents in the collaborating team observe the actions taken by the peers which enable them to learn about the environment faster by leveraging both their own experiences (using DQN) and modelling the behavior of the team mate by using CNN (social/observation learning). This way, CDQN also reduces sample complexity. Figure 4(b) shows the fraction of times (averaged over 50 experiments) the final actions of the agents in collaborating team are based on the observation gathered from the action of the peer. We plot this for every game instance. As it can be seen from this figure, in a few initial games, the action based on the observation from the peer is taken more frequently while in the later games this frequency decreases. Initially collaborating agents learn to rely on social/observation based learning as the DQN models take longer time to train. However, as the game progresses, the agents' self-learned DQN models improve, inducing it to slowly switch to the DQN models for a greater fraction of time.

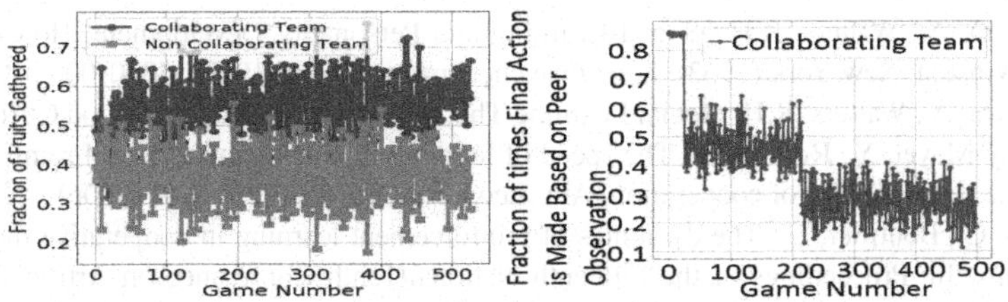

(a) (b)

Fig. 4. (a) Fraction of fruits gathered by Collaborating Team and Non-Collaborating Team in each game (b) Fraction of time agents in Collaborating Team played the CNN based action (trained based on peer's actions)

5 Conclusion

Proposed CRL method was shown to increase both the learning rate of the agents and the sustained levels of cooperation as compared to traditional RL. RL methods are expensive for agents, in terms of the large amounts of samples required to explore the state-action space and the loss of reward during the exploration stage. CRL is inspired from social learning theory, where agents leverage observations of others' actions (and indirectly, their polices) as an inexpensive way to learn their policies thereby overcoming the drawbacks of RL. Children pay attention to people around them and encode their behavior. At a later time they may mimic the behavior they have observed. As they become adults, they rely more on their own experiences and reinforcements (positive or negative) to take their actions. In the same way, we have observed that the agents that use CRL, initially choose actions obtained from the observational learning more often. Subsequently, as the RL/DQN models improve, they slowly reduce the frequency of choosing actions from observational learning. To validate the proposed methodology we compared experiments on agents playing a game of iterated PGG and play a game of FGG. We have shown that when agents collaborate with each other, they increase the cooperation among themselves. We also show that when agents collaborate with each other their cooperation level goes up faster due to reduction in the exploration phase of learning.

References

1. Abbeel, P., Ng, A.Y.: Apprenticeship learning via inverse reinforcement learning. In: Proceedings of the Twenty-First International Conference on Machine Learning, ICML 2004, p. 1. ACM, New York (2004)
2. Andreoni, J., Harbaugh, W., Vesterlund, L.: The carrot or the stick: rewards, punishments, and cooperation. Am. Econ. Rev. **93**(3), 893–902 (2003)
3. Atkeson, C.G., Schaal, S.: Robot learning from demonstration. In: Proceedings of the Fourteenth International Conference on Machine Learning, ICML 1997, pp. 12–20. Morgan Kaufmann Publishers Inc., San Francisco (1997)
4. Axelrod, R., Hamilton, W.: The evolution of cooperation. Biosystems **211**(1–2), 1390–1396 (1996)

5. Bandura, A., Walters, R.H.: Social Learning and Personality Development. Holt Rinehart and Winston, New York (1963). https://psycnet.apa.org/record/1963-35030-000
6. Bandura, A., Walters, R.H.: Social Learning Theory. Prentice-Hall, Englewood Cliffs (1977)
7. Bereby-Meyer, Y., Roth, A.E.: The speed of learning in noisy games: partial reinforcement and the sustainability of cooperation. Am. Econ. Rev. **96**(4), 1029–1042 (2006)
8. Claus, C., Boutilier, C.: The dynamics of reinforcement learning in cooperative multiagent systems. In: Proceedings of the Fifteenth National/Tenth Conference on Artificial Intelligence/Innovative Applications of Artificial Intelligence, pp. 746–752 (1998)
9. Engelmore, R.: Prisoner's dilemma-recollections and observations. In: Rapoport, A. (ed.) Game Theory as a Theory of a Conflict Resolution, pp. 17–34. Springer, Dordrecht (1978). https://doi.org/10.1007/978-94-010-2161-6_2
10. Fehr, E., Gachter, S.: Cooperation and punishment in public goods experiments. Am. Econ. Rev. **90**(4), 980–994 (2000)
11. Fu, F., Hauert, C., Nowa, M.A., Wang, L.: Reputation-based partner choice promotes cooperation in social networks. Phys. Rev. E **78**, 026117 (2008)
12. Gunnthorsdottir, A., Rapoport, A.: Embedding social dilemmas in intergroup competition reduces free-riding. Organ. Beha. Hum. Decis. Processes **101**(2), 184–199 (2006)
13. Hu, J., Wellman, M.P.: Multiagent reinforcement learning: theoretical framework and an algorithm. In: Proceedings of the Fifteenth International Conference on Machine Learning, ICML 1998, pp. 242–250. Morgan Kaufmann Publishers Inc., San Francisco (1998)
14. Lange, P.A.V., Joireman, J., Parks, C.D., Dijk, E.V.: The psychology of social dilemmas: a review. Organ. Behav. Hum. Decis. Processes **120**(2), 125–141 (2013)
15. Ledyard, J.: A survey of experimental research. In: Kagel, J.H., Roth, A.E. (eds.) The Handbook of Experimental Economics. Princeton University Press, Princeton (1995)
16. Leibo, J.Z., Zambaldi, V., Lanctot, M., Marecki, J., Graepel, T.: Multi-agent reinforcement learning in sequential social dilemmas. In: Proceedings of the 16th Conference on Autonomous Agents and Multiagent Systems, pp. 464–473 (2017)
17. Mnih, V., et al.: Playing Atari with deep reinforcement learning. In: NIPS Deep Learning Workshop 2013 (2013)
18. Nowak, M.A., Signmund, K.: Evolution of indirect reciprocity. In: Proceedings of the National Academy of Sciences, pp. 1291–1298 (2005)
19. Rand, D.G., Arbesman, S., Christakis, N.A.: Dynamic social networks promote cooperation in experiments with humans. In: Proceedings of the National Academy of Sciences, pp. 19193–19198 (2011)
20. Sandholm, T.W., Crites, R.H.: Multiagent reinforcement learning in the iterated prisoner's dilemma. Biosystems **37**(1–2), 147–166 (1996)
21. van Veelen, M., Garcia, J., Rand, D.G., Nowak, M.A.: Direct reciprocity in structured populations. Proc. Natl. Acad. Sci. **109**, 9929–9934 (2012)
22. Wunder, M., Littman, M., Babes, M.: Classes of multiagent q-learning dynamics with greedy exploration. In: Proceedings of the 27th International Conference on Machine Learning, ICML 2010 (2010)
23. Yang, Y., Luo, R., Li, M., Zhou, M., Zhang, W., Wang, J.: Mean field multi-agent reinforcement learning. In: Proceedings of the 35th International Conference on Machine Learning, pp. 5571–5580 (2018)
24. Zhou, L., Yang, P., Chen, C., Gao, Y.: Multiagent reinforcement learning with sparse interactions by negotiation and knowledge transfer. IEEE Trans. Cybern. **47**(5), 1238–1250 (2017)

SigTran: Signature Vectors for Detecting Illicit Activities in Blockchain Transaction Networks

Farimah Poursafaei[1,3](✉), Reihaneh Rabbany[2,3], and Zeljko Zilic[1]

[1] ECE, McGill University, Montreal, Canada
zeljko.zilic@mcgill.ca
[2] CSC, McGill University, Montreal, Canada
rrabba@cs.mcgill.ca
[3] Mila, Montreal, Canada
farimah.poursafaei@mila.quebec

Abstract. Cryptocurrency networks have evolved into multi-billion-dollar havens for a variety of disputable financial activities, including phishing, ponzi schemes, money-laundering, and ransomware. In this paper, we propose an *efficient* graph-based method, SigTran, for detecting illicit nodes on blockchain networks. SigTran first generates a graph based on the transaction records from blockchain. It then represents the nodes based on their structural and transactional characteristics. These node representations *accurately* differentiate nodes involved in illicit activities. SigTran is *generic* and can be applied to records extracted from different networks. SigTran achieves an F_1 score of 0.92 on Bitcoin and 0.94 on Ethereum, which outperforms the state-of-the-art performance on these benchmarks obtained by much more complex, platform-dependent models.

1 Introduction

Blockchain-based cryptocurrencies, such as Bitcoin and Ethereum, have taken a considerable share of financial market [11]. Malicious users increasingly exploit these platforms to undermine legal control or conduct illicit activities [4,5,11,17]. In particular, billions of dollars attained through illegal activities such as drug smuggling and human trafficking are laundered smoothly through blockchain-based cryptocurrencies exploiting their pseudonymity [21]. Given the openness and transparency of blockchain [19], it is of paramount importance to mine this data for detecting such illicit activities.

Although the recent machine learning advances have enhanced the exploration of large-scale complex networks, the performance of these methods relies on the quality of the data representations and extracted features [14]. Likewise, in blockchain networks with high anonymity and large number of participants with diverse transactional patterns, any illicit node detection method is effective only when the extracted characteristics of the data efficiently distinguish the illicit and licit components of the network. Hence, developing effective illicit node detection method depends heavily on the efficiency of the data representations and extracted features. Considering that network topologies can reflect the roles of the different nodes, graph representation learning methods have been potentially conceived as great means for capturing neighborhood similarities and community detection [22]. Additionally, machine learning analysis on

K. Karlapalem et al. (Eds.): PAKDD 2021, LNAI 12712, pp. 27–39, 2021.
https://doi.org/10.1007/978-3-030-75762-5_3

large networks is becoming viable due to the efficiency, scalability and ease of use of graph representation learning methods [15,22].

Driven by the need to enhance the security of the blockchain through transaction network analysis, and by recent advances in graph representation learning, we propose an efficient graph-based method SIGTRAN for detecting illicit nodes in the transaction network of blockchain-based cryptocurrencies. SIGTRAN can be applied for warning honest parties against transferring assets to illicit nodes. In addition to providing a trading ledger for cryptocurrencies, blockchain can be perceived as a network that analyzing its dynamic properties enhances our understanding of the interactions within the network [9,26]. SIGTRAN first constructs a graph based on the extracted transactions from the blockchain ledger considering integral characteristics of the blockchain network. Then it extracts structural, transactional and higher-order features for graph nodes; these features strengthen the ability to classify the illicit nodes. SIGTRAN shows superior performance compared to the previous platform-dependant state-of-the-arts (SOTAs), while it is generic and applicable to different blockchain network models. Particularly, SIGTRAN achieves an F_1 score of 0.92 and 0.94 for detecting illicit nodes in Bitcoin and Ethereum network, respectively. Moreover, SIGTRAN is scalable and simpler compared to much more complex SOTA contenders. In short, SIGTRAN is:

- **Generic:** SIGTRAN is platform independent and applicable to different blockchain networks, unlike current contenders.
- **Accurate:** SIGTRAN outperforms much more complex SOTA methods on the platforms they are designed for.
- **Reproducible:** we use publicly available datasets, and the code for our method and scripts to reproduce the results is available at: https://github.com/fpour/SigTran.

2 Background and Related Work

The increasingly huge amount of data being appended to the blockchain ledger makes the manual analysis of the transactions impossible. Multiple works proposed different machine learning techniques for detection of the illicit entities on cryptocurrency networks [12,13,16,21,28,31]. Specifically, [12,16,28] investigate supervised detection methods for de-anonymizing and classifying illegitimate activities on the Bitcoin network. In parallel, Farrugia et al. [13] focus on examining the transaction histories on the Ethereum aiming to detect illicit accounts through a supervised classification approach.

Lorenz et al. [21] address the problem of tracking money laundering activities on the Bitcoin network. Concentrating on the scarcity of the labeled data on a crypocurrency network such as Bitcoin, Lorenz et al. [21] argue against the unsupervised anomaly detection methods for the detection of illicit patterns on the Bitcoin transaction network. Instead, they propose an active learning solution as effective as its supervised counterpart using a relatively small labeled dataset.

Previous work often focuses on analyzing the transactions on the cryptocurrency networks with *platform-dependent* features, e.g. works in [13,21,25,27]. There are also several *task-dependent* studies investigating a specific type of illicit activity, for example [6,10] focus on ponzi schemes, which are illicit investments. Here, we propose a method that is *independent of the platform and/or task* and looks at basic structural

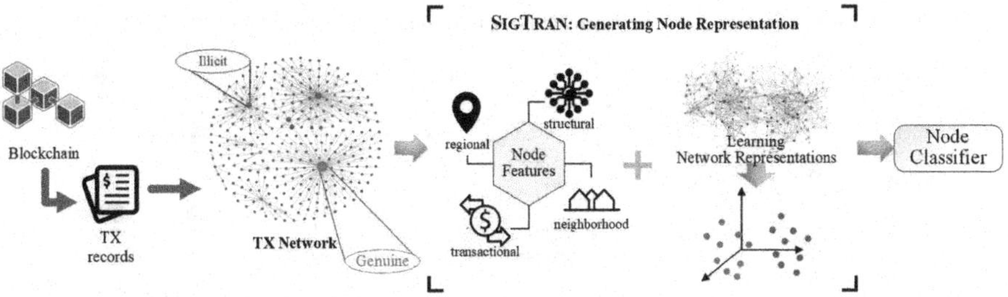

Fig. 1. Overview of SIGTRAN to detect illicit nodes on a blockchain network.

features of the underlying transaction networks. This is inline with more recent graph based methods which we summarize below.

Graph-based anomaly detection approaches have emerged recently as a promising solution for analyzing the growing data of the blockchain ledger, for both Bitcoin [18], and Ethereum [20,29–31]. In particular, Wu et al. [30] model the transaction network of Ethereum as a multi-edge directed graph where edges represent transactions among Ethereum public addresses, and are assigned weights based on the amount of transactions and also a timestamp. They propose *trans2vec* that is based on adopting a random walk-based graph representation embedding method and is specifically designed for the Ethereum network. Weber et al. [28] model the Bitcoin transaction network as a directed acyclic graph where the nodes represent the transactions and the edges illustrate the flow of cryptocurrency among them. They construct a set of features for the transactions of the network based on the publicly available information. Then, they apply Graph Convolutional Networks to detect illicit nodes. For node features, they consider local information (e.g. transactions fee, average Bitcoin sent/received), and other aggregated features representing characteristics of the neighborhood of the nodes. In the experiment section we show that SIGTRAN is more accurate than these two platform-dependent SOTAs.

More generally, graph representation learning has became commonplace in network analysis with superior performance in a wide-range of real-world tasks. As a pioneering work, *node2vec* [15] explores the neighborhood of each node through truncated random walks which are used for achieving representations that preserve the local neighborhood of the nodes. *RiWalk* [22] mainly focuses on learning the structural node representations through coupling a role identification procedure and a network embedding step. For a more detailed survey of network embedding methods, see [14].

3 Proposed Method

Given the publicity of the transaction ledger, our aim is to detect illicit activities on a blockchain-based cryptocurrency network. We formulate the problem as a node classification task in the transaction graph. Specifically, given the transaction records of a set of blockchain nodes, we devise the transaction graph and investigate the authenticity of different nodes by predicting the probability of each being involved in an illegitimate activities such as phishing, scam, malware, etc. We propose an efficient feature vector generation approach for nodes in these networks which demonstrates node activity

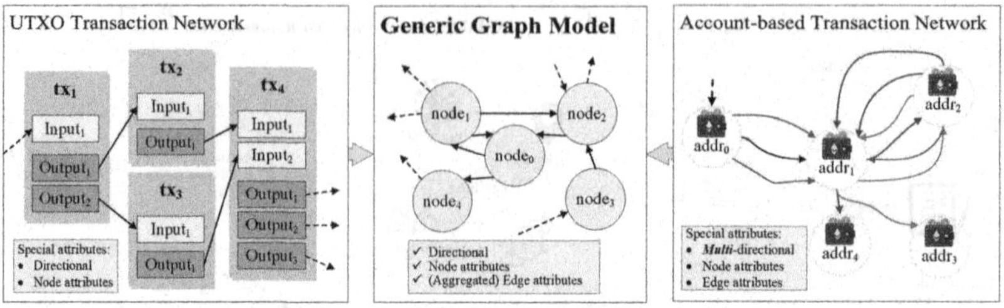

Fig. 2. SIGTRAN creates a generic graph model based on the transaction networks.

signatures which can be used to distinguish illicit nodes. An overview of SIGTRAN framework is illustrated in Fig. 1. SIGTRAN extracts the transaction (*TX*) history from the blockchain ledger and constructs a transaction network from those records. To generate node representations, it then extracts a set of useful features which are fused with the corresponding node representations produced by a node embedding method. The final representations are then classified to detect illicit nodes. These steps are explained in detail in the following.

3.1 Transactions History Retrieval

The required transaction records can be obtained directly from the blockchain public ledger of the target cryptocurrency. For instance, for Bitcoin or Ethereum, we can use the client software of these peer-to-peer networks to pull down the blockchain data in binary format which is converted to human-readable formats like CSV via an appropriate parser. As an example, for converting the binary records of the Bitcoin and Ethereum ledger, *SoChain* [1] and [3] can be employed respectively. The transaction records contain details such as timestamp, amount sent or received, incoming and outgoing addresses, and other related information. Different authoritative websites (such as *EtherScamDB* [2] for Ethereum network) helps in gathering a list of illicit nodes on the blockchain network. Together transaction records and the information about the authenticity of the network nodes constitute the dataset required.

3.2 Network Construction

A cryptocurrency transactions network is modeled as a graph demonstrating the interactions among participants of the network. We model a blockhchain network as a graph $G = (V, E)$, where V represents the set of nodes and E expresses the set of edges. Nodes and edges could have extra attributes, such as labels for nodes, and amount and timestamp of transaction for edges. Essentially, blockchain networks can be classified into two categories: (a) unspent transaction output *(UTXO)* model where the nodes specify the transactions, and the edges denote the flow of the cryptocurrency among nodes. Bitcoin, Dash, Z-Cash, and Litecoin are cyrptocurrencies based on the UTXO model [26], and (b) *account-based* model, where the account addresses are considered as the nodes and the transactions among addresses as the edges of the graph. Ethereum network is based on the account-based model. Considering the different categories of blockchain networks, we construct a generic graph model, as illustrated in Fig. 2, to which the instances of both the UTXO as well as the account-based network models are easily convertible. In the generic graph, the nodes specify the network entities in which we are interested to investigate their authenticity, while the edges denote the interactions among the nodes. The generated graph model entails any features associated with the nodes, whereas multiple edges between any two nodes with the same direction are aggregated into a single edge. It is noteworthy that based on the underlying blockchain network (i.e. UTXO or account-based), nodes and edges of the generic graph can have different intuitions. Particularly, if the graph is constructed based on an UTXO blockchain, the nodes represent cryprocurrency transactions which may belong to licit or illicit categories of real entities. However, if the graph is constructed based on an account-based blockchain, each node represents either an illicit or licit address. In both cases, node representations and classification are applied incognizant of the underlying blockchain model.

3.3 SIGTRAN

After modeling the blockchain transactions network as a graph, we need to develop proper representations for the nodes. This consists of a set of carefully crafted features which are fused with learned node representations, explained below respectively.

SIGTRAN-Feature Extraction. For each node u, we gain a diverse set of features consisting of four main categories as follows. It is important to note that the features of the nodes (e.g., labels) and edges (e.g., amount and timestamp) of the original network are preserved in the constructed generic model, since we employ these attributes for extracting the features of the nodes.

- *Structural features* consist of in-degree ($D_{in}(u) = \sum_{v \in N_u} |e_{vu}|$), out-degree ($D_{out}(u) = \sum_{v \in N_u} |e_{uv}|$), and total degree ($D_{tot}(u) = D_{in}(u) + D_{out}(u)$) of node u. As there may exist multiple edges between two nodes, $|e_{vu}|$ determines the number of edges from v to u, and N_u consists of all first-order neighbors of node u.
- *Transactional features* investigate the characteristics related to the amount and time interval of the transactions. Indeed, blockchain specific information of the transaction network is mainly enriched in this set of features. Each edge e_{uv} from u to v is associated with a set of attributes including the amount and time interval of the transactions from node u to node v. For obtaining transactional features, we consider a set of aggregation functions, G, which includes *summation*, *maximum*, *minimum*, *average*, *standard deviation*, and *entropy* operations over an arbitrary given distribution x as follows:

$$G = \{\sum(x), \max(x), \min(x), \overline{x}, \sigma(x), H(x)\} \tag{1}$$

With the set of aggregation functions G, transactional features of node u are defined as:

$$tx_u^{amnt} = \{g(e_u^a) \mid g \in G, e_u^a \subseteq \{e_{uv}^a, e_{vu}^a\}\}, tx_u^{freq} = \{g(e_u^\tau) \mid g \in G, e_u^\tau \subseteq \{e_{uv}^\tau, e_{vu}^\tau\}\}$$

where e_u^a denotes the amount related to (in/out) edges of node u. Similarly, e_u^τ denotes the time interval related to (in/out) edges of node u.

- *Regional features* are defined with regard to the ego network of a node. We consider the *egonet* of node u ($S_u = (V_u, E_u)$) as a subgraph of the original graph consisting of u and its first-order neighbors (i.e. N_u), with all the edges amongst these nodes. As an example, considering the generic graph model in Fig. 2, the egonet of $node_0$ consists of $\{node_1, node_2, node_4\}$. Having the definition of the egonet in mind, we consider the number of edges of S_u as one of the regional features of node u. Besides, the in-degree, out-degree, and total degree of S_u are considered as the other regional features according to $D_{in}(S_u) = |\{e_{wv} \in E \mid w \notin V_u, v \in V_u\}|$, $D_{out}(S_u) = |\{e_{wv} \in E \mid w \in V_u, v \notin V_u\}|$, and $D_{tot}(S_u) = D_{in}(S_u)+D_{out}(S_u)$, where $V_u = u \cup N_u$.
- *Neighborhood features* analyze the aggregated characteristics of neighbors of node u. Considering the aggregation functions in (1), the neighborhood features of node u are defined as: $D_{in}(N_u) = \{g(D_{in}(v)) \mid g \in G, v \in N_u\}$, $D_{out}(N_u) = \{g(D_{out}(v)) \mid g \in G, v \in N_u\}$, and $D_{tot}(N_u) = \{g(D_{tot}(v)) \mid g \in G, v \in N_u\}$.

Network Representation Learning. In order to learn node representations which fuse topological perspectives of the nodes in a cryptocurrency transaction network, SIGTRAN combines the extracted features explained in above (which are obtained focusing on the specific characteristics of the cryptocurrency networks such as amount and time interval of transactions) with the node representations that are learned automatically through a network embedding procedure. For retrieving more efficient node representations, we exploit a common network embedding method for learning the features of the nodes in the generic graph model. Then, we fuse the extracted features with the node embeddings in an effective manner so that the ultimate node representations effectively demonstrate the fundamental characteristics of the nodes. For fusing the extracted features and the node embeddigns, we investigate two approaches explained in the following subsections.

RiWalk-Enhanced. In this approach, we focus on the fact that nodes with different functionalities have different roles in a network, and the structure of the network can be investigated for gleaning these roles [22]. Hence, we consider the SIGTRAN-features as powerful indicators of similarity among nodes, and decouple the node embedding procedure into two steps. First, we identify the top ten SIGTRAN-features with the highest importance in detecting the illicit nodes and retrieve the values of those features for each node u as $\mathbf{f}_u^*$. We then relabel each neighbor of node u such as v according to the function $\phi(v) = h(\mathbf{f}_\mathbf{u}^*) \oplus h(\mathbf{f}_\mathbf{v}^*) \oplus d_{uv}$. Here, d_{uv} denotes the shortest path length from u to v, $\oplus$ is the concatenation operation, and $h(x)$ is defined as $h(x) = \lfloor log_2(x+1) \rfloor$. The new labels which are generated based on the node features roughly indicates the role of the nodes (thus, *Ri: Role identification*). Thereafter, the second step consists of a random-walk-based network embedding method for learning the node representations. Specifically, we generate several random walks starting from each node, then merge the random walks to construct a corpus and adopt the Skip-Gram model with negative sampling of *word2vec* [24] to learn the node representations.

SIGTRAN. In this approach, we consider the fusion of the SIGTRAN-features and automatically generated node embeddings through a concatenation procedure. Particularly, we apply a random-walk-based node embedding method such as node2vec [15] and for each node u obtain its embedding as $\mathbf{e}_\mathbf{u}^*$. Then, we generate the final representations by concatenating the SIGTRAN-features $\mathbf{f}_\mathbf{u}^*$ with the node embeddings for each node (i.e., $\mathbf{e}_\mathbf{u}^* \oplus \mathbf{f}_\mathbf{u}^*$) intending to achieve accurate node representations.

3.4 Node Classification

The generated node representations can then be used in the downstream task for classification of the illicit and genuine nodes. The illicit node detection task is akin to the common task of fraud detection and anti-money laundering applications. We simply employ *Logistic Regression* for the classification task because of its widespread adoption in similar tasks as well as its high interpretability [7,8,25,28]. This simple choice enables us to better compare the effect of different embedding techniques.

Table 1. Statistics of the investigated Blockchain-based Cryptocurrency Networks.

Dataset	Nodes	Edges	Illicit nodes
Bitcoin	203,769	234,355	4,545
Ethereum	2,973,489	13,551,303	1,165

4 Experiments

This section evaluates SIGTRAN experimentally. We present the datasets in Sect. 4.1, baseline methods in Sect. 4.2, and discuss the results in Sect. 4.3.

4.1 Dataset Description

We investigated two real-world transaction datasets consisting of the most widely adopted cryptocurrencies: (a) Bitcoin blockchain network which is the largest cryptocurrency system based on UTXO model, and (b) Ethereum that support smart contracts, holds the second largest cryptocurrency, and provides an account-based model.

We employed Bitcoin transactions dataset shared by Weber et al. [28] in which 21% of the transactions are labeled as *licit* (corresponding to different legitimate categories such as exchanges, miners, wallet provider, etc.), 2% as *illicit* (corresponding to different categories of illegitimate activities such as scams, malware, ponzi scheme, ransomeware, etc.), and there are no labels for the rest of the transactions. In addition to the transaction records, the Bitcoin dataset consists of a set of handcrafted features representing the characteristics of the considered transactions. Since the dataset is fully anonymized, we could only generate *structural*, *regional*, and *neighborhood* features for the nodes of the Bitcoin graph. We combined SIGTRAN-features with the initial attributes available in the dataset to form the node features of the Bitcoin network. In addition, we investigated the Ethereum transactions data shared by Wu et al. [30]. This dataset consists of Ethereum transaction records for a set of addresses consisting of licit addresses as well as illicit addresses reported to be involved in phishing and scam activities. The statistical details of the Bitcoin and Ethereum dataset are shown in Table 1.

4.2 Baseline Methods

Several SOTA methods were evaluated and compared.

- *node2vec* [15] is a random-walk-based node representation method which employs biased random walks to explore the neighborhood of the nodes with the consideration of local and global network similarities. Default parameters of the node2vec are set in line with the typical values mentioned in the paper [15]: context size $k = 10$, embedding size $d = 64$, walk length $l = 5$, and number of walk per node $r = 20$. We have also considered setting $p = 0.25$ and $q = 4$ to better exploit the structural equivalency of the nodes according to the discussion in the paper [15].
- *RiWalk* [22] is another random-walk-based node embedding methods which focuses on learning structural node representations through decoupling the role identification and the network embedding procedures [22]. We considered the *RiWalk-WL* which aims to imitate the neighborhood aggregation notion of the Weisfeiler-Lehman graph kernels, and captures fine-grained connection similarity patterns.

Table 2. Bitcoin performance: SIGTRAN significantly outperforms baselines on the Bitcoin dataset. The last three rows are introduced in this paper.

	Training Set Performance					*Test Set Performance*				
Algorithm	Precision	Recall	F_1	Accuracy	AUC	Precision	Recall	F_1	Accuracy	AUC
Weber et al. [28]	0.910	0.933	0.921	0.921	0.980	0.901	0.929	0.915	0.913	0.976
node2vec	0.626	0.300	0.405	0.560	0.580	0.627	0.312	0.415	0.563	0.580
RiWalk	0.556	0.348	0.426	0.535	0.556	0.549	0.343	0.421	0.530	0.547
RiWalk-enhanced	0.584	0.478	0.518	0.573	0.620	0.582	0.486	0.522	0.573	0.619
SIGTRAN-Features	0.911	0.936	0.924	0.923	0.980	**0.905**	0.926	0.915	0.914	0.976
SIGTRAN	0.898	0.940	0.919	0.917	0.978	0.890	**0.947**	**0.918**	**0.915**	**0.976**

Table 3. Ethereum performance: SIGTRAN significantly outperforms baselines on the Ethereum dataset. The last three rows are introduced in this paper.

	Training Set Performance					*Test Set Performance*				
Algorithm	Precision	Recall	F_1	Accuracy	AUC	Precision	Recall	F_1	Accuracy	AUC
trans2vec [30]	0.912	0.909	0.910	0.911	0.966	0.919	0.894	0.906	0.908	0.967
node2vec	0.908	0.935	0.921	0.920	0.970	0.917	0.907	0.912	0.912	0.964
RiWalk	0.922	0.772	0.840	0.852	0.904	0.931	0.764	0.838	0.853	0.894
RiWalk-enhanced	0.921	0.846	0.882	0.887	0.908	0.928	0.832	0.877	0.884	0.899
SIGTRAN-Features	0.926	0.945	0.935	0.934	0.962	0.923	0.926	0.925	0.925	0.958
SIGTRAN	0.946	0.953	0.949	0.949	0.983	**0.944**	**0.940**	**0.942**	**0.942**	**0.976**

We also compared the performance of SIGTRAN with methods specifically designed for Bitcoin or Ethereum network.

- Bitcoin: we considered the method proposed by *Weber et al.* [28] as the baseline.
- Ethereum: we considered phishing scams detection method by Wu et al. [30] denoted as trans2vec as the baseline method. To make a fair comparison, we set the default parameters of trans2vec inline with the parameters of the node2vec.

4.3 Performance Evaluation

To evaluate the performance of SIGTRAN, we considered the illicit nodes as the target of the detection approach and randomly selected an equal number of genuine nodes to form our set of anchor nodes. We extracted the first-order neighbors of all the anchor nodes and all edges among these nodes to construct a subgraph for investigation. Random selection of genuine nodes was repeated for 50 times, thus 50 different subgraphs were examined and the average performance was reported. Logistic regression with *L1* regularization was implemented in Scikit-learn Python package as the node classifier. The performance evaluation results for the Bitcoin and Ethereum network are illustrated in Table 2 and Table 3, respectively. To investigate the importance of SIGTRAN-*features*, both tables also report the performance of the classification tasks when only SIGTRAN-features were used as the node representations.

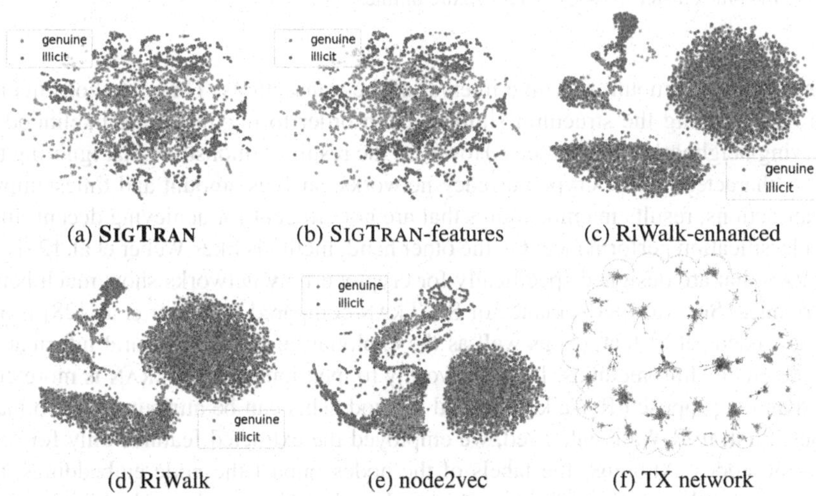

Fig. 3. Bitcoin Embeddings: SIGTRAN better separate illicit (red) and genuine (blue) transactions in Bitcoin network (plotted in (f)) compared to other baselines. (Color figure online)

Bitcoin. Considering the results of illicit node detection on Bitcoin network in Table 2, it can be observed that node embedding methods namely node2vec and RiWalk did not generate efficient node representations. Therefore, the classification task had very low performance in detecting illicit nodes. The poor performance of node2vec and RiWalk is due to the fact that these methods are not specifically dealing with the intrinsic characteristics of financial networks, such as having multiple edges among nodes, or being

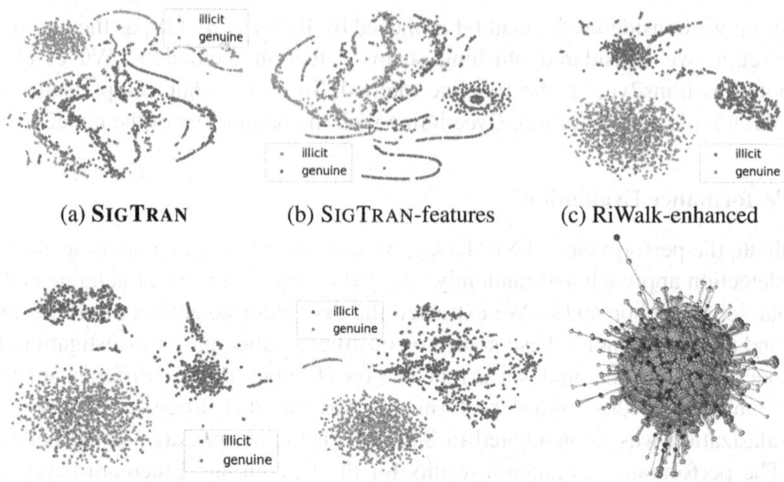

(a) SIGTRAN (b) SIGTRAN-features (c) RiWalk-enhanced

(d) RiWalk (e) node2vec (f) TX network

Fig. 4. Ethereum Embeddings: SIGTRAN better separate illicit (red) and genuine (blue) accounts in Ethereum network (plotted in (f)) compared to other baselines. Notice the red nodes mixed in the blue cluster in (c–e). (Color figure online)

dependent on the amount and time interval of the transactions. These methods mainly focus on exploiting the structural similarities in order to maximize the likelihood of preserving neighborhoods of nodes. However, the results demonstrate that ignoring the specific characteristics of cryptocurrency networks, such as amount and timestamp of the transactions, results in embeddings that are not efficient for achieving decent illicit node classification performance. On the other hand, methods likes Weber et al. [28] and SIGTRAN that are designed specifically for cryptocurreny networks show much better performance. Superior performance of SIGTRAN compared to Weber et al. [28] is due to its extended set of features as well as the exploitation of the structural information via node embedding methods. It is noteworthy to mention that SIGTRAN is more efficient than the proposed RiWalk-enhanced method. This can be attributed to two main reasons. First, in RiWalk-enhanced, we employed the extracted features only for relabeling the nodes. Although the labels of the nodes impact the node embeddings, the exact values of the extracted features do not directly influence the embeddings values which are later used for the node classification task. Moreover, it should be noted that the new labels combine the extracted features of the anchor and neighbor nodes as well as their shortest path distance. Thus, modified values of the extracted features are used for labeling. However, it is noteworthy that RiWalk-enhanced outperforms its counterpart RiWalk, which underlines the importance of fusing the extracted features with the node embeddings in terms of improving the performance of the node classification task. For a qualitative comparison of the different embedding methods, we have depicted the t-SNE [23] transformations of different node representations methods for one of the subgraphs of the Bitcoin network in Fig. 3. According to Fig. 3, it can be observed that the embeddings produced by SIGTRAN shape more separable distributions.

Ethereum. For the Ethereum dataset as shown in Table 3, it can be observed that SIGTRAN demonstrates considerably better performance than the other methods. Although trans2vec and node2vec demonstrate high performance, the superior performance of SIGTRAN underlines its efficiency in employing the native characteristics of the cryptocurrency networks as well as structural information obtained by the node embedding methods. Besides, we can observe that the extracted features improved the performance of the RiWalk-enhanced compared to RiWalk. Due to the fact that SIGTRAN better incorporates the extracted features with the network structural embeddings, it achieves the most decent performance on the Ethereum network as well. We have also depicted t-SNE [23] transformations of different node embedding methods for a subgraph of the Ethereum network in Fig. 4. Considering Fig. 4, it is observable that embeddings obtained by SIGTRAN show considerable distinction between illicit and licit nodes, while for example in Fig. 4e, there are several illicit nodes (marked with red) in the licit cluster (marked with blue).

5 Conclusions

We propose SIGTRAN that extracts signature vectors for detecting illicit activities in blockchain network. Our proposed SIGTRAN transforms the blockchain network into a simple graph and then extracts carefully designed features which explain structural, transactional, regional, and neighborhood features of the nodes. These features are then combined with generic node representations which encode the roles of the nodes in a given graph. SIGTRAN should be considered as a simple and strong baseline when developing more complex models. Our proposed SIGTRAN baseline is:

- **Accurate:** SIGTRAN outperforms state-of-the-art alternatives in detecting illicit activities in blockchain transaction records.
- **Generic:** SIGTRAN is platform independent and we apply it to blockchain data extracted from both Bitcoin and Ethereum.

Reproducibility. The code and data are available at https://github.com/fpour/SigTran.

References

1. Bitcoin block explorer and API. https://sochain.com/. Accessed 15 Sept 2020
2. Ethereum scam database. https://etherscamdb.info/scams. Accessed 14 May 2020
3. Github - blockchain-etl/ethereum-etl: Python scripts for ETL (extract, transform and load) jobs for Ethereum blocks, transactions, ERC20/ERC721 tokens, transfers, receipts, logs, contracts, internal transactions. Data is available in Google BigQuery https://goo.gl/oy5bcq. https://github.com/blockchain-etl/ethereum-etl. Accessed 15 Sept 2020
4. Atzei, N., Bartoletti, M., Cimoli, T.: A survey of attacks on Ethereum smart contracts (SoK). In: Maffei, M., Ryan, M. (eds.) POST 2017. LNCS, vol. 10204, pp. 164–186. Springer, Heidelberg (2017). https://doi.org/10.1007/978-3-662-54455-6_8
5. Badawi, E., Jourdan, G.V.: Cryptocurrencies emerging threats and defensive mechanisms: a systematic literature review. IEEE Access **8**, 200021–200037 (2020)

6. Bartoletti, M., Pes, B., Serusi, S.: Data mining for detecting Bitcoin Ponzi schemes. In: 2018 Crypto Valley Conference on Blockchain Technology (CVCBT), pp. 75–84. IEEE (2018)
7. Bhattacharyya, S., Jha, S., Tharakunnel, K., Westland, J.C.: Data mining for credit card fraud: a comparative study. Decis. Support Syst. **50**(3), 602–613 (2011)
8. Carneiro, N., Figueira, G., Costa, M.: A data mining based system for credit-card fraud detection in e-tail. Decis. Support Syst. **95**, 91–101 (2017)
9. Chen, T., et al.: Understanding Ethereum via graph analysis. In: IEEE INFOCOM 2018-IEEE Conference on Computer Communications, pp. 1484–1492. IEEE (2018)
10. Chen, W., Zheng, Z., Ngai, E.C.H., Zheng, P., Zhou, Y.: Exploiting blockchain data to detect smart Ponzi schemes on Ethereum. IEEE Access **7**, 37575–37586 (2019)
11. Conti, M., Kumar, E.S., Lal, C., Ruj, S.: A survey on security and privacy issues of Bitcoin. IEEE Commun. Surv. Tutor. **20**(4), 3416–3452 (2018)
12. Dey, S.: Securing majority-attack in blockchain using machine learning and algorithmic game theory: a proof of work. arXiv preprint arXiv:1806.05477 (2018)
13. Farrugia, S., Ellul, J., Azzopardi, G.: Detection of illicit accounts over the Ethereum blockchain. Expert Syst. Appl **150**, 113318 (2020)
14. Goyal, P., Ferrara, E.: Graph embedding techniques, applications, and performance: a survey. Knowl. Based Syst. **151**, 78–94 (2018)
15. Grover, A., Leskovec, J.: node2vec: scalable feature learning for networks. In: Proceedings of the 22nd ACM SIGKDD International Conference on Knowledge Discovery and Data Mining, pp. 855–864 (2016)
16. Harlev, M.A., Sun Yin, H., Langenheldt, K.C., Mukkamala, R., Vatrapu, R.: Breaking bad: de-anonymising entity types on the Bitcoin blockchain using supervised machine learning. In: Proceedings of the 51st Hawaii International Conference on System Sciences (2018)
17. Howell, B.E., Potgieter, P.H.: Industry self-regulation of cryptocurrency exchanges (2019)
18. Hu, Y., Seneviratne, S., Thilakarathna, K., Fukuda, K., Seneviratne, A.: Characterizing and detecting money laundering activities on the Bitcoin network. arXiv preprint arXiv:1912.12060 (2019)
19. Huang, H., Kong, W., Zhou, S., Zheng, Z., Guo, S.: A survey of state-of-the-art on blockchains: theories, modelings, and tools. arXiv preprint arXiv:2007.03520 (2020)
20. Lin, D., Wu, J., Yuan, Q., Zheng, Z.: Modeling and understanding Ethereum transaction records via a complex network approach. IEEE Trans. Circ. Syst. II Express Briefs **67**, 2737–2741 (2020)
21. Lorenz, J., Silva, M.I., Aparício, D., Ascensão, J.T., Bizarro, P.: Machine learning methods to detect money laundering in the Bitcoin blockchain in the presence of label scarcity. arXiv preprint arXiv:2005.14635 (2020)
22. Ma, X., Qin, G., Qiu, Z., Zheng, M., Wang, Z.: RiWalk: fast structural node embedding via role identification. In: 2019 IEEE International Conference on Data Mining (ICDM), pp. 478–487. IEEE (2019)
23. van der Maaten, L., Hinton, G.: Visualizing data using t-SNE. J. Mach. Learn. Res. **9**, 2579–2605 (2008)
24. Mikolov, T., Sutskever, I., Chen, K., Corrado, G.S., Dean, J.: Distributed representations of words and phrases and their compositionality. In: Advances in Neural Information Processing Systems 26, pp. 3111–3119 (2013)
25. Monamo, P.M., Marivate, V., Twala, B.: A multifaceted approach to Bitcoin fraud detection: global and local outliers. In: 2016 15th IEEE International Conference on Machine Learning and Applications (ICMLA), pp. 188–194. IEEE (2016)
26. Motamed, A.P., Bahrak, B.: Quantitative analysis of cryptocurrencies transaction graph. Appl. Netw. Sci. **4**(1), 1–21 (2019)
27. Pham, T., Lee, S.: Anomaly detection in the Bitcoin system-a network perspective. arXiv preprint arXiv:1611.03942 (2016)

28. Weber, M., et al.: Anti-money laundering in Bitcoin: experimenting with graph convolutional networks for financial forensics. arXiv preprint arXiv:1908.02591 (2019)
29. Wu, J., Lin, D., Zheng, Z., Yuan, Q.: T-edge: temporal weighted multidigraph embedding for Ethereum transaction network analysis. arXiv preprint arXiv:1905.08038 (2019)
30. Wu, J., et al.: Who are the phishers? Phishing scam detection on Ethereum via network embedding. IEEE Trans. Syst. Man Cybern. Syst. (2020)
31. Yuan, Z., Yuan, Q., Wu, J.: Phishing detection on Ethereum via learning representation of transaction subgraphs. In: Zheng, Z., Dai, H.-N., Fu, X., Chen, B. (eds.) BlockSys 2020. CCIS, vol. 1267, pp. 178–191. Springer, Singapore (2020). https://doi.org/10.1007/978-981-15-9213-3_14

VOA*: Fast Angle-Based Outlier Detection over High-Dimensional Data Streams

Vijdan Khalique[1(✉)] and Hiroyuki Kitagawa[2]

[1] Graduate School of Systems and Information Engineering, University of Tsukuba, Tsukuba, Japan
khalique.vijdan@kde.cs.tsukuba.ac.jp
[2] Center for Computational Sciences, University of Tsukuba, Tsukuba, Japan
kitagawa@cs.tsukuba.ac.jp

Abstract. Outlier detection in the high-dimensional data stream is a challenging data mining task. In high-dimensional data, the distance-based measures of outlierness become less effective and unreliable. Angle-based outlier detection ABOD technique was proposed as a more suitable scheme for high-dimensional data. However, ABOD is designed for static datasets and its naive application on a sliding window over data streams will result in poor performance. In this research, we propose two incremental algorithms for fast outlier detection based on an outlier threshold value in high-dimensional data streams: *IncrementalVOA* and VOA^*. *IncrementalVOA* is a basic incremental algorithm for computing outlier factor of each data point in each window. VOA^* enhances the incremental computation by using a bound-based pruning method and a retrospect-based incremental computation technique. The effectiveness and efficiency of the proposed algorithms are experimentally evaluated on synthetic and real world datasets where VOA^* outperformed other methods.

Keywords: Outlier detection · Angle-based outlier detection · High-dimensional data · Data streams

1 Introduction

Outlier detection is a data mining task of identifying data points which are different from the majority of data [6]. Such data points are indicative of some unexpected behavior in the process generating the data. As a result, outlier detection has many important real-world applications such as detecting financial frauds, hazardous roads [9] and network intrusion, and monitoring surveillance videos [13]. For example, in the case of surveillance videos [13], it is required to analyze video frames from cameras where each frame is represented as features in high-dimensional space. The outlier detection algorithm receives featurized video frames as input and detects the ones that are different from the majority.

Most of the existing outlier detection algorithm [2,14,19,21] consider distance-based measure of outlierness. These approaches are ineffective in

K. Karlapalem et al. (Eds.): PAKDD 2021, LNAI 12712, pp. 40–52, 2021.
https://doi.org/10.1007/978-3-030-75762-5_4

high-dimensional data because distance between the farthest and nearest points becomes insignificant due to the curse of dimensionality [7].

For high-dimensional data streams, the outlier detection task becomes more challenging due to high arrival rate and the unbounded nature of the data stream. Therefore, outlier detection is done over a sliding window which logically bounds the data in terms of a time interval (time-based window) or the number of data points (count-based window). Thus, for each window, the algorithm is required to calculate outlier factors of newly arriving data and update the existing ones.

One of the effective outlier detection methods in high-dimensional data is angle-based outlier detection (ABOD) proposed in [12]. ABOD is based on the observation that a data point in the cluster is surrounded by many other data points compared to the one which is located away from the cluster. Consequently, a data point within a cluster will have higher variance of angles than the ones located away from the cluster. Moreover, a point at the border of the cluster will have variance of angles larger than the point away from the cluster and smaller than the point within the cluster. Accordingly, ABOD uses variances of angles as the measure of outlierness of each data point. Furthermore, angle between a point and pairs of points is more stable in high-dimensional spaces, which makes ABOD more suitable for high-dimensional datasets. However, ABOD uses distance based weighting of angles on which Pham et el. in [15] argued that it is less meaningful due to the curse of dimensionality. As a result, they considered calculating pure variances of angles (VOA) and proposed *FastVOA* which uses random projection based method to approximate VOA.

This paper presents fast angle-based outlier detection methods for high-dimensional data streams based on accurate and unbiased calculation of variances of angles (VOA). The main contributions of this paper are: (1) an incremental angle-based outlier detection algorithm (*IncrementalVOA*) for detecting the outliers in high-dimensional data streams, (2) an improved incremental algorithm VOA^*, which uses bound-based pruning and retrospect-based technique to speed-up the outlier detection, and (3) experiments on synthetic and real world datasets to show the accuracy and performance of the proposed algorithms.

2 Related Work

There are several techniques for detecting outliers from static and streaming data with low and high dimensionality. They define the outlier data points based on observations e.g. distance-based [8,10,11], density-based [2,16] and clustering-based [1,3,4]. Methods based on these observations are also proposed for data streams [17,18,21]. However, all of these techniques use some kind of distance-based measure to determine neighbors, cluster or density. In contrast, the angle-based approach called Angle-based Outlier Detection (ABOD) was proposed by Kriegel et al. in [12] based on the observation discussed in the previous section. Therefore, the variance of angles can be used as the outlier factor and data points with least outlier factors are the outliers in the given dataset. Kreigel et al. further proposed a k-nearest neighbor-based approximation technique *FastABOD* for approximating the angle-based outlier score to improve

Table 1. Notations and Data Structures

Symbols	Description
n, d	Window size, number of dimensions
θ_{pqr}	Angle between vectors $q-p$ and $q-r$
p, p'	A new data point and expired data point at timestamp t
Δ_q	The displacement in terms of angle for q between p' and p
$soa_q, sosa_q$	Sum of angles and sum of square of angles of point q at t_{last}
t_{last}	Timestamp when soa and $sosa$ were last updated
$\mathbb{H}_t$	Set of n tuples $\{(pid, t_{last}, soa, sosa)\}$
$\mathbb{E}_t$	Set of $(n-2)$ expired points of timestamps $t-2n+1$ to $t-n$
Data Structures	Description
$SUM_\theta[n]$	Array of sum of angles θ_{apb}s of points in W_t
$SUM_{\theta^2}[n]$	Array of sum of squared angles θ^2_{apb}s of points in W_t
$VOA_t[n]$	Array of VOA_ts of points in W_t
$SUM_\theta^{(UB/LB)}[n]$	Array of UB or LB of sum of θ_{apb}s of points in W_t
$SUM_{\theta^2}^{(UB/LB)}[n]$	Array of UB or LB of sum of θ^2_{apb}s of points in W_t
$VOA_t{}^{(UB/LB)}[n]$	Array of UB and LB of VOA_ts of points in W_t
$\mathbb{O}_t$	Array of outliers in W_t

the performance. Another approximation technique called *FastVOA* [15] used random projection and AMS sketches to approximate variances of unweighted angles. For data streams, *DSABOD* and *FastDSABOD* were proposed in [20], which are incremental versions of *ABOD* and *FastABOD*. As they use distance-based weighting of angles, the resulting outlier factors may become unreliable in high-dimensional data.

3 Problem Definition

This section formally defines the problem of angle-based outlier detection over streaming data. We consider the exact and unweighted angles for calculating VOA as in [15]. In addition, we formally define data stream, sliding window and angle-based outlier factor before defining the problem definition.

Definition 1. *(Data Stream) A data stream is an infinite sequence of data points:* $p_1, ..., p_{t-1}, p_t, p_{t+1}, ...,$ *where* $p_t \in \mathbb{R}^d$ *arrives at timestamp t.*

A sliding window over an infinite data stream maintains a finite set of data points. We consider a sliding window of a fix size n. The notion of *slide* is that the window slides at each timestamp and a new data point is added and an oldest one is removed.

Definition 2. *(Sliding Window) Given a timestamp t, a sliding window* W_t *of size n is defined as a set of n data points* $W_t = \{p_{t-n+1}, p_{t-n+2} ..., p_t\}$.

Algorithm 1: IncrementalVOA

Data: p (new point) and p' (expired point) at timestamp t
Result: $\mathbb{O}_t$

Populate: SUM_θ, $SUM_{\theta 2}$ at timestamp $t-1$
for $q \in W_t \setminus \{p\}$ **do**
  for $r \in W_t \setminus \{p, q\}$ **do**
    Calculate: $\theta_{p'qr}$ and θ_{pqr}
    $SUM_\theta[q] \leftarrow SUM_\theta[q] - \theta_{p'qr} + \theta_{pqr}$, $SUM_{\theta 2}[q] \leftarrow SUM_{\theta 2}[q] - \theta^2_{p'qr} + \theta^2_{pqr}$
  end
  Calculate: $MOA_1[q]$ and $MOA_2[q]$ from $SUM_\theta[q]$ and $SUM_{\theta 2}[q]$
  $VOA_t[q] \leftarrow MOA_2(q) - (MOA_1(q))^2$
  if $VOA_t[q] \leq \tau$ **then**
    insert in $\mathbb{O}_t$
  end
end

For simplicity, we consider a simple count-based sliding window which slides at each timestamp t where a new data point is included and an oldest data point is removed. However, more general definitions may be used for defining the sliding window. The proposed methods are applicable in case of time-based window or a different notion of slide in which the slide size is more than 1.

The angle-based outlier factor in streaming environment is defined as follows:

Definition 3. *(VOA_t) Given a window W_t of size n and a data point $p \in W_t$, let θ_{apb} denote the angle between the vectors $a - p$ and $b - p$ for a random pair of points $a, b \in W_t \setminus \{p\}$. The outlier factor of p is defined as follows: $VOA_t[p] = Var_t[\theta_{apb}] = MOA_2[p] - (MOA_1[p])^2$ where MOA_1 and MOA_2 are as follows: $MOA_1[p] = \frac{\sum_{a,b \in W_t \setminus \{p\}} \theta_{apb}}{\frac{1}{2}(n-1)(n-2)}$, $MOA_2[p] = \frac{\sum_{a,b \in W_t \setminus \{p\}} \theta^2_{apb}}{\frac{1}{2}(n-1)(n-2)}$.*

Now, the angle-based outlier detection problem in the streaming environment can be defined as follows:

Definition 4. *(Angle-based outlier detection over streaming data) Given an outlier threshold τ, the angle-based outlier detection over streaming data is the problem of continuously finding the set of outliers $\mathbb{O}_t = \{p \in W_t | VOA_t[p] \leq \tau\}$ for each window W_t.*

4 Proposed Methods

This research proposes two algorithms for angle-based outlier detection for high-dimensional data streams called *IncrementalVOA* and VOA^*. *IncrementalVOA* is a basic incremental algorithm. While, VOA^* uses bound-based pruning to obtain $\mathbb{X}_t \subset W_t$ and applies a retrospect-based incremental algorithm *RetroIncrementaVOA* for updating VOA_ts. Table 1 shows the description of symbols and data structures used in the proposed algorithms.

4.1 IncrementalVOA

IncrementalVOA (*Algorithm* 1) is a basic incremental algorithm which calculates outlier factor of $p \in W_t$ and updates $n-1$ points in the window W_t. Overall, the window W_n is filled with the initial n points and VOA_t of each data point is calculated by considering all pairs of data points. For W_t, where $t > n$, $VOA_t[q]$) of a new data point p is calculated according to *Definition* 3 and if $VOA_t(p) \leq \tau$ then it is inserted in the list $\mathbb{O}_t$. *IncrementalVOA* maintains SUM_θ and SUM_{θ^2} for all points in W_t. It updates remaining $n-1$ data points $q \in W_t \setminus \{p\}$ by using p and p', and calculating the old angle between vectors $q-p'$ and $q-r$ and the new angle between vectors $q-p$ and $q-r$ (line 4 and 5). Finally, $VOA_t[q]$ is calculated (line 7–8), and q is inserted in $\mathbb{O}_t$ if it is an outlier (line 9–11). Therefore, *IncrementalVOA* avoids calculating VOA_t for all point every time the window slides.

4.2 VOA*

In data streams, the difference between a new and an expired data point can be small. In this case, outlier factors of many points in W_t are not much affected by the window slide. Suppose there are n objects being monitored, where each object produces d-dimensional data points. All the objects produce data one-by-one in a sequence and a sliding window of size n contains the data from all n objects at each timestamp. For example, n surveillance cameras mounted on different locations produce d-dimensional representation of frames where the monitoring system receives data from one location at-a-time in a specific sequence. The difference among n data points in W_t can be large due to different locations of mounted cameras. However, the expired data point corresponds with the same camera which has produced the new data point at the given time. If the monitoring cycle is not too long, the new data point is likely to be similar to the expired data point. Typically, such kind of situation may occur in the real world, where the new and expired data points are similar. VOA^* can deal with such kind of data streams well by using bound-based pruning and retrospect-based methods which are explained in detail in the following subsections. The overall VOA^* algorithm is given in *Algorithm* 2.

Bound-Based Pruning. We need to check VOAs of data points $q \in W_t \setminus \{p\}$ when p arrives and p' expires. The angle $\theta_{pqp'}$ is denoted as Δ_q because it shows displacement between p' and p in terms of angle. When Δ_q is added to and subtracted from $\theta_{p'qr}$ (where r is another random point in W_t), an upper bound $\theta_{pqr}^{(UB)}$ and a lower bound $\theta_{pqr}^{(LB)}$ can be obtained which indicates the maximum and minimum possible value of the new angle θ_{pqr}. Hence, upper and lower bounds can be given as: $\theta_{pqr}^{(UB)} = \theta_{p'qr} + \Delta_q$ and $\theta_{pqr}^{(LB)} = \theta_{p'qr} - \Delta_q$. In order to calculate $VOA_t[q]$, SUM_θ and SUM_{θ^2} for q are required. For incrementally computing $SUM_\theta[q]$, it can be written as: $SUM_\theta[q] = SUM_\theta[q] + \theta_{pqr} - \theta_{p'qr}$.

Algorithm 2: *VOA**

```
Data: p (new point) and p' (expired point) at timestamp t
Result: O_t
for q ∈ W_t \ {p} do
    calculate: Δ_q
    calculate: LB and UB of SUM_θ and SUM_θ2 of q using eqs. (1)(2)(3)(4)
    calculate: VOA_t^(LB)[q] and VOA_t^(UB)[q] using eqs. (5)(6)
    if VOA_t^(UB)[q] ≤ τ then
        insert in O_t
    end
    if VOA_t^(LB)[q] ≤ τ < VOA_t^(UB)[q] then
        voa ← RetroIncrementalVOA(q, E_t, H_t)
        if voa ≤ τ then
            insert q in O_t
        end
    end
end
```

As there are $n-1$ data points in $W_t \setminus \{p\}$ and Δ_q bounds $|\theta_{pqr} - \theta_{p'qr}|$, we can calculate $SUM_\theta^{(UB)}$ and $SUM_\theta^{(LB)}$ as following:

$$SUM_\theta^{(UB)}[q] = SUM_\theta{}^{(UB)}[q] + (n-1)\Delta_q \tag{1}$$

$$SUM_\theta^{(LB)}[q] = SUM_\theta{}^{(LB)}[q] - (n-1)\Delta_q \tag{2}$$

Similarly, incremental update of $SUM_{\theta^2}[q]$ can be written as: $SUM_{\theta^2}[q] = SUM_{\theta^2}[q] + \theta_{pqr}^2 - \theta_{pqr'}^2$. It can be further expanded as: $SUM_{\theta^2}[q] + (\theta_{pqr} - \theta_{p'qr})(\theta_{pqr} + \theta_{p'qr})$ where Δ_q bounds $|\theta_{pqr} - \theta_{p'qr}|$ and $(\theta_{pqr} + \theta_{p'qr})$ can take the maximum value of 2π. Hence, $SUM_{\theta^2}^{(UB)}$ and $SUM_{\theta^2}^{(LB)}$ can be calculated as following:

$$SUM_{\theta^2}^{(UB)}[q] = SUM_{\theta^2}^{(UB)}[q] + 2\pi\Delta_q \tag{3}$$

$$SUM_{\theta^2}^{(LB)}[q] = SUM_{\theta^2}^{(LB)}[q] - 2\pi\Delta_q \tag{4}$$

Given the $SUM_{\theta^2}^{(UB)}$ and $SUM_{\theta^2}^{(LB)}$, $VOA_t^{(UB)}$ and $VOA_t^{(LB)}$ can be calculated as:

$$VOA_t^{(UB)}[q] = \frac{SUM_{\theta^2}^{(UB)}[q]}{\frac{1}{2}(n-1)(n-2)} - \left(\frac{SUM_\theta^{(LB)}[q]}{\frac{1}{2}(n-1)(n-2)}\right)^2 \tag{5}$$

$$VOA_t^{(LB)}[q] = \frac{SUM_{\theta^2}^{(LB)}[q]}{\frac{1}{2}(n-1)(n-2)} - \left(\frac{SUM_\theta^{(UB)}[q]}{\frac{1}{2}(n-1)(n-2)}\right)^2 \tag{6}$$

$VOA_t^{(UB)}$ and $VOA_t^{(LB)}$ represent the maximum and minimum value of outlier factor of a given data point q which can be used for pruning the data points

Algorithm 3: RetroIncrementalVOA

Data: $q, \mathbb{E}_t, \mathbb{H}_t$
Result: $voaValue$ of q

Get $pid, t_{last}, soa_q, sosa_q$ of q from $\mathbb{H}_t$, $soa \leftarrow soa_q, sosa \leftarrow sosa_q$
for $i = t_{last} + 1 \rightarrow t$ **do**
  $j \leftarrow i - n$, $p \leftarrow p_i \in W_t$, $p' \leftarrow p_j \in \mathbb{E}_t$
  for $k = i - n + 1 \rightarrow i$ *except* t_{last} **do**
    $r \leftarrow p_k \in (W_t \cup \mathbb{E}_t)$
    Calculate: θ_{pqr} and $\theta_{p'qr}$
    $soa \leftarrow soa + \theta_{pqr} - \theta_{p'qr}, sosa \leftarrow sosa + \theta^2_{pqr} - \theta^2_{p'qr}$
  end
end
$SUM^{(UB)}_{\theta}[q] \leftarrow SUM^{(LB)}_{\theta}[q] \leftarrow soa$, $SUM^{(UB)}_{\theta^2}[q] \leftarrow SUM^{(LB)}_{\theta^2}[q] \leftarrow sosa$
Calculate: $MOA_1[q]$ and $MOA_2[q]$ from soa and $sosa$
$voaValue \leftarrow MOA_2[q] - (MOA_1[q])^2$
$updateHistory(\mathbb{H}_t)$
return $voaValue$

and obtaining a small subset $\mathbb{X}_t \subset (W_t \setminus \{p\})$ effectively. Consequently, outlier factors of smaller number of data points in $\mathbb{X}_t$ are recalculated. Given the threshold τ, $VOA^{(UB)}_t[q]$ and $VOA^{(LB)}_t[q]$ of $q \in \mathbb{X}_t$, one of the cases holds: *(A)* if $VOA^{(UB)}_t[q] \leq \tau$ then q is outlier, *(B)* if $VOA^{(LB)}_t[q] > \tau$ then q is inlier or, *(C)* if $VOA^{(LB)}_t[q] \leq \tau < VOA^{(UB)}_t[q]$, recalculate $VOA_t[q]$.

A factor that effects the pruning and the size of $\mathbb{X}_t$ is the difference between p and p' in terms of their locations in the data space. If the difference is not significant then the value of Δ_q is small. This is the best case for *VOA** as given an appropriate value of τ, a large number of points can be pruned resulting in a small fraction of points in $\mathbb{X}_t$. Consequently, the overall processing time for updating outlier factors of $n-1$ data points in W_t can be significantly reduced. In the worst case, the difference between p and p' is large which will result in more data points satisfying the *Case C* and the size of $\mathbb{X}_t$ will increase.

Retrospect-Based Incremental Algorithm. In order to compute $VOA_t[q]$ ($q \in \mathbb{X}_t$), the naïve way is to calculate it from the scratch by considering all pairs of points in the window. *VOA** (in line 8–13 of *Algorithm* 2) invokes *RetroIncrementalVOA* (Algorithm 3) to further improve the performance by incrementally computing VOA_ts of data points in $\mathbb{X}_t$. It maintains the most recently expired data points and the history of data points in the window to update the outlier factors from the time when the VOA_ts of the data points were last calculated.

RetroIncrementalVOA maintains the history of each data point as a tuple $(pid, t_{last}, soa, sosa)$ in a set $\mathbb{H}_t$. It also stores $n-2$ most recently expired data points in $\mathbb{E}_t$. $\mathbb{E}_t$ and $\mathbb{H}_t$ are updated at each timestamp where for the newly arriving point, $\mathbb{H}_t$ is updated and the expired data point is inserted in $\mathbb{E}_t$ and its history is removed from $\mathbb{H}_t$. For the 1st window ($t = n$), $\mathbb{H}_n$ is populated with $pid, t_{last}, soa, sosa$ where $t_{last} = n$ for all data points. soa and $sosa$ are stored so that they can be used for incrementally computing the outlier factor. Given a point $q \in \mathbb{X}_t$, it is passed to $RetroIncrementalVOA$ for incremental computation of $VOA_t[q]$ along with $\mathbb{H}_t$ and $\mathbb{E}_t$. $RetroIncrementalVOA$ searches q in $\mathbb{H}_t$ and finds out in which window soa_q and $sosa_q$ were last calculated and then it updates soa_q and $sosa_q$ from the last updated window ($t_{last} + 1$) up to the current window (t). The algorithm considers the entire history of a data point to incrementally update its outlier factor and it fetches the expired data points in $\mathbb{E}_t$ with timestamps greater than or equal to $t_{last} - n + 2$. Hence, only small fraction of points have to be processed for updating soa_q and $sosa_q$. Additionally, if q satisfies *Case C* in two closely located windows, the processing time to update $VOA_t[q]$ will be reduced significantly because few new points have arrived since the previous update. Furthermore, the size of $\mathbb{H}_t$ is n and $\mathbb{E}_t$ is $n - 2$, which does not require large memory space.

5 Experiments and Results

All algorithms are implemented on Java 1.8. We conducted experiments on 3.6 GHz Intel Core i7-7700 CPU with 32 GB RAM and 64-bit Windows 10 Pro installed. In the experiments, the performance and accuracy of the proposed method is evaluated on synthetic and real world datasets.

5.1 Accuracy

The accuracy is evaluated on three real world dataset - Isolet, Multiple Feature (Mfeat) and Human Activity Recognition (HAR) used for classification and machine learning [5]. Isolet is a 617 dimensional dataset of pronunciation of English alphabets. We selected B, C, D, E as inlier and L as outlier classes. Mfeat is a 649 dimensional dataset of handwritten digits (0–9) in which we considered 6 and 9 as inlier and 0 as outlier classes. HAR is a sensor data of 561 dimensions representing 6 activities (1. walking, 2. walking upstairs, 3. walking downstairs, 4. sitting, 5. standing and 6. laying) where 1, 2, 3 were chosen as inlier and 6 as outlier classes. We randomly chose 20 data objects from the selected outlier classes in each dataset. The accuracy of VOA^* is compared with *DSABOD* and *FastDSABOD* proposed in [20]. Accuracy evaluation of VOA and ABOD in [15] was based on ranking of detected outliers. However, VOA^* detects outliers according to a threshold value. Therefore, we have evaluated the accuracy on various thresholds and reported the results as precision-recall graphs in Fig. 1. All experiments are done with window size $n = 1000$, slide size $m = 1$ and dimensions corresponding with the dimensionality of the real world datasets. It may be noted that VOA^*, *IncrementalVOA*, and *SimpleVOA**, give the same results, therefore, they have same accuracies. Consequently, we do not compare their accuracies with other methods. Furthermore, we omit the comparison of VOA^*, *DSABOD* and *FastDSABOD* on synthetic datasets as they all show almost perfect results.

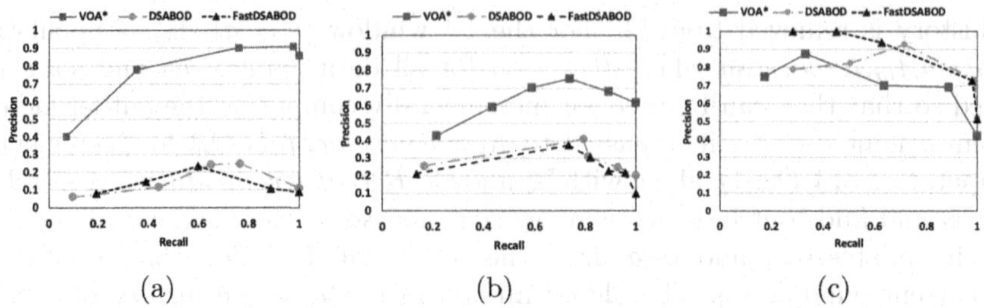

Fig. 1. Accuracy evaluation on real world datasets. (a) Mfeat (b) HAR (c) Isolet

Overall, VOA^* outperforms *DSABOD* and *FastDSABOD* on Mfeat and HAR datasets as shown in Fig. 1(a) and Fig. 1(b). VOA^* shows consistent accuracy on all the datasets. As *DSABOD* and *FastDSABOD* consider distance-based weighting, such weighting becomes less meaningful in high-dimensional space. As a result, they were not able to detect outliers effectively. VOA^* is more than 80% and 50% effective on Mfeat and HAR respectively. However, on Isolet (Fig. 1(c)) it shows comparable accuracy.

5.2 Performance

We evaluated the performance of the proposed method after validating its accuracy by considering the following parameters on the synthetic datasets: i) the increase in the deviation between the new and expired point, ii) window size n, iii) dimensions d and iv) slide size m. We compared the performance with *NaiveVOA* which calculates VOA_ts of data points in each window W_t according to Definition 3 and *SimpleVOA** which uses bound-based pruning without retrospect-based incremental algorithm. *SimpleVOA** naïvely calculates the outlier factors of data points in $\mathbb{X}_t$ unlike the VOA^*. All algorithms were executed 20 times, and their performances were recorded as the average CPU time per window. Furthermore, the CPU time for the 1st window (i.e. W_n) is not considered for *IncrementalVOA*, *SimpleVOA** and VOA^*.

Synthetic Datasets. The high-dimensional synthetic datasets were generated using the same approach as adopted in previous researches on angle-based outlier detection [12,15,20]. In each synthetic dataset, 20 outliers are introduced randomly from uniform distribution over the whole data space. The dataset size corresponds with the window size n and after n points are fed to the window of size n, data points from the same dataset are input again in the same sequence at each timestamp in order to simulate a data stream. Furthermore, we control the difference between the new and expired data point. At each timestamp, in order to increase the difference between new point p and expired point p', we add a random value (called *deviation*) in p. The deviation is drawn randomly from a normal distribution. If the standard deviation of the normal distribution is increased, the value of random deviation increases. It can be expressed as: $p = p' + \eta$ where $\eta = (x_1, x_2, ..., x_d)$, $x_i \sim N(0, \sigma)$ and $\sigma \in \mathbb{R}$. The parameter settings of experiments for synthetic datasets are given in Table 2 where the underlined values are the default settings. The experimental results for performance on synthetic data are shown in Fig. 2 and Fig. 4.

Table 2. Experimental Parameters

Parameters	Values
Window size (n)	500, <u>1000</u>, 2000, 3000, 4000, 5000
Dimensions (d)	<u>50</u>, 100, 200, 400, 600, 800
Slide size (m)	<u>1</u>, 1%, 5%, 10%, 15%, 20%
Deviation	<u>100</u>, 500, 1000, 2000, 3000
Threshold τ	<u>0.02</u>

Figure 2 shows the comparison of *VOA**, *SimpleVOA**, *NaiveVOA* and *IncrementalVOA* with default parameter settings on log scale. *VOA** performs almost 500x faster than *NaiveVOA* and 5x faster than *IncrementalVOA* at *deviation* = 100. At *deviation* = 3000, *VOA** runs 100x times faster than *SimpleVOA**. The performance of *VOA** deteriorate when the difference between expired and new point (deviation) increases and smaller number of points are pruned. However, the CPU time is still less than or equal to *IncrementalVOA*. Similarly, the effect of increasing window size, dimensions and slide size are shown in Fig. 4 with default settings where only relevant parameter is changed in each experiment. We have omitted *SimpleVOA** and *NaiveVOA* in these experiments for clear exposition. In summary, it can be seen that the CPU time increases with increasing n, d and m but *VOA**'s performance remains better than *IncrementalVOA*. This is primarily due to its bound-based pruning method which prunes most of the points when deviation is small.

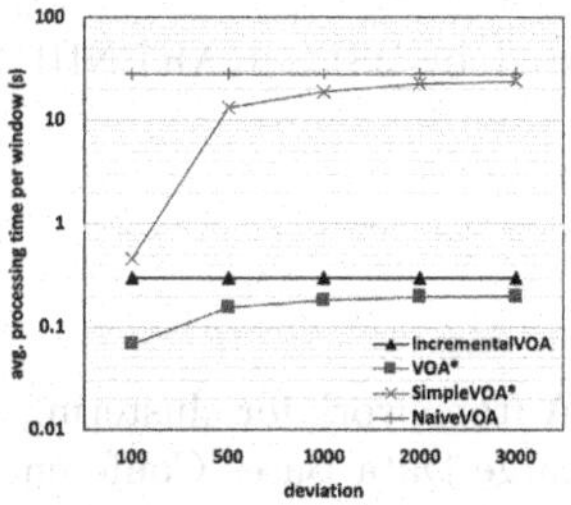

Fig. 2. Effect of varying deviation

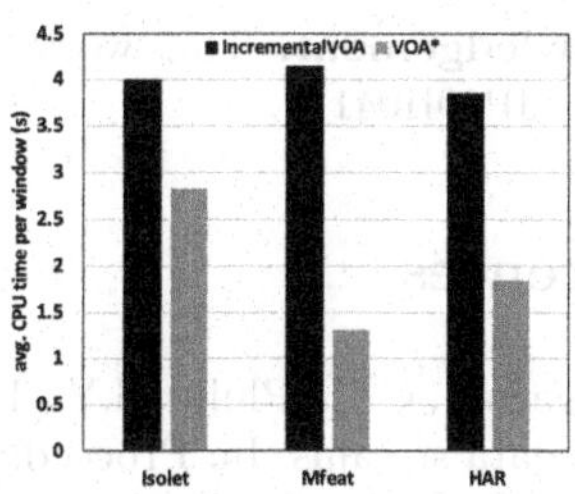

Fig. 3. CPU time on real world datasets

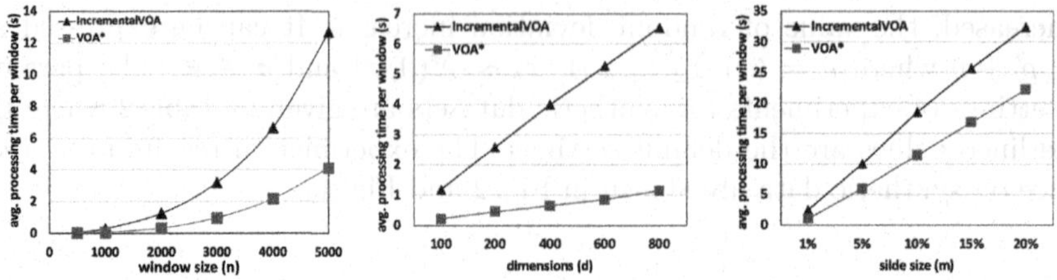

(a) (b) (c)

Fig. 4. Effect of increasing: (a) window size (b) dimensions (c) slide size

Real-World Datasets. The performance is also evaluated on the real world datasets - Isolet, Mfeat and HAR with threshold values 0.03, 0.07 and 0.024 respectively. Figure 3 shows the results of performance of VOA^* and *IncrementalVOA*. Generally, VOA^* outperformed *IncrementalVOA* on all datasets and performed more than 2x faster than *IncrementalVOA* on Mfeat and HAR and on Isolet it performed almost 1.5x faster.

6 Conclusion

This paper proposes two algorithms VOA^* and *IncrementalVOA* for incrementally detecting angle-based outliers for a sliding window over a high-dimensional data stream given an outlier threshold τ. *IncrementalVOA* is the basic incremental algorithm and VOA^* further enhances the incremental computation by using bound-based pruning and retrospect-based methods. The experimental results on synthetic and real world datasets showed that VOA^* performs much faster than *IncrementalVOA*, *NaiveVOA* and *SimpleVOA**. Moreover, VOA^* has better accuracy of outlier detection than distance weighted angle-based outlier factor calculated by *DSABOD* and *FastDSABOD*. In the future, we aim to extend this work for maintaining the list of top-k outliers in high-dimensional data streams.

Acknowledgement. This work was partly supported by JSPS KAKENHI Grant Number JP19H04114.

References

1. Aggarwal, C.C., Philip, S.Y., Han, J., Wang, J.: A framework for clustering evolving data streams. In: Proceedings of 2003 Very Large Data Bases Conference, pp. 81–92. Elsevier (2003)
2. Breunig, M.M., Kriegel, H.P., Ng, R.T., Sander, J.: LOF: identifying density-based local outliers. In: Proceedings of 2000 ACM SIGMOD International Conference on Management of Data, pp. 93–104 (2000)

3. Cao, F., Estert, M., Qian, W., Zhou, A.: Density-based clustering over an evolving data stream with noise. In: Proceedings of 2006 SIAM International Conference on Data Mining, pp. 328–339. SIAM (2006)
4. Chandola, V., Banerjee, A., Kumar, V.: Anomaly detection: a survey. ACM Comput. Surv. **41**(3), 1–58 (2009)
5. Dua, D., Graff, C.: UCI machine learning repository (2017). http://archive.ics.uci.edu/ml
6. Hawkins, D.M.: Identification of Outliers, vol. 11. Springer, Heidelberg (1980). https://doi.org/10.1007/978-94-015-3994-4
7. Hinneburg, A., Aggarwal, C.C., Keim, D.A.: What is the nearest neighbor in high dimensional spaces? In: Proceedings of 26th International Conference on Very Large Data Bases, pp. 506–515 (2000)
8. Ishida, K., Kitagawa, H.: Detecting current outliers: continuous outlier detection over time-series data streams. In: Bhowmick, S.S., Küng, J., Wagner, R. (eds.) DEXA 2008. LNCS, vol. 5181, pp. 255–268. Springer, Heidelberg (2008). https://doi.org/10.1007/978-3-540-85654-2_26
9. Kieu, T., Yang, B., Jensen, C.S.: Outlier detection for multidimensional time series using deep neural networks. In: Proceedings of 2018 19th IEEE International Conference on Mobile Data Management, pp. 125–134. IEEE (2018)
10. Knorr, E.M., Ng, R.T., Tucakov, V.: Distance-based outliers: algorithms and applications. VLDB J. **8**(3–4), 237–253 (2000)
11. Kontaki, M., Gounaris, A., Papadopoulos, A.N., Tsichlas, K., Manolopoulos, Y.: Continuous monitoring of distance-based outliers over data streams. In: Proceedings of 2011 IEEE 27th International Conference on Data Engineering, pp. 135–146. IEEE (2011)
12. Kriegel, H.P., Schubert, M., Zimek, A.: Angle-based outlier detection in high-dimensional data. In: Proceedings of 14th ACM SIGKDD International Conference on Knowledge Discovery and Data Mining, pp. 444–452 (2008)
13. Mahadevan, V., Li, W., Bhalodia, V., Vasconcelos, N.: Anomaly detection in crowded scenes. In: Proceedings of 2010 IEEE Computer Society Conference on Computer Vision and Pattern Recognition, pp. 1975–1981. IEEE (2010)
14. Papadimitriou, S., Kitagawa, H., Gibbons, P.B., Faloutsos, C.: LOCI: fast outlier detection using the local correlation integral. In: Proceedings of 19th International Conference on Data Engineering, pp. 315–326. IEEE (2003)
15. Pham, N., Pagh, R.: A near-linear time approximation algorithm for angle-based outlier detection in high-dimensional data. In: Proceedings of 18th ACM SIGKDD International Conference on Knowledge Discovery and Data Mining, pp. 877–885 (2012)
16. Pokrajac, D., Lazarevic, A., Latecki, L.J.: Incremental local outlier detection for data streams. In: 2007 IEEE Symposium on Computational Intelligence and Data Mining, pp. 504–515. IEEE (2007)
17. Salehi, M., Leckie, C., Bezdek, J.C., Vaithianathan, T., Zhang, X.: Fast memory efficient local outlier detection in data streams. IEEE Trans. Knowl. Data Eng. **28**(12), 3246–3260 (2016)
18. Shaikh, S.A., Kitagawa, H.: Continuous outlier detection on uncertain data streams. In: Proceedings of 2014 IEEE Ninth International Conference on Intelligent Sensors, Sensor Networks and Information Processing, pp. 1–7. IEEE (2014)
19. Tran, L., Fan, L., Shahabi, C.: Distance-based outlier detection in data streams. PVLDB Endow. **9**(12), 1089–1100 (2016)

20. Ye, H., Kitagawa, H., Xiao, J.: Continuous angle-based outlier detection on high-dimensional data streams. In: Proceedings of 19th International Database Engineering and Applications Symposium, pp. 162–167 (2015)
21. Yoon, S., Lee, J.G., Lee, B.S.: NETS: extremely fast outlier detection from a data stream via set-based processing. PVLDB Endow. **12**(11), 1303–1315 (2019)

Learning Probabilistic Latent Structure for Outlier Detection from Multi-view Data

Zhen Wang[1], Ji Zhang[2(✉)], Yizheng Chen[1], Chenhao Lu[1], Jerry Chun-Wei Lin[3], Jing Xiao[4], and Rage Uday Kiran[5]

[1] Zhejiang Lab, Hangzhou, China
wangzhen@zhejianglab.com
[2] University of Southern Queensland, Toowoomba, Australia
[3] Western Norway University of Applied Sciences, Bergen, Norway
[4] South China Normal University, Guangzhou, China
[5] University of Aizu, Aizu Wakamatsu, Fukushima, Japan

Abstract. Mining anomalous objects from multi-view data is a challenging issue as data collected from diverse sources have more complicated distributions and exhibit inconsistently heterogeneous properties. Existing multi-view outlier detection approaches mainly focus on transduction, which becomes very costly when new data points are introduced by an input stream. Besides, the existing detection methods use either the pairwise multiplication of cross-view data vectors to quantify outlier scores or the predicted joint probability to measure anomalousness, which are less extensible to support more sources. To resolve these challenges, we propose in this paper a Bayesian probabilistic model for finding multi-view outliers in an inductive learning manner. Specifically, we first follow the probabilistic projection method of latent variable for exploring the structural correlation and consistency of different views. Then, we seek for a promising Bayesian treatment for the generative model to approach the issue of selecting the optimal dimensionality. Next, we explore a variational approximation approach to estimate model parameters and achieve model selection. The outlier score for every sample can then be predicted by analyzing mutual distances of its representations across different views in the latent subspace. Finally, we benchmark the efficacy of the proposed method by conducting comprehensive experiments.

Keywords: Multi-view outlier detection · Probabilistic modeling · Subspace learning · Variational approximation

1 Introduction

As a fundamental data mining task, outlier detection (a.k.a anomaly detection in many scenarios) aims to identify anomalous and rare items, events or observations which raise suspicions by differing from the majority of the data. It has been widely applied in many fields, such as Web spam detection [29], fraud transaction detection [4], cybersecurity and malicious insider attack detection [26]. Over

K. Karlapalem et al. (Eds.): PAKDD 2021, LNAI 12712, pp. 53–65, 2021.
https://doi.org/10.1007/978-3-030-75762-5_5

the past decades, vast amounts of effective outlier detection algorithms have been developed, including distance-based approach [23], information-theoretic algorithm [34], density-based approach [22], reference-based approach [24] and clustering-based approach [14]. These methods were designed for single-view data. However, in many scenarios, data are often collected from diverse domains and are describable by multiple feature sets, each called one particular view. The traditional detection methods only examine a single view of instances and cannot discover multi-view anomalies, i.e., instances that exhibit inconsistent behaviors across different views. Thus, how to effectively leverage multiple views of data to detect anomalies, also called multi-view outlier detection [11], becomes an interesting and useful direction. Detecting outliers from multi-view data is a challenging problem due to: 1) the multi-view data usually have more complicated characteristics than the single-view data. 2) the data points may exhibit inconsistently heterogeneous properties across different views.

To tackle the problem, a number of multi-view outlier detection methods have been proposed in the literature. HOrizontal Anomaly Detection (HOAD) [11] pioneers this work. In HOAD, the author first constructs a combined similarity graph based on the similarity matrices, and computes the key eigenvectors of the graph Laplacian of the combined matrix. Then, outliers are identified by computing cosine distance between the components of these eigenvectors. This idea is further studied by [21] and [18] for different application tasks. Another successful group of methods is developed from the perspective of data representation [16,17,37]. They assume that a normal sample usually serves as a good contributor in representing the other normal samples while the outliers do not. By calculating the representation coefficients in low-rank matrix recovery, the multi-view outliers can be identified. In addition, Iwata *et al.* [13] utilize a sophisticated statistical machine learning algorithm to detect anomalies. A deep neural networks-based method has been developed for detecting multi-view outliers in [15] where they design a neural networks model to learn the latent intact representation for multi-view data.

Those detection methods are all unsupervised and transductive. In some applications, however, one can often obtain plenty of labeled normal data. In these cases, it is natural to hypothesize that these data will enable the detector to better capture normal behaviors than unlabelled data. In addition, transduction does not build a predictive model. If a new data point is added to the testing dataset, then they will have to re-run the algorithm (e.g. clustering or matrix factorization) from the beginning for predicting the labels, which can become more computationally costly. Wang *et al.* propose the first inductive multi-view semi-supervised anomaly detection (IMVSAD) approach in [32](here we stick on the definition of semi-supervised outlier detection in [1,3,4]). They design a probabilistic generative model for multi-view normal data and estimate model parameters by maximizing the likelihood of sample data with the EM algorithm. Then, the outlier score of an instance can be derived by computing its joint probability of all views. Though their multi-view detector achieves some competitive results in their evaluation, major limitations still exist. First, the number of latent components needs to be specified (often by cross-validation) before performing their model. Second, maximum likelihood estimates can be heavily biased for small samples, and the optimality properties may not be applicable to small samples. Lastly, the outlier score computation of their model can be computationally expensive when faced with the increasing number of views, because they have to calculate the inverse of the large covariance matrix of joint probability.

To deal with the issue and improve the work, we develop a Bayesian detector model based on variational inference in this paper. We show how it obtains the appropriate model dimensionality by introducing continuous hyper-parameters and the suitable prior distributions to sparsify their columns for automatically determining the dimension of latent factor. Also, the proposed Bayesian treatment version can cope with per-view noise which often corrupt the observations across different views at different levels [7]. Furthermore, we define a new outlier score measurement by using the latent representations to detect outliers in the subspace, which avoids the expensive computation of the inverse of a large matrix or comparing different views in a pairwise manner. Additionally, we conduct more comprehensive experiments and evaluate the model's detection performance on a wide variety of data sets.

The rest of this paper is organized as follows. In Sect. 2, we introduce the related work in multi-view outlier detection. In Sect. 3, we present the proposed model, including its notation setting, the problem formulation, model specification and inference, and outlier score measurement. In Sect. 4, we present experimental studies, results and discussion. In Sect. 5, we conclude the presented study and highlight future work.

2 Related Work

In general, our work is related to the research work on outlier detection and multi-view subspace modeling.

Outlier Detection. Many efforts have been made to tackle the problem of outlier detection over recent decades [20,31,38]. To detect abnormal activities on graph data, e.g. spot mis-behavioral IP sources in the traffic network, Tong *et al.* [31] propose a non-negative residual matrix factorization framework to handle the problem. Liu *et al.* [19] propose a SVDD-based learning framework to identify outliers in imperfect data by assigning likelihood values to input data. In [9], Du *et al.* design an outlier detection method by combining discriminative information learning and metric learning. All these methods were designed for single-view data. To date, a few methods have been developed to handle the outlier detection problem in multi-view data settings. Some of them try to find the samples that have inconsistent cross-view cluster memberships [11,18,21]. For example, Marcos *et al.* [21] propose a multi-view detection method via Affinity Propagation by comparing inconsistent affinity vectors in multiple views. To identify class outlier and attribute outlier simultaneously, $l_{2,1}$-norm induced error terms are integrated into low-rank subspace learning [17] and k-means clustering [37]. Besides, Li *et al.* [16] raise the problem of detecting the third type outlier (class-attribute outlier). Wang *et al.* [33] offer a novel hierarchical generative model to find multi-view outliers under a semi-supervised detection scenario via

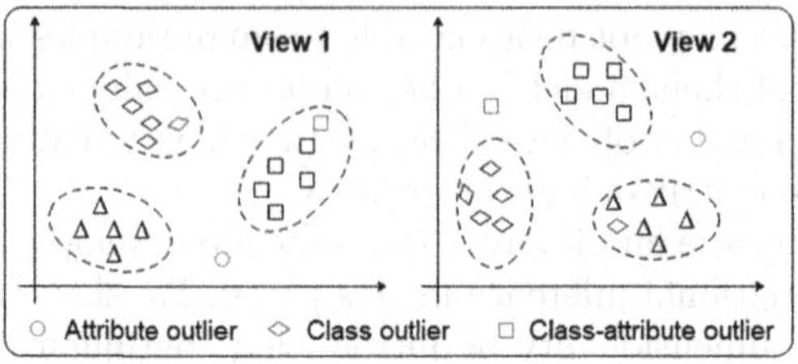

Fig. 1. Illustration of three types of outliers in a multi-view setting.

inductive learning. Sheng *et al.* [27] propose a nearest neighbour-based outlier measurement criterion to estimate the set of normal instances explicitly.

Multi-view Subspace Learning. Our approach is also related to subspace modeling. Subspace learning-based approaches aim to obtain a latent subspace shared by multiple views by assuming that observed views are generated from this latent space [36]. Canonical Correlation Analysis is an early and classical method for dimensionality reduction by learning the shared latent subspace jointly from two views [12]. It was further developed to conduct multi-view clustering [5], nonlinear subspace learning [10] and sparse formulation [2,30]. In contrast to Canonical Correlation Analysis, Diethe *et al.* [8] generalize Fisher's discriminant analysis to explore the latent subspace spanned by multi-view data in a supervised setting. Xia *et al.* [35] develop a multi-view spectral embedding algorithm which encodes multi-view features simultaneously to achieve a more meaningful low-dimensional embedding. Quadrianto *et al.* [25] study multi-view metric learning by constructing an accurate metric to precisely measure the dissimilarity between different examples associated with multiple views. The learned latent subspace can be used to infer another view from the observation view. Shon *et al.* [28] propose the shared GPLVM that can handle multiple observation spaces, where each set of observations is parameterized by a different set of kernels. Chen *et al.* [6] present a generic large-margin learning framework which is based on an undirected latent space Markov network for discovering predictive latent subspace representations shared by structured multi-view data.

3 Methodology

In this section, we first review the problem definition and explain notational conventions used throughout the paper, then we revisit the multi-view data generation assumption. Based on that, we illustrate the proposed detection model together with its estimation and the outlier score computation.

3.1 Problem Definition

Following the definition used in [16], there are three kinds of outliers in multi-view setting. As shown in Fig. 1, a *class-outlier* is an outlier that exhibits

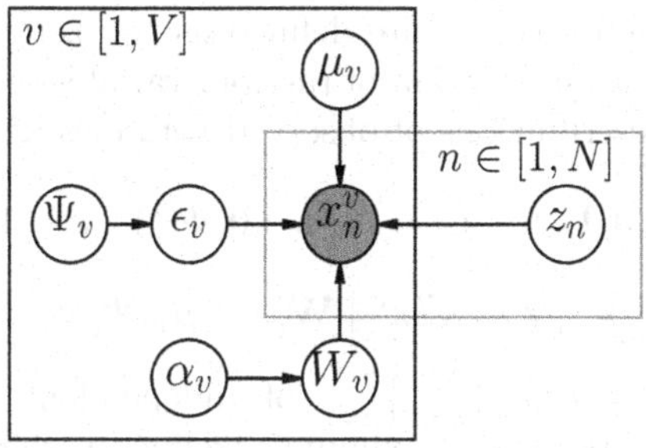

Fig. 2. The graphical representation of proposed Bayesian model.

inconsistent characteristics (e.g., cluster membership) across different views. An *attribute-outlier* is an outlier that exhibits consistent abnormal behaviours (e.g. far away from the majority) in each view. A *class-attribute-outlier* is an outlier that exhibits class outlier characteristics in some views while shows attribute outlier properties in the other views. Suppose we are given a dataset $\mathcal{D}$ which consists of N instances, denoted by $n = 1, ..., N$, described by V views with each view $v = 1, 2, ..., V$. The feature representation of instance n under view v is $\mathbf{x}_n^v \in \mathbb{R}^{d^v}$, where d^v is the dimensionality of view v. $X^v = [\mathbf{x}_1^v, \mathbf{x}_2^v, ..., \mathbf{x}_N^v] \in \mathbb{R}^{d^v \times N}$ is sample set observed in view v. In this way, the whole dataset is denoted as $\mathcal{D} = \{\mathbf{X}^1, \mathbf{X}^2, ..., \mathbf{X}^V\}$. Then, the multi-view outlier detection method computes a outlier score for each instance and compares it to a threshold $\hat{\tau}$ for finding the outlier in multi-view setting.

To link the multiple views $\mathbf{x}^1, \mathbf{x}^2, ...,$ and $\mathbf{x}^V$ together, a common latent variable $\mathbf{z}$ has been introduced in [32], and the task is to learn these common structure or the correspondence between observed views and the unobserved space. To explore it, authors in [32] propose a probabilistic model which describes the generative process of multi-view instances whose views are linked via a single, reduced-dimensionality latent variable space. Specifically, $\mathbf{x}^1, \mathbf{x}^2, ...,$ and $\mathbf{x}^V$ are generated from $\mathbf{z}$ by first choosing a value for the latent variable $\mathbf{z}$ and then sampling observed variables conditioned on this latent value. The $d^1, d^2, ..., d^V$-dimensional observed vectors $\mathbf{x}_n^1, \mathbf{x}_n^2, ..., \mathbf{x}_n^V$ are defined by a linear transformation governed by the matrix $\mathbf{W}_v \in \mathbb{R}^{d^v \times m}$ of the latent vector $\mathbf{z} \in \mathbb{R}^m$ plus a projection noise $\boldsymbol{\epsilon}_v \in \mathbb{R}^{d^v}$ and the data offset $\boldsymbol{\mu}_v \in \mathbb{R}^{d^v}$, so that

$$\mathbf{x}_n^v = \mathbf{W}_v \mathbf{z}_n + \boldsymbol{\mu}_v + \boldsymbol{\epsilon}_v \quad v = 1, 2, ..., V \tag{1}$$

3.2 Multi-view Bayesian Outlier Detector Formulation

In this section, we propose our Bayesian probabilistic model for capturing view-consistency patterns of multi-view normal data and detecting multi-view outliers in an inductive learning manner. We start with in the following the introduction of the specific probabilistic formulation. The distributions over the latent variable $\mathbf{z}$ and projection noise $\boldsymbol{\epsilon}_v$ are defined as two zero mean Gaussians. By the property of affine transformation of random variable, combining (1) and (2) gives the conditional distributions of observed variables $\mathbf{x}^v$ as (3)

$$\mathbf{z} \sim \mathcal{N}(\mathbf{z} \mid \mathbf{0}, \mathbf{I}_m); \quad \boldsymbol{\epsilon}_v \sim \mathcal{N}(\boldsymbol{\epsilon}_v \mid \mathbf{0}, \boldsymbol{\Psi}_v^{-1}) \quad v = 1, 2, ..., V \tag{2}$$

$$\mathbf{x}^v \mid \mathbf{z} \sim \mathcal{N}(\mathbf{x}^v \mid \mathbf{W}_v \mathbf{z} + \boldsymbol{\mu}_v, \boldsymbol{\Psi}_v^{-1}) \tag{3}$$

Given a dataset $\mathcal{D} = \left\{\{\mathbf{x}_n^v\}_{n=1}^N\right\}_{v=1}^V$ of independent and identically distributed observed points, the maximum likelihood solution of parameters $\{\mathbf{W}_v, \boldsymbol{\mu}_v, \boldsymbol{\Psi}_v^{-1}\}_{v=1}^V$ can be found by an EM algorithm [32]. However, the maximum-likelihood solution may lead to overfitting to small data sets. It is also not easy in practice to identify how many latent components they are since employing cross-validation to do so is computationally costly. Therefore, a Bayesian model selection is offered here, and we extend to a Bayesian generative model by introducing suitable prior distributions. In particular, we adopt ARD prior over $\mathbf{W}_v$ to automatically sparsify its columns

$$\mathbf{W}_v \mid \boldsymbol{\alpha}_v \sim \prod_{i=1}^{d^v} \mathcal{N}\left(\mathbf{w}_{vi} \mid \mathbf{0}, \left(diag(\boldsymbol{\alpha}_v)\right)^{-1}\right); \quad \boldsymbol{\alpha}_v \sim \prod_{j=1}^{m} \mathcal{G}(\alpha_{vj} \mid a_\alpha, b_\alpha) \tag{4}$$

where $\mathbf{w}_{vi} \in \mathbb{R}^m$ is the i_{th} row of $\mathbf{W}_v$, Gamma prior $\boldsymbol{\alpha}_v$ controls the magnitude of $\mathbf{W}_v$. If certain α_{vj} is large, the jth column of $\mathbf{W}_v$ will tend to take value zero and become little importance. We parameterize the distributions over $\boldsymbol{\mu}_v$ and $\boldsymbol{\Psi}_v$ by defining

$$\boldsymbol{\mu}_v \sim \mathcal{N}(\boldsymbol{\mu}_v \mid \mathbf{0}, \beta_v^{-1}\mathbf{I}_{d^v}); \quad \boldsymbol{\Psi}_v \sim \mathcal{W}(\boldsymbol{\Psi}_v \mid \mathbf{K}_v^{-1}, \nu_v) \tag{5}$$

where $\mathcal{W}(\boldsymbol{\Psi}_v|\mathbf{K}_v^{-1}, \nu_v) \propto |\boldsymbol{\Psi}_v|^{\frac{(\nu_v - d^v - 1)}{2}} \exp\left(-\frac{1}{2}Tr(\mathbf{K}_v\boldsymbol{\Psi}_v)\right)$ denotes the Wishart distribution. Since we have no further knowledge about the hyperparameters of priors, we choose broad ones by setting $a_\alpha = b_\alpha = \beta_v = 10^{-3}$, $\mathbf{K}_v = 10^{-3}\mathbf{I}_{d^v}$ and $m = \min\{d^v - 1; v = 1, \ldots, V\}$. Figure 2 summarizes the full Bayesian model and the inter-dependencies between the model parameters over a data set of N instances. Arrows represent conditional dependencies between random variables.

3.3 Multi-view Bayesian Outlier Detector Inference

In Sect. 3.2, we seek for a promising Bayesian treatment for the generative model to avoid discrete model selection and instead uses continuous hyper-parameters to automatically determine the appropriate dimensionality. In this section, we further explore a variational approximation approach to estimate model parameters and achieve model selection by learning from multi-view data. From Fig. 2, the joint probability of the data $\mathcal{D}$, latent components $\mathbf{Z} = \{\mathbf{z}_1, \ldots, \mathbf{z}_N\}$, and parameters $\boldsymbol{\Theta} = \{\{\mathbf{W}_v, \boldsymbol{\alpha}_v, \boldsymbol{\mu}_v, \boldsymbol{\Psi}_v\}_{v=1}^V\}$ can be written as

$$\begin{aligned} p(\mathbf{X}^1, \ldots, \mathbf{X}^V, \mathbf{Z}, \boldsymbol{\Theta}) = &\prod_{v=1}^V p(\mathbf{W}_v|\boldsymbol{\alpha}_v) p(\boldsymbol{\alpha}_v) p(\boldsymbol{\mu}_v) p(\boldsymbol{\Psi}_v) \\ &\times \prod_{n=1}^N \prod_{v=1}^V p(\mathbf{x}_n^v|\mathbf{z}_n, \mathbf{W}_v, \boldsymbol{\mu}_v, \boldsymbol{\Psi}_v) p(\mathbf{z}_n) \end{aligned} \tag{6}$$

Our goal is to learn the posterior distributions over all unknowns given data $\mathcal{D}$. It is analytically intractable to derive the posterior distribution $p(\mathbf{Z}, \boldsymbol{\Theta}|\mathcal{D})$ from Eq. (6) by marginalizing the joint distribution. Here, we apply variational inference for approximating the posterior by a factorized distribution

$$q(\mathbf{Z}, \boldsymbol{\Theta}) = \prod_{n=1}^{N} q(\mathbf{z}_n) \times \prod_{v=1}^{V} \left(q(\boldsymbol{\Psi}_v)q(\boldsymbol{\mu}_v) \prod_{j=1}^{m} q(\alpha_{vj}) \prod_{i=1}^{d^v} q(\mathbf{w}_{vi})\right) \quad (7)$$

The distribution q is found by maximizing the lower bound $\mathcal{L}_{q(\mathbf{Z},\boldsymbol{\Theta})}$

$$= \log p(\mathcal{D}) - KL(q(\mathbf{Z}, \boldsymbol{\Theta})||p(\mathbf{Z}, \boldsymbol{\Theta}|\mathcal{D})) = \iiint q(\mathbf{Z}, \boldsymbol{\Theta}) \log \frac{p(\mathcal{D}, \mathbf{Z}, \boldsymbol{\Theta})}{q(\mathbf{Z}, \boldsymbol{\Theta})} d\mathbf{Z} d\boldsymbol{\Theta} \quad (8)$$

Since $\log p(\mathcal{D})$ is constant, maximizing the low bound is equivalent to minimizing the KL divergence between $q(\mathbf{Z}, \boldsymbol{\Theta})$ and $p(\mathbf{Z}, \boldsymbol{\Theta}|\mathcal{D})$ through adjusting the approximate posterior. We substitute factor distributions in Eq. (7) into (8) and then dissect out the dependence on one of the factors $q_l(\boldsymbol{\Omega}_l)$. Maximizing $\mathcal{L}$ gives the result $\log q_l(\boldsymbol{\Omega}_l) =$

$$\mathbb{E}_{q_k(\boldsymbol{\Omega}_k), k \neq l}\left[\ln \prod_n p(\mathbf{z}_n) \prod_v p(\mathbf{x}_n^v|\mathbf{z}_n) + \ln \prod_v p(\mathbf{W}_v|\boldsymbol{\alpha}_v)p(\boldsymbol{\alpha}_v)p(\boldsymbol{\mu}_v)p(\boldsymbol{\Psi}_v)\right] + c \quad (9)$$

where $\boldsymbol{\Omega} = \{\{\mathbf{z}_n\}_{n=1}^{N}, \{\mathbf{W}_v, \{\alpha_{vj}\}_{j=1}^{m}, \boldsymbol{\mu}_v, \boldsymbol{\Psi}_v\}_{v=1}^{V}\}$ refers all latent components and parameters of model, $\mathbb{E}_{k \neq l}[\cdot]$ represents an expectation *w.r.t.* distribution $q_k(\boldsymbol{\Omega}_k)$ for all $k = 1, \ldots, |\boldsymbol{\Omega}|$ and $k \neq l$. Combining (9), (6) with the distributions defined in Sect. 3.2, we obtain the following factor distributions in turn.

$$q(\mathbf{z}_n) = \mathcal{N}(\mathbf{z}_n|\boldsymbol{\mu}_{\mathbf{z}_n}, \boldsymbol{\Sigma}_{\mathbf{z}_n}); \quad q(\boldsymbol{\Psi}_v) = \mathcal{W}(\boldsymbol{\Psi}_v|\hat{\mathbf{K}}_v^{-1}, \hat{\nu}_v) \quad (10)$$

$$q(\boldsymbol{\mu}_v) = \mathcal{N}(\boldsymbol{\mu}_v|\boldsymbol{\mu}_{\boldsymbol{\mu}_v}, \boldsymbol{\Sigma}_{\boldsymbol{\mu}_v}); \quad q(\alpha_{vj}) = \mathcal{G}(\alpha_{vj}|\hat{a}_\alpha, \hat{b}_\alpha) \quad (11)$$

$$q(\mathbf{w}_{vi}) = \mathcal{N}(\mathbf{w}_{vi}|\boldsymbol{\mu}_{\mathbf{w}_{vi}}, \boldsymbol{\Sigma}_{\mathbf{w}_{vi}}) \qquad i = 1, \ldots, d^v \quad (12)$$

where $n = 1, \ldots, N$, $v = 1, \ldots, V$, $j = 1, \ldots, m$,

$$\boldsymbol{\Sigma}_{\mathbf{z}_n} = \left(\sum_v \langle \mathbf{W}_v^T \boldsymbol{\Psi}_v \mathbf{W}_v \rangle + \mathbf{I}_m\right)^{-1}; \; \boldsymbol{\mu}_{\mathbf{z}_n} = \boldsymbol{\Sigma}_{\mathbf{z}_n}\left[\sum_v \langle \mathbf{W}_v^T \rangle \langle \boldsymbol{\Psi}_v \rangle (\mathbf{x}_n^v - \langle \boldsymbol{\mu}_v \rangle)\right] \quad (13)$$

$$\begin{aligned} \hat{\mathbf{K}}_v = \mathbf{K}_v + \sum_n \Big(&\mathbf{x}_n^v \mathbf{x}_n^{vT} + \langle \boldsymbol{\mu}_v \boldsymbol{\mu}_v^T \rangle - \langle \boldsymbol{\mu}_v \rangle \mathbf{x}_n^{vT} - \mathbf{x}_n^v \langle \boldsymbol{\mu}_v^T \rangle - \langle \mathbf{W}_v \rangle \langle \mathbf{z}_n \rangle \mathbf{x}_n^{vT} \\ &- \mathbf{x}_n^v \langle \mathbf{z}_n^T \rangle \langle \mathbf{W}_v^T \rangle + \langle \mathbf{W}_v \rangle \langle \mathbf{z}_n \rangle \langle \boldsymbol{\mu}_v^T \rangle + \langle \boldsymbol{\mu}_v \rangle \langle \mathbf{z}_n^T \rangle \langle \mathbf{W}_v^T \rangle + \langle \mathbf{W}_v \rangle \langle \mathbf{z}_n \mathbf{z}_n^T \rangle \langle \mathbf{W}_v^T \rangle \end{aligned} \quad (14)$$

$$\begin{aligned} &+ diag\big([Tr(\boldsymbol{\Sigma}_{\mathbf{w}_{v1}} \langle \mathbf{z}_n \mathbf{z}_n^T \rangle), \ldots, Tr(\boldsymbol{\Sigma}_{\mathbf{w}_{vd^v}} \langle \mathbf{z}_n \mathbf{z}_n^T \rangle)]\big)\Big) \\ &\boldsymbol{\Sigma}_v = \left(N \langle \boldsymbol{\Psi}_v \rangle + \beta_v \mathbf{I}_{d^v}\right)^{-1}; \quad \boldsymbol{\mu}_v = \boldsymbol{\Sigma}_v \sum_n \langle \boldsymbol{\Psi}_v \rangle (\mathbf{x}_n^v - \langle \mathbf{W}_v \rangle \langle \mathbf{z}_n \rangle) \end{aligned} \quad (15)$$

$$\hat{a} = a + \frac{d^v}{2}; \quad \hat{b} = b + \frac{\langle ||\mathbf{W}_{v,:j}||^2 \rangle}{2}; \quad \hat{\nu}_v = \nu_v + N \quad (16)$$

$$\boldsymbol{\Sigma}_{\mathbf{w}_{vi}} = \left(diag(\langle \boldsymbol{\alpha}_v \rangle) + \sum_n \langle \mathbf{z}_n \mathbf{z}_n^T \rangle \langle \boldsymbol{\Psi}_v \rangle_{ii}\right)^{-1} \tag{17}$$

$$\boldsymbol{\mu}_{\mathbf{w}_{vi}} = \boldsymbol{\Sigma}_{\mathbf{w}_{vi}} \sum_n \left(\langle \mathbf{z}_n \rangle \langle \boldsymbol{\Psi}_v \rangle_{,:i}{}^T (\mathbf{x}_n^v - \langle \boldsymbol{\mu}_v \rangle) - \langle \mathbf{z}_n \mathbf{z}_n^T \rangle \sum_{i'=1, i' \neq i}^{d^v} \langle \boldsymbol{\Psi}_v \rangle_{i'i} \langle \mathbf{w}_{vi'} \rangle \right) \tag{18}$$

where $M_{,:i}$ and $M_{i'i}$ represent the ith column and the (i', i) entry of matrix M respectively, and $\langle \cdot \rangle$ denotes the expectation. The set of equations (13–18) represent a set of consistency conditions for the maximum of the lower bound subject to the factorization constraint. We will seek a consistent solution by first initializing all of the factors $q_k(\boldsymbol{\Omega}_k)$ appropriately and then cycling through the factors and replacing each in turn with a revised estimate given by the right-hand side of equations evaluated using the current estimates for all of the other factors. We monitor the convergence of the variational optimization by evaluating the lower bound $\mathcal{L}(q(\mathbf{Z}, \boldsymbol{\Theta}))$ during the re-estimation of these factors. To summarize Sect. (3.2–3.3), we seek for a promising Bayesian treatment for the generative model to capture view-consistency patterns of multi-view normal data and approach the issue of selecting the optimal dimensionality by exploring a variational approximation approach.

3.4 Multi-view Outlier Score Estimation

After inferring the general Bayesian generative model of multi-view data from the observed training cases, we in this section characterize the new outlier score measurement proposed for every new arriving sample by analyzing the mutual distances of its representations across different views in the latent subspace. With optimal solution of factors in the $q(\mathbf{Z}, \boldsymbol{\Theta})$, we know the posterior distribution of the latent variable is expressed, from the Bayes' theorem, in the form

$$p(\mathbf{z}|\mathbf{x}^v) \propto p(\mathbf{x}^v|\mathbf{z})p(\mathbf{z}) = \int p(\mathbf{x}^v|\mathbf{z})p(\mathbf{z})d\mathbf{z} = \mathcal{N}(\mathbf{z}|\mathbf{L}^{-1}\mathbf{W}_v^T \boldsymbol{\Psi}_v (\mathbf{x}^v - \boldsymbol{\mu}_v), \mathbf{L}^{-1}) \tag{19}$$

by integrating out $\mathbf{z}$, it gives a joint (multivariate) Gaussian for $\mathbf{z}|\mathbf{x}^v$, where $\mathbf{L} = \mathbf{W}_v^T \boldsymbol{\Psi}_v \mathbf{W}_v + \mathbf{I}_m$. According to our generative assumption, all views of an normal example are generated from shared latent vectors. In other words, the distances between all latent vectors projected back from each observed view of the nominal should be shortened as more as possible. Thus it is feasible to use the difference of latent vectors to measure abnormality of the testing example. By this insight, we formulate the outlier score $s(\mathbf{x}_i)$ of an instance $\mathbf{x}_i = [\mathbf{x}_i^1; \mathbf{x}_i^2; \ldots; \mathbf{x}_i^V]$

$$s(\mathbf{x}_i) \propto \sum_{v=1}^{V} \|\mathbf{z}_i^v - \bar{\mathbf{z}}_i\| \tag{20}$$

where $\bar{\mathbf{z}}_i$ is the mean of representations of the i-th sample from all views in the latent subspace. The goal of our outlier score is to measure to what extent a sample's structure differs in different views in the latent space. The output of our proposed method is a ranked list of anomalies based on their outlier scores.

4 Experimental Evaluations

In this section, extensive experiments are conducted to demonstrate the effectiveness of our proposed method.

4.1 Experiment Setup

The performance of our method is evaluated on public Outlier Detection Datasets (ODDS)[1] and WebKB[2] dataset. For ODDS datasets (namely *thyroid, annthyroid, forestcover, vowels, pima* and *vertebral*), we synthesize multiple views by randomly splitting the features (often adopted in previous works, e.g. [16]), where each feature belongs to only one view. The real multi-view WebKB dataset contains webpages collected from four universities, including Cornell, Texas, Washington and Wisconsin. Without loss of generality, we report the detection results on one representative subset only *Cornell* in our experiments while similar results are obtained on the other three subsets. The Cornell subset contains 195 webpages over five labels. Each webpage is described by four views: content, inbound link, outbound link and cites. To generate three types of multi-view outliers, we follow the strategy presented in previous works (e.g. [16]) for a fair comparison. After the outlier generation stage, we evenly split all normal instances into two parts for keeping the normal as a clear majority in the test set, and use one of them as the training set to train the proposed model. Then, we verify the outlier detection performance on the test set which consists of the remaining normal data and the generated outliers.

Table 1. AUC values (mean ± std) on six ODDS datasets with outlier ratio 5%.

	Data sets	HOAD[11]	AP[21]	PLVM[13]	LDSR[16]	IMVSAD[32]	Proposed
Attribute Outlier	Thyroid	.5202 ± .0864	.6737 ± .1164	.8989 ± .0091	.9751 ± .0074	.9957 ± .0009	.9877 ± .0056
	Annthyroid	.5078 ± .0724	.5747 ± .0669	.8904 ± .0363	.9876 ± .0022	.9991 ± .0003	.9977 ± .0017
	Forestcover	.6801 ± .0866	.6774 ± .0739	.4870 ± .0126	.9983 ± .0005	.9990 ± .0020	.9985 ± .0078
	Vowels	.8540 ± .0691	.7062 ± .1125	.5481 ± .0067	.9181 ± .0153	.9991 ± .0022	.9900 ± .0084
	Pima	.5921 ± .0768	.9376 ± .0293	.9086 ± .0083	.9858 ± .0057	.9891 ± .0053	.9993 ± .0059
	Vertebral	.8338 ± .0972	.8586 ± .0604	.7564 ± .0061	.9793 ± .0200	.9998 ± .0004	.9986 ± .0036
Class Outlier	Thyroid	.5393 ± .0303	.5847 ± .0227	.5676 ± .0093	.8631 ± .0217	.8773 ± .0160	.8744 ± .0205
	Annthyroid	.5849 ± .0348	.5265 ± .0350	.4087 ± .0176	.7128 ± .0418	.7196 ± .0363	.7160 ± .0427
	Forestcover	.6872 ± .0337	.7906 ± .0332	.6035 ± .0044	.7551 ± .0293	.8883 ± .0137	.8669 ± .0192
	Vowels	.3818 ± .0384	.7520 ± .0513	.5479 ± .0282	.9245 ± .0173	.9650 ± .0163	.9670 ± .0152
	Pima	.5557 ± .0310	.5659 ± .0365	.5425 ± .0138	.5924 ± .0543	.6585 ± .0333	.6707 ± .0322
	Vertebral	.5209 ± .0812	.5272 ± .0449	.4444 ± .0416	.6070 ± .0568	.9782 ± .0065	.9848 ± .1526
Class-Attribute Outlier	Thyroid	.4934 ± .0270	.6380 ± .0723	.7122 ± .0191	.9344 ± .0179	.9914 ± .0034	.9863 ± .0075
	Annthyroid	.4976 ± .0311	.5647 ± .0819	.8933 ± .0134	.9122 ± .0220	.9936 ± .0021	.9896 ± .0096
	Forestcover	.4342 ± .0468	.8054 ± .0373	.8184 ± .0087	.9845 ± .0049	.9923 ± .0065	.9832 ± .0089
	Vowels	.5994 ± .1342	.8511 ± .0713	.6390 ± .0223	.9642 ± .0064	.9864 ± .0078	.9772 ± .0071
	Pima	.4181 ± .0260	.7916 ± .0555	.8249 ± .0063	.9315 ± .0146	.9646 ± .0080	.9719 ± .0120
	Vertebral	.7386 ± .0700	.7277 ± .0524	.6913 ± .0261	.9185 ± .0371	.9994 ± .0010	.9937 ± .0045

[1] https://odds.cs.stonybrook.edu.

[2] http://lig-membres.imag.fr/grimal/data.html.

We compared the proposed model with several state-of-the-art multi-view outlier detection methods: HOAD [11], AP [21], Probabilistic Latent Variable Model (PLVM) [13], Latent Discriminant Subspace Representation (LDSR) [16], and IMVSAD [32]. ROC curve and AUC score are used as the performance evaluation metrics in the experiments.

4.2 Experiment Results

For reducing the impact of fluctuations caused by the process of data construction, we repeat the random outlier generation procedure 20 times on each dataset, and at each time, we perturb 5% of the data in that procedure. We average their performance and report AUC results (mean ± std). From Table 1, we can observe that 1) inductive detectors (including the IMVSAD and the proposed one) consistently outperform other transductive counterparts (including HOAD, AP, PLVM and LDSR) on all six ODDS datasets for all kinds of multi-view outliers. The superiority of inductive detectors is expected as they leverage the semi-supervised outlier detection technique that can, to the highest degree, capture the nature and property of normal instances. This, in turn, can help the learned models to better distinguish whether the test instance is normal or not, thus improving the detection performance; 2) among inductive detectors, our proposed model achieves similarly competing detecting results as the IMVSAD while our proposed model avoids the time-consuming cross-validation procedure for model selection in the IMVSAD. Our Bayesian solution offers a significant advantage in allowing the accurate components of the model to be determined and the precise latent subspace to be constructed automatically; 3) the proposed model works better than IMVSAD on pima and vertebral datasets (especially for identifying multi-view class outliers), which falls in line with our previous remark that MLE relatively biases to small samples while Bayesian estimation can mitigate the problem by creating a statistical model to link data to parameters and formulating prior information about parameters. Figure 4 shows the ROC curves of all compared methods on the WebKB dataset with outlier ratios of 5% (left) and 10% (right). Likewise, we can observe that our approach achieves a higher AUC than its competitors, which demonstrates the effectiveness of our Bayesian detection method.

To investigate how the number of outliers affects the performance of different models, we experiment on data corrupted by progressively higher percentages of outliers. Figure 3 shows the variation of AUCs on dataset *pima* with outlier ratios of 2%, 5%, 10%, 15%, 20%, 25% and 30% for three types of outliers. We see that, in general, the performance decreases for all the methods as the outlier rate increases. Yet, our proposed method is comparatively more robust and outperforms other compared ones under all the different outlier ratio settings investigated.

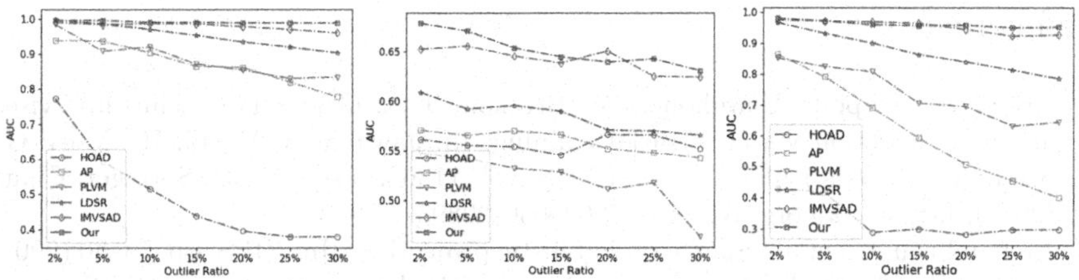

(a) Attribute Outlier (b) Class Outlier (c) Class-Attribute Outlier

Fig. 3. The variation curves of AUC W.R.T outlier ratio.

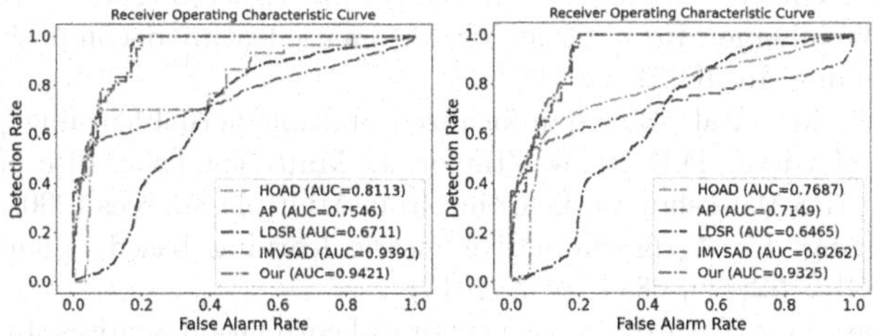

Fig. 4. ROC curves of compared methods on WebKB dataset.

5 Conclusions

In this work, we investigate the inductive learning framework for multi-view outlier detection. Through exploiting the correlation for multi-view data in the subspace and learning a predictive subspace shared by multiple views, we propose a novel solution to detecting outliers in multi-view data. The proposed model avoids extra computational cost which the preceding unsupervised multi-view outlier detectors often suffer. Also, it achieves the automatic specification of the dimensionality through a Bayesian reformulation of multi-view probabilistic projection. Furthermore, the proposed method defines a new outlier score measurement to detect outliers in the latent subspace. Thanks to the avoidance of calculating outlier scores by permuting all view pairs, our measurement can be easily extensible to more views. Experimental results on multiple datasets demonstrate its promising effectiveness. Our future work will focus on detecting multi-view outliers from more complex data types including temporal and dynamically changing data.

Acknowledgement. The authors would like to thank the support from Zhejiang Lab (111007-PI2001) and Zhejiang Provincial Natural Science Foundation (LZ21F030001).

References

1. Akcay, S., Atapour-Abarghouei, A., Breckon, T.P.: GANomaly: semi-supervised anomaly detection via adversarial training. In: Jawahar, C.V., Li, H., Mori, G., Schindler, K. (eds.) ACCV 2018. LNCS, vol. 11363, pp. 622–637. Springer, Cham (2019). https://doi.org/10.1007/978-3-030-20893-6_39
2. Archambeau, C., et al.: Sparse probabilistic projections. In: NIPS, pp. 73–80 (2009)
3. Chalapathy, R., et al.: Deep learning for anomaly detection: a survey. CoRR (2019)
4. Chandola, V., Banerjee, A., Kumar, V.: Anomaly detection: a survey. ACM Comput. Surv. (CSUR) **41**(3), 1–58 (2009)
5. Chaudhuri, K., Kakade, S.M., Livescu, K., Sridharan, K.: Multi-view clustering via canonical correlation analysis. In: ICML, pp. 129–136 (2009)
6. Chen, M., Weinberger, K.Q., Chen, Y.: Automatic feature decomposition for single view co-training. In: ICML (2011)
7. Christoudias, M., et al.: Bayesian localized multiple kernel learning (2009)
8. Diethe, T., Hardoon, D.R., Shawe-Taylor, J.: Multiview fisher discriminant analysis. In: NeurIPS Workshop on Learning from Multiple Sources (2008)
9. Du, B., Zhang, L.: A discriminative metric learning based anomaly detection method. TGRS **52**(11), 6844–6857 (2014)
10. Fyfe, C., Lai, P.L.: ICA using kernel canonical correlation analysis. In: Proceedings of International Workshop on Independent Component Analysis and Blind Signal Separation (2000)
11. Gao, J., et al.: A spectral framework for detecting inconsistency across multi-source object relationships. In: ICDM, pp. 1050–1055. IEEE (2011)
12. Hotelling, H.: Relations between two sets of variates. In: Kotz, S., Johnson, N.L. (eds.) Breakthroughs in Statistics. Springer Series in Statistics (Perspectives in Statistics), pp. 162–190. Springer, Heidelberg (1992). https://doi.org/10.1007/978-1-4612-4380-9_14
13. Iwata, T., Yamada, M.: Multi-view anomaly detection via robust probabilistic latent variable models. In: NeurIPS, pp. 1136–1144 (2016)
14. Izakian, H., Pedrycz, W.: Anomaly detection in time series data using a fuzzy c-means clustering. In: IFSA/NAFIPS, pp. 1513–1518. IEEE (2013)
15. Ji, Y.X., et al.: Multi-view outlier detection in deep intact space. In: ICDM, pp. 1132–1137. IEEE (2019)
16. Li, K., Li, S., Ding, Z., Zhang, W., Fu, Y.: Latent discriminant subspace representations for multi-view outlier detection. In: AAAI (2018)
17. Li, S., Shao, M., Fu, Y.: Multi-view low-rank analysis for outlier detection. In: SDM, pp. 748–756. SIAM (2015)
18. Liu, A.Y., Lam, D.N.: Using consensus clustering for multi-view anomaly detection. In: SPW, pp. 117–124. IEEE (2012)
19. Liu, B., Xiao, Y., Philip, S.Y., Hao, Z., Cao, L.: An efficient approach for outlier detection with imperfect data labels. TKDE **26**(7), 1602–1616 (2013)
20. Liu, F.T., et al.: Isolation-based anomaly detection. TKDD **6**(1), 1–39 (2012)
21. Marcos Alvarez, A., Yamada, M., Kimura, A., Iwata, T.: Clustering-based anomaly detection in multi-view data. In: CIKM, pp. 1545–1548. ACM (2013)
22. Na, G.S., Kim, D., Yu, H.: DILOF: Effective and memory efficient local outlier detection in data streams. In: KDD, pp. 1993–2002 (2018)
23. Pang, G., Cao, L., Liu, H.: Learning representations of ultrahigh-dimensional data for random distance-based outlier detection. In: KDD, pp. 2041–2050 (2018)

24. Pei, Y., Zaiane, O.R., Gao, Y.: An efficient reference-based approach to outlier detection in large datasets. In: ICDM, pp. 478–487. IEEE (2006)
25. Quadrianto, N., Lampert, C.H.: Learning multi-view neighborhood preserving projections. In: ICML (2011)
26. Senator, T.E., et al.: Detecting insider threats in a real corporate database of computer usage activity. In: KDD, pp. 1393–1401 (2013)
27. Sheng, X.R., Zhan, D.C., Lu, S., Jiang, Y.: Multi-view anomaly detection: neighborhood in locality matters. In: AAAI, vol. 33, pp. 4894–4901 (2019)
28. Shon, A., Grochow, K., Hertzmann, A., Rao, R.P.: Learning shared latent structure for image synthesis and robotic imitation. In: NeurIPS, pp. 1233–1240 (2006)
29. Spirin, N., Han, J.: Survey on web spam detection: principles and algorithms. ACM SIGKDD Explor. Newsl. **13**(2), 50–64 (2012)
30. Suo, X., et al.: Sparse canonical correlation analysis. CoRR (2017)
31. Tong, H., Lin, C.Y.: Non-negative residual matrix factorization with application to graph anomaly detection. In: SDM, pp. 143–153. SIAM (2011)
32. Wang, Z., et al.: Inductive multi-view semi-supervised anomaly detection via probabilistic modeling. In: ICBK, pp. 257–264. IEEE (2019)
33. Wang, Z., Lan, C.: Towards a hierarchical Bayesian model of multi-view anomaly detection. In: IJCAI (2020)
34. Wu, S., Wang, S.: Information-theoretic outlier detection for large-scale categorical data. TKDE **25**(3), 589–602 (2011)
35. Xia, T., et al.: Multiview spectral embedding. IEEE Trans. Syst. Man Cybern. Part B (Cybern.) **40**(6), 1438–1446 (2010)
36. Xu, C., Tao, D., Xu, C.: A survey on multi-view learning. CoRR (2013)
37. Zhao, H., Fu, Y.: Dual-regularized multi-view outlier detection. In: IJCAI (2015)
38. Zimek, A., Gaudet, M., Campello, R.J., Sander, J.: Subsampling for efficient and effective unsupervised outlier detection ensembles. In: KDD, pp. 428–436 (2013)

GLAD-PAW: Graph-Based Log Anomaly Detection by Position Aware Weighted Graph Attention Network

Yi Wan, Yilin Liu, Dong Wang, and Yujin Wen(✉)

AI Application and Research Center, Huawei Technologies, Shenzhen, China
{wanyi20,wangdong153,wenyujin}@huawei.com

Abstract. Anomaly detection is a crucial and challenging subject that has been studied within diverse research areas. In this work, we focus on log data (especially computer system logs) which is a valuable source to investigate system status and detect system abnormality. In order to capture transition pattern and position information of records in logs simultaneously, we transfer log files to session graphs and formulate the log anomaly detection problem as a graph classification task. Specifically, we propose GLAD-PAW, a graph-based log anomaly detection model utilizing a new position aware weighted graph attention layer (PAWGAT) and a global attention readout function to learn embeddings of records and session graphs. Extensive experimental studies demonstrate that our proposed model outperforms existing log anomaly detection methods including both statistical and deep learning approaches.

Keywords: Anomaly detection · Graph neural network · Log data

1 Introduction

Computer systems are ubiquitous among modern industry and applications. With increasing complexity of computations and large volume data, these systems can be vulnerable to attacks and abnormalities. Even a single service anomaly can impact millions of users' experience. Accurate and timely anomaly detection is a crucial task in monitoring system executions and assists in building trustworthy computer systems.

Computer systems routinely generate logs to record notable events and executions indicating what has happened. These widely available logs are valuable sources to investigate system status and detect anomalies. As summarized in [6], log-based anomaly detection framework mainly contains four steps: log parsing, session creation, feature representation learning, and anomaly detection. The purpose of log parsing is to extract a group of log templates (referred to as "events"), whereby raw logs can be transferred into structured texts. For example, as shown in Fig. 1, logs L_2 and L_3 are alike and both can be parsed as

Y. Wan and Y. Liu— Equal contribution.

K. Karlapalem et al. (Eds.): PAKDD 2021, LNAI 12712, pp. 66–77, 2021.
https://doi.org/10.1007/978-3-030-75762-5_6

Logs:
L1: 081109 203518 35 INFO dfs.FSNamesystem: BLOCK* NameSystem. allocateBlock: /mnt/hadoop /mapred/system/job_200811092030_0001/job.jar. blk_-1608999687919862906
L2: 081109 203519 143 INFO dfs.DataNode$DataXceiver: Receiving block blk_-1608999687919862906 src: /10.250. 10.6:40524 dest: /10.250.10.6:50010
L3: 081109 203519 145 INFO dfs.DataNode$DataXceiver: Receiving block blk_-1608999687919862906 src: /10.250. 14.224:42420 dest: /10.250.14.224:50010
L4: 081109 203519 145 INFO dfs.DataNode$Packet Responder: PacketResponder 1 for block blk_-1608999 687919862906 terminating
L5: 081109 203519 145 INFO dfs.DataNode$Packet Responder: PacketResponder 2 for block blk_-1608999 687919862906 terminating

block_id:
blk_-1608999687919862906
Templates:
v1: BLOCK* NameSystem.allocateBlock: <*> <*>
logs: L1
v2: Receiving block <*> src: <*> dest: <*>
logs: L2, L3
v3: PacketResponder <*> for block <*> <*>
logs: L4, L5

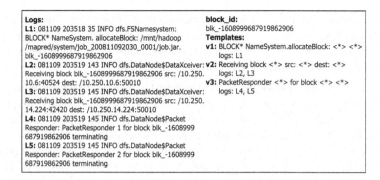

Fig. 1. Example of log entries and their corresponding templates.

"Receiving block $< * >$ src: $< * >$ dest: $< * >$". In session creation step, parsed logs are usually grouped by one of the three following windows: fixed window, sliding window, and session window. In Fig. 1, we group logs based on block_id (session window), and denote templates as v_1, v_2, v_3. Then we obtain an event session $(v_1, v_2, v_2, v_3, v_3)$ with block_id 1608999687919862906. The sessions are further encoded into numerical feature vectors and fed into anomaly detection models for judgment. In this paper, we focus on the feature representation learning and anomaly detection steps. As for log parsing, we apply a tree based log parsing algorithm called DRAIN [5] in all experiments.

Existing approaches for log anomaly detection can be classified into two categories: 1) quantitative detection models, and 2) sequential detection models. The first type of approaches utilizes quantitative information of log events that reflects underlying relationships between records. When such relationships are broken, the input logs would be marked as anomalies. Some representative models include PCA [19], Isolation Forest [11], Invariant Mining [12] and Log Clustering [10]. The second type of approaches assumes that log sequences follow some intrinsic patterns. A log sequence is treated as sequential anomaly if it deviates from normal patterns. For example, Spectrogram [15] utilizes mixture-of-Markov-Chains; DeepLog [3] implements the Long Short-Term Memory (LSTM); LogAnomaly [13] adds attention mechanism and frequency features to DeepLog.

Even though successes have been achieved in certain scenarios, the models above still have some drawbacks: statistical models fail to include the order of log sequences; Markov Chain based models assume that every log event is influenced only by its recent N predecessors rather than the entire history, and with the increase of N, the model computational complexity will grow exponentially; RNN based models fail to explicitly capture fine-grained relationships between events and are not always ideal in dealing with long-term dependency.

To address aforementioned drawbacks, we propose GLAD-PAW, a graph neural network (GNNs)-based log anomaly detection model regarding log events as nodes and interactions between log events as edges. GNNs are proposed to combine the feature information and the graph structure to learn better representations on graphs via propagation and aggregation. The recent success in node classification, link prediction and graph classification tasks of GNNs suggests potential approaches to implementing graph structure in anomaly detection.

The main contributions of this work are as follows.

- To the best of our knowledge, we are the first to investigate log-based anomaly detection task based on graph neural network models. Specifically, we propose a novel log anomaly detection model (GLAD-PAW) which performs anomaly detection task by transferring it into graph classification problem.
- In GLAD-PAW, a novel PAWGAT graph layer is proposed, which is a position aware weighted graph attention layer. This layer serves as the node feature encoder by adding position information of each node and learning to assign different weights to different neighbors, which effectively utilizes information between events in sessions.
- We evaluate GLAD-PAW through extensive experiments on two benchmark log datasets and achieve state-of-the-art results.

2 Related Work

2.1 Log-Based Anomaly Detection

Logs are valuable sources to investigate system status and detect system anomalies. Tremendous efforts have been made to perform log-based anomaly detection. Statistical models utilize quantitative information of log events to detect anomalies. Principal component analysis (PCA) [19] constructs normal and abnormal subspaces, and computes the squared distance of the projection of event frequency vectors onto the anomaly subspace. Invariant Mining method [12] tries to estimate the invariant space via singular value decomposition (SVD) on input matrix. Log Clustering [10] clusters the logs to ease log-based problem identification and utilizes a knowledge base to check if the log sequences occurred before. Apart from these distance-based methods, [11] proposes Isolation Forest that constructs isolation trees assuming that anomalies are more susceptible to isolation. Different from statistical methods above, mixture-of-Markov-Chains model [15] is proposed by learning a proper distribution on overlapping n-grams, which takes into account sequential information of logs. In recent years, deep learning methods have been developed to incorporate sequential information and temporal dependency in anomaly detection. In general, these approaches aim to capture the intrinsic pattern of the system under normal executions and treat deviations as anomalies. For example, DeepLog [3] models log sequences as natural language and utilize Long Short-Term Memory (LSTM) [7] architecture to learn log patterns from normal executions. LogAnomaly [13] adds attention mechanism and frequency features to DeepLog.

2.2 Graph Neural Networks

Graph neural networks (GNNs) are proposed to combine the feature information and the graph structure to learn better representations on graphs via propagation and aggregation. Recently, GNNs have been applied on graph level tasks, e.g., graph classification, which requires global graph information including both graph structures and node attributes. Its objective is to train a model that generates the corresponding label for every input graph, in which the key is to learn the graph level feature representation. [4] proposed a readout mechanism that globally combines the representations of graph nodes. [14] proposed a similar approach by introducing a virtual node that bidirectionally connects to all existing nodes in the graph to represent the graph level feature. SortPool [21] sorts node embeddings according to their structural roles within a graph and feeds the sorted embeddings to the next convolutional layer. DiffPool [20] and SAGPool [1] both consider a hierarchical representations of graphs and can be combined with various graph neural network architectures in an end-to-end fashion. EdgePool [2] is proposed based on edge contraction, which performs the task by choosing edges instead of nodes and pooling the connected nodes.

3 The Proposed Model

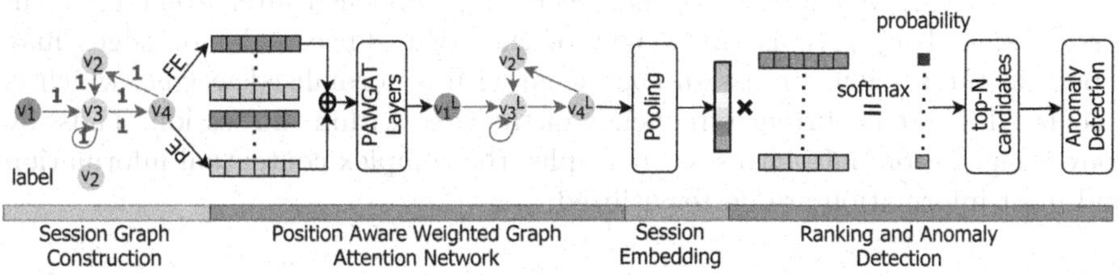

Fig. 2. The overall architecture of GLAD-PAW with L PAWGAT layers. FE and PE denote event embeddings and position embeddings, and $\times$ refers to the inner product of two vectors.

In this section, we introduce the proposed GLAD-PAW model, as shown in Fig. 2. We first formulate the problem of session-based log anomaly detection, then explain how to construct graphs from logs. Following the construction of session graphs, we describe PAWGAT layer and our GLAD-PAW model in details. In the following content, we denote vectors with bold lowercase letters, matrices with bold uppercase letters.

3.1 Problem Statement

The objective of log-based anomaly detection is to predict anomalies in log sessions based on historical sequential records without labels, i.e., an unsupervised anomaly detection task. In this work, we divide this task into two steps: next event prediction and anomaly detection. For a given log file, after log parsing, let $V = \{v_1, v_2, \ldots, v_{|V|}\}$ denote the set of all unique events after log parsing,

s denote an anonymous session, $v_{s,i} \in V$ $(i = 1, 2, \ldots, n)$ represent the ith event in session s, and thereby s can be represented by $s = (v_{s,1}, v_{s,2}, \ldots, v_{s,n})$ in time order.

The first step aims to predict the last event $v_{s,n}$ based on the first $(n-1)$ events for each session. To be exact, our model first generates probability vector $\hat{\mathbf{y}} = (\hat{y}_1, \hat{y}_2, \ldots, \hat{y}_{|V|})$ for all possible events, where $\hat{y}_i$ is the prediction score of the corresponding item. The events with top-N greatest values in $\hat{\mathbf{y}}$ will be the candidate events. In the second step, if the truth label is not included in top-N candidates predicted in the first step, such session will be labelled as abnormal.

3.2 Graph Construction

As shown in Fig. 2, at first, each session is converted into a weighted directed session graph, $G_s = (V_s, E_s)$, $G_s \in \mathcal{G}$, where $\mathcal{G}$ is the set of all session graphs. For example, given a session $s = (v_1, v_3, v_4, v_2, v_3, v_3, v_2)$, the session graph and corresponding truth label are shown in Fig. 2. In session graph G_s, the node set V_s is composed by unique events in s with initial feature vector $\mathbf{h}_v \in \mathbb{R}^d$, where d is the dimension of the latent space. The edge set E_s contains all directed edges $(v_{s,i-1}, v_{s,i}, w_{s,(i-1)i})$ indicating that event $v_{s,i}$ happened after event $v_{s,i-1}$ in session s, and $w_{s,(i-1)i}$ is the weight of the edge. Since nodes or edges may repeat in the session, we assign each edge with a normalized weight, which is calculated as the occurrence frequency of the edge within the session. Thus, by converting session data into session graphs, the complex contextual information and prior information are better utilized.

3.3 Position Aware Weighted Graph Attention Layer

Position Aware Node Embedding. In the aforementioned graph construction method, repeated events appearing at different positions in the session are regarded as a single node on the graph, which will lead to loss of position information in sessions. To tackle this problem, inspired by [17], we record the last occurrence position of each item in a session graph and inject a learnable embedding $\mathbf{h}_p \in \mathbb{R}^d$ for each position, with $p \in \{1, 2, \ldots, \ell\}$ and ℓ being the maximum number of unique events in all sessions. For example, session $s = (v_1, v_3, v_4, v_2, v_3, v_3, v_2)$ contains four unique events v_1, v_2, v_3, v_4, whose initial position can be represented as 4, 2, 1, 3 according to the reverse order of their last occurrences in s. Then we map their position into d dimensional embeddings to match the feature embeddings.

Intuitively, the reason for choosing the last occurrence position is that actions close to the current moment often have greater impact on the next possible events. With position embedding, for item $v_{s,i}$ in session s, the position aware node embedding could be formulated as:

$$\mathbf{h}_{s,i} = \mathbf{h}_{v_{s,i}} + \mathbf{h}_{p_{s,i}}, \tag{1}$$

where $\mathbf{h}_{v_{s,i}}$, $\mathbf{h}_{p_{s,i}}$ are feature embeddings and position embeddings for ith node in session s.

Weighted Graph Attention Layer. The input to our weighted graph attention layer is session graph node embedding vectors $\mathbf{h}_s = [\mathbf{h}_{s,1}, \mathbf{h}_{s,2}, \ldots, \mathbf{h}_{s,n}]^\top$, where $\mathbf{h}_{s,i} \in \mathbb{R}^d$ is obtained by Eq. (1) and $\cdot^\top$ represents transposition. The layer produces a new set of node features $\mathbf{h}'_s = [\mathbf{h}'_{s,1}, \mathbf{h}'_{s,2}, \ldots, \mathbf{h}'_{s,n}]^\top$ for each session graph as its output, where $\mathbf{h}'_{s,i} \in \mathbb{R}^d$. In order to obtain sufficient expressive ability to transform the input features into higher order features, we first apply a shared linear transformation parameterized by a weight matrix $\mathbf{W} \in \mathbb{R}^{d' \times d}$ to every node, where d' is the latent dimensionality of graph layer. Then for every node $v_{s,i}$ in session s, a self-attention mechanism is applied to aggregate information from its first-order neighbors $\mathcal{N}(i)$. In this work, the size of the session graph is constrained by window size, which should not be too large in practice. Thus, we take the entire neighborhood of nodes without sampling strategy. The self-attention mechanism a: $\mathbb{R}^{d'} \times \mathbb{R}^{d'} \times \mathbb{R} \rightarrow \mathbb{R}$ computes attention coefficients as:

$$e_{s,ij} = a(\mathbf{W}\mathbf{h}_{s,i}, \mathbf{W}\mathbf{h}_{s,j}, w_{s,ij}) \tag{2}$$

which implies the importance of node j's features to node i. To compare the importance of different neighbors directly, we normalize e_{ij} using the softmax function across all nodes $j \in \mathcal{N}(i)$:

$$\alpha_s = \text{softmax}(e_{s,ij}) = \frac{\exp(e_{s,ij})}{\sum_{k \in \mathcal{N}(i)} \exp(e_{s,ik})}. \tag{3}$$

The choice of attention mechanism varies. In this work, we use a single MLP layer with a shared weight vector $\mathbf{a} \in \mathbb{R}^{2d'+1}$, and apply activation function LeakyReLU nonlinearity (with negative slope 0.2). Expanded fully, the attention coefficients can be computed as follows:

$$\alpha_{s,ij} = \frac{\exp\left(\text{LeakyReLU}\left(\mathbf{a}^\top [\mathbf{W}\mathbf{h}_{s,i} || \mathbf{W}\mathbf{h}_{s,j} || w_{s,ij}]\right)\right)}{\sum_{k \in \mathcal{N}(i)} \exp\left(\text{LeakyReLU}\left(\mathbf{a}^\top [\mathbf{W}\mathbf{h}_{s,i} || \mathbf{W}\mathbf{h}_{s,k} || w_{s,ik}]\right)\right)}, \tag{4}$$

where $||$ is the concatenation operation.

For every node i in session graph G_s, once obtained all attention coefficients of its neighbors computed as Eq. (4), we aggregate the features of neighbors into a single vector. Here, we choose a linear combination aggregator (with nonlinearity, σ) which summarizes attention coefficients as:

$$\mathbf{h}'_{s,i} = \sigma\left(\sum_{j \in \mathcal{N}(i)} \alpha_{s,ij} \mathbf{W}\mathbf{h}_{s,j}\right). \tag{5}$$

To stabilize the training process of the self-attention layers, we apply the multi-head setting as suggested in previous work [16]. Specifically, we extend our attention mechanism processed in Eq. (4) by employing K independent heads, and concatenating the results of all K heads as the output feature representation:

$$\mathbf{h}'_{s,i} = ||_{k=1}^{K} \sigma\left(\sum_{j \in \mathcal{N}_i} \alpha_{s,ij}^k \mathbf{W}^k \mathbf{h}_{s,j}\right), \tag{6}$$

where $\alpha^k_{s,ij}$ are normalized attention coefficients computed by the k-th attention mechanism shown in Eq. (4), and $\mathbf{W}^k$ is k-th input linear transformation weight matrix. In this setting, we finally get output features, $\mathbf{h}'_{s,i} \in \mathbb{R}^{Kd'}$.

3.4 Session Embeddings

By stacking L PAWGAT layers, we can obtain the embedding vectors $\mathbf{h}'^L_s$ of all nodes involved in session s. The next step is to use a readout function to generate a representation of session graph $\mathcal{G}_s$. Common approaches to this problem include simply summing up or averaging all node embeddings in final layer, which cannot provide sufficient model capacity for learning a representative session embedding for the session graph. In contrast, we choose the readout function as in [9] to define a graph level representation vector:

$$\mathbf{h}_{\mathcal{G}_s} = \tanh\left(\sum_{i=1}^{|V_s|} \operatorname{softmax}\left(f(\mathbf{h}'^L_{s,i})\right) \odot \tanh\left(g(\mathbf{h}'^L_{s,i})\right)\right), \tag{7}$$

where f and g are neural networks applied on $\mathbf{h}'^L_{s,i}$. Actually, $\operatorname{softmax}\left(f(\mathbf{h}'^L_{s,v})\right)$ acts as a soft attention mechanism in [18] that decides the importance of nodes in session graph $\mathcal{G}_s$ relevant to current task.

3.5 Prediction and Anomaly Detection

After obtaining the session graph embedding $\mathbf{h}_{\mathcal{G}_s} \in \mathbb{R}^d$, we can calculate a prediction score for every event $v_i \in V$ by multiplying its initial node embedding $\mathbf{h}_{v_i} \in \mathbb{R}^d$ $(i = 1, \cdots, |V|)$ with $\mathbf{h}_{\mathcal{G}_s}$:

$$\hat{z}_i = \mathbf{h}_{v_i}^\top \mathbf{h}_{\mathcal{G}_s}. \tag{8}$$

Then we apply a softmax function to transform the prediction score vector into the probability distribution form $\hat{\mathbf{y}}$

$$\hat{\mathbf{y}} = \operatorname{softmax}(\hat{\mathbf{z}}). \tag{9}$$

where $\hat{\mathbf{z}} = (\hat{z}_1, \ldots, \hat{z}_{|V|})$ denotes score vector over all events in set V and $\hat{\mathbf{y}} = (\hat{y}_1, \ldots, \hat{y}_{|V|})$ represents the probability vector of nodes appearing to be the next event in session s. Since we formulate our next event prediction task as a graph level classification problem, we train our model by minimizing cross-entropy loss function between $\hat{\mathbf{y}}$ and true one-hot encoded label $\mathbf{y}$. For the top-N candidates, we simply pick the highest N probabilities over all events based on $\hat{\mathbf{y}}$. If the truth label is not included in top-N candidates, a session will be labelled as abnormal.

4 Experiments

In this section, we present our experimental setup and results. Our experiments are designed to answer the following research questions:

- **RQ1:** Does GLAD-PAW achieve the state-of-the-art performance?
- **RQ2:** How does PAWGAT layer work for anomaly detection problem compared with other graph layers (including ablation test)?
- **RQ3:** How do the key hyperparameters affect model performance, such as the graph layer and number of heads?

4.1 Datasets

We evaluate our work on two representative benchmark datasets:

HDFS. HDFS dataset is generated and collected from Amazon EC2 platform through running Hadoop-based map-reduce jobs. It contains messages about blocks when they are allocated, replicated, written, or deleted. Each block is assigned with a unique ID in the log, treated as log sessions.

BGL. BGL dataset is Blue Gene/L supercomputer system logs, which consists of 128K processors and is deployed at Lawrence Livermore National Laboratory (LLNL). We generate sessions from BGL dataset by sliding window with window size 10 and step 1. A session is labeled as anomalous if it contains at least one abnormal entry.

We randomly take 80% of normal sessions as training data and the rest 20% normal sessions along with all abnormal sessions as test data. Statistics of both datasets are shown in Table 1.

Table 1. Statistics of datasets.

Dataset	Duration	Entries	Anomalies	Events
HDFS	38.7 h	11,197,954	16,838 (blocks)	29
BGL	7 month	4,747,963	348,460 (logs)/391,936 (sessions)	380

4.2 Baslines and Evaluation Metrics

In order to prove the advantage our proposed model, we compare GLAD-PAW with six representative baseline methods: Isolation Forest [11], PCA [19], Invariant Mining [12], Log Clustering [10], Deeplog [3], and LogAnomaly [13].

Since anomalies are rare in real life, instead of using Accuracy (very likely to achieve values close to 1), we use Precision, Recall and F_1-measure as the evaluation metrics in our experiments. It is worth to note that, BGL dataset contains a number of out-of-vocabulary (OOV) events, which are events that have never seen in the training set. Therefore, we report the OOV/non-OOV results for BGL in Table 2.

4.3 Experimental Setup

We apply GLAD-PAW using two PAWGAT layers with each layer containing four heads. The hidden sizes of the node feature vectors are set to 128 for all layers. All parameters of our model are initialized using a Gaussian distribution with mean 0 and variance 0.1^2. For optimizer, we choose Adam optimizer with learning rate 0.001 and the linear schedule decay rate strategy. We repeat all the methods five times and choose the average results as the final results. The experiments are conducted on a Linux server with Intel(R) Xeon(R) 2.40GHz CPU and Tesla V100 GPU, running with Python 3.6, PyTorch 1.5.0.

4.4 RQ1: Comparison with Baseline Models

GLAD-PAW utilizes multiple layers of PAWGAT to capture transition pattern and position information of log records within session graphs, and applies an attention based readout function that generates graph level representations to detect anomalies. Experimental results on HDFS and BGL are shown in Table 2. It can be seen that the proposed GLAD-PAW model achieves the state-of-the-art performance on benchmark datasets - it outperforms all the baseline methods with an average F_1 score of 0.984 on the HDFS dataset, and 0.979 on the BGL dataset. We also substitute the readout function with a mean pooling layer (GLAD-MEAN). This variant also performs better than any other models except GLAD-PAW, which demonstrates the efficacy of PAWGAT layer and the attention based readout function.

The statistical baseline models perform much worse than neural network based baselines and our proposed model. These statistical models only consider the frequency features and ignore transition orders, which lead to failure of capturing reasonable representations for events and sessions in logs. The implementation of neural networks, e.g., DeepLog, LogAnomaly, GLAD-MEAN have largely improved the performances. DeepLog is the first model to apply LSTM as an encoder unit in log-based anomaly detection task. Though LSTM is a good match for sequence prediction, it treats every input event as equally important, leading to its main drawback. LogAnomaly incorporates an attention mechanism with LSTM units and adds frequency information into consideration, which outperforms DeepLog on both datasets. However, it still suffers from the same drawbacks of LSTM units.

4.5 RQ2: Comparison with Other GNN Layers

As mentioned above, many different types of GNN layers can be used to learn node embeddings. To prove the efficacy of PAWGAT, we substitute all graph layers with GCN [8], GGNN [9] and GAT [16]. Besides, we also perform ablation studies by dropping position information (WGAT) and weight information (PAGAT) of PAWGAT. In Fig. 3 (a), results of above five GNN layers are shown with Precision, Recall and F_1-measure on HDFS dataset.

Table 2. Evaluation on benchmark logs.

Model	HDFS			BGL		
	Precision	Recall	F_1	Precision	Recall	F_1
Isolation Forest	0.950	0.756	0.842	0.483	0.468	0.475
PCA	0.974	0.445	0.611	0.180	0.054	0.083
Invariant Mining	0.975	0.633	0.768	0.320	0.455	0.376
Log Clustering	0.498	0.530	0.513	0.249	0.045	0.077
DeepLog	0.970	0.907	0.934	0.971/0.862	**0.969/0.851**	0.970/0.856
LogAnomaly	**0.980**	0.905	0.941	0.988/0.934	0.964/ 0.828	0.976/0.878
GLAD-MEAN	0.977	0.952	0.964	**0.996/0.980**	0.962/0.822	**0.979/0.894**
GLAD-PAW	0.978	**0.991**	**0.984**	**0.996**/0.978	0.962/0.819	**0.979**/0.891

GLAD-PAW with PAWGAT layer achieves the best performance among all involved GNN layers on HDFS dataset. From Fig. 3 (a), graph layers with attention mechanism are more powerful than GCN and GGNN, since this mechanism specifies the importance of each node's neighbors. From the comparison between GAT, WGAT, PAGAT and PAWGAT, we can conclude that both position embeddings and weights between edges play important roles in conveying the representation and structural information between events within session graphs.

4.6 RQ3: Parameter Sensitivity

To answer RQ3, we examine two key parameters, number of heads and number of graph layers, on how they influence the performance on HDFS dataset. The experimental results on different number of heads in set $\{1, 2, 4, 8\}$ and graph layers in set $\{1, 2, 3, 4\}$ are shown in Fig. 3(b). It can be seen that the best performance is achieved by stacking 2 PAWGAT layers with 4 heads. It should also be noted that simpler parameter settings (e.g., 1 layer, 1 head) bring about worse representation ability whereas complicated settings can result in over-fitting or over-smoothing for graphs, and both will impair the overall performance of the model.

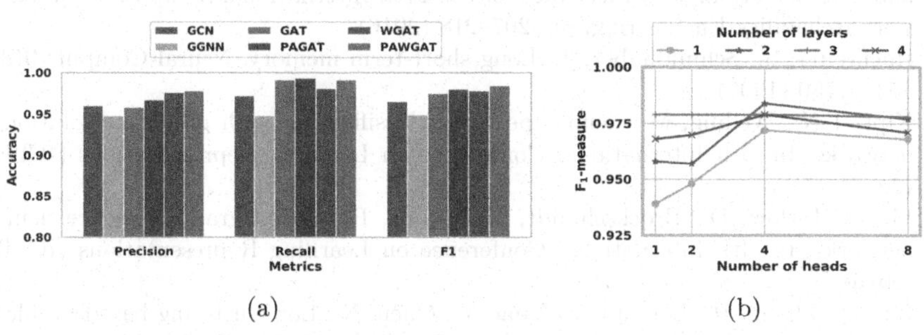

(a) (b)

Fig. 3. (a) A comparison of anomaly detection models using different GNN layers on HDFS dataset. (b) A comparison of F_1-measure under different parameter settings (number of heads and graph layers) on HDFS dataset.

5 Conclusion

In this work, we propose a novel architecture GLAD-PAW for log-based anomaly detection that incorporates graph models in representing log sessions. Specifically, GLAD-PAW constructs directed session graphs from log records and applies a novel position aware weighted graph attention network (PAWGAT) to generate informative representations for all events. PAWGAT contains position and weight information which can better capture local contextual feature and transition pattern within log sessions. Moreover, after stackings of graph layers, an attention based pooling layer assists in learning the global information for each log session. Extensive evaluation on large system logs have clearly demonstrated the superior effectiveness of GLAD-PAW compared with previous methods.

Future work include but are not limited to incorporating other types of graph classification mechanisms into GLAD-PAW to improve the effectiveness and efficiency, and applying our model to different domains (e.g., operation system or database failure detection).

References

1. Cangea, C., Veličković, P., Jovanović, N., Kipf, T., Liò, P.: Towards sparse hierarchical graph classifiers. arXiv preprint arXiv:1811.01287 (2018)
2. Diehl, F.: Edge contraction pooling for graph neural networks. arXiv preprint arXiv:1905.10990 (2019)
3. Du, M., Li, F., Zheng, G., Srikumar, V.: Deeplog: Anomaly detection and diagnosis from system logs through deep learning. In: Proceedings of the 2017 ACM SIGSAC Conference on Computer and Communications Security, pp. 1285–1298 (2017)
4. Gilmer, J., Schoenholz, S.S., Riley, P.F., Vinyals, O., Dahl, G.E.: Neural message passing for quantum chemistry. In: Proceedings of the 34th International Conference on Machine Learning-Volume 70, pp. 1263–1272 (2017)
5. He, P., Zhu, J., Zheng, Z., Lyu, M.R.: Drain: An online log parsing approach with fixed depth tree. In: Proceedings of the IEEE International Conference on Web Services, pp. 33–40 (2017)
6. He, S., Zhu, J., He, P., Lyu, M.R.: Experience report: System log analysis for anomaly detection. In: Proceedings of the 27th International Symposium on Software Reliability Engineering, pp. 207–218 (2016)
7. Hochreiter, S., Schmidhuber, J.: Long short-term memory. Neural Comput. **9**(8), 1735–1780 (1997)
8. Kipf, T.N., Welling, M.: Semi-supervised classification with graph convolutional networks. In: 5th International Conference on Learning Representations (ICLR) (2017)
9. Li, Y., Tarlow, D., Brockschmidt, M., Zemel, R.: Gated graph sequence neural networks. In: 4th International Conference on Learning Representations (ICLR) (2016)
10. Lin, Q., Zhang, H., Lou, J.G., Zhang, Y., Chen, X.: Log clustering based problem identification for online service systems. In: 2016 IEEE/ACM 38th International Conference on Software Engineering Companion (ICSE-C), pp. 102–111. IEEE (2016)

11. Liu, F.T., Ting, K.M., Zhou, Z.H.: Isolation forest. In: 2008 Eighth IEEE International Conference on Data Mining, pp. 413–422. IEEE (2008)
12. Lou, J.G., Fu, Q., Yang, S., Xu, Y., Li, J.: Mining invariants from console logs for system problem detection. In: USENIX Annual Technical Conference, pp. 1–14 (2010)
13. Meng, W., et al.: Loganomaly: unsupervised detection of sequential and quantitative anomalies in unstructured logs. In: Proceedings of the 28th International Joint Conference on Artificial Intelligence, pp. 4739–4745 (2019)
14. Pham, T., Tran, T., Dam, H., Venkatesh, S.: Graph classification via deep learning with virtual nodes. arXiv preprint arXiv:1708.04357 (2017)
15. Song, Y., Keromytis, A.D., Stolfo, S.J.: Spectrogram: a mixture-of-markov-chains model for anomaly detection in web traffic. In: NDSS (2009)
16. Veličković, P., Cucurull, G., Casanova, A., Romero, A., Lio, P., Bengio, Y.: Graph attention networks. In: 6th International Conference on Learning Representations (ICLR) (2018)
17. Wang, J., Xu, Q., Lei, J., Lin, C., Xiao, B.: Pa-ggan: session-based recommendation with position-aware gated graph attention network. In: 2020 IEEE International Conference on Multimedia and Expo (ICME), pp. 1–6 (2020)
18. Xu, K., et al.: Show, attend and tell: neural image caption generation with visual attention. In: Proceedings of the 32nd International Conference on Machine Learning, vol. 37, pp. 2048–2057 (2015)
19. Xu, W., Huang, L., Fox, A., Patterson, D., Jordan, M.: Largescale system problem detection by mining console logs. In: Proceedings of SOSP 2009 (2009)
20. Ying, Z., You, J., Morris, C., Ren, X., Hamilton, W., Leskovec, J.: Hierarchical graph representation learning with differentiable pooling. In: Advances in Neural Information Processing Systems, pp. 4800–4810 (2018)
21. Zhang, M., Cui, Z., Neumann, M., Chen, Y.: An end-to-end deep learning architecture for graph classification. In: AAAI, vol. 18, pp. 4438–4445 (2018)

CubeFlow: Money Laundering Detection with Coupled Tensors

Xiaobing Sun[1], Jiabao Zhang[1,2], Qiming Zhao[1], Shenghua Liu[1,2(✉)], Jinglei Chen[3], Ruoyu Zhuang[3], Huawei Shen[1,2], and Xueqi Cheng[1,2]

[1] CAS Key Laboratory of Network Data Science and Technology, Institute of Computing Technology, Chinese Academy of Sciences, Beijing, China
qmzhao@cqu.edu.cn, {liushnghua,shenhuawei,cxq}@ict.ac.cn
[2] University of Chinese Academy of Sciences, Beijing, China
zhangjiabao18@mails.ucas.edu.cn
[3] China Construction Bank Fintech, Beijing, China

Abstract. Money laundering (ML) is the behavior to conceal the source of money achieved by illegitimate activities, and always be a fast process involving frequent and chained transactions. How can we detect ML and fraudulent activity in large scale attributed transaction data (i.e. tensors)? Most existing methods detect dense blocks in a graph or a tensor, which do not consider the fact that money are frequently transferred through middle accounts. CubeFlow proposed in this paper is a scalable, flow-based approach to spot fraud from a mass of transactions by modeling them as two coupled tensors and applying a novel multi-attribute metric which can reveal the transfer chains accurately. Extensive experiments show CubeFlow outperforms state-of-the-art baselines in ML behavior detection in both synthetic and real data.

1 Introduction

Given a large amount of real-world transferring records, including a pair of accounts, some transaction attributes (e.g. time, types), and volume of money, how can we detect money laundering (ML) accurately in a scalable way? One of common ML processes disperses dirty money into different *source* accounts, transfers them through many *middle* accounts to *destination* accounts for gathering in a fast way. Thus the key problem for ML detection are:

Informal Problem 1. ***Given*** *a large amount of candidates of source, middle, and destination accounts, and the transferring records, which can be formalized as two coupled tensors with entries of (source candidates, middle candidates, time, $\cdots$), and (middle candidates, destination candidates, time, $\cdots$),* ***how to find*** *the accounts in such a ML process accurately and efficiently.*

X. Sun, J. Zhang and Q. Zhao—Contribute equally.
The work was done when Xiaobing Sun and Qiming Zhao were visiting students at ICT CAS, who are separately from NanKai University and Chongqing University.

K. Karlapalem et al. (Eds.): PAKDD 2021, LNAI 12712, pp. 78–90, 2021.
https://doi.org/10.1007/978-3-030-75762-5_7

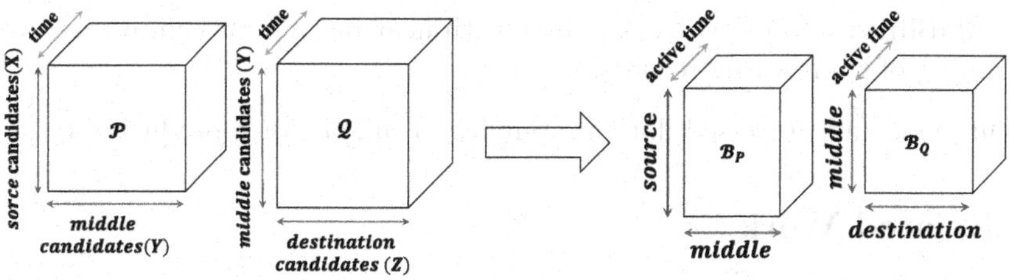

Fig. 1. An example of ML detection. Two coupled input tensors indicate a money flow from X to Y to Z. Modes X, Y and Z denote the candidates of source, middle and destination accounts. The purpose of detection is to find two dense coupled blocks in original coupled tensors, i.e., catching fraudsters involving in two-step ML activities.

Figure 1 shows an example of ML detection with two coupled tensors indicating a flow from source to middle to destination accounts. Those candidates can be pre-selected by existing feature-based models or in an empirical way. For example, in a bank, we can let source candidates simply be external accounts with more money transferring into the bank than out of the bank, destination candidates be the opposite ones, and middle candidates be the inner accounts.

Most existing dense subtensor detection methods [6,18,19] have been used for tensor fraud detection, but only can deal with one independent tensor. Therefore, they exploit exactly single-step transfers, but do not account for "transfer chains". Such methods based on graph's density [5,14,17], have the same problem and even not be able to leverage multi-attributes. Although, FlowScope [12] designed for dense and multi-step flow, it fails to take into account some important properties (e.g. time) because of the limits by the graph.

Therefore, we propose CubeFlow, a ML detection method with coupled tensors. CubeFlow not only considers the flow of funds (from sources, through middle accounts, to destinations), but also can combine some attributes, such as transferring time to model fraudsters' highly frequent transfers. We define a novel multi-attribute metric for fraudulent transferring flows of ML. CubeFlow considers the suspicious in-and-out balance for middle accounts within short time intervals and detects the chain of fraudulent transfers accurately. The experiments on real-world datasets show that CubeFlow detects various adversarial injections and real ML fraudsters both with high accuracy and robustness.

In summary, the main advantages of our work are:

- **Multi-attribute metric for money-laundering flow:** We propose a novel multi-attribute metric for detecting dense transferring flows in coupled tensors, which measures the anomalousness of typical two-step laundering with suspicious in-and-out balance for middle accounts within many short time intervals.
- **Effectiveness and robustness:** CubeFlow outperforms baselines under various injection densities by descending the amount of money or ascending # of accounts on real-world datasets. And CubeFlow shows its robustness in different proportions of accounts with better accuracy.

- **Scalability:** CubeFlow is scalable, with near-linear time complexity in the number of transferring records.

Our code and processed data are publicly available for reproducibility [1].

2 Related Work

Our work pulls from two main different fields of research: (i) domain-specific ML detection methods; and (ii) general anomaly detection methods in graphs and tensors.

Money Laundering Detection. The most classical approaches for Anti-ML are those of rule based classification relying heavily on expertise. Khan et al. [9] used Bayesian network designed with guidance of the rules to assign risk scores to transactions. The system proposed by [10] monitored ongoing transactions and assessed their degree of anomaly. However, rule based algorithms are easy to be evaded by fraudsters. To involve more attributes and handle the high-dimensional data, machine learning models such as SVM [21], decision trees [22] and neural networks [15] are applied, while these methods are focused on isolated transaction level. Stavarache et al. [20] proposed a deep learning based method trained for Anti-ML tasks using customer-to-customer relations. However, these algorithms detect the ML activities in supervised or semi-supervised manners, suffering from imbalanced class and lack of adaptability and interpretability.

General-Purpose Anomaly Detection in Graphs and Tensors. Graphs (i.e. tensors) provide a powerful mechanism to capture interrelated associations between data objects [1], and there have been many graph-based techniques developed for discovering structural anomalies. SpokEn [17] studied patterns in eigenvectors, and was applied for anomaly detection in [8] later. CatchSync [7] exploited two of the tell-tale signs created by fraudsters. And many existing methods rely on graph (i.e. tensor)'s density, e.g., Fraudar [5] proposed a suspiciousness measure on the density, HoloScope [13,14] considered temporal spikes and hyperbolic topology and SpecGreedy [3] proposed a unified framework based on the graph spectral properties. D-Cube [19], M-Zoom [18] and CrossSpot [6] adopted greedy approximation algorithms to detect dense subtensors, while CP Decomposition (CPD) [11] focused on tensor decomposition methods. However, these methods are designed for general-purpose anomaly detection tasks, which not take the flow across multiple nodes into account.

3 Problem Formulation

In general money laundering (ML) scenario, fraudsters transfer money from source accounts to destination accounts through several middle accounts in order to cover up the true source of funds. Here, we summarize three typical characteristics of money laundering:

[1] https://github.com/BGT-M/spartan2-tutorials/blob/master/CubeFlow.ipynb.

Table 1. Notations and symbols

Symbol	Definition
$\mathcal{P}(X, Y, A_3, \ldots, A_N, V)$	Relation representing the tensor which stands money trans from X to Y
$\mathcal{Q}(Y, Z, A_3, \ldots, A_N, V)$	Relation representing the tensor which stands money trans from Y to Z
N	Number of mode attributes in $\mathcal{P}$ (or $\mathcal{Q}$)
A_n	n-th mode attribute name in $\mathcal{P}$ (or $\mathcal{Q}$)
V	measure attribute (e.g. money) in $\mathcal{P}$ (or $\mathcal{Q}$)
$\mathcal{B}_\mathbf{P}$ (or $\mathcal{B}_\mathbf{Q}$)	a block(i.e. subtensor) in $\mathcal{P}$ (or $\mathcal{Q}$)
$\mathbf{P}_x$, $\mathbf{P}_y$, $\mathbf{Q}_z$,$\mathbf{P}_{a_n}$	set of distinct values of X, Y, Z, A_n in $\mathcal{P}$ (or $\mathcal{Q}$)
$\mathbf{B}_x$,$\mathbf{B}_y$, $\mathbf{B}_z$, $\mathbf{B}_{a_n}$	set of distinct values of X, Y, Z, A_n in $\mathcal{B}_\mathbf{P}$ (or $\mathcal{B}_\mathbf{Q}$)
$M_{x,y,a_3,\ldots,a_N}(\mathcal{B}_\mathbf{P})$	attribute-value mass of $(x, y, a_3, \ldots, a_N)$ in $(X, Y, A_3, \ldots, A_N)$
$M_{y,z,a_3,\ldots,a_N}(\mathcal{B}_\mathbf{Q})$	attribute-value mass of $(y, z, a_3, \ldots, a_N)$ in $(Y, Z, A_3, \ldots, A_N)$
ω_i	Weighted assigned to a node in priority tree
$g(\mathcal{B}_\mathbf{P}, \mathcal{B}_\mathbf{Q})$	Metric of ML anomalousness
$[3, N]$	$\{3, 4, \ldots, N\}$

Density: In ML activities, a high volume of funds needs to be transferred from source to destination accounts with limited number of middle accounts. Due to the risk of detection, fraudsters tend to use shorter time and fewer trading channels in the process of ML which will create a high-volume and dense subtensor of transfers.

Zero Out Middle Accounts: The role of middle accounts can be regarded as a bridge in ML: only a small amount of balance will be kept in these accounts for the sake of camouflage, most of the received money will be transferred out. This is because the less balances retained, the less losses will be incurred if these accounts are detected or frozen.

Fast in and Fast Out: To reduce banks' attention, "dirty money" is always divided into multiple parts and transferred through the middle accounts one by one. The "transfer", which means a part of fund is transferred in and out of a middle account, is usually done within a very short time interval. This is because the sooner the transfer is done, the more benefits fraudsters will get.

Algorithms which focus on individual transfers, e.g. feature-based approaches, can be easily evaded by adversaries by keeping each individual transfer looks normal. Instead, our goal is to detect dense blocks in tensors composed of source, middle, destination accounts and other multi-attributes, as follows:

Informal Problem 2 (ML Detection with Coupling Tensor). *Given two money transfer tensors* $\mathcal{P}(X, Y, A_3, \ldots, A_N, V)$ *and* $\mathcal{Q}(Y, Z, A_3, \ldots, A_N, V)$, *with attributes of source, middle and destination candidates as* X, Y, Z, *other coupling attributes (e.g. time) as* A_n, *and a nonnegative measure attribute (e.g. volume of money) as* V.

Find: *two dense blocks (i.e. subtensor) of $\boldsymbol{\mathcal{P}}$ and $\boldsymbol{\mathcal{Q}}$.*
Such that:

- *it maximizes density.*
- *for each middle account, the money transfers satisfy zero-out and fast-in-and-fast-out characteristics.*

Symbols used in the paper are listed in Table 1. As common in other literature, we denote tensors and modes of tensors by boldface calligraphic letters(e.g. $\boldsymbol{\mathcal{P}}$) and capital letters(e.g. X, Y, Z, A_n) individually. For the possible values of different modes, boldface uppercase letters (e.g. $\mathbf{P}_x, \mathbf{P}_y, \mathbf{Q}_z, \mathbf{P}_{a_n}$) are used in this paper. Since $\boldsymbol{\mathcal{P}}(X, Y, A_3, \ldots, A_N, V)$ and $\boldsymbol{\mathcal{Q}}(Y, Z, A_3, \ldots, A_N, V)$ are coupled tensors sharing the same sets of modes Y, A_n, we have $\mathbf{P}_y = \mathbf{Q}_y$ and $\mathbf{P}_{a_n} = \mathbf{Q}_{a_n}$. Our targets, the dense blocks (i.e. subtensor) of $\boldsymbol{\mathcal{P}}$ and $\boldsymbol{\mathcal{Q}}$, are represented by $\boldsymbol{\mathcal{B}}_\mathbf{P}$ and $\boldsymbol{\mathcal{B}}_\mathbf{Q}$. Similarly, the mode's possible values in these blocks are written as $\mathbf{B}_x, \mathbf{B}_y, \mathbf{B}_z, \mathbf{B}_{a_n}$. An entry $(x, y, a_3, \ldots, a_N)$ indicates that account x transfers money to account y when other modes are equal to $a_3, \ldots, a_N$ (e.g. during a_3 time-bin), and $M_{x,y,a_3,\ldots,a_N}(\boldsymbol{\mathcal{B}}_\mathbf{P})$ is the total amount of money on the subtensor.

4 Proposed Method

4.1 Proposed Metric

First, we give the concept of fiber: *A fiber of a tensor $\boldsymbol{\mathcal{P}}$ is a vector obtained by fixing all but one $\boldsymbol{\mathcal{P}}$'s indices.* For example, in ML process with A_3 representing transaction timestamp, total money transferred from source accounts X into a middle account y at time-bin a_3 is the mass of fiber $\boldsymbol{\mathcal{P}}(X, y, a_3, V)$ which can be denoted by $M_{:,y,a_3}(\boldsymbol{\mathcal{B}}_\mathbf{P})$, while total money out of the middle account can be denoted by $M_{y,:,a_3}(\boldsymbol{\mathcal{B}}_\mathbf{Q})$.

In general form, we can define the minimum and maximum value between total amount of money transferred into and out of a middle account $y \in \mathbf{B}_y$ with other attributes equal to $a_3 \in \mathbf{B}_{a_3}, \ldots, a_N \in \mathbf{B}_{a_N}$:

$$f_{y,a_3,\ldots,a_N}(\boldsymbol{\mathcal{B}}_\mathbf{P}, \boldsymbol{\mathcal{B}}_\mathbf{Q}) = min\{M_{:,y,a_3,\ldots,a_N}(\boldsymbol{\mathcal{B}}_\mathbf{P}), M_{y,:,a_3,\ldots,a_N}(\boldsymbol{\mathcal{B}}_\mathbf{Q})\} \quad (1)$$

$$q_{y,a_3,\ldots,a_N}(\boldsymbol{\mathcal{B}}_\mathbf{P}, \boldsymbol{\mathcal{B}}_\mathbf{Q}) = max\{M_{:,y,a_3,\ldots,a_N}(\boldsymbol{\mathcal{B}}_\mathbf{P}), M_{y,:,a_3,\ldots,a_N}(\boldsymbol{\mathcal{B}}_\mathbf{Q})\} \quad (2)$$

Then we can define the difference between the maximum and minimum value:

$$r_{y,a_3,\ldots,a_N}(\boldsymbol{\mathcal{B}}_\mathbf{P}, \boldsymbol{\mathcal{B}}_\mathbf{Q}) = q_{y,a_3,\ldots,a_N}(\boldsymbol{\mathcal{B}}_\mathbf{P}, \boldsymbol{\mathcal{B}}_\mathbf{Q}) - f_{y,a_3,\ldots,a_N}(\boldsymbol{\mathcal{B}}_\mathbf{P}, \boldsymbol{\mathcal{B}}_\mathbf{Q}) \quad (3)$$

Next, our ML metric is defined as follows for spotting multi-attribute money-laundering flow:

Definition 1. *(Anomalousness of coupled blocks of ML) The anomalousness of a flow from a set of nodes* $\mathbf{B}_x$*, through the inner accounts* $\mathbf{B}_y$*, to another subset* $\mathbf{B}_z$*, where other attribute values are* $a_n \in \mathbf{B}_{a_n}$*:*

$$\begin{aligned} g(\boldsymbol{\mathcal{B}}_\mathbf{P}, \boldsymbol{\mathcal{B}}_\mathbf{Q}) &= \frac{\sum_{y \in \mathbf{B}_y, a_i \in \mathbf{B}_{a_i}} ((1-\alpha) f_{y,a_3,\ldots,a_N}(\boldsymbol{\mathcal{B}}_\mathbf{P}, \boldsymbol{\mathcal{B}}_\mathbf{Q}) - \alpha \cdot r_{y,a_3,\ldots,a_N}(\boldsymbol{\mathcal{B}}_\mathbf{P}, \boldsymbol{\mathcal{B}}_\mathbf{Q}))}{\sum_{i=3}^{N}(|\mathbf{B}_{a_i}|) + |\mathbf{B}_x| + |\mathbf{B}_y| + |\mathbf{B}_z|} \\ &= \frac{\sum_{y \in \mathbf{B}_y, a_i \in \mathbf{B}_{a_i}} (f_{y,a_3,\ldots,a_N}(\boldsymbol{\mathcal{B}}_\mathbf{P}, \boldsymbol{\mathcal{B}}_\mathbf{Q}) - \alpha q_{y,a_3,\ldots,a_N}(\boldsymbol{\mathcal{B}}_\mathbf{P}, \boldsymbol{\mathcal{B}}_\mathbf{Q}))}{\sum_{i=3}^{N}(|\mathbf{B}_{a_i}|) + |\mathbf{B}_x| + |\mathbf{B}_y| + |\mathbf{B}_z|} \end{aligned} \tag{4}$$

Intuitively, $f_{y,a_3,\ldots,a_N}(\boldsymbol{\mathcal{B}}_\mathbf{P})$ is the maximum possible flow that could go through middle account $y \in \mathbf{B}_y$ when other attributes are $a_n \in \mathbf{B}_{a_n}$. $r_{y,a_3,\ldots,a_N}(\boldsymbol{\mathcal{B}}_\mathbf{P}, \boldsymbol{\mathcal{B}}_\mathbf{Q})$ is the absolute value of "remaining money" in account y after transfer, i.e., retention or deficit, which can be regarded as a penalty for ML, since fraudsters prefer to keep small account balance at any situations. When we set A_3 as time dimension, we consider the "remaining money" in each time bin which will catch the trait of fast in and fast out during ML. We define α as the coefficient of imbalance cost rate in the range of 0 to 1.

4.2 Proposed Algorithm: CubeFlow

We use a near-greedy algorithm CubeFlow, to find two dense blocks $\boldsymbol{\mathcal{B}}_\mathbf{P}$ and $\boldsymbol{\mathcal{B}}_\mathbf{Q}$ maximizing the objective $g(\boldsymbol{\mathcal{B}}_\mathbf{P}, \boldsymbol{\mathcal{B}}_\mathbf{Q})$ in (4).

To develop an efficient algorithm for our metric, we unfold the tensor $\boldsymbol{\mathcal{P}}$ on mode-X and $\boldsymbol{\mathcal{Q}}$ on mode-Z. For example, a tensor unfolding of $\boldsymbol{\mathcal{P}} \in \mathbb{R}^{|\mathbf{B}_x| \times |\mathbf{B}_y| \times |\mathbf{B}_{a_3}|}$ on mode-X will produce a $|\mathbf{B}_x| \times (|\mathbf{B}_y| \times |\mathbf{B}_{a_3}|)$ matrix.

For clarity, we define the index set $\mathbf{I}$, whose size equals to the number of columns of matrix:

$$\mathbf{I} = \mathbf{B}_y \bowtie \mathbf{B}_{a_3} \bowtie \ldots \bowtie \mathbf{B}_{a_N} \tag{5}$$

where $\bowtie$ denotes Cartesian product. Therefore, the denominator of (4) can be approximated by $|\mathbf{B}_x| + |\mathbf{I}| + |\mathbf{B}_z|$.

First, we build a priority tree for entries in $\boldsymbol{\mathcal{B}}_\mathbf{P}$ and $\boldsymbol{\mathcal{B}}_\mathbf{Q}$. The weight (i.e. priority) assigned to index i is defined as:

$$\omega_i(\boldsymbol{\mathcal{B}}_\mathbf{P}, \boldsymbol{\mathcal{B}}_\mathbf{Q}) = \begin{cases} f_i(\boldsymbol{\mathcal{B}}_\mathbf{P}, \boldsymbol{\mathcal{B}}_\mathbf{Q}) - \alpha q_i(\boldsymbol{\mathcal{B}}_\mathbf{P}, \boldsymbol{\mathcal{B}}_\mathbf{Q}), & \text{if } i \in \mathbf{I} \\ M_{i,:,:,\ldots,:}(\boldsymbol{\mathcal{B}}_\mathbf{P}), & \text{if } i \in \mathbf{B}_x \\ M_{:,i,:,\ldots,:}(\boldsymbol{\mathcal{B}}_\mathbf{Q}), & \text{if } i \in \mathbf{B}_z \end{cases} \tag{6}$$

The algorithm is described in Algorithm 1. After building the priority tree, we perform the near greedy optimization: block $\boldsymbol{\mathcal{B}}_\mathbf{P}$ and $\boldsymbol{\mathcal{B}}_\mathbf{Q}$ start with whole tensor $\boldsymbol{\mathcal{P}}$ and $\boldsymbol{\mathcal{Q}}$. Let we denote $\mathbf{I} \cup \mathbf{B}_x \cup \mathbf{B}_z$ as $\mathbf{S}$. In every iteration, we remove the node v in $\mathbf{S}$ with minimum weight in the tree, approximately maximizing objective (4); and then we update the weight of all its neighbors. The iteration is repeated until one of node sets $\mathbf{B}_x, \mathbf{B}_z, \mathbf{I}$ is empty. Finally, two dense blocks $\hat{\boldsymbol{\mathcal{B}}}_\mathbf{P}, \hat{\boldsymbol{\mathcal{B}}}_\mathbf{Q}$ that we have seen with the largest value $g(\hat{\boldsymbol{\mathcal{B}}}_\mathbf{P}, \hat{\boldsymbol{\mathcal{B}}}_\mathbf{Q})$ are returned.

Algorithm 1: CubeFlow

Input: relation $\mathcal{P}$, relation $\mathcal{Q}$
Output: dense block $\mathcal{B}_\mathbf{P}$, dense block $\mathcal{B}_\mathbf{Q}$

$\mathcal{B}_\mathbf{P} \leftarrow \mathcal{P}$, $\mathcal{B}_\mathbf{Q} \leftarrow \mathcal{Q}$;
$\mathbf{S} \leftarrow \mathbf{I} \cup \mathbf{B}_x \cup \mathbf{B}_z$;
$\omega_i \leftarrow$ calculate node weight as Eq. (6) ;
$T \leftarrow$ build priority tree for $\mathcal{B}_\mathbf{P}$ and $\mathcal{B}_\mathbf{Q}$ with $\omega_i(\mathcal{B}_\mathbf{P}, \mathcal{B}_\mathbf{Q})$;
while $\mathbf{B}_x, \mathbf{B}_z$ *and* $\mathbf{I}$ *is not empty* **do**
  $v \leftarrow$ find the minimum weighted node in T;
  $\mathbf{S} \leftarrow \mathbf{S} \backslash \{v\}$;
  update priorities in T for all neighbors of v;
  $g(\mathcal{B}_\mathbf{P}, \mathcal{B}_\mathbf{Q}) \leftarrow$ calculate as Eq. (4);
end
return $\hat{\mathcal{B}}_\mathbf{P}, \hat{\mathcal{B}}_\mathbf{Q}$ that maximizes $g(\mathcal{B}_\mathbf{P}, \mathcal{B}_\mathbf{Q})$ seen during the loop.

5 Experiments

We design experiments to answer the following questions:

- **Q1. Effectiveness:** How early and accurate does our method detect synthetic ML behavior comparing to the baselines?
- **Q2. Performance on Real-World Data:** How early and accurate does our CubeFlow detect real-world ML activity comparing to the baselines?
- **Q3. Performance on 4-Mode Tensor:** How accurate does CubeFlow compare to the baselines dealing with multi-mode data?
- **Q4. Scalability:** Does our method scale linearly with the number of edges?

Table 2. Dataset description

Name	Volume	# Tuples
3-mode bank transfer record (from_acct, to_acct, time, money)		
CBank	$491295 \times 561699 \times 576$	$2.94M$
	$561699 \times 1370249 \times 576$	$2.60M$
CFD-3	$2030 \times 2351 \times 728$	$0.12M$
	$2351 \times 7001 \times 730$	$0.27M$
4-mode bank transfer record (from_acct, to_acct, time, k_symbol, money)		
CFD-4	$2030 \times 2351 \times 728 \times 7$	$0.12M$
	$2351 \times 7001 \times 730 \times 8$	$0.27M$

5.1 Experimental Setting

Machine: We ran all experiments on a machine with 2.7 GHZ Intel Xeon E7-8837 CPUs and 512 GB memory.

Data: Table 2 lists data used in our paper. CBank data is a real-world transferring data from an anonymous bank under an NDA agreement. Czech Financial Data (CFD) is an anonymous transferring data of Czech bank released for Discovery Challenge in [16]. We model CFD data as two 3-mode tensors consisting of entries (a_1, a_2, t, m), which means that account a_1 transfers the amount of money, m, to account a_2 at time t. Specifically, we divide the account whose money transferring into it is much larger than out of the account into X, on the contrary, into Z, and the rest into Y. Note that it does not mean that X, Y, Z have to be disjoint, while this preprocessing helps speed up our algorithm. We also model CFD data as two 4-mode tensors having an additional dimension *k_Symbol* (characterization of transaction, e.g., insurance payment and payment of statement). And we call two CFD data as CFD-3 and CFD-4 resp.

Implementations: We implement CubeFlow in Python, CP Decomposition(CPD) [11] in Matlab and run the open source code of D-Cube [19], M-Zoom [18] and CrossSpot [6]. We use the sparse tensor format for efficient space utility. Besides, the length of time bins of CBank and CFD are 20 min and 3 days respectively, and the value of α is 0.8 as default.

5.2 Q1.Effectiveness

To verify the effectiveness of CubeFlow, we inject ML activities as follows: fraudulent accounts are randomly chosen as the tripartite groups, denoted by B_x, B_y and B_z. The fraudulent edges between each group are randomly generated with probability p. We use Dirichlet distribution (the value of scaling parameter is 100) to generate the amount of money for each edge. And for each account in B_y, the amount of money received from B_x and that of transferred to B_z are almost the same. Actually, we can regard the remaining money of accounts in B_y as camouflage, with amount of money conforms to a random distribution ranging from 0 to 100000 (less than 1% of injected amount of money). To satisfy the trait "Fast in and Fast out", we randomly choose the time from one time bin for all edges connected with the same middle account.

The Influence of the Amount of Money: In this experiment, the number of B_x, B_y, B_z is 5, 10 and 5 resp, and we increase the amount of injected money laundered step by step while fixing other conditions. As shown in Fig. 2(a), CubeFlow detects the ML behavior earliest and accurately, and the methods based on bipartite graph are unable to catch suspicious tripartite dense flows in the tensor.

The Influence of the Number of Fraudulent Accounts: Another possible case is that fraudsters may employ as many as people to launder money, making the ML behavior much harder to detect. In this experiment, we increase the number of fraudsters step by step at a fixed ratio (5 : 10 : 5) while keeping the amount of laundering money and other conditions unchanged. As Fig. 2(b) shown, our method achieves the best results.

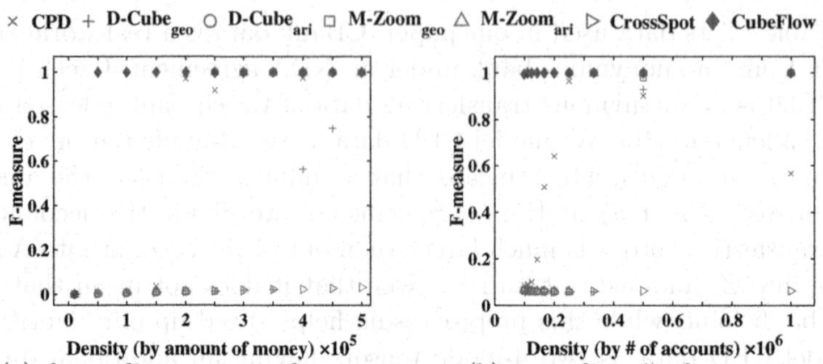

(a) descending amount of money (b) ascending # of accounts

Fig. 2. CubeFlow outperforms baselines under different injected densities by descending amount of money (a) or ascending # of accounts (b) on CFD-3 data.

Table 3. Experimental results on CFD-3 data with different injection ratios of accounts

X:Y:Z	CubeFlow	$D\text{-}Cube_{geo}$	$D\text{-}Cube_{ari}$	CPD	CrossSpot	$M\text{-}Zoom_{geo}$	$M\text{-}Zoom_{ari}$
5:10:5	**0.940**	0.189	0.553	0.641	0.015	0.455	0.553
10:10:10	**0.940**	0.427	0.653	0.647	0.024	0.555	0.555
10:5:10	**0.970**	0.652	0.652	0.725	0.020	0.652	0.652

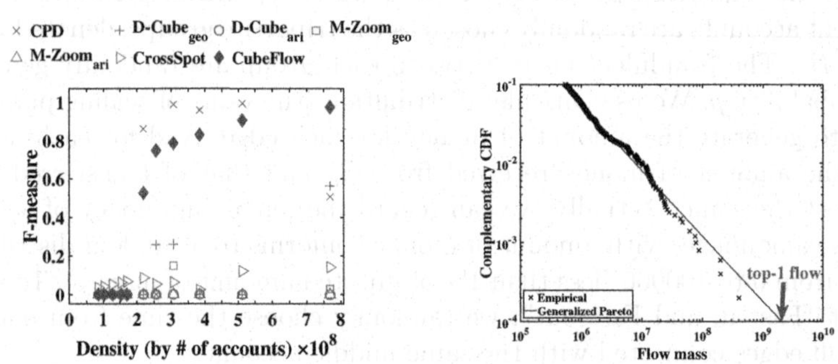

(a) CBank data with ground-truth (b) mass distribution of flows

Fig. 3. CubeFlow performs best in real CBank data. (a) CubeFlow detects earliest (less money being laundered) and accurately in ground-truth community. (b) The GP distribution closely fits mass distributions of real flows: Black crosses indicate the empirical mass distribution for flows with same size with top-1 flow detected by CubeFlow, in the form of its complementary CDF (i.e. CCDF).

Robustness with Different Injection Ratios of Accounts: To verify the robustness of our method, we randomly pick B_x, B_y and B_z under three ratios as presented in Table 3. The metric for comparison is **FAUC**: the areas under curve of F-measure as in Fig. 2. We normalize the density in horizontal axis to scale FAUC between 0 and 1, and higher FAUC indicates better performance.

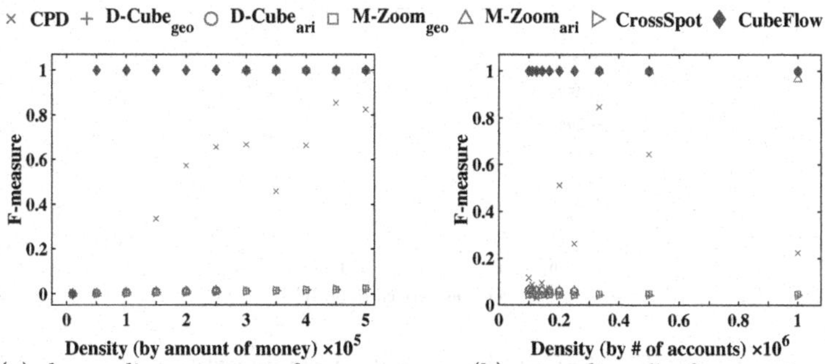

(a) descending amount of money (b) ascending # of accounts

Fig. 4. CUBEFLOW outperforms the baselines under different adversarial densities by descending the amount of money (a) or ascending # of accounts (b) on CFD-4 data.

And we can see from Table 3, CUBEFLOW achieves far better performance than other baselines under all settings, indicating earlier and more accurate detection for more fraudulent accounts.

5.3 Q2. Performance on Real-World Data

CBank data contains labeled ML activity: based on the X to Y to Z schema, the number of each type of accounts is 4, 12, 2. To test how accurately and early we can detect the fraudsters in CBank data, we first scale down the percentage of dirty money laundered from source accounts to destination accounts, then gradually increase the volume of money laundering linearly back to the actual value in the data. Figure 3(a) shows that CUBEFLOW can the catch the ML behaviors earliest, exhibiting our method's utility in detecting real-world ML activities. Note that although CPD works well at some densities, it fluctuate greatly, indicating that CPD is not very suitable for ML behavior detection.

Flow Surprisingness Estimation with Extreme Value Theory: Inspired by [4], we use Generalized Pareto (GP) Distribution, a commonly used probability distribution within extreme value theory, to estimate the extreme **tail** of a distribution without making strong assumptions about the distribution itself. GP distributions exhibit heavy-tailed decay (i.e. power law tails), which can approximate the tails of almost any distribution, with error approaching zero [2].

Specifically, we estimate tail of GP distribution via sampling. Given a flow corresponding to two blocks, $\boldsymbol{\mathcal{B}}_{\mathbf{P}}$, $\boldsymbol{\mathcal{B}}_{\mathbf{Q}}$, with total mass $M(\boldsymbol{\mathcal{B}}_{\mathbf{P}}) + M(\boldsymbol{\mathcal{B}}_{\mathbf{Q}})$, we sample 5000 uniformly random flows from data with same size. For $\epsilon = 0.1$, we fit a GP distribution using maximum likelihood to the largest ϵN masses. The surprisingness of flow is the **CDF** of this GP distribution, evaluated at its mass. As shown in Fig. 3(b), masses of sampled flows follow a GP distribution and tail measure score (i.e. CDF) of top-1 flow detected by CUBEFLOW is very close to 1 (pointed by red arrow), indicating that this activity is quite extreme(i.e. rare) in CBank data.

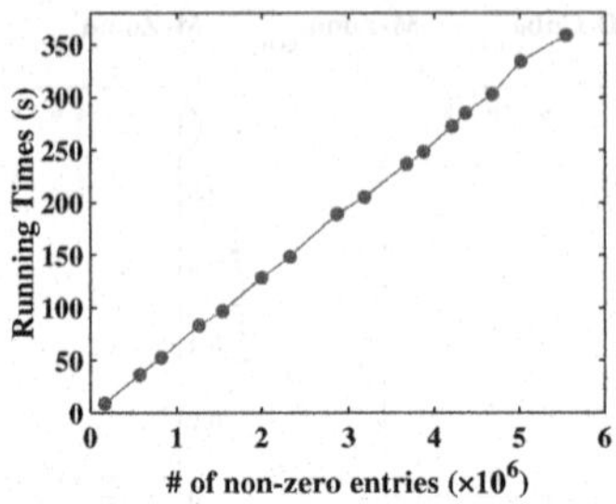

Fig. 5. CubeFlow scales near linearly

5.4 Q3. Performance on 4-mode Tensor

To evaluate the performance of CubeFlow dealing with multi-mode data, we conducted similar experiments on CFD-4 data. Figure 4(a)–4(b) show that our method takes significant advantages over the baselines as our method achieves excellent performance far earlier.

5.5 Q4. Scalability

Scalability: We demonstrate the linearly scalability with of CubeFlow by measuring how rapidly its update time increases as a tensor grows. As Fig. 5 shown, CubeFlow scales linearly with the size of non-zero entries.

6 Conclusion

In this paper, we propose a money laundering detection method, CubeFlow, which is a scalable, flow-based approach to spot the fraud in big attributed transaction tensors. We model the problem with two coupled tensors and propose a novel multi-attribute metric which can utilize different characteristics of money-laundering flow. Experiments based on different data have demonstrated the effectiveness and robustness of CubeFlow's utility as it outperforms state-of-the-art baselines. The source code is opened for reproducibility.

Acknowledgements. This paper is partially supported by the National Science Foundation of China under Grant No.91746301, 61772498, U1911401, 61872206, 61802370. This paper is also supported by the Strategic Priority Research Program of the Chinese Academy of Sciences, Grant No. XDA19020400 and 2020 Tencent Wechat Rhino-Bird Focused Research Program.

References

1. Akoglu, L., Tong, H., Koutra, D.: Graph based anomaly detection and description: a survey. Data Mining Knowl. Discov. **29**(3), 626–688 (2014). https://doi.org/10.1007/s10618-014-0365-y
2. Balkema, A.A., De Haan, L.: Residual life time at great age. Annals of Probability (1974)
3. Feng, W., Liu, S., Danai, K., Shen, H., Cheng, X.: Specgreedy: unified dense subgraph detection. In: European Conference on Machine Learning and Principles and Practice of Knowledge Discovery in Databases (ECML-PKDD) (2020)
4. Hooi, B., Shin, K., Lamba, H., Faloutsos, C.: Telltail: fast scoring and detection of dense subgraphs. In: AAAI (2020)
5. Hooi, B., Song, H.A., Beutel, A., Shah, N., Shin, K., Faloutsos, C.: Fraudar: bounding graph fraud in the face of camouflage. In: SIGKDD. ACM (2016)
6. Jiang, M., Beutel, A., Cui, P., Hooi, B., Yang, S., Faloutsos, C.: A general suspiciousness metric for dense blocks in multimodal data. In: ICDM (2015)
7. Jiang, M., Cui, P., Beutel, A., Faloutsos, C., Yang, S.: Catchsync: catching synchronized behavior in large directed graphs. In: SIGKDD. ACM (2014)
8. Jiang, M., Cui, P., Beutel, A., Faloutsos, C., Yang, S.: Inferring strange behavior from connectivity pattern in social networks. In: Tseng, V.S., Ho, T.B., Zhou, Z.-H., Chen, A.L.P., Kao, H.-Y. (eds.) PAKDD 2014. LNCS (LNAI), vol. 8443, pp. 126–138. Springer, Cham (2014). https://doi.org/10.1007/978-3-319-06608-0_11
9. Khan, N.S., Larik, A.S., Rajput, Q., Haider, S.: A Bayesian approach for suspicious financial activity reporting. Int. J. Comput. Appl. **35**, 181–187 (2013)
10. Khanuja, H.K., Adane, D.S.: Forensic analysis for monitoring database transactions. In: Mauri, J.L., Thampi, S.M., Rawat, D.B., Jin, D. (eds.) SSCC 2014. CCIS, vol. 467, pp. 201–210. Springer, Heidelberg (2014). https://doi.org/10.1007/978-3-662-44966-0_19
11. Kolda, T., Bader, B.: Tensor decompositions and applications. SIAM Review (2009)
12. Li, X., et al.: Flowscope: spotting money laundering based on graphs. In: AAAI (2020)
13. Liu, S., Hooi, B., Faloutsos, C.: A contrast metric for fraud detection in rich graphs. IEEE Trans. Knowl. Data Eng. **31**, 2235–2248 (2019)
14. Liu, S., Hooi, B., Faloutsos, C.: Holoscope: topology-and-spike aware fraud detection. In: CIKM. ACM (2017)
15. Lv, L.T., Ji, N., Zhang, J.L.: A RBF neural network model for anti-money laundering. In: ICWAPR. IEEE (2008)
16. Lütkebohle, I.: Bworld robot control software. https://data.world/lpetrocelli/czech-financial-dataset-real-anonymized-transactions/. Accessed 2 Nov 2018
17. Prakash, B.A., Sridharan, A., Seshadri, M., Machiraju, S., Faloutsos, C.: Eigen-Spokes: surprising patterns and scalable community chipping in large graphs. In: Zaki, M.J., Yu, J.X., Ravindran, B., Pudi, V. (eds.) PAKDD 2010. LNCS (LNAI), vol. 6119, pp. 435–448. Springer, Heidelberg (2010). https://doi.org/10.1007/978-3-642-13672-6_42
18. Shin, K., Hooi, B., Faloutsos, C.: M-Zoom: fast dense-block detection in tensors with quality guarantees. In: Frasconi, P., Landwehr, N., Manco, G., Vreeken, J. (eds.) ECML PKDD 2016. LNCS (LNAI), vol. 9851, pp. 264–280. Springer, Cham (2016). https://doi.org/10.1007/978-3-319-46128-1_17

19. Shin, K., Hooi, B., Kim, J., Faloutsos, C.: D-cube: dense-block detection in terabyte-scale tensors. In: WSDM. ACM (2017)
20. Stavarache, L.L., Narbutis, D., Suzumura, T., Harishankar, R., Žaltauskas, A.: Exploring multi-banking customer-to-customer relations in aml context with poincar\'e embeddings. arXiv preprint arXiv:1912.07701 (2019)
21. Tang, J., Yin, J.: Developing an intelligent data discriminating system of anti-money laundering based on SVM. In: ICMLC. IEEE (2005)
22. Wang, S.N., Yang, J.G.: A money laundering risk evaluation method based on decision tree. In: ICMLC. IEEE (2007)

Unsupervised Boosting-Based Autoencoder Ensembles for Outlier Detection

Hamed Sarvari[1], Carlotta Domeniconi[1], Bardh Prenkaj[2], and Giovanni Stilo[3(✉)]

[1] George Mason University, Fairfax, VA, USA
{hsarvari,cdomenic}@gmu.edu
[2] Sapienza University of Rome, Rome, Italy
prenkaj@di.uniroma1.it
[3] University of L'Aquila, L'Aquila, Italy
giovanni.stilo@univaq.it

Abstract. Autoencoders have been recently applied to outlier detection. However, neural networks are known to be vulnerable to overfitting, and therefore have limited potential in the unsupervised outlier detection setting. The majority of existing deep learning methods for anomaly detection is sensitive to contamination of the training data to anomalous instances. To overcome the aforementioned limitations we develop a Boosting-based Autoencoder Ensemble approach (BAE). BAE is an unsupervised ensemble method that, similarly to boosting, builds an adaptive cascade of autoencoders to achieve improved and robust results. BAE trains the autoencoder components sequentially by performing a weighted sampling of the data, aimed at reducing the amount of outliers used during training, and at injecting diversity in the ensemble. We perform extensive experiments and show that the proposed methodology outperforms state-of-the-art approaches under a variety of conditions.

1 Introduction

Outlier (or anomaly) detection is the process of automatically identifying irregularity in the data. An outlier is an observation that deviates drastically from the given norm or average of the data. This is a widely accepted definition of an anomaly. Nevertheless, outlier detection, being intrinsically unsupervised, is an under-specified, and thus *ill-posed* problem. This makes the task challenging.

Several anomaly detection techniques have been introduced in the literature, e.g. distance-based [11] , density-based [1], and subspace-based methods [8]. Neural networks, and specifically autoencoders, have also been used for outlier detection [6,7]. An autoencoder is a multi-layer symmetric neural network whose goal

Giovanni Stilo—His work is partially supported by Territori Aperti a project funded by Fondo Territori Lavoro e Conoscenza CGIL CISL UIL and by SoBigData-PlusPlus H2020-INFRAIA-2019-1 EU project, contract number 871042.

K. Karlapalem et al. (Eds.): PAKDD 2021, LNAI 12712, pp. 91–103, 2021.
https://doi.org/10.1007/978-3-030-75762-5_8

is to reconstruct the data provided in input. To achieve this goal, an autoencoder learns a new reduced representation of the input data that minimizes the reconstruction error. When using an autoencoder for outlier detection, the reconstruction error indicates the level of outlierness of the corresponding input.

As also discussed in [6], while deep neural networks have shown great promise in recent years when trained on very large datasets, they are prone to overfitting when data is limited due to their model complexity. For this reason, autoencoders are not popular for outlier detection, where data availability is an issue.

The authors in [6] have attempted to tackle the aforementioned challenges by generating randomly connected autoencoders, instead of fully connected, thus reducing the number of parameters to be tuned for each model. As such, they obtain an ensemble of autoencoders, whose outlier scores are combined to achieve the ultimate scoring values. The resulting approach is called RandNet.

Ensembles can address the ill-posed nature of outlier detection, and they have been deployed with success to boost the performance in classification and clustering, and to some extent in outlier discovery as well [6,12,15]. The aggregation step of ensembles can filter out spurious findings of individual learners, which are due to the specific learning biases being induced, and thus can achieve a consensus that is more robust than the individual components. To be effective, though, an ensemble must attain a good trade-off between the accuracy and the diversity of its components.

In this paper we focus on both autoencoders and the ensemble methodology to design an effective approach to outlier detection, and propose a **B**oosting-based **A**utoencoder **E**nsemble method (BAE). Unlike RandNet, which trains the autoencoders independently of one another, our approach uses fully connected autoencoders, and aims at achieving diversity by performing an adaptive weighted sampling, similarly to boosting but for an unsupervised scenario. We train an ensemble of autoencoders in sequence, where the data sampled to train a given component depend on the reconstruction errors obtained by the previous one; specifically, the larger the error, the smaller the weight assigned to the corresponding data point is. As such, our adaptive weighted sampling progressively forces the autoencoders to focus on inliers, and thus to learn representations which lead to large reconstruction errors for outliers. This process facilitates the generation of components with accurate outlier scores. We observe that, unlike standard supervised boosting, our weighting scheme is not prone to overfitting because we proportionally assign more weight to the inliers, which are the "easy to learn" fraction of the data.

Overall the advantage achieved by BAE is twofold. First, the progressive reduction of outliers enables the autoencoders to learn better representations of "normal" data, which also results in accurate outlier scores. Second, each autoencoder is exposed to a different set of outliers, thus promoting diversity among them. Our experimental results show that this dual effect indeed results in a good accuracy-diversity trade-off, and ultimately to a competitive or superior performance against state-of-the-art competitors.

2 Related Work

Various outlier detection techniques have been proposed in the literature, ranging from distance-based [11], density-based [1], to subspace-based [8] methods. A survey of anomaly detection methods can be found in [5]. Neural networks, as a powerful learning tool, have also been used for outlier detection, and autoencoders are the fundamental architecture being deployed. Hawkins [7] used autoencoders for the first time to address the anomaly detection task. Deep autoencoders, though, are known to be prone to over-fitting when limited data are available [6]. A survey on different neural network based methods used for the anomaly discovery task can be found in [4].

Generative Adversarial Networks (GANs) have been recently applied to anomaly detection. In particular, Liu et al. [10] introduced SO-GAAL and MO-GAAL, where the generator produces potential outliers and the discriminator learns to separate real data from outliers.

Outlier detection, as an unsupervised learning task, is a challenging task and susceptible to over-fitting. The ensemble learning methodology has been used in the literature to make the outlier detection process more robust and reliable [12,15]. For example, HiCS [8] is an ensemble strategy which finds high contrast subspace projections of the data, and aggregates outlier scores in each of those subspaces. HiCS shows its potential when outliers do not appear in the full space and are hidden in some subspace projections of the data. Isolation forest [9] builds an ensemble of trees and assigns larger outlier scores to the instances that have shorter average path lengths on the tree structure. CARE [13] is a sequential ensemble approach that uses a two-phase aggregation of results in each iteration to reach the final detection scores. Ensembles can dodge the tendency of autoencoders to overfit data [6]. Only a few ensemble-based outlier detection methods using autoencoders have been proposed. RandNet [6] uses an ensemble of autoencoders with randomly dropped connections.

An ensemble approach consists of two main steps: generating individual ensemble members, and then combining their predictions. A common approach to build the consensus is to aggregate all the predictions equally weighted. A drawback of this approach is that poor components can negatively impact the effectiveness of the ensemble. In order to address this issue, selective outlier ensembles have been introduced. For example, DivE [17] and SelectV [12] are greedy strategies based on a (pseudo ground-truth) target ranking, and a graph-based selective strategy for outlier ensembles is given in [15]. BoostSelect [2] improves the performance of SelectV and DivE. Like SelectV, BoostSelect chooses the components that progressively improve the correlation with the target vector. In addition, BoostSelect favors components that identify new outliers, not yet spotted by previously selected components.

3 Methodology

We build an ensemble of autoencoders inspired by boosting. In boosting, weak learners are trained sequentially to build stronger predictors. Likewise, we train a

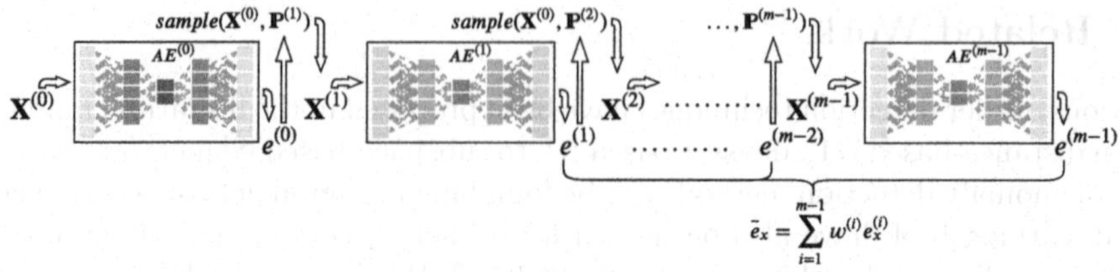

Fig. 1. BAE overview

sequence of autoencoders on different data sub-samples, thus building a boosted ensemble (BAE) for anomaly detection. At each step of the cascade, we sample data from the original collection using a probability distribution defined in terms of reconstruction errors. Finally, the components are combined into a weighted sum that represents the final anomaly score ranking of the boosted model. Like boosting, BAE generates different training samples at each iteration. This enhances the variability across the learners. Unlike the original boosting strategy, though, our sampling strategy favors the inliers (the "easy to learn" fraction of the data), thus reducing the risk of overfitting the learned models to the outliers. Moreover, unlike boosting, our approach is fully unsupervised.

When detecting anomalies, it's common to use the reconstruction error of an autoencoder as indicator of the outlierness level of each data point. The rationale behind this approach is that autoencoders represent an approximated identity function. Therefore, the model is compelled to learn the common characteristics of the data. As a consequence, the learnt data representation typically does not characterize outliers well, thus generating higher errors for the latter.

An overview of BAE is given in Fig. 1. We refer to each step of the boosting cascade as an iteration $i \in [0, m)$ of BAE which produces a component of the ensemble. The number of iterations, i.e. the size of the ensemble m, is a parameter of BAE. $\mathrm{AE}^{(i)}$ is the autoencoder trained on the sample data $\mathbf{X}^{(i)}$ generated in the i-th iteration. BAE iteratively maintains a distribution over the data to sample new training sets. At each iteration $i \in [0, m)$, instances are sampled with replacement from the original set $\mathbf{X}^{(0)}$ according to a probability function $\mathbb{P}^{(i)}$. The sampled instances compose the new dataset $\mathbf{X}^{(i)}$. The probability function $\mathbb{P}^{(i)}$ is defined using the reconstruction error achieved by the autoencoder of the previous iteration $(i-1)$ on the original dataset $\mathbf{X}^{(0)}$. Our goal is to progressively reduce the percentage of outliers in $\mathbf{X}^{(i)}$. To this end, we define a distribution that assigns weights to data which are inversely proportional to the corresponding reconstruction error. Specifically, the function $\mathbb{P}^{(i)}$ assigns higher probability to data points with lower reconstruction error:

$$\mathbb{P}_x^{(i)} = \frac{1/e_x^{(i-1)}}{\sum 1/e_x^{(i-1)}}$$

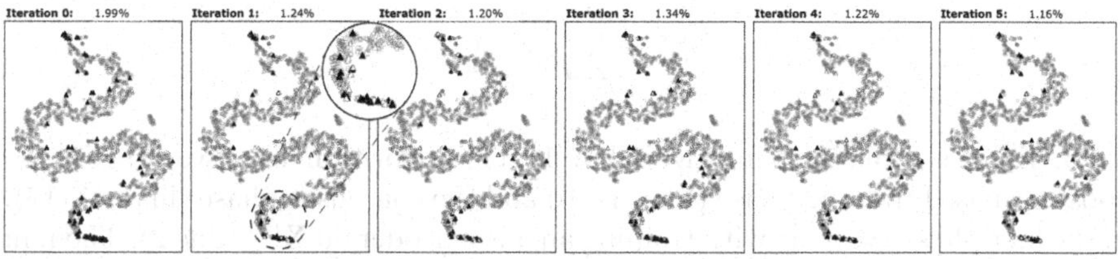

Fig. 2. Percentage of outliers (reported above each sub-image) in successive samples $\mathbf{X}^{(i)}$ generated by BAE for PageBlocks. The triangles are the outliers (the white ones are discarded via weighted sampling). The red circles are the inliers. (Color figure online)

$\mathbb{P}_x^{(i)}$ is the probability that $x \in \mathbf{X}^{(0)}$ is selected, and thus added to $\mathbf{X}^{(i)}$. The reconstruction error $e_x^{(i-1)}$ of x is computed using the autoencoder $\mathrm{AE}^{(i-1)}$:

$$e_x^{(i-1)} = \left(||x - AE^{(i-1)}(x)||_2\right)^2$$

By sampling points according to $\mathbb{P}^{(i)}$, we reduce the percentage of outliers which are included in the training data in successive iterations of BAE. Figure 2 shows this process on the PageBlocks dataset. Each sub-figure plots the points in $\mathbf{X}^{(0)}$ while highlighting the outliers sampled at each iteration. The red circles are inliers and the triangles are outliers[1]. The filled triangles correspond to the outliers which are sampled, and the unfilled triangles are the outliers which are not sampled. For each iteration we compute the percentage of sampled outliers w.r.t. the total number of instances in $\mathbf{X}^{(0)}$ (see value on top of each sub-figure). The Figure shows an overall declining trend of this percentage through the iterations, thereby confirming the effect of our weighted sampling technique.

The progressive sifting of outliers in the sequence of training data built by BAE enables the autoencoders to learn encoders tuned to the inliers, leading to lower reconstruction errors for the inliers and larger ones for the outliers. As such, the reconstruction error of an instance becomes a measure of its outlierness. Like boosting, we assign an importance factor to the autoencoders. The importance (or weight) $w^{(i)}$ of $\mathrm{AE}^{(i)}$ depends on the reconstruction error that $\mathrm{AE}^{(i)}$ achieves on its training data $\mathbf{X}^{(i)}$: the smaller the error is, the higher its importance will be. Specifically, we define $w^{(i)}$ as follows (we discard the error of $\mathrm{AE}^{(0)}$):

$$w^{(i)} = \frac{1/\sum_{x \in \mathbf{X}^{(i)}} e_x^{(i)}}{\sum_{i=1}^{m-1} \left(1/\sum_{x \in \mathbf{X}^{(i)}} e_x^{(i)}\right)}$$

The contribution of $\mathrm{AE}^{(i)}$ to the final anomaly score of $x \in \mathbf{X}^{(0)}$ is proportional to its estimated importance $w^{(i)}$. We compute the consensus outlier score $\bar{e}_x$ of x by aggregating the errors $e_x^{(i)}$, $i = 1, \ldots, m-1$, each weighted by $w^{(i)}$:

[1] We use two colors—black and white—to highlight the sampling of outliers. The inliers are represented without a color variation.

$$\bar{e}_x = \sum_{i=1}^{m-1} w^{(i)} e_x^{(i)}$$

BAE[2] is summarized in Algorithm 1. The algorithm has two stages: the boosting-based training phase (lines 1–10) and the consensus phase (lines 11–17). In the first phase, we start with training an autoencoder on $\mathbf{X}^{(0)}$ (line 2). Then, in each iteration (lines 3–10), we compute the resulting reconstruction errors (line 5) and the probability distribution for points in $\mathbf{X}^{(0)}$ (line 6); we then sample the training data (line 8), and train the autoencoder (line 9). During the consensus phase, we compute the importance of each autoencoder of the ensemble (lines 12–14), and finally we aggregate the contribution of each autoencoder to obtain the consensus anomaly scores (lines 15–17).

Algorithm 1. BAE: Boosting-based Autoencoder Ensembles

Require: $\mathbf{X}^{(0)}, m$
1: — *Boosting-based training phase* —
2: Train $\text{AE}^{(0)}$ on $\mathbf{X}^{(0)}$
3: **for** $i = 1$ to $m-1$ **do**
4:     **for** $x \in \mathbf{X}^{(0)}$ **do**
5:         $e_x^{(i-1)} \leftarrow \left(||x - \text{AE}^{(i-1)}(x)||_2 \right)^2$
6:         $\mathbb{P}_x^{(i)} \leftarrow \frac{1/e_x^{(i-1)}}{\sum 1/e_x^{(i-1)}}$
7:     **end for**
8:     $\mathbf{X}^{(i)} \leftarrow \text{sample}(\mathbf{X}^{(0)}, \mathbb{P}^{(i)})$
9:     Train $\text{AE}^{(i)}$ on $\mathbf{X}^{(i)}$
10: **end for**
11: — *Consensus phase* —
12: **for** $i = 1$ to $m-1$ **do**
13:     $w^{(i)} \leftarrow \frac{1/\sum_{x \in \mathbf{X}^{(i)}} e_x^{(i)}}{\sum_{i=1}^{m-1} \left(1/\sum_{x \in \mathbf{X}^{(i)}} e_x^{(i)} \right)}$
14: **end for**
15: **for** $x \in \mathbf{X}^{(0)}$ **do**
16:     $\bar{e}_x \leftarrow \sum_{i=1}^{m-1} w^{(i)} e_x^{(i)}$
17: **end for**
18: **return** $\bar{e}$

3.1 Architecture and Training of Autoencoders

The base learners of our ensemble are fully-connected autoencoders. An autoencoder is a neural network with a symmetric design, where the number of input and output neurons is equal to the data dimensionality, which we denote as d. The complete architecture depends on the number of hidden layers (corresponding to the depth l), the number of neurons in each hidden layer, and the choice

[2] The code of BAE is available at https://gitlab.com/bardhp95/bae.

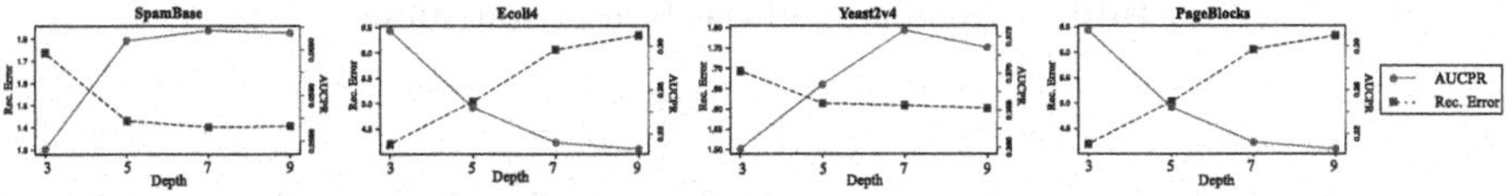

Fig. 3. Relationship between the average reconstruction error of the training samples across all ensemble components and performance (AUCPR) for different depths of the autoencoders.

of activation functions. To set the number of neurons in the hidden layers, we followed the strategy proposed in [6]. Hence, we use a Sigmoid activation function for the first (hidden) layer and for the output layer. For all the other layers we use the Rectified Linear (ReLU) activation function.

To determine the number of layers l, we ran BAE on each dataset for different values of $l \in \{3, 5, 7, 9\}$. Figure 3 shows the reconstruction error and the AUCPR for increasing l values. The two measures are negatively correlated (see Sect. 5 for details). Based on these results, we designed an unsupervised optimization strategy to select the depth l that gives the minimum reconstruction error for each dataset. Specifically, we choose the value l that minimizes the following optimization function:

$$\text{argmin}_{l \in \mathcal{L}} \left\{ \frac{1}{m-1} \sum_{i=1}^{m-1} \sum_{x \in \mathbf{X}^{(i)}} \left(||x - \text{AE}_l^{(i)}(x)||_2 \right)^2 \right\}$$

where $\text{AE}_l^{(i)}$ is the autoencoder of depth l trained at iteration i, and $\mathcal{L}$ is the set of tested l values.

4 Experiments

Methods: To assess the effectiveness of BAE, we compare its performance against baseline and ensemble techniques. Specifically, the outlier detection algorithms are LOF [1], a single autoencoder with nine layers (in short, SAE9), Hawkins [7], one-class SVM [16], and MO-GAAL [10]. The ensemble techniques are HiCS [8], RandNet [6], BoostSelect [2], CARE [13], and iForest [9].

LOF is a well-known density-based outlier detection method. For each run of LOF, the value of the MinPts parameter is randomly chosen from the set $\{3, 5, 7, 9\}$. The autoencoder structure of SAE9 is the same as the components of BAE, as discussed in Sect. 3. For one-class SVM (OCSVM) we use an RBF kernel with length-scale parameter set to 1; we randomly select the soft margin parameter in the interval $[0.25, 0.75]$ for each of the 30 runs, and set the tolerance for the stopping criterion equal to 0.001.

For Hawkins, we set the number of layers to 5, as suggested by the authors. For HiCS, we use 50 Monte-Carlo iterations, $\alpha = 0.05$, and a cutoff of 500 for the number of candidates in the subspaces. We use LOF with 10 neighbors as base outlier detector algorithm. For RandNet we use 7 layers, and we set the

Table 1. Summary of the dataset statistics.

	ALOI	Ecoli4	Glass	KDDCup99	Lympho	PageBlocks	Pima	SatImage	Shuttle	SpamBase	Stamps	WDBC	Wilt	WPBC	Yeast05679v4	Yeast2v4	MNIST0	MNIST1	MNIST2	MNIST3	MNIST4	MNIST5	MNIST6	MNIST7	MNIST8	MNIST9
Instances	49,534	336	214	48,113	148	4,982	510	1,105	1,013	2,579	315	367	4,655	198	528	514	7,534	8,499	7,621	7,770	7,456	6,950	7,508	7,921	7,457	7,589
Attributes	27	7	7	40	18	10	8	36	9	57	9	30	5	34	8	8	784	784	784	784	784	784	784	784	784	784
Outliers %	3	6	4	0.4	4	2	2	3	1	2	2	2.45	2	2.4	10	10	8	7	8	8	8	9	8	8	8	8

structure parameter to 0.5. We set the growth ratio equal to 1 and we use 20 different ensemble components. For iForest, we set the number of base detectors to 100 and the percentage of outliers as in the original paper. We ran CARE, using both LOF and kNN as base detectors, by setting the number of neighbors count to 5 and the maximum number of iterations to 15, as reported in [13]. We report only the results of CARE-kNN since it outperformed[3] CARE-LOF.

BoostSelect is a selective outlier detection method introduced in Sect. 2. To generate the ensemble components of BoostSelect, we apply 9 neighborhood-based outlier detection algorithms [2], namely: KNN, KNNW, LOF, SimplifiedLOF, LoOP, LDOF, KDEOS, INFLO, and ODIN. We ran each algorithm 20 times with a different neighborhood size k randomly choosen from the set {1,2, ... ,100}. This gives us a total of 180 components.

All the aforementioned runs were made by using the ELKI data mining software[4]. BoostSelect's parameters are consistent with those in the original paper (drop rate $d \in \{0.25, 0.5, 0.75\}$ and threshold $t \in \{0.05, 0.1, 0.15\}$); each of the nine possible combinations of parameters was used the same number of times.

To evaluate the effect of the selective strategy of BoostSelect, we consider an additional "Naive" ensemble baseline, called Naive180, which averages the 180 individual outlier score rankings produced by the components of the ensemble, without applying any selection procedure. MO-GAAL [10] is a state-of-the-art GAN-based method for unsupervised outlier detection. We use the PyOD[5] implementation of MO-GAAL with the default parameters.

Setting of BAE parameters: We set the ensemble size to 20. We tested the sensitivity of BAE under different ensemble sizes, and observed that the performance is stable for sizes beyond 20. We train each autoencoder $AE^{(i)}$ for 50 epochs. We stop earlier if the reconstruction errors converge, that is when the difference between the average reconstruction error at the end of two consecutive epochs is lower than 10^{-4}. We use ADAM as the optimizer of each $AE^{(i)}$ with learning rate $lr = 10^{-3}$ and weight decay equal to 10^{-5}. We train BAE multiple times on various depths. Hence, for each dataset, we select the optimal depth on $\mathcal{L} = \{3, 5, 7, 9\}$ following the unsupervised strategy discussed in Sect. 3.

Datasets: Data designed for classification are often used to assess the feasibility of outlier detection approaches. A common approach in outlier detection is to

[3] By 2.5% on average.

[4] https://elki-project.github.io/.

[5] https://pyod.readthedocs.io/.

use labeled data from the UCI Machine Learning Repository[6]. Instances from the majority class, or from multiple large classes, give the inliers, and instances from the minority classes (or down-sampled classes) constitute the outliers [3]. For ALOI, KDDCup99, Lymphography, Shuttle, SpamBase, WDBC, Wilt, and WPBC we follow the conversion process used in [3] (we used the normalized and without duplicate datasets[7] version#1). Ecoli4, Pima, Segment0, Yeast2v4, and Yeast05679v4 were generated according to the Keel framework approach[8]. In SatImage, the points of the majority class are used as inliers, and 1% of the rest of the data is sampled to obtain the outliers. We also test the approaches on MNIST, a large collection of images of handwritten digits. We generate a separate dataset for each digit, where the images (represented as a vector with 784 dimensions) of the chosen digit are the inliers. The outliers are obtained by uniformly sampling 1% of the other digits[9].

We found that HiCS does not converge with the larger datasets (i.e. KDD-Cup99 and MNIST)[10]. We solved this problem by reducing the dimensionality for the aforementioned datasets and let HiCS successfully run. We used TSNE with three components for the reduction of KDDCup99, and PCA with ten components for the reduction of the MNIST datasets. We then used the reduced versions of KDDCup99 and MNIST for HiCS only. Table 1 reports a summary of the statistics of the datasets used in our experiments.

Performance Evaluation: We use the area under the precision-recall curve (AUCPR) - often called average precision - to quantify the performance of each method. This measure is known to be more effective than the area under the ROC curve (AUCROC) when dealing with imbalanced datasets [14].

Results: Table 2 shows the performance achieved by all the tested methods. We run each technique 30 times, and report the average AUCPR values. Statistical significance is evaluated using a one-way ANOVA with a post-hoc Tukey HSD test with a p-value threshold equal to 0.01. For each dataset, the technique with the best performance score, and also any other method that is not statistically significantly inferior to it, is bold-faced.

The results indicate that BAE achieves a statistically superior performance (tied with other competitor in some cases) in 15 out of the 26 datasets. On several datasets, including some of the larger ones, i.e. KDDCup99, and the majority of the MNIST data, BAE wins by a large margin across all competing techniques. Overall, BAE is the most robust performer across all datasets, with an average AUCPR score of 41.6%. CARE and RandNet come second and third. BAE clearly demonstrates a superior performance against RandNet; in fact, its performance is always superior to that of RandNet, except in a few cases, where

[6] https://archive.ics.uci.edu/ml/datasets.php.

[7] Data from: https://www.dbs.ifi.lmu.de/research/outlier-evaluation/DAMI/.

[8] https://sci2s.ugr.es/keel/imbalanced.php.

[9] Used data available at: https://figshare.com/articles/dataset/MNIST_dataset_for_Outliers_Detection_-_MNIST4OD_/9954986.

[10] The ELKI project implementation of HiCS produces a warning message and the execution does not reach a converging state.

Table 2. Average AUCPR values in percentage (over 30 runs) for all methods.

Dataset	LOF	SAE9	OCSVM	Hawkins	Hics	RandNet	MOGAAL	BoostSelect	Naive180	CARE-knn	IF	BAE
ALOI	**13.3**	4.5	4.2	4.0	3.9	4.5	3.6	11.6	11.4	4.7	3.3	4.5
Ecoli4	5.3	8.4	12.3	11.5	4.8	9.5	5.4	12.9	12.9	**18.5**	**17.5**	**16.6**
Glass	15.9	9.2	7.6	10.5	7.8	9.2	10.8	14.2	13.2	**17.6**	**16.6**	12.3
KDD99	3.1	20.7	34.8	20.1	0.5	15.8	6.8	3.1	6.2	9.4	37.7	**40.8**
Lympho	47.0	76.2	80.0	23.7	4.0	85.0	80.7	**92.8**	**98.2**	**97.8**	**98.3**	**97.0**
PageBlocks	25.1	19.5	23.8	29.3	15.2	21.0	20.2	33.4	**40.6**	17.4	24.5	30.6
Pima	1.9	3.1	2.8	3.1	3.2	3.1	**5.3**	3.8	3.6	4.0	4.3	3.4
SatImage	21.1	63.7	52.4	29.4	2.3	59.9	**79.1**	71.3	65.6	55.6	**80.9**	67.5
Shuttle	11.7	2.1	2.2	3.9	**23.0**	2.1	1.0	12.3	12.0	3.3	7.3	2.1
SpamBase	7.1	5.8	6.6	6.0	2.0	5.8	9.0	**15.4**	14.2	13.5	**16.2**	5.9
Stamps	9.9	13.8	11.8	14.4	**21.4**	12.9	13.0	**20.1**	15.6	15.5	13.3	13.8
WDBC	**56.3**	49.7	**59.7**	41.2	28.7	**59.2**	**57.4**	**60.4**	**62.2**	**53.8**	**65.9**	**56.6**
Wilt	7.3	1.2	1.4	1.75	**16.1**	1.2	1.4	1.7	1.9	2.0	1.6	1.3
WPBC	21.2	22	22.7	21.5	25.0	22.1	**28.7**	23.0	23.6	22.4	23.0	23.9
Yeast05679v4	13.6	10.3	**15.4**	11.2	**15.6**	10.9	13.2	**15.0**	**14.4**	**16.2**	14.7	**17.8**
Yeast2v4	14.6	24.0	**35.8**	23.4	15.7	25.7	20.3	30.6	29.1	30.4	31.7	**37.1**
MNIST0	18.9	66.1	49.6	46.3	19.1	71.3	43.5	46.0	41.0	**78.2**	46.9	**78.9**
MNIST1	14.3	92.1	90.0	**93.7**	9.3	**94.7**	52.5	86.6	60.9	**96.5**	92.2	**96.1**
MNIST2	17.3	42.7	27.5	37.0	13.9	43.6	27.4	33.9	29.2	39.1	23.7	**47.9**
MNIST3	19.9	45.6	37.4	45.1	18.9	47.4	38.0	37.8	33.2	50.5	33.8	**55.4**
MNIST4	24.2	57.7	46.8	51.6	15.6	67.6	40.9	57.6	49.5	58.8	33.8	**71.2**
MNIST5	28.1	**49.7**	23.3	**48.8**	17.0	**49.5**	18.3	**48.9**	**41.9**	**45.1**	23.2	**48.4**
MNIST6	23.4	67.5	42.8	53.3	18.4	70.6	35.7	61.4	51.2	64.1	34.8	**73.8**
MNIST7	18.3	66.9	52.8	59.7	17.5	69.8	34.6	52.3	47.0	71.3	54.0	**74.8**
MNIST8	17.7	32.8	27.5	40.3	15.4	39.4	**47.1**	33.9	29.7	32.6	21.0	39.4
MNIST9	28.2	60.8	41.5	50.6	16.9	**64.8**	32.0	**63.2**	56.1	55.7	37.6	**63.9**
Average	18.6	35.2	31.3	30.4	13.5	37.2	27.9	35.7	32.6	37.5	32.6	*41.6*

both BAE and RandNet are in the same statistical significance tier. CARE shows a superior performance in eight datasets, the largest number among the competitors. We observe that Naive180 performs similarly to BoostSelect in most cases, and it's slightly inferior to BoostSelect on average. This means that the performance achieved by BoostSelect mostly stems from the use of a large number of outlier detection algorithms and components to form the ensemble.

As discussed in [6], the performance of GAN-based and autoencoder-based methods for unsupervised outlier detection is affected by the contamination of training data with outliers. BAE, as shown in Fig. 4, successfully isolates a considerable portion of outliers, and therefore achieves a superior performance compared to other deep methods.

All methods, except HiCS, perform poorly on Shuttle, Stamps, and Wilt. This may be due to the nature of the underlying distributions of inliers and outliers in these data. For example, Shuttle is generated as specified in Zhang et al. [18], where data of six different classes are labeled as inliers, and data from the remaining class are defined as outliers. For Wilt, the data points correspond to image segments; samples representing "diseased trees" are labelled as outliers, and samples corresponding to "all other land covers" are labelled as inliers. Inliers of different classes belong to different distributions, possibly spanning a variety of subspaces. In this scenario, learning a single global repre-

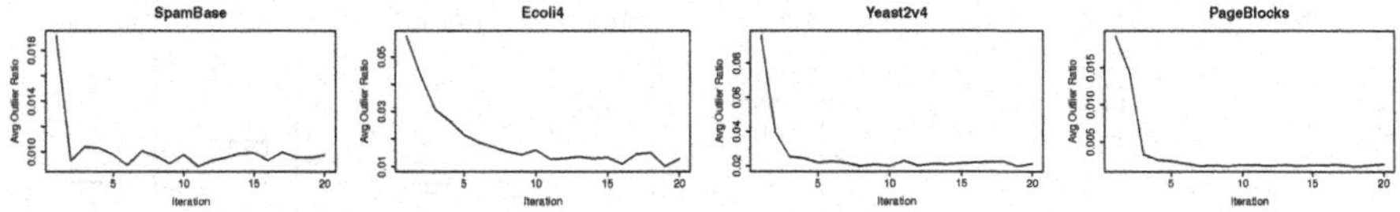

Fig. 4. Average outlier ratio in the training sample across iterations of BAE for some datasets.

sentation (as GAN-based methods and autoencoders do) may not capture well the complex nature of inliers, thus compromising the ability to detect outliers. Since HiCS discovers local subspaces to which data belong, it can fare better in these scenarios. We will perform further investigation on this analysis.

5 Discussion

In this section, we evaluate the impact of the used parameters, the implemented strategies, and discuss some of the analyses that guide our choices.

Depth of the BAE Components: The performance of BAE is a function of the depth of the autoencoders used to build the ensemble. As discussed in Sect. 3, we train BAE on multiple depths of base autoencoders. The optimal depth is then chosen based on the average reconstruction error achieved by each ensemble. In fact, we choose as optimal depth the one corresponding to the number of layers of the ensemble that generates the minimum average reconstruction error. Figure 3 demonstrates the inverse relationship between ensemble performance, in AUCPR, and average reconstruction error on training samples. Due to space limitations we show these plots only for some datasets.

Outlier ratio variation: One of the main objectives of BAE is to reduce the number of outliers in the training samples through a sequence of autoencoders. In Fig. 4, we plot the ratio of outliers sampled as part of the training sets $\mathbf{X}^{(i)}$ in successive iterations i of BAE. The observed decreasing trends confirm the efficacy of BAE in reducing the overall number of outliers in the different subsamples. Due to the page limit, we only showcase the decreasing outlier ratio trends for some datasets (similar results were obtained on all datasets).

Ensemble efficacy: In order to assess the effectiveness of BAE as an ensemble-based method, we constructed the box plots of the performance of BAE's components and compare them with those of a single autoencoder, i.e. SAE9. In Fig. 5, the blue diamond shows the performance of BAE and the red circle marks the performance of SAE9. To have meaningful box plots, we use the scores from a single run. Therefore, the AUCPR values can be slightly different from those reported in Table 2. We observe that the performance of BAE is either comparable or superior to the median AUCPR, thus confirming the effectiveness of our ensemble approach.

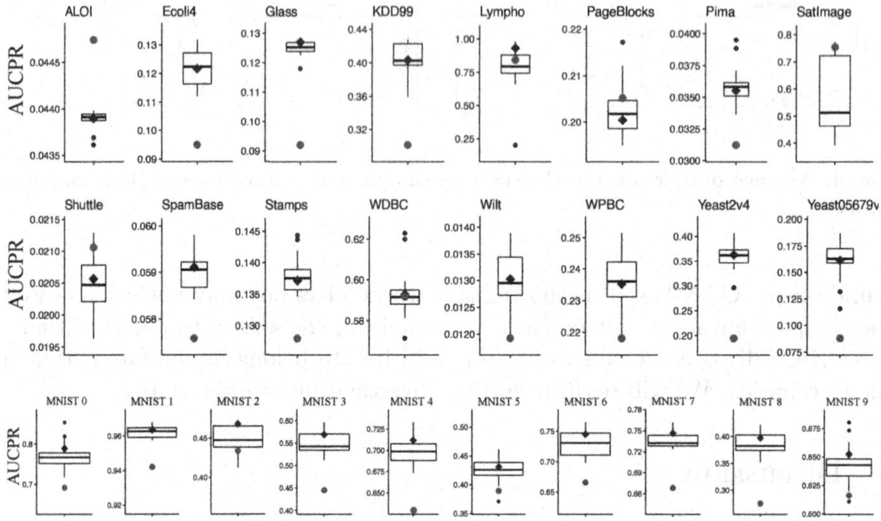

Fig. 5. Box plots of the ensemble performance of BAE for a single run. The blue diamond marks the performance of the ensemble consensus, and the red circle marks the performance of SAE9. (Color figure online)

6 Conclusion

We introduced a boosting-based outlier ensemble approach. BAE trains a sequence of autoencoders by performing a weighted sampling of the data with the aim of progressively reducing the number of outliers in the training. This enables the autoencoders to learn better representations of the inliers, which results in improved outlier scores. Furthermore, each autoencoder is exposed to a different set of outliers, thus promoting their diversity. Our experimental results show that BAE outperforms state-of-the-art competitors.

References

1. Breunig, M.M., Kriegel, H.P., Ng, R.T., Sander, J.: Lof: identifying density-based local outliers. In: ACM Sigmod Record, vol. 29, pp. 93–104. ACM (2000)
2. Campos, G.O., Zimek, A., Meira, W.: An unsupervised boosting strategy for outlier detection ensembles. In: Pacific-Asia Conference on Knowledge Discovery and Data Mining, pp. 564–576. Springer (2018)
3. Campos, G.O., Zimek, A., Sander, J., Campello, R.J., Micenková, B., Schubert, E., Assent, I., Houle, M.E.: On the evaluation of unsupervised outlier detection: measures, datasets, and an empirical study. Data Min. Knowl. Disc. **30**(4), 891–927 (2016)
4. Chalapathy, R., Chawla, S.: Deep learning for anomaly detection: a survey. arXiv:1901.03407 (2019)
5. Chandola, V., Banerjee, A., Kumar, V.: Anomaly detection: a survey. ACM Comput. Surv. **41**(3) (2009)

6. Chen, J., Sathe, S., Aggarwal, C., Turaga, D.: Outlier detection with autoencoder ensembles. In: Proceedings of the 2017 SIAM International Conference on Data Mining, pp. 90–98. SIAM (2017)
7. Hawkins, S., He, H., Williams, G., Baxter, R.: Outlier detection using replicator neural networks. In: International Conference on Data Warehousing and Knowledge Discovery, pp. 170–180. Springer (2002)
8. Keller, F., Muller, E., Bohm, K.: Hics: high contrast subspaces for density-based outlier ranking. In: 2012 IEEE 28th international conference on data engineering, pp. 1037–1048. IEEE (2012)
9. Liu, F.T., Ting, K.M., Zhou, Z.H.: Isolation forest. In: 2008 Eighth IEEE International Conference on Data Mining, pp. 413–422. IEEE (2008)
10. Liu, Y., Li, Z., Zhou, C., Jiang, Y., Sun, J., Wang, M., He, X.: Generative adversarial active learning for unsupervised outlier detection. IEEE Trans. Knowl. Data Eng. **38**(8), 1517–1528 (2019)
11. Ramaswamy, S., Rastogi, R., Shim, K.: Efficient algorithms for mining outliers from large data sets. In: ACM Sigmod Record, vol. 29, pp. 427–438. ACM (2000)
12. Rayana, S., Akoglu, L.: Less is more: Building selective anomaly ensembles with application to event detection in temporal graphs. In: Proceedings of the 2015 SIAM International Conference on Data Mining, pp. 622–630. SIAM (2015)
13. Rayana, S., Zhong, W., Akoglu, L.: Sequential ensemble learning for outlier detection: a bias-variance perspective. In: 2016 IEEE 16th International Conference on Data Mining (ICDM), pp. 1167–1172. IEEE (2016)
14. Saito, T., Rehmsmeier, M.: The precision-recall plot is more informative than the roc plot when evaluating binary classifiers on imbalanced datasets. PloS one **10**(3) (2015)
15. Sarvari, H., Domeniconi, C., Stilo, G.: Graph-based selective outlier ensembles. In: Proceedings of the 34th ACM/SIGAPP Symposium on Applied Computing, pp. 518–525. ACM (2019)
16. Schölkopf, B., Platt, J.C., Shawe-Taylor, J., Smola, A.J., Williamson, R.C.: Estimating the support of a high-dimensional distribution. Neural comput. **13**(7) (2001)
17. Schubert, R., Wojdanowski, R., Zimek, A., Kriegel, H.P.: On evaluation of outlier rankings and outlier scores. In: Proceedings of the SIAM International Conference on Data Mining, pp. 1047–1058. SIAM (2012)
18. Zhang, K., Hutter, M., Jin, H.: A new local distance-based outlier detection approach for scattered real-world data. In: Pacific-Asia Conference on Knowledge Discovery and Data Mining. pp. 813–822. Springer (2009)

Unsupervised Domain Adaptation for 3D Medical Image with High Efficiency

Chufu Deng, Kuilin Li, and Zhiguang Chen(✉)

Sun Yat-Sen University, Guangzhou, China
{dengchf3,liklin3}@mail2.sysu.edu.cn, chenzhg29@mail.sysu.edu.cn

Abstract. Domain adaptation is a fundamental problem in the 3D medical image process. The current methods mainly cut the 3D image into 2D slices and then use 2D CNN for processing, which may ignore the inter-slice information of the 3D medical image. Methods based on 3D CNN can capture the inter-slice information but lead to extensive memory consumption and lots of training time. In this paper, we aim to model the inter-slice information in 2D CNN to realize the unsupervised 3D medical image's domain adaptation without additional cost from 3D convolutions. To better capture the inter-slice information, we train the model from the adjacent (local) slices and the global slice sequence perspective. We first propose the Slice Subtract Module method (SSM), which can easily embed into 2D CNN and model the adjacent slices by introducing very limited extra computation cost. We then train the adaptation model with a distance consistency loss supervised by the LSTM component to model the global slice sequence. Extensive experiments on BraTS2019 and Chaos datasets show that our method can effectively improve the quality of domain adaptation and achieve state-of-the-art accuracy on segmentation tasks with little computation increased while remaining parameters unchanged.

Keywords: Domain adaptation · 3D medical image · Slice sequence

1 Introduction

The advanced deep convolutional neural network (CNN) has achieved a significant leap in many recognition tasks including classification, super-resolution, and semantic segmentation. However, these successful algorithms mostly rely on large-scale annotated datasets. In reality, medical images' segmentation label requires experienced doctors to spend lots of time annotating each pixel, resulting in many open medical datasets labeled little or not. When we directly apply the well-trained supervised deep learning model to the target dataset with little labels, due to the huge gap in the distribution of the training dataset and the test dataset, also known as domain shift, the model's powerful capability is difficult to reproduce. To make the model perform well on the target dataset, some studies have proposed domain adaptation methods to reduce domain shift between the source dataset and the target dataset.

K. Karlapalem et al. (Eds.): PAKDD 2021, LNAI 12712, pp. 104–116, 2021.
https://doi.org/10.1007/978-3-030-75762-5_9

In the field of medical image processing, domain adaptation is widespread. Unlike the natural images usually obtained by optical cameras [1], the medical image uses various imaging methods (e.g., MRI, CT) to capture different physical characteristics. These different imaging methods cause domain shift among different modalities, making the model trained on one modality (e.g., MRI) difficult to apply to another modality (e.g., CT). Jia et al. [2] transfers CT images to the Cone-Beam CT (CBCT) images, and enables task-driven adversarial learning by combining adaptation network and CBCT segmentor; Carolina et al. [3] realizes the domain adaptation between human embryonic stem cell-derived cardiomyocytes and adult cardiomyocytes by adding a loss function that penalizes the Maximum Mean Difference in the feature space. These methods achieve satisfactory results in domain adaptation by designing new losses or combining specific target tasks, but suffer from the lack of 3D image's inter-slice information. And these information may help to improve the accuracy [4]. Thus in this paper, we mainly discuss how to adapt the 3D medical image efficiently.

There are three main aspects for applying deep learning to 3D medical image processing: (1) Cut the 3D image into 2D slices [5], then use common 2D models to process each 2D slice, finally, concatenate the results of each 2D slice into a 3D form. Although these solutions make full use of the entire 2D slices information, they ignore the relationship among slices. (2) Use 3D CNN to directly process the 3D images [6]. Since the cost of parameters and computation of 3D CNN itself is expensive, directly using 3D CNN on the original high-resolution 3D image may lead to insufficient memory and lots of training time. (3) In common practice, 3D medical images are usually expressed as 2D slice sequences, and the combination of CNN and RNN is used respectively responsible for intra-slice and inter-slice context information. However, these methods [7,8] often need multi RNNs to achieve better accuracy.

To adapt the 3D medical image efficiently, in this paper, we take both the adjacent slices and the global slice sequence information into consideration. For the adjacent slices, we propose a universal method Slice Subtract Module method (SSM) with high efficiency and powerful performance inspired by Temporal Shift Module (TSM) [9]. Specifically, TSM shifts part of the channels in the temporal dimension, making it convenient to exchange information between adjacent frames. Considering the distribution of differences between the adjacent slices is relatively uniform and regular in 3D medical images, using shift operation may be difficult to capture the changed parts between adjacent slices. Thus, we designed the SSM for the medical image, which mainly uses the subtract operation between slices to highlight the changed parts. It can be conveniently inserted into 2D CNN neural network to achieve slice sequence modeling with little computation increased while remaining parameters unchanged. These advantages enable SSM to achieve the performance of 3D CNN while maintaining the complexity of 2D CNN. For the preservation of the global slice sequence information, we introduce a LSTM component to extract and embed contextual information of slice sequence into the process of the domain adaptation. A series of experiments verify that our method is effective.

In summary, this paper's main concern is how to effectively use the sequence information among 2D slices of 3D medical images during the process of domain adaptation while reducing the cost of computation and parameters. For this purpose, the method we designed has the following contributions:

- Compared to 3D CNN, The designed SSM can be easily embedded into 2D CNN with much smaller cost but competitive accuracy.
- A distance consistency loss based on LSTM is added to the domain adaptation model to realize the global slice sequence information preservation.
- The proposed domain adaptation model is applied to the segmentor to achieve the target domain's segmentation labels with state-of-the-art accuracy.

2 Related Work

2.1 Domain Adaptation

Domain adaptation learning theory was first proposed in the field of natural language processing. Early common domain adaptation methods include instance-based domain adaptation [10], feature representation based domain adaptation [11], and classifier based domain adaptation [12]. With the development of deep learning, some researches [13] directly used neural networks to learn domain invariant space, and some studies [14,15] projected the target domain to the source domain through domain transformation network. Since the generation adversarial network (GAN) [16] was proposed, there have been many studies [17] uses adversarial loss to replace the traditional Maximum Mean Difference loss. These GAN-based methods optimized the adversarial objective function of the domain classifier to minimize the difference distance between domains, and achieved state-of-the-art accuracy. With the proposal of CycleGAN [18], unsupervised domain transformation has developed rapidly. Many CycleGAN based methods [19,20] used unpaired data to achieve satisfactory adaptation results on different tasks, reducing the dependence on the dataset with paired data.

2.2 3D Medical Image Processing

Many researchers have achieved satisfactory results in various medical image processing tasks by proposing new deep learning models currently. On the ISBI cell tracking challenge in 2015, Olaf et al. [21] proposed the well-known U-Net architecture and won the first prize, leading a new trend in 2D medical image segmentation. In biomedical imaging, 3D images often consist of highly anisotropic dimensions [4], that is, the scale of each voxel in depth can be much larger than that in the slice plane. At present, more and more researches are devoted to using sequential information to improve performance. Since the methods based on 2D CNN often suffer from missing 3D contextual information, Xia et al. [22] trained deep networks from different dimensions and fused multi-dimensions information at the final stage. Chen et al. [7] introduced RNNs to capture sequential information to achieve higher accuracy. Cicek et al. [23] proposed a 3D version of

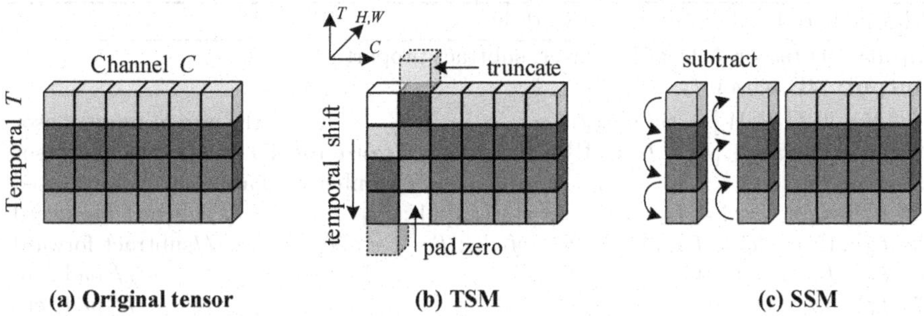

Fig. 1. The proposed SSM method. (a) is the original tensor, (b) is the TSM, and (c) is our SSM.

U-Net to implement 3D image segmentation. Considering the expensive costs of directly training a 3D CNN, Mohammad et al. [24] propose a novel two-pathway architecture and split the 3D image into small patches for training. In this paper, we propose a novel way to incorporate slice sequence information into 2D CNN. Compared to the above methods, our method can retain the well-designed 2D network architectures unchanged by using residual block, while enjoying better accuracy, computation and parameters trade-off.

3 Methods

3.1 Slice Subtract Module (SSM)

Like video files, there is a sequential relationship among slices in 3D medical images. Generally, a video file can be expressed as $A \in \mathbb{R}^{N\times T\times C\times H\times W}$, where N denotes the batch size, T denotes the time dimension, C denotes the number of channels, and H and W denotes the spatial resolution. 3D convolution operation can model T, H, and W, while 2D convolution operation can not model T, thus 2D CNN will not be sensitive to time series. As shown in Fig. 1, to enable 2D CNN to model slice sequence, TSM shifts part of the channels in the time dimension from two directions, forward and backward. After the shift operation, a certain frame will include both the information from past and future frames. we specifically design the SSM for 3D medical images based on TSM, which can capture the subtle differences and highlight the changed part between slices. Different from TSM, SSM replaces the shift operation with the subtract operation. Specifically, along the slice sequence, we use part of the channels of the current slice to subtract the corresponding channels of the previous slice, and use another part of the channels of the current slice to subtract the corresponding channels of the next slice. Referring to TSM, since shift all channels will cause training latency and performance degradation due to large data movement and worse spatial modeling ability, we use 1/8 channels to subtract forward and 1/8 channels to subtract backward. To easily embed into any 2D CNN with the structure changed minimum, we insert the SSM into the residual block. The

Algorithm 1. Slice Subtract Module

Input: 4D tensor f_i from last layer, subtract proportion p.
Output: 4D tensor f_o.

```
[N × T, C, H, W] = getsize(f_i)                       // get the size of input tensor
f_m = resize(f_i, [N, T, C, H, W])                    // resize the 4D tensor into 5D tensor
c = C/p                                               // number of channels to be processed
f_m[:, :T − 1, :c] = f_m[:, :T − 1, :c] − f_m[:, 1 :, :c]     // subtract backward
f_m[:, 1 :, c : 2c] = f_m[:, 1 :, c : 2c] − f_m[:, :T − 1, c : 2c]   //subtract forward
f_m[:, T − 1, :c] = 0                                 // pad zero
f_m[:, 0, c : 2c] = 0                                 // pad zero
f_m = Conv2d(f_m)
f_o = f_i + f_m
```

details are shown in Algorithm 1. In 3D CNN, the 3D image is input into the neural network in the form of a 5D tensor with a size of $[N, T, C, H, W]$. Since we perform adaptation for 3D images and our adapter is implemented using 2D CNN, we resize the 5D tensor into a 4D tensor with a size of $[N \times T, C, H, W]$. Compared to 3D CNN, SSM can achieve a satisfactory result with faster speed and lower parameter cost. Compared to TSM, SSM can achieve better result with a small amount of computation increased, and the parameter cost remains unchanged.

3.2 LSTM Module

Long Short-Term Memory (LSTM) was initially used in natural language processing to model text with temporal relationships, and then complete tasks such as translation, sentiment analysis, and human-computer dialogue. And the Bi-directional LSTM is proposed to model the contextual text better and improve the model's performance. Recently, some studies have applied LSTM to video analysis and prediction, which provides a new idea for 3D medical image processing [8]. Inspired by these, we add the BiLSTM structure to the domain adaptation model to preserve the global slice sequence. Giving a 3D medical image x with the size of $T \times C \times H \times W$, we split the T into three parts T_l, T_m and T_r with the ratio of 1:6:1. Let x_l, x_m and x_r denote the 3D images with T_l, T_m and T_r slices. We use x_m as our BiLSTM's input to predict the x_l and x_r. Since the BiLSTM modules do not need to rely on labels, we first use source data and target data to train the corresponding BiLSTM modules. The loss function is as follows:

$$\mathcal{L}(BiLSTM) = \mathbb{E}_x[\|BiLSTM(x_m) - [x_l, x_r]\|_2^2], \tag{1}$$

Then, during the process of domain adaptation, we apply BiLSTM modules to the images generated by the adaptation model to provide distance consistency loss, which encourages the adaptation model to pay attention to the preservation of the global slice sequence information.

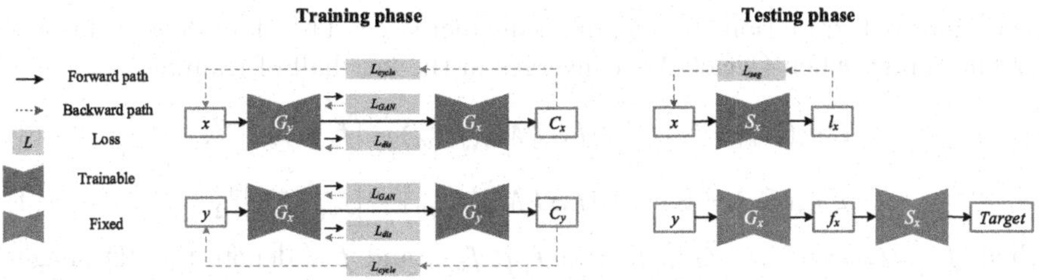

Fig. 2. The proposed 3D medical image domain adaptation framework. **Training phase:** The two generators with SSM are jointly trained together with the two discriminators, under the help of the pre-trained and fixed BiLSTM module. **Testing phase:** The segmentor is trained on source domain data under the supervision of $\mathcal{L}_{seg}$. Then, the target domain data is input into the generator with the f_x as output, and the f_x is input into the segmentor to get the target label.

3.3 Training of the Domain Adaptation

The objective of this study is to implement the domain adaptation for 3D medical images. Our method is based on CycleGAN, and the adaptation task is divided into two phases: the training phase and the testing phase. The overall process is shown in the Fig. 2. Let x denote the source domain 3D data, y denote the target domain 3D data. Generator G_x means to transfer the input data into source domain data, and generator G_y means to transfer the input data into target domain data. Given the two domain's unpaired data, we can not directly train the generator through supervision loss. CycleGan takes a different way. It introduces two training process: $G_x(G_y(x)) = C_x$ and $G_y(G_x(x)) = C_y$, so that the generators can be trained by calculating the supervision loss of C_x and x, C_y and y. In this training phase, we embed the SSM into the CycleGAN generators, and the embedding method is shown in Sect. 3.1. This process involves the following losses:

Cyclegan Loss. We adopt pixel-level mean square error loss to force the reconstructed sample $G_x(G_y(x))$ and $G_y(G_x(x))$ to be identical to their inputs x and y,

$$\mathcal{L}_{cycle}(G_x, G_y) = \mathbb{E}_{x,y}[\|x - G_x(G_y(x))\|_2^2 + \|y - G_y(G_x(y))\|_2^2], \tag{2}$$

and the widely used adversarial loss $\mathcal{L}_{GAN}$ [16] is adopted to train the discriminator D_x and D_y, and the generator G_x and G_y.

Distance Consistency Loss. In the training phase, we pre-trained the BiLSTM modules as described in Sect. 3.2. Then we add the trained BiLSTM module to the CycleGAN to provide the distance consistency loss to the generators. Due to the low quality of the generated images in the early training stage, the sequence information cannot be extracted by the BiLSTM modules, resulting in performance degradation and training time increased. Therefore we add this loss for fine-tuning when CycleGAN converges in the last half of training.

$$\mathcal{L}_{dis}(G_x) = \mathbb{E}_{f_x}[\|BiLSTM_x(f_{x,m}) - [f_{x,l}, f_{x,r}]\|_2^2, \tag{3}$$

$$\mathcal{L}_{dis}(G_y) = \mathbb{E}_{f_y}[\|BiLSTM_y(f_{y,m}) - [f_{y,l}, f_{y,r}]\|_2^2, \tag{4}$$

where $f_x = G_x(y)$, $f_y = G_y(x)$. And $f_{x,l}$, $f_{x,m}$ and $f_{x,r}$ denote the 3D images with T_l, T_m and T_r slices described in Sect. 3.2.

Seg Loss. Segmentation is a common and important task for medical image processing. Therefore, in the testing phase, we use the segmentation results to verify our method's effectiveness. Since our domain adaptation method is for 3D images, we use the traditional 3D U-net [23] as the segmentor, and the following losses are designed,

$$\mathcal{L}_{seg}(S_x) = \mathbb{E}_x[L_{CE}(S_x(x), l_x)] \, . \tag{5}$$

where l_x is the segmentation label, $L_{CE}(\cdot)$ represents the cross-entropy loss based on the real and expected outputs of the segmentor S_x. Combined with the generators trained in the training phase, we input the f_x generated by the target domain y into the segmentor to obtain the target domain labels we need.

4 Experiments

4.1 Dataset and Experimental Settings

BRATS2019[1] is an open dataset with four registered modalities, i.e., T1, T2, T1c and Flair. The training set contains 274 subjects, where each subject has registered four-modal 3D MRI with the size of $155 \times 240 \times 240$ and a tumor segmentation label. The lesion segmentation label include the enhancing region of the tumor (ET), the tumor core (TC) which is the union of necrotic, non-enhancing and enhancing regions of the tumor, and the whole tumor (WT) which is the union of the TC and the peritumoral edematous region. Chaos[2] is an open dataset with unregistered CT, MRI T1, and MRI T2. Chaos is divided into training set with abdominal segmentation labels and test set without the labels. The training set contains 20 3D samples with different number of slices and different sizes, and we resize them into 256×256 in our experiments. Since the label is not available in both the BraTS2019 and Chaos's test sets, we don't use the test sets in our experiment. Instead, We perform 5-fold cross-validation on the training set in all the experiments. We use Min-Max Normalization to preprocess our input data.

We perform experiments on a single NVIDIA Tesla V100 GPU. Our adapter's input is $[1 \times 16, 1, 240, 240]$ on BraTS2019, and $[1 \times 16, 1, 256, 256]$ on Chaos. Our segmentor is a 3D U-net, the input of segmentor is $[1, 48, 1, 240, 240]$ on

[1] https://www.med.upenn.edu/cbica/brats-2019/.

[2] https://chaos.grand-challenge.org/Data/.

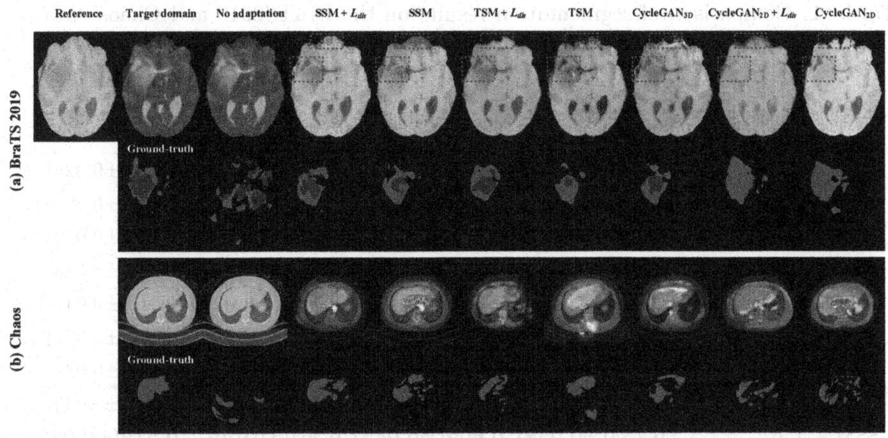

Fig. 3. The segmentation results of domain adaptation models on (a) BraTS2019 and (b) Chaos. The first row is the target domain image and the corresponding adaptation results, and the Reference denotes the registered source domain image. The second row is the ground-truth segmentation label and the segmentation results of different models.

BraTS2019, and [1, 16, 1, 256, 256] on Chaos since there are not too many slices in Chaos samples. In the testing phase, the adapter would adapt the slices from the same 3D images and combine them into a 3D subject as the segmentor's input. We use Adam optimizer to optimize our model, with the initial learning rate of 2e-3, and the loss weights of $\mathcal{L}_{cycle}$, $\mathcal{L}_{GAN}$, and $\mathcal{L}_{dis}$ are [1, 1, 0.2] in the training phase. We use the Dice score to evaluate segmentation results.

4.2 Segmentation Evaluation

Here we use the segmentation results to show the proposed method's effectiveness. In BraTS2019, we let MRI T1 as source domain and MRI T2 as target domain. In Chaos we let MRI T2 as source domain and CT as target domain. CycleGAN$_{2D}$ is CycleGAN using 2D CNN, CycleGAN$_{3D}$ is CycleGAN using 3D CNN. And the TSM, SSM use CycleGAN$_{2D}$ as backbones. As shown in Fig. 3, the segmentation result is inferior when the segmentor trained in the source domain is directly used to test on the target domain without adaptation. When we perform adaptation, the result was significantly improved. Table 1 shows the dice score of the models on BarTS2019 and Chaos. It can be seen that the TSM and SSM methods are more effective than CycleGAN$_{2D}$ since they consider the relationship between slices. And SSM is better than TSM by catching the difference between slices effectively. Compared to models without $\mathcal{L}_{dis}$'s constraint, models with $\mathcal{L}_{dis}$ do help to improve the adaptation results by considering the global sequence information. And the complete model SSM + $\mathcal{L}_{dis}$ outperforms other models in both BraTS2019 and Chaos datasets.

Table 1. Comparison of segmentation results on the BraTS2019 and Chaos dataset. The dice score is shown in the form of mean(±std).

Method	BraTS2019			Chaos
	WT	TC	ET	
On source	0.823(±0.048)	0.747(±0.034)	0.624(±0.017)	0.683(±0.024)
No adaptation	0.097(±0.008)	0.084(±0.009)	0.006(±0.004)	0.106(±0.009)
CycleGAN$_{2D}$	0.643(±0.062)	0.534(±0.043)	0.304(±0.015)	0.438(±0.030)
CycleGAN$_{2D}$ + $\mathcal{L}_{dis}$	0.651(±0.043)	0.537(±0.055)	0.312(±0.021)	0.453(±0.033)
CycleGAN$_{3D}$	0.712(±0.055)	0.584(±0.058)	**0.366(±0.018)**	0.511(±0.041)
TSM	0.684(±0.058)	0.553(±0.054)	0.331(±0.020)	0.477(±0.027)
TSM + $\mathcal{L}_{dis}$	0.697(±0.047)	0.582(±0.070)	0.351(±0.019)	0.497(±0.029)
SSM	0.704(±0.056)	0.571(±0.053)	0.353(±0.017)	0.496(±0.030)
SSM + $\mathcal{L}_{dis}$	**0.734(±0.058)**	**0.609(±0.045)**	0.364(±0.020)	**0.519(±0.032)**

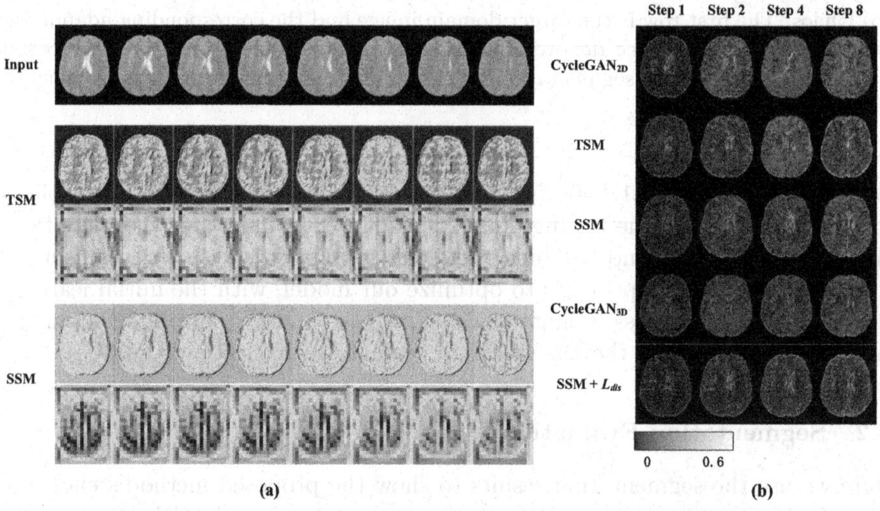

Fig. 4. The visual results of the slices difference on BraTS2019. (a) The feature maps are the output of TSM and SSM in the down-sample convolution layers (the first and the last down-sample convolution layer in our encoder). (b) The step means the distance between slices.

4.3 Slices Difference Evaluation

To further figure out the impact of TSM and SSM on the slice sequence in the domain adaptation process. In our down-sample convolution layers, we visualize some feature maps of TSM and SSM. As shown in Fig. 4.(a), the input are consecutive slices, and we can see that the adjacent slices are similar. The feature maps are the average value of all the channels in different slices. We can see that TSM may not provide much additional information as the network go deeper, and the SSM focuses on the changed areas instead.

Table 2. The MAE scores of slices difference. The step means the distance between slices.

Method	Step 1	Step 2	Step 4	Step 8
CycleGAN_{2D}	0.019	0.028	0.039	0.058
CycleGAN_{3D}	0.016	0.020	0.032	0.039
TSM	0.016	0.023	0.039	0.047
SSM	**0.014**	0.022	0.040	0.046
SSM + $\mathcal{L}_{dis}$	0.015	**0.019**	**0.025**	**0.036**

Furthermore, we attempt to calculate the slice differences between registered source images and adapted target images with the following formulate:

$$f(x, y) = MAE((x_i - x_j) - (y_i - y_j)) \quad (6)$$

where x means the source domain image, y means the target domain image after adaptation, i, j means the ith and j^{th} slice of x and y. We first use $x_i - x_j$ to obtain the variations between the source domain slices and use $y_i - y_j$ to obtain the variations between the adapted target domain slices. Then we use Mean Absolute Error (MAE) to capture the differences of these variations. In our experiments, we set the distance between slices to 1, 2, 4, and 8, which means $j - i = 1, 2, 4, 8$. The MAE scores are shown in Table 2, and the visual results are shown in Fig. 4.(b). We can see that the difference rises as the step size increases in all models. Without the slice sequence constraint, CycleGAN_{2D} performs worse than others. From step 1, we can see that the SSM does help to reduce the difference. With the $\mathcal{L}_{dis}$ constrain, even if the step is long, SSM + $\mathcal{L}_{dis}$ can maintain a relatively small difference.

4.4 Cost Evaluation

In addition to analyzing our adaptation model's effectiveness, we also make comparisons for parameters and computational cost. In these experiments, We do not consider the segmentor, which is shared by all adaptation models in cost measurement. We show the accuracy, FLOPs, and the number of parameters trade-off in Fig. 5. We can see that TSM does not increase the number of parameters compared to CycleGAN_{2D} but only increases the shift operation. Compared to TSM, SSM only increases the computation cost of channel subtraction without parameter number increase either. SSM can achieve very competitive performance at high efficiency: its performance is closer to 3D CNN while enjoying fast inference with high efficiency and low computational cost. And SSM + $\mathcal{L}_{dis}$ achieve the state-of-the-art results by adding little parameters and FLOPs. We also compared the effect of the number of slices on model performance. CycleGAN_{2D} cannot get much performance improvement from the increase in the number of slices. And Other models have more performance improvements.

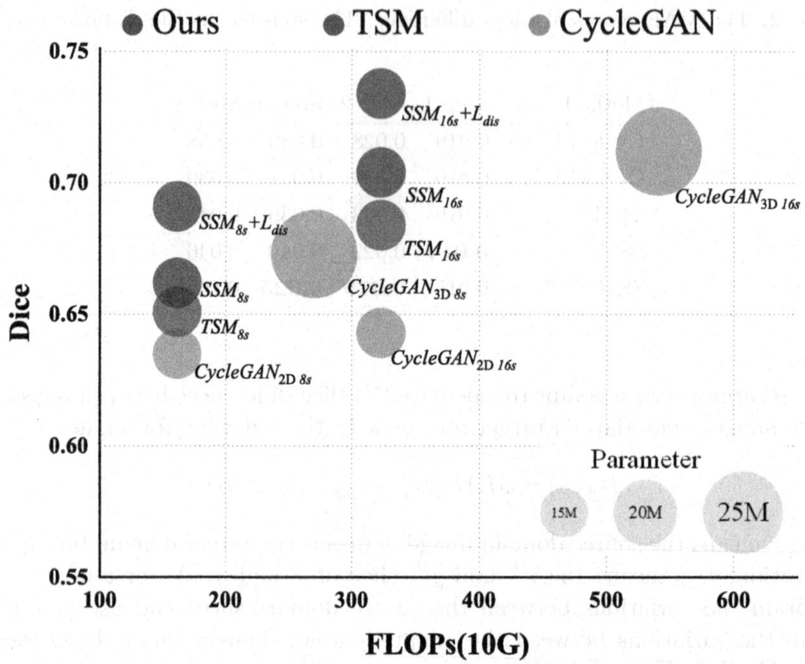

Fig. 5. The accuracy, FLOPs, and the number of parameters of different models. The x-axis is the FLOPs. The y-axis is the dice score of the whole tumor in BraTS2019. The area of the circles indicates the number of parameters. The $8s$, $16s$ means the slice number T is 8 and 16.

5 Conclusion

The paper proposed a new unsupervised domain adaptation framework which can be used to effectively adapt the 3D medical image. The effectiveness of the framework is built on the novel SSM for adjacent slices and the novel distance loss terms based on the BiLSTM modules for the global slice sequence. The effectiveness of the domain adaptor have been verified by the extensive experiments on MRI and CT images from different human organs. Particularly, we also analyze the framework's cost and practicality, and our framework shows high efficiency and accuracy, as confirmed in multiple experiments.

Acknowledgement. This work was supported by the National Key R&D Program of China under Grant NO.2018YFB0204303, Nature Science Foundation of China under Grant NO.U1811461, the Guangdong Natural Science Foundation under Grant NO.2018B030312002, and the Program for Guangdong Introducing Innovative and Entrepreneurial Teams under Grant NO.2016ZT06D211.

References

1. Dou, Q., Ouyang, C., Chen, C., Chen, H., Heng, P.-A.: Pnp-adanet: plug-and-play adversarial domain adaptation network with a benchmark at cross-modality cardiac segmentation. IEEE Access **7**, 99065–99076 (2019)
2. Jia, X., Wang, S., Liang, X., Balagopal, A., Nguyen, D., Yang,, M., Wang, Z., Ji, J., Qian, X., Jiang, S.: Cone-beam computed tomography (cbct) segmentation by adversarial learning domain adaptation. In: Proceedings of the Medical Image Computing and Computer Assisted Intervention, pp. 567–575 (2019)
3. Pacheco, C., Vidal, R.: An unsupervised domain adaptation approach to classification of stem cell-derived cardiomyocytes. In: Proceedings of Medical Image Computing and Computer Assisted Intervention, pp. 806–814 (2019)
4. Lee, K., Zlateski, A., Ashwin, V., Seung, H.S.: Recursive training of 2d–3d convolutional networks for neuronal boundary prediction. In: Advances in Neural Information Processing Systems, pp. 3573–3581 (2015)
5. Fripp, J., Crozier, S., Warfield, S.K., Ourselin, S.: Automatic segmentation and quantitative analysis of the articular cartilages from magnetic resonance images of the knee. IEEE Trans. Med. Imaging **29**(1), 55–64 (2010)
6. Lai, M.: Deep learning for medical image segmentation. arXiv: preprint (2015)
7. Chen, J., Yang, L., Zhang, Y., Alber, M., Chen, D.Z.: Combining fully convolutional and recurrent neural networks for 3d biomedical image segmentation. In: Advances in Neural Information Processing Systems, pp. 3036–3044 (2016)
8. Stollenga, M.F., Byeon, W., Liwicki, M., Schmidhube, J.: Parallel multi-dimensional lSTM, with application to fast biomedical volumetric image segmentation. In: Advances in Neural Information Processing Systems, pp. 2998–3006 (2015)
9. Lin, J., Gan, C., Han, S.: Tsm: temporal shift module for efficient video understanding. In: Proceedings of the International Conference on Computer Vision, pp. 7082–7092 (2019)
10. Zadrozny, B.: Learning and evaluating classifiers under sample selection bias. In: Proceedings of the Twenty-First International Conference on Machine Learning, pp. 114–121 (2004)
11. Blitzer, J., Mcdonald, R., Pereira, F.: Domain adaptation with structural correspondence learning. In: Proceedings of the 2006 Conference on Empirical Methods in Natural Language Processing, pp. 120–128 (2006)
12. Pan, S.J., Yang, Q.: A survey on transfer learning. IEEE Trans. Knowl. Data Eng. **22**(10), 1345–1359 (2010)
13. Long, M., Zhu, H., Wang, J., Jordan, M.I.: Unsupervised domain adaptation with residual transfer networks. In: Proceedings of the 30th International Conference on Neural Information Processing Systems, pp. 136–144 (2016)
14. Isola, P., Zhu, J., Zhou, T., Efros, A.A.: Image-to-image translation with conditional adversarial networks. In: Proceedings of the Computer Vision and Pattern Recognition, pp. 5967–5976 (2016)
15. Liu, M., Breuel, T.M., Kautz, J.: Unsupervised image-to-image translation networks. Computer, Vision and Pattern Recognition. arXiv (2017)
16. Goodfellow, I., Pouget-Abadie, J., Mirza, M., Xu, B., Warde-Farley, D., Ozair, S., Courville, A., Bengio, Y.: Generative adversarial nets. In: Proceedings of the Neural Information Processing Systems, pp. 2672–2680 (2014)

17. Luo Y., Zheng, L., Guan, T., Yu, J., Yang, Y.: Taking a closer look at domain shift: category-level adversaries for semantics consistent domain adaptation. In: Proceedings of Conference on Computer Vision and Pattern Recognition, pp. 2502–2511 (2019)
18. Zhu, J., Park, T., Isola, P., Efros, A.A.: Unpaired image-to-image translation using cycle-consistent adversarial networks. In: Proceedings of International Conference on Computer Vision, pp. 2242–2251 (2017)
19. Royer, A., Bousmalis, K., Gouws, S., Bertsch, F., Mosseri, I., Cole, F., Murphy, K.: XGAN: unsupervised Image-to-Image Translation for many-to-many Mappings, pp. 33–49 (2020)
20. Anoosheh, A, Agustsson, E., Timofte, R., Van Gool, L.: Combogan: unrestrained scalability for image domain translation. In: Proceedings of the Computer Vision and Pattern Recognition Workshops, pp. 896–8967 (2018)
21. Ronneberger, O., Fischer, P., Brox, T.: U-net: convolutional networks for biomedical image segmentation. In: Proceedings of the International Conference on Medical Image Computing and Computer Assisted Intervention, pp. 234–241 (2015)
22. Xia, Y., Xie, L., Liu, F., Zhu, Z., Fishman, E.K., Yuille, A.L.: Bridging the gap between 2d and 3d organ segmentation with volumetric fusion net. In: Proceedings of Medical Image Computing and Computer Assisted Intervention, pp. 445–453 (2018)
23. Çiçek, Ö., Abdulkadir, A., Lienkamp, S.S., Brox, T., Ronneberger, O.: 3d u-net: learning dense volumetric segmentation from sparse annotation. In: Proceedings of Medical Image Computing and Computer-Assisted Intervention, pp. 424–432 (2016)
24. Havaei, M., Davy, A., Warde-Farley, D., Biard, A., Courville, A., Bengio, Y., Pal, C., Jodoin, P.-M., Larochelle, H.: Brain tumor segmentation with deep neural networks. Med. Image Anal. **35**, 18–31 (2017)

A Hierarchical Structure-Aware Embedding Method for Predicting Phenotype-Gene Associations

Lin Wang[1], Mingming Liu[1], Wenqian He[1], Xu Jin[1], Maoqiang Xie[1](✉), and Yalou Huang[2]

[1] College of Software, NanKai University, TianJin 300071, China
{wangl_2019,hewenqian,jinx_2018}@mail.nankai.edu.cn,
{liumingming,xiemq}@nankai.edu.cn

[2] TianJin International Joint Academy of Biomedicine, TianJin, China
huangyl@nankai.edu.cn

Abstract. Identifying potential causal genes for disease phenotypes is essential for disease treatment and facilitates drug development. Inspired by existing random-walk based embedding methods and the hierarchical structure of Human Phenotype Ontology (HPO), this work presents a Hierarchical Structure-Aware Embedding Method (HSAEM) for predicting phenotype-gene associations, which explicitly incorporates node type information and node individual difference into random walks. Unlike existing meta-path-guided heterogeneous network embedding techniques, HSAEM estimates an individual jumping probability for each node learned from hierarchical structures of phenotypes and different node influences among genes. The jumping probability guides the current node to select either a heterogeneous neighborhood or a homogeneous neighborhood as the next node, when performing random walks over the heterogeneous network including HPO, phenotype-gene and Protein-Protein Interaction (PPI) networks. The generated node sequences are then fed into a heterogeneous SkipGram model to perform node representations. By defining the individual jumping probability based on hierarchical structure, HSAEM can effectively capture co-occurrence of nodes in the heterogeneous network. HSAEM yields its extraordinary performance not only in the statistical evaluation metrics compared to baselines but also in the practical effectiveness of prioritizing causal genes for Parkinson's Disease.

Keywords: Network embedding · Heterogeneous network · Hierarchical structure-aware · Phenotype-gene association prediction

1 Introduction

Recognizing potential pathogenic genes for disease phenotypes is important for understanding the mechanism of disease and developing new targeted treatment methods [4]. Comparing with in vivo or biochemical experimental methods, which consumes a lot of manpower and financial resources, computational approaches can efficiently identify potential pathogenic genes.

L. Wang and M. Liu—Equal contribution.

K. Karlapalem et al. (Eds.): PAKDD 2021, LNAI 12712, pp. 117–128, 2021.
https://doi.org/10.1007/978-3-030-75762-5_10

To understand biological information and predict the associations effectively, some methods follow the "guilt-by-association" principle [23] with the assumption that genes related to the same or similar diseases tend to accumulate in the same neighborhood of the molecular network. PRINCE [21] uses known pathogenic genes to transform the disease phenotype into a initial query vector represented by genes and uses label propagation to sequence the possible pathogenic genes of a specified disease. RWRH [13] ranks phenotypes and genes simultaneously and selects the most similar phenotype to the query disease. Besides, BiRW [24] performs random walks on a constructed network to capture circular bigraphs patterns in networks for unveiling phenotype-gene association. These methods always rely on known phenotype-gene associations but sparseness problem of phenotype-gene association matrices impedes the methods to achieve good results on link prediction.

Recently, network embedding methods have gained increasing popularity in link prediction, since they can not only learn highly interpretable and expressive representations for each node but also alleviate the data sparsity problem. With the thought of Word2vec [12] that the word neighborhood can be identified by co-occurrence rate, random-walk based methods are exploited to generate random paths over networks, where the nodes in network are viewed as words and random paths are viewed as sentences. Deepwalk [15] and Node2vec [9] apply the idea to homogeneous networks. For heterogeneous networks, Metapath2vec [6] formalizes meta-path based random walks to construct the heterogeneous neighborhood of a node and propose a heterogeneous SkipGram model to perform node representations. Since the choice of meta-paths highly affect the quality of the learnt node representations, meta-path based embedding method often require prior knowledge or heuristics. To solve the problem, JU&ST [10] propose JUmp and STay strategies to select the next node without any meta-path, which can balance between homogeneous and heterogeneous edges. In addition, these methods can be also extended for bioinformatic tasks, such as miRNAs-diseases association prediction [14] and drug-target interaction prediction [26]. However, these methods only pay attention to the node type information in heterogeneous networks and ignore the different contributions on transferring information among nodes, especially for the situation that a hierarchical structure exists in the data. The key problem is how to design a random walk strategy to select "neighborhood" for nodes in a heterogeneous network.

To tackle the problem, we proposed an innovative algorithm called HSAEM (A **H**ierarchical **S**tructure-**A**ware **E**mbedding **M**ethod) for predicting phenotype-gene associations. HSAEM performs random walks simultaneously on Human Phenotype Ontology(HPO), phenotype-gene and Protein-Protein Interaction(PPI) networks to generate node sequences. Then a heterogeneous SkipGram model is utilized to learn interpretable features from the sequences. The contributions of our work are as follows:

- HSAEM benefits capturing the hierarchical structure of phenotypes and quantifying an individual jumping probability for each phenotype nodes, which can effectively capture co-occurrence between gene nodes and phenotype nodes in the heterogeneous network.
- Establishing a hierarchy for PPI network by sorting the node influence of gene nodes and attaching higher jumping probabilities to higher-level gene nodes, which ensures

the gene nodes that are strongly related to phenotypes more inclined to select heterogeneous neighbors during random walk.
- The approach can model both node type and individual difference information in the network. Comprehensive experiments clearly validate that HSAEM outperforms state-of-the-art methods, which proves the effectiveness of predicting phenotype-gene associations.

2 Method

Our heterogeneous network is composed of a PPI network (green lines between green circle nodes), the Human Phenotype Ontology (HPO) (orange lines between orange square nodes), and a phenotype-gene association network (blue lines between orange square nodes and green circle nodes) in Fig. 1(a). HSAEM consists of three steps. Firstly, random walk is performed over the input heterogeneous network to generate several random node sequences according to individual jumping probability. Secondly, heterogeneous SkipGram is used to learn meaningful representations of nodes by feeding the node sequences generated in the first step into it. Finally, the inner product of node representations are used to predict phenotype-gene associations. An illustration of the HSAEM procedure is illustrated in Fig. 1.

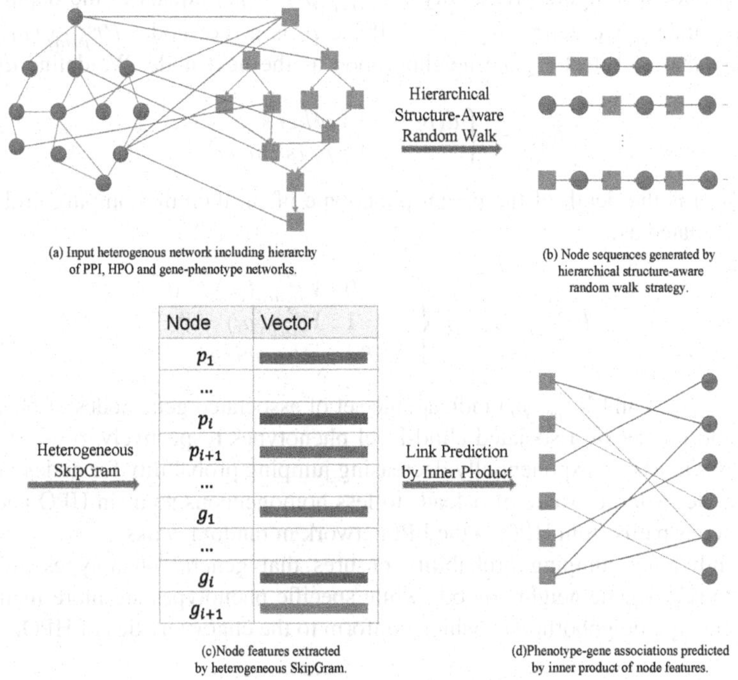
Hierarchical
Structure-Aware
Random Walk
(a) Input heterogenous network including hierarchy of PPI, HPO and gene-phenotype networks.
(b) Node sequences generated by hierarchical structure-aware random walk strategy.
Node
Vector
p_1
...
p_i
p_{i+1}
...
g_1
...
g_i
g_{i+1}
Heterogeneous
SkipGram
Link Prediction
by Inner Product
(c)Node features extracted by heterogeneous SkipGram.
(d)Phenotype-gene associations predicted by inner product of node features.

Fig. 1. Illustration of the HSAEM procedure.

2.1 Hierarchical Structure-Aware Random Walk

Random walks are performed on the network to obtain fixed-length node sequences, in each step of which, there are two options to select the next node: selecting a homogeneous neighborhood in the current network or jumping to another network to select a heterogeneous neighborhood. The probability of jumping to another network for current node v is referred to as the *jumping probability* of node v, otherwise defined as *staying probability*. In this section, we introduce the jumping probability in details.

(a) Jumping Probability for Phenotypes in HPO
Phenotype ontology is organized as a directed acyclic graph (DAG) with hierarchical structure, which serves as a standardized vocabulary of phenotypic abnormalities that have been seen in human disease [18]. The higher level phenotypes locate, the more general they are; the deeper level phenotypes located, the more specific. Therefore, the phenotypes in the phenotype ontology have large individual differences. There are mainly the following two aspects: 1) For phenotype-gene association, general phenotypes in higher levels are associated with more genes, while specific phenotypes in deeper levels are associated with fewer genes; 2) For phenotype-phenotype association, general phenotypes are loosely associated in higher lever, while specific phenotypes are tightly associated in deeper levels.

To better capture the characteristics in phenotype data, we introduce variable $d_(p_i)$ and an individual jumping probability $Pr^G_{jump}(p_i)$. $d(p_i)$ equals to the depth of the phenotype node i, we define $d(p_i) = 1$ if the p_i is a root node. $Pr^G_{jump}(p_i)$ is the probability that p_i select a gene neighborhood as the next node. We define $d(p_i)$ as follows:

$$d(p_i) = \begin{cases} d(p_k) + 1 : others \\ 1 : p_i \; is \; the \; root \end{cases} \tag{1}$$

where $d(p_k)$ is the depth of the parent phenotype of p_i, then the jumping probability can be calculated as:

$$Pr^G_{jump}(p_i) = \begin{cases} 0 : V^G_{jump}(p_i) = \emptyset \\ 1 : V^P_{stay}(p_i) = \emptyset \\ \alpha^{d(p_i)} : otherwise \end{cases} \tag{2}$$

where $V^G_{jump}(p_i)$ and $V^P_{stay}(p_i)$ indicate the set of associated gene nodes of phenotype node p_i and the set of associated child-level phenotypes respectively. $\alpha \in [0, 1]$ is a decay factor to assign exponentially decreasing jumping probability for nodes far from the root node. A larger value of α leads to less homogeneous paths in HPO and more heterogeneous paths from HPO to the PPI network in random walks.

The individual jumping probability ensures that general phenotypes are more inclined to select gene neighborhoods while specific phenotypes are more inclined to select phenotype neighborhoods, which conform to the characteristics of HPO.

(b) Jumping Probability for Genes in PPI Network
The gene nodes in the PPI network also have obvious individual differences on transferring information, such as the degree of gene nodes(max degree 572, min degree 2) and the number of heterogeneous edges with phenotype nodes(max 565, min 3). For better measuring individual differences among gene nodes, the Jaccard Index are extended to quantify the node influence(NI) [7] of gene nodes as follow:

$$NI(g_j) = \sum_{p_i \in leaf} \frac{|Ng(g_j) \cap Ng(p_i)|}{|Ng(g_j) \cup Ng(p_i)|}, \tag{3}$$

where $Ng(g_j)$ and $Ng(p_i)$ denote the set of associated gene nodes of gene g_j and phenotype p_i respectively. If p_i and g_j share many common neighbours, they will be more probably influenced with each other. Since a higher-level node in the phenotype ontology is a collection of deeper nodes, the gene set associated with deeper-level phenotypes is usually a subset of the one associated with higher-level phenotype. Therefore, only phenotypes in leaf nodes of phenotype ontology are included when calculating public genes. Then, we construct a hierarchy for the PPI network by sorting the node influence of genes, which follows two principles: 1) nodes with higher influence index are located in the higher level; 2) nodes directly linked to nodes with high influence are located in the higher level. $d(g_j)$ equals to depth of the gene node j where $d(g_j) = 1$ if g_j with the max NI. We define $d(g_j)$ as follows:

$$d(g_j) = \begin{cases} d(g_k) + 1 : NI(g_j) \le NI(g_k) \\ \lfloor (d(g_k) + 1)\frac{NI(g_j)}{NI(g_k)} \rfloor : NI(g_j) > NI(g_k) \\ 1 : g_j \text{ with the max } NI \end{cases} \tag{4}$$

where $d(g_k)$ is the depth of a neighbor of g_j with masked depth.

With the assumption that a gene with higher quantified influence has more probability to interact with phenotypes and this node should have larger probability to select a phenotype neighborhood, for current gene g_j, the jumping probability are defined similarly as follow:

$$Pr^P_{jump}(g_j) = \begin{cases} 0 : V^P_{jump}(g_j) = \emptyset \\ 1 : V^G_{stay}(g_j) = \emptyset \\ \beta^{d(g_j)} : otherwise \end{cases} \tag{5}$$

where $V^P_{jump}(g_j)$ and $V^G_{stay}(g_j)$ indicate the set of associated phenotype nodes of gene g_j and the set of associated gene nodes respectively. $\beta \in [0, 1]$ controls the individual jumping probability of gene node g_j, which ensures the higher-level gene nodes are more inclined to select a phenotype neighborhood. A larger value of β leads to less homogeneous paths in the PPI network and more heterogeneous paths from PPI to the HPO network.

2.2 Node Embedding Learning with SkipGram

Based on the node sequences generated by random walks, SkipGram model [5] can be applied to learn vector representation for each node in the heterogeneous network, which maximizes the co-occurrence rate of the nodes appearing within a context window of length w in node sequences. Moreover, a negative sampling strategy is utilized to minimize the co-occurrence rate of the center node v and a randomly sampled negative node u that not appears in the set of random walks. For our heterogeneous network, we adopt the heterogeneous SkipGram model [6], considering different node types of the neighborhood nodes.

In detail, for the pair of nodes (c_t, v), the SkipGram model maximizes the following objective function:

$$\log \sigma(\vec{c_t} \cdot \vec{v}) + \gamma \cdot \mathbb{E}_{u_t \sim P_t(u_t)}[log\sigma(-\vec{u_t} \cdot \vec{v})] \qquad (6)$$

where $log\sigma(x) = log(\frac{1}{1+e^{-x}})$ is the logistic sigmoid function, c_t is the neighbor set of v, $\gamma \in \mathbb{Z}^+$ refers to the number of negative samples. The sampling distribution $P_t(u)$ is specified by the node type t(phenotype or gene) of the neighbor c_t. Asynchronous Stochastic Gradient Descent (ASGD) can be used to learn the node vector representations of phenotypes and genes effectively.

2.3 Prediction of Phenotype-Gene Associations

The task of phenotype-gene prediction is considered as a ranking problem over a set of candidate phenotypes $p_i \in P$. The proposed model aims to return the top ranked candidate genes by measuring the relevance scores between phenotypes and genes. Specifically, given a phenotype-gene pair, the relevance scores is calculated by the inner product of their vectors $r = \boldsymbol{p_i^T g_j}$, where $\boldsymbol{p_i}$ and $\boldsymbol{g_j}$ are the vector representations of phenotypes and genes respectively and r will serve as the score for ranking.

3 Experiments

3.1 Data Preparation

We downloaded the human gene–phenotype associations from the Human Phenotype Ontology project, consisting of 6253 phenotypes and 18,533 genes in September 2016. The human protein–protein interaction (PPI) network was obtained from the Human Protein Reference Database (HPRD) [16]. All phenotypes are formed as a sixteen-level hierarchical structure where each phenotype (except the one in leaf and root node) is associated with several child-level and parent-level phenotypes. The PPI network contains 145,856 binary interactions between the 18,533 genes. In the experiments, the set of genes by intersection of the datasets were used. After removing isolated genes and the genes not associated with root phenotype node of HPO, there remain 141,530 associations between 6,253 phenotypes and 2,354 genes as well as 7,589 and 12,820 associations between these genes and phenotypes respectively. A more recent version of the HPO, from January 2020, was used to measure the performance of the models for predicting new associations.

3.2 Baselines and Results

(a) Baselines

In this section, we compare HSAEM with state-of-the-art methods to demonstrate the superior performance of our method. PRINCE [21], RWRH [13], BIRW [24] are classical methods used to predict the phenotype-gene and disease-gene interactions. These network-based methods use basic label propagation, i.e. random walk on networks to prioritize phenotype genes in different ways. OGL [11] and GC^2NMF [25] are the methods that take hierarchical structure of phenotype ontology into consideration in the prediction of phenotype-gene interaction. OGL captures the hierarchical structures in the phenotype ontology structure that aggregate phenotypes in the same annotation paths to root by utilizing ontology-guided group Lasso. GC^2NMF measures discriminatory similarity according to phenotype levels and introduce weighted graph constraint into NMF.

Deepwalk [15], Metapath2vec [6] and JU&ST [10] are random-walk based embedding methods with different random walk strategies. Note that DeepWalk are originally designed for homogeneous networks, which is applied to the heterogeneous network by ignoring the schema of the network and treating all nodes and edges equally as for a homogeneous network. For Metapath2vec, we need to specify the meta-path scheme ("PGP") to guide random walks. These embedding methods can learn node representations for phenotypes and genes. Then the inner product of phenotype and gene vectors as the relevance score is used to predict phenotype-gene associations.

The baselines are iterated by 50 times, and the hyper parameters for baselines are tuned according to their literature. For all random-walk based embedding methods, we use the same parameters listed below: (1) The number of walks per node n: 10; (2) The walk length l: 100; (3) The embedding dimension d: 128; (4) The window size w: 8.

(b) Phenotype-Gene Association Prediction in Cross-Validation

In the experiment, 10-fold cross-validation is adopted to evaluate our method. 90% of the phenotype-gene associations are randomly selected as train set and the remaining 10% associations are used as test set. Since the number of unobserved samples is far larger than positive ones, a balanced sampling strategy is adopted to randomly select equal number of unobserved samples as negative samples. The experiment for each method is repeated 10 times independently and the average result as the final result. For a fair and comprehensive comparison, AUC (Area Under Receiver Operating Characteristic Curve), AUPR (Area Under Precision-Recall Curve), F-Score (F-measure) and MCC (Matthews Correlation Coefficient) are chosen as evaluation metrics.

Table 1 shows the performances on phenotype-gene association prediction. It is demonstrated that HSAEM has a remarkable advantage over other phenotype-gene association prediction methods, which proves the network embedding method is suitable for the task of phenotype-gene association prediction and has advantages over traditional random walk or label propagation methods. Comparing with OGL and GC^2NMF, HSAEM makes more effective use of the hierarchical structure of HPO. OGL captures the ontology structure by a group Lasso smoothness among the phenotypes on the same phenotype paths, introducing a bias towards more general phenotypes in higher levels. GC^2NMF applies the hierarchical structure of the phenotypes to

Table 1. Performance of HSAEM and baselines in cross-validation

Method	AUC	AUPR	F-Score	MCC
PRINCE($\alpha = 0.01$, $\beta = -20$)	0.7671	0.8445	0.7492	0.4751
RWRH($\gamma = \lambda = 0.9$, $\eta = 0.5$)	0.9147	0.9188	0.8613	0.7969
BIRW($\alpha = 0.9$, $m = n = 2$)	0.8762	0.8925	0.8158	0.6849
OGL($\beta = 10^5$,$\gamma = 10^6$)	0.8639	0.8902	0.8351	0.6332
GC^{2}NMF($\lambda_1 = 0.1$, $\lambda_2 = 10$)	0.7083	0.8254	0.7181	0.3627
Deepwalk	0.8939	0.9287	0.8708	0.6955
Metapath2vec	0.8435	0.8968	0.8140	0.6004
JU& ST($\alpha = 0.5$)	0.9357	0.9634	0.9264	0.7924
HSAEM	**0.9512**	**0.9736**	**0.9434**	**0.8281**

Table 2. Performance of Predicting New Associations

Method	AUC	AUPR	F-Score	MCC
PRINCE	0.5726	0.5064	0.4369	0.1176
RWRH	0.6933	0.6018	0.5251	0.2502
BIRW	0.6989	0.6277	0.5585	0.3044
OGL	0.6472	0.6035	0.5077	0.2495
GC^{2}NMF	0.5396	0.4358	0.3781	0.0552
Deepwalk	0.7163	0.6188	0.5359	0.2717
Metapath2vec	0.6696	0.6084	0.5290	0.2530
JU&ST	0.7418	0.6689	0.5875	0.3343
HSAEM	**0.7529**	**0.6846**	**0.6003**	**0.3515**

the calculation of the similarity between phenotypes, which cannot capture heterogeneous links directly. In addition, HSAEM achieves much better prediction performance than the other random-walk based embedding methods, which illustrate the hierarchical structure-aware random walk strategy can more effectively discover the co-occurrence among nodes.

(c) Predicting New Phenotype-Gene Associations

To further demonstrate that HSAEM is not overfitting the training data, we evaluated the models by predicting a set of new associations that were added into HPO between September 2016 and January 2020. In the experiment, all the models were trained using the data from September 2016, and the trained model was used to predict the 151,043 new associations. An equal number of unobserved samples are selected randomly as negative samples. The experiment for each method is repeated 10 times independently and the average performances are reported in Table 2.

With help of hierachical structure-aware random walk strategy, HSAEM consistently outperformed the other methods on all evaluation metrics. The performance confirms that the distinctive designs of individual jumping probability for phenotype and gene nodes is helpful for improving the prediction.

3.3 Effectiveness of Hierarchical Structure-Aware Random Walk

To study the contribution of our proposed hierarchical structure-aware random walk strategy with individual jumping probability, we also test the performance of random walk with fixed jumping probability Pr^G_{jump} and Pr^P_{jump} for all phenotype nodes and gene nodes, respectively. By tuning Pr^G_{jump} and Pr^P_{jump} from 0.1 to 1, Fig. 2 reveals the AUC of the method under different combined configurations Pr^G_{jump} and Pr^P_{jump}. It is demonstrated that the method with fixed jumping probability can achieve the best AUC (0.9409) when Pr^G_{jump} and Pr^P_{jump} are 0.3 and 0.7, which is lower than the AUC (0.9514) of HSAEM that utilizes individual jumping probability ($\alpha = 0.6, \beta = 0.8$) for different gene and phenotype nodes.

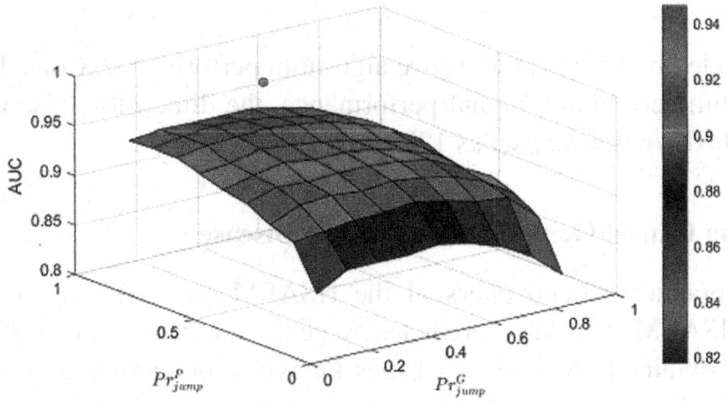

Fig. 2. Performance on fixed jumping probability.

The experimental result suggests that homogeneous edges linking phenotype nodes in HPO and heterogeneous edges linking nodes across PPI network and HPO play the primary role in learning node embeddings for phenotypes and genes. HSAEM assigns a higher jumping probability to more general phenotypes in higher levels, which can alleviate the bias caused by the inconsistency of the associated genes of them. Futhermore, assigning a higher jumping probability to the gene nodes with higher node influence can capture heterogeneous links more effectively. This comparison further corroborates the effectiveness and rationality of individual jumping probability.

3.4 Analysis on Different Parameter Settings in SkipGram

The effect of the dimension of node representations is tested by setting other parameters to their best ones, and the result is shown in Fig. 3(a). According to the result, the value of AUC increases as the dimension gets larger but the change becomes insignificant after the dimension exceeds 128.

The impact of the size of context windows in SkipGram is also analyzed by setting it to the range from 1 to 10. The results presented in Fig. 3(b) indicates that the optimal result is obtained when the window size reaches 8, which reveals that capturing

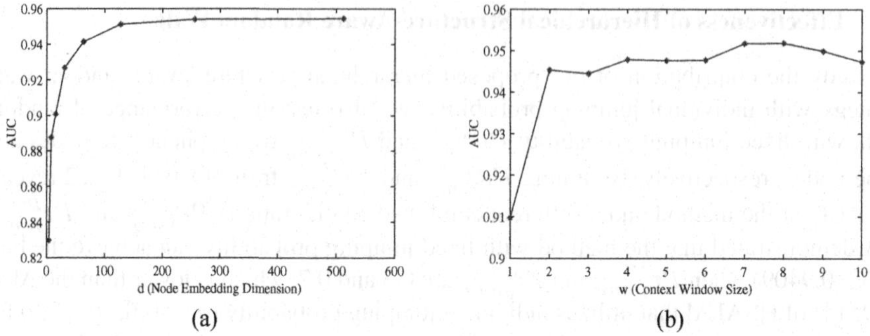

Fig. 3. a) Impact of node embedding dimension d; b) context window size w

higher-order node proximity can improve algorithm performance. Considering the balance between time consumption and performance, the dimension of embeddings and the size of context window are set as 128 and 8 respectively.

3.5 Predicting Causal Genes for Parkinson's Disease

To verify the practical effectiveness of the HSAEM on predicting phenotype-gene associations, HSAEM is used to predict new genes for Parkinsonism (HP:0001300). In the data we obtained, there are 27 genes known to be associated with Parkinsonism(Parkinson's Disease) on HPO till January 2020. Apart from the known 27 disease genes for Parkinson's disease, the top-10 ranked predicted genes of HSAEM are searched in literature. The results shown in Table 3 demonstrate that there are 8 out of 10 interactions verified by some literature, thus showing the practicality of our proposed method.

Table 3. Top-10 predicted genes for Parkinsonism and the references

Rank	Gene	Score	Ref
1	PDYN	0.9565	[22]
2	GBA2	0.9469	[8]
3	MAPK12	0.9231	[2]
4	MCEE	0.9083	[1]
5	TRPM7	0.8755	[19]
6	SLC19A3	0.7496	-
7	C12ORF65	0.7300	[17]
8	TRIM63	0.6988	[3]
9	ATG2A	0.6785	-
10	NDUFB5	0.5493	[20]

4 Conclusion

The paper propose a novel random-walk based embedding method called HSAEM to predict phenotype-gene associations. It benefits from the hierarchical structure-aware random-walk strategy with the distinctive designs of individual jumping probability for phenotype and gene nodes. The experimental results show that HSAEM outperforms than other state-of-the-art methods. The prediction of causal genes for Parkinson's disease proves its practical effectiveness, thus it is well believed it will be utilized in identifying causing genes for rare studied diseases in the future.

Acknowledgements. This work is supported by the Natural Science Foundation of Tianjin (No. 18JCYBJC15700), the National Natural Science Foundation of China (No. 81171407) and National Key R&D Program of China(2018YFB0204304).

References

1. Andréasson, M., Zetterström, R.H., von Döbeln, U., Wedell, A., Svenningsson, P.: MCEE mutations in an adult patient with Parkinson's disease, dementia, stroke and elevated levels of methylmalonic acid. Int. J. Mol. Sci. **20**(11), 2631 (2019)
2. Bohush, A., Niewiadomska, G., Filipek, A.: Role of mitogen activated protein kinase signaling in Parkinson's disease. Int. J. Mol. Sci. **19**(10) (2018)
3. Bonne, G., Rivier, F., Hamroun, D.: The 2019 version of the gene table of neuromuscular disorders (nuclear genome). Neuromuscul. Disord. **28**(12), 1031–1063 (2018)
4. Botstein, D., Risch, N.: Discovering genotypes underlying human phenotypes: past successes for mendelian disease, future approaches for complex disease. Nat. Genet. **33**(3), 228–237 (2003)
5. Cheng, W., Greaves, C., Warren, M.: From n-gram to skipgram to concgram. Int. J. Corpus Linguist. **11**(4), 411–433 (2006)
6. Dong, Y., Chawla, N.V., Swami, A.: Metapath2vec: scalable representation learning for heterogeneous networks. In: Proceedings of the 23rd ACM SIGKDD International Conference on Knowledge Discovery and Data Mining, pp. 135–144 (2017)
7. Estrada, E.: Generalized walks-based centrality measures for complex biological networks. J. Theor. Biol. **263**(4), 556–565 (2010)
8. Franco, R., Sánchez-Arias, J.A., Navarro, G., Lanciego, J.L.: Glucocerebrosidase mutations and synucleinopathies. potential role of sterylglucosides and relevance of studying both GBA1 and GBA2 genes. Front. Neuroanat. **12**, 52 (2018)
9. Grover, A., Leskovec, J.: Node2vec: scalable feature learning for networks. In: Proceedings of the 22nd ACM SIGKDD International Conference on Knowledge Discovery and Data Mining, pp. 855–864 (2016)
10. Hussein, R., Yang, D., Cudré-Mauroux, P.: Are meta-paths necessary? Revisiting heterogeneous graph embeddings. In: Proceedings of the 27th ACM International Conference on Information and Knowledge Management, pp. 437–446 (2018)
11. Kim, S., Xing, E.P., et al.: Tree-guided group lasso for multi-response regression with structured sparsity, with an application to eqtl mapping. Ann. Appl. Stat. **6**(3), 1095–1117 (2012)
12. Le, Q., Mikolov, T.: Distributed representations of sentences and documents. In: Proceedings of the 31st International Conference on Machine Learning, pp. 1188–1196 (2014)
13. Li, Y., Patra, J.C.: Genome-wide inferring gene-phenotype relationship by walking on the heterogeneous network. Bioinformatics **26**(9), 1219–1224 (2010)

14. Luo, Y., et al.: A network integration approach for drug-target interaction prediction and computational drug repositioning from heterogeneous information. Nat. Commun. **8**(1), 1–13 (2017)
15. Perozzi, B., Al-Rfou, R., Skiena, S.: Deepwalk: online learning of social representations. In: Proceedings of the 20th ACM SIGKDD International Conference on Knowledge Discovery and Data Mining pp. 701–710 (2014)
16. Petegrosso, R., Park, S., Hwang, T.H., Kuang, R.: Transfer learning across ontologies for phenome-genome association prediction. Bioinformatics **33**(4), 529–536 (2017)
17. Pyle, A., Ramesh, V., Bartsakoulia, M., Boczonadi, V., Horvath, R.: Behr's syndrome is typically associated with disturbed mitochondrial translation and mutations in the c12orf65 gene. J. Neuromuscul. Dis. **1**(1), 55–63 (2014)
18. Robinson, P.N., Mundlos, S.: The human phenotype ontology. Clin. Genet. **77**(6), 525–534 (2010)
19. Sun, Y., Sukumaran, P., Schaar, A., Singh, B.B.: TRPM7 and its role in neurodegenerative diseases. Channels **9**(5), 253–261 (2015)
20. Talebi, R., Ahmadi, A., Afraz, F., Abdoli, R.: Parkinson's disease and lactoferrin: analysis of dependent protein networks. Gene Rep. **4**, 177–183 (2016)
21. Vanunu, O., Magger, O., Ruppin, E., Shlomi, T., Sharan, R.: Associating genes and protein complexes with disease via network propagation. PLoS Comput. Biol. **6**(1), e1000641 (2010)
22. Westin, J.E., Andersson, M., Lundblad, M., Cenci, M.A.: Persistent changes in striatal gene expression induced by long-term L-DOPA treatment in a rat model of Parkinson's disease. Eur. J. Neurosci. **14**(7), 1171–1176 (2010)
23. Wolfe, C.J., Kohane, I.S., Butte, A.J.: Systematic survey reveals general applicability of "guilt-by-association" within gene coexpression networks. BMC Bioinformat. **6**(1), 1–10 (2005)
24. Xie, M., Xu, Y., Zhang, Y., Hwang, T., Kuang, R.: Network-based phenome-genome association prediction by bi-random walk. PloS One **10**(5), e0125138 (2015)
25. Zhang, Y., Wang, Y., Liu, J., Huang, Y., Xie, M.: Weighted graph constraint and group centric non-negative matrix factorization for gene-phenotype association prediction. In: Proceedings of the 22nd IEEE Symposium on Computers and Communications, pp. 943–950 (2017)
26. Zong, N., Kim, H., Ngo, V., Harismendy, O.: Deep mining heterogeneous networks of biomedical linked data to predict novel drug-target associations. Bioinformatics **33**(15), 2337–2344 (2017)

Autonomous Vehicle Path Prediction Using Conditional Variational Autoencoder Networks

D. N. Jagadish(✉), Arun Chauhan, and Lakshman Mahto

Indian Institute of Information Technology Dharwad, Dharwad 580009, Karnataka, India

Abstract. Path prediction of autonomous vehicles is an essential requirement under any given traffic scenario. Trajectory of several agent vehicles in the vicinity of ego vehicle, at least for a short future, is needed to be predicted in order to decide upon the maneuver of the ego vehicle. We explore variational autoencoder networks to obtain multimodal trajectories of agent vehicles. In our work, we condition the network on past trajectories of agents and traffic scenes as well. The latent space representation of traffic scenes is achieved by using another variational autoencoder network. The performance of the proposed networks is compared against a residual baseline model.

Keywords: Autonomous vehicle path prediction · Conditional variational autoencoder · Deep learning

1 Introduction

Smart city road traffic could comprise various categories of vehicles. The agents could be cyclists, pedestrians and motor vehicles in several forms. Few of the agents may as well possess intelligence to be a capable autonomous vehicle (AV). The agents maneuver on the road guided by locality and traffic signs. However, each of these agents present independent movement in the vicinity of the AV. For the AV to maneuver towards its destination, the traffic scenario and locality are to be known, and the surrounding agents' short future trajectories are to be predicted. AV following path of higher prediction probability can avoid dangerous situations ahead and thereby promises a safer, accident free and optimum maneuverability.

Several motion models exist in the literature for predicting agents' trajectory. Physics-based models represent an AV as dynamic entities governed by the laws of physics bearing no dependencies with the agents nearby [1]. These models assume a constant speed or acceleration and orientation for the vehicles. These models are limited to predict very short-term motion of agents as they rely on low level motion details. They lack to anticipate changes in motion under varying context. Trajectories that vehicle drivers usually intend to perform are considered for maneuver-based motion models [2]. These models consider agents as independent maneuvering entities. The motion prediction is carried by identifying the maneuver intention of a driver. As similar to the previous model categories, these models also suffer from the inaccuracies, especially at traffic junctions, as the agents' dependencies are considered. Agents are maneuvering

K. Karlapalem et al. (Eds.): PAKDD 2021, LNAI 12712, pp. 129–139, 2021.
https://doi.org/10.1007/978-3-030-75762-5_11

entities interacting with each other in the interaction-aware motion models. In trajectory prototypes methods [3], trajectories ending up with unavoidable collisions are penalized. Therefore the methods can yield safe trajectories over hazardous ones. Dynamic Bayesian Networks are frequently utilized ones among the interaction-aware motion models. Asymmetric coupled Hidden Markov Models exploit pairwise dependencies between agents [4]. As traffic rules can impose interactions among agents, such contexts are taken advantage in [5]. The work in [6] focuses on exploiting the mutual influences of agents. The causal dependencies exhibited between the agents are taken into account as a function of local situational context.

Agent detection and probable trajectory prediction are the primary tasks for an AV. Pedestrians are quite often the vulnerable road users. Their behavior prediction is accomplished when computer vision techniques are utilized [7, 8]. Classification and building of localized traffic scenarios are made possible using acquired time series data from sensors, while the decision making is aided from understanding agents' dynamics and their near future behavior [9]. The significant non-trivial tasks under agents' trajectory predictions constitute learning the following. The interdependencies between the agents in any traffic scenario. The traffic rules and road geometry that are influencing the agents maneuver. Multimodal trajectories of the agents resembling more than one possible future path given an observation of the past.

In addition to the above mentioned intrinsic challenges, several practical limitations exist while implementing the prediction model. To mention a few, the AV using on-board sensors may only partially observe the surrounding environment and therefore limited in the range, suffer from occlusion and noise from sensors, and on-board computational resources could turn out to be insufficient. The AV under the drive collecting sensor data and whose path prediction is the ultimate goal is often referred to as ego vehicle (EV). Most implementation works consider the availability of an unobstructed top-down view of the surrounding environment. EV mounted with surveillance cameras can capture the view [10]. Such an arrangement is not cost effective. Nevertheless, such a dataset [11] comprising EV translation and orientation, agents' relative translation from the EV and yaws, and traffic lights status along with semantic maps of the road segments is available for exploration using deep learning techniques. Shown in Fig. 1 is a sample semantic map highlighting agents' location, their intended paths and the traffic lights status.

In this work, we make use of conditional variational autoencoder (CVAE) networks for predicting multimodal trajectories of traffic agents. We introduce variants in CVAE, whose constituents range from simple dense layers when data represents 2D coordinates to convolutional and upsampling layers when data is represented as a bird's eye view (BEV) image. Remainder of the paper provides insight into the related works in the application domain, proposed methods and experimental results.

2 Related Works

Interaction-aware based motion models have been a popular choice. Attributes, such as the agent's relative position, speed, direction and acceleration to EV are considered in [12–14] to determine the behavior. The agents are treated as isolated entities even though other agents surround the target agent. The temporal dependencies are captured well

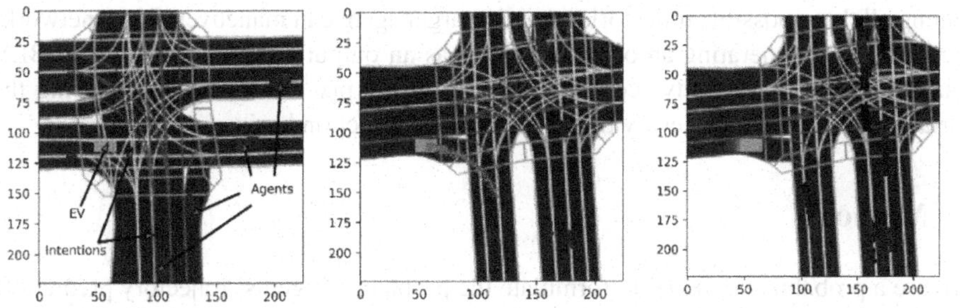

(a) (b) (c)

Fig. 1. Images from the dataset. (a) Semantic map of a road segment describing a traffic scenario. (b) Target agent, green in color, surrounded by other agents, shown in blue, at a time instant. Trace in pink is the ground truth trajectory of the target agent. (c) 2D RGB color image of a data tensor, formed by fusion of agents past trajectories. (Color figure online)

enough with simple recurrent neural networks [15]. However, these works fall short in accounting for the agents' interdependencies. The models under a dense traffic scenario produce erroneous and unsafe results. This shortcoming is overcome by feeding the track history of agents in vicinity of the target agent in [16–19]. The selection of number of surrounding agent vehicles to take into account is a factor largely depending on the direction, distance and lane. Multiple RNNs, based on a gated recurrent unit (GRU) [20] of a long short-term memory (LSTM) [21], are made use to obtain predictions and include agent interdependencies. Environmental factors could play a role impacting the target agent's behavior, which the models fail to include. Under dense traffic scenarios, few of the agent vehicles in the vicinity of the target agent may get occluded as the EV might lose the line of sight. Agent bounding boxes, traffic light conditions and driver intended lanes are color coded when used as a BEV image in [22, 23] (refer Fig. 1a). Environment in the BEV image is segmented in [24]. Traffic scenes encoded as image-like data are well suited for exploring spatial relationships of agents using convolutional neural networks (CNN) layers. The outcome of such an encoding has been very motivating. Further, the temporal characteristics in the data are learnt using 3D convolutions [25]. 2D convolutional layers in cascade extract the spatial features. Moving forward, the advantages of CNNs and RNNs are retained by forming hybrid networks [24, 26]. The prediction network output is complex in nature. So a conditional generative model is proposed in [24]. To this end, a CVAE network is utilized for predicting trajectories [27]. The network may predict maneuver intentions of agents or multimodal or unimodal future paths. The maneuver intentions are a sort of high-level understanding of the traffic scenario. They only help to figure out the probability that the agent moves on a straight lane or takes a right turn or turn left at an intersection of traffic roads. Instead of predicting mere intentions, the network predicting the trajectory is more useful as agents' behavior in future is learnt. The network when fed with past agents' movement, semantic map and traffic conditions can output several possible maneuver paths for the target agent. Hence the trajectories distribution is multimodal. A unimodal trajectory predictor network on the other hand is limited to predict a highest likelihood trajectory among all those possible trajectories that the target agent can maneuver. Some networks are capable of generating an occupancy map as an output in correspondence to BEV image [28, 29]. Probability score is attached to each pixel in the map describing the certainty of occupancy at every timestamp in the prediction horizon.

3 Methods

We use a probabilistic model to formulate the problem of agents' trajectory prediction. Let the position of an ith agent at a timestamp t be represented by $x^{i}{}_{t}$. N denotes total agents in the surrounding. The predicted trajectory of all the agents in the horizon length k is defined as:

$$X_{\text{aN}} = \left\{x_t^i, x_{t+1}^i, x_{t+2}^i, \ldots, x_{t+k}^i\right\} \quad \forall i \in N \tag{1}$$

The observations recorded by the EV are denoted by O_{EV}. Given the observations we compute the conditional distribution $p(x_{\text{aN}}|o_{\text{EV}})$ in the proposed models. We introduce a few of CVAE based neural networks to attain the objective of path prediction. CVAEs are conditional generative models for structured output prediction. The encoder part of the network models the prior distribution. Upon re-parameterization, the decoder part of the network generates output by learning the input distribution.

We introduce two variants in CVAEs in perspective of data representation the networks can handle. In CVAE-I, the encoder part of the network receives future trajectories of the agents X_{aN} to represent input as a latent vector z. We intend to generate a distribution $p_\theta(\text{z}|x_{\text{aN}})$ conditioned on c at the bottleneck layer. Vector c is nothing but represents O_{EV} of the EV. The network on training completion can yield the trajectories X_{aN} at its output due to conditioned probability distribution $p_\theta(x_{\text{aN}}|z, c)$ learnt by the generator. CVAE-II differs from the previous by added conditioning. The traffic scene captured as a BEV image is used for this purpose.

The models are trained to maximize the conditional marginal log-likelihood. The backpropagation algorithm is implemented using stochastic gradient variational Bayes method to reduce the cost function. The empirical lower bound is:

$$\mathcal{L}(x_{aN}; \theta, \varphi) = KL(q_\varphi(\text{z}|x_{aN}, y)||p_\theta(\text{z}|x_{aN}, y)) + \frac{1}{L}\sum\nolimits_{l=1}^{L} \log p_\theta\left(y|x_{aN}, z^{(l)}\right) \tag{2}$$

where y is the network output, $z^{(l)}$ is a Gaussian latent variable, and L is the number of samples. Once the networks are trained, for an observation of agents' motion O_{EV}, we sample the stochastic latent space and condition the decoder on vector c, to get multimodal path predictions of the agents.

CVAE-I. We select a simple network for learning the data distribution and Fig. 2a shows the network. The input to the network X_{aN}, is the prediction of an agent, described by 50 position coordinates. The input is encoded by utilizing just two dense layers, dense 1 and dense 2 having nodes 50 and 16 respectively. Since the reconstruction is all about predicting 2-dimension coordinates of an agent, we proceed to have a latent space variable in 2-dimension. The network is conditioned on c, the agent's 10 history and a

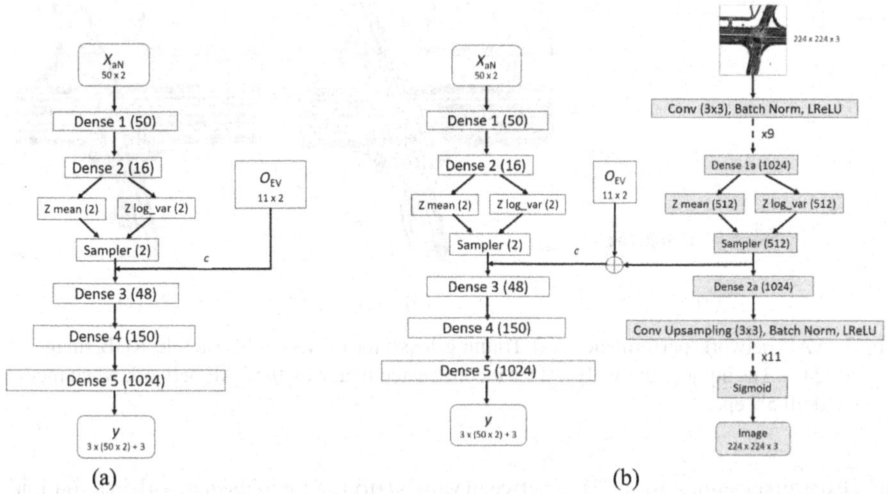

Fig. 2. CVAE networks. (a) CVAE-I network conditioned on agent's history and (b) CVAE-II network conditioned on both agent's history and latent space variable of a variational autoencoder.

present position. The decoder uses three dense layers, dense 3, 4 and 5 for reconstruction. The network output is multimodal having the ability to predict 03 probable trajectories y_1, y_2 and y_3, and their respective probabilities p_1, p_2 and p_3. The network is trained by backpropogation during which we minimize the sum of negative multi-log-likelihood and KL divergence losses.

CVAE-II. As we compare the previous network architecture, we argue that a traffic context aware network architecture should outperform it. Hence we bring in additional context aware information while conditioning the CVAE. The network is shown in Fig. 2b. The surrounding agents' density, locations, lanes and semantic view all impact the target agent's behavior. The BEV image of the traffic scene, available with the dataset as $224 \times 224 \times 25$ tensor (describing 10 historic placement of all agents on a semantic map), is transformed to an RGB image of $224 \times 224 \times 3$ pixels size is encoded using a variational autoencoder (VAE) network. The latent space variable of this VAE (512×2 in size) and O_{EV} of the agent (11×2 in size) conditions the CVAE-II network, which in all aspects is similar to CVAE-I. Once again as with CVAE-I, we train the CVAE-II network by backpropogation during which we minimize the sum of negative multi-log-likelihood and KL divergence losses.

The VAE has a sequence of 09 2-dimension convolutional layers (kernel size 3×3, stride $= 2$ or 1 and padding $= 1$) with batch normalization and leaky ReLU activation layers to follow. As the input image is complex enough, we choose the latent space to be composed of 512 Gaussian distributions, wherein we are interested to know their mean and log variance values. The generator part of the VAE, has a sequence of 11 convolutional upsampling layers (kernel size 3×3 or 4×4, stride 2 or 1 and padding $= 1$) with batch normalization and leaky ReLU activation layers to follow. The bilinear interpolation using convolutional upsampling layers are a better choice than transpose layers as they avoid checker board effects on reconstructed images. As the fed in RGB

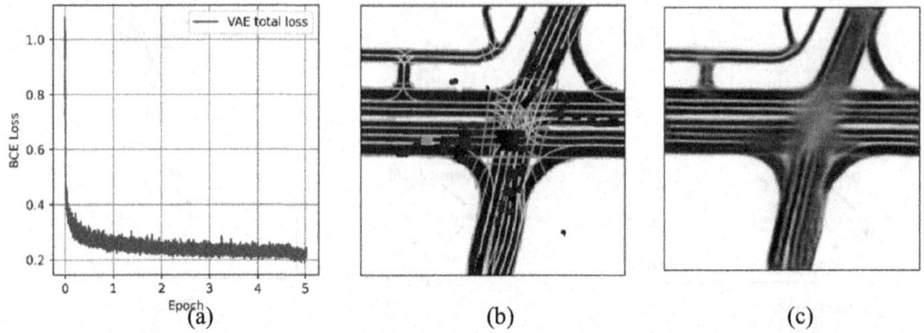

Fig. 3. VAE network performance. (a) Training loss plot of VAE. (b) Sample RGB image of $224 \times 224 \times 3$ as input to the VAE and (c) reconstructed image by the VAE network when model is trained till 5th epoch.

image has pixel values normalized between values 0 to 1, we end the network in a sigmoid function. We have chosen binary cross entropy loss as a cost function for reconstruction and KL divergence loss for distribution matching (refer (2)).

Figure 3a shows the training loss plot. The VAE network is trained till 05 epochs. Binary cross entropy loss per pixel value is shown on y-axis. The loss approaches 0.2. The training dataset is described in Sect. 4. A sample image shown in Fig. 3b is provided as input after training and as a response the reconstructed image is shown in Fig. 3c. On comparison, it can be noticed that the VAE network is able to reconstruct back the semantic map, lanes and agents' density and thus proving a capable enough distribution capturing in latent space for CVAE-II network conditioning.

4 Experiment

We introduce the dataset used for experimenting the proposed networks, the configurations of the CVAE networks while implementation and the results obtained in the following subsections. As the dataset is very recently introduced, and much work is currently not available, we implement a baseline residual deep neural network for comparison with our networks.

Dataset. Around one thousand hours of traffic agent movement data is recorded in the dataset presented in [11]. EV is mounted with LiDAR and driven by human drivers on pre-declared road segments. While moving around, EV captures self and other agents' translation, yaw, rotation and extent. Many such EVs on road segments cover up to 16,000 miles of distance travel. Since road segments are pre-declared, cameras fixed at high altitudes capture the required road semantic view. Roughly 15,000 semantic map annotations are included in the dataset. The preprocessed data is presented as a collection of scenes, wherein each scene on an average holds 248.36 frames. The frames are sampled at a rate of 10 fps and placed in a chronological order.

The data can be rendered as an image so that convolutional layer based deep neural networks can be benefitted. The target agent is translated to the location of interest in the image, and so are the other agents placed relative to this position. All agents are color coded with bounding box dimensions proportional to their extent. Figure 1b reveals a sample image of size 224 × 224 pixels. Target agent of interest is translated to (64, 112). On a semantic map the image includes details of driver intention lanes, ground truth path of the target agent and traffic light condition. Collection of images in a chronological order with target agent centered at (64, 112) are fused. The tensor so created includes past and present movement of all agents over the road segment. To train our proposed CVAE networks, past 10 images having specific target agent centered are fused. The ground truth paths are taken to compute cost function error. Figure 1c shows the 2D view of the 224 × 224 × 25 tensor. The trace of the agents' bounding box resembles agents' movement within a timeframe.

For our work, we use 0.15 million 224 × 224 × 25 size tensors for training of networks and 0.05 million such tensors for validation purposes.

Residual Network. To serve as a baseline network to compare the performance of CVAEs we make use of resnet50 [30]. The early convolutional layer of the network segregates temporal information from the data. The series of convolutional layers in the later part of the network captures spatial information contained in the input data tensor, the size being (224 × 224) pixels × 25 channels. The network has multiple conv and identity blocks, with both having internal skip connections. The skip connection does help the network to overcome the vanishing gradient problem. Every identity block has 3 (conv + batch norm + activation) layers, whereas a conv block has a (conv + batch norm) layer in the skip connection path. The network outputs agent's multimodal positions in 2D with their probability scores.

Configuration. The following are chosen to train all the networks. Each image pixel size is 0.5 m × 0.5m , history_frames = 10, future_frames = 50, batch_size = 32, optimizer = Adam with learning_rate = 1e-03, b–tch normalization layer having ε = 1e–05 and momentum = 0.1, reconstruction cost function = negative multi-log likelihood, multimodal output to have 03 probable trajectories per target agent. To compare distributions we resort to KL divergence loss. For CVAE networks, the latent space dimension is 02, whereas for VAE of Fig. 2b the latent space dimension is 512. Table 1 provides total count on trainable parameters.

Results. We train the CVAE variants and residual networks for 10 epochs. A plot of training and validation errors for the networks, in negative multi-log likelihood, averaged over 1,000 iterations is shown in Fig. 4a. Networks' performance summarized in Table 1 shows the lowest error during training and validation. The CVAE-II network manages a low training and validation error. The CVAE-I network is not bad either given its simplicity in architecture, with as few as 479,683 trainable weights. Limitation in capturing input features due to reduced network complexity resulted in a high error. The same simplistic network, when conditioned with traffic scene context, outperforms the huge baseline network. It is to be noted that the CVAE-II network will only use around 50% of its size while drawing predictions and hence is at least 6X smaller in size

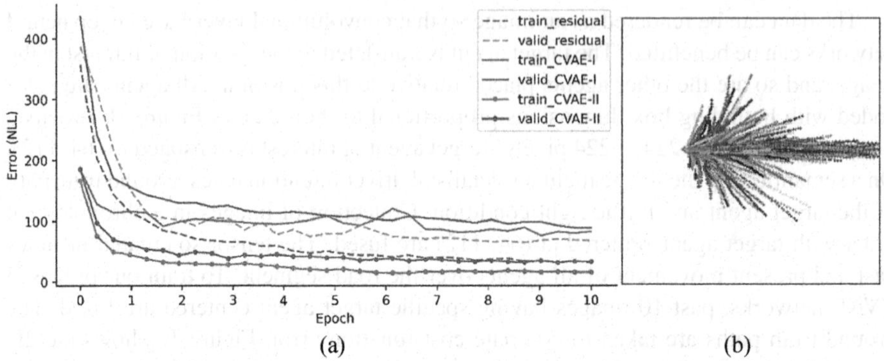

Fig. 4. Networks performance. (a) Training and validation plots of CVAEs and baseline residual network. (b) Plot of predicted trajectories by the CVAE-II network over a few randomly selected inputs from validation dataset.

Table 1. Training and validation performance of the networks.

Network	Total parameters	Training loss (NLL)	Validation loss (NLL)
Residual	35,260,119	31.37	64.9
CVAE-I	479,683	76.12	88.82
CVAE-II	504,259 + 10,403,251	15.49	31.76

when compared to the baseline network. Prediction trajectories by the network over a few randomly selected input samples from validation dataset is shown in Fig. 4b.

Figure 5 displays multimodal output trajectories on semantic maps by all the implemented networks for a few of the input images from the dataset. For a reference to compare, the first row is the ground truth trajectory. Subfigure titles are data sample indices. Second row images show output trajectories predicted by the residual network. The predicted path probabilities are mentioned within the image. The same are shown for the CVAE-I and CVAE-II networks in the third row and last row of the figure, respectively. The prediction trajectories are highlighted against the background semantic maps in all the subfigures. On a close observation, it is quite evident that the path predictions of the CVAE-II network are a close match to the ground truths.

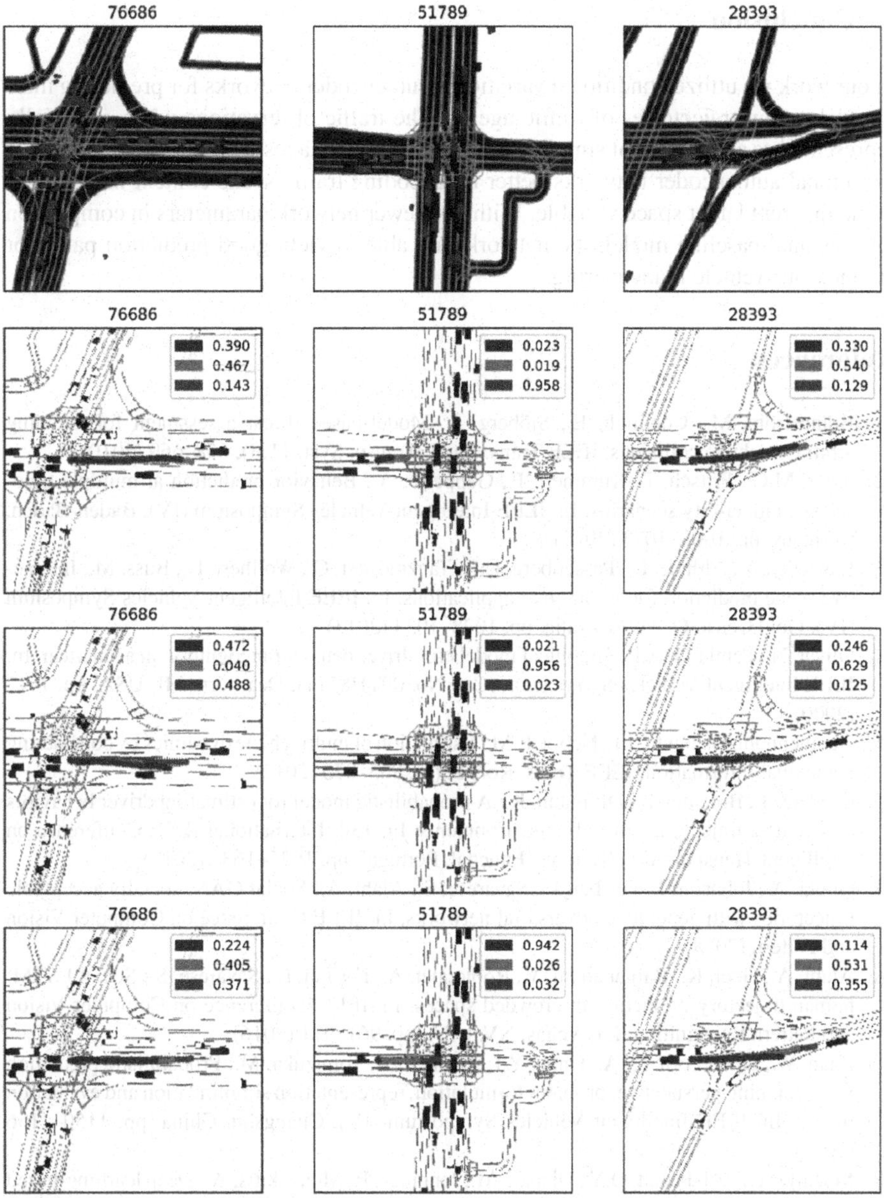

Fig. 5. Predictions by the networks on a few dataset samples. First row: Ground truth trajectory (shown in pink color). In the second column the target agent movement has stopped; Second row: Residual baseline network prediction trajectories; Third row: CVAE-I network prediction trajectories and last row: CVAE-II network prediction trajectories. (Probability scores of the trajectories are shown as legends within subfigures).

5 Conclusion

In our work we utilize conditional variational autoencoder networks for predicting multimodal output trajectories of traffic agents. The traffic observations when essentially represented as a sequence of simplified bird's eye view images will help to condition the variational autoencoder networks better by encoding traffic scene context information in the inherent latent space variable. With just fewer network parameters in comparison to a residual baseline model, the networks are able to yield good prediction paths for autonomous vehicle maneuvering.

References

1. Brännström, M., Coelingh, E., Sjöberg, J.: Model-based threat assessment for avoiding arbitrary vehicle collisions. IEEE Trans. Intell. Transp. Syst. **11**(3), 658–669 (2010)
2. Ortiz, M.G., Fritsch, J., Kummert, F., Gepperth, A.: Behavior prediction at multiple timescales in inner-city scenarios. In: IEEE Intelligent Vehicles Symposium (IV), Baden-Baden, Germany, pp. 1068–1073 (2011)
3. Lawitzky, A., Althoff, D., Passenberg, C.F., Tanzmeister, G., Wollherr, D., Buss, M.: Interactive scene prediction for automotive applications. In: IEEE Intelligent Vehicles Symposium (IV), Gold Coast, QLD, Australia, pp. 1028–1033 (2013)
4. Oliver, N., Pentland, A.P.: Graphical models for driver behavior recognition in a smartcar. In: IEEE Intelligent Vehicles Symposium (Cat. No. 00TH8511), Dearborn, MI, USA, pp. 7–12 (2000)
5. Agamennoni, G., Nieto, J.I., Nebot, E.M.: Estimation of multivehicle dynamics by considering contextual information. IEEE Trans. Rob. **28**(4), 855–870 (2012)
6. Gindele, T., Brechtel, S., Dillmann, R.: A probabilistic model for estimating driver behaviors and vehicle trajectories in traffic environments. In: 13th International IEEE Conference on Intelligent Transportation Systems, Funchal, Portugal, pp. 1625–1631 (2010)
7. Gupta, A., Johnson, J., Fei-Fei, L., Savarese, S., Alahi, A.: Social GAN: socially acceptable trajectories with generative adversarial networks. In: IEEE Conference on Computer Vision and Pattern (2018)
8. Alahi, A., Goel, K., Ramanathan, V., Robicquet, A., Fei-Fei, L., Savarese, S.: Social LSTM: human trajectory prediction in crowded spaces. In: IEEE Conference on Computer Vision and Pattern Recognition, Las Vegas, NV, USA, pp. 961–971 (2016)
9. Zhan, W., La de Fortelle, A., Chen, Y.T., Chan, C.Y., Tomizuka, M.: Probabilistic prediction from planning perspective: problem formulation, representation simplification and evaluation metric. In: IEEE Intelligent Vehicles Symposium (IV), Changshu, China, pp. 1150–1156 (2018)
10. Mozaffari, S., Al-Jarrah, O.Y., Dianati, M., Jennings, P., Mouzakitis, A.: Deep learning-based vehicle behavior prediction for autonomous driving applications: a review. IEEE IEEE Trans. Intell. Transp. Syst. (2020). https://doi.org/10.1109/TITS.2020.3012034
11. Houston, J., et al.: One thousand and one hours: self-driving motion prediction dataset. arXiv preprint, arXiv:2006.14480 (2020)
12. Zyner, A., Worrall, S., Nebot, E.: A recurrent neural network solution for predicting driver intention at unsignalized intersections. IEEE Robot. Autom. Lett. **3**(3), 1759–1764 (2018)
13. Zyner, A., Worrall, S., Ward, J., Nebot, E.: Long short-term memory for driver intent prediction. In: IEEE Intelligent Vehicles Symposium (IV), Los Angeles, CA, USA, pp. 1484–1489 (2017)

14. Xin, L., Wang, P., Chan, C.Y., Chen, J., Li, S.E., Cheng, B.: Intention-aware long horizon trajectory pre-diction of surrounding vehicles using dual LSTM networks. In: 21st International Conference on Intelligent Transportation Systems (ITSC), Maui, HI, USA, pp. 1441–1446 (2018)
15. Hammer, B.: On the approximation capability of recurrent neural networks. Neurocomputing **31**(1–4), 107–123 (2000)
16. Deo, N., Trivedi, M.M.: Multi-modal trajectory prediction of surrounding vehicles with maneuver based LSTMs. In: IEEE Intelligent Vehicles Symposium (IV), Changshu, China, pp. 1179–1184 (2018)
17. Phillips, D.J., Wheeler, T.A., Kochenderfer, M.J.: Generalizable intention prediction of human drivers at intersections. In: IEEE Intelligent Vehicles Symposium (IV), Los Angeles, CA, USA, pp. 1665–1670 (2017)
18. Dai, S., Li, L., Li, Z.: Modeling vehicle interactions via modified LSTM models for trajectory prediction. IEEE Access **7**, 38287–38296 (2019)
19. Hu, Y., Zhan, W., Tomizuka, M.: Probabilistic prediction of vehicle semantic intention and motion. In: IEEE Intelligent Vehicles Symposium (IV), Changshu, China, pp. 307–313 (2018)
20. Cho, K., et al.: Learning phrase representations using RNN encoder-decoder for statistical machine translation. arXiv preprint, arXiv:1406.1078 (2014)
21. Hochreiter, S., Schmidhuber, J.: Long short-term memory. Neural Comput. **9**(8), 1735–1780 (1997)
22. Cui, H., et al.: Multi-modal trajectory predictions for autonomous driving using deep convolutional networks. In: International Conference on Robotics and Automation (ICRA), Montreal, Canada, pp. 2090–2096 (2019)
23. Djuric, N., et al.: Motion prediction of traffic actors for autonomous driving using deep convolutional networks. arXiv preprint, arXiv:1808.05819, 2 (2018)
24. Lee, N., Choi, W., Vernaza, P., Choy, C.B., Torr, P.H., Chandraker, M.: Desire: distant future prediction in dynamic scenes with interacting agents. In: IEEE Conference on Computer Vision and Pattern Recognition, Honolulu, HI, USA, pp. 336–345 (2017)
25. Luo, W., Yang, B., Urtasun, R.: Fast and furious: real time end-to-end 3D detection, tracking and motion fore-casting with a single convolutional net. In: IEEE Conference on Computer Vision and Pattern Recognition, Salt Lake City, US, pp. 3569–3577 (2018)
26. Zhao, T., et al.: Multi-agent tensor fusion for contextual trajectory prediction. In: IEEE Conference on Computer Vision and Pattern Recognition, California, US, pp. 12126–12134 (2019)
27. Jagadish, D.N., Chauhan, A., Mahto, L: Deep learning techniques for autonomous vehicle path prediction. In: AAAI Workshop on AI for Urban Mobility. Vancouver, Canada (2021)
28. Hoermann, S., Bach, M., Dietmayer, K.: Dynamic occupancy grid prediction for urban autonomous driving: A deep learning approach with fully automatic labeling. In: IEEE International Conference on Robotics and Automation (ICRA), Brisbane, Australia, pp. 2056–2063 (2018)
29. Schreiber, M., Hoermann, S., Dietmayer, K.: Long-term occupancy grid prediction using recurrent neural networks. In: International Conference on Robotics and Automation (ICRA), Montreal, Canada, pp. 9299–9305 (2019)
30. He, K., Zhang, X., Ren, S., Sun, J.: Deep residual learning for image recognition. In: IEEE conference on computer vision and pattern recognition, Las Vegas, US, pp. 770–778 (2016)

Heterogeneous Graph Attention Network for Small and Medium-Sized Enterprises Bankruptcy Prediction

Yizhen Zheng[1(✉)], Vincent C. S. Lee[1], Zonghan Wu[2], and Shirui Pan[1]

[1] Department of Data Science and AI, Faculty of IT, Monash University, Melbourne, Australia
yzhe0006@student.monash.edu, {vincent.cs.lee,shirui.pan}@monash.edu
[2] University of Technology Sydney, Ultimo, Australia
zonghan.wu-3@student.uts.edu.au

Abstract. Credit assessment for Small and Medium-sized Enterprises (SMEs) is of great interest to financial institutions such as commercial banks and Peer-to-Peer lending platforms. Effective credit rating modeling can help them make loan-granted decisions while limiting their risk exposure. Despite a substantial amount of research being conducted in this domain, there are three existing issues. Firstly, many of them are mainly developed based on financial statements, which usually are not publicly-accessible for SMEs. Secondly, they always neglect the rich relational information embodied in financial networks. Finally, existing graph-neural-network-based (GNN) approaches for credit assessment are only applicable to homogeneous networks. To address these issues, we propose a heterogeneous-attention-network-based model (HAT) to facilitate SMEs bankruptcy prediction using publicly-accessible data. Specifically, our model has two major components: a heterogeneous neighborhood encoding layer and a triple attention output layer. While the first layer can encapsulate target nodes' heterogeneous neighborhood information to address the graph heterogeneity, the latter can generate the prediction by considering the importance of different metapath-based neighbors, metapaths, and networks. Extensive experiments in a real-world dataset demonstrate the effectiveness of our model compared with baselines.

Keywords: Bankruptcy prediction · Financial network · Heterogeneous graph · Graph neural networks · Graph attention networks

1 Introduction

Representing about 90% of the business entities and 50% of employment worldwide, small and medium-sized enterprises (SMEs) play an essential role in the economy [1]. However, many of them face huge financing obstacles, especially when applying for loans from commercial banks, due to their financial status's

K. Karlapalem et al. (Eds.): PAKDD 2021, LNAI 12712, pp. 140–151, 2021.
https://doi.org/10.1007/978-3-030-75762-5_12

opacity. One of the reasons is that SMEs are usually not listed, and thus, they do not have the accountability to publish their financial reports and statements regularly. Therefore, there is a dearth of public SMEs' financial data, and it poses a challenge on credit risk assessment tasks for them. The other reason is that even though financial professionals in banks can get access to SMEs' financial statements, in practice, SMEs may intentionally beautify their statements and commit accounting fraud to meet the issuer's criteria. As a result, credit issuers bear the significant risk of uncertainty in lending to SMEs and thus are reluctant to accept their financing applications.

Most existing credit risk assessing methods are based on traditional machine learning algorithms using structured financial statement data [5,11]. However, it is difficult to obtain SMEs' financial statements, which poses a significant challenge for SMEs' credit risk assessment. Also, these traditional machine learning approaches cannot learn the rich relational information embodied in financial graphs. Financial graphs provide rich relational information, which is valuable in inferring a firm's credit condition since its financial status can be affected by its connections, such as its shareholders and executives. In recent years, some studies are attempting to apply GNN-based models to exploiting financial networks. However, these methods are developed on homogeneous financial networks such as company-to-company guarantee networks [4,13]. This kind of network has only one type of node. There are no GNN-based approaches tailored for heterogeneous financial networks which contain multiple node types and edge types. Though there are some general heterogeneous information network (HIN) embedding methods, such as HAN, HetGNN, and MAGNN [6,17,19], which can learn node embeddings on heterogeneous graphs, these methods cannot be directly applied for credit risk assessment tasks. First, these models can only handle node information from a single source or graph, whereas for credit risk assessing a node's information may need to be retrieved and processed from multiple heterogeneous financial graphs. Second, these methods cannot effectively capture the information in target nodes' heterogeneous neighborhoods. This is because they only use a transformation matrix to project the features of different types of nodes into the same vector space to address nodes' heterogeneity instead of sufficiently learning the heterogeneous neighborhood for each node.

We propose a new heterogeneous graph attention network-based approach (HAT) for SMEs' practical risk assessment by using publicly accessible data to address the issues mentioned above. This model can adequately utilize the features of different types of nodes and relations from multiple heterogeneous data sources to predict companies' bankruptcy. Specifically, we first use a random walk with a restart mechanism (RWR) to find a fixed-size of strongly correlated type-specific neighbors for each heterogeneous neighbor type for target nodes. Then, we use the attention mechanism to aggregate heterogeneous type-specific neighbor's features and concatenate the output with target nodes' features to obtain target nodes' heterogeneous neighborhood-level embeddings. In this way, information from all heterogeneous neighbors can be embedded with target nodes' representation. Following, with target nodes' heterogeneous neighborhood-level embeddings, we use the triple attention output layer to train an end-to-end SMEs' bankruptcy predictor. Using triple-level attention, our model can consider the importance of different metapath-based neighbors, metapaths, and graph data sources. To evaluate our proposed model's performance, we build a real-world dataset by collecting SMEs' public data from multiple open data sources. Then, we use this data to construct two heterogeneous financial graphs: a shareholder network and a board member and executive network. The effectiveness of our proposed model has been verified through extensive experiments. In a nutshell, the main contributions of this paper can be summarized as follows:

1. To the best of our knowledge, this is the first attempt to approach bankruptcy prediction using heterogeneous graph neural network techniques. Our work opens a way to develop the company's credit risk evaluation solutions considering rich information embodied in HINs.
2. We propose HAT, a heterogeneous attention network, which can effectively encode information of multiple types of node features and relations from multiple graph data sources to evaluate SMEs' credit risk.
3. We have conducted extensive experiments on a self-collected real-world dataset. The result demonstrates and verifies the effectiveness of our method.

2 Related Work

Bankruptcy Prediction. Recently, researchers have developed machine learning-based bankruptcy prediction models using financial ratio data or privately-owned data provided by financial institutions. For example, Mai et al. [11] collect accounting data for 11827 US-listed firms to train a deep learning model with an average embedding layer to forecast bankruptcy. Chen et al. [3] exploited the ensemble proportion learning model with SVM based on four public bankruptcy datasets on the UCI repository. However, these methods relied heavily on financial ratios, which can be distorted with accounting beautifying techniques such as "window dressing". Also, this data is usually not publicly accessible for SMEs. Though some studies conducted their research on public bankruptcy datasets, their number is very limited and always out-of-date. Furthermore, these studies have not exploited financial graphs, whose rich relational information can provide valuable clues when inferring firms' credit status.

Graph Neural Networks. Exploiting graph-structured data with deep learning, graph neural networks (GNN), can effectively generate a low-dimensional vector representation for each network node [16,18,20]. These latent representations can then be used for various downstream tasks, such as node classification and link prediction [9]. Extending the graph convolutional networks (GCNs) with masked self-attention layers, Velickovic et al. [15] proposed Graph attention networks (GAT) to consider neighbor nodes' weights, rather than treating neighbors' nodes equally important. In recent years, some pioneering studies are implementing GNN-based approaches in financial networks. For instance,

Cheng et al. [4] proposed a high-order graph attention representation method to infer the systematic credit risk based on company-to-company guarantee networks. Shumovskaia et al. [13] developed a recursive-neural-network-based GNN model to explore bank-client transactional networks. However, there is no study applying the GNN-based approach in financial HINs. This paper introduces a novel model, HAT, to learn node embeddings in financial HINs effectively.

3 Problem Formulation

Financial Heterogeneous Graph. This paper considers two financial networks, including shareholder networks, representing firms' shareholding structure and board member networks, showing firms' board members and executives as heterogeneous graphs. Figure 1 and Fig. 2 show examples for these two financial graphs. For a heterogeneous graph $G = (V, E)$, V and E represent nodes and edges in the graph, respectively. Here, node V includes two types of nodes: companies C and individuals I. The edges consist of company-individual edges E_{ci} and company-company edges E_{cc}.

Problem Statement. Given a set of heterogeneous financial graphs $\{g_1, g_2..., g_n\}$, in which each graph is defined as $G = (V, E)$, we aim to predict whether a firm will go bankrupt by learning a global node embedding $\mathbf{h}_c$ for each firm node c and mapping it to a probability distribution.

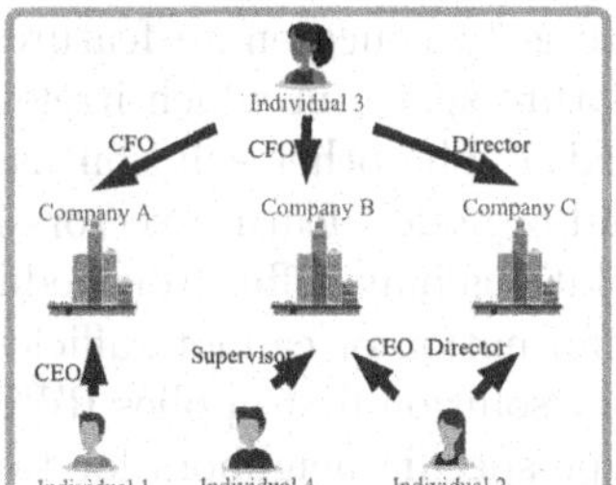

Fig. 1. An example of board member and executives' networks

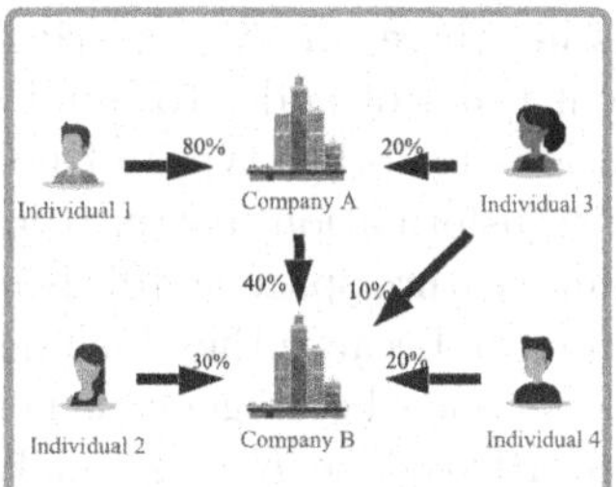

Fig. 2. An example of shareholder networks

4 Methodology of Our Proposed Model

In this section, we describe HAT for heterogeneous graph embedding learning in financial networks. As shown in Fig. 3, HAT consists of two main components: a heterogeneous neighborhood encoding layer and a triple attention output layer. While the heterogeneous neighborhood encoding layer can sufficiently exploit the heterogeneous neighborhood for target nodes, the triple attention and output layer can generate the final embedding and prediction results with three levels of attention: node-level, metapath-level, and network-level.

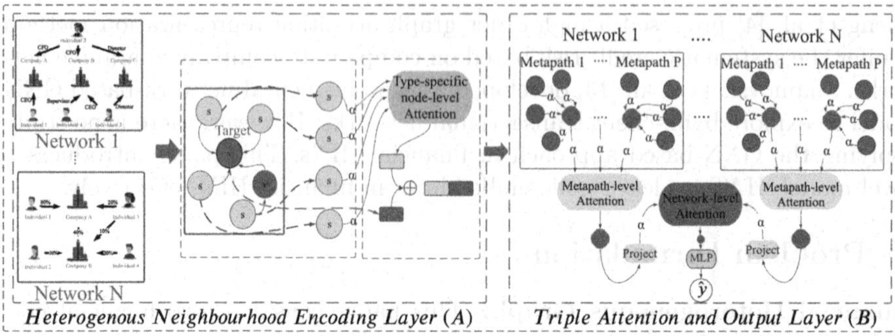

Fig. 3. The architecture of HAT for bankruptcy prediction

4.1 Heterogeneous Neighborhood Encoding Layer

A heterogeneous neighborhood encoding layer can address the heterogeneity of a heterogeneous graph. This layer first uses the RWR method to find each heterogeneous type top K type-specific neighbors for target nodes and aggregates them to get type-specific neighborhood embeddings. Then, it concatenates these type-specific neighborhood embeddings with target nodes' own features to obtain the all-type neighborhood embeddings.

It is challenging to solve heterogeneous graphs' heterogeneity since different types of nodes can have their features lie in different feature spaces. To solve this issue, there are two existing solutions. One is to concatenate features of different types of nodes to build a new large feature space, in which irrelevant dimensions for other types of nodes are assigned 0. The other solution uses a linear transformation matrix to project each kind of node's feature vectors into the same feature space [8,10]. However, these solutions may suffer from reduced performance because they include more noisy information or cannot sufficiently encode a node's heterogeneous neighborhood. Our solution first applies RWR to explore a fixed-size K of strongly correlated type-specific neighbors for target nodes. Here K is a tunable parameter. As shown in Fig. 3(A), given a node v and its $s \in S$ type neighbors, the two steps of this process are presented below:

1. Step-1: Sampling a fixed-length sequence based on RWR for the node v: We start a random walk from a node v. According to a probability p, the walk decides whether it will move to one of the current node's neighbors or restart from node v iteratively until it has collected a sequence with a preset fixed number of nodes, defined as RWR (v).
2. Step-2: Finding top K neighbors for v: Based on s type neighbors occurrence frequency on RWR (v), we select top K s type neighbors for v.

We can then aggregate these top K collected nodes' features to encode v's s type neighborhoods. If a target node has multiple heterogeneous neighbor types, we can repeat the two-step process for each neighbor type to generate each type neighborhood representation. Since each type-specific neighbor may contribute differently to a target node, it is necessary to consider their weight in the aggregation process. Therefore, we use an attention layer in heterogeneous neighbor nodes' aggregation. Given a fixed number K and a heterogeneous neighbor node type $s \in S$, for a target node, it has a set of type-specific neighbors features h_1^s, h_2^s, $\ldots h_K^s$. With these features, we can use an attention layer to learn the weight of each type-specific neighbor α_1^s, α_2^s, $\ldots \alpha_K^s$ and this process can be formulated as follows:

$$\{\alpha_1^s, \alpha_2^s, \ldots \alpha_K^s\} = att_{heterneigh}\{h_1^s, h_2^s, \ldots h_K^s\}. \quad (1)$$

In this equation, $att_{heterneigh}$ means the self-attention-based [14] deep neural network for type-based neighbors' aggregation. Specifically, we first use a one-layer MLP to transform these features into a heterogeneous type-specific node-level attention vector. Then, we can get the importance of each heterogeneous type-based neighbor, denoted as e_k^s, where $\{k|k \in Z, 0 \leq k \leq K\}$:

$$e_k^s = q_H^T \cdot tanh(W_H \cdot h_k^s + b_H), \quad (2)$$

where q_H is the parameterized attention vector, W_H is the weight matrix, and b_H is the bias vector. All of these parameters are learnable. Then, by using a softmax layer, we can normalize e_k^s to get the coefficient α_k^s:

$$\alpha_k^s = softmax(e_k^s) = \frac{exp(e_k^s)}{\sum_{k=1}^{K} exp(e_k^s)}, \quad (3)$$

here α_k^s represents the relative importance of a type-specific neighbor. With α_k^s, we can now polymerize the features of top K type-based neighbors for target nodes to obtain the learned embedding of node i's type s neighborhood. This process is presented as follows:

$$Z_i^s = \sum_{k=1}^{K} \alpha_k^s \cdot h_k^s, \quad (4)$$

where Z_i^s is the representation of node i's type s neighborhood. Finally, to obtain the embedding of node i's all type neighborhood, including node i's self-generated features, we need to aggregate the learning embedding of all Z_i^s with node i's own feature as follows:

$$Z_i^S = (\|_{s \in S} Z_i^s) \parallel h_i, \quad (5)$$

where $\|$ means concatenation, Z_i^S represents the all type neighborhood embedding for the node i, and h_i means the target node i's own feature. To better understand this aggregating and concatenation process, we present a graphical illustration in Fig. 4. From this figure, we can see node $c1$ has two types of heterogeneous neighbors: nn and ii. Through neighbor sampling, we can get the nn type and ii type top K neighbors for the node $c1$. Here $\{\alpha_1^{s1}, \alpha_2^{s1}, .., \alpha_4^{s1}\}$ and $\{\alpha_1^{s2}, \alpha_2^{s2}, .., \alpha_4^{s2}\}$ are two sets of computed attention coefficient for $c1$'s ii type neighbors and nn type neighbors respectively. After aggregation, node $c1$ gets its ii type neighborhood representation Z_{c1}^{s1}, and nn Z_{c1}^{s2}. Finally, by aggregating these two representations with c1's own feature h_{c1}. Node $c1$ get its all type neighborhood embedding Z_{c1}^S.

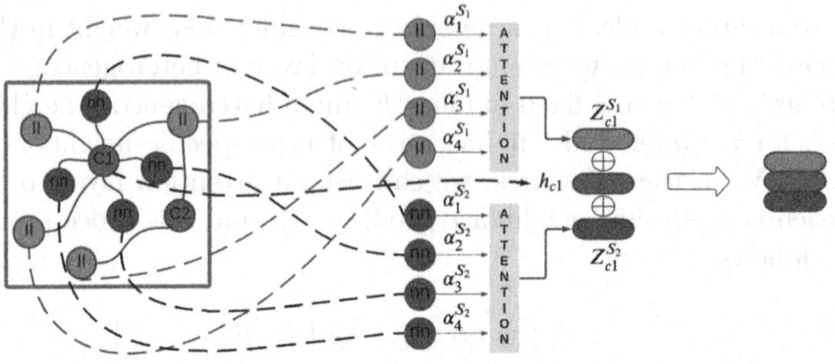

Fig. 4. Graphical illustration for the heterogeneous neighborhood encoding layer

4.2 Triple Attention and Output Layer

With the all-type neighborhood embeddings, the triple attention and output layer can generate the final embeddings for target nodes using triple-level attention: metapath-based node level, metapath level, and network level. The architecture of the triple-level attention layer is presented in Fig. 3(B).

Metapath-Based Node-Level Attention. For the metapath-based node-level attention, we adopt the multi-head graph attention operation [15], which is an attention mechanism for homogeneous graph representation learning. This operation is applicable here since, within a metapath, a target node is within a homogeneous graph with its same type neighbors. By using this operation, the model can learn the weight of different metapath-based neighbors.

A heterogeneous network has several metapaths denoted as $m \in M$ based on a target node type. Given a specific node i with type s, the node-level attention learns the weights of its metapath-based neighbor $j \in N_i^m$, where N_i^m represents a set of nodes including both node i's metapath-based neighbors and node i itself. Then, we compute the node-level attention value α_{ij}^m by using the all-type neighborhood embeddings Z_i^S and Z_j^S. With the normalized importance value α_{ij}^m, we can weight sum all the nodes in the set N_i^m to get the metapath-level embedding for node i. This process is formulated as follows:

$$\alpha_{ij}^m = \frac{exp(\sigma(q_m^T \cdot [Z_i^S \parallel Z_j^S]))}{\sum_{o \in N_i^m} exp(\sigma(q_m^T \cdot [Z_i^S \parallel Z_o^S]))}, \tag{6}$$

$$Z_i^m = \sigma(\sum_{j \in N_i^m} \alpha_{ij}^m \cdot Z_j^S), \tag{7}$$

where Z_i^m denotes the m metapath-based embedding for node i. To stabilize this learning process, multi-head attention can be used to improve the representation ability. Specifically, we can repeat the node-level attention L times to get L embeddings for node i. Then, the metapath-specific embedding can be generated by concatenating these embeddings:

$$Z_i^m = \|_{l=1}^{L} \sigma(\sum_{j \in N_i^m} \alpha_{ij}^m \cdot Z_j^S). \tag{8}$$

After learning the embedding for each metapath m, we can get a set of Z_i^m for the node i. This set of embedding will be aggregated through the metapath-level attention to get the network-level embedding for i.

Metapath-Level Attention. We introduce a self-attention-based metapath-level attention [14] to compute the weights for different metapaths. Given a specific node i with type s, it has several metapath-specific embeddings Z_i^m for several meta paths $m \in M$. With this set of semantic-specific embeddings, we can calculate the normalized metapath level attention α_m^P. Please note that, here α_m^P is shared by all nodes, since there is some similar connection between a metapath and a type of nodes. Then, we fuse metapath-specific embeddings of i by weighted summing all Z_i^m with α_m^P, to create the network-level embedding:

$$\alpha_m^P = \frac{exp(\frac{1}{|V_s|}\sum_{i \in V_s} q_O^T tanh(W_O \cdot Z_i^m + b_O))}{\sum_{m \in M} exp(\frac{1}{|V_s|}\sum_{i \in V_s} q_O^T tanh(W_O \cdot Z_i^m + b_O))}, \tag{9}$$

$$Z_i^g = \sum_{m \in M} \alpha_m^P \cdot Z_i^m, \tag{10}$$

here Z_i^g is the network-specific embedding for the node i, q_O is a parameterized vector, W_O is a weight matrix, and b_O is a bias vector. With all Z_i^g, we can polymerize them to get node i's final embedding via network-level attention.

Network-Level Attention. Network-level attention can address the heterogeneity of different networks and determine their importance. Firstly, we project network-level embeddings to the same vector space with a linear transformation layer. Given a node i with type s, and $g \in G$, where G is the set of graphs input, the process of projection is shown below:

$$Z_i^{g'} = \theta_s \cdot Z_i^g, \tag{11}$$

where $Z_i^{g'}$ is the projected network-level embeddings, Z_i^g is the original network-level embeddings, and θ_s is the transformation matrix for the node type s. Like metapath-level attention, network-level attention applies the attention mechanism to distill the network semantic information from multiple graphs. Given a node i with type s, and a network $g \in G$, we can use the transformed embedding $Z_i^{g'}$ to compute the normalized network-level attention α_g. Then, by weighted sum all $Z_i^{g'}$, we can get the final embedding for node i. This process is formulated as below:

$$\alpha_n = \frac{exp(\frac{1}{|V_s|}\sum_{i \in V_s} q_R^T \cdot tanh(W_R \cdot Z_i^{g'} + b_R))}{\sum_{g \in G} exp(\frac{1}{|V_s|}\sum_{i \in V_s} q_R^T \cdot tanh(W_R \cdot Z_i^{g'} + b_R))}, \tag{12}$$

$$Z_i = \sum_{g \in G} \alpha_g \cdot Z_i^{g'}, \tag{13}$$

here Z_i is the final embedding for node i, q_R is a parameterized vector, W_R is a weight matrix, and b_R is a bias vector. Z_i can be transformed to the desired output dimension for further tasks.

4.3 Training

We consider bankruptcy prediction as a binary classification task. The output layer predicts the labels of companies based on its final embedding Z_i. To obtain the output, we feed the embedding to a softmax layer for classification as follows:

$$Z_i = softmax(Z_i) \tag{14}$$

Then, we can optimize the model parameters by minimizing the cross-entropy loss over training data with the L2-norm:

$$L = -\sum_{i \in V_s} Y_i log(C \cdot Z_i) + \eta \parallel \theta \parallel_2 \tag{15}$$

where C is the parameter of the classifier, $i \in V_s$ represents a node with the type s, Y_i is the ground truth label for node i, Z_i is i's node embedding, θ is the model parameters, and η represents the regularization factor. With node labels' guidance, we adopt the gradient descent method to process the backpropagation and learn the node embeddings for target nodes.

5 Experiment

Datasets. We collect data from multiple Chinese government open data sources and build a real-world dataset containing a board member network and a shareholder network for 13489 companies. In the dataset, 3566 companies are labeled as bankrupt. The experiment source code and collected dataset are available in our Github repository[1].

Experiment Methods. We select five baselines to verify the effectiveness of our proposed method.

- **Logistic Regression (LR).** LR is a statistical model powered by a shallow neural network. It is widely adopted in bankruptcy prediction tasks [7].
- **Support Vector Machine (SVM).** SVM is a machine learning technique, which builds hyperplanes in a high dimensional space for classification and regression tasks. It is widely adopted in bankruptcy prediction tasks [2].

[1] https://github.com/hetergraphforbankruptcypredict/HAT.

Table 1. Experiment result

Method	Accuracy	Macro-F1	Recall	Precision
LR	76.03	62.83	61.63	68.51
SVM	75.89	62.45	61.31	68.25
DeepWalk	60.30	50.97	51.22	51.09
GAT	76.50	62.10	60.98	69.95
HAN	77.06	63.07	61.75	71.22
HAT	**78.31**	**64.32**	**62.72**	**74.86**

Table 2. Variants result

Method	Accuracy	Macro-F1	Recall	Precision
HAT_{mean}	78.08	63.41	61.99	74.80
HAT_{max}	77.61	64.42	62.82	72.23
HAT_{min}	78.22	64.51	62.89	74.30
HAT_{dual}	78.26	**66.18**	**64.40**	73.18
HAT	**78.31**	64.32	62.72	**74.86**

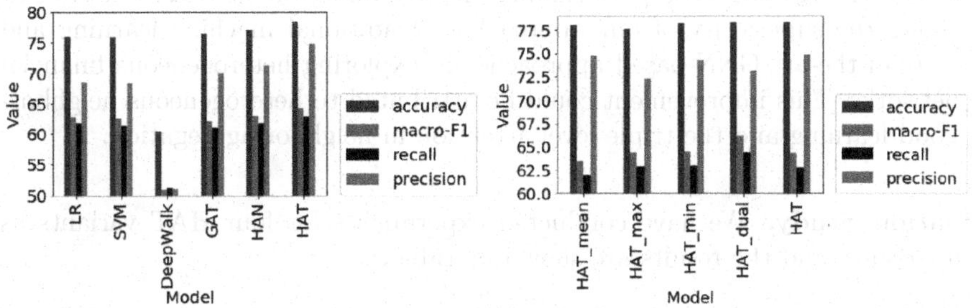

Fig. 5. Comparison of baselines and variants of HAT.

- **DeepWalk (DW).** DW is a random-walk-based graph embedding method for homogeneous graphs. In the experiment, we ignore the heterogeneous graph's heterogeneity and process the DW through the whole graph [12].
- **GAT.** GAT is a semi-supervised attention-based GNN approach for homogeneous graphs. Here we only consider the best metapath [15].
- **HAN.** HAN is a state-of-the-art heterogeneous graph neural network with dual-level attention. It is a general heterogeneous GNN method [17].

Evaluation Metrics. We have selected four metrics: Micro-F1, Macro-F1, Recall, and Precision to evaluate the performance of our model.

Implementation Details. In the experiment, we consider the data split based on the bankruptcy date of company nodes. While the training set includes all nodes that went bankrupt before 2019, the validation and test set include nodes that went bankrupt after 2019. This is reasonable because the model's goal is to predict companies' bankruptcy. For active company nodes, we distribute them according to the ratio 68:16:16 for training, validation, and test set, respectively. This is because the training set contains 68% of all bankruptcy nodes in our dataset, while the remaining bankruptcy nodes are distributed equally for the other two sets. Thus, with this ratio, the data distribution in all three sets would be similar.

Results and Analysis. The performance of HAT and baseline methods is presented in Table 1 and Fig. 5. From this table, we have the following observations:

- Surprisingly, GNN-based method GAT only obtain similar results to LR and SVM, the reason may be that GAT only uses one metapath-based homogeneous graph as input, which cannot provide sufficient graph structural and relational information to advance the model training.
- HAN outperforms the other four baselines, which is reasonable because HAN can learn additional useful signals provided in HINs.
- HAT achieves the best performance in all evaluation metrics. This result shows the superiority of our approach to traditional machine learning and state-of-the-art GNN-based approaches in exploring heterogeneous financial networks. This improvement could be attributed to heterogeneous neighborhood learning and the triple-level attention in neighbor aggregation.

Ablation Study. We have conducted experiments for four HAT variants as comparisons, and the results are shown in Table 2.

- **HAT**$_{mean}$**:** Compared with HAN, HAT$_{mean}$ has the heterogeneous neighborhood encoding layer and use mean pooling in neighbors aggregation. Also, it applies the triple-level attention instead of the dual-level attention.
- **HAT**$_{max}$**:** Compared with HAT$_{mean}$, HAT$_{max}$ applys max pooling in the heterogeneous neighborhood encoding layer.
- **HAT**$_{min}$**:** Compared with HAT$_{mean}$, HAT$_{min}$ adopts min pooling in the heterogeneous neighborhood encoding layer.
- **HAT**$_{dual}$**:** Compared with HAT, HAT$_{dual}$ has no network-level attention.
- **HAT:** HAT employs an attention mechanism in the heterogeneous neighborhood encoding layer.

From Table 2 and Fig. 5, we can see that all HAT variants outperform HAN in all evaluation metrics. This result demonstrates the heterogeneous neighborhood encoding layer's effectiveness, which can encapsulate the heterogeneous neighborhood information to boost the model performance. Also, while HAT$_{dual}$ achieves the best performance for accuracy and precision, HAT obtains the highest results for Macro-F1 and Recall. It is hard to tell which one is better between HAT$_{dual}$ and HAT in the scenario in which heterogeneous graph inputs have the same set of node types. However, suppose graph inputs have a different set of node types. In that case, only HAT can address the heterogeneity of different networks since it has additional network-level attention to project network-level embeddings. Therefore, HAT is preferable due to its comprehensiveness in handling graph inputs.

6 Conclusion

This paper develops a novel heterogeneous GNN-based method HAT for SMEs' bankruptcy prediction using public data from multiple sources. Our model can effectively learn the heterogeneous neighborhood for target nodes and generate node embeddings considering the weights of metapath-based neighbors, metapaths, and networks, with a triple attention output layer. The results of extensive experiments have shown that our model outperforms five baselines in exploring financial HINs. For future works, we plan to apply our method to other scenarios, e.g., social and e-commerce networks.

References

1. Small and Medium Enterprises (SME) finance (2020). https://www.worldbank.org/en/topic/smefinance. Accessed 4 Nov 2020
2. Chaudhuri, A., De, K.: Fuzzy support vector machine for bankruptcy prediction. Appl. Soft Comput. **11**(2), 2472–2486 (2011)
3. Chen, Z., Chen, W., Shi, Y.: Ensemble learning with label proportions for bankruptcy prediction. Expert Syst. Appl. **146**, 113115 (2020)
4. Cheng, D., Zhang, Y., Yang, F., Tu, Y., Niu, Z., Zhang, L.: A dynamic default prediction framework for networked-guarantee loans. In: CIKM (2019)
5. Erdogan, B.E.: Prediction of bankruptcy using support vector machines: an application to bank bankruptcy. J. Stat. Comput. Simul. **83**(8), 1543–1555 (2013)
6. Fu, X., Zhang, J., Meng, Z., King, I.: MAGNN: metapath aggregated graph neural network for heterogeneous graph embedding. In: WWW 2020 (2020)
7. Hauser, R.P., Booth, D.: Predicting bankruptcy with robust logistic regression. J. Data Sci. **9**(4), 565–584 (2011)
8. Huang, Q., Yu, J., Wu, J., Wang, B.: Heterogeneous graph attention networks for early detection of rumors on Twitter. In: IJCNN (2020)
9. Kipf, T.N., Welling, M.: Semi-supervised classification with graph convolutional networks. In: ICLR (2016)
10. Linmei, H., Yang, T., Shi, C., Ji, H., Li, X.: Heterogeneous graph attention networks for semi-supervised short text classification. In: EMNLP-IJCNLP (2019)
11. Mai, F., Tian, S., Lee, C., Ma, L.: Deep learning models for bankruptcy prediction using textual disclosures. Eur. J. Oper. Res. **274**(2), 743–758 (2019)
12. Perozzi, B., Al-Rfou, R., Skiena, S.: DeepWalk: online learning of social representations. In: KDD (2014)
13. Shumovskaia, V., Fedyanin, K., Sukharev, I., Berestnev, D., Panov, M.: Linking bank clients using graph neural networks powered by rich transactional data (2020)
14. Vaswani, A., et al.: Attention is all you need. In: NeurIPS (2017)
15. Veličković, P., Cucurull, G., Casanova, A., Romero, A., Lio, P., Bengio, Y.: Graph attention networks. In: ICLR (2017)
16. Wang, H., Zhou, C., Chen, X., Wu, J., Pan, S., Wang, J.: Graph stochastic neural networks for semi-supervised learning. In: NeurIPS (2020)
17. Wang, X., et al.: Heterogeneous graph attention network. In: WWW (2019)
18. Wu, Z., Pan, S., Chen, F., Long, G., Zhang, C., Philip, S.Y.: A comprehensive survey on graph neural networks. In: IEEE Transactions on Neural Networks and Learning Systems (2020)
19. Zhang, C., Song, D., Huang, C., Swami, A., Chawla, N.V.: Heterogeneous graph neural network. In: SIGKDD (2019)
20. Zhu, S., Pan, S., Zhou, C., Wu, J., Cao, Y., Wang, B.: Graph geometry interaction learning. In: NeurIPS (2020)

Algorithm Selection as Superset Learning: Constructing Algorithm Selectors from Imprecise Performance Data

Jonas Hanselle(✉), Alexander Tornede, Marcel Wever, and Eyke Hüllermeier

Heinz Nixdorf Institut, Department of Computer Science, Paderborn University, Paderborn, Germany
{jonas.hanselle,alexander.tornede,marcel.wever,eyke}@upb.de

Abstract. Algorithm selection refers to the task of automatically selecting the most suitable algorithm for solving an instance of a computational problem from a set of candidate algorithms. Here, suitability is typically measured in terms of the algorithms' runtimes. To allow the selection of algorithms on new problem instances, machine learning models are trained on previously observed performance data and then used to predict the algorithms' performances. Due to the computational effort, the execution of such algorithms is often prematurely terminated, which leads to *right-censored* observations representing a lower bound on the actual runtime. While simply neglecting these censored samples leads to overly optimistic models, imputing them with precise though hypothetical values, such as the commonly used penalized average runtime, is a rather arbitrary and biased approach. In this paper, we propose a simple regression method based on so-called superset learning, in which right-censored runtime data are explicitly incorporated in terms of interval-valued observations, offering an intuitive and efficient approach to handling censored data. Benchmarking on publicly available algorithm performance data, we demonstrate that it outperforms the aforementioned naïve ways of dealing with censored samples and is competitive to established methods for censored regression in the field of algorithm selection.

Keywords: Algorithm selection · Superset learning · Censored data

1 Introduction

Per-instance Algorithm Selection (AS) denotes the problem of recommending an algorithm that appears to be most suitable for a given instance of a problem class. The suitability is assessed with respect to a specific performance criterion, such as solution quality or runtime. The latter is of special interest when dealing with computationally hard problems, such as the Boolean satisfiability problem (SAT) [25,26], the traveling salesperson problem (TSP) [19], or constraint satisfaction problems (CSP) [18], just to name a few. For this kind of problems, different

K. Karlapalem et al. (Eds.): PAKDD 2021, LNAI 12712, pp. 152–163, 2021.
https://doi.org/10.1007/978-3-030-75762-5_13

solvers or optimizers have been developed, which build on different heuristics to exploit certain structures inherent to the problem instances. As the heuristics are complementary in some sense, choosing the solver or optimizer on a per-instance basis can drastically improve the overall performance [13].

A common approach to AS is the use of machine learning methods to predict the runtime of algorithms on unseen problem instances. One major challenge for such approaches concerns the training data, parts of which are usually censored. The censoring has its root in the way the training data is generated. To assess the true runtime of an algorithm for a problem instance, it is simply run on that particular instance. However, for specifically hard instances, the algorithms may take days, weeks, or even years before returning a solution. To keep the computational effort reasonable, a cutoff time is set at which the execution of still running algorithms is aborted. In the commonly used AS benchmarking suite ASlib [2], up to 70% of the instances in the training data are censored (cf. Table 1).

To deal with censored data, several approaches consider them as missing values and apply imputation techniques to replace these values [21,26]. Alternatively, techniques from the field of survival analysis have been proposed [5,6,24]. While the latter approach appropriately captures the information provided by a censored observation, namely that the runtime exceeds the cutoff time C, the former turns it into unduly precise information and comes with the risk of incorporating a bias in the learning process.

In this paper, we propose to consider AS as a so-called superset learning problem [9], where the learner induces a (precise) predictive model from possibly imprecise training data. More concretely, we learn a regression model for runtime prediction from training instances that are either precisely labeled or labeled with an interval of the form (C, ∞). This is not only a just representation of the (weak albeit non-void) information that the true runtime exceeds C, but, as will be seen, also offers an efficient way for handling censored data. In our experimental evaluation, we show that methods based on superset learning can induce (linear) models that outperform naïve strategies for dealing with censored data in AS.

2 The Per-instance Algorithm Selection Problem

The per-instance algorithm selection problem, first introduced in [20], is essentially a recommendation problem where the sought recommendation is an algorithm to be used on a specific instance of an algorithmic problem class (such as SAT). More formally, given a problem class (instance space) $\mathcal{I}$ and a set of algorithms $\mathcal{A}$ able to solve instances from this class, the goal is to learn a mapping $s : \mathcal{I} \longrightarrow \mathcal{A}$, called *algorithm selector*. Provided a problem instance $i \in \mathcal{I}$ and a performance measure $m : \mathcal{I} \times \mathcal{A} \longrightarrow \mathbb{R}$, such a selector s should ideally return the algorithm performing best in terms of this measure. This ideal mapping s^* is called the *virtual best solver* (VBS), aka. *oracle*, and formally defined as

$$s^*(i) := \arg\min_{a \in \mathcal{A}} \mathbb{E}\left[m(i, a)\right] \tag{1}$$

for all instances $i \in \mathcal{I}$, where the expectation accounts for the potential randomness in the application of the algorithm. While conceptually simple to define, the computation cannot be performed through an exhaustive enumeration, as m is usually costly to evaluate and often even requires running the algorithm on the particular instance at hand, like in the case of runtime as a measure. For the remainder of the paper, we will assume that runtime is used as the performance measure. In contrast to the oracle, the *single best solver* (SBS) is the strategy that always selects the algorithm which is best on average across all instances, and accordingly can be seen as a natural baseline.

Closely related to AS is the problem of *algorithm scheduling* [17], where the recommendation target is a schedule of algorithms (instead of a single algorithm) assumed to be executed sequentially for a specific amount of time, until a solution is either found or the schedule is over. Also closely related is the *extreme algorithm selection* problem [22], where the recommendation target is still a single algorithm, but the set of algorithms $\mathcal{A}$ to choose from is assumed to be extremely large, i.e., in the hundreds to thousands compared to tens in the standard AS problem. Finally, *meta-algorithm selection* [23] refers to the problem of choosing among the many existing algorithm selection approaches and can be formalized as an algorithm selection problem itself.

2.1 Common Algorithm Selection Solutions

To circumvent the problem of costly evaluations of the performance measure m, the majority of existing AS approaches in one way or another learns a surrogate $\widehat{m} : \mathcal{I} \times \mathcal{A} \longrightarrow \mathbb{R}$ of the original performance measure, with the property of being cheap to evaluate. This allows for the exhaustive enumeration in (1), and hence to determine an algorithm selector $s : \mathcal{I} \longrightarrow \mathcal{A}$ with

$$s(i) := \arg\min_{a \in \mathcal{A}} \widehat{m}(i, a). \qquad (2)$$

For the purpose of representing instances, we assume the existence of a feature function $f : \mathcal{I} \longrightarrow \mathbb{R}^d$ mapping instances to d-dimensional feature vectors. Such features should comprise properties of the instance that are potentially relevant to determine the best algorithm. For SAT instances, examples of such features include the number of variables or the number of clauses in a formula. The computation of such features takes time, which is important to consider when the measure m to optimize is (related to) runtime. Moreover, learning such surrogate models requires training instances $\mathcal{I}_D \subset \mathcal{I}$ for which the performance value $m(i, a) \in \mathbb{R}$ is available for some algorithms $a \in \mathcal{A}$.

One of the simplest instantiations of the surrogate framework described above is to learn a performance predictor $\widehat{m}_a : \mathcal{I} \longrightarrow \mathbb{R}$ separately for each algorithm $a \in \mathcal{A}$ as done in Satzilla'07 [26]. In a later version of Satzilla, these models were replaced by cost-sensitive decision forests $\widehat{m}_{a,a'} : \mathcal{I} \longrightarrow \{0, 1\}$ for each pair of algorithms $a \neq a' \in \mathcal{A}$. Here, predictions are pairwise comparisons suggesting which of two candidate algorithms will most likely perform better on a given instance, and the final selection is determined by majority voting. In fact,

Satzilla'11 can be seen as a cost-sensitive version of the all-pairs decomposition scheme for solving multi-class classification problems, leading to the more general application of multi-class classification approaches $\widehat{m} : \mathcal{I} \longrightarrow \mathcal{A}$, where each algorithm $a \in \mathcal{A}$ is considered a class [24].

Moreover, to estimate the unknown performance of an instance/algorithm pair, instance-based approaches such as SUNNY [1] or ISAC [12] rely on similar instances for which evaluations are available, and make use of k-nearest neighbor or clustering techniques. In a more recent work [8], hybrid ranking and regression models are successfully applied to the problem of algorithm selection, with the aim to benefit from both perspectives (ranking and regression) on the problem.

2.2 The Influence of Censored Data

Recall that we assume a set of training instances $\mathcal{I}_D \subset \mathcal{I}$ with potentially missing values for the performance measure. That is, the performance values $m(i, a) \in \mathbb{R}$ are available for *some* but rarely for all algorithms $a \in \mathcal{A}$. This is because the training data generation for algorithm selection is usually performed under the constraint of a timeout. When an algorithm $a \in \mathcal{A}$ is run on an instance $i \in \mathcal{I}_D$ and does not terminate before a given *cutoff* time C, its execution is simply stopped at this point, as some algorithms can take extremely long to solve specific instances of combinatorial optimization problems [7]. In such a case, no precise performance value, i.e. runtime $m(i, a)$, can be recorded. Instead, the true runtime is only known to exceed the cutoff C, i.e. $m(i, a) > C$. In other words, the observation is *right-censored* [15]. The well-known ASlib [2] benchmark contains scenarios where over 70% of the data is censored. Obviously, since these data points make up a significant amount of available data, they need to be treated with care.

The simplest approach for dealing with censored data points is to completely ignore them when training surrogate models $\widehat{m}$. This strategy does not only waste important information, but also bears the danger of learning overoptimistic models, as potentially very long runtimes are not considered during the training. A slightly more advanced strategy is to impute the censored points either by the cutoff time C or a multiple thereof, motivated by the penalized average runtime (PARN) score. However, once again, this strategy can lead to strongly biased models, as in cases with many censored data points the majority of the training data is deliberately distorted. A more advanced imputation strategy has been proposed in [21] in the context of algorithm configuration (AC) [4,10]. It iteratively estimates truncated normal distributions over the censored data based on generalizations made by models over the uncensored data, and finally replaces each censored sample by the mean of the corresponding distribution. While offering a theoretical foundation, this approach makes a strong assumption on the distribution of the runtimes of the algorithms. Moreover, it is computationally expensive, as the process involves fitting the requested model repeatedly before returning a final model.

An ideal algorithm selection technique would directly incorporate censored samples in the training process, instead of artificially removing any imprecision in a preprocessing step. This is accomplished by AS techniques leveraging concepts from *survival analysis*, which were first introduced in [5,6] for the setting of algorithm scheduling and recently improved by Run2Survive [24]. The latter is an algorithm for learning runtime distributions conditioned on an instance, which are in turn used in a decision-theoretic framework to tailor the selection of an algorithm towards a specific performance measure. Such methods do nevertheless rely on distributional assumptions.

Therefore, in this paper, we adopt another approach based on so-called *superset learning*, a problem setting that has recently attracted increasing attention in the machine learning literature [9]. Superset learning is an extension of standard supervised learning, which deals with the problem of learning predictive models from possibly imprecise training data characterized in the form of *sets* of candidate values. In principle, every method for supervised learning, whether parametric or non-parametric, can be extended to the setting of superset learning. Moreover, imprecision can be modeled in a rather flexible way and is not restricted to censoring.

3 Superset Learning with Right-Censored Data

In supervised learning, training data is supposed to be given in the form of tuples $(\boldsymbol{x}_n, y_n) \in \mathcal{X} \times \mathcal{Y}$, where $\mathcal{X}$ is the instance and $\mathcal{Y}$ the output space, and the task is to learn the dependency between instances and outcomes. More specifically, given a hypothesis space $\mathcal{H} \subset \mathcal{Y}^{\mathcal{X}}$ to choose from, the learner seeks to find an accurate predictive model in the form of a risk-minimizing hypothesis

$$h^* \in \operatorname*{argmin}_{h \in \mathcal{H}} \int_{(\boldsymbol{x},y) \in \mathcal{X} \times \mathcal{Y}} L(h(\boldsymbol{x}), y) \, \mathrm{d}\, \mathbb{P}(\boldsymbol{x}, y) \,, \tag{3}$$

where $L : \mathcal{Y} \times \mathcal{Y} \rightarrow \mathbb{R}$ is a loss function and $\mathbb{P}$ an unknown joint probability distribution characterizing the data-generating process. In superset learning, the values of the target variables, i.e., the outcomes y_n (and possibly also the instances $\boldsymbol{x}_n$, though this is of less interest for our purpose), are not necessarily observed precisely. Instead, they are characterized in terms of sets $Y_n \ni y_n$ known to cover the true observation y_n. For example, in the case of classification, Y_n is a (finite) set of candidate classes. In regression, where $\mathcal{Y} = \mathbb{R}$ is infinite, imprecise observations are typically modeled in the form of intervals.

Several methods for learning from imprecise observations of that kind have been proposed in the literature. Here, we adopt the approach of generalized loss minimization based on the so-called *optimistic superset loss* [11], which has also been proposed under the name *infimum loss* [3]. The optimistic superset loss (OSL) is a generalization of the original loss L in (3) that compares $h(\boldsymbol{x}) = \hat{y}$ with the set-valued target in an "optimistic" way:

$$L^*(\hat{y}, Y) = \inf_{y \in Y} L(\hat{y}, y) \tag{4}$$

Learning is then accomplished by using L^* in place of L, for example by minimizing the generalized empirical risk (or a regularized version thereof) on the training data $\mathcal{D} = \{(\boldsymbol{x}_n, Y_n)\}_{n=1}^N$:

$$h^* \in \operatorname*{argmin}_{h \in \mathcal{H}} \frac{1}{N} \sum_{n=1}^{N} L^*(h(\boldsymbol{x}_n), Y_n)$$

This way of learning from imprecise data is essentially motivated by the idea of data disambiguation, i.e., of figuring out the true observations y_n among the candidates suggested by the set-valued observations Y_n. We refer to [9] for a more detailed explanation and to [3] for a theoretical foundation of this approach.

In our setting, hypotheses h correspond to algorithm-specific surrogate models $\widehat{m}_a : \mathcal{I} \to \mathbb{R}$ that estimate the runtime of algorithm $a \in \mathcal{A}$ on problem instance $i \in \mathcal{I}$. More specifically, we seek to learn a function $\mathbb{R}^d \to \mathbb{R}$ that maps instance feature vectors to algorithm runtime estimates. Here, precise values of the target variable correspond to true runtimes, i.e., the exact time it took an algorithm to solve a problem instance. However, these values are normally not measured precisely, and there are various reasons for why runtimes might better be characterized in the form of intervals $[l, u] \subset \mathbb{R}$. For example, the actual runtime might have been slowed down due to another process running in parallel, so that only a rough estimate could be derived.

Here, we are specifically interested in imprecision due to censoring, namely due to terminating an algorithm upon exceeding the cutoff time C. As we only know a lower bound on the ground truth runtime, in this case, the observations are right-censored. This information can be modeled in terms of the imprecise observation $Y = (C, \infty)$. As a consequence, the OSL in (4) simplifies as follows[1]:

$$L^*(\hat{y}, Y) = \begin{cases} L(\hat{y}, y) & \text{if } Y = \{y\} \\ L(\hat{y}, C) & \text{if } Y = (C, \infty) \text{ and } \hat{y} < C \\ 0 & \text{if } Y = (C, \infty) \text{ and } \hat{y} \geq C \end{cases} , \tag{5}$$

where the first case corresponds to precise observations and the second case to underestimated right-censored data points. In the third case, the runtime is correctly predicted to exceed C, whence the loss is 0. In principle, the OSL L^* can be instantiated with any suitable regression loss $L : \mathcal{Y} \times \mathcal{Y} \to \mathbb{R}$. For this paper, we choose the commonly used L_2-loss, also known as the *least squares error* $L(\hat{y}, y) = L_2(\hat{y}, y) := (y - \hat{y})^2$. An illustration of the OSL instantiated with this loss function is given in Fig. 1.

In spite of its simplicity, the superset learning approach has several properties that make it appealing for dealing with right-censored data in the algorithm selection setting. Whereas the established iterative imputation scheme by Schmee and Hahn (SCHMEE&HAHN) [21] assumes that the censored samples

[1] Here, we make the assumptions that $L(y, y) = 0$ for all $y \in \mathcal{Y}$, $L(y, \cdot)$ is monotone decreasing on $(-\infty, y)$ and monotone increasing on (y, ∞), which hold for all reasonable loss functions.

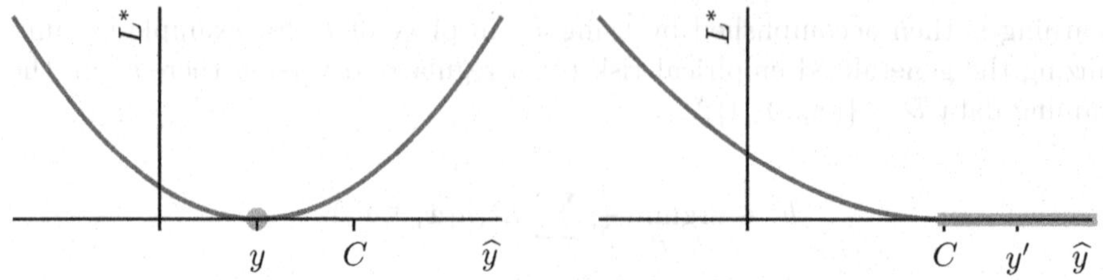

Fig. 1. Illustration of the OSL instantiated with the L_2-loss. The left example shows a precise observation $y \leq C$ for which we simply penalize the prediction with the squared distance to the actual observation. The right example shows a censored observation of a timed out algorithm run. As $y' > C$, the actual runtime y' is unknown and we only know the lower bound C. Thus, we have an imprecise observation $Y' = (C, \infty)$, for which we do not impose a penalty if the prediction $\hat{y}$ lies in the interval, i.e. if $\hat{y} \in Y'$. If $\hat{y} \notin Y'$, we penalize with the squared distance to the interval's lower bound C.

are drawn from a truncated normal distribution, the superset learning approach does not make any assumptions about the distribution of censored data. Another significant disadvantage of the iterative method is its computational cost. While SCHMEE&HAHN needs to iteratively refit a regression model multiple times until convergence (or a maximum number of iterations is reached), the superset learning method only needs a single pass for integrating censored samples into the regression model. Also intuitively, the approach is well-suited for the task of algorithm selection: To select a well-performing algorithm for a new problem instance, there is no need for precisely estimating the runtime of poorly performing algorithms. Still, to avoid selecting these algorithms, they need to be identified, which is reflected by enforcing predictions above the cutoff time C.

A toy example of precise and censored data points as well as the resulting linear regression models when using the L_2-loss and its OSL extension is given in Fig. 2. The example illustrates why imputing censored samples with a constant leads to strongly biased models. The superset learning approach circumvents this problem by treating the censored samples as imprecise (interval-valued) observations, for which the loss is 0 if the prediction lies in the interval.

In the scope of this paper, we consider the class of linear models for the algorithm-specific runtime surrogates $\widehat{m}_a$. We minimize the proposed optimistic superset loss as depicted in (5) using standard gradient-based optimization.

4 Experimental Evaluation

For conducting the experimental evaluation of the proposed superset learning approach to algorithm selection, we use the *ASlib* benchmark [2]. This benchmark library consists of AS *scenarios*, which are collections of problem instances from various algorithmic problem domains such as SAT or TSP, characterized

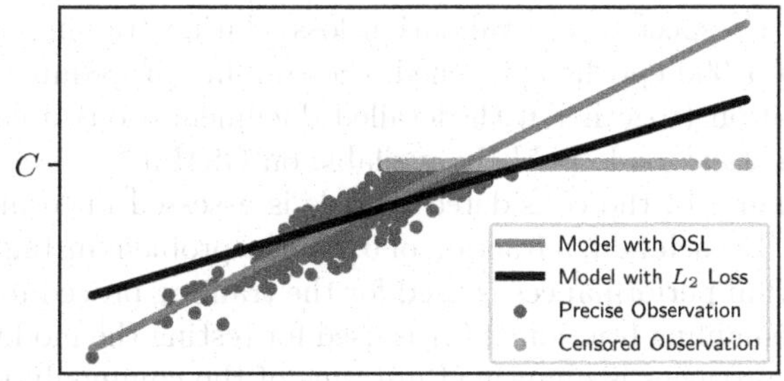

Fig. 2. Toy example of data with a linear dependency between feature and target and Gaussian centered noise. Observations above the cutoff C have been censored. The two lines indicate linear regression models, the green one has been trained using the OSL, the black one using the standard L_2-loss for which the censored observations were imputed with C. (Color figure online)

in terms of feature vectors as well as the performances achieved by all candidate algorithms on the respective problem instance. These scenarios entail significant amounts of censored algorithm runs; an overview of all scenarios is given in Table 1.

Table 1. Overview of the *ASlib* scenarios with the corresponding number of instances (#I), unsolved instances (#U), algorithms (#A), instance features (#F), the cutoff (C) as well as the fraction of censored algorithm runs (%C).

	ASP-POTASSCO	BNSL-2016	CPMP-2015	CSP-2010	CSP-MZN-2013	CSP-Minizinc-Time-2016	MAXSAT-PMS-2016	MAXSAT-WPMS-2016	MAXSAT12-PMS	MAXSAT15-PMS-INDU	MIP-2016	PROTEUS-2014	QBF-2011
#I	1294	1179	527	2024	4642	100	601	630	876	601	218	4021	1368
#U	82	0	0	253	944	17	45	89	129	44	0	456	314
#A	11	8	4	2	11	20	19	18	6	29	5	22	5
#F	138	86	22	86	155	95	37	37	37	37	143	198	46
C	600	7200	3600	5000	1800	1200	1800	1800	2100	1800	7200	3600	3600
%C	20.1	28.1	27.6	19.6	70.2	50	39.4	57.9	41.2	48.9	20	60.2	54.7

	QBF-2014	QBF-2016	SAT03-16_INDU	SAT11-HAND	SAT11-INDU	SAT11-RAND	SAT12-ALL	SAT12-HAND	SAT12-INDU	SAT12-RAND	SAT15-INDU	TSP-LION2015
#I	1254	825	2000	296	300	600	1614	767	1167	1362	300	3106
#U	241	55	269	77	47	108	20	229	209	322	17	0
#A	14	24	10	15	18	9	31	31	31	31	28	4
#F	46	46	483	115	115	115	115	115	115	115	54	122
C	900	1800	5000	5000	5000	5000	1200	1200	1200	1200	3600	3600
%C	55.6	36.2	24.7	60.7	33.3	47.5	53.9	66.7	50.4	73.5	23.5	9.6

4.1 Experimental Setup

In this work, we consider linear models for modeling the performance surrogate $\widehat{m}$. These models are trained using the *Adam* optimizer [14], minimizing the OSL instantiated with the L_2-loss. We chose a learning rate of 0.05 and a batch size of 32. For regularization, validation-based early stopping was employed: Before the training procedure starts, a fraction of 0.3 of the training data is put aside as validation data. Then, during training, this validation data is used to compute the validation loss periodically. If this loss increases for 16 consecutive checks, training is terminated and the model parameters are set to the best observed parameters (with respect to the validation loss) during training. After a maximum number of 5,000 epochs is reached, the training procedure is terminated. The implementation (together with detailed documentation) used for the conducted experiments is made publicly available on GitHub[2].

The performance of the considered models is assessed in terms of a 10-fold cross-validation. In each fold, a fraction of 0.9 of the problem instances and corresponding algorithm performances is used for the training procedure as described above and the remaining fraction of 0.1 is used for testing the model. The performance of the approaches is quantified in terms of the commonly used *penalized average runtime* with a penalty factor of 10 (PAR10). This measure simply averages the runtimes achieved by the selected algorithms over the problem instances in a scenario, while penalizing algorithm runs that timed out with a constant $10 \cdot C$.

4.2 Baseline Approaches

We compare the proposed algorithm selection technique based on superset learning with the commonly used strategies for dealing with censored samples. Using the same model configuration as described in the previous section, we consider the IGNORE strategy, in which we neglect training instances of censored runs as well as the naïve imputation schemes CLIP and PAR10, in which we impute the censored data points with the cutoff C or $10 \cdot C$, respectively, for training regression models using the L_2-loss. As a more elaborate solution, we compare our method against the iterative imputation scheme SCHMEE&HAHN [21].

4.3 Results

In the following, we discuss the results of the conducted experiments, which are summarized in Table 2. A • next to a baseline result indicates that the SUPERSET approach was significantly better, while a ∘ indicates that it was significantly worse on the corresponding scenario according to a Wilcoxon signed-rank test with $p = 0.05$. Per scenario, the best achieved result is highlighted in bold and the rank of each result is given in parentheses. We observe that the IGNORE strategy is the worst strategy among all considered approaches. As the *ASlib* scenarios include up to 73.5% of censored algorithm runs, this comes at no surprise, as large parts of the training data are simply discarded. The approaches CLIP and PAR10 that impute censored observations with constants yield better results than ignoring censored runs. Here, the approach CLIP that imputes censored runs with the exact algorithm cutoff yields better results than imputing with $10 \cdot C$ as done for PAR10.

Due to the computational expense of iteratively refitting the regression models, the majority of the experimental runs for the SCHMEE&HAHN baseline method did not manage to finish within a given time frame of 96 CPU days. While it is evident that the superset learning approach has a clear advantage with

[2] https://github.com/JonasHanselle/Superset_Learning_AS.

Table 2. PAR10 scores of the proposed approach as well as all considered baselines for all considered scenarios.

Scenario	CLIP	IGNORE	PAR10	SCHMEE&HAHN	SUPERSET
ASP-POTASSCO	**267.16** (1)	785.49 (5) •	285.82 (3)	312.05 (4) •	267.75 (2)
BNSL-2016	**2333.59** (1)	10221.65 (5) •	5829.51 (4) •	2350.71 (2)	2392.99 (3)
CPMP-2015	6223.74 (3)	11490.41 (5) •	6683.52 (4)	**5754.33** (1)	6020.67 (2)
CSP-2010	**723.25** (1)	989.05 (4)	1081.57 (5) •	814.91 (3)	726.05 (2)
CSP-MZN-2013	1384.45 (2)	12466.45 (5) •	1460.69 (3) •	1582.99 (4) •	**1337.03** (1)
CSP-Minizinc-Time-2016	2417.4 (2)	7416.88 (5) •	3314.75 (3)	3394.89 (4)	**2417.29** (1)
MAXSAT-PMS-2016	605.77 (3)	8793.23 (5) •	2474.75 (4) •	563.13 (2)	**542.93** (1)
MAXSAT-WPMS-2016	2391.88 (3) •	14633.54 (5) •	4173.15 (4) •	2124.11 (2)	**2100.98** (1)
MAXSAT12-PMS	961.97 (3) •	2141.06 (5) •	1313.61 (4)	866.79 (2)	**814.62** (1)
MAXSAT15-PMS-INDU	**905.35** (1)	16085.16 (5) •	2658.04 (4) •	1116.16 (3) •	930.83 (2)
MIP-2016	**9403.18** (2) ○	20182.02 (5) •	10097.0 (4)	9784.38 (3)	**9403.18** (1)
PROTEUS-2014	5198.77 (2)	20111.7 (5) •	5623.13 (4)	5508.62 (3)	**5095.73** (1)
QBF-2011	**3601.4** (1)	20402.16 (5) •	6343.31 (4) •	3886.31 (3)	3771.58 (2)
QBF-2014	1706.53 (4) •	4345.37 (5) •	1666.69 (3) •	1469.3 (2)	**1455.48** (1)
QBF-2016	2552.29 (2)	6431.72 (5) •	2997.55 (4)	2766.28 (3)	**2531.79** (1)
SAT03-16_INDU	**3528.8** (1) ○	7254.0 (5) •	3999.94 (4)	3836.84 (3)	3581.81 (2)
SAT11-HAND	11831.89 (3) ○	24563.68 (5) •	11960.84 (4)	**11399.28** (1)	11831.89 (2)
SAT11-INDU	**7107.54** (2) ○	12992.03 (5) •	8419.32 (4)	7326.67 (3)	**7107.54** (1)
SAT11-RAND	**3376.17** (1)	16395.14 (5) •	4193.79 (4)	3929.21 (3)	3380.65 (2)
SAT12-ALL	**1635.69** (1)	8439.57 (5) •	1643.47 (3)	2073.78 (4) •	1639.68 (2)
SAT12-HAND	2612.79 (4)	8046.59 (5) •	2587.83 (3)	2554.14 (2)	**2551.2** (1)
SAT12-INDU	1857.48 (2)	11204.39 (5) •	3411.23 (4) •	2123.33 (3) •	**1840.33** (1)
SAT12-RAND	**568.54** (1)	10576.49 (5) •	1075.56 (4) •	665.15 (3)	659.1 (2)
SAT15-INDU	6133.29 (2)	11637.43 (5) •	8575.57 (4)	7548.6 (3) •	**6126.67** (1)
TSP-LION2015	**485.22** (1)	9585.78 (5) •	743.14 (4) •	650.07 (3)	496.57 (2)
Avg. rank	1.96	4.96	3.80	2.76	**1.52**
Avg. PAR10	3192.57	11087.64	4104.55	3376.08	**3160.97**

respect to runtime over iterative imputation methods such as SCHMEE&HAHN, it also exhibits the strongest predictive accuracy for the task of algorithm selection in the conducted experiments[3].

5 Conclusion

In this paper, we proposed to consider algorithm selection within the framework of superset learning to handle censored data in a proper way. Instead of replacing the censored labels with more or less arbitrary precise numbers, or leaving out these data points entirely, the labels are considered as imprecise observations in the form of intervals (ranging from the cutoff time to infinity). Invoking a principle of generalized loss minimization, a regression model is then built to predict (precise) runtimes of algorithms for unseen problem instances.

[3] Note that for the TSP-LION2015 scenario, only 9 out of 10 folds were evaluated using the SCHMEE&HAHN imputation due to technical issues.

Our experimental evaluation reveals that working with these imprecisely annotated data, the generalization performance of the fitted models is significantly better compared to naïve strategies for dealing with censored data. Moreover, we showed superset learning to be competitive or superior to the imputation technique proposed in [21]. Instead of repeatedly rebuilding the model until convergence of the predicted values, each model needs to be fit only once. As the number of such models equals the number of algorithms to choose from, superset learning has a clear advantage over SCHMEE & HAHN in terms of runtime complexity.

Interesting directions for future work include integrating superset learning with a hybrid regression & ranking approach [8], or to consider the decision which algorithm to choose according to the predicted runtimes in terms of a stacked classifier instead of simply taking the algorithm with minimum predicted runtime [16]. Besides, as already said, superset learning is a quite versatile approach that allows for modeling imprecision in a flexible way. This might also be beneficial in the context of algorithm selection, where censoring is certainly not the only source of imprecision.

Acknowledgments. This work was supported by the German Federal Ministry of Economic Affairs and Energy (BMWi) within the "Innovationswettbewerb Künstliche Intelligenz" and the German Research Foundation (DFG) within the Collaborative Research Center "On-The-Fly Computing" (SFB 901/3 project no. 160364472). The authors also gratefully acknowledge support of this project through computing time provided by the Paderborn Center for Parallel Computing (PC^2).

References

1. Amadini, R., Gabbrielli, M., Mauro, J.: SUNNY: a lazy portfolio approach for constraint solving. Theor. Pract. Log. Prog. **14**(4–5), 509–524 (2014)
2. Bischl, B., et al.: ASlib: a benchmark library for algorithm selection. Artif. Intell. **237**, 41–58 (2016)
3. Cabannes, V., Rudi, A., Bach, F.: Structured prediction with partial labelling through the infimum loss. In: Proceedings of ICML, International Conference on Machine Learning (2020)
4. Eggensperger, K., Lindauer, M., Hoos, H.H., Hutter, F., Leyton-Brown, K.: Efficient benchmarking of algorithm configurators via model-based surrogates. Mach. Learn. **107**(1), 15–41 (2017). https://doi.org/10.1007/s10994-017-5683-z
5. Gagliolo, M., Legrand, C.: Algorithm survival analysis. In: Bartz-Beielstein, T., Chiarandini, M., Paquete, L., Preuss, M. (eds.) Experimental Methods for the Analysis of Optimization Algorithms, pp. 161–184. Springer, Heidelberg (2010). https://doi.org/10.1007/978-3-642-02538-9_7
6. Gagliolo, M., Schmidhuber, J.: Leaning dynamic algorithm portfolios. Ann. Math. Artif. Intell. **47**, 295–328 (2006)
7. Gomes, C.P., Selman, B., Crato, N.: Heavy-tailed distributions in combinatorial search. In: Smolka, G. (ed.) CP 1997. LNCS, vol. 1330, pp. 121–135. Springer, Heidelberg (1997). https://doi.org/10.1007/BFb0017434
8. Hanselle, J., Tornede, A., Wever, M., Hüllermeier, E.: Hybrid ranking and regression for algorithm selection. In: Schmid, U., Klügl, F., Wolter, D. (eds.) KI 2020. LNCS (LNAI), vol. 12325, pp. 59–72. Springer, Cham (2020). https://doi.org/10.1007/978-3-030-58285-2_5

9. Hüllermeier, E., Cheng, W.: Superset learning based on generalized loss minimization. In: Appice, A., Rodrigues, P.P., Santos Costa, V., Gama, J., Jorge, A., Soares, C. (eds.) ECML PKDD 2015. LNCS (LNAI), vol. 9285, pp. 260–275. Springer, Cham (2015). https://doi.org/10.1007/978-3-319-23525-7_16
10. Hutter, F., Hoos, H.H., Leyton-Brown, K.: Bayesian optimization with censored response data. In: NIPS Workshop on Bayesian Optimization, Sequential Experimental Design, and Bandits (2011)
11. Hüllermeier, E.: Learning from imprecise and fuzzy observations: data disambiguation through generalized loss minimization. Int. J. Approx. Reason. **55**(7), 1519–1534 (2014). Special issue: Harnessing the information contained in low-quality data sources
12. Kadioglu, S., Malitsky, Y., Sellmann, M., Tierney, K.: ISAC - instance-specific algorithm configuration. In: ECAI (2010)
13. Kerschke, P., Hoos, H.H., Neumann, F., Trautmann, H.: Automated algorithm selection: survey and perspectives. Evol. Comput. **27**(1), 3–45 (2019)
14. Kingma, D.P., Ba, J.: Adam: a method for stochastic optimization. In: 3rd International Conference on Learning Representations, ICLR 2015, San Diego, CA, USA, 7–9 May 2015. Conference Track Proceedings (2015)
15. Kleinbaum, D.G., Klein, M.: Survival Analysis. Survival Analysis, vol. 3. Springer, New York (2012). https://doi.org/10.1007/978-1-4419-6646-9
16. Kotthoff, L.: Hybrid regression-classification models for algorithm selection. In: ECAI, pp. 480–485 (2012)
17. Lindauer, M., Bergdoll, R.-D., Hutter, F.: An empirical study of per-instance algorithm scheduling. In: Festa, P., Sellmann, M., Vanschoren, J. (eds.) LION 2016. LNCS, vol. 10079, pp. 253–259. Springer, Cham (2016). https://doi.org/10.1007/978-3-319-50349-3_20
18. Lobjois, L., Lemaître, M., et al.: Branch and bound algorithm selection by performance prediction. In: AAAI/IAAI, pp. 353–358 (1998)
19. Pihera, J., Musliu, N.: Application of machine learning to algorithm selection for TSP. In: 26th IEEE International Conference on Tools with Artificial Intelligence, ICTAI 2014, Limassol, Cyprus, 10–12 November 2014, pp. 47–54. IEEE Computer Society (2014)
20. Rice, J.R.: The algorithm selection problem. Adv. Comput. **15**, 65–118 (1976)
21. Schmee, J., Hahn, G.J.: A simple method for regression analysis with censored data. Technometrics **21**(4), 417–432 (1979)
22. Tornede, A., Wever, M., Hüllermeier, E.: Extreme algorithm selection with dyadic feature representation. In: Appice, A., Tsoumakas, G., Manolopoulos, Y., Matwin, S. (eds.) DS 2020. LNCS (LNAI), vol. 12323, pp. 309–324. Springer, Cham (2020). https://doi.org/10.1007/978-3-030-61527-7_21
23. Tornede, A., Wever, M., Hüllermeier, E.: Towards meta-algorithm selection. In: Workshop on Meta-Learning (MetaLearn 2020) @ NeurIPS 2020 (2020)
24. Tornede, A., Wever, M., Werner, S., Mohr, F., Hüllermeier, E.: Run2survive: a decision-theoretic approach to algorithm selection based on survival analysis. In: Asian Conference on Machine Learning, pp. 737–752. PMLR (2020)
25. Xu, L., Hutter, F., Hoos, H., Leyton-Brown, K.: Hydra-mip: automated algorithm configuration and selection for mixed integer programming. In: RCRA Workshop @ IJCAI (2011)
26. Xu, L., Hutter, F., Hoos, H.H., Leyton-Brown, K.: The design and analysis of an algorithm portfolio for SAT. In: Bessière, C. (ed.) CP 2007. LNCS, vol. 4741, pp. 712–727. Springer, Heidelberg (2007). https://doi.org/10.1007/978-3-540-74970-7_50

Sim2Real for Metagenomes: Accelerating Animal Diagnostics with Adversarial Co-training

Vineela Indla, Vennela Indla, Sai Narayanan, Akhilesh Ramachandran(✉), Arunkumar Bagavathi, Vishalini Laguduva Ramnath, and Sathyanarayanan N. Aakur(✉)

Oklahoma State University, Stillwater, OK 74048, USA
{rakhile,saakurn}@okstate.edu

Abstract. Machine learning models have made great strides in many fields of research, including bioinformatics and animal diagnostics. Recently, attention has shifted to detecting pathogens from genome sequences for disease diagnostics with computational models. While there has been tremendous progress, it has primarily been driven by large amounts of annotated data, which is expensive and hard to obtain. Hence, there is a need to develop models that can leverage low-cost, *synthetic* genome sequences to help tackle complex metagenome classification problems for diagnostics. In this paper, we present one of the first *sim2real* approaches to help multi-task deep learning models learn robust feature representations from synthetic metagenome sequences and transfer the learning to predict pathogen sequences in real data. Extensive experiments show that our model can successfully leverage synthetic and real genome sequences to obtain 80% accuracy on metagenome sequence classification. Additionally, we show that our proposed model obtains *76%* accuracy, with limited real metagenome sequences.

Keywords: Sim2Real · Adversarial co-training · Pathogen detection

1 Introduction

Machine learning has driven great advancements in processing genome sequences for pathogenic potential prediction [4], disease gene prediction [10] and genome assembly [15]. Given such rapid and encouraging progress, research has recently shifted focus towards genome sequence classification [4,20] for detecting and diagnosing infectious diseases in humans and animals from clinical metagenome data. Traditional approaches for metagenome analysis are largely driven by highly skilled personnel and specialized bioinformatics pipelines to collect and analyze enormous amounts of DNA sequences (Metagenomes) collected from clinical specimens such as tissue biopsies. Even though highly accurate, traditional methods are time-consuming and can be very expensive.

K. Karlapalem et al. (Eds.): PAKDD 2021, LNAI 12712, pp. 164–175, 2021.
https://doi.org/10.1007/978-3-030-75762-5_14

Machine learning-based frameworks [4,20] have been used to accelerate this process by creating efficient pipelines to classify genome sequence reads. However, success has mostly been driven by supervised machine learning models, which require large amounts of real-world annotated data to learn existing patterns and generalize assumptions to make accurate predictions. Large amounts of annotated data can be hard to obtain and are expensive to annotate. Additionally, metagenome data can contain sensitive biological information and bring security and privacy issues into consideration. *Synthetic data*, on the other hand, offers an exciting alternative to annotated, real-world metagenome sequences due to its low-cost construction and richness in labels. Hence, there is a need for developing frameworks that can learn useful patterns from low-cost and easily accessible simulated data.

In this work, we tackle the problem of metagenome sequence classification of real-world clinical sequences using the idea of *sim2real* transfer, a promising research direction that leverages synthetic data to learn robust representations for real-world applications. Specifically, we focus on classifying metagenome sequences to diagnose the Bovine Respiratory Disease Complex (BRDC) - a multi-etiologic disease caused by multiple bacterial and viral agents. We analyze the viability of using synthetic genome sequences to learn efficient feature representations that can help classify real metagenome sequences from clinical experiments by navigating various challenges such as noise, measurement errors, and intra-class variation. We propose a novel adversarial noise-based co-training approach for deep learning models trained in a multi-task learning setting. While adversarial noise-based learning [6] has focused mainly on increasing model robustness to synthetic noise, we use adversarial co-training to learn efficient representations from synthetic data for metagenome sequence classification.

The **contributions** of our work are three-fold: (i) we introduce the first *sim2real* framework for metagenome sequence classification for single pathogen detection, (ii) present a novel adversarial noise-based co-training procedure for deep neural networks to learn efficient feature representations that transfer to both simulated and real metagenome sequences, and (iii) perform extensive experiments over multiple features and baselines to help assess the viability of using synthetic data for complex diagnostic tasks using both traditional machine learning and deep learning approaches.

2 Related Work

ML for Metagenome Classification: Machine learning based metagenome classification with traditional k-mer frequencies are effective in several problems like chromatin accessibility prediction [18], bacteria detection [5], and pathogen prediction [4]. Higher-order embeddings learned with deep learning and machine learning models have also shown great success in metagenome sequence prediction and classification by capturing simple nucleotide representations [4], hierarchical feature representations [21], long-term dependencies in metagenome sequences [18]. Graph-based representation for genomics research is also prominent in genomic sub-compartment prediction [2], disease gene predictions [10],

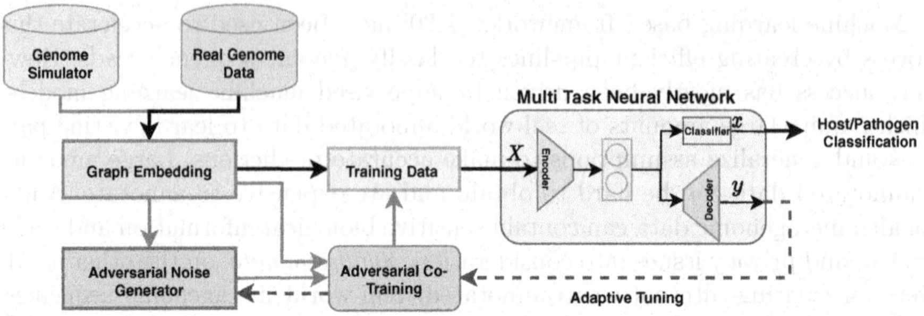

Fig. 1. Overall architecture for the proposed *sim2real* approach: a novel sim2real framework that trains a multi-task neural network with adversarial noise using both synthetic and real genome sequences for host/pathogen classification.

genome assembly [15]. In this study, we use representations learned from a special type of genome graph called a *De Bruijn graph* [15] and use deep learning to extract robust metagenome representations. **Sim2Real:** With growing trends in deep learning and easy accessibility of simulated data, *sim2real* approaches have been explored in natural language processing (NLP) [17], robotics [23], computer vision [14,16], and image processing [8]. Despite their promise, *sim2real* approaches have a disadvantage of not able to transfer the simulated learning exactly to the actual real-world setup [3], which state-of-the-art methods are trying to overcome and close the gap [11] in all domains. The existing deep learning models [4] for metagenome classification does not handle unknown real data in making predictions. To overcome these limitations, we propose an adversarial co-training-based multi-task learning framework to help transfer simulated metagenome learning to the real-world setup.

3 Sim2Real Metagenome Sequence Classification

We introduce a *sim2real* approach, illustrated in Fig. 1, for robust metagenome representation learning. In this section, we discuss i) A graph-based method to map raw genome sequences to high-dimensional features, ii) Issues of learning from synthetic genome sequences, and iii) Our novel adversarial co-training multi-task deep neural network architecture.

3.1 Feature Representation Using De Bruijn Graphs

We first construct a feature representation from raw genome sequences by representing the nucleotide base-pair structure as a series of characters. Each character ($\{A, T, C, G\}$) represents a nucleotide base in the genome of the subject (*bovine lung sample*, in our case). From this sequence, we construct a De Bruijn graph - a directed graph structure commonly used in bioinformatics for *de novo* assembly of sequences into a whole genome [15]. The constructed De Bruijn graph's

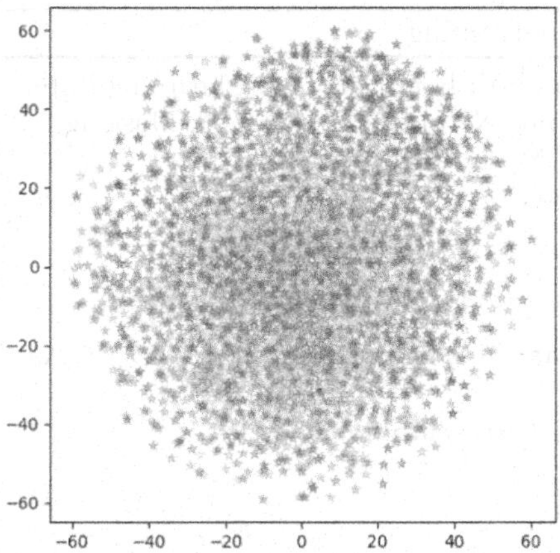

(a) 3-*mer* De Bruijn graph embeddings

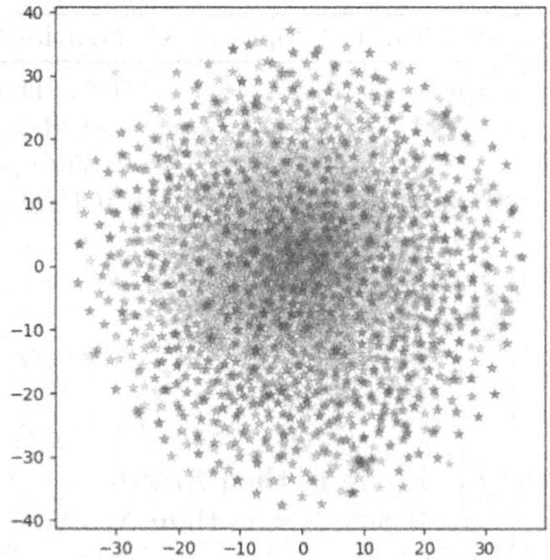

(b) 6-*mer* De Bruijn graph embeddings

Fig. 2. t-SNE visualization of *node2vec* embeddings. Host samples are indicated in green and red for synthetic and real data, respectively. Pathogen sequences are visualized in yellow and **blue** for synthetic and real data, respectively. (Color figure online)

structure captures the overlap between the sequential occurrence of nucleotide base-pairs in the genome sequence, with each node representing a k-*mer*. Each k-*mer* is a subsequence of length k constructed from the genome sequence. It helps break each sequence read into smaller chunks, and they capture complex structures beyond simple base-pair interactions. The number of nodes in the De Bruijn graph ($\mathcal{G} = (\mathcal{V}, \mathcal{E})$) is 4^k, conditioned on the parameter k. Hence larger values of k increase the number of nodes while reducing the number of edges, leading to more sparse graphs. Hence the structure of the De Bruijn graph can vary greatly between species. If the given subject is made of g genome sequences, we construct g De Bruijn graphs, one from each sequence. To exploit the local neighborhood structure, we use *node2vec* [7] to construct a continuous-valued feature embedding from each constructed De Bruijn graphs. This intuition was empirically evaluated in prior works [20], where *node2vec* outperformed algorithms that captured the global structural properties such as *graph2vec* [19] and various graph kernels such as random walks [12].

3.2 The Problem with Using *Pure* Synthetic Data

While *node2vec* representations performed well on simulated genome sequences [20], there are some challenges in using them to augment training data for classifying *real* genome sequences captured by sequencing technologies such as MinION or Illumina. First, real-world sequencing approaches can have errors in reading nucleotides from the collected samples, significantly altering the constructed graph structure. Some common modes of errors include substitution, insertion, and deletion errors, which can alter the constructed graph structure. This effect is particularly acute in low-cost genome sequencing models such as MinION

Algorithm 1. Proposed Adversarial Co-Training

Input : Synthetic Data $\tilde{X}, \tilde{Y}$; Real Data $\hat{X}, \hat{Y}$; Learning Rate η; Mask Rate: m_r; Network Parameters θ; Masking Probability p_1; Noise Rate: p_2

Output : Learnt Parameters θ

Randomly initialize θ; $m = m_r$

repeat

  for $\tilde{X}_i \in \tilde{X}$ **do**

    $M_i = \tilde{X}_i^{\setminus u:v}$ `// Mask` m `random indices` $(u : v)$ `of` $\tilde{X}_i$

    $N_i \sim N(\mu_{\tilde{X}}, \sigma_{\tilde{X}})$

    $t_1 \leftarrow UniformSample(0, 1)$; $t_2 \leftarrow UniformSample(0, 1)$

    if $t_1 < p_1$ **then** $\tilde{X}_i = M_i$;

    if $p_1 < t_2 < p_2$ **then** $\tilde{X}_i = \tilde{X}_i + N_i$;

  end

  $X = \{\tilde{X}, \hat{X}\}$; $Y = \{\tilde{Y}, \hat{Y}\}$

  Forward Propagation: $\{x, y\} = f(X; \theta)$

  Compute Loss : $\mathcal{L} = \lambda_1 \mathcal{L}_{CE}(y, Y) + \lambda_2 \mathcal{L}_{recons}(x, X)$

  Back Propagation : $\theta = \theta - \eta \nabla_\theta$

  if *AdaptiveLearning*($\mathcal{L}_{recons}$) **then**

    $\eta = \eta/10; m = m_r + 0.1; p_2 = p_2 + 0.1; p_1 = p_1 + 0.1;$

  else $m = m_r; p_1 = 0; p_2 = 0;$

until *model converges*;

(our choice of sequencing), where error rates can be as high as 38.2% [13]. Second, simulated host sequences are *pure*, i.e., there are only genome sequence samples from the host and the target pathogen. However, in real-world sequences, there can exist random concentrations of non-host, pathogen, and non-pathogen sequences found in real-world samples. Hence, using synthetic data *as is*, can result in graph embeddings that are more homogeneous and not reflective of the real genome sequences found "in the wild".

We experimentally verify these challenges by visualizing the initial embeddings constructed from De Bruijn graphs from simulated and real genome sequences. Figure 2 shows the *node2vec* embeddings of both simulated and real data for (a) 3-*mer* and (b) 6-*mer* configurations. As can be seen, the simulated data (in green and yellow) are more homogeneous (closely populated), while the real genome data (in red and blue) has more variation. As k increases, this effect is less pronounced and allows for better transfer between simulated and real data. Empirically, in Sect. 4.2, we can see that as k increases, the gap between models trained with and without synthetic data increases.

3.3 Adversarial Co-training: A Deep Learning Baseline

To overcome these challenges, we propose the use of adversarial co-training to enable deep learning models to use simulated genome sequences effectively. Traditional adversarial training makes a deep neural network more robust to small perturbations in the input image. On the other hand, we use the idea of adversarial noise to help make synthetic data more realistic to mimic the variations found in real-world genome sequences (see Sect. 3.2). We use the term adversarial *co-training* to refer to the process of generating noisy synthetic data using adversarial noise functions. Inspired by the success of multi-task learning for metagenome sequence classification [20], we use a deep learning model with two output networks - one for classification and one for the reconstruction of the input feature vectors. A notable difference is that we use undersampling to ensure a balanced split of classes and pre-train with synthetic data *with* the proposed adversarial training before co-training with a fraction of real genome sequences. The final training objective is the sum of the classification loss ($\mathcal{L}_{CE}$: unweighted cross-entropy) and the reconstruction loss ($\mathcal{L}_{recons}$) and is given by

$$\mathcal{L} = \lambda_1 \mathcal{L}_{CE}(y, Y) + \lambda_2 \sum_{i=1}^{n} \|X - x\|_{\ell_1}^2 \tag{1}$$

where n is the dimension of the input embedding and x, y refer to the multi-task neural network predictions. X and Y are the input data generated through the adversarial co-training process. The reconstruction loss $\mathcal{L}_{recons}$ (the second term) is defined as the sum-squared error between the reconstructed features and the actual unperturbed input features. Note that although the objective function is similar to [20], there are 2 key differences: (i) the inputs are perturbed using adversarial noise, and (ii) we use a sum-squared error rather than the L-2 norm since the goal is not perfect reconstruction, but to detect and remove noise.

We generate the noise-perturbed input feature vectors using the adversarial co-training process outlined in Algorithm 1. The algorithm inputs are a neural network function f with parameters θ; a mixture of real and synthetic data, an initial learning rate η and default adversarial noise parameters such as masking rate, masking probability, and noise rate. We use a mix of both *additive* and *multiplicative* functions to augment the pure synthetic genome sequences with adversarial noise. The former refers to the addition of a noise vector to the input features and is the most common form of adversarial noise explored in current literature. The latter refers to the *multiplication* of crafted noise vectors with the input data to form a feature vector with almost imperceptible noise. The idea of *masking* is used to introduce the multiplicative noise into the feature vectors, i.e., a random portion of the input vector is set to an arbitrary value (0 in our case). The additive noise is used to mimic the substitution errors caused by real-world genome sequence reads, and the multiplicative noise is used to model the insertion and deletion errors.

We use an adaptive learning mechanism, inspired by Aakur *et al.* [1], to control the level of masking and addition of noise to the input features to ensure that the learning process converges. Specifically, we monitor the reconstruction loss $\mathcal{L}_{recons}$ using a low pass filter, whose threshold is the running average of the reconstruction loss. As the reconstruction loss begins to increase, i.e., the model is not able to successfully separate the noise and original signal, the learning rate is reduced. The unperturbed features are presented in the next training iteration.

Table 1. Dataset characteristics (number of metagenome sequences) used for training and evaluation of the proposed approach. Real-world host sequences are undersampled such that the ratio of host-pathogen sequences is approximately 3:2.

Data split	Synthetic data		Real data		
	Host	Pathogen	# of specimens	Host	Pathogen
Training	1000	1000	7	2100	1494
Validation	500	500	2	225	102
Test	–	–	4	400	251

If the reconstruction loss is low, then the amount of noise added increases by increasing the probability of adversarial noise. This training process ensures that the model learns relevant features from the perturbed input vectors and does not overfit the training data. Combined with the multi-task learning objective, the adversarial co-training adds regularization to the network and ensures that representations learned from both synthetic and real data are robust.

3.4 Implementation Details

Network Structure: The multi-task deep learning model has three components: an encoder network, a decoder (reconstruction) network, and a classifier network. The encoder network has five (5) dense layers, intersperses with a dropout layer. The initial dense layer is set to have 256 neurons, with each subsequent layer's capacity increasing by 0.5%. The dropout probability is set to 0.5. The encoded feature dimension is set to 4096. The decoder network (for reconstruction) consists of three (3) dense layers to rescale the encoded features back to the input vector's dimension. The classification network uses a dense layer to produce the metagenome classification. We set the initial learning rate to be 4×10^{-5}, and we first perform a *cold start*, i.e., for five epochs, the learning rate is set to be 4×10^{-12} and then increased. The masking probability m_r is set to be 30%. The masking probability (p_1) and noise rate (p_2) are set to be 0.5 and 0.75, respectively. Finally, the modulating parameters λ_1 and λ_2 (from Eq. 1) are set to be 1 and 5, respectively. All hyperparameters are found using a grid search and kept constant for all experiments.

4 Experimental Evaluation

4.1 Data and Baselines

We use a mixture of simulated and real genome sequences to train and evaluate our approach. Table 1 presents the statistics of simulated and real datasets used in all our experiments.

Synthetic Data. Inspired by prior work from Narayanan *et al.* [20], we use the ART simulator [9] to generate synthetic genome data. The NCBI nucleotide

Table 2. Quantitative evaluation using both simulated and varying amounts of real training data on two k-mer configurations. All experiments use real data for evaluation.

Training data	Accuracy (%)									
	3-mer					6-mer				
	LR	SVM	NN	DL	Ours	LR	SVM	NN	DL	Ours
Synthetic Only	61.14	60.52	61.60	61.44	**62.21**	41.94	48.54	50.23	47.62	**61.59**
Real Only	68.48	69.46	65.84	**69.74**	69.59	**80.00**	74.38	76.41	78.49	79.41
Synthetic + Real (5%)	48.54	39.11	38.56	60.81	**69.28**	47.46	53.54	51.98	67.12	**70.81**
Synthetic + Real (25%)	63.01	51.40	38.56	65.44	**71.43**	56.65	57.29	57.45	70.05	**76.77**
Synthetic + Real (50%)	63.90	54.99	40.52	66.51	**70.20**	58.64	60.79	61.47	74.81	**78.49**
Synthetic + Real (100%)	64.54	67.83	65.49	70.05	**71.74**	67.86	66.94	65.09	79.42	**80.03**

database [24] is used to obtain reference genomes for both the host (bovine) and pathogen (*Mannheimia haemolytica*). The sequence output from the ART simulator is simulated for the host genome with a minimum Q-score (Quality Score) of 28 and a maximum Q score of 35, and varying quantities of the pathogen genome reads were added to the generated bovine genome. The resulting genome sequences are used to simulate metagenome datasets from bovine lung samples.

Real-World Metagenome Data. To obtain real-world genome data, we extracted metagenome sequences from thirteen BRDC suspect lung specimens at the Oklahoma Animal Disease Diagnostic Laboratory. DNA sequences are extracted from lung samples using the DNeasy Blood and Tissue Kit. Sequencing libraries are prepared from the extracted DNA using the Ligation Sequencing Kit and the Rapid Barcoding Kit. Prepared libraries are sequenced using MinION (R9.4 Flow cells), and sequences with an average Q-score of more than seven are used in the final genome. Pathogens in the metagenome sequence data are identified using a modified version of the bioinformatics pipeline [25].

Baselines. In addition to our proposed approach in Sect. 3.3, we train and evaluate several traditional machine learning and deep learning baselines. The traditional machine learning baselines include Support Vector Machines (SVM), Logistic Regression (LR), and feed-forward neural network (NN) with two hidden layers with 512 neurons each. We also compare against the multi-task learning-based approach (DL) [20] as a deep learning baseline. All baselines were trained with both synthetic and real data, *without adversarial co-training.*

4.2 Quantitative Evaluation

We quantitatively evaluate all models on two commonly used k-*mer* configurations for De Bruijn graph construction and summarize the results in Table 2. It can be seen that all models trained purely on synthetic data, *without any labeled real metagenome sequences*, transferred well to the real data. The proposed approach obtains a best performance of 62.2% on the 3-*mer* configuration and 61.59% on the 6-*mer* configuration. Other baselines do not transfer well

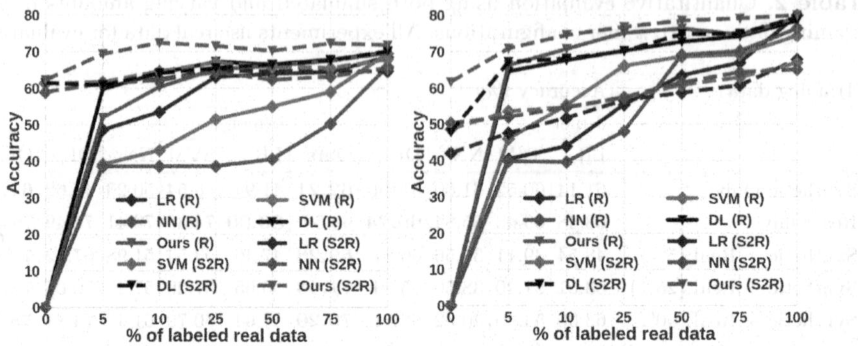

(a) 3-mer features (b) 6-mer features

Fig. 3. Performance of models trained with various degree of labeled real data (S2R) with models trained with pure real data (R), in (a) 3-mer and (b) 6-mer configurations.

to the real-world genome sequences while obtaining significantly better performance when trained purely on real-world sequences. These results back up our intuition that *pure* synthetic data does not quite model real-world characteristics (Sect. 3.2) and hence requires a smarter treatment. Using the adversarial co-training (ours) approach improves the performance significantly, with complete simulated and real data for training outperforms models trained with pure real data (∼2% gain for 3-mer and ∼0.03% gain for 6-mer). But our goal is to use limited real data for model training to set a benchmark using the *sim2real* approach. As given in Table 2, our proposed *sim2real* approach with complete synthetic data along with 50% of random real data training obtains performance within ∼ ±1% change compared to best a model trained with pure real data and within ∼ ±2% compared to the model trained with 100% synthetic and real data.

Learning with Limited Data. We further evaluate the effectiveness of using machine learning algorithms for genome classification with limited labeled data. We add varied amount of real-world metagenome sequences available for training to the complete synthetic data and present results of both 3-*mer* and 6-*mer* configurations in Fig. 3. It can be seen that for both configurations, the lack of training data (both synthetic and real) greatly impacts the performance of all models and that this effect is more pronounced in the 6-*mer* configuration. Significantly, it can be seen that we outperform all baselines on both configurations with only 5% of labeled real data to outperform *all baselines*.

Robustness to k-mer Configuration. Given the drastic difference in classification performance of machine learning models between the 3-*mer* and 6-*mer* configurations, we evaluate the robustness of the machine learning baselines for multiple k-*mer* configurations by varying the value of k. Results are summarized in Fig. 4, where we construct node2vec embedding for both synthetic and real data for different k-*mer* configurations obtained by varying k from 2 to 10. It can be seen that our approach generally performs well across all configurations

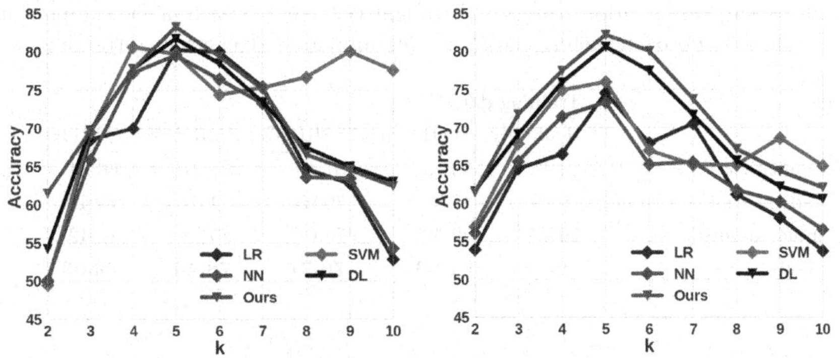

(a) Training with only real data (b) Training with synthetic and real data

Fig. 4. Robustness evaluation of all (a) models trained with only real data and models trained with both synthetic and real data when different k-mer configurations are used.

in the presence of synthetic data, which was the intended training environment. Other baselines do not quite perform consistently across different configurations, except for the SVM baseline, which has the most consistent performance of all machine learning models for varying k-*mer* configurations. An interesting point to note is that our approach gains as much as 82% classification accuracy for 5-*mer* configuration, a significant jump from the two standard configurations ($k \in \{3, 6\}$). This could be attributed to the fact that tetranucleotide biases from phylogenetically similar species are very similar between closely related species [22]. Given that the host and pathogen are drastically different, the classification performance is greater and provides a promising future direction of research to detect multiple pathogens.

Ablation Studies. We perform ablation studies to evaluate the different aspects of the system to ascertain each component's contribution. We conduct all evaluations on the 6-*mer* configuration and summarize results in Table 3. It can be seen that multi-task learning provides a significant jump. Still, the most significant factor is the use of adversarial co-training of multi-task learning models. We also experiment with different sources of noise, such as from a uniform distribution, but the gains are minimal compared to the use of Gaussian noise. These results indicate that the use of adversarial co-training provides a significantly more efficient way to use synthetic data compared to a naïve usage and hence offers an exciting avenue of research to improve machine learning-driven genome sequence classification.

Table 3. Results on ablation studies to highlight the effect of each of the components in the proposed approach such as multi-task learning and adversarial co-training.

Model	Training data				
	Real only	Synth. only	S2R(25%)	S2R(50%)	S2R(100%)
W/o multitask	76.41	50.23	57.45	61.47	66.94
W/o adversarial co-training	78.49	47.62	70.05	74.81	79.42
W/ uniform noise	76.47	59.32	74.04	75.27	75.42
Full model	**79.41**	**61.59**	**76.77**	**78.49**	**80.03**

5 Discussion and Future Work

In this work, we presented the first adversarial co-training approach to perform efficient *sim2real* learning to leverage simulated synthetic data for genome sequence classification. Extensive experiments on synthetic and real genome sequence data show that the proposed approach can learn robust features for effective genome sequence classification. We hope that this work inspires the exploration of low-cost, synthetic data to train and use machine learning models for metagenome classification. We aim to explore this approach further to extend this pipeline to detect multiple pathogens in real genome sequences to perform end-to-end animal diagnosis *with limited or no clinical data for training.*

Acknowledgement. This research was supported in part by the US Department of Agriculture (USDA) grants AP20VSD and B000C011.

References

1. Aakur, S.N., Sarkar, S.: A perceptual prediction framework for self supervised event segmentation. In: Proceedings of the IEEE Conference on Computer Vision and Pattern Recognition, pp. 1197–1206 (2019)
2. Ashoor, H., et al.: Graph embedding and unsupervised learning predict genomic sub-compartments from hic chromatin interaction data. Nat. Commun. **11**(1), 1–11 (2020)
3. Baker, B., et al.: Emergent tool use from multi-agent autocurricula. arXiv preprint arXiv:1909.07528 (2019)
4. Bartoszewicz, J.M., Seidel, A., Rentzsch, R., Renard, B.Y.: DeePaC: predicting pathogenic potential of novel DNA with reverse-complement neural networks. Bioinformatics **36**(1), 81–89 (2020)
5. Fiannaca, A., et al.: Deep learning models for bacteria taxonomic classification of metagenomic data. BMC Bioinform. **19**(7), 198 (2018)
6. Goodfellow, I.J., Shlens, J., Szegedy, C.: Explaining and harnessing adversarial examples. arXiv preprint arXiv:1412.6572 (2014)
7. Grover, A., Leskovec, J.: node2vec: scalable feature learning for networks. In: Proceedings of the 22nd ACM SIGKDD International Conference on Knowledge Discovery and Data Mining, pp. 855–864 (2016)

8. He, K., Zhang, X., Ren, S., Sun, J.: Deep residual learning for image recognition. In: Proceedings of the IEEE Conference on Computer Vision and Pattern Recognition, pp. 770–778 (2016)
9. Huang, W., Li, L., Myers, J.R., Marth, G.T.: Art: a next-generation sequencing read simulator. Bioinformatics **28**(4), 593–594 (2012)
10. Hwang, S., Kim, C.Y., Yang, S., Kim, E., Hart, T., Marcotte, E.M., Lee, I.: Humannet v2: human gene networks for disease research. Nucleic Acids Res. **47**(D1), D573–D580 (2019)
11. Kadian, A., et al.: Sim2real predictivity: does evaluation in simulation predict real-world performance? IEEE Robot. Autom. Lett. **5**(4), 6670–6677 (2020)
12. Kang, U., Tong, H., Sun, J.: Fast random walk graph kernel. In: Proceedings of the 2012 SIAM International Conference on Data Mining, pp. 828–838. SIAM (2012)
13. Laver, T., et al.: Assessing the performance of the oxford nanopore technologies minion. Biomol. Detect. Quantif. **3**, 1–8 (2015)
14. Li, X., et al.: Online adaptation for consistent mesh reconstruction in the wild. In: Advances in Neural Information Processing Systems, 33 (2020)
15. Lin, Y., Yuan, J., Kolmogorov, M., Shen, M.W., Chaisson, M., Pevzner, P.A.: Assembly of long error-prone reads using de Bruijn graphs. Proc. Nat. Acad. Sci. **113**(52), E8396–E8405 (2016)
16. Lu, J., Yang, J., Batra, D., Parikh, D.: Hierarchical question-image co-attention for visual question answering. In: Advances in Neural Information Processing Systems, 29, pp. 289–297 (2016)
17. Marzoev, A., Madden, S., Kaashoek, M.F., Cafarella, M., Andreas, J.: Unnatural language processing: bridging the gap between synthetic and natural language data. arXiv preprint arXiv:2004.13645 (2020)
18. Min, X., Zeng, W., Chen, N., Chen, T., Jiang, R.: Chromatin accessibility prediction via convolutional long short-term memory networks with k-mer embedding. Bioinform. **33**(14), i92–i101 (2017)
19. Narayanan, A., Chandramohan, M., Venkatesan, R., Chen, L., Liu, Y., Jaiswal, S.: graph2vec: learning distributed representations of graphs. arXiv preprint arXiv:1707.05005 (2017)
20. Narayanan, S., Ramachandran, A., Aakur, S.N., Bagavathi, A.: Genome sequence classification for animal diagnostics with graph representations and deep neural networks. arXiv preprint arXiv:2007.12791 (2020)
21. Nguyen, T.H., Chevaleyre, Y., Prifti, E., Sokolovska, N., Zucker, J.D.: Deep learning for metagenomic data: using 2D embeddings and convolutional neural networks. arXiv preprint arXiv:1712.00244 (2017)
22. Perry, S.C., Beiko, R.G.: Distinguishing microbial genome fragments based on their composition: evolutionary and comparative genomic perspectives. Genome Biol. Evol. **2**, 117–131 (2010)
23. Sadeghi, F., Toshev, A., Jang, E., Levine, S.: Sim2Real viewpoint invariant visual servoing by recurrent control. In: Proceedings of the IEEE Conference on Computer Vision and Pattern Recognition (CVPR), (June 2018)
24. Sherry, S.T., et al.: dbSNP: the NCBI database of genetic variation. Nucleic Acids Res. **29**(1), 308–311 (2001)
25. Stobbe, A.H., et al.: E-probe Diagnostic Nucleic acid Analysis (edna): a theoretical approach for handling of next generation sequencing data for diagnostics. J. Microbiol. Methods **94**(3), 356–366 (2013)

Attack Is the Best Defense: A Multi-Mode Poisoning PUF Against Machine Learning Attacks

Chia-Chih Lin(✉) and Ming-Syan Chen

National Taiwan University, Taipei, Taiwan
cclin@arbor.ee.ntu.edu.tw, mschen@ntu.edu.tw

Abstract. Resistance to modeling attacks is an important issue for Physical Unclonable Functions (PUFs). Deep learning, the state-of-the-art modeling attack, has recently been shown to be able to break many newly developed PUFs. Since then, many more complex PUF structures or challenge obfuscations have been proposed to resist deep learning attacks. However, the proposed methods typically focus on increasing the nonlinearity of PUF structure and challenge-response mapping. In this paper, we explore another direction with a multi-mode poisoning approach for a classic PUF (MMP PUF) in which each working mode is a simple add-on function for a classic PUF. By dividing the original challenge space for each working mode, the proposed MMP PUF generates a multi-modal challenge-response dataset that poisons machine learning algorithms. To validate the idea, we design two working mode types, challenge shift and response flip, as examples with widely-used delay-based Arbiter PUF. Experimental results show that our approach respectively achieves 74.37%, 68.08%, and 50.09% accuracy for dual-mode shift, quad-mode circular shift and dual-mode flip with deep learning models trained on over 3 million challenge-response pairs.

Keywords: Physical Unclonable Function (PUF) · Modeling attacks · Machine learning · Deep neural network · Poisoning attack

1 Introduction

The physical unclonable function (PUF) is essentially a circuit that uses intrinsic random physical characteristics to provide unique tags for each single chip. During the authentication process, PUF receives a sequence of bits, also known as challenges, from a server and generates a sequence of bits, so-called responses, to be transmitted to the server. A challenge and a corresponding response generated by a PUF is called a challenges-response pair (CRP). PUF has been considered as the key technology for the hardware security, especially for various applications of embedded systems [2]. Replicating PUF instances is hard for attackers because the randomness introduced by the manufacturing process is generally unpredictable. Despite the promise of being unclonable, the majority

K. Karlapalem et al. (Eds.): PAKDD 2021, LNAI 12712, pp. 176–187, 2021.
https://doi.org/10.1007/978-3-030-75762-5_15

of PUFs are vulnerable to machine learning (ML) based modeling attacks [15]. Moreover, with recent rapid advances in deep learning techniques, deep learning (DL) attacks have become a new powerful threat that pushes the boundary of modeling attacks to a new level [7,17].

This study explores another adversarial direction and proposes a novel PUF design paradigm which combines multiple working modes with an existing PUF instance, as shown in Fig. 1. We refer to the proposed Multi-Mode Poisoning PUF as MMP PUF. A working mode is an add-on function working with a classic PUF instance. It can be a challenge obfuscation function that transforms the challenge before it is applied to the PUF, or a post processing function that alters the output of the PUF. When the MMP PUF receives a challenge, only one of the modes is responsible for generating a response. By definition, the proposed MMP PUF is a form of Controlled PUF [5]. However, the MMP PUF will generate a noisy and multi-modal dataset by rapidly switching between the working modes with equal likelihood. The behavior of the MMP PUF is essentially a data poisoning attack [3] raised by PUF to significantly reduce the training accuracy of an eavesdropper. This multi-mode approach has some advantages such as: (1) The MMP PUF transfers the effort of introducing nonlinearity to the mode selection such that it does not matter whether each of the working modes or PUF itself is vulnerable to a modeling attack. In other words, a working mode should be very simple and lightweight to meet the requirements of IoT devices. (2) The MMP PUF can achieve a nearly ideal performance, a $\approx 50\%$ accuracy for predicting a 1-bit response, by designing working modes that maximize the entropy of the response. (3) Since the working modes are fully digital without feedback structures, the reliability of the multi-mode approach depends on which PUF instance is used. (4) The PUF instance of MMP PUF can be replaced by different PUF instances because the working modes are solely add-on functions. To verify the feasibility of the proposed approach, we evaluated MMP PUF through two working mode types, namely challenge shift and response flip, using the widely-used classic Arbiter PUF. Moreover, we provide a guideline of working mode design for customized extensions based on our experimental results. The contributions of this paper are summarized as follows:

- To our best knowledge, we are the first to propose a multi-mode PUF design paradigm which generates a multi-modal CRP dataset against ML attack.
- The MMP PUF is a general and flexible framework that one can design customized modes and a selection mechanism for required constraints.
- We verified the approach using two types of working modes and Arbiter PUF. The result shows that even if a naive mode design for a dual-mode PUF, the accuracy of the DL attack reduces to 74.37%. Besides, an ideal working mode design can obtain a 50.09% accuracy. Note that 50% is the perfect performance for predicting a 1-bit response.
- We provide a design guideline for MMP PUF based on analysis of experimental results.

The rest of this paper is arranged as follows: Sect. 2 describes the preliminary work. Section 3 defines working modes and the mode selection mechanism.

Two types of the working modes are also introduced as the examples of design in this section. Section 4 presents an evaluation of the proposed multi-mode PUF and experimental results. Suggestions for designing working modes and the mode selection mechanism are also discussed based on the experimental results. Finally, conclusions are drawn in Sect. 5.

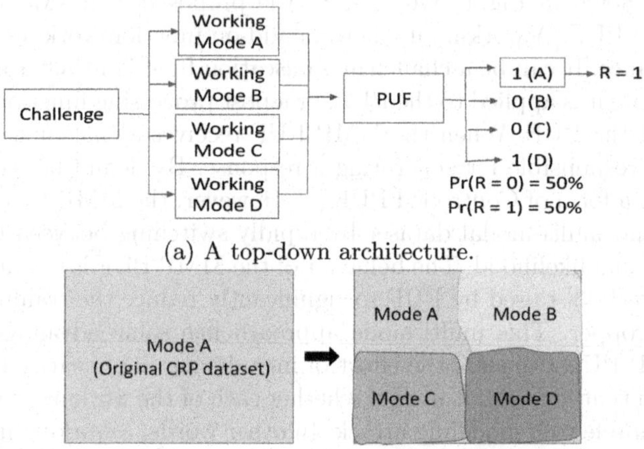

(a) A top-down architecture.

(b) Multi-Mode in the challenge-response space.

Fig. 1. Multi-mode PUF approach. (a) A top-down architecture (b) An illustration of MMP PUF in the challenge-response space.

2 Related Work

Several works have shown Arbiter PUF(APUF), XOR APUF and lightweight secure PUF (LSPUF) can be broken by traditional machine learning (ML) algorithms [12,17]. With the growth of deep learning techniques, newly developed PUFs such as multiplexer PUF [16], interpose PUF (IPUF) [13] and double Arbiter PUF (DAPUF) [9] have also been broken by deep learning models [7,17]. To overcome an ML attack, several complex PUF structures have recently been proposed to introduce structural nonlinearity. Ma et al. proposed a Multi-PUF on FPGA which combines weak PUFs with a Strong PUF to reduce the predictability of CRP mapping [8]. Awano et al. proposed an Ising PUF, a lattice-like structure consisting of small Arbiter PUFs, to increase learning complexity [1]. Another direction is to obfuscate challenges before they are applied to the PUF. Zalivaka et al. [20] used a multiple input signature register (MISR) to modify input challenges. Mispan et al. proposed a challenge permutation and substitution method to disorder the output transition probability of Arbiter PUF [11]. Nevertheless, most approaches did not consider DL based modeling attacks. In addition, these approaches typically focus on increasing nonlinearity of PUF structure or challenge-response mapping, which requires non-negligible additional area and computation resources. In contrast, the MMP PUF explores a different direction based on the hardness of learning from a poisoned dataset [6] that can be cost-efficient without compromising performance against DL attacks.

Similar to our concept of using multiple modes, a dual-mode PUF was proposed by Wang et al. to resist machine learning attacks [18]. The proposed PUF works with a ring oscillator (RO) PUF and a bistable ring (BR) PUF using different numbers of inverters. However, their approach is only available for a reconfigurable RO PUF. On the contrary, although we evaluated the multi-mode approach with Arbiter PUFs, the approach can be further generalized to other PUFs. Wang et al. also proposed an approach where the response of a PUF is determined to be inverted by a function [19]. Their approach can be considered as a dual-mode case of our proposed architecture. Explicitly, we show that the multi-mode approach with challenge transformations also achieves similar results in this work. We argue that the MMP PUF is a more general and flexible framework that allows one to explore challenge and response transformations for additional customized design such as a PUF with multiple-bit output or obfuscated circuit against Physical attacks.

3 Multi-Mode Poisoning PUF

The MMP PUF consists of two parts, namely working modes and the mode selection mechanism. An ideal working mode design partitions the original challenge-response into disjoint spaces, whereas a naive design leaves learnable intersections to ML algorithms. On the other hand, the mode selection is also critical to guarantee the security level. In addition, an imbalanced mode selection might expose the vulnerability to an adversary. We discuss working mode design and the mode selection in Sects. 3.1 and 3.2. Finally, the DL attack considered in this work is described in Sect. 3.3.

3.1 Working Mode Design

A working mode is defined as a simple combinational logic circuit that combines with a classic PUF to generate a response. As mentioned previously, it can be a challenge obfuscation that transforms challenges before applying them to the PUF, a post processing function which changes the response generated by the PUF, or a combination of smaller working modes. Given a challenge, only one working mode is evoked and each working mode determines one response. However, each single working mode should be satisfied with the quality metrics of PUF such as uniformity and uniqueness. That is, the property of the original PUF instance should be retained.

The ideal case of multiple working modes design is to maximize the uncertainty of the response of MMP PUF. To this end, responses generated by each mode need to be uniformly distributed over possible outcomes. More specifically, for an n-mode MMP PUF that generates a 1-bit response, the ideal case given a challenge would be: $n/2$ modes generate '1' and the rest of $n/2$ modes generate '0' such that the entropy of the MMP PUF equals the maximum value of 1, i.e.,

$$H = -\sum_{i=0}^{1} p_i \log p_i = 1 \tag{1}$$

where p_i is the probability of outcome i and $p_0 = p_1 = \frac{1}{2}$. Suppose the attacker does not know which mode is active. The prediction accuracy would be close to 50% if the selection of either modes is equally likely. In contrast to the ideal case, even if the working modes are naively designed, the prediction accuracy is also reduced significantly for the MMP PUF. In this section, we first discussed the expected performance of the naive n-mode MMP PUF, and then proposed two working mode types as an example of naive and ideal n-mode design.

Since we assume that all working modes follow the uniformity property, the probability of response $R_i = 0$ is 0.5 for each mode i, that is, $p = P\{R_i = 0\} = P\{R_i = 1\} = 0.5$. Assuming all the working modes work independently and have an equal likelihood of being selected, the probability of possible outcomes of naive n-mode can be approximated with a Binomial distribution:

$$B(n,x) = \sum_{x=0}^{n} \binom{n}{x} p^x (1-p)^{n-x} \tag{2}$$

Hence, if an adversary does not know the exact mode given a challenge, the maximum probability of the correct prediction will be:

$$B(n,x) = \sum_{x=0}^{n} \binom{n}{x} p^x (1-p)^{n-x} max(q, 1-q) \tag{3}$$

where q is the probability of being "1" or "0" of n outcomes generated by each mode.

Intuitively, increasing n leads to reduced prediction accuracy for the adversary. However, increasing the number of working modes implies using additional area, which violates the original lightweight intention. In addition, Eq. (3) indicates the prediction accuracy finally converges as the number of working modes increases linearly. We explore the tradeoff between number of working modes and prediction accuracy in Sect. 4.2.

To verify MMP PUF, two types of working modes are designed as naive and ideal examples in this work. The first one considered as a naive design is an n-bit right circular shift which rearranges the input challenge by n-bit shifting before applying it to a k-bit PUF where $n = 0, 1, ...k - 1$. A dual-mode n-bit shifting can be designed as follows: the first working mode right shifts the challenge by n bits before applying to the PUF whereas the second working mode remains unchanged and directly applies the challenge to the PUF. Similarly, a quad-mode n-bit circular shift can be designed as $(n \times i)$-bit right shifting where i is the $i-th$ mode and $i = 0, 1, ...3$. Although the ideal case mentioned above might also be achieved by the challenge permutation technique, however, the permutation is not trivial to design. A simple way to demonstrate the ideal design is to directly flip the response. Therefore, the second type of working mode is a flip mode which flips the output of the PUF. Hence, a dual-mode flip can be designed as follows: The first mode flips the response after applying a challenge to the PUF whereas the second mode remains unchanged.

3.2 Hiding the Working Modes

As the first barrier, the working mode selection of the MMP PUF should be robust to ML attacks. In addition, the probability of each working mode should be equal whenever a set of challenges is collected. More specifically, suppose k challenges are generated for a n-mode PUF, the number of challenges assigned for each mode should be close to $\frac{k}{n}$ to prevent one of the working modes from being exposed to the attacker. Intuitively, a corresponding working mode can be determined by XORing n challenge bits from fixed, selected indices for each challenge as the trigger signal in [19]. However, such design is not secure for Chosen Challenge Attack (CCA) [12]. Instead, we used a mask as a pseudo-random selector to mitigate this problem. Figure 2 depicts an example of the proposed working mode selection which rearranges the challenge to generate a mask and then applies a bitwise AND to obtain the masked challenge bits to decide which challenge bits are selected as the input of XOR operations.

For a dual-mode PUF, mode selection is simply determined by the outcomes of XOR operation since the probability of the outcomes is equally likely. Similarly, for a quad-mode case, challenge bit indices can be divided into two sets and the working modes can be determined two by two. For instance, suppose a k-bit challenge is divided into n parts. Then, one can use $(k/n) \times i$ indices for mode 1 and 2, and use $(k/n) \times i + 1$ indices for the rest of two modes where $i = 0, 1, ..., k-1$. The XOR operations used in the mode selection mechanism can also be combined with AND and OR operations. In that case, one can customize a mode selection mechanism with imbalanced switching probability for specific design requirements. However, note that since each of the working modes is not ML resistant, imbalanced working modes might reduce the learning complexity of MMP PUF. An attacker can simply hack the MMP PUF by precisely predicting the majority mode and almost randomly guessing for the rest of modes. More precisely, given a set of CRPs which consists of $\alpha\%$ of a majority mode and $(1-\alpha)\%$ of minority modes, the expected prediction accuracy can be approximated by:

$$\alpha + 0.5 \times (1 - \alpha) \tag{4}$$

where $0 < \alpha < 1$.

3.3 Deep Learning Attack

We use a deep neural network (DNN) adopted from [7,17] as the model of ML-based modeling attack. The DNN model is a fully-connected network consisting of an input layer that takes parity vectors as input, several hidden layers with

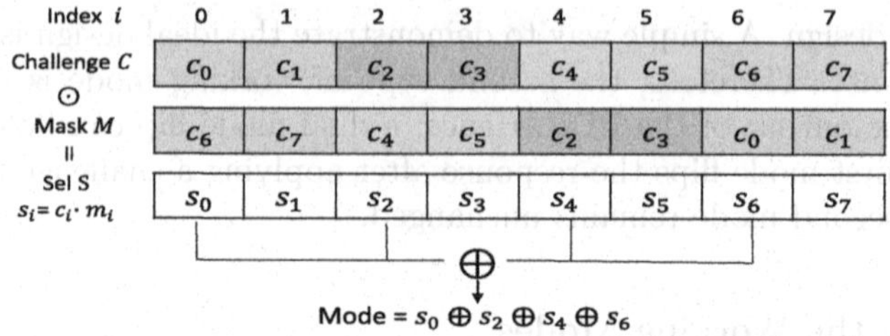

Fig. 2. An example of mode selection mechanism using 8-bit PUF.

ReLU activation functions, and 1 neuron output with a sigmoid activation function. To avoid overfitting, a dropout layer is used before the output layer. The loss function is a binary cross entropy loss and an Adam optimizer is used. The number of hidden layers and the corresponding number of neurons are selected based on the scale of the CRP set and the PUF instance. For the multi-mode approach, 1–9 hidden layers and 15–1000 neurons were evaluated in our experiments. One might ask whether the considered DNN structure is too simple. Nonetheless, since there are no local features or sequence relation in challenge-response data, the convolutional layer or recurrence network does not make sense in this case. Another question is whether it is possible to separately train a controller model and a model for each working mode to break our approach. We argue that if solving the mode selection is hard for an adversary, the learning problem of one mode of the MMP PUF is the same as learning a target function from a very noisy dataset, which is a fundamentally hard problem in the machine learning field.

4 Experimental Result

The experimental results not only demonstrate the performance of MMP PUF under ML attack, but also provide a guideline for someone who would like to extend the MMP PUF. Our findings are summarized as follows: (1) While the ideal design of the working mode achieves a nearly perfect performance ($\approx 50\%$ accuracy), a naive design also reduces the prediction accuracy significantly. (2) For a naive mode design, an ideal performance can be approximated by increasing the number of modes. (3) Imbalanced mode selection reduces the security level of MMP PUF seriously (4) We evaluated the Strict Avalanche Criteria (SAC) property of the MMP PUF to investigate predictability under CCA attack. Our result shows that the proposed masked mode selection achieves better SAC property than the unmasked one.

4.1 Experimental Setup

Our experimental setup is generally followed recent work in [7]. A threat model where an adversary can apply arbitrary challenges to the PUF is considered.

An arbiter PUF constructed by a linear additive model can be used to evaluate the performance of the modeling attack [14]. In the experiments, challenges were generated randomly. A 64-bit Arbiter PUF was simulated by python as the PUF instance of the MMP PUF. The simulated PUF instance and challenge transformation generally followed [14] where the weight vector ω followed a normal distribution with $\mu = 0$ and $\sigma = 1$. Afterward, noise-free CRPs were generated and split into training and testing sets with a ratio of 9:1 to train the DNN models. We use Keras [4] to implement the DNN models in Sect. 3.3.

4.2 Performance of Working Modes Design

We first investigated the performance of the dual-mode shift where mode 1 generated responses from the original PUF and mode 2 shifted challenges by 8-bits before applying the challenges to the PUF. Figure 3a compares the performance of mode 1, mode 2, and the dual-mode approach. The 0% shifted data indicates all challenges are applied to the PUF directly, which is the case for mode 1. Similarly, 100% shifted data indicates all responses are generated by mode 2; the 50% shift represents that responses are generated by the dual-mode approach. As shown in Fig. 3a, mode 2 generally outperforms mode 1 when the number of CRPs is small. This confirms that the challenge permutation technique increases the learning complexity [10,11]. However, using either mode 1 or mode 2 alone leaves the system vulnerable to DL attacks as the number of CRPs grows. In contrast to the dual-mode APUF, the prediction accuracy is reduced significantly. This implies robustness to machine learning attack is not the sole design concern for each of the working modes. Moreover, the accuracy of the dual-mode design is close to 75%, which is the same as the approximated value from Eq. (3). This finding implies the accuracy is highly related to the entropy of the modes given a challenge.

Flip mode aims to maximize the uncertainty of the MMP PUF to achieve an ideal performance of 50% prediction accuracy. For the dual-flip APUF, mode 1 directly generates the response from the PUF and mode 2 flips the output of the PUF. Since the original APUF is well known to be vulnerable to modeling attacks, flip mode alone is also vulnerable to modeling attacks. Therefore, instead of comparing the two modes used in isolation, we directly compared the performance of dual-mode shift and dual-mode flip in Fig. 3b. The performance of both modes finally converges to the approximated value in Eq. (3), which again indicates the performance of the multi-mode approach highly depends on the uncertainty of the outcomes generated by each mode.

4.3 Impact of Increasing the Number of Working Modes

Although designing working modes naively obtains acceptable performance, one might ask if the prediction accuracy is close to 50% when multiple working modes are used. To investigate the tradeoff between the number of working modes and predictability, the number of modes was varied from 2 to 9 using n-mode shift with a 64-bit Arbiter PUF. For each working mode in the n-mode

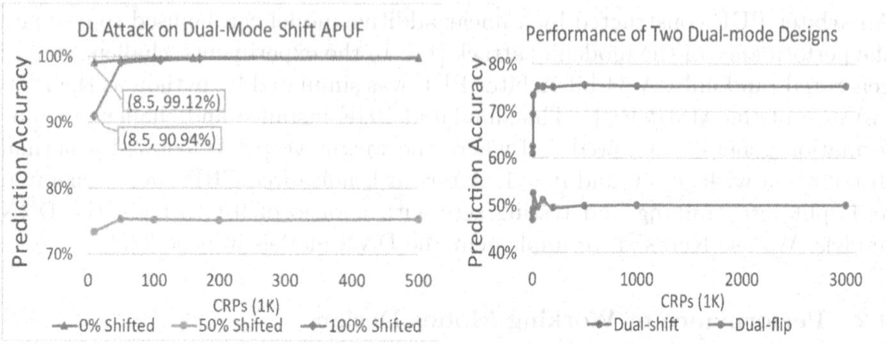

(a) DL attack on dual-mode shift APUF. (b) Performance of two types of modes

Fig. 3. Deep learning attack on dual-mode APUF

shift approaches, challenges were shifted by $4 \times k$ bits where $k = 0, 1 ... n - 1$. Furthermore, all modes were required to have equal likelihood of being selected to eliminate the impact of imbalanced data. We compared the prediction accuracy to the approximation based on Eq. (3) in Fig. 4. As shown in Fig. 4, increasing the number of working modes increases the learning complexity for the modeling attack, but the reduced accuracy is saturated as the number of modes increases incrementally. Since increasing the number of modes also increases hardware implementation complexity, using more than 4 modes is not considered cost-efficient.

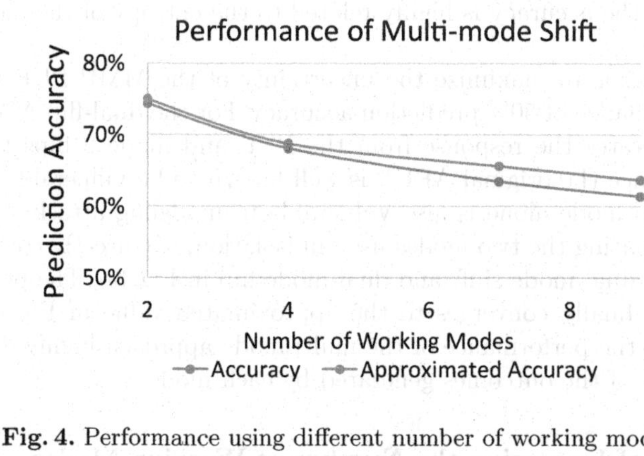

Fig. 4. Performance using different number of working modes

Table 1. Performance of each working modes for imbalanced dual-mode PUFs

Proportion	Non-shift	Shift	Overall Acc.	Approx. Acc.
50%/50%	75.01%	73.93%	74.46%	75%
75%/25%	97.32%	49.65%	85.83%	87.5%
25%/75%	49.19%	96.94%	85.43%	87.5%

4.4 Impact of Imbalanced Working Modes

We compared dual-mode shifts with different proportions of modes to explore the impact of imbalanced working modes in Table 1. The overall performance of imbalanced working modes is generally lower than that of the balanced cases. Moreover, the performances are almost the same for the 25% and 75% cases where the proportion between the majority and minority modes is 3 to 1. Both of the majority modes in the 25% and 75% cases achieve almost 100% accuracy but the results for the minority modes are nearly random guesses. This indicates the DNN model exhibits bias that tends to over-classify the majority mode in the imbalanced cases. In contrast, the model learns the overall output probability of all modes in the balanced cases. Therefore, a better strategy is to select the working modes with equal likelihood while each of the working modes is vulnerable to modeling attacks.

4.5 Impact of Output Transition Probability

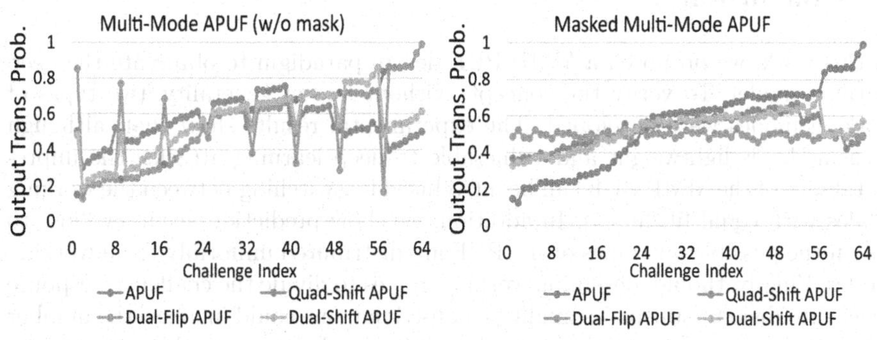

(a) Unmasked MMP APUF (b) Masked MMP APUF

Fig. 5. Output Transition Probability of Multi-Mode Poisoning APUF with and without a mask for the mode selection

A PUF instance with k-bit challenge and 1-bit response is said to satisfy strict avalanche criteria (SAC) if its output transition probability is 0.5 whenever any challenge bit is flipped. The SAC property is widely used to investigate the predictability of a PUF [11,12]. We followed the algorithm described in [11] to estimate the output transition probability of the proposed n-mode design. We evaluated the proposed dual-mode and quad-mode MMP PUFs. Besides, the mode selection which directly XORs fixed selected challenge bits (referred to as unmasked case) is compared with the masked version proposed in Sect. 3.2. The indices of bits are chosen by $n * i$, where $n = 8$ and $i = 0, 1...7$. Then the selected bits are XORed for the mode selection. In Quad-mode cases, the selected bits are partitioned into two sets; each set determines one of two possible outcomes thus 2×2 possible modes are determined. For the unmasked cases, the bits are selected from the challenge directly. For masked cases, the bits are selected from the masked challenge described in Sect. 3.2. We reported the result in Fig. 5.

As shown in Fig. 5a and b, there is a linear pattern in Arbiter PUF where the output transition probability is proportional to the challenge bit index. This corresponds to previous findings [11,12] that the pattern leaks information, resulting in increased vulnerability to machine learning algorithms. The output transition probability of each n-mode approach generally follows the probability of their corresponding PUF instances in unmasked case in Fig. 5a. Interestingly, we observed significant fluctuations at all challenge bit indices used in the unmasked mode selection, especially for the dual-flip case. These fluctuations expose potential risks of predictability if an adversary explicitly collects CRPs with the same pattern (1-bit complemented) used in the SAC test. By contrast, our masked mode selection alleviates the problem to achieve a better SAC property in all cases. The result in Fig. 5 indicates the design of the masked mode selection mechanism is better than the unmasked one.

5 Conclusion

In this work, we proposed a MMP PUF design paradigm to obfuscate the deep learning model. To verify the concept, without loss of generality, two types of working modes were developed. The experimental results show that, although each modes is lightweight and vulnerable to deep learning attacks, the unpredictability of the MMP PUF can be introduced by switching between the working modes with equal likelihood. In addition, we show prediction accuracy close to 50% if the possible outcomes of the PUF are distributed uniformly to all working modes. Finally, though designing working modes to divide the challenge-response space into disjoint subspaces might be non-trivial, one could increase the number of modes to approximate the desired security level. However, there is a trade-off between performance and hardware cost. While our current mode selection is defined heuristically and adopted to all PUF instances for ease of exposition, the mode selection can be further extended to provide a strong security guarantee.

References

1. Awano, H., Sato, T.: Ising-PUF: a machine learning attack resistant PUF featuring lattice like arrangement of arbiter-PUFs. In: Proceedings of DATE, pp. 1447–1452 (2018)
2. Babaei, A., Schiele, G.: Physical unclonable functions in the internet of things: state of the art and open challenges. Sensors **19**, 3208 (2019)
3. Biggio, B., Nelson, B., Laskov, P.: Poisoning attacks against support vector machines. In: Proceedings of ICML, pp. 1467–1474 (2012)
4. Chollet, F., et al.: Keras (2015). https://keras.io
5. Gassend, B., Clarke, D., van Dijk, M., Devadas, S.: Controlled physical random functions. In: Proceedings of ACSAC, pp. 149–160 (2002)
6. Kearns, M., Li, M.: Learning in the presence of malicious errors. In: Proceedings of STOC, pp. 267–280 (1988)
7. Khalafalla, M., Gebotys, C.: PUFs deep attacks: enhanced modeling attacks using deep learning techniques to break the security of double arbiter PUFs. In: Proceedings of DATE, pp. 204–209 (2019)
8. Ma, Q., Gu, C., Hanley, N., Wang, C., Liu, W., O'Neill, M.: A machine learning attack resistant multi-PUF design on FPGA. In: Proceedings of ASP-DAC, pp. 97–104 (2018)
9. Machida, T., Yamamoto, D., Iwamoto, M., Sakiyama, K.: A new arbiter PUF for enhancing unpredictability on FPGA. Sci. World J. **2015** (2015)
10. Majzoobi, M., Koushanfar, F., Potkonjak, M.: Lightweight secure PUFs. In: Proceedings of IEEE/ACM ICCAD, pp. 670–673 (2008)
11. Mispan, M.S., Su, H., Zwolinski, M., Halak, B.: Cost-efficient design for modeling attacks resistant PUFs. In: Proceedings of DATE, pp. 467–472 (2018)
12. Nguyen, P.H., Sahoo, D.P., Chakraborty, R.S., Mukhopadhyay, D.: Security analysis of arbiter PUF and its lightweight compositions under predictability test. ACM TODAES **22**(2), December 2016
13. Nguyen, P.H., Sahoo, D.P., Jin, C., Mahmood, K., Ruhrmair, U., van Dijk,M.: The interpose PUF: secure PUF design against state-of-the-art machine learning attacks. In: TCHES, vol. 2019, no. 4, pp. 243–290, August 2019
14. Rührmair, U., et al.: PUF modeling attacks on simulated and silicon data. IEEE TIFS **8**(11), 1876–1891 (2013)
15. Rührmair, U., Sehnke, F., Sölter, J., Dror, G., Devadas, S., Schmidhuber, J.: Modeling attacks on physical unclonable functions. In: Proceedings of CCS, pp. 237–249. ACM (2010)
16. Sahoo, D.P., Mukhopadhyay, D., Chakraborty, R.S., Nguyen, P.H.: A multiplexer-based arbiter PUF composition with enhanced reliability and security. IEEE TC **67**(3), 403–417 (2018)
17. Santikellur, P., Bhattacharyay, A., Chakraborty, R.S.: Deep learning based model building attacks on arbiter PUF compositions. IACR Cryptol. ePrint Arch. **2019**, 566 (2019)
18. Wang, Q., Gao, M., Qu, G.: A machine learning attack resistant dual-mode PUF. In: Proceedings of ACM GLSVLSI, pp. 177–182 (2018)
19. Wang, S., Chen, Y., Li, K.S.: Adversarial attack against modeling attack on PUFs. In: Proceedings of ACM/IEEE DAC, pp. 1–6 (2019)
20. Zalivaka, S.S., Ivaniuk, A.A., Chang, C.: Low-cost fortification of arbiter PUF against modeling attack. In: Proceedings of IEEE ISCAS, pp. 1–4 (2017)

Combining Exogenous and Endogenous Signals with a Semi-supervised Co-attention Network for Early Detection of COVID-19 Fake Tweets

Rachit Bansal[1](✉), William Scott Paka[2], Nidhi[3], Shubhashis Sengupta[3], and Tanmoy Chakraborty[2]

[1] Delhi Technological University, Delhi, India
rachitbansal_2k18ee152@dtu.ac.in
[2] IIIT-Delhi, Delhi, India
{william18026,tanmoy}@iiitd.ac.in
[3] Accenture Labs, Delhi, India
{nidhi.sultan,shubhashis.sengupta}@accenture.com

Abstract. Fake tweets are observed to be ever-increasing, demanding immediate countermeasures to combat their spread. During COVID-19, tweets with misinformation should be flagged and neutralised in their early stages to mitigate the damages. Most of the existing methods for early detection of fake news assume to have *enough propagation information* for *large labelled tweets* – which may not be an ideal setting for cases like COVID-19 where both aspects are largely absent. In this work, we present ENDEMIC, a novel early detection model which leverages exogenous and endogenous signals related to tweets, while learning on limited labelled data. We first develop a novel dataset, called ECTF for early COVID-19 Twitter fake news, with additional behavioural test-sets to validate early detection. We build a heterogeneous graph with follower-followee, user-tweet, and tweet-retweet connections and train a graph embedding model to aggregate propagation information. Graph embeddings and contextual features constitute endogenous, while time-relative web-scraped information constitutes exogenous signals. ENDEMIC is trained in a semi-supervised fashion, overcoming the challenge of limited labelled data. We propose a co-attention mechanism to fuse signal representations optimally. Experimental results on ECTF, *PolitiFact*, and *GossipCop* show that ENDEMIC is highly reliable in detecting early fake tweets, outperforming nine state-of-the-art methods significantly.

1 Introduction

Over the past couple of years, several social networking platforms have seen drastic increase in the number of users and their activities online [1]. Due to the lockdown situations and work from home conditions during COVID-19 pandemic, the screen time on social media platforms is at an all time high. Twitter is one such micro-blogging platform where users share opinions and even rely on news updates. Twitter users exposed

R. Bansal and W. S. Paka—Equal Contribution. The work was done when Rachit was an intern at IIIT-Delhi.

K. Karlapalem et al. (Eds.): PAKDD 2021, LNAI 12712, pp. 188–200, 2021.
https://doi.org/10.1007/978-3-030-75762-5_16

to unverified information and opinions of others often get influenced and become contributors to further spreading. The characteristics of fake news spreading faster farther and deeper than genuine news is well-studied [12]. Fake news on health during COVID-19 might endanger people's lives as a few of them call for action. Although our proposed method is highly generalised, we choose to specifically focus on the ongoing pandemic situation of COVID-19 as it is timely and needs scaleable solutions.

State-of-the-art fake news detection models for Twitter have proved useful when trained on a sufficient amounts of labelled data. Any emerging fake news in its initial stages could not be detected by such models due to the lack of corresponding labelled data. Moreover, these models tend to fail when the fake news is not represented largely in the training set. By the time the fake news is detected, it has spread and caused damages to a wide range of users. Detecting a fake news in an early stage of its spread gives us the advantage of flagging it early. A few state-of-the-art models for early fake news detection use propagation based methods [9], i.e., they are based on how a particular news event (fake/genuine) spreads (both wider and deeper); retweet chain in Twitter is one such example. These models work with the propagation chains and require sufficient historical data (retweet/reply cascade) of each tweet, which are very hard to collect within limited time.

In this work, we design `ENDEMIC`, a novel approach to detect fake news in its early stage. To this end, we developed an Early COVID-19 Twitter Fake news dataset, called `ECTF` with additional test set (early-test) for early detection. We collected vast amount of COVID-19 tweets and label them using trusted information, transformer models and human annotation. We finally curate a standard training dataset, composed of numerous rumour clusters, while simulating a scenario of emerging fake news through early-test set. Next, we extract exogenous signals in form of most suitable stances from relevant external web domains, relative to the time the tweet is posted. Unlike existing studies which extract propagation paths per tweet, we create a massive heterogeneous graph with follower-followee, tweet-retweet and tweet-user connections and obtain the node representations in an unsupervised manner. The graph embeddings (both users and tweets) constitute the major component of the endogenous signals (within Twitter). The time-variant contextual tweet and user features are used to provide additional context, and a few of them are masked to validate the model for early detection. Lastly, to overcome the challenge of limited labelled dataset, we setup the whole model in a semi-supervised fashion, learning in an adversarial setting to utilise the vast unlabelled data.

In Summary, fake news on Twitter is an inevitable threat especially during COVID-19 pandemic, where inaccurate or deviating medical information could be harmful. As a result, a timely model which can detect fake news in its early stages is important due to the current conditions of emerging fake news. We propose `ENDEMIC`, a semi-supervised co-attention network which utilises both exogenous and endogenous signals. Experimental results show that `ENDEMIC` outperforms nine state-of-the-art models in the task of early fake tweet detection. `ENDEMIC` produces 93.7% accuracy, on `ECTF` for fake tweet detection and 91.8% accuracy for early detection, outperforming baselines significantly. We also show the generalisation of `ENDEMIC` by showing its efficacy on two publicly available fake news datasets, *PolitiFact* and *GossipCop*; `ENDEMIC` achieves 91.2% and 84.7& accuracy on the two datasets, respectively.

In particular, our major contributions are as follows:

- We introduce ECTF, an Early COVID-19 Twitter Fake news dataset with additional early-test set to evaluate the performance of fake tweet detection models for early detection.
- As propagation paths for tweets might not be available in all scenarios, we build connections of follower-followee, tweet-retweet and user-tweet network and extract representations upon learning the graph, constituting of our endogenous signals.
- We use exogenous signals which informs the model on the realities of information available on the web at tweet time, and helps in learning from weak external signals.
- Adding an effort towards early detection, time variant features are masked at test time for further evaluation on early detection.
- We further show the generalisation of ENDEMIC by presenting its superior performance on two other general fake news datasets .

Reproducibility: The code and the datasets are public at: https://github.com/LCS2-IIITD/ENDEMIC.

2 Related Work

Our work focuses majorly on early detection of fake tweets. As our model involves techniques such as graphs and semi-supervised learning, we present related studies pertaining to our model.

Fake News Detection. Proliferation of fake news over the Internet has given rise to many research, and hence there exist an abundance of literature. Early studies on fake news relied on linguistic features of texts to detect if the content is fake [3]. Wang et al. [27] used textual embeddings and weak labels with reinforcement learning to detect fake news. They leveraged an annotator component to obtain fresh and high-quality labelled samples to overcome the challenge of limited labelled data. Recent approaches have started exploring directed graph and Weisfeiler-Lehman Kernel based model for social media dataset that use similarity between different graph kernels [21]. A recent survey [31] shows that there are four approaches to detect fake news: (i) knowledge-based methods, where the content is verified with known facts, (ii) style-based methods, by analysing the writing style of the content, (iii) propagation methods, based on how a particular news event spreads, and (iv) source-based methods, by verification of credibility of sources. As studies show, each of these methods used individually is not enough to build an efficient classifier [16].

Semi-supervised Learning. Semi-supervised models have been used often in the past to leverage vast unlabelled datasets in various fields. Helmstetter et al. [8] explored weakly supervised learning for fake news detection which automatically collects large scale noisy dataset to aid the classification task. Yu et al. [29] used constrained semi-supervised learning for social media spammer detection. Tensor embeddings are used to design a semi-supervised model for content based fake news detection [6]. A few studies leveraged variational auto-encoders in the form of sequence-to-sequence modelling on text classification and sequential labelling [7]. Nigam et al. [18] classified the text using a combination of Naive Bayes and Expectation Maximisation algorithms and demonstrated substantial performance improvements. Miyato et al. [15] utilised adversarial and virtual adversarial training to the text domain by applying perturbations to the word embeddings. Chen et al. [4] introduced MixText that combines labelled, unlabelled and augmented data for the task of text classification.

Early Detection. Early fake news detection methods detect fake news at an initial stage where the news has not yet been popularised. There exist very limited studies on early detection. Liu et al. [9] built an early detection model using propagation paths of news spreading as a multivariate time series and then training a recurrent and convolution classifier on user features. Rosenfeld et al. used graph kernels on Twitter cascades to capture intricate details of the data-set for fake news detection without feeding user identity and time for an early detection model [21]. Shu et al. used content engagement and cleaned labelled dataset for early detection with deep neural network [24].

3 Our Proposed Dataset: `ECTF`

As there is no publicly available COVID-19 Twitter dataset particularly for *early detection*, we attempt to develop our own dataset, `ECTF`, specifically crafting it for early detection. We expand on `CTF`, a general COVID-19 Twitter fake news dataset, proposed by Paka et al. [19]. `CTF` was formed using multiple sources- unlabelled Twitter datasets that are publicly released [2,22,26], hydration of tweets using predefined hashtags, and governmental health organisations and fact checking websites for verified news. They considered statements released by the latter to be true, and applied Sentence-BERT [20] and RoBERTa [11] to convert these tweets and verified news into contextual embeddings, pairwise cosine similarity is then computed to assign a label 'fake' or 'genuine'. This way, `CTF` composes of $72,578$ labelled and $2,59,469$ unlabelled tweets, partially verified manually.

We took a sample of labelled and unlabelled tweets from `CTF`, forming our train set. A **'general-test'** set is created by randomly sampling from the remaining labelled and unlabelled tweets. The training dataset, which contains a wide variety of rumour clusters, is kept constant. We identified small rumour clusters in the labelled dataset and use those to form additional test set, called **'early-test'** set, to perform behavioural analysis of our algorithm for early detection. These small rumour clusters contain the fake news that are not popularised yet (having chance of getting popular), simulating early stages of fake news events. We extracted more tweets belonging to these rumour clusters, while keeping an upper limit on the time since they were posted. This ensures that the tweets belonging to this early-test set are in their early stages.

We refer to the complete dataset composed of the train and test sets as **`ECTF`**, **E**arly **C**OVID-19 **T**witter **F**ake news dataset.

4 Our Proposed Methodology: `ENDEMIC`

Here we present our model, called **`ENDEMIC`** (**E**xogenous and e**ND**ogenous signals with s**EMI**-supervised **C**o-attention network). Figure 1 represents the model

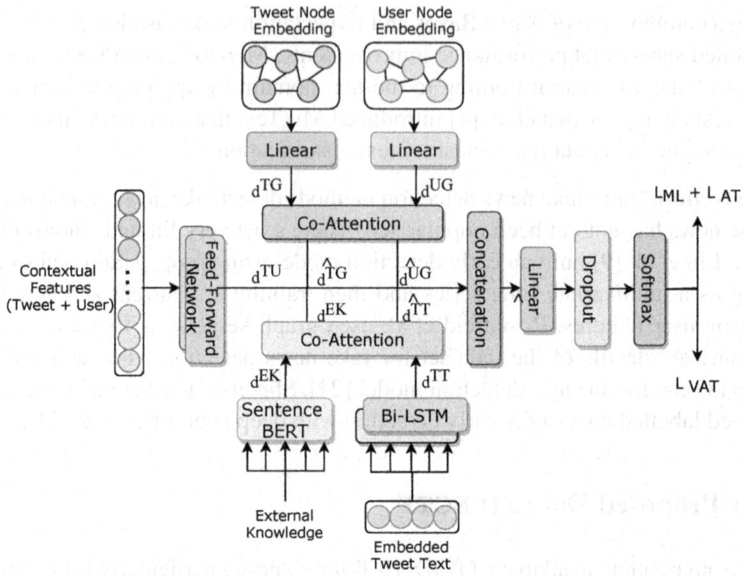

Fig. 1. A schematic architecture of ENDEMIC. The encoded interpolations of external knowledge (d^{EK}), tweet text (d^{TT}), contextual features (d^{TU}), and tweet (d^{TG}) and user node embeddings (d^{UG}) are shown. $\hat{\cdot}$ represents an output from co-attention.

architecture. Here we show various input features, each of which passes through separate modules before being concatenated. Our approach towards early detection relies on simulating the early stages of the news event testing with time invariant features.

4.1 Exogenous Signals

Traditional supervised fake news detection methods are bounded by the knowledge present in the data they are trained on [30]. This makes them incapable of classifying tweets on information the model is not trained on. This is particularly challenging for domains where large amount of new theories and their consequent studies arrive rapidly within small duration. Very often, the general stance of such news also changes radically over time. Just as a human expert, classification models too need to be well-aware of current information in a domain in order to be reliable and efficient. To address this problem, we make use of exogenous signals through *external knowledge*.

We curate exogenous content using web-scraped articles from various sources. In order to simulate the early stages of detection, this knowledge base is collected differently for training and testing instances. We build the external knowledge for training as a static knowledge base composed of various sources for each input. We hypothesise that collecting this external knowledge in accordance with the time the post was made simulates an early detection scenario, consequently making the model adapt to such scenarios better where the information about an emerging fake news will be limited.

Using this, the model learns the weak signals that are present to perform a classification. In case of testing, the external knowledge is scraped for every test instance at the time of testing itself. The dynamic nature of building the external knowledge base during testing ensures that the model makes use of latest sources to make the prediction.

We perform a web scraping with tweets as queries using Google Web API. Since a large amount of information in form of web pages are available per query, and the content per page could also be excessively large, selecting the right content is vital. We do so by firstly tokenizing each web page $x^{EK,i}$ into sentences. The jth sentence of $x^{EK,i}$, denoted by $x_j^{EK,i}$, is encoded using Sentence-BERT [20] to obtain its contextual representation, $d_j^{EK,i} \in \mathbb{R}^K$, where K is the dimension of the encoding. This representation is then compared with the representation of the tweet text, $x^{TT} \rightarrow d_x^{TT} \in \mathbb{R}^K$, encoded using the same Sentence-BERT model. This comparison is made using cosine similarity. If the cosine similarity between these two encodings, $cos(d_j^{EK,i}||d_x^{TT})$, is greater than a threshold ϵ (set as 0.8), then the sentence $x_j^{EK,i}$ is added to the set of input external knowledge for that particular tweet. The same representation, $d_j^{EK,i}$ is used as the corresponding input to the model for all sentences belonging to the tweet. This process is done for the entire set of input queries, until we obtain 50 such sentences for each, with the amount of phrases per web-source being limited to 10. Thus, the net external knowledge input to ENDEMIC during training is obtained by concatenating the encoding for each input in 2D fashion and is given by $d^{EK} \in \mathbb{R}^{n \times 50 \times K}$, where n is the input size. For most of our experiments, we keep $K = 512$.

4.2 Endogenous Signals

Input Tweet Embedding. The original tweet text x_i^{TT} of sequence length N (say) is first represented by a one-hot vector using a vocabulary of size V. A word embedding look-up table transforms these one-hot vectors into a dense tensor. This embedding vector is further encoded using a Bidirectional LSTM. A final state output $d \in \mathbb{R}^{K/2}$ is obtained at both the forward and backward layers, which are then concatenated to give a 2D vector corresponding to each text input. The final representation of the tweet text input to the model is, thus, $d^{TT} \in \mathbb{R}^{n \times N \times K}$.

Graph Embeddings. Extracting the propagation chain of retweets has been proven effective by the existing early detection models [9,14]. However, these methods limit the training as for each training or test sample, the entire set of retweet chains needs to be extracted which is often computationally expensive. Here we build a heterogeneous graph $\mathcal{G}(\mathcal{V}, \mathcal{E})$ as follows: the set of nodes V can be users or tweets, and edges E are formed in the following ways: two users are connected via follower-followee link, a user is connected to her posted tweet, two tweets are connected if one is a retweet of another. We keep $\mathcal{G}$ as undirected intentionally ignoring the direction of some edge types (follower-followee) in order to maintain uniformity. The formed heterogeneous graph contains around $51M$ nodes and $70M$ edges. Such huge connections, when added into one graph, form many disconnected clusters. In our graph, we observe one giant cluster with almost millions of nodes and edges, which stands dominating compared to other small (and disconnected) clusters.

We obtain the embedding of the graph using GraphSAGE [17] in an unsupervised fashion due to its capability to scale and learn large graphs easily. We label each node with its characteristics such as parent, tweet, retweet, user, fake tweet. The generalisability of GraphSAGE helps in extracting the embeddings of unseen users and tweets. We use a teleportation probability of 0.3 for the graph to randomly jump across nodes, which helps with clusters having less number of connections or being disconnected from the rest. In this work, we show that using the embeddings of both users and tweets in combination with co-attention leads to better supervision and learning of the model.

We represent the tweet and user graph embeddings as d^{TG} and $d^{UG} \in \mathbb{R}^{n \times G}$ respectively, where G represents the embedding dimension, and is kept as $G = 768$, for most of our experiments.

Contextual Features. Social media platforms like Twitter offer a variety of additional features that can play a crucial role in identifying the general sentiment and stance on a post by the users. Moreover, some features of user could also be used as indicative measures of their social role and responsibility, in general. Therefore, we use a variety of such tweet and user features to provide additional context regarding the input tweet and the corresponding users who interact with it. Some of the tweet features used are *number of favourites, number of retweets, PageRank reliability score of domains mentioned in the tweet, text harmonic sentiment*[1], etc. Some of the user features include *follower and followee counts*, *verified status*, and *number of tweets made by the user*.

Note that majority of these features continually change over time and can be regarded as time-variant and accordingly, the inferences drawn also would change over time. Therefore, for early detection, it is vital not to rely too heavily on such features. For instance, the number of likes and retweets for a tweet changes over time. Similarly, such additional features for a new user cannot be expected to give a proper indicative measure of a user's tendency of believing, spreading, or curbing misinformation. And masking the time-variance during evaluation is better explained in Sect. 5.3.

Throughout this study, we represent these contextual tweet and user features as $x^{TF} \in \mathbb{R}^{n \times N_{TF}}$ and $x^{TF} \in \mathbb{R}^{n \times N_{UF}}$, respectively, where N_{TF} and N_{UF} indicate the number of such features. As shown in Fig. 1, these input features are concatenated and passed across a common feed-forward network (FFN), which interpolates $x_i^{TF} \oplus x_i^{UF} \in \mathbb{R}^{1 \times (N_{TF}+N_{UF})}$ to $d_i^{TU} \in \mathbb{R}^C$, where C is the output dimension of FFN.

4.3 Connecting Components and Training

Co-Attention. In order to jointly attend and reason about various interpolated inputs, we use the parallel co-attention mechanism [13] (Fig. 2). As shown in Fig. 1, this is done at two places of ENDEMIC, namely, to co-attend between external knowledge $d_i^{EK} \in \mathbb{R}^{50 \times K}$ and tweet text $d^{TT} \in \mathbb{R}^{N \times K}$ in 2D, and tweet $d^{TG} \in \mathbb{R}^{1 \times G}$ and user $d^{UG} \in \mathbb{R}^{1 \times G}$ graph embeddings in 1D. The same process is followed for two; therefore, to unify the notation, we use $d^A \in \mathbb{R}^{X \times Z}$ and $d^B \in \mathbb{R}^{Y \times Z}$ to explain the mechanism used.

[1] Obtained from the textblob tool in Python.

Firstly, an affinity matrix $C \in \mathbb{R}^{Y \times X}$ is obtained as, $C = tanh(d^A W_b d^{B^\top})$. Here $W_b \in \mathbb{R}^{Z,Z}$ represents the learnable weight matrix. Further the corresponding attention maps between A and B are calculated as follows:

$$H^A = tanh(W_A d^{A^\top} + (W_B d^{B^\top})C), \quad H^B = tanh(W_B d^{B^\top} + (W_A d^{A^\top})C^\top)$$

where, W_A, $W_B \in \mathbb{R}^{k \times Z}$ again represent the learnable weight matrices. Further, to compute the attention probabilities of each element in A with each element of B, we use,

$$a^A = Softmax(w_{hA}{}^\top H^A), \quad a^B = Softmax(w_{hB}{}^\top H^B)$$

where, $w_{hA}, w_{hB} \in \mathbb{R}^k$ represent the weight parameters, while $a^A \in \mathbb{R}^X$ and $a^B \in \mathbb{R}^Y$ represent the resultant attention probabilities. Finally, the net attention vectors between A and B are computed as a weighted sum of the corresponding features, i.e.,

$$\widehat{A} = \sum_{i=1}^{X} a_i^A d_i^A, \quad \widehat{B} = \sum_{i=1}^{Y} a_i^B d_i^B \tag{1}$$

$\widehat{A}$ and $\widehat{B} \in \mathbb{R}^{1 \times Z}$ are the learned feature vectors obtained through co-attention. This, for instance, represents how each representation in the tweet graph embeddings attends to the user graph embeddings, when A represents tweet graph embeddings (TG), and B represents user graph embeddings (UG).

The interpolations of the various inputs, obtained through the separate smaller pipelines are combined using a unified head pipeline in the model architecture. Considering a single input, firstly, the representations from the two co-attention layers, $\widehat{d^{EK}}$ and $\widehat{d^{TT}} \in \mathbb{R}^{1 \times K}$, and $\widehat{d^{TG}}$ & $\widehat{d^{UG}} \in \mathbb{R}^{1 \times G}$, are concatenated along with $d^{TU} \in \mathbb{R}^{1 \times C}$.

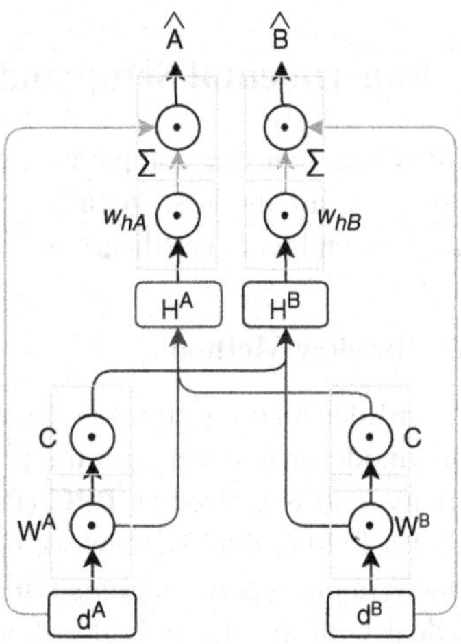

Fig. 2. The co-attention mechanism.

$$d = \widehat{d^{EK}} \oplus \widehat{d^{TT}} \oplus \widehat{d^{TG}} \oplus \widehat{d^{UG}} \oplus d^{TU}$$

This net representation, $d \in \mathbb{R}^{2K+2G+C}$, then passes onto a dropout ($p_{drop} = 0.2$) regularised feed-forward layer with an output size of 2, and finally a Softmax function, to give the probability per output class.

Training. In order to overcome the limitations produced by scarce labelled data, we use the Virtual Adversarial Loss (VAT) across both labelled and unlabelled data. VAT introduces a small perturbation in the input embedding and computes the loss using weak labels. Maximum Likelihood (ML) loss and Adversarial Training (AT) losses are further used to train `ENDEMIC` on the labelled data [15]. These additional losses allow `ENDEMIC` to be more robust, and the abundant unlabelled data being used this way allow it to form better understanding and representation of the domain-specific texts.

Table 1. Features used and performance comparison on **general-test** dataset of ECTF (TG: Tweet Graph, UG: User Graph, EK: External Knowledge, UL: Unlabelled Data).

Model	Feature used				Performance			
	TG	UG	EK	UL	Accuracy	Precision	Recall	F1 Score
HAN					0.865	0.685	0.865	0.762
MixText				✓	0.875	0.835	0.843	0.847
GCAN		✓			0.852	0.817	0.820	0.813
CSI	✓		✓		0.873	0.805	0.915	0.854
dEFEND	✓		✓		0.890	0.830	0.891	0.862
CNN-MWSS				✓	0.805	0.785	0.810	0.790
RoBERTa-MWSS				✓	0.825	0.815	0.852	0.840
FNED				✓	0.911	0.907	0.920	0.915
PPC	✓				0.861	0.860	0.832	0.845
ENDEMIC	✓	✓	✓	✓	**0.937**	**0.922**	**0.943**	**0.932**

5 Experimental Setup and Results

Here we present our comparative analysis by empirically comparing ENDEMIC with state-of-the-art models, including techniques for general fake news detection, early detection, and text classification.

5.1 Baseline Methods

We consider nine baselines as follows. Liu and Wu [9] introduced a GRU and CNN-based model which relies on the propagation path (PP) of a news source to classify its veracity, and is termed as **PPC** (Propagation Path Classification). **FNED** [10] makes use of a PU-learning framework to perform weak classification, and is claimed to be suitable in data scarce settings with large amount of unlabelled data, as is the scenario we deal with in this study. Further, Shu et al. [25] proposed to learn from Multiple-sources of weak Social Supervision (**MWSS**), on top of a classifier such as CNN and RoBERTa [11]. They too relied on weak labelling, but constructed them through social engagements. Both **dEFEND** and **CSI** deployed an RNN-based framework for classifying news; the former makes use of user comments with co-attention, while the latter relies on a three-step process of *capturing, scoring* and *integrating* the general user response. Furthermore, **GCAN** [14] presents a dual co-attention mechanism over source and retweet tweet representations, relying on interaction and propagation for detecting fake news. Finally, we also make some interesting observations by employing text classification models, namely **MixText** [4] and **HAN** [28], as additional baselines.

5.2 Evaluating on General-Test Set

Table 1 shows the performance comparison of ENDEMIC compared to the baselines on the **general-test** set of ECTF. We observe that all the features (graph based, external

Table 2. Comparing model performance on **early-test** set of ECTF.

Model	Performance				ΔAcc
	Acc.	Prec.	Rec.	F1	
HAN	0.850	0.692	0.841	0.751	1.5%
MixText	0.865	0.832	0.851	0.844	1.0%
GCAN	0.809	0.800	0.785	0.795	4.3%
CSI	0.808	0.821	0.795	0.815	6.5%
dEFEND	0.800	0.790	0.815	0.790	9.0%
MWSS	0.798	0.800	0.812	0.805	2.7%
FNED	0.901	0.887	0.913	0.895	1.0%
PPC	0.831	0.805	0.797	0.810	3%
ENDEMIC	**0.918**	**0.910**	**0.923**	**0.920**	1.9%

Table 3. Performance comparison for **mask-detect** on ECTF. The masked features and corresponding metric scores are shown.

Model	Time-variant features	Performance				ΔAcc
		Acc.	Prec.	Rec.	F1	
HAN	–	0.858	0.705	0.850	0.755	1.2%
MixText	–	0.860	0.815	0.833	0.820	1.0&
GCAN	Tweet propagation	0.792	0.812	0.775	0.805	6.0%
CSI	User response	0.783	0.701	0.745	0.725	9.0%
dEFEND	User comments	0.768	0.750	0.804	0.774	12.2%
MWSS	–	0.810	0.800	0.835	0.844	1.5%
FNED	–	0.892	0.890	0.925	0.905	1.9%
PPC	Propagation paths	0.855	0.810	0.790	0.805	0.6%
ENDEMIC	Contextual features	**0.928**	**0.920**	**0.930**	**0.927**	0.9%

knowledge and unlabelled data) play a major role in determining the corresponding performance of detecting fake tweets. While general fake news detection models like dEFEND and CSI strive to integrate more such features, early detection models like FNED and PPC tend to rely more on the time-invariant features like text alone. The effect produced by the absence of one over the other is apparent from Table 1. In general, ENDEMIC shows a benchmark performance of 0.937 accuracy, outperforming all the baselines across all the evaluation measures significantly.

5.3 Evaluating on Early Detection

Early-Test. Table 2 shows the comparative results on the specially curated evaluation set (**early-test**) for early detection of fake news. Even though the change in accuracy (*ΔAcc*) of ENDEMIC as compared to its performance on general fake news (as shown in Table 1) is not least among the rest of the models, it can be seen that it still comfortably outperforms all other baselines. Interestingly, the general purpose text classifiers, which are not particularly designed for fake news detection, show a relatively lesser *ΔAcc*, while dEFEND [5] suffers from the largest difference of 9%. We attribute this to the heavy reliance of these models on time-variant features which provide critically less context in early stages. At the same time, as shown in Table 1, only text-based and closely related non-time-variant features are not enough to reliably detect fake news. Thus, the set of input features used by ENDEMIC optimises the trade-off between reliability and time-variance.

Masking Time Variance. To further verify our model's prowess in detecting early fake news, we introduce a unique masking approach to be applied upon the early-test evaluation set. For this technique, the tweets belonging to the smaller rumour clusters (as defined in Sect. 3) are further simulated as actual early tweets for a model, by masking all time-variant features used by the model. We call this approach **'mask-detect'**.

Most of the existing techniques for fake news detection make use of some particular input features which are time-variant. In case of ENDEMIC, these are the additional

Table 4. Performance comparison on two general-domain datasets to check the generalisability of the models.

Model	PolitiFact				GossipCop			
	Accuracy	Precision	Recall	F1 Score	Accuracy	Precision	Recall	F1 Score
HAN	0.863	0.730	0.855	0.790	0.800	0.685	0.775	0.770
MixText	0.870	0.840	0.860	0.853	0.785	0.805	0.795	0.813
GCAN	0.867	0.855	0.880	0.875	0.790	0.805	0.795	0.803
CSI	0.827	0.847	0.897	0.871	0.772	0.732	0.638	0.682
dEFEND	0.904	0.902	**0.956**	**0.928**	0.808	0.729	0.782	0.755
CNN-MWSS	0.823	0.811	0.830	0.820	0.770	0.793	0.815	0.795
RoBERTa-MWSS	0.825	0.833	0.795	0.805	0.803	0.815	0.810	0.807
FNED	0.907	**0.910**	0.922	0.916	0.832	0.830	0.825	0.830
PPC	0.885	0.887	0.870	0.880	0.791	0.805	0.832	0.811
ENDEMIC	**0.912**	**0.910**	0.904	0.915	**0.847**	**0.835**	**0.822**	**0.840**

contextual tweet features and user features, which rapidly change over time. When such features are masked, there is effectively no way to distinguish an old tweet from a new one. Therefore, we perform mask-detect by replacing the numerical values of the relevant time-variant features with a common masking token. These features are different for each model. Therefore, we first identify such features and then perform the masking. Table 3 shows the features masked for the various fake news detection models and their corresponding performance.

5.4 Evaluating on General-Domain Fake News

Although we tested our model on COVID-19, ENDEMIC is highly generalised to other domains. And to prove the generalisability, we also evaluate ENDEMIC on the FakeNewsNet [23] datasets – *PolitiFact* and *GossipCop*. Both of these datasets are obtained from the respective sources and contain labelled textual news content along with the social context. Evaluating on these datasets allows us to validate ENDEMIC's generalisation across domains outside of COVID-19 and beyond ECTF.

Table 4 shows that ENDEMIC outperforms all baselines with more than 0.5% accuracy on *PolitiFact* and 1.5% on *GossipCop*. The performance of ENDEMIC is highly comparable compared to other baselines w.r.t. other evaluation metrics. These results clearly establish that ENDEMIC generalises well on any fake news domain and dataset.

6 Conclusion

In this work, we introduced the task of early detection of COVID-19 fake tweets. We developed a COVID-19 dataset with additional test set to evaluate models for early detection. We took measures to simulate the early stages of the fake news events by extracting the external knowledge relative to the time the tweet is posted. Our proposed model, ENDEMIC overcomes the challenges of limited dataset in a semi-supervised fashion. We adhere to co-attention as a better information fusion tested on time invariant features. ENDEMIC outperformed nine baselines in both the tasks of general and early-stage fake news detection. Experimental results compared to nine state-of-the-art models show that ENDEMIC is highly capable for early detection task. We also showed the generalisability of ENDEMIC on two other publicly available fake news dataset which are not specific to COVID-19.

Acknowledgements. The work was partially supported by Accenture Labs, SPARC (MHRD) and CAI, IIIT-Delhi. T. Chakraborty would like to thank the support of the Ramanujan Fellowship.

References

1. Bruna, J., Zaremba, W., Szlam, A., LeCun, Y.: Spectral networks and locally connected networks on graphs. arXiv preprint arXiv:1312.6203 (2013)
2. Carlson: Coronavirus tweets, tweets (json) for coronavirus on Kaggle (2020). https://www.kaggle.com/carlsonhoo/coronavirus-tweets. Accessed 2020
3. Castillo, C., Mendoza, M., Poblete, B.: Information credibility on Twitter. In: WWW, pp. 675–684 (2011)
4. Chen, J., Yang, Z., Yang, D.: Mixtext: linguistically-informed interpolation of hidden space for semi-supervised text classification. arXiv preprint arXiv:2004.12239 (2020)
5. Cui, L., Shu, K., Wang, S., Lee, D., Liu, H.: dEFEND: a system for explainable fake news detection. In: CIKM, pp. 2961–2964 (2019)
6. Guacho, G.B., Abdali, S., Shah, N., Papalexakis, E.E.: Semi-supervised content-based detection of misinformation via tensor embeddings. In: ASONAM, pp. 322–325. IEEE (2018)
7. Gururangan, S., Dang, T., Card, D., Smith, N.A.: Variational pretraining for semi-supervised text classification. arXiv preprint arXiv:1906.02242 (2019)
8. Helmstetter, S., Paulheim, H.: Weakly supervised learning for fake news detection on Twitter. In: ASONAM, pp. 274–277. IEEE (2018)
9. Liu, Y., Wu, Y.: Early detection of fake news on social media through propagation path classification with recurrent and convolutional networks. In: AAAI, pp. 354–361 (2018)
10. Liu, Y., Wu, Y.F.B.: FNED: a deep network for fake news early detection on social media. ACM Trans. Inf. Syst. **38**(3), 1–33 (2020)
11. Liu, Y., et al.: RoBERTa: a robustly optimized BERT pretraining approach. arXiv preprint arXiv:1907.11692 (2019)
12. Lohr, S.: It's true: false news spreads faster and wider. And humans are to blame. The New York Times 8 (2018)
13. Lu, J., Yang, J., Batra, D., Parikh, D.: Hierarchical question-image co-attention for visual question answering. In: Lee, D., Sugiyama, M., Luxburg, U., Guyon, I., Garnett, R. (eds.) NIPS, vol. 29, pp. 289–297. Curran Associates, Inc. (2016)
14. Lu, Y.J., Li, C.T.: GCAN: graph-aware co-attention networks for explainable fake news detection on social media. In: ACL, pp. 505–514, July 2020
15. Miyato, T., Dai, A.M., Goodfellow, I.: Adversarial training methods for semi-supervised text classification. arXiv preprint arXiv:1605.07725 (2016)
16. Monti, F., Frasca, F., Eynard, D., Mannion, D., Bronstein, M.M.: Fake news detection on social media using geometric deep learning. arXiv preprint arXiv:1902.06673 (2019)
17. Niepert, M., Ahmed, M., Kutzkov, K.: Learning convolutional neural networks for graphs. In: ICML, pp. 2014–2023 (2016)

18. Nigam, K., McCallum, A.K., Thrun, S., Mitchell, T.: Text classification from labeled and unlabeled documents using EM. Mach. Learn. **39**(2–3), 103–134 (2000)
19. Paka, W.S., Bansal, R., Kaushik, A., Sengupta, S., Chakraborty, T.: Cross-SEAN: a cross-stitch semi-supervised neural attention model for COVID-19 fake news detection. arXiv preprint arXiv:2102.08924 (2021)
20. Reimers, N., Gurevych, I.: Sentence-BERT: sentence embeddings using Siamese BERT-networks. In: EMNLP-IJCNLP, Hong Kong, China, pp. 3982–3992 (2019)
21. Rosenfeld, N., Szanto, A., Parkes, D.C.: A kernel of truth: determining rumor veracity on Twitter by diffusion pattern alone. In: The Web Conference, pp. 1018–1028 (2020)
22. Smith, S.: Coronavirus (covid19) Tweets - early April (2020). https://www.kaggle.com/smid80/coronavirus-covid19-tweets-early-april. Accessed 2020
23. Shu, K., Mahudeswaran, D., Wang, S., Lee, D., Liu, H.: FakeNewsNet: a data repository with news content, social context and dynamic information for studying fake news on social media. arXiv preprint arXiv:1809.01286 8 (2018)
24. Shu, K., Wang, S., Liu, H.: Understanding user profiles on social media for fake news detection. In: MIPR, pp. 430–435. IEEE (2018)
25. Shu, K., et al.: Leveraging multi-source weak social supervision for early detection of fake news. arXiv preprint arXiv:2004.01732 (2020)
26. Celin, S.: COVID-19 tweets afternoon 31.03.2020 (2020). https://www.kaggle.com/svencelin/covid19-tweets-afternoon-31032020. Accessed 2020
27. Wang, Y., et al.: Weak supervision for fake news detection via reinforcement learning. In: AAAI, vol. 34, pp. 516–523 (2020)
28. Yang, Z., Yang, D., Dyer, C., He, X., Smola, A., Hovy, E.: Hierarchical attention networks for document classification. In: NAACL, pp. 1480–1489 (2016)
29. Yu, D., Chen, N., Jiang, F., Fu, B., Qin, A.: Constrained NMF-based semi-supervised learning for social media spammer detection. Knowl. Based Syst. **125**, 64–73 (2017)
30. Zhou, X., Jain, A., Phoha, V.V., Zafarani, R.: Fake news early detection: A theory-driven model. Digital Threats Res. Pract. **1**(2) (2020)
31. Zhou, X., Zafarani, R.: A survey of fake news: fundamental theories, detection methods, and opportunities. ACM CSUR **53**(5), 1–40 (2020)

TLife-LSTM: Forecasting Future COVID-19 Progression with Topological Signatures of Atmospheric Conditions

Ignacio Segovia-Dominguez[1,2](✉), Zhiwei Zhen[1], Rishabh Wagh[1], Huikyo Lee[2], and Yulia R. Gel[1]

[1] The University of Texas at Dallas, Richardson, TX 75080, USA
ignacio.segoviadominguez@utdallas.edu
[2] Jet Propulsion Laboratory, California Institute of Technology, Pasadena, CA 91109, USA

Abstract. Understanding the impact of atmospheric conditions on SARS-CoV2 is critical to model COVID-19 dynamics and sheds a light on the future spread around the world. Furthermore, geographic distributions of expected clinical severity of COVID-19 may be closely linked to prior history of respiratory diseases and changes in humidity, temperature, and air quality. In this context, we postulate that by tracking topological features of atmospheric conditions over time, we can provide a quantifiable structural distribution of atmospheric changes that are likely to be related to COVID-19 dynamics. As such, we apply the machinery of persistence homology on time series of graphs to extract topological signatures and to follow geographical changes in relative humidity and temperature. We develop an integrative machine learning framework named Topological Lifespan LSTM (TLife-LSTM) and test its predictive capabilities on forecasting the dynamics of SARS-CoV2 cases. We validate our framework using the number of confirmed cases and hospitalization rates recorded in the states of Washington and California in the USA. Our results demonstrate the predictive potential of TLife-LSTM in forecasting the dynamics of COVID-19 and modeling its complex spatio-temporal spread dynamics.

Keywords: Dynamic networks · COVID-19 · Topological Data Analysis · Long Short Term Memory · Environmental factors · Clinical severity

1 Introduction

Nowadays, there is an ever-increasing spike of interest in enhancing our understanding of hidden mechanisms behind transmission of SARS-CoV2 (i.e., the virus that causes COVID-19) and its potential response to atmospheric conditions including temperature and relative humidity [4,16,23]. Understanding the impact of atmospheric conditions on COVID-19 trajectory and associated mortality is urgent and critical, not only in terms of efficiently responding to the

K. Karlapalem et al. (Eds.): PAKDD 2021, LNAI 12712, pp. 201–212, 2021.
https://doi.org/10.1007/978-3-030-75762-5_17

current pandemic (e.g., preparing an adequate health care response in areas with expected higher clinical coronavirus severity), but also in terms of forecasting impending hotspots and potential next-wave occurrences.

However, as shown in many recent biosurveillance studies [2,20], the non-trivial relationship between the spatio-temporal dynamics of atmospheric data and disease transmission may not be captured well by conventional metrics based on Euclidean distances. This phenomenon can be partially explained by a sophisticated spatio-temporal dependence structure of the atmospheric conditions. A number of recent results on (re)emerging infectious have introduced the concepts of topological data analysis (TDA) into modeling the spread of climate-sensitive viruses. In particular, TDA and, specifically, persistent homology have been employed by [11] for analysis of influenza-like illness during the 2008–2013 flu seasons in Portugal and Italy. Most recently, [19,25] show the explanatory and predictive power of TDA for analysis of Zika spread. The obtained results show that topological descriptors of spatio-temporal dynamics of atmospheric data tend to improve forecasting of infectious diseases. These findings are largely due to the fact that topological descriptors allow for capturing higher-order dependencies among atmospheric variables that otherwise might be unassessable via conventional spatio-temporal modeling approaches based on geographical proximity assessed via Euclidean distance.

In this paper, we develop a novel predictive deep learning (DL) platform for COVID-19 spatio-temporal spread, coupled with topological information on atmospheric conditions. The key idea of the new approach is based on the integration of the most essential (or *persistent*) topological descriptors of temperature and humidity, into the Long Short Term Memory (LSTM) model. The new Topological Lifespan Long Short Term Memory (TLife-LSTM) approach allows us to track and forecast COVID-19 spread, while accounting for important local and global variability of epidemiological factors and its complex dynamic interplay with environmental and socio-demographic variables.

The significance of our paper can be summarized as follows:

- To the best of our knowledge, this is the first paper, systematically addressing complex nonlinear relationships between atmospheric conditions and COVID-19 dynamics.
- The proposed TLife-LSTM model, coupled with TDA, allows for efficient and mathematically rigorous integration of hidden factors behind COVID-19 progression which are otherwise inaccessible with conventional predictive approaches, based on Euclidean distances.
- Our case studies indicate that the new TLife-LSTM approach delivers a highly competitive performance for COVID-19 spatio-temporal forecasting in the states of California and Washington on a county-level basis. These findings suggest that TLife-LSTM and, more generally, biosurveillance tools based on topological DL might be the most promising predictive tools for COVID-19 dynamics under the scenarios of limited and noisy epidemiological data records.

2 Related Work

Throughout the last year numerous studies have found that DL and, particularly, LSTM exhibit predictive utility for COVID-19 spread at the country level and varying forecasting horizons e.g. 2, 4, 6, 8, 10, 12 and 14 days [1,3,5,27]. Similar biosurveillance performance in terms of Root Mean Square Error (RMSE) and Mean Absolute Error (MAE), have been also obtained for various Recurrent Neural Networks (RNN) architectures [24,27]. Hence, despite requiring higher historical epidemiological records, DL tools nowadays are viewed as one of the most promising modeling approaches to address short- and medium-term prediction of COVID-19 dynamics [22].

Topological Data Analysis (TDA) provides a rigorous theoretical background to explore topological and geometric features of complex geospatial data. TDA has been proven useful in a broad range of data science applications [8,15,21], including biosurveillance [11,19,25], COVID-19 spread visualization [10,12,18] and COVID-19 therapy analysis [9]. However, to the best of our knowledge, there yet exists no *predictive models for COVID-19, harnessing the power of TDA*. Even more, there are, still, few research studies incorporating atmospheric conditions as inputs of RNNs to forecasting COVID-19 spread. Our research makes contributions in both areas and creates connections between TDA and DL through adding topological signatures, i.e. covariates, into a LSTM architecture.

3 Problem Statement

Our primary goal is to construct a RNN model which harnesses the power of TDA to forecast COVID-19 dynamics at various spatial resolutions. We exploit the coordinate-free PH method to extract topological features from environmental variables and to summarize the most essential topological signatures of atmospheric conditions.

In this project we use two distinctive data types: 1) the number of COVID-19 confirmed cases/hospitalizations from US states at county-level, and 2) daily records of environmental variables from weather stations. Our goal is to model the COVID-19 dynamics and predict its spread and mortality at county-level in US states, while accounting for complex relationships between atmospheric conditions and current pandemic trends.

4 Background

We start from providing a background on the key concepts employed in this project, namely, TDA and LSTM neural networks.

4.1 Preliminaries on Topological Data Analysis

The fundamental idea of TDA is that the observed data $\mathbb{X}$ represent a discrete sample from some metric space and, due to sampling, the underlying structure of this space has been lost. The main goal is then to systematically retrieve the lost underlying structural properties, by quantifying dynamics of topological properties exhibited by the data, assessed through multiple user-defined (dis)similarity scales [6,7,14]. Such (dis)similarity measure can be, for instance, similarity of temperature values or homogeneity of COVID-19 counts in various counties. Hence, given this idea, the derived TDA summaries are expected to be inherently robust to missing data, to matching multiple spatio-temporal data resolutions, and other types of uncertainties in the observed epidemiological information.

The TDA approach is implemented in the three main steps (see the toy example in Fig. 1):

1. We associate $\mathbb{X}$ with some filtration of $\mathbb{X}$: $\mathbb{X}_1 \subseteq \mathbb{X}_2 \subseteq \ldots \subseteq \mathbb{X}_k = \mathbb{X}$.
2. We then monitor the evolution of various pattern occurrences (e.g., cycles, cavities, and more generally k-dimensional holes) in nested sequence of subsets. To make this process efficient and systematic, we equip $\mathbb{X}$ with certain combinatorial objects, e.g., simplicial complexes. Formally, a simplicial complex is defined as a collection $\mathcal{C}$ of finite subsets of $\mathcal{G}$ such that if $\sigma \in \mathcal{C}$ then $\tau \in \mathcal{C}$ for all $\tau \subseteq \sigma$. The basic unit of simplicial complexes is called the *simplex*, and if $|\sigma| = m+1$ then σ is called an m-simplex. Here we employ a Vietoris–Rips (VR) complex which is one of the most widely used complexes due its computational cost benefits [8,14]. Filtration of $\mathbb{X}_1 \subseteq \mathbb{X}_2 \subseteq \ldots$ is then associated with filtration of VR complexes $VR_1 \subseteq VR_2 \subseteq \ldots$.
3. We track the index of VR complex t_b when each topological feature is first recorded (i.e., born) and the index of VR complex t_d when this feature is last seen (i.e., died). Then lifespan of this feature is $t_d - t_b$. The extracted topological characteristics with a longer lifespan are called *persistent* and are likelier connected to some fundamental mechanisms behind the underlying system organization and dynamics, e.g., higher-order interactions between disease transmissibility and changes in atmospheric variables. In turn, topological features with a shorter lifespan are referred to as *topological noise*.

We then use distributions of the recorded lifespans of topological features as inherent signatures, characterizing the observed data $\mathbb{X}$ and integrate these topological descriptors into our TLife-LSTM model.

4.2 Long Short Term Memory

RNNs and, particularly, Long Short Term Memory (LSTM) approaches have been successfully used to model time series and other time-dependent data [26]. The LSTM architecture addresses the problem of the gradient instability of predecessors and adds extra flexibility due to the memory storage and forget gates. Each LSTM unit contains three transition functions: input gate i_t, output gate o_t and forget gate f_t;

$$i_t = \sigma(W_i \cdot [h_{t-1}, x_t] + b_i) \quad (1)$$

$$o_t = \sigma(W_o \cdot [h_{t-1}, x_t] + b_o) \quad (2)$$

$$f_t = \sigma(W_f \cdot [h_{t-1}, x_t] + b_f) \quad (3)$$

where σ represents the sigmoid function. The gates' information and the output vector h_{t-1} of the hidden layer obtained at step $t-1$ serve as input to the memory cell c_t, Eq. (5). It allows to recursively use vector output h_t to extract patterns from complex time series.

$$g_t = \tanh(W_g \cdot [h_{t-1}, x_t] + b_g) \tag{4}$$

$$c_t = f_t \odot c_{t-1} + i_t \odot g_t \tag{5}$$

$$h_t = o_t \odot \tanh c_t \tag{6}$$

During the training stage, LSTM neural networks learn the weight matrices W_i, W_o, W_f and W_g, while the performance is affected by bias vectors b_i, b_o, b_f and b_g. Most of variants of LSTM architecture performs similarly well in large scale studies, see [17].

5 The Proposed Topological Lifespan Long Short Term Memory (TLife-LSTM) Approach

We now introduce a new DL framework to capture topological patterns in the observed spatio-temporal data and to model complex relationships in time series. Details are divided into two sections. First, we explain the construction of dynamic networks and posterior extraction of n-dimensional features. Then, we expand on the general methodology and provide specifics on the neural network architecture.

5.1 Topological Features of Environmental Dynamic Networks

Topological features, as defined in Sect. 4.1, describe the shape structure of underlying data which are invariant under continuous transformations such as twisting, bending, and stretching. In this study, we primarily focus on lifespans of such topological features computed on environmental dynamic networks.

Definition 1. *Let $\mathcal{G} = (V, E, \omega)$ be a weighted graph, where V is a set of vertices (e.g., weather stations), $E = \{e_1, e_2, \ldots\} \subseteq V \times V$ is a set of edges, and $\omega = \omega(e) : E \to \mathbb{Z}^+$ for all $e \in E$ are edge weights. (Here $\omega(e)$ may represent a level of similarity exhibited by atmospheric conditions at two weather stations).*

In particular, our biosurveillance methodology is inspired by the recent advancements of applying PH on dynamic networks. Let $\mathbb{G} = \{\mathcal{G}_t\}_{t=1}^{T} = \{\mathcal{G}_1, \ldots, \mathcal{G}_T\}$ be a sequence of time-evolving weighted networks observed over index-time t, with $1 \le t \le T < \infty$. Then, the objective is to compute and summarize persistent topological features at each $\mathcal{G}_t$; particularly lifespans as defined in Sect. 4.1. Hence, PH emerges as a topological technique to track changes in dynamic networks.

Meteorologists monitor atmospheric conditions using ground-based weather stations and satellites. Ground-based stations collect observations at different time resolutions. Let $O^{(i)}$ be historical-meteorological data, i.e., time series observations, collected from station $S^{(i)} = \{\text{latitude}, \text{longitude}\}$ such that $O^{(i)}$ is a time-ordered sample where each element contains λ observed measures of

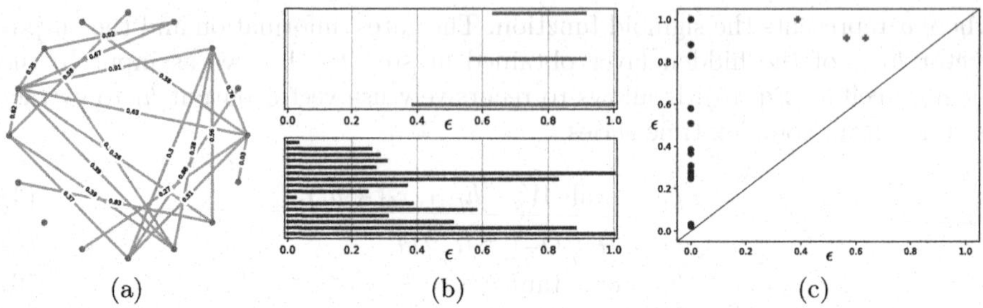

Fig. 1. Example of applying persistent homology (PH) in a weighted graph. (a) Weighted graph. (b) Barcode, birth and death of topological summaries. Blue: 0-dimensional features. Red: 1-dimensional features. (c) Persistence diagram. (Color figure online)

weather conditions, i.e. $O_t^{(i)} = \{O_{t1}^{(i)}, O_{t2}^{(i)}, \ldots, O_{t\lambda}^{(i)}\}$, and indexed by time t. Given a set of stations $\{S^{(i)}\}_{i=1}^{M}$ sited on the area of study (California and Washington), we construct a dynamic network $\mathbb{G} = \{\mathcal{G}_t\}_{t=1}^{T} = \{\mathcal{G}_1, \ldots, \mathcal{G}_T\}$ using each station as vertex $v_i \in V$. Each undirected graph $\mathcal{G}_t$ contains edge weights $\omega_t(e)$ which vary over time according with weather measurements.

Our goal is to build time series of graphs that reflect the underlying connections of atmospheric conditions across local regions. That is, attaching weighted edges as a function of temporal-varying temperature or relative humidity provides a natural way to track weather changes between regions and link these with the trend in COVID-19 dynamics.

Algorithm 1. Topological Feature Extraction from Environmental Dynamic Networks

---

INPUT: Weather Conditions $\{O^{(i)}\}_{i=1}^{M}$; Station Locations $\{S^{(i)}\}_{i=1}^{M}$
OUTPUT: Topological summaries

for $i \leftarrow 1 : M-1$ **do**
  for $j \leftarrow i+1 : M$ **do**
    Compute $L_{ij} = Norm(S^{(i)} - S^{(j)})$
for $t \leftarrow 1 : T$ **do**
  for $h \leftarrow 1 : \lambda$ **do**
    for $i \leftarrow 1 : M-1$ **do**
      for $j \leftarrow i+1 : M$ **do**
        Compute $D_{th}^{(ij)} = O_{th}^{(i)} - O_{th}^{(j)}$
    Standardize matrix D_{th}
  Compute $D_t = \frac{1}{2\lambda} \sum_{h=1}^{\lambda} D_{th} + \frac{1}{2} L$
  Compute $\omega_t(e) = D_t/\text{max element in } D_t$
  Generate $\mathcal{G}_t$ based on $\omega_t(e)$
Apply PH on dynamic networks $\mathbb{G} = \{\mathcal{G}_t\}_{t=1}^{T}$ for 0, 1, 2 and 3 dimensions
Calculate mean, total sum, variance of features' lifespan, and the number of topological features obtained through PH in each dimension

---

Algorithm 1 describes our methodology to create each weighted graph $\mathcal{G}_t$, based on atmospheric conditions, $O_t^{(i)}$, and its corresponding feature extraction. We propose to use the TDA tools, particularly, persistent homology (PH) to track qualitative features that persist across multiple scales. First, we generate a lower triangular matrix L to represent the distance between stations. Then for each weather measurement, we use D_{th} to record the difference in O_t between stations. For D_{th}, we take all low-triangular elements as a sample to do the standardization and generate a new matrix D_t based on the atmospheric connections D_{th} and the distance L. To ensure that weights in the dynamic networks are between 0 and 1, we divide D_t by its maximum element. Finally, we generate our dynamic networks and use the VR complex to get one persistence diagram on each $\mathcal{G}_t$. Based on features' lifespans, we compute the mean, total sum, and variance of the lifespan of topological features, and use these topological signatures as input for our RNN architecture.

5.2 Topological Machine Learning Methodology: TLife-LSTM

Let $\mathbf{Y} = \{Y_i\}_{i=1}^N$ be a multivariate time series, where $Y_i \in \mathbf{Y}$ and each Y_{it} is a sequence of outcomes from random variables Z_{it}, $t = 1, 2, \ldots, T$ indexed by time t. Hence, historical data of the $i - esim$ element Y_i is a time-ordered sample of observations $Y_{i1}, Y_{i2}, \ldots, Y_{iT}$. Given the partition of $\mathbf{Y} = \{\mathbf{A}, \mathbf{B}\}$ such that $\mathbf{A} = \{Y_1, Y_2, \ldots, Y_\tau\}$ and $\mathbf{B} = \{Y_{\tau+1}, Y_{\tau+2}, \ldots, Y_T\}$, vectors $Y_j \in \mathbf{A}$ are formed by historical data of COVID-19 progression, whilst vectors $Y_k \in \mathbf{B}$ are topological signatures from an environmental dynamic network $\mathbb{G}$.

Given multivariate time-series dataset $\mathbf{Y} = \{Y_i\}_{i=1}^N$, we train a LSTM model to capture complex relationships between variables. Notice that input comes from two different sources: 1) historical data of COVID-19 dynamics, and 2) persistent signatures of atmospheric conditions. We then create a suitable RNN architecture to handle these data. Figure 2a shows a graphical representation of the main modules in our RNN architecture. First, observations are received

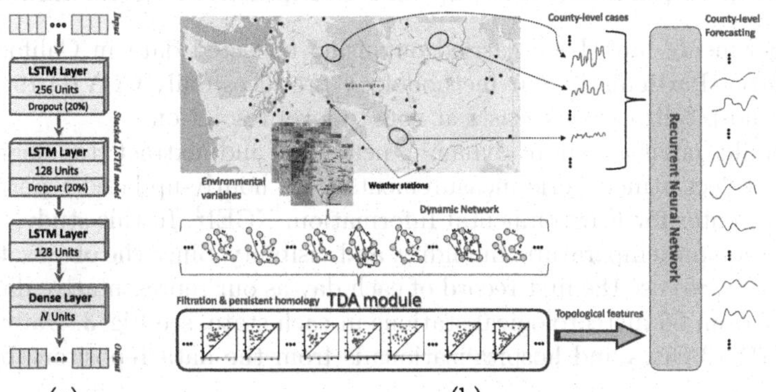

Fig. 2. Topological LSTM (a) RNN architecture. (b) Topological Lifespan LSTM.

as real vectors and passed through a series of three-stacked LSTM layers with 256, 128, and 128 units, respectively. To avoid overfitting, we follow the dropout regularization technique, i.e., randomly dropping units in the training step. Next, a densely connected layer accounts for additional relationship and prepares the data-flow to produce N outputs. In the current study, each output corresponds to each local region, e.g., counties in the two US states.

Figure 2b depicts our proposed Topological Lifespan LSTM (TLife-LSTM). The key idea is to assemble the computed topological summaries along with the time series of confirmed COVID-19 cases. Hence, we complement conventional biosurveillance information with the topological summaries, in order to enrich the input data in our deep learning phase. Although there have been various approaches using neural networks to predict COVID-19 dynamics, our methodology is unique as it integrates atmospheric conditions in its learning step. Furthermore, incorporation of topological signatures aims to ensure that the learning phase only relates distinctive and persistent features of atmospheric conditions along with intrinsic dynamics of COVID-19 progression. Hence, with TLife-LSTM, we extract higher-order dependence properties among temperature, relative humidity, and COVID-19 spread which are otherwise inaccessible with more conventional methods based on the analysis of geographic proximity.

6 Experiments

To assess the predictive capabilities of the proposed approach, we present experimental forecasts of COVID-19 dynamics: 1) the number of confirmed cases (**Cases**), and 2) the number of hospitalizations (**Hospi**). To evaluate the performance of our forecasts, we compare results using common evaluation metrics and present insightful visualizations at county-level. All experiments are run using Keras and Tensorflow, source codes[1] are published online to encourage reproducibility and replicability.

6.1 Data, Experimental Setup and Comparative Performance

Our experiments have been carried out using collected data in California and Washington. Particularly, our methodology produces daily COVID-19 progression and hospitalization forecasts at county-level resolution.

To build environmental dynamic networks and extract topological features, we select atmospheric measurements from hourly-updated reports of the National Center for Environmental Information, NCEI[2]. In this study, we focus our attention on temperature, humidity, and visibility. Since the observations are made hourly, we use the first record of each day as our representative daily measurement from 66 meteorological stations in each state, see Fig. 3. Daily records on COVID-19 cases and hospitalizations is from the data repository by Johns

[1] Available at Source codes (repository).
[2] Available at https://www.ncei.noaa.gov.

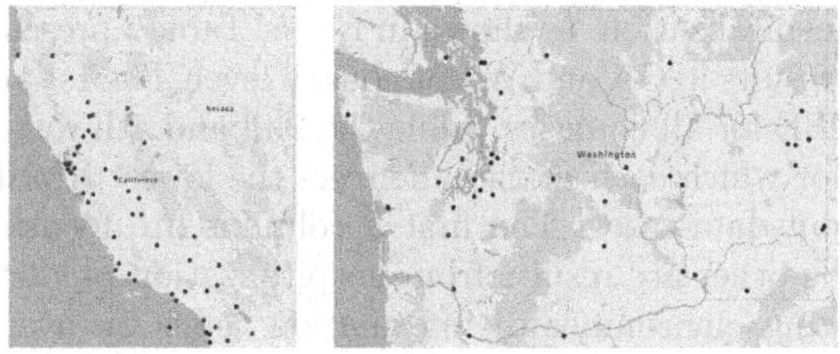

(a) (b)

Fig. 3. Land-based selected meteorological stations. (a) California state. (b) Washington state.

Hopkins University[3], see [13], which includes aggregated data sources from the World Health Organization, European Center for Disease Prevention and Control, and US Center for Disease Control and Prevention. For information about the COVID-19 disease progression, and additional modeling resources, we use the MIDAS online portal for COVID-19 modeling research[4].

We split the datasets into a training set, from April 15 to August 31, and a test set, the whole month of September. We train our RNN architecture to produce daily predictions for the next 30 days, i.e., one month ahead of forecasting, and use RMSE as a metric to assess the predictive capability of the derived forecasts.

To verify the value added by the topological descriptors, we perform predictions on three different input-data-cases: a) using only the number of confirmed cases (LSTM), b) using the number of confirmed cases plus environmental variables (LSTM+EV), and c) using the number of confirmed cases plus topological summaries (TLife-LSTM). Notice that case (a) corresponds to the traditional application of LSTM on COVID-19 data as in literature, and case (b) is equal to use LSTM with environmental variables. As Fig. 4 show, forecasting performance of new daily COVID-19 cases at a county level tend to benefit from integrating topological summaries of atmospheric conditions. Similarly, the topological features extracted from environmental dynamic networks also tend to enhance the

Table 1. RMSE Results in California (CA) and Washington (WA). Performance comparison.

Statistics	LSTM **Cases**	LSTM+EV **Cases**	TLife-LSTM **Cases**	LSTM **Hospi**	LSTM+EV **Hospi**	TLife-LSTM **Hospi**
CA (means)	64.5662	58.7211	60.5828	13.8351	19.3277	12.1730
CA (freq)	16	22	21	16	9	31
WA (means)	17.2531	11.4251	17.10648	3.1329	2.4663	3.0746
WA (freq)	4	32	4	11	15	11

[3] Available at https://github.com/CSSEGISandData/COVID-19.
[4] Available at https://midasnetwork.us/covid-19/.

forecasts of the hospitalization, as shown in Fig. 5. Table 1 presents the summary of the prediction results in CA and WA at county level. The 1st and 3rd row show the means of RMSE for all counties, while the 2nd and 4th rows show the numbers of counties for which each method delivers the lowest RMSE. The columns represent the input-data-cases. The first 3 columns are focusing on the cases prediction and the other are results from hospitalization prediction. As Table 1 suggests, integrating environmental information tends on average to improve the forecasting performance of the number of cases and hospitalizations in both states. While we tend to believe that the impact of environmental variables on COVID-19 transmission is likely to be only indirect, the impact of environmental information on future hospitalization appears to be much more profound. This phenomenon is likely to be linked to the connection of atmospheric variables and multiple pre-existing health conditions, e.g., respiratory and cardiovascular diseases, that result in elevating risks of COVID-19 severity. As such, the important future extensions of this analysis include integration of various sociodemographic variables (e.g., population, health conditions, and median age) into TLife-LSTM.

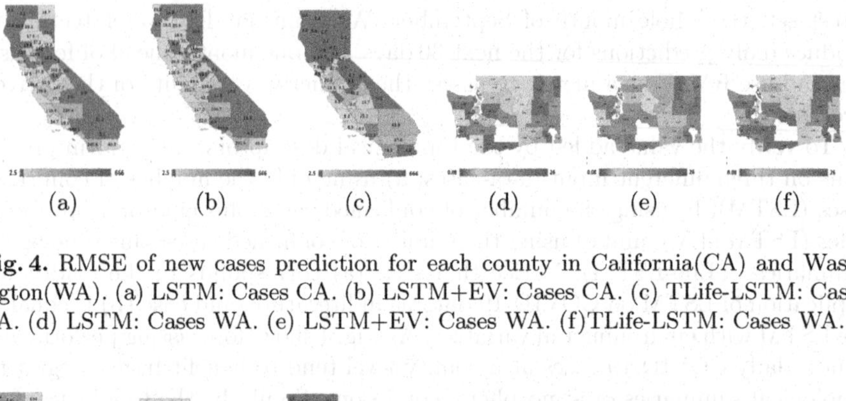

(a) (b) (c) (d) (e) (f)

Fig. 4. RMSE of new cases prediction for each county in California(CA) and Washington(WA). (a) LSTM: Cases CA. (b) LSTM+EV: Cases CA. (c) TLife-LSTM: Cases CA. (d) LSTM: Cases WA. (e) LSTM+EV: Cases WA. (f)TLife-LSTM: Cases WA.

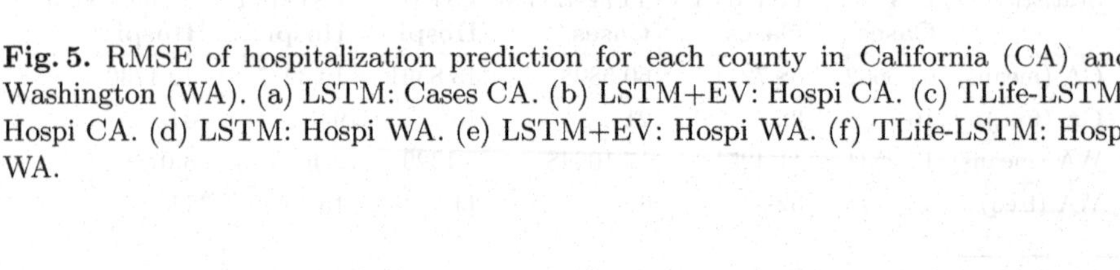

(a) (b) (c) (d) (e) (f)

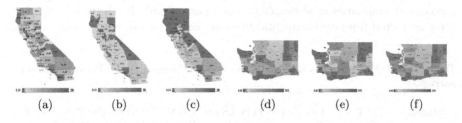

Fig. 5. RMSE of hospitalization prediction for each county in California (CA) and Washington (WA). (a) LSTM: Cases CA. (b) LSTM+EV: Hospi CA. (c) TLife-LSTM: Hospi CA. (d) LSTM: Hospi WA. (e) LSTM+EV: Hospi WA. (f) TLife-LSTM: Hospi WA.

7 Conclusions and Future Scope

We have developed a new topological machine learning model TLife-LSTM to forecast COVID-19 dynamics, with a focus on using environmental and topological features along with time series of the historical COVID-19 records. Our experiments have indicated that TLife-LSTM yields more accurate forecasts of future number of cases and hospitalizations rates in the states of California and Washington on a county-level basis. As a result, TLife-LSTM indicates the impact of atmospheric conditions on COVID-19 progression and associated mortality; critical to efficiently respond to the current pandemic and forecast impeding hotspots and potential waves. In the future, we plan to introduce socio-demographic variables to our model and encode multi-parameter persistent characteristics of epidemiological and atmospheric conditions.

Acknowledgments. This work has been supported in part by grants NSF DMS 2027793 and NASA 20-RRNES20-0021. Huikyo Lee's research was carried out at the Jet Propulsion Laboratory, California Institute of Technology, under a contract with the National Aeronautics and Space Administration (80NM0018D0004).

References

1. Alazab, M., Awajan, A., Mesleh, A., Abraham, A., Jatana, V., Alhyari, S.: COVID-19 prediction and detection using deep learning. Int. J. Comput. Inf. Syst. Ind. Manage. Appl. **12**, 168–181 (2020)
2. de Ángel Solá, D.E., Wang, L., Vázquez, M., Méndez Lázaro, P.A.: Weathering the pandemic: how the Caribbean Basin can use viral and environmental patterns to predict, prepare and respond to COVID-19. J. Med. Virol. **92**(9), pp. 1460–1468 (2020)
3. Arora, P., Kumar, H., Panigrahi, B.: Prediction and analysis of COVID-19 positive cases using deep learning models: a descriptive case study of India. Chaos Solitons Fractals **139**, 110017 (2020). https://doi.org/10.1016/j.chaos.2020.110017
4. Berumen, J., et al.: Trends of SARS-Cov-2 infection in 67 countries: role of climate zone, temperature, humidity and curve behavior of cumulative frequency on duplication time. medRxiv (2020)
5. Bouhamed, H.: COVID-19 cases and recovery previsions with Deep Learning nested sequence prediction models with Long Short-Term Memory (LSTM) architecture. Int. J. Sci. Res. Comput. Sci. Eng. **8**, 10–15 (2020)
6. Carlsson, G.: Topology and data. BAMS **46**(2), 255–308 (2009)
7. Carlsson, G.: Persistent homology and applied homotopy theory. In: Handbook of Homotopy Theory. CRC Press, Boca Raton (2019)
8. Chazal, F., Michel, B.: An introduction to topological data analysis: fundamental and practical aspects for data scientists. arxiv:1710.04019 (2017)
9. Chen, J., Gao, K., Wang, R., Nguyen, D.D., Wei, G.W.: Review of COVID-19 antibody therapies. Annu. Rev. Biophys. **50** (2020)
10. Chen, Y., Volic, I.: Topological data analysis model for the spread of the coronavirus. arXiv:2008.05989 (2020)
11. Costa, J.P., Škraba, P.: A topological data analysis approach to epidemiology. In: European Conference of Complexity Science (2014)

12. Dlotko, P., Rudkin, S.: Visualising the evolution of English COVID-19 cases with topological data analysis ball mapper. arXiv:2004.03282 (2020)
13. Dong, E., Du, H., Gardner, L.: An interactive web-based dashboard to track COVID-19 in real time. Lancet Infect. Dis. **20**(5), 533–534 (2020). https://doi.org/10.1016/S1473-3099(20)30120-1
14. Edelsbrunner, H., Harer, J.: Persistent homology - a survey. Contemp. Math. **453**, 257–282 (2008)
15. Falk, M., et al.: Topological data analysis made easy with the topology toolkit, what is new? (2020)
16. Franch-Pardo, I., Napoletano, B.M., Rosete-Verges, F., Billa, L.: Spatial analysis and GIS in the study of COVID-19. A review. Sci. Total Environ. **739**, 140033 (2020)
17. Greff, K., Srivastava, R.K., Koutník, J., Steunebrink, B.R., Schmidhuber, J.: LSTM: a search space odyssey. IEEE Trans. Neural Netw. Learn. Syst. **28**(10), 2222–2232 (2017). https://doi.org/10.1109/TNNLS.2016.2582924
18. Johnson, L., Schieberl, L.: Topological visualization of COVID-19 spread in California, Florida, and New York (2020)
19. Lo, D., Park, B.: Modeling the spread of the Zika virus using topological data analysis. PLoS One **13**(2), e0192120 (2018)
20. Metcalf, C.J.E., et al.: Identifying climate drivers of infectious disease dynamics: recent advances and challenges ahead. Proc. R. Soc. B Biol. Sci. **284**(1860), 20170901 (2017)
21. Otter, N., Porter, M.A., Tillmann, U., Grindrod, P., Harrington, H.A.: A roadmap for the computation of persistent homology. EPJ Data Sci. **6**(1), 1–38 (2017). https://doi.org/10.1140/epjds/s13688-017-0109-5
22. Ramchandani, A., Fan, C., Mostafavi, A.: DeepCOVIDNet: an interpretable deep learning model for predictive surveillance of COVID-19 using heterogeneous features and their interactions. IEEE Access **8**, 159915–159930 (2020). https://doi.org/10.1109/ACCESS.2020.3019989
23. Rouen, A., Adda, J., Roy, O., Rogers, E., Lévy, P.: COVID-19: relationship between atmospheric temperature and daily new cases growth rate. Epidemiol. Infect. **148** (2020)
24. Shahid, F., Zameer, A.: Predictions for COVID-19 with deep learning models of LSTM, GRU, and Bi-LSTM. Chaos, Solitons Fractals **140**, 110212 (2020)
25. Soliman, M., Lyubchich, V., Gel, Y.: Ensemble forecasting of the Zika space-time spread with topological data analysis. Environmetrics **31**(7), e2629 (2020). https://doi.org/10.1002/env.2629
26. Yu, Y., Si, X., Hu, C., Zhang, J.: A review of recurrent neural networks: LSTM cells and network architectures. Neural Comput. **31**(7), 1235–1270 (2019)
27. Zeroual, A., Harrou, F., Abdelkader, D., Sun, Y.: Deep learning methods for forecasting COVID-19 time-series data: a comparative study. Chaos Solitons Fractals **140**, 110121 (2020). https://doi.org/10.1016/j.chaos.2020.110121

Lifelong Learning Based Disease Diagnosis on Clinical Notes

Zifeng Wang[1(✉)], Yifan Yang[2], Rui Wen[2], Xi Chen[2], Shao-Lun Huang[1], and Yefeng Zheng[2]

[1] Tsinghua-Berkeley Shenzhen Institute, Tsinghua University, Beijing, China
wangzf18@mails.tsinghua.edu.cn
[2] Jarvis Lab, Tencent, Shenzhen, China

Abstract. Current deep learning based disease diagnosis systems usually fall short in catastrophic forgetting, i.e., directly fine-tuning the disease diagnosis model on new tasks usually leads to abrupt decay of performance on previous tasks. What is worse, the trained diagnosis system would be fixed once deployed but collecting training data that covers enough diseases is infeasible, which inspires us to develop a lifelong learning diagnosis system. In this work, we propose to adopt attention to combine medical entities and context, embedding episodic memory and consolidation to retain knowledge, such that the learned model is capable of adapting to sequential disease-diagnosis tasks. Moreover, we establish a new benchmark, named Jarvis-40, which contains clinical notes collected from various hospitals. Experiments show that the proposed method can achieve state-of-the-art performance on the proposed benchmark. Code is available at https://github.com/yifyang/LifelongLearningDiseaseDiagnosis.

Keywords: Lifelong learning · Disease diagnosis · Clinical notes

1 Introduction

An automatic disease diagnosis on clinical notes benefits medical practices for two aspects: **(1)** primary care providers, who are responsible for coordinating patient care among specialists and other care levels, can refer to this diagnosis result to decide which enhanced healthcare services are needed; **(2)** the web-based healthcare service system offers self-diagnosis service for new users based on their chief complaints [19,20]. Most of previous works in deep learning based disease diagnosis focus on **individual disease risk prediction**, e.g., DoctorAI [3], RETAIN [4] and Dipole [12], all of them are based on the sequential historical visits of individuals. Nevertheless, it is not realistic to collect a large amount of data with a long history of clinical visits as those in public MIMIC-III or CPRD datasets, especially when the system is deployed online where the prior information of users is usually unavailable. On the contrary, we work on disease diagnosis system that can make decisions purely based on clinical notes.

Y. Yang—Equal contribution

K. Karlapalem et al. (Eds.): PAKDD 2021, LNAI 12712, pp. 213–224, 2021.
https://doi.org/10.1007/978-3-030-75762-5_18

On the other hand, modern deep learning models are notoriously known to be **catastrophic forgetting**, i.e., neural networks would overwrite the previously learned knowledge when receiving new samples [15]. Defying catastrophic forgetting has been of interest to researchers recently for several reasons [9]: Firstly, diagnosis systems would be fixed once developed and deployed but it is difficult to collect enough training samples initially, which requires the space for incremental learning of the existing system; Secondly, due to the stakes of privacy and security of medical data, the availability of the clinical notes is often granted for a certain period and/or at a certain location. In most scenarios, we are only allowed to maintain a tiny copy of the data. As far as we know, lifelong learning (or continual learning) disease diagnosis has not been explored until [10]. However, [10] only considered specific skin disease classification on images. In this work, we thoroughly investigate lifelong learning disease diagnosis based on clinical notes. Our method aims at handling four major challenges.

1. The model should be trained in the lack of patient visit history and adapt to existing medical scenarios.
2. Considering the importance of medical entity knowledge for disease diagnosis, previously learned knowledge should be combined with contexts and transferred as much as possible when facing new tasks.
3. We identify that the approaches used in Computer Vision tasks usually confront severe performance deterioration in Natural Language Processing (NLP) tasks, hence the learned model should retain the previous knowledge and maintain the performance on NLP tasks.
4. Since excessive computational complexity is a common problem of many state-of-the-art continual learning methods, the proposed method should be efficient enough for its practical use.

We built a system based on clinical notes instead of patients' historical data. For the second aim, we design a knowledge guided attention mechanism combining entities and their types with context as medical domain knowledge. Likewise, we adopt episodic memory embedding and consolidation to retain the previous knowledge and realize continual learning diagnosis. Since external knowledge can be obtained from contexts, our method outperforms baselines with smaller memory size (i.e., our method is more efficient and have greater potential to be applied in practical scenario). Our main contributions are:

1. We propose a new continual learning framework in medical fields. To the best of our knowledge, we are the first to investigate continual learning in medical fields that handles disease diagnosis grounded on clinical notes.
2. We propose an approach to utilize medical domain knowledge for disease diagnosis by leveraging both context and medical entity features, in order to transfer knowledge to new stages.
3. We propose a novel method named embedding episodic memory and consolidation (E^2MC) to prevent catastrophic forgetting on disease diagnosis tasks.

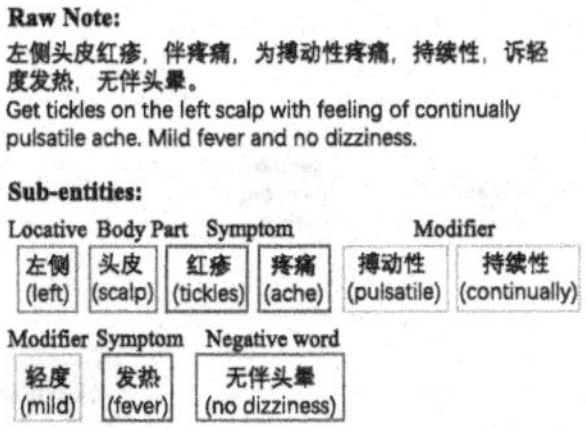

Fig. 1. An example of the raw note and the extracted sub-entities based on medical knowledge.

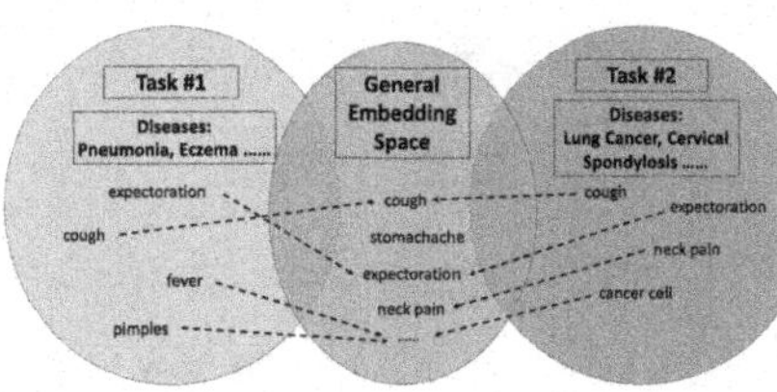

Fig. 2. Transfer of entity knowledge along different tasks.

4. We introduce a novel lifelong learning disease diagnosis benchmark, called JARVIS-40, from large scale real-world clinical notes with labeled diagnosis results by professional clinicians. We hope this work would inspire more research in future to adopt state-of-the-art continual learning techniques in medical applications based on clinical notes.

2 Methodology

In this section, we present the main technique of the proposed E^{2}MC framework. We first present the definition of the lifelong learning problem. Then, we introduce the proposed continual learning disease diagnosis benchmark, named JARVIS-40. In addition, we detail main body of the architecture, including the context & sub-entity encoder, knowledge fusion module, and the embedding episodic memory & consolidation module.

2.1 Problem Definition and Nomenclature

There is a stream of tasks $\{\mathcal{T}_1, \mathcal{T}_2, \ldots, \mathcal{T}_K\}$, where the total number of tasks K is not restricted certainly in lifelong learning tasks. Each $\mathcal{T}_k$ is a supervised task, e.g., classification problem which consists of a pair of data and the corresponding label $\mathcal{T}_k = (\mathcal{X}_k, \mathcal{Y}_k)$. Moreover, suppose class sets between tasks are not overlapped, i.e., $\mathcal{Y}_k \cap \mathcal{Y}_{k'} = \emptyset$, $\forall k, k' \in [K]$, the total number of classes increases with new tasks coming in. Denote the accumulated label set on the k-th task by $\mathcal{Y}_{:k}$, we have $|\mathcal{Y}_{:K}| = \sum_{k=1}^{K-1} |\mathcal{Y}_{:k}|$ where $|\mathcal{Y}|$ denotes cardinaility of the label set $\mathcal{Y}$. In this scenario, the aim of lifelong learning is to learn a classification model f. At the step k, the model is optimized only on $\mathcal{T}_k$, but should still maintain its performance on the previous $k-1$ tasks. The metrics for evaluating continual learning methods are then defined by the accuracy on the first task $\mathcal{T}_1$ as $\text{acc}_{f,1}$, and the average accuracy on all seen tasks: $\text{ACC}_{\text{avg}}^K = \frac{1}{K}\sum_{k=1}^K \text{acc}_{f,k}$.

Furthermore, storing a memory set $\mathcal{M}_k$ of each seen task $\mathcal{T}_k$ can enable better performance [16,17]. We store a subset with memory size B for each task. The total memory $\mathcal{M}_{:k} = \mathcal{M}_1 \bigcup \mathcal{M}_2 \cdots \bigcup \mathcal{M}_{k-1}$, and $|\mathcal{M}_{:k}| = (k-1) * B$. They are either reused as model inputs for rehearsal, or to constrain the optimization of the new task loss thus defying forgetting.

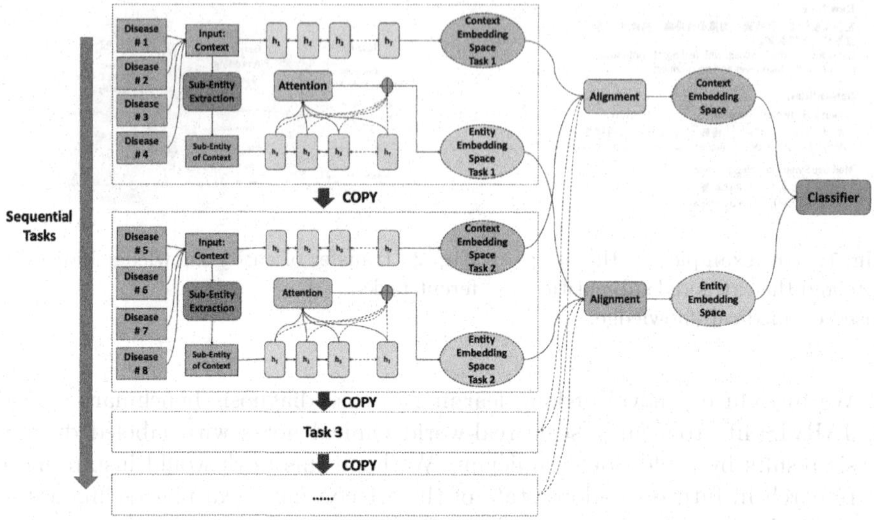

Fig. 3. The overall architecture of the proposed continual learning based disease diagnosis model.

2.2 Lifelong Learning Diagnosis Benchmark: Jarvis-40

There has been no public dataset about clinical notes and the corresponding diagnosis results in Chinese. We create a clinical-note-based disease diagnosis benchmark named Jarvis-40, collected from several hospitals in China. Each sample contains the patient's chief complaints and clinical notes written by a clinician with the diagnosis which may be more than one disease. We divide the 40 classes into 10 disjoint tasks. This data was obtained via our contracts with hospitals, then de-identified and verified manually to ensure that no private information would be emitted.

2.3 Overall Framework

We design a novel architecture for lifelong learning disease diagnosis, as shown in Fig. 3. At each step (i.e., task), clinical notes of different diseases are fed to the model. With a sub-entity extraction module, the original context and sub-entities constitute two channels. After that, two bidirectional long short-term memory (BiLSTM) modules process the input context and the extracted sub-entities respectively. Before going to the alignment layer, the sub-entity embeddings are fused by attention with the context embedding, forming two different embedding spaces for each task. When the training of a task is done, the fused sub-entity embedding and context embedding are aligned to a general embedding space and concatenated as the input for the final classifier.

Context Encoder
The context (clinical note) is tokenized into T tokens in *character* level: $\boldsymbol{x}^c = (x_1^c, x_2^c, \ldots, x_T^c)$, where their initial word embeddings are set with fastText pre-trained embedding [7]: $\boldsymbol{h}_t^{c,\text{init}} \in \mathbb{R}^d, t = 1, \ldots, T$. After that, we adopt bidirectional LSTM (BiLSTM) [5] which consists of both forward and backward networks to process the text: $\overrightarrow{\boldsymbol{h}_t^c} = \text{LSTM}(x_t^c, \overrightarrow{\boldsymbol{h}_{t-1}^c})$, $\overleftarrow{\boldsymbol{h}_t^c} = \text{LSTM}(x_t^c, \overleftarrow{\boldsymbol{h}_{t-1}^c})$. The final embedding is obtained by the concatenation of both right and left embeddings $\boldsymbol{h}_t^c = (\overrightarrow{\boldsymbol{h}_t^c}; \overleftarrow{\boldsymbol{h}_t^c})$. An aggregation function is used to obtain the aggregated context embedding by $\boldsymbol{h}^c = \phi_{\text{agg}}(\boldsymbol{h}_1^c, \ldots, \boldsymbol{h}_T^c)$, where ϕ_{agg} can be *concatenation*, *max pooling*, or *average pooling*. Moreover, to mitigate the sentence embedding distortion when training on new tasks [18], we utilize an alignment layer operated as

$$\boldsymbol{z}^c = \mathcal{W}_{\text{align}}^{c\top} \boldsymbol{h}^c = \mathcal{W}_{\text{align}}^{c\top} f_{\text{enc}}^c(\boldsymbol{x}^c), \tag{1}$$

which projects the aggregated context embedding $\boldsymbol{h}^c$ to the embedding space of the current task; $f_{\text{enc}}^c(\boldsymbol{x}^c)$ here concludes the whole encoding process from the input example $\boldsymbol{x}^c$ to $\boldsymbol{h}^c$.

Sub-entity Encoder and Knowledge Fusion
Knowledge like sub-entities is utilized as shown in Fig. 2, where we believe it should be transferred and maintained along sequence of disease-prediction tasks (i.e., various diseases). For instance, when a clinician makes decisions, he/she usually pays more attention on those important entities, e.g., the major symptoms. However, obtaining sub-entities is not an easy task. The extracted entities by a named entity recognition (NER) model trained on general corpus might be inaccurate or not aligned with medical knowledge, e.g., *subarachnoid hemorrhage* might be split into *cobweb*, *membrane*, *inferior vena* and *bleed.*[1] To solve these two challenges, we implement a medical NER model to extract medical entities from raw notes, and propose an attention mechanism to fuse medical knowledge with the context embeddings.

Sub-entity Extraction. The prior knowledge obtained from external medical resources often provides rich information that powers downstream tasks [1]. We utilize BERT-LSTM-CRF [6] as the NER model. As there is no available public medical corpus for NER in Chinese, we collected more than 50,000 medical descriptions, labeled 14 entity types including doctor, disease, symptom, medication and treatment and 11 sub-entity types of symptom entity including body part, negative word, conjunction word and feature word. We then trained a BERT-LSTM-CRF model on this labeled corpus for NER task. Figure 1 shows extracted sub-entities by the NER model from one clinical note.

Attentive Encoder. Referring to the example in Fig. 1, a clinician might focus on the symptoms like *rash* to decide the disease range roughly in dermatosis, while the entity *two days* only implies the disease severity. And, he/she notices that the rash is on *scalp*, which strengthens the belief in dermatosis. Similarly, we

[1] In Chinese, subarachnoid homorrhage is "蛛网膜下腔出血", which can be split into "蛛网", "膜", "下腔", "出血" by an NER model considering general semantics.

propose a knowledge fusion method that leverages medical sub-entities and fuses this information with the context to obtain the knowledge-augmented embeddings.

Suppose there are M extracted sub-entities: $\boldsymbol{x}^s = (x_1^s, \ldots, x_M^s)$. Similarly, each sub-entity is assigned a pre-trained embedding $\boldsymbol{h}_m^{s,\text{init}}, m = 1, \ldots, M$. This sequence of sub-entity embeddings is then processed by another BiLSTM, yielding their representations $\boldsymbol{h}_m^s$. In order to pay more attention to important sub-entities, we introduce *context towards sub-entity attention* to enhance the crucial sub-entities in sub-entity encoding:

$$u_m = \frac{\boldsymbol{h}_m^{s\top} \boldsymbol{h}^c}{\|\boldsymbol{h}_m^s\|_2 \|\boldsymbol{h}^c\|_2}, \quad a_m = \frac{\exp(u_m)}{\sum_{j=1}^{M} \exp(u_j)}, \quad \boldsymbol{h}^s = \sum_{m=1}^{M} a_m \boldsymbol{h}_m^s. \tag{2}$$

The obtained attention map $\boldsymbol{a} = (a_1, \ldots, a_M)$ has the same length as the input sequence of sub-entities, which implies the importance of each sub-entity to the task. The knowledge-augmented entity embedding $\boldsymbol{h}^s$ encodes medical entities and the corresponding importance by a weighted summation. Then, we adopt an alignment layer to get the final sub-entity embedding as $\boldsymbol{z}^s = \mathcal{W}_{\text{align}}^{s\top} \boldsymbol{h}^s = \mathcal{W}_{\text{align}}^{s\top} f_{\text{enc}}^s(\boldsymbol{x}^s)$, where $f_{\text{enc}}^s(\boldsymbol{x}^s)$ has the similar definition as $f_{\text{enc}}^c(\boldsymbol{x}^c)$ in Eq. (1). We concatenate $\boldsymbol{z}^s$ with the context embedding $\boldsymbol{z}^c$ in Eq. (1) to build the input embedding $\boldsymbol{z}$ for the classifier, which could be a one linear layer plus a softmax layer to predict the class of this note: $p(y = l|\boldsymbol{x}; \mathcal{W}_{\text{clf}}) = \frac{\exp(\mathcal{W}_{\text{clf},l}^\top \boldsymbol{z})}{\sum_{l=1}^{|\mathcal{Y}|} \exp(\mathcal{W}_{\text{clf},l}^\top \boldsymbol{z})}$. Here, $\boldsymbol{x} = (\boldsymbol{x}^s, \boldsymbol{x}^c)$ denotes the input encompassing both character and sub-entity level tokens from the same clinical note.

Embedding Episodic Memory and Consolidation (E²MC)

During the training on task $\mathcal{T}_k$, aside from sampling a training batch $\mathcal{B}_{\text{train}}^k \subset \mathcal{T}_k$, we also retrieve a replay batch from the memory set $\mathcal{M}_{:k}$ as $\mathcal{B}_{\text{replay}}^k \subset \mathcal{M}_{:k}$. For each example $(x, y) \in \mathcal{B}^k = \mathcal{B}_{\text{train}}^k \bigcup \mathcal{B}_{\text{replay}}^k$, we try to regularize the context embedding $\boldsymbol{z}^c$ and the sub-entity embeddings $\boldsymbol{z}^s$, respectively, rather than model parameters. Denote the context encoder learned at the stage k by $\boldsymbol{z}_k^c = \mathcal{W}_{\text{align}}^{c\top} f_{\text{enc}}^{c,k}(\boldsymbol{x}^c)$, the regularization term is defined between $\boldsymbol{z}_k^c$ and $\boldsymbol{z}_{k-1}^c$ as $\Omega(\boldsymbol{x}^c) = \|\boldsymbol{z}_k^c - \boldsymbol{z}_{k-1}^c\|_2^2$. And the regularization term on the sub-entity encoder can be built by $\Omega(\boldsymbol{x}^s) = \|\boldsymbol{z}_k^s - \boldsymbol{z}_{k-1}^s\|_2^2$. The full objective function is as follows

$$\begin{aligned} \mathcal{L}(\Theta) = \mathcal{L}_{\text{xent}}(\Theta) + \mathcal{L}_{\text{reg}}(\Theta) = \sum_{(x_i, y_i) \in \mathcal{B}^k} \ell(\mathcal{W}_{\text{clf}}^\top \boldsymbol{z}_i, y_i) + \alpha \Omega(\boldsymbol{x}_i^c) + \beta \Omega(\boldsymbol{x}_i^s) \\ \text{where} \quad \ell(\mathcal{W}_{\text{clf}}^\top \boldsymbol{z}_i, y_i) = -\log p(y_i|\boldsymbol{x}_i; \mathcal{W}_{\text{clf}}). \end{aligned} \tag{3}$$

Here, α and β are hyperparameters to control the regularization strength; $\ell(\mathcal{W}_{\text{clf}}^\top \boldsymbol{z}_i, y_i)$ is defined by negative log likelihood (also called cross entropy) between the groundtruth and the prediction, which is the most common loss for classification tasks; $\Theta = \{\Theta_{\text{emb}}^c, \Theta_{\text{emb}}^s, \Theta_{\text{LSTM}}^c, \Theta_{\text{LSTM}}^s, \mathcal{W}_{\text{align}}^c, \mathcal{W}_{\text{align}}^s, \mathcal{W}_{\text{clf}}\}$ contains all trainable parameters in this model, notably Θ_{emb}^c and Θ_{emb}^s are word embeddings in character level and in sub-entity level, respectively. Please note that we propose to optimize on the objective function in Eq. (3) by two steps. Firstly, we freeze the alignment layer to optimize on the negative log likelihood

$$\min_{\Theta \backslash \{\mathcal{W}_{\text{align}}^c, \mathcal{W}_{\text{align}}^s\}} \mathcal{L}_{\text{xent}}(\Theta) = \sum_{(x_i, y_i) \in \mathcal{B}^k} \ell(\mathcal{W}_{\text{clf}}^{\top} z_i, y_i), \tag{4}$$

which is for learning new tasks; Secondly, we unfreeze the alignment layers with all other parameters fixed to optimize on the consolidation terms

$$\min_{\mathcal{W}_{\text{align}}^c, \mathcal{W}_{\text{align}}^s} \mathcal{L}_{\text{reg}}(\Theta) = \sum_{(x_i, y_i) \in \mathcal{B}^k} \alpha \Omega(\boldsymbol{x}_i^c) + \beta \Omega(\boldsymbol{x}_i^s), \tag{5}$$

which aims to retain the knowledge by regularizing on the embedding space.

3 Experiments

3.1 Compared Baselines

We compare our methods with the following baselines on Jarvis-40:

- **Fine-tuning**. This is the naive method that directly finetune the model on new tasks, which serves as a lower bound of all continual learning methods.
- **Multi-Task**. This is a strategy that remembers models are trained sequentially but all the data of seen tasks would be utilized to refresh the learned knowledge, which serves as an upper bound for all continual learning methods.
- **EWC** [8]. Elastic weight consolidation proposed to constrain the learning speed of important parameters when training on new tasks, thus retaining performance on previous tasks.
- **GEM** [11]. Gradient episodic memory aims at projecting the gradients closer to the gradients of previous tasks, by means of constrained optimization.
- **AGEM** [2]. Average GEM tries to alleviate computational complexity of GEM by regularizing gradients only on those computed on the randomly retrieved examples.
- **MBPA++** [14]. Memory-based parameter adaptation is an episodic memory model that performs sparse experience replay and local adaptation to defy catastrophic forgetting.

3.2 Dataset

Our experiments are conducted on both Jarvis-40_{small} and Jarvis-40_{full}[2] in which there are 43,600 and 199,882 clinical notes respectively, training set and test set are randomly divided. After performing sub-entity extraction, we obtain around 2,600 different types of sub-entities. We split the raw data into 10 disjoint tasks randomly where each task contains $40/10 = 4$ classes of diseases.[3] As described

[2] For the sake of privacy, we are only permitted by hospitals to release Jarvis-40_{small}. All the data released has been manually desensitized.

[3] Different diseases can be classified in various ways (e.g., specialties and severity). Therefore, it is natural to split the whole set into disjoint subsets (i.e., tasks).

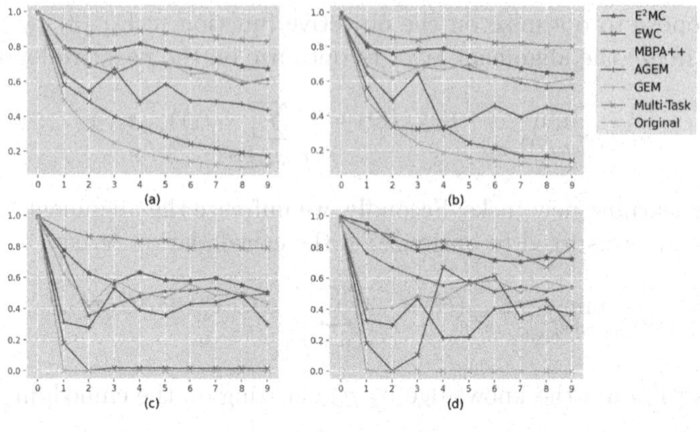

Fig. 4. Diagnosis accuracy of all methods. x-axis denotes the stage ranging from one to ten, y-axis denotes accuracy. (a), (b) are the average accuracy on Jarvis-40_{full} and Jarvis-40_{small}, respectively; (c), (d) are the first task accuracy on Jarvis-40_{full} and Jarvis-40_{small}, respectively.

in Sect. 2.1, we adopt the accuracy on the first task $\text{acc}_{f,1}$, and the average accuracy on all seen tasks $\text{ACC}^K_{\text{avg}}$ as metrics to evaluate these methods.

We also testified our method on another practical dataset to investigate a common application scenario (Medical Referral) in China. We collected another 112,000 clinical notes (not included in Jarvis-40) from four hospitals, where we pick five typical diseases respectively. These hospitals vary in functions, e.g., the hospital # 1 often receives common patients but lacks data of severer diseases hence we found cough and fever are representative diseases; the hospital # 4 often receives relatively severer diseases like lung cancer but lacks data of common diseases. In this situation, the model should incrementally learn from new hospital notes where disease distribution varies with disease severity. The results of this experiment are shown in Sect. 3.7.

3.3 Experimental Protocol

As mentioned in Sect. 2.3 and Sect. 2.3, both context and sub-entity encoders utilize BiLSTM with same settings. As shown by Eq. (4) and Eq. (5), the training is done in two phase. For the first phase training on BiLSTM model, we pick learning rate $1e^{-3}$, batch size 50; for the second phase training on context alignment model and entity alignment model, learning rates are set as $1e^{-4}$ and $2e^{-5}$ respectively, and the batch size is set 32. Memory size is set as 128 for all methods with replay module except for EWC which has memory size 1024 on Jarvis-40_{small} and 2048 on Jarvis-40_{full}.

3.4 Overall Results on Benchmark Javis-40

Figure 4 shows the overall performances on Jarvis-40. Results of average accuracy on all tasks are shown by Fig. 4 (a) and (b). It could be observed that (1) when

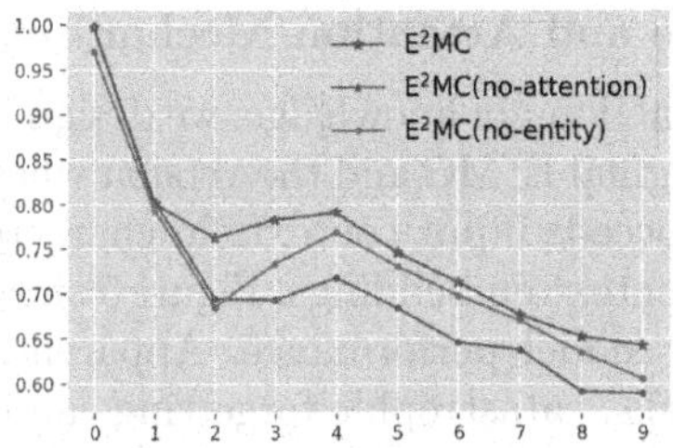

Fig. 5. Ablation study for demonstrating the benefits led by external medical domain knowledge, where it illustrates the average accuracy on Jarvis-40_{small}.

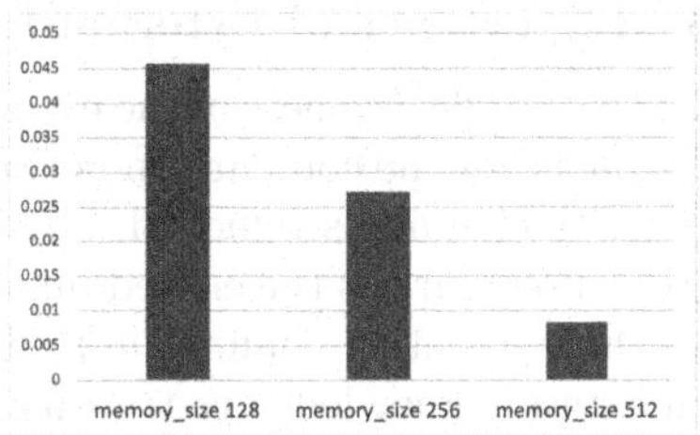

Fig. 6. The relative accuracy improvement of E^2MC over the corresponding no-entity baseline with different memory sizes.

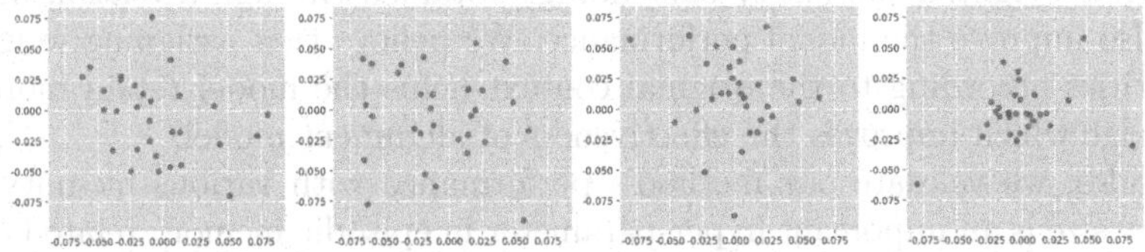

Fig. 7. Embedding visualization of entity embeddings trained on the medical referral data, where each point represents one unique entity. Training goes from early stage on the left figures to last stage on the right figures.

the number of tasks increases, all methods including Multi-Task decay to some extent. It implies the inevitable performance decay when the task complexity increase; (2) our method consistently outperforms all other baselines. It indicates the benefits brought by external medical domain knowledge and further shows that embedding episodic memory as well as consolidation contribute to a better ability of defying forgetting; (3) the tendency of methods without gradient or weight consolidation like MBPA++, Multi-Task and our method is similar. This tendency demonstrates the interference and transfer between tasks.

Figure 4 (c) and (d) demonstrate the model accuracy on the first task of Jarvis-40_{full} and Jarvis-40_{small}, respectively. We see that (1) our method E^2MC outperforms all the baselines. In particular, on Jarvis-40_{full}, our method performance is very close to the Multi-Task method, meanwhile, it significantly outperforms other methods; (2) the traditional parameter regularization method performs poorly. The results of EWC decay dramatically when receiving new tasks and do not yield improvement over the vanilla method; (3) all methods perform better on the small dataset and they are also closer to the Multi-Task upper bound. The reason behind is that the model is be trained on less data at each task which cause less forgetting.

3.5 Importance of External Knowledge and Attention Mechanism

To show the effectiveness of the external medical domain knowledge, we design an experiment for the comparison between the original E^2MC and the version whose sub-entity channel is removed, i.e., it only proceeds input contexts in character level and regularizes the embedding space of context embeddings. Figure 5 shows that the sub-entity channel indeed improves model performance. Apparently, with external knowledge to transfer among tasks, catastrophic forgetting can be further alleviated.

Likewise, in order to identify how the attention mechanism helps the whole system, we compare the original E^2MC and the version without attention mechanism. Under such setting, the alignment module for sub-entity channel directly works on each sub-entity and all sub-entities share same weight. It is obvious in Fig. 5 that weighted sub-entity embeddings obtained through attention mechanism also improve the model performance. We believe that assigning weight to sub-entities according to the original context helps the model retain more key knowledge which improves the effectiveness of alignment model.

Besides, we validate our method's performance with various memory size. Memory size is an important hyper-parameter in episodic memory related methods, The results are shown in Fig. 6. We identify that our method obtains larger improvement than the baseline with smaller memory size. This shows how knowledge strengthens the information richness of attentive embedding episodic memory. With this advancement, our model can generally perform well with less memory consumption.

3.6 Analysis of Entity Embeddings and Visualization

To explore how the embedding consolidation layer works, we show a visualization of entity embeddings trained on the practical data in Fig. 7. The high dimensional embeddings are reduced by t-SNE [13] to 2D space for visualization. A clear tendency is that embedding space becomes more and more squeezed. That is, with more and more tasks seen, those embeddings are projected into an aggregated space such that the performance decay will be alleviated. We also analyze the effect of projection with average cosine similarity between entity embeddings in a certain sample. As shown in Fig. 8, entity embeddings after alignment module is obviously closer to each other (i.e., more squeezed) at each step compared with results from no-alignment model.

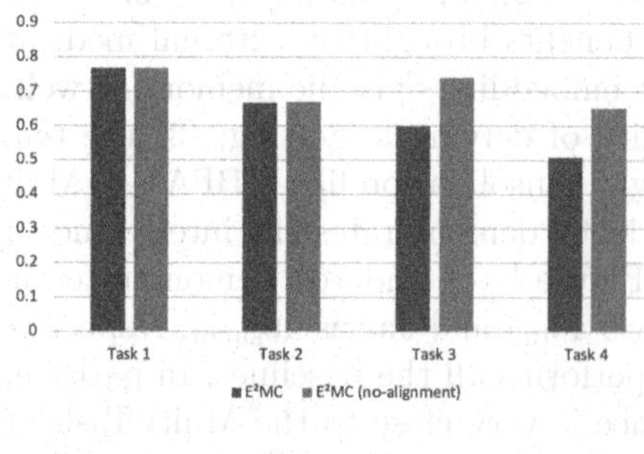

Fig. 8. The changes in aggregation degree of entity embeddings over a sequence of tasks, demonstrating the influence of alignment module. The average cosine similarity among all entities in a sample of Task 1 is used to represent aggregation degree.

Table 1. Accuracy of models on the practical data. Best ones (except the multi-task method) are in bold. FT, MT, E^2MC(\e.) denote fine-tuning, multi-task, and E^2MC(no entity), respectively.

Stage	FT	EWC [8]	GEM [11]	AGEM [2]	MBPA++ [14]	E^2MC(\e.)	E^2MC	MT
1	0.4936	0.4930	0.6958	0.5944	0.4025	0.7902	**0.7912**	0.8996
2	0.3742	0.4317	0.6715	0.5931	0.5311	0.7521	**0.7582**	0.8517
3	0.2406	0.2456	0.6997	0.6661	0.5788	**0.7677**	0.7632	0.8771
4	0.1894	0.2855	0.6574	0.5014	0.6154	0.7516	**0.7824**	0.8752

3.7 Experiment Based on Medical Referral

Results on the practical data mentioned in Sect. 3.2 are shown in Table 1. It can be seen that our method outperforms other baselines significantly. Considering the composition of diseases in each task varies dramatically in this experiment, we prove that our method indeed improves the ability of transferring and maintaining knowledge. And our method has a greater potential to be implemented in medical referral scenario in the future.

4 Conclusion

In this paper, we propose a new framework E^2MC for lifelong learning disease diagnosis on clinical notes. To the best of our knowledge, it is the first time that continual learning is introduced to medical diagnosis on clinical notes. Compared with existing methods in the NLP literature, our method requires less memory while still obtains superior results. As privacy and security of patients data arises more and more concern, it would be expected that continual learning in medical applications becomes increasingly important. The code and the dataset will be releases later.

References

1. Cao, S., Qian, B., Yin, C., Li, X., Wei, J., Zheng, Q., Davidson, I.: Knowledge guided short-text classification for healthcare applications. In: IEEE International Conference on Data Mining, pp. 31–40. IEEE (2017)
2. Chaudhry, A., Ranzato, M., Rohrbach, M., Elhoseiny, M.: Efficient lifelong learning with a-gem. In: International Conference on Learning Representations (2018)
3. Choi, E., Bahadori, M.T., Schuetz, A., Stewart, W.F., Sun, J.: Doctor AI: predicting clinical events via recurrent neural networks. In: Machine Learning for Healthcare Conference, pp. 301–318 (2016)
4. Choi, E., Bahadori, M.T., Sun, J., Kulas, J., Schuetz, A., Stewart, W.: RETAIN: an interpretable predictive model for healthcare using reverse time attention mechanism. In: Advances in Neural Information Processing Systems, pp. 3504–3512 (2016)

5. Hao, Y., et al.: An end-to-end model for question answering over knowledge base with cross-attention combining global knowledge. In: Annual Meeting of the Association for Computational Linguistics, pp. 221–231 (2017)
6. Huang, Z., Xu, W., Yu, K.: Bidirectional lstm-crf models for sequence tagging. arXiv preprint arXiv:1508.01991 (2015)
7. Joulin, A., Grave, E., Bojanowski, P., Mikolov, T.: Bag of tricks for efficient text classification. arXiv preprint arXiv:1607.01759 (2016)
8. Kirkpatrick, J., et al.: Overcoming catastrophic forgetting in neural networks. Proc. Natl. Acad. Sci. **114**(13), 3521–3526 (2017)
9. Lee, C.S., Lee, A.Y.: Clinical applications of continual learning machine learning. Lancet Digit Health **2**(6), e279–e281 (2020)
10. Li, Z., Zhong, C., Wang, R., Zheng, W.-S.: Continual learning of new diseases with dual distillation and ensemble strategy. In: Martel, A.L., et al. (eds.) MICCAI 2020. LNCS, vol. 12261, pp. 169–178. Springer, Cham (2020). https://doi.org/10.1007/978-3-030-59710-8_17
11. Lopez-Paz, D., Ranzato, M.: Gradient episodic memory for continual learning. In: Advances in Neural Information Processing Systems, pp. 6467–6476 (2017)
12. Ma, F., Chitta, R., Zhou, J., You, Q., Sun, T., Gao, J.: Dipole: diagnosis prediction in healthcare via attention-based bidirectional recurrent neural networks. In: ACM SIGKDD International Conference on Knowledge Discovery and Data Mining, pp. 1903–1911 (2017)
13. Maaten, L.V.D., Hinton, G.: Visualizing data using t-sne. J. Mach. Learn. Res. **9**, 2579–2605 (2008)
14. de Masson d'Autume, C., Ruder, S., Kong, L., Yogatama, D.: Episodic memory in lifelong language learning. In: Advances in Neural Information Processing Systems, pp. 13143–13152 (2019)
15. McClelland, J.L., McNaughton, B.L., O'Reilly, R.C.: Why there are complementary learning systems in the hippocampus and neocortex: insights from the successes and failures of connectionist models of learning and memory. Psychol. Rev. **102**(3), 419–457 (1995)
16. Robins, A.: Catastrophic forgetting, rehearsal and pseudorehearsal. Connect. Sci. **7**(2), 123–146 (1995)
17. Silver, D.L., Mercer, R.E.: The task rehearsal method of life-long learning: overcoming impoverished data. In: Cohen, R., Spencer, B. (eds.) AI 2002. LNCS (LNAI), vol. 2338, pp. 90–101. Springer, Heidelberg (2002). https://doi.org/10.1007/3-540-47922-8_8
18. Wang, H., Xiong, W., Yu, M., Guo, X., Chang, S., Wang, W.Y.: Sentence embedding alignment for lifelong relation extraction. In: International Conference of the North American Chapter of the Association for Computational Linguistics, pp. 796–806 (2019)
19. Wang, Z., et al.: Online disease self-diagnosis with inductive heterogeneous graph convolutional networks. In: Proceedings of The Web Conference 2021 (2021)
20. Zhang, X., et al.: Learning robust patient representations from multi-modal electronic health records: a supervised deep learning approach. In: Proceedings of the SIAM International Conference on Data Mining (2021)

GrabQC: Graph Based Query Contextualization for Automated ICD Coding

Jeshuren Chelladurai[1,2(✉)], Sudarsun Santhiappan[1], and Balaraman Ravindran[1,2]

[1] Department of Computer Science and Engineering, Indian Institute of Technology Madras, Chennai, India
{jeshuren,sudarsun,ravi}@cse.iitm.ac.in
[2] Robert Bosch Centre for Data Science and AI, Chennai, India

Abstract. Automated medical coding is a process of codifying clinical notes to appropriate diagnosis and procedure codes automatically from the standard taxonomies such as ICD (International Classification of Diseases) and CPT (Current Procedure Terminology). The manual coding process involves the identification of entities from the clinical notes followed by querying a commercial or non-commercial medical codes Information Retrieval (IR) system that follows the Centre for Medicare and Medicaid Services (CMS) guidelines. We propose to automate this manual process by automatically constructing a query for the IR system using the entities auto-extracted from the clinical notes. We propose **GrabQC**, a **Gra**ph based **Q**uery **C**ontextualization method that automatically extracts queries from the clinical text, contextualizes the queries using a Graph Neural Network (GNN) model and obtains the ICD Codes using an external IR system. We also propose a method for labelling the dataset for training the model. We perform experiments on two datasets of clinical text in three different setups to assert the effectiveness of our approach. The experimental results show that our proposed method is better than the compared baselines in all three settings.

Keywords: Automated medical coding · ICD10 · Query contextualization · Healthcare data analytics · Revenue Cycle Management

1 Introduction

Automated medical coding is a research direction of great interest to the healthcare industry [3,22], especially in the Revenue Cycle Management (RCM) space, as a solution to the traditional human-powered coding limitations. Medical coding is a process of codifying diagnoses, conditions, symptoms, procedures, techniques, the equipment described in a medical chart or a clinical note of a patient. The codifying process involves mapping the medical concepts in context to one or more accurate codes from the standard taxonomies such as ICD (International Classification of Diseases) and CPT (Current Procedure Terminology).

K. Karlapalem et al. (Eds.): PAKDD 2021, LNAI 12712, pp. 225–237, 2021.
https://doi.org/10.1007/978-3-030-75762-5_19

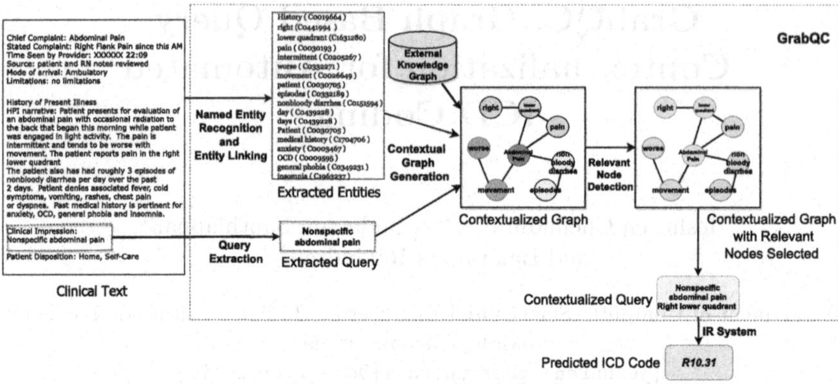

Fig. 1. The figure describes the overall pipeline of our proposed method.

The ICD taxonomy is a hierarchy of diagnostic codes maintained by the World Health Organisation. Medical coders are trained professionals who study a medical chart and assign appropriate codes based on their interpretation. The most significant drawbacks in manual coding are its Turn-Around Time (TAT), typically 24–48 h, and the inability to scale to large volumes of data. Automatic medical coding addresses both problems by applying AI and Natural Language Understanding (NLU) by mimicking and automating the manual coding process.

The problem of automating the assignment of ICD codes to clinical notes is challenging due to several factors, such as the lack of consistent document structure, variability in physicians' writing style, choice of vocabularies to represent a medical concept, non-explicit narratives, typographic and OCR conversion errors. Several approaches to solve these problems are available in the literature such as methods based on Deep Learning [2,14,18,22,23], Knowledge Bases [19] and Extractive Text Summarisation [5]. The most recent works treat this problem as a multi-label classification problem and solve them using various deep learning model architectures. Although approaches based on deep learning greatly reduce the manual labor required in the feature engineering process, there are certain challenges in applying these to the medical coding task:

- Lack of explainability
- Requirement of large amounts of labelled data
- Large label space (ICD9 ∼14,000 codes, ICD10 ∼72,000 codes)

There have been several attempts to address these challenges, for e.g., using an attention mechanism [10,14,21], transfer learning [23] and extreme classification [1,23]. However, much more needs to be done to develop truly satisfactory deployable systems. In this work, we focus on the question of operating with large label space. We study how a medical coder arrives at the final ICD codes for a given clinical note or medical chart. We observe that the medical coders study entities from different subsections of a document, such as Chief Complaint, Procedure, Impression, Diagnosis, Findings, etc., to construct evidence for every ICD code. We also observe that the medical coders use several commercial and non-commercial Information Retrieval (IR) tools such as Optum EncoderPro, AAPC Codify and open tools such as CDC's ICD-CM-10 Browser Tool for associating the entities to relevant ICD codes. We propose a solution for automated ICD coding by automatically constructing a contextually enhanced text query containing entities from the clinical notes along with an existing information retrieval system. Figure 1 depicts our method, which extracts an accurate entities based query, with which the relevant ICD codes could be fetched by querying an ICD IR system. Our method provides explainability to the retrieved codes in terms of the contextual entities, usually lacking in the end-to-end DL based solutions.

We propose GrabQC, a Graph-based Query contextualization method, alongside an existing Information Retrieval system to automatically assign ICD codes to clinical notes and medical records. The overall architecture of our proposed method, which consists of four essential modules, is shown in Fig. 1. The first module (Sect. 3.1) extracts all the data elements (entities) from the clinical notes along with their respective types such as condition, body part, symptom, drug, technique, procedure, etc. The second module (Sect. 3.2) extracts the primary diagnosis available typically under the Chief Complaint section of a medical chart. The third module (Sect. 3.3) constructs a graph from the entities enriched by a pre-constructed external Knowledge Base. The fourth module (Sect. 3.4) prunes the constructed graph based on relevance to the clinical note concepts. We then construct the contextualized query for the integrated IR system to fetch the relevant ICD codes.

The main contributions in this work are as follows:

- *GrabQC*, a Graph-based Query Contextualization Module to extract and generate contextually enriched queries from clinical notes to query an IR system.
- A Graph Neural Network (GNN) model to filter relevant nodes in a graph to contextualize the concepts.
- A distant supervised method to generate labelled dataset for training the GNN model for relevant node detection.

The rest of this paper is organised as follows. In Sect. 2, we provide a summary of the other approaches in the literature that solve the problem of ICD coding. We describe the proposed GrabQC module in Sect. 3. We present our experimental setup and the results in Sect. 4 and 5 respectively. We finally give concluding remarks and possible future directions of our work in Sect. 6.

2 Related Work

Automatic ICD coding is a challenging problem, and works in literature have focused on solving this problem using two broad paradigms [17], Information Retrieval (IR) and Machine Learning (ML). Our current work is a hybrid approach that uses both paradigms. We first use ML to contextualize the query based on the extracted entities, and then use the contextualized query for inquiring the IR system. Our work is a special case of query expansion, where we use the source document to expand by contextualizing the extracted query.

Information Retrieval: Rizzo et al. [17] presented an incremental approach to the ICD coding problem through top-K retrieval. They also used the Transfer Learning techniques to expand the skewed dataset. The authors also established experimentally that the performance of ML methods decreases as the number of classes increases. Park et al. [16] used IR to map free-text disease descriptions to ICD-10 codes.

Machine/Deep Learning: Automatic ICD coding problem has also been solved by framing it as a multi-label classification problem [2,4,8,10,14,18,21,22]. The past works can be broadly classified as belonging to one of the three categories, ML feature engineering [9], Deep representation learning architectures [2,14,18,21] and addition of additional information, such as the hierarchy of ICD codes [11], label descriptions [14,23]. There are also research works that solve these problem on other forms of clinical texts, such as the Death certificates [8] and Radiology reports [4].

Graph based DL for Automated ICD Coding: Very few works have used Graphs in solving the problem of ICD coding. Graph convolutional neural networks [7] was used for regularization of the model to capture the hierarchical relationships among the codes. Fei et al. [19] proposed a method which uses graph representation learning to incorporate information from the Freebase Knowledge Base about the diseases in the clinical text. The learned representation was used along with a Multi-CNN representation of the clinical texts to predict the ICD codes.

3 GrabQC - Our Proposed Approach

Automatic ICD coding is the task of mapping a clinical note $\mathcal{D}$ into a subset of the ICD codes $\mathcal{Y}$. We propose GrabQC, a graph-based Query Contextualization method to solve this problem using a hybrid of Deep Learning and Information Retrieval approaches. We will now discuss this proposed method in detail.

3.1 Named Entity Recognition and Entity Linking

A semi-structured clinical note $\mathcal{D}$ consists of a set of subsections $\mathcal{T}$, where each section comprises one or more sentences. The initial step is to identify interesting entities, which would serve as building blocks of the contextualized graph. We use a Named Entity Recognition (NER) module to extract a set of entities, $\mathcal{E} = \{e_1, e_2, \ldots, e_m\}$ where e_i can be a single word or a span of words from $\mathcal{D}$. To obtain more information about the extracted entities, we need to associate them with a knowledge base $\mathcal{B}$. An entity linker takes as input the sentence and the extracted entity and outputs a linked entity from the Knowledge base $\mathcal{B}$. The extracted entity set $\mathcal{E}$ is linked with $\mathcal{B}$ using the entity linker to produce the set of linked entities $\mathcal{L} = \{l_1, l_2, \ldots, l_m\}$.

3.2 Query Extraction

The physician lists the diagnoses of the patient in a pre-defined subsection of the clinical note. This subsection can be extracted using rules and regex-based methods. Each item in the extracted list of diagnoses constitute the initial query set $\mathcal{Q} = \{q_1, q_2, \ldots, q_n\}$, where n is the number of queries/diagnoses. We also extract entities from the queries in $\mathcal{Q}$ and link them with the $\mathcal{B}$ to produce a set of linked entities $\mathcal{P}_i$, for each q_i.

3.3 Generation of Contextual Graph

From the methods described in Sect. 3.1 and 3.2, we convert the clinical note $\mathcal{D}$ into a set of queries $\mathcal{Q}$, set of linked entities $\mathcal{L}$ and set of query-linked entities $\mathcal{P}_i, \forall q_i \in \mathcal{Q}$. We generate a Contextual Graph $\mathcal{G}_i$, for each query $q_i \in \mathcal{Q}$. We construct a set of matched entities V_i, with the entities in $\mathcal{P}_i$ that have a simple path with $\mathcal{L}$ in a Knowledge Base ($\mathcal{B}$). We construct the edge set of $\mathcal{G}_i$, E_i by adding an edge (p_i, l_j), if the path $(p_i, \cdots, l_j) \in \mathcal{B}$. For every matched entity, we add an edge between the matched entity and other linked entities in the same sentence. Repeating this process, we obtain the contextual graph $\mathcal{G}_i = (V_i, E_i), \forall q_i \in \mathcal{Q}$.

3.4 Relevant Node Detection

We frame the problem of detecting relevant nodes in a Contextual Graph ($\mathcal{G}$) as a supervised node classification problem. Let $\mathcal{G}$ be the contextual graph of the query q, obtained from Sect. 3.3. Here $\mathcal{G} = (V, E)$, where $V = \{n_1, n_2, \cdots, n_N\}$ is the set of nodes and $E \subseteq \{V \times V\}$ is the set of edges. The structure of the $\mathcal{G}$ is denoted using an adjacency matrix, $A \in \mathbb{R}^{N \times N}$, where $A_{ij} = 1$ if the nodes i and j are connected in the $\mathcal{G}$, 0 otherwise. Given $\mathcal{G}$ and query q, we train a neural network which passes $\mathcal{G}$ through a Graph Neural Network(GNN) model to compute the hidden representation for all the nodes in the $\mathcal{G}$. We also obtain the representation for q by averaging over the words' embeddings in the query and passing it through a dense layer. We then concatenate the individual nodes' GNN representations with the query to obtain the final representation for each node-query pair. Finally, these node-query pair representations are passed to a softmax layer to classify them as either relevant or non-relevant. We then obtain the contextual query q_c by concatenating the query with the predicted relevant nodes' words. We will now describe each of these elements in more detail.

At the base layer of the model we have d-dimensional pre-trained word embeddings for each word (v_i) in the node of the contextual graph. The d-dimensional embeddings (x_i) of the N nodes in the $\mathcal{G}$ are vertically stacked to form the initial feature matrix, $\mathbf{X}^{(0)} = [x_1; x_2; \cdots; x_N] \in \mathbb{R}^{N \times d}$. The node representations at each layer are obtained using the Graph Neural Network as described by Morris et al., [13] with $\mathbf{W}_1^{(\ell)}$ and $\mathbf{W}_2^{(\ell)}$ as the parameters in each layer. We use this variant of the GNN rather than the widely-used Graph Convolutional Networks [7] because it preserves the central node information and omits neighbourhood normalization. To obtain the hidden representation at each layer, we perform,

$$\mathbf{X}^{(\ell+1)} = \mathbf{W}_1^{(\ell+1)}\mathbf{X}^{(\ell)} + \mathbf{W}_2^{(\ell+1)}A\mathbf{X}^{(\ell)} \tag{1}$$

To obtain the hidden representation for the query q, we average the d-dimensional pre-trained word embeddings(w_i) of the words present in q. Then we pass the averaged word embedding through a dense layer, with $\mathbf{W}_q, \mathbf{b}_q$ as its learnable parameters. The hidden representation of the query ($\mathbf{x}^q$) is given by,

$$\mathbf{x}^q = \mathbf{W}_q \left(\frac{1}{|w|} \sum_{w_i \in w} w_i \right) + \mathbf{b}_q \tag{2}$$

The input to the final classification layer is the combined representation of the node in the final layer (k) of the GNN and the query's hidden representation. The final prediction of the model whether a given node $n_i \in \mathcal{G}$ is relevant for the query q is given by,

$$Pr(n_i = \mathit{relevant} \mid q) = \mathbf{sigmoid}(\mathbf{W}_c[\mathbf{x}_i^k; \mathbf{x}_q] + \mathbf{b}_c) \tag{3}$$

$$\hat{y}_i = \mathbb{I}\left[Pr(n_i = \mathit{relevant} \mid q) \geq \tau\right] \tag{4}$$

Here $\mathbb{I}$ is an indicator function that outputs 1 if the score is greater than the threshold (τ), 0 otherwise. We train the model by optimizing the Weighted Binary Cross-Entropy Loss with L_2-regularization using the Adam Optimizer [6].

The contextual query q_c is given by,

$$q_c = \{q \cup v_i \mid \hat{y}_i = 1, 1 \leq i \leq N\} \tag{5}$$

3.5 Distant Supervised Dataset Creation

The training of the model described in Sect. 3.4 requires labels for each node in the contextual graph $\mathcal{G}$. Gathering relevance labels for each node in the $\mathcal{G}$ is challenging and time-consuming, as the available datasets have ICD codes only for each clinical chart. Hence, we use distant supervision [12] for annotating the nodes in the $\mathcal{G}$. We chose to use the descriptions of the ICD codes[1] for providing supervision to the labelling process.

Let $\mathcal{D}$ be the clinical note and $L = \{l_1, \cdots, l_m\}$ be the manually annotated ICD-codes for $\mathcal{D}$. Let the descriptions for the ICD codes in Y be $B = \{b_1, \cdots, b_m\}$. We use the methods described in Sect. 3.3, to convert $\mathcal{D}$ into $\mathcal{Q} = \{q_1, \cdots, q_n\}$ and $\mathcal{G} = \{\mathcal{G}_1, \cdots, \mathcal{G}_n\}$, where $\mathcal{G}_i = (V_i, E_i)$. We need to assign a relevancy label $y_j, \forall v_j \in V_i$. Let W_j be the set of words in node $v_j \in V_i$ and l_k be the label corresponding to q_i. We label $v_j \in V_i$ to be relevant ($y_j = 1$), if the Jaccard similarity of W_j and b_k is greater than a threshold(t). Otherwise, we label v_i as non-relevant ($y_j = 0$). Repeating this process for all the nodes, we obtain the node-labelled $\mathcal{G}_i$.

[1] https://www.cms.gov/Medicare/Coding/ICD10/2018-ICD-10-CM-and-GEMs.

4 Experimental Setup

In this section, we perform experiments to answer the following questions,

- How does the proposed relevant node detection model perform in identifying the relevant nodes in the contextualized graph? (Sect. 5.1)
- Is the query's performance generated by GrabQC better than the initial query on the same IR system? (Sect. 5.2)
- How does the proposed method compare against other machine-learning-based methods in the literature? (Sect. 5.3)

4.1 Dataset Description

We use anonymized versions of two clinical text datasets, a subset of the publicly available MIMIC-III dataset and a proprietary dataset (EM-clinical notes). The MIMIC-III dataset is a collection of discharge summaries of patients from the Intensive Care Unit (ICU). At the same time, the EM-clinical notes (referred to as *EM*) are Electronic Health Records (EHRs) of patients visiting the Emergency Room in a hospital. Both the datasets are manually coded with the corresponding ICD diagnosis codes. The choice of dataset from different specialties help in understanding the performance of our method across them. The subset of MIMIC-III (referred to as *M3s*) dataset consists of discharge summaries. The number of ICD codes and the number of extracted queries had a difference of less than 2. M3s dataset contains 7166 discharge summaries with 1934 unique ICD-9 codes, while the EM dataset has 7991 EHRs with 2027 ICD-10 codes. Since ICD-10 is the latest version used commercially and lacks a public source ICD-10 coded clinical notes dataset, we use a proprietary dataset to test our method.

4.2 Setup and Baselines

We follow three experimental setups to assess the performance of the proposed method. For all the experiments, we have split the datasets into train/dev/test splits. We have trained our model on the training set by tuning the hyper-parameters in the dev set. We report the results on the test set.

- Initially, to validate our proposed Relevant Node Detection model, we compare it against other variants of Graph Neural Network. We use the same architecture of the model, but only change the GNN layers with Graph Convolutional Neural Network [7], and Graph Attention Network [20] for comparison. We use macro-average precision, recall, and F1-score as the metrics for evaluation.
- To compare the performance of the GrabQC generated contextual query, we use an Information Retrieval (IR) based evaluation. We keep the IR system as constant and compare the performance by providing the various baseline queries. We use Recall@k as the evaluation metric, as we are concerned with the method's ability to retrieve the specific code from the large pool of labels. The retrieved codes can be pruned further using rules or business logic.

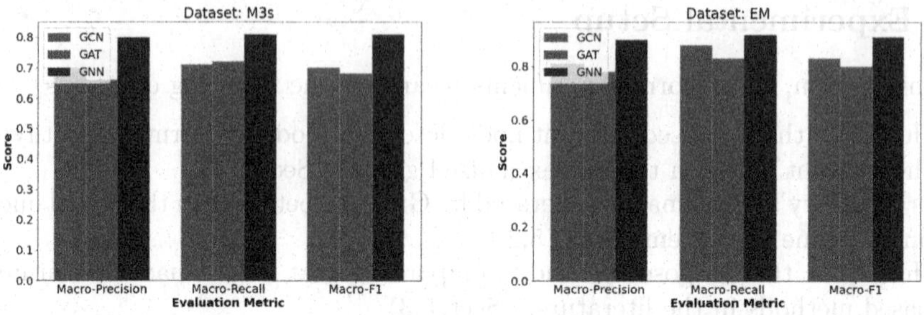

(a) M3s Dataset (b) EM Dataset

Fig. 2. Results of the various Graph Neural Network architectures on the test set of the datasets

- As GrabQC involves machine-learning for generating the contextual query, we compare the performance against other ML-based methods used for autocoding. We compare against other Neural network architectures (CNN, Bi-GRU) and two popular methods (CAML, MultiResCNN) in the literature [10,14]. We use label macro-average precision, recall, F1-score, and also Recall@k as the evaluation metrics. We use macro-average as it reflects the performance on a large label set and emphasises rare-label prediction [14].

4.3 Implementation Details

We use UMLS[2] and SNOMED[3] as Knowledge Base ($\mathcal{B}$) for the entity linking (Sect. 3.1) and path extraction. We use SciSpacy [15] for Named Entity Recognition and Entity Linking to UMLS. We use ElasticSearch[4] in its default settings as our IR system. We index the IR system by the ICD codes and their corresponding description. There are certain keywords (**External Keywords**) that occur in the ICD code descriptions but never occur in the clinical notes (e.g., without, unspecified). We add the Top 4 occurring external keywords as nodes to the $\mathcal{G}$ and link them with the linked entities in the query. We use PyTorch for implementing the relevant node detection model.

5 Results and Discussion

5.1 Performance of the Relevant Node Detection Model

In the Contextual Graph, the query's entities are the centre nodes that link the other entities. Hence the information of the centre node is essential for classifying

[2] https://www.nlm.nih.gov/research/umls/index.html.
[3] https://www.snomed.org/.
[4] https://www.elastic.co/.

Table 1. Results of the different queries along with the contextual query generated by GrabQC. Here, Normal refers to the query (q) extracted from the clinical text; Contextual Graph refers to the $\mathcal{G}$, External refers to the concatenation of the described query with the external keywords described in Sect. 4.3. The best score, with the external keywords, is marked in **bold*** and without the external keywords is marked in **bold**.

Dataset	Query Type	Recall @ 1	Recall @ 8	Recall @ 15
M3s	Normal	0.2098	0.5830	0.6684
	Contextual Graph	0.1911	0.5540	0.6404
	GrabQC	**0.2130**	**0.5978**	**0.6767**
	Normal + External	0.3000	0.5732	0.6242
	Contextual Graph + External	0.2792	0.5839	0.6550
	GrabQC + External	**0.3129***	**0.6230***	**0.6876***
EM	Normal	**0.2562**	**0.5999**	0.6541
	Contextual Graph (CG)	0.2474	0.5888	0.6462
	GrabQC	0.2525	0.5996	**0.6547**
	Normal + External	0.2752	0.5539	0.5999
	CG + External	0.2472	0.5941	0.6504
	GrabQC + External	**0.3437***	**0.6465***	**0.6872***

whether the node is relevant or not. Since the GNN model captures that information explicitly by skip connections and without normalizing on the neighbors, it performs better than the other compared methods seen in Fig. 2.

5.2 Query-Level Comparison

From Table 1, it is evident that the performance of the system changes on the addition of keywords from the source document. The system's performance degrades when we pass $\mathcal{G}$ as the query. This behavior validates the fact that there are many non-relevant terms in the $\mathcal{G}$ that makes the IR system provide noisy labels. When GrabQC filters the relevant terms, the performance increases substantially and is better than the extracted query. Also, the addition of external keywords significantly improves the performance, as the IR system is indexed using the ICD-code descriptions. Our method GrabQC again filters relevant terms from the external keywords, which is evident from the increase in performance compared to the standard query with external keywords.

5.3 Comparison with Machine Learning Models

From Table 2, we observe that the macro-average performance of the methods is deficient. The deficiency is due to the high number of labels in the ICD coding task. Our method of using the contextual query generated by GrabQC performs better than the other compared methods. Complex models like CAML and MultiResCNN are unable to perform well in this case due to a large number of labels

Table 2. Experimental results of the proposed method along with the other machine learning baselines on the test set. The best scores are marked in **bold**.

Dataset	Method	Macro			Recall	
		Precision	Recall	F1-score	@8	@15
M3s	CNN	0.0310	0.0273	0.0291	0.5177	0.5783
	Bi-GRU	0.0222	0.0104	0.0141	0.4578	0.5079
	CAML [14]	0.0093	0.0067	0.0078	0.2383	0.3064
	MultiResCNN [10]	0.0118	0.0071	0.0089	0.4701	0.5415
	GrabQC	0.1164	0.1153	0.1027	0.5978	0.6767
	GrabQC + External	**0.1334**	**0.1193**	**0.1129**	**0.6230**	**0.6876**
EM	CNN	0.0711	0.0538	0.0613	0.6459	0.6771
	Bi-GRU	0.0755	0.0431	0.0549	0.6333	0.6731
	CAML [14]	0.0267	0.0158	0.0199	0.1410	0.2040
	MultiResCNN [10]	0.0288	0.0176	0.0218	0.1410	0.2057
	GrabQC	0.1333	0.1266	0.1226	0.5996	0.6547
	GrabQC + External	**0.1670**	**0.1567**	**0.1534**	**0.6465**	**0.6872**

and a lesser number of training samples. Even with a limited amount of training data, our method can perform significantly better than the other baselines.

5.4 Analysis

We analyze the effect of the number of layers in the Graph Neural Network on the performance of GrabQC. The results of this analysis are tabulated in Table 3. We observe that choosing the number of layers as 3 gave us the best result in both the datasets.

Table 3. Macro-average F1-score on the test set of the datasets obtained by varying the number of layers in the Graph Neural Network.

# layers in GNN	M3s	EM
1	0.89	0.78
2	0.90	0.80
3	**0.91**	**0.80**
4	0.90	0.79

Table 4. Examples of GrabQC contextual queries generated for the queries extracted from the clinical note.

Normal Query (q)	Contextual Query (q_c)
Sciatica	Sciatica pain right lower back
Acute knee pain	Acute knee pain left
Laceration	Lip Laceration
Strain of lumbar region	Strain of lumbar region lower back pain
History of breast cancer	History of breast cancer left breast carcinoma

Table 4 tabulates the contextual query generated by GrabQC. From the examples, we see that the GrabQC method can contextualize the standard query. GrabQC can associate the extracted query with the region of the diagnosis, the laterality, and associated conditions. The contextual query can also serve as an explanation for the ICD codes returned by the IR system.

6 Conclusion

We proposed an automation process that expedites the current medical coding practice in the industry by decreasing the turn-around time. Our method mimics the manual coding process that provides full explainability. We proposed **GrabQC**, an automatic way of generating a query from clinical notes using a graph neural network-based query contextualization module. We successfully demonstrated the effectiveness of the proposed method in three settings. As a next step, we plan to use Reinforcement Learning (RL) for automatically contextualizing the query, without the need for relevance labels. The proposed contextualization method can also be extended as hard attention for DL-based methods.

Acknowledgements. The authors would like to thank Buddi.AI for funding this research work through their project RB1920CS200BUDD008156.

References

1. Bhatia, K., Dahiya, K., Jain, H., Mittal, A., Prabhu, Y., Varma, M.: The extreme classification repository: multi-label datasets and code (2016). http://manikvarma.org/downloads/XC/XMLRepository.html
2. Duarte, F., Martins, B., Pinto, C.S., Silva, M.J.: Deep neural models for icd-10 coding of death certificates and autopsy reports in free-text. J. Biomed. Inform. **80**, 64 – 77 (2018). https://doi.org/10.1016/j.jbi.2018.02.011, http://www.sciencedirect.com/science/article/pii/S1532046418300303
3. Farkas, R., Szarvas, G.: Automatic construction of rule-based icd-9-cm coding systems. BMC bioinformatics 9 (Suppl 3), S10 (2008). https://doi.org/10.1186/1471-2105-9-S3-S10
4. Karimi, S., Dai, X., Hassanzadeh, H., Nguyen, A.: Automatic diagnosis coding of radiology reports: a comparison of deep learning and conventional classification methods (2017). https://doi.org/10.18653/v1/W17-2342
5. Kavuluru, R., Han, S., Harris, D.: Unsupervised extraction of diagnosis codes from EMRS using knowledge-based and extractive text summarization techniques. In: Zaïane, O.R., Zilles, S. (eds.) AI 2013. LNCS (LNAI), vol. 7884, pp. 77–88. Springer, Heidelberg (2013). https://doi.org/10.1007/978-3-642-38457-8_7
6. Kingma, D., Ba, J.: Adam: a method for stochastic optimization. In: International Conference on Learning Representations (2014)
7. Kipf, T.N., Welling, M.: Semi-supervised classification with graph convolutional networks. In: International Conference on Learning Representations (ICLR) (2017)

8. Koopman, B., Zuccon, G., Nguyen, A., Bergheim, A., Grayson, N.: Automatic icd-10 classification of cancers from free-text deathcertificates. Int. J. Med. Inform 84 (2015). https://doi.org/10.1016/j.ijmedinf.2015.08.004
9. Larkey, L.S., Croft, W.B.: Combining classifiers in text categorization. In: Proceedings of the 19th Annual International ACM SIGIR Conference on Research and Development in Information Retrieval. SIGIR 1996, pp. 289–297. Association for Computing Machinery, New York (1996). https://doi.org/10.1145/243199.243276, https://doi.org/10.1145/243199.243276
10. Li, F., Yu, H.: ICD coding from clinical text using multi-filter residual convolutional neural network. In: Proceedings of the Thirty-Fourth AAAI Conference on Artificial Intelligence (2020)
11. Lima, L., Laender, A., Ribeiro-neto, B.: A hierarchical approach to the automatic categorization of medical documents, pp. 132–139 (1998). https://doi.org/10.1145/288627.288649
12. Mintz, M., Bills, S., Snow, R., Jurafsky, D.: Distant supervision for relation extraction without labeled data. In: Proceedings of the Joint Conference of the 47th Annual Meeting of the ACL and the 4th International Joint Conference on Natural Language Processing of the AFNLP: Volume 2. ACL 2009, vol. 2, pp. 1003–1011. Association for Computational Linguistics, USA (2009)
13. Morris, C., et al.: Weisfeiler and leman go neural: Higher-order graph neural networks. In: Proceedings of the AAAI Conference on Artificial Intelligence, vol. 33, no. 01, pp. 4602–4609, July 2019. https://doi.org/10.1609/aaai.v33i01.33014602, https://ojs.aaai.org/index.php/AAAI/article/view/4384
14. Mullenbach, J., Wiegreffe, S., Duke, J., Sun, J., Eisenstein, J.: Explainable prediction of medical codes from clinical text. In: Proceedings of the 2018 Conference of the North American Chapter of the Association for Computational Linguistics: Human Language Technologies, Volume 1 (Long Papers), pp. 1101–1111. Association for Computational Linguistics, New Orleans, Louisiana, June 2018. https://doi.org/10.18653/v1/N18-1100, https://www.aclweb.org/anthology/N18-1100
15. Neumann, M., King, D., Beltagy, I., Ammar, W.: ScispaCy: fast and robust models for biomedical natural language processing. In: Proceedings of the 18th BioNLP Workshop and Shared Task, pp. 319–327. Association for Computational Linguistics, Florence, Italy, August 2019. https://doi.org/10.18653/v1/W19-5034, https://www.aclweb.org/anthology/W19-5034
16. Park, H., et al.: An information retrieval approach to icd-10 classification. Stud. Health Technol. Inform. **264**, 1564–1565 (2019). https://doi.org/10.3233/SHTI190536
17. Rizzo, S.G., Montesi, D., Fabbri, A., Marchesini, G.: ICD code retrieval: novel approach for assisted disease classification. In: Ashish, N., Ambite, J.-L. (eds.) DILS 2015. LNCS, vol. 9162, pp. 147–161. Springer, Cham (2015). https://doi.org/10.1007/978-3-319-21843-4_12
18. Shi, H., Xie, P., Hu, Z., Zhang, M., Xing, E.: Towards automated ICD coding using deep learning (2017)
19. Teng, F., Yang, W., Chen, L., Huang, L., Xu, Q.: Explainable prediction of medical codes with knowledge graphs. Front. Bioeng. Biotechnol. **8**, 867 (2020). https://doi.org/10.3389/fbioe.2020.00867, https://www.frontiersin.org/article/10.3389/fbioe.2020.00867
20. Veličković, P., Cucurull, G., Casanova, A., Romero, A., Liò, P., Bengio, Y.: Graph Attention Networks. In: International Conference on Learning Representations (2018). https://openreview.net/forum?id=rJXMpikCZ, accepted as poster

21. Vu, T., Nguyen, D.Q., Nguyen, A.: A label attention model for ICD coding from clinical text, pp. 3307–3313 (2020). https://doi.org/10.24963/ijcai.2020/457
22. Xie, P., Xing, E.: A neural architecture for automated ICD coding, pp. 1066–1076 (2018). https://doi.org/10.18653/v1/P18-1098
23. Zhang, Z., Liu, J., Razavian, N.: Bert-xml: Large scale automated ICD coding using Bert pretraining (2020)

Deep Gaussian Mixture Model on Multiple Interpretable Features of Fetal Heart Rate for Pregnancy Wellness

Yan Kong[1], Bin Xu[1,2](✉), Bowen Zhao[2], and Ji Qi[2]

[1] Global Innovation Exchange Institute, Tsinghua University, Beijing, China
xubin@tsinghua.edu.cn

[2] Department of Computer Science and Technology, Tsinghua University, Beijing, China

Abstract. About 1% of the newborns have disorders of cardiac rhythm and conduction that result in heart failure, brain damage and even sudden newborn death. To ensure the safety of pregnancy, the electronic fetal monitoring (EFM) technique is widely used in obstetrics. However, the existing automatic diagnosis methods suffer from two main problems: insufficient features and low interpretability. In order to improve the interpretability and effect of the method, we propose a novel fully interpretable method. We first propose an iterative local linear regression (ILLR) method of linear complexity, which calculates over all local ranges of the fetal heart rate (FHR) and generates local gradients and coefficients of determination, that are used as indicators of intensity and typicality of fetal heart activity. Then, we elaborate the methodology of extraction of dozens of features by interpretable methods. Finally, we propose an interpretable deep Gaussian mixture model that can automatically select multiple features, which is composed of a mixture model based on Gaussian model weighted by features and a regression model. We conduct cross validation experiments on the full benchmark intrapartum database CTU-UHB, which shows that our method obtains significant improvements of 5.61% accuracy over state-of-the-art baselines.

Keywords: Electronic fetal monitoring · Cardiotocography · Iterative local linear regression · Deep Gaussian mixture model

1 Introduction

Electronic fetal monitoring (EFM), also known as cardiotocogram (CTG) is a technology of detecting whether a fetus is appearing signs of distress during the labor. This technique is widely used in clinics to ensure the safety of pregnancy. According to statistics, about 1% of the newborns are found to have disorders of cardiac rhythm and conduction that shown to result in heart failure, brain damage and even sudden newborn death [20]. As shown in Fig. 1, electronic fetal monitoring technology includes several steps: obtaining monitoring record data,

K. Karlapalem et al. (Eds.): PAKDD 2021, LNAI 12712, pp. 238–250, 2021.
https://doi.org/10.1007/978-3-030-75762-5_20

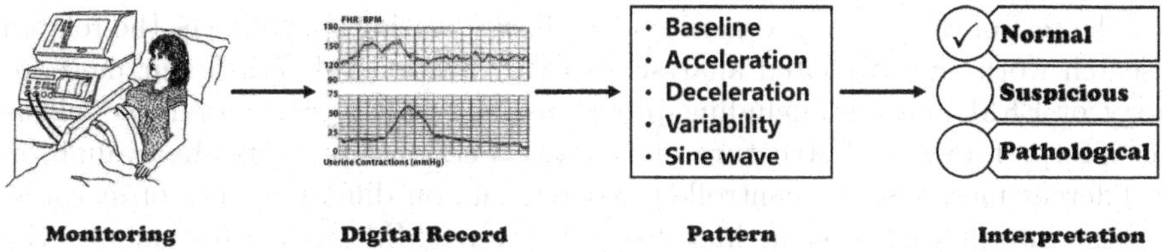

Fig. 1. The overview of the Electronic Fetal Monitoring technology.

identifying FHR patterns and making EFM interpretation. The interpretation standard of EFM is different between different countries and regions of the world, like FIGO [2], ACOG [1], and NICE [15]. At the moment, the main problem about EFM is that it is often difficult for doctors to agree on a result according to the same guideline [7]. As a result, the low-level hospitals are still using the non-stress testing (NST) or contraction stress testing (CST) as evaluation methods, whose specifities are shown no more than 50% [10,11]. However, this paper proposes the view that the use of automatic diagnosis system can better solve the problem of inconsistent diagnosis results, and furthermore improve the overall medical level of developing areas as well as regional medical equity.

There are some problems of the automatic analysis technology with EFM. The existing feature studies on EFM are far from enough [14]. Due to this limitation, current feature based methods are not effective in actual. And the existing uninterpretable methods can not reach good effect, either.

The study of this paper focuses on an effective analytical method of EFM records by using interpretable methods, including data preprocessing and feature extraction of various scopes, and an interpretable model with the ability of multiple feature learning and high sensitivity on small data. In order to verify the effect of our method, we carry out a verification of our complete classification method on the open-access intrapartum CTG database named CTU-UHB [4].

The contributions of this paper are as follows.

- We propose a framework of new interpretable methods to acquire an automated analysis of EFM diagnosis with high efficiency. Our first innovation is the ILLR method that can calculate the least squares linear regression and coefficient of determination over N local ranges of signal of length N with O(N) complexity both in time and space. The second is the methodology of feature extraction of FHR records, which can obtain 77 FHR features in various scopes that are closer to the clinical application than existing methods. Third one is, we propose an deep Gaussian mixture model (DGMM) that learn multiple features comprehensively. The model is completely interpretable, which is very in line with the preference of health and medical application.
- We evaluated our framework by cross validation on CTU-UHB dataset, and the result shows that our methods obtain significant improvements of 5.61% accuracy over the state-of-the-art baseline.

The rest of this paper is organized as follows. Section 2 introduces the related research work on automated analysis of FHR. Section 3 describes our methodology of EFM analysis including preprocessing, feature extraction as well as the interpretation and structure of DGMM. Section 4 presents the evaluation of different methods, the controlled experements on different types of features, and the interpretation of the knowledge of the DGMM. Section 5 concludes the researches of FHR analysis methods in this paper.

2 Related Work

There are many researches on automatic analysis of FHR. [8,9] researched the approaches to determining the baseline, accelerations and decelerations. There are also some works on feature analysis [12,16,18]. [13] researched the analysis methods of FHR and put forward that computer-based EFM will be improved by better signal quality, more additional information, and more clinical data.

There have been efforts on the public intrapartum dataset CTU-UHB [4], as this can be the standard benchmark data that include discrete signal data, available biochemical marks as well as some more general clinical features. The CTG dataset contains 552 discrete records in which 46 ones are cesarean, and there is no data on inherited congenital diseases. Researches on this dataset can be divided into two paradigm.

In the first paradigm, researchers are exploiting the features in the frequency domain using wave transform. [6] proposed short time Fourier transform (STFT) to obtain the visual representation of time-frequency information. In addition, [3] and [22] proposed continuous wavelet transform (CWT) for the same purpose. However, they both have the test leakage problem since the training and test data come from similar images generated by the wavelet of different scales. [21] proposed image-based time-frequency (IBTF) analysis and LS-SVM. Besides, [19] used fast Fourier transform (FFT) to check the energy in different frequency ranges but did not get good enough result.

As for researches beyond frequency domain methods, [5] researched on feature based artificial neural network (ANN). They labeled the data by specialists for supervised learning, but only used those of certain results. Therefore, the effect is not guaranteed in general cases. [14] have used 54 features to do classification experiments by least squares support vector machines (LS-SVMs). But they came to the conclusion that using only 3 features can work best. It owes to the low comprehensiveness of their model to learn complicated features.

3 Approach

As shown in Fig. 2, our interpretable framework includes 3 steps: data preprocessing, feature extraction and classification by the DGMM model.

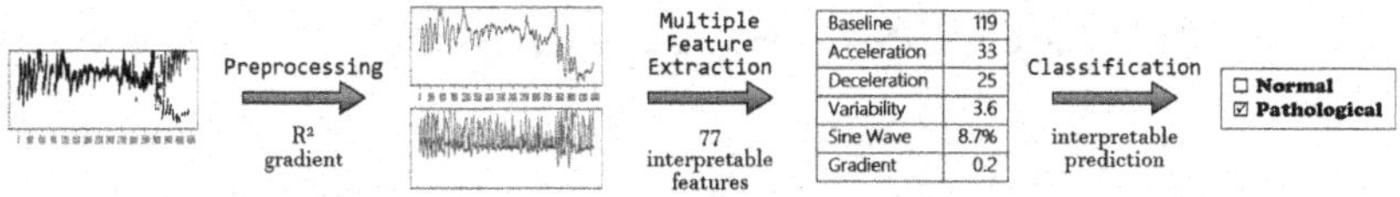

Fig. 2. The outline of our analysis framework of EFM

3.1 Data Preprocessing

The current EFM technique produces two types of discrete digital signals, namely fetal heart rate (FHR) and uterine contraction (UC). Our method only requires FHR traces for adaptability. In the preprocessing stage, we first reset the abnormal data of a period less than 15 s. Then, the ILLR method is used to perform local linear regression and data fitting.

Iterative Local Linear Regression. The ILLR method is based on least squares linear regression. This paper will give the steps and needed equations for the method. It can be proved this method can perform linear regression and linear fitting both of N times on an FHR trace of length N by O(N) complexity.

For 2-dimensional data of length N, $p_i = (x_i, y_i), i = 1, \ldots, N$, given a range parameter r, we propose a function summation operator $S_i(f)$ to denote a summation operation for function f on the points close to p_i as

$$S_i(f) = \sum_j f(j),\ \left(|x_j - x_i| \leqslant r\right), \tag{1}$$

where f could be subscript operations, like $S_i(xy) = \sum_j x_j y_j,\ (|x_j - x_i| \leqslant r)$. And we specially define n as $S_i(1)$, which means the number of the points close to p_i. Consequently, the least squares solution of linear function $\hat{y} = \hat{a}_i x + \hat{b}_i$ of domain $\{p_j \,|\, |x_j - x_i| \leqslant r\}$ is equivalent to

$$\hat{a}_i = \frac{n \cdot S_i(xy) - S_i(x) \cdot S_i(y)}{n \cdot S_i(x^2) - S_i(x)^2}, \hat{b}_i = \bar{y} - \hat{a}_i \bar{x}, \tag{2}$$

where $\bar{x} = \frac{S_i(x)}{n}, \bar{y} = \frac{S_i(y)}{n}$. And the coefficient of decision is equivalent to $R_i^2 = 1 - \frac{SSE}{SST}$. It can be deduced $SSE = S_i\left((y - \hat{y})^2\right) = S_i\left((y - \bar{y} - (\hat{a}_i \cdot (x - \bar{x})))^2\right) = \frac{1}{n}\left(n \cdot S_i(y^2) - S_i(y)^2 + \left(\left(n \cdot S_i(x^2) - S_i(x)^2\right)\hat{a}_i - 2 \cdot (n \cdot S_i(xy) - S_i(x) \cdot S_i(y))\right) \cdot \hat{a}_i\right)$, and $SST = S_i\left((y - \bar{y})^2\right) = \frac{1}{n} \cdot \left(n \cdot S_i(y^2) - S_i(y)^2\right)$. Hence,

$$R_i^2 = 1 - \frac{SSE}{SST} = \frac{2\left(n \cdot S_i(xy) - S_i(x) \cdot S_i(y)\right) - \hat{a}_i\left(n \cdot S_i(x^2) - S_i(x)^2\right)}{n \cdot S_i(y^2) - S_i(y)^2} \hat{a}_i. \tag{3}$$

At last, we can refit the data with the following formula.

$$Y(x) = \frac{\sum_k R_k^2 \cdot \left(\hat{a}_k \cdot x_i + \hat{b}_k\right)}{\sum_k R_k^2}, k \in \{k \,|\, |x_k - x_i| \leqslant r\} \tag{4}$$

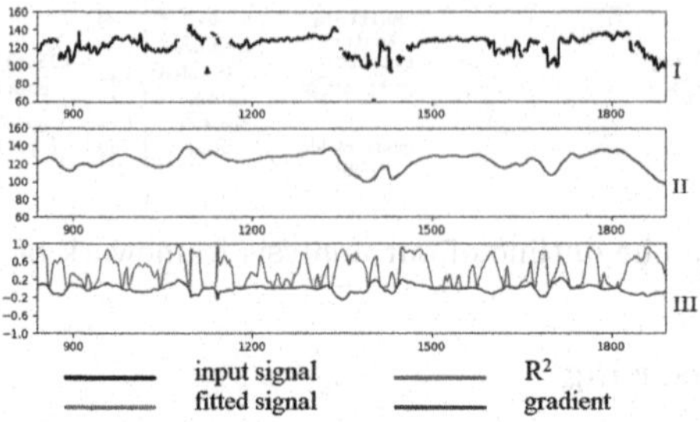

Fig. 3. Signal processing with ILLR method

According to the above formula, as the central point p_i changes, only such values need to be maintained to calculate the regression and refit the data: $S(1), S(x), S(y), S(x^2), S(xy), S(y^2)$. If we sort the points $p_i = (x_i, y_i), i = 1, \ldots, N$ in order of x_i, and calculate iteratively over ordered central point p_i, the total complexity of time and space will be O(N).

The ILLR method produce 2 kind of features. One refers to $\hat{a}_i$, which is called "gradient" in this paper. The other refers to R^2, namely determination coefficient. We will introduce the significance of them later. And by the refit function, we can change the signal frequency, the missing data can also be completed.

3.2 Extraction of Multiple Interpretable Features

In our method, totally 77 features of diversified scopes are extracted from each FHR trace as shown in Table 2. And the feature extraction method is as follows.

Overall Features. In the first stage, we apply ILLR method on the whole FHR with the regression parameter $r = 15\,(s)$, and acquire the result as shown in Fig. 3. To the knowledge of this paper, the FHR is influenced by vagus nerve, atrioventricular status [17], blood oxygen saturation, and other factors. And we use the gradients in representation of their influence.[1] On the other hand, the R^2, called coefficient of determination, is referred as the typicality of fetal heart activity. To the clinical knowledge, an overall reference to the FHR is needed, that is, whether the FHR activity is good and typical. For this reason, we abstracted 2 kinds of new features, which are the intensity and typicality of fetal heart activity as above.

Baseline. In the second stage, we determine the baseline of FHR, which is variable over time. We pick out the moderate signal parts or the low point. Then,

[1] According to the principle of signal processing, the relationship between the factors and heart rate is like control signal and output signal. The 2-norm of the gradient divided by the length is the ratio of energy to time, namely the power, which reflect the intensity of fetal heart activity. But this paper is not in favor of obtaining the so-called control signal from the first order difference of the FHR.

we enumerate the combinations of them and use scoring method to select the best one. The score items include prior probability, part length, value distribution of the parts, and the difference between each part and the surrounding signals. Finally, we smooth the selected parts and join them with quintic S-curve.

Accelerations and Decelerations. For the third stage, we detect accelerations (acc) and decelerations (dec). We developed a complex boundary decision algorithm because the real cases are very complicated, like shape overlapping and length change from seconds to minutes. For a specific acceleration/deceleration, we define the following features. "Time" is the duration time. "Peak" is the maximum difference between the FHR and baseline in beats per minute (bpm). "Area" is the integral area with respect to baseline. "Rank" is defined as the average rank in Table 1. "Reference" is defined as $rank/10 * (1 - miss\_rate)$. These indices are closer to the clinical application than original features, as the area index can reflect the blood oxygen compensation ability and distress degree of the fetus by experience.

Table 1. Ranks of accelerations and decelerations

Rank			1	2	3	4	5	6	7	8	9	10	12	15	18
Acc	area	$(bpm \cdot s)$	150	200	300	450	600	750	900	1050	1200	1400	1800	-	-
	peak	(bpm)	8	10	15	17.5	20	23.3	26.7	30	35	40	50	-	-
	time	(s)	10	12	15	30	60	90	120	160	200	240	360	-	-
Dec	area	$(bpm \cdot s)$	100	200	250	375	500	550	600	650	700	750	850	1000	-
	peak	(bpm)	-	10	15	17.5	20	21.7	23.3	25	26.2	27.5	30	32.5	35
	time	(s)	7.5	15	20	25	30	36	42	48	54	60	72	90	-

Variability. We determine the long-term variability (LTV) and short-term variability (STV) by range analysis on data of 60 s and 2.5 s. We also calculate baseline variability on the data without accelerations and decelerations and determine the type by the proportion of variation range.

Sine Wave. To detect sine waves, we first split FHR into 60-s segments with the overlap of 30 s. Then, we estimate a sinusoidal function for each segment and calculate the normalized square loss of the function to the FHR. Once there is a sine wave detected on a segment with the loss not more than 0.2 times the amplitude, we combine the adjacent wave segments of little change in the amplitude and wavelength. If a sine wave with 10 continuous periods exists, the FHR is of sine wave pattern.

3.3 Interpretable Deep Gaussian Mixture Model

The mixture model, also known as the hybrid model, is composed of several different models, which stands for the existence of submodels in a large model. As

Table 2. Features with indices of amounts. acc: accelaration, dec: deceleration

Type	Feature scope	Concrete content
Acc&Dec 35	reference 2	total reference coefficient of all acc/dec
	area 2	total area of all acc/dec (*bpm·s*)
	time 2	total time of all acc/dec (*s*)
	peak 2	the maximum peak of all acc/dec (*bpm*)
	rank 7	the merged rank and average rank of all acc/dec, and the minimum, maximum and median of the merged rank of accelerations for every 20 min
	number 8	the numbers of acc/dec in 3 different scales, and the minimum and maximum numbers of accelerations for every 20 min
	subtype 12	the quite real and total numbers of specific types: abrupt acceleration, prolonged acceleration, variable deceleration (VD), late or early deceleration (LD or ED), prolonged deceleration and severe deceleration
Top 16	acceleration 8	area (*bpm·s*), duration (*s*), peak (*bpm*) and rank of the first and second top acc/dec
	deceleration 8	
ILLR 6	gradient 3	the 1-norm and 2-norm both divided by data length, the standard deviation
	R^2 3	
Other 20	raw signal 3	
	full time 1	the full time of FHR record (*s*)
	quality 1	1 - the miss rate of geometric length of the data
	baseline 3	whether the global baseline is decided, the amount of 10-minute baseline periods, the average baseline (*bpm*)
	tachycardia 3	the specific area (*bpm·s*), time (*s*), and the average value obtained by dividing area by time (*bpm*)
	bradycardia 3	
	variability 4	long-term variability (*bpm*), short-term variability (*bpm*), baseline variability (*bpm*), the type of baseline variability (1 4: absent, minimal, moderate, marked)
	sine wave 2	whether the FHR is a sine wave pattern, the time of sine wave (*s*)

the most suitable atomic model, the Gaussian model is robust enough to tackle with the complex realistic features that always conforms to the Gaussian distribution on the whole. In addition, the DGMM also learns the uniform weights of features to balance the information comprehensiveness, this further reduces over fitting and improves versatility.

Theoretical Basis of Mixture Model Based on Gaussian Model. For a data vector $\boldsymbol{d} = (v_1, v_2, \ldots, v_P)$, supposing that each dimension is independent of the others, we define the relative probability function of the tth Gaussian model, $g\left(\boldsymbol{d}, M^{(t)}\right)$, as formula 5.

$$\prod_{i=1}^{P} \frac{1}{\sigma_i^{(t)}} \cdot e^{-\frac{\left(v_i-\mu_i^{(t)}\right)^2}{2\left(\sigma_i^{(t)}\right)^2}} = \exp\left(\sum_{i=1}^{P}\left(-\log\left(\sigma_i^{(t)}\right)-\frac{1}{2}\left(\frac{v_i-\mu_i^{(t)}}{\sigma_i^{(t)}}\right)^2\right)\right) \tag{5}$$

In avoid of exponential explosion that make the power of the exponent become 0, the exponent should be extracted and applied to the *softmax* function, as shown in formula 6.

$$\boldsymbol{PD}(\boldsymbol{d}) = softmax\left(\log\left(g\left(\boldsymbol{d}, M^{(1)}\right)\right), \ldots, \log\left(g\left(\boldsymbol{d}, M^{(T)}\right)\right)\right), \tag{6}$$
$$\text{where } softmax\left(e_1, ..., e_n\right) = \frac{1}{\sum_i \exp\left(e_i\right)} \exp\left(e_1, ..., e_n\right)$$

Weighting Features. The over fitting and homogenization problems are common for the network of ordinary GMM. As a solution, we apply a non-negative weight to each feature for relaxation, $\boldsymbol{w} = (w_1, \ldots, w_P), w_1, \ldots, w_P \geqslant 0$. And we introduce the prior probability of the submodels $\exp(\boldsymbol{P}) = \exp\left(p^{(1)}, \ldots, p^{(T)}\right)$. Consequently, we derive the final formula 7 from $\log g\left(\boldsymbol{d}, M^{(t)}\right)$ as follow.

$$\sum_{i=1}^{P} w_i\left(-log\left(\sigma_i^{(t)}\right)-\frac{1}{2}\left(\frac{v_i-\mu_i^{(t)}}{\sigma_i^{(t)}}\right)^2\right)+p^{(t)} \tag{7}$$

Topology of Deep Gaussian Mixture Model. As is shown in Fig. 4, the whole model consists of two parts. The first part is a mixture model involving feature learning that outputs $\boldsymbol{PD}(\boldsymbol{d})$ as mentioned above. The rest is a linear regression model with one convolution layer with parameter $\boldsymbol{W}_1 = \left(u^{(1)}, u^{(2)}, \ldots, u^{(T)}\right)$, and outputs the result $R(\boldsymbol{d}) = \int \boldsymbol{PD}(\boldsymbol{d}) * \boldsymbol{W}_1$. In addition, for initial start-up, MLE parameters can be used.

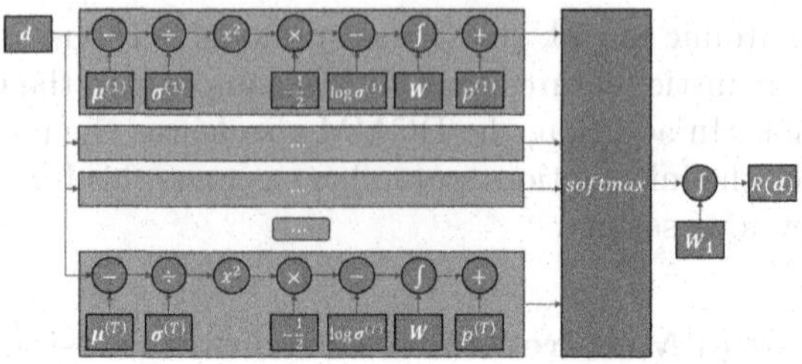

Fig. 4. The structure of deep Gaussian mixture model

4 Experiments

4.1 Dataset

We conduct experiments on the standard benchmark dataset CTU-UHB [4] to evaluate and compare our proposed method with different model parameter settings T. The database consists of 552 records of intrapartum data, each including the raw signal of FHR and UC for 30 to 90 min. The data are divided into 2 types: normal and pathological. We take the data with pH < 7.15 as pathological, and otherwise as normal. Eventually, 447 normal data and 105 pathological data are obtained.

4.2 Settings

We first evaluate our method by 10-fold cross-validation for a more reliable comparison with other methods. The feature dimension of DGMM is 108, including 77 original features, and normalized features that linearly related to time length. To compare with DGMM, DNN is used to test the same features, whose structure contains an input layer of size 77, three hidden full connection layers of size 128, 64 and 16, and an output layer of size 2. The result demonstrate the DGMM is much better than DNN on the same features, and reflect that the problem of overfitting will occur as the larger T is leading to better performance. The baseline methods compared with our method include ANN [5], LS-SVMs [14] and IBTF with LS-SVM [21]. Furthermore, an ablation study was conducted on all the features introduced above with different feature types, in which the model parameter was fixed with $T = 32$. In addition, we attempt to explain the DGMM's knowledge of key features learned from the data set in order of their uniform weight $\boldsymbol{w}$ of the model.

To measure the performances, the confusion matrix is used for each test. It has four prognostic indices: True Positive (TP), True Negative (TN), False Positive (FP), and False Negative (FN). The positive and negative that represent the predicted result stand for pathological and normal respectively, and true or false represents the correctness of the prediction. The performance criteria is shown in Table 3.

Table 3. The performance criteria

Criterion	Accuracy (Acc)	Sensitivity (Se)	Specificity (Sp)	Quality index (QI)
Formulation	$\frac{TP+TN}{TP+TN+FP+FN}$	$\frac{TP}{TP+FN}$	$\frac{TN}{TN+FP}$	$\sqrt{Se \cdot Sp}$

Table 4. The results of our methods and previous methods

Method	ACC	Se	Sp	QI
DNN	67.57	42.86	73.38	56.08
DGMM (T = 16)	90.22	76.19	**93.51**	84.41
DGMM (T = 24)	90.22	**79.05**	92.84	85.67
DGMM (T = 32)	**90.58**	**79.05**	93.29	**85.87**
DGMM (T = 40)	89.86	**79.05**	92.39	85.46
ANN	84.97	74.46	89.61	81.68
LS-SVMs	76.94	68.48	77.68	72.94
IBTF	65.41	63.45	65.88	64.04

Table 5. The results of ablation study

Ablation type	ACC	Se	Sp	QI
None	90.58	79.05	93.29	85.87
Acc&Dec	71.01	54.29	74.94	63.78
Top	86.23	70.48	89.93	79.61
ILLR	83.51	58.10	89.49	72.10
Other	68.66	41.90	74.94	56.04
ILLR+Other	69.20	28.57	78.75	47.43

The results are obtained by 10-fold cross validation, based on DGMM (T = 32).

4.3 Results

Table 4 presents comparisons between the proposed model and baseline models. Note that the ANN method is tested on manually selected simple data in different stages, so we objectively calculate the average score from their report. In the comparison, our method shows a significant improvement in sensitivity, because DGMM can effectively learn feature relevance and global distribution regardless of the common error of true data, as if doctors are paying more attention to comprehensive interpretation of data. So the classification boundary is more general and human-like.

Table 5 further displays the results obtained by ablation experiments. The ablation group of *acc&dec* and of *other* both have a very low result, and the features in these groups are almost all features mentioned in obstetric guidelines. The ablation group of *top* shows a little change, because they have less chance to play a special role. And the controlled group of ILLR proposed in this paper shows a 7.07% decrease of accuracy, which shows the importance of the ILLR method.

4.4 Interpretation of DGMM

We present some of the features in weight order in Table 6. And the EFM type of each submodel can be explained according to the ACOG guideline, as follows. Model 4 reflects scarce LD and ED, but frequent prolonged decelerations, and straight baseline (very low variation power as shown by gradient), such is a very serious pathological pattern. Besides, it also implies frequent accelerations and long-term tachycardia. Model 5 shows low STV and long sinusoidal waveform (18% of 1.5 h). Model 6 means acceleration and deceleration scarcity, which is

Table 6. The feature expectation of submodel for DGMM (T = 8)

Feature	$M^{(1)}$	$M^{(2)}$	$M^{(3)}$	$M^{(4)}$	$M^{(5)}$	$M^{(6)}$	$M^{(7)}$	$M^{(8)}$
LD or ED per 20 min	0.80	0.83	0.79	**0.17**	0.81	**0.12**	0.79	**−0.01**
Variability type *	2.45	2.97	2.15	3.04	2.47	2.27	2.18	2.24
Prolonged decelerations per 20 min	0.49	0.75	0.57	**1.22**	0.01	0.16	0.23	0.71
Minimum accelerations in 20 min	2.45	1.78	1.44	1.65	2.95	**0.28**	2.21	**0.58**
Accelerations per 20 min	2.18	2.13	1.24	2.33	1.49	1.05	1.38	1.07
Sine wave proportion	−0.44	0.08	0.05	−0.06	**0.18**	0.04	**0.21**	0.05
Normalized 2-norm of gradient	0.29	0.37	0.36	**0.06**	0.26	0.28	**0.51**	0.36
Tachycardia area per 20 min	−0.27	0.71	0.82	**3.50**	−0.39	0.48	0.94	**1.83**
Average tachycardia heart rate	3.16	5.43	7.46	4.59	7.09	6.13	1.64	**6.80**
EFM type **	1.83	1.55	1.42	−1.54	−1.82	−2.02	−2.07	−2.12

*: (1: absent, 2: minimal, 3: moderate, 4: marked)
**: (1: normal, -1: pathological)

also a bad pattern. Model 7 is long sinusoidal waveform and excessive variation power. Model 8 includes acceleration and deceleration scarcity as well as model 6, but also tachycardia that of both large area and high heart rate.

4.5 Discussion

Our method shows the best accuracy and quality index in contrast to other methods and is completely interpretable. In the methods of many researches, the researchers mostly cut off the partial data to improve the effect, because of the huge variability of features of FHR trace for an extended period. This also reflects the features we are dealing with are more complex and the classification work is more difficult. But our method can adapt to the changes of data length, and can also follow the baseline changes of FHR, so we solved the problem of processing long FHR traces with good effect.

5 Conclusion

In this paper, we propose a new function summation operator $S_i(f)$ that can be fast iterated, then we propose and prove the ILLR method that can calculate the least squares linear regression and coefficient of determination over N local ranges of signal of length N with the cost of O(N). And we put forward the idea of using the gradient and determination coefficient as the indicators of intensity and typicality of fetal heart activity.

We propose a completely interpretable methodology for preprocessing and extracting 77 different features from tens of scopes. And our experiments show the advantages of DGMM in multiple feature learning, as if doctors pay more attention to the comprehensive interpretation of data.

According to our controlled experiments, the features obtained by ILLR method work for at least 7% accuracy. And our experiments highlight the significance of the features related to accelerations and decelerations by showing them accounting for nearly 20% accuracy of the results. This paper also draws a rare conclusion that the top shape and second top shape of accelerations and decelerations have a little meaning in the analysis of FHR, too.

References

1. Arnold, K.C., Flint, C.J.: Acog practice bulletin no 106: intrapartum fetal heart rate monitoring: nomenclature, interpretation, and general management principles. Obstetr. Gynecol. **114**(1), 192–202 (2017)
2. Ayres-De-Campos, D., Spong, C.Y., Chandraharan, E.: Figo consensus guidelines on intrapartum fetal monitoring: cardiotocography. Int. J. Gynecol. Obstetr. (2015)
3. Bursa, M., Lhotska, L.: The use of convolutional neural networks in biomedical data processing. In: International Conference on Information Technology in Bio- and Medical Informatics (2017)
4. Chudáek, V., Spilka, J., Bura, M., Jank, P., Lhotská, L.: Open access intrapartum CTG database. BMC Pregn. Childbirth **14**(1), 16–16 (2014)
5. Comert, Z., Kocamaz, A.F.: Evaluation of fetal distress diagnosis during delivery stages based on linear and nonlinear features of fetal heart rate for neural network community. Int. J. Comput. Appl. **156**(4), 26–31 (2016)
6. Comert, Z., Kocamaz, A.F.: Fetal hypoxia detection based on deep convolutional neural network with transfer learning approach (2018)
7. Da, G., Luis, M., Costa-Santos, C., Ugwumadu, A., Schnettler, W.: Agreement and accuracy using the figo, acog and nice cardiotocography interpretation guidelines. Acta Obstetricia et Gynecologica Scandinavica: Official Publication of the Nordisk Forening for Obstetrik och Gynekologi (2017)
8. De Laulnoit, A.H., et al.: Automated fetal heart rate analysis for baseline determination and acceleration/deceleration detection: a comparison of 11 methods versus expert consensus. Biomed. Sig. Process. Control **49**, 113–123 (2019)
9. De Laulnoit, A.H., Boudet, S., Demailly, R., Peyrodie, L., Beuscart, R., De Laulnoit, D.H.: Baseline fetal heart rate analysis: Eleven automatic methods versus expert consensus 2016, pp. 3576–3581 (2016)
10. Devoe, L.D.: Nonstress testing and contraction stress testing. Obstetr. Gynecol. Clin. North Am. **26**(4), 535–556 (1999)
11. Devoe, L.D.: Nonstress and contraction stress testing. The Global Library of Women's Medicine (2009)
12. Doret, M., Spilka, J., Chudacek, V., Goncalves, P., Abry, P.: Fractal analysis and hurst parameter for intrapartum fetal heart rate variability analysis: a versatile alternative to frequency bands and lf/hf ratio. PLOS ONE **10**(8) (2015)

13. Georgieva, A., Abry, P., Chudacek, V., Djuric, P.M., Frasch, M.G., Kok, R., Lear, C.A., Lemmens, S.N., Nunes, I., Papageorghiou, A.T., et al.: Computer-based intrapartum fetal monitoring and beyond: a review of the 2nd workshop on signal processing and monitoring in labor (october 2017, oxford, uk). Acta Obstetricia et Gynecologica Scandinavica **98**(9), 1207–1217 (2019)
14. Georgoulas, G., Karvelis, P., Spilka, J., Chudáček, V., Stylios, C.D., Lhotská, L.: Investigating ph based evaluation of fetal heart rate (fhr) recordings. Health and Technology **7**(1) (2017)
15. Guideline, N.: Intrapartum care for healthy women and babies (2014)
16. Magenes, G., Signorini, M., Ferrario, M., Lunghi, F.: 2CTG2: a new system for the antepartum analysis of fetal heart rate, pp. 781–784 (2007). https://doi.org/10.1007/978-3-540-73044-6_203
17. Mohajer, M.P., Sahota, D.S., Reed, N.N., James, D.K.: Atrioventricular block during fetal heart rate decelerations. Arch. Dis. Childhood-Fetal Neonatal Ed. **72**(1), F51–F53 (1995)
18. Parer, J.T., Ikeda, T.: A framework for standardized management of intrapartum fetal heart rate patterns. Am. J. Obstetr. Gynecol. **197**(1), 26-e1 (2007)
19. Rotariu, C., Pasarică, A., Costin, H., Nemescu, D.: Spectral analysis of fetal heart rate variability associated with fetal acidosis and base deficit values. In: International Conference on Development & Application Systems (2014)
20. Southall, D., et al.: Prospective study of fetal heart rate and rhythm patterns. Arch. Disease Childhood **55**(7), 506–511 (1980)
21. Zafer, C., Fatih, K.A., Velappan, S.: Prognostic model based on image-based time-frequency features and genetic algorithm for fetal hypoxia assessment. Comput. Biol. Med. pp. S0010482518301458 (2018)
22. Zhao, Z., Deng, Y., Zhang, Y., Zhang, Y., Shao, L.: Deepfhr: intelligent prediction of fetal acidemia using fetal heart rate signals based on convolutional neural network. BMC Med. Inform. Dec. Making **19**(1) (2019)

Adverse Drug Events Detection, Extraction and Normalization from Online Comments of Chinese Patent Medicines

Zi Chai[1,2] and Xiaojun Wan[1,2](✉)

[1] Wangxuan Institue of Computer Technology, Peking University, Beijing, China
{chaizi,wanxiaojun}@pku.edu.cn
[2] The MOE Key Laboratory of Computational Linguistics, Peking University, Beijing, China

Abstract. Chinese Patent Medicines (CPMs) are welcomed by many people around the world, but the lack of information about their Adverse Drug Reactions (ADRs) is a big issue for drug safety. To get this information, we need to analyze from a number of real-world Adverse Drug Events (ADEs). However, current surveillance systems can only capture a small portion of them and there is a significant time lag in processing the reported data. With the rapid growth of E-commerce in recent years, quantities of patient-oriented user comments are posted on social media in real-time, making it of great value to automatically discover ADEs. To this end, we build a dataset containing 17K patient-oriented user posts about CPMs and further propose a new model that jointly performs ADE detection, extraction and normalization. Different from most previous works dealing with these tasks independently, we show how multi-task learning helps tasks to facilitate each other. To better deal with colloquial expressions and confusing statements in user comments, we leverage standard ADR-terms as prior knowledge as well as finding clues from other related comments. Experimental results show that our model outperforms previous work by a substantial margin.

Keywords: Adverse Drug Event (ADE) · ADE detection · ADE extraction · ADE normalization · Chinese patent medicines

1 Introduction

An Adverse Drug Event (ADE) refers to any undesirable experience associated with the real-world use of a medical product. By analyzing if there is a causative relationship between the event and medicine from a number of ADEs, we can dicover information about Adverse Drug Reactions (ADRs). This information will be further written in the instructions of medical products.

Chinese Patent Medicines (CPMs) are produced under the guidance of modern pharmacy theory using traditional Chinese medicines as raw materials, and they are welcomed by a number of people around the world. However, the ADEs

K. Karlapalem et al. (Eds.): PAKDD 2021, LNAI 12712, pp. 251–262, 2021.
https://doi.org/10.1007/978-3-030-75762-5_21

of many CPMs are not fully discovered, and thus many of their instructions use "still uncertain (尚不明确)" to describe the side effects. This is a big issue since health problems caused by CPMs have taken up over 17% of all the reported health issues in recent years[1]. To discover ADEs caused by CPMs, clinical trials are widely performed. However, they are far from enough and the results in ideal circumstances can be different in actual use [12]. Although many spontaneous reporting systems are used for post-marketing surveillance, they often suffer from under-reporting problems [2,6] as well as a time lag in processing reported data[22]. With the growth of E-commerce, many people are willing to buy medicines online and post some comments describing their drug use. These online drugstores can be viewed as a special type of social media where large amounts of patient-oriented user posts are generated in real-time, making it possible for automatically harvesting ADEs. To this end, we build **the first** dataset about CPMs from user posts and perform data-driven ADE discovery.

Prior works about ADE discovery mainly focused on three typical tasks. *ADE detection* is a binary classification task aiming to pick out user posts that mention certain ADEs. *ADE extraction* is a Named Entity Recognition (NER) task that extracts text-segments describing ADEs. *ADE normalization* performs Named Entity Normalization (NEN) that align each ADE entity to a standard ADR-term. Table 1 shows examples of the three tasks. It is worth noticing that these tasks are **strongly related** with each other, e.g., if a comment is classified as "not containing any ADE", the result of ADE extraction should be "None".

Table 1. Typical comments from our data (mentioned ADEs are marked out) and the expected outputs for each task.

	Comment	Classify	Extract	Normalize
1	After taking the pills, my **stomach burns**. (吃了药有胃在灼烧的感觉。)	1	stomach burns (胃在灼烧)	gastritis (胃炎)
2	Hope it works for my headache. (希望可以治好我的头疼。)	0	None	None
3	Used for rhinitis, but my **head hurt** and I **felt dizzy** after taking it. (用来治鼻炎，但吃完头疼，晕乎乎。)	1	head hurt (头疼) felt dizzy (晕乎乎)	gastritis (头痛) dizziness (头晕)

There are two major challenges in dealing with online comments. First, **colloquial expressions** are widely appeared, e.g., in *comment 1*, "stomach burns" is the colloquial expression of "stomachache". Second, there are many **confusing statements** describing the *patient's symptoms* instead of ADEs, e.g., *comment-2*, "headache" is a symptom but it may directly used by other comments to describe ADEs. As a result, an ADE mining model should understand context dependencies, especially when dealing with *comment-3*, where confusing statements "rhinitis" and colloquial-expressed ADEs "head hurt" are mixed together.

[1] http://www.nmpa.gov.cn/.

Based on the intuition that three tasks are closely related, we propose a model using multi-task learning to deal with all of them. It is composed by an encoder, a primary decoder, and an enhancing decoder. The encoder is mainly used for ADE-detection, and we regard both ADE extraction and normalization as sequence labeling tasks performed by the two decoders. To deal with colloquial expressions and confusing statements, our model leverages standard ADE-terms as prior knowledge, and can refer to other comments to find useful clues to better understand colloquial expressions. The main contributions of this paper are:

- We build the first dataset about CPMs containing 17K user comments from social media and further perform automatic ADE mining.
- We propose a model that jointly perform ADE detection, extraction and normalization. It deals with the colloquial expressions and confusing statements challenges by leveraging standard ADR-terms as prior knowledge and referring to other related comments to find useful clues.
- We perform extensive experiments to show that our model outperforms previous work by a substantial margin.

Resources for this paper are available at https://github.com/ChaiZ-pku.

2 Related Work

2.1 ADE Mining in Social Media

Data on social media have gained much attention for ADE mining because of their high volume and availability. User posts from blogs, forums and online communities are widely used by prior works, but comments from online drugstores were **little used** before. As far as we know, our dataset is the first for CPMs. Compared with prior datasets coming from biochemical literature, comments in our dataset is much shorter. Besides, users tend to use colloquial expressions while literatures written by experts are quite formal.

The ADE detection task aims to filter ADE-unrelated user posts. A commonly used method is data-driven supervised learning. Traditional machine learning methods [11,19] are widely used based on carefully designed features. In recent years, deep-learning models based on distributed representation have taken up the mainstream [1,9,24]. Since these methods require a large number of human-labeled data which are not easy to acquire, unsupervised [17] and semi-supervised classifiers [15] are getting more and more attention.

The ADE extraction task can be viewed as a specific NER task. Traditionally, this is performed by lexicon-based approaches [18] matching inputs with standard ADE-lexicons. In recent years, most studies regard it as a sequence labeling problem performed by statistical learning methods [8,18,29], especially deep learning methods [3,25,27]. The ADE normalization task can be viewed as a certain type of NEN task [23], which aims at determining a standard ADR-term for each specific ADE entity. Most prior works train a multi-class classifier after ADE extraction [5,21] to perform NEN. In recent years, using sequence labeling to directly perform ADE normalization from original user posts [30] is also adopted to avoid error-cascade.

2.2 Multi-task Learning in ADE Discovery

As mentioned above, different tasks about ADE discorery are closely related, and thus some researches [4,7] combine ADE detection and extraction together, others [16,30,30] try to perform ADE extraction with normalization. However, the multi-task learning is achieved by simply sharing low-level representations, e.g., using the same LSTM or CNN architecture to extract features for both of the two tasks and extraction respectively. Our method is **the first** to combine all these tasks together as far as we know. Besides, we introduce an encoder with two decoders to make these tasks **deeply related**.

3 Dataset

We chose several large-scale online drugstores to build our dataset, including Jingodng, Yiyao, Xunyiwenyao, Jinxiang and Jianke online Drugstores[2]. We chose 1k CPMs which have the most related user comments and crawled 186K user posts. After that, we performed data cleaning by discarding comments with URLs (most of them are useless advertisements), filtering out non-textual components (links to images or videos, labels about post time, etc.), removing comments which are less than three (Chinese) characters, and removing duplicated comments on the same CPM. Finally, we got 71K comments in total.

After data collection, we hired three native speakers to pick out ADE-entities in each comment (the definition of an ADE-entity is described in Sect. 4.1). To further guarantee the annotation quality, we hired another project manager who daily examined 10% of the annotations from each annotator and provided feedbacks. The annotation was considered valid only when the accuracy of examined results surpasses 90%. In this way, we discovered 5.2K comments containing 1.1K ADE-entities, and all of them were regarded as "ADE-comments" (the rest 66K comments were regarded as "Other-comments"). As we can see, only a small portion of user comments mention certain ADEs, which is not surprising since in most cases, CPMs indeed work and will not cause ADEs. Besides, there are many noise data in online comments. For ADE normalization, we put all the discovered ADE-entities into an ADE-set. For each entity, annotators were asked to normalize it into a standard ADR-term from the (Chinese version of) World Health Organization Adverse Reactions Terminology (WHOART). Finally, there were 108 related ADR-terms, and for each term, there were 10 different expressions on average. This reflects the challenge of colloquial expression.

We used all the 5.2K ADE-comments for our experiments. To make our task more challenging, we divided the rest 66K Other-comments into two subsets: if an Other-comment contains any item in our ADE-set, i.e., there are confusing statements, it is regarded as a "confusing Other-comment", else a "simple Other-comment". Finally, we got 3.8K "confusing Other-comments". Using all ADE-comments, confusing Other-comments and 8K ramdonly sampled "simple Other-comments", we built the final version of dataset for experiments.

[2] http://mall.jd.com/index-1000015441.html, http://www.111.com.cn, http://yao.xywy.com/, http://www.jinxiang.com/, https://www.jianke.com/.

4 Model

4.1 Problem Formulation

In this section, we formalize the three tasks. We regard each comment as a token-sequence and further define the concept of *ADE-entity* as follows: (1) an ADE-entity is a continuous sub-sequence from the comment, (2) an ADE-entity only describes one ADE, and it has no intersection with other ADE-entities in the same comment, (3) so long as an ADE-entity contains enough information to reveal an ADE, it should be as short as possible. The detection task predicts whether a comment contains any ADE-entity; the extraction task extracts all ADE-entities given an input comment; and the normalization aligns each extracted ADE-entity with an ADR-term. We regard the detection task as a binary classification task, and view both extraction and normalization tasks as sequence labeling tasks. Figure 1 shows the target outputs of each task given *comment-1* in Table 1 as input. Different from most prior works regarding ADE normalization as a multi-class classification task after ADE extraction, we directly perform normalization from input comments. This can not only avoid error-cascade, but also make it more convenient to combine normalization and extraction tasks.

Comment:	吃	了	药	有	胃	在	灼	烧	的	感	觉	。
ADE extraction:	O	O	O	O	B	I	I	I	O	O	O	O
ADE normalization:	O	O	O	O	T_{03}	T_{03}	T_{03}	T_{03}	O	O	O	O
ADE detection:	ADE-comment.											

Fig. 1. Target outputs of each task given the first comment in Table 1 as input. Since the ADE-entity "stomach burns (胃在灼烧)" corresponds with the 3rd term "gastritis (胃炎)" in the term-dictionary, the label for ADE normalization is T_{03}.

4.2 Encoder and Primary Decoder

The architecture of our model is shown in Fig. 2(a) composed by an encoder, a primary decoder, and an enhancing decoder. Our encoder and primary dedcoder takes the form of Transformer [26], which use multi-head self-attention to capture text dependencies.

As mentioned above, the ADE detection task is regarded as a binary classification task. It is performed by using a TextCNN model [13] based on the outputs of the encoder. As we can see, the attention between encoder and primary decoder explicitly combines ADE detection tasks with the other two tasks.

4.3 Enhancing Decoder

To deal with challenges caused by colloquial expressions and confusing statements, we use the enhancing decoder introducing extra information from standard ADR-terms as well as referring to other related comments. First, we introduce our info fusion block. It aims to use n queries $\boldsymbol{Q} \in R^{n \times d}$ to find useful

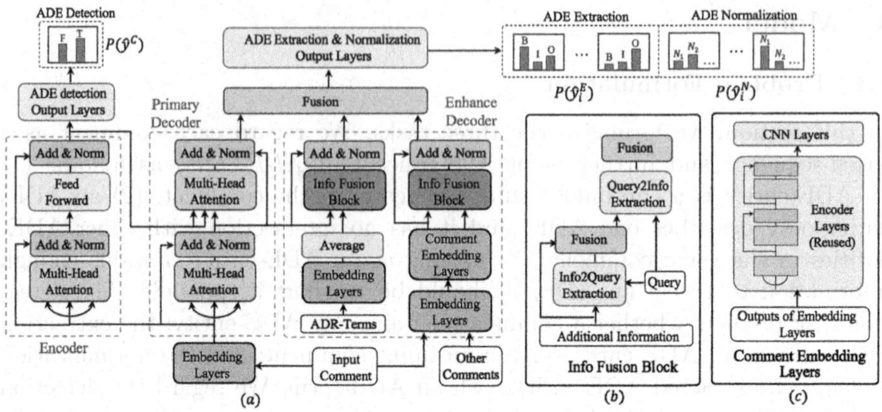

Fig. 2. An illustration of our model. (a) the whole architecture, (b) an info fusion block, (c) comment embedding layers.

clues from a memory, $M \in R^{m \times d}$ storing m slices of prior knowledge. To make queries and memories closely interacted, we adopt the coattention scenario [28] composed by a prior2query attention and a reverse query2prior attention.

Standard ADR-terms are the final outputs of our ADE normalization task. They can also be viewed as the results of ADE extraction (described in a formal way). As a result, these terms are suitable for acting as prior knowledge. For each ADR-term which is a sequence of tokens, we first reuse the embedding layers to get a sequence of vector tokens, and average this sequence as the final term-representation. These representations are further sent to our info fusion block as memories providing extra knowledge.

Since there are always many comments on the same medicine product and they are naturally related (e.g., the same medical product may cause similar ADEs), our enhancing decoder also refers to other user posts. For an input comment, we look up all other comments towards the same medical product and choose k of them which are most similar to the input comment. To measure the similarity between two comments, we use the Jaccard index similarity on their bag-of-token representations. After getting the k comments, we use comment embedding layers, as illustrated in Fig. 2(c), to embed them into a memory containing k vectors. The comment embedding layer first reuses the encoder to convert an input comment into a sequence of vectors, and then adopt the TextCNN model to get a single vector represtntation. The intuition for choosing TextCNN is based on the observation that most colloquial expressions and confusing statements are phrase-level challenges.

4.4 Output Layers

The ADE extraction and normalization tasks are both regarded as sequence labeling problems. For the i-th token in the input comment, we denote the corresponding output of primary decoder as $e_i^{\text{prim}} \in R^{d \times 1}$. For enhancing decoder, extra information extracted from ADE-terms, other comments are denoted as $e_i^{\text{term}}, e_i^{\text{comm}} \in= R^{d\times 1}$, respectively. Our model first fuse them by:

$$\begin{aligned} o_i^{\text{fuse}} &= \text{Concat}(e_i^{\text{prim}}, e_i^{\text{term}}, e_i^{\text{comm}}) \cdot w^o \\ o_i &= \text{LayerNorm}(o_i^{\text{fuse}} + FF(o_i^{\text{fuse}})) \end{aligned} \tag{1}$$

where FF is means feed-forward layer and w^o is a parameter.

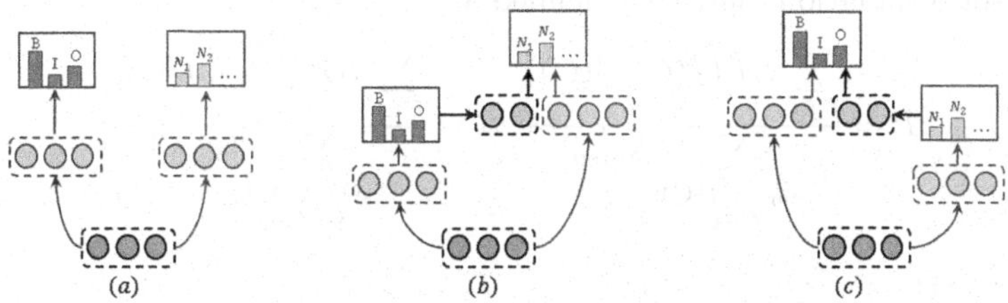

Fig. 3. Comparisons of: (a) Parallel, (b) Hierarchical-EN, (c) Hierarchical-NE.

We denote the prediction of our extraction, normalization task by $\hat{y}_i^e$, $\hat{y}_i^n$ respectively, and use three scenarios to get these labels:

- Parallel, which predicts $\hat{y}_i^e$, $\hat{y}_i^n$ independently based on o_i.
- Hierarchical-EN, which first predicts $\hat{y}_i^e$, and then the predicted label is embedded into a vector to further predict $\hat{y}_i^n$, as shown in Fig. 3(b).
- Hierarchical-NE, which is symmetry to Hierarchical-EN, i.e., the extraction task is based on the results of normalization, as shown in Fig. 3(c).

The cross-entropy loss function is used by each task, and the total loss of our model is a weighted sum of each loss.

5 Experiments

5.1 Evaluation Metric

For both ADE extraction and normalization tasks, we introduce two metrics:

Strict Evaluation. We denote a set of comments as $\mathcal{C} = \{C\}$. For any predicted entity, it is regarded as "correct" only if *exactly the same* with the corresponding gold-standard annotation. In this way, we can computs the precision and recall values.

Soft Evaluation. When a model nearly finds the correct answer, i.e., extracting "my stomach burns" instead of "stomach burns", directly regarding the prediction as "incorrect" is kind of arbitrary. For this reason, we use soft evaluations raised by [10]. We first define the span coverage between entities:

$$sp(s_{\text{pred}}, s_{\text{gold}}) = \frac{|s_{\text{pred}} \cap s_{\text{gold}}|}{|s_{\text{gold}}|} \tag{2}$$

where $|\cdot|$ returns the number of tokens in an entity and $\cap$ returns the set of tokens that two segments have in common (the two tokens should also appear in the same position). Based on the span coverage between two segments, we can further define the span coverage between two sets:

$$SP(\hat{f}(C), f(C)) = \sum_{s_{\text{pred}}, s_{\text{gold}} \in \hat{f}(C)} sp(s_{\text{pred}}, s_{\text{gold}}) \tag{3}$$

The soft evaluation metrics is defined as:

$$P = \frac{\sum_{C \in \mathcal{C}} SP(\hat{f}(C), f(C))}{\sum_{\substack{s_{\text{pred}} \in \hat{f}(C) \\ C \in \mathcal{C}}} |s_{\text{pred}}|}, R = \frac{\sum_{C \in \mathcal{C}} SP(\hat{f}(C), f(C))}{\sum_{\substack{s_{\text{gold}} \in f(C) \\ C \in \mathcal{C}}} |s_{\text{gold}}|} \tag{4}$$

5.2 Baselines and Scenarios

In this section, we introduce baseline models for each task.

ADE Detection. We chose four typical baseline methods. First, we used SVM and Random Forest. Following the conventions, we implemented the two models based on tf-idf and Part-of-Speech (POS) features. Second, we implemented RNN and CNN models. We used the CRNN, RCNN model in [9] and the ADE-CNN model is similar to [15] as strong baseline models.

ADE Extraction and Normalization. We first chose some traditional statistical baselines. An HMM model similar with [31] is used. As MaxEnt [20] was used for ADE extraction by [8], it was also used as a baseline model. The CRF model [14] was widely adopted by sequence labeling, and thus we use a model similar with [29] for ADE extraction. We also implemented some deep learning models. We adopted [3] which used an BiLSTM model, as well as the transformer architecture [26]. CRF-BiLSTM is powerful for many sequence labeling tasks, [25] used this model for ADE detection and [27] further combined character and word embeddings to extract ADEs in electronic health records. We implemented such a model as our strong baseline.

Baselines using Multi-task Learning. As mentioned above, performing multi-task learning has become a new trend in ADE discovery. For this reason, we used such models in recent years as strong baselines. First, Gupta2018 [7] combined ADE detection and extraction task by sharing the same LSTM at the bottom of the model. Second, Zhao2019 [30] first combined ADE extraction and normalization task by regarding both of them as sequence labeling problems. Besides, explicit feedbacks were introduced between the two tasks.

5.3 Results for ADE Detection

The results for ADE detection task are shown in Table 2. As we can see, each model can get rather high-level performances. For our model, the results of ADE detection is not affected by which scenario the output layers for ADE extraction and normalization takes. So we simply show the results under "parallel" scenario. As we can see, using multi-task learning does not shown significant high performance, but this does not bother considering that the performances are already good enough.

Table 2. Results for the ADE detection task.

Model	F1	P	R	Model	F1	P	R
SVM	91.34	93.87	88.95	Random Forest	88.36	90.80	86.05
CRNN	92.58	93.02	92.15	RCNN	92.10	92.32	92.89
ADE-CNN	**93.68**	**95.09**	92.30	Gupta2018	92.48	92.82	92.13
Ours	92.79	91.13	**94.52**				

5.4 Results for ADE Extraction

The results for ADE extraction are shown in Table 3. When it comes to strict evaluations, our model based on the hierarchical-NE scenario gets the highest performances. Besides, Gupta2018 and Zhao2019 that leverage multi-task learning are better than the rest baselines. This illustrates the importance of using multi-task learning. When it comes to soft evaluations, our model still gets the highest F1. Although it does not reach the best precision and recall values, it maintains high-level performances. Besides, models using multi-task learning can reach better performances.

Table 3. Results for the ADE extraction task. In each column, we underline the best results of all baselines and set the best result of the whole column in bold.

Model	Strict Evaluatoin			Soft evaluation		
	F1	P	R	F1	P	R
HMM	60.04	53.03	69.20	65.80	56.00	**79.76**
MaxEnt	62.42	57.89	67.72	66.45	62.56	70.88
CRF	69.92	73.92	66.33	77.24	91.97	66.59
BiLSTM	69.98	79.02	62.79	73.81	92.52	61.40
Transformer	71.80	78.98	65.82	75.29	91.72	63.85
CRF-LSTM	71.86	74.95	69.02	79.63	**93.53**	69.33
Gupta2018	72.19	76.70	68.19	78.06	91.69	67.96
Zhao2019	72.47	75.09	70.03	78.53	90.72	69.21
Ours (Parallel)	74.05	79.65	69.19	79.01	91.84	69.32
Ours (Hierarchical-NE)	**75.51**	**80.15**	**71.38**	**80.54**	92.39	71.37
Ours (Hierarchical-EN)	74.07	77.78	**71.38**	79.50	90.90	70.63

When we compare models that use multi-task learning, we can find that Zhao2019, which combines ADE extraction and alignment task always performs better than Gupta2018, which combines ADE extraction and detection task. This illustrates that the extraction and normalization tasks are more related and thus easier to benefit from each other. Besides, deep learning models (especially the CRF-LSTM model) always perform better than traditional machine learning methods. By the way, getting a high recall value is much harder than precision for all models.

5.5 Results of ADE Normalization

The results of ADE normalization are shown in Table 4. Different from ADE extraction which aims to discover as much information about ADE-entities as we can, the goal of ADE normalization is precisely aligning ADR-terms to ADE-entities. To this end, we only adopt the strict evaluation metrics. From normalization results, we can find that our model based on the hierarchical-NE scenario gets the best recall and F1 values. Among all baselines, Zhao2019 which combines ADE normalization and extraction gets the best performances.

Table 4. Results for the ADE normalization task. In each column, we underline the best results of all baselines, and set the best result of the whole column in bold.

Model	F1	P	R	Model	F1	P	R
HMM	54.80	58.89	51.24	MaxEnt	59.47	51.85	69.71
CRF	62.91	69.71	57.31				
BiLSTM	63.33	84.67	50.58	Transformer	64.30	86.27	51.24
CRF-BiLSTM	67.01	<u>**91.38**</u>	52.91				
Gupta2018	63.84	85.84	50.82	Zhao2019	<u>68.18</u>	64.44	<u>72.37</u>
Ours (Parallel)	67.59	63.15	72.71				
Ours (Hierarchical-NE)	69.24	65.96	72.87	Ours (Hierarchical-EN)	**70.07**	66.37	**74.21**

6 Conclusion and Future Work

In this paper, we studied how to use comments from online drugstores to detect ADEs caused by CPMs by jointly performing ADE detection, extraction and normalization tasks. We build the first dataset about CPMs containing 17K patient-oriented online comments, and further view the three tasks in a joint manner. We also tried to deal with colloquial expressions and confusing statements by introducing extra information, i.e., leveraging prior knowledge from ADR-terms as well as referring to other related comments. Experiment results prove that our model outperforms previous work by a substantial margin.

For future work, it would be interesting to enlarge the dataset and put our model in real-world use. In this way, we can compare its performance with real-world surveillance systems. We will also apply our model to other similar tasks to test its robustness.

References

1. Akhtyamova, L., Ignatov, A., Cardiff, J.: A large-scale CNN ensemble for medication safety analysis. In: Frasincar, F., Ittoo, A., Nguyen, L.M., Métais, E. (eds.) NLDB 2017. LNCS, vol. 10260, pp. 247–253. Springer, Cham (2017). https://doi.org/10.1007/978-3-319-59569-6_29
2. Bates, D.W., Evans, R.S., Murff, H., Stetson, P.D., Pizziferri, L., Hripcsak, G.: Detecting adverse events using information technology. J. Am. Med. Inform. Assoc. **10**(2), 115–128 (2003)
3. Cocos, A., Fiks, A.G., Masino, A.J.: Deep learning for pharmacovigilance: recurrent neural network architectures for labeling adverse drug reactions in twitter posts. J. Am. Med. Inform. Assoc. **24**(4), 813–821 (2017)
4. Crichton, G., Pyysalo, S., Chiu, B., Korhonen, A.: A neural network multi-task learning approach to biomedical named entity recognition. BMC Bioinform. **18**(1), 368 (2017)
5. Doan, S., Xu, H.: Recognizing medication related entities in hospital discharge summaries using support vector machine. In: Proceedings of the 23rd International Conference on Computational Linguistics: Posters, pp. 259–266. Association for Computational Linguistics (2010)
6. Drazen, J.M., et al.: Adverse drug event reporting: the roles of consumers and health-care professionals: workshop summary. National Academy Press (2007)
7. Gupta, S., Gupta, M., Varma, V., Pawar, S., Ramrakhiyani, N., Palshikar, G.K.: Multi-task learning for extraction of adverse drug reaction mentions from tweets. In: Pasi, G., Piwowarski, B., Azzopardi, L., Hanbury, A. (eds.) ECIR 2018. LNCS, vol. 10772, pp. 59–71. Springer, Cham (2018). https://doi.org/10.1007/978-3-319-76941-7_5
8. Gurulingappa, H., et al.: Development of a benchmark corpus to support the automatic extraction of drug-related adverse effects from medical case reports. J. Biomed. Inform. **45**(5), 885–892 (2012)
9. Huynh, T., He, Y., Willis, A., Rüger, S.: Adverse drug reaction classification with deep neural networks. In: Proceedings of COLING 2016, International Conference on Computational Linguistics: Technical Papers, pp. 877–887 (2016)
10. Johansson, R., Moschitti, A.: Syntactic and semantic structure for opinion expression detection. In: Proceedings of the Fourteenth Conference on Computational Natural Language Learning, pp. 67–76. Association for Computational Linguistics (2010)
11. Jonnagaddala, J., Jue, T.R., Dai, H.J.: Binary classification of twitter posts for adverse drug reactions. In: Proceedings of the Social Media Mining Shared Task Workshop at the Pacific Symposium on Biocomputing, Big Island, HI, USA, pp. 4–8 (2016)
12. Jüni, P., Altman, D.G., Egger, M.: Assessing the quality of controlled clinical trials. BMJ **323**(7303), 42–46 (2001)
13. Kim, Y.: Convolutional neural networks for sentence classification. arXiv preprint arXiv:1408.5882 (2014)
14. Lafferty, J., McCallum, A., Pereira, F.C.: Conditional random fields: Probabilistic models for segmenting and labeling sequence data (2001)
15. Lee, K., et al.: Adverse drug event detection in tweets with semi-supervised convolutional neural networks. In: Proceedings of the 26th International Conference on World Wide Web, pp. 705–714. International World Wide Web Conferences Steering Committee (2017)

16. Lou, Y., Zhang, Y., Qian, T., Li, F., Xiong, S., Ji, D.: A transition-based joint model for disease named entity recognition and normalization. Bioinformatics **33**(15), 2363–2371 (2017)
17. Ma, F., et al.: Unsupervised discovery of drug side-effects from heterogeneous data sources. In: Proceedings of the 23rd ACM SIGKDD International Conference on Knowledge Discovery and Data Mining, pp. 967–976. ACM (2017)
18. Nikfarjam, A., Sarker, A., O'Connor, K., Ginn, R., Gonzalez, G.: Pharmacovigilance from social media: mining adverse drug reaction mentions using sequence labeling with word embedding cluster features. J. Am. Med. Inform. Assoc. **22**(3), 671–681 (2015)
19. Plachouras, V., Leidner, J.L., Garrow, A.G.: Quantifying self-reported adverse drug events on twitter: signal and topic analysis. In: Proceedings of the 7th International Conference on Social Media & Society, p. 6. ACM (2016)
20. Ratnaparkhi, A.: A maximum entropy model for part-of-speech tagging. In: Conference on Empirical Methods in Natural Language Processing (1996)
21. Sahu, S.K., Anand, A.: Recurrent neural network models for disease name recognition using domain invariant features. arXiv preprint arXiv:1606.09371 (2016)
22. Sampathkumar, H., Chen, X.W., Luo, B.: Mining adverse drug reactions from online healthcare forums using hidden markov model. BMC Med. Inform. Decis. Mak. **14**(1), 91 (2014)
23. Schiefer, C.: Neural biomedical named entity normalization
24. Tafti, A.P., et al.: Adverse drug event discovery using biomedical literature: a big data neural network adventure. JMIR Med. Inform. **5**(4) (2017)
25. Tutubalina, E., Nikolenko, S.: Combination of deep recurrent neural networks and conditional random fields for extracting adverse drug reactions from user reviews. J. Healthcare Eng. 2017 (2017)
26. Vaswani, A., et al.: Attention is all you need. In: Advances in Neural Information Processing Systems, pp. 5998–6008 (2017)
27. Wunnava, S., Qin, X., Kakar, T., Rundensteiner, E.A., Kong, X.: Bidirectional LSTM-CRF for adverse drug event tagging in electronic health records. Proc. Mach. Learn. Res. **90**, 48–56 (2018)
28. Xiong, C., Zhong, V., Socher, R.: Dynamic coattention networks for question answering. arXiv preprint arXiv:1611.01604 (2016)
29. Yates, A., Goharian, N., Frieder, O.: Extracting adverse drug reactions from social media. AAAI **15**, 2460–2467 (2015)
30. Zhao, S., Liu, T., Zhao, S., Wang, F.: A neural multi-task learning framework to jointly model medical named entity recognition and normalization. In: Proceedings of the AAAI Conference on Artificial Intelligence, vol. 33, pp. 817–824 (2019)
31. Zhou, G., Su, J.: Named entity recognition using an hmm-based chunk tagger. In: proceedings of the 40th Annual Meeting on Association for Computational Linguistics, pp. 473–480. Association for Computational Linguistics (2002)

Adaptive Graph Co-Attention Networks for Traffic Forecasting

Boyu Li, Ting Guo(✉), Yang Wang, Amir H. Gandomi, and Fang Chen

Data Science Institute, University of Technology Sydney, Ultimo, Australia
{boyu.li,ting.guo,yang.wang,Amirhossein.Gandomi,fang.chen}@uts.edu.au

Abstract. Traffic forecasting has remained a challenging topic in the field of transportation, due to the time-varying traffic patterns and complicated spatial dependencies on road networks. To address such challenges, we propose an adaptive graph co-attention network (AGCAN) to predict traffic conditions on a road network graph. In our model, an adaptive graph modelling method is adopted to learn a dynamic relational graph in which the links can capture the dynamic spatial correlations of traffic patterns among nodes, even though the adjacent nodes may not be physically connected. Besides, we propose a novel co-attention network targeting long- and short-term traffic patterns. The long-term graph attention module is used to derive periodic patterns from historical data, while the short-term graph attention module is employed to respond to sudden traffic changes, like car accidents and special events. To minimize the loss generated during the learning process, we adopt an encoder-decoder architecture, where both the encoder and decoder consist of novel hierarchical spatio-temporal attention blocks to model the impact of potential factors on traffic conditions. Overall, the experimental results on two real-world traffic prediction tasks demonstrate the superiority of AGCAN.

Keywords: Graph Neural Network · Traffic forecasting · Attention mechanism · Spatio-temporal · ITSs

1 Introduction

Intelligent Transportation Systems (ITSs) are gaining increasing attention in the planning of urban areas as they can significantly improve the operation level of cities by aiding in traffic management [1,16]. The prediction of traffic conditions is a vital component of advanced ITSs, which aims to influence travel behavior, reduce traffic congestion, improve mobility and enhance air quality [2,7]. The goal of traffic forecasting is to predict the traffic conditions (e.g., traffic speed) of future time-steps given the historical traffic data [26]. However, the predictions become particularly challenging due to short-term (e.g., accidents and special events) and long-term (e.g., peak/off-peak and weekday/weekend) traffic patterns, and the uncertainty of traffic conditions caused by the natural spatial dependencies of the road networks.

K. Karlapalem et al. (Eds.): PAKDD 2021, LNAI 12712, pp. 263–276, 2021.
https://doi.org/10.1007/978-3-030-75762-5_22

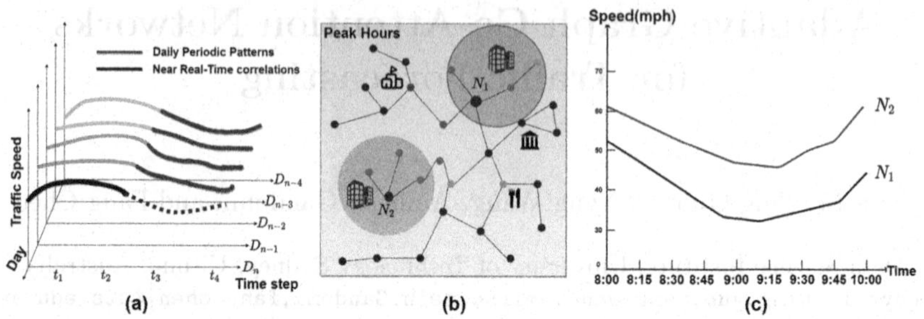

Fig. 1. Cross-time and cross-region information of traffic network. (a) The current traffic condition (the red dash curve) has high correlations with the daily periodic patterns and the near real-time traffic conditions. (b, c) The nonadjacent locations (N_1 and N_2) may have similar traffic conditions at the same time steps. (Color figure online)

Traffic forecasting, which can be considered as a typical spatial and temporal problem, has been studied for decades [3]. Recently, data-driven methods have become more popular, especially with the rise of deep learning modelling techniques. To form traffic information networks, various sensors are deployed at key traffic intersections to capture traffic data over time. Previous studies in deep learning methods, such as Convolutional Neural Networks (CNNs) and Recurrent Neural Network (RNNs), have achieved acceptable results from both spatial and temporal perspectives [10,11,17,23]. With the recent growth of interest in studies on Graph Neural Networks (GNNs), the incredible potential of applying GNNs to traffic forecasting tasks has been discovered [4,9,22]. Traffic data from entire traffic network can be sampled as traffic graph signals. In traffic networks, each node represents a sensor and the signals of the node are the traffic data records. A novel Spatial-Temporal GCN model (STGCN) was developed to learn both the spatial and temporal features from neighbors [25]. This method has been proven to be very effective in comparison with previous traditional deep learning methods. To balance the information derived from spatial and temporal domains, the authors proposed Diffusion Convolutional Recurrent Neural Network (DCRNN), which applied a diffusion process with STGCN to predict traffic flows [8]. Meanwhile, the attention mechanism has proven to be a simple but effective method that can achieve great success in sequence data processing [6,20,21]. Therefore, researchers have started to consider combining the attention mechanism and GNNs to attain better performance of traffic predictions and enhanced SOTA results on traffic benchmark datasets [15,27]. However, the existing GNN-based models still lack satisfactory progress for traffic forecasting, mainly due to the following two challenges.

- **Lack of cross-region information extraction.** While existing methods only collect spatial information from neighbors, some of this data are not always true as road attributes (like road width and direction) may cause differences. Meanwhile, two nodes belonging to different regions which are not connected may have similar traffic patterns (refer to the example shown in Fig. 1(b) and (c)). Thus, extracting useful information from both neighbors and cross-region nodes is a big challenge.

- **Lack of cross-time information integration.** An important feature of traffic data for prediction tasks is the daily periodic patterns (i.e., long-term traffic patterns). The traffic conditions are always similar during weekday peak hours for a given location. In contrast, near real-time traffic information (i.e., short-term patterns) is a key indicator of emergencies, like car incidents, special events, road congestion, etc. Therefore, integrating useful information from both long- and short-term patterns for a better prediction is another challenge (as shown in Fig. 1(a)).

To address the aforementioned challenges, we propose an Adaptive Graph Co-Attention Network (AGCAN) to predict the traffic conditions on a given road network over defined time steps. Although we focus on traffic speed predictions for illustration purposes, our model could be applied to predict other numerical traffic data. The contributions of this work are summarized as follows:

- We introduce an adaptive graph attention learning module to capture the cross-region spatial dependencies with the dynamic trend. The generated adaptive graph can help GNNs collect useful information from a larger range of areas.
- We propose a long- and short-term co-attention network which learns both temporal periodic patterns and near real-time traffic patterns for better prediction results. This attention mechanism builds direct relationships between historical and future time steps to alleviate the problem of error propagation.
- We evaluate our adaptive graph co-attention network (AGCAN) on two real-world traffic datasets and observed improvements over state-of-the-art baseline methods for both long-term (60 min) and short-term (15–30 min) predictions.

2 Preliminaries

Traffic forecasting is a typical time-series prediction problem, i.e., predicting the most likely traffic measurements (e.g., speed and traffic volume) in the subsequent Q time steps given the previous P traffic observations as:

$$F^* = \{X^*_{t+1}, X^*_{t+2}, \cdots, X^*_{t+Q} \mid H\} \tag{1}$$

where $H = \{X_{t-P+1}, X_{t-P+2}, \ldots, X_t\} \in \mathbb{R}^{P \times N \times C}$ is the observations of historical P time steps. $X_t \in \mathbb{R}^{N \times C}$ is the traffic condition at time step t. $[X_t]_{i\cdot}$ means the historical traffic signal of Location i at time step t, which can be long- or short-term traffic signals. C is the number of traffic conditions of interest (e.g., traffic volume, traffic speed, etc.). N refers to N locations of interest (e.g., traffic sensors or road segments).

In this work, we define the traffic network on a graph and focus on structured traffic time series [25]. The observation H is not independent but is linked by pairwise connections in the graph. Therefore, H can be regarded as a graph signal that is defined on an undirected graph (or directed one) $\mathcal{G} = (\mathcal{V}, \mathcal{E}, \mathcal{A})$, where $\mathcal{V}$ is the set of interested locations, $\mathcal{E}$ is corresponding edge set, and $\mathcal{A} \in \mathbb{R}^{N \times N}$ is the adjacency matrix with weights $\mathcal{A}_{ij}$ measured by the road network distance between node v_i and v_j.

3 Adaptive Graph Co-Attention Network

As discussed in Sect. 1, the study is aimed to extract and integrate useful cross-region and cross-time information from a given traffic network. To capture the cross-region spatial dependencies with the dynamic trend, an adaptive graph attention learning module is developed. Besides the actual connections between neighboring traffic nodes considered in the original traffic network (i.e., physical links), we introduce dynamic connections (i.e., cross-region links) to capture the potential correlations between areas with similar attributes (e.g., business or school districts) that may be separated by long distances. To learn the cross-time dependencies, we propose a co-attention network to take both long- and short-term traffic signals into consideration. The co-attention network follows the encoder-decoder architecture, where the encoder encodes the input traffic features and the decoder predicts the output sequence. Long-term and short-term temporal information will be processed in parallel, and a hierarchical attention block is developed to operate spatio-temporal signals under a hierarchy of successively higher-level units. The framework of our proposed AGCAN model is shown in Fig. 2.

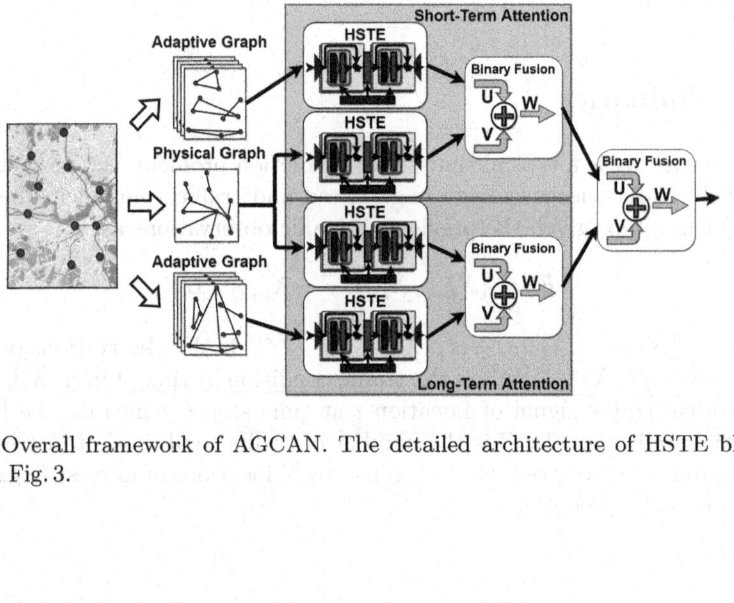

Fig. 2. Overall framework of AGCAN. The detailed architecture of HSTE blocks is given in Fig. 3.

3.1 Adaptive Graph Learning

The cross-region spatial correlations widely exist in the real-world traffic network with a dynamic trend. For instance, traffic congestion always occurs in business districts during peak hour and in school districts at the end of school. The cross-region correlations can be represented as cross-region links over the traffic network. If the traffic patterns are similar at a time step, there is a hyper link between the two nodes. The graph composed of functional link that changes over time is denoted an *Adaptive Graph*. To distinguish it from the original graph with physical links, we specify the physical graph and adaptive graph as:

Physical Graph is constructed based on the connectivity of the actual traffic network, represented as $\mathcal{G}_{\mathcal{P}} = (\mathcal{V}, \mathcal{E}_{\mathcal{P}}, \mathcal{A}_{\mathcal{P}})$. The adjacency matrix $\mathcal{A}_{\mathcal{P}}$ shows the spatial dependencies between traffic locations by the physical distances $d(\mathcal{V}_i, \mathcal{V}_j)$. We compute the pairwise traffic network distances between nodes and build the adjacency matrix using the threshold Gaussian kernel [8,18]:

$$[\mathcal{A}_{\mathcal{P}}]_{ij} = \begin{cases} \exp\left(-\frac{\mathrm{d}(\mathcal{V}_i, \mathcal{V}_j)^2}{\sigma^2}\right), \mathrm{d}\left(\mathcal{V}_i, \mathcal{V}_j\right) \geq \rho; \\ 0, \qquad \text{otherwise} . \end{cases} \tag{2}$$

where σ is the standard deviation of distances, and ρ is the threshold. As the interested locations are fixed, the physical graph is static.

Adaptive Graph is a dynamic graph that is constructed based on the dynamic correlations (cross-region links) between any nodes of the traffic network at time step t, represented as $\mathcal{G}_{\mathcal{F}}{}^t = (\mathcal{V}, \mathcal{E}_{\mathcal{F}}{}^t, \mathcal{A}_{\mathcal{F}}{}^t)$. To dynamically capture the cross-region spatial dependencies, we introduce a self-adaptive method to generate $\mathcal{A}_{\mathcal{F}}{}^t$. The attention mechanism is adopted here to learn the signal similarity among nodes. It is unrealistic to calculate the similarity of one target node and the rest of the nodes for each time step. Hence, we pre-train a set of cross-region clusters over the traffic network. The nodes in the same clusters have coarse similarity in some traits (e.g., similar traffic patterns or functional areas). Only the nodes in the same cluster are considered for message passing. The clusters can be obtained based on regional attributes or the correlations of historical traffic signals. In this paper, we use K-NN algorithm to generate the set of clusters, represented as $\mathcal{M}$. $\mathcal{M}_{ij} = 1$ means $\mathcal{V}_i$ and $\mathcal{V}_j$ are in the same cluster, and $\mathcal{M}_{ij} = 0$ means otherwise. We adopt an attention block to capture the dynamic spatial similarity for a given time step t:

$$\alpha_{ij}^t = \frac{\exp(\langle \mathscr{F}_\alpha([X_t]_{i\cdot}), \mathscr{F}_\alpha([X_t]_{j\cdot})\rangle / \sqrt{C})}{\sum_{m \in \{m|\mathcal{M}_{ij}=1\}} \exp(\langle \mathscr{F}_\alpha([X_t]_{i\cdot}), \mathscr{F}_\alpha([X_t]_{m\cdot})\rangle / \sqrt{C})} \tag{3}$$

where $\mathscr{F}_\alpha(x) = \mathbf{ReLU}(x\mathbf{W} + \mathbf{b})$ is a nonlinear projection of x with learnable parameters $\mathbf{W}$ and $\mathbf{b}$, and $\langle \cdot, \cdot \rangle$ denotes the inner product operator. Therefore, the adjacency matrix of $\mathcal{G}_{\mathcal{F}}{}^t$ is:

$$[\mathcal{A_F}^t]_{ij} = \begin{cases} \alpha_{ij}^t, & \mathcal{M}_{ij} = 1; \\ 0, & \text{otherwise} . \end{cases} \tag{4}$$

The attention of $\mathcal{A_F}^t$ is integrated into our co-attention model. All the learnable parameters will be trained and updated consistently with other modules.

3.2 Hierarchical Spatio-Temporal Embedding

To model the dynamic spatio-temporal embedding, we adopt an encoder-decoder architecture, represented as **HSTE**, with novel hierarchical spatio-temporal attention blocks (H-Att). Both the encoder and decoder contain $\mathcal{L}$ H-Att blocks with residual connections. In each H-Att block, the temporal unit is stacked upon the spatial unit. The motivation is that the spatial dependencies captured by applying spatial attention can reduce the data perturbation and greatly help the temporal attention unit to better capture the long- and short-term traffic patterns. We also incorporate the graph structure and signal information into multi-attention mechanisms through spatio-temporal embedding. The modules are detailed as follows.

Spatio-Temporal Embedding. The idea of Spatio-Temporal Embedding (ST embedding) was proposed by Zheng, et al. [27]. The *node2vec* approach [5] is applied on the adjacency matrix ($\mathcal{A_P}$ or $\mathcal{A_F}^t$ in our case) to learn the vertex representations to preserve the graph structure. The obtained spatial embedding of node $\mathcal{V}_i$ is represented as $e_i^S \in \mathbb{R}^C$. To obtain the time-variant vertex representations, the spatial embedding and the temporal input ($e_i^T \in \mathbb{R}^C$) are fused by ST embedding, as shown in Fig. 3, and defined as $e_i^{STE} = e_i^S + e_i^T$, where e_i^T can be either long-term or short-term traffic signals. The ST embedding contains both the graph structure and time information and is used in spatial and temporal attention mechanisms, as shown in Fig. 3.

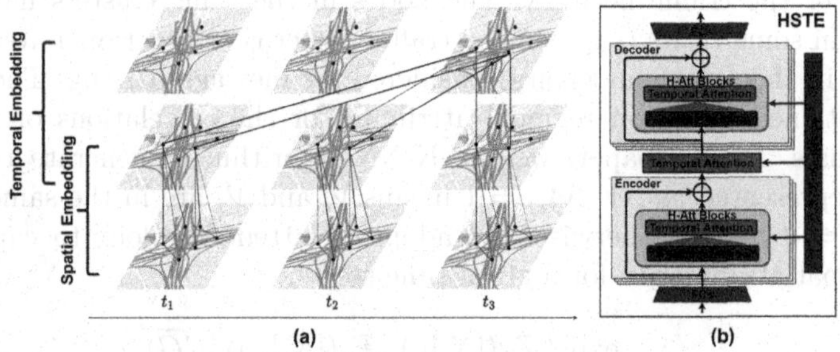

Fig. 3. The architecture of the HSTE block. (a) The spatial and temporal correlations are extracted and integrated hierarchically. (b) The encoder-decoder architecture with spatio-temporal embedding in HSTE block.

H-Att Blocks. The spatial and temporal attention units of an H-Att block are various classic attention mechanisms [21]. For the spatial attention units, the attention context is defined on the spatial neighbors with either physical or cross-region links. For the temporal attention units, the context is defined in the long-/short-term historical time steps. The output of spatial attention units will be fed to the temporal attention units for hierarchical learning, as shown in Fig. 3. Formally, the two units can be defined as:

Weighted Spatial Attention To model the traffic conditions of a location impacted by other locations, we design a weighted spatial attention mechanism by considering the graph structure ($\mathcal{G}_{\mathcal{P}}$ or $\mathcal{G}_{\mathcal{F}}$):

$$\beta_{ij}^{(k)t} = \frac{\exp(\langle \mathscr{F}_{\beta_1}^{(k)}([H_t^{(l-1)}]_{i\cdot}), \mathscr{F}_{\beta_1}^{(k)}([H_t^{(l-1)}]_{j\cdot})\rangle/\sqrt{D}) \cdot \mathcal{A}_{ij}}{\sum_{a\in\{a|\mathcal{A}_{ia}\neq 1\}} \exp(\langle \mathscr{F}_{\beta_1}^{(k)}([H_t^{(l-1)}]_{i\cdot}), \mathscr{F}_{\beta_1}^{(k)}([H_t^{(l-1)}]_{a\cdot})\rangle/\sqrt{D}) \cdot \mathcal{A}_{ia}} \tag{5}$$

$$[H_t^{(l)}]_{i\cdot} = \|_{k=1}^{K} \left\{ \sum_{j\in\{j|\mathcal{A}_{ij}\neq 1\}} \beta_{ij}^{(k)t} \cdot \mathscr{F}_{\beta_2}^{(k)}\left([H_t^{(l-1)}]_{j\cdot}\right) \right\} \tag{6}$$

where $H^{(l-1)} \in \mathbb{R}^{P\times N\times D}$ is the input of the lth block and the hidden embedding of $\mathcal{V}_i$ at time step t is represented as $[H_t^{(l)}]_{i\cdot}$. $\mathcal{A}$ could be $\mathcal{A}_{\mathcal{P}}$ or $\mathcal{A}_{\mathcal{F}}$ depending on which module the weighted spatial attention belongs to. $\mathscr{F}_{\beta_1}^{(k)}(\cdot)$ and $\mathscr{F}_{\beta_2}^{(k)}(\cdot)$ are two different nonlinear projections, and $||$ is the concatenation operation.

Temporal Attention. To model the traffic condition at a location correlated with its previous observations, we adopt a temporal attention mechanism to dynamically model the nonlinear correlations between different time steps. In our AGCAN model, we considered long- and short-term correlations which can be modelled by applying the same temporal attention unit:

$$\gamma_{ij}^{(k)v} = \frac{\exp(\langle \mathscr{F}_{\gamma_1}^{(k)}([H_i^{(l-1)}]_{v\cdot}), \mathscr{F}_{\gamma_1}^{(k)}([H_j^{(l-1)}]_{v\cdot})\rangle/\sqrt{D})}{\sum_{a\in\mathcal{N}_i} \exp(\langle \mathscr{F}_{\gamma_1}^{(k)}([H_i^{(l-1)}]_{v\cdot}), \mathscr{F}_{\gamma_1}^{(k)}([H_a^{(l-1)}]_{v\cdot})\rangle/\sqrt{D})} \tag{7}$$

$$[H_i^{(l)}]_{v\cdot} = \|_{k=1}^{K} \left\{ \sum_{j\in\mathcal{N}_i} \gamma_{ij}^{(k)v} \cdot \mathscr{F}_{\gamma_2}^{(k)}\left([H_j^{(l-1)}]_{v\cdot}\right) \right\} \tag{8}$$

The output of Eq. 8 is a predictive embedding of vertex v for time step i inference from historical traffic patterns. $\mathcal{N}_i$ denotes a set of time steps before time step i. Information from time steps earlier than the target step to enable causality is considered for short-term learning, while the segments on the past few days at the same time period are considered for long-term learning.

Hierarchical Structure The weighted spatial attention and temporal attention units constitute an H-Att block (as shown in Fig. 3). The lower-level spatial attention unit is responsible for collecting information from spatial neighbors and reducing the data perturbation. In comparison, the upper-level temporal attention unit learns the traffic patterns based on its previous observations.

Encoder-Decoder. Following previous works [8,27], HSTE adopts an encoder-decoder architecture for hierarchical spatio-temporal embedding (as shown in Fig. 3). Before entering into the encoder, the historical observation $H \in \mathbb{R}^{P \times N \times C}$ is transformed into $H^{(0)} \in \mathbb{R}^{P \times N \times D}$ using fully-connected layers. Then, $H^{(0)}$ is fed into the encoder with $\mathcal{L}$ H-Att blocks with residual connections and produces an output $H^{(L)} \in \mathbb{R}^{P \times N \times D}$. Following the encoder, a prediction attention unit is added to convert the encoded feature $H^{(L)}$ to generate the future sequence representation $H^{(L+1)} \in \mathbb{R}^{Q \times N \times D}$. The prediction attention unit has the same structure as the temporal attention unit (Eqs. 7 and 8). Next, the decoder stacks $\mathcal{L}$ H-Att blocks on $H^{(L+1)}$ and produces the output $H^{(2L+1)} \in \mathbb{R}^{Q \times N \times D}$. Finally, the fully-connected layers produce the Q time steps ahead of the prediction $F^* \in \mathbb{R}^{Q \times N \times C}$.

3.3 Long- and Short-Term Co-Attention Network

As shown in Fig. 2, AGCAN is a co-attention structure containing long- and short-term attentions that process concurrently to capture daily periodic patterns and near real-time changes in traffic networks. Both long- and short-term attention modules contain two HSTE blocks in which the cross-region and neighborhood spatio-temporal information are collected and processed respectively. The inputs of long- and short-term attention modules are defined as follows. Suppose the sampling frequency is g times per day and that the current time is t_0 and the size of predicting time window is T_p. As shown in Fig. 2, we intercept two time-series segments of lengths T_S and T_L along the time axis as the input of the short-term and long-term inputs respectively, where T_S and T_L are all integer multiples of T_p.

Short-Term Input. $H^S = \{X_{t_0-T_S+1}, X_{t_0-T_S+2}, \cdots, X_{t_0}\} \in \mathbb{R}^{T_S \times N \times C}$, a segment of historical time series, is directly adjacent to the predicting period, as represented by the purple curve in Fig. 1. Intuitively, the formation and dispersion of traffic congestion are gradual. Hence, the near real-time traffic flows inevitably have an influence on future traffic flows.

Long-Term Input. $H^L = \{X_{t_0-(T_L/T_p)g+1}, \cdots, X_{t_0-(T_L/T_p)g+T_p}, X_{t_0-(T_L/T_p-1)g+1}, \cdots, X_{t_0-(T_L/T_p-1)g+T_p}, \cdots, X_{t_0-g+T_p}\} \in \mathbb{R}^{T_L \times N \times C}$ consists of the segments from the past few days at the same time period as the prediction windows, as indicated by the blue lines in Fig. 1. As the travel modes are totally different for weekdays and weekends, we use the past few weekdays to predict future weekdays, while the past few weekends are used for weekend prediction. As the traffic patterns always repeat due to the regular daily routines of individuals, the purpose of long-term attention is to learn the daily periodic patterns.

Binary Fusion. To adaptively fuse the future sequence representations generated from different modules/blocks, we adopt a binary fusion derived from

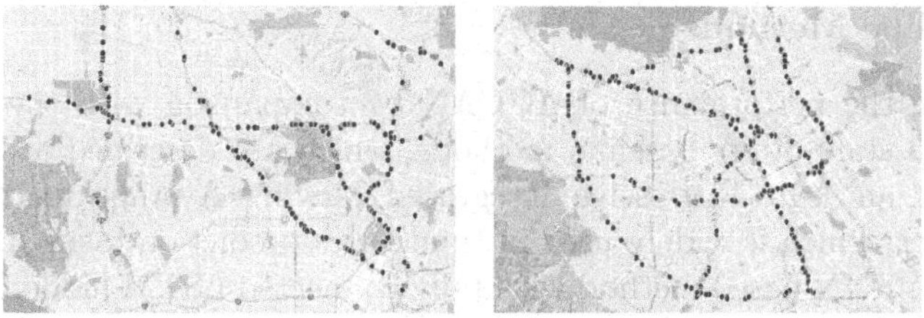

(a) METR-LA (b) PEMS-BAY

Fig. 4. Sensor distribution of (a) METR-LA (Los Angles) and (b) PEMS-BAY (Bay Area) datasets.

Recurrent Neural Networks (RNNs) [12]. The inputs of a binary fusion can be represented by H_1 and H_2 and are fused as:

$$H_{BF} = (H_1\mathbf{U} + H_2\mathbf{V} + \mathbf{b}_1)\mathbf{W} + \mathbf{b}_2 \quad (9)$$

The outputs of AGCAN are the binary-fused future sequence representations from both long- and short-term attentions (as shown in Fig. 2). AGCAN can be trained end-to-end via back-propagation by minimizing the mean absolute error (MAE) between predicted values and ground truths, as follows:

$$\mathcal{L}(\Theta) = \frac{1}{Q}\sum_{t_0}^{t_{0+Q}} \left|F^* - \hat{F}^*\right| \quad (10)$$

where Θ denotes all learnable parameters in AGCAN.

4 Experiment

4.1 Datasets

To validate the performance of AGCAN in real-world applications, we choose two benchmark traffic datasets, **PEMS-BAY** and **METR-LA**. There are total 325 sensors located in PEMS-BAY and 207 in METR-LA, shown in Fig. 4. The details of these datasets are provided in Table 1. The traffic speeds (mph) are collected by those sensors every 5 min.

Table 1. Benchmark dataset statistics

Dataset	# Nodes	# Edges	Time Range	# Time Steps
PEMS-BAY	325	2369	2017/01/01 ~ 2017/06/30	52116
METR-LA	207	1515	2012/03/01 ~ 2012/06/27	34272

4.2 Baseline Methods

We evaluate the performance of AGCAN by comparison with the 7 different classical and state-of-art baseline methods, which are described as follows: (1) **ARIMA** is an Auto-Regressive Integrated Moving Average method with a Kalman filter, which is widely used in time series prediction tasks [13]. (2) **FC-LSTM** is an RNN-based method with fully-connected LSTM hidden layers [19]. (3) **WaveNet** is a convolution network architecture for time series data [14]. (4) **STGCN**, a spatial-temporal graph conventional network, combines both the spatial and temporal convolution [25]. (5) **DCRNN**, a diffusion convolutions recurrent neural network, adopts an encoder-decoder architecture and diffusion convolution to help capture the complex spatial information [8]. (6) **Graph WaveNet** combines the graph convolution with dilated casual convolutions [24]. (7) **GMAN** adopts an encoder-decoder architecture that contains multi-attention mechanisms to capture the spatio-temporal correlations of traffic networks [27]. Herein, we employ the recommend setting for ARIMA and FC-LSTM from a previous work [8]. For other methods, we use the original parameters from their origin work.

4.3 Experiment Settings

For a fair comparison, the construction of the physical adjacency matrix for AGCAN and other graph-based methods are the same as shown in Eq. 2. The two datasets are spilt into 70% for training, 20% for validation, and 10% for testing. We adopt three standard metrics for performance evaluations in the traffic prediction tasks: *Mean Absolute Error* (MAE), *Mean Squared Error* (RMSE), *Mean Absolute Percentage Error* (MAPE).

For the prediction tasks, we use the previous $P = 12$ time steps to predict the future $Q = 12$ time steps traffic speeds. For each HSTE in our model (as shown in Fig. 2), we adopt 4 H-Att blocks in both the encoder and decoder, where each block contains $K = 6$ attention heads.

4.4 Experimental Results

Forecasting Performance Comparison. Table 2 shows the comparison of different methods for 15 min, 30 min, and 1 h ahead predictions on two datasets. The results show that AGCAN achieved state-of-art performance in the three evaluation metrics. The experimental observations are as follows: (1) Traditional traffic prediction methods (ARMIA, FC-LSTM) show relatively poor performance in comparison because they only consider the temporal correlations. With the introduction of spatial dependencies, DCRNN and other graph-based models make significant progress. (2) Our proposed model achieves more improvements in both short and long-term predictions compared to the latest GNN-based methods, such as Graph WaveNet and GMANm, which not consider the cross-region and cross-time correlations (the effect of different components of AGCAN are presented in the following section). To compare the performance of spatio-temporal methods (DCRNN, Graph WaveNet, GMAN and AGCAN), we show their predictions during peak hours, as shown in Fig. 5. It is easy to observe that the forecasting deviations exist between 9:00 and 13:00 for GMAN and Graph WaveNet (Fig. 5(a)) and between 9:00 and 11:00 for DCRNN (Fig. 5(b)). The robustness of AGCAN in predicting complicated traffic conditions is due to the extraction and integration of information from the cross-region and cross-time domains.

Table 2. Performance of ASGAN and other models

Data	Model	15 min			30 min			60 min		
		MAE	RMSE	MAPE	MAE	RMSE	MAPE	MAE	RMSE	MAPE
METR-LA	ARIMA	3.99	8.21	9.60%	5.15	10.45	12.70%	6.90	13.23	17.40%
	FC-LSTM	3.44	6.30	9.60%	3.77	7.23	10.90%	4.37	8.69	13.20%
	WaveNet	2.99	5.89	8.04%	3.59	7.28	10.25%	4.45	8.93	13.62%
	STGCN	2.88	5.74	7.62%	3.47	7.24	9.57%	4.59	9.4	12.70%
	DCRNN	2.77	5.38	7.30%	3.15	6.45	8.80%	4.59	9.40	12.70%
	Graph WaveNet	2.69	5.15	6.90%	3.07	6.22	8.37%	3.53	7.37	10.01%
	GMAN	2.77	5.44	7.26%	3.10	6.34	8.53%	3.44	7.21	9.99%
	AGCAN	**2.63**	**5.08**	**6.72%**	**3.02**	**6.23**	**8.17%**	**3.39**	**7.19**	**9.72%**
PEMS-BAY	ARIMA	1.62	3.30	3.50%	2.33	4.76	5.40%	3.38	6.50	8.30%
	FC-LSTM	2.05	4.19	4.80%	2.20	4.55	5.20%	2.37	4.96	5.70%
	WaveNet	1.39	3.01	2.91%	1.83	4.21	4.16%	2.35	5.43	5.87%
	STGCN	1.36	2.96	2.90%	1.81	4.27	4.17%	2.49	5.69	5.79%
	DCRNN	1.38	2.95	2.90%	1.74	3.97	3.90%	2.07	4.74	4.90%
	Graph WaveNet	**1.30**	2.74	2.73%	1.63	3.70	3.67%	1.95	4.52	4.63%
	GMAN	1.34	2.82	2.81%	1.62	3.72	**3.63%**	1.86	4.32	**4.31%**
	AGCAN	**1.30**	**2.73**	**2.68%**	**1.60**	**3.66**	3.65%	**1.82**	**4.30**	**4.31%**

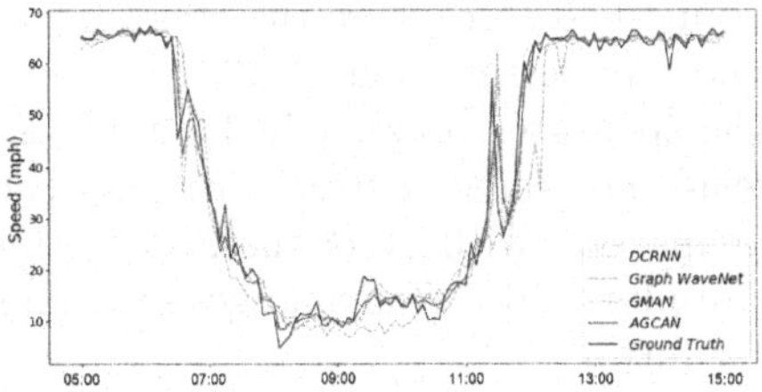

(a) Morning Peak.

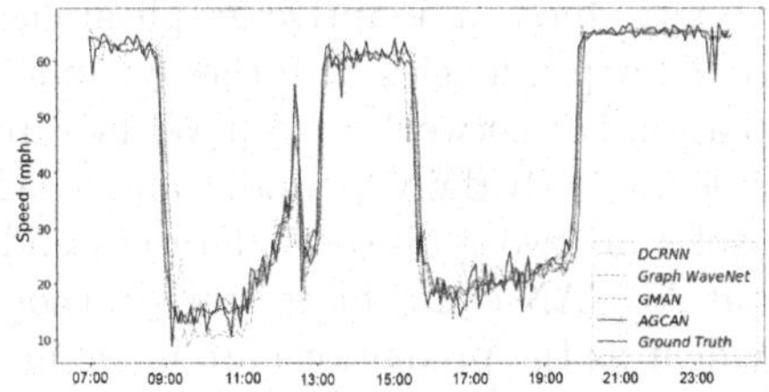

(b) Morning & Evening Peak.

Fig. 5. Speed predictions with spatio-temporal methods in the peak hours of two datasets.

Contribution of Each Component. To better understand the contribution of each component in our model, we test the performance of four variants by separating the different components separately, i.e., $\text{AGCAN}_{\mathcal{A}_{\mathcal{P}},\text{H}^{\text{S}}}$, $\text{AGCAN}_{\mathcal{A}_{\mathcal{F}},\text{H}^{\text{S}}}$,

$\mathrm{AGCAN}_{\mathcal{A}_\mathcal{P},\mathrm{H^L}}$, and $\mathrm{AGCAN}_{\mathcal{A}_\mathcal{F},\mathrm{H^L}}$. $\mathrm{AGCAN}_{\mathcal{A}_\mathcal{P},\mathrm{H^S}}$ refers to we only consider the physical graph ($\mathcal{G}_\mathcal{P}$) and short-term inputs (H^S) in this variant. Figure 6 shows the MAE in each prediction step of AGCAN and the four variants. We observe that AGCAN consistently outperforms its variants, indicating the usefulness of the combinations of cross-region and cross-time information in modeling the complicated spatio-temporal correlations. Moreover, even though $\mathrm{AGCAN}_{*,\mathrm{H^L}}$ performed worse than $\mathrm{AGCAN}_{*,\mathrm{H^S}}$, meaning the short-term temporal dependency is more informative than the long-term input, the combination of the two can greatly improve the prediction accuracy.

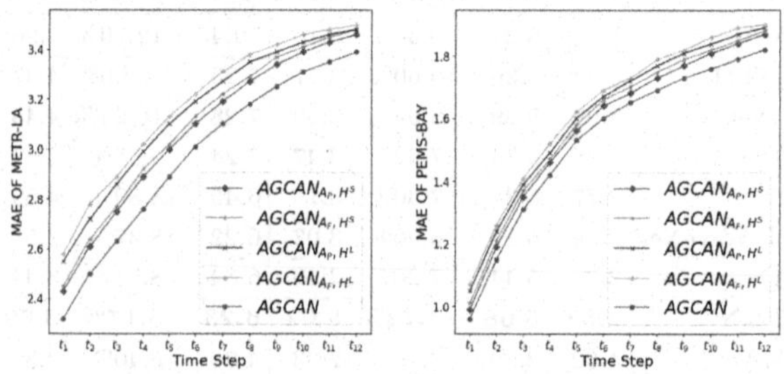

(a) METR-LA. (b) PEMS-BAY.

Fig. 6. MAE of each prediction step for AGCAN and its variants.

4.5 Conclusions

In this paper, we propose an adaptive graph co-attention networks (AGCAN) to predict the traffic conditions on a given road network over time steps ahead. We introduce an adaptive graph modelling method to capture the cross-region spatial dependencies with the dynamic trend. We design a long- and short-term co-attention network with novel hierarchical spatio-temporal embedding blocks to learn both daily periodic and near real-time traffic patterns. Experiment results on two real-world datasets and comparison with other methods verify that AGCAN can achieve state-of-the-art results, and the advantages are more evident as the predictions are made into far future.

References

1. An, S.H., Lee, B.H., Shin, D.R.: A survey of intelligent transportation systems. In: 2011 Third International Conference on Computational Intelligence, Communication Systems and Networks, pp. 332–337. IEEE (2011)
2. Bolshinsky, E., Friedman, R.: Traffic flow forecast survey. Technical report, Computer Science Department, Technion (2012)
3. Cascetta, E.: Transportation Systems Engineering: Theory and Methods, vol. 49. Springer, Cham (2013)

4. Ermagun, A., Levinson, D.: Spatiotemporal traffic forecasting: review and proposed directions. Transp. Rev. **38**(6), 786–814 (2018)
5. Grover, A., Leskovec, J.: node2vec: Scalable feature learning for networks. In: Proceedings of the 22nd ACM SIGKDD International Conference on Knowledge Discovery and Data Mining, pp. 855–864 (2016)
6. Jaderberg, M., Simonyan, K., Zisserman, A., et al.: Spatial transformer networks. In: Advances in Neural Information Processing Systems, pp. 2017–2025 (2015)
7. Lana, I., Del Ser, J., Velez, M., Vlahogianni, E.I.: Road traffic forecasting: recent advances and new challenges. IEEE Intell. Transp. Syst. Mag. **10**(2), 93–109 (2018)
8. Li, Y., Yu, R., Shahabi, C., Liu, Y.: Diffusion convolutional recurrent neural network: data-driven traffic forecasting. arXiv preprint arXiv:1707.01926 (2017)
9. Lu, H., Huang, D., Song, Y., Jiang, D., Zhou, T., Qin, J.: St-trafficnet: a spatial-temporal deep learning network for traffic forecasting. Electronics **9**(9), 1474 (2020)
10. Lv, Z., Xu, J., Zheng, K., Yin, H., Zhao, P., Zhou, X.: LC-RNN: a deep learning model for traffic speed prediction. In: IJCAI, pp. 3470–3476 (2018)
11. Ma, X., Dai, Z., He, Z., Ma, J., Wang, Y., Wang, Y.: Learning traffic as images: a deep convolutional neural network for large-scale transportation network speed prediction. Sensors **17**(4), 818 (2017)
12. Maas, A., Le, Q.V., O'neil, T.M., Vinyals, O., Nguyen, P., Ng, A.Y.: Recurrent neural networks for noise reduction in robust ASR (2012)
13. Makridakis, S., Hibon, M.: ARMA models and the box-Jenkins methodology. J. Forecast. **16**(3), 147–163 (1997)
14. Oord, A.V.D., et al.: Wavenet: A generative model for raw audio. arXiv preprint arXiv:1609.03499 (2016)
15. Park, C., Lee, C., Bahng, H., Kim, K., Jin, S., Ko, S., Choo, J., et al.: Stgrat: A spatio-temporal graph attention network for traffic forecasting. arXiv preprint arXiv:1911.13181 (2019)
16. Qi, L.: Research on intelligent transportation system technologies and applications. In: 2008 Workshop on Power Electronics and Intelligent Transportation System, pp. 529–531. IEEE (2008)
17. Shi, X., Chen, Z., Wang, H., Yeung, D.Y., Wong, W.K., Woo, W.C.: Convolutional LSTM network: a machine learning approach for precipitation nowcasting. In: Advances in Neural Information Processing Systems, vol. 28, pp. 802–810 (2015)
18. Shuman, D.I., Narang, S.K., Frossard, P., Ortega, A., Vandergheynst, P.: The emerging field of signal processing on graphs: Extending high-dimensional data analysis to networks and other irregular domains. IEEE Signal Process. Mag. **30**(3), 83–98 (2013)
19. Sutskever, I., Vinyals, O., Le, Q.V.: Sequence to sequence learning with neural networks. In: Advances in Neural Information Processing Systems, pp. 3104–3112 (2014)
20. Thekumparampil, K.K., Wang, C., Oh, S., Li, L.J.: Attention-based graph neural network for semi-supervised learning. arXiv preprint arXiv:1803.03735 (2018)
21. Vaswani, A., et al.: Attention is all you need. In: Advances in Neural Information Processing Systems, pp. 5998–6008 (2017)
22. Wei, L., et al.: Dual graph for traffic forecasting. IEEE Access (2019)
23. Wu, Y., Tan, H.: Short-term traffic flow forecasting with spatial-temporal correlation in a hybrid deep learning framework. arXiv preprint arXiv:1612.01022 (2016)
24. Wu, Z., Pan, S., Long, G., Jiang, J., Zhang, C.: Graph wavenet for deep spatial-temporal graph modeling. arXiv preprint arXiv:1906.00121 (2019)
25. Yu, B., Yin, H., Zhu, Z.: Spatio-temporal graph convolutional networks: a deep learning framework for traffic forecasting. arXiv preprint arXiv:1709.04875 (2017)

26. Zambrano-Martinez, J.L., Calafate, C.T., Soler, D., Cano, J.C., Manzoni, P.: Modeling and characterization of traffic flows in urban environments. Sensors **18**(7), 2020 (2018)
27. Zheng, C., Fan, X., Wang, C., Qi, J.: GMAN: a graph multi-attention network for traffic prediction. In: Proceedings of the AAAI Conference on Artificial Intelligence, vol. 34, pp. 1234–1241 (2020)

Dual-Stage Bayesian Sequence to Sequence Embeddings for Energy Demand Forecasting

Frances Cameron-Muller(✉), Dilusha Weeraddana, Raghav Chalapathy, and Nguyen Lu Dang Khoa

Data61-The Commonwealth Scientific and Industrial Research Organisation (CSIRO), Eveleigh, Australia

francesc-m@bigpond.com, {dilusha.weeraddana, raghav.chalapathy,khoa.nguyen}@data61.csiro.au

Abstract. Bayesian methods provide a robust framework to model uncertainty estimation. However, conventional Bayesian models are hard to tune and often fail to scale over high dimensional data. We propose a novel Dual-Stage Bayesian Sequence to Sequence (DBS2S) model by extending Sequence to Sequence (S2S) deep learning architecture with exogenous variable input to capture uncertainty and forecast energy consumption accurately. DBS2S model is trained with a two-stage S2S encoder-decoder network and benefits from the feature representation capability of S2S to capture complex multimodal posterior distributions within embeddings. We evaluate the proposed model for probabilistic energy consumption forecasting using four real-world public datasets and achieve improved prediction accuracy up to 64% in terms of mean absolute percentage error over existing, state-of-the-art Bayesian neural networks. Additionally, probabilistic prediction intervals forecasted by DBS2S is utilized to detect outliers and flag diverse consumer behavior.

Keywords: Energy demand forecasting · Bayesian neural networks · Prediction uncertainty · Time series · Anomaly detection

1 Introduction

Accurate Energy Demand Forecasting (EDF) plays a pivotal role in improving the reliability and efficiency of the grid [20]. According to the research reported in [15], the predictive analytics could save a significant amount of operational cost even by 1% improvement in the load forecast accuracy.

However, EDF is becoming increasingly challenging with the rising population and growing expectations for comfort and convenience [23]. Consumer energy consumption behaviour vastly differs between households of different compositions and is heavily influenced by various exogenous features such as time of day, day of the week and daily weather conditions [31]. Furthermore, energy flow between renewable sources such as solar energy generation plant to the electric grid, are subject to uncertainty due to various factors [28]. Hence, it is crucial for energy companies and utility operators to forecast not only the "point" forecast but also the uncertainty associated with the energy

K. Karlapalem et al. (Eds.): PAKDD 2021, LNAI 12712, pp. 277–289, 2021.
https://doi.org/10.1007/978-3-030-75762-5_23

forecast. In particular, probabilistic load forecasting is critical for energy authorities in many decision-making processes, such as demand response scheduling.

Bayesian methods have the natural ability to automatically quantify the uncertainty of the inference through the posterior distribution. These models have been extended to provide a principled framework for uncertainty quantification for EDF [29]. Moreover, deep learning has attracted much attention in recent years for time series forecasting due to major attributes, i.e., unsupervised feature learning, strong statistical flexibility and computational scalability [10]. However, conventional deep learning methods may not be the best choice for EDF due to the fact that they lack the ability to quantify uncertainty inherently.

Hence Bayesian deep learning was developed from Bayesian probability theory and constructs learning models based on a deep learning framework, which provides outstanding results while incorporating uncertainty [4]. S2S models are class of deep learning models employed to encode complex long-term sequential relationships for energy time-series data.

Motivated by the ability of S2S models to successfully learn long-range temporal dependencies, we propose a novel DBS2S model, to forecast energy demand along with probabilistic uncertainty estimation. Additionally, the proposed model can jointly produce rich feature representations while also inferring complex multimodal posterior distributions. The proposed DBS2S model captures three types of uncertainty namely: Aleatoric (uncertainty inherent in the observation noise), Epistemic (ignorance about the correct model that generated the data) and model misspecification [4]. Uncertainty induced due to unseen samples which differ from the training data set is popularly known as *model misspecification* [32]. All three aspects of uncertainty are captured in DBS2S model by adopting a dual-stage training procedure inspired by the staged training proposed by Zhu et al. [32] (Uber's model). The first training stage uses an encoder-decoder network to learn embeddings of the multivariate time series, which are then passed as input for the second stage of training comprising of another encoder-decoder network to perform inference.

To the best of our knowledge, this is the first attempt to utilize a DBS2S encoder-decoder network for EDF.

The main contributions of this paper are the following:

- We propose a novel DBS2S model to accommodate exogenous variables and capture long-term temporal relationships within energy data to forecast energy consumption accurately and quantify uncertainty.
- We evaluate the proposed model for probabilistic energy consumption forecasting using diverse four real-world public datasets.
- We utilize the probabilistic prediction intervals forecasted by DBS2S to detect outliers and flag diverse consumer behavior.

2 Related Works

The three most common techniques adopted to perform EDF are traditional time series methods, soft computing methods and hybrid models [10]. Traditional time series methods for EDF include econometric and Auto-Regressive Integrated Moving Average

(ARIMA) models [1]. Soft computing methods include neural networks, support vector machines, tree-based models, genetic algorithms and fuzzy logic [16]. Hybrid models mostly combine an artificial neural network (ANNs) with another soft computing method such as support vector machines [25], support vector regression [5] or particle swarm optimization [21] to forecast energy demand. Soft computing and hybrid methods have been shown to outperform traditional time series methods [22]. In the broader field of machine learning, recent years have witnessed a rapid proliferation of deep sequential networks with unprecedented results in EDF [10, 13]. The main advantage of deep sequential and LSTM based S2S models is the ability to extract hidden non-linear features and high-level invariant structures automatically, thus avoiding the laborious feature engineering. S2S models are popular since they enable end-to-end modeling with the capability to effectively model multivariate input data and extract non-linear patterns over longer input sequences [2, 24]. Forecasting the uncertainty in future energy demand using Bayesian Neural Networks (BNNs) have gained attention in EDF. BNNs are categorised by the posterior distribution approximation method the model implements. Approximation is commonly achieved either through variational inference or sampling. Variational inference estimates the distribution through optimisation by postulating a family of posteriors and selecting the proposed posterior that is closest to the exact posterior distribution by minimising the Kullback-Leibler divergence [4, 7, 11]. However, the variational inference methods significantly increase the number of model parameters that lead to scalability issues for large datasets for EDF. Additionally, variational inference methods also require the specification of the loss function and training method for each case due to their unique optimization problems [32]. Bayesian inference through sampling methods most commonly belong to the Markov Chain Monte Carlo family of algorithms [18] and approximate the posterior distribution through sampling the weights in the parameter space in proportion to a specific probability. Popular sampling methods implemented in BNNs include Hamiltonian Markov Chain Monte Carlo (MCMC) [19], Metropolis Hasting Algorithm [17], No-U-Turn Sampler [14] and Monte Carlo Dropout [8, 9]. Extensive experimentation into the choice of sampling method was conducted before Monte Carlo Dropout was selected as the sampling method employed in both networks in our model due to its empirical superiority.

The closest related work similar to our proposed model is by Uber [32] serving as inspiration for the dual-staged training procedure we modified and adopted. Our training method deviates from Uber's model, in terms of the prediction network architecture and handling of exogenous features. Additionally, our work builds upon Uber's model by applying the method to a different context (EDF) and testing the robustness of the proposed model. This was achieved by comparing performance on four diverse public datasets against other state-of-art BNNs that also produce probabilistic uncertainty estimation. The performance of Uber's model was only compared to basic baseline models and deep learning methods that lack principled uncertainty estimation. The inclusion of an additional training stage before inference is performed to extend the model to capture the often overlooked source of uncertainty, model misspecification, along with epistemic and aleatoric uncertainty [32].

3 Method

3.1 Uncertainty Estimation

The goal of the DBS2S model $f^{\hat{W}}(\cdot)$ trained with parameters $\hat{W}$ is to produce a prediction $\hat{y}_t = f^{\hat{W}}(x_t)$ for every new sample x_t and accurately estimate η_t, the prediction standard error, at each time step t. From this, a $\alpha\%$ prediction interval for energy demand for every future time step t can be constructed by $[\bar{y}_t - z_{\frac{\alpha}{2}}\eta_t,\ \bar{y}_t + z_{\frac{\alpha}{2}}\eta_t]$ where $z_{\frac{\alpha}{2}}$ is the z value of the desired confidence level α. Prediction intervals are a vital tool in the energy production planning procedure of utilities and electric grid operators [30]. This section will illustrate how the decomposition of the prediction uncertainty into three parts is derived and how each aspect of the uncertainty is captured.

Derivation of the Decomposition of Prediction Uncertainty. Consider a BNN denoted by $f^W(\cdot)$ where W is the set of weight parameters with an assumed Gaussian prior $W \sim N(0,I)$ in accordance with common practices. For the data generating distribution $p(y_t|f^W(\cdot))$, in regression we assume that $y_t|W \sim N(f^W(x_t), \sigma^2)$ with some noise level σ. Bayesian inference for time series regression given a set of T number of observations $X = (x_1, \ldots, x_T)$ and $Y = (y_1, \ldots, y_T)$ aims to find the optimal posterior distribution over the model parameters $p(W|X,Y)$ necessary for the construction of the prediction distribution. For each new sample x_t, the prediction distribution is given by the following integral

$$p(y_t|x_t) = \int_W p(y_t|f^W(x_t))p(W|X,Y)\,dW \tag{1}$$

By the law of total variance, the variance of the prediction distribution, η^2, can be decomposed into

$$Var(y_t|x_t) = Var[E(y_t|W,x_t)] + E[Var(y_t|W,x_t)] = Var(f^W(x_t)) + \sigma^2 \tag{2}$$

$$\eta^2 = Var(f^W(x_t)) + \sigma^2 \tag{3}$$

Hence, the prediction variance in Eq. 2 can be decomposed into (i) $Var(f^W(x))$, the epistemic uncertainty which reflects the ignorance over the set of model parameters W and (ii) σ^2, the aleatoric uncertainty that describes the inherent noise level during data generating process. An underlying assumption of Eq. 2 is that predicted output series y, is generated by the same procedure. However, energy demand is highly variable and dependent on numerous exogenous features does not follow this assumption. It is expected that there will be new samples that don't follow the patterns of the training data reflecting constant changes in society, consumer usage behaviour, population and climate. Therefore, we include a third aspect of prediction uncertainty called model misspecification, denoted by μ, that captures the uncertainty when the assumption is no longer complied [32]. Thus, model misspecification must be included in Eq. 2 to more accurately define prediction uncertainty as

$$\eta^2 = Var(f^W(x)) + \sigma^2 + \mu \tag{4}$$

As discussed with further detail in the next section, epistemic uncertainty and model misspecification are captured simultaneously in our model design and so Eq. 5 can be expressed more concisely by

$$\psi_t = Var(f^W(x))_t + \mu \tag{5}$$

$$\eta_t^2 = \psi_t + \sigma^2 \tag{6}$$

Handling the Three Aspects of Uncertainty. The first aspect of the variance, epistemic uncertainty is approximated in the second training stage through Monte Carlo Dropout [8,9] within the prediction network. By randomly diminishing some of the weights in the parameter space W to zero, the model's output can be viewed as a random sample from the posterior prediction distribution. Monte Carlo techniques can be utilized after generating multiple random samples to characterize the prediction distribution through its mean and standard deviation. The second aspect of variance, the aleatoric uncertainty of the data is captured using a held-out validation set, separate from the training and test set. The variance of the model's predictions on the validation set is used as an estimate of the inherent noise in the data and approximation for the aleatoric uncertainty for the test set. The third aspect of variance, the model misspecification is handled in the first stage of training. A latent embedding space of the training data is fit by the encoder part of the encoder-decoder network in the first training stage and passed to the prediction network to perform inference in the following stage. Model misspecification is alleviated by producing forecasts from the learned embeddings of the training data instead of the raw training data.

The overall prediction uncertainty is estimated by combining the three components of uncertainty and illustrated in Algorithm 1.

Algorithm 1. Uncertainty Estimation

INPUT: test set $X_T = (x_1, \ldots, x_T)$, validation set $X_V = (x_1, \ldots, x_V)$, Bayesian S2S model with dropout applied after each hidden layer $f(\cdot)$, number of iterations B
OUTPUT: prediction $\hat{y}_T = (\hat{y}_1, \ldots, \hat{y}_T)$, prediction uncertainty $\eta_T = (\eta_1, \ldots, \eta_T)$

Step 1: Predict the test set and produce point forecasts $\hat{y}_T$ and uncertainty estimates ψ_T
for $b = 1$ to B **do**
  for x_t in the test set X_T **do**
    $\hat{y}_{t,b} \leftarrow f(x_{t,b})$
$\bar{y}_T \leftarrow \frac{1}{B} \sum_{b=1}^{B} \hat{y}_{T,b}$
$\psi_T \leftarrow \frac{1}{B} \sum_{b=1}^{B} (\hat{y}_{T,b} - \bar{y}_T)^2$

Step 2: Use validation set to estimate aleatoric uncertainty σ^2
for x_v in validation set X_V **do**
  $\hat{y}_v \leftarrow f(x_v)$
$\bar{y}_V \leftarrow \frac{1}{V} \sum_{v=1}^{V} \hat{y}_v$
$\sigma^2 \leftarrow \frac{1}{V} \sum_{v=1}^{V} (\hat{y}_v - \bar{y}_V)^2$

Step 3: Estimate prediction standard error η_T (Eq. 3)
for x_t in the test set X_T **do**
  $\eta_t \leftarrow \sqrt{\psi_t + \sigma^2}$
return $\hat{y}_T, \eta_T$

3.2 Model Design

Our proposed model, as illustrated in Fig. 1 has a dual-stage training procedure, an embeddings and inference network to perform EDF and uncertainty estimation respectively. Firstly, we learn embeddings from the historical load and exogenous features (i.e., weather and calendar features) using the S2S encoder. Secondly, the learned embeddings are fed to the second S2S model to perform inference. The optimal model architecture was chosen empirically through extensive experimentation to achieve best results. Unlike Uber's model [32], DBS2S model incorporates exogenous feature inputs within historical load to learn long-term sequential dependencies.

Table 1 summarises the performance of all Bayesian S2S models with various configurations of model architectures and handling of exogenous features tested. Table 1 illustrates the optimal results achieved using a second S2S architecture for the inference network instead of a simple multi-layer perceptron (MLP) employed by Uber's model and passing the exogenous features as input to the first of the two networks along with the historical load. Monte Carlo Dropout was employed to achieve probabilistic uncertainty estimation in both embedding and inference networks.

Table 1. Performance of the Bayesian S2S models with various configurations of model architectures and handling of exogenous features tested.

Model			Sydney		Newcastle		Appliances		Power	
1st network	2nd network	Feature handling	SMAPE	95% PI	SMAPE	95% PI	SMAPE	95% PI	SMAPE	95% PI
LSTM	MLP	1st	12.75	71.25	12.42	73.78	5.54	93.77	36.05	84.09
CNN	LSTM	1st	26.89	**96.48**	28.15	**94.37**	5.74	**95.25**	35.60	85.04
LSTM	MLP	2nd	28.63	52.66	14.44	82.43	8.32	80.98	57.49	76.26
LSTM	Deep MLP	1st	27.06	89.47	12.17	99.84	7.78	89.18	53.70	68.34
LSTM	MLP	1st	10.59	92.26	9.67	94.30	5.22	94.59	36.61	84.47
LSTM	Deep MLP	1st	9.164	97.65	10.80	95.52	5.38	95.08	37.09	84.96
LSTM	CNN	1st	10.60	92.26	9.94	94.01	5.24	95.41	36.80	83.62
LSTM	Deep CNN	1st	27.40	0.00	28.71	0.00	5.48	92.95	35.62	82.37
Deep CNN	LSTM	1st	34.26	75.96	32.01	67.32	7.63	85.08	36.64	**86.61**
LSTM	LSTM	1st	**9.07**	99.99	**8.88**	99.67	**5.16**	94.43	**35.07**	84.47

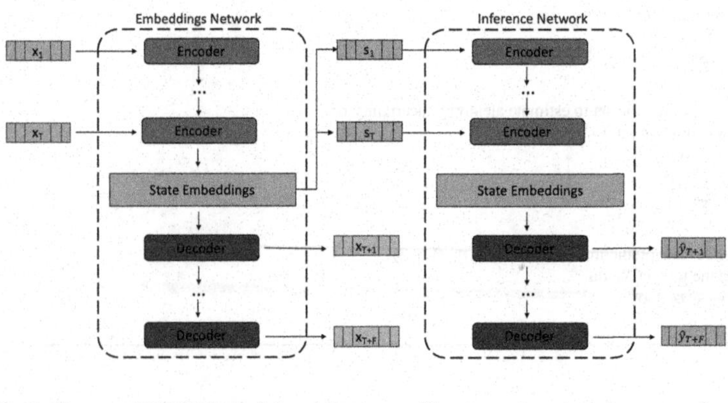

Fig. 1. Proposed DBS2S model architecture with a two-stage training procedure.

4 Experimental Settings

4.1 Description of the Datasets

The proposed model performance is evaluated using four public energy consumption datasets. Two datasets are sourced from Ausgrid containing half-hourly aggregated energy consumption of around 150 solar homes for two major Australian cities, Sydney and Newcastle [3]. The Ausgrid datasets are two years long from 2011–2013. For both these datasets, we used the first year as the training set, the following month for the validation set, and the final eleven months as the test set. The other two datasets were sourced from the public UCI Machine Learning Repository. The third dataset, the 'Appliances' dataset, was re-indexed to consist of the hourly energy consumption of a single household in Belgium over the span of 5 months [26]. The final dataset, the Individual Household Electric Power Consumption 'Power' dataset, was re-indexed to contain the hourly electric power consumption of a single household in France over the span of 4 years [27]. For both of these datasets, the first 70% of the data was used for the training set, the following 10% for the validation set and the final 20% was set aside as test set.

4.2 Feature Engineering and Selection

The input feature set consists of exogenous calendar features such as day of the week, hour of the day and month to capture the daily and seasonal dynamics in consumer behaviour patterns. The strong cyclic daily pattern in energy consumption is shown in Fig. 2 with plot (a) illustrating the daily usage pattern for weekday and weekends for the Ausgrid Sydney test dataset. Energy consumption's strong seasonality due to the effects of weather and holiday periods is demonstrated in Fig. 2(b) as the daily usage pattern varies between the seasons. Weather features including the temperature and humidity were also included where available to account for the obvious effect of climate on energy demand. Another associated exogenous feature, 'Global Intensity', was included for the 'Power' dataset where weather was missing.

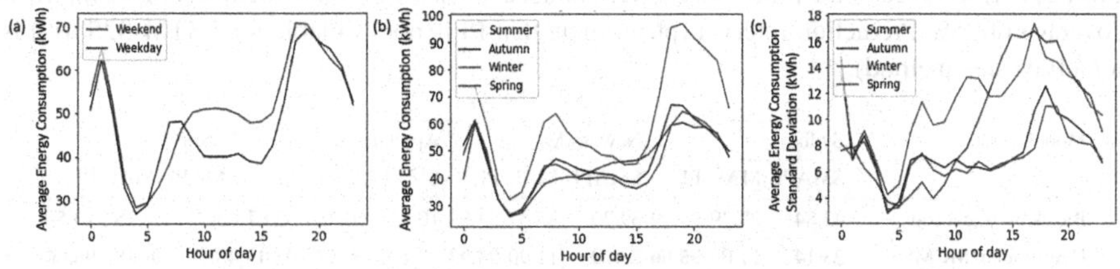

Fig. 2. Plot (a) shows the average energy consumption for each hour of the day for weekdays and weekends in the Ausgrid Sydney test dataset. Plot (b) shows the average energy consumption across the day for the four seasons and plot (c) shows the standard deviation. Similar behaviour was observed in remaining datasets.

4.3 Network Settings

The encoder-decoder networks used in both stages of training have the same architecture for the best performing model. The DBS2S encoder-decoder network is constructed with 2-layer LSTM cells, with 128 and 32 hidden states, respectively. Samples were constructed using a sliding window with step size of 1. Each sliding window contained 48 lags of each feature in the multivariate time series. Each sample is scaled before being passed to the model. The best model configuration considers input samples that contain one- or two-days' of historical data to forecast the energy consumption for next step ahead.

5 Experimental Results

We use three standard performance measures: Mean Absolute Error (MAE), Mean Absolute Percentage Error (MAPE), and Symmetric Mean Absolute Percentage Error (SMAPE).

The performance of the proposed model was evaluated in comparison to four widely adopted state-of-art S2S and BNN models: LSTM-S2S [13], BNN with Bayes by Backprop [4], BNN with Hamiltonian MCMC [19], Uber's Model [32].

5.1 Prediction Performance

The experimental results for the proposed DBS2S model are compared with the baseline methods as illustrated in Fig. 3(a). The detailed tabular results are recorded for reference in Table 2. Figure 3(a) illustrates our proposed model outperformed all four baseline models for EDF on all four datasets when compared to the above-mentioned three performance metrics. This is testament to the strength of our model to produce accurate point forecasts for energy demand and its robustness. Therefore, the proposed DBS2S framework successfully models the diverse and unique sequential dynamics of all four datasets.

Table 2. Performance of the proposed DBS2S model against the baseline methods. PI = Empirical Coverage of 95% Prediction Interval (please note that PI is not available for LSTM-S2S as it is not a Bayesian method).

Model	Sydney			Newcastle			Appliances			Power		
	SMAPE	MAE	PI	SMAPE	MAE	PI	SMAPE	MAE	PI	SMAPE	MAE	PI
Bayes by Backprop	42.84	27.79	95.79	39.70	35.57	23.89	10.79	0.48	32.67	65.74	0.75	10.55
Hamiltonian MCMC	23.14	11.06	**95.66**	23.10	11.90	**94.93**	8.11	0.37	93.43	48.98	0.48	**96.66**
Uber's model	10.43	5.06	93.09	9.67	4.98	94.30	7.19	0.33	90.16	61.80	0.60	49.24
LSTM-S2S	10.97	5.48	NA	11.32	6.17	NA	6.06	0.28	NA	36.35	0.36	NA
DBS2S model (proposed)	**9.07**	**4.42**	99.99	**8.88**	**4.55**	99.67	**5.16**	**0.24**	**94.43**	**35.07**	**0.35**	84.47

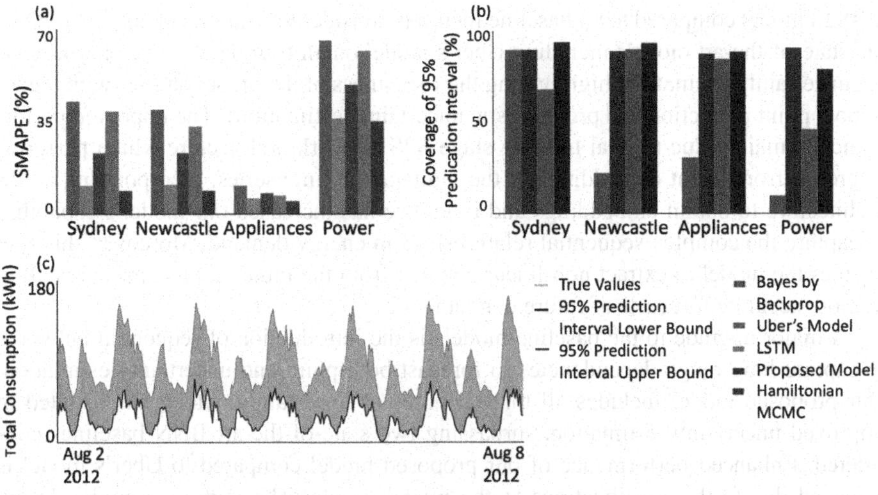

Fig. 3. (a) SMAPE of the baseline models and the proposed model for each dataset, (b) Empirical Coverage of the 95% prediction intervals of the baseline models and our model for each dataset and (c) Prediction interval produced by the model for the first week of the test set of the Sydney dataset.

5.2 Uncertainty Estimation

The quality of uncertainty estimation for each model was evaluated by comparing the empirical coverage of the 95% prediction interval as illustrated in Fig. 3(b) and Table 2. The baseline LSTM-S2S model, is not a Bayesian model and thus does not produce a probabilistic prediction distribution. Therefore, LSTM-S2S model can not be compared in this setting. An example of the 95% prediction intervals produced by the proposed Bayesian S2S model is shown in Fig. 3(c), presenting the model's output for the first week in the test set of the 'Sydney' dataset. Figure 3(b) illustrates the strong results achieved by our model in-terms of uncertainty estimation with respect to the baseline models. Our model outperformed the baseline models: BNN with Bayes by Backprop and Uber's Model, and on all four datasets achieving 95% prediction intervals with empirical coverages closer to 95%. We observed that our model also outperforms the final baseline model, BNN with Hamiltonian MCMC, in uncertainty estimation when the dataset doesn't contain strong seasonality but is slightly outperformed when the data suffers from seasonality. Detail into the results, strengths and weaknesses of the model is discussed in the next section.

5.3 Summary

Accurate point prediction for EDF is a significant strength of our proposed DBS2S model. The probabilistic nature of the uncertainty estimation achieved by incorporating Bayesian inference within our model is a key advantage of the proposed model over non-Bayesian LSTMs. Furthermore, Table 2 illustrate the superiority of the proposed

DBS2S model compared to all baseline methods considered. Our model outperforms all the state-of-the-art models including Uber's model on all four datasets except one case for uncertainty estimation, highlighting the robustness of the model output (with regard to both point prediction and probabilistic uncertainty estimation). The improved performance is mainly due to dual training stage S2S network architecture which performs inference from latent embeddings of the multivariate time series. Incorporating a S2S architecture to obtain embeddings and inference has increased our model's capability to capture the complex sequential relationships in energy demand. Moreover, this also enables the model to extract non-linear patterns from the input feature space, resulting in more accurate forecasts of future demand.

A major upgrade to the baseline models is the introduction of sequential networks to capture long-range dependencies to forecast both point and uncertainty estimation. Our proposed model includes all three sources of uncertainty, which has resulted in improved uncertainty estimation, surpassing two state-of-the-art BNN baselines considered. Enhanced performance of our proposed model compared to Uber's model is achieved due to the modifications to the model design. The improved model design for uncertainty estimation is achieved by learning exogenous features embeddings and sequential dependencies within the historical load. The results suggest that uncertainty from model misspecification might also arise from the exogenous features not complying with the assumption of Eq. 2. This adaptation along with changing the prediction network to another encoder-decoder have shown to improve the level of uncertainty estimation. We observed that our model performs well in uncertainty estimation in comparison to the final baseline model, BNN with Hamiltonian MCMC, when the data doesn't suffer from strong seasonality such as "Appliances" dataset that only spans 5 months, but doesn't perform as well for heavily seasonal datasets. This limitation of our model is due to the use of a validation set to estimate the aleatoric uncertainty. It is evident from Fig. 2(c), the standard deviation of energy consumption being significantly higher in winter for the two Ausgrid datasets than the other seasons. The results produced feature validation data samples from the month of July, with higher seasonal variation, resulted in the overall prediction standard error being overestimated and the 95% prediction interval being relatively wider. However, when the validation set was changed from July to March for the "Newcastle" dataset, the coverage of the prediction interval converged to the ideal 95% dropping from 99.67% to 98.49%, illustrating that this potential limitation can be overcome by careful selection of the validation set. Hence, an interesting path for further work would be to transform the data to remove the seasonality before passing it to our model. Furthermore, choosing a k-fold cross validation approach to selectively choose a validation set to tune the model would render the model more robust to seasonal variation.

5.4 Detecting Anomalies in Energy Consumption

Energy authorities, are interested in identifying anomalous consumption behaviors, for effective power supply management [12]. All household demands larger than two standard deviations are flagged as outliers similar to formulations by Chou et al. [6]. Table 3 demonstrates that extending uncertainty estimation to anomaly detection provides a promising and useful application. For instance, anomalous instances identified within Ausgrid-Sydney data is indicative of diverse consumer behaviour.

Table 3. Anomaly detection results table

	Sydney	Newcastle	Appliances	Power
Balanced accuracy	0.98	0.73	0.68	0.56
AUC	0.50	0.57	0.60	0.74

6 Conclusion

Energy authorities are interested in obtaining demand forecasting with uncertainty estimation since the "point" predictions can often be inaccurate, and the planning process must accommodate the errors. The reasons for inaccuracy in forecasted demand include the chaotic nature of weather conditions, penetration of reproducible energy sources, and other associated operational factors. Hence energy demand forecasting with accurate uncertainty estimation is vital for managing energy sources efficiently. To the best of our knowledge, this is the first recorded attempt to investigate a dual-stage sequence to sequence framework to learn embeddings and jointly produce future energy demand along with uncertainty estimation. We present comprehensive experimental results highlighting our proposed approach's robustness on four diverse public energy datasets over state-of-the-art baseline methods. Additionally, the application of probabilistic prediction intervals to anomaly detection is shown to flag diverse consumer behavior. Ultimately, we believe this work, at the intersection of probabilistic forecasting and deep learning will lead to development of more accurate time-series prediction models.

References

1. Ahmadi, S., Hossien Fakehi, A., Haddadi, M., Iranmanesh, S.H., et al.: A hybrid stochastic model based Bayesian approach for long term energy demand managements. Energy Strategy Rev. **28**, 100462 (2020)
2. Assaad, M., Boné, R., Cardot, H.: A new boosting algorithm for improved time-series forecasting with recurrent neural networks. Inf. Fusion **9**(1), 41–55 (2008)
3. Ausgrid: Solar home electricity data (2011–2013). https://www.ausgrid.com.au/Industry/Our-Research/Data-to-share/Solar-home-electricity-data
4. Blundell, C., Cornebise, J., Kavukcuoglu, K., Wierstra, D.: Weight uncertainty in neural networks. arXiv preprint arXiv:1505.05424 (2015)
5. Casteleiro-Roca, J.L., et al.: Short-term energy demand forecast in hotels using hybrid intelligent modeling. Sensors **19**(11), 2485 (2019)
6. Chou, J.S., Telaga, A.S.: Real-time detection of anomalous power consumption. Renew. Sustain. Energy Rev. **33**, 400–411 (2014)
7. Fortunato, M., Blundell, C., Vinyals, O.: Bayesian recurrent neural networks. arXiv preprint arXiv:1704.02798 (2017)

8. Gal, Y., Ghahramani, Z.: Dropout as a Bayesian approximation: representing model uncertainty in deep learning. In: International Conference on Machine Learning, pp. 1050–1059 (2016)
9. Gal, Y., Hron, J., Kendall, A.: Concrete dropout. In: Advances in Neural Information Processing Systems, pp. 3581–3590 (2017)
10. Ghalehkhondabi, I., Ardjmand, E., Weckman, G.R., Young, W.A.: An overview of energy demand forecasting methods published in 2005–2015. Energy Syst. **8**(2), 411–447 (2017)
11. Hernández-Lobato, J.M., Adams, R.: Probabilistic backpropagation for scalable learning of Bayesian neural networks. In: International Conference on Machine Learning, pp. 1861–1869 (2015)
12. Himeur, Y., Alsalemi, A., Bensaali, F., Amira, A.: Robust event-based non-intrusive appliance recognition using multi-scale wavelet packet tree and ensemble bagging tree. Appl. Energy **267**, 114877 (2020)
13. Hochreiter, S., Schmidhuber, J.: Long short-term memory. Neural Comput. **9**(8), 1735–1780 (1997)
14. Hoffman, M.D., Gelman, A.: The no-u-turn sampler: adaptively setting path lengths in Hamiltonian Monte Carlo. J. Mach. Learn. Res. **15**(1), 1593–1623 (2014)
15. Hong, T.: Crystal ball lessons in predictive analytics. EnergyBiz Mag. **12**(2), 35–37 (2015)
16. Kaytez, F., Taplamacioglu, M.C., Cam, E., Hardalac, F.: Forecasting electricity consumption: a comparison of regression analysis, neural networks and least squares support vector machines. Int. J. Electr. Power Energy Syst. **67**, 431–438 (2015)
17. Mullachery, V., Khera, A., Husain, A.: Bayesian neural networks. arXiv preprint arXiv:1801.07710 (2018)
18. Neal, R.M.: Bayesian Learning for Neural Networks, vol. 118. Springer, New York (2012). https://doi.org/10.1007/978-1-4612-0745-0
19. Neal, R.M., et al.: MCMC using Hamiltonian dynamics. Handbook of Markov chain Monte Carlo **2**(11), 2 (2011)
20. Raza, M.Q., Khosravi, A.: A review on artificial intelligence based load demand forecasting techniques for smart grid and buildings. Renew. Sustain. Energy Rev. **50**, 1352–1372 (2015)
21. Raza, M.Q., Nadarajah, M., Hung, D.Q., Baharudin, Z.: An intelligent hybrid short-term load forecasting model for smart power grids. Sustain. Urban Areas **31**, 264–275 (2017)
22. Srinivasan, D.: Energy demand prediction using GMDH networks. Neurocomputing **72**(1–3), 625–629 (2008)
23. Tang, T., Bhamra, T.: Changing energy consumption behaviour through sustainable product design. In: DS 48: Proceedings DESIGN 2008, the 10th International Design Conference, Dubrovnik, Croatia (2008)
24. Tokgöz, A., Ünal, G.: A RNN based time series approach for forecasting Turkish electricity load. In: 2018 26th Signal Processing and Communications Applications Conference (SIU), pp. 1–4. IEEE (2018)
25. Torabi, M., Hashemi, S., Saybani, M.R., Shamshirband, S., Mosavi, A.: A hybrid clustering and classification technique for forecasting short-term energy consumption. Environ. Progress Sustain. Energy **38**(1), 66–76 (2019)
26. UCI Machine Learning Repository: Appliance energy prediction (2006). https://archive.ics.uci.edu/ml/datasets/Appliances+energy+prediction
27. UCI Machine Learning Repository: Individual household electric power consumption (2010). https://archive.ics.uci.edu/ml/datasets/Individual+household+electric+power+consumption
28. Valibeygi, A., Habib, A.H., de Callafon, R.A.: Robust power scheduling for microgrids with uncertainty in renewable energy generation. In: 2019 IEEE Power & Energy Society Innovative Smart Grid Technologies Conference (ISGT), pp. 1–5. IEEE (2019)

29. Dong, Y., Ifrim, G., Mladenić, D., Saunders, C., Van Hoecke, S. (eds.): ECML PKDD 2020. LNCS (LNAI), vol. 12461. Springer, Cham (2021). https://doi.org/10.1007/978-3-030-67670-4
30. Wijaya, T.K., Sinn, M., Chen, B.: Forecasting uncertainty in electricity demand. In: Workshops at the Twenty-Ninth AAAI Conference on Artificial Intelligence (2015)
31. Zhou, K., Yang, S.: Understanding household energy consumption behavior: the contribution of energy big data analytics. Renew. Sustain. Energy Rev. **56**, 810–819 (2016)
32. Zhu, L., Laptev, N.: Deep and confident prediction for time series at Uber. In: 2017 IEEE International Conference on Data Mining Workshops (ICDMW), pp. 103–110. IEEE (2017)

AA-LSTM: An Adversarial Autoencoder Joint Model for Prediction of Equipment Remaining Useful Life

Dong Zhu, Chengkun Wu(✉), Chuanfu Xu, and Zhenghua Wang

Institute for Quantum Information and State Key Laboratory of High Performance Computing, College of Computer Science and Technology, National University of Defense Technology, Changsha, China
{zhudong,chengkun_wu,xuchuanfu,zhhwang}@nudt.edu.cn

Abstract. Remaining Useful Life (RUL) prediction of equipment can estimate the time when equipment reaches the safe operating limit, which is essential for strategy formulation to reduce the possibility of loss due to unexpected shutdowns. This paper proposes a novel RUL prediction model named AA-LSTM. We use a Bi-LSTM-based autoencoder to extract degradation information contained in the time series data. Meanwhile, a generative adversarial network is used to assist the autoencoder in extracting abstract representation, and then a predictor estimates the RUL based on the abstract representation learned by the autoencoder. AA-LSTM is an end-to-end model, which jointly optimizes autoencoder, generative adversarial network, and predictor. This training mechanism improves the model's feature extraction and prediction capabilities for time series. We validate AA-LSTM on turbine engine datasets, and its performance outperforms state-of-the-art methods, especially on datasets with complex working conditions.

Keywords: Remaining useful life · Time series · Adversarial autoencoder · Feature extraction

1 Introduction

Remaining Useful Life (RUL) prediction is one of the main tasks in Prognostic and Health Management (PHM) [1], which estimates the remaining useful time for specified equipment before reaching safe operating limits. Managers can schedule maintenance plans according to RULs in advance, thus preventing potentially serious failures.

Current research on RUL prediction is mainly based on data-driven approaches. Machine learning methods have been used for RUL predictions, including Support Vector Machines (SVM) [2], Random Forests (RF), Hidden Markov Models (HMM) [3], etc. However, these methods also require an abundance of domain knowledge to craft effective features, which are usually high-dimensional. Recently, deep learning, which can automatically extract abstract

K. Karlapalem et al. (Eds.): PAKDD 2021, LNAI 12712, pp. 290–301, 2021.
https://doi.org/10.1007/978-3-030-75762-5_24

feature representations through very deep hidden layers, has been widely used in RUL prediction.

Recurrent Neural Networks (RNN) are preferred for RUL prediction due to the temporal property of the data collected by the sensors. Long Short Term Memory networks (LSTM) [4] is a widely employed type of recurrent neural networks. Zheng et al. [5] used a two-layer LSTM network followed by a full connectivity layer for engine RUL prediction, and the model takes raw data as input. Listou et al. [6] added an RBM layer to the initial pre-training phase to extract degradation-related features from raw data in an unsupervised manner and use it for LSTM prediction. Huang et al. [7] used bidirectional LSTM (Bi-LSTM) for RUL prediction, which outplays LSTM in capturing historical information.

Convolutional Neural Networks (CNN) have been used in many areas such as computer vision and speech recognition [8]. To note, CNNs perform equally well on serial data. Li et al. [9] used a deep 1D convolutional network to predict the RUL of the engine without a conventional pooling layer. It used multiple convolutional layers to fully learn the high-level abstract representation. Wen et al. [10] added residual blocks to traditional CNNs to overcome the problem of gradient vanish or explosion.

Hybrid neural networks can improve the ability of RNN prediction compared to a single network. CNN can capture spatial and local features, while LSTM can make full use of historical information. The combination can be performed in two modes. One is the serial mode, where the CNN is used to extract spatial features first, and then the LSTM network is connected [11–13]. The other is the parallel mode, where the CNN and LSTM learn different feature representations respectively, then the extracted features are fused for RUL prediction by the full connection layer [14,15].

In this paper, we propose a novel end-to-end model named AA-LSTM for RUL prediction. Firstly, the Bi-LSTM encoder transforms the input time series into a latent code vector, and then the decoder reconstructs the time series from the latent code vector, making it as close to the original time series. Meanwhile, a discriminator is used to discriminate the distribution of the latent code vector and make it match the normal distribution. In addition, a designed RUL predictor used the latent code vector to predict the remaining life of each input. Through the proposed joint training mechanism, the losses of the three parts are optimized jointly and constitute an end-to-end model. Finally, the performance of the proposed model is validated on a turbofan engine dataset.

2 Proposed Approach

An illustrative framework of the proposed AA-LSTM is shown in Fig. 1. AA-LSTM contains three stages. In the first stage, the input time series data is encoded into a latent vector and reconstructed by the decoder. In the second stage, the discriminator and the encoder are trained adversarially each other so that the latent vector is close to a normal distribution. In the third stage, the RUL predictor predicts the final RUL of the sequence based on the latent vector.

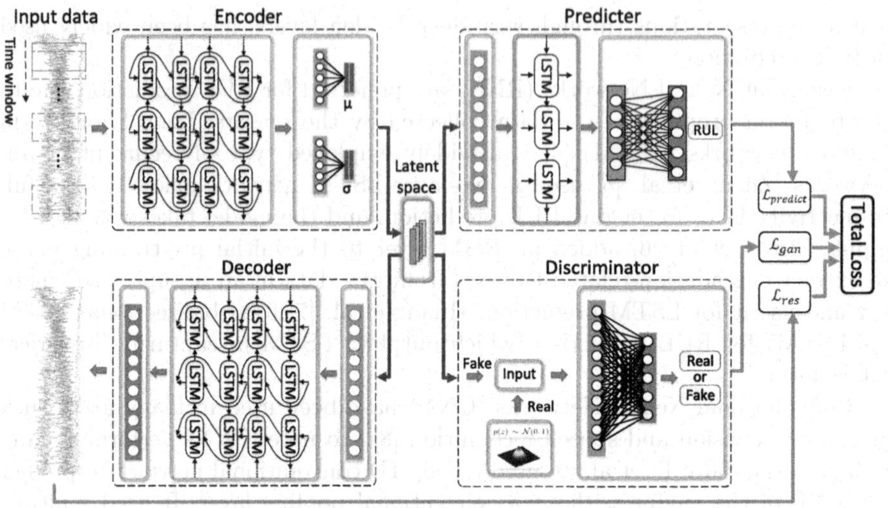

Fig. 1. Overall structure of the proposed model.

2.1 Encoder Model

The encoder model is depicted in Fig. 1. The raw sensor data is divided into multiple fragments by the time window. Each fragment can be represented as $X = [x_1, x_2, ..., x_t] \in \mathbb{R}^{t\times d}$, where t represents the time step size, d represents the feature dimension, and $x_t \in \mathbb{R}^{1\times d}$ represents the vector of each time step t.

The time series x is embedded in the latent code vector z by the encoder. The distribution of encoder can be expressed as $q\left(z\,|x\right)$. Meanwhile, the encoder needs to learn the aggregated posterior distribution $q\left(z\right) = \int q(z|x)p(x)dx$.

Bi-LSTM is used to capture the long-term dependence of the input sensor time sequence from two opposite directions. LSTM can overcome the problem of long-term learning difficulties for RNN [4]. The core idea of the LSTM is to use non-linear gating units to control the information flow of neural units [16].

The forget gate f_t determines what historical information in the unit state is forgotten, the input gating i_t controls the updating of the information, and the candidate cell state $\widetilde{S}_t$ ensures that the value range is between -1 and 1. Besides, the output gating o_t determines the information to be output by the storage unit. These gating units can be expressed as follows,

$$f_t = \sigma(W_f x_t + R_f h_{t-1} + b_f) \tag{1}$$

$$i_t = \sigma(W_i x_t + R_i h_{t-1} + b_i) \tag{2}$$

$$\widetilde{S}_t = \tanh(W_s x_t + R_s h_{t-1} + b_s) \tag{3}$$

$$o_t = \sigma(W_o x_t + R_o h_{t-1} + b_o) \tag{4}$$

where W represents the input weight, R represents the cyclic weight, and b represents the bias weight.

Next, the current time step cell state S_t can be updated to:

$$S_t = f_t \otimes S_{t-1} + i_t \otimes \widetilde{S}_t \tag{5}$$

The current time forward hidden state $\overrightarrow{h_t}$ can be expressed as:

$$\overrightarrow{h_t} = o_t \otimes \tanh(S_t) \tag{6}$$

Similarly, the backward hidden state $\overleftarrow{h_t}$ can be calculated by the opposite time step direction.

Finally, the historical and future information of the sequential input can be evaluated at the same time, and the hidden state of the current time step in the Bi-LSTM layer can be expressed as:

$$h_t = \overrightarrow{h_t} \oplus \overleftarrow{h_t} \tag{7}$$

The sensor sequence is embedded as the latent code vector z through a two-layer Bi-LSTM network, and each layer contains 64 units. In addition, the 2 fully connected layers respectively fit the mean and variance of z, making it roughly follow a standard normal distribution. The reparameterization trick [17] is used to obtain the final latent code vector z of the encoder.

2.2 Decoder Model

The distribution of decoder can be expressed as $p(x|z)$, which reconstructs the latent code vector z as $\widetilde{x}$. The decoder uses the distribution of z to learn the data distribution $p_d(x) = \int p(x|z)p(z)dz$.

The structure of the decoder is shown in Fig. 1. Similarly, the decoder infers the hidden representation of z through a 2-layer Bi-LSTM network with the same structure as the encoder. Before being fed into Bi-LSTM, z is expanded into a three-dimensional vector by the RepeatVector layer to meet the input requirements of Bi-LSTM. In addition, the TimeDistributed layer ensures that the output of the Bi-LSTM layer has the same shape as the raw input time series. The reconstruction loss can be expressed as:

$$\mathcal{L}_{res} = \frac{1}{n}\sum_{i=1}^{n} |x_i - \widetilde{x}_i|^2 \tag{8}$$

2.3 Discriminator Model

To ensure the generation ability of the autoencoder, latent code vector z is imposed with a arbitrary prior distribution $p(z)$, and encoder must make $q(z)$ match $p(z)$.

Generally, the similarity between $q(z)$ and $p(z)$ is measured by Kullback-Leibler (KL) divergence, such as variational autoencoder (VAE) [17]. This is equivalent to adding a regularization term to encoder, which can prevent encoder from obtaining a meaningless distribution. In our model, KL divergence is replaced by adversarial training procedure [18].

Generative Adversarial Network (GAN) is a training procedure proposed by Goodfellow [19]. A generator G and a discriminator D construct a mini-max adversarial game, in which G samples from random noise space and generates fake sample x' to trick the discriminator D. The discriminator needs to learn whether the data generated by G is real or fake [18]. The optimization process of GAN is that G and D reach a nash equilibrium.

In our model, the encoder can be regarded as the generator of GAN, which maps the input x to the latent code vector z, hoping to match a certain prior distribution $p(z)$. Following the original VAE [17], we let $p(z)$ be the standard normal distribution $p(z) \sim \mathcal{N}(0,1)$. At the same time, a discriminator is added to discriminate whether z is derived from $q(z)$ (fake samples), which is computed by the encoder, or from the prior distribution $p(z)$ (real samples). The discriminator adds a regularization constraint to the encoder, encouraging $q(z)$ to math $p(z)$. The loss of adversarial training procedure can be expressed as:

$$\mathcal{L}_{adversarial} = \min_G \max_D \mathbb{E}_{z \sim p(z)} \left[\log D\left(p(z)\right)\right] + \mathbb{E}_{x \sim p(x)} \left[\log\left(1 - D\left(q(z)\right)\right)\right] \quad (9)$$

The structure of the discriminator is shown in Fig. 1. The discriminator is composed of 2 fully connected layers, the number of neurons is 512 and 256 respectively, and the output layer uses the sigmoid activation function.

2.4 RUL Predictor

The latent code vector z contains the degradation information, and the predictor needs to infer the RUL of the current sequence according to z.

An LSTM layer is used to discover hidden features in z. Since the z is a 2-dimensional vector, the sequence information is compressed, and the RepeatVector layer needs to be used to extend the time step before being input to the LSTM layer. Then, two fully connected layers integrate the features learned by LSTM and map them to RUL.

Let y_i be the actual RUL value and $\widetilde{y_i}$ be the estimated value, the loss of the RUL predictor can be expressed as:

$$\mathcal{L}_{predict} = \frac{1}{n}\sum_{i=1}^{n} \left|y_i - \widetilde{y_i}\right|^2 \quad (10)$$

2.5 Training Process

Algorithm 1 describes the joint training process of AA-LSTM. The three parts of the entire model are jointly trained using Adam to optimize parameters. In the time series reconstruction phase, the encoder maps the input sequence to z, which obeys the posterior distribution $q(z)$, and the decoder reconstructs it to $\widetilde{x}$, updating the encoder and decoder to minimize the reconstruction loss $\mathcal{L}_{res}$. In the adversarial training phase, $\widetilde{z}$ is sampled from the prior distribution $p(z)$, and the discriminator is updated to minimize the adversarial loss $\mathcal{L}_{adversarial}$. Then update the generator (the encoder of the autoencoder), to improve the ability of the confusion discriminator. In the prediction phase, the RUL value is calculated according to z, and the predictor is updated to minimize the prediction loss $\mathcal{L}_{predict}$.

Let the parameters α, β, and γ be the weight values. The overall loss of AA-LSTM can be expressed as:

$$\mathcal{L} = \alpha\mathcal{L}_{res} + \beta\mathcal{L}_{adversarial} + \gamma\mathcal{L}_{predict} \tag{11}$$

Algorithm 1. The joint training process of AA-LSTM.

for each training iterations **do**
  Sample $\left\{x^{(i)}\right\}_{i=1}^{m}$ a batch from the time series data.
  Generate the latent code vectors $\left\{z^{(i)}\right\}_{i=1}^{m} \sim q(z)$ from the encoder.
  Reconstruct the time series $\left\{\widetilde{x}^{(i)}\right\}_{i=1}^{m}$ from the decoder.
  Sample $\left\{\widetilde{z}^{(i)}\right\}_{i=1}^{m}$ a batch from the prior distribution $p(z)$.
  Predict the RUL $\left\{\widetilde{y}^{(i)}\right\}_{i=1}^{m}$ of the $\left\{z^{(i)}\right\}_{i=1}^{m}$.
  Update the encoder and the decoder to decrease $\mathcal{L}_{res}$.
  Update discriminator to decrease $\mathcal{L}_{adversarial}$.
  Update predictor to decrease $\mathcal{L}_{predict}$.
  Update encoder to increase $D(z)$ and $D(\widetilde{z})$.
end for

3 Experiment

3.1 Introduction to Turbofan Data Set

C-MAPSS is a benchmark dataset in the field of RUL, and NASA uses the Commercial Modular Aero Propulsion System Simulation Tool to simulate engines in the 90,000 lb thrust level [20].

C-MAPSS is divided into 4 data subsets, each with different operation modes and failure modes. Table 1 gives a summary of C-MAPSS. Each unit has 26 dimensions, including the engine number, the operating cycle number, 3 operation settings, and monitoring values for 21 sensors. Each data subset contains a train set and a test set. The train set contains the entire life cycle data of each engine from the initial state to the failure, and in the test set only data for a certain period before failure. See [20] for a detailed description of the C-MAPSS dataset.

Table 1. Basic introduction of C-MAPSS dataset.

Dataset	FD001	FD002	FD003	FD004
Fault modes	1	1	2	2
Operating conditions	1	6	1	6
Train units	100	260	100	249
Test units	100	259	100	248
Train minimum cycles	128	156	145	128
Test minimum cycles	31	21	38	19

3.2 Data Preprocessing

Feature Selection. Not all sensors help predict the target. Some sensor values are constant throughout the life cycle of the engine, so they can be removed to reduce the dimensionality of the data, which leaves sensors 2, 3, 4, 7, 8, 9, 11, 12, 13, 14, 15, 17, 20 and 21.

Data Normalization. For selected sensors, min-max normalization is used to normalize the values of all sensors to $[-1, 1]$. In addition, different operating conditions affect the sensor values. FD002 and FD004 have six different operating conditions, given by three operating variables. To eliminate this effect, we use K-means [21] to divide the three operating variables into six classes. The normalization formula is as follows:

$$x^{i,j}_{(c)} = \frac{2(x^{i,j}_{(c)} - \min x^{j}_{(c)})}{\max x^{j}_{(c)} - \min x^{j}_{(c)}} - 1 \tag{12}$$

Where $x^{i,j}_{(c)}$ denotes the raw data, c represents the operating conditions, i represents the i-th data point, j represents the j-th sensor, and $\max x^{j}_{(c)}$ and $\min x^{j}_{(c)}$ denote the maximum and minimum values of the j-th sensor, respectively.

Remaining Useful Life Label. For the training set, the RUL for each sample can be expressed as the maximum number of cycles run for that engine minus the current last number of cycles. The research of [22] showed that it is more reasonable to estimate RUL using piece-wise linear function, which can reduce the loss of the target and improve the prediction performance of the model. This piece-wise linear function can be expressed as the equation:

$$kink(x) = \begin{cases} 125 & if\ x \geq R_{max} \\ x & otherwise \end{cases} \tag{13}$$

3.3 Performance Metrics

The performance of the AA-LSTM is evaluated using the RMSE and a scoring function. Let $RUL_i{}'$ be the predicted value of the i-th engine, and the error value $d_i = RUL_i{}' - RUL_i$, then the RMSE of N engines can be defined as:

$$RMSE = \sqrt{\frac{1}{N}\sum_{i=1}^{N} {d_i}^2} \tag{14}$$

The scoring function used in this paper is derived from the evaluation metrics in the PHM08 data competition and the formula for the scoring function is:

$$Score = \begin{cases} \sum_{i=1}^{N}\left(e^{-\frac{d_i}{13}} - 1\right), d_i < 0 \\ \sum_{i=1}^{N}(e^{-\frac{d_i}{10}} - 1), d_i \geq 0 \end{cases} \tag{15}$$

3.4 Results

Validation Results of the RUL on the Test Set. In this section, we will present the performance of AA-LSTM by validating it on the four validation sets of C-MAPSS. In order to reduce the effect of randomness on the model results, 10 identical experiments were performed. The key parameters were set as follows: batch = 128, $\alpha = 1$, $\beta = 0.001$, and $\gamma = 0.001$. The training epochs for the four subsets were 30, 15, 30, and 25.

The results for each subset on the test set are shown in the Fig. 2 and Table 2. On subsets FD001 and FD003, the metrics RMSE and score are relatively lower, mainly because these two datasets are simpler and easier to predict. FD002 and FD004 contain multiple operating states and failure modes, making it more difficult to predict the RUL, especially on FD004.

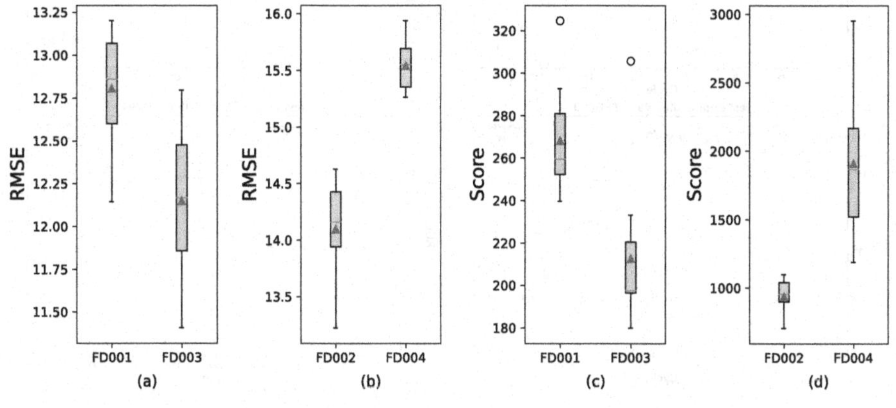

Fig. 2. RMSE and score values of proposed model on test dataset.

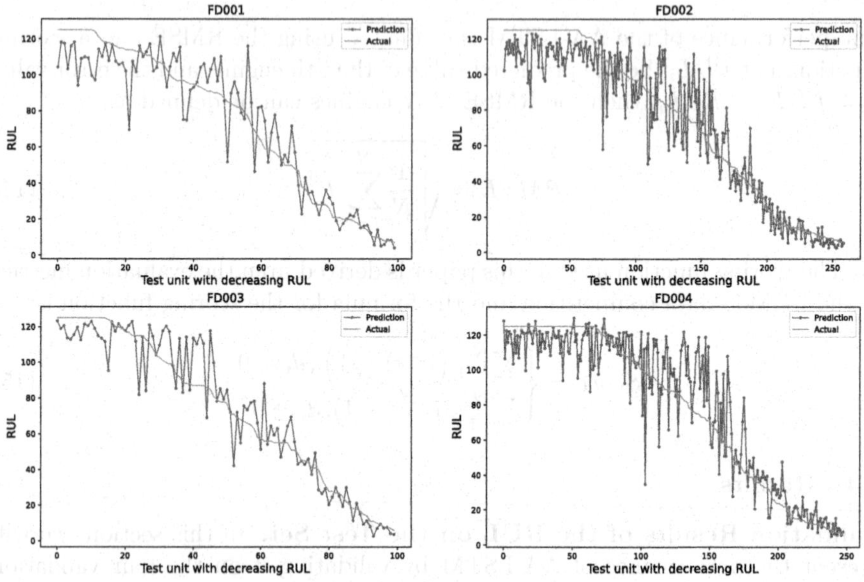

Fig. 3. The prediction for each testing engine units in test dataset.

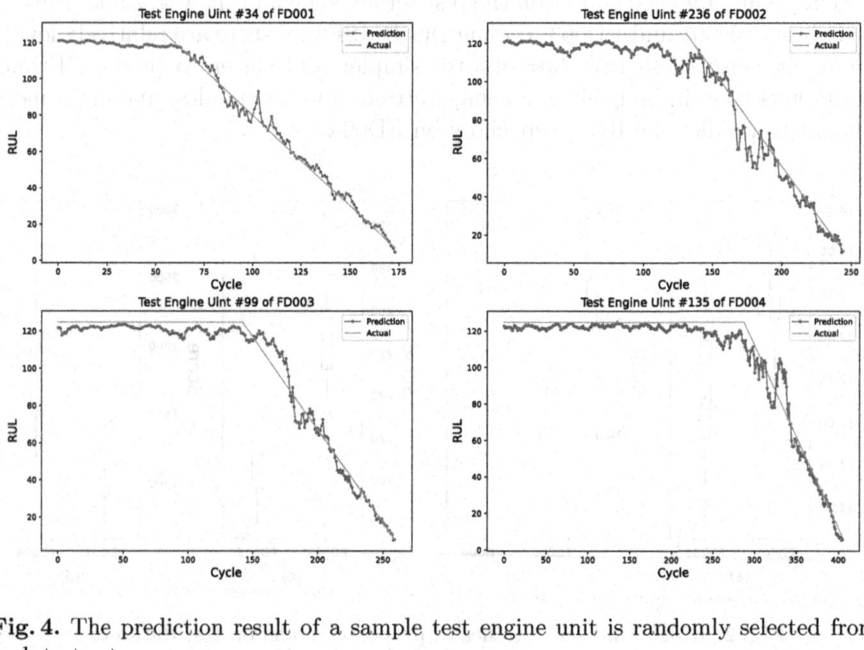

Fig. 4. The prediction result of a sample test engine unit is randomly selected from each test set.

Figure 3 shows the deviation between the true RUL and the predicted value of each test unit in the 4 data sets. To facilitate the observation of the predictive performance of the model on the degradation process, the test units in each data subset are arranged in descending order according to the true RUL values. We can find that the prediction effects of FD001 and FD003 are relatively good. When the value of RUL_{max} is near, the predicted value is very close to RUL_{max}. At the end of life, that is, when the equipment is close to damage, the model has seen a lot more historical cycles. Therefore, the prediction error is pretty small, which has positive significance. Because AA-LSTM can accurately predict the life of the machine that is about to be damaged, it can help managers formulate warranty and maintenance strategies in a timely manner, and reduce the loss caused by shutdown maintenance.

Furthermore, a test unit is randomly selected from each subset of the test set. Based on the RUL value provided by the test set at the last moment, its label for each life cycle is calculated and then its entire life cycle is predicted using the proposed model. The results are shown in Fig. 4. It can be found that the model predicts the RUL of the device throughout its entire life cycle well, especially when the equipment is close to damage, with particularly small prediction errors.

Table 2. Evaluation metrics of different methods on C-MAPSS dataset

	FD001		FD002		FD003		FD004	
	RMSE	Score	RMSE	Score	RMSE	Score	RMSE	Score
RF [23]	17.91	480	29.59	70457	20.27	711	31.12	46568
GB [23]	15.67	474	29.09	87280	16.84	577	29.01	17818
LSTM [5]	16.14	338	24.49	4450	16.18	852	28.17	5550
LSTMBS [24]	14.89	481	26.86	7982	15.11	493	27.11	5200
DCNN [9]	<u>12.61</u>	274	22.36	10412	<u>12.64</u>	284	23.31	12466
AGCNN [13]	**12.42**	**226**	<u>19.43</u>	<u>1492</u>	13.39	<u>227</u>	<u>21.5</u>	<u>3392</u>
AA-LSTM	12.81	<u>268</u>	**14.1**	**940**	**12.15**	**213**	**15.54**	**1911**
IMP	/	/	27.43%	37.00%	3.88%	6.17%	27.72%	43.66%

Discussion. At present, many researchers have conducted RUL prediction studies based on C-MPASS. To further evaluate the performance of AA-LSTM, recent state-of-the-art results are compared and the results are presented in the Table 2, where IMP indicates the percentage of improvement compared to the current optimal results. We can see that AA-LSTM is slightly weaker than the model using CNN on the simplest dataset, FD001, in terms of RMSE and score. On FD003, the RMSE is improved by 3.88% compared to DCNN, 6.17% compared to AGCNN [13]. On FD002, AA-LSTM has a 27.43% improvement in RMSE and a 37% increase in score. On FD004, the most complex dataset, the performance improvement is most pronounced, with a 27.72% improvement on RMSE and a 43.66% improvement on score.

According to what was presented in the previous subsection, both FD001 and FD003 have only one failure mode, FD001 has only one operating condition, while FD003 has six operating conditions. This shows that AA-LSTM is slightly weaker than the current CNN model on a particularly simple data set. However, on a data set with complex working conditions, AA-LSTM performs better and predicts RUL more accurately. This is of practical significance because, in real industrial scenarios, the operating conditions of the equipment are complex, and AA-LSTM is more adaptive to actual working conditions.

4 Conclusion

In this paper, we proposed a novel end-to-end RUL prediction model named AA-LSTM. The adversarial mechanism is added to the training of the Bi-LSTM-based autoencoder, so that the distribution learned by the encoder conforms to the prior distribution, and further enhances the learning ability of the autoencoder. Subsequently, a predictor predicts the RUL value based on the latent code vector. The three modules are trained jointly and the loss of the model is optimized simultaneously, which improves the extraction of degenerate information from the time series and the prediction of RUL. The proposed method has been validated on the benchmark dataset C-MAPSS. The overall performance of AA-LSTM outplays state-of-the-art models, especially on data sets with complex operating conditions and failure modes.

Acknowledgments. This work is jointly funded by the National Science Foundation of China (U1811462), the National Key R&D project by Ministry of Science and Technology of China (2018YFB1003203), and the open fund from the State Key Laboratory of High Performance Computing (No. 201901-11).

References

1. Xia, T., Dong, Y., Xiao, L., Du, S., Pan, E., Xi, L.: Recent advances in prognostics and health management for advanced manufacturing paradigms. Reliab. Eng. Syst. Safety **178**, 255–268 (2018)
2. Khelif, R., Chebel-Morello, B., Malinowski, S., Laajili, E., Fnaiech, F., Zerhouni, N.: Direct remaining useful life estimation based on support vector regression. IEEE Trans. Ind. Electron. **64**(3), 2276–2285 (2017)
3. Zhu, K., Liu, T.: Online tool wear monitoring via hidden semi-Markov model with dependent durations. IEEE Trans. Ind. Inform. **14**, 69–78 (2018)
4. Hochreiter, S., Schmidhuber, J.: Long short-term memory. Neural Comput. **9**(8), 1735–1780 (1997)
5. Zheng, S., Ristovski, K., Farahat, A., Gupta, C.: Long short-term memory network for remaining useful life estimation. In: 2017 IEEE International Conference on Prognostics and Health Management (ICPHM) (2017)
6. Listou Ellefsen, A., Bjørlykhaug, E., Æsøy., Ushakov, S., Zhang, H.: Remaining useful life predictions for turbofan engine degradation using semi-supervised deep architecture. Reliab. Eng. Syst. Safety **183**, 240–251 (2019)

7. Huang, C.G., Huang, H.Z., Li, Y.F.: A bidirectional LSTM prognostics method under multiple operational conditions. IEEE Trans. Ind. Electron. **66**, 8792–8802 (2019)
8. Wang, Q., et al.: Deep image clustering using convolutional autoencoder embedding with inception-like block. In: 2018 25th IEEE International Conference on Image Processing (ICIP), pp. 2356–2360 (2018)
9. Li, X., Ding, Q., Sun, J.Q.: Remaining useful life estimation in prognostics using deep convolution neural networks. Reliab. Eng. Syst. Safety **172**, 1–11 (2018)
10. Long, W., Yan, D., Liang, G.: A new ensemble residual convolutional neural network for remaining useful life estimation. Math. Biosci. En. MBE **16**(2), 862–880 (2019)
11. Hong, C.W., Lee, K., Ko, M.S., Kim, J.K., Oh, K., Hur, K.: Multivariate time series forecasting for remaining useful life of turbofan engine using deep-stacked neural network and correlation analysis. In: 2020 IEEE International Conference on Big Data and Smart Computing (BigComp) (2020)
12. Xia, T., Song, Y., Zheng, Y., Pan, E., Xi, L.: An ensemble framework based on convolutional bi-directional LSTM with multiple time windows for remaining useful life estimation. Comput. Ind. **115**, 103182 (2020)
13. Liu, H., Liu, Z., Jia, W., Lin, X.: Remaining useful life prediction using a novel feature-attention based end-to-end approach. IEEE Trans. Ind. Inform. **PP**(99), 1 (2020)
14. Zhang, W., Jin, F., Zhang, G., Zhao, B., Hou, Y.: Aero-engine remaining useful life estimation based on 1-dimensional FCN-LSTM neural networks. In: 2019 Chinese Control Conference (CCC), pp. 4913–4918 (2019)
15. Al-Dulaimi, A., Zabihi, S., Asif, A., Mohammadi, A.: Hybrid deep neural network model for remaining useful life estimation. In: ICASSP 2019–2019 IEEE International Conference on Acoustics, Speech and Signal Processing (ICASSP) (2019)
16. Sutskever, I., Vinyals, O., Le, Q.V.: Sequence to sequence learning with neural networks. In: Advances in Neural Information Processing Systems (2014)
17. Kingma, D.P., Welling, M.: Auto-encoding variational bayes. arXiv preprint arXiv:1312.6114 (2013)
18. Makhzani, A., Shlens, J., Jaitly, N., Goodfellow, I.: Adversarial autoencoders. In: ICLR (2016)
19. Goodfellow, I., et al.: Generative adversarial nets. In: Advances in Neural Information Processing Systems, pp. 2672–2680 (2014)
20. Saxena, A., Goebel, K., Simon, D., Eklund, N.: Damage propagation modeling for aircraft engine run-to-failure simulation. In: 2008 International Conference on Prognostics and Health Management, pp. 1–9 (2008)
21. Liu, X., et al.: Multiple kernel k-means with incomplete kernels. IEEE Trans. Pattern Anal. Machine Intell. **42**, 1191–1204 (2017)
22. Heimes, F.O.: Recurrent neural networks for remaining useful life estimation. In: 2008 International Conference on Prognostics and Health Management (2008)
23. Zhang, C., Lim, P., Qin, A.K., Tan, K.C.: Multiobjective deep belief networks ensemble for remaining useful life estimation in prognostics. IEEE Trans. Neural Netw. Learn Syst. **28**(10), 2306–2318 (2017)
24. Liao, Y., Zhang, L., Liu, C.: Uncertainty prediction of remaining useful life using long short-term memory network based on bootstrap method. In: 2018 IEEE International Conference on Prognostics and Health Management (ICPHM), pp. 1–8 (2018)

Data Mining of Specialized Data

Analyzing Topic Transitions in Text-Based Social Cascades Using Dual-Network Hawkes Process

Jayesh Choudhari[1(✉)], Srikanta Bedathur[2], Indrajit Bhattacharya[3], and Anirban Dasgupta[4]

[1] University of Warwick, Coventry, UK
choudhari.jayesh@alumni.iitgn.ac.in
[2] Indian Institute of Technology Delhi, New Delhi, India
srikanta@cse.iitd.ac.in
[3] TCS Research, Kolkata, India
b.indrajit@tcs.com
[4] Indian Institute of Technology Gandhinagar, Ahmedabad, India
anirbandg@iitgn.ac.in

Abstract. We address the problem of modeling bursty diffusion of text-based events over a social network of user nodes. The purpose is to recover, disentangle and analyze overlapping social conversations from the perspective of user-topic preferences, user-user connection strengths and, importantly, topic transitions. For this, we propose a Dual-Network Hawkes Process (DNHP), which executes over a graph whose nodes are user-topic pairs, and closeness of nodes is captured using topic-topic, a user-user, and user-topic interactions. No existing Hawkes Process model captures such multiple interactions simultaneously. Additionally, unlike existing Hawkes Process based models, where event times are generated first, and event topics are conditioned on the event times, the DNHP is more faithful to the underlying social process by making the event times depend on interacting (user, topic) pairs. We develop a Gibbs sampling algorithm for estimating the three network parameters that allows evidence to flow between the parameter spaces. Using experiments over large real collection of tweets by US politicians, we show that the DNHP generalizes better than state of the art models, and also provides interesting insights about user and topic transitions.

Keywords: Network Hawkes process · Generative models · Gibbs sampling

This project has received funding from the Engineering and Physical Sciences Research Council, UK (EPSRC) under Grant Ref: EP/S03353X/1, CISCO University grant, and Google India AI-ML award.

Electronic supplementary material The online version of this chapter (https://doi.org/10.1007/978-3-030-75762-5_25) contains supplementary material, which is available to authorized users.

K. Karlapalem et al. (Eds.): PAKDD 2021, LNAI 12712, pp. 305–319, 2021.
https://doi.org/10.1007/978-3-030-75762-5_25

1 Introduction

We address the problem of modeling text-based information cascades, generated over a social network. Observed data on social media is a tangle of multiple overlapping conversations, each propagating from users to their connections, with the rate depending on connection strengths between the users and the conversation topics. The individual conversations, their paths and topics are not directly observed and needs to be recovered. Additionally, individual conversations involve topic shifts, according to the preferences of the users [1]. Our goal is to analyze the user connection strengths, their topic preferences, and the topic-transition patterns from such social conversations.

There exists a number of models that uses a variety of Hawkes processes to model such cascades [1,8,10]. None of these capture user-user, user-topic and topic-topic interactions simultaneously. Additionally, in these models, the content does not influence the response rate. This is a significant disconnect with the underlying social process, where the rate of response for a user depends on the user and topic of the 'parent' post, as well as the (possibly different) topic that gets triggered for the responding user. As a result, two related and important questions are yet unexplored– (1) *how to decompose the overall responsiveness for a pair of users and a pair of topics*, and (2) *how to incorporate the influence of topics on the event rate?*. For example, in the US context, our model should be able to capture a higher response rate for a user passionate about healthcare engaging with another passionate about politics, than for the same user engaging with another talking about gun violence.

In this paper, we address these two issues by extending the Network Hawkes Process [10] which executes over a one-dimensional network over users, to propose a *Dual-Network Hawkes Process* (DNHP) which unfolds over a two-dimensional space of user-topic pairs. Individual events now trigger for a user-topic pair. Each such event spawns a new Poisson process for every other user-topic pair in the neighborhood, whose rate is determined by the two (user, topic) pairs. For tractability and generalization, we decompose this 4-dimensional interaction into *three* interaction matrices. These represent the connection strengths between (a) the pair of users, (b) the pair of topics, and (c) the responding user-topic pair. This decomposition leads to significant parameter sharing between individual point processes. Thus, in addition to being closer to the generation of real-life topical information cascades, the Dual-Network Hawkes Process promises significantly better generalization based on limited training data via parameter sharing.

Using the model, we address the task of recovering the user-user, user-topic and topic-topic connection strengths, along with recovering the latent topic and parent (or triggering) event for each event. A significant challenge for parameter estimation is that the user-user and topic-topic weights are intrinsically coupled in our model and cannot be integrated out analytically. We address the coupling issue by showing that the posterior distribution of the user-user (topic-topic) weights is conditionally Gamma distributed given the topic-topic (user-user) weights. Based on the conditional distributions, we propose a Gibbs sampling

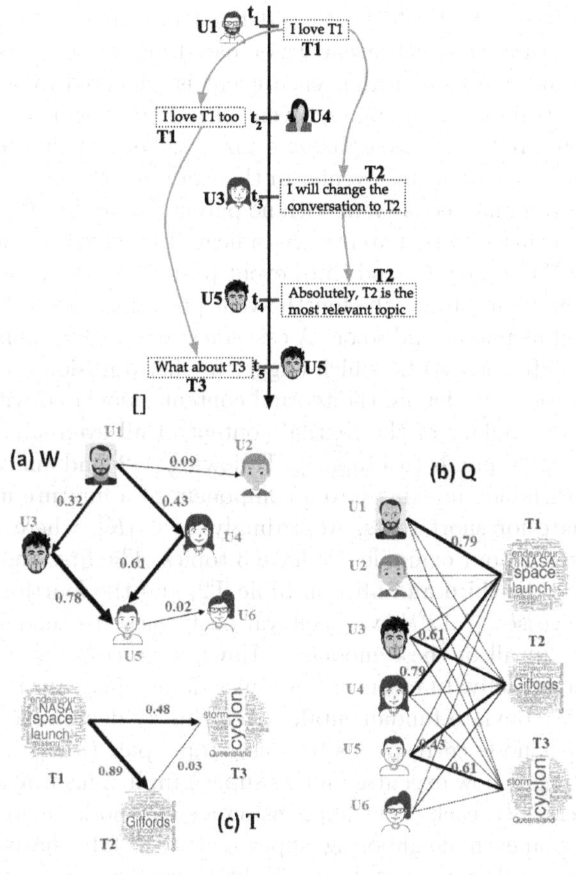

Fig. 1. Illustration of DNHP (A) Event generation process, (B) Model Parameters

based inference algorithm for these tasks. In our inference algorithm, the update equations for the user-user and topic-topic weights become coupled, thereby allowing the flow of evidence between them.

We perform extensive experiments over a large real collection of tweets by US politicians. We show that by being more faithful to the underlying process, or model generalizes much better over held-out tweets compared to state of the art baselines. Further, we report revealing insights involving users groups, topics and their interactions, demonstrating the analytical abilities of our model.

2 Dual-Network Hawkes Process

We consider text based cascades generated by a set of users $U = \{1, 2, \ldots, n\}$, connected by edges $\mathcal{E}$. Let $E = \{e\}$ be the set of all events, which may be tweets or social media posts, created by the users U. The example in Fig. 1(b) shows a toy collection of 5 events. Each event e is defined as a tuple $e = (t_e, c_e, d_e, \eta_e, z_e)$, where, t_e is the time at which event was created and, $c_e \in U$, is the user who created this event. We assume that each event is triggered by a unique parent event. Let $z_e \in E$ indicate the parent event of e. Events which are triggered by some other event are termed as *diffusion* events, and events that happen on their own are termed as *spontaneous* events. In the example, the first event posted by $U1$ at time t_1 is a spontaneous event with no parent, while the others are diffusion events. For the other events, parents are indicated by arrows. The second event posted by user $U4$ at time t_2, and third event posted by user $U3$ at time t_3 have the first event as their parent, the fourth event posted by user $U5$ at time t_4 has the second event as parent and so on. A cascade starts with a spontaneous event, which triggers diffusion events, which trigger further diffusion events, leading to a cascade. We use d_e to denote the textual content associated with event e. Let $\mathscr{V}$ denote the vocabulary of the textual content of all events, i.e. $d_e \subset \mathscr{V}$. We assume that d_e corresponds to a topic η_e. Following [1,2] and unlike [8], we model η_e as discrete variable, indexing into a component of a mixture model, which is more appropriate for short texts. Accordingly, $\eta_e \in [K]$, where K denotes the number of topics. In our example, we have 3 topics. The first and second events are on topic T1, the third and fifth on topic T2, and the fourth on topic T3.

Hawkes processes [7,12] have been variously used to model such cascade events [1,8,10]. In all of these models, a Hawkes Process executes over a network of user nodes only. In essence, the topics do not play a role in the Hawkes process itself. We deviate fundamentally from this by defining a super-graph $\mathcal{G}$, where each super-node corresponds to a user-topic pair (u, k), $u \in U$, $k \in [K]$. The Hawkes Process now executes on this super-graph. This is also illustrated in Fig. 1(A). Specifically, each event happens on a super-node (u, v), and spawns a Poisson Process on each 'neighboring' super-node (v, k'). In the example, according to the super-node representation, the first event happens at $(U1, T1)$, the second at $(U4, T1)$ and so on. Each event spans events on each neighboring super-node. We define two super-nodes to be neighbors is there corresponding users are neighbors in the social graph. The graph in Fig. 1(B)(a) shows the social graph for our example. As a result of this, the first event at $(U1, T1)$ will trigger Poisson Processes at super-nodes with users $U2$ (i.e. $(U2, T1), (U2, T2), (U2, T3)$), $U3$ (i.e. $(U3, T1), (U3, T2), (U3, T3)$), and $U4$ (i.e. $(U4, T1), (U4, T2), (U4, T3)$). The rate of each Poisson Process, for example that triggered $(U4, T1)$, is determined by the 'closeness' of the super-node pair. We discuss this in more detail later in this section. We call this the Dual-Network Hawkes Process (DNHP), because the process executes on a two-dimensional network, unlike those based on the Network Hawkes Process which have a one-dimensional network. Once the DNHP has generated events until some time horizon T, the textual content d_e of each event at super-node (c_e, η_e) is generated independently according to the distribution associated with its topic η_e. We first describe the generation of the super-node (c_e, η_e) and time t_e of each event, and then that of the textual content d_e.

Modeling Time and Topic: In this phase, the time t_e, user c_e, parent z_e, and topic η_e) for each event is generated using the Multivariate Hawkes Process (MHP) on graph $\mathcal{G}$. We follow the general process of existing models [1,8,10,13], but replace user nodes with user-topic super-nodes. In the following, when we refer to a pair (u, k), we will assume $u \in U$, and $k \in [K]$.

Let $\mathcal{H}_{t-}$ denote the set of all events generated prior to time t. Then, following the definition of the Hawkes Process, the intensity function $\lambda_{(v,k)}(t)$ for super-node (v, k) is given by the superposition of the base intensity $\mu_{(v,k)}(t)$ of (v, k) and the impulse responses of historical events $e \in \mathcal{H}_{t-}$ at super-nodes (c_e, η_e) at time t_e: $\lambda_{(v,k)}(t) = \mu_{(v,k)}(t) + \sum_{e \in \mathcal{H}_{t-}} h_{(c_e,\eta_e),(v,k)}(t - t_e)$. The base intensity for node (v, k) is defined as $\mu_{(v,k)}(t) = \mu_v(t) \times \mu_k(t)$, where, $\mu_v(t)$ is base intensity associated with user v, and $\mu_k(t)$ is the base intensity for topic k.

In the context of super-nodes, the parameterization of the impulse response $h_{(u,k),(v,k')}$ becomes a challenge. The naive 4-dimensional parameterization is unlikely to have enough data for confident estimation, while complete factorization with four 1-dimensional parameters is overly biased. We propose its decomposition into three factors: $h_{(u,k),(v,k')}(\Delta t) = W_{u,v} \mathcal{T}_{k,k'} Q_{v,k'} f(\Delta t)$. These three factors form the parameters of our model. Here, $W_{u,v}$ captures user-user preference, $\mathcal{T}_{k,k'}$ captures topic-topic interaction, and $Q_{u,k}$ user-topic preference. We believe that this captures the most important interactions in the data, while providing generalization ability.

Figure 1(B) illustrates this parameterization. Figure 1(B)(a) shows parameter $W_{u,v}$. User pairs (U3, U5) have the strongest connection, indicating the U5 responds to U3 with the quickest rate, followed by (U4, U5), etc. Note that this parameterization is directional. Figure 1(B)(b) shows parameter $Q_{u,k}$. Here, (U1, T1) has the strongest connection, indicating that user U1 posts on topic T1 with the quickest rate, followed by the others. Figure 1(B)(c) shows parameter $\mathcal{T}_{,}kk'$. This shows that topic transitions happen from T1 to T2 with the quickest rate, while those from T2 to T3 happen much slower. Note that this parameter is also directional. The overall rate of the process induced at (U2, T1) by the event at (U1, T1) is determined by the product of the factors $W_{U1,U2}$, $Q_{U2,T1}$ and $\mathcal{T}_{T1,T1}$. Finally, $f(\Delta t)$ is the time-kernel term. We model the time-kernel term $f(\Delta t)$ using a simple exponential function i.e. $\exp(\Delta t)$.

To generate events with the intensity function $\lambda_{(u,k)}(t)$, we follow the level-wise generation of events [13]. Let, Π_0 be the level 0 events which are generated with the base intensity of the nodes $(v, k) \in \mathcal{G}$, i.e. $\mu_{(u,k)}(t)$. In our example, this generates the first event (U1, T1) with time-stamp t1. Then, the events at level $\ell > 0$ are at each super-node (u', k') are generated as per the following non-homogeneous Poisson process: $\Pi_\ell \sim Poisson\left(\sum_{(t_e,(c_e,\eta_e),z_e) \in \Pi_{\ell-1}} h_{(c_e,\eta),(u',k')}(t - t_e)\right)$.

Influencing happens only on neighboring super-nodes (u', k') for $(c_e, \eta_e), e \in \Pi_{\ell-1}$. Recall that two super-nodes (u, k) and (u', k') are neighbors if the corresponding users u and u' are neighbors in the social network. Imagine our example set of events in Fig. 1(A) being generated using the parameterization in Fig. 1(B)

Algorithm 1. DNHP Generative Model

for all $u \in U$ **do**
  for all $v \in \mathcal{N}_u$ **do** Sample $W_{u,v} \sim Gamma(\alpha_1', \beta_2')$ ▷ User-User Influence
  for all $k \in [K]$ **do** Sample $Q_{u,k} \sim Gamma(\alpha_3', \beta_3')$ ▷ User-Topic Preference
for all $k \in [K]$ **do**
  Sample $\zeta_k \sim Dirichlet_K(\alpha)$ ▷ Topic-word Distribution
  for all $l \in [K]$ **do** Sample $T_{k,l} \sim Gamma(\alpha_2', \beta_2')$ ▷ Topic-Topic Interaction
Generate $(t_e, (c_e, \eta_e), z_e)$ for each event as described in Section 2
for all $e \in E$ **do**
  Sample $N_{d_e} \sim Poisson(\lambda)$; Sample N_{d_e} words from ζ_{η_e} ▷ #words to sample

according to the level-wise generation process. Here, $\Pi_0 = \{(U1, T1, t1)\}$, $\Pi_1 = \{(U4, T1, t2), (U3, T2, t3)\}$, and $\Pi_2 = \{(U5, T3, t4), (U5, T2, t5)\}$.

Modeling Documents: Once the events are generated on super-nodes, generation of each document d_e for event e happens conditioned only on the topic η_e of the super-node, using a distribution over words ζ_{η_e} specific to topic η_e. The words in the first and second events are generated i.i.d. from ζ_{T1}, those in the third event from ζ_{T2}, and so on. The complete generative process for DNHP is presented in Algorithm 1.

Additionally, in the supplementary[1], we show that DNHP is stable in the sense that it does not generates infinite number of events in finite time horizon because of the recurrent nature of the Hawkes processes.

Approximate Posterior Inference: The latent variables associated with each event in case of DNHP are the parent ($z_e = \{0(spontaneous), e'(triggering\ event)\}$), and the topic ($\eta_e$) of the event. Along with these latent variables, the model parameters, namely, the user-user influence matrix W, the user-topic preference matrix Q, the topic-topic interaction matrix $\mathcal{T}$, and base rates for each user and each topic need to be estimated. As the exact inference is intractable, we perform inference using an iterative Gibbs sampling algorithm. The posterior distributions for all the parameters and the pseudo-code for the Gibbs sampling algorithm is described in the supplementary (See Footnote 1).

3 Experiments and Results

In this section we validate the strengths of DNHP empirically. Here we describe the models compared, datasets used, and the evaluation tasks and results.

3.1 Models Evaluated

(1) HMHP: HMHP [1] is a model for the generation of text-based cascades. HMHP incorporates user temporal dynamics, user-topic preferences, user-user

[1] https://drive.google.com/file/d/1V7bpwGPuUdX114Mevh5ww0gdam-QO5JG/view.

influence, along with topical interactions. However, in HMHP, similar to the other models in the literature, the generation of time stamp of an event is independent of the topic associated with the event. Additionally, HMHP does not capture user-topic preferences.
(2) NHWKS & NHLDA: Network Hawkes [10] jointly models event time stamps and the user-user network strengths. As opposed to HMHP and DNHP, NHWKS does not model text content. Therefore, we define a simple extension of NHWKS that additionally generated topic labels and content for events, following up on the NHWKS generative process. Specifically, we use an LDA mixture model, that assigns a topic to each event by sampling from a prior categorical distribution, and then draws the words of that event i.i.d by sampling from the word-distribution specific to that topic. We call this Network Hawkes LDA (NHLDA).
(3) DNHP: This is our model with Gibbs sampling based inference algorithm.

3.2 Datasets

(1) Real Dataset: The dataset that we consider here, denoted as USPol (US Politics Twitter Dataset), is a set of roughly $370K$ tweets extracted for 151 users who are members of the US Congress[2]. The tweets were extracted in July 2018 using the *Twitter API* and each tweet consists of time stamp, the user-id, and the tweet-text(content). The total vocabulary size here is roughly $33k$ after removing rare tokens. Ground truth information about parent events is not available for this dataset. Also, we do not consider retweets explicitly. Note that retweets have same topic as that of the original tweet, and retweets form only a small fraction of the parent-child relations that we intend to infer.

(2) Semi-synthetic Dataset: As for the USPol dataset the gold standard for topics and parents is not available, we also generate a semi-synthetic dataset called SynthUSPol using the DNHP generative model, with the same user-user graph as USPol. Here, $\mu_u = \mu_k = 0.003, K = 25, |\mathscr{V}| = 5000$, and the topic-word distributions are sampled from $Dirichlet(0.01)$. For each event e, the number of words in document d_e is sampled from a Poisson distribution with mean 10 to mimic tweets. Using this configuration, we generate 3 different samples of roughly $370K$ events each. For this dataset, due to space constraints, we only report parent identification performance in Table 1. Note that the user-user weight estimates depend directly and only on the identified parents.

3.3 Evaluation Tasks and Results

We evaluate the performance of the models based on following tasks:

(A) ***Cascade Reconstruction Performance:*** For the parent identification task the evaluation metrics used are the accuracy and the recall. Accuracy is defined as the percentage of events for which the correct parent is identified. And,

[2] https://bit.ly/2ufvRWR.

given a ranked list of the predicted parents for each event, recall is calculated by considering the top 1, 3, 5 and 7 predicted parents.

(B) ***Generalization Performance:*** We compare the performance of the models using Log-Likelihood ($\mathcal{LL}$) of the held-out test set. We perform this task on the semi-synthetic dataset `SynthUSPol` and also on the real dataset `USPol`. For each event e in the `Test` set the observed variables are the time t_e, the creator-id c_e, and the words/content d_e, while the parent z_e and the topic η_e are latent.

The calculation of the log-likelihood of the test data involves a significant computational challenge. Let $\mathcal{X}$ and $\mathcal{Y}$ denote the set of events in the `Train` and `Test` sets respectively. As per DNHP, the total log-likelihood $\mathcal{LL}(DNHP)$ of the test set $\mathcal{Y}$ is given as described in Fig. 2. Here, the summations over $e' \in E$, over $\eta_{e'}$, and over η_e are for the marginalization over the candidate set of parents, topic of parent event, and topic of event e respectively. The expression for HMHP requires a similar summation.

$$\mathcal{LL}(DNHP) = \sum_{e \in \mathcal{Y}} (\log P(t_e, c_e, w_e)) = \sum_{e \in \mathcal{Y}} \sum_{\substack{e' \in E \\ c_e \in \mathcal{N}(c_{e'}) \\ t_{e'} < t_e}} \sum_{\eta_{e'}} \sum_{\eta_e} (\exp(-W_{c_{e'},c_e} \mathcal{T}_{\eta_{e'},\eta_e} Q_{c_e,\eta_e})$$
$$\times W_{c_{e'},c_e} \mathcal{T}_{\eta_{e'},\eta_e} Q_{c_e,\eta_e} \exp(\Delta t_e)) \times P(w_e|\eta_e) + \sum_{\eta_e} \exp(\mu_{c_e} \mu_{\eta_e} T) \mu_{c_e} \mu_{\eta_e} \times P(w_e|\eta_e)$$

Fig. 2. DNHP total log-likelihood

In general, when the candidate parents are not in the training set, the parent event also has latent variables. Observe the summation in Fig. 2 over candidate parent's topic $\eta_{e'}$. Therefore, calculating $\mathcal{LL}$ for all the events $e \in \mathcal{Y}$ involves recursively enumerating and summing over all possible test cascades. We avoid this summation by assuming that the parent event for each test event is in the training set, and create our test sets accordingly.

For the semi-synthetic dataset `SynthUSPol`, this is simple, since we know the actual cascades. We take the events at a specific level as the `Test` set, and the events at all previous levels as the `Train` set. However, for the real dataset `USPol`, the true cascade structure is unknown. So we use some heuristics to ensure that the events in the `Test` set are very likely to have parents in the `Train` set. We also design controls for the `Test` set size. We process events sequentially. Each event $e \in E$ is added to the `Test` set $\mathcal{Y}$ if and only if at most p_{test} fraction of its candidate parents are already in the `Test` set $\mathcal{Y}$. This ensures that $1 - p_{test}$ fraction of its candidate parents are still in the `Train` set $\mathcal{X}$. Note that increasing (decreasing) p_{test} results in increasing (decreasing) the test set size, and decreasing (increasing) the train set size. To study the effects of increasing training data size without reducing the test size, we use an additional parameter $0 \leq p_{data} \leq 1$ to decide whether to include an event in our experiments at all. Specifically, we first randomly include each event in the dataset with probability p_{data}, and then the `Train` and `Test` split is performed.

3.4 Results

Parent Identification: Table 1 presents the results for this task. For both the models, recall improves significantly as we consider more candidates predicted parents. The accuracy and recall@1 for the DNHP is ~20% better than that of the HMHP model. In summary, DNHP outperforms the HMHP model with respect to the reconstruction performance for the synthetic data.

Generalization Performance: The generalization performance for the models is evaluated on the basis of their ability to estimate the heldout $\mathcal{LL}$. Tables 2 and 3 present the heldout $\mathcal{LL}$ of time and content for DNHP and HMHP, and $\mathcal{LL}$ of time by for DNHP and NHWKS for the SynthUSPol and USPol datasets respectively.

(1) SynthUSPol Dataset: The results are averaged over 3 independent samples each of size ~$370K$ The size of the Train set upto 3rd last level and upto 2nd last level is ~$170K$ and ~$340K$ respectively. In general, with more training data, for both models $\mathcal{LL}$ improves. Overall, DNHP outperforms NHWKS in explaining the time stamps by benefiting from estimating topic-topic parameters given the text, and in turn using those to better estimate the user-user parameters. In the same way, by better estimating both of these parameters using coupled updates, DNHP outperforms HMHP in explaining time stamps and textual content together.

Table 1. Parent Identification task on SynthUSPol Dataset. (Avg. over 5 samples)

Parent identification		
	DNHP	HMHP
Accuracy	**0.47**	0.40
Recall@1	**0.48**	0.40
Recall@3	**0.75**	0.68
Recall@5	**0.84**	0.79

Table 2. Avg. $\mathcal{LL}$ - SynthUSPol data ($K = 25$)

Average log-likelihood of time & content		
Test on	DNHP	HMHP
2nd last level	**−58.66**	−59.31
Last level	**−57.6**	−58.48
Average log-likelihood of time		
Test on	DNHP	NHWKS
2nd last level	**−2.56**	−2.76
Last level	**−2.32**	−2.48

(2) USPol Dataset: The rows indicate the Train (Test) when the events are selected w.p. p_{data} set as 0.5, 0.7, and 1.0 (which is the complete dataset). Then the Train-Test split is performed with p_{test} set as 0.3 *and* 0.5, which indicate the maximum fraction of candidate parents for each event in the Test set.

Observe that as expected all the models, DNHP, HMHP and NHWKS, get better at estimating the $\mathcal{LL}$ with increase in the size of the dataset. However, DNHP outperforms the competitors by a significant margin A crucial point to note is that the gap between DNHP and the two baselines HMHP and NHWKS is larger when the training dataset size is smaller. This agrees with our understanding of parameter sharing leading to better generalization given limited volumes of training data.

Table 3. Average Log-Likelihood of Time+Content and Time with $K = 100$ for the real dataset USPol. (The Train(Test) sizes mentioned are approximate)

$p_{test} = 0.3$					
Train (Test)	Time + Content			Time	
	DNHP	HMHP	NHLDA	DNHP	NHWKS
$114K(70K)$	**−82.11**	−96.51	−96.72	**−8.03**	−24.46
$177K(100K)$	**−79.09**	−87.32	−87.63	**−7.07**	−16.32
$240K(130K)$	**−77.03**	−80.71	−81.00	**−6.34**	−10.27
$p_{test} = 0.5$					
Train (Test)	Time + Content			Time	
	DNHP	HMHP	NHLDA	DNHP	NHWKS
$86K(98K)$	**−83.37**	−96.15	−105.56	**−8.09**	−23.57
$133K(144K)$	**−80.45**	−87.78	−93.18	**−7.21**	−16.02
$179K(190K)$	**−78.27**	−81.96	−85.90	**−6.53**	−10.90

This demonstrates that DNHP has already learned the parameters efficiently with the smaller dataset size, using flow of evidence between the parameters.

3.5 Analytical Insight from USPol Dataset

In order to extract analytical insights from the USPol dataset, we first fit the model using $K = 100$ topics. For ease of understanding, these 100 topics were further manually annotated by one of the following 8 topics– {Politics, Climate, Social, Defence, Guns, Economy, Healthcare, Technology, Guns}[3] by looking at the set of top words in the topic. Henceforth we refer to these as the topics. Each user was tagged as either Democrat (D) or Republican (R) (based on their Wiki. page). We then extract insights by considering the set of values $\{W_{uv}\mathcal{T}_{k,k'}Q_{v,k'}\}$ for every pair of users (u, v) such that v follows u and (k, k') is a topic-pair.

Figure 3 shows the heat-maps obtained taking various marginalizations over the four tuple (u, k, v, k'). The heatmap in Fig. 3A represents the matrix obtained by $\sum_{u,v} W_{uv}\mathcal{T}_{k,k'}Q_{v,k'}$, and hence estimates rate of a parent child topic pair (k, k'). It is instructive to observe that there are off-diagonal transitions (e.g. Politics → Social and Economy → Politics, Social → Politics) that have higher value than some of the diagonal entries, indicating how the conversations evolve across topics. Figure 3B indicates the aggregated user-user rates across parties obtained by aggregating across all topic-pairs and over all users in the same party. The heatmap clearly indicates that Democrat user have a higher aggregated rate, irrespective of the party affiliation of the child tweet's user. Figure 3C and 3D show two different views of the user-topic rate, where Fig. 3C includes spontaneous posts too, but Fig. 3D includes only replies. Certain topics are equally

[3] Open sourced along with the rest of the data.

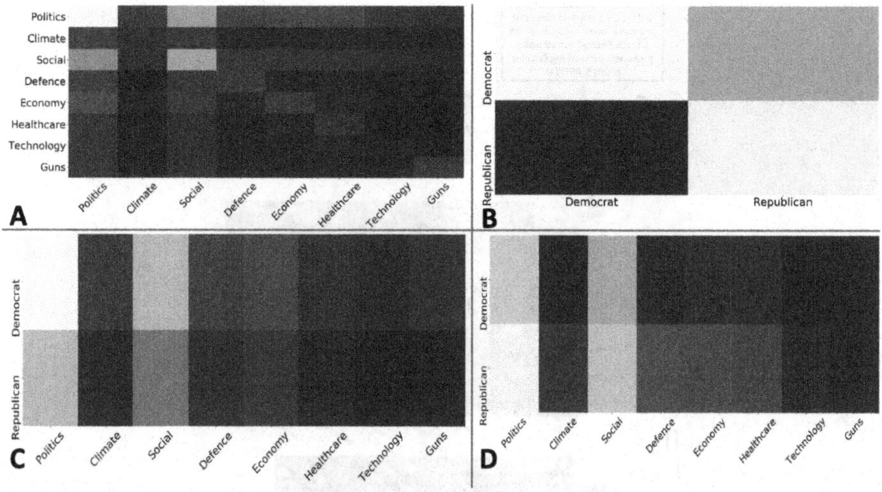

Fig. 3. (A)Topic-topic transition ($\sum_{u,v} W_{u,v}\mathcal{T}_{k,k'}Q_{v,k'}$), (B) User-User Transition ($\sum_{k,k'} W_{u,v}\mathcal{T}_{k,k'}Q_{v,k'}$), (C) (Source) User-(Source) Topic Emission ($\sum_{v,k'} W_{u,v}$ $\mathcal{T}_{k,k'}Q_{v,k'}$), and (D) (Destination)User transiting to (Destination) Topic ($\sum_{u,k} W_{u,v}$ $\mathcal{T}_{k,k'}Q_{v,k'}$)

prominent in both, but there are topics (e.g. Economy, Healthcare) that have a higher rate for the reply tweets than in the source ones.

Drill-down Analysis: To identify interesting topical interactions and parent-child tweet examples we drill down further following two *top-down* approaches:

1) Topics to Users Interaction: Fig. 4(a) explains pictorially the first approach. We start with the matrix $\sum_{u,v} W_{u,v}\mathcal{T}_{k,k'}Q_{v,k'}$ (a *topic*×*topic* matrix), and identify some asymmetric topic pairs. In Fig. 4(a)(A) the (*Economy, Healthcare*) pair is chosen for drilling down further. For this selected *topic-topic* pair we the find the aggregated user-user interaction rate. In the corresponding (*user* × *user*) matrix, (obtained by fixing the topic pairs in the set $\{W_{u,v}\mathcal{T}_{k,k'}Q_{v,k'}\}$), we identify the cells which corresponds to users with different affiliations. In Fig. 4(a)(B) the (*Democrat, Republican*) pair is chosen. We then extract some sample interactions between these users and present as anecdotes in Fig. 4(a)(C).

2) Users to Topics Interaction: Fig. 4(b) – Here we start with the (*user* × *user*) matrix defined by $\sum_{k,k'} W_{u,v}\mathcal{T}_{k,k'}Q_{v,k'}$ (matrix in Fig. 4(b)(A)). We then follow a similar process as in the previous case, i.e. we identify the cell which corresponds to users with different affiliations then calculate the aggregate rate of interaction for all topic-pairs. This gives a (*topic*×*topic*) matrix restricted to the users from matrix Fig. 4(b)(A). In this topic-topic interaction matrix we identify dominant cells with asymmetric topics (namely, (*Social, Politics*) cell in matrix in Fig. 4(b)(B)) and then identify anecdotal parent-child tweet pairs. We note that both the (finer grained) topic assignments, as well as the relation among the tweet-pairs looks reasonable.

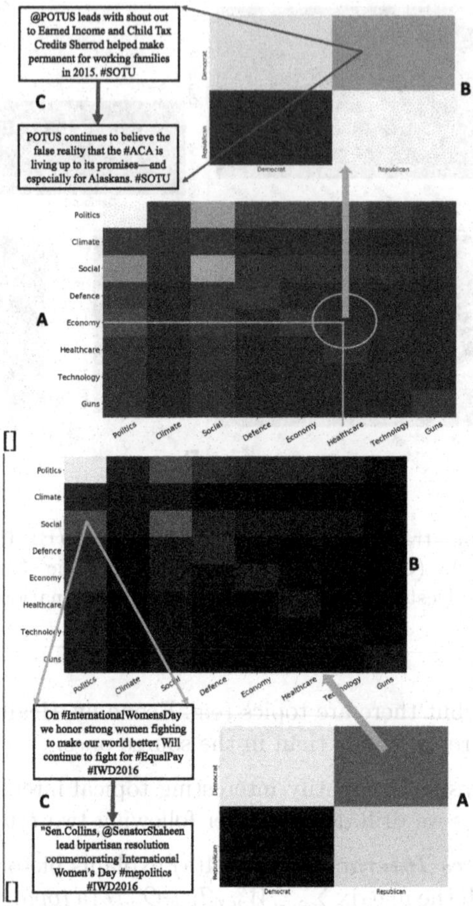

Fig. 4. (a) User-User transition for a specific Topic-Topic transition, (b) Topic-Topic transition for a specific User-User transition (darker the color smaller the interaction intensity)

Finally, in Table 4, we show some additional examples of parent child tweet pairs that correspond to different topics and also users with different political affiliations. In each row, the topics of the tweets are annotated in bold. Observe that the conversation transitions naturally from one topic to another. This is difficult to capture for other state-of-the-art models.

4 Related Work

Recently, there has been a spate of research work in inferring information diffusion networks. The network reconstruction task can be based on just the event times ([4–6, 10, 14, 16]), where the content of the events is not considered. Dirichlet Hawkes Process (DHP) [2] is one of the models that uses the content and time information, but the tasks performed are not related to network inference or cascade reconstruction. Similar to our model the DHP, as well is a mixture model and assigns single topic to each event, but it does not have any notion of parent event or topical interactions. Li et al. [9] investigate the problem of inferring branching structures of heterogeneous MHPs using two levels of Bayesian non-parametric processes. However, they do not consider latent topics or topical interactions between events.

The recent models such as HTM [8], and HMHP [1] show that using the content information can be profitable and can given better estimates for the network inference tasks as well the cascade inference task. HMHP model is the closest model to our model, which considers topical interactions as well. However, in both HMHP and HTM [8], the event times are not conditioned on even time stamps. Instead, the topics are generated conditioned on users and parent events. While all of these capture interactions between users, only HMHP and HTM captures interactions between topics. None of these models capture interactions between users, between topics and between users and topics together.

Following a different line of research, recently there has been effort in using Recurrent Neural Networks (RNN) to model the intensity of point processes [3, 11, 15]. These look to replace pre-defined temporal decay functions with positive functions of time that are learnt from data. So far, these have not considered latent marks, such as topics, or topic-topic interactions.

Table 4. Example parent-child tweet pairs with different topics and different political affiliations for users

Parent Tweet	Child Tweet
(Media): Joined Cheryl Tan & Don Roberts on WAVY News this morning, to discuss #Syria & where I stand	**(Foreign):** #AlQueda positioned to take #Syria if US action ousts Assad. What msg are we sending our troops? "Fight'em in Iraq support'em in Syria"
(Politics): Every American should be free to live & work according to their beliefs w/out fear of punishment by govt #Notmybossbusiness	**(Women's Rights):** Women's private health decisions are between her & her doctor, not her boss. #NotMyBossBusiness
(Foreign): 50 years of isolating Cuba had failed to promote democracy, setting us back. Thats why we restored diplomatic relations. @POTUS #SOTU	**(Politics):** Mr. President you've done enough, now its our time to repair damage you've done & make this country great again#FinalSOTU #SOTU
(House Proceedings): @POTUS delivered vision for expanding opportunity. Let's build a future where anyone who works hard & plays by the rules can succeed #SOTU	**(Foreign)** Would like to hear from @POTUS how he plans to get our U.S. sailors in Iranian custody back. So far nothing #outoftouch

Conclusions. We proposed the Dual-Network Hawkes process to address the problem of reconstructing and analyzing text-based social cascades by capturing user-topic, user-user and topic-topic interactions. DNHP executes on a graph with nodes as user-topic pairs, the event times being determined by both the posting and reacting pairs users and topics. We show that DNHP fits real social data better than state-of-the-art baselines for text-based cascades by using a large collection of US apolitical tweets; the model also reveals interesting insights about social interactions at various levels of granularity.

References

1. Choudhari, J., Dasgupta, A., Bhattacharya, I., Bedathur, S.: Discovering topical interactions in text-based cascades using hidden Markov Hawkes processes. In: ICDM (2018)
2. Du, N., Farajtabar, M., Ahmed, A., Smola, A., Song, L.: Dirichlet-Hawkes processes with applications to clustering continuous-time document streams. In: SIGKDD (2015)
3. Du, N., Dai, H., Trivedi, R., Upadhyay, U., Gomez-Rodriguez, M., Song, L.: Recurrent marked temporal point processes: Embedding event history to vector. In: SIGKDD (2016)
4. Gomez-Rodriguez, M., Leskovec, J., Balduzzi, D., Schölkopf, B.: Uncovering the structure and temporal dynamics of information propagation. Netw. Sci. **2**(1), 26–65 (2014)
5. Gomez-Rodriguez, M., Leskovec, J., Krause, A.: Inferring networks of diffusion and influence. ACM Trans. Knowl. Discovery from Data (TKDD) **5**(4), 1–37 (2012)
6. Gomez-Rodriguez, M., Leskovec, J., Schölkopf, B.: Modeling information propagation with survival theory. In: International Conference on Machine Learning, pp. 666–674 (2013)
7. Hawkes, A.: Spectra of some self-exciting and mutually exciting point processes. Biometrika **58**(1), 83–90 (1971)
8. He, X., Rekatsinas, T., Foulds, J., Getoor, L., Liu, Y.: Hawkestopic: a joint model for network inference and topic modeling from text-based cascades. In: ICML (2015)
9. Li, H., Li, H., Bhowmick, S.S.: BRUNCH: branching structure inference of hybrid multivariate hawkes processes with application to social media. In: Lauw, H.W., Wong, R.C.-W., Ntoulas, A., Lim, E.-P., Ng, S.-K., Pan, S.J. (eds.) PAKDD 2020. LNCS (LNAI), vol. 12084, pp. 553–566. Springer, Cham (2020). https://doi.org/10.1007/978-3-030-47426-3_43
10. Linderman, S., Adams, R.: Discovering latent network structure in point process data. In: ICML (2014)
11. Mei, H., Eisner, J.M.: The neural Hawkes process: a neurally self-modulating multivariate point process. In: Advances in Neural Information Processing Systems, pp. 6754–6764 (2017)
12. Rizoiu, M., Lee, Y., Mishra, S., Xie, L.: A tutorial on hawkes processes for events in social media. In: arXiv (2017)
13. Simma, A., Jordan, M.I.: Modeling events with cascades of poisson processes. In: Proceedings of the Twenty-Sixth Conference on Uncertainty in Artificial Intelligence, pp. 546–555 (2010)

14. Wang, S., Hu, X., Yu, P., Li, Z.: Mmrate: Inferring multi-aspect diffusion networks with multi-pattern cascades. In: SIGKDD (2014)
15. Xiao, S., Yan, J., Yang, X., Zha, H., Chu, S.M.: Modeling the intensity function of point process via recurrent neural networks. In: AAAI (2017)
16. Yang, S.H., Zha, H.: Mixture of mutually exciting processes for viral diffusion. In: International Conference on Machine Learning, pp. 1–9 (2013)

HiPaR: Hierarchical Pattern-Aided Regression

Luis Galárraga[1(✉)], Olivier Pelgrin[2], and Alexandre Termier[3]

[1] Inria, Rennes, France
luis.galarraga@inria.fr
[2] Aalborg University, Aalborg, Denmark
olivier@cs.aau.dk
[3] University of Rennes I, Rennes, France
alexandre.termier@irisa.fr

Abstract. We introduce HiPaR, a novel pattern-aided regression method for data with both categorical and numerical attributes. HiPaR mines hybrid rules of the form $p \Rightarrow y = f(X)$ where p is the characterization of a data region and $f(X)$ is a linear regression model on a variable of interest y. The novelty of the method lies in the combination of an enumerative approach to explore the space of regions and efficient heuristics that guide the search. Such a strategy provides more flexibility when selecting a small set of jointly accurate and human-readable hybrid rules that explain the entire dataset. As our experiments shows, HiPaR mines fewer rules than existing pattern-based regression methods while still attaining state-of-the-art prediction performance.

Keywords: Linear regression · Rule mining

1 Introduction

In the golden age of data, regression analysis is of great utility in absolutely all domains. As the need for interpretable methods becomes compelling, a body of literature focuses on pattern-aided linear regression [3,11–13,16]. Such approaches mine *hybrid rules*, i.e., statements such as *property-type* = "*cottage*" ⇒ *price* $= \alpha + \beta \times$ *rooms* $+ \gamma \times$ *surface*, where the left-hand side is a logical expression on categorical features, and the right-hand side is a linear model valid only for the points satisfying the given condition, for instance, $\{x^1, x^2, x^3\}$ in Table 1.

The advantage of models based on hybrid rules is that they can boost the accuracy of linear functions and still deliver human-readable statements. In contrast, they usually lag behind black-box methods such as gradient boosting trees [7] or random forests [2] in terms of accuracy. Indeed, our experiments show that existing pattern-aided regression methods struggle at providing satisfactory performance and interpretability simultaneously: accuracy is achieved by mining numerous and long rules, whereas compactness comes at the expense of

K. Karlapalem et al. (Eds.): PAKDD 2021, LNAI 12712, pp. 320–332, 2021.
https://doi.org/10.1007/978-3-030-75762-5_26

Table 1. Toy example for the prediction of real estate prices. The symbols "*", "+", and "-" denote high, medium, and low prices respectively.

Id	Property-type	State	Rooms	Surface	Price
x^1	Cottage	Very good	5	120	510k (*)
x^2	Cottage	Very good	3	55	410k (*)
x^3	Cottage	Excellent	3	50	350k (+)
x^4	Apartment	Excellent	5	85	320k (+)
x^5	Apartment	Good	4	52	140k (-)
x^6	Apartment	Good	3	45	125k (-)

prediction power. The goal of this work is to reach a sweet spot where accuracy is attainable with few hybrid rules that can be easily grasped by a human user.

Finding such a good set of hybrid rules is hard, because the search space of possible conditions (the left-hand side of the rules) is huge. Methods such as regression trees (RT) resort to a greedy approach that refines rules with the best condition at a given stage. This is illustrated by Fig. 1 that shows two splitting steps of an RT on the example in Table 1. Regions with a high goodness of fit spanning a partition, e.g., the dashed region on the left, will never be explored even if they have short descriptions (here *state* = "*excellent*"). They may only be described by two patterns (*ptype* = "*cottage*" $\wedge$ *state* $\neq$ "*v.good*" and *ptype* $\neq$ "*cottage*" $\wedge$ *state* $\neq$ "*good*"), which is less interpretable.

Our main contribution lies in a novel strategy to explore the space of hybrid rules. Such a strategy is hierarchical, as depicted in Fig. 2, and looks "everywhere" for (potentially overlapping) regions with a high goodness of fit. This is followed by a selection phase that picks a small set of rules with good joint prediction performance. Our experiments suggest that our method, called HiPaR, strikes an interesting performance/intepretability balance, achieving as much error reduction as the best interpretable approaches but with one order of magnitude fewer atomic elements (conditions) to examine for the analyst. Before elaborating on our approach, we introduce relevant concepts and related work.

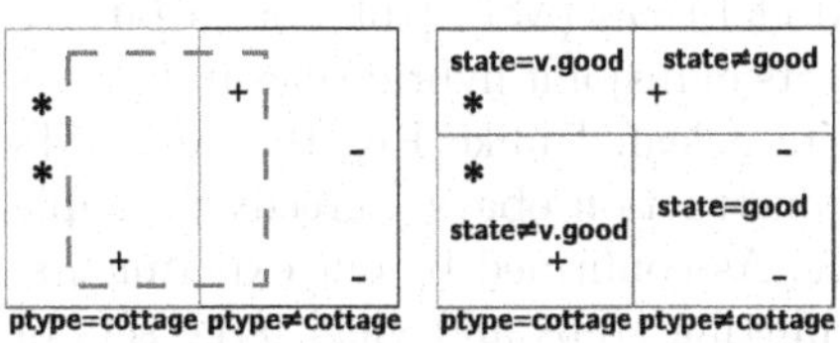

Fig. 1. Two exploration steps of an RT.

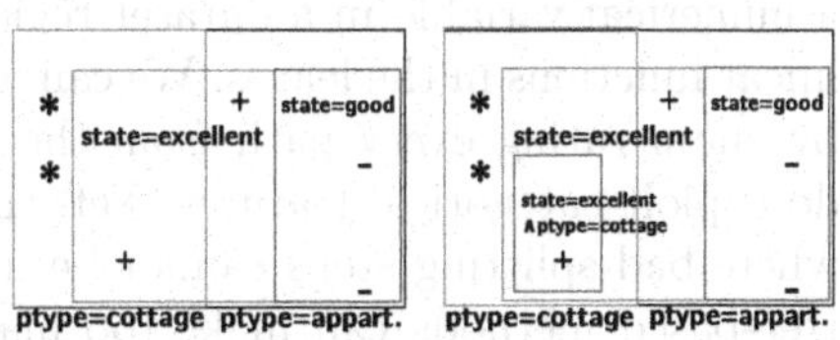

Fig. 2. Our hierarchical exploration.

2 Preliminaries and Notation

Datasets. A *dataset* $D = \{x^1, \ldots, x^n\} \subseteq V^{|A|}$ is a set of $|A|$-dimensional points or observations, where A is a set of attributes and each component of x^i, denoted by x^i_a, is associated to an attribute $a \in A$ (with domain V_a), e.g., x^1_{state} = "*very good*" in Table 1. We use the terms *attribute*, *feature*, and *variable* interchangeably. A *categorical* attribute holds elements for which partial and total orders are meaningless, e.g., property types as in Table 1. A *numerical* attribute takes numeric values and represents measurable quantities such as prices or temperature measures. They are the target of regression analysis.

Patterns. A pattern is a characterization of a dataset region (subset of points). An example is p : *property-type* = "*cottage*" ∧*surface* $\in (-\infty, 60]$ with associated region $D_p = \{x^2, x^3\}$ in Table 1. In this work we focus on conjunctive patterns on non-negated conditions. These conditions take the form $a_i = v$ for categorical attributes or $a_j \in I$ for discretized numerical attributes, where I is an interval in $\mathbb{R}$. If p is a pattern, its support and relative support in a dataset D are defined as $s_D(p) = |D_p|$ and $\bar{s}_D(p) = \frac{s_D(p)}{|D|}$, respectively. When D is implicit, we write $s(p)$ and $\bar{s}(p)$ for the sake of brevity. A pattern p is *frequent* in D if $s_D(p) \geq \theta$ for a given user-defined threshold θ. A pattern p is *closed* if no longer pattern can describe the same region. As each region can be described by a single closed pattern, we define the closure operator $\mathbf{cl}(p)$ so that $\mathbf{cl}$ returns D_p's closed pattern. For instance, given p : *state* = "*good*" characterizing the region $\{x^5, x^6\}$ in Table 1, $\mathbf{cl}(p)$ is *state* = "*good*" ∧*property-type* = "*apartment*", because this is the maximal pattern that still describes $\{x^5, x^6\}$.

3 Related Work

Classical Methods. Piecewise Regression (PR) [13] learns hybrid rules of the form $z \in [v_1, v_2] \Rightarrow y = f(X)$ $(v_1, v_2 \in \mathbb{R})$, and is among the first approaches for pattern-aided regression. PR is outmatched by more recent methods due to its limitations, e.g., it defines conditions on *only one numerical* attribute. A regression tree [11] (RT) is a decision tree whose leaves predict the average of a numerical variable in a dataset region. Model trees (MT) [16] store arbitrary linear functions in the leaves. We can mine sets of disjoint hybrid rules from trees by enumerating every path from the root to a leaf. Unlike PR, RTs and MTs do exploit categorical features. Yet, their construction obeys a greedy principle where bad splitting steps cannot be undone. As confirmed by our experiments, tree-based methods can make too many splitting steps and yield large sets of rules, even though comparable performance is attainable with fewer rules.

Contrast Pattern-Aided Regression. The authors of [3] propose CPXR, a method that relies on contrast pattern mining to find hybrid rules on the regions of a dataset where a global linear model performs poorly—called the LE (large error) region. A subsequent iterative process then selects a small set of those rules with low overlap and good error reduction w.r.t. the global model. Despite its emphasis on error reduction, CPXR has some limitations: regions spanning outside the region LE are disregarded, and the search is restricted to the class of contrast patterns. While the latter decision keeps the search space under control, our experiments show that exploring the (larger) class of closed patterns allows for a significant gain in prediction accuracy with a reasonable penalty in runtime.

Related Paradigms. The problem of finding data regions with a high goodness of fit w.r.t. a target variable is similar to the problems of subgroup discovery [9] (SD) and exceptional model mining (EMM) [4,5]. In their general formulation, these approaches report subgroups—data regions in our jargon—where the behavior of one or multiple target variables deviates notably from the norm. There exist plenty of SD and EMM approaches [5,10] tailored for different description languages, types of variables and notions of "exceptionality". Finding subgroups correlated with a variable can be framed as an SD or EMM task, nonetheless, these paradigms are concerned with reporting *individually exceptional* subgroups. For this reason, these methods are often greedy and resort to strategies such as beam search. Conversely, we search for *jointly exceptional* sets of rules that explain the whole dataset, and jointly achieve good performance. Indeed, our evaluation shows that an SD-like selection for hybrid rules yields lower performance gains than the strategy proposed in this paper.

4 HiPaR

We now describe our pattern-aided regression method called HiPaR, summarized in Algorithm 1. HiPaR mines hybrid rules of the form $p \Rightarrow y = f_p(A'_{num})$ for a pattern p characterizing a region $D_p \subseteq D$, and a target variable y on a dataset D with categorical and numerical attributes A_{num}, A_{cat} ($A'_{num} = A_{num} \setminus \{y\}$). The patterns define a containment hierarchy that guides HiPaR's search and provides flexibility to the subsequent rule selection phase. This hierarchy is rooted at the empty pattern $\top$ that represents the entire dataset. After learning a global linear model $\top \Rightarrow y = f_\top(A'_{num})$ (called the *default model*), HiPaR operates in three stages detailed next.

Algorithm 1: HiPaR

Input: a dataset: $\mathcal{D}$ with attributes A_{cat}, A_{num}; target variable: $y \in A_{num}$ with $A'_{num} = A_{num} \setminus \{y\}$; support threshold: θ

Output: a set R of hybrid rules $p \Rightarrow y = f_p(A'_{num})$

1 Learn default hybrid rule $r_\top : \top \Rightarrow y = f_\top(A'_{num})$ from D

2 $C :=$ *hipar-init*(D, y, θ)

3 $\mathcal{R} :=$ *hipar-candidates-enum*$(\mathcal{D}, r_\top, C, \theta)$

4 **return** *hipar-rules-selection*$(\mathcal{R} \cup \{r_\top\})$

4.1 Initialization

The initialization phase (line 2 in Algorithm 1) bootstraps HiPaR's hierarchical search. We describe the initialization routine *hipar-init* in Algorithm 2. The

Algorithm 2: hipar-init

Input: a dataset: D with attributes A_{cat}, A_{num}; target variable: $y \in A_{num}$; support threshold: θ

Output: a set of frequent patterns of size 1

1 $C_{cat} := \bigcup_{a \in A_{cat}} \{c : a = v \mid s_D(c) \geq \theta\}$

2 $C_{num} := \emptyset$

3 **for** $a \in A_{num}$ **do**

4 $\quad C_{num} := C_{num} \cup \{c : (a \in I) \in discr(a, D, y) \mid s_D(c) \geq \theta\}$

5 **return** $C_{cat} \cup C_{num}$

procedure computes frequent patterns of the form $a = v$ for categorical attributes (line 1), and $a \in I$ for numerical attributes (lines 2–4), where I is an interval of the form $(-\infty, \alpha)$, $[\alpha, \beta]$, or (β, ∞). The intervals are calculated by discretizing the numerical attributes w.r.t. the values of the target variable [6]. This way, HiPaR minimizes the variance of y for the points that match a pattern.

4.2 Candidates Enumeration

The patterns computed in the initialization step are the starting point to explore the hierarchy of regions and learn candidate rules. These regions are characterized by closed patterns on categorical and discretized numerical variables. Our preference for closed patterns is based on two reasons. First, a region is characterized by a unique closed pattern, which prevents us from visiting the same region multiple times. Second, closed patterns compile the maximal set of conditions that portray a region. This expressivity may come handy when analysts need to identify *all* the attributes that correlate with the target variable.

Inspired on the LCM algorithm [15] for closed itemset mining, the routine *hipar-candidates-enumeration* in Algorithm 3 takes as input a hybrid rule $p \Rightarrow y = f_p(A'_{num})$ learnt on the region D_p, and returns a set of hybrid rules defined on closed descendants of p. Those descendants are visited in a depth-first hierarchical manner. Lines 1 and 2 initialize the procedure by generating a working copy C' of the conditions used to refine p (loop in lines 5–18). At each iteration in line 5, Algorithm 3 refines the parent pattern p with a condition $c' \in C'$ and removes c' from the working copy (line 7). After this refinement step, the routine proceeds in multiple phases described below.

Pruning. In line 8, Algorithm 3 enforces thresholds on support and interclass variance (iv) for the newly refined pattern $\hat{p} = p \wedge c'$. The support threshold θ serves two purposes. First, it prevents us from learning rules on extremely small subsets and incurring overfitting. Second, it allows for pruning, and hence lower runtime. This is crucial since HiPaR's search space is exponential in the number of conditions. The threshold on interclass variance—an exceptionality metric for patterns proposed in [14]—prunes regions with a low goodness of fit. We highlight the heuristic nature of using iv here, since it lacks the anti-monotonicity of support. That said, thresholding on *iv* proves effective at keeping the size of the search space under control with no impact on prediction performance[1].

Closure Computation. If a refinement $\hat{p} = p \wedge c'$ passes the test in line 8, HiPaR computes its corresponding closed pattern p' in line 9. The closure operator may add further conditions besides c', hence line 10 excludes those conditions for future refinements[2] of p. Next, line 11 guarantees that no path in the search space is explored more than once. This is achieved by verifying whether pattern p is the leftmost parent of p'. This ensures that the hierarchy rooted at the node *ptype* = "*cottage*" ∧*ptype* = "*excellent*" is explored only once, in this case when visited from its leftmost parent *ptype* = "*cottage*".

Algorithm 3: hipar-candidates-enum

Input: a dataset: D with attributes A_{cat}, A_{num}; parent hybrid rule: $r_p : p \Rightarrow y = f_p(A'_{num})$; patterns of size 1: C; support threshold: θ
Output: a set $\mathcal{R}$ of candidate hybrid rules $p \Rightarrow y = f_p(A'_{num})$

1 $\mathcal{R} := \emptyset$
2 $C' := C$
3 $C_n := \{c \in C \mid c : a \in I \wedge a \in A_{num}\}$
4 $\nu :=$ k-th percentile of iv_D in C_n
5 **for** $c' \in C'$ **do**
6     $\hat{p} := p \wedge c'$
7     $C' := C' \setminus \{c'\}$
8   **if** $s_D(\hat{p}) \geq \theta \wedge iv_D(\hat{p}) > \nu$ **then**
9         $p' = \mathbf{cl}(\hat{p})$
10         $C' := C' \setminus p'$
11   **if** *p is the left-most parent of p'* **then**
12   Learn $r_{p'} : p' \Rightarrow y = f_{p'}(A'_{num})$ on $D_{p'}$
13   **if** $m(r_{p'}) < m(r_{p^*}) \; \forall p^* : p^*$ *is parent of* p' **then**
14                 $\mathcal{R} = \mathcal{R} \cup \{r_{p'}\}$
15                 $C'_n := \emptyset$
16   **for** $a \in A'_{num} \setminus attrs(p')$ **do**
17                     $C'_n := \{c \in discr(a, D_{p'}, y) \mid s_D(c) \geq \theta\} \cup C'_n$
18                 $\mathcal{R} := \mathcal{R} \cup$ *hipar-candidates-enum*$(D, r_{p'}, (C' \setminus C_n) \cup C'_n, \theta)$
19 **return** $\mathcal{R}$

Learning a Regression Model. In line 12, Algorithm 3 learns a hybrid rule $p' \Rightarrow y = f_{p'}(A'_{num})$ from the data points that match p'. Before being accepted as a candidate (line 14), this new rule must pass a test in the spirit of Occam's razor (line 13): if a hybrid rule defined on region $D_{p'}$ does not predict strictly better than all the hybrid rules defined in the super-regions of $D_{p'}$, then the rule is redundant, because we can obtain good predictions with more general

---

[1] We set ν empirically to the k = 85-th percentile of iv_D (line 4 in Algorithm 3). Lower percentiles did not yield better performance in our experimental datasets.

[2] Our abuse of notation treats p as a set of conditions.

and simpler models. Performance is defined in terms of an error metric m (e.g., RMSE). This requirement makes our search diverge from a pure DFS.

DFS Exploration. The final stage of *hipar-candidates-enum* discretizes the numerical variables not yet discretized in p' (lines 16–17)[3] and uses those conditions to explore the descendants of p' recursively (line 18). We remark that this recursive step could be carried out regardless of whether p' passed or not the test in line 13: Error metrics are generally not anti-monotonic, thus the region of a rejected candidate may still contain sub-regions that yield more accurate hybrid rules. Those regions, however, are more numerous, have a lower support, and are described by longer patterns. Hence, recursive steps become less appealing as we descend in the hierarchy.

4.3 Rule Selection

The set $\mathcal{R}$ of candidate rules $r_p : p \Rightarrow y = f_p(A'_{num})$ returned by the enumeration phase is likely to be too large for presentation to a human user. Thus, HiPaR carries out a selection process (line 4 in Algorithm 1) that picks a subset (of rules) of minimal size, minimal joint error, and maximal coverage. We formulate these multi-objective desiderata as an integer linear program (ILP):

$$\begin{aligned} \min \; & \sum_{r_p \in \mathcal{R}} -\alpha_p \cdot z_p + \sum_{r_p, r_q \in \mathcal{R}, p \neq q} (\omega \cdot \mathcal{J}(p,q) \cdot (\alpha_p + \alpha_q)) \cdot z_{pq} \\ \text{s.t.} \; & \sum_{r_p \in \mathcal{R}} z_p \geq 1 \qquad \forall r_p, r_q \in \mathcal{R}, p \neq q : z_p + z_q - 2z_{pq} \leq 1 \\ & \forall r_p, r_q \in \mathcal{R}, p \neq q : z_p, z_{pq} \in \{0,1\} \end{aligned} \tag{1}$$

Each variable z_p can take either 1 or 0 depending on whether the rule $r_p \in \mathcal{R}$ is selected or not. The first constraint guarantees a non-empty set of rules. The term $\alpha_p = \bar{s}(p)^{\sigma} \times \bar{e}(r_p)^{-1}$ is the support-to-error trade-off of the rule. Its terms $\bar{e}(r_p) \in [0,1]$ and $\bar{s}(p) \in [0,1]$ correspond to the normalized error and normalized support of rule r_p. The term $\bar{e}(r_p)$ is computed by dividing r_p's error by the total error of all the rules in $\mathcal{R}$ ($\bar{s}(p)$ is computed the same way). The objective function rewards rules with small error defined on regions of large support. This latter property accounts for our maximal coverage desideratum and protects us from overfitting. The support bias $\sigma \in \mathbb{R}_{\geq 0}$ is a meta-parameter that controls the importance of support in rule selection.

Due to our hierarchical exploration, the second term in the objective function penalizes overlapping rules. The overlap is measured via the Jaccard coefficient $\mathcal{J}$ on the regions of every pair of rules. If two rules r_p, r_q are selected, i.e., z_p and z_q are set to 1, the second family of constraints enforces the variable z_{pq} to be 1 and pay a penalty proportional to $\omega \times (\alpha_p + \alpha_q)$ times the degree of overlap between D_p and D_q in the objective function. The overlap bias $\omega \in \mathbb{R}_{\geq 0}$ controls the magnitude of the penalty. Values closer to 0 make the solver tolerate

[3] *attrs*(p) returns the set of attributes present in a pattern.

overlaps and choose more rules ($\omega = 0$ chooses all rules). Large values favor sets of non-overlapping rules (as in regression trees). The solution to Eq. 1 is a set of hybrid rules that can be used as a prediction model for the target variable y.

4.4 Prediction with HiPaR

A rule $r : p \Rightarrow y = f_p(A'_{num})$ *is relevant to* or *covers* a data point $\hat{x}$ if the condition defined by p evaluates to true on $\hat{x}$. If $\hat{x}$ is not covered by any rule, HIPAR uses the default regression model $r_\top$ to produce a prediction. Otherwise, HIPAR returns a weighted sum of the predictions of all the rules relevant to $\hat{x}$. The weight of a rule r is equals the inverse of its error in training normalized by the inverse of the error of all the rules that cover $\hat{x}$.

5 Evaluation

We evaluate HIPAR in two rounds of experiments along three axes, namely prediction power, interpretability, and runtime. The first round compares HIPAR with state-of-the-art regression methods (Sect. 5.2). The second round provides an anecdotal evaluation that shows the utility of HIPAR's rules for data analysis (Sect. 5.3). Section 5.1 serves as a preamble that describes our experimental setup.

5.1 Experimental Setup

We provide the most important aspects of the experimental settings and refer the reader to our complete technical report [1] for details (Sect. 5.1).

HIPAR's Settings. Unless explicitly stated, we set $\theta = 0.02$ (minimum support threshold) and $\sigma = \omega = 1.0$ (support bias and overlap biases) for HIPAR. Our choices are based on a study of the impact of HIPAR's hyper-parameters in the runtime and quality of the resulting rule sets [1]. Linear functions are learned using sparse (regularized) linear regression (e.g., LASSO).

Competitors. We compare HIPAR to multiple regression methods comprising[4]:
- The pattern-based methods, CPXR [3], regression trees (RT) [11] and model trees (MT) [16].
- Three accurate black-box methods, namely, random forests (RF) [2], gradient boosting trees (GBT) [7], and rule fit (Rfit) [8].
- HIPAR with no subsequent rule selection (called HIPAR$_f$), and HIPAR with a rule selection in the spirit of subgroup discovery (HIPAR$_{sd}$) where the top q rules with the best support-to-error trade-off are reported. The parameter q is set to the average number of rules output by HIPAR in cross-validation.
- Two hybrid methods resulting from the combination of HIPAR's enumeration with Rfit's rule selection (HIPAR+Rfit), and Rfit's rule generation with HIPAR's rule selection (Rfit+HIPAR).

[4] Hyper-parameters were tuned using Hyperopt. For CPXR we set $\theta = 0.02$ as in [3].

Datasets. We test HiPaR and its competitors on 17 datasets, 7 of them used in [3]: *abalone*, *cpu*, *houses*, *mpg2001*, *servo*, *strikes*, and *yatch*[5]. In addition, our anecdotal evaluation in Sect. 5.3 relies on the results presented in [4] on the datasets *giffen* and *wine2*. The other datasets were downloaded from `kaggle.com`, namely *cb_nanotubes*, *fuel_consumption*, *healthcare*, *optical*, *wine*, *concrete*, *beer_consumption* and *admission*. The datasets match the keyword "regression" in Kaggle's search engine (as of 2019), and define meaningful regression tasks.

Metrics. In line with [3], we evaluate prediction accuracy in terms of the average RMSE reduction w.r.t. a baseline non-regularized linear model in 10-fold cross-validation. Since our goal is to mine human-readable rules, we also evaluate the "interpretability" of the rule sets. Given the subjective and context-dependent nature of this notion—and the diversity of domains of our datasets— we use output complexity as a proxy for interpretability. This is defined as the number of elements in a rule set. An element is either a condition on a categorical attribute or a numerical variable with a non-zero coefficient in a linear function.

5.2 Comparison with the State of the Art

Accuracy and Complexity Evaluation. Figure 3 depicts the mean RMSE reduction for the different methods for all executions in our experimental datasets. We first note that the black-box methods (in blue) rank higher than the pattern-based methods. The tree-based approaches usually achieve good performance, however this comes at the expense of complex sets of rules as suggested in Fig. 5. If we limit the number of leaves in the trees using HiPaR[6]—denoted by RT_H and MT_H—, the results are mixed: a positive impact on MTs, and a slight drop in performance for RTs. This suggests that the greedy exploration of the tree-based methods may sometimes lead to unnecessary splitting steps where good performance is actually attainable with fewer rules.

We observe that HiPaR's median error reduction is comparable to MT and HiPaR_f. Yet, HiPaR outputs one order of magnitude fewer elements (Fig. 5), thanks to our rule selection step. Besides, HiPaR's behavior yields fewer extreme values, outperforms HiPaR_{sd}, and is comparable to RF.

We finally observe that CPXR and HiPaR+Rfit lie at the bottom of the ranking in Fig. 3. As for CPXR, the method is too selective and reports only 1.42 rules on average in contrast to HiPaR and MT that find on average 8.81 and 23.92 rules, respectively. This is also reflected by the low variance of the reductions compared to other methods. We highlight the large variance of HiPaR+Rfit. While it can achieve high positive error reductions, its rule extraction is not designed for subsequent filtering, because Rfit reports weak estimators (trees) that become accurate only in combination, for instance, by aggregating their answers or as features for a linear regressor.

[5] These are the only publicly available datasets used in [3].

[6] By setting this limit to the avg. number of rules found by HiPaR in cross-validation.

All in all, Fig. 5 suggests that HiPaR offers an interesting trade-off between model complexity and prediction accuracy. This makes it appealing for situations where users need to inspect the correlations that explain the data, or for tuning other methods as we did for RTs and MTs.

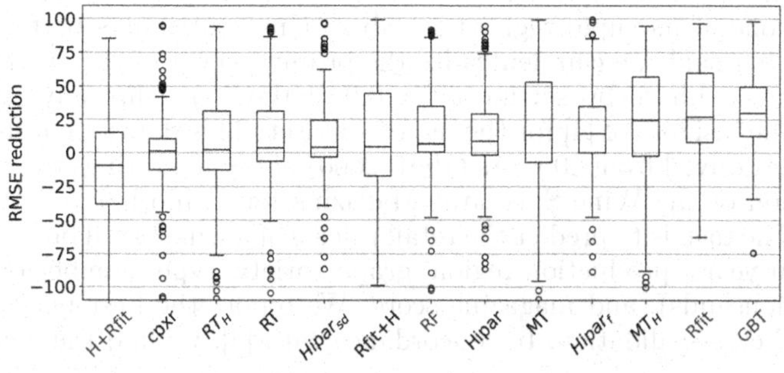

Fig. 3. Mean RMSE reduction in cross-validation. Black-box methods are in blue (Color figure online)

Runtime Evaluation. Figure 4 depicts the average runtime of a fold of cross-validation for the different regression methods. We advise the reader to take these results with a grain of salt. RT, GBT, and RF are by far the most performant algorithms partly because they count on a highly optimized scikit-learn implementation. They are followed by Rfit and the hybrid methods, which combine Rfit and HiPaR. We observe HiPaR is slower than its variants HiPaR$_f$ and HiPaR$_{sd}$ because it adds a more sophisticated rule selection that can take on average 46% of the total runtime (97% for *optical*, 0.26% for *carbon_nanotubes*). Finally, we highlight that MT is one order of magnitude slower than HiPaR despite its best-first-search implementation.

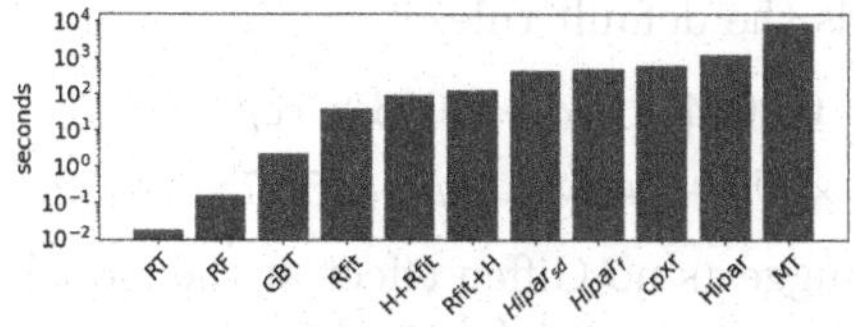

Fig. 4. Average training time.

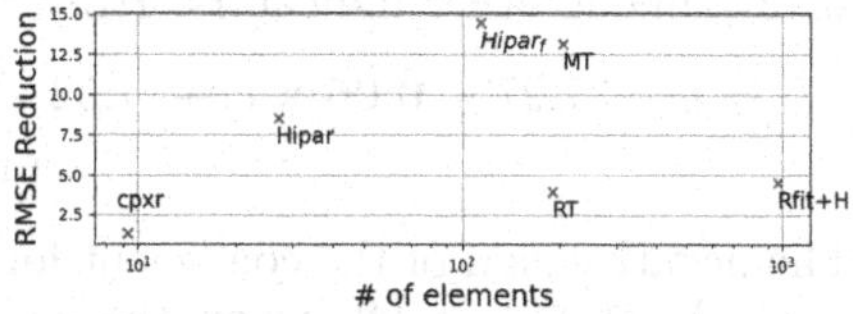

Fig. 5. Complexity accuracy trade-off.

5.3 Anecdotal Evaluation

We illustrate the utility of HiPaR at finding interpretable rules on two use cases used in the evaluation of the EMM approach presented in [4]. In this work, the authors introduce the Cook's distance between the coefficients of the default model and the coefficients of the local models as a measure of exceptionality for regions—referred as subgroups. A subgroup with a large Cook's distance is cohesive and its slope vector deviates considerably from the slope vector of the bulk of the data (w.r.t. a target variable). We emphasize that HIPAR's goal is different from EMM's: The former looks for compact sets of accurate rules, whereas the latter searches for individually exceptional regions. In this spirit, nothing prevents HIPAR from pruning an exceptional region according to EMM if one of its super-regions or sub-regions contributes better to reduce the error. That said, we can neutralize the pruning effect of the overlap penalty in the selection phase by setting $\omega = 0.0$ (HIPAR$_f$). This way HIPAR can reproduce the insights of [4] for the *wine2* dataset. This dataset consists of 9600 observations derived from 10 years (1991–2000) of tasting ratings reported in the online version of the Wine Spectator Magazine for California and Washington red wines. The task is to predict the retail price y of a wine based on features such as its age in years, production region, grape variety, type, number of produced cases (in thousands), and magazine score. We report the best performing rule set in 5-fold cross-validation. In concordance with [4], we find the default rule:

$$\top \Rightarrow y = -189.69 - 0.0002 \times \mathit{cases} + 2.39 \times \mathit{score} + 5.08 \times \mathit{age}.$$

As pointed out in [4], non-varietal wines, i.e., those produced from several grape varieties, tend to have a higher price, and this price is more sensitive to score and age. HIPAR$_f$ ($\theta = 0.05$) found 69 rules including the rule supporting this finding (support 7%):

$$\mathit{variety} = \text{``}\mathit{non\text{-}varietal}\text{''} \Rightarrow y = -349.78 - 0.003 \times \mathit{cases} + 4.20 \times \mathit{score} + 7.97 \times \mathit{age}$$

HIPAR could also detect the so called *Giffen effect*, observed when, contrary to common sense, the price-demand curve exhibits an upward slope. We observe this phenomenon by running HIPAR$_f$ on the *giffen* dataset that contains records of the consumption habits of households in the Chinese province of Hunan at different stages of the implementation of a subsidy on staple foodstuffs. The target variable y is the percent change in household consumption of rice, which is predicted via other attributes such as the change in price (*cp*), the household size (*hs*), the income per capita (*ipc*), the calorie consumption per capita (*ccpc*), the share of calories coming from (a) fats (*shf*), and (b) staple foodstuffs (*shs*, *shs2*), among other indicators. HIPAR finds the default rule:

$$\top \Rightarrow y = 37.27 - \mathbf{0.06} \times \mathit{cp} + 1.52 \times \mathit{hs} + 0.0004 \times \mathit{ipc} + 0.003 \times \mathit{ccpc}$$
$$-146.28 \times \mathit{shf} + 54.13 \times \mathit{shs} - 156.78 \times \mathit{shs2}$$

The negative sign of the coefficient for *cp* suggests no Giffen effect at the global level. As stated in [4], when the subsidy was removed (characterized by the condition $\mathit{round} = 3$), the Giffen effect was also not observed in affluent and very poor households. It rather concerned those moderately poor who, despite the surge in the price of rice, increased their consumption at the expense of other sources of calories. Such households can be characterized by intervals in the income and calories per capita (*ipc*, *ccpc*), or by their share of calories from staple foodstuffs (*sh*, *sh2*). This is confirmed by the hybrid rule (support 4%):

$$\mathit{round} = 3 \wedge \mathit{cpc} \in [1898, 2480) \wedge \mathit{sh2} \in [0.7093, \infty) \Rightarrow y = 42.88$$
$$+\mathbf{1.17} \times \mathit{cp} + 1.08 \times \mathit{hs} - 0.005 \times \mathit{ipc} + 0.018 \times \mathit{ccpc}$$
$$-7.42 \times \mathit{shf} - 3.08 \times \mathit{shs} - 114.21 \times \mathit{shs2}.$$

The positive coefficient associated to *cp* shows the Giffen effect for households with moderate calories per capita (calories share from staples higher than 0.7093). The latter condition aligns with the findings of [4] that suggests that households with higher values for *sh* were more prone to this phenomenon.

6 Conclusions and Outlook

We have presented HiPaR, a novel pattern-aided regression method. HiPaR mines compact sets of accurate hybrid rules thanks to (1) a hierarchical exploration of the search space of data regions, and (2) a selection strategy that optimizes for small sets of rules with joint low prediction error and good coverage. HiPaR mines fewer rules than state-of-the-art methods at comparable performance. As future work, we envision to extend the rule language bias to allow for negated conditions as in RT and MT, parallelize the enumeration phase, and integrate further criteria in the search, e.g., p-values for linear coefficients. A natural follow-up is to mine hybrid classification rules. HiPaR's source code and experimental data are available at https://gitlab.inria.fr/lgalarra/hipar.

References

1. HiPaR: hierarchical pattern-aided regression. Technical report. https://arxiv.org/abs/2102.12370
2. Breiman, L.: Random forests. Machine Learn. **45**(1), 5–32 (2001)
3. Dong, G., Taslimitehrani, V.: Pattern-aided regression modeling and prediction model analysis. IEEE Trans. Knowl. Data Eng. **27**(9), 2452–2465 (2015)
4. Duivesteijn, W., Feelders, A., Knobbe, A.: Different slopes for different folks: mining for exceptional regression models with Cook's distance. In: ACM SIGKDD (2012)
5. Duivesteijn, W., Feelders, A.J., Knobbe, A.: Exceptional model mining. Data Min. Knowl. Disc. **30**(1), 47–98 (2015). https://doi.org/10.1007/s10618-015-0403-4
6. Fayyad, U.M., Irani, K.B.: Multi-interval discretization of continuous-valued attributes for classification learning. In: IJCAI (1993)
7. Friedman, J.H.: Greedy function approximation: a gradient boosting machine. Ann. Stat. **29**(5), 1189–1232 (2001)
8. Friedman, J.H., Popescu, B.E.: Predictive learning via rule ensembles. Ann. Appl. Stat. **2**(3), 916–954 (2008)
9. Grosskreutz, H., Rüping, S.: On subgroup discovery in numerical domains. Data Min. Knowl. Disc. **19**(2), 210–226 (2009)
10. Herrera, F., Carmona, C.J., González, P., del Jesus, M.J.: An overview on subgroup discovery: foundations and applications. Knowl. Inf. Syst. **29**(3), 495–525 (2011)
11. Kramer, S.: Structural regression trees. In: AAAI (1996)
12. Malerba, D., Esposito, F., Ceci, M., Appice, A.: Top-down induction of model trees with regression and splitting nodes. IEEE Trans. Pattern Anal. Mach. Intell. **26**(5), 612–625 (2004)
13. McGee, V.E., Carleton, W.T.: Piecewise regression. J. Am. Stat. Assoc. **65**(331), 1109–1124 (1970)

14. Morishita, S., Sese, J.: Traversing itemset lattices with statistical metric pruning. In: SIGMOD/PODS (2000)
15. Uno, T., Asai, T., Uchida, Y., Arimura, H.: LCM: an efficient algorithm for enumerating frequent closed item sets. In: FIMI (2003)
16. Wang, Y., Witten, I.H.: Inducing model trees for continuous classes. In: ECML Poster Papers (1997)

Improved Topology Extraction Using Discriminative Parameter Mining of Logs

Atri Mandal[1(✉)], Saranya Gupta[2], Shivali Agarwal[1], and Prateeti Mohapatra[1]

[1] IBM Research, Bangalore, India
{atri.mandal,shivaaga,pramoh01}@in.ibm.com
[2] CUHK, Hong Kong, Hong Kong SAR

Abstract. Analytics on log data from various sources like application, middleware and infrastructure plays a very important role in troubleshooting of distributed applications. The existing tools for log analytics work by mining log templates and template sequences. The template sequences are then used to derive the application control flow or topology. In this work, we show how the use of parameters in logs enables discovery of a more accurate application topology, thereby aiding troubleshooting. An accurate application topology information helps in better correlation of logs at runtime, enabling troubleshooting tasks like anomaly detection, fault localization and root cause analysis to be more accurate.

To this end, we propose a novel log template mining approach which uses parameter mining combined with fuzzy clustering on historical runtime logs to mine better quality templates. We also leverage parameter flows between log templates using a novel *discriminative parameter mining* approach for better topology extraction. In our method we do not assume any source code instrumentation or application specific assumptions like presence of transaction identifiers. We demonstrate the effectiveness of our approach in mining templates and application topology using real world as well as simulated data.

Keywords: Observability · Application topology · Parameter mining · Root cause analysis

1 Introduction

IT Operational Analytics (ITOA) aims at providing insights into the operational health of IT components deployed in a cloud datacenter. A very challenging problem in this domain entails troubleshooting a distributed application deployed on the cloud by analyzing large volumes of machine-data like logs, collected from the different layers of the software stack like application, middleware and infrastructure. The execution traces obtained from the logs can be used to find out the precise location and root cause of a fault in a service. A lot of research is happening in the area of automating tasks like topology inference, anomaly detection, fault localization using log analytics.

K. Karlapalem et al. (Eds.): PAKDD 2021, LNAI 12712, pp. 333–345, 2021.
https://doi.org/10.1007/978-3-030-75762-5_27

While a lot of ground has been covered by the wide range of log analytics tools and techniques [1,2] proposed in literature, there are still some key areas that are open for exploration. One of the open questions is the role of parameters present in the log data for getting more informative templates and leveraging them for troubleshooting and diagnostics tasks. Another open challenge is how to keep application topology up-to-date automatically which is ever evolving due to agile development. In absence of a proper topology, the log correlations at runtime will not be accurate having a negative impact on downstream troubleshooting tasks like fault localization.

In this paper, we present a system for analyzing execution logs of microservices in cloud-native applications based on insights obtained from parameters embedded in the log lines. We do not depend on any kind of code instrumentation and assume that source code is not available to us. The main contributions of this paper are as follows:

1. **More accurate template and parameter extraction:** Existing state-of-the-art methods on template mining are not well equipped to handle log lines with a long chain of parameters as they rely solely on the textual similarity between log lines [3–6]. We improve on this technique by using a more parameter focused approach. Specifically, we use a novel *frequency histogram-based soft boundary detection* technique for better distinction between invariant and parameter words. We also employ a novel *fuzzy clustering* method for getting more fine-grained and well-formatted templates.
2. **Discriminative parameter mining for better application topology extraction:** In addition to extracting topology at inter-microservice level, our system also discovers the application control flow within individual microservices. To discover the application control flow, we use a method which not only relies on the temporal correlation of templates but also looks at specific parameter positions within the templates to infer the relation between them. Our system can mine the application traces in interleaved logs *without relying on transaction identifiers.*

We present a comprehensive study of the efficacy of our approach using four real applications and different types of simulated workloads. Our study shows that the proposed technique outperforms the state of the art in template mining with good margin on most datasets. Our topology extraction works specially well in presence of highly interleaved logs generated due to high number of concurrent microservices in an application. The rest of the paper is organized as follows: Sect. 2 describes the related work. Section 3 talks about the implementation of the system and the different stages of the pipeline. In Sect. 4 we present our experimental results and we conclude in Sect. 5.

2 Related Work

In recent years, with the growing popularity of cloud and Kubernetes platforms, there has a been a lot of interest in building effective log monitoring and observability systems for services deployed as cloud-native applications on Kubernetes.

For example, Kiali [7] provides a service mesh for cloud-native applications which requires istio proxy [8] sidecar injection into the application. OpenTelemetry [9] based methods can be used to collect traces, metrics and logs but they require code instrumentation. APM tools like AppDynamics [10] and DynaTrace [11] can dynamically create topology maps to visualize the performance of applications. These tools, however, provide visibility only at the service level. To get insights at a finer granualarity level one has to manually analyze the logs.

Many recent works deal with the control flow mining of application from logs [12–14]. However, most of these works assume the presence of global identifiers such as transaction ids in the log data for handling interleaved logs. This may not be practical in client scenarios as such knowledge might be missing or difficult to tag in complex systems containing a lot of microservices. The work in [2] is the closest in this area which mines the execution control flow from interleaved logs using temporal correlation. This work rely heavily on templates being accurately mined. [2] uses a novel method for inferring templates that relies on a combination of text and temporal vicinity signals but ignores parameters completely. Existing implementations of template mining approaches rely primarily on textual similarity based clustering and regular expression based matching for identifying invariants and parameters [3–6]. Such a rigid approach can result in inaccuracies in the log parsing and clustering when log lines include long parameter chains (e.g. parameters resulting from object dumps), misclassification of terms (such as when a term appears as both an invariant and a parameter), etc.

3 System Overview

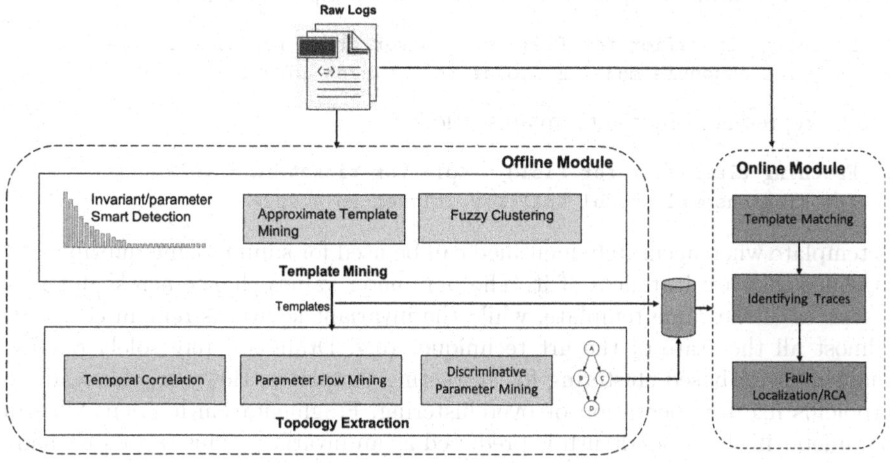

Fig. 1. System overview

We define application topology of a distributed cloud-native application as a graph $G = (V, E)$, where $v \in V$ are nodes denoted by log templates, and $e \in E$ are edges that capture the control flow between the templates s.t. templates belong to microservices in an application. The graph in Fig. 3 shows a sample topology spanning two microservices. We now describe the proposed system that mines the complete application topology including calls between the constituent microservices. Figure 1 outlines the different stages of our pipeline. The focus of our paper is the offline module which does template and topology mining. For sake of completeness, we show how the output of the offline module feeds into the runtime (online) module which does trace analysis, fault localization and other troubleshooting tasks. Our method of mining topology builds upon the OASIS approach described in [2]. OASIS uses only temporal correlations in log templates to construct the topology, and does not capture parameters. As a result, OASIS finds it hard to disambiguate edges that have a branching probability below a threshold (say 5%). Such edges cannot be distinguished from noise caused by random interleaving of logs. Another shortcoming of using only temporal correlations is finding spurious edges caused by frequent interleavings of unrelated templates. Our approach addresses the shortcomings of OASIS by mining parameters along with the templates. So the final topology mined by our approach not only consists of edges but also the important parameters that flow across the edges. We now provide details of the key components in the proposed system.

3.1 Template Mining

A template is an abstraction of a print statement in source code, which manifests itself in raw logs with different embedded parameter values in different executions. We represent templates as a set of invariant keywords and parameters (denoted by parameter placeholder $\langle p_i \rangle$). For example the following log lines:

```
1. Fetching classifier for field = resolver-group for ticket-id = 100
2. Invoking ensemble model E-0x67cf for  ticket-id = 100
```

can be represented by the templates below:

```
1. Fetching classifier for field = <p1> for ticket-id = <p2>
2. Invoking ensemble model <p1> for  ticket-id = <p2>
```

A template when accurately identified can be used for summarizing multiple raw log lines that are instances of it. The parameter values change across different instances of the same template, while the invariant keywords remain constant. Almost all the state of the art techniques (e.g. Drain [6]) rely solely on textual similarity-based clustering for extracting templates. However, this leads to problems like fragmentation or overclustering. Fragmentation happens when a parameter in the groundtruth is predicted as an invariant. This results in many mined templates mapping to a single groundtruth template. The reverse happens when an invariant in groundtruth is predicted as a parameter [2]. To overcome these limitations we use a novel template mining method which focuses on better extraction of parameters.

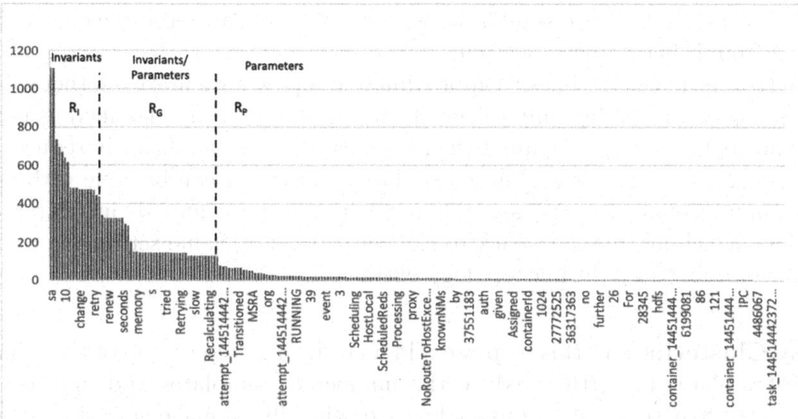

Fig. 2. Frequency histogram based soft boundary detection

Approximate Template Mining: As our intention is to match the print pattern as closely as possible we make our best attempt at segregating invariants and parameters from the log line upfront and create approximate templates. This enables us to get fairly accurate templates in just one linear pass over the log file thereby vastly reducing the number of log lines under consideration (in some cases by more than 10000x) for the subsequent clustering step which is an $O(n^2)$ operation. We make use of two key insights in differentiating between invariants and parameters. First, we notice that by its presence in every single manifestation of a template, the invariant words have much higher frequencies than parameter words. Second, there is an inherent difference in the structure of an invariant (primarily English dictionary words, technical domain specific words, etc.) and a parameter (IP addresses, server names, port numbers, usernames, user ids, timestamps, etc.). Using these two key insights we differentiate between an invariant and parameter with high confidence.

Frequency Histogram Based Approach with Soft Invariant/Parameter Boundary Detection: We first compute a frequency histogram of all tokens in the log file. Then we use two separate cutoff frequencies, one near the high-frequency range and another near the low-frequency range as shown in the Fig. 2. The cutoffs are determined empirically. This divides the histogram into three distinct regions viz. R_I containing mostly invariants, R_P containing mostly parameters and a grey area R_G which contains both invariants and parameter. To determine the type of each token in R_G, we use a word embedding technique viz. FastText [15] as it is more suited to support out-of-vocabulary (OOV) words. Specifically, we train FastText embeddings on the log datasets. Using these trained embeddings, we obtain the k nearest neighbors (k is odd) for each token in R_G. The nearest neighbors are selected from labeled tokens in R_I and R_P. We select the token type using majority voting. Note that we need not be restricted to labeled tokens from the same dataset - as the features of invariant words and parameters are similar across datasets. We can reuse the embeddings trained from other datasets as well.

To transform the log lines to approximate templates we make another scan of the log records, retaining only tokens in the log line that are classified as invariants and replacing the remaining with a wildcard. This significantly reduces the clustering space so the next step of text-based clustering can be done with much less computational overhead as explained earlier. The wildcards are parameter positions which are mapped back to the corresponding value set during parameter flow mining step during topology extraction.

Fuzzy Clustering: In this step, we add a clustering step on top of the approximate templates to further reduce the number of templates and improve the quality. We first sort the templates lexicographically to make sure similar template patterns (which may eventually fall in the same cluster) come close to one another. After this, we make a single pass over the approximate templates and only compare adjacent templates (T_i and T_{i+1}) in the sorted order. Templates that have high similarity score keep getting clustered till a steep decline is observed between a pair. Using this method we can infer the cluster membership in a single linear scan and avoid a pairwise similarity computation which has $O(n^2)$ complexity. For templates T_i and T_{i+1}, we, first, create an array of tokens R and S to represent each template respectively. We then compute the similarity between the token sets R and S using a variant of the Jaro-Winkler [16] distance metric. The reason for using this metric is to give higher preference to similarity in the beginning of the strings as the invariant words are more likely to appear in the beginning. Given two sets R and S, the similarity score of templates T_i and T_{i+1} is given by the following similarity equation:

$$Score(R, S) = \frac{FuzzyJW(R, S)}{|R| + |S| - FuzzyJW(R, S)} \tag{1}$$

Here, *FuzzyJW* represents the similarity measure that is used to calculate the similarity score (normalized to lie in range 0–1) between token sets R and S. The score indicates to what extent T_i matches T_{i+1}.

3.2 Topology Extraction

To mine the application topology, we need to mine relationships between the templates mined in the previous step. Figure 3 shows an example of the ground truth application topology (in black colored arrows and nodes) for a proprietary machine-learning based application used for automatic classification of helpdesk email tickets [17], henceforth referred to as MLApp. This application is known to have some rare edges like $T2 \rightarrow T3$ and $T12 \rightarrow T14$ which manifest less than 5% times in logs. Our goal is to extract the topology faithfully from logs as shown in the snapshot on the right side. As we had mentioned before, we build upon OASIS [2] which has two main shortcomings of i) missing out on low-frequency but relevant edges and ii) mining spurious edges. Figure 3 shows the case of

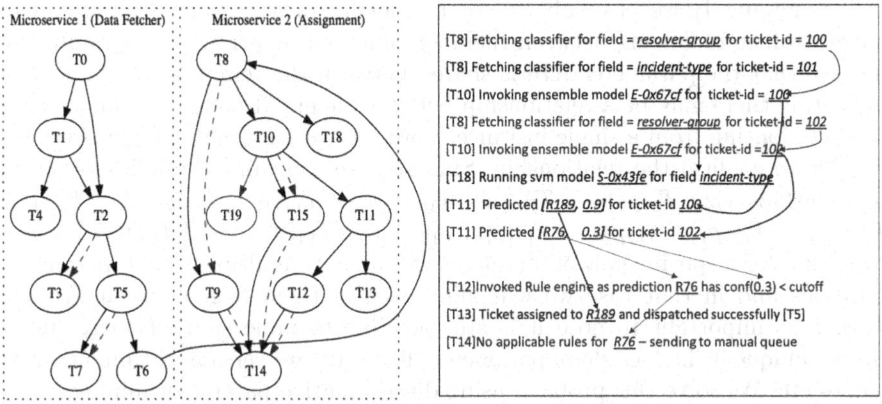

Fig. 3. Figure explaining parameter flows in application (Color figure online)

missing edges by OASIS on the ground truth topology through red-dotted lines; the red solid line and the red colored node shows the spurious edge suggested by OASIS. We address both these shortcoming using the concept of *parameter flows* and *discriminative parameters* as explained in the section below. The first step of temporal correlations is pretty standard and we leverage OASIS for it without any threshold based filtering.

Parameter Flow Mining. Now we come to one of the primary insights of this paper which uses information contained in parameters of log records to establish a connection between templates and mine robust edges for the application topology. We first convert each log record into a template event with template and parameter annotations. The wildcards, that were introduced during template mining, are replaced with the parameter value set. We also chunk the log record into time windows as before and mine shared parameter positions between each pair of templates appearing in the same time window. In mathematical terms, let $\mathrm{T}_a(\mathrm{p}_i)$ denote that p_i is the ith parameter position in templates T_a and $\mathrm{val}(\mathrm{p}_i)$ denote the parameter value set at parameter position i. Then, we mine relationships of the form:

$$\exists(T_a(p_i), T_b(p_j)), val(p_i) = val(p_j) \Longrightarrow T_a \rightarrow T_b \tag{2}$$

where $T_a \rightarrow T_b$ denotes an edge between the two templates if the precondition is satisfied. We shall use $T_a(p_i) \rightarrow T_b(p_j)$ when we need to convey additionally the parameter positions in the templates that lead to the edge. To understand how this works we can focus on the log snippet given in Fig. 3. There are 3 interleaved traces in this snippet as indicated by the different colours viz. ($T8 \rightarrow T18$), ($T8 \rightarrow T10 \rightarrow T11 \rightarrow T13$) and ($T8 \rightarrow T10 \rightarrow T11 \rightarrow$T12$\rightarrow T14$). We can also see that the log lines do not have clearly demarcated transaction-ids making it difficult to mine the template relations using simple techniques like process mining. However we observe that almost all of the sequences in the traces have parameters flowing between them as indicated by arrows. For example we observe that the parameter $R76$ is shared between the templates $T12$ and $T14$ indicating there may be a relationship between them. Although we cannot infer this relationship from a single instance, if we look at sufficient time bins we will be able to confirm the relationship. Similarly we can mine the following other relationships viz.: $T_8(p_2) \rightarrow T_{10}(p_2)$, $T_{10}(p_2) \rightarrow T_{11}(p_3)$, $T_{11}(p_1) \rightarrow T_{13}(p_1)$, $T_{11}(p_1) \rightarrow T_{12}(p_1)$, $T_{11}(p_2) \rightarrow T_{12}(p_2)$, $T_{12}(p_1) \rightarrow T_{14}(p_1)$ and $T_8(p_1) \rightarrow T_{18}(p_2)$. Note that for a specific pair of templates we can have multiple shared parameter positions and in that case we will mine each of them (e.g. between T_{11} and T_{12}). It is important to note it is still possible to mine spurious edges using this technique if the logs share parameters like software versions, common server names etc. We solve this problem using discriminative parameter mining.

Discriminative Parameter Mining. To refine the parameter flow space we use certain heuristics. Firstly, we eliminate those parameters whose value is same across a large number of templates in the same time window. Also, we remove those parameters which are not consistently shared across time windows and therefore show no temporal correlation. The parameters which are retained after the refinement process are the **discriminative parameters** which help in establishing relationships between a pair of templates. This is a key step which helps us confirm the sequence between template pairs even though it may not be apparent using temporal correlation. Using this technique we can mine rare edges like $T2 \rightarrow T3$ or $T12 \rightarrow T14$ (these edges have $< 5\%$ probability which is below the typical thresholds).

It is important to note that a template may have any number of parameters but not all of them may be discriminative. Also a template can have more than one discriminative parameter positions. For example in the case of $T8$ both parameter positions are discriminative viz. p_1(field) for the pair $(T8, T18)$ and p_2(ticket-id) for the pair $(T8, T10)$. *The edges obtained after applying all the above steps give the application topology.*

3.3 Online Module Usecase

Once the template relationships are established using parameter flows it is trivial to infer execution traces from logs. For example in Fig. 3 the relation $T_8(p_1) \rightarrow T_{18}(p_2)$ helps us identify the connection between the log trace indicated in black. It is important to note that we cannot connect these log lines simply using the application topology on the left without the parameter relationships. In the figure all the traces can be identified using the template relationships derived above using parameter information. The above example clearly shows the power of discriminative parameters in identifying individual execution traces in interleaved logs. This property can be utilised effectively to detect system anomalies as well as to localize software faults with high accuracy and reliability.

4 Experimental Results

To evaluate the overall effectiveness of the system, we tested with both real and synthetic datasets. Although our real datasets are sufficiently large for evaluation of templates, there is no ground truth available for topology. To obtain ground truth we implemented a custom simulator which can generate logs having characteristics of application logs of micro-services deployed on the cloud. The simulator uses a **Topology Generator** to construct a hypothetical micro-service control flow having all fundamental building blocks like forks, merges, detour paths etc. The simulator also uses a **Template Generator** to generate print patterns for each node in the topology graph. We also simulate parameter flows by replicating parameter values across topology edges based on configuration parameters. Finally the simulator uses a **Log Generator** to generate execution logs using the hypothetical topology and the templates generated in the previous steps. Interleaving is simulated by having multiple instances of the simulator run in parallel.

Along with the synthetic dataset, we used 4 real-world applications viz. Watson Assistant(WA), OpenStack [18], MLApp and Train Ticket [19]. For WA and MLApp we used real client logs from production. The OpenStack and TrainTicket applications were deployed as cloud-native applications in a Redhat Openshift cluster and load generators were used to generate logs. For WA only we use 2 separate client datasets referred to as WA-1 and WA-2 respectively. The ground truth was manually curated for each of the real-world applications.

4.1 Invariant/Parameter Identification from Logs

In the first experiment, we evaluate how well the invariants and parameters are identified. This is crucial for the proper functioning of the remaining stages of our algorithm. To evaluate this we use a simple metric for accuracy viz. we evaluate how many tokens (both invariants and parameters) in the groundtruth were correctly reported by the template miner. For example, in the following case, we will report an accuracy of 8/10 i.e. 0.8.

```
[Log line] NETWORK [conn4774] end connection 127.0.0.1:59924 (3 connections
now open)
[Mined template] NETWORK <P> end connection 127.0.0.1:<P> (<P> <P> now open)
[Ground truth] NETWORK <P> end connection <P>:<P> (<P> connections now open)
```

Our results clearly indicate that the state-of-the-art template mining techniques often confuse invariants and parameters and have a lot of scope for improvement. The performance is noticeably poor for WA and OpenStack datasets as both these datasets contain log lines with long chain of parameters.

4.2 Template Extraction

Table 2 shows the results of our approach in template mining. To calculate precision, we only consider template patterns which can be uniquely mapped to a singleton pattern in the ground truth.

Table 1. Invariant/parameter identification

Dataset	Log lines	OASIS	Drain (vanilla)	Fuzzy mining (our approach)
WA-1	1.1m	58.9	47.5	**93.5**
WA-2	1.8m	49.2	34.5	**84.2**
MLApp	200K	90.8	88.9	**95.6**
OpenStack	1m	26.5	32.0	**74.3**
Train-Ticket	1m	66.1	62.6	**94.9**
Synthetic	1m	93.5	91.4	**98.4**

The very poor precision for WA (by OASIS and Drain) can be explained by the presence of long parameter chains which results in heavy fragmentation. Surprisingly, on OpenStack which also has long parameters, OASIS performed best and outperformed our approach on precision. On deeper analysis, we found that OpenStack had a lot of very short templates constituting mainly of invariants while WA templates usually had a long sequence of invariants. Also, there is a poor segregation of invariants/parameters in OpenStack but all parameters come towards the end in WA which can be handled well by Fuzzy mining. These conditions led to fragmentation in Fuzzy mining and overclustering in OASIS which explains the precision and recall results. It is interesting to note the contrasting result in Table 1.

Table 2. Template extraction: P(precision), R(recall), GT(groundtruth)

Dataset	Log lines	GT templates	OASIS (P/R)	Drain (P/R)	Fuzzy (P/R)
WA-1	1.1m	148	7.1/67.9	22.6/89.2	**77.5/92.6**
WA-2	1.8m	173	5.1/68.2	15.9/90.8	**75.8/92.5**
MLApp	200K	43	88.1/86.0	69.5/95.3	**93.3/97.6**
OpenStack	1m	45	**91.7**/73.3	69.0/84.4	81.25/**86.7**
Train-Ticket	1m	24	45.5/62.5	54.5/75.0	**96.15/82.1**
Synthetic	1m	140	79.6/80.7	82.9/86.4	**95.2/98.6**

4.3 Application Topology Discovery

Table 3 shows the results of our approach in discovering application topology on the datasets for which ground truth was available. For our experiments, we kept the number of concurrent threads for each micro-service to 1 (t = 1), varied the number (n) of concurrently running microservices from 5 to 50. We also parameterized the size of each topology (s = 50) with a standard deviation ($\delta_s = 7$). The values of s and $\delta_s = 7$ were empirically chosen by observing the values in typical cloud-native deployments. The results were evaluated against 4 baseline approaches viz. sequence-mining (SM-vanilla), sequence mining with transitive reduction of edges (SM-tred), Process Mining (PM) and OASIS. It can be seen that for the synthetic dataset the precision decreases gradually with more concurrent micro-services with most baselines because spurious edges are mined. Temporal correlation (OASIS) works reasonably well but misses out some rare edges.

Our approach works very well for the MLApp application compared to traditional techniques as there are a lot of flow parameters. *For example for the section of the topology shown in Fig.* 3 *our approach was able to mine all the groundtruth edges accurately.* Other methods like OASIS ended up mining spurious edges ($T10 \rightarrow T19$) and also missed valid rare edges ($T2 \rightarrow T3$,$T5 \rightarrow T7$, $T10 \rightarrow T15$, $T15 \rightarrow T14$, $T9 \rightarrow T14$, $T12 \rightarrow T14$). The PM approach suffers from low recall as the transaction ids are not present in most datasets including MLApp.

Table 3. Topology extraction

Method name	Synthetic dataset (t = 1, s = 50, δ_s = 7)			MLApp
	n = 5	n = 10	n = 50	
SM-vanilla	6.5/87.7	3/89	1.6/88	7.5/71.9
SM-tred	49/37.4	45.1/37.3	37.3/35	55.1/59.0
PM	61.9/53	24/29.3	15/17	63.0/27.2
OASIS	98.0/92.7	93.2/82.3	88.6/77	85.3/77.8
Fuzzy mining	**100/98.5**	**97.5/97.5**	**93.2/96.0**	**95.6/90.3**

4.4 Observations

It can be seen that our approach performs well in challenging conditions like high concurrency and absence of transaction-ids. We observed during evaluation that the high precision of our approach leads to very efficient execution of online tasks like trace mining on fresh event logs. We also observed some weaknesses of our approach during evaluation. Firstly if there are no flow parameters in logs then our approach doesn't work and we'll fare no better than traditional techniques. Secondly, in some cases, our parameter position based approach does not work as the parameters are more complex and span multiple positions and have variable lengths. To summarize, our technique is very amenable to real world applications where logging standards are adhered and performs better than baselines most of the time.

5 Conclusion and Future Work

We proposed a novel log template and discriminative parameter mining approach to extract more accurate topology from historical logs of distributed application deployed in cloud environment. We outperformed the baseline techniques on most datasets. As part of future work, we would like to make parameter position identification more robust and use advanced machine learning to identify flows when pairs of neighboring templates do not share any discriminative parameters.

References

1. Gupta, M., Mandal, A., Dasgupta, G., Serebrenik, A.: Runtime monitoring in continuous deployment by differencing execution behavior model. In: Pahl, C., Vukovic, M., Yin, J., Yu, Q. (eds.) ICSOC 2018. LNCS, vol. 11236, pp. 812–827. Springer, Cham (2018). https://doi.org/10.1007/978-3-030-03596-9_58
2. Nandi, A., Mandal, A., Atreja, S., Dasgupta, G.B., Bhattacharya, S.: Anomaly detection using program control flow graph mining from execution logs. In: Proceedings of the 22nd ACM SIGKDD International Conference on Knowledge Discovery and Data Mining, pp. 215–224 (2016)
3. Vaarandi, R.: A data clustering algorithm for mining patterns from event logs. In: Proceedings of the 3rd IEEE Workshop on IP Operations Management (IPOM 2003) (IEEE Cat. No.03EX764), pp. 119–126 (2003)
4. Vaarandi, R., Pihelgas, M.: Logcluster - a data clustering and pattern mining algorithm for event logs. In: 2015 11th International Conference on Network and Service Management (CNSM), pp. 1–7 (2015)
5. Vaarandi, R.: Mining event logs with SLCT and loghound. In: NOMS 2008–2008 IEEE Network Operations and Management Symposium, pp. 1071–1074 (2008)
6. He, P., Zhu, J., Zheng, Z., Lyu, M.R.: Drain: An online log parsing approach with fixed depth tree. In 2017 IEEE International Conference on Web Services (ICWS), pp. 33–40 (2017)
7. Kiali. https://github.com/kiali/kiali
8. Istio service mesh. https://istio.io/
9. Opentelemetry - an observability framework for cloud-native software. https://opentelemetry.io/
10. Appdynamics. https://www.appdynamics.com/
11. Dynatrace. https://www.dynatrace.com/
12. Sigelman, B.H., et al.: Dapper, a large-scale distributed systems tracing infrastructure. Technical report, Google Inc, (2010)
13. Barham, P., Isaacs, R., Mortier, R., Narayanan, D.: Magpie: online modelling and performance-aware systems, pp. 85–90 (2003)
14. Fonseca, R., Porter, G., Katz, R.H., Shenker, S., Stoica, I.: X-trace: A pervasive network tracing framework. In: NSDI 2007, USA, p. 20. USENIX Association (2007)
15. Bojanowski, P., Grave, E., Joulin, A., Mikolov, T.: Enriching word vectors with subword information. arXiv preprint arXiv:1607.04606 (2016)
16. Winkler, W.E.: Overview of record linkage and current research directions. Technical report, BUREAU OF THE CENSUS (2006)

17. Mandal, A., et al.: Automated dispatch of helpdesk email tickets: Pushing the limits with AI. In: The Thirty-Third AAAI Conference on Artificial Intelligence, AAAI 2019, Honolulu, Hawaii, USA, 27 January–1 February 2019, pp. 9381–9388. AAAI Press (2019)
18. Openstack: Open-source software for creating public and private clouds
19. Zhou, X., et al.: Latent error prediction and fault localization for microservice applications by learning from system trace logs. In: Dumas, M., Pfahl, S., Apel, S., Russo, A. (eds.) Proceedings of the ACM Joint Meeting on European Software Engineering Conference and Symposium on the Foundations of Software Engineering, ESEC/SIGSOFT FSE 2019, Tallinn, Estonia, 26–30 August 2019, pp. 683–694. ACM (2019)

Back to Prior Knowledge: Joint Event Causality Extraction via Convolutional Semantic Infusion

Zijian Wang, Hao Wang(✉), Xiangfeng Luo(✉), and Jianqi Gao

School of Computer Engineering and Science, Shanghai University, Shanghai 200444, China
{zijianwang,wang-hao,luoxf,gjqss}@shu.edu.cn

Abstract. Joint event and causality extraction is a challenging yet essential task in information retrieval and data mining. Recently, pre-trained language models (e.g., BERT) yield state-of-the-art results and dominate in a variety of NLP tasks. However, these models are incapable of imposing external knowledge in domain-specific extraction. Considering the prior knowledge of frequent n-grams that represent cause/effect events may benefit both event and causality extraction, in this paper, we propose convolutional knowledge infusion for frequent n-grams with different windows of length within a joint extraction framework. Knowledge infusion during convolutional filter initialization does not only help the model capture both intra-event (i.e., features in an event cluster) and inter-event (i.e., associations across event clusters) features but also boost training convergence. Experimental results on the benchmark datasets show that our model significantly outperforms the strong BERT+CSNN baseline.

Keywords: Causality extraction · Prior knowledge · Semantic infusion

1 Introduction

Joint event and causality extraction from natural language text is a challenging task in knowledge discovery [1], discourse understanding [12], and machine comprehension [19]. Formally, joint event causality extraction is defined as a procedure of extracting an event triplet consisting of a cause event $e1$, an effect event $e2$ with the underlying causality relationship. Due to the complexity and ambiguity of natural languages, event causality extraction remains a hard problem in case no high-quality training dataset is not available like the financial domain.

Conventional methods commonly treat event causality extraction as a two-phase process in a pipeline manner, divided into event mining and relation inference [22]. However, the pipeline approach suffers from error propagation. Incorporating the current neural model with the prior knowledge in existing domain-specific knowledge bases is complicated (Fig. 1).

Recently, neural-based extraction methods have become a majority of related tasks in NLP, relation classification [18], relation extraction [14] and sequence

K. Karlapalem et al. (Eds.): PAKDD 2021, LNAI 12712, pp. 346–357, 2021.
https://doi.org/10.1007/978-3-030-75762-5_28

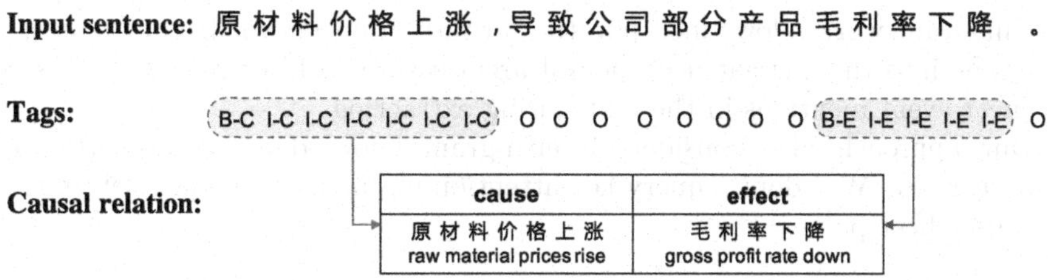

Fig. 1. An example of joint event causality extraction from raw financial text. It extracts cause and effect events with the underlying cause-effect relation simultaneously.

tagging [10]. Among these methods, pre-trained language models [4] (e.g., BERT) dominate the state-of-the-art results on a wide range of NLP tasks. BERT provides a masked language model (MLM) supporting fine-tuning to achieve better performance on a new dataset while reducing serious feature engineering efforts. However, there are still drawbacks when applying such pre-trained language models for event causality extraction as follows:

- The small size of available datasets is the main bottleneck of fine-tuning BERT-based models to satisfactory performance.
- Even if BERT trained on a large-scale corpus contains commonsense knowledge, it may not be sufficient for specific-domain such as financial area.
- Domain-specific frequency analysis of n-grams should be essential prior knowledge to recognize events. In contrast, these important hints have not been fully emphasized in the current neural architecture.

To tackle these issues, in this paper, we propose a novel joint method for event causality extraction. We first pass the input text through the BERT encoding layer to generate token representations. Secondly, to infuse the knowledge like frequent cause/effect n-grams on different scales, we utilize multiple convolutional filters simultaneously (i.e., infusing intra-event knowledge). The weights of these filters are manually initialized with the centroid vectors of the cause/effect n-gram clusters. After that, we link the cause and effect events using the key-query attention mechanism to alleviate incorrect cause-effect pair candidates (i.e., infusing inter-event knowledge). Finally, we predict target labels given the contextual representation by combing bidirectional long short-term memory (LSTM) with the conditional random field (CRF). Empirical results show that our model fusing advantages of both intra- and inter-event knowledge significantly improve event causality extraction and obtain state-of-the-art performance.

The contributions of this paper can be summarized as follows:

1. We propose a novel joint framework of event causality extraction based on recent advances of the pre-trained language models, taking account of frequent domain-specific and event-relevant n-grams given statistic analysis.

2. This framework allows incorporating intra-n-gram knowledge of frequent n-grams into the current deep neural architecture to filter potential cause or effect event mentions in the text during extracting.
3. Our approach also considers inter-n-gram knowledge of cause-effect co-occurrence. We adopt a query-key attention mechanism to pairwisely extract cause-effect pairs.

2 Related Work

The methods for event or causality extraction in the literature fall into three categories: rule-based, machine learning, and neural network. The rule-based methods employ linguistic resources or NLP toolkits to perform pattern matching. These methods often have low adaptability across domains and require extensive in-domain knowledge to deal with the domain generalisation problem. The second category methods are mainly based on machine learning techniques, requiring considerable human effort and additional time cost in feature engineering. These methods heavily rely on manual selection of textual feature sets. The third category methods are depending on the neural network. In this section, we survey those methods and figure out the problems existing in those methods.

2.1 Pattern Matching

Numerous rule-based methods have been dedicated to event causality mining. Early works predominantly perform syntactic pattern matching, where the patterns or templates are handcrafted for the specific-domain texts. For instance, Grishman et al. [6] perform syntactic and semantic analysis to extract the temporal and causal relation networks from texts. Kontos et al. [11] match expository text with structural patterns to detect causal events. Girju et al. [5] validate acquired patterns in a semi-supervised way. They check whether to express a causal relationship based on constraints of nouns and verbs. As a result, those methods cannot generalize to a variety of domains.

2.2 Machine Learning

The paradigm of automatic causal extraction dates back to machine learning techniques using trigger words combined with decision trees [17] to extract causal relations. Sorgente et al. [19] first extract the candidates of causal event pairs with pre-defined templates and then use a Bayesian classifier to filter non-causal pairs. Zhao et al. [22] compute the similarity of syntactic dependency structures to integrate causal connectives. Indeed, these methods suffer from data sparsity and require professional annotators.

2.3 Neural Network

Due to the powerful deep neural representations, neural networks can effectively extract implicit causal relations. In recent years, the adoption of deep learning techniques for causality extraction has become a popular choice for researchers. Methods of event extraction can be roughly divided into the template argument filling approach or sequence labelling approach. For example, Chen et al. [3] propose to process extraction as a serial execution of relation extraction and event extraction. This method splits the task into two phases. They first extract the event elements, then put them into a pre-defined template consisting of event elements. By sharing features during extraction, it effectively avoids missing the vital information.

Other models obey a sequential labelling manner. Fu et al. [9] propose to treat the causality extraction problem as a sequential labelling problem. Martinez et al. [16] employ an LSTM model to make contextual reasoning for predicting event relation. Jin et al. [10] propose a cascaded network model capturing local phrasal features as well as non-local dependencies to help cause and effect extraction. Despite their success, those methods ignore the fact of domain-specific extraction. On the one hand, an n-gram representing cause or effect event should frequently appear in the text. On the other hand, the events involving causality have a higher probability of co-occurrence.

3 Our Model

Figure 2 shows the overall architecture of the model. It is composed of four modules: 1) We first pass the input sentence through the BERT encoding layer and the alignment layer successively to output the BERT representation for each token. 2) A convolutional knowledge infusion layer is created to capture useful n-gram patterns to focus on domain-relevant n-grams at the beginning of the training phase. 3) We characterize inter-associations among cause and effect using a key-query attention mechanism. 4) Temporal dependencies are established utilising the LSTM combined with CRF layers, which significantly improves the performance of causality extraction.

3.1 BERT Encoder

The BERT is one of the pre-trained language models, which contains multi-layers of the bidirectional transformer, designed to jointly condition on both left and right context. BERT supports fine-tuning for a wide range of tasks without substantial task-specific architecture modifications. We omit the exhaustive background description of the architecture of BERT in this paper.

In our model, we use the BERT encoder to model sentences that contain the event mentions via computing a context-aware representation for each token. We take the packed sentences $[CLS, S, SEP]$ as the input. "[CLS]" is inserted as the first token of each input sequence. "[SEP]" denotes the end of the sentence. For each token s_i in S, the BERT input can be represented as:

$$s_i = [s_i^{tok} \oplus s_i^{pos} \oplus s_i^{seg}], \tag{1}$$

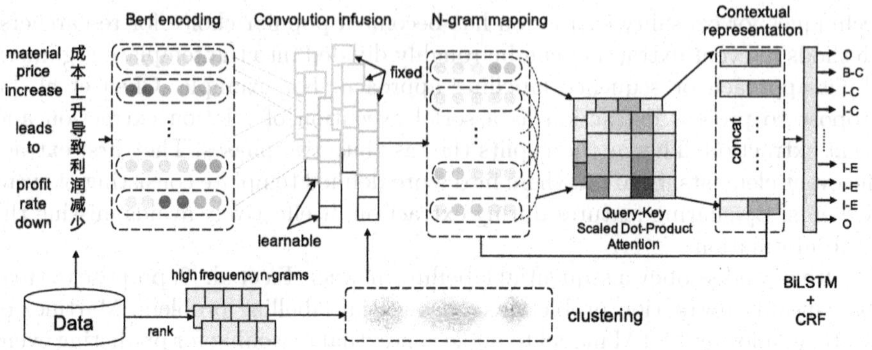

Fig. 2. Illustration of the model architecture. Intra- and inter-event feature are extracted using convolutional knowledge infusion and key-query attention mechanism. This model allows to extract cause and effect events associated with a event causal relation at the same time.

where s_i^{tok}, s_i^{pos}, and s_i^{seg} represent token, position, and segment embeddings for i, respectively. In this regard, a sentence is expressed as a vector $H \in \mathbb{R}^{l \times e}$, where l is the length of sentence and e is the size of embedding dimension. h_i is a vector standing for word embedding of the i-th word in the sentence.

$$h_N^i = BERT([CLS, s_1, \ldots, s_i, \ldots, s_l, SEP]), \quad (2)$$

where h_N^i represents hidden states of the last layer generated by BERT. We use them as the context-aware representation for each token.

3.2 Convolutional Semantic Infusion

In natural language processing, the convolutional operation can be understood as a process in which a sliding window continuously captures semantic features in a sentence of specific length. The convolution operation will extract semantic features similar to the convolution filter (i.e., the vectors similar to the convolution filter should share a higher weight). In our model, the convolution layer aims to enhance event-relevant features by embedding causal patterns into feature maps. Considering that word embedding can be initialized using pre-trained word vectors, it also should be possible to initialize the convolutional filter (i.e., kernel) with the vector related to frequent event n-grams. Therefore, inspired by [13], we launch unsupervised clustering to collect n-grams mentioning the similar cause or effect events in the text into the same clusters. Thus, these clusters can elegantly represent cause/effect semantic. Then, we compute the centroid vector of the cluster given all cause/effect event n-grams in the cluster and initialize convolution filters with these vectors. We only fix a part of weight in the filter using the cluster semantic vector, which allows our model to learn more features by itself during training. The details are given below:

Ngram Collection: Sentences can be segmented into particular chunks according to the size of the window in the convolution operation, e.g., "我喜欢苹果 (I like apples)" are split into four bigrams "我喜,喜欢,欢苹,苹果" when n = 2, step = 1). This method takes advantage over standard word representations to obtain ampler semantic information, which motivates us to recognize the event phrase boundary in this example.

The texts in a different domain often have different length for event n-grams, which also dominates convolutional window size. At the same time, the n-gram length is a kind of vital prior knowledge in event extraction. However, it is often ignored by conventional character- or word-based methods, resulting in errors of event boundary recognition. Therefore, we count the event length in the dataset. We find the number of the events in the most frequent length accounted for 27% among all examples in the Financial dataset. Motivated by this phenomenon, we associate the n-gram length feature as well as n-gram clusters by applying the convolution filters with various window sizes.

Intuitively, "profit drop" should be more important than the location n-gram like "New York" in most cases in the financial domain. Thus, we distill the relevant n-grams in a cluster with the centroid vector. In practical, we collect n-grams only from the training data. For its simplicity and effectiveness, we sort the n-grams using Naïve Bayes to obtain scores. The ranking score r of the n-gram w is calculated as follows:

$$r = \frac{(p_c^w + b) / \| p_c \|_1}{(p_e^w + b) / \| p_e \|_1}, \tag{3}$$

where c denotes cause event and e indicates effect event. p_c^w is the number of sentences that contains n-grams w in class c. $\|p_c\|_1$ is the number of n-grams related to cause c, p_e^w is the number of sentences that contain n-grams w related to e, $\|p_e\|_1$ is the number of n-grams in e, and b is a smoothing parameter. We select the top $n\%$ n-gram vectors by scoring n-grams using Formula 3.

Filter Initialization: First of all, we embed each n-gram into a vector representation using BERT. Since filters in CNNs are insufficient on the amount to account all selected n-grams, we only extract the cluster vectors to represent the generalized cause/effect features. We perform k-means clustering to obtain clusters. Here, the cluster vector is defined as the centroid vector of the cluster. Considering that non-causal n-grams exist in the sentence, we purposely leave some blank. We do not fully fill the convolution filters with centroid vectors and randomly initialize the rest weights of the filters.

$$c_i^n = f(W \cdot h_{i:i+n-1} + b) \tag{4}$$

Namely, W is the weight of initialized filter, b is a bias term and f is a nonlinear function such as sigmoid, ReLU. As a single cluster is often not enough to describe the overall information, we need to cluster with different length scales. To capture the features of causal events at different scales, we use parallel convolution operation with varying windows. For example, convolution window sizes of n_0, n_1, n_2 are chosen when there are three convolution layers. For example, when length equals to 4, the model tends to extract "股价下跌" (share prices fell) and "销量减少" (sales reduced) in the Financial dataset. The convolution output can be computed as the concatenation of sub-spaces:

$$c_i = [c_i^{n_0} \oplus c_i^{n_1} \oplus c_i^{n_2}] \tag{5}$$

3.3 Query-Key Attention

Query-key attention is a special attention mechanism according to a sequence to compute its representation. It has been successfully applied in many NLP tasks, such as machine translation and language understanding. In our model, the query-key attention mechanism is employed to mine inter associations between cause and effect events. After the multi-scaled convolution operation, we obtain the feature mapping for each token. Instead of max-pooling, we use the multi-head attention mechanism to check whether exist a cause-effect relation. Following Vaswani et al. [21], we split the encoded representations of the sequence into homogeneous sub-vectors, called heads. The input consists of queries $Q \in \mathbb{R}^{t \times d}$, keys $K \in \mathbb{R}^{t \times d}$ and values $V \in \mathbb{R}^{t \times d}$, while d is the dimension size. The mathematical formulation is shown below:

$$Attention\,(Q, K, V) = softmax\left(\frac{QK^T}{\sqrt{d}}\right)V, \tag{6}$$

$$H_i = Attention\left(QW_i^Q, KW_i^K, KW_i^V\right), \tag{7}$$

$$H_{\text{head}} = ([H_1 \oplus H_2 \oplus \cdots \oplus H_h])\,W, \tag{8}$$

where h is the number of heads. Query, key, and value matrices dimension is d/h. We perform the attention in parallel and concatenate the output values of h heads. The parameter matrices of i-th linear projections $W_i^Q \in \mathbb{R}^{n \times (\frac{d}{h})}$, $W_i^K \in \mathbb{R}^{n \times (\frac{d}{h})}$, $W_i^V \in \mathbb{R}^{n \times (\frac{d}{h})}$. In addition, the outputs of CNN and attention structure are cascaded to output token-wise contextual representations.

3.4 BiLSTM+CRF

Long Short Term Memory (LSTM) is a particular Recurrent Neural Networks (RNN) that overcomes the vanishing and exploding gradient problems of traditional RNN models. Through the specially designed gate structure of LSTM, the model can optionally keep context information. Considering that the text field has obvious temporal characteristics, we use BiLSTM to model the temporal dependencies of tokens. Conditional random field (CRF) can obtain tags in the global optimal chain given input sequence, taking the correlation tags between neighbour tokens into consideration. Therefore, for the sentence $S = \{s_1, s_2, s_3, ..., s_n\}$ along with a path of tags sequence $y = \{y_1, y_2, y_3, ..., y_n\}$, CRF scores the outputs using the following formula:

$$score(S, y) = \sum_{i=1}^{n+1} A_{y_{i-1}, y_i} + \sum_{i=1}^{n} P_{i, y_i} \tag{9}$$

3.5 Training Objective and Loss Function

The goal of our training is to minimize the loss function. The mathematical formulation is defined as below:

$$E = log \sum_{y \in Y} exp^{s(y)} - score(s, y), \tag{10}$$

where Y is the set of all possible tagging sequences for an input sentence.

4 Experiment

4.1 Datasets

We conduct our experiments on three datasets. The first dataset is the Chinese Emergency Corpus (CEC). CEC is an event ontology corpus[1] publicly available. It consists of six event categories: outbreak, earthquake, fire, traffic accident, terrorist attack, and food poisoning. We extract 1,026 sentences mention event and causality from this dataset. Since there are few publicly available datasets, to make the fair comparison, we also conduct experiments on two in-house datasets, one called "Financial", which is built based on Chinese Web Encyclopedias, such as Jinrongjie[2] and Hexun[3]. This dataset contains a large number of financial articles, including cause-effect mentions. The Financial dataset is divided into a training set (1,900 instances), validation set (200 instances), and test set (170 instances). Because few English datasets are publicly available for event causality extraction, we re-annotated the SemEval-2010 task 8 dataset [7] and evaluate our model on this dataset. Finally, we obtain 1,003 causality instances from this dataset (Table 1).

Table 1. Statistic details of datasets, including training, development and test sets.

Statistics	CEC	Fiancial	SemEval2010
Average sentence length	31.14	57.94	18.54
Mean distance between causal events	10.24	13.49	5.33
Mode of cause event length and proportion	2(41%)	4(31%)	1(85%)
Mode of effect event length and proportion	4(27%)	4(23%)	1(91%)
Average value of cause event length	4.03	6.06	0.96
Average value of effect event length	5.41	6.46	0.98

[1] https://github.com/shijiebei2009/CEC-Corpus.
[2] http://www.jrj.com.cn/.
[3] http://www.hexun.com/.

Table 2. Average F1-scores (%) of joint event causality extraction using different models. Boldface indicates scores better than the baseline system. The lengths of high-frequency cause/effect events are 2, 4, and 1 for the CEC, Financial and SemEval2010 dataset, respectively.

Model		CEC	Financial	SemEval2010
IDCNN+CRF	[20]	68.26	71.81	68.59
BiLSTM+CRF	[8]	68.74	74.75	73.20
CNN+BiLSTM+CRF	[15]	71.68	74.31	74.20
CSNN	[10]	70.61	74.59	73.71
BERT+CSNN (baseline)		74.61	76.23	75.69
CISAN		72.49	75.99	74.20
BERT+CISAN (unigram)		**75.26**	76.07	**77.65**
BERT+CISAN (bigram)		**75.93**	**76.70**	**77.26**
BERT+CISAN (trigram)		74.45	**76.29**	**77.14**
BERT+CISAN (quagram)		**75.27**	**77.09**	**77.35**

4.2 Experimental Settings

We use the pre-trained uncased BERT model. We set the model hyper-parameters according to [4]. For all datasets, we set the maximum length of sentences to 100, the size of the batch to 8, the learning rate for Adam to 1×10^{-5}, the number of training epochs to 100. To prevent over-fitting, we set the dropout rate to 0.5. It is worth mentioning that the lengths for n-grams are assigned through our preliminary statistics. For n-gram clustering, we set the number of clusters equal to the dimension of convolution filters; both are 100.

4.3 Results and Analysis

We make a comparison between our model and previous models and also conduct additional ablation experiments to prove the effectiveness of our model. Table 2 reports the results on CEC, Financial and SemEval2010 datasets. Following previous works, we use the standard F1-score as the evaluation metrics.

IDCNN+CRF [20]: A modified CNN model, using Iterated Dilated Convolutional Neural Networks, which permits fixed-depth convolutions to run in parallel across documents to speedup while retaining accuracy comparable to the BiLSTM-CRF.

BiLSTM+CRF: This is a classic sequential labeling model [8], which models context information using Bi-LSTM and learns sequential tags using CRF.

CNN+BiLSTM+CRF: This model [2] uses CNN to enhance BiLSTM+CRF to capture local n-gram dependencies.

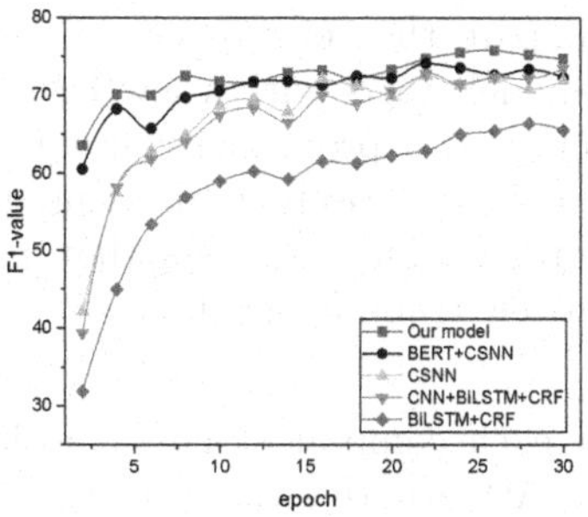

Fig. 3. F1-score on test dataset of Financial w.r.t training epoch.

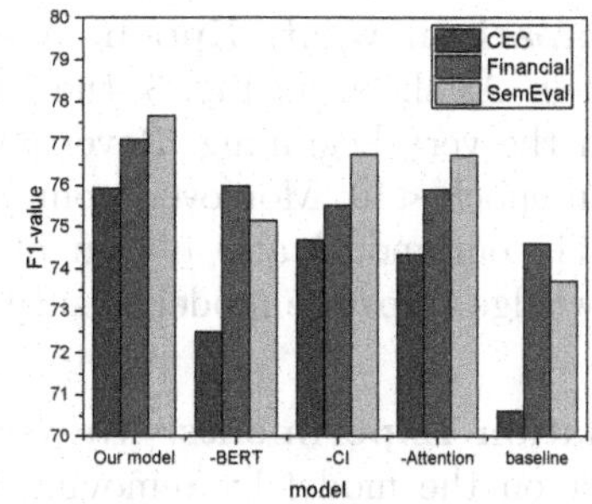

Fig. 4. Ablation analysis of our model in three datasets.

CSNN: The variation [10] modifies the basic architecture of BiLSTM+CRF with additional CNN layer to address n-gram inside features, and the self-attention mechanism to address n-gram outside features.

CISAN: Our model with convolutional semantic-infused filters and key-query attention, using GloVe word embeddings.

BERT+CSNN: BERT serves as the encoder for CSNN, replacing the GloVe embedding layer.

BERT+CISAN: Our model utilizes convolutional semantic-infused filters, key-query attention incorporating with the BERT encoder.

4.4 Effectiveness of Semantic Infusion

To verify that the effectiveness of BERT and convolution semantic infusion both contribute to extraction, we construct another strong baseline BERT+CSNN. The result on BERT+CISAN produces a decent f1 score, which is 1.32%, 0.86%, and 1.96% higher than BERT+CSNN on CEC, Financial, and SemEval2010, to show the effectiveness of n-grams and knowledge produced by the pre-trained model.

As described, the length of n-grams serves as a determining hyper-parameter, which reflects the frequency information of event length in our model. To prove this hypothesis, we assign the different value of lengths to verify the effectiveness and investigate the influence on the results. Table 2 indicates that the best length of high-frequency cause/effect n-grams for the CEC, Financial and SemEval2010 dataset are 2, 4 and 1, respectively. Intuitively, the causal events in Financial are very likely to be described as a phrase whose length = 4 like "股价下跌(Share prices fell)" and "销量减少 (sales reduced)". Thus, We think that convolution is more sensitive to such events. When using our method with the length of 4 for n-grams, we get the best results compared to other lengths, which demonstrates the effectiveness of semantic infusion.

Comparison w.r.t. Epoch. We further compare the convergence speed of models. As shown in Fig. 3, the NON-BERT model starts with a low F1 score from the very beginning. Reversely, our model gains a promising consequence when epoch <10. Moreover, compared with the strong baseline represented by BERT, our model also obtain obvious advantages. This suggests that prior knowledge helps the model to surpass others at few training epochs.

Ablation Experiments. We verify whether each component has a positive effect on the model by removing (BERT, CI, Attention) layer in order. The results of experiments launched on three datasets are shown in Fig. 4. The contribution of BERT layer is most significant for the reason that BERT has sufficient pre-training and enriched with rich semantic knowledge. Moreover, the model is promoted by extra gains of CI and attention layers. The CI layer incorporating knowledge related to cause/effect event improves the F1 score. The attention layer applied is also efficient in removing off the noise. In conclusion, all these layers simultaneously contribute to obtaining SOTA performance.

5 Conclusion

This paper introduces a novel n-gram-based neural solution for joint casual relation and event extraction. The model looks up high-frequency cause/effect n-grams in the sentence by partially encoding the causal features into convolution filters. In practical, we model the associations of n-gram pairs to find the potential cause-effect pairs. As a result, our proposal achieves significant improvements compared with baseline. This work shows the feasibility of further enhancing intra- and inter- knowledge around n-grams in guiding neural extractor. In the future, a potential direction is to adopt our method to few-shot learning of coarse-to-fine causality extraction. We are also interested in multiple cause-effect pairs extraction rather than only extracting one cause-effect event pair per time.

References

1. Asghar, N.: Automatic extraction of causal relations from natural language texts: a comprehensive survey. arXiv preprint arXiv:1605.07895 (2016)
2. Chen, T., Xu, R., He, Y., Wang, X.: Improving sentiment analysis via sentence type classification using BiLSTM-CRF and CNN. Expert Syst. Appl. **72**, 221–230 (2017)
3. Chen, Y., Xu, L., Liu, K., Zeng, D., Zhao, J.: Event extraction via dynamic multi-pooling convolutional neural networks. In: Proceedings of the 53rd Annual Meeting of the Association for Computational Linguistics and the 7th International Joint Conference on Natural Language Processing (Volume 1: Long Papers), pp. 167–176. Association for Computational Linguistics (July 2015)
4. Devlin, J., Chang, M.W., Lee, K., Toutanova, K.: Bert: pre-training of deep bidirectional transformers for language understanding. arXiv preprint arXiv:1810.04805 (2018)

5. Girju, R., Moldovan, D.I., et al.: Text mining for causal relations. In: FLAIRS Conference, pp. 360–364 (2002)
6. Grishman, R.: Domain modeling for language analysis. Tech. rep., New York Univ. NY (1988)
7. Hendrickx, I., et al.: Semeval-2010 task 8: multi-way classification of semantic relations between pairs of nominals. In: Proceedings of the 5th International Workshop on Semantic Evaluation, pp. 33–38. Association for Computational Linguistics (2010)
8. Huang, Z., Xu, W., Yu, K.: Bidirectional LSTM-CRF models for sequence tagging. arXiv preprint arXiv:1508.01991 (2015)
9. Jian, F., Zong-Tian, L., Wei, L., Wen, Z.: Event causal relation extraction based on cascaded conditional random fields. Pattern Recognit. Artif. Intell. **24**(4), 567–573 (2011)
10. Jin, X., Wang, X., Luo, X., Huang, S., Gu, S.: Inter-sentence and implicit causality extraction from Chinese corpus. In: Lauw, H.W., Wong, R.C.-W., Ntoulas, A., Lim, E.-P., Ng, S.-K., Pan, S.J. (eds.) PAKDD 2020. LNCS (LNAI), vol. 12084, pp. 739–751. Springer, Cham (2020). https://doi.org/10.1007/978-3-030-47426-3_57
11. Kontos, J., Sidiropoulou, M.: On the acquisition of causal knowledge from scientific texts with attribute grammars. Int. J. Appl. Expert Syst. **4**(1), 31–48 (1991)
12. Li, P., Mao, K.: Knowledge-oriented convolutional neural network for causal relation extraction from natural language texts. Expert Syst. Appl. **115**, 512–523 (2019)
13. Li, S., Zhao, Z., Liu, T., Hu, R., Du, X.: Initializing convolutional filters with semantic features for text classification. In: Proceedings of the 2017 Conference on Empirical Methods in Natural Language Processing, pp. 1884–1889 (2017)
14. Lin, Y., Shen, S., Liu, Z., Luan, H., Sun, M.: Neural relation extraction with selective attention over instances. In: Proceedings of the 54th Annual Meeting of the Association for Computational Linguistics (Volume 1: Long Papers), pp. 2124–2133 (2016)
15. Ma, X., Hovy, E.: End-to-end sequence labeling via bi-directional LSTM-CNNs-CRF. arXiv preprint arXiv:1603.01354 (2016)
16. Martínez, E., Shwartz, V., Gurevych, I., Dagan, I.: Neural disambiguation of causal lexical markers based on context. In: IWCS 2017–12th International Conference on Computational Semantics-Short papers (2017)
17. Riaz, M., Girju, R.: Another look at causality: discovering scenario-specific contingency relationships with no supervision. In: 2010 IEEE Fourth International Conference on Semantic Computing, pp. 361–368. IEEE (2010)
18. Shen, Y., Huang, X.J.: Attention-based convolutional neural network for semantic relation extraction. In: Proceedings of COLING 2016, the 26th International Conference on Computational Linguistics: Technical Papers, pp. 2526–2536 (2016)
19. Sorgente, A., Vettigli, G., Mele, F.: Automatic extraction of cause-effect relations in natural language text. DART@ AI* IA **2013**, 37–48 (2013)
20. Strubell, E., Verga, P., Belanger, D., McCallum, A.: Fast and accurate entity recognition with iterated dilated convolutions. arXiv preprint arXiv:1702.02098 (2017)
21. Vaswani, A., et al.: Attention is all you need. In: Advances in Neural Information Processing Systems, pp. 5998–6008 (2017)
22. Zhao, S., Liu, T., Zhao, S., Chen, Y., Nie, J.Y.: Event causality extraction based on connectives analysis. Neurocomputing **173**, 1943–1950 (2016)

A k-MCST Based Algorithm for Discovering Core-Periphery Structures in Graphs

Susheela Polepalli(✉) and Raj Bhatnagar(✉)

University of Cincinnati, Cincinnati, OH 45221, USA
polepasa@mail.uc.edu, bhatnark@ucmail.uc.edu

Abstract. Core-periphery structures are examples of meso-scale characteristics of graphs. Most existing algorithms for core-periphery (CP) structures work by first finding the dense cores of a network and then discovering the peripheral nodes around them. Our algorithm presented here seeks to query a graph to return the CP structures centered around any selected query node. Our algorithm significantly reduces the computational complexity of repeatedly querying the CP structures from a network. Our algorithm repeatedly extracts minimum cost spanning trees (MCSTs), first from the original network, and then successively from the residual networks. From the union of these MCSTs, our algorithm efficiently answers the queries for CP structures around nodes. We validate our algorithm on example networks taken from two domains.

Keywords: Graph mining · Core-periphery structures · Community detection

1 Introduction

The underlying data for a number of domains can be viewed in the form of graph structures. This representation denotes entities as nodes, and strengths of relationships among entities as edge labels. Most of the existing community detection algorithms are based on one of the following main ideas: minimum cut method, hierarchical clustering, Girvan–Newman algorithm [12], and modularity maximization method. In all these algorithms a common theme is to identify densely connected subgraphs of the complete graph.

Communities having core-periphery (CP) structures are observed in many domains. They occur in the form of a densely connected core set of nodes, surrounded by a layer of relatively loosely connected peripheral nodes, as introduced by Borgatti and Everett in [2]. A number of approaches have attempted to find CP structures using graph-based ideas of overlapping communities [4], connected k-plexes [1], and random walk signatures [9]. There are quite a few applications in social and scientific networks that require us to repeatedly query the networks for CP community structures around specific individual nodes of the network. We may want to ask for the social group around a single individual in a social network or for the closest proteins to a specific protein in a biological network.

K. Karlapalem et al. (Eds.): PAKDD 2021, LNAI 12712, pp. 358–370, 2021.
https://doi.org/10.1007/978-3-030-75762-5_29

These query nodes may or may not be parts of dense clusters of the complete network. To address this need we reduce the size of the graph, by dropping weaker edges, so that repeated query tasks are made computationally efficient.

To construct the CP structure from the reduced graph, we initialize the core set by including in it all the immediate 1-hop neighbors of the query node. We compute the edge density of this initial core and set it as the initial threshold for edge density. Neighbors of the nodes that are already in the core are then successively added to the core such that these node additions do not reduce the edge density. We then adds layers of peripheral nodes that follow a similar closure property on edge density, and the density thresholds decrease successively from one peripheral layer to the next. Our algorithm uses the clustering coefficient of each node, if needed, to identify nodes that have high connectivity around them.

2 Related Work

The work by Borgatti and Everett in [2] identified the CP structures as dense blocks of 1's, connected to sparser blocks, in graphs' adjacency matrices. Exploiting the properties of overlapping communities, J. Yang and J. Leskovec used the idea of overlapping tiles [4] to identify CP structures. The algorithm *ClusterONE-CP* [3] also works with overlapping communities discovered by the clustering algorithm *ClusterONE* and analyzes the core and overlapping sections of communities to discover the CP structures. Silva et al. [8] have defined core coefficients for nodes using closeness centrality and community modularity metrics. Most of these algorithms seek to first identify densely connected cores and then find nodes that are peripheral to these cores. One problem with algorithms based on centrality and affinity metrics for nodes and edges is that each core community may have a different average value for these centrality metrics. Some of the algorithms will not be able to work with such variations of average densities among cores across the whole graph. In our algorithm we achieve independence from affinity variations by considering a few most significant of the minimum cost spanning trees (MCSTs) extracted from the target graph. Each MCST tends to select the highest affinity edges for each neighborhood. The concept of Merged MCSTs was introduced in [7] where it was shown that the K-merged MCST neighborhood graphs perform better than the k-nearest neighbor graphs for capturing the notion of distance (or affinity) between pairs of nodes in a graph. Two rounds of MCSTs were also used in *2MSTclus* algorithm [5] to extract separate but touching clusters. Use of k-MCSTs by our algorithm has, therefore, precedence and also justification for their use in effectively capturing inter-node similarities in graph clustering algorithms.

3 Our Approach

The first phase of our algorithm constructs the k-MCST graph from the original graph. This is done by first determining the MCST of the graph and then dropping from the graph all those edges that are included in the MCST. We again find the MCST from the residual graph and repeat this process until we have obtained k MCSTs. If the graph is not connected and has multiple components, we can obtain and work with the Minimum Cost Spanning Forests (MCSFs). We then merge the first k MCSTs to form the graph within which we will look for the CP structures for query nodes. The resulting graph has the properties [7] that it is k-connected, and contains only the highest affinity edges. This ensures that each node is connected to at least k other nodes, and these k edges have the strongest affinity in the neighborhood of this node.

3.1 Definitions

Structural Density ρ: We introduce a metric called structural density for a cluster of nodes, defined as: $\rho = \frac{Number\ of\ edges\ in\ the\ cluster}{Number\ of\ possible\ edges\ in\ the\ cluster}$. When a set of nodes is included in a cluster, all edges from the k-MCST that connect these nodes among themselves are also included in the cluster.

Core-Periphery Structure: A core-periphery network for a query node N, derived from a k-MCST neighborhood graph, consists of a set of nodes constituting the core C, surrounded by its sets of nodes for peripheral layers (P_1 U P_2 U ... P_N). Let us say a core C is a set of nodes, and p is a node that is not in C but has a direct edge to at least one node in C. Then considering all those edges that can be included in C from the k-MCST, a core set of nodes C has the following property: For each nodes p connected to C, the density $\rho(C) \geq \rho(C \cup \{p\})$. That is, C is that closed set of nodes to which no other immediate neighboring node can be added without reducing its structural density.

We add a caveat here about a limiting situation for the above definition of a core set. It is possible that the affinity values of the query node to its nearest neighbors are the smallest in a very large neighborhood around it. This will result in very large cores for such nodes, and in extreme cases the entire graph may be included as core for a node. In real graphs such nodes are rare. To account for such query nodes we may place a limit on the growth of a core set.

We consider a number of peripheral layers of nodes that surround the core. Each successive layer's affinity to the core nodes is weaker than that of the preceding layer. A node p belongs to the jth peripheral layer P_j when it has a direct edge to at least one of the nodes in core set C and the following are satisfied: $for\ j == 1 : (1^{st}\ peripheral\ layer) : \rho(C) > \rho(C \cup \{p\}) \geq P_j\_threshold$ and for $j > 1 : P_{j-1}\_threshold > \rho(C \cup \{p\}) \geq P_j\_threshold$.

3.2 Algorithm

Our methodology has the following two main phases. *Phase 1:* Construction of k-MCST graph from the original graph. We also compute the clustering coefficient of each node in the k-MCST graph to aid in choosing the nodes in the denser parts of the graph, if needed. *Phase 2:* Construction of core-periphery structures for query nodes guided by the structural density metric

Phase 1: We construct the k-MCST graph from the original graph as follows [7]. Let G = (V, E) denote the complete graph where the label for each edge represents the inverse of the affinity between its two nodes. $MCST_1$ denotes the set of edges of this graph's MCST, that is, $MCST_1 = MCST(V, E)$. The second MCST is computed from the graph that results after removing all edges of $MCST_1$ from the original graph. That is, $\mathrm{MCST}_2 = \mathrm{MCST}(\mathrm{V}, (\mathrm{E} - \text{Edges-of-}\mathrm{MCST}_1))$. Similarly, MCST_i denotes the set of edges of the MCST of G with edges of $\sum_{j=1}^{i-1} MCST_{\mathrm{j}}$ removed from the original graph. We define the k-MCST neighborhood graph as: $k\text{-MCST} = (V, \sum_{i=1}^{k} MCST_{\mathrm{i}})$. Typically, a large number of weaker edges are discarded in this process. For the next phase of CP construction we no longer consider the edge labels (affinity values) and work only with the edge densities of the subgraphs.

Phase 2: Construction of CP Structures. We take the query node N and grow its core subgraph around it. The main idea of constructing a core around a seed node is to identify that closed set of nodes and edges, whose structural density can not be either increased or maintained the same by adding any new node. The main steps for growing the core are:

1. Take the query node N and add to it all its immediate neighbors, and also all the edges from k-MCST that connect them. This is the starting core set in the core growth process. We then compute the structural density ρ_{temp} for this core, and set it as the initial density threshold.
2. We now consider all those immediate 1-hop neighbor nodes of the nodes in the core, that are not yet included in the core, and call this set L. We test each member of L, one by one, to see if its addition to the core set increases or decreases the value ρ_{temp}. All those nodes from L that, individually, do not decrease ρ_{temp}, when potentially added to the current core, are marked for addition to the core. After all nodes in L have been individually tested, the marked nodes are added to the core. A new structural density is computed for this new set of nodes. If this new density value is larger than rho_{temp} then rho_{temp} is set to this new higher value; else, the old value of rho_{temp} is retained.
3. Step-2 above is performed repeatedly until there is no node in L that gets marked for addition to the core. In this case the core growth process is stopped.

For any set of nodes L the process of adding nodes to the core is independent of the order in which individual nodes in L are tested for their effect on structural density.

In effect, a core set of nodes includes (i) the query node, (ii) all its 1-hop neighbors, and (iii) all those nodes that are m hops away from the query node, and when added to the core, maintain or exceed the edge density of the core formed from nodes that are at least m-1 hops away. This results in cores that bring in densely connected neighboring nodes irrespective of the number of hops they may be away from the query node.

Defining Periphery Thresholds: We define the density thresholds for the peripheral layers in terms of the density of the core. The first peripheral layer threshold for a given core C is: $Periphery_1threshold = \rho_{\mathrm{core}}(1 - r)$. Here r is a parameter by which the density threshold is reduced for each successive peripheral layer from that of the previous layer. That is, if r is set to 0.1 then the density threshold for the first peripheral layer is 0.9 times the density of the core. The density thresholds for other outer peripheral layers are obtained by successively reducing the threshold by the parameter $(1 - r)$. $Periphery_j threshold = \rho_{core}(1 - j \times r)$ where j is the peripheral layer number.

Identifying Peripheral Nodes: We essentially repeat the same process that is used for core growth, but now use the lower density threshold of peripheral layer (l-1), to add nodes to the first peripheral layer (l). Step-2 and step-3 of the core growth process above are repeated but with the lower threshold. Once the iterations of these steps are completed for a layer, we lower the threshold to that for the next outer peripheral layer and repeat the process. From our results we have seen that it is possible for a node three hops away from the query node to be included in the core, but a node that is two hops away to be in the peripheral layer. Details of this algorithm can be found in [14].

3.3 Complexity Analysis

In the first phase, the neighborhood graph is constructed by merging k MCSTs. For generating k MCSTs, the time complexity is of the order of $\mathcal{O}(k(E \log n))$, where E is the number of edges and n is the number of nodes in the graph. For identifying nodes that are in the denser regions of the graph we compute the clustering coefficient of each node in the k-MCST. The complexity of computing the clustering coefficients, using some good data structures to represent the graphs is, either $\mathcal{O}(m^2/n)$ or $\mathcal{O}(m^{1.48})$ [10] where m is the number of edges in the graph and n is the number of nodes in the graph. Having significantly reduced the number of edges in the k-MCST compared to the original graph, we have made the task of computing clustering coefficients much less compute intensive. In the second phase, as each node is connected to at least k nearest nodes, the time complexity for growing a core and build peripheries for a given query node is $\mathcal{O}(k^{number-of-layers})$. Since we expect to query the k-MCST multiple times, complexity of this repetitive task will be significantly reduced. The number of nearest neighbors of a node in the k-MCST is, on average, equal to $2 * k$, but will be much higher in the complete, unreduced graph. So, the complexity of our algorithm is significantly reduced for the operations of computing clustering coefficients and for repeatedly computing the CP structures from a k-MCST (Fig. 1).

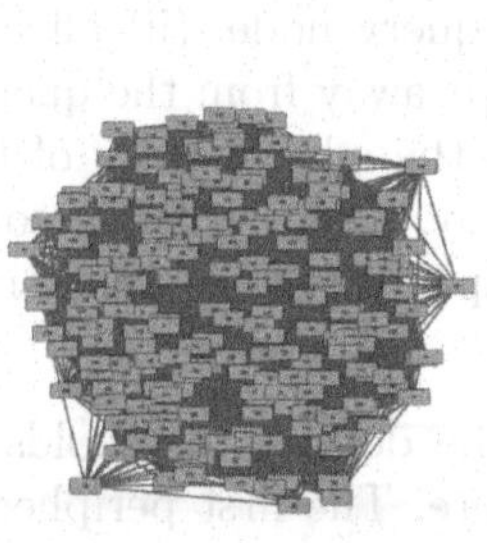

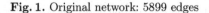

Fig. 1. Original network: 5899 edges

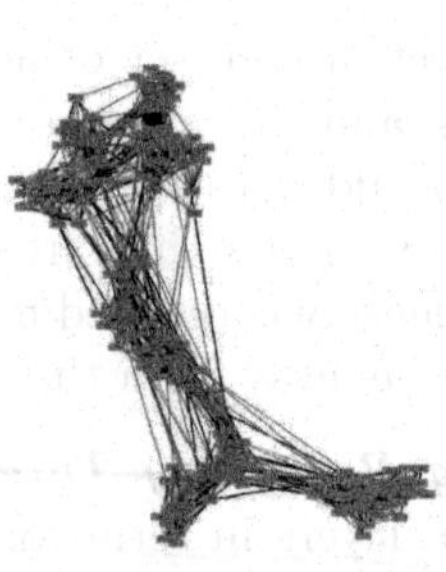

Fig. 2. 7-MCST network: 1645 edges

4 Experiments and Results on Real World Data Sets

4.1 Primary School Data Set

This data set is taken from a study of contact networks from a primary school conducted by SocioPatterns [11]. It describes interactions among students and teachers in a primary school for one day. Nodes represent students and teachers. Edges represent cumulative interaction time spent between two people, measured in seconds. Each node is also labeled by the class and gender of the person. The network has 236 nodes and 5899 labeled edges.

k-MCST Generation: The edge weights here describe the amount of time spent between two people. Since distance or cost for an edge is inverse of the time spent, for our MCST algorithm we replace each edge label with the inverse of the time spent between pairs of people. From this network, we constructed a merged k-MCST graph for various values of k, and the one for $k = 7$ is shown in Fig. 2. The number of edges in this 7-MCST network is only 27.8% of those in the original network. This significantly reduces the complexity of computing the clustering coefficients of nodes and also the CP structures.

CP Network for Node with Highest Clustering Coefficient: Among all the students in the 7-MCST network, student #52 has the highest clustering coefficient (0.92). The core-periphery structure for this student, obtained by our algorithm, is shown in Fig. 6. The periphery thresholds are defined using the reduction factor of $r = 5\%$.

The core for student #52 contains 10 students, (Core(52) is shown in yellow nodes). There are a total of 31 other nodes that have at least one edge incident on at least one of the core nodes. These 31 nodes are tested for placement in the next peripheral layer. The core exhibits a high structural density (0.97) and is surrounded by peripheral layers in decreasing order of structural density. Figure 3 shows the structural density of the (core+node) pairs when each of these 31 nodes is tested for placement in the next peripheral layer. It is clear from this plot that the nodes in the outer peripheral layers have decreasing connectivity to the core.

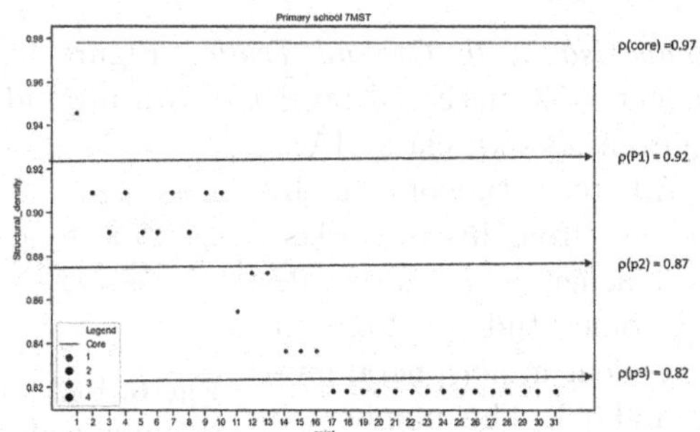

Fig. 3. Density values when each of 31 neighbors is added to the core

CP Network for a Node with Low Clustering Coefficient: We constructed the core-periphery network for student #125 who has a very low clustering coefficient (0.19) in the 7-MCST network. As this student is very sparsely connected to his immediate neighbors, the structural density of his core social network is very low. The algorithm then finds a large number of other sparsely connected students who meet this low threshold and become members of his core and periphery sets. That is, his core community consists of a large number of students, each of whom is only lowly connected to others in the core community. This core community has its peripheral layers at decreasing levels of edge densities, and they are also highly populated due to low density thresholds (Figs. 4 and 5).

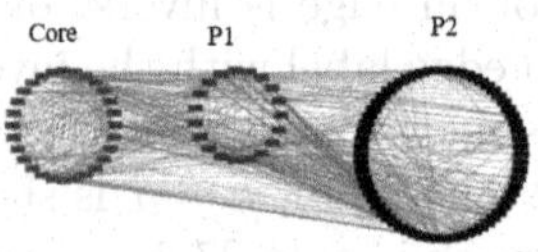

Fig. 4. CP structure of student #125

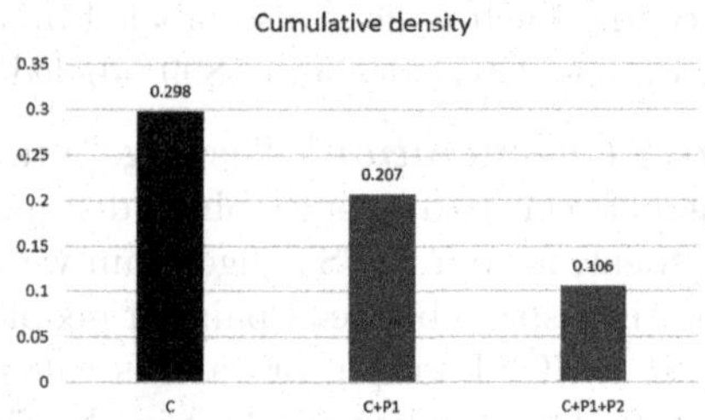

Fig. 5. Cumulative density of CP structure of student #125

This shows that a community for a query node with high clustering coefficient will include fewer but highly connected students, and when the clustering coefficient of the query node is low, we are likely to get a loosely connected and highly populated community.

Comparison with Ground Truth: Figure 6 here shows the CP structure for student #52, with red border drawn around nodes representing students who are from his own class (4A).

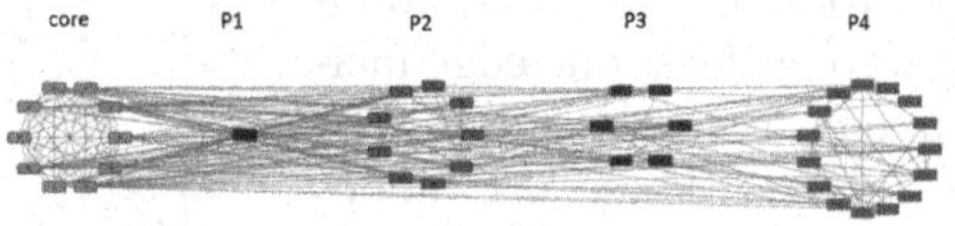

Fig. 6. Red bordered students are in the same class as query node #52 (Color figure online)

All ten students in his core are from his own class, his teacher is in layer P1, and some students from his class occur in peripheral layer P2 and p3. The numbers in Table 1 show the distribution of classes the students come from in each layer of the CP structure. This is very much in line with the expectation that students interact more with their own classmates and have fewer interactions with students from other classes. The teachers were also included in the CP structure and the CP structure explains the interaction of the students with their class teacher. The teacher is in the first peripheral layer and this says that the student #52 interacts more with his friends and less with his teacher.

Similarly, as we analyze the class distribution of students in the CP structure of student #125 shown in Fig. 7 and Table 2, it is seen that the student's core includes students mostly his own class (1A), and many other students from 2nd, 3rd and 4th grades appear in his peripheral layers.

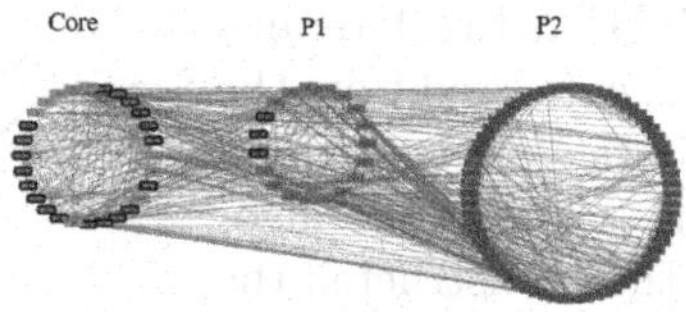

Fig. 7. Green-bordered students are in the same class as student #125 (Color figure online)

Though the strength of connectivity with each student is low, the structural density of his core and peripheries is also lower, but he has his own CP structure in the sense of communities for other students.

Table 1. Distribution of nodes in CP structure of student #52

Node class	Core	P1	P2	P3	P4
4A	10	0	7	4	0
4B	0	0	2	1	4
5A	0	0	0	0	2
5B	0	0	0	1	5
3B	0	0	0	0	2
2B	0	0	0	0	2
Teachers	0	1	0	0	0

Table 2. Distribution of nodes in CP structure of student #125

Node class	Core	P1	P2
1A	24	2	0
1B	2	9	12
2A	2	3	13
2B	4	7	13
3A	0	1	7
4A	0	0	2
4B	1	0	13
Teachers	1	1	2

Our algorithm is thus able to construct meaningful CP structure for any node, from denser or sparser parts of the graph. This is in contrast to traditional algorithms that find their cores in only the dense parts of the graphs.

4.2 Airport Network

This data set describes passenger traffic through a network of airports in year 2010. The data is taken from the Bureau of Transportation Statistics (BTS).

The data set includes 1574 airports as nodes and 28236 edges which are labeled with the number of passengers traveling between pairs of airports. In this data set, each edge represents the number of passengers who traveled between two airports. The affinity between two airports is determined by the number of passengers traveling between them.

A 5-MCST network retains 5396 edges, extracted from the original data set containing 28236 edges. Using our algorithm we then constructed the core-periphery network for the CVG airport as the query node. CVG airport is an international airport located in Hebron, Kentucky. The cities in the core set when CVG airport is used as the query are shown in Table 3. The set of cities in the core for the same query airport CVG has fewer members when we use the 3-MCST network. This is shown in Table 4. The effect of k on the CP structures has been described in the next subsection. The core and periphery cities are plotted on the map in Fig. 8 and its properties are illustrated in Table 5. The yellow, pink and blue nodes represent the core, periphery-1 and periphery-2 respectively.

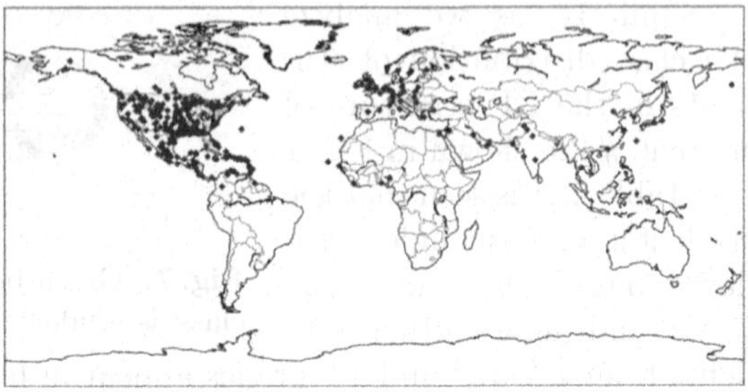

Fig. 8. Core and periphery cities of CVG airport (Color figure online)

Table 3. Core cities for CVG airport on 5-MCST network

Montreal, Canada	Chicago, IL
Atlanta, GA	Cincinnati, OH
Cleveland, OH	Charlotte, NC
Knoxville, TN	Asheville, NC
South Bend, IN	Accra, Ghana
Dallas/Fort Worth, TX	Evansville, IN
Fort Wayne, IN	Ashland, WV
Salt Lake City, UT	Detroit, MI
Orlando, Florida	

Table 4. Core cities of CVG airport on 3-MCST network

Montreal, Canada	Chicago, IL
Atlanta, GA	Cincinnati, OH
Cleveland, OH	Charlotte, NC
Knoxville, TN	Asheville, NC

Table 5. Properties of each layer in core-periphery network of CVG airport

Layer	#airports	ρ(Layer)	Avg. # passengers/day (mean edge weight)	Avg. # passengers/day from core to each periphery layer
Core	16	0.35	633	–
P1	30	0.03	793	713
P2	408	0.007	167	267

4.3 Effect of Varying the Parameter *k*

The number of MCSTs included in the reduced graph, k, has significant impact on the nature of CP structure obtained. Tables 6, and 7 below show the variation in the characteristics of CP structures obtained for $k = 3$, 5, 7, and 9 for both our data sets. An increase in the value of k results in more edges getting included in the k-MCST graph. This causes cores to have higher edge densities, and some nodes that were in peripheries for lower k's increase their connectivity to the cores and gets pulled into the core sets. We see that the choice of k affects the sizes of the core sets and also the average affinity values within the nodes of the core. Therefore, a choice of k must be made for each CP structure to identify the cores of desired size and optimal average affinity value. A solution for this is to maintain k-MCSTs for multiple values of k, process the query on all of them, and choose the CP structure with optimal affinity values.

Table 6. Primary school data set: variation of CP structure of student 52

k	#edges(k-MCST)	ρ(core)	Avg core edge wght	Core list
3	705	0.57	14.11	34, 38, 47, 51, 52, 119, 176, 196
5	1175	0.82	12.6	34, 38, 47, 51, 52, 119, 176, 196
7	1645	0.93	11.97	34, 37, 38, 47, 49, 51, 52, 119, 176, 196
9	2115	0.95	9.33	34, 37, 38, 46, 47, 49, 50, 51, 52, 53, 54, 93, 119, 176, 185, 196

4.4 Validation of Use of *k*-MCSTs for CP Structures

One assumption we have made is that the CP structures discovered from the reduced k-MCST graphs are as valid as those discovered from the complete graphs. To validate this assumption we apply our algorithm to the original graph and also to its reduced k-MCST graph and compare the resulting CP structures.

Table 7. Airport data set: variation in CP structure

k	#edges in kMCST	ρ(core)	Avg # passengers/day in core cities	# core nodes
3	3822	0.36	529	8
5	5396	0.35	686	15
7	6675	0.35	1084	27
9	7755	0.35	893	38

We compute two different types of ratios from average affinity values corresponding to the edges in: (i) Core vs. (Core + P1), (ii) Core vs. (Core + P1

+ P2), and (iii) Core vs. (Core + P1 + P2 + P3), etc. The ratios show how much relatively weaker the peripheries are compared to the core's own affinities. The precise descriptions of the two ratios used are given in the following paragraphs. These ratios are expected to decline as more and more peripheries are included in the denominator. The individual ratio values represent the relative distribution of the affinities across a core and its peripheries. The trend of these ratios, along increasing inclusion of peripheries, reflects the way the peripheral strengths decline as we move away from the core.

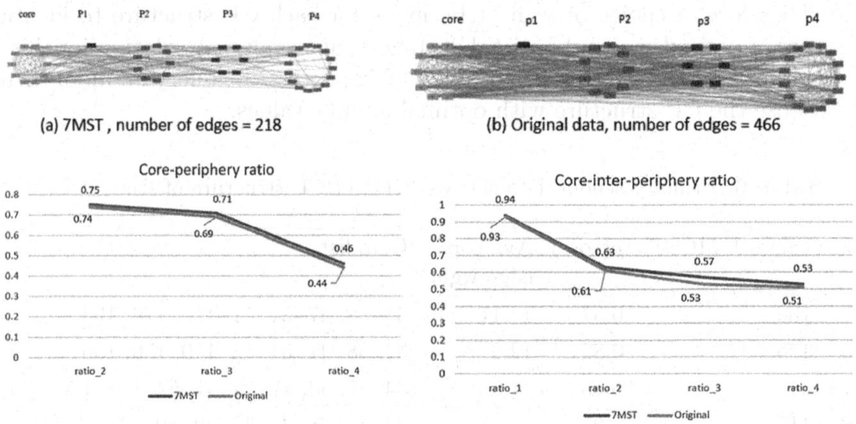

Fig. 9. Validation of core-periphery structure of student 52 (primary school)

We run our algorithm on the original complete graph and also on the reduced k-MCST graph to construct CP structures. If these ratios show very similar values and very similar declining trend, then we can infer that the average affinities of edges included in cores and peripheries are also very similar. Similar values and trends of the ratios will, therefore, show that the assumption of working with the k-MCSTs is justifiable. The names and precise descriptions of the two ratios that we have used to characterize the CP-structures are: (i) Core-periphery ratio, and (ii) Core-inter-periphery ratio. Given a core and its n periphery layers, we define core-periphery ratio up to a particular periphery level k as: Core-periphery_ratio$_k$ $= \frac{W_{\text{in}}(Core)}{W_{\text{in}}(Core)+\sum_{k=1}^{n} W_{\text{in}}(Periphery_k)}$ where $1 \leq \text{k} \leq \text{n}$, and $W_{in}(set\ of\ edges)$ is the sum of the weights in *set of edges*. That is, we sum the edge weights for all edges included exclusively in the core, and then separately in individual peripheral layers. The core-periphery ratio is defined as the ratio of sum of core's internal edge weights to the core and peripheries' internal edge weights only. This measure depicts the cohesiveness of core and peripheries as new outer peripheral layers are added. As seen in Fig. 9 and Fig. 10 CP structures constructed from the k-MCST and the complete graph show very similar values and decreasing trends. For the school data set the individual ratios are almost identical. For the airport data, the trend is the same but the individual values differ because in this case, during the core generation phase, some very weak edges got included in the core. We can thus see that the core-periphery networks generated using our approach are not significantly affected by dropping the weaker edges and retaining only the k-MCST graphs.

Now we consider the second metric for characterizing the CP structures. Core exhibits properties of cohesiveness by having high intra-core affinity values and low affinity values for the edges that connect the core nodes to peripheral nodes. To study the relative weights of intra-core and inter-core-periphery edges we introduce the measure called core-inter-periphery ratio. This is computed as: $\text{Core-inter-periphery\_ratio}_{\text{k}} = \frac{W_{\text{in}}(Core)}{W_{\text{in}}(Core)+\sum_{k=1}^{n} Interedge(Core, Periphery_{\text{k}})}$

where $1 \leq \text{k} \leq \text{n}$, $W_{in}(set\ of\ edges)$ is the sum of the weights in the *set of edges*, and $Interedge(C, P_{\text{j}})$ is the sum of edge-weights connecting Core to P_{j}. As periphery layers are added, we compute the successive core-inter-periphery ratio and examine its trend. As seen in Figs. 9 and 11, the CP structures obtained using k-MCST and the complete graph show very similar values and similar decreasing trends.

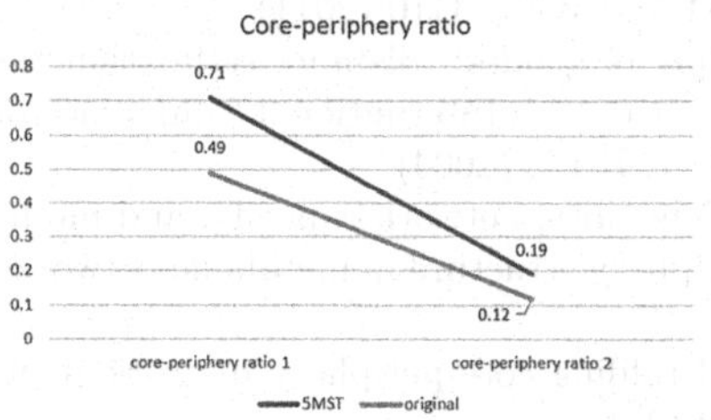

Fig. 10. Core-periphery ratio validation

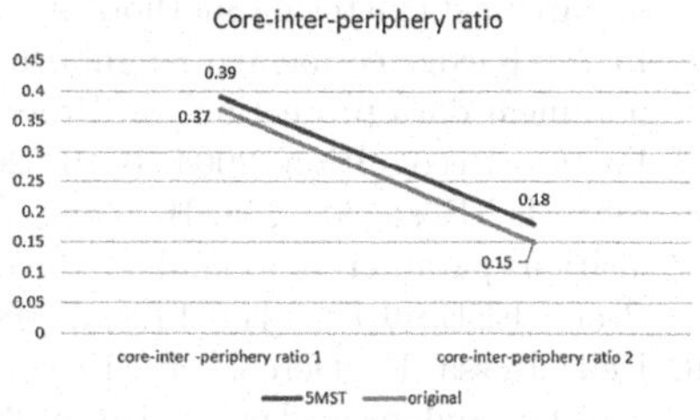

Fig. 11. Core-inter periphery ratio validation.

This shows that the CP structures obtained from the k-MCSTs have essentially the same characteristics as those that would have been obtained from the complete graph. Details of this algorithm can be found in [14].

5 Conclusion

In this paper we have presented a new algorithm for constructing core-periphery structures for individual query nodes from an edge-weighted undirected graph. Our algorithm is tailored for applications where we may want to repeatedly query a very large graph for CP structures surrounding individual nodes. Our algorithm reduces the computational complexity for each CP query by significantly reducing the size of the graph to be processed, by retaining only the strongest affinity edges. We have demonstrated the effectiveness of our algorithm by executing it on a social network and on an airport passenger traffic network. The ground truths from these domains validate the CP structures produced by our algorithm. Using two different validation measures, we have shown that CP structures obtained from k-MCSTs and the original complete graphs shows very similar trends in terms of structural and edge weight densities among the cores and their peripheries.

References

1. Luo, F., et al.: Core and periphery structures in protein interaction networks. In: BMC Bioinformatics, vol. 10, no. 4. BioMed Central (2009)
2. Borgatti, S.P., Everett, M.G.: Models of core/periphery structures. Soc. Netw. **21**(4), 375–395 (2000)
3. Sardana, D., Raj, B.: Core periphery structures in weighted graphs using greedy growth. In: 2016 IEEE/WIC/ACM International Conference on Web Intelligence (WI). IEEE (2016)
4. Yang, J., Leskovec, J.: Overlapping communities explain core-periphery organization of networks. Proc. IEEE **102**(12), 1892–1902 (2014)
5. Zhong, C., Miao, D., Wang, R.: A graph-theoretical clustering method based on two rounds of minimum spanning trees. Pattern Recogn. **43**(3), 752–766 (2010)
6. Zhong, C., Miao, D., Fränti, P.: Minimum spanning tree based split-and-merge: a hierarchical clustering method. Inf. Sci. **181**(16), 3397–3410 (2011)
7. Li, Y.: K-edge connected neighborhood graph for geodesic distance estimation and nonlinear data projection. In: Proceedings of the 17th International Conference on Pattern Recognition, 2004, ICPR 2004, vol. 1. IEEE (2004)
8. Silva, D., Rosa, M., Ma, H., Zeng, A.-P.: Centrality, network capacity, and modularity as parameters to analyze the core-periphery structure in metabolic networks. Proc. IEEE **96**(8), 1411–1420 (2008)
9. Della Rossa, F., Dercole, F., Piccardi, C.: Profiling core-periphery network structure by random walkers. Sci Rep. **3**(1), 1–8 (2013)
10. Schank, T., Wagner, D.: Approximating clustering coefficient and transitivity. J. Graph Algorithms Appl. **9**(2), 265–275 (2005)
11. Stehlé, J., et al.: High-resolution measurements of face-to-face contact patterns in a primary school. PloS One **6**(8), e23176 (2011)
12. Newman, M.E.J.: Fast algorithm for detecting community structure in networks. Phys. Rev. E **69**(6), 066133 (2004)
13. Ailem, M., Role, F., Nadif, M.: Graph modularity maximization as an effective method for co-clustering text data. Knowl.-Based Syst. **109**, 160–173 (2016)
14. Polepalli, S.: Discovery of core-periphery structures in networks using k-MSTs. Diss. University of Cincinnati (2019)

Detecting Sequentially Novel Classes with Stable Generalization Ability

Da-Wei Zhou[1], Yang Yang[2], and De-Chuan Zhan[1(✉)]

[1] National Key Laboratory for Novel Software Technology, Nanjing University, Nanjing 210023, China
zhoudw@lamda.nju.edu.cn, zhandc@nju.edu.cn

[2] Nanjing University of Science and Technology, Nanjing 210023, China
yyang@njust.edu.cn

Abstract. Recognizing object of unseen novel classes is of great importance in real-world incremental data scenarios. Additionally, novel classes arise frequently in data stream mining, *e.g.*, new topics in opinion monitoring, and novel protein families in protein sequence classification. Conducting streaming novel class detection is a complex problem composed of detection and model update. However, when updating the model solely with detected novel instances, it concentrates more on the novel patterns than known ones; thus the detection ability of the model may degrade consequently. This would exert harmful affections to further classification as the data evolving. To this end, in this paper, we consider the accuracy of the detection along with the robustness of model updating, and propose DEtecting Sequentially Novel clAsses with Stable generalization Ability (DesNasa). Specifically, DesNasa utilizes a prototypical network to reflect the structure information between scattered prototypes for novel class detection. Furthermore, the well-designed data augmentation method can help the model learning novel patterns robustly without degrading detection ability. Experiments on various datasets successfully validate the effectiveness of our proposed method.

Keywords: Novel class detection · Data stream classification · Open world learning

1 Introduction

In our dynamically evolving world, data is often with stream format, which may only be available temporarily for storage constraints or privacy issues. To tackle this, many incremental algorithms are proposed, *e.g.*, open-ended object detection [16], incremental image classification [17] and online video classification [21]. While these methods ignore an essential issue of streaming data, namely the emergence of unseen category [25], *e.g.*, in online opinion monitoring, new topics often emerge as news happens [13]; in protein sequence classification, new types of proteins would arise as nature evolves [23]. Under such circumstances, the problem can be decomposed into three sub-problems: detecting novel classes,

K. Karlapalem et al. (Eds.): PAKDD 2021, LNAI 12712, pp. 371–382, 2021.
https://doi.org/10.1007/978-3-030-75762-5_30

classifying known instances, and updating model in the data stream [14]. As a result, traditional anomaly detection methods cannot handle this situation. They only stress one of the three tasks as novelty detection, thus triggering the development in streaming sequentially novel class detection.

There is much research regarding efficiently detecting novel classes with model updating. [6] detects outliers having strong cohesion among themselves, and uses ensemble of clustering models for classification. [3] utilizes nearest neighbor ensemble to deal with problems of different geometric distances. [13] assumes the label of stream data is available after some time delay and adopts clustering method for novelty detection and known classification. [14] utilizes completely-random trees with growing mechanism for model update, and [15] approximates original information by a dynamic low-dimensional structure via matrix sketching. However, great hidden trouble often occurs when deploying these methods with sequentially new classes, namely *detection ability degrading* [6]. Since these methods are designed for detecting and model update in a few periods, they ignore the ability of forecasting in the long future. That is, solely updating the model with limited novel instances incorporates imbalanced confusion into the current model. The performance of the model shall suffer a decline in the following periods, which harms both the accuracy of classifying known classes and the ability to detect the unknown novel classes. As a result, after several times of updating, error accumulates and the models will fail to predict. These two problems are concurrent, and impose the challenge of detection efficiency and detection stability to exploit these methods for developing algorithms.

To solve the co-occurred problems in class incremental learning, we propose detecting sequentially novel classes with stable generalization ability (DesNasa), which can identify the emergency new classes and learn novel patterns simultaneously without degrading detection ability. Specifically, DesNasa utilizes a prototypical network with intra-class compactness, aiming to acquire the ability of learning distance function over input space and the task space. Thus, the framework can consider information from the scattered prototypes and incoming instance for new class detection. Moreover, the prototypes can help in maintaining forecasting ability when learning novel patterns by mixed replay. Consequently, we can detect and learn novel patterns more efficiently and meanwhile maintain stable performance during the data stream evolution.

2 Related Work

Three key sub-problems, *i.e.,* detecting emerging new classes, classifying known instances, and updating models with detected novel patterns form the current problem. Traditional anomaly detection methods [12,22] mainly focus on identifying anomaly or outlier under static environment. However, these methods need the whole dataset to process, which is not suitable for streaming format data in the dynamic environment. Consequently, several online novelty detection methods have been proposed. [2] extends novelty detection to dynamic streams, while the detection component and the update component differs in these methods, which results in accumulating error and hard to tune parameters. On the other hand, these methods are designed for novelty detection, which is a binary classification problem. While in streaming novel class detection, a significant difference lies in that anomaly instances and known instances should be assigned with certain class labels [14]. The model should have the ability to classify known instances and update with data stream, thus making the above methods incapable of stream novel class detection scenarios. To solve the three sub-problems in a unified framework, many works try to introduce additional information in stream mining. [13] used a clustering approach ECSMiner which tackles the novel class detection and classification problems by introducing time constraints for delayed classification. LACU [4] needs an unlabeled dataset to help classification. [14] utilizes completely-random trees with growing mechanism, and [3] studied nearest neighbor ensemble for different geometric distances. These methods need additional information such as saving relatively large subsets of the original data or obtaining an extra unlabeled dataset. However, heuristically saving subsets consumes huge memory and complex select cost.

Other lines of study involve open-set recognition [5] and active learning [8]. Open-set recognition aims to conduct classification on known class instances and reject unknown ones for *pre-collected* datasets. However, it differs with the current scenario that the open-set model cannot update incrementally, thus these methods are incapable of handling data streams. Active learning is designed for achieving greater performance with fewer chosen labeled training instances. In the current problem, only novel instances are chosen to update the model, which can be seen as a particular case of active learning. Nevertheless, our target is not only learning novel patterns but also maintaining the knowledge of former classes, which is outside the scope of active learning.

3 Preliminaries

Notations. Given the initial model $\mathcal{M}$ pretrained on previous data, which contains k classes, we have the streaming data as: $S = \{(\mathbf{x}_t, y_t)\}_{t=1}^{\infty}$, where $\mathbf{x}_t \in R^d$ and $y_t \in Y = \{1, 2, \ldots, k, k+1, \ldots, K\}$ with $K \gg k$, t is the timestamp. The goal is to utilize $\mathcal{M}$ as a detector for emerging new class and a classifier for known class [3,14]. Then $\mathcal{M}$ is updated timely such that it maintains accurate predictions for known and emerging new classes on streaming data S.

Problem Overview. The goal is to detect novel candidates and classify known instances in the data stream. Stream data are fed into the model, and it scores each instance with the probability of being a novel class. Instances predicted as known classes would be passed to the classifier for prediction, while those predicted as novel instances are saved in the buffer $\mathcal{B}$. After the prediction of every period, the model will request the true label of instances in the buffer to reduce query cost, and remove instances belonging to existing classes. *Only true-novel instances in the buffer will be used to update the model* [14]. The left part of Fig. 1 shows the framework. More novel classes will arise, and the model can learn to classify with time.

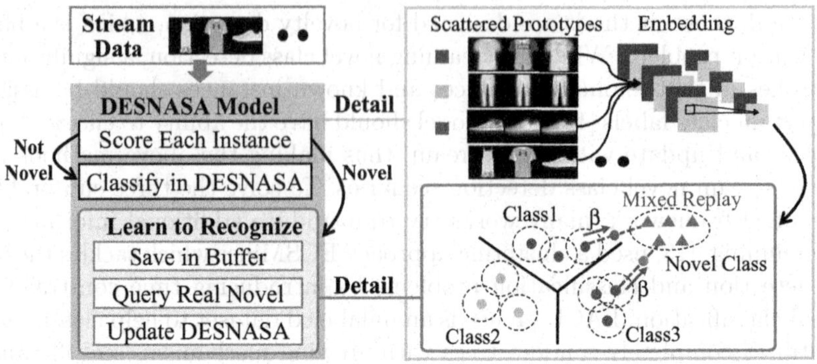

Fig. 1. Illustration of the framework. Stream data are fed into the model, and DESNASA scores each instance of being novel. Instances predicted as novel are saved in the buffer. Only real-novel instances in the buffer will be used to update the model. Specifically, it assigns probability by the normalized distance between instance and prototypes in embedding space. In the update stage, prototypes and novel instances are mixed for alleviating detection ability decay.

4 Proposed Method

4.1 The Framework of DESNASA

Since the stream data is dynamic and ever-changing, a proper way to model each class is to pick prototypical samples. To this end, we design a prototypical deep model, which embeds scattered class centers and arrange probability of belonging to a known class from distance in the embedding space. In detail, DESNASA utilizes a deep network as embedding module $\mathcal{E}(\cdot)$. Besides, the class centers in the data steam can be maintained to represent every class, leading to the class center set C. Then for class k, the mean of all instances belong to this class can be calculated as: $\mathbf{c}_k = \frac{1}{|\Omega_k|} \sum_{\mathbf{x} \in \Omega_k} \mathbf{x}$. In consideration of the property of data stream, we can maintain class centers through an incremental way:

$$\hat{\mathbf{c}}_k = \mathbf{c}_k + \frac{\mathbf{x} - \mathbf{c}_k}{|\Omega_k| + 1} . \quad (1)$$

In order to obtain more diverse class centers, we introduce scattered class centers with instance picking probability p:

$$\hat{\mathbf{c}}_{k,p} = \mathbf{c}_{k,p} + \mathbb{I}(\gamma < p) \frac{\mathbf{x} - \mathbf{c}_{k,p}}{|\Omega_k| + 1} , \quad (2)$$

where $\mathbb{I}(\cdot)$ is the indicator function, and γ is randomly sampled from $\mathcal{U}(0, 1)$. We define the picking probability of each prototype by dividing $0 \sim 1$ into P equal parts, where P is the number of prototypes maintained for each class. That is, prototypes with higher p will be calculated with more instances from this class, thus different prototypes of same class would reveal diversity from each other.

We denote scattered prototypes as C_{ij}, where $i \in \{1, 2, \cdots, k\}$ represents class index and $j \in \{1, 2, \cdots, P\}$ stands for prototype index.

After calculating the scattered prototypes, the embedding module $\mathcal{E}(\cdot)$ will then produce feature maps of $\mathcal{E}(C_{ij})$ and $\mathcal{E}(\mathbf{x})$ in the embedding space. We assume that the distance in the embedding space can be used to measure the similarity between samples and prototypes, *i.e.*, $p(\mathbf{x} \in C_{ij} \mid \mathbf{x}) \propto -\|\mathcal{E}(\mathbf{x}) - \mathcal{E}(C_{ij})\|_2^2$. Thus, by normalizing the distance of all known prototypes, we are able to acquire the probability of $\mathbf{x}$ belonging to prototype C_{ij} by:

$$p(\mathbf{x} \in C_{ij} \mid \mathbf{x}) = \frac{e^{-d(\mathcal{E}(\mathbf{x}), \mathcal{E}(C_{ij}))/\tau}}{\sum_{m=1}^{k} \sum_{n=1}^{P} e^{-d(\mathcal{E}(\mathbf{x}), \mathcal{E}(C_{mn}))/\tau}}, \tag{3}$$

where $d(\mathcal{E}(\mathbf{x}), \mathcal{E}(C_{ij})) = \|\mathcal{E}(\mathbf{x}) - \mathcal{E}(C_{ij})\|_2^2$ represents the distance between $\mathcal{E}(\mathbf{x})$ and $\mathcal{E}(C_{ij})$, τ corresponds to the temperature which controls the smoothness of probability assignment. We further get the probability of $p(y \mid \mathbf{x})$ by summing the probability belonging to prototypes of same class up:

$$p(y \mid \mathbf{x}) = \sum_{j=1}^{P} p(\mathbf{x} \in C_{yj} \mid \mathbf{x}). \tag{4}$$

Considering the probability calculated from embedding space, we then optimize the embedding module by defining distance-based cross entropy:

$$Loss = -\log p(y \mid \mathbf{x}) = -\log(\sum_{j=1}^{P} p(\mathbf{x} \in C_{yj} \mid \mathbf{x})). \tag{5}$$

During the training process, we optimize the model by minimizing loss function Eq. 5. Then for instances from known class, the model maximizes the probability of $\mathbf{x}$ being associated with a prototype among prototype set C. Thus it decreases the distance between incoming instances with the centroids from the genuine class of the samples. We then introduce the novel class detection algorithm.

4.2 Novel Class Detection

Classical novel class detection methods base on predict uncertainty or structure information [3,14], and score each instance with the probability of being novel. These methods will then manually set fixed threshold to determine the decision boundary between unknown and known. However, the fixed threshold differs in different tasks with different input distribution and thus hard to tune. In this section, we develop a suitable strategy to set a dynamic threshold, which can utilize the predicted uncertainty with structure information for better separation.

During the stream data deployment, for every instance in the stream S, we should determine if it belongs to a known class or a novel class. Considering the probability arranged by embedding distance in Eq. 3. Instances with higher probability are those closer to corresponding prototypes in the embedding

Algorithm 1. Novel Class Detection

Input: Data stream : $S = \{\mathbf{x}_t\}_{t=1}^{\infty}$
Output: Class label of each $\mathbf{x}_t$

Empty buffer $\mathcal{B} \leftarrow \varnothing$;
repeat
  Get a test instance $\mathbf{x}_t$;
  Calculate the embedding of instance $\mathcal{E}(\mathbf{x}_t)$;
  Calculate the embeddings of prototypes $\mathcal{E}(C_{ij}), \forall i = 1, 2, \cdots, k; \forall j = 1, 2, \cdots, P$;
  Calculate normalized probability $p(y \mid \mathbf{x}_t) \leftarrow$ Eq.4 ;
  Calculate the confidence of predicted instance $Conf(\mathbf{x}_t) = \max p(y \mid \mathbf{x}_t)$;
  if $Conf(\mathbf{x}_t) <$ dynamic threshold dt **then**
    Save $\mathbf{x}_t$ in the buffer; $\mathcal{B} \leftarrow \mathcal{B} \cup \{\mathbf{x}_t\}$;
    $y_t = k + 1$, Predict as a novel class item;
    Update dummy class center $C_{novel} \leftarrow$ Eq.1;
    Update dynamic threshold $dt \leftarrow$ Eq.7;
  else
    $y_t = \operatorname*{argmax}_{y=1,\dots,k} p(y \mid \mathbf{x})$, predict as a known class;
  end if
until Stream is empty

space. This indicates that the confidence of the model is equivalent to the output highest probability $\max p(y \mid \mathbf{x})$ of Eq. 4. For a known instance, we have the scattered prototypes maintained in the training process, and the corresponding confidence should be relatively high. But for an unseen novel instance, we have no prior about what this class is like, and the arranged probability shall separate between all known prototypes; thus the model would not indicate relatively high confidence than a known instance. We can design the detection formula of the model:

$$f(\mathbf{x}) = \begin{cases} \operatorname*{argmax}_{y=1,\dots,k} p(y \mid \mathbf{x}) & \text{if } \max p(y \mid \mathbf{x}) > \text{ threshold} \\ k+1 & \text{otherwise} \end{cases} . \tag{6}$$

Note that the *threshold* is designed to tell apart known instances from novel classes, which is hard to tune between different tasks. We then design a *dummy class* method for convenient threshold implementation. In detail, once we detect novel-like candidates in the test stream, we then save them in the novelty buffer $\mathcal{B}$. Buffered instances will form the dummy class to estimate novel class distribution by forming novel prototypes with Eq. 1. The dynamic threshold dt is updated by:

$$\hat{dt} = \lambda * dt + (1 - \lambda) * (\max p(y \mid C_{dummy}) + m), \tag{7}$$

where λ is a trade-off parameter close to 1, and m is a positive margin, which aims at making the threshold converge to the confidence of novel prototype and keep it slightly higher. And m can be estimated with $\eta std(\max p(y \mid C_{dummy}))$. Considering when initially define the threshold, it may not well stand for the cutoff between known classes and novel class. Each time the model detects a novel candidate, the dynamic threshold will be updated through Eq. 7. Thus it

Algorithm 2. Learn to Recognize

Input: Predicted novel instance buffer: $\mathcal{B} = \{(\mathbf{x}_j, \mathbf{y}_j)\}_{j=1}^{m}$
Output: Updated model : $\mathcal{M}$

```
1: Request true labels from B, and remove instances belonging to known classes.
2: for each novel instance (x_j, y_j) ∈ B do
3:     Update novel class center ← Eq.2
4:     Sample a known class prototype (x_i, y_i)
5:     Mix known class center and novel instance ← Eq.8
6:     Calculate the embedding of the mixed instance E(x̃)
7:     Calculate the embeddings of prototypes E(C_ij), ∀i = 1, 2, ··· , k; ∀j = 1, 2, ··· , P;
8:     Calculate the normalized probability p(y | x̃) ← Eq.4
9:     Calculate distance-based cross-entropy L_j ← Eq. 5
10:    Obtain the derivative ∂L_j/∂Θ, update model;
11: end for
```

steps closer towards the ideal cutoff between novel dummy prototypes and known prototypes. The algorithm for novel class detection is shown in Algorithm 1.

4.3 Learn to Recognize

Another essential part of DesNasa is to learn the pattern of novel class without degrading detection ability. That is, how to update the model with only true-novel instances in the buffer. Traditional methods based on structure information may fail for unable to depict the novel class distribution. In contrast, DNN-based methods may suffer severe imbalanced learning due to pure novel instances. While in our model, we solve the problem by learning *mixed replay*.

For our model, a straightforward way to learn the novel patterns is to conduct stochastic gradient descent on these pure-novel instances. While the performance of the model will degrade when novel instances incorporating. This is caused by the imbalanced learning phase in model updating, which harms the prior knowledge, and thus makes the model incapable of forecasting as data evolves. The model should concentrate equally on both former and current distributions.

Considering DesNasa maintains the prototypes in the data stream, which stand for the data from the former distribution. A proper way is to retrain them when updating the model with novel instances. That is, to train with the prototype along with novel instances. Thus the model can optimize for novel patterns and meanwhile preserving former classes knowledge. However, a fatal problem of overfitting often occurs when conducting many times of retraining on limited prototypes [7]. To solve this problem, we design mixed replay which combines scattered class centers along with novel instances to form mixed instance:

$$\tilde{\mathbf{x}} = \beta \mathbf{x}_i + (1 - \beta)\mathbf{x}_j; \quad \tilde{\mathbf{y}} = \beta \mathbf{y}_i + (1 - \beta)\mathbf{y}_j \ , \tag{8}$$

where $(\mathbf{x}_i, \mathbf{y}_i)$ is a known prototype, and $(\mathbf{x}_j, \mathbf{y}_j)$ is one novel class instance, $\mathbf{y}_i, \mathbf{y}_j$ are corresponding one-hot label encodings, and $\beta \in [0, 1]$, we utilize the interpolation of one known class center and one novel instance to force the model concentrate on knowledge learned before. [24] proved Eq. 8 establishes a linear

Table 1. Prediction results (mean ± standard deviation) of different compared methods on simulated streams. The best performance is bolded.

Methods	MNIST		Fashion-MNIST		HAR		CIFAR10	
	F-Measure	Accuracy	F-Measure	Accuracy	F-Measure	Accuracy	F-Measure	Accuracy
iForest+KNN	0.697 ± 0.04	0.687 ± 0.05	0.623 ± 0.03	0.679 ± 0.06	0.678 ± 0.02	0.696 ± 0.02	0.627 ± 0.04	0.666 ± 0.03
ODIN	0.752 ± 0.05	0.787 ± 0.05	0.744 ± 0.04	0.729 ± 0.03	0.776 ± 0.02	0.783 ± 0.01	0.801 ± 0.03	0.796 ± 0.03
LACU-SVM	0.699 ± 0.06	0.705 ± 0.08	0.633 ± 0.06	0.677 ± 0.04	0.691 ± 0.05	0.752 ± 0.03	0.644 ± 0.02	0.652 ± 0.04
ECSMiner	0.714 ± 0.06	0.737 ± 0.09	0.700 ± 0.02	0.746 ± 0.03	0.660 ± 0.04	0.682 ± 0.06	0.727 ± 0.05	0.731 ± 0.05
SENCForest	0.747 ± 0.04	0.763 ± 0.03	0.779 ± 0.02	0.781 ± 0.06	0.762 ± 0.06	0.782 ± 0.04	0.766 ± 0.03	0.717 ± 0.03
SENNE	0.807 ± 0.04	0.789 ± 0.04	0.794 ± 0.06	0.783 ± 0.05	0.811 ± 0.02	0.808 ± 0.03	0.798 ± 0.03	0.770 ± 0.02
CPL	0.747 ± 0.06	0.755 ± 0.06	0.766 ± 0.08	0.750 ± 0.07	0.744 ± 0.04	0.785 ± 0.06	0.764 ± 0.03	0.768 ± 0.03
DesNasa	**0.830 ± 0.03**	**0.829 ± 0.03**	**0.816 ± 0.03**	**0.812 ± 0.02**	**0.819 ± 0.03**	**0.824 ± 0.02**	**0.862 ± 0.03**	**0.860 ± 0.03**

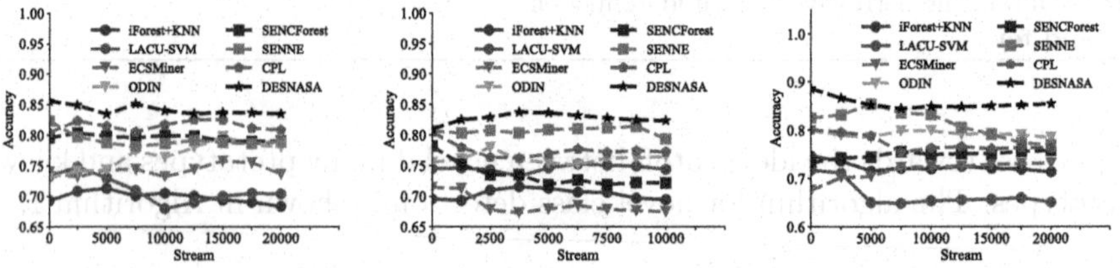

(a) MNIST (b) HAR (c) CIFAR10

Fig. 2. Accuracy curve of different methods over sequential stages.

relationship between data augmentation and the supervision signal, and thus leads to a strong regularizer that improves generalization. Considering the phenomenon that solely replays over instances from former distribution shall cause overfitting, Mixed replay offers a simple way to overcome this drawback.

Indeed, after many rounds of training with mixed replay $(\tilde{\mathbf{x}}, \tilde{\mathbf{y}})$, the model acquires the pattern of novel classes and meanwhile consolidates prior knowledge. The guideline for learning to recognize novel class is shown in Algorithm 2.

5 Experiments

Datasets and Configurations. We provide the empirical results and performance comparison of DesNasa. In particular, we test with 4 benchmark datasets, *i.e.,* MNIST [10], Fashion-MNIST [19], CIFAR10 [9], and HAR [1], and 1 real-world incremental dataset, *i.e.,* NYTimes [14], which consists of 35,000 latest news crawled with the New York Times API. Each news item is classified into six categories, namely, 'Arts', 'Business Day', 'Sports', 'U.S.', 'Technology' and 'World'. Each instance is converted into a 100 dimension vector with word2vec.

Following the typical setting as [3,14], to simulate emerging new classes in the data stream, we assume that training set D with two known classes is available at the initial stage. Instances of these two known classes and an emerging new class appear in the first period of the data stream with uniform distribution. In the second period, instances of these three classes seen in the first period and

Table 2. Prediction results (mean ± standard deviation) of different compared methods on novel class forecasting, which stands for performance of the model during next period when updating with different compared methods.

Methods	MNIST		Fashion-MNIST		HAR		CIFAR10	
	F-Measure	Accuracy	F-Measure	Accuracy	F-Measure	Accuracy	F-Measure	Accuracy
SGD	0.310 ± 0.06	0.386 ± 0.05	0.619 ± 0.05	0.656 ± 0.05	0.626 ± 0.09	0.676 ± 0.07	0.671 ± 0.05	0.705 ± 0.04
Selective Replay	0.410 ± 0.04	0.463 ± 0.03	0.604 ± 0.07	0.648 ± 0.08	0.640 ± 0.07	0.680 ± 0.04	0.681 ± 0.06	0.707 ± 0.05
Distill Replay	0.496 ± 0.08	0.552 ± 0.06	0.666 ± 0.09	0.693 ± 0.08	0.656 ± 0.05	0.691 ± 0.04	0.772 ± 0.02	0.768 ± 0.02
Generative Replay	0.401 ± 0.09	0.468 ± 0.08	0.498 ± 0.06	0.557 ± 0.07	0.581 ± 0.07	0.609 ± 0.09	0.680 ± 0.02	0.729 ± 0.02
Mixed Replay	**0.832 ± 0.03**	**0.831 ± 0.03**	**0.834 ± 0.07**	**0.835 ± 0.05**	**0.792 ± 0.06**	**0.797 ± 0.05**	**0.863 ± 0.02**	**0.862 ± 0.02**

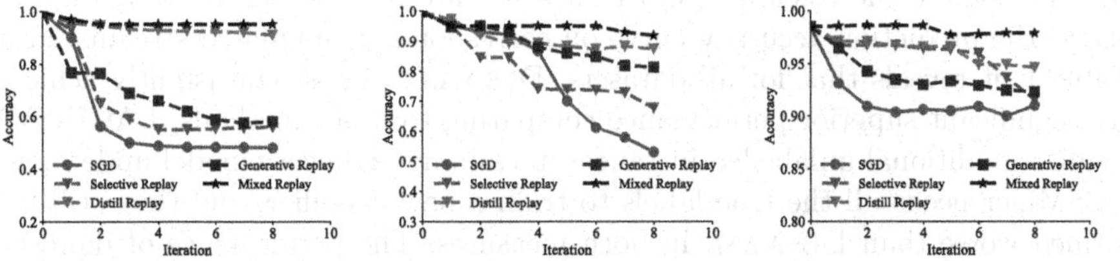

(a) MNIST (b) HAR (c) CIFAR10

Fig. 3. The test accuracy on initial classes when model update with different methods.

another emerging new class appear with uniform distribution. Instances appear one at a time, and the model should make a prediction for every instance before processing the next. Each dataset is used to simulate a data stream over five trails, and both mean and standard variance of the performance are reported. Considering novel classes may arise simultaneously in a single period, while some compared methods cannot handle it. We simulate one novel class to arrive in a single period for a fair comparison.

Compared Methods. We first compare to a common-used novelty detection method iForest [12] and deep anomaly detection algorithm ODIN [11]. We also compare to the state-of-the-art methods which combine novel class detection with model update: LACU-SVM [4], ECSMiner [13], SENC-Forest [14], SENNE [3]. Convolutional prototype learning method CPL [20] is also compared. We also conduct experiments to evaluate the forecasting ability of the model. We adopt the *same* embedding module and detection rules for novel class detection, and 4 major methods, *i.e.*, SGD, Selective Replay [7], Generative Replay [18] and Distill Replay [17] are implemented for the model update process: **SGD**: update model with only novel instances; **Selective Replay**: save former instances with reservoir sampling, and retrain them with the novel instance at update stage; **Generative Replay**: training generative model for former data distribution, and rehearsal generated data with novel instance to update; **Distill Replay**: consider knowledge distillation loss of former instance when updating the model. Note that the latter three methods are initially proposed to overcoming forgetting of former knowledge in incremental learning [21], which can be viewed as a similar problem as the detection ability decline in our paper.

F-Measure and Accuracy are recorded to evaluate the detection and classification performance, respectively [3,14]. F-measure evaluates the effect of novel class detection, which is calculated at several intervals and use the average as the final evaluation. Accuracy= $\frac{A_{\text{novel}}+A_{\text{known}}}{m}$, where m is the total number of stream instances, A_{novel} is the number of novel instances identified correctly, A_{known} is the number of known class instances classified correctly.

Real-World Stream Data Classification. The stream prediction results, which calculate the Accuracy and F-measure are reported in Table 1. Figure 2 shows the prediction accuracy curve on corresponding simulated streams. From Table 1, it reveals that for all datasets, DesNasa almost consistently achieves the significant superior performance comparing to other methods. LACU-SVM requires additional unlabelled instances in training and every model update, and ECSMiner needs all the true labels to train a new classifier, and they still performed worse than DesNasa in both measures. The performance of iForest is dissatisfactory on the dataset with high dimensions for it is tree-based, which only uses a few dimensions for tree building. SENNE needs to store huge amounts of data/ensemble models, but our DesNasa still outperforms it in all datasets. ODIN and CPL utilize the same DNN backbone as DesNasa , while they reveal unsatisfactory results for the fixed threshold is hard to tune, but DesNasa can dynamically adjusting threshold with data stream evolving.

Detection Ability Maintenance. We also conduct experiments about detection ability maintenance. Table 2 and Fig. 3 shows the result of *same base model* (embedding and detection rules) updating with different methods. In Fig. 3, after updating with other compared methods, prior knowledge of the model will be disturbed, which results in the degrading of detection ability. Additionally, Table 2 shows the corresponding prediction performance of the *next* period when updating with these methods, which reveals that the forecasting ability will be degraded with other methods. In comparison, the model updated with mixed replay suffers the least disturbance in model updating, and can achieve the best performance in the next-period prediction.

Stream News Identification. In this part, we conduct experiments in a real-world text stream, *i.e.,* NYTimes. News categorization for a news stream is an important issue, where a new topic of news may arise due to a newly occurred event. News data are categorized into six classes, and we treat two of them as known classes at the initial stage and simulate others to appear sequentially in the later stages. Figure 4 shows the experiment results. We can infer from Fig. 4(a) that in a real-world text stream, DesNasa can still achieve the best performance among all compared methods. Figure 4(b) indicates the mixed replay in DesNasa can help the model dealing with sequentially novel classes with the least detection ability degrading.

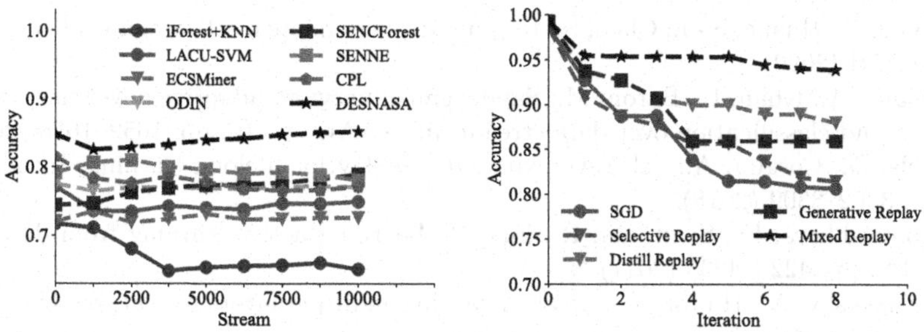

Fig. 4. Accuracy curve on NYTimes. Left: prediction accuracy with data evolves. Right: test accuracy on initial known classes when model update with different methods.

6 Conclusion

In many real-world incremental data scenarios, data arise with novel classes. To solve novel class detection and model extension problem, we propose a novel framework to detect sequentially novel classes with stable generalization ability (DesNasa). It can efficiently detect the novel class and extend the model with the least detection ability degrade. More specifically, DesNasa can detect novel class instances by utilizing the structure information of a prototypical network. Furthermore, it can learn to recognize through data augmentation between prototypes and novel instances without degrading detection ability. Experiments on image and text datasets successfully validate DesNasa stabilizes the performance of the regularly updated model.

Acknowledgments. This research was supported by NSFC (61773198, 61632004, 61921006, 62006118), NSFC-NRF Joint Research Project under Grant 61861146001, Collaborative Innovation Center of Novel Software Technology and Industrialization, CCF- Baidu Open Fund (CCF-BAIDU OF2020011), Baidu TIC Open Fund, Natural Science Foundation of Jiangsu Province of China under Grant (BK20200460) and Nanjing University Innovation Program for PhD candidate (CXYJ21-53). De-Chuan Zhan is the corresponding author.

References

1. Aggarwal, C.C.: A survey of stream clustering algorithms. In: Data Clustering, pp. 231–258. Chapman and Hall/CRC (2018)
2. Ahmad, S., Lavin, A., Purdy, S., Agha, Z.: Unsupervised real-time anomaly detection for streaming data. Neurocomputing **262**, 134–147 (2017)
3. Cai, X.-Q., Zhao, P., Ting, K.-M., Mu, X., Jiang, Y.: Nearest neighbor ensembles: an effective method for difficult problems in streaming classification with emerging new classes. In: ICDM, pp. 970–975. IEEE (2019)
4. Da, Q., Yu, Y., Zhou, Z.-H.: Learning with augmented class by exploiting unlabeled data. In: AAAI, pp. 1760–1766 (2014)

5. Geng, C., Huang, S.-J., Chen, S.: Recent advances in open set recognition: a survey. TPAMI (2020)
6. Haque, A., Khan, L., Baron, M.: Sand: semi-supervised adaptive novel class detection and classification over data stream. In: AAAI, vol. 16, pp. 1652–1658 (2016)
7. Isele, D., Cosgun, A.: Selective experience replay for lifelong learning. In: AAAI, pp. 3302–3309 (2018)
8. Konyushkova, K., Sznitman, R., Fua, P.: Learning active learning from data. In: NIPS, pp. 4225–4235 (2017)
9. Krizhevsky, A., Hinton, G., et al.: Learning multiple layers of features from tiny images. Technical report, Citeseer (2009)
10. LeCun, Y., Cortes, C., Burges, C.J.: Mnist handwritten digit database. AT&T Labs, 2:18 (2010). http://yann.lecun.com/exdb/mnist
11. Liang, S., Li, Y., Srikant, R.: Enhancing the reliability of out-of-distribution image detection in neural networks. In: ICLR (2018)
12. Liu, F.T., Ting, K.M., Zhou, Z.-H.: Isolation forest. In: ICDM, pp. 413–422 (2008)
13. Masud, M., Gao, J., Khan, L., Han, J., Thuraisingham, B.M.: Classification and novel class detection in concept-drifting data streams under time constraints. TKDE **23**(6), 859–874 (2010)
14. Mu, X., Ting, K.M., Zhou, Z.-H.: Classification under streaming emerging new classes: a solution using completely-random trees. TKDE **29**(8), 1605–1618 (2017)
15. Mu, X., Zhu, F., Du, J., Lim, E.-P., Zhou, Z.-H.: Streaming classification with emerging new class by class matrix sketching. In: AAAI, pp. 2373–2379 (2017)
16. Perez-Rua, J.-M., Zhu, X., Hospedales, T.M., Xiang, T.: Incremental few-shot object detection. In: CVPR, pp. 13846–13855 (2020)
17. Rebuffi, S.-A., Kolesnikov, A., Sperl, G., Lampert, C.H.: icarl: incremental classifier and representation learning. In: CVPR, pp. 2001–2010 (2017)
18. Shin, H., Lee, J.K., Kim, J., Kim, J.: Continual learning with deep generative replay. In: NeurIPS, pp. 2990–2999 (2017)
19. Xiao, H., Rasul, K., Vollgraf, R.: Fashion-Mnist: a novel image dataset for benchmarking machine learning algorithms. preprint arXiv:1708.07747 (2017)
20. Yang, H.-M., Zhang, X.-Y., Yin, F., Liu, C.-L.: Robust classification with convolutional prototype learning. In: CVPR, pp. 3474–3482 (2018)
21. Yang, Y., Zhou, D.-W., Zhan, D.-C., Xiong, H., Jiang, Y.: Adaptive deep models for incremental learning: considering capacity scalability and sustainability. In: SIGKDD, pp. 74–82 (2019)
22. Yang, Y., Zhang, J., Carbonell, J., Jin, C.: Topic-conditioned novelty detection. In: SIGKDD, pp. 688–693 (2002)
23. Zhang, D., Liu, Y., Si, L.: Serendipitous learning: learning beyond the predefined label space. In: SIGKDD, pp. 1343–1351 (2011)
24. Zhang, H., Cisse, M., Dauphin, Y.N., Lopez-Paz, D.: mixup: beyond empirical risk minimization. In: ICLR (2018)
25. Zhou, Z.-H.: Learnware: on the future of machine learning. Front. Comput. Sci. **10**(4), 589–590 (2016)

Learning-Based Dynamic Graph Stream Sketch

Ding Li, Wenzhong Li(✉), Yizhou Chen, Mingkai Lin, and Sanglu Lu

State Key Laboratory for Novel Software Technology, Nanjing University, Nanjing 210023, China
liding@smail.nju.edu.cn, lwz@nju.edu.cn

Abstract. A graph stream is a kind of dynamic graph representation that consists of a consecutive sequence of edges where each edge is represented by two endpoints and a weight. Graph stream is widely applied in many application scenarios to describe the relationships in social networks, communication networks, academic collaboration networks, etc. Graph sketch mechanisms were proposed to summarize large-scale graphs by compact data structures with hash functions to support fast queries in a graph stream. However, the existing graph sketches use fixed-size memory and inevitably suffer from dramatic performance drops after a massive number of edge updates. In this paper, we propose a novel Dynamic Graph Sketch (DGS) mechanism, which is able to adaptively extend graph sketch size to mitigate the performance degradation caused by memory overload. The proposed DGS mechanism incorporates deep neural network structures with graph sketch to actively detect the query errors, and dynamically expand the memory size and hash space of a graph sketch to keep the error below a pre-defined threshold. We conducted extensive experiments on three real-world graph stream datasets, which show that DGS outperforms the state-of-the-arts with regard to the accuracy of different kinds of graph queries.

Keywords: Sketch · Data stream · Graph stream

1 Introduction

A *graph stream* [4,13] is a consecutive sequence of items, where each item represents a graph edge. Each edge is usually denoted by a tuple consisting of two endpoints and a weight. Nowadays graph stream is ubiquitously applied to describe the relationships in social networks, communication networks, academic

This work was partially supported by the National Key R&D Program of China (Grant No. 2018YFB1004704), the National Natural Science Foundation of China (Grant Nos. 61972196, 61832008, 61832005), the Key R&D Program of Jiangsu Province, China (Grant No. BE2018116), the science and technology project from State Grid Corporation of China (Contract No. SGJSXT00XTJS2100049), and the Collaborative Innovation Center of Novel Software Technology and Industrialization.

K. Karlapalem et al. (Eds.): PAKDD 2021, LNAI 12712, pp. 383–394, 2021.
https://doi.org/10.1007/978-3-030-75762-5_31

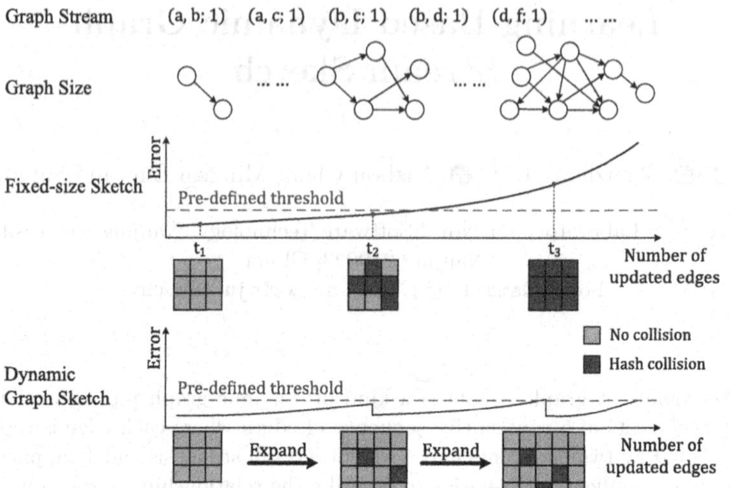

Fig. 1. This figure shows the change of query error with the increasing number of nodes and edge updates using fixed-size sketch and dynamic graph sketch, respectively.

collaboration networks, etc. For example, a graph stream can be used to describe temporal-spatial varying network traffics, or represent a dynamic social network with increasing number of users and their social interactions. It plays an important role since the graph topology is indispensable to many data analysis tasks in these applications. Apparently, traditional data structures such as adjacency matrix and the adjacency list cannot be directly adopted to store a graph stream due to its large volume and high dynamicity.

Sketch is a compact data structure for data stream summarization. Traditional sketches such as CM-sketch [2] and CU-sketch [3] were designed to summarize a stream of isolated data items, which are not suitable for graph stream summarization due to the lack of ability to capture the connections between items and to answer graph topology queries. To address the problem, Tang et al. proposed a *graph sketch* called TCM [13]. It summarized a graph stream by a matrix where each edge was mapped to a bucket of the matrix using a hash function, and the edge weight was recorded in the corresponding bucket. Since the graph sketch recorded not only the edge weight, but also the connections of a graph, it was able to answer graph topology queries such as reachability queries and subgraph queries. Several variants of graph sketch were proposed to improve the query accuracy and efficiency [4,8].

However, the major drawback of the existing graph sketches is that they construct a graph summarization based on a pre-defined fixed-size matrix which are not expandable. Many real-world applications are dealing with dynamic graph streams, i.e., the population (the nodes) of a social network may increase rapidly and the interaction between users (the edges) may change dynamically. Applying the fixed-size graph sketch on the dynamic graph streams inevitably suffers from poor accuracy of query results after a large number of edge updates. We illustrate this problem in Fig. 1. The figure shows the change of the average relative error of edge queries with the number of edge updates in TCM. As can be seen, TCM suffers from serious performance degradation during the edge updating process. At the beginning, hash collisions are rare and the query error is relatively low. With more and more edge updates, hash collision occurs in every buckets of TCM and the memory is overloaded. This leads to the fact that the average relative error of edge queries becomes very high. Thus, it is desirable for a graph sketch to have the ability of incremental expansion to avoid rapid performance degradation and keep the relative error below a threshold.

In this paper, we propose a novel Dynamic Graph Sketch (DGS), which is able to adaptively extend its size and redistribute hash collisions to mitigate the performance degradation caused by overloaded memory. In contrast to the existing graph sketches that use fixed-size memory, DGS is able to keep the query accuracy always at a high level, no matter how many edges have been updated. Figure 1 presents an example how DGS works and how the performance of DGS changes with the number of edge updates. As shown in the figure, DGS periodically predict its current query error. If the error exceeds a pre-defined threshold, it will expand its size and simultaneously redistribute hash collisions to reduce the query error. Then, the expanded sketch continues to record the subsequent edges. In this way, the query error can be always limited under the pre-defined threshold.

The main results and contributions of this paper are summarized as follows:

- We propose Dynamic Graph Sketch (DGS), a novel mechanism for graph stream summarization. It is able to adaptively expand its size to mitigate the performance degradation during the edge updating process, keeping the query accuracy always at a high level. To the best of our knowledge, we are the first to propose the dynamic graph sketch which uses adaptively incremental memory.
- We integrate deep learning techniques into graph sketch design. Specifically, we introduce a deep neural network (DNN) to predict a graph sketch's query error, and based on which we design a convolutional neural network (CNN) to aid expanding a small-size graph sketch to a large-size one and simultaneously mitigate hash collisions to improve graph query accuracy.
- We conduct extensive experiments on three real-world graph streams to evaluate the effectiveness of our proposed algorithm. The experimental results show that DGS outperforms state-of-the-art graph sketches in terms of different kinds of graph queries.

2 Related Work

Data stream sketches are designed to summarize the data streams that are modeled as isolated items. Thus, they cannot answer graph topology queries. C-sketch [1] utilized multiple hash tables to store a data stream and was able to estimate the frequencies of all the items. However, it suffered from both overestimation and underestimation. Cormode et al. improved the work of [1] and proposed CM-sketch [2] which only suffered from overestimation. Estan et al. designed CU-sketch [3] to improve CM-sketch's query accuracy at the cost of not supporting item deletions. More recently, Tang et al. proposed MV-sketch [12] which tracked candidate heavy items inside the sketch data structure via the idea of majority voting. Liu et al. proposed SF-sketch [9] which consisted of both a large sketch and a small sketch to upgrade the query accuracy. Besides designing novel data structures, some researchers proposed to utilize the power of machine learning to optimize the performance of sketch. Later, Hsu et al. [6] and Zhang et al. [14] proposed Learned Count-Min algorithm which combined the traditional CM-sketch with a heavy hitter oracle to improve efficiency.

In contrast to data stream sketches, graph sketches are able to summarize graph streams, keeping the topology of a graph and thus supporting more kinds of queries including 1-hop precursor/successor query, reachability query, etc. Zhao et al. designed gSketch [15], which utilized CM-sketch to support edge query and aggregate subgraph query for a graph stream. Tang et al. proposed TCM [13], which adopted an adjacency matrix to store the compressed graph stream. gMatrix [8] was similar to TCM, and it used reversible hash functions to generate graph sketches. More recently, Gou et al. proposed GSS [4], which was the state-of-the-art for graph stream summarization. To improve query accuracy, GSS consisted of not only an adjacency matrix, but also an adjacency list buffer, which can avoid edges collisions to some extent and improve the query accuracy.

In summary, all the existing sketches use fixed-size memory and thus suffer from serious performance degradation after a certain number of edge updates. To the best of our knowledge, the dynamic graph sketch expansion problem has not been well addressed in the past.

3 Preliminaries

In this section, we introduce some preliminaries about graph stream summarization.

Definition 1 (Graph stream). *A graph stream is a consecutive sequence of items $S = \{e_1, e_2, \ldots, e_n\}$, where each item $e_i = (s, d, t, \omega)$ denotes a directed edge from node s to node d arriving at timestamp t with weight ω. Note that an edge e_i can be updated multiple times at different timestamps.*

A graph stream S forms a dynamic directed graph $G = (V, E)$ that changes with every arrival of an edge update, where V denotes the node set and E denotes the edge set of the graph. Since an edge e_i may appear multiple times in the graph stream S, the weight of e_i is computed by an *aggregation function* based on all the edge weights that share the same endpoints. Common aggregation functions include $min(\cdot)$, $max(\cdot)$, $average(\cdot)$, $sum(\cdot)$, etc. In the rest of this paper, we adopt $sum(\cdot)$ as the default aggregation function to introduce our method.

Definition 2 (Graph sketch [13]**).** *A graph sketch is a graph* $K = (V_K, E_K)$ *whose size is smaller than the original graph* $G = (V, E)$ *formed by a given graph stream. Specifically,* $|V_K| \leq |V|$ *and* $|E_K| \leq |E|$*. A hash function* $H(\cdot)$ *is used to map each node in* V *to a node in* V_K*, and edge* (s, d) *in* E *is mapped to edge* $(H(s), H(d))$ *in* E_K*.*

The graph sketch is usually implemented by an adjacency matrix M, and each element $M[i][j]$ in the adjacency matrix is usually called a *counter*.

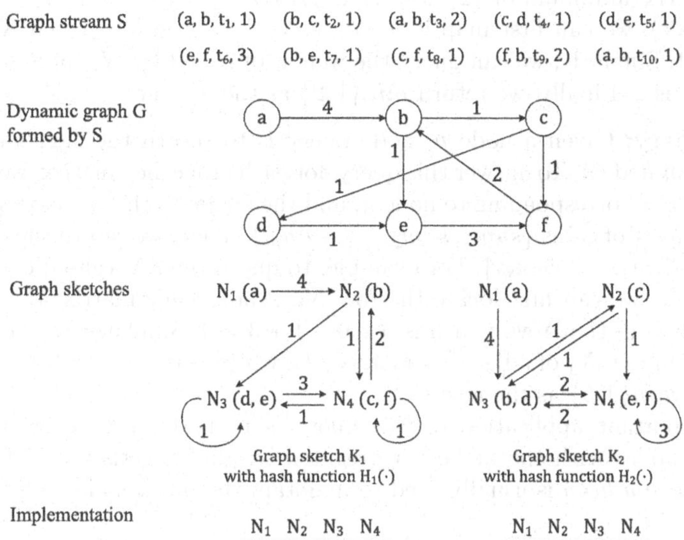

	N_1	N_2	N_3	N_4
N_1	0	4	0	0
N_2	0	0	1	1
N_3	0	0	1	3
N_4	0	2	1	1

	N_1	N_2	N_3	N_4
N_1	0	0	4	0
N_2	0	0	1	0
N_3	0	1	0	2
N_4	0	0	2	3

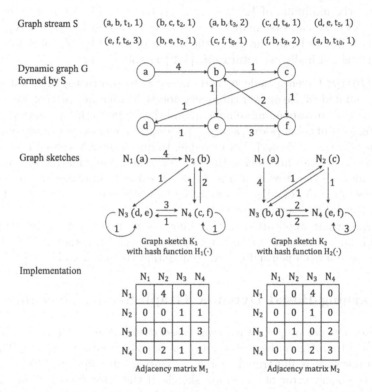

Fig. 2. An example of using two graph sketches to summarize a graph stream.

Usually, several graph sketches will be simultaneously used to record a graph stream to reduce the query error. The hash functions of these sketches are different and mutually independent. In order to better illustrate how to use a set of graph sketches to summarize a graph stream, we give an example which is presented in Fig. 2. As shown in the figure, two graph sketches with different hash functions are used to summarize graph stream S. For each edge in the graph stream, the graph sketches conduct an *edge update* as follows.

Edge Update: To record an edge $e_i = (s, d, t, \omega)$ of the graph stream, the graph sketch first calculates the hash values $(H(s), H(d))$. Then, it locates the corresponding position $M[H(s)][H(d)]$ in the adjacency matrix, and adds the value in that position by ω. For example, to record the edge $(b, c, t_2, 1)$, graph sketch K_1 first calculates the hash values $(H_1(b), H_1(c)) = (N_2, N_4)$, and then the value in $M_1[N_2][N_4]$ is added by 1. Similarly, for graph sketch K_2, the value in $M_2[H_2(b)][H_2(c)]$ is added by 1.

Graph sketches usually support two basic queries: *edge query* and *node query.*

Edge Query: Given an edge e_i, edge query is to return the weight of e_i. To answer this query, we can first query the weight of the edge that e_i is mapped to in all the graph sketches, obtaining a set of weights $\{\omega_1, \omega_2, \ldots, \omega_m\}$. Then, we return the minimum of $\{\omega_1, \omega_2, \ldots, \omega_m\}$. For example, to query the weight of edge (b, e), we can first map it to edge (N_2, N_3) of graph sketch K_1, whose weight is 1. Similarly, we can query the weight of edge (N_3, N_4) of graph sketch K_2, which is 2. Finally, we return $min\{1, 2\}$ as the answer.

Node Query: Given a node n, node query is to return the aggregated edge weight *from* node n. To answer this query, for each adjacency matrix, we can first locate the row corresponding to node n, and then sum up the values in that row, obtaining a set of sums $\{sum_1, sum_2, \ldots, sum_m\}$. Then, we return the minimum of $\{sum_1, sum_2, \ldots, sum_m\}$. For example, to query the aggregated edge weight *from* node e, we can first locate the row N_3 of adjacency matrix M_1, and sum up the values in that row, which is $0 + 0 + 1 + 3 = 4$. Similarly, we can sum up the values in row N_4 of adjacency matrix M_2, which is $0 + 0 + 2 + 3 = 5$. Finally, we return $min\{4, 5\}$ as the answer.

An important application of this query is to find *top-k heavy nodes*, i.e. the top-k nodes with the highest aggregated weight. Together with the graph sketches, a *min-heap* is usually used to maintain the top-k nodes [13].

4 Learning-Based Dynamic Graph Sketch Mechanism

In this section, we propose a learning-based dynamic graph sketch mechanism called DGS to mitigate the performance problem of fixed-sized sketch. The proposed framework is illustrated in Fig. 3. During edge updates, DGS actively detects the query error of the graph sketch. If the error exceeds a pre-defined threshold, it adaptively expands the sketch size and hash space to reduce hash collisions and improve query accuracy. The framework consists of two major components: a *learning-based error detection* module and a *CNN-based sketch expansion* module, which are introduced in detail as follows.

4.1 Learning-Based Error Detection

Given a set of graph sketches formed after a certain number of edge updates, we periodically detect its current query error to decide whether to expand it. Since it is infeasible to calculate the precise error by exhausting all possible node queries and edge queries, we construct a learning-based prediction model to estimate the average relative error of edge queries.

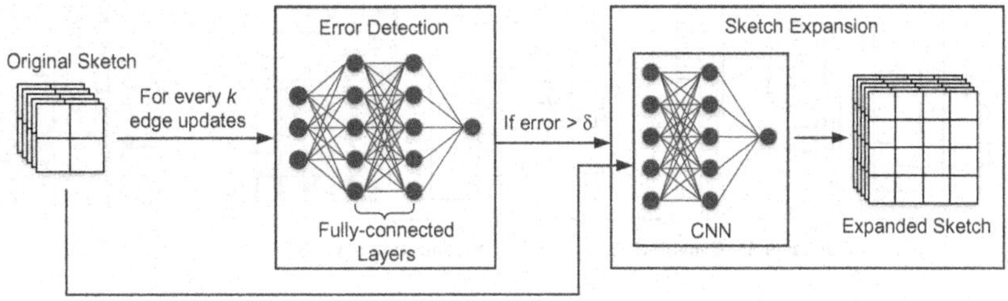

Fig. 3. The framework of dynamic graph sketch (DGS).

The proposed prediction model is illustrated in Fig. 3. It is a neural network consisting of two fully-connected layers. It takes a set of graph sketches as input, flattens all the counters of the graph sketches to a real-value vector, and feeds the vector into the neural network, which outputs the predicted relative error.

The error detection model can be trained based on historical data. For example, we can use a small percentage of the graph stream edge updates to form several graph sketches at different timestamps, and test the edge query errors of those graph sketches to form a labeled dataset, which can be used to train the prediction model.

4.2 CNN-Based Graph Sketch Expansion

Given a set of graph sketches whose current query error is detected to be higher than a pre-defined threshold, we introduce a DNN-based graph sketch expansion module to expand its size and reduce the hash collision and query error.

Convolutional Neural Network (CNN) Architecture. Inspired by the fact that CNN is expert in image super-resolution [5,11] and that images are usually represented by matrices, we design a deep convolutional neural network for graph sketch expansion. The basic idea is to design a CNN model to map low-resolution matrices (smaller-size graph sketches) to high-resolution ones (larger-size graph sketches) while keeping the similarities between them.

The structure of the proposed deep convolutional neural network is presented in Fig. 4. It includes three convolutional layers and a reshaping layer. A graph sketch with size $w \times w$ first goes through three convolutional layers, and the last convolutional layer outputs 4 matrices with size $w \times w$. To form a larger graph sketch, the reshaping layer reshapes these 4 matrices, obtaining a larger matrix with size $2w \times 2w$.

Similar to the training of the error prediction model in Sect. 4.1, this network can be trained using historical data as well. We use a small percentage of edge updates to form several small-size sketches and corresponding large-size sketches at different timestamps. A small-size sketch together with its corresponding large-size sketch at the same timestamp can be used as one training sample to train the network.

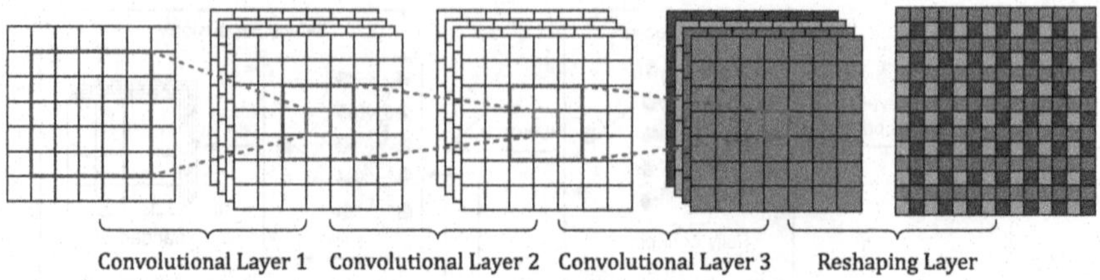

Fig. 4. The proposed CNN structure for graph sketch expansion.

Reshape of Counter Values. After graph sketch expansion, we can obtain a set of large-size graph sketches. However, although the expanded graph sketches have the similar shape as the original ones, their weights (counters) maybe not conform with the definition of graph sketch. To better illustrate this problem, we give an example illustrated in Fig. 5. The original graph sketch is a 2×2 matrix with hash function $H(x) = x \ mod \ 2$. After expansion, the matrix size is 4×4 with hash function $H'(x) = x \ mod \ 4$ (the hash space is expanded both horizontally and vertically). Therefore an edge originally mapped to cell (0, 0) can be now mapped to four cells (0, 0), (0, 2), (2, 0), (2, 2) in the expanded graph sketch. The sum of the weights in the four cells are 20 + 30 + 30 + 40 = 120 which is unequal to the original weight 12. According to the definition of graph sketch, the weights of the matrix should be the counter of the graph stream edges, so we need to reshape the weight of the expanded graph sketch with a normalization trick.

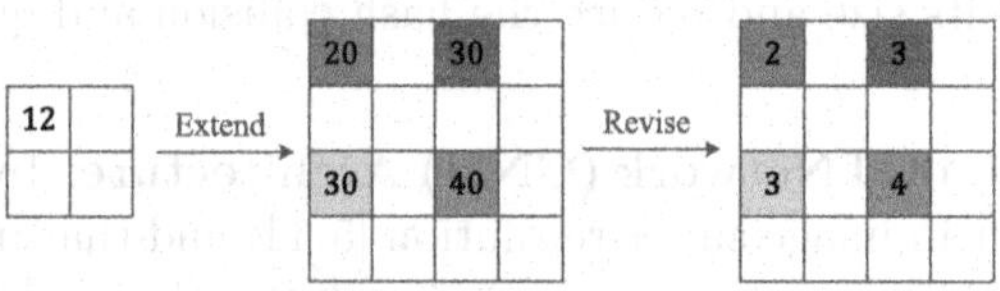

Fig. 5. Reshape of counter values.

Generally speaking, denoted by $K^{original}$ the original graph sketch of size $w \times w$ and $K^{expanded}$ the expanded graph sketch of size $2w \times 2w$, we adopt the following normalized formula to reshape the counter values in the expanded graph sketch:

$$K_{i,j}^{expanded} \leftarrow K_{H(i),H(j)}^{original} \times \frac{K_{i,j}^{expanded}}{sum} \tag{1}$$

where *sum* represents sum of the four counters expanded from the original one which can be calculated by:

$$sum = K_{H(i),H(j)}^{expanded} + K_{H(i),H(j)+w}^{expanded} + K_{H(i)+w,H(j)}^{expanded} + K_{H(i)+w,H(j)+w}^{expanded} \tag{2}$$

5 Performance Evaluation

5.1 Experimental Environment

We conduct extensive experiments to validate the effectiveness of the proposed dynamic graph sketch (DGS). We compare our method with two state-of-the-art graph sketches: TCM [13] and GSS [4]. All experiments were performed on a desktop with Intel Core i7-7700 processors (4 cores, 8 threads), 8 GB of memory, and NVIDIA GeForce GTX 1050 GPU. All sketches except GSS were implemented in Java. For GSS, we used the C++ source code provided on the Github[1]. For fair comparison, we disabled the buffer list of GSS since it does not limit the memory usage. We used the PyTorch library [10] to implement our proposed learning-based prediction and expansion model.

The experiments are based on three real-world graph stream datasets.

- **lkml-reply[2]:** The first dataset is a collection of communication records in the network of the Linux kernel mailing list. It contains 63,399 email addresses (nodes) and 1,096,440 communication records (edges).
- **prosper-loans[3]:** The second dataset is loans between members of the peer-to-peer lending network at Prosper.com. Nodes represent members; edges represent loans and are directed from lenders to borrowers. The dataset contains 89,269 nodes and 3,394,979 edges.
- **facebook-wosn-wall[4]:** The third data set is the directed network of a small subset of posts to other user's wall on Facebook. The nodes of the network are Facebook users, and each directed edge represents one post, linking the users writing a post to the users whose wall the post is written on. The dataset contains 46,952 nodes and 876,993 edges.

We adopt the following performance metrics in our experiments.

- **Average relative error (ARE):** measures the accuracy of the reported weights in edge queries. Given a query q, the *relative error* $RE(q)$ is defined as: $RE(q) = \frac{|\hat{f}(q) - f(q)|}{f(q)}$ where $\hat{f}(q)$ denotes the estimated answer of query q, and $f(q)$ denotes the real answer of query q. Given a set of queries $Q = \{q_1, q_2, \ldots, q_n\}$, the ARE is calculated by $ARE(Q) = \frac{\sum_{i=1}^{n} RE(q_i)}{n}$.
- **Intersection accuracy (IA)** [13]**:** measures the accuracy of the top-k heavy nodes reported by a sketch. Let X be the set of reported top-k heavy nodes, and Y be the set of ground truth top-k heavy nodes. The IA is formulated as $IA = \frac{|X \cap Y|}{k}$.
- **Normalized discounted cumulative gain (NDCG)** [7]**:** measures the quality of a ranking. Given a ranking list of heavy nodes reported by a sketch, *discounted cumulative gain* ($DCG@k$) is defined as $DCG@k =$

[1] https://github.com/Puppy95/Graph-Stream-Sketch.
[2] http://konect.cc/networks/lkml-reply.
[3] http://konect.cc/networks/prosper-loans.
[4] http://konect.cc/networks/facebook-wosn-wall.

$\sum_{i=1}^{k} \frac{r(i)}{log_2(i+1)}$, where $r(i)$ denotes the relative score of the ith node in the ranking list and it belongs to $\{0, 1\}$. If the ith node in the ranking list is indeed a heavy node, $r(i)$ will be 1; otherwise, it will be 0. Using the definition above, $NDCG@k$ is formulated as $NDCG@k = \frac{DCG@k}{IDCG@k}$, where $IDCG@k$ represents the $DCG@k$ of an ideal ranking list obtained by sorting the nodes in the ranking list in descending order with respect to their relative scores. Thus, NDCG ranges in $[0, 1]$, which can be used to evaluate the ability to find top-k heavy nodes of graph sketches (the higher the better).

5.2 Numerical Results

We analyze the performance for edge query and node query of different sketches.

Edge Query. Figure 6 shows the change of edge query's ARE in TCM, GSS, and our proposed DGS. As can be seen, with the increase of updated edges, the ARE of edge queries in fixed-size sketches (TCM and GSS) grows drastically. For high compression ratio (compression ratio = 1/160) in TCM and GSS, the ARE of edge queries in dataset *lkml-reply* is 2.032 and 5.691 after 25% edges updates, and it grows to 14.022 and 28.714 after 75% edge updates, respectively. The performance of TCM (compression ratio = 1/40) and GSS (compression ratio = 1/40) performs better, but they also suffer from high ARE from 75% to 100% edge updates. In contrast, the ARE of the proposed DGS is not sensitive to the number of edge updates, and it outperforms the other algorithms significantly when the percentage of edge updates is larger than 50%. It verifies the effectiveness of variable-size graph sketch for edge query.

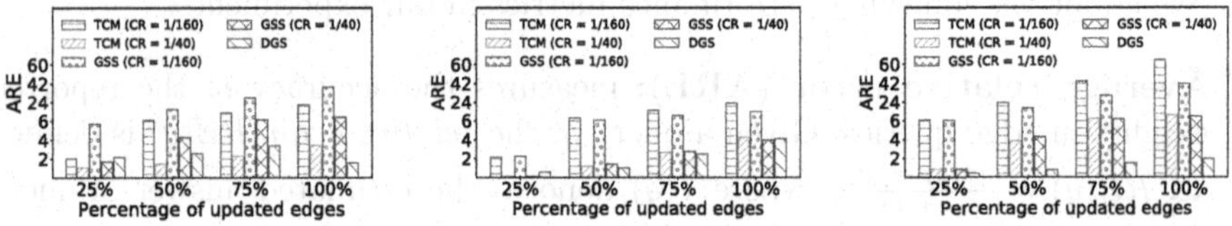

(a) lkml-reply (b) prosper-loans (c) facebook-wosn-wall

Fig. 6. The ARE of edge queries.

Node Query. We evaluate the ability to find top-k heavy nodes of DGS as well as TCM. We do not conduct this experiment on GSS since GSS does not support heavy node query. The results are shown in Fig. 7 and Fig. 8. As shown in Fig. 7, when the size of TCM is small, i.e., with compression ratio 1/160, the intersection accuracy falls down to 24% in some cases. For larger sketch size, i.e., TCM(1/40), the performance improved. In the figures, TCM(ideal) means we set a sufficient large memory size for TCM, which is as large as the final size of DGS. Although

TCM(ideal) achieves the best accuracy in finding top-k nodes, we emphasize that it is hard to set a proper size for TCM at initialization since the size of the incoming graph stream is unknown. As shown in Fig. 7, DGS achieves an intersection accuracy of 85%, 85%, and 80% in the task of finding top-20 heavy nodes on data set *lkml-reply*, *prosper-loans*, and *facebook-wosn-wall* respectively, which outperforms other situations significantly, and it performs very close to the TCM(ideal) situation. This illustrates that our proposed dynamic graph sketch expansion algorithm is effective and able to keep the performance always at a high level.

We also calculate the NDCG based on the result list of top-k heavy node query. The results are shown in Fig. 8. Similarly, the NDCG is worse for predefined fixed-size TCM(1/160) and TCM(1/40). Again, DGS outperforms other situations and it performs very close to TCM(ideal).

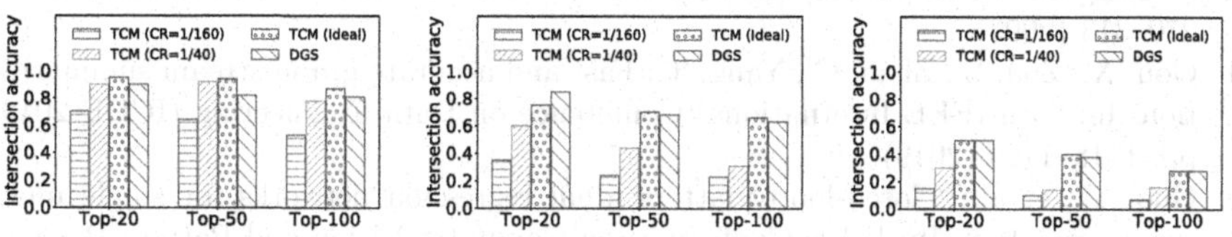

(a) lkml-reply (b) prosper-loans (c) facebook-wosn-wall

Fig. 7. Heavy node query (intersection accuracy)

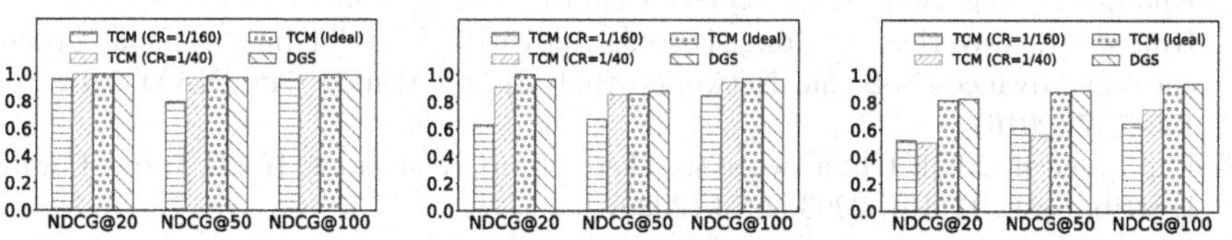

(a) lkml-reply (b) prosper-loans (c) facebook-wosn-wall

Fig. 8. Heavy node query (NDCG)

6 Conclusion

In this paper, we proposed *Dynamic Graph Sketch (DGS)*, a novel framework for large-scale graph stream summarization. Unlike conventional graph sketches that used fixed-sized memory, DGS adopted an expandable-size design to avoid performance drop after a large number of edge updates. DGS introduced a deep neural network to actively predict the query error of the graph sketch. If the predicted error exceeded a pre-defined threshold, it used a convolutional neural network to learn to expand its memory size and hash space. Extensive experiments based on three real-world graph streams showed that the proposed method is able to achieve high accuracy for different kinds of graph queries compared to the state-of-the-arts.

References

1. Charikar, M., Chen, K., Farach-Colton, M.: Finding frequent items in data streams. In: Widmayer, P., Eidenbenz, S., Triguero, F., Morales, R., Conejo, R., Hennessy, M. (eds.) ICALP 2002. LNCS, vol. 2380, pp. 693–703. Springer, Heidelberg (2002). https://doi.org/10.1007/3-540-45465-9_59
2. Cormode, G., Muthukrishnan, S.: An improved data stream summary: the count-min sketch and its applications. J. Algorithms **55**(1), 58–75 (2005)
3. Estan, C., Varghese, G.: New directions in traffic measurement and accounting: focusing on the elephants, ignoring the mice. ACM Trans. Comput. Syst. **21**(3), 270–313 (2003)
4. Gou, X., Zou, L., Zhao, C., Yang, T.: Fast and accurate graph stream summarization. In: 35th IEEE International Conference on Data Engineering (ICDE 2019), pp. 1118–1129 (2019)
5. Guo, Y., et al.: Closed-loop matters: dual regression networks for single image super-resolution. In: IEEE Conference on Computer Vision and Pattern Recognition (CVPR), pp. 5406–5415 (2020)
6. Hsu, C., Indyk, P., Katabi, D., Vakilian, A.: Learning-based frequency estimation algorithms. In: 7th International Conference on Learning Representations (ICLR 2019) (2019)
7. Järvelin, K., Kekäläinen, J.: Cumulated gain-based evaluation of IR techniques. ACM Trans. Inf. Syst. **20**(4), 422–446 (2002)
8. Khan, A., Aggarwal, C.C.: Query-friendly compression of graph streams. In: Kumar, R., Caverlee, J., Tong, H. (eds.) 2016 IEEE/ACM International Conference on Advances in Social Networks Analysis and Mining (ASONAM 2016), pp. 130–137 (2016)
9. Liu, L., et al.: Sf-sketch: a two-stage sketch for data streams. IEEE Trans. Parallel Distrib. Syst. **31**(10), 2263–2276 (2020)
10. Paszke, A., et al.: Pytorch: an imperative style, high-performance deep learning library. In: Annual Conference on Neural Information Processing Systems (NeurIPS 2019), pp. 8024–8035 (2019)
11. Shi, W., et al.: Real-time single image and video super-resolution using an efficient sub-pixel convolutional neural network. In: IEEE Conference on Computer Vision and Pattern Recognition (CVPR), pp. 1874–1883 (2016)
12. Tang, L., Huang, Q., Lee, P.P.C.: Mv-sketch: a fast and compact invertible sketch for heavy flow detection in network data streams. In: IEEE Conference on Computer Communications (INFOCOM 2019), pp. 2026–2034 (2019)
13. Tang, N., Chen, Q., Mitra, P.: Graph stream summarization: from big bang to big crunch. In: Proceedings of the 2016 International Conference on Management of Data (SIGMOD 2016), pp. 1481–1496 (2016)
14. Zhang, M., Wang, H., Li, J., Gao, H.: Learned sketches for frequency estimation. Inf. Sci. **507**, 365–385 (2020)
15. Zhao, P., Aggarwal, C.C., Wang, M.: gSketch: on query estimation in graph streams. Proc. VLDB Endow. **5**(3), 193–204 (2011)

Discovering Dense Correlated Subgraphs in Dynamic Networks

Giulia Preti[1(✉)], Polina Rozenshtein[2], Aristides Gionis[3], and Yannis Velegrakis[4]

[1] ISI Foundation, Turin, Italy
giulia.preti@isi.it
[2] Amazon, Tokyo, Japan
prrozens@amazon.co.jp
[3] KTH Royal Institute of Technology, Stockholm, Sweden
argioni@kth.se
[4] University of Trento and Utrecht University, Utrecht, The Netherlands
i.velegrakis@uu.nl

Abstract. Given a dynamic network, where edges appear and disappear over time, we are interested in finding sets of edges that have similar temporal behavior and form a dense subgraph. Formally, we define the problem as the enumeration of the maximal subgraphs that satisfy specific density and similarity thresholds. To measure the similarity of the temporal behavior, we use the correlation between the binary time series that represent the activity of the edges. For the density, we study two variants based on the average degree. For these problem variants we enumerate the maximal subgraphs and compute a compact subset of subgraphs that have limited overlap. We propose an approximate algorithm that scales well with the size of the network, while achieving a high accuracy. We evaluate our framework on both real and synthetic datasets. The results of the synthetic data demonstrate the high accuracy of the approximation and show the scalability of the framework.

1 Introduction

A popular graph-mining task is discovering dense subgraphs, i.e., densely connected portions of the graph. Finding dense subgraphs was well studied in computer science and data-mining communities with many real-world applications.

Many highly dynamic real-life applications are modeled by continuously changing graphs. In some cases, the nodes and edges may evolve in a convergent manner and display correlated behavior. These groups of correlated elements, especially when they are topologically close, can represent regions of interest in the network. This work is focused on the discovery of such patterns, i.e., on the *correlated dense subgraphs* in dynamic networks. We consider graphs with edges that appear and disappear as time passes. Our goal is to identify sets of edges

P. Rozenshtein—Work done while at Aalto University, Finland and IDS, NUS, Singapore.

K. Karlapalem et al. (Eds.): PAKDD 2021, LNAI 12712, pp. 395–407, 2021.
https://doi.org/10.1007/978-3-030-75762-5_32

that show a similar behavior in terms of their presence in the graph, and at the same time, are densely connected. Previous works [5,6] considered a similar problem, but limited to the time intervals and hard network partition. In this work we propose a general framework for finding dense correlated subgraphs, which can work with any temporal and spatial measure. Given specific density and correlation thresholds, we enumerate all maximal (the output set does not contain graphs, which are subgraphs of one another) subgraphs that satisfy the thresholds. Furthermore, since outputting a large number of highly overlapping subgraphs is not practical, we produce a manageable and informative set of highly diverse subgraphs.

Our main contributions are: **(i)** We introduce and formally define the generic problem of detecting a set of dense and correlated subgraphs in dynamic networks (Sect. 2), and explain how it differs from other similar works (Sect. 5); **(ii)** We propose two different measures to compute the density of a group of edges that change over time, which are based on the average-degree density [7], and a measure to compute their correlation, based on the Pearson correlation (Sect. 2); **(iii)** We develop an exact solution, called ExCoDe, for enumerating all the subgraphs that satisfy given density and correlation thresholds. We also propose an approximate solution that scales well with the size of the network, and at the same time achieves high accuracy (Sect. 3); **(iv)** We study the problem of identifying a more compact and diverse subset of results. We extend our framework to extract a set of subgraphs with a pairwise overlap less than a specified threshold (Sect. 3); **(v)** We evaluate our framework on both real and synthetic datasets, confirming the correctness of the exact solution, the high accuracy of the approximate one, the scalability of the framework, and the applicability of the solution on networks of different nature (Sect. 4).

The extended version of the paper can be found at http://arxiv.org/abs/2103.00451.

2 Problem Statement

A *dynamic network* is a graph that models data that change over time. It is represented as a sequence of static graphs (*snapshots* of the network).

Dynamic Network. Let $T \subseteq \mathcal{T}$ be a set of time instances over a domain $\mathcal{T}$. A *dynamic network* $D = (V, E)$ is a sequence of graphs $G_i = (V, E_i)$ with $i \in T$, referred to as snapshots of the network, where V is a set of vertices, $E_i \subseteq V \times V$ is a set of edges between vertices. The set E denotes the union of the edges in the snapshots, i.e., $E = \cup_{i \in T} E_i$.

We assume that all the snapshots share the same set of nodes. If a node does not interact in a snapshot, then it is present as a singleton.

Given a graph $G = (V, E)$, a *subgraph* H of G is a graph $H = (V_H, E_H)$, such that $V_H \subseteq V$ and $E_H \subseteq E$. In static graphs, the density of a subgraph is traditionally computed as the average degree of its nodes [7]:

Density. The *density* of a (static) graph $G = (V, E)$ is the average degree of its nodes, i.e., $\rho(G) = 2|E|/|V|$.

In the case of a dynamic network D, the edges of a subgraph H may not exist in all the snapshots, meaning that the density may be different in each snapshot. Therefore, we propose two approaches to aggregate the density values. Let $G_i(H) = (V_H, E_H \cap E_i)$ denote the subgraph induced by H in the snapshot i. The *minimum density*, denoted as ρ_m, is the minimum density of any subgraph induced by H across the snapshots of D; while the *average density*, denoted as ρ_a, is the average density among these induced subgraphs. In particular,

$$\rho_m(H) = \min_{i \in T} \rho(G_i(H)), \quad \rho_a(H) = \frac{1}{|T|} \sum_{i \in T} \rho(G_i(H)). \tag{1}$$

Given a density threshold δ, a subgraph H is called δ-dense if $\rho_m(G) \geq \delta$ or $\rho_a(G) \geq \delta$, respectively.

These intuitive definitions are too strict for those practical situations where an interesting event or anomaly exhibits itself only in a small number of snapshots of the network [2]. To account for such situations, we introduce the notion of *activity* and say that a subgraph H is *active* at time t if at least k edges of H exist in t, i.e., $|E_t \cap E_H| \geq k$. Then, we relax our density definitions and compute the minimum and average density of H by aggregating only over the snapshots where H is active. Let T_H^k denote the subset of snapshots H is active, i.e., $T_H^k = \{t \mid t \in T \text{ and } |E_t \cap E_H| \geq k\}$. We redefine Eq. 1 as follows:

$$\rho_m^k(H) = \min_{i \in T_H^k} \rho(G_i(H)), \quad \rho_a^k(H) = \frac{1}{|T_H^k|} \sum_{i \in T_H^k} \rho(G_i(H)). \tag{2}$$

If T_H^k is empty, then both $\rho_m^k(H)$ and $\rho_a^k(H)$ are set to 0. We use the notation ρ^k to refer collectively at ρ_m^k and ρ_a^k.

We say that a subgraph is *correlated* if its edges are pairwise correlated. We therefore represent every edge as a time series over the snapshots, and measure the correlation between two edges as the Pearson correlation between the time series. Pearson correlation is widely used to detect associations between time series [8]. However, our framework can work with any other correlation measure. Let $\mathbf{t}(e)$ denote the time series of the edge e, where each coordinate is set to $t_i(e) = 1$ if e appears in the snapshot i, and thus $t_i(e) = 0$ otherwise.

Edge Correlation. Let $D = (V,E)$ be a dynamic network, $e_1, e_2 \in E$ be two edges with respective time series $\mathbf{t}(e_1) = \{t_1(e_1), \ldots, t_T(e_1)\}$, $\mathbf{t}(e_2) = \{t_1(e_2), \ldots, t_T(e_2)\}$, and $\bar{t}(e) = \frac{1}{|T|} \sum_{i=1}^{T} t_i(e)$. The correlation between e_1 and e_2, denoted as $c(e_1, e_2)$, is the Pearson correlation between $t(e_1)$ and $t(e_2)$, i.e.,

$$c(e_1, e_2) = \frac{\sum_{i=1}^{T} (t_i(e_1) - \bar{t}(e_1))(t_i(e_2) - \bar{t}(e_2))}{\sqrt{\sum_{i=1}^{T} (t_i(e_1) - \bar{t}(e_1))^2} \sqrt{\sum_{i=1}^{T} (t_i(e_2) - \bar{t}(e_2))^2}}.$$

Given a correlation threshold σ, the edges e_1 and e_2 are considered correlated if $c(e_1, e_2) \geq \sigma$.

We define the correlation of a subgraph H as the minimum pairwise correlation between its edges, i.e., $c_m(H) = \min_{e_i \neq e_j \in E_H} c(e_i, e_j)$, and say that H is σ-correlated if $c_m(H) \geq \sigma$.

Our goal is to identify all the dense and correlated subgraphs in a dynamic network. However, since a dense correlated subgraph may contain dense correlated substructures due to the nature of the density and correlation measures used, we restrict our attention to the *maximal* subgraphs. Thus, given a dynamic network D, a density threshold δ, and a correlation threshold σ, we want to find all the subgraphs H that are δ-dense and σ-correlated, and are not a strict subset of another δ-dense σ-correlated subgraph.

As it is often the case with problems that enumerate a complete set of solutions that satisfy given constraints, the answer set could potentially be very large and contain solutions with a large degree of overlap. To counter this effect, we further focus on reporting only the *diverse* subgraphs, which are subgraphs that differ from one another and are representative of the whole answer set. To measure the similarity between subgraphs, we use the Jaccard similarity between their edge sets, i.e., the Jaccard similarity between the graph $G'=(V', E')$ and $G''=(V'', E'')$, denoted as $J(G', G'')$, is $J(G', G'')=|E' \cap E''|/|E' \cup E''|$. Then, we require that the pairwise similarities between subgraphs in the answer set are lower than a given similarity threshold ϵ. This is in line with previous work that has aimed at finding a diverse collection of dense subgraphs [9].

Diverse Dense Correlated Subgraphs Problem [DiCorDiS]**.** Given a dynamic network D, a density threshold δ, a correlation threshold σ, and a similarity threshold ϵ, find a collection $\mathcal{S}$ of maximal and diverse subgraphs such that for each $H \in \mathcal{S}$, H is δ-dense and σ-correlated, and for each distinct $H, H' \in \mathcal{S}$, $J(H, H') \leq \epsilon$.

3 Solution

To solve DiCorDiS, we propose a two-step approach, called ExCoDe (**Ex**tract **Co**rrelated **D**ense **E**dges). It first identifies maximal sets of correlated edges, and then extracts subsets of edges that form a dense subgraph according the density measures ρ_m^k or ρ_a^k. The correlation of a set of edges is computed as c_m.

Given the dynamic network $D = (V, E)$ we create a *correlation graph* $\mathcal{G} = (E, \mathcal{E})$, such that the vertex set of $\mathcal{G}$ is the edge set E of D, and the edges of $\mathcal{G}$ are the pairs $(e_1, e_2) \in E \times E$ that have correlation $c(e_1, e_2) \geq \sigma$. It is easy to see that a *maximal clique* in the correlation graph $\mathcal{G}$ corresponds to a maximal set of correlated edges in D.

The flow of ExCoDe is illustrated in Algorithm 1. Starting from the dynamic network $D = (V, E)$ the algorithm first creates the correlation graph $\mathcal{G}$ by adding a meta-edge between two edges of D if their correlation is greater than σ. Then it enumerates all the maximal cliques in $\mathcal{G}$. This collection of maximal cliques in $\mathcal{G}$ corresponds to a collection $\mathcal{C}$ of maximal correlated edge sets in D. Finally, FindDiverseDenseEdges examines each connected component in $\mathcal{C}$ (by using

Algorithm 1. ExCoDe

Input: Dynamic network $D = (V, E)$, Density function ρ^k
Input: Thresholds: Correlation σ, Density δ, Size s_M
Input: Thresholds: Edges-per-snapshot k, Similarity ϵ
Output: Diverse dense correlated maximal subgraphs $\mathcal{S}$

$\mathcal{G} \leftarrow$ CreateCorrelationGraph(G, σ)
$\mathcal{C} \leftarrow$ FindMaximalCliques$(\mathcal{G})$
$\mathcal{S} \leftarrow$ FindDiverseDenseEdges$(D, \mathcal{C}, \rho^k, \delta, k, s_M, \epsilon)$
return $\mathcal{S}$

Algorithm 2. FindDiverseDenseEdges

Input: Dynamic Network $D = (V, E)$
Input: Set of maximal cliques $\mathcal{C}$, Density function ρ^k
Input: Thresholds: Density δ, Size s_M, Edges-per-snapshot k, Similarity ϵ
Output: Set of diverse dense maximal subgraphs $\mathcal{S}$

$\mathcal{S} \leftarrow \emptyset$; $\mathcal{P} \leftarrow \emptyset$
$CC \leftarrow$ extractCC$(\mathcal{C})$
for each $X \in CC$ **do**
  if $X.size < s_M$ **and** isMaximal$(X, \mathcal{S} \cup \mathcal{P})$ **and** isDiverse$(X, \mathcal{S})$ **then**
    $(flag, R) \leftarrow$ isDense$(D, X, k, \rho^k, \delta)$
    add X to $\mathcal{S}$ **if** $flag = 1$
    add R to $\mathcal{P}$ **if** $flag = 0$
for each $X \in \mathcal{P}$ **do**
  add X to $\mathcal{S}$ **if** isMaximal$(X, \mathcal{S})$ **and** isDiverse(X, $\mathcal{S}$)
return $\mathcal{S}$

either the density measure ρ^k_m or ρ^k_a) to identify those constituting dense subgraphs in D, retaining only a subset of pairwise dissimilar subgraphs according to the similarity threshold ϵ. Next we describe the key elements of Algorithm 1, while the detailed descriptions of all the subroutines are in the supplementary materials.

Creation of the Correlation Graph. The correlation graph $\mathcal{G}$ can be built exactly, by computing the correlation $c(e_1, e_2)$ between each pair of edges $e_1, e_2 \in E$ and retaining those pairs satisfying $c(e_1, e_2) \geq \sigma$. However, when D is large, comparing each pair of edges is prohibitively expensive, and thus we propose an approximate solution based on *min-wise hashing* [3]. Here we exploit the fact that a strong correlation between two edges implies a high Jaccard similarity of the sets of snapshots where the edges appear. We use min-wise hashing to identify sets of candidate correlated edges. Specifically, we use a variant of the TAPER algorithm [19].

Enumeration of the Maximal Cliques. After the creation of the correlation graph $\mathcal{G}$, the maximal groups of correlated edges are enumerated by identifying the maximal cliques in $\mathcal{G}$. To this aim, we use our implementation of the *GP* algorithm of Wang et al. [17].

Algorithm 3. IsDense

```
Input: Dynamic Network D = (V, E)
Input: A set of edges X, Density function ρ^k
Input: Thresholds: Density δ, Edges-per-snapshot k
Output: (1, ∅) if X is dense; (0, R) if X contains the dense subsets R; (−1, ∅) o.w.
K ← kEdgeSnapshots(X, k)
if ρ^k(X, K, δ) then return (1, ∅)
if containsDense(X, K, k) = ∅ then return (−1, ∅)
return (0, extractDense(X, K, k))

function containsDense(X, K, k)
    while ρ^k(X, K, δ/2) = false do
        if X = ∅ or K = ∅ then return false
        max ← getMaxDeg(X)
        n ← getMinDegNode(X)
        if max < δ/2 then return false
        X ← X \ adj(n)
        K ← kEdgeSnapshots(X, k)
    return true
function extractDense(X, K, k)
    R ← ∅; Q ← {(X, K)}
    while Q ≠ ∅ do
        extract (Y, K') from Q
        if ρ^k(Y, K', δ) then
            R ← R ∪ {Y}
        else if K' ≠ ∅ then
            max ← getMaxDeg(Y)
            N ← getMinDegNodes(Y)
            if max < δ then continue
            for each n ∈ N do
                Y ← Y \ adj(n)
                K' ← kEdgeSnapshots(Y, k)
                add (Y, K') to Q if Y ≠ ∅
    return R
```

Discovery of the Dense Subgraphs. The goal of this step is to find connected groups of edges that form a dense subgraph, using either ρ_m^k or ρ_a^k as density function ρ^k. FindDiverseDenseEdges receives in input a set of maximal cliques $\mathcal{C}$, each of which represents a maximal group of correlated edges. Since some of the edges in a clique may not be connected in the network D, the algorithm extracts all the distinct connected components from the cliques (calling subroutine extractCC), before computing the density values. To allow a faster discovery of the maximal groups of dense edges, the connected components are sorted in descending order of their size and processed iteratively. If no larger or similar dense set of the current candidate X has been discovered yet (line 4), and if the size of X does not exceed the threshold s_M, the density of

X is computed by ISDENSE (line 5) calling subroutine ISAVGDENSE (for $\rho_a^k(X)$) or subroutine ISMINDENSE (for $\rho_m^k(X)$). Both subroutines iterate through the snapshots where at least k edges of X are present, but then the former aggregates the density values, while the latter keeps track of the minimum density. The latter allows for early stopping, if it encounters a snapshot where the density is below the threshold. However, thanks to the optimizations described in the next paragraph, the implementation of ISAVGDENSE is more efficient than that of ISMINDENSE, and thus we call the latter only when the former returns **true**, given that the average is an upper bound to the minimum.

When the density of the subgraph H induced by X is above the threshold δ, X is inserted in the result set $\mathcal{S}$ (line 6). Otherwise, some subset $X' \subseteq X$ may satisfy the condition $\rho_a^k(H') \geq \delta$. Since examining all the subsets of X is costly, we use Procedure CONTAINSDENSE, which is based on a 2-approximation algorithm for the densest subgraph problem [7], to prune the search space. In details, Procedure CONTAINSDENSE iteratively removes the vertex with lowest degree from the subgraph H, until it becomes empty or its density is greater than $\delta/2$. Every time a vertex is removed, its outgoing edges are removed as well (line 11), and thus the set of valid snapshots K must be updated (line 12). If K becomes empty, any subset of X has zero density, and thus the algorithm returns **false** (line 7). If the maximum value of density calculated during the execution of this algorithm is below the threshold $\delta/2$, it holds that X cannot contain a subset X' with density above δ [7], and thus CONTAINSDENSE returns **false**. Therefore, EXTRACTDENSE, which extracts all the dense subsets in the set X, is invoked (line 4) only when CONTAINSDENSE returns **true**.

When Procedure CONTAINSDENSE returns **true**, EXTRACTDENSE iteratively searches for all the dense subsets in X. At each iteration, a subset of edges Y is extracted from the queue Q and its density is checked. If Y is not dense but the set of valid snapshots is not empty (line 20), a new candidate is created for each vertex n with lowest degree in the subgraph induced by Y. These candidates are then inserted into Q. On the other hand, when Y is dense, it is inserted into the result set R. At the end of Algorithm 2, the maximal subsets in the set $\mathcal{P}$, which contains the elements of all the R sets computed during the search, are checked for similarity with the subsets already in $\mathcal{S}$. Those with Jaccard similarity below ϵ with any subsets in $\mathcal{S}$ are finally added to $\mathcal{S}$ (lines 8–9).

Computing Average Density Efficiently. The average density $\rho_a(H)$ of a subnetwork $H = (V_H, E_H)$ in a dynamic network D can be computed via the *summary graph* of D defined as the *static graph* $\mathcal{R} = (V, E, \sigma)$ where V is the set of vertices of D, E is the union of the edges E_i of all the snapshots of D, and $\sigma : E \mapsto \mathbb{R}$ is a weighting function that assigns, to each edge $e \in E$, a value equal to its average appearance over all the snapshots of D, i.e., $\sigma(e) = 1/|T| \sum_{i \in T} t_i(e)$. The following proposition ensures that $\rho_a(H)$ is equivalent to the weighted density of H in the summary graph $\mathcal{R}$, which is defined as $w\rho(H) = 2\sum_{e \in E_H} \sigma(e)/|V_H|$.

Proposition 1. *Given a dynamic network D, its summary graph $\mathcal{R}$, and a subnetwork H, it holds that $\rho_a(H) = w\rho(H)$.*

Proof.

$$\rho_a(H) = \frac{1}{|T|}\sum_{t\in T}\rho(G_t(H)) = \frac{1}{|T|}\sum_{i\in T}\left(\frac{2|E_H \cap E_t|}{|V_H|}\right)$$

$$= \frac{2}{|V_H|}\frac{1}{|T|}\sum_{i\in T}\sum_{e\in E_H} t_i(e) = \frac{2}{|V_H|}\sum_{e\in E_H}\sigma(e) = w\rho(H). \quad \square$$

The weighted density of H in the summary graph $\mathcal{R}$ can be calculated significantly faster than its average density in the dynamic network D, since the former is obtained by summing the appearances of the edges of H defined by σ, while the latter is obtained by constructing the subgraph induced by E_H in each snapshot, computing the average node degree of each induced subgraph, and taking the average among those values. Thus, Proposition 1 allows us to improve the efficiency of our algorithm when using ρ_a (and ρ_a^k) density function.

ExCoDe Complexity. The exact construction of $\mathcal{G}$ takes $\mathcal{O}(|E|^2)$, as it requires the computation of all the pairwise edge correlations. The approximate solution creates $h \cdot r$ hash values for the edges in $\mathcal{O}(h \cdot r \cdot |E|)$ and compares only the edges that share at least one hash code. Even though the worst-case time complexity is still $\mathcal{O}(|E|^2)$ (every pair of edges share some hash code), practically, the actual number of comparisons is much smaller than $|E|^2$. The time complexity of the maximal clique enumeration is $\mathcal{O}\left(|E| \cdot \kappa(\mathcal{G})\right)$, where $\kappa(\mathcal{G})$ is the number of cliques in $\mathcal{G}$. The computation of the connected components in the maximal cliques takes $\mathcal{O}(|E| \cdot \kappa(\mathcal{G}))$, as it requires a visit of the network D for each clique. In the worst case, each edge of the network belongs to a different connected component, and thus Algorithm 2 must iterate $|E|$ times. At each iteration, it calls Procedure isDense to compute the density of the current set of edges X if its size is lower than s_M. Procedure isDense calculates the average node degree of each subgraph induced by X in all the snapshots where at least k edges of X are present (at most $|T|$), and thus its time is bounded by $\mathcal{O}(s_M \cdot |T|)$. When X is not dense, the algorithm further calls Procedure containsDense and Procedure extractDense. The former runs in s_M, since it removes at least one edge from X at each iteration; while the latter must process all the subsets of X in the worst case (2^{s_M}). The complexity of Algorithm 2 is therefore $\mathcal{O}(\kappa(\mathcal{G}) \cdot |E| + |E|(s_M \cdot |T| + s_M + 2^{s_M})) = \mathcal{O}(|E|(\kappa(\mathcal{G}) + s_M|T| + 2^{s_M}))$, which is also the complexity of Algorithm 1.

4 Experimental Evaluation

We evaluate the performance of our exact and approximate solutions in terms of accuracy and execution time. We also integrated our solution into a tool demonstrated at ICDMW19 [14]. More experimental results can be found in the supplementary materials, due to space limitations.

The datasets considered are 3 real networks and 6 randomly-generated networks, the characteristics of which are shown in Tables 1 and 2, respectively. They report the number of vertices $|V|$, edges $|E|$, and snapshots $|T|$; the average node degree $d_a(G)$; the average node degree per snapshot $d_a(G_i)$; and the

Table 1. Real datasets

Dataset	$\|V\|$	$\|E\|$	$\|T\|$	$d_a(G)$	$d_a(G_i)$	$c_a(e)$
HAGGLE	274	2K	90	15.5	5.2	5.4
TWITTER-S	767	2K	2K	6.2	3	121.1
TWITTER-M	1.2K	7K	2K	12.1	3.2	86.4
TWITTER-L	1.3K	10K	2K	15.2	3.3	68.6
MOBILE-S	5K	42K	48	15.3	4.9	3.8
MOBILE-M	5K	80K	48	28.6	6.4	3.6
MOBILE-L	5K	118K	48	41.4	7.5	3.6

Table 2. Synthetic datasets

Dataset	$\|V\|$	$\|E\|$	$\|T\|$	p_{in}	p_{out}	*Independent*			*Correlated*	
						$d_a(G)$	$d_a(G_i)$	$a_a(e)$	$d_a(G_i)$	$c_a(e)$
GAUSSIAN-1-7-1	100	1059	100	0.7	0.1	21.1	10.6	50	10.32	48.2
GAUSSIAN-2-7-1	200	3029	100	0.7	0.1	30.2	15.1	50.1	15.1	50.1
GAUSSIAN-3-7-1	300	6070	100	0.7	0.1	40.4	20.2	50	20.3	50.3
GAUSSIAN-1-7-3	100	1825	100	0.7	0.3	36.5	18.2	50	18.2	49.9
GAUSSIAN-2-7-3	200	6828	100	0.7	0.3	68.2	34	49.9	33.8	49.6
GAUSSIAN-3-7-3	300	14723	100	0.7	0.3	98.1	49	49.9	48.9	49.8

Table 3. Minimum (F_m) and average (F_a) F-score, running time. Worst case values in parenthesis. "-" for runs longer than 2 days.

Dataset	CIFORAGER			EXCODE		
	F_m	F_a	$t(min)$	F_m	F_a	$t(sec)$
GAUSSIAN-1-7-1	0	(.07) .03	(19) .2	(.98) 1	(.99) 1	(1.1) .71
GAUSSIAN-2-7-1	0	(.03) .01	(365) 4	(.00) 1	(.95) 1	(2.2) 1.6
GAUSSIAN-3-7-1	0	(-) .007	(-) 44	(.98) 1	(.99) 1	(8.0) 4.5
GAUSSIAN-1-7-3	0	(.02) .01	(78) .9	(.98) 1	(.99) 1	(1.2) 1
GAUSSIAN-2-7-3	0	(-) .003	(-) 332	(.97) 1	(.99) 1	(10) 5.5
GAUSSIAN-3-7-3	-	(-) -	(-) -	(.98) 1	(.99) 1	(51) 23

average number of appearances of an edge in the snapshots $c_a(e)$. HAGGLE [4] is a human-contact network, TWITTER [12] is a hashtag co-occurrence network created using tweets collected from 2011 to 2016, and MOBILE [11] is a network modeling calls between users made available by Telecom Italia. The GAUSSIAN-X-Y-Z are synthetic networks generated using the *gaussian random partition graph* generator in the Python NetworkX library[1]. A graph is obtained by partitioning

[1] https://tinyurl.com/y5sezq73.

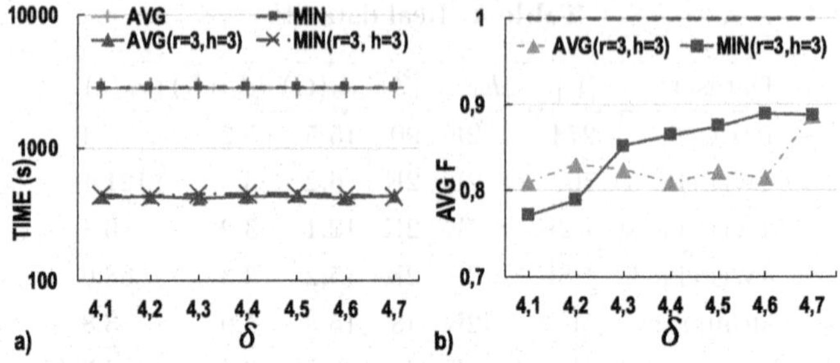

Fig. 1. Running time and average F score for varying δ

the set of n nodes into k groups each of size drawn from a normal distribution $\mathcal{N}(s, s/v)$, and then adding intra-cluster edges with probability p_{in} and inter-cluster edges with probability p_{out}.

We implemented our algorithms in Java 1.8, and run the experiments on a 24-Core (2.40 GHz) Intel Xeon E5-2440 with 188 Gb RAM with Linux 3.13, limiting the amount of memory available to 150 Gb. In addition, we implemented a Java version of ciForager [6], which is the approach most related to ours. For the synthetic datasets we report results based on 100 runs.

Effectiveness of the Exact Solution. We tested the effectiveness of our exact algorithm in detecting the actual dense groups of correlated edges in the synthetic networks Gaussian-x-7-1 and Gaussian-x-7-3. The correlation threshold σ is set to 0.8; the density threshold δ is equal to the minimum average degree among the actual dense groups, namely 2; and the maximum size s_M is set to ∞ to ensure we do not miss any dense group. We measured the accuracy in terms of the Jaccard similarity between the groups of edges discovered $\mathcal{S}$ and the dense groups in the ground-truth $\mathcal{G}$. First, for each group $H \in \mathcal{S}$, we computed the Jaccard similarity with its closest dense group in $\mathcal{G}$, and then calculated minimum and average precision as P_a and P_m:

$$P_a = \frac{1}{|\mathcal{S}|} \sum_{H \in \mathcal{S}} \max_{J \in \mathcal{G}} \text{Jacc}(H, J) \qquad P_m = \min_{H \in \mathcal{S}} \max_{J \in \mathcal{G}} \text{Jacc}(H, J)$$

Then, for each dense group $J \in \mathcal{G}$, we computed the Jaccard similarity with its closest group in $\mathcal{S}$, and calculated minimum and average recall R_a and R_m:

$$R_a = \frac{1}{|\mathcal{G}|} \sum_{J \in \mathcal{G}} \max_{H \in \mathcal{S}} \text{Jacc}(H, J) \qquad R_m = \min_{J \in \mathcal{G}} \max_{H \in \mathcal{S}} \text{Jacc}(H, J)$$

Finally, we report the average and minimum F-score, as: $F_a = 2(P_a \cdot R_a)/(P_a + R_a)$ and $F_m = 2(P_m \cdot R_m)(P_m + R_m)$.

As shown in Table 3, for each synthetic network ExCoDe obtained both $F_a = 1$ and $F_m = 1$, meaning that the algorithm correctly identified all the dense groups despite the extra edges added between the groups in the networks. We achieved lower scores only when using correlation thresholds $\sigma \leq 0.2$ for the smallest network, and $\sigma \leq 0.3$ for the others. In these cases it is more likely that some inter-group edges have a correlation greater than σ, and therefore the algorithm discovers sets of edges that are supersets of the actual dense groups. Nonetheless, the F_a score is always greater than 0.94, while the F_m score is lower than 0.97 only for network GAUSSIAN-2-7-1.

In addition, Table 3 compares our approach with CIFORAGER [6], the closest competitor. CIFORAGER creates a partition of edges to optimize temporal (aggregated over temporal windows) correlation and spatial (all-pairs path) distance of each partition. We run the code with the default parameters: window length $w_l = 10$, window overlap size $w_i = 1$, clustering similarity threshold 0.25, region similarity threshold 0.2. The window length is the size of the window used to segment the sequence of graph snapshots into overlapping subsequences, while the overlap size indicates how much the subsequences overlap. The clustering threshold decides which edges to be grouped together, and the region threshold determines how to merge the groups of edges found in different windows.

Since the output of CIFORAGER is an edge partition, and thus contains edges that are not part of any dense group, the average and minimum precisions P_a and P_m are always low, and, as a consequence, the F_a and F_m are always lower than those of EXCODE. Performance wise, CIFORAGER is expensive since it computes the temporal distance for almost each pair of edges and for each window. In contrast, our algorithm was able to terminate in less than a minute with every configuration and network tested.

Efficiency of the Approximate Solution. Figure 1 shows the performance and running time of EXCODE *approximate* to find the (0.9)-correlated δ-dense subgraphs in the MOBILE-M network, using both ρ_a^k and ρ_m^k, and varying δ. As we can see, the approximate solution is one order of magnitude faster than the exact algorithm, and yet achieves a F_a score of at least 0.8 for the ρ_a^k (AVG) case, and 0.77 for the ρ_m^k (MIN). We observed a similar behavior also in the other networks. As an example, in the TWITTER samples we obtained the highest F_a score at high density values, while a minimum of 0.63 at low density values.

5 Related Work

Our problem is close to dense subgraph mining in dynamic networks. Works in this field aim at retrieving the highest-scoring temporal subgraph [13], the densest temporally compact subgraph [15], or the group of nodes most densely connected in all the snapshots [16]. Although they can be adapted to retrieve multiple subgraphs, the detected subgraphs are non-overlapping, and with edges that are not temporally correlated. The enumeration of dense structures has been studied in the context of frequent subgraph mining [1], and top densest subgraph mining [9]. When the input is a dynamic network, these groups represent subgraphs that persist over time; however, in general they are not temporally correlated. In anomaly and fraud detection, other measures have been considered, together with the density, with the goal of finding interesting regions in the snapshots of a dynamic network [2]. All such works focus on the statistically significant structures, while our interest is on dense groups of edges with a similar behavior over time. A notion of correlation has been used in approaches that characterize the event dynamics by the number of articular labels in the vicinity of spacial reference nodes [10], or that compute a decay factor [18]. In contrast to our work, both these approaches retrieve only anomalous nodes. The works closest to ours are CStag [5] and its incremental version ciForager [6], which find regions of correlated temporal change in dynamic graphs. However, they partition the edges into L regions, meaning that each edge is a part of the output, and hence the output can be very large and contain a number of low quality graphs. In contrast, we enumerate only the subgraphs with large density and high pairwise edge correlation.

6 Conclusions

We studied the problem of finding maximal dense correlated subgraphs in dynamic networks. We proposed two measures to compute the density of a subgraph that changes over time, and one to assess the temporal correlation of its edges. We described a framework that uses those measures to identify such subgraphs for given density and correlation thresholds. We experimentally demonstrated the limitations of the existing solutions and provided an approximate solution that runs in an order of magnitude faster, yet achieving a good solution quality.

Acknowledgments. Aristides Gionis and Giulia Preti are supported by EC H2020 RIA project "SoBigData++" (871042). Aristides Gionis is supported by three Academy of Finland projects (286211, 313927, 317085), the ERC Advanced Grant REBOUND (834862), the Wallenberg AI, Autonomous Systems and Software Program (WASP).

References

1. Abdelhamid, E., Canim, M., Sadoghi, M., Bhattacharjee, B., Chang, Y.C., Kalnis, P.: Incremental frequent subgraph mining on large evolving graphs. TKDE **29**(12), 2710–2723 (2017)
2. Akoglu, L., Tong, H., Koutra, D.: Graph based anomaly detection and description: a survey. Data Min. Knowl Disc. **29**(3), 626–688 (2014). https://doi.org/10.1007/s10618-014-0365-y
3. Broder, A.Z., Charikar, M., Frieze, A.M., Mitzenmacher, M.: Min-wise independent permutations. JCSS **60**(3), 630–659 (2000)
4. Chaintreau, A., Hui, P., Crowcroft, J., Diot, C., Gass, R., Scott, J.: Impact of human mobility on opportunistic forwarding algorithms. Trans. Mob. Comput. **6**(6), 606–620 (2007)
5. Chan, J., Bailey, J., Leckie, C.: Discovering correlated spatio-temporal changes in evolving graphs. KAIS **16**(1), 53–96 (2008)
6. Chan, J., Bailey, J., Leckie, C., Houle, M.: ciForager: incrementally discovering regions of correlated change in evolving graphs. TKDD (2012)
7. Charikar, M.: Greedy approximation algorithms for finding dense components in a graph. In: APPROX (2000)

8. Fu, T.C.: A review on time series data mining. Eng. Appl. Artif. Intell. **24**(1), 164–181 (2011)
9. Galbrun, E., Gionis, A., Tatti, N.: Top-k overlapping densest subgraphs. Data Min. Knowl. Disc. **30**(5), 1134–1165 (2016)
10. Guan, Z., Yan, X., Kaplan, L.M.: Measuring two-event structural correlations on graphs. PVLDB **5**(11) (2012)
11. Italia, T.: Telecommunications - tn to tn (2015). https://doi.org/10.7910/DVN/KCRS61
12. Lahoti, P., Garimella, K., Gionis, A.: Joint non-negative matrix factorization for learning ideological leaning on twitter. In: WSDM (2018)
13. Ma, S., Hu, R., Wang, L., Lin, X., Huai, J.: Fast computation of dense temporal subgraphs. In: ICDE (2017)
14. Preti, G., Rozenshtein, P., Gionis, A., Velegrakis, Y.: Excode: a tool for discovering and visualizing regions of correlation in dynamic networks. In: ICDMW (2019)
15. Rozenshtein, P., Tatti, N., Gionis, A.: Finding dynamic dense subgraphs. TKDD **11**(3), 1–30 (2017)
16. Semertzidis, K., Pitoura, E., Terzi, E., Tsaparas, P.: Finding lasting dense subgraphs. Data Min. Knowl. Disc. **33**(5), 1417–1445 (2018). https://doi.org/10.1007/s10618-018-0602-x
17. Wang, Z., et al.: Parallelizing maximal clique and k-plex enumeration over graph data. J. Parallel Distr. Comput. **106**, 79–91 (2017)
18. Yu, W., Aggarwal, C.C., Ma, S., Wang, H.: On anomalous hotspot discovery in graph streams. In: ICDM (2013)
19. Zhang, J., Feigenbaum, J.: Finding highly correlated pairs efficiently with powerful pruning. In: CIKM (2006)

Fake News Detection with Heterogenous Deep Graph Convolutional Network

Zhezhou Kang[1,2], Yanan Cao[1,2(✉)], Yanmin Shang[1,2], Tao Liang[1,2], Hengzhu Tang[1,2], and Lingling Tong[3(✉)]

[1] Institute of Information Engineering, Chinese Academy of Sciences, Beijing, China
{kangzhezhou,caoyanan,shangyanmin,liangtao0305,tanghengzhu}@iie.ac.cn
[2] School of Cyber Security, University of Chinese Academy of Sciences, Beijing, China
[3] National Computer Network Emergency Response Technical Team/Coordination Center of China, Beijing, China
tongling300@sina.com

Abstract. Fake news detection is a challenging problem due to its tremendous real-world political and social impacts. Previous works judged the authenticity of news mainly based on the content of a single news, which is generally not effective because the fake news is often written to mislead users by mimicking the true news. This paper innovatively utilizes the connection between multiple news, such as their relevance in time, content, topic and source, to detect fake news. We construct a heterogeneous graph with different types of nodes and edges, which is named as News Detection Graph (NDG), to integrate various information of multiple news. In order to learn deep representation of news nodes, we propose a Heterogenous Deep Convolutional Network (HDGCN) which utilizes a wider receptive field, a neighbor sampling strategy and a hierarchical attention mechanism. Extensive experiments carried on two real-world datasets demonstrated the effectiveness of our work in solving the fake news detection problem.

Keywords: Fake news detection · Heterogenous graph · Graph convolution network

1 Introduction

In recent years, fake news spreading on social media becomes a serious concern. Researchers devoted great efforts to the task of fake news detection. Most previous methods are mainly based on the analysis of news content, such as mining the lexical features [11], syntactic features [5], deep syntax and rhetorical structure [3], or more recently, writing styles [11,17]. However, it is difficult to effectively identify the fake news with highly imitated content only relying on the news content. Recent research focuses on the utilization of social information in news websites. These works add additional social context features, including

K. Karlapalem et al. (Eds.): PAKDD 2021, LNAI 12712, pp. 408–420, 2021.
https://doi.org/10.1007/978-3-030-75762-5_33

user demographics [9], social network structure [13], propagation patterns [20] and user reactions [19]. These methods give a lot of innovative ideas to detect fake news. However, all existing works judge the authority of news mainly based on the features of a single news, which is not effective because the information provided by single news is limited and may deviate from the facts.

To tackle the above problems, we introduce several additional information which not only help to establish the interaction between multiple news but also help to provide useful information to detect fake news. This information includes news sources, user reviews, news release time and news domain. (1) *News source* mainly reflects the reliability of the content. For example, news published by BBC is more likely to be true news. On the contrary, the news published by some unknown websites is mostly unconfirmed. (2) *User reviews* contain a lot of information that has never appeared in the reports. For some news, users' reviews directly reflect the authenticity of the news. (3) *Time* and *Domain* are important attributes of news. For example, news with similar release times or similar content can be different reports about the same event or relevant events. We use these similarity features to increase the information interaction between different news and reduce repeated judgments to improve the performance and efficiency. For example, if we know the authenticity of a piece of news, we can quickly judge the authenticity of other news according to the similarity.

To better integrate different types of information and relationships between news, we introduce a heterogeneous graph which is named as News Detection Graph (NDG). NDG contains four types of nodes, which are news nodes, domain nodes, review nodes and source nodes. We link the news nodes with similar topic, content, source and time respectively to establish the relationships between multiple news and we also link the news nodes with its reviews. Then, we aim to learn the embedding of the news nodes and judge whether they are fake news. In order to properly embed the NDG, we propose a Heterogenous Deep Graph Convolutional Network (HDGCN) which makes three improvements on the basis of HGCN [8]. (1) In NDG, some news nodes are directly connected, while others are indirectly connected through domain or source nodes. In order to capture the relationship between news nodes with different distances, we expand the range of neighbors in graph convolution to increase the receptive field of HGCN. (2) With the increase of receptive field, the number of aggregated neighbor nodes increases sharply, which leads to the problem of "neighbor explosion" and over smoothing. So, we introduce the neighbor sampling strategy to control the number of aggregated nodes. Then we deepen the graph model to repeat sampling in each layer, which not only retains node information in the graph, but also relieves the "neighbor explosion" and over smoothing problem. (3) In order to obtain more useful information in the graph, we introduce a novel hierarchical attention mechanism which can capture the different contributions of different relationships, different node types and different layers to node embedding.

Above all, our contributions are listed as follows: (1) Unlike previous works which just focus on the information of a single news, we use multiple news to jointly judge the authenticity of news. (2) We propose NDG which introduces additional information to construct multiple relationships between news nodes. (3) We propose HDGCN with a wider receptive field, a neighbor sampling strategy and a novel hierarchical attention mechanism to learn better node embedding. (4) Experimental results show that the proposed NDG-HDGCN framework achieves good performance on two large scale real-world datasets compared to several state-of-the-art methods.

2 Methodology

2.1 NDG Construction

NDG is formally defined as $G = \{V, E\}$, where V and E represent the set of nodes and edges respectively. The construction of NDG includes nodes initialization and edges construction. Figure 1(F) clearly illustrates the overall structure of NDG.

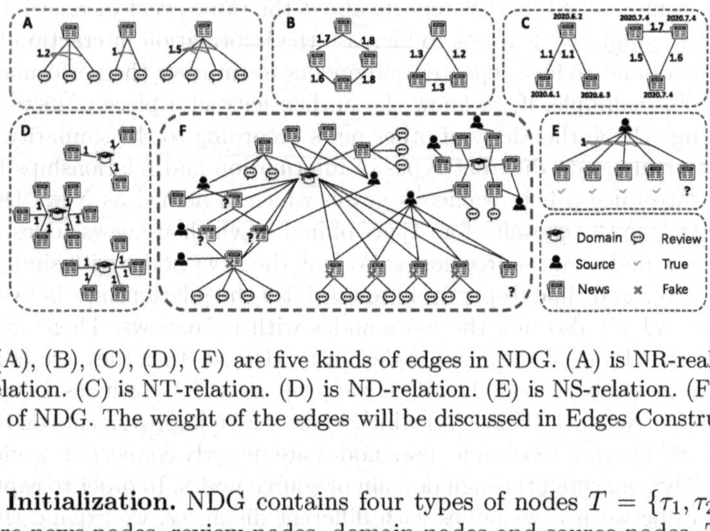

Fig. 1. (A), (B), (C), (D), (F) are five kinds of edges in NDG. (A) is NR-realtion. (B) is NN-relation. (C) is NT-relation. (D) is ND-relation. (E) is NS-relation. (F) is a toy example of NDG. The weight of the edges will be discussed in Edges Construction.

Nodes Initialization. NDG contains four types of nodes $T = \{\tau_1, \tau_2, \tau_3, \tau_4\}$, which are news nodes, review nodes, domain nodes and source nodes. The purpose of node initialization is to obtain a reasonable initial representation of each type of node. We first use Glove [15] to learn the contextualized representation for each word in news content and user reviews. For news nodes and review nodes, we use Bi-LSTM [6] to obtain content features as the initial embeddings. For source nodes, characteristics like source quality, writing styles, are mainly reflected in the published news and user reviews. So, the initial embeddings are obtained by summing all the related news embeddings and review embeddings. For domain nodes, we sum the embeddings of the top 100 high-frequency words in each domain as the initial embeddings. The dimension of the initialization vector is 100. Initialization process is shown in Fig. 3(B).

Edges Construction. NDG contains five types of edges, which represent five types of relations, namely ND-relation, NR-relation, NS-relation, NT-relation and NN-relation. The weight of each kind of edge is denoted as Φ.

- ND-relation links news nodes and domain nodes. If news content involves a certain domain, there is an edge between them (Fig. 1(D)), and the weight of the edge is 1 ($\Phi = 1$). Adding this edge makes the news in the same field closely connected, while news in different domain is sparse.
- NR-relation links news nodes with review nodes. If news has a review, then there is an edge between them (Fig. 1(A)). It has been proved by related researches [19] that user reviews have rich information from the crowd on social media, which is helpful for fake news detection. The weight of the edge should reflect the users' attitude (support, neutrality, opposition) to the news content. According to this assumption, when user support or oppose the news, the edge weight should be high and when it is neutral, the edge weight should be low. Furthermore, when the support rate is near 0 and 1, the weight of the edge should quickly approach the maximum value. So, we set the weight of this edge as Eq. 1. a_1, a_2, a_3 are coefficients of equations. *upvote* is the proportion of users who like the news and $upvote = \frac{\#like}{\#like+\#dislike}$.

$$\Phi = a_1 * upvote^2 - a_2 * upvote + a_3 \tag{1}$$

- NS-relation links news nodes and the related source node (Fig. 1(E)). We build this edge to measure the reliability of news source and make the news with same source connected. If the news source is very authoritative, then the connected news is likely to be true, and vice versa. This conclusion has been proved by Vydiswaran's group [22]. The weight of the edge is 1 ($\Phi = 1$).
- NT-relation links two news nodes if they are released at a similar time (Fig. 1(C)). Here we roughly link the news released within one day. For simplicity, we use a linear function to calculate the edge weight which is expressed as Eq. 2. Here t is the time interval between two news. k and b are artificial coefficients. According to our setting, the closer the release time of two news is, the greater the edge weight between them is.

$$\Phi = k * |t| + b \tag{2}$$

- NN-relation links two news nodes if the semantic similarity of them exceeds a certain threshold (Fig. 1(B)), denoted by θ. We use cosine similarity to calculate the distance between two news nodes. Here the similarity score is denoted by γ. When γ increase, Φ should increase rapidly. Specially, we truncate the edge weight to a max value ϕ_{max}, which is a artificial coefficient. The weight of NN-edge is defined in Eq. 3.

$$\Phi = -log(1 - \gamma) \tag{3}$$

2.2 HDGCN

To obtain more accurate embeddings of different types of nodes in NDG. We proposed HDGCN includes deep strategy and hierarchical attention strategy.

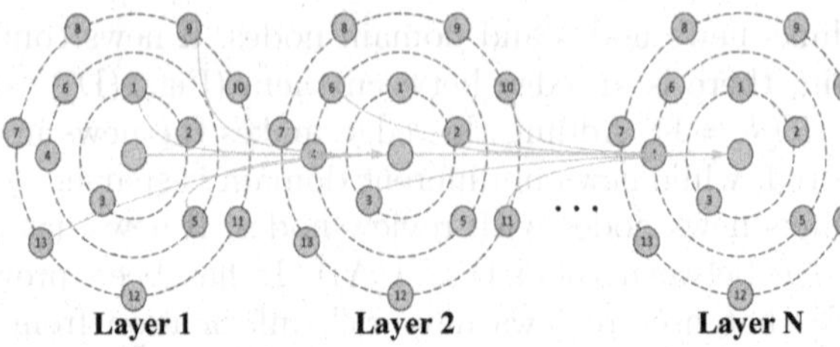

Fig. 2. An example of Deep Strategy.

- ***Deep Strategy.*** Deep strategy consists of two steps. We first extend the neighborhood of our model to d_{max} hops and take all the nodes in d_{max} hops as candidate neighbor nodes $\mathbb{N}'$.
  Then we introduce a parameter p to denote the probability of nodes being retained. Each node in $\mathbb{N}'$ can be selected as a neighbor node with the probability p. The neighbor nodes set is denoted by $\mathbb{N}$. The maximum number of $\mathbb{N}$ is n_{max}, and nodes exceeding n_{max} will be ignored. For each layer, we repeat the neighbor sampling process and then we aggregate the selected nodes $\mathbb{N}$ to the central node. In this way, each layer has an output representation aggregated by different sampling nodes. This makes us alleviate the problem of "neighbor explosion" and over smoothing. The structure of deep strategy can be seen in Fig. 2.

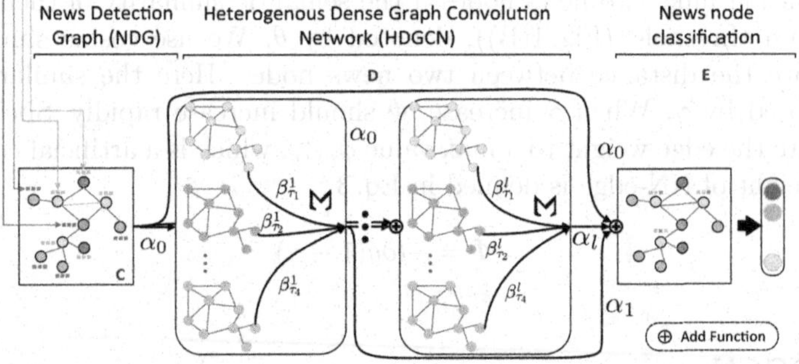

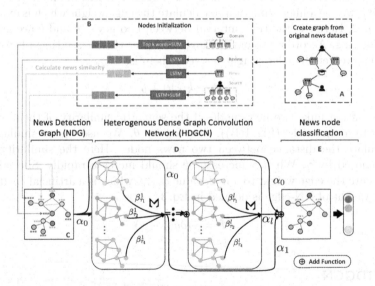

Fig. 3. Nodes initialization and example of Heterogenous Deep GCN. α_i is the attention score of ith layer. β^l_τ is the attention score of type τ in lth layer.

- ***Hierarchical Attention Strategy.*** Our model contains three levels of attention. First of all, our graph is a weighted graph. Different relationships in the graph correspond to different weights. Therefore, when we aggregate nodes, we first focus on the influence of different relationships on nodes embeddings.

Then we extend the work [8] to embed NDG and we also pay attention to different types of nodes. The heterogeneous embedding strategies are as follows:

$$\begin{aligned} H^{(l+1)} &= f(\beta^{(l)}, A^{(l)}, H^{(l)}, W^{(l)}) \\ &= \sigma(\sum_{\tau \in T} \beta_\tau^{(l)} * \widetilde{A_\tau} \cdot H_\tau^{(l)} \cdot W_\tau^{(l)}) \end{aligned} \tag{4}$$

where $\widetilde{A_\tau} \in \mathbb{R}^{|V| \cdot |V_\tau|}$ is the submatrix of $\widetilde{A}$. $\widetilde{A}$ represents the symmetric normalized adjacency matrix and $\widetilde{A} = M^{-\frac{1}{2}} \cdot A \cdot M^{-\frac{1}{2}}$, here A is the weighted adjacent matrix of G. The rows of $\widetilde{A_\tau}$ represent all nodes in G, and the columns of $\widetilde{A_\tau}$ represent all nodes of type τ. Each row of $\widetilde{A_\tau}$ represents all neighbor nodes of type τ for one node in G. $\beta_\tau^{(l)}$ is the attention weight of type τ in layer l. σ is Relu active function. The hidden representation in layer $l+1$ of nodes $H^{(l+1)}$ is obtained by aggregating the hidden representations of each type of nodes $H_\tau^{(l)}$ with different transformation matrix $W_\tau^{(l)} \in \mathbb{R}^{q^{(l)} \cdot q^{(l+1)}}$ in layer L. $W_\tau^{(l)}$ needs to be learned. Initially, $H_\tau^{(0)} = X_\tau$, and X_τ is the initialization vectors of the nodes of type τ , which can be obtained in the NDG construction.

At last, we add layer attention to HDGCN, which is, we connect all the layers in the graph convolution network with different attention score (Eq. 5). Because the output of each layer is aggregated by different sampling nodes, and the depth of feature extraction of each layer is different, we hope to give different attention weights to different layers. Each layer in the convolution network accepts the output of all the layers in front of it as input. That is to say, H^{l+1} consists of the graph convolution transitions from all previous layers.

$$H^{(l+1)} = T(\alpha_{l+1} * f(H^{(l)}, W^{(l)}), \alpha_l * H^{(l)}, ..., \alpha_0 * H^{(0)}) \tag{5}$$

Here, T is a connection function which can be add function or concatenation function. α_i is the attention score of layer i which learns the importance of this layer. f function is defined in Eq. 4.

2.3 Optimazation Objective

The final representation H^L of news nodes in NDG is fed to softmax classifier based on MLP to obtain category probability matrix Z of all categories. W' is parameter matrix of MLP. Finally, the cross-entropy loss is used as the optimization objective function for fake news detection.

$$Z = softmax(H^{(L)} * W') \tag{6}$$

$$Loss = -\sum_{i \in D_{train}} \sum_{j=1}^{C} Y_{ij} * log Z_{ij} \tag{7}$$

where C is the number of news categories (real or fake, etc.). D_{train} is the training set of news node. Y is the label of news data.

3 Experiment

In this section, we first introduce two real-world datasets (Fakeddit [12] and Weibo [10]), then we will introduce our parameter setting and comparison models, and finally analyze the performance of the proposed framework NDG-HDGCN.

3.1 Dataset

Fakeddit. [12] A large-scale benchmark dataset from Reddit. This dataset can support news classification of different granularities, including 2-class, 3-class and 6-class. In this paper, we sample 40000 news samples from Fakeddit data. We divide NDG into training set, dev set and test set according to the ratio of 3:1:1.

Weibo. [10] The dataset from the largest Chinese social media. It is a two-category dataset. Following previous work [24], we randomly select 10% instances as the dev set, and split the rest for training and testing set with a ratio of 3:1.

The statistic information of both datasets is shown in Table 1.

Table 1. Statistics of NDG

		News node	Domain node	Review node	Author node	NN-relation	NS-relation	NR-relation	NT-relation	ND-relation
Fakeddit NDG	Train	24000	3308	137564	16447	29840	24000	137564	11834	24000
	Test	8000	627	56241	5350	8376	8000	56241	3798	8000
Weibo NDG	Train	3148	50	23773	225	4170	3148	23773	1231	3148
	Test	1049	50	8039	112	1093	1049	8039	633	1049

3.2 Parameter Setting and Evaluation Metrics

In NDG, we use different formulas to calculate different kinds of edge weights. For these edges, we limit the weight to 1–2. For NN-relation, we set the coefficient $a_1 = 4, a_2 = 4, a_3 = 2$ (Fig. 4(a)). For NR-relation, we set the coefficient $k = -\frac{1}{24}, b = 2$ (Fig. 4(b)). For NT-relation, we set the threshold $\theta = 0.9$ and the truncate value $\psi_{max} = 2$ (Fig. 4(c)). For our model, we set the maximum number of neighbor nodes $n_{max} = 10$ and the maximum distance between the central node and its neighbors $d_{max} = 2$.

The parameters are update by Adam algorithm and the learning rate is initialized as 1e-3. We select the best parameter configuration based on performance on the dev set and evaluate the configuration on the test set. We compare our method with baselines of fake news detection in terms of accuracy and F1 score.

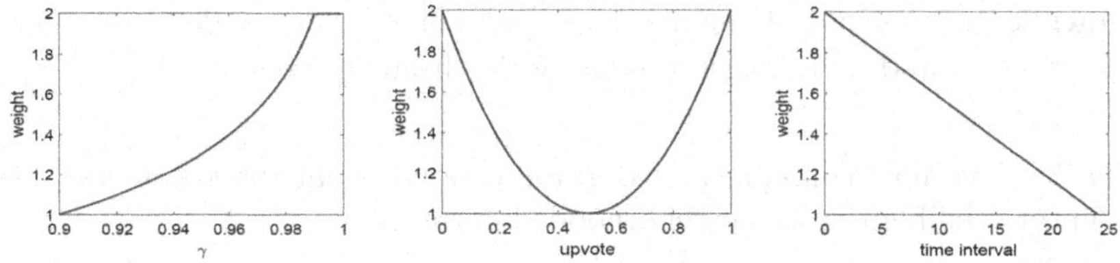

(a) NN-relation (b) NR-relation (c) NT-relation

Fig. 4. The weight of different edges.

3.3 Baselines

We compare our NDG-HDGCN with multiple outstanding embedding methods which can divided into text-based models and graph-based models. Specially, CSI and GLAN are state-of-the-art models in text-based and graph-based method respectively. Models are list as follow:

Bi-LSTM: A widely used traditional RNN-based model that learns temporal-linguistic patterns from news content and user reviews.

TextCNN: The most commonly used CNN model for text analysis.

InferSent [2] **:** A model that once trained on a high-quality language inference task. It aims at providing a universal sentence representation.

Bert [4]**:** A pre-train model which achieves state-of-the-art results on many natural language processing tasks.
CSI [18]**:** A text-based method using hybrid deep learning model to obtain news representation from news content, users' response and news source.

Table 2. Accuracy and F1-score on Fakeddit and Weibo datasets

Basic model	Fakeddit (Acc, F1)			Weibo (Acc, F1)
	2-class	3-class	6-class	2-class
Bi-LSTM	0.839, 0.811	0.821, 0.791	0.719, 0.731	0.903, 0.872
TextCNN	0.834, 0.832	0.779, 0.731	0.698, 0.721	0.870, 0.823
InferSent [2]	0.876, 0.818	0.821, 0.789	0.749, 0.787	0.911, 0.922
Bert [4]	0.886, 0.835	0.851, 0.818	0.813, 0.812	0.936, 0.915
CSI [18]	-,-	-,-	-,-	0.953, 0.954
Deepwalk [16]	0.875, 0.843	0.841, 0.796	0.760, 0.737	0.906, 0.874
GCN [7]	0.892, 0.879	0.853, 0.812	0.799, 0.763	0.931, 0.943
GAT [21]	0.881, 0.877	0.832, 0.834	0.772, 0.781	0.911, 0.917
GLAN [24]	-,-	-,-	-,-	0.946, 0.946
HDGCN	**0.906, 0.885**	**0.877, 0.858**	**0.823, 0.832**	**0.961, 0.960**

Deepwalk [16]**:** A classical homogeneous graph embedding model. Node embeddings are generated by taking n random walks from the center node.

GCN [7]**:** The most commonly used graph convolutional network model which can capture the high order neighborhoods information.

GAT [21]**:** An model that leverages the spatial information of a node by learning different weights for different neighbors using a self-attention mechanism.
GLAN [24]**:** GLAN is a novel heterogeneous graph-based method using global-local attention for rumor detection.

3.4 Performance Comparison

Table 2 shows the results of overall performance. From these tables, we can observe that our method outperforms all baselines on two datasets. More specially:

Our model performs better than the text-based model Bert and CSI. Among the text-based fake news detection methods, Bert and CSI are the most competitive. However, these methods only focus on the information of a single news. They treat each piece of news as an independent sample, which makes them lose a lot of information. In contrast, our NDG enables us to obtain additional information from multiple news and we also make better use of various types of information than text-based methods.

Among the graph-based methods, HDGCN is better than homogeneous graph-based method GCN, GAT and heterogeneous graph-based method GLAN. Compared with GCN and GAT, HDGCN benefits from the weighted graph and it can learn heterogeneous information from different types of nodes and edges. Furthermore, HDGCN is a deep graph model which makes it have stronger node representation ability. For GLAN, it only uses content information and user information, and it constructs a message propagation graph. The structural information GLAN learned is mainly the mode of information dissemination. For us, our graph is built around news content. Social context information is also added to NDG. Most importantly, NDG constructs the relationships between different news which makes it possible to detect fake news with other time-related or content-related news as a supplement.

3.5 Ablation Studies

In order to better understand the contribution of different components to the performance, we run several ablation studies on both datasets.
Deep strategy we remove the deep strategy from a 10 layers HDGCN. Table 3 shows that when we remove deep strategy, F1 and ACC will decrease by 3%–5%, which indicates that our deep strategy is useful for deepening graph network model.

Table 3. Effect of HDGCN

	Fakeddit (Acc, F1)			Weibo (Acc, F1)
	2-class	3-class	6-class	
(-) deep strategy	−0.05, −0.03	−0.04, −0.03	−0.04, −0.04	−0.04, −0.05
(-) edge weight	−0.05, −0.05	−0.04, −0.03	—0.05, −0.05	−0.06, −0.05
(-) nodes type attention	−0.02, −0.01	−0.01, −0.02	−0.02, −0.02	−0.03, −0.02
(-) layer attention	−0.02, −0.02	−0.02, −0.02	−0.02, −0.03	−0.02, −0.03
(-) hierarchical attention	−0.06, −0.07	−0.05, −0.06	−0.07, −0.07	−0.09, −0.08

Table 4. Effect of types of nodes and edges

Nodes/Edges	Fakeddit (Acc, F1)			Weibo (Acc, F1)
	2-class	3-class	6-class	
(-) domain	−0.03,−0.03	−0.03,−0.02	−0.03,−0.02	−0.03,−0.04
(-) review	−0.04,−0.05	−0.04,−0.03	−0.04,−0.04	−0.05,−0.06
(-) author	−0.03,−0.01	−0.03,−0.02	−0.04,−0.02	−0.03,−0.03
(-) NT-relation	−0.02,−0.01	−0.01,−0.01	−0.02,−0.02	−0.01,−0.01
(-) NN-relation	−0.04,−0.04	−0.04,−0.03	−0.03,−0.02	−0.04,−0.04

Hierarchical attention strategy We remove edge weight, nodes type attention, layer attention respectively, and at last, we remove them all. Table 3 shows that edge weight is most effective component in our attention mechanism. It shows that carefully designed edge weights are useful in graph representation learning. Furthermore, we can see that these three kinds of attention weights can jointly improve the model effect.

Nodes and Edges We remove each type of nodes and edges in NDG and Table 4 shows that F1 and ACC will decrease by 3%–6%. However, the metrics decrease slightly when removing NT-relations. The reason may be that the number of NT-relation is so small that it has little effect on the results.

3.6 Further Analysis

The Number of Convolution Layers. We analyze the influence of the number of convolution layers on HDGCN. Figure 5(a) show that with the increase of the number of layers, the model results remain stable and slightly improved.

Selection of p Probability. We analyze p with a 10-layer HDGCN on both datasets. Figure 5(b) shows that the performance of our model achieves the best result when p is in 0.6–0.8. When p exceed 0.8, the performance begins to decline. Because duplicate neighbor nodes are aggregated, which can not eliminate the over smoothing problem. When p is less than 0.6, the number of neighbor nodes is insufficient. So the experimental results are not good neither.

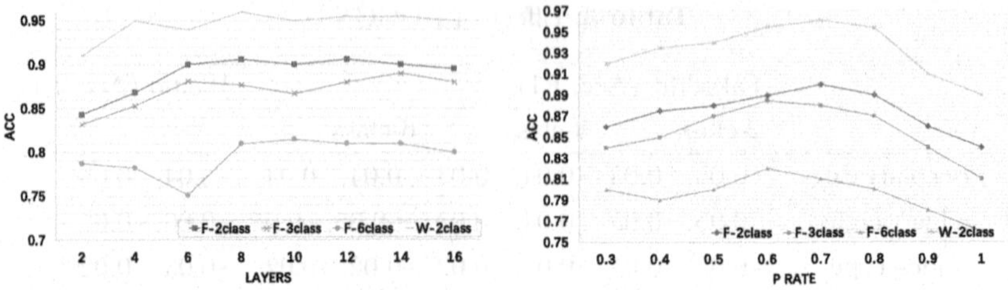

(a) Analysis of the number of layers (b) Analysis of p probability

Fig. 5. Analysis on both datasets. F: Fakeddit, W: Weibo.

4 Related Work

Most of the previous methods are mainly based on news content. These methods focus on mining the lexical features [11], syntactic features [5], deep syntax and rhetorical structure [3], or more recently, writing styles [11,17]. Despite the success of existing content based fake news detection methods, however, fake news is becoming more and more deceptive. Recent research advancements aggregate users' social engagement on news pieces to help infer which articles are fake. These works are mainly based on social context features, including user demographics [1,9], social network structure [13], propagation patterns [20] and user reactions [19]. Other works are knowledge-based methods. They uses external sources [14,23]to fact-checking claims in news content. These methods give a lot of innovative ideas to detect fake news, and all of them have improved the existing methods to a certain extent.

5 Conclusion

In this paper, we present a novel graph-base fake news detecting framework NDG-HDGCN. NDG integrates various types of information and constructs the relationships between multiple news. Different from the existing methods, we are the first work to detect fake news by using multiple news. HDGCN improves the existing GCN models and makes deep node embedding possible. Experiments show that our framework can train a deep graph model on both news datasets.

Acknowledgements. This research is supported by the National Key Research and Development Program of China (No. 2018YFB1004703) and National Natural Science Foundation of China (No. U1936110, NO61902394).

References

1. Chen, W., et al.: Semi-supervised user profiling with heterogeneous graph attention networks. IJCAI. **19**, 2116–2122 (2019)

2. Conneau, A., Kiela, D., Schwenk, H., Barrault, L., Bordes, A.: Supervised learning of universal sentence representations from natural language inference data. In: Conference on Empirical Methods in Natural Language Processing (2017)
3. Conroy, N.J., Rubin, V.L., Chen, Y.: Automatic deception detection: methods for finding fake news. In: ASIS&T2015 (2015)
4. Devlin, J., Chang, M.W., Lee, K., Toutanova, K.: Bert: pre-training of deep bidirectional transformers for language understanding. arXiv preprint arXiv:1810.04805 (2018)
5. Feng, S., Banerjee, R., Choi, Y.: Syntactic stylometry for deception detection. In: Meeting of the Association for Computational Linguistics: Short Papers (2012)
6. Graves, A., Schmidhuber, J.: Framewise phoneme classification with bidirectional LSTM and other neural network architectures. Neural Netw. **18**(5–6), 602–610 (2005)
7. Kipf, T.N., Welling, M.: Semi-supervised classification with graph convolutional networks. arXiv preprint arXiv:1609.02907 (2016)
8. Linmei, H., Yang, T., Shi, C., Ji, H., Li, X.: Heterogeneous graph attention networks for semi-supervised short text classification. In: Proceedings of the 2019 Conference on Empirical Methods in Natural Language Processing and the 9th International Joint Conference on Natural Language Processing (EMNLP-IJCNLP) (2019)
9. Long, Y.: Fake news detection through multi-perspective speaker profiles. Association for Computational Linguistics (2017)
10. Ma, J., Gao, W., Mitra, P., Kwon, S., Cha, M.: Detecting rumors from microblogs with recurrent neural networks. In: International Joint Conference on Artificial Intelligence (2016)
11. Markowitz, D.M., Hancock, J.T., Daniele, F.: Linguistic traces of a scientific fraud: the case of diederik stapel. Plos One **9**(8), e105937 (2014)
12. Nakamura, K., Levy, S., Wang, W.Y.: r/fakeddit: a new multimodal benchmark dataset for fine-grained fake news detection. arXiv preprint arXiv:1911.03854 (2019)
13. Nguyen, V.H., Sugiyama, K., Nakov, P., Kan, M.Y.: Fang: leveraging social context for fake news detection using graph representation. In: Proceedings of the 29th ACM International Conference on Information & Knowledge Management, pp. 1165–1174 (2020)
14. Pan, J.Z., Pavlova, S., Li, C., Li, N., Li, Y., Liu, J.: Content based fake news detection using knowledge graphs. In: Vrandečić, D., et al. (eds.) ISWC 2018. LNCS, vol. 11136, pp. 669–683. Springer, Cham (2018). https://doi.org/10.1007/978-3-030-00671-6_39
15. Pennington, J., Socher, R., Manning, C.: Glove: global vectors for word representation. In: Conference on Empirical Methods in Natural Language Processing (2014)
16. Perozzi, B., Al-Rfou, R., Skiena, S.: Deepwalk: online learning of social representations. In: Proceedings of the 20th ACM SIGKDD international conference on Knowledge discovery and data mining, pp. 701–710 (2014)
17. Potthast, M., Kiesel, J., Reinartz, K., Bevendorff, J., Stein, B.: A stylometric inquiry into hyperpartisan and fake news. arXiv preprint arXiv:1702.05638 (2017)
18. Ruchansky, N., Seo, S., Liu, Y.: Csi: a hybrid deep model for fake news detection. In: Proceedings of the 2017 ACM on Conference on Information and Knowledge Management, pp. 797–806 (2017)
19. Shu, K., Cui, L., Wang, S., Lee, D., Liu, H.: Defend: Explainable fake news detection. In: KDD (2019)

20. Shu, K., Mahudeswaran, D., Wang, S., Liu, H.: Hierarchical propagation networks for fake news detection: investigation and exploitation. In: Proceedings of the International AAAI Conference on Web and Social Media, Vol. 14, pp. 626–637 (2019)
21. Velikovi, P., Cucurull, G., Casanova, A., Romero, A., Liò, P., Bengio, Y.: Graph attention networks. arXiv preprint arXiv:1710.10903 (2017)
22. Vydiswaran, V.G.V., Zhai, C.X., Roth, D.: Content-driven trust propagation framework. In: ACM Sigkdd International Conference on Knowledge Discovery & Data Mining (2011)
23. Yu, B., Zhang, Z., Liu, T., Wang, B., Li, Q.: Beyond word attention: using segment attention in neural relation extraction. In: Twenty-Eighth İnternational Joint Conference on Artificial Intelligence IJCAI-19 (2019)
24. Yuan, C., Ma, Q., Zhou, W., Han, J., Hu, S.: Jointly embedding the local and global relations of heterogeneous graph for rumor detection. In: 2019 IEEE International Conference on Data Mining (ICDM) (2020)

Incrementally Finding the Vertices Absent from the Maximum Independent Sets

Xiaochen Liu, Weiguo Zheng(✉), Zhenyi Chen, Zhenying He, and X. Sean Wang

Fudan University, Shanghai, China
{xiaochenliu18,zhengweiguo,zhenyichen20, zhenying,xywangCS}@fudan.edu.cn

Abstract. A vertex v in a graph G is called an *absent vertex* if it is not in any maximum independent set of G. Absent vertex discovery is useful in various scenarios. For example, if G depicts a wireless communication interference graph, the existence of absent vertices in G may indicate network throughput bottlenecks. However, finding all the absent vertices is hard since it is at least as difficult as finding all the maximum independent sets, which is NP-hard. This paper focuses on a method that finds the absent vertices incrementally, in the hope of finding many such vertices quickly in the early incremental stages. The method iteratively invokes two polynomial-time algorithms to find the 'easy' absent vertices, and then the expensive exact maximum independent set solver to find the 'difficult' ones. At each iteration, the Mirror theorem is used to find extra absent vertices, and then all the absent vertices found so far are removed from the graph before going into the next iteration, until all absent vertices are found. Experimental results show that the above method can find most absent vertices much earlier than the baseline brute-force method on several widely-used datasets, showing its effectiveness.

Keywords: Absent vertex · Maximum independent set · Reduction rules

1 Introduction

Since graph representation can depict complex relationships among various objects, it has been widely used to model combinatorial optimization problems. One of the well-known problems regards the maximum independent set (MIS for short), which is a maximum subgraph such that no two vertices in the subgraph have an edge between them. MIS arises widely in various applications, such as mining investment from stock market interaction graphs [4], detecting malicious user in voting pools [3], and assigning channels in a wireless network [12,13]. In this paper, we study a special group of vertices concerning MIS, where each

K. Karlapalem et al. (Eds.): PAKDD 2021, LNAI 12712, pp. 421–433, 2021.
https://doi.org/10.1007/978-3-030-75762-5_34

vertex in the group is absent from all the maximum independent sets. This kind of vertex has special properties in practical applications. For instance, let us consider the scenario of channel assignment in a wireless network below.

In wireless networks, the common channel shared by devices will lead to interference between multiple transmitters in the same area [12]. Hence, channel management is a major challenge, which is more so in the fifth-generation (5G) network with the broad implementation of ultra-dense networks [7]. In the binary interference model, a conflict graph is constructed, where a vertex represents a possible communication between two devices and an edge represents an interference relationship between two communications. An MIS of the conflict graph describes the largest set of interferer allowed by the MAC protocol on a particular channel [11]. In the ideal situation, each time slice's channel assignment should be selected from one of the maximum independent sets. However, if some vertices are not included in any maximum independent set, their corresponding communication will be blocked under any maximizing throughput schedule. Fortunately, with the technologies of directional antennas and smart antennas, it is possible to reduce interferences [1], which is equivalent to reduce edges in the conflict graph, to put these vertices back in some MIS. The first step to implementing such optimization is finding the vertices not included in any MIS (called *absent vertices* in this paper).

The maximum independent set is related to the vertex cover problem, where a vertex subset C of $V(G)$ is a vertex cover iff each edge in $E(G)$ has at least one endpoint in C. It has been proven [15] that C is a minimum vertex cover of G iff $V(G)\backslash C$ is a maximum independent set of G. Thus computing the absent vertices is equivalent to computing the vertices included in all minimum vertex covers. Similarly, the absent vertices are also excluded in all maximum cliques of G's complement. In short, finding absent vertices contributes to enumerating maximum independent sets, maximum cliques, and minimum vertex covers of a graph, which have applications in graph coloring problem [6] and bioinformatics research [9]. When solving the above three problems, the absent vertices can be removed to simplify the graph structure, that can be helpful to the process.

It is costly to find all the absent vertices since it is at least as difficult as to find all MIS. In practice, however, finding a part of the absent vertices is useful. Even removing a part of the absent vertices can simplify the aforementioned problems. In the above wireless communication example, any absent vertex points to existing bottlenecks in the wireless network.

Hence, we turn to find the absent vertices incrementally and try to find many absent vertices as early as possible. We can do this because there are 'easy' absent vertices that polynomial-time algorithms can find. Also, given some absent vertices, we may be able to locate other absent vertices using Mirror theorem [10]. When no 'easy' absent vertices can be found, we then resort to the expensive algorithm to find some 'difficult' ones. Furthermore, removing absent vertices does not impact the absent vertices in the rest of the graph. So we remove known absent vertices to simplify the graph and repeat the process. Experiments show that this iterative process is effective.

Hence, the main contributions of this paper are as follows.

- We define the absent vertex, which is useful for combinatorial optimization problems. However, we prove that finding all the absent vertices is NP-Hard.
- We propose an incremental method to find all the absent vertices. The basic building blocks for the incremental method are integrative framework, two polynomial-time algorithms, the Mirror theorem, and the exact MIS solver.
- We conduct experiments on some widely-used datasets, analyzing absent vertices' degree distribution. As shown in the experimental result, our approach finds most of the absent vertices much earlier than a baseline method adapted from the exact MIS solver. For example, in the dblp-2011 dataset, our approach can find 93.9% absent vertices within 512 s, while the baseline method can only find 0.3%.

The rest of this paper is organized as follows. We formally define the absent vertex in Sect. 2. In Sect. 3 and 4, we give the baseline method, mirror reduction, our framework, and the two polynomial-time algorithms. We describe our experiment integrative results in Sect. 5. Finally, we conclude in Sect. 6.

2 Preliminary

This section presents the preliminary concepts and formal problem statement. Without loss of generality, we focus on *unweighted undirected graphs*. Let $G = (V, E)$ denote a graph consisting of the vertex set V and edge set E. We denote the one-hop neighbors of a vertex v by $N(v) = u \in V|(u, v) \in E$, and the closed one-hop neighbors by $N[v] = N(v) \cup \{v\}$.

Definition 1 *(Independent Set). A vertex subset I ($I \subseteq V(G)$) is an independent set of G (denoted by IS(G)) if no edge exists between any two vertices u and v in G (i.e., $\forall u, v \in I$, $(u, v) \notin E(G)$).*

Definition 2 *(Maximum Independent Set). An independent set I, $I \in IS(G)$, is a maximum independent set (denoted by MIS(G)) if it is not smaller than any other independent set. For such vertex set I, its cardinality is called the independence number of G, denoted by $\alpha(G)$.*

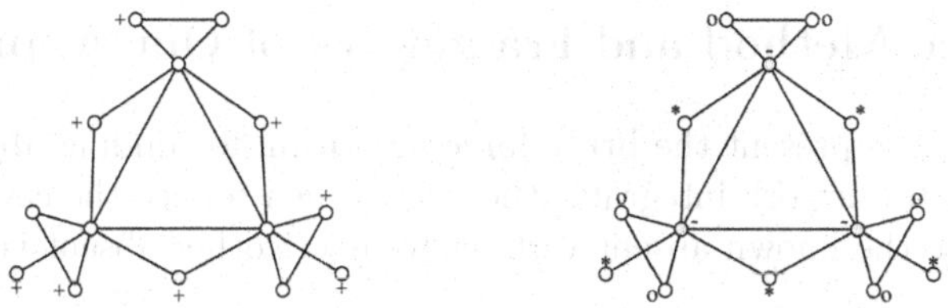

(a) independent set (b) classified vertices

Fig. 1. An example of independent set and vertex categories

The Fig. 1(a) shows an example of maximum independent with $\alpha(G) = 8$, where '+' sign marked vertices constitute an $MIS(G)$. Note that there may exist more than one vertex subset that is $MIS(G)$.

Definition 3 *(Vertex Categories). The vertices in G are classified into following two categories according to whether they belong to an $MIS(G)$ or not.*

- *Positive vertex. A vertex u is called a positive vertex if it belongs to at least one maximum independent set. Let V^+ denote the set of positive vertices in G. V^+ can be further divided into two groups:*
  - *Exact vertex. A vertex u is called an exact vertex if it is contained in all maximum independent sets of G. Let V^* denote the set of exact vertices.*
  - *Optional vertex. The other vertices in V^+ are called optional vertices (denoted by V^o), i.e., $V^o = V^+ \setminus V^*$.*
- *Absent vertex. The vertices that do not belong to any maximum independent set are called absent vertices (denoted by V^-), i.e., $V^- = V \setminus V^+$.*

In the example shown in Fig. 1(b), the vertices marked with $-$, o and $*$ correspond to V^-, V^o and V^*, respectively.

Problem Statement: *Given a graph G, the goal is to find its absent vertices.*

Theorem 1. *Finding all vertices in V^- is NP-Hard.*

Proof. Assume that there exists a polynomial algorithm that can find all absent vertices. Hence, for finding a maximum independent set in a graph, we can iteratively apply the following process: first, we use such a polynomial algorithm to find the absent vertices of G_t; second, we randomly pick a vertex u not in the absent ones to join the maintaining independent set; after that, we delete u and its one-hop neighbor in the graph, i.e., $G_{t+1} = G_t[V(G_t) \setminus N[u]]$. The G_0 is initialized with G. In each round, the randomly picked vertex is a positive vertex of G_t, which assures that the choice of this vertex can lead to a maximum independent of G_t, since the picked vertex is in a maximum independent set of G_t. Because the polynomial algorithm can be applied $O(|V(G)|)$ times at most, the time complexity of finding a maximum independent set is also polynomial. However, finding a maximum independent set is NP-Hard. Under the assumption that $NP \neq P$, such a polynomial algorithm does not exist.

3 Baseline Method and Framework of Our Approach

In this section, we present the brute force algorithm for finding absent vertices. Moreover, our framework integrates the ways which reduce the candidate set of V^- or leverage the known absent vertices to assert other absent vertices.

3.1 Baseline Brute-Force Method

Intuitively, we can enumerate all maximum independent sets and find the vertices absent from the sets. However, the number of enumerated sets is huge.

Algorithm 1: *ClassifyVertices*$(G(V, E), u)$

Input: u: the vertex to be classified
Output: V^*, V^o or V^-.

1 $G' \leftarrow newGraph(V(G) \setminus \{u\}, E(G))$
2 **if** `Calculate_MIS_Number(`G'`)`$< \alpha(G)$ **then**
3   return V^*
4 **else**
5     $G'' \leftarrow newGraph(V(G) \setminus Neighbors(u), E(G))$
6   **if** `Calculate_MIS_Number(`G''`)`$< \alpha(G) - 1$ **then**
7   return V^-
8   **else**
9   return V^o

Moreover, the enumeration process is time-consuming since finding one maximum independent set is NP-hard. We establish the following theorems to improve the exhaustive enumeration.

Theorem 2. *A vertex u belongs to V^* if the independent number of the graph obtained by removing u is less than $\alpha(G)$.*

Proof. Since the definition of *Exact Vertex* is the vertex contained in each maximum independent set, the independent set excluding u cannot be the maximum independent set, which means its cardinality is less than $\alpha(G)$.

Theorem 3. *A vertex u belongs to V^-, if the independent number of the graph obtained by removing $N[u]$ less than $\alpha(G) - 1$.*

Proof. If $u \in V^+$, there exists an $MIS(G)$, I, containing u and excluding $N(u)$. Hence, for the graph obtained by removing $N[u]$, $I \setminus \{u\}$ is one of its maximum independent sets, which means its independent number is $\alpha(G) - 1$. However, the real $\alpha(G')$ is less than $\alpha(G) - 1$. Hence, $u \notin V^+$ and $u \in V^-$.

By using Theorem 2 and Theorem 3, we devise Algorithm 1 to determine the category that a vertex belongs to. The Algorithm solves the independent number of the graphs, removing u and $N[u]$, to judge whether vertex u meets the criteria of being an exact vertex and an absent vertex in line 2 and line 5, respectively. For the vertex, neither V^* nor V^-, it is an optional vertex.

3.2 Framework of Our Approach

During the computation of MIS in Algorithm 1, various MISs would be found. For the vertices absent from the found MISs, they might be members of V^-. As for the vertices, whose status fluctuates between chosen and unchosen among known $MIS(G)$, we assert them as optional vertices. Besides utilizing the known MISs, Mirror theorem can leverage the known absent vertices for finding other absent ones. In this subsection, we extend the *Mirror Reduction* to verify other absent vertices through known absent vertices and propose our framework.

Algorithm 2: *IntegrativeFramework($G(V,E)$)*

Input: $G(V,E)$: an unweighted undirected graphs
Output: V^-: absent vertices
1 $V^- \leftarrow \emptyset$
2 $(P, N, \alpha(G)) \leftarrow$ `MIS_Solver`(G)
  `// The candidate sets for exact vertcies and absent vertcies.`
3 **foreach** $v \in N$ **do**
4     $V^-, N \leftarrow$ `PolynomialAlgorithm`(V, E, N)
5   **if** $v \notin V^-$ **then**
6   **if** `check_V_minus`$(G, v, \alpha(G)) ==$ *True* **then**
7             $V^- \leftarrow V^- \cup \{v\}$
8   **foreach** $u \in Mirrors(v)$ **do**
9                 $V^- \leftarrow V^- \cup \{u\}$; $V(G) \leftarrow V(G) \setminus \{u\}$
10             $V(G) \leftarrow V(G) \setminus \{v\}$
11   **else**
              `// Utilize the found MIS to refine the candidate set.`
12             $N \leftarrow$ `updateCandidates` (N)
13 return V^-

Mirror Reduction: As shown in a previous study [10], the choice between some pairwise vertices can be found easily. We utilize the theorem to leverage known absent vertices for finding the others.

Definition 4 *(Mirror). A mirror of v is a vertex u, u is one of v's two-hop neighbors, that $N(v) \setminus N(u)$ constructs a clique (or an empty set).*

Theorem 4. *For a vertex v, $v \in V^-$, its mirrors belong to V^- as well.*

Proof. For an absent vertex u, there are at least two of its neighbors in each $MIS(G)$. (If not, there will exist a $MIS(G)$ containing it or one of its neighbors. In the former case, it contradicts V^-'s definition. For the latter case, we can swap u with its neighbor to compose another $MIS(G)$, which contradicts the absent assumption.) Only one of the vertices in a clique could be taken into an independent set. There is at least one vertex in $N(u) \cap N(v)$, where v is a mirror of u, is in every $MIS(G)$. As a result, mirrors of u are also absent vertices.

Then, whenever a vertex is confirmed to be V^-, its mirror will be asserted as a subset of V^-. To find a vertex's mirrors, we firstly enumerate its two-hop neighbors, where there are at most Δ^2 vertices (the Δ denotes the maximum degree in the graph). For a vertex v and one of its two-hop neighbors, u, we check whether $N(v) \setminus N(u)$ construct a clique. The examination of clique structure costs $O(\Delta^2)$. Hence, the overall process's time complexity is $O(\Delta^4)$.

We propose these methods to cut off the computation cost by shrinking the candidate sets size for V^- and expanding the known V^- set. Finally, we construct a synthetical framework by assembling these modules into the basic one, as shown in Algorithm 2. In the beginning, We invoke the MIS solver to initialize the candidate sets for exact vertices and absent vertices in line 2. Then, we enumerate each v from the candidate set for absent vertices. We will first try if the polynomial-time algorithms can find some absent vertices during the enumeration. The polynomial-time algorithms illustrated in the next section are combined with mirror reduction in line 4. If polynomial-time algorithms and mirror reduction cannot process vertex v, we utilize the Theorem 3 to check whether v is an absent vertex in line 6. If v is an absent vertex, we apply mirror reduction to find other absent vertices in line 8–9 and temporarily remove the found absent ones. If not, we get one another $MIS(G)$, which shrinks the candidate set of V^- in line 12.

4 Polynomial Algorithms

In this section, we propose two heuristic algorithms to find some absent vertices that can enhance the performance of the incremental process in polynomial time.

4.1 Extended Domination Reduction

Lemma 1 (Dominance Reduction [10]). *For two adjacent vertices u and v, if $N(u) \setminus v \subseteq N(v)$, there exists a maximum independent set excluding v. We denote this relationship as $\langle u, v \rangle$, i.e., u dominates v. Besides that, $D(v)$, is used to denote the set of vertices dominating v.*

Lemma 2 (Triangle Count [8]). *A vertex u dominates its neighbor v iff $\Delta(u, v) = d(u) - 1$, where $\Delta(u, v)$ denotes the number of triangles containing edge (u, v) and $d(u)$ denotes the degree of vertex u.*

Theorem 5. *Vertex u_1 is an absent vertex, if there are two vertices u_2 and u_3 holding that: 1. $u_2 \in N(u_1)$ and $u_3 \in N(u_1)$; 2. $(u_2, u_3) \notin E$; 3. u_2 dominates u_1 and u_3 dominates u_1.*

Proof. Assume that there exists an MIS M having the vertex u_1. According to the definition of independent set, $N(u_1) \cap M = \emptyset$, which means that $u_2, u_3 \notin M$. However, $(M \setminus \{u_1\} \cup \{u_2, u_3\})$ is also an independent set, which is larger than the assumed maximum independent set M. It contradicts with the assumption. Therefore, the assumption is invalid, and there does not exists such a MIS M.

Theorem 5 inspires us a way to find absent vertices by finding the triple satisfying conditions above. We denote such way as Extended Domination Reduction (shorted as **EDR**). Finding such triple is equivalent to find a vertex u such that $D(u)$ cannot constitute a clique and $|D(u)| \geq 2$. We construct an auxiliary graph to find these triples efficiently.

We initialize the auxiliary graph G_a with empty. There are tree vertex sets in G_a, V_{check}, $V_{suspend}$ and V_{rest}, storing the probable absent vertices, the possible vertices, and other vertices, respectively.

Lemma 3. *If u_3 dominates u_2 and u_2 dominates u_1, then u_3 dominates u_1.*

Proof. With known dominations, $N(u_3) \setminus u_2 \subset N(u_2)$, and $N(u_2) \setminus u_1 \subset N(u_1)$. In addition, $u_3 \in N(u_2)$. Hence, $(u_1, u_3) \in E(G)$, u_3 dominates u_1.

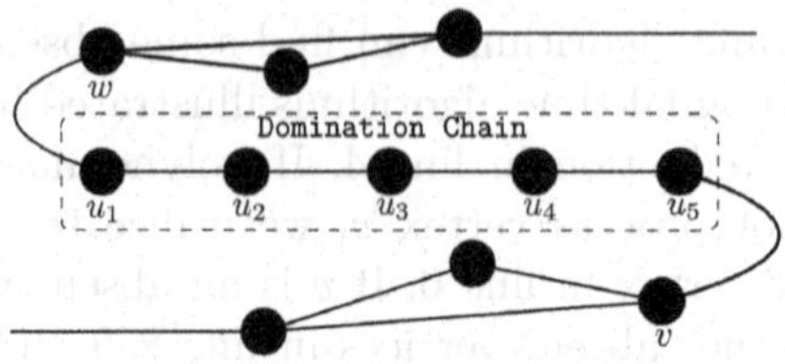

Fig. 2. Domination chain

With Lemma 3, there exists a topological order for verifying whether a vertex satisfying Theorem 5. When u dominates v and v dominates w, u must dominate w and $N[u] \subset N[v] \subset N[w]$. If $D(w)$ forms a clique, $D(v)$ can form one as well. Therefore, $D(v)$ have the higher priority to be checked as a clique than w.

Whenever a domination relationship, $\langle u, v \rangle$, is found, the following processes are taken: (1). project these two vertices into G_a; (2). add the u's counter, cnt_u by one, which records u's topological order with respect to Lemma 3; (3). if v is outside V_{check} and its counter is 0, delete its related edges in G_a and put v into the V_{check}. Otherwise, move v to $V_{suspend}$; (4). If u is in V_{check}, move it to $V_{suspend}$. Otherwise, put u to V_{rest}; (5). When it is first time to be put in V_{rest} or $V_{suspend}$, add the edges $\{(u,x)|x \in V_{rest} \cup V_{suspend}, (u,x) \in E(G)\}$ to $E(G_a)$. (6). Add the edge (u, v) to $E(G_a)$.

For the particular case that u and v dominate each other, let the vertex having a smaller identifier dominate the another one.

As for the clique structure examination, it is taken after the addition process. If $v \in V_{check}$, we need to check whether $\Delta_{G_a}(u, v)$ equals to $degree_{G_a}\{v\}$. If not, vertex v is an absent vertex. Then, v is removed from auxiliary graph. And for each vertex u, where u dominates v and is in $V_{suspend}$, the counter cnt_u should be decreased by 1. Once cnt_u is decreased to 0, u is moved back to V_{check}, since no vertex has the higher topological order than u now. At this time, edges linking to u without the meaning of domination in G_a should be removed as well.

The time complexity of using an auxiliary graph is $O(m \times \Delta)$. First of all, $O(m \times \Delta)$ is counted for initializing triangle counts to find out domination relations in G. With known domination relations, an auxiliary graph is built. Each vertex will iterate its edges in G for deleting or adding operations for at most two times when the vertex is added to V_{rest}, converted from $V_{suspend}$ to V_{check} and finally removed from G_a. Therefore, each edge in $E(G)$ will be traversed four times at most, and $O(\Delta)$ for each edge addition and deletion. In conclusion, the total time complexity is $O(m \times \Delta)$.

4.2 Chain-Chased Reduction

Definition 5. *(Domination Chain). Domination Chain (denoted by* **D-Chain***) is a path composed of vertices, of which length is odd, linking two terminal vertices, v and w, which is dominated by one of its neighbors, respectively. Each vertex, excluding terminal vertices, are numbered from one end of the path. The vertices, whose number are odd (even), are* ***odd vertices (even vertices)****.*

Figure 2 provides an example of D-Chain, where $\{u_i\}$ composes a D-chain, $\{u_1, u_3, u_5\}$ is the set of odd vertices and $\{u_2, u_4\}$ are even vertices.

Lemma 4. *For one D-Chain, if the degree of each vertex in the path is two, all odd vertices and even vertices will belong to V^* and V^-, respectively. The terminal vertices belong to V^-.*

Proof. Firstly, a path, having the odd length and no branch, has the sole maximum independent set composed of all odd vertices. If not all odd vertices are chosen, we must gain at least one extra vertex from other parts in the graph G. Hence, at least one terminal vertex should be chosen. Otherwise, the rest part of the graph will not change. However, when a terminal vertex is chosen, the terminal vertex's neighbor, which dominates it, cannot be chosen. As proven by domination reduction, this situation is not better than choosing its domination neighbor, and there does not exist such extra vertex.

Lemma 5. *Given two graph $G(V, E)$ and $G'(V, E_p \cup E)$, where $E_p = \{(u_x, u_y)| u_x \in V_G^-, u_y \in V(G)\}$, these graphs' maximum independent sets are the same.*

Proof. First of all, $\alpha(G) \geq \alpha(G')$. And if $\alpha(G) = \alpha(G')$, the maximum independent set of G' is also one of $MIS(G)$. Excluding the absent vertices, the rest part of G is identical to G'. Because of the definition of V^-, G's maximum independent sets are applicable to G'. Hence, $\alpha(G) = \alpha(G')$ and $MIS(G) = MIS(G')$.

With Lemma 5, we can extend Lemma 4's application scope. The even vertices having extra branches to other paths take no effect on vertices' classification since the graph with extra branches can be reduced to the situation of Lemma 4 after deleting the edges from absent vertices.

In our implementation of finding D-Chain, we take depth-first-search (DFS) starting with one domination relationship (u, v), regarding v as a terminal vertex in a D-Chain. During the search process, we assume the odd vertices and even vertices are members of V^* and V^-, respectively. When the search process reaches a vertex w that is dominated by other vertex or goes back to the v, we check whether the traversed path is satisfied with the Lemma 4 after deleting the extra branches from the even vertices. A threshold θ is set to limit the search tree's depth. In the experimental phase, we set θ to 5, so that the time complexity for a single trial is $O(\Delta^5)$.

5 Experimental Results

We evaluate the performance of our integrative algorithm in several real-world datasets. The experimental settings and evaluation results are listed as follows.

Datasets. We evaluate our algorithms on real graphs downloaded from the Laboratory for Web Algorithmics[5] and Stanford Network Analysis Platform [14]. Statistics of these graphs are shown in Table 1. The first four columns provide the graph's name, vertex number, edge number, and independent number (the last five columns are explained below).

Table 1. Statistics of real graphs and absent vertices found by Syn&E&D

Graphs	#Vertices	#Edges	Independence Number	$\|V^-\|$	#Found Absent Vertices			
					Brute	Mirror	EDR	D-Chain
AstroPh	18,772	198,050	6,760	7,128	490	99	6,419	120
Epinions	75,879	405,740	53,599	14,078	1,224	240	12,516	98
Email	265,214	364,481	246,898	4,498	49	16	4,431	2
dblp-2011	932,803	3,353,337	434,074	218,077	8,578	2,387	202,815	4,297
in-2004	1,382,790	13,591,223	896,673	271,928	55,763	48,859	160,271	7,035
wiki-Talk	2,394,385	4,659,565	2,338,222	32,686	840	256	31,576	14

Integrative Algorithms with Different Modules. We conduct extensive empirical studies through real graph data to evaluate the efficiency and effectiveness of our developed framework and heuristic methods for classifying vertices. We use the VCSolver [2] as the basic module for computing maximum independent sets. We implement the following algorithms[1] and evaluate their performance.

- **Brute**: enumerate each vertex, utilizing Theorem 2 and Theorem 3to find the absent vertices (see Sect. 3.1).
- **Syn**: integrate the theorems about mirror reduction and LP-relexation to infer absent vertices from the known information (see Algorithm 2).
- **Syn&EDR**: integrate EDR to integrative framework (see Sect. 4.1).
- **Syn&D-Chain**: integrate D-Chain to integrative framework (see Sect. 4.2)
- **Syn&E&D**: integrate EDR and D-Chain to integrative framework.

Experiments are conducted on a Ubuntu 18.04 server with an Intel(R) Xeon(R) Gold 5215 @ 2.50 GHz and 64 GB RAM. Algorithms are implemented in Java.

Eval-I: Evaluating Algorithms' Efficiency in Finding Absent Vertices. We evaluate the efficiency in several small datasets, since it is very expensive to run the MIS solver on large graphs, where the result is shown in Fig. 3(a), Fig. 3(b) and Fig. 3(c). The efficiency gap between the **Brute** and **Syn&E&D** keeps a level of 2x. The D-Chain Reduction achieves a slight acceleration, but the effectiveness of EDR is outstanding.

Eval-II: Evaluating Incremental Process in Given Time Budget. We set the time budget to 512 seconds to compare the aforementioned algorithms' performances. We report the relationship between the running time and found absent vertices percentage for each algorithm. The total number of absent vertices is calculated by **Syn&E&D** within a day. We also classify absent vertices according to the method of finding them out, which is reported in the last five columns of Table 1.

[1] We have shared the codes at https://github.com/Kelukin/Absent-Vertices.

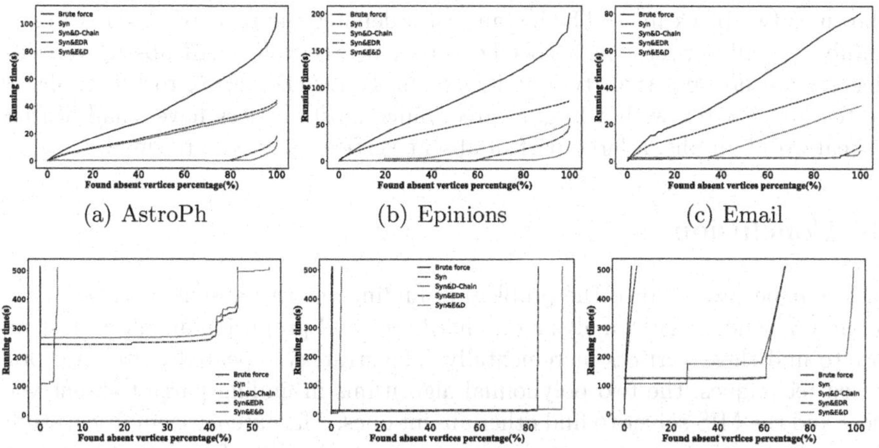

(a) AstroPh (b) Epinions (c) Email

(d) in-2004 (e) dblp-2011 (f) wiki-Talk

Fig. 3. Incremental process on different datasets

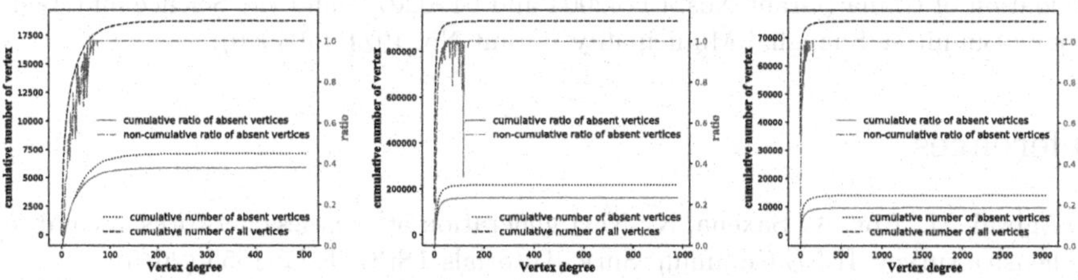

(a) AstroPh (b) dblp-2011 (c) Epinions

Fig. 4. Degree distribution of the absent vertices

As shown in the Fig. 3, the **Syn&E&D** outperforms the competitors in all cases. Since the index used in **Syn&E&D** is more complex than others, the time cost in the initialization phase is higher, which causes the **Syn&E&D** finds the first absent vertex much slower than the **Brute** and the **Syn** in Fig. 3(d) and Fig. 3(f). The two polynomial algorithms can find batches of absent vertices in little time, especially the EDR (supported by Table 1). When no polynomial algorithm is applicable, **Syn&E&D** uses the MIS solver and Mirror reduction to find other absent ones. With more absent vertices are found, the graph structure is simplified as well. Hence, the polynomial might be applicable again and boost the incremental process as shown in Fig. 3(d).

Eval-III: Evaluating Degree Distribution of Absent Vertices. We study the degree distribution of absent vertices and select three representative distribution, as shown in Fig. 4. The two red lines represent two ratios, the accumulated ratio and the instant ratio. The accumulated ratio corresponds to the ratio of absent vertices' accumulated number to all vertices', where the two numbers are shown as two black lines. The instant ratio denotes the ratio of absent vertices' number to all vertices' in a specific vertex degree value. All absent vertices' degrees are not less than two. It is very likely for the vertex to be an absent vertex for a vertex with a large degree. Since most vertices have small degrees in real-world graphs, a large part of absent vertices' degrees are small.

6 Conclusion

In this paper, we studied the problem of finding vertices absent from the maximum independent sets, analyzing its hardness, and proposing an effective strategy to find these vertices incrementally. The strategy repeatedly uses two proposed techniques: the two polynomial algorithms in finding part of absent vertices and the MIS solver to find other absent ones. The extensive empirical study demonstrates that our proposed strategy can find most of the absent vertices much earlier than the baseline brute-force method on several datasets.

Acknowledgements. This paper was supported by the National Natural Science Foundation of China (Grant Nos. 61732004 and 61902074) and the Science and Technology Committee Shanghai Municipality (Grant No. 19ZR1404900).

References

1. Agiwal, M., Roy, A., Saxena, N.: Next generation 5G wireless networks: a comprehensive survey. IEEE Commun. Surv. Tutorials **18**(3), 1617–1655 (2016)
2. Akiba, T., Iwata, Y.: Branch-and-reduce exponential/fpt algorithms in practice: a case study of vertex cover. Theor. Comput. Sci. **609**, 211–225 (2016)
3. Araújo, F., Farinha, J., Domingues, P., Silaghi, G.C., Kondo, D.: A maximum independent set approach for collusion detection in voting pools. J. Parallel Distrib. Comput. **71**(10), 1356–1366 (2011)
4. Boginski, V., Butenko, S., Pardalos, P.M.: Mining market data: a network approach. Comput. Oper. Res. **33**(11), 3171–3184 (2006)
5. Boldi, P., Vigna, S.: 004, May). The webgraph framework I: compression techniques. In: ACM WWW (2004)
6. Byskov, J.M.: Enumerating maximal independent sets with applications to graph colouring. Oper. Res. Lett. **32**(6), 547–556 (2004)
7. Cao, J., et al.: A neural network based conflict-graph construction approach for ultra-dense networks. In: IEEE Globecom Workshops (2018)
8. Chang, L., Li, W., Zhang, W.: Computing a near-maximum independent set in linear time by reducing-peeling. In: ACM SIGMOD, pp. 1181–1196 (2017)
9. Eblen, J.D., Phillips, C.A., Rogers, G.L., Langston, M.A.: The maximum clique enumeration problem: algorithms, applications, and implementations. BMC Bioinform. **13**(S–10), S5 (2012)
10. Fomin, F.V., Grandoni, F., Kratsch, D.: A measure & conquer approach for the analysis of exact algorithms. J. ACM **56**(5), 25:1-25:32 (2009)
11. Friend, D.: Cognitive networks: foundations to applications. Ph.D. thesis, Virginia Tech, Blacksburg, VA, USA (2009)

12. Hoefer, M., Kesselheim, T., Vöcking, B.: Approximation algorithms for secondary spectrum auctions. ACM Trans. Internet Techn. **14**(2–3), 16:1-16:24 (2014)
13. Joo, C., Lin, X., Ryu, J., Shroff, N.B.: Distributed greedy approximation to maximum weighted independent set for scheduling with fading channels. IEEE/ACM Trans. Netw. **24**(3), 1476–1488 (2016)
14. Leskovec, J., Krevl, A.: SNAP datasets: stanford large network dataset collection. http://snap.stanford.edu/data (2014)
15. Skiena, S.: The Algorithm Design Manual. Second Edition, Springer (2008)

Neighbours and Kinsmen: Hateful Users Detection with Graph Neural Network

Shu Li[1,2,3,4](✉), Nayyar A. Zaidi[1], Qingyun Liu[2,3], and Gang Li[5]

[1] School of Information Technology, Deakin University, Geelong 3216, Australia
{shul,nayyar.zaidi}@deakin.edu.au

[2] Institute of Information Engineering, Chinese Academy of Sciences, Beijing, China

[3] National Engineering Laboratory of Information Security Technologies, Beijing, China
liuqingyun@iie.ac.cn

[4] School of Cyber Security, University of Chinese Academy of Sciences, Beijing, China

[5] Centre for Cyber Security Research and Innovation, Deakin University, Geelong, VIC 3216, Australia
gang.li@deakin.edu.au

Abstract. With a massive rise of user-generated web content on social media, the amount of hate speech is also increasing. Countering online hate speech is a critical yet challenging task. Previous research has primarily focused on hateful content detection. In this study, we shift the attention from hateful content detection towards hateful users detection. Note, hateful users detection can benefit from users' tweets, profiles, social relationships, but the real benefit is that it can be aided by Graph Neural Networks (GNN). Typical Graph Neural Networks, such as `GraphSAGE`, only considers local neighbourhood information and samples the neighbourhood uniformly, thus they lack the ability to capture long-range relationships or to differentiate neighbours of a node. In this paper, we present `HateGNN` – a GNN-based method to address these two limitations. Our proposed method relies on the notion of latent neighbourhood, as well as systematic sampling of the neighbourhood nodes. The experimental results demonstrate that `HateGNN` outperforms state-of-the-art baselines in the task of detecting hateful users. We also provide a detailed analysis to demonstrate the efficacy of the proposed method.

Keywords: Hateful users · GNNs · Biased sampling · Latent connections

1 Introduction

The proliferation of social media enables people to freely express their opinions online. However, it also becomes the breeding ground of hate speech that is described as abusive language, cyberbullying, discrimination, racism, sexism, threats, or toxicity [2]. The stark increase of hateful content on the internet has

K. Karlapalem et al. (Eds.): PAKDD 2021, LNAI 12712, pp. 434–446, 2021.
https://doi.org/10.1007/978-3-030-75762-5_35

resulted in the emergence of conflict and hate [9]. Thus, hate speech classification has become a topic of growing interest for industry and academia. Over the past few years, a variety of models and methods on hate speech detection formulated it as a text classification task [13]. Current methods exploit the text representation as character n-grams or TF-IDF, and then resort to machine learning techniques, such as *Logistic Regression*, *SVM*, *Decision Trees*, and *Random Forests*. Recently, deep learning methods, such as *Recurrent Neural Networks* [10,14] and *Convolutional Neural Networks* [3], have been popularized in natural language processing to analyse online content.

Despite existing efforts in this area, hate speech detection remains a challenge. First, the state-of-the-art models oversimplify the problem, such as considering only tweets with hate-related words [1]. These methods rely entirely on textual (i.e., lexical and semantic) features [8], and are not aware of user and community information. Second, the state-of-the-art hate speech classifiers are vulnerable to extremely simple, model-agnostic attacks [7] [1]. These realistic attacks reduce the detection recall by nearly 50% in some cases. Therefore, hate speech classifiers depending only on text detection are not robust against adversaries who deliberately mislead the classifiers.

Fortunately, the textual contents are not the only information that can be used to study hate speech in the social network. There is information that is often linked to a profile representing a person or an organization. Investigating such information presents plenty of opportunities to explore a richer feature space that can be helpful in identifying online hate speech. Moreover, it is more natural that users' profile is considered when detecting hate speech, rather than just considering isolated tweets. In addition, directly identifying (and controlling) hateful users who are intentionally propagating the hate speech is an important and effective measure for countering online hate speech. Therefore, in this study we shift the attention from hate speech toward hateful users.

Since hateful users detection can benefit from users' tweets, profiles and social relationships – it can be aided by *Graph Neural Networks* (GNNs). Specifically, users in the social network are nodes in the graph and the social relationship can be regarded as edges. Each node contains abundant property information such as user's profile, user's tweets. GNNs can be naturally applied to such node classification task by employing deep neural networks to aggregate feature information of neighbouring nodes. However, detecting hateful users based on GNNs has three main challenges.

- It is likely to be ineffective to aggregate neighbouring information for nodes that have no or too few relations with other nodes in the graph.
- A neighbourhood of a node is defined as the set of all neighbours which are one or more hops away. Typically, only features of the neighbourhood are aggregated. We conjecture that there might be nodes which could be very similar to the node in question, but are not in the neighbourhood of the

[1] The model-agnostic attacks include diluting the hateful signal, obfuscating hateful tokens through character level perturbations, or injecting non-hate distractor.

node. The existing methods are not able to aggregate such high-similarity (non-neighbourhood) nodes.
- The current GNNs, especially the spatial-based methods like `GraphSAGE` [4], sample all neighbours equally when aggregating their information. It does not consider the fact that different neighbours may influence the node differently.

To address the limitations above, we propose a framework – `HateGNN`, for hateful user detection in the social network. `HateGNN` has following salient features:

- To address the first and second challenge described above, apart from the existing (explicit) social graph, we create a latent graph based on the node property information. This will help drastically for nodes which have no or too few neighbours. Moreover, the latent graph has the ability to capture the important features from distant but informative nodes.
- To address the third challenge, a bias strategy is applied to sample neighbours (not only immediate neighbours but latent neighbours) for differentiating the influences of the neighbours. We have proposed a sampling strategy to help choose the most informative features.
- Once the neighbours are selected, we aggregate the social and latent neighbourhoods to compute the final node embeddings.

We claim that the final embeddings obtained with `HateGNN` are much powerful than existing state-of-the-art methods. We back-up this claim by conducting experiments on two public datasets of hate speech. The results demonstrate the superior performance of `HateGNN` over state-of-the-art baselines. Moreover, we separately validate the efficacy of the salient features of `HateGNN`. We also provide detailed analyses on how various parameters (e.g., the size of sampling neighbours set, the similarity, the node properties) impact the model performance.

2 Related Work

Hate speech detection has been a popular research topic for decades. However, most of the existing literature focused on hate speech detection towards textual contents [13], and they attempted to adopt various typical classification algorithms [3,10,14]. Research on hateful users is still relatively under explored. The research of [11] characterized the hateful users on Twitter, and it showed that hateful users differ from normal ones in terms of their word usage, activity patterns and network structure. However, the work of [11] mainly focused on the analysis of hateful users rather than the detection model.

GNNs have been widely used to learn node embeddings. GNNs encode nodes into vectors by aggregating feature information from node's local neighbourhood via neural networks. To reduce the computational costs and improve performance, several recent studies have attempted to use different ways of neighbourhood aggregation. *Graph Convolutional Network* (GCN) uses a graph convolutional layer to encapsulate each node's hidden representation by summing "message" from all one-hop neighbours [6]. *Graph Attention Network* (GAT) [12] assigns different importance to different neighbours by utilising self-attention mechanism, and then combines their impacts to generate node embeddings. As a general inductive framework, `GraphSAGE` [4] is able to efficiently generate node embeddings for previously unseen data by sampling and aggregating features from a node's local neighbourhood. However, existing neighbourhood aggregation methods are not able to aggregate nodes that are similar but are far away from each other (i.e., not in immediate neighbourhood of each other). Moreover, these methods overlook the fact that different neighbours can influence a node in different ways. As discussed, we will address these issues with our proposed `HateGNN` method.

3 Problem Definition

In this section, we introduce the notions of social graph and latent graph, and then formally define the problem of *Hateful Users Detection with Graph Neural Network* (`HateGNN`). The social network, such as Twitter, is represented by a huge graph, in which every user is represented by a node and the follower/followee relationships between nodes are represented by an edge. Apart from the follower-followee network, the tweeting/retweeting network, representing the flow of information, can also be represented by a retweet graph, with users as the nodes and edges implying that one user has retweeted the post of another user. We define a social graph as:

Definition 1 (Social Graph). *A social graph is graph that can be created based on existing (explicit) connections between entities, and is parametrised as:* $\mathcal{G}^o = (\mathbf{V}, \mathbf{E}_{\mathcal{G}^o}, \mathbf{P})$, *where* $\mathbf{V}$ *are the nodes,* $\mathbf{E}_{\mathcal{G}^o}$ *correspond to the relationships of nodes, and* $\mathbf{P}$ *denotes the properties of nodes.*

The social graph defined above is fundamental to describe the relationship of the users, in which the node representation can be efficiently improved by aggregating features from its neighbours. However, it is unable to capture the long-distance dependencies among nodes with similar properties or topology structures (e.g. core, betweenness, and community bridges) when they are far away in the social graph. In order to solve the above problems, we propose a latent graph as:

Definition 2 (Latent Graph). *A latent graph is parametrised as:* $\mathcal{G}^l = (\mathbf{V}, \mathbf{E}_{\mathcal{G}^l}, \mathbf{P}, \lambda_{\mathcal{G}^l})$ *with nodes* $\mathbf{V}$ *and edges* $\mathbf{E}_{\mathcal{G}^l}$, *where the edges between two nodes* $u, v \in \mathbf{V}$ *denotes their similarity exceeds a certain threshold* $\lambda_{\mathcal{G}^l}$. $\mathbf{P}$ *represents the property of nodes as the same meaning in the* $\mathcal{G}^o$.

The problem of `HateGNN` is then formulated as follows. Given a set of users and their tweets in the social network, `HateGNN` aims to design a model $\mathcal{M}$ to learn the embedding of user, denoted as $\mathbf{x}_v$, and the function $\mathcal{F} : \mathbf{x}_v \rightarrow \Sigma$ that assigns the label information to users, so as to detect whether one user is hateful or not in the social network.

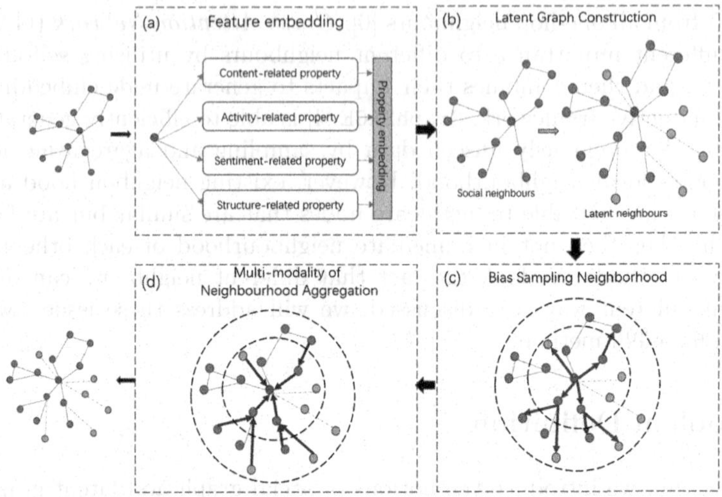

Fig. 1. Pictorial illustration of the of HateGNN framework.

4 The HateGNN Framework

In this section, we will describe our proposed HateGNN method, and illustrate it in Fig. 1. First, the social graph $\mathcal{G}^o$ is obtained[2], where the nodes are the users, and each node has its property (e.g. users' tweets). Second, when the similarity of property vector between two nodes exceeds the threshold $\lambda_{\mathcal{G}^l}$, they are linked by a latent edge. These latent edges and the corresponding nodes form the latent graph $\mathcal{G}^l$. Third, a biased neighbourhood sampling strategy is implemented. Neighbours (both social and latent) that are more similar to the processed node, have higher priority to be sampled until a fixed-size set of neighbours $N^s_{(v)}$ is obtained. Finally, after choosing neighbours $N^s_{(v)}$, we aggregate their property information to obtain node embedding $\mathbf{x}_v$ by multiplying the weight matrices that are trained with neural networks.

4.1 Latent Graph Construction

Let us discuss the creation of latent graph. Note, for hateful user detection, the properties of node $v \in \mathbf{V}$, could be content-related, activity-related, or sentiment-related, etc. In addition, the topology structure for each node, could also be considered as a property[3]. The properties used in this work are detailed in Table 1. We obtain vector representation for each property (e.g. one-hot and

[2] The follower-followee or tweeting/retweeting relationship can be obtained conveniently by using Application Programming Interface (API) provided by the social network.

[3] The topology structure refers to centrality measurements for v in $\mathcal{G}^o$.

Table 1. List of various nodes' properties used in `HateGNN`.

Property type	Property values
Content-related property	users' tweets
Activity-related property	the number of tweets, retweet, follower, followees, favourites, hashtags, quote, URLs, mentions per tweet in average, and average and median time interval between tweets
Sentiment-related property	sentiment of tweets, and bad-words usage
Structure-related property	betweenness, eigenvector, indegree, outdegree, and the above property for the 1-neighbourhood of a user

labels encoders, `GloVe` for word embedding) and get the initial feature of node $v \in \mathbf{V}$, denoted as $\mathbf{z}_v$. For $u, v \in \mathbf{V}$, their similarity is defined based on as:

$$S(v,u) = \text{PearsonSimilarity}(\mathbf{z}_v, \mathbf{z}_u). \tag{1}$$

When the similarity between two nodes exceeds the threshold $\lambda_{\mathcal{G}^l}$, they are linked by a latent edge, which finally creates a latent graph $\mathcal{G}^l$.

4.2 Biased Sampling Neighbourhood

When the graph have high-degree nodes (i.e., the nodes have a large number of neighbours), considering all neighbours for aggregation is usually inefficient and unnecessary [5]. Given that a node's neighbours in graph have no natural ordering (e.g., sentences, images), `GraphSAGE` [4] proposed to uniformly sample a fixed-size set of neighbours, which outperforms strong GNNs models. Our proposed framework `HateGNN` improves `GraphSAGE` by deriving a set of sampled neighbours based on their similarity. The intuition is that similar neighbours (similar in any type of properties listed in Table 1) could consolidate and enhance the node embedding results.

Algorithm 1 describes the overall procedure of our sampling process. The operations from step 4 to 12 calculate the node similarity to sample neighbours. Then, for each node $v \in \mathbf{V}$, it aggregates the representations of its sampled neighbourhood, $\{\mathbf{h}_u^{k-1}, \forall u \in N_{(v)}^s\}$, and then concatenates the node's current representation, $\mathbf{h}_v^{k-1}$. This concatenated vector is fed through a fully connected layer with non-linear activation function σ. Finally, we get the final representations output at depth K, denoted as $\mathbf{x}_v = \mathbf{h}_v^k, \forall v \in \mathbf{V}$. Specifically, the similarity threshold $\lambda_{\mathcal{G}^l}$ in the latent graph could be set equal to the bias parameter η in the `HateGNN` model.

4.3 Multi-modality of Neighbourhood Aggregation

The neighbourhoods $N(v) = \left\{N_o(v), N_l(v)\right\}$ of node v includes its neighbourhood in both the social graph and the latent graph. The social-neighbourhood $N_o(v)$ consists of the set of v's adjacent nodes in the social graph $\mathcal{G}^o$, and the

Algorithm 1. `HateGNN` using Biased Sampling Neighbourhood

Input: Graph $\mathcal{G} = (\mathbf{V}, \mathbf{E}, \mathbf{P})$; input features $\{\mathbf{z_v}, \forall v \in \mathbf{V}\}$; depth K;
Weight matrices $\mathbf{W}^k, \forall k \in \{1..., K\}$; non-linearity σ;
Mean aggregator functions $\text{AGGREGATE}_k^{mean}$, $\forall k \in \{1..., K\}$;
Neighbourhood $N(v)$;
The sampled neighbourhood $N_{(v)}^s$, the size $|N_{(v)}^s|$;
The size of neighbours to be sampled β; The bias parameter η;
Output: Vector representations $\mathbf{x}_v$ for all $v \in \mathbf{V}$;
1: $\mathbf{h}_v^0 \leftarrow \mathbf{z_v}, \forall v \in \mathbf{V}$;
2: **for** $\{k = 1...K\}$ **do**
3:   **for** $v \in \mathbf{V}$ **do**
4:     **for** $u \in N_{(v)}$ **do**
5:       **if** $|N_{(v)}^s| \leq \beta$ **then**
6:         **if** $S(v, u) \geq \eta$ using Equation (1) **then**
7:           $N_{(v)}^s \leftarrow u$;
8:         **end if**
9:       else
10:         **break;**
11:       **end if**
12:     **end for**
13:     $\mathbf{h}_{N_{(v)}^s}^k \leftarrow \text{AGGREGATE}_k^{mean}\left(\{\mathbf{h}_u^{k-1}, \forall u \in N_{(v)}^s\}\right)$;
14:     $\mathbf{h}_v^k \leftarrow \sigma\left(\mathbf{W}^k \cdot \text{CONCAT}\ (\mathbf{h}_v^{k-1}, \mathbf{h}_{N_{(v)}^s}^k)\right)$;
15:   **end for**
16:   $\mathbf{h}_v^k \leftarrow \mathbf{h}_v^k / ||\mathbf{h}_v^k||_2, \forall v \in \mathbf{V}$;
17: **end for**
18: $\mathbf{x}_v \leftarrow \mathbf{h}_v^K, \forall v \in \mathbf{V}$;

latent-neighbourhood $N_l(v)$ are those whose similarity to node v are higher than a parameter $\lambda_{\mathcal{G}^l}$. In aggregation process, we combine the social neighbourhood and the latent neighbourhood to generate the node embedding. The motivation is that different types of neighbours will make different contributions to the final node representation. For the social-neighbourhood, it denotes the effect of user' social nature. In comparison to this explicit relationship, the latent-neighbourhood indicates the long-range dependencies with the node, which is invisible and cannot be captured directly. Thus the step 13 in `HateGNN` could be updated with Eq. (2), in which $N_{o_{(v)}}^s$ and $N_{l_{(v)}}^s$ are the sampled neighbours from the social graph and the latent graph, respectively, that is:

$$\mathbf{h}_{N_{(v)}^s}^k \leftarrow \text{AGGREGATE}_k^{mean}\left(\left\{\mathbf{h}_u^{k-1}, \forall u \in \{N_{o_{(v)}}^s \cup N_{l_{(v)}}^s\}\right\}\right). \tag{2}$$

4.4 Model Training

`HateGNN` is not attempting to learn the embedding results for all nodes in a graph, but to learn a mapping that generates embedding for each node. Depending on

Table 2. Datasets used in the Experiments

Data	Users	Edges
`HateUser5K`	100,386	retweet edges: 2,286,592
`Tweet9K`	1,448	follower-followee edges: 3,471

the dataset with node labels for hateful user detection, we train the model in a semi-supervised learning paradigm. With the labelled nodes, we train `HateGNN` by minimizing the cross entropy via back-propagation and gradient descent. Thus, the loss function is calculated as:

$$\mathcal{L} = \sum_{v \in \mathbf{V}} \Big(y_v \log p_v + (1 - y_v) \log(1 - p_v) \Big), \quad \text{where} \quad p_v = \sigma(\mathbf{w}^T \mathbf{x}_v + b).$$

5 Experiments

In this section, we conduct an empirical evaluation of our proposed method `HateGNN`, with the aim of answering the following research questions:

RQ1: How does `HateGNN` perform vs. the baselines for hateful users detection?
RQ2: How do the components of `HateGNN` (latent neighbourhood, bias-sampling, multi-modality of neighbourhood aggregation) affect the model performance?
RQ3: How do various parameters, e.g., the size of sampling neighbours set, the similarity, the node properties, impact performance of the model?

We compared `HateGNN` with five baselines, including `AdaBoost`, `GradBoost`, `GCN` [6], `GAT` [12] and `GraphSAGE` [4], and we evaluated their performances using two widely-used datasets in the hate speech domain, for which the statistics are summarized in Table 2.

HateUser5K contains a network of 100k users, out of which about 5k were annotated to be either hateful or not. Hateful users are those who endorse any type of hate speech (e.g., abusive language, discrimination, racism). Each user has several activity-related, content-related, and structure-related properties, as shown in Table 1. If one user has retweeted another users, such retweet connection is represented as the social graph in this dataset.

Tweet9K is the dataset of online tweets, which contains 16,907 tweet IDs and their labels. It was collected from *Twitter* by [13], and was annotated as sexism, racism, both or neither by recruited experts. By using the *Tweepy* library, we retrieved the tweets and also collected the follower-followee information for the users as the edges in the social graph. Since some users have now been suspended, only 9,755 tweets of 1,448 users were acquired.

5.1 Comparisons with Baselines (RQ1)

We compared our model with other baselines (`AdaBoost, GradBoost, GCN, GAT, GraphSAGE`) in Table 3, and used `Accuracy, F1-score, Area under the`

ROC Curve (AUC), to evaluate the performance of classification. It is noteworthy that HateGNN outperforms all baselines on both datasets. In terms of accuracy, our model leads to a performance improvement of over 5% on HateUser5K, and over 2% on Tweet9K, which is very encouraging.

5.2 Performance Analysis (RQ2)

To answer **RQ2**, we design experiments to evaluate the efficacy of each component of HateGNN. The performance is reported in Table 4, where the best results are highlighted in bold.

Table 3. Comparison of HateGNN with baselines and state-of-the art GraphSage.

Dataset	Methods	Accuracy	F1-score	AUC
HateUser5K	AdaBoost	0.6894 ± 0.0132	0.3724 ± 0.0137	0.8499 ± 0.0182
	GradBoost	0.8389 ± 0.0104	0.5043 ± 0.0217	0.8768 ± 0.0086
	GCN	0.8543 ± 0.0254	0.5397 ± 0.0127	0.8716 ± 0.0376
	GAT	0.8643 ± 0.0302	0.4578 ± 0.0212	0.7900 ± 0.0197
	GraphSAGE	0.8904 ± 0.0372	0.6355 ± 0.0880	0.9392 ± 0.0334
	HateGNN	**0.9509 ± 0.0354**	**0.7987 ± 0.1385**	**0.9649 ± 0.0442**
Tweet9K	AdaBoost	0.4475 ± 0.0168	0.4891 ± 0.0112	0.7567 ± 0.0290
	GradBoost	0.7342 ± 0.0280	0.5919 ± 0.0404	0.7813 ± 0.0376
	GCN	0.6897 ± 0.1880	0.5048 ± 0.0314	0.7003 ± 0.0323
	GAT	0.7023 ± 0.0820	0.4991 ± 0.0124	0.6813 ± 0.0536
	GraphSAGE	0.8598 ± 0.0974	0.7889 ± 0.1302	0.9243 ± 0.0694
	HateGNN	**0.8715 ± 0.1052**	**0.8062 ± 0.1434**	**0.9244 ± 0.0843**

Table 4. The effects of each component of HateGNN.

Dataset	Graph	Methods	Accuracy	F1-score	AUC
HateUser5K	Social Graph	GraphSAGE	0.8922 ± 0.0339	0.6399 ± 0.0789	0.9396 ± 0.0346
		HateGNN	**0.9276 ± 0.0364**	**0.7264 ± 0.1196**	**0.9509 ± 0.0437**
	Latent Graph	GraphSAGE	0.8892 ± 0.0341	0.6337 ± 0.0787	0.9398 ± 0.0326
		HateGNN	**0.9237 ± 0.0408**	**0.7302 ± 0.1252**	**0.9570 ± 0.0336**
	Social + Latent	GraphSAGE	0.9008 ± 0.0356	0.6652 ± 0.0886	0.9458 ± 0.0339
		HateGNN	**0.9260 ± 0.0322**	**0.7204 ± 0.1031**	**0.9548 ± 0.0359**
Tweet9K	Social Graph	GraphSAGE	0.8345 ± 0.1128	0.7812 ± 0.1362	**0.9005 ± 0.0959**
		HateGNN	**0.8429 ± 0.1118**	**0.7856 ± 0.1439**	0.8938 ± 0.1044
	Latent Graph	GraphSAGE	0.8598 ± 0.0974	0.7889 ± 0.1302	**0.9243 ± 0.0694**
		HateGNN	**0.8626 ± 0.0974**	**0.7926 ± 0.1328**	0.9205 ± 0.0825
	Social + Latent	GraphSAGE	0.8660 ± 0.0970	0.7974 ± 0.1322	**0.9273 ± 0.0734**
		HateGNN	**0.8715 ± 0.1052**	**0.8062 ± 0.1434**	0.9244 ± 0.0843

The Efficacy of the Latent Neighbourhood. It can be seen from Table 4 by focusing on results reported individually on social graph and latent graph – that, the performance on latent graph is comparable with that on social graph. It is encouraging to see that the performance on latent graph is even better than that on the social graph of dataset Tweet9K, thus, demonstrating the efficacy of the latent connections for this problem. Please note that the individual reported results on social graph and latent graph, do not include biased sampling of the neighbours.

The Efficacy of the Biased Sampling Strategy. We conducted the experiments on social graph, latent graph and the (social+latent) graph, respectively. The results in Table 4 show that HateGNN generally outperforms GraphSAGE. This confirms that the biased sampling strategy helps learn the node embedding from the neighbours.

The Effect of Multi-modality of Neighbourhood Aggregation. According to Table 4, for HateUser5K dataset, the performances of HateGNN on the (social+latent) graph is slightly superior than the results on latent graph in Accuracy and on social graph in AUC. In terms of Tweet9K, HateGNN performs better on the (social+latent) graph than on individually latent graph or social graph. Moreover, GraphSAGE trained on the combination of the social and latent graph has a better performance than GraphSAGE on either on HateUser5K or Tweet9K, demonstrating the efficacy of multi-modality neighbourhood.

Table 5. The comparative analysis of the sampling size.

Methods	Sampling size	Accuracy	F1-score	AUC
GraphSAGE	$S_1 = 5$; $S_2 = 1$	0.8904 ± 0.0372	0.6355 ± 0.0880	0.9392 ± 0.0334
	$S_1 = 10$; $S_2 = 5$	0.8944 ± 0.0363	0.6448 ± 0.0875	0.9418 ± 0.035
	$S_1 = 25$; $S_2 = 10$	0.8936 ± 0.0365	0.6423 ± 0.0890	0.9427 ± 0.0340
	$S_1 = 100$; $S_2 = 50$	0.8968 ± 0.0357	**0.6516 ± 0.0879**	0.9440 ± 0.0349
	$S_1 = 500$; $S_2 = 100$	**0.8972 ± 0.03466**	0.6493 ± 0.0876	**0.9447 ± 0.0344**
HateGNN	$S_1 = 5$; $S_2 = 1$	**0.9260 ± 0.0322**	**0.7204 ± 0.1031**	**0.9548 ± 0.0359**
	$S_1 = 10$; $S_2 = 5$	0.9155 ± 0.0348	0.6983 ± 0.0970	0.9500 ± 0.0355
	$S_1 = 25$; $S_2 = 10$	0.9103 ± 0.0355	0.6844 ± 0.0980	0.9488 ± 0.0359
	$S_1 = 100$; $S_2 = 50$	0.9063 ± 0.0296	0.6729 ± 0.0768	0.9496 ± 0.0330
	$S_1 = 500$; $S_2 = 100$	0.9016 ± 0.0344	0.6599 ± 0.0889	0.9474 ± 0.0334

5.3 Parameter Analysis (RQ3)

How do various parameters, e.g., the size of sampling neighbours set, the similarity measure and the node properties impact the model performance? To answer **RQ3**, we discuss these questions in this section. Due to space constraints, we only present results on HateUser5K dataset, however, a similar pattern of results was observed on Tweet9K.

The Setting of the Sampling Size. In this section, we probe the influence of the size of sampling neighbours set on the model performance. [4] found that setting the depth of neighbourhood $K = 2$ provided a consistent boost in accuracy. Thus, we set the default value for K. Because of the memory limit and the run-time requirements, we only adjust the neighbourhood sample sizes S_1 and S_2 from $\{5, 1\}$ to $\{500, 100\}$. The results are presented in Table 5. For random sampling of `GraphSAGE`, increasing the neighbourhood sample size basically obtained no more than 1% performance improvement. For biased sampling strategy of `HateGNN`, a small sampling size achieved the best performance, showing that learning node embedding from a small number of sampling neighbours is able to maintain promising results. It is encouraging to see that a small sample size for `HateGNN` leads to much better accuracy that was achieved with `GraphSage` with much larger sample.

The Similarity Measure. We have to use some forms of similarity measure to calculate the similarity among nodes, and then conduct latent graph construction and biased sampling. In this experiment, we compare two similarity measures, namely *Spearman* and *Pearson*, as presented in Table 6. Compared with random sampling, `HateGNN` model trained with *Spearman* or *Pearson* similarity measures, has much better performance. Secondly, *Pearson*-based similarity measure leads to better performance than *Spearman*. This is the reason, we present *Pearson* as the default option, and all the results presented in this work are based on *Pearson* measure.

The Effect of Similarity and Dissimilarity. In Sect. 5.2, we demonstrated the effectiveness of biased sampling. Here we discuss why we can not do biased sampling with dissimilarity measures? In this experiment, we compared `HateGNN` model performance by biased sampling neighbours according to similarity and dissimilarity (Table 7). Dissimilarity-based sampling strategy is unable to achieve further performance improvement, even worse than `GraphSAGE` with random

Table 6. The impact of various similarity approaches.

Neighbourhood	Similarity	Accuracy	F1-score	AUC
Social Neighbourhood	Random	0.8922 ± 0.0339	0.6399 ± 0.0789	0.9396 ± 0.0346
	Spearman	0.9163 ± 0.0335	0.7009 ± 0.0961	0.9505 ± 0.0361
	Pearson	0.9276 ± 0.0364	0.7264 ± 0.1196	0.9509 ± 0.0437
Latent Neighbourhood	Random	0.8892 ± 0.0341	0.6337 ± 0.0787	0.9398 ± 0.0326
	Spearman	0.9139 ± 0.0360	0.6965 ± 0.1063	0.9512 ± 0.0373
	Pearson	0.9237 ± 0.0408	0.7302 ± 0.1252	0.9570 ± 0.0336
Social + Latent	Random	0.8904 ± 0.0372	0.6355 ± 0.0880	0.9392 ± 0.0334
	Spearman	0.9177 ± 0.0349	0.6994 ± 0.1059	0.9509 ± 0.0377
	Pearson	0.9260 ± 0.0322	0.7204 ± 0.1031	0.9548 ± 0.0359

Table 7. The effect of the use of similarity or dissimilarity measure.

Bias	Accuracy	F1-score	AUC
Dissimilarity	0.8853 ± 0.0412	0.6313 ± 0.0951	0.9333 ± 0.0327
Random	0.8904 ± 0.0372	0.6355 ± 0.0880	0.9392 ± 0.0334
Similarity	**0.9260 ± 0.0322**	**0.7204 ± 0.1031**	**0.9548 ± 0.0359**

sampling. It is demonstrated that nodes embedding cannot benefit from the dissimilar neighbours in this task.

The Impacts of the Nodes' Properties. As collecting content-related and activity-related properties are relatively easy (i.e. only the user itself is involved), we have only utilized these two properties in this study. Here, we will explore the impact of more nodes' properties on `HateGNN` model. It can be seen from Table 8, that by adding activity-related property, the relative improvements are no more than 1% in three evaluation metrics. But when using all properties, `HateGNN` model achieves further performance improvement, with the relative improvements are about 3%, 9%, 2% in `Accuracy`, `F1-score`, `AUC`, respectively. This shows the benefit of utilizing more nodes' properties in the model for this task.

Table 8. The impact of nodes' properties

Property	Accuracy	F1-score	AUC
Content (300d)	0.9195 ± 0.0389	0.7084 ± 0.1204	0.9480 ± 0.0438
Content+Activity(320d)	0.9260 ± 0.0322	0.7204 ± 0.1031	0.9548 ± 0.0359
Content+Activity+ Sentiment+Structure(1028d)	**0.9509 ± 0.0354**	**0.7987 ± 0.1385**	**0.9649 ± 0.0442**

6 Conclusions

In this paper, we develop a sophisticated framework for hateful users detection in the social network – `HateGNN`, which not only exploits the explicit social graph, but also builds a latent graph. In addition, it has an effective neighbour sampling technique that can choose the most informative features from neighbours. On two standard hate-speech detection datasets, the proposed model leads to better performance than existing state of the art methods such as `GraphSAGE`, etc. In the future, we will investigate the application of `HateGNN` on even larger datasets, together with biased sampling weight learning based on multi-view node properties. It is important to note that formulation of `HateGNN` is general, and though we have constrained ourselves to hate-speech detection problem in this work, the application of `HateGNN` to general graphs is straight-forward and is currently under-progress.

Acknowledgment. This work is supported by Scientific Research Guiding Project (Grant No. Y9W0013401), Key Technical Talents Project of CAS (Grant No. Y8YY041101) and National Natural Science Fund of China (Project No. 71871090).

References

1. Arango, A., Pérez, J., Poblete, B.: Hate speech detection is not as easy as you may think: a closer look at model validation. In: COLING (2019)
2. Fortuna, P., Nunes, S.: A survey on automatic detection of hate speech in text. ACM Comput. Surv. **51**(4), 1–30 (2018)
3. Georgakopoulos, S.V., Tasoulis, S.K., Vrahatis, A.G., Plagianakos, V.P.: Convolutional neural networks for toxic comment classification. In: Proceedings of the 10th Hellenic Conference on Artificial Intelligence, SETN (2018)
4. Hamilton, W.L., Ying, Z., Leskovec, J.: Inductive representation learning on large graphs. In: NIPS, pp. 1024–1034 (2017)
5. Hou, Y., Chen, H., Li, C., Cheng, J., Yang, M.C.: A representation learning framework for property graphs. In: SIGKDD, pp. 65–73. ACM (2019)
6. Kipf, T.N., Welling, M.: Semi-supervised classification with graph convolutional networks. In: ICLR (2016)
7. Kurita, K., Belova, A., Anastasopoulos, A.: Towards robust toxic content classification. arXiv preprint arXiv:1912.06872 (2019)
8. Mishra, P., Del Tredici, M., Yannakoudakis, H., Shutova, E.: Author profiling for abuse detection. In: COLING (2018)
9. Petulla, S., Kupperman, T., Schneider, J.: Hate crimes spurned by group-based hatred (2018)
10. Pitsilis, G.K., Ramampiaro, H., Langseth, H.: Effective hate-speech detection in twitter data using recurrent neural networks. Appl. Intell. **48**(12), 4730–4742 (2018)
11. Ribeiro, M.H., Calais, P.H., Santos, Y.A., Almeida, V.A., Meira Jr, W.: Characterizing and detecting hateful users on twitter. In: ICWSM (2018)
12. Velickovic, P., Cucurull, G., Casanova, A., Romero, A., Liò, P., Bengio, Y.: Graph attention networks. In: ICLR (2018)
13. Waseem, Z., Hovy, D.: Hateful symbols or hateful people? Predictive features for hate speech detection on twitter. In: SRW@HLT-NAACL, pp. 88–93 (2016)
14. Zhou, P., Qi, Z., Zheng, S., Xu, J., Bao, H., Xu, B.: Text classification improved by integrating bidirectional LSTM with two-dimensional max pooling. In: COLING (2016)

Graph Neural Networks for Soft Semi-Supervised Learning on Hypergraphs

Naganand Yadati[1(✉)], Tingran Gao[2], Shahab Asoodeh[3], Partha Talukdar[1], and Anand Louis[1]

[1] Indian Institute of Science, Bangalore, Karnataka 560012, India
{naganand,ppt,anandl}@iisc.ac.in
[2] University of Chicago, Chicago, IL 60637, USA
tingrangao@galton.uchicago.edu
[3] Harvard University, Cambridge, MA 02138, USA
shahab@seas.harvard.edu

Abstract. Graph-based semi-supervised learning (SSL) assigns labels to initially unlabelled vertices in a graph. Graph neural networks (GNNs), esp. graph convolutional networks (GCNs), are at the core of the current-state-of-the art models for graph-based SSL problems. GCNs have recently been extended to undirected hypergraphs in which relationships go beyond pairwise associations. There is a need to extend GCNs to directed hypergraphs which represent more expressively many real-world data sets such as co-authorship networks and recommendation networks. Furthermore, labels of interest in these applications are most naturally represented by probability distributions. Motivated by these needs, in this paper, we propose a novel GNN-based method for directed hypergraphs, called Directed Hypergraph Network (DHN) for semi-supervised learning of probability distributions (Soft SSL). A key contribution of this paper is to establish generalisation error bounds for GNN-based soft SSL. In fact, our theoretical analysis is quite general and has straightforward applicability to DHN as well as to existing hypergraph methods. We demonstrate the effectiveness of our method through detailed experimentation on real-world datasets. We have made the code available.

Keywords: Graph neural network · Hypergraph · Soft SSL

1 Introduction

In the last decade, deep learning models have been successfully embraced in many different fields and have been shown to achieve excellent performance on a vast range of applications. Graph Convolutional Networks (GCNs) [16] have been recently proposed as an adaptation of a particular deep learning model (i.e., convolutional neural networks) to enable handling of graph-structured data. GCN has been shown to be effective especially in semi-supervised learning on attributed graphs. GCNs have inspired the current state-of-the art models for graph-based SSL [24,28].

K. Karlapalem et al. (Eds.): PAKDD 2021, LNAI 12712, pp. 447–458, 2021.
https://doi.org/10.1007/978-3-030-75762-5_36

While graphs are powerful data representations for pairwise relationships, hypergraphs provide more flexible data representations for relationships beyond pairwise associations. A hypergraph relaxes the notion of an edge (commonly called hyperedge) to contain more than two vertices. Real-world datasets such as co-authorship networks, recommendation networks, email communication networks, protein-protein interaction networks, etc. can be flexibly modelled by hypergraphs. For example, in a co-authoship network, a document (hyperedge) can be co-authored by more than two authors (vertices). The existence of such relationships naturally motivates the problem of hypergraph-based semi-supervised learning (SSL).

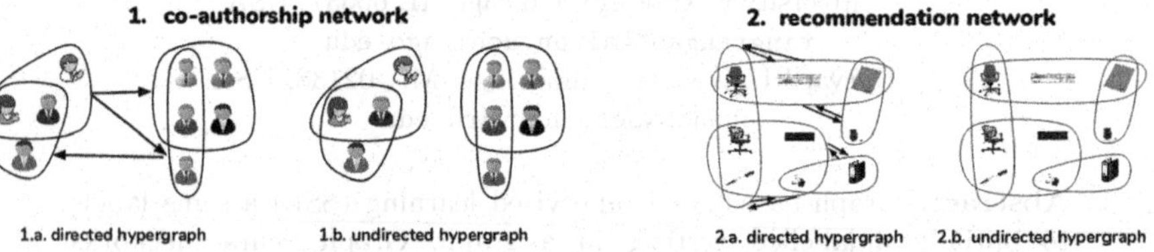

Fig. 1. (Best seen in colour) Examples of real-world networks modelled as directed hypergraphs and undirected hypergraphs. To the left is 1. co-authorship network in which vertices are authors, and hyperedges are collaborations (documents). 1.a. shows the network modelled as a directed hypergraph in which directions are citations among documents. 1.b. shows the undirected version in which the citation relationships are absent. To the right is 2. recommendation network in which vertices are products, and each hyperedge contains all products bought by a user. 2.a. shows the directed hypergraph in which directions represent user similarity (two-way) and 2.b. shows the undirected version. We are interested in semi-supervised vertex classification of probability distributions in these networks. The vertex labels in the examples are research topic interests for co-authorship and product ratings for recommendation networks. (Color figure online)

There exist GNN-based methods for semi-supervised vertex classification on undirected hypergraphs [7,29]. However, these methods do not work for directed hypergraphs. Directed hypergraphs encode additional relationships as illustrated in Fig. 1. For example, in a co-authorship network, documents (hyperedges) are related by directed citation relationships. Motivated by this, our focus in this paper is on semi-supervised vertex classification on the more powerful directed hypergraphs which encode an additional layer of relationships (Fig. 1).

Furthermore, vertex labels in these applications involving directed hypergraphs e.g. research interests of authors in co-authorship networks, product ratings of products in recommendation networks, etc. are most naturally represented by probability distributions (soft labels). Following up on a prior work [20] that generalised label propagation to graph-based SSL of probability distributions (graph-based *soft* SSL) and motivated by the fact that directed hypergraphs and soft-labels occur *simultaneously* in real-world, we make the following contributions.

- We explore GNNs for (hyper)graph-based soft SSL. We propose DHN (Directed Hypergraph Network), a novel GNN-based method for directed hypergraphs. DHN can be applied for soft SSL using existing tools from optimal transportation.
- Our second contribution is to provide generalisation error bounds for GNN-based soft SSL. Our effort in this direction has lead to deriving generalisation error bounds for GNNs within the framework of algorithmic stability. We establish that such models, which use filters with bounded eigenvalues independent of graph size, can satisfy the strong notion of uniform stability and thus are generalisable. In particular, the algorithmic stability of GNNs depends on the largest absolute eigenvalue of the graph convolution filter . Our analysis is quite general and the error bounds can be easily established for DHN and existing hypergraph neural methods.
- We demonstrate DHN's effectiveness through detailed experimentation on real-world data. In particular, we demonstrate superiority over state-of-the-art hypergraph-based neural networks such as HGNN [7] and HyperGCN [29]. We provide new empirical benchmarks for soft-SSL. We have made the code available to foster reproducible research.

We have made the code and supplementary pdf available at https://drive.google.com/file/d/1DNJXIqKdpWqrimLQU3yZxUtYUYBXh1Oe/view?usp=sharing.

2 Related Work

Geometric deep learning is an umbrella phrase for emerging techniques attempting to generalise (structured) deep neural networks to non-Euclidean domains such as graphs and manifolds. GCN [16] and their various extensions are the current state-of-the art for graph-based SSL [28] and graph-based unsupervised learning [12] problems. The reader is referred to recent books [13,18] on this topic. Recently, graph-based deep models (also message-passing neural networks [11]) have been analysed theoretically [25].

Learning on Hypergraphs: Hypergraph is a combinatorial structure consisting of vertices and hyperedges, where each hyperedge is allowed to connect any number of vertices, thus generalizing graphs. This additional flexibility facilitates the capture of higher order interactions among objects; applications have been found in many fields such as computer vision, network clustering, folksonomies, cellular networks, and community detection.

The seminal work on hypergraphs [32] introduced the popular clique expansion [7] of a hypergraph. Hypergraph neural networks (HGNN) [7] use the clique expansion while HyperGCN [29] uses the mediator-based Laplacian to extend GCNs to hypergraphs. Another line of work uses the mathematically appealing tensor methods but they are limited to uniform hypergraphs. Recent developments work for arbitrary hypergraphs and fully exploit the hypergraph structure [2,15,30].

Graph-Based Soft SSL: Researchers have shown that using unlabelled data during training can improve label prediction significantly [22]. While most methods assume that labels of interest are numerical or categorical variables, other works "soften" this assumption and handle "soft labels" such as histograms [4,23]. One way of propagating histograms is to minimise the Kullback-Leibler (KL) divergence [21]. Recent studies have replaced the metric-agnostic KL divergence with metric-aware Wasserstein distance (interactions between histogram bins) for graphs [20] and hypergraphs [10].

Embeddings in Wasserstein Space: There exist at least a couple of recent works that embed Gaussian distributions in the Wasserstein space [19]. Inspired by a recent work [8], in this work, we focus on embedding input data as a discrete probability distrirbution on a fixed support set. The Wasserstein distance and its gradient require the solution of a linear program [27] and are costly to compute. A popular efficient approximation is the Sinkhorn divergence [5] in which the underlying problem is regularised and is computed efficiently by a fixed-point iteration.

In all the papers that we have discussed above, the proposed methods are either restricted to graphs or undirected hypergraphs and do not work for directed hypergraphs. Also, none of the GNN-based methods discusses soft SSL. Our contributions are precisely to address these limitations.

3 Method

In this section, we first describe soft SSL on directed hypergraphs and then propose DHN (Directed Hypergraph Network) for the problem.

3.1 Directed Hypergraph

A directed hypergraph [9] is an ordered pair $\mathcal{H} = (V, E_d)$ where $V = \{v_1, \cdots, v_n\}$ is a set of n vertices and $E_d = \{(t_1, h_1), \cdots, (t_m, h_m)\} \subseteq 2^V \times 2^V$ is a set of m directed hyperedges. Each element in E_d is an ordered pair (t, h) where $t \subseteq V$ is the *tail* and $h \subseteq V$ is the *head* with $t \neq \emptyset$, $h \neq \emptyset$. Denote the set of all undirected hyperedges by E i.e., $E = \bigcup_{(t,h)\in E_d} \Big(t \cup h\Big)$. Denote $I \in \{0,1\}^{|V|\times|E|}$ to be the incidence matrix of E i.e. $I(v, e) = 1$ if $v \in e$ and 0 otherwise.

3.2 Soft SSL on Directed Hypergraphs

We consider the problem of predicting probability distributions for the vertices in $\mathcal{H} = (V, E_d)$ given a typically small subset $V_k \subseteq V$ of vertices with known distributions. In this work, we are concerned with discrete distributions modelled on a metric space i.e. an ordered pair (M, C) in which M is a set and $C : M \times M \to \mathbb{R}$ is the cost function (metric) associated with the set. Furthermore, we assume that we are provided with a feature matrix, $X_V \in \mathbb{R}^{n\times D_V}$, in which each vertex $v \in V$ is represented by a D_V-dimensional feature vector x_v (here $n = |V|$). We are also provided with a hyperedge feature matrix $X_E \in \mathbb{R}^{m \times D_E}$ with $x_e, e \in E$ as D_E-dimensional feature representations (here $m = |E_d|$).

Our objective is to learn a labelling function $Z = \phi(\mathcal{H}, X_V, X_E)$ that maps each vertex to a probability distribution in the space of discrete probability distributions $\mathcal{P}_F(M)$ on F atoms (F is number of histogram bins) defined on the metric space (M, C). The cost function C can be represented by a non-negative symmetric matrix of size $F \times F$. Note that each row of $Z \in [0,1]^{n \times F}$ maps each vertex $v \in V$ to a probability distribution $Z_v \in [0,1]^F$. The function h is going to be trained on a supervised loss, L ,w.r.t to the vertices in V_k so that the trained h can be used to predict distributions of all the vertices in $V \setminus V_k$. We now give an example application and then the details of the labelling function h followed by the supervised loss L.

Example Application: Predicting topic distributions of authors in co-authorship networks can be posed as a soft SSL problem on directed hypergraphs. V represents the set of authors, E the set of all collaborations (documents), E_d the citation relationships among the documents, F the number of possible research interests of authors (Machine Learning, Theoretical Computer Science, etc.), X_V and X_E any available features on the authors and documents respectively (e.g. text attributes).

3.3 Directed Hypergraph Network (DHN)

Hypergraphs contain hyperedges in which relationships can go beyond pairwise and hence are challenging to deal with. A flexible way to embed vertices of a hypergraph is to "approximate" the hypergraph by a suitable graph and then apply traditional graph-based methods on the vertices. Two notable candidates of ϕ are Hypergraph neural network (HGNN) [7] and Hypergraph Convolutional Network (HyperGCN) [29]. HGNN uses the clique expansion of the hypergraph [32] while HyperGCN uses the mediator-based Laplacian [2] to approximate the input hypergraph. However, they are restricted to undirected hyperedges and also cannot exploit the hyperedge feature matrix X_E.

A key idea of our approach is to treat each hyperedge $e \in E$ as a vertex of the graph $\mathcal{G} = (E, E_d)$. We then pass $\mathcal{G}$ through a graph neural network to obtain $H_E = f_{GNN}(\mathcal{G}, X_E)$ so that the initial features, X_E, are refined to H_E. We then propose the layer-wise propagation rule of DHN as:

$$H_V^{(t+1)} = \sigma\left(\left[H_V^{(t)},\ \ I \cdot H_E^{(t)} \cdot \Theta^{(t)}\right]\right), \quad t = 0, \cdots, \tau - 1 \tag{1}$$

where $[\cdot, \cdot]$ denotes concatenation, t is the time step, I is the incidence matrix, $H_E^{(t+1)} = \sigma_1\left(I^T H_V^{(t)}\right)$ for $t = 1, \cdots, \tau - 1$, $H_E^{(0)} = f_{GNN}(\mathcal{G}, X_E)$, σ and σ_1 are non-linear activation functions, and τ is the total number of propagation steps with $H_V^{(0)} = X_V$. Note that the labelling function $Z = \phi(\mathcal{H}, X_V, X_E) = \text{softmax}\left(H_V^{\tau}\right)$ where softmax is applied row-wise.

3.4 The Supervised Loss L

A crucial observation here is that because of the softmax layer, the output of h is (already) inherently a probability distribution. For each vertex $v \in V_k$, the predicted distribution Z_v and the (known) true distribution Y_v must be "close" to each other. A natural way to compare probability distributions is to use the KL-divergence between Y_v and Z_v. However, KL-divergence cannot exploit the metric space (M, C) and suffers from stability issues [3]. In this work, we use the more stable Wasserstein distance to exploit the metric space [10].

$$L = \sum_{v \in V_k} W_p\Big(Z_v, Y_v\Big), \quad W_p(\mu, \nu) = \left(\inf_{\pi \in \Pi(\mu,\nu)} \int_{M \times M} C(x_1, x_2)^p d\pi(x_1, x_2)\right)^{\frac{1}{p}}.$$

For discrete distributions, W_p is the solution of a linear program. For practical purposes, we compute the regularised distance using the Sinkhorn algorithm. Please see the supplementary material for more details.

Optimisation: We call DHN optimised with the Wasserstein loss as Soft-DHN. All parameters are learned using stochastic gradient descent (SGD). Please see the supplementary material for time complexity.

4 Theoretical Analysis: Generalisation Error

A key contribution of this chapter is to provide generalisation error bounds for (hyper)graph-based soft-SSL. Our effort to derive these bounds has lead us to a more powerful outcome, namely, proving generalisation error bounds for a one-layer GNN by extending the results of a traditional GCN [25] to the soft SSL setting with Wasserstein loss. The main novelty is to generalise the error bounds to the learning problem "valued in the Wasserstein space." The main challenge is that the Wasserstein space is an abstract metric space without linear structure.

The section is organised as follows. We first introduce all the notations needed (ego-graph view, semi-supervised learning setting, etc.). We then give single layer and SGD bounds using the notations. We finally give the main result (proposition 1) which states that a GNN trained with Wasserstein loss has the same generalisation error bound as the traditional GCN (trained with cross entropy).

Let $G = (V, E)$ be a graph with $|V| = n$. We consider a one-layer GNN

$$f(X, \Theta) = \sigma(KX\Theta) \tag{2}$$

where $X \in \mathbb{R}^{n \times d}$ is the feature matrix (n is the number of vertices in a graph, d is the dimension of the feature vectors), $K = g(L_G)$ is a graph filter (typically symmetrically normalised adjacency with self loops, and $L_G \in \mathbb{R}^{n \times n}$ is the graph Laplacian), and $\Theta \in \mathbb{R}^{d \times F}$ is the set of parameters. We note that our proposed DHN falls under this formulation in special circumstances. Specifically if the non-linearity σ_1 in Eq. 1 is removed we get the kernel $K = II^T$ (also known as the clique expansion of the hypergraph [32].) The non-linearity σ in Eq. 2 is the softmax function acting on each row of the product $KX\Theta \in \mathbb{R}^{n\times F}$; the output is of dimension $n \times F$, where each output row is a discrete probability distribution, i.e., $f(X,\Theta) \geq 0$ and $f(X,\Theta)\mathbf{1}_F = \mathbf{1}_n$ where $\mathbf{1}_F = (1,\ldots,1)^F \in \mathbb{R}^F$, and similarly for $\mathbf{1}_n$. Without loss of generality, we assume $d = 1$. Note that in order for the output to be nontrivial probability distributions, we must assume $F > 1$.

We adopt an ego-graph view [25] to simplify our discussion for local behavior of the soft GCN at a particular vertex. Whenever no confusion arises, we identify a vertices x and χ in the graph G with their respective D-dimensional feature vectors. Thus the output of f at $x \in V$ is $f(x,\Theta) = \sigma\left(\sum_{\chi\in\mathcal{N}(x)} K_{x\chi}\chi\Theta\right) = \sigma\left(\left(\sum_{\chi\in\mathcal{N}(x)} K_{x\chi}\chi\right)\cdot\Theta\right)$ where $\mathcal{N}(x)$ denotes for the one-hop neighborhood of x with respect to the adjacency relation defined by matrix K, and $K_{x\chi} \in \mathbb{R}$ stands for the entry in $K \in \mathbb{R}^{n\times n}$ that describes the adjacency relation between vertices x and χ. Let $E_x := \sum_{\chi\in\mathcal{N}(x)} K_{x\chi}\chi \in \mathbb{R}$ so that $f(x,\Theta) = \sigma(E_x \cdot \Theta)$.

We consider the supervised learning setting, and learn GNN from the training set $\{z_i = (x_i, y_i), i = 1,\ldots,m\}$ sampled i.i.d. from the product space $V \times \mathcal{P}_F$ with respect to probability distribution $\mathcal{D}$ on this product space, where $\mathcal{P}_F$ is the space of discrete probability distributions on F atoms. The output of softmax lies in $\mathcal{P}_F$, which is a convex cone. For any new data $z = (x,y) \sim \mathcal{D}$, we evaluate the performance of GNN f using a Wasserstein cost $\ell(f(\cdot,\Theta),z) = \ell(f(\cdot,\Theta),(x,y)) = W(f(x,\Theta),y)$.

Here the Wasserstein cost is defined with respect to a cost function penalizing moving masses across bins. Since we are working only with histograms in GNN, we shall use a cost function $C \in \mathbb{R}^{F\times F}$ that is defined for pairs of histogram bins. The transport problem is a linear program with $z = f(x,\Theta)$.

4.1 Assumptions/Notations

To avoid unnecessary technical complications, assume the histogram admits a geometric realisation over the one-dimensional Euclidean space, such that the ith bin is placed at location $b_i \in \mathbb{R}$, and set $C_{ij} := |b_i - b_j|, \quad \forall 1 \leq i,j \leq F$. Without loss of generality, we assume $b_1 \leq b_2 \leq \cdots \leq b_F$, and write $h_i := b_{i+1} - b_i \geq 0$ for all $i = 1,\ldots,F-1$. Denote the diameter of the support by $D := \max_{1\leq i,j\leq F}|b_i - b_j| = b_F - b_1$. We take the Wasserstein cost as the Wasserstein-1 distance: $W(\mu,\nu) := W_1(\mu,\nu)$. In this particular one-dimensional setting, we have a particularly simple form for the cost function:

$$W_1(\mu,\nu) = \int_0^1 \left|F_\mu^{-1}(s) - F_\nu^{-1}(s)\right| \mathrm{d}s = \int_{-\infty}^{\infty} |F_\mu(t) - F_\nu(t)| \,\mathrm{d}t \qquad (3)$$

where $F_\mu : \mathbb{R} \to [0,1]$, $F_\nu : \mathbb{R} \to [0,1]$ are the cumulative distribution functions of μ, ν, respectively; F_μ^{-1}, F_ν^{-1} are the *generalized inverses* of F_μ and F_ν, respectively, defined as (similar for F_ν^{-1})

$$F_\mu^{-1}(t) := \inf\{b \in \mathbb{R} : F_\mu(b) > t\}, \qquad \forall t \in [0,1]. \qquad (4)$$

This characterisation is seen in any standard literature on optimal transport, e.g., [26].

4.2 Definitions: Generalisation and Empirical Erros

Let the learning algorithm A_S on a dataset S be a function from ζ^m to $(\mathcal{Y})^X$. Where $\mathcal{X}$ is the input Hilber space, $\mathcal{Y}$ is the output Hilbert space, $\zeta = \mathcal{X} \times \mathcal{Y}$. The training set of datapoints, labels is $S = \{z_1 = (x_1, y_1), \cdots, z_k = (x_k, y_k))\}$. Let the loss function be $\ell : \zeta^m \times \zeta \to \mathbb{R}$. Then the generalisation error or risk $R(A_S)$ is deined as $R(A_S) := \mathbb{E}\Big[\ell(A_S, z)\Big] = \int \ell(A_S, z)p(z)dz$ where $p(z)$ is the probability of seeing the sample $z \in S$. The empirical error, on the other hand, is $R_{\text{emp}}(A_S) := \frac{1}{k}\sum_{j=1}^{k} \ell(A_S, z_j)$

4.3 Extension to GNNs on Hypergraphs

We note that our proposed DHN falls under this formulation in special circumstances. In particular, if the non-linearities in Eq. 1 are removed, our DHN can be seen as $K = IA^2$ where $f_{GNN}(\mathcal{G}, X_E)$ is the simple graph convolution operator [28] and A is the symmetrically normalised adjacency (with self loops) of the graph $\mathcal{G}$. Also, the analysis can be extended to exisiting hypergraph GNNs such as hypergraph neural network (HGNN [7]) where $K = II^T$ (also known as the clique expansion [32] of the hypergraph). Hence, our theoretical analysis is quite general that can be easily appplied to DHN and existing hypergraph neural methods. The full proof is given in the supplementary material.

Theorem 1. *Let A_S be a one-layer GNN algorithm (of Eq. 2) equipped with the graph convolutional filter $g(L_G)$ and trained on a dataset S for T iterations. Let the loss and activation functions be Lipschitz-continuous and smooth. Then the following expected generalisation gap holds with probability at least $1 - \delta$, $\delta \in \{0, 1\}$:*

$$\mathbb{E}_{\text{SGD}}\Big[R(A_S) - R_{\text{emp}}(A_S)\Big] \leq \frac{1}{m}O\Big((\lambda_G^{max})^{2T}\Big) + \Big(O\Big((\lambda_G^{max})^{2T}\Big) + B\Big)\sqrt{\frac{\log\frac{1}{\delta}}{2m}} \quad (5)$$

where the expectation $\mathbb{E}_{SGD}$ is taken over the randomness inherent in SGD, m is the no. training samples, and B is a constant which depends on the loss function. Our theorem states that GNN trained with the Wasserstein loss enjoys the same generalisation error bound as the traditional GCN (trained with cross entropy). We establish that such models, which use filters with bounded eigenvalues independent of graph size, can satisfy the strong notion of uniform stability and thus is generalisable.

5 Experiments

We conducted experiments on 5 real-world directed hypergraphs (four are co-authorship datasets and one is a recommendation dataset). Statistics of the datasets are in the supplementary. Vertex labels in all our datasets are discrete distributions seen in the real world (not synthetic in any way). For example, in

Table 1. Results on real-world directed hypergraphs. We report 100× mean squared errors (lower is better) over 10 different train-test splits. All reported numbers are to be multiplied by 0.01 to get the actual numbers. Please see Sect. 5 for details.

Method	Cora	DBLP	ACM	Amazon	arXiv
KL-MLP	8.94 ± 0.16	7.72 ± 0.14	8.47 ± 0.15	6.81 ± 0.16	10.87 ± 0.25
OT-MLP	7.45 ± 0.35	7.53 ± 0.18	7.85 ± 0.26	6.78 ± 0.24	10.01 ± 0.23
KLR-MLP	8.05 ± 0.22	7.35 ± 0.18	7.82 ± 0.29	6.74 ± 0.15	–
OTR-MLP	6.57 ± 0.43	7.24 ± 0.18	6.77 ± 0.32	6.72 ± 0.23	–
KL-HGNN	7.86 ± 0.25	7.17 ± 0.12	7.23 ± 0.19	6.71 ± 0.19	9.95 ± 0.25
KL-HyperGCN	7.95 ± 0.27	7.15 ± 0.17	7.53 ± 0.21	6.69 ± 0.17	9.99 ± 0.23
Soft-HGNN	5.97 ± 0.37	6.18 ± 0.37	6.02 ± 0.37	6.63 ± 0.39	8.61 ± 0.49
Soft-HyperGCN	6.02 ± 0.32	6.21 ± 0.35	6.04 ± 0.32	$\mathbf{6.61 \pm 0.30}$	8.60 ± 0.47
Soft-HGAT	5.83 ± 0.39	6.05 ± 0.33	5.94 ± 0.39	6.62 ± 0.45	8.47 ± 0.46
Soft-SAGNN	5.69 ± 0.32	6.07 ± 0.47	5.82 ± 0.41	$\mathbf{6.59 \pm 0.32}$	8.29 ± 0.27
Soft-HNHN	5.64 ± 0.37	6.11 ± 0.39	5.88 ± 0.27	6.64 ± 0.36	8.34 ± 0.35
Soft-DHGCN	5.62 ± 0.35	6.06 ± 0.45	5.84 ± 0.39	6.67 ± 0.33	8.31 ± 0.29
KL-DHN (ours)	7.04 ± 0.24	6.97 ± 0.22	7.16 ± 0.24	6.65 ± 0.17	9.34 ± 0.32
Soft-DHN (ours)	$\mathbf{4.87 \pm 0.40}$	$\mathbf{5.65 \pm 0.42}$	$\mathbf{5.12 \pm 0.34}$	$\mathbf{6.55 \pm 0.33}$	$\mathbf{7.69 \pm 0.36}$

a co-authorship dataset, an author that has published 7 papers in vision conferences, and 3 in NLP conferences, is assigned the soft label [0.7, 0.3] (this is neither multi-label nor single label). In fact, more than 80% vertices in all our datasets have such proper soft labels. For more details on dataset construction, please see Sect. 6 of the supplementary material. Our focus is on predicting true probability distributions. Existing hypergraph neural methods such as HGNN [7], and HyperGCN [29] were originally designed for multi-class SSL (Hard SSL). However, we adapted them for soft SSL by training them with KL and Wasserstein losses.

5.1 Experimental Setup

We take extensive measures to ensure fairness of comparison with baselines. Inspired by the experimental setups of prior related works [16,17], we tune hyperparameters using the Cora citation network dataset alone and use the optimal hyperparameters for all the other datasets. We hyperparameterise the cost matrix (base metric of the Wasserstein distance) as $C_{ii} = 1, i = 1, \cdots, F$ and $C_{ij} = \eta, i \neq j$. The cost matrix C is an $F \times F$ matrix (F is the number of histogram bins) with ones on the diagonal and a hyperparameter η elsewhere. We could have used a matrix of all ηs. But it is no different from a matrix of all ones from the optimisation perspective and so we used the above more general matrix. Details of hyperparameter tuning and optimal hyperparameters of all methods (including baselines) are in the suplementary material. We use the mean squared error (MSE) between true and predicted distributions on the test set of vertices. Table 1 shows MSEs on the test split for all the five datasets.

5.2 Baselines

We used both Wasserstein distance and KL divergence to train different models. As already noted, we used the Sinkhorn algorithm to compute the (regularised) Wasserstein distance. We compared DHN with the following baselines:

- **KL-MLP**: We used a simple multi-layer perceptron (MLP) on the features of the vertices and trained it using KL-divergence
- **OT-MLP**: We trained another MLP with the Wasserstein distance as the loss function. Note that this baseline and the previous baseline do not use the structure (graph/hypergraph)
- **KLR-MLP**: We regularised an MLP with explicit KL-divergence-based regularisation that uses the structure (graph/hypergraph) [21].
- **OTR-MLP**: We regularised an MLP with explicit Wasserestein-distance-based regularisation that uses the structure (graph/hypergraph) [20]. For hypergraphs we used the clique expansion of the hypergraph [10].
- **KL-HGNN/KL-HyperGCN**: We trained the different GCN-based methods on hypergraphs with KL divergence loss function on the labelled vertices.
- **Soft-HGNN/Soft-HyperGCN**: We trained the different GCN-based methods on hypergraphs with the Wasserstein distance as the loss function.
- **Soft-Hyper-Atten**: This model is a generalisation of graph attention to hypergraphs [1] We trained it with the Wasserstein distance.
- **Soft-Hyper-SAGNN**: We trained the recently proposed self-attention hypergraph-based method [31] with the Wasserstein distance loss.
- **Soft-HNHN**: We also used this very recent model [6] as a baseline (with Wasserstein loss). HNHN uses hypereges as neurons and computes hyperedge representations.
- **Soft-DHGCN**: This baselines [14] uses separate incidence matrices for tail and head. We trained it with Wasserstein loss.

5.3 Discussion

We used a simple one-layer architecture for our proposed DHN and a 2-hop simplified GCN [28] as the GNN model on the graph $\mathcal{G} = (E, E_d)$ i.e.

$$Z = softmax(I \cdot H_E \cdot \Theta_1), \quad H_E = A^2 X_E \Theta_2$$

where A is the symmetically normalised adjacency (with self loops) of the graph $\mathcal{G}$. We demonstrate that this simple model is effective enough through an ablation study in Sect. 5.1 of the supplementary. Please see Sect. 5 of the supplementary for more experiments on arXiv. Our results on real-world datasets demonstrate strong performances across all the datasets esp. on the co-authorships networks. Specifically, we observe that Soft models (that use the Wasserstein loss) are almost always superior to their counterparts that use the KL divergence as the loss function. This is because the Soft models can exploit the distance matrix C while KL-divergence does not. Moreover, our proposed DHN outperforms several hypergraph baselines. This is because they do not exploit the rich structural information in the directed hyperedges (connections among hyperedges) while our DHN does exploit them. Though baselines such as HNHN [6] compute representations for hyperedges, they do not exploit dependencies between them. The DHGCN baseline [14] does exploit such relationships. However, it does not treat the relationships between hyperedges as separate edges. The benefits of doing so of a separate graph are evident in the table where we report 2 hops of a GNN run on this graph. We also experimented on standard graph node-classification datasets such as Cora, Citeseer, and Pubmed by treating the class label as one-hot distribution. We used the Soft variants of GCN [16], Simple GCN [28], and GAT [24]. We achieved competitive results as in Sect. 5 (Table 3) of the supplementary.

6 Conclusion

We have proposed DHN, a novel method for soft SSL on directed hypergraphs. DHN can effectively propagate histograms to unknown vertices by integrating vertex features, directed hyperedges and undirected hypergraph structure. We have established generalisation bounds for DHN within the framework of algorithmic stability. Specifically we modified the "gradient" in Wasserstein space to satisfy Lipschitz condition required in the stability framework. DHN is effective compared to SOTA baselines.

References

1. Bai, S., Zhang, F., Torr, P.H.S.: Hypergraph convolution and hypergraph attention. Pattern Recogn. **110**, 107637 (2021)
2. Chan, T.H., Liang, Z.: Generalizing the hypergraph laplacian via a diffusion process with mediators. In: COCOON (2018)
3. Chen, Y., Ye, J., Li, J.: A distance for HMMs based on aggregated wasserstein metric and state registration. In: Leibe, B., Matas, J., Sebe, N., Welling, M. (eds.) ECCV 2016. LNCS, vol. 9910, pp. 451–466. Springer, Cham (2016). https://doi.org/10.1007/978-3-319-46466-4_27
4. Corduneanu, A., Jaakkola, T.S.: Distributed information regularization on graphs. In: NIPS, pp. 297–304. MIT Press (2005)
5. Cuturi, M.: Sinkhorn distances: lightspeed computation of optimal transport. In: NIPS. Curran Associates, Inc. (2013)
6. Dong, Y., Sawin, W., Bengio, Y.: HNHN: hypergraph networks with hyperedge neurons. arXiv preprint arXiv:2006.12278 (2020)
7. Feng, Y., You, H., Zhang, Z., Ji, R., Gao, Y.: Hypergraph neural networks. In: AAAI (2019)
8. Frogner, C., Mirzazadeh, F., Solomon, J.: Learning entropic wasserstein embeddings. In: ICLR (2019)
9. Gallo, G., Longo, G., Pallottino, S., Nguyen, S.: Directed hypergraphs and applications. Discrete Appl. Math. **42**(2–3), 177–201 (1993)
10. Gao, T., Asoodeh, S., Huang, Y., Evans, J.: Wasserstein soft label propagation on hypergraphs: algorithm and generalization error bounds. In: AAAI (2019)

11. Gilmer, J., Schoenholz, S.S., Riley, P.F., Vinyals, O., Dahl, G.E.: Neural message passing for quantum chemistry. In: ICML (2017)
12. Hamilton, W., Ying, Z., Leskovec, J.: Inductive representation learning on large graphs. In: NIPS. Curran Associates, Inc. (2017)
13. Hamilton, W.L.: Graph representation learning. Synth. Lect. Artif. Intell. Mach. Learn. **14**(3), 1–159 (2020)
14. Han, J., Cheng, B., Wang, X.: Two-phase hypergraph based reasoning with dynamic relations for multi-hop KBQA. In: Proceedings of the Twenty-Ninth International Joint Conference on Artificial Intelligence (IJCAI), pp. 3615–3621 (2020)
15. Hein, M., Setzer, S., Jost, L., Rangapuram, S.S.: The total variation on hypergraphs - learning on hypergraphs revisited. In: NIPS. Curran Associates, Inc. (2013)
16. Kipf, T.N., Welling, M.: Semi-supervised classification with graph convolutional networks. In: ICLR (2017)
17. Liao, R., Zhao, Z., Urtasun, R., Zemel, R.S.: Lanczosnet: multi-scale deep graph convolutional networks. In: ICLR (2019)
18. Ma, Y., Tang, J.: Deep learning on graphs. Cambridge University Press (2020)
19. Muzellec, B., Cuturi, M.: Generalizing point embeddings using the wasserstein space of elliptical distributions. In: NeurIPS. Curran Associates, Inc. (2018)
20. Solomon, J., Rustamov, R., Guibas, L., Butscher, A.: Wasserstein propagation for semi-supervised learning. In: ICML (2014)
21. Subramanya, A., Bilmes, J.: Semi-supervised learning with measure propagation. J. Mach. Learn. Res. **12**, 3311–3370 (2011)
22. Subramanya, A., Talukdar, P.P.: Graph-based semi-supervised learning. Synth. Lect. Artif. Intell. Mach. Learn. **8**(4), 1–125 (2014)
23. Tsuda, K.: Propagating distributions on a hypergraph by dual information regularization. In: ICML (2005)
24. Veličković, P., Cucurull, G., Casanova, A., Romero, A., Liò, P., Bengio, Y.: Graph attention networks. In: ICLR (2018)
25. Verma, S., Zhang, Z.L.: Stability and generalization of graph convolutional neural networks. In: KDD (2019)
26. Villani, C.: Topics in Optimal Transportation Theory (2003)
27. Villani, C.: Optimal Transport - Old and New. vol. 338. Springer-Verlag (2008)
28. Wu, F., Souza, A., Zhang, T., Fifty, C., Yu, T., Weinberger, K.: Simplifying graph convolutional networks. In: ICML (2019)
29. Yadati, N., Nimishakavi, M., Yadav, P., Nitin, V., Louis, A., Talukdar, P.: HyperGCN: a new method of training graph convolutional networks on hypergraphs. In: NeurIPS. Curran Associates, Inc. (2019)
30. Zhang, C., Hu, S., Tang, Z.G., Chan, T.H.H.: Re-revisiting learning on hypergraphs: confidence interval and subgradient method. In: ICML (2017)
31. Zhang, R., Zou, Y., Ma, J.: Hyper-sagnn: a self-attention based graph neural network for hypergraphs. In: International Conference on Learning Representations (ICLR) (2020)
32. Zhou, D., Huang, J., Schölkopf, B.: Learning with hypergraphs: clustering, classification, and embedding. In: Schölkopf, B., Platt, J.C., Hoffman, T. (eds.) NIPS. MIT Press (2007)

A Meta-path Based Graph Convolutional Network with Multi-scale Semantic Extractions for Heterogeneous Event Classification

Haiyang Wang, Xin Song, Yujia Liu, ChenGuang Chen, and Bin Zhou(✉)

National University of Defense Technology, ChangSha, China
{wanghaiyang19,songxin,liuyujia,cchenguang,binzhou}@nudt.edu.cn

Abstract. Heterogeneous social events modeling in large and noisy data sources is an important task for applications such as international situation assessment and disaster relief. Accurate and interpretable classification can help human analysts better understand global social dynamics and make quick and accurate decisions. However, achieving these goals is challenging due to several factors: (i) it is not easy to model different types of objects and relations in heterogeneous events in an unified manner, (ii) it is difficult to extract different semantic dependences at different scales among word sequences, and (iii) it is hard to accurately learn the subtle difference between events. Recently, graph neural networks have demonstrated advantages in learning complex and heterogeneous data. In this paper, we design a social event modeling method based on a Heterogeneous Information Network (HIN) and meta-path to calculate the similarity of events. In order to extract different semantic dependence, we propose a multi-scales semantic feature extraction framework. We present a Local Extrema Graph convolution Network (LEGCN) to expand the difference of various types events achieving accurate classification. We conduct extensive experiments on multiple real-world datasets and show that the proposed method exhibits comparable or even superior performance and also provides convincing interpretation capabilities.

Keywords: Heterogeneous events classification · Multi-scale semantic · Local Extrema Graph Convolutional Network

1 Introduction

Social events such as protest, fight, or cooperation have a major impact on the society. Classification of social events from large and heterogeneous open source media is a challenging problem. Previous event classification approach mainly treats social event as homogeneous words/elements co-occurrence graph [8]. Despite the compelling results achieved, these methods are ambiguous in

K. Karlapalem et al. (Eds.): PAKDD 2021, LNAI 12712, pp. 459–471, 2021.
https://doi.org/10.1007/978-3-030-75762-5_37

event modeling and do not fully consider the different types of elements contained in the event, such as actors, time, location etc. However, the heterogeneous elements and their relationships in modeling events are crucial to achieve interpretable and accurate classification.

Currently, events in the real-world are often extracted from formal reports or news articles. Those events can be structed as graphs with multi-types of nodes and edges, known as heterogeneous information network (HIN) [11]. In HIN, certain types of objects connected by sequences of relations are often called *meta-path*. Meta-path plays an important role in capturing the semantic relevance between objects in HIN. Taking the social events data shown in Fig. 1 as an example, it contains three types of nodes including event, actor and time. A relation between two events can be revealed by the meta-path Event-Actor-Event (EAE) which describes the co-actor relation. Meanwhile, Event-Time-Event (ETE) means that they happened at the same time. Thus, we can obtain more abundant and detailed semantic information by applying HIN and meta-path. In this paper, we model social events based on HIN and meta-path to obtain the similarity among them.

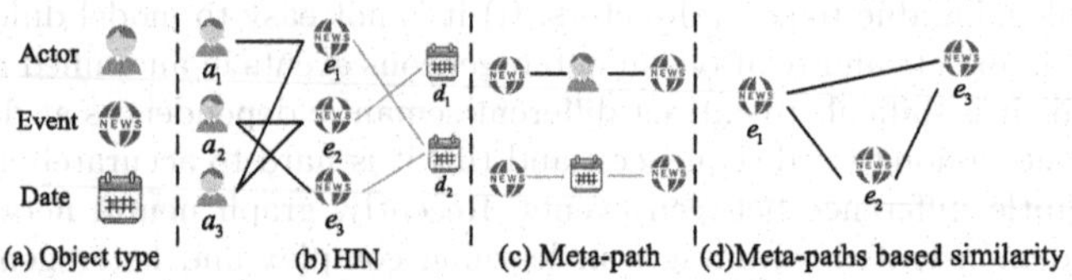

Fig. 1. An illustrative example of a HIN (social events). (a) Three types of nodes. (b) A HIN consists three types of nodes and two types of connections. (c) Two meta-paths involved in social events. (d) Similar events e_1, e_2, e_3 based on meta-path

Event objects, event category, and content are always closely related. It is vital to extract an apposite and comprehensive feature representation from content. Existing methods usually encode words of input text and feed them as input to some recurrent neural networks (e.g. RNN, GRU, BiLSTM, etc.) to get semantic dependencies between adjacent words. Besides, attention mechanism is widely used to extract important words for enhancing semantic feature representation. For instance, a protest category event described as "*Iraqi protesters vow to continue regardless of government offers*". We can get the importance score of every word in the sentence by attention mechanism. Obviously, *protesters* and *government* will be given a high score to help the classification model classify it into protest category. Hence, we construct the global context feature representation based on semantic dependencies and important score of words.

In addition to the global context information, local semantic dependences of the sentence is also vital. In the above example, Although these three words (*protesters, continue, government*) are not adjacent and have the same spacing, their combination is meaningful. It means the protesters will continue their demonstrations against the government. This is helpful for classifying events in Iraq in the next few days. Thus, it is vital to capture semantics in different distance scales. In this paper, we employ the multi-scale dilated convolution to obtain the local feature representation. Since we use news titles as the description of the event, we use three scales of dilated convolutions with intervals of 0, 1, and 2 words, taking into account the limitation of its length. Base on the method, we extract the local semantic feature representation.

Finally, we construct a graph-based classification model. A node in the graph represents a news title of an event. The node feature is a concatenation of global context feature as well as local semantic feature. A edge weight is the similarity between events calculated by the meta-path based methods. To perform heterogeneous events classification, we utilize a novel **L**ocal **E**xtrema **G**raph **C**onvolutional **N**etwork model, namely LEGCN. LEGCN uses the difference operators to expand the feature difference between different categories of events and shrink the same ones. Our contributions are summarized as follows:

- We introduce **Meta-paths based Similarity Measure** by modeling the events as Heterogeneous Information Network, which helps to obtain a more accurate similarity measure with abundant semantics between events.
- We design a new **Multi-scale Semantic Feature Extraction** framework for integrating global context information and local semantic dependencies to get a comprehensive feature representation of events.
- We propose a novel **LEGCN** for heterogeneous events classification. We evaluate the proposed method with state-of-the-art models on multiple real-world datasets and demonstrate its strengths and interpretability.

2 Related Work

In this section, we review the research process of HIN and meta-path. We also summarize some recent work on graph neural networks.

2.1 Meta-paths in HIN

A heterogeneous information network considers multiple objects and their relation. It fully demonstrates the complex interactive relationship among various objects in social events. Originally, Fouss et al. [3] designed a similarity metric ECTD with nice properties and interpretation to model heterogeneous data. Unfortunately, due to the lack of path constraints, ECTD cannot capture the subtle semantics of heterogeneous networks. Considering the differences in semantics in the meta-paths made up of different types of objects in a HIN, important researches have been explored using meta-paths to mine similar peer entities from largescale heterogeneous information networks, such as PathSim [12], HeteSim [10], and PathRank [5]. In addition, Zhang et al. [16] proposed a network embedding algorithm MetaGraph2Vec based on meta-graphs to capture semantic information between distant nodes and embed multiple types of nodes, aiming to solve the problem of data-sparse.

2.2 Graph Neural Networks

In the past few years, due to the fact of many real-world data can be represented as graphs, Graph Neural Networks (GNNs) has received continuous attentions [1]. A lot of previous research create various Graph Neural Networks model to adapt to diverse graph-structured data. GNNs model can be constructed by the spectral or non-spectral approaches. Spectral methods usually use Fourier transformation and graph Laplacian to define convolution operation. And Non-spectral methods define convolution through a local neighborhood around nodes in the graph. In the non-spectral methods, a simplified layer-wise graph neural network model which called graph convolutional networks (GCN) achieves state-of-the-art classification results on several benchmark graph datasets [4]. Moreover, many GCN variants have been proposed in recent years, such as Fast-GCN [2], TextGCN [14] and GAT [13]. For NLP tasks, GCN and subsequent variants have shown superior performance in semantic role labeling, machine translation, event detection, relation classification and text classification.

3 Methodology

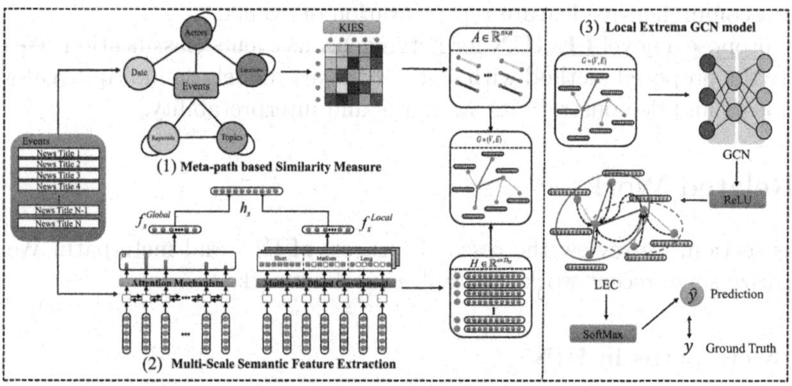

Fig. 2. System framework of our model. We construct graph based on news titles and feed them into LEGCN.

3.1 Problem Definition

We formulate the heterogeneous event classification problem as a graph-based node classification. We have N description of events ($S = s_1, s_2, \cdots, s_N$) in total. In our paper, we use news titles as the description. Each event news title is a node. Similarity exists between nodes, and each node has semantics. Thus, in order to construct a graph, we make use of the similarity as edge weight and the semantics feature as node embedding. Base on the graph, local extrema GCN is developed to give the label(y) of a certain type of nodes.

Figure 2 provides an overview of the proposed method. The key objectives are (1) to calculate edge weight by using a meta-path based similarity measure; (2) to capture the semantic feature by applying attention mechanism and multi-scale dilated convolution; (3) to classify nodes by employing the LEGCN.

3.2 Edge: Meta-path Based Similarity Measure

To obtain the adjacency matrix, we design a meta-paths based similarity measure. We first construct a HIN based on events. For hierarchical topic structures and the affiliation relationship between keywords and topics, we employ the JoSH model [7]. And We use the manually organized synonyms to add synonym relationship among keywords. Actors, Date and Locations can be obtained from our datasets. Then we calculate the similarity of events based on the meta-paths between different entity types.

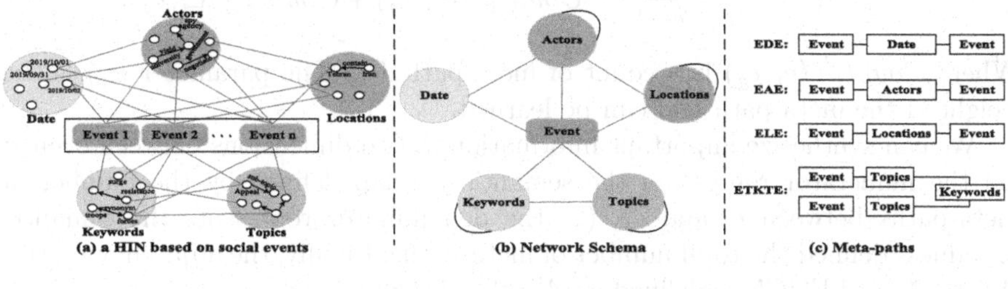

Fig. 3. Illustration of events as a HIN.

Definition 1. *Heterogeneous Information Network (HIN) is a directed graph* $G = (V, E)$ *with the object type mapping* $\Phi : V \rightarrow A$ *and the link mapping* $\Psi : E \rightarrow R$*, while the types of objects* $|A| > 1$ *or relations* $|R| > 1$.

For example, Fig. 3(a) shows a HIN of events. It includes five different object types and different types of relations between them. The event *is initiated by* the Actors and *occurs* at different times. In addition, there are connections within object of the same type. There are *sub-topics* within topics, and there are *synonyms* and *antonyms* within keywords. To better understand the complex object and link types in the HIN, we introduce its network schema and meta-path..

Definition 2. *Network schema is a meta template for a HIN. It is a directed graph defined over object types* A*, with edges as relations from* R*, denoted as* $T_G = (A, R)$.

Definition 3. *A meta-path* P *is a sequence of relations defined between different object types on* T_G*. The form of* P *is* $A_1 R_1 \rightarrow A_2 R_2 \rightarrow \cdots \rightarrow A_{l+1} R_{l+1}$*. It defines a composite relation* $R = R_1 \bullet R_2 \bullet \cdots \bullet R_{l+1}$ *between type* $A_1, A_2, \cdots, A_{l+1}$*, where* $\bullet$ *denotes the composition operator and* $l + 1$ *is the length of* P.

Since the meta-paths express different semantics, the similarity of events calculated based on the meta-path is accurate. Figure 3(c) shows four different types of meta-paths based on Fig. 3(b). For example, the meta-path EAE means two events are similar if they have the same actors; the ETKTE means two events are similar if their topics are same keywords. Note that if there is a relation between entities of the same type owned by two events, they can also be considered similar. For example, *e1-Location (tehran) (-location(Iran))-e2*, Tehran is the capital of Iran. So, we also think e_1 and e_2 are similar. We enumerated 25 different meta-paths to include abundant semantic information.

In a HIN, given a collection of meaningful meta-paths $\{P_m|m = 1, 2, 3, \ldots M\}$, we measure the similarity between two events by the knowledgeable meta-paths instances based social event similarity ($KIES$) method [8], the $KIES$ between two events e_i and e_j is defined as:

$$KIES\left(e_i, e_j\right) = \sum_{m=1}^{M} w_m \frac{2 \times Count_{P_m}\left(e_i, e_j\right)}{Count_{P_m}\left(e_i, e_i\right) + Count_{P_m}\left(e_j, e_j\right)} \tag{1}$$

Where $Count_{P_m}(e_i, e_j)$ is a count of meta-path P_m. The parameter w_m is the weight of the meta-path and can be learned.

We can synthesize important information in two dimensions between events: (1) the numerator represents the semantic overlap defined by the number of meta-paths between e_i and e_j; (2) the denominator represents the semantic broadness defined the total number of meta-paths. Finally, the adjacency matrix $A \in \mathbb{R}^{N \times N}$of LEGCN is defined as $A[i, j] = KIES(e_i, e_j)$.

3.3 Node Representation: Multi-scale Semantic Feature Extraction

As for a news title s, we apply GloVe embeddings to transform each word into a real-value vector ($X \in \mathbb{R}^{D_w \times L_s}$), where D_w is the dimension of the word embedding and L_s is the sentence length. Secondly, we use BiLSTM and attention mechanism to capture the global context feature of titles. Then, we apply the multi-scale dilated convolution [6] on X to capture the vital local semantic dependence at different levels of granularity. Finally, we concatenate two different features as the node feature representation.

Global Context Feature Extraction. We use the i-th step as an example to describe global context feature extraction mechanism. It first uses a BiLSTM for initial word embedding matrix (X), and the hidden vector is calculated based on previous hidden vector ($\overrightarrow{h_i}$ or $\overleftarrow{h_i}$) and current input word embedding x_i:

$$\overrightarrow{h_i} = LSTM\left(x_i, \overrightarrow{h_{i-1}}\right), \; \overleftarrow{h_i} = LSTM\left(x_i, \overleftarrow{h_{i-1}}\right) \tag{2}$$

We can obtain $h_i^{lstm} = \left[\overrightarrow{h_i}; \overleftarrow{h_i}\right] \in \mathbb{R}^{2D_h}$ by concatenating the two hidden vectors, D_h is the dimension of the hidden state. Moreover, we apply the attention mechanism to calculate weight of each word. It first computes the importance score($Score_i$) of each word in the title. $Score_i$ is calculated based on sentence vector s_v and the final hidden vector of BiLSTM h_i^{lstm}:

$$Score_i = v_s^T \tanh\left(W^s s_v + W^w h_i^{lstm} + b^s\right) + b^v \tag{3}$$

where W^s, W^i ,v, b^s and b^v are trainable parameters. s_v is trained by the Doc2vec for each news title. Attention weight α_i is computed as:

$$\alpha_i = \frac{\exp\left(Score_i\right)}{\sum_{j=1}^{L_s} \exp\left(Score_j\right)} \tag{4}$$

The final vector representation is the weight sum of hidden vectors:

$$f_s^{Global} = \sum_{i=1}^{L_s} \alpha_i h_i^{lstm} \tag{5}$$

Local Semantic Feature Extraction. In addition to global contextual features, the local semantic dependence between words with different intervals in a sentence is also very important. Recently, the convolutional layer is used to capture important local features. Yu demonstrated the effectiveness of Dilated Convolution to extract local patterns on images [15]. It consists of multiple convolutional layers with the same filter and stride size but different dilation rates. Formally, the dilated convolution is applied to input vector with defined gaps. Inspired by this, we apply 2D CNN filters with different dilation rates on X, defined as:

$$d[i,j] = \sum_{l_r=1}^{D_w} \sum_{l_c=1}^{L_c} X[i + l_r, j + k \times l_c] \times \mathbf{c}[l_r, l_c] \tag{6}$$

where d is the output feature vector, c represents the convolutional filter of dimension $D_w \times L_c$, k is the dilation rate. We define K filters with three different dilations rates $k_s, k_m, k_l (k_l > k_m > k_s)$ for short-distance, medium-distance and long-distance semantic dependences. Then, we concatenate the multiple filter vectors and apply a nonlinear layer to it. Finally, we get the final convolution output, denoted as f_s^{Local} for sentence s.

Feature Concatenation. For sentence s, we learn the global context feature (f_s^{Global}) by using the BiLSTM and attention mechanism, as well as the Local semantic feature (f_s^{Local}) by using the multi-scale dilated convolution. We combine these two features as the node feature representation $h_s = [f_s^{Global}; f_s^{Local}]$. As for events, we can obtain the feature matrix $H \in \mathbb{R}^{N \times D_H}$.

3.4 Local Extrema GCN Model

Given the constructed graph, we introduce our proposed model named Local Extrema Graph Convolution Network (LEGCN). It consists of a graph convolutional layer and a local extrema convolution layer.

Graph Convolution Network Layer. We use Graph Convolution Network (GCN) [4] capture the discriminative spatial features for node classification. GCN is defined as:

$$H^{(l+1)} = \sigma\left(\tilde{D}^{-\frac{1}{2}}\tilde{A}\tilde{D}^{\frac{1}{2}}H^{(l)}W^{(l)}\right) \tag{7}$$

where $\tilde{A} = A + I_N$ for self-loops. $\tilde{D}$ is the degree matrix and $\tilde{D}_{ii} = \sum_j \tilde{A}_{ij}$. σ is a non-linear activation function. l denotes the layer number, $H^{(l)}$is the matrix passing by layers. $W^{(l)}$ are parameters of the GCN layer.

Local Extrema Convolution Layer. In this paper, we use a novel Local Extrema Convolution (LEC) layer to capture local extrema information. Recently, LEC is used in graph classification tasks [9]. It plays an important role in finding local extrema nodes to generate pooled graphs. In our model, the LEC is used in node classification task. It can calculate the difference between a node and its neighbor nodes. In details, LEC uses the difference operator to update the node feature representation. For a node, LEC can make the difference between it and its neighbors of the same type smaller, and the different types larger. The local extrema convolutional layer is defined as:

$$h_i^{(l+1)} = \sigma\left(h_i^{(l)}W_1^{(l)} + \sum_{j\in\mathcal{N}(i)} a_{i,j}\left(h_i^{(l)}W_2^{(l)} - h_j^{(l)}W_3^{(l)}\right)\right) \tag{8}$$

where $\mathcal{N}(i)$ denotes the neighborhood of the ith node in the graph. $W_{1,2,3}^{(l)}$ are learnable parameters of the l layer. $a_{i,j}$ is the edge weight between node i and node j.

Optimization. Finally, we get $Z = sigmoid(LEC(A, GCN(A, H)))$. And loss function is defined as the cross-entropy error over all labeled events.

4 Experiments

In this section, we evaluate the proposed model on the heterogeneous event classification tasks. Specifically, we want to determine the following key questions: (1) Can the meta-path based similarity measure accurately compared to other methods? (2) Can our model provide interpretable results? (3) Can LEGCN model achieve excellent classification results compared to other baselines?

4.1 Datasets and Settings

The experimental evaluation is performed on the Global Database of Events, Language, and Tone event data(**GDELT**)[1]. It contains political events to evaluate national and international crises or events. These news events are encoded

[1] https://www.gdeltproject.org/.

with 20 categories such as Cooperate, Protest, Fight etc. These events contain a lot of entity and relationship information. Therefore, GDELT is very suitable as a data source for heterogeneous events classification. In this paper, we use two ways to label data: (1) The events are divided into protest or non-protest; (2) Same as the GDELT categories. We select GDELT data for five countries (**Iran**, **Iraq**, **Saudi Arabia**, **Syria**, **Turkey**) from July 1, 2019, to June 30, 2020.

For our model, we set the word and sentence embedding size as 100. We tuned other parameters and set the learning rate as 0.02. We randomly selected 20% of the training set as the validation set. Model parameters can be trained via back-propagation and optimized by the Adam algorithm. We trained LEGCN for a maximum of 200 epochs in 2 labels datasets and 2000 epochs in 20 labels datasets. As for baselines, we use the default parameter settings in their papers.

4.2 Comparative Methods

We compareLEGCN with multiple SOTA classification models: (1) **CNN:** we convert the texts as 1-by-S-by-C images (height 1, width S, and C channels). Then, we train a CNN model based on that. (2) **TextGCN:** the TextGCN model defined in [14], which uses two GCN layers and one-hot vector to represent word or document. (3) **GAT:** we employ the GAT model [13] which contains two GAT layers. And we use the Doc2vec to learn the feature for every event.

We conduct ablation tests to investigate the effects of different modules: (1) without KIES (**-KIES**): we only use cosine similarity and remove KIES. (2) without BiLSTM and attention mechanism (**-BA**): removing the global context feature. (3) without multi-scale dilated convolutional module (**-MSD**): removing the local semantic feature. (4) without LEC (**-LEC**): we only use one GCN layer and remove the local extrema convolution layer.

Table 1. Test Accuracy on heterogeneous events classification task.

Method	Iran		Iraq		Saudi Arabia		Syria		Turkey	
	2	20	2	20	2	20	2	20	2	20
CNN	0.7763	0.5715	0.7568	0.5758	0.7750	0.5676	0.7845	0.5980	0.7807	0.5764
TextGCN	0.8223	0.7482	0.8195	0.7189	0.8302	0.7376	0.8265	0.7163	0.8453	0.7238
GAT	0.8209	0.7325	0.8240	0.6695	0.8368	0.7340	0.8365	0.6845	0.8203	0.6911
LEGCN	**0.9566**	**0.8659**	**0.9733**	**0.8628**	**0.9679**	**0.8848**	**0.9569**	**0.8697**	**0.9707**	**0.8674**
-KIES	0.8467	0.7259	0.8557	0.7014	0.8513	0.6952	0.8295	0.7174	0.8439	0.7353
-BA	0.9132	0.8092	0.8989	0.8035	0.9124	0.8225	0.8934	0.8047	0.8942	0.8286
-MSD	0.8857	0.8034	0.8952	0.8053	0.8935	0.8056	0.9165	0.7903	0.9117	0.7917
-LEC	0.8531	0.7702	0.8753	0.7856	0.8694	0.7752	0.8918	0.7804	0.8879	0.7631

4.3 Results and Evaluation

Classification Performance. Table 1 presents the test accuracy of each model. LEGCN performs significantly outperforms all other baselines on five datasets.

It shows that the effectiveness of the proposed method especially on 20 labels datasets. The difference of accuracy across different datasets may be due the different distribution of event categories in each country. Most of the graph-based methods show relatively good performance. The reason may be the superior message passing function of the graph model. More in-depth performance analysis, CNN have poor performance because many semantic features are lost in the process of conversion. Text GCN and GAT model have comparable performance. However, TextGCN is more stable. In order to analyze the effect of each component in our model, we perform the ablation tests on all datasets.

Ablation Tests. We observe that the performance of variant versions without BA or MSD reduce slightly. However, when the KIES module or LEC module is removed, the performance reduce greatly. Without the KIES module, the similarity between nodes may not be accurately calculated. Lack of the LEC layer, nodes feature cannot be well learned. KISE and LEC play a relatively important role in our model, but to achieve higher accuracy, MSD and BA are also essential.

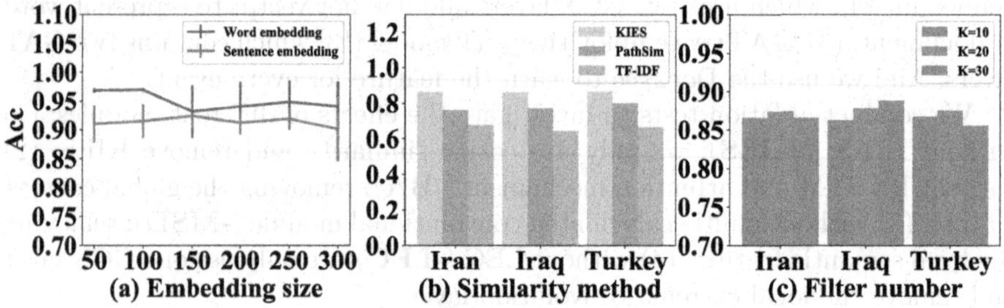

Fig. 4. Sensitivity analysis

4.4 Sensitivity Analysis

We investigate how the performance varies with hyperparameters.

- *Embedding Dimension.* To test if our model is sensitive to the length of word or sentence embedding size, we evaluate different embedding dimension from 50 to 300. As shown in Fig. 4(a), we can see that test accuracy fluctuates slightly with the embedding dimensions. High dimensional embeddings not improve performances of classification and may cost more training time.
- *Similarity Method.* We replace KIES with PathSim and TF-IDF. As shown in Fig. 4(b), KIES performs better in all datasets. It is because KIES and PathSim consider the semantic of different entities. Besides, KIES quantifies the weights of different meta-paths.

- *Filter Number.* We perform sensitivity analysis on different numbers of filters and the results are shown in Fig. 4 (c). In both the Iran and Turkey datasets, the proposed model achieves the best performance when K = 20. More filters tend to reduce the classification power of the model.

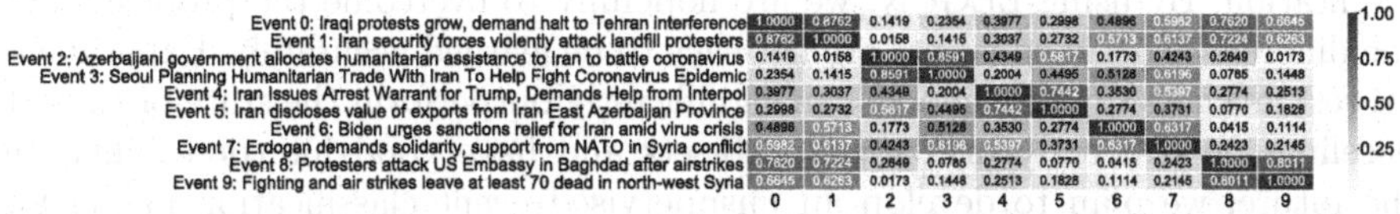

Fig. 5. The heatmap of the similarity between events. There 10 events of five types that occurred in Iran, including Protest (0, 1), Cooperation (2, 3), Make public statement (4, 5), Demand (6, 7), and Fight (8, 9).

4.5 Interpretability

In this section, we analyze the interpretability of our model.

- *KIES.* The value of KIES ranges from 0 to 1. Higher KIES means two events are more similar. From Fig. 5 we can see that: (i) the event similarity matrix from KIES is symmetric; (ii) the KIES of two events is higher if they are of the same category; (iii) the KIES is significantly different if two events belong to different categories, which is beneficial to events classification.
- *LEGCN.* we provide a t-SNE visualization of the computed feature representations of the LEGCN on Syria datasets in Fig. 6 In order to get a more intuitive display, we only show 5 classes of events. Points of different colors are the learned feature in different event classes. It verifies the discriminative feature learned from the LEGCN model.

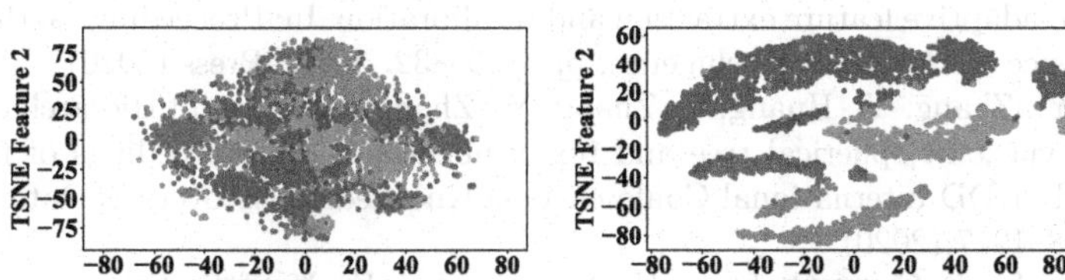

(a) Before LEGCN (b) After LEGCN

Fig. 6. The t-SNE visualization of test events feature representation in SA.

5 Conclusions

In this paper, we introduce a meta-path based social event similarity measure, and propose a graph-based model LEGCN for heterogeneous event classification. The model takes the similarity between events as edge weight. A multi-scale semantic feature extraction framework is proposed to learn event feature representation. By using LEGCN, we are hopefully to overcome the problem that the differences between events are not easy to be learned accurately. Experimental results show that our model can significantly outperform the state-of-the-art baselines on five real-world social datasets and have good interpretability. In the future, we plan to develop an unsupervised event classification model for classification on large-scale unlabeled event data.

Acknowledgements. This work was supported by the National Key Research and Development Program of China NO. 2017YFB0803301, and Postgraduate Scientific Research Innovation Project of Hunan Province CX20200015.

References

1. Cai, H., Zheng, V.W., Chang, K.C.C.: A comprehensive survey of graph embedding: problems, techniques, and applications. IEEE Trans. Knowl. Data Eng. **30**(9), 1616–1637 (2018)
2. Chen, J., Ma, T., Xiao, C.: FastGCN: fast learning with graph convolutional networks via importance sampling. arXiv preprint arXiv:1801.10247 (2018)
3. Fouss, F., Pirotte, A., Renders, J.M., Saerens, M.: Random-walk computation of similarities between nodes of a graph with application to collaborative recommendation. IEEE Trans. Knowl. Data Eng. **19**(3), 355–369 (2007)
4. Kipf, T.N., Welling, M.: Semi-supervised classification with graph convolutional networks. arXiv preprint arXiv:1609.02907 (2016)
5. Lee, S., Park, S., Kahng, M., Lee, S.G.: PathRank: a novel node ranking measure on a heterogeneous graph for recommender systems. In: Proceedings of the 21st ACM International Conference on Information and Knowledge Management, pp. 1637–1641 (2012)
6. Ma, L., et al.: AdaCare: explainable clinical health status representation learning via scale-adaptive feature extraction and recalibration. In: Proceedings of the AAAI Conference on Artificial Intelligence, pp. 825–832. AAAI Press (2020)
7. Meng, Y., Zhang, Y., Huang, J., Zhang, Y., Zhang, C., Han, J.: Hierarchical topic mining via joint spherical tree and text embedding. In: Proceedings of the 26th ACM SIGKDD International Conference on Knowledge Discovery & Data Mining, pp. 1908–1917 (2020)
8. Peng, H., Li, J., Gong, Q., Song, Y., Ning, Y., Lai, K., Yu, P.S.: Fine-grained event categorization with heterogeneous graph convolutional networks. arXiv preprint arXiv:1906.04580 (2019)
9. Ranjan, E., Sanyal, S., Talukdar, P.P.: Asap: adaptive structure aware pooling for learning hierarchical graph representations. In: AAAI, pp. 5470–5477 (2020)
10. Shi, C., Kong, X., Huang, Y., Philip, S.Y., Wu, B.: HeteSIM: a general framework for relevance measure in heterogeneous networks. IEEE Trans. Knowl. Data Eng. **26**(10), 2479–2492 (2014)

11. Shi, C., Li, Y., Zhang, J., Sun, Y., Philip, S.Y.: A survey of heterogeneous information network analysis. IEEE Trans. Knowl. Data Eng. **29**(1), 17–37 (2016)
12. Sun, Y., Han, J., Yan, X., Yu, P.S., Wu, T.: PathSIM: meta path-based top-k similarity search in heterogeneous information networks. Proc. VLDB Endow. **4**(11), 992–1003 (2011)
13. Veličković, P., Cucurull, G., Casanova, A., Romero, A., Lio, P., Bengio, Y.: Graph attention networks. arXiv preprint arXiv:1710.10903 (2017)
14. Yao, L., Mao, C., Luo, Y.: Graph convolutional networks for text classification. In: Proceedings of the AAAI Conference on Artificial Intelligence, vol. 33, pp. 7370–7377 (2019)
15. Yu, F., Koltun, V.: Multi-scale context aggregation by dilated convolutions. arXiv preprint arXiv:1511.07122 (2015)
16. Zhang, M., Wang, J., Wang, W.: HeteRank: a general similarity measure in heterogeneous information networks by integrating multi-type relationships. Inf. Sci. **453**, 389–407 (2018)

Noise-Enhanced Unsupervised Link Prediction

Reyhaneh Abdolazimi(✉) and Reza Zafarani

Data Lab, EECS Department, Syracuse University, Syracuse, USA
{rabdolaz,reza}@data.syr.edu

Abstract. Link prediction has attracted attention from multiple research areas. Although several – mostly unsupervised – link prediction methods have been proposed, improving them is still under study. In several fields of science, noise is used as an advantage to improve information processing, inspiring us to also investigate noise enhancement in link prediction. In this research, we study link prediction from a data preprocessing point of view by introducing a noise-enhanced link prediction framework that improves the links predicted by current link prediction heuristics. The framework proposes three noise methods to help predict better links. Theoretical explanation and extensive experiments on synthetic and real-world datasets show that our framework helps improve current link prediction methods.

Keywords: Link prediction · Noise-enhanced methods · Graph algorithms

1 Introduction

Link prediction is a fundamental problem in graph mining, aiming to predict the existence of a link between two nodes in a graph. Link prediction has two main tasks: (1) predicting the links that will be added to a graph in the future, and (2) identifying missing links in an observed graph. Both tasks have important applications such as in identifying interactions between proteins in bioinformatics, building recommender systems, suggesting friends in social networks, and the like.

With link prediction being useful in many applications, a large number of link prediction algorithms have been proposed, which are different in aspects such as performance (e.g., accuracy) and computational complexity. An alternative to designing a new algorithm for improving link prediction can be to modify the input data (input graph) to such an algorithm. A common approach to change data is to add noise. While noise is often redundant, and research often tries to remove or reduce its effects, it has been shown to be invaluable in many areas of science, especially in nonlinear information processing systems [3]. Noise enhancement has long been used in physical systems as *stochastic resonance* and

K. Karlapalem et al. (Eds.): PAKDD 2021, LNAI 12712, pp. 472–487, 2021.
https://doi.org/10.1007/978-3-030-75762-5_38

has also shown promise in areas such as stochastic optimization, image processing, and machine learning [2,3,15]. Such benefits of adding noise have motivated us to explore the possibility of enhancing link prediction by adding noise. Adding noise introduces an extra step to the existing algorithms. This extra *noise injection* step introduces some level of randomization to link prediction algorithms. A natural approach to introduce noise in a network is to add edges as it allows one to systematically compare the predicted links in noisy and noiseless networks.

(a) original graph (b) noisy train graph

Fig. 1. Original graph before (Figure a) and after (Figure b) adding noise (dashed red edge) using the same link prediction algorithm (Adamic/Adar method). Adding noise edge (2, 6) in Figure (b) improves link prediction performance, as observed by the %23 and %250 increase in the values of link prediction quantitative criteria (*ROC* and *Average Precision*, here). (Color figure online)

To provide some intuition on how adding noise can improve link prediction, we provide an example. Consider the graph in Fig. 1a with 8 nodes and 9 edges. We can split this graph into two subgraphs: (I) "train subgraph" with all 8 nodes and all the 8 black edges, and (II) "test subgraph" with 2 nodes (3 and 6) and one blue edge. We can predict the links in this graph using the Adamic/Adar link prediction method [7], and evaluate them using metrics such as *Average Precision (AP)* and *ROC*. Here, we obtain AP and ROC values of 0.14 and 0.76 respectively. Next, we add a single noise edge (2, 6) to the train graph (the dashed red line) to get the *noisy train graph* in Fig. 1b. The same Adamic/Adar method can be applied to predict links. The added noise not only increases the accuracy of the predicted links (%250 increase in AP and %23 increase in ROC), but also yields a better ranking on the edges predicted (%48 increase in Kendall's Tau).

Noise-Enhanced Link Prediction. In this paper, we investigate noise-enhanced link prediction. We propose a simple framework to improve the predicted links in a network by adding noise, as outlined in Algorithm 1. Our approach first divides the original graph into two subgraphs: (1) train subgraph, and (2) test subgraph and then applies a link prediction algorithm on the train subgraph. After that, as our approach is iterative (to account for noise randomness), in each iteration, the algorithm builds a noisy network by adding noise (edges) to the train subgraph, and predict the links in this noisy network. It then evaluates the predicted links using some evaluation criteria on the original test subgraph. It iterates a few times and returns the best set of predicted links, e.g., with the best score based on some evaluation metric, as the noise-enhanced links. Our goal is to detect better links (in terms of some evaluation metric)

Algorithm 1. Noise-Enhanced Link Prediction

Input: Graph G, Noise Injection Method `Noise`, Link Prediction method `LP`, Evaluation Metric `Eval`, `Iterations`.
Output: Noise-enhanced ranked links L_{best}
$G_{train}, G_{test} \leftarrow$ `Divide`(G)
$L_0 \leftarrow$ `LP`(G_{train})
$E_0 \leftarrow$ `Eval`$(G_{train}, G_{test}, L_0)$
$E_{best} \leftarrow E_0$, $L_{best} \leftarrow L_0$
for $i = 1$ **to Iterations do**
    $\tilde{G}_{train_i} \leftarrow$ `Noise`(G_{train})
    $L_i \leftarrow$ `LP`$(\tilde{G}_{train_i})$
    $E_i \leftarrow$ `Eval`$(\tilde{G}_{train_i}, G_{test}, L_i)$
    if E_i improves compared to E_{best} **then**
        $E_{best} \leftarrow E_i$, $L_{best} \leftarrow L_i$
return L_{best}

while adding limited noise, i.e., without extremely increasing the link prediction execution time. In particular, we aim to answer two questions:

Q1. Does adding noise improve the performance of link prediction algorithms? If yes, by how much?
Q2. Adding noise increases the cost of predicting links. Are the cost acceptable relative to the performance improvements in predicting links? What is the trade-off?

By addressing these questions at a high level, our framework makes the following contributions:

1. We introduce noise-enhanced link prediction, a framework that relies on adding noise to improve existing link prediction algorithms;
2. We propose three methods to add noise to a graph which can be translated as a preprocessing step to improve current link prediction algorithms (Sect. 3);
3. We provide a theoretical foundation for noise-enhanced link prediction by showing that the suggested noise injection methods improve link prediction measures (Sect. 5); and
4. We evaluate our framework on several real-world and synthetic networks using well-established link prediction methods. Our results show that noise helps predict better links in networks compared to the predicted links in the original graphs (Sect. 6).

2 Literature Review

To our best knowledge, there is no past research on adding noise to link prediction methods. Here, we briefly review (and relate to our work) both (I) link prediction and (II) noise-enhanced algorithms.

Link Prediction. There are two main groups of link prediction methods: unsupervised and supervised. Here, we focus on unsupervised methods. These methods attempt to predict links by assigning scores to all node pairs based on network structure. Unsupervised methods can be grouped into:

I. *Neighborhood-based Methods* assume nodes x and y are likely to link in the future if they share common neighbors. Let $\Gamma(v)$ denote the neighbors (adjacent nodes) of node v. Examples include:
 - *Common neighbors*, used in many applications [7], is formulated as $|\Gamma(x) \cap \Gamma(y)|$.
 - *Jaccard's coefficient* is the probability of selecting a common neighbor of a pair of nodes x and y among all the neighbors of nodes x and y, formulated as $\frac{|\Gamma(x) \cap \Gamma(y)|}{|\Gamma(x) \cup \Gamma(y)|}$.
 - *Adamic/Adar* is a well-established measure for link prediction [7]. Adamic/Adar measure is defined as $\sum_{z \in \Gamma(x) \cap \Gamma(y)} \frac{1}{\log |\Gamma(z)|}$, giving more weight to adjacent nodes with less neighbors.
 - *Preferential Attachment* measure assumes the probability of a new link between nodes x and y is proportional to the number of neighbors of x and y: $|\Gamma(x)| \cdot |\Gamma(y)|$

II. *Path-based Methods* consider all paths in the graph including the shortest path.
 - *Katz* counts all paths between two nodes in the graph. The paths are exponentially damped by their lengths giving more weights to shorter paths. It is measured as $\sum_{l=1}^{\infty} \beta^l \cdot |path_{x,y}^{<l>}|$, where $path_{x,y}^{<l>}$ is the set of all paths from x to y, and β is the damping parameter ($\beta, l > 0$).
 - *Hitting time* is the expected number of steps needed for a random walk to start from node x and reach node y ($H_{x,y}$). The normalized version of hitting time in undirected graphs is $NHT(x, y) = H_{x,y} \cdot \pi_y + H_{y,x} \cdot \pi_x$ where π is the stationary probability of the respective node.
 - *Rooted PageRank* is a modified version of Pagerank. In Pagerank, the score between nodes x and y can be calculated as the stationary probability of y in a random walk that moves to a random neighbor with probability β, returning to x with probability $1 - \beta$. Let D be a diagonal degree matrix, and $N = D^{-1}A$ be the normalized adjacency matrix, then $RPR(x, y) = (1 - \beta)(1 - \beta N)^{-1}$
 - *SimRank* assumes the similarity of nodes depends on the similarity of nodes that they are connected to. From a random walk viewpoint, the SimRank score measures how soon two random walkers meet at a special node if they both start from node x to y [9].

As our goal is to add noise to the link prediction process, we experiment with well-known approaches from each unsupervised link prediction category as we will discuss in our experiments.

Noise-Enhanced Systems. Noise enhances performance in many areas [3]. We review some here:

I. *Stochastic Resonance (SR)* is observed when increasing random noise leads to an increase in the signal detection performance [10]. SR is frequently used in noise-enhanced information systems with examples in biological, physical, and engineered systems [4,11].

II. *Image Processing* also benefits from noise enhancement. Adding noise to images before thresholding can improve the human brain's ability to perceive noisy visual patterns [17]. Noise can also improve image segmentation [6] and image resizing detection.
III. *Signal Detection.* Noise can help signals' detection. For example, for detecting a constant signal in a Gaussian mixture noise background, some white Gaussian noise can improve the performance of the sign detector [5]. Additive noise can also help more efficiently detect a weak sinusoid signal [21].
IV. *Optimization.* Randomization helps finding optimal or near-optimal solutions in search algorithms, when searching for an optimum is likely to get trapped in local minima. For example, the randomization in Genetic Algorithms [18] helps avoid self-similarity in the population [3]. Mutation is similar to adding noise and often a suitable mutation rate can result in performance improvement.
V. *Machine Learning.* Noise decreases the convergence time in many clustering and competitive learning algorithms [15]. It also reduces the convergence time of backpropagation algorithm while training convolutional neural network [2]. This happens as backpropagation and some clustering algorithms such as k-means which are special cases of Expectation-Maximization (EM) algorithm [16], improves by noise enhancement.
VI. *Graph algorithms.* Noise can also improve graph algorithms by modifying the input data of such algorithms. For example, it has been shown that noise can help improve current community detection methods [1] in terms of objective functions and similarity to ground-truth communities.

3 Noise Injection Methods

Based on Algorithm 1, our framework has three steps: adding noisy edges to the graph, predicting links using link prediction methods, and evaluating the predicted links via performance metrics. To analyze noise enhancement in link prediction systematically, we experiment with various link prediction methods and evaluation criteria. We propose three general ways to add noise to graphs.

Our noise injection methods focus on nodes with a high degree as links are more probable to form around higher degree nodes. This will be theoretically justified later in Sect. 5.

Therefore, we implement the following steps in the suggested noise methods : (1) sorting nodes based on their degrees, (2) choosing the top p percent of sorted nodes as *candidates*, and (3) adding edges within candidates. For adding each edge, we select a pair of nodes (edge endpoints) from candidates. The proposed methods differ in how these pairs of nodes are chosen.

I. Random Noise (`Random`). Edge endpoints are randomly selected from the candidates. Before adding a noise edge, we check its existence in the graph. If we select all nodes as candidates, `Random` simply connects nodes irrespective of their degree.

II. Weighted Noise (`Weighted`). Each node (edge endpoint) v_i is selected with probability $P_{\texttt{Weighted}}(v_i)$ that depends on its degree among degrees of other candidate nodes:

$$P_{\texttt{Weighted}}(v_i) = \frac{d_i}{\sum_{i=1}^{n} d_i},$$

where d_i refers to the degree of node i, and n is the number of candidate nodes.

III. Frequency Noise (`Frequency`). We select nodes based on the degree distribution of the candidates, where nodes with more frequent degrees are less likely to be selected:

$$P_{\texttt{Frequency}}(v_i) = \frac{1 - f_{d_i}/n}{f_{d_i} \times \sum_{d=1}^{k}(1 - f_d/n)}, \tag{1}$$

where f_d is the frequency of degree d inside candidates. This method is inspired by the observation that in most real-world networks, the degrees follow a power law distribution.

Time Complexity. Our introduced noise methods consist of the following time complexities in a graph with $|V|$ nodes and $|E|$ edges: Calculating nodes' degrees in $O(|E|)$, sorting nodes based on their degrees in $O(|V|\log|V|)$, choosing *candidates* in $O(1)$, computing node probabilities in $O(|V|)$, and adding noise edges based on node probabilities in $O(|E|)$ (The maximum number of noise edges is at most $|E|$). Hence, the final time complexity due to adding noise is $\max(O|E|, O(|V|\log|V|))$.

Example 1. Consider the graph shown in Fig. 2 with 6 nodes $\{a, b, c, d, e, f\}$ with degrees {1, 4, 2, 2, 1, 2} and 6 edges. First, nodes are sorted based on their degree $\{b, c, d, f, a, e\}$, and then the top %80 of these sorted nodes are chosen as candidates $\{b, c, d, f\}$.

Finally, pairs of nodes are selected from candidates as follows:

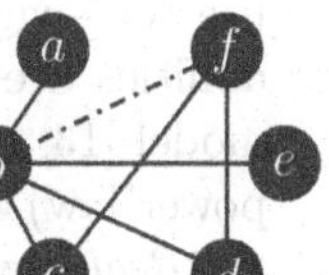

Fig. 2. Sample graph

`I. Random`. Randomly connects pairs of nodes, e.g., it may add a noise edge between nodes b and f (dashed line in Fig. 2).

`II. Weighted`. Nodes with higher degrees have higher probability to be selected. The probabilities $P_{\texttt{Weighted}}(v_i)$ for $v_i \in \{b, c, d, f\}$ are $\{0.4, 0.2, 0.2, 0.2\}$.

`III. Frequency`. Candidates with less frequent degrees are more probable to be selected. So node b with degree 4 and $f_4 = 1$ is more likely ($P_{\texttt{Frequency}}(b) = 0.75$) to be chosen as a source or destination node. The frequencies of other candidates are $P_{\texttt{Frequency}}(c) = P_{\texttt{Frequency}}(d) = P_{\texttt{Frequency}}(f) = 1/12$.

Table 1. Synthetic datasets statistics

Graph model	Graph size (n)	Parameters
Random (n, p)	$1,000$	$p \in \{0.001, 0.003, 0.006, 0.007\}$
Small-world (n, k, p)	$1,000$	$p \in \{0.0001, 0.001, 0.01, 0.1, 1\}$, $k = 10$
Configuration $(deg - seq)$	$1,000$	powerlaw $deg - seq$

Table 2. Real-world datasets

(a) Real-world Datasets Statistics.

Type	Network	$\|V\| = n$	$\|E\| = m$	Time Range	Avg. Degree	Clustering Coeff.
Social Network	UCIrvine Messages [14]	1899	15587	2008/03-2008/07	16.4	0.109
	Bitcoin	3783	14124	2010/11-2016/01	7.46	0.176
	Internet growth [12]	25526	52412	2004/01-2007/07	4.1	0.213
	Fb Wall posts [19]	46952	876993	2004/11-2009/01	37.3	0.107

(b) Real-world Datasets details.

Network	Train Time	Test Time	#Sample Nodes	#Sample Edges
UCIrvine Messages	2008/03-2008/07	2008/08-2008/10	1899	15587
Bitcoin	2010/11-2013/12	2014/01-2016/01	3783	14124
Internet growth	2004/01-2006/12	2007/01-2007/12	8000	20310
Fb Wall posts	2004/11-2008/10	2008/11-2009/01	8000	50174

4 Experimental Setup

In this section, we describe the datasets, the proportion of noisy edges added, data preparation, candidate size, link prediction methods, performance metrics, and evaluation metrics.

I. *Datasets.* Noise impact on link prediction is studied in both synthetic and real-world networks:
(1) Synthetic Networks. To better investigate the performance of link prediction after adding noise, we evaluated our framework on synthetic graphs generated by three network models: (1) random graphs; (2) small-world model; and (3) configuration model. The properties of these three models are provided in Table 1. For random graphs, we used the *Erdős-Rényi* graph model with $n = 1,000$ nodes and edge formation probabilities equal to $\{1/n, \log n/n, 2\log n/n\}$=$\{0.001, 0.003, 0.006, 0.007\}$. To create small-world graphs, we use the Watts-Strogatz model [20]. For its parameters, we use the suggestions provided by the authors [20], i.e., edge rewiring probabilities are $\{0.0001, 0.001, 0.01, 0.1, 1\}$. Finally, we use the configuration model [13] to create random graphs with specific degree sequences (we use power law).
(2) Real-world Networks: Our framework is also evaluated on real-world networks. For systematic analysis, we use four real-world networks. Table 2a provides the statistics of these networks.
II. *Data Preparation.* To ensure the data is ready for our experiments, some processing steps are taken: (1) We select data samples for each dataset. For sampling, we sample 8,000 nodes and all their connected edges (induced subgraph) using Breath-First Search. For datasets with nodes less than $8,000$ nodes, we leave them as they are. The specific sample numbers are shown in Table 2b; (2) We split the input data into training and testing sets. This division is performed based on timestamps with a ratio of training data: test data $= 90\% : 10\%$ as shown in Table 2b.

III. *Noise Proportion.* The amount of added noise (e percent) depends on the number of edges in the graph. For example, if there are 5,000 edges in the graph, we can add %10 of current edges (500 edges) as noise edges to the graph. We change e values from 1% to 10% with 1% increments.
IV. *Candidates Size.* Based on Sect. 3, the top p percent of sorted nodes are selected as candidates. We vary p from 10% to 100% with 10% increments. When $p = 100\%$, candidates contain all nodes.
V. *Link Prediction Methods.* We select four widely used unsupervised link prediction measures: (1) Common neighbor, Adamic-Adar, and preferential attachment from neighborhood-based methods, and (2) Katz from path-based methods. All selected methods have shown great performance in predicting future or missing links [7].
VI. *Performance Metrics.* Two groups of quantitative criteria can be considered for evaluating predicted links [8]: (1) *fixed-threshold* metrics, which depend on several types of thresholds, and (2) *threshold curves*, which are used when the data distribution is highly imbalanced. To evaluate the predicted links, we choose two criteria from each of the aforementioned groups; accuracy and F_1-score from fixed threshold metrics, and Receiver Operation Characteristics (ROC) and Precision-Recall (PR) from threshold curves. The Average Precision score is also used in the experiments to summarize the Precision-Recall curve. For calculating the above criteria, we use a cut-off rank k to return the top-ranked results. We vary k from 20% to 100% of test edges with 20% increments. We also evaluate *Kendall's* τ coefficient to measure the ordinal association between the ranked predicted links and test edges.
VII. *Evaluation Metrics.* To assess noise enhancement, we measure the following for each link prediction performance metric:

- `Expected First Success (EFS)` is the expected number of times that we require to add noise to the graph to ensure that we improve the predicted links at least once. For example, if we predict better links in 45 tests out of 100 tests, the expected first success is 3 as $\frac{100}{45} \simeq 2.22$. Formally, for performance metric m:

$$\texttt{EFS}_m = \left\lceil \frac{\text{number of test}}{\text{successful tests}} \right\rceil$$

- `Relative performance Improvement (RPI)` is the relative performance metric value improvement after noise enhancement:

$$\texttt{RPI}_m = \frac{m_{\text{noise-enhanced}} - m_{\text{original}}}{m_{\text{original}}} \times 100$$

Before performing the experiments, we show that these link prediction similarity-based scores can in theory be improved after adding noise.

5 Theoretical Analysis

Although empirical studies [7] show that the similarity-based measures for link prediction work well on different graphs, but there should be still ways to improve the scores. In this section, we show that all similarity-based scores can be improved by adding noisy edges between high degree nodes. To simplify, consider graph G with $|v| = n$ nodes, E edges, and two nodes i and j. Let $\Gamma(i)$ denote the set of neighbors of i and $\Gamma(j)$ the set of neighbors of j. For ease of presenting the proofs and without loss of generality, denote HIGHDEGREE as the set of nodes with higher degrees compared to that of other nodes (this can be formalized). Let $i \sim j$ denote nodes i and j are linked.

Theorem 1 (Common Neighbor Score Change). *Connecting two* HIGHDEGREE *nodes can increase common neighbor score the most.*

Proof. There are two ways to look at this. First, consider adding edges between pairs of nodes to increase the common neighbor score. How we can increase $CN(i, j)$? This increase is possible by (1) connecting neighbors x of i to j, i.e., $x \sim j$, such that $x \in \Gamma(i)$; or (2) neighbors y of j to i, i.e., $y \sim i$, $y \in \Gamma(j)$. How can we form edges to increase common neighbors of more nodes? If we select neighbors x and y such that $x, y \in$ HIGHDEGREE, more nodes find common neighbor with j and i. Similarly, if $j, i \in$ HIGHDEGREE, we will also increase the common neighbor of more nodes (i.e., i and j now act as the common neighbors). Therefore, connecting two HIGHDEGREE nodes, $x \sim j$ and $y \sim i$, increases common neighbor scores among more node pairs in G.

Secondly, we can show this property relatively compared to edges between lower degree nodes. Consider $i, j \in$ HIGHDEGREE and $i', j' \notin$ HIGHDEGREE , then $i \sim j$ increases common neighbor scores more compared to $i' \sim j'$. This is because the edge $i \sim j$ increases score for all $CN(k, j)$, $k \in \Gamma(i)$ and $CN(i, l)$, $l \in \Gamma(j)$. Similarly, edge $i' \sim j'$ increases $CN(k', j')$, $k' \in \Gamma(i')$ and $CN(i', l')$, $l' \in \Gamma(j')$. Since $|\Gamma(i)| > |\Gamma(i')|$ and $|\Gamma(j)| > |\Gamma(j')|$, $i \sim j$ can increase the common neighbor score between more nodes in G compared to $i' \sim j'$.

Theorem 2 (Adamic/Adar Score Change). *Connecting two* HIGHDEGREE *nodes can increase the Adamic/Adar score the most.*

Proof. Adamic/Adar score sums up the degree of common neighbors by giving more weight to low degree nodes and less weight to high degree nodes. Consider connecting pairs of nodes that increase Adamic/Adar score. How can connecting two nodes increase $AA(i, j)$ most? This is feasible by (1) increasing $CN(i, j)$; or (2) minimizing the degree of common neighbor nodes ($\min\{deg(k)|\forall k \in \Gamma(i) \cap \Gamma(j)\}$). As it is described in Theorem 1, connecting two HIGHDEGREE nodes increases common neighbor scores among more node pairs in G. For the second part, decreasing the degrees of common neighbor nodes is not considered in this paper as we only add edges. As adding edges decreases the weight of the common neighbor node in calculating Adamic/Adar score, how we can handle this trade-off to finally increase $AA(i, j)$?

We consider the worst-case scenario for $AA(i,j)$. If $x \in \Gamma(i)$ or $x \in \Gamma(j)$, then $x \sim k, k \notin \{i,j\}$ decreases $AA(i,j)$ by $(\frac{1}{\log(|\Gamma(x)|)} - \frac{1}{\log(|\Gamma(x)|+1)})$. This loss can be minimized if $x \in$ HIGHDEGREE since changes in logarithm value of $\Gamma(x)$ have less impact on $AA(i,j)$ while $x \in$ HIGHDEGREE compared to when $x \notin$ HIGHDEGREE. Similarly, if $k \in$ HIGHDEGREE, this weight difference will be too small. Therefore, if the gain resulted by the new common neighbors is higher than the weight loss due to current common neighbors, $AA(i,j)$ will be increased, which shows the power of connecting two high degree nodes to improve Adamic/Adar score.

Theorem 3 (Preferential Attachment Score Change). *Connecting two* HIGHDEGREE *nodes can increase the preferential attachment score the most.*

Proof. Preferential attachment selects the pair of nodes with the maximum value of $PA(i,j) = |\Gamma(i)| \cdot |\Gamma(j)|$, which indirectly implies $PA(i,j)$ mostly chooses nodes $\{i,j\} \in$ HIGHDEGREE. First, consider connecting node pairs that increase $PA(i,j)$. How does a noisy edge increase $PA(i,j)$? The gain of adding noisy edges can be calculated in two situations; 1) if we only add a noise edge $x \sim i$ ($x \notin \{i,j\}$), the gain is $|\Gamma(i)+1| \cdot |\Gamma(j)| - |\Gamma(i)| \cdot |\Gamma(j)| = |\Gamma(j)|$, and 2) if we add both noise edges $x \sim i$ and $y \sim j$ ($x, y \notin \{i,j\}$), where the gain will be $|\Gamma(i)+1| \cdot |\Gamma(j)+1| - |\Gamma(i)| \cdot |\Gamma(j)| = |\Gamma(i)| + |\Gamma(j)| + 1$. So, more neighbors in the noisy edges' endpoints lead to more gains in $PA(i,j)$, and if both $\{i,j\} \in$ HIGHDEGREE, $PA(i,j)$ can be increased more compared to the situation $\{i,j\} \notin$ HIGHDEGREE.

The mentioned gains can be extended to all sources and destinations of noisy edges, e.g., for x and y, if $\{x,y\} \in$ HIGHDEGREE, their chance of being selected by preferential attachment will be increased. Therefore, noise can also improve $PA(i,j)$ by connecting two HIGHDEGREE nodes.

Theorem 4 (Katz Score Change). *linking two* HIGHDEGREE *nodes increases Katz score the most.*

Proof. $Katz(i,j)$ counts all paths between two nodes and gives more weights to shorter paths. How we can increase $Katz(i,j)$? This increase is possible by (1) increasing the number of paths between i and j, or (2) making the path between i and j shorter.

Adding a random edge can create paths that did not exist before. This random edge can have the highest effect on $Katz(i,j)$ if it creates a path between a large number of nodes. As nodes in HIGHDEGREE act as hubs in the graph and they are connected to a large number of nodes, connecting two of them ($x \sim y$) make a bridge between $\Gamma(x)$ and $\Gamma(y)$. As a result, not only the lengths of paths between many nodes will be decreased, but also the number of paths between them will be increased. So, connecting two HIGHDEGREE nodes can also increase Katz score.

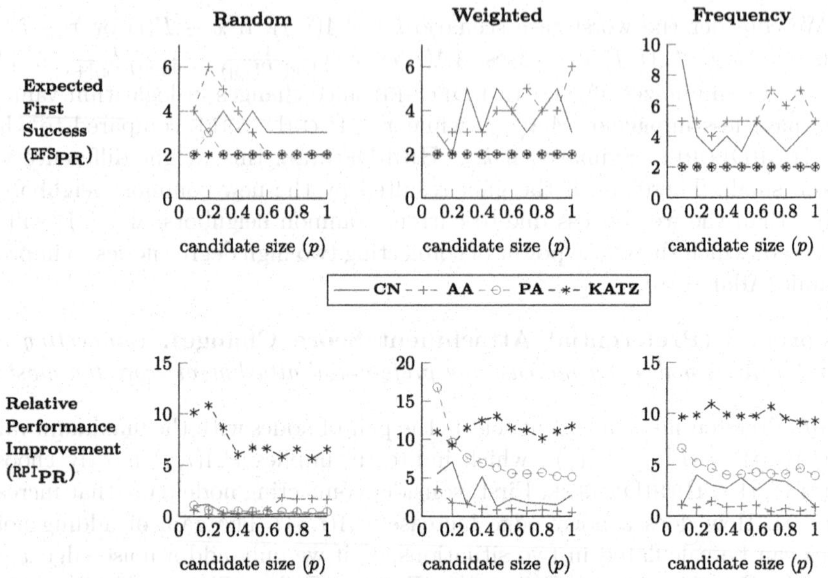

Fig. 3. Impact of all three proposed noise methods on area under precision-recall (PR) curve of the predicted links on Bitcoin. The average EFS = 3 and RPI = 4.2 for all link prediction measures.It shows we only needs to add noise three times to Bitcoin to predict better links in terms of PR.

6 Experimental Analysis

We evaluate the impact of adding noise on link prediction in both real-world and synthetic networks.

6.1 Noise-Enhanced Link Prediction in Real-World Networks

For each network and each candidate size p, we enhance the network using three noise methods by adding different proportions of noise e and measure both evaluation metrics (EFS and RPI) for all performance metrics. As e varies from 1% to 10% with 1% increments, and the experiments are done 10 times for each e (to assess stability of the results), 100 experiments are done for each p. By taking the average of EFS and RPI values of these 100 tries, the results of adding different proportions of noise for each candidate size p can be summarized by a number. The results of both evaluation metrics (RPI and EFS) on different candidate sizes, for each dataset and performance metric, can be summarized using 6 plots as shown as an example in Fig. 3. The figure shows the impact of all three proposed noise methods on the area under precision-recall (PR) curve of the predicted links on Bitcoin dataset. As shown in the figure, EFS is on average 3 for all link prediction measures, so on average, one only needs to add noise three times to Bitcoin to predict better links in terms of PR. The average RPI(= 4.2) for PR curve is also shown in Fig. 3. It is interesting that Katz works very well with noise in Fig. 3 with Low EFS (= 2) values and high RPI values.

For space reasons, we summarize all results in Table 3, which provides the average EFS and RPI of noise-enhanced link prediction methods on all real-world networks. For each network, there are six rows, one for each performance metric. For each network, noise method, and link prediction method, we provide both EFS and RPI. We summarize the findings in Table 3 as follows:

- Noise-enhanced link prediction has an average EFS = 17 and RPI = 9.4 over all networks, noise methods, link prediction methods, and performance metrics, implying noise yields improvements.
- Weighted Noise is the best noise option for Bitcoin (EFS = 3 and RPI = 3.5), and Internet growth (EFS = 2 and RPI = 16). Frequency Noise performs the best on UCIrvine Messages (EFS = 4 and RPI = 29) and Random Noise is the best noise choice for Fb Wall posts (EFS = 4 and RPI = 2.3). In overall, Weighted is the first noise priority to apply on a real-world social network; and
- Each link prediction method works best with a specific type of noise: (1) Common neighbor improves more with Frequency Noise and Weighted Noise, where for Frequency: (EFS = 5, RPI = 2.8), and for Weighted: (EFS = 5, RPI = 1.8). These numbers are the average values for the area under curve of ROC, accuracy, F_1-score, Precision-Recall, average-precision, and Kendal's tau; (2) Adamic/Adar improves more with Frequency Noise and Random Noise, where for Frequency: (EFS = 4, RPI = 7.1), and for Random: (EFS = 7, RPI = 7.2); (3) Preferential-Attachments improves more with Random Noise and Weighted Noise, where for Random: (EFS = 3, RPI = 10.7), and for Weighted: (EFS = 7, RPI = 11.9); (4) Katz improves best with Weighted Noise, where for Weighted: (EFS = 3, RPI = 23.1).

Table 3. Expected First Success (EFS) and Relative Performance Improvement (RPI) of noise-enhanced link prediction methods on 4 real-world networks. These numbers show that noise-enhanced link prediction has an average EFS = 17 and RPI = 9.4 over all networks, noise methods, link prediction methods, and performance metrics.

Network Dataset	Performance Metric	Random Noise								Weighted Noise								Frequency Noise							
		CommonNeighbor		*Adamic/Adar*		*PreferentialAttach*		*Katz*		*CommonNeighbor*		*Adamic/Adar*		*PreferentialAttach*		*Katz*		*CommonNeighbor*		*Adamic/Adar*		*PreferentialAttach*		*Katz*	
		EFS	RPI	EFS	RPI	EFS	RPI	EFS	RPI	EFS	RPI	EFS	RPI	EFS	RPI	EFS	RPI	EFS	RPI	EFS	RPI	EFS	RPI	EFS	RPI
IrvineMessages	ROC-AUC	5	1.7	3	4.3	4	2	3	7.3	2	9.7	2	14.5	2	4.2	5	13.1	2	9.5	2	15	2	2.5	5	35.5
	AveragePrecision	4	3.4	3	5.8	5	1.2	3	41.32	29		2	33.9	2	14.2	6	62	2	28	2	35.7	2	17.9	5	93.1
	Acurracy	11	8.3	10	4.4	17	6.2	32	2.2	8	11.9	8	11.8	9	12.6	15	5.3	4	28	5	20.5	6	10.9	8	11.8
	PR-AUC	4	3.4	3	5.6	5	1.1	3	54.3	2	29	2	33.8	2	13.9	6	77	2	30	2	36.9	2	17.8	5	122.7
	F_1-score	11	8.3	19	4.4	17	6.2	32	2.2	8	12	8	11.7	9	12.5	15	5.2	4	27	5	20.4	6	10.8	8	11.8
	kendall's tau	4	95	3	34.6	3	3	3	5.9	2	576	2	80	2	7.7	6	7.4	2	670	2	84.7	2	0.1	5	4.2
Bitcoin	ROC-AUC	6	0.5	3	0.7	2	2.3	2	1.5	2	3	3	1.4	2	3.4	2	3.7	3	1.8	3	1	2	3.8	2	4.8
	AveragePrecision	3	0.4	3	1	2	0.6	2	7.7	4	1.8	3	1.5	2	7.7	2	12.3	5	1.3	4	1.2	2	4.7	2	10.8
	Acurracy	6	0.5	4	0.9	6	0.7	4	0.8	5	0.9	4	1.2	3	3	3	1.7	6	0.7	7	0.7	3	3.2	6	0.8
	PR-AUC	3	0.4	3	0.9	2	0.5	2	7.3	3	3	4	1.2	2	7.4	2	11.2	4	2.6	5	0.8	2	4.5	2	9.8
	F_1-score	6	0.51	4	0.8	6	0.7	4	0.7	5	0.88	4	1.2	3	2.9	3	1.6	6	0.74	7	0.6	3	3.1	6	0.8
	kendall's tau	2	5.8	3	3.8	2	88.4	2	5.3	2	19.4	3	7.4	2	125	2	12.6	3	10	3	4.8	2	155.9	2	16.3
Internet-growth	ROC-AUC	3	0.4	2	1.1	2	2.4	2	0.8	2	6.5	2	5.4	2	5	7	0.5	2	3.9	2	4.5	2	5.3	9	0.4
	AveragePrecision	2	0.45	2	0.9	2	1	2	5.8	2	35	1	44	2	32.9	3	9.2	2	16.5	2	20.7	2	10	4	6.3
	Acurracy	8	0.37	4	1	3	2.4	5	0.5	2	8.3	2	8.2	2	10.4	2	7	2	10.3	2	9.8	2	6.7	2	6
	PR-AUC	3	0.26	2	0.8	2	1	2	5.2	2	34	1	43	2	32.7	3	8.3	2	16	2	21.1	2	9.9	4	5.8
	F_1-score	8	0.36	4	1	3	2.3	5	0.5	2	7.9	2	7.9	2	9.9	2	6.7	2	10	2	9.4	2	6.4	2	5.7
	kendall's tau	3	8.1	2	24	2	27	2	24	2	96	2	162	2	75147	7	6.4	2	103	2	157	2	116	11	2.5
Facebook	ROC-AUC	2	1.2	3	0.7	2	5.8	2	1.2	3	1.4	3	0.8	2	6.8	3	2.9	3	1.6	4	0.6	2	13.9	2	13.8
	AveragePrecision	2	0.8	3	0.5	2	7.6	2	7.5	5	2.1	4	0.4	2	13.8	3	5.7	7	3.4	15	0.7	2	127	7	5.8
	Acurracy	9	0.21	6	0.43	6	2.9	9	0.26	151	0.01	11	0.24	4	5.2	23	0.12	334	0.009	239	0.01	2	30	455	0.008
	PR-AUC	2	0.85	3	0.48	2	6.5	2	5.7	4	1.3	5	0.42	2	11.7	3	4.4	7	2.8	15	0.78	2	145	8	4.1
	F_1-score	9	0.2	6	0.42	6	2.9	9	0.25	151	0.01	11	0.23	4	5.1	23	0.11	334	0.009	239	0.01	2	20	455	0.007
	kendall's tau	2	7.6	3	7	2	34.6	2	5.2	4	7.1	3	8	2	42	2	11.7	4	7.5	6	3.8	2	223	2	54

Table 4. Expected First Success (EFS) and Relative Performance Improvement (RPI) of noise-enhanced link prediction on networks generated by the configuration model with $n = 1,000$ nodes. Link prediction methods perform well on configuration graphs after adding noise with an average EFS = 3 and RPI = 4.21.

Network Dataset	Performance Metric	Random Noise: *CommonNeighbor* EFS	*CommonNeighbor* RPI	*Adamic/Adar* EFS	*Adamic/Adar* RPI	*PreferentialAttach* EFS	*PreferentialAttach* RPI	*Katz* EFS	*Katz* RPI
$n = 1000$	ROC-AUC	3	1	2	2	4	0.83	2	3.85
	AveragePrecision	3	1.23	2	3.36	5	0.33	2	7.16
	Acurracy	3	2.19	3	2.11	3	1.83	3	2.23
	PR-AUC	2	1.73	2	4.41	5	0.43	2	6.9
	F_1-score	3	1.92	3	1.83	3	1.52	3	1.96
	kendall's tau	3	3.23	2	7.37	4	2.81	2	4.4

Network Dataset	Performance Metric	Weighted Noise: *CommonNeighbor* EFS	*CommonNeighbor* RPI	*Adamic/Adar* EFS	*Adamic/Adar* RPI	*PreferentialAttach* EFS	*PreferentialAttach* RPI	*Katz* EFS	*Katz* RPI
$n = 1000$	ROC-AUC	2	2.7	2	2.57	3	0.84	2	5.7
	AveragePrecision	2	6.7	2	7.08	2	2.89	2	11.7
	Acurracy	3	2.33	2	3.77	2	3.24	2	4
	PR-AUC	2	7.94	2	8.73	2	3.4	2	11.28
	F_1-score	3	2	2	3.25	2	2.7	2	3.56
	kendall's tau	2	8.01	2	9.91	3	3.2	2	6.73

Network Dataset	Performance Metric	Frequency Noise: *CommonNeighbor* EFS	*CommonNeighbor* RPI	*Adamic/Adar* EFS	*Adamic/Adar* RPI	*PreferentialAttach* EFS	*PreferentialAttach* RPI	*Katz* EFS	*Katz* RPI
$n = 1000$	ROC-AUC	3	2	2	2.4	3	0.94	2	7
	AveragePrecision	2	4.57	2	6.78	2	3.4	2	10.91
	Acurracy	3	2.36	3	2.68	2	2.79	3	2.46
	PR-AUC	2	5.49	2	8.57	2	3.99	2	10.46
	F_1-score	3	2.04	3	2.31	2	2.3	3	2.16
	kendall's tau	3	4.79	2	8.83	3	3.41	2	9.3

Table 5. Expected First Success (EFS) and Relative Performance Improvement (RPI) of noise-enhanced link prediction methods on random networks with edge formation probability $p \in \{0.006, 0.007\}$. The noise-enhanced link prediction obtains the average EFS = 71 and RPI = 19.2 over all measures and noise methods on random graph models with $n = 1000$ nodes.

Network Dataset	Performance Metric	Random Noise: *CommonNeighbor* EFS	*CommonNeighbor* RPI	*Adamic/Adar* EFS	*Adamic/Adar* RPI	*PreferentialAttach* EFS	*PreferentialAttach* RPI	*Katz* EFS	*Katz* RPI
$p = 0.006$	ROC-AUC	23	5E-5	23	1.59			6	5.21
	AveragePrecision	23	4E-5	56	1.93			6	53.83
	Acurracy	23	3E-5	77	0.8			63	1.1
	PR-AUC	23	2E-5	56	2.61			6	55.39
	F_1-score	23	6E-5	77	0.8			63	1.1
	kendall's tau			19	1.24			6	32.15
$p = 0.007$	ROC-AUC	5	6.03	59	1.08			2	59.92
	AveragePrecision	36	1.76	53	0.49			2	107.19
	Acurracy	51	1.75	91	1.1			36	2.9
	PR-AUC	46	1.28	51	0.76			2	116.44
	F_1-score	51	1.74	91	1.1			36	2.59
	kendall's tau	5	203.18	59	0.1			2	52.59

Network Dataset	Performance Metric	Weighted Noise: *CommonNeighbor* EFS	*CommonNeighbor* RPI	*Adamic/Adar* EFS	*Adamic/Adar* RPI	*PreferentialAttach* EFS	*PreferentialAttach* RPI	*Katz* EFS	*Katz* RPI
$p = 0.006$	ROC-AUC	30	4E-3	72	0.2			7	4.24
	AveragePrecision	30	3E-5	72	0.52			7	70.83
	Acurracy	30	2E-5	72	0.4			59	0.7
	PR-AUC	30	1E-5	72	0.96			7	95.51
	F_1-score	30	4E-5	72	0.4			59	0.7
	kendall's tau			126	0.25			8	26.02
$p = 0.007$	ROC-AUC	91	0.21	251	1.32			2	67.66
	AveragePrecision	21	12.38	143	0.37			2	90.34
	Acurracy	21	4.3	143	0.7			44	2.6
	PR-AUC	20	126.51	126	0.65			2	95.55
	F_1-score	21	4.28	143	0.7			44	2.59
	kendall's tau	112	3.19	251	0.64			2	62.47

Network Dataset	Performance Metric	Frequency Noise: *CommonNeighbor* EFS	*CommonNeighbor* RPI	*Adamic/Adar* EFS	*Adamic/Adar* RPI	*PreferentialAttach* EFS	*PreferentialAttach* RPI	*Katz* EFS	*Katz* RPI
$p = 0.006$	ROC-AUC	28	3E-3	112	0.21			16	1.9
	AveragePrecision	28	3E-5	201	0.45			16	15.31
	Acurracy	28	2E-5	201	0.2			201	0.2
	PR-AUC	28	1E-5	201	0.64			16	15.77
	F_1-score	28	4E-5	201	0.2			201	0.2
	kendall's tau			167	0.26	-	-	17	12.12
$p = 0.007$	ROC-AUC	9	2.65	251	0.03			2	91.82
	AveragePrecision	48	0.48	251	0			2	204.64
	Acurracy	77	1.2	501	0.2			51	2
	PR-AUC	72	0.28	167	0.04			2	237.87
	F_1-score	77	1.2	501	0.2			51	1.99
	kendall's tau	9	21.12	251	0.02			2	67.58

Table 6. Expected First Success (EFS) and Relative Performance Improvement (RPI) of noise-enhanced link prediction methods on small-world networks with edge rewiring probability $p = 1$. The average (EFS, RPI) of the introduced framework is (5,54) for all measures and noise methods in small-world graphs with $n = 1000$ nodes.

Network Dataset	Performance Metric	Random Noise: *Adamic/Adar* EFS	*Adamic/Adar* RPI	*CommonNeighbor* EFS	*CommonNeighbor* RPI	*Katz* EFS	*Katz* RPI	*PreferentialAttach* EFS	*PreferentialAttach* RPI
$p = 1$	ROC-AUC	4	2.2	4	9.82	3	39.33	2	44.51
	AveragePrecision	5	11.03	4	27.36	3	38.67	2	285
	Acurracy	6	3.82	6	12.3	3	9.5	5	19.2
	PR-AUC	6	134.6	4	25.52	3	45.6	2	370
	F_1-score	6	3.8	6	12.24	3	9.4	5	19.09
	kendall's tau	8	49	4	68	9	11.6	2	231

Network Dataset	Performance Metric	Weighted Noise: *Adamic/Adar* EFS	*Adamic/Adar* RPI	*CommonNeighbor* EFS	*CommonNeighbor* RPI	*Katz* EFS	*Katz* RPI	*PreferentialAttach* EFS	*PreferentialAttach* RPI
$p = 1$	ROC-AUC	21	2.3	4	9.9	3	42.94	2	44.93
	AveragePrecision	5	26.43	4	24.37	3	42.48	2	194
	Acurracy	5	13.55	6	10.4	3	13	4	21.5
	PR-AUC	5	264	4	21.97	3	54.35	2	235
	F_1-score	5	13.49	6	10.35	3	12.96	4	21.38
	kendall's tau	9	33	4	74.8	8	13.1	2	209

Network Dataset	Performance Metric	Frequency Noise: *Adamic/Adar* EFS	*Adamic/Adar* RPI	*CommonNeighbor* EFS	*CommonNeighbor* RPI	*Katz* EFS	*Katz* RPI	*PreferentialAttach* EFS	*PreferentialAttach* RPI
$p = 1$	ROC-AUC	8	3.54	4	11.87	3	0.03	2	42.33
	AveragePrecision	4	40.44	6	19.88	3	2.E-4	2	192
	Acurracy	5	16.1	8	8.35	3	1E-4	5	19.55
	PR-AUC	4	60.31	6	19.7	3	1E-4	2	194
	F_1-score	5	16.8	8	8.3	3	2E-4	5	19.4
	kendall's tau	10	47.7	4	74.8	-	-	2	209

6.2 Noise-Enhanced Link Prediction in Synthetic Networks

This section evaluates noise-enhanced link prediction on synthetic graphs generated by different graph models to address the following questions: (1) which link prediction measure works best under noise?; (2) which noise-enhanced measure performs the best for each network model? (3) Which type of network model in general yields better results? Which one has the worst results? To answer these questions, we provided the results in two parts: 1) Tables including statistics on EFS and RPI values after adding noise; and 2) Tables qualitatively summarizing the results of all measures and noise methods. We explain these two parts in detail.

Quantitative Tables. For space reasons, we summarize synthetic networks' results in Tables 4, 5, and 6, which provides the average EFS and RPI of noise-enhanced link prediction methods on configuration model, random model, and small-world model graphs. For each network, noise method, and link prediction method, we provide both EFS and RPI of each performance metric. Gray cells indicate that the specific link prediction method did not improve with the specific noise method. We summarize the findings in Table 4, 5, 6 for each model as follows:

I. Configuration Model. Table 4 provides the results of noise-enhanced link prediction as follows:

- Noise-enhanced link prediction obtains an average $\texttt{EFS} = 3$ and $\texttt{RPI} = 4.21$ over configuration model networks, link prediction methods, noise methods, and performance metrics, indicating that noise is always able to predict better links in configuration model graphs.
- Although Weighted Noise is the best noise method for improving the link prediction in configuration model (average $\texttt{EFS} = 3$ and $\texttt{RPI} = 5.18$), `Frequency` and `Random` also perform very well (`Random`: $\texttt{EFS} = 3$ and $\texttt{RPI} = 2.78$, and `Frequency` $\texttt{EFS} = 3$ and $\texttt{RPI} = 4.68$).
- Katz works better under noise ($\texttt{EFS} = 3$, $\texttt{RPI} = 6.23$). Noise can also improve preferential attachment ($\texttt{EFS} = 4, \texttt{RPI} = 2.27$), but works worse than other measures. Adamic/Adar and common neighbor also show good performance after adding noise ($\texttt{EFS} = 3, \texttt{RPI} = 3.46$).

II. Random Graphs. Table 5 represents the results of our noise-enhanced framework as follows:

- For random graphs, our framework obtains the average $\texttt{EFS} = 71$ and $\texttt{RPI} = 19.2$ over all measures and noise methods when the edge formation probability $p \in \{006, 0.007\}$. When $p \in \{0.001, 0.003\}$, the noise-enhanced framework poorly works.
- Random Noise works the best with link prediction measures ($\texttt{EFS} = 37, \texttt{RPI} = 20$), and Frequency Noise works the worst for link prediction measures ($\texttt{EFS} = 113, \texttt{RPI} = 18$).
- Katz shows the best performance after adding noise ($\texttt{EFS} = 30, \texttt{RPI} = 46$) and Adamic/Adar shows the worst ($\texttt{EFS} = 147, \texttt{RPI} = 0.6$). Common neighbors also improves ($\texttt{EFS} = 36, \texttt{RPI} = 10.9$).

III. Small-world model. Table 6 represents the results of noise-enhanced framework as follows:

- The average (`EFS`, `RPI`) is (5,54) when edge rewiring probability p equals 1 in small-world graphs for all measures and noise methods.
- When $0.01 <= p <= 0.1$ (sweet spot), only Frequency Noise can perform well (`EFS`, `RPI`)=(2, 0.03). `Weighted` and `Random` perform poorly. For $p = 1$, all the noise methods work well (`Random`=($\texttt{EFS} = 5, \texttt{RPI} = 61$), `Weighted`=($\texttt{EFS} = 6, \texttt{RPI} = 58$)), `Frequency`=($\texttt{EFS} = 5, \texttt{RPI} = 41$)).
- When $0.01 <= p <= 0.1$ (sweet spot), Adamic/Adar works the best after adding noise ($\texttt{EFS} = 2, \texttt{RPI} = 0.08$), and Katz works the worst ($\texttt{EFS} = 2, \texttt{RPI} = 0.007$). Common neighbors and preferential attachment with the average ($\texttt{EFS} = 2, \texttt{RPI} = 0.06$) also perform well with noise-enhancement. For $p = 1$, Katz performs the best after noise with the average $\texttt{EFS} = 3, \texttt{RPI} = 132$.

Qualitative Table. Tables 7a and 7b summarize all statistics and figures. Table 7a answers the question on the type of noise that works best under each network model. As shown in Table 4, the configuration model works very well with all types of noise; noise should be ordered as `Weighted`, `Frequency`, and `Random` for best performance. The noise method performing best for small-world graphs is different for edge rewiring probability $p = 0.1$ and $p = 1$. When $p <= 0.1$, only `Frequency` performs well and using `Weighted`, and `Random` is not recommended. For $p = 1$, best methods can be ranked as `Random`, `Frequency`, `Weighted`. Note that `Frequency`, and `Weighted` are both our second priority as they are not very different. For random graphs, the noise ranking is: `Random`, `Weighted`, and `Frequency`. Table 7b answers the question on the type of noise-enhanced link prediction measures that work best under each type of graph. Although all measures work well for the configuration model, Katz works slightly better. Link prediction measures' ranking on noisy Random graphs is Katz, Common neighbor, and Adamic/Adar. Katz also works the best on noisy small-world graphs with $p = 1$, but performs the worst when $p < 0.1$. According to this table, Katz works well on all noisy graph models except for small-world networks' sweet-spot.

Table 7. Synthetic networks results

(a) Which type of noise works best for each network model?

Models	Parameters	Noise Priority Ranking		
Random(n,p)	all p	Random(1st)	**Weighted(2nd)**	Frequency(3rd)
Small-world	$p <= 0.1$	Frequency(1st)	**Weighted(2nd)**	Random(3rd)
	$p = 1$	Random(1st)	**Frequency(2nd)**	**Weighted(2nd)**
Configuration	Powerlaw	Weighted(1st)	**Frequency(2nd)**	Random(3rd)

(b) Which noise-enhanced link prediction measures works best for each type of graph?

Models	Parameters	Noise-enhanced link prediction methods Ranking			
Random(n,p)	all p	Katz(1st)	**CN(2nd)**	AA(3rd)	-
Small-world	$p <= 0.1$	AA(1st)	**CN(2nd)**	PA(3rd)	Katz(4th)
	$p = 1$	Katz(1st)	**PA(2nd)**	CN(3rd)	AA(4th)
Configuration	Powerlaw	Katz(1st)	**AA(2nd)**	CN(3rd)	PA(4th)

7 Conclusions

We proposed a framework to improve link prediction by adding noise to networks. The noise-enhanced framework adds a preprocessing phase before applying the existing link prediction methods to modify the input network. We introduced three noise methods to add noisy edges to the graph by considering nodes with high degrees as noisy edges' endpoints. Both theoretical and experimental results show that our framework can help the current link prediction methods predict better links in terms of performance metrics in both real-world and synthetic datasets.

References

1. Abdolazimi, R., Jin, S., Zafarani, R.: Noise-enhanced community detection. In: Proceedings of the 31st ACM Conference on Hypertext and Social Media, pp. 271–280 (2020)

2. Audhkhasi, K., Osoba, O., Kosko, B.: Noise-enhanced convolutional neural networks. Neural Netw. **78**, 15–23 (2016)
3. Chen, H., Varshney, L.R., Varshney, P.K.: Noise-enhanced information systems. PIEEE (2014)
4. Gammaitoni, L., Hänggi, P., Jung, P., Marchesoni, F.: Stochastic resonance. Rev. Modern Phys **70**(1), 223 (1998)
5. Kay, S.: Can detectability be improved by adding noise? IEEE Signal Proc. Lett. **7**(1), 8–10 (2000)
6. Krishna, O., Jha, R.K., Tiwari, A.K., Soni, B.: Noise induced segmentation of noisy color image. In: 2013 NCC, pp. 1–5, February 2013
7. Liben-Nowell, D., Kleinberg, J.: The link-prediction problem for social networks. J. Am. Soc. Inform. Sci. Technol. **58**(7), 1019–1031 (2007)
8. Lichtnwalter, R., Chawla, N.V.: Link prediction: fair and effective evaluation. In: 2012 IEEE/ACM ASONAM, pp. 376–383. IEEE (2012)
9. Lü, L., Zhou, T.: Link prediction in complex networks: a survey. Phys. A **390**(6), 1150–1170 (2011)
10. McDonnell, M.D., Abbott, D.: What is stochastic resonance? definitions, misconceptions, debates, and its relevance to biology. PLoS Comp. Bio. **5**(5) (2009)
11. McDonnell, M.D., Ward, L.M.: The benefits of noise in neural systems: bridging theory and experiment. Nat. Rev. Neurosci. **12**(7), 415 (2011)
12. Mislove, A.: Online social networks: measurement, analysis, and applications to distributed information systems. Ph.D. thesis, Rice University, Department of Computer Science, May 2009
13. Newman, M.E.: The structure and function of complex networks. SIAM Rev. **45**(2), 167–256 (2003)
14. Opsahl, T., Panzarasa, P.: Clustering in weighted networks. Soc. Netw. **31**(2), 155–163 (2009)
15. Osoba, O., Kosko, B.: Noise-enhanced clustering and competitive learning algorithms. Neural Netw. **37**, 132–140 (2013)
16. Osoba, O., Mitaim, S., Kosko, B.: The noisy expectation-maximization algorithm. Fluct. Noise Lett. **12**(03), 1350012 (2013)
17. Simonotto, E., Riani, M., Seife, C., Roberts, M., Twitty, J., Moss, F.: Visual perception of stochastic resonance. Phys. Rev. Lett. **78** (1997)
18. Tang, K.S., Man, K.F., Kwong, S., He, Q.: Genetic algorithms and their applications. IEEE Signal Process. Mag. **13**(6), 22–37 (1996)
19. Viswanath, B., Mislove, A., Cha, M., Gummadi, K.P.: On the evolution of user interaction in Facebook. In: Proceedings of the 2nd ACM SIGCOMM Workshop on Social Networks (WOSN 2009), August 2009
20. Watts, D.J., Strogatz, S.H.: Collective dynamics of 'small-world' networks. Nature **393**(6684), 440 (1998)
21. Zozor, S., Amblard, P.O.: On the use of stochastic resonance in sine detection. Signal Proc. **82**(3) (2002)

Weak Supervision Network Embedding for Constrained Graph Learning

Ting Guo[1], Xingquan Zhu[2](✉), Yang Wang[1], and Fang Chen[1]

[1] Data Science Institute, University of Technology Sydney, Sydney, Australia
{ting.guo,yang.wang,fang.chen}@uts.edu.au

[2] Department of Computer and Electrical Engineering and Computer Science, Florida Atlantic University, Boca Raton, USA
xzhu3@fau.edu

Abstract. Constrained learning, a weakly supervised learning task, aims to incorporate domain constraints to learn models without requiring labels for each instance. Because weak supervision knowledge is useful and easy to obtain, constrained learning outperforms unsupervised learning in performance and is preferable than supervised learning in terms of labeling costs. To date, constrained learning, especially constrained clustering, has been extensively studied, but was primarily focused on data in the Euclidean space. In this paper, we propose a weak supervision network embedding (WSNE) for constrained learning of graphs. Because no label is available for individual nodes, we propose a new loss function to quantify the constraint-based loss, and integrate this loss in a graph convolutional neural network (GCN) and variational graph auto-encoder (VGAE) combined framework to jointly model graph structures and node attributes. The joint optimization allows WSNE to learn embedding not only preserving network topology and content, but also satisfying the constraints. Experiments show that WSNE outperforms baselines for constrained graph learning tasks, including constrained graph clustering and constrained graph classification.

Keywords: Weak supervision · GNN · Network embedding

1 Introduction

Graph-structured data are becoming increasingly common in many real-world applications, such as social networks, citation networks, knowledge graphs, telecommunication networks, and biological networks [4]. In graph-structured data, nodes represent individual entities, and edges represent relationships and interactions between entities. For example, in citation networks, each document denotes a node, and a citation link between two documents is treated as an edge. Learning node embeddings is one of the most important and active research topics in network feature representation learning. Many models [24] have been proposed to embed each node into a continuous vector space. Because embedding

K. Karlapalem et al. (Eds.): PAKDD 2021, LNAI 12712, pp. 488–500, 2021.
https://doi.org/10.1007/978-3-030-75762-5_39

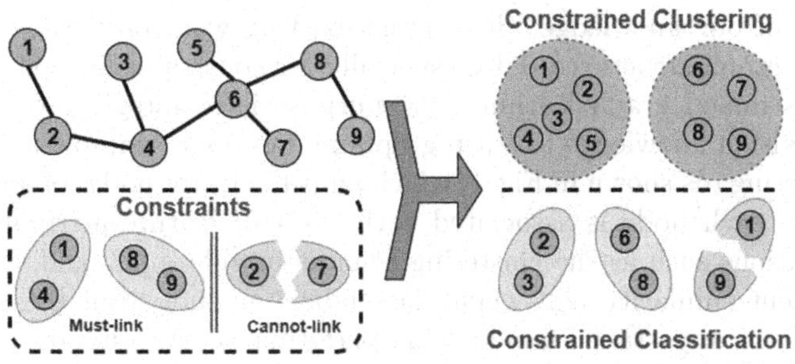

Fig. 1. Constrained graph learning. The weak supervision, must-link and cannot-link, specifies pairwise node constraints. Constrained graph clustering clusters node into groups, and constrained classification classifies a pair of nodes as either same-group or different-group.

vectors preserve topology and node content information, they can be directly used in downstream tasks, such as clustering or classification.

Depending on the availability of node labels, existing node embedding methods fall into two categories: (1) *Strong-supervision:* labels are provided to individual nodes (or a portion of nodes) for finding embeddings with maximum separability. (2) *Non-supervision:* no labels are provided to the learning task, so embedding aims to find a mapping to an output space without labeled responses. For strong-supervision methods, they require labeled training data, which often imply high labeling costs and obstacles. This bottleneck effect manifests itself in various ways, including the insufficient quantity of labeled data, insufficient subject-matter expertise to label data, and insufficient time to label and prepare data. On the other hand, for non-supervision methods, because no labels are given to the learning algorithms to differentiate samples, it is difficult for them to find structures satisfying users' requirements.

In addition to the above strong- *vs.* non-supervision scenarios, *Weak supervision* in machine learning provides a new setting where no labels are provided for individual instances, but some noisy, low-quality, conditional constraints over unlabeled data are available for model learning [1]. Weak supervision learning is intended to decrease labeling costs and increase the efficiency of human efforts expended in hand-labeling data while makes the outputs usable and comprehensive to specific problems. For example, pairwise constraints, *must-link* and *cannot-link*, are a means of weak supervision that constrain a pair of data points to belong to the same cluster (must-link) or different clusters (cannot-link). By integrating such pairwise constraints, constrained clustering is able to learn cluster structures much better than pure clustering methods [2]. A variety of studies have attempted to learn models using weak supervision [5,12,13,26], some works [25] have also recently advanced deep learning to weak supervision, but existing methods are primarily focused on data in the Euclidean space. While both weakly supervised learning and graph data have been extensively studied, to the best of our knowledge, there is no existing work on weakly supervised learning for graph-structured data, especially for constrained graph learning.

For constrained graph learning, the purpose is to integrate constraints as weak supervision knowledge to learn graph models. One example is constrained graph clustering (as shown in Fig. 1, which aims to cluster nodes of an attributed graph where each node is associated with a set of feature attributes. In specific applications such as the clustering of faces in videos [23] and the assessing of interpatient similarity [22] when class labels are not available, constraints are particularly important for enhancing performance. Another example is constrained graph classification (as shown in Fig. 1), which learns a binary classifier to identify whether two given nodes belong to the same group or not. Like recommendation systems, the predicted pairwise constraints (pairwise association rules) are very important in suggesting relevant items to users (as known as pairwise preference learning) [8].

The above observations motivate our research to propose a weak supervision node embedding model (WSNE) for constrained graph learning. We consider pairwise node constraints as weak supervision, and the main idea is to learn optimal node embedding by simultaneously integrating constraint loss and graph reconstruction loss in a deep graph convolution network (GCN) and variational graph auto-encoder (VGAE) combined framework. Different from existing strong-supervision graph embedding methods (including supervised/semi-supervised), our approach can utilize both constrained and unconstrained nodes to derive a high-order embedding evaluation criterion. This evaluation criterion estimates the effectiveness of node embedding based upon the high-level distances of must-link/cannot-link among constrained nodes and the average squared distances between unconstrained nodes. The GCN framework is used to learn a target node's representation by propagating neighbour information in an iterative manner until a stable fixed point is reached. And the VGAE framework is used to learn latent node representations through reconstructing graph topology information such as the graph adjacency matrix. To make better use of the constraints in graph learning, we also propose a novel topology optimization to fully utilize the potential constraint information during the message passing process in GCN as the given network topology may induce a performance degradation if it is directly employed in classification/clustering tasks.

In summary, the main contribution of the paper, compared to existing methods in the field, is threefold:

- We formulate a new weak-supervision network embedding task to utilize weak supervision knowledge to find effective latent vector space.
- We propose a new pairwise constraint evaluation criterion to efficiently evaluate the quality of embedding vector on constrained and unconstrained graph data.
- We develop a new constrained topology optimization method for graph convolution layers which take must-link and cannot-link into consideration.

2 Problem Definition

An undirected connected attributed graph $\mathcal{G} = \{\mathcal{V}, \mathcal{E}, \mathcal{A}, \mathcal{X}\}$ consists of a set of nodes $\mathcal{V}$ with $|\mathcal{V}| = n$, a set of edges $\mathcal{E}$ with $|\mathcal{E}| = m$, the adjacency matrix $\mathcal{A}$, and node attribute matrix $\mathcal{X}$. If there is an edge between node i and node j, the entry $\mathcal{A}_{ij}$ denotes the weight of the edge; otherwise, $\mathcal{A}_{ij} = 0$. For unweighted graphs, we simply set $\mathcal{A}_{ij} = 1$.

For each node, its content (features) is represented as a vector $x \in \mathbb{R}^n$, where x_i denotes feature values of node i (Node attributes, node content, and node features are equivalent terms in this paper). Therefore, $\mathcal{X} \in \mathbb{R}^{n \times d}$ denotes the node attribute matrix of the graph, and the columns of $\mathcal{X}$ are the d features of the graph.

Network embedding aims to embed a graph $\mathcal{G}$ in a low-dimensional space $\mathcal{Z} \in \mathbb{R}^{n \times m}$, where $m << d$ and the columns of $\mathcal{Z}$ are the m embedded signals of the graph. In the context of constrained graph learning, we consider two pairwise node-level constraints as weak supervision knowledge:

- **Must-link**: Two nodes belong to be in the same cluster/group, *i.e.* $\mathcal{C}^+ = \{< i, j > | \mathcal{C}_i = \mathcal{C}_j\}$, where $\mathcal{C}_k$ means the cluster that node k belongs to.
- **Cannot-link**: Two nodes do not belong to the same cluster/group, *i.e.* $\mathcal{C}^- = \{< i, j > | \mathcal{C}_i \neq \mathcal{C}_j\}$.

Given graph $\mathcal{G}$ and weak supervision constraints $\{\mathcal{C}^+, \mathcal{C}^-\}$, constrained graph learning **aims** to solve learning sub-tasks as follows:

- **Constrained graph clustering:** incorporate pairwise node constraints to cluster nodes into different groups.
- **Constrained graph classification:** incorporate constraints $\{\mathcal{C}^+, \mathcal{C}^-\}$ to learn a binary classifier to classify a pair of nodes as either same-group or different-group.

For constrained graph learning, our theme is to impose constraints on network embedding process to learn a discriminative representation for each node. The pairwise constraints $\{\mathcal{C}^+, \mathcal{C}^-\}$ define transitive binary relations over the nodes. Consequently, when making use of constraints, we take a transitive closure over the constraints. The full set of derived constraints is then presented to the learning algorithm.

3 CEL: Constraint Embedding Loss

Because class labels are not available for individual nodes, we have to design a new approach to utilize constraints $\{\mathcal{C}^+, \mathcal{C}^-\}$ for node embedding learning. In this section, we address the weak supervision problem discussed in Sect. 1 by finding optimal node embedding of weak supervision from both must-link and cannot-link constraints. From the perspective of weak supervision, we assume that the optimal node embeddings should have the following properties: (1) *Must-link distance*: The pairwise must-link nodes should be close to each other in the node embedding space; (2) *Cannot-link distance*: The pairwise cannot-link nodes should be far away from each other in the node embedding space; (3) *Separability*: unconstrained nodes should be able to be separated from each other in the node embedding space.

Intuitively, (1) and (2) only consider the constraints from constrained pairs and tend to optimize the node embeddings based on the constraints. This is motivated by the commonly observed phenomenon [17] that nodes close to each other tend to share common content information. Note (3) incorporates the distribution of unconstrained nodes, and tends to optimize the node embeddings that can separate nodes far from each other. It is similar to the PCA's assumption, which is expressed as the average squared distance between unlabelled samples.

Based upon the above properties, we derive a new evaluation criterion $\mathcal{L}_{CEL}(\mathcal{Z})$, for a given node embedding matrix $\mathcal{Z}$ as follow:

$$\begin{aligned}\mathcal{L}_{CEL}(\mathcal{Z}) = & \frac{\gamma^{+}}{2|\mathcal{C}^{+}|}\sum_{<i,j>\in\mathcal{C}^{+}}(\mathcal{Z}_{i\cdot}-\mathcal{Z}_{j\cdot})^2 - \frac{\gamma^{-}}{2|\mathcal{C}^{-}|}\sum_{<i,j>\in\mathcal{C}^{-}}(\mathcal{Z}_{i\cdot}-\mathcal{Z}_{j\cdot})^2 \\ & - \frac{1}{2|\mathcal{C}^{u}|^2}\sum_{<i,j>\in\mathcal{C}^{u}}(\mathcal{Z}_{i\cdot}-\mathcal{Z}_{j\cdot})^2\end{aligned} \tag{1}$$

In Eq. (1), $\mathcal{C}^u = \{< i,j > \mid < i,\cdot >\notin \mathcal{C}^+ \cup \mathcal{C}^- \ and \ < j,\cdot >\notin \mathcal{C}^+ \cup \mathcal{C}^-\}$ is the pairwise unconstraint sets. γ^+ and γ^- are two parameters, which control the weights of the three types of constraints. Based on the experiments, we found that applying feature binarization after 0–1 scaling on $\mathcal{Z}$ for calculating $\mathcal{L}_{CEL}$ can accelerate the convergence during the training process. By defining a pairwised constraint matrix $P = [P_{ij}]^{n\times n}$ as

$$P_{ij} = \begin{cases} \gamma^+/|\mathcal{C}^+| & if \ < i,j >\in \mathcal{C}^+ \\ -\gamma^-/|\mathcal{C}^-| & if \ < i,j >\in \mathcal{C}^- \\ -1/2|\mathcal{C}^u|^2 & if \ < i,j >\in \mathcal{C}^u \\ 0 & otherwise \end{cases} \tag{2}$$

We can then rewrite the $\mathcal{L}_{CEL}(\mathcal{Z})$ in Eq. 1 as follow:

$$\mathcal{L}_{CEL}(\mathcal{Z}) = \frac{1}{2}\sum_{i,j}(\mathcal{Z}_{i\cdot}-\mathcal{Z}_{j\cdot})^2 P_{ij} = tr(\mathcal{Z}^\top(D^p - P)\mathcal{Z}) = tr(\mathcal{Z}^\top L^p \mathcal{Z}) \tag{3}$$

where $tr(\cdot)$ is the trace of a matrix, D^p is the diagnal matrix whose entries are column sums of P, i.e. $D^p_{ii} = \sum_j P_{ij}$. $L^p = D^p - P$ is a Laplacian matrix.

Therefore, the framework to optimize the node embeddings for graph learning by considering constraints is to minimize $\mathcal{L}_{CEL}(\mathcal{Z})$ during the training process.

4 WSNE for Constrained Graph Learning

The above constraint embedding loss $\mathcal{L}_{CEL}$ allows us to quantify embedding loss without knowing labels of individual nodes. In this paper, we incorporate

$\mathcal{L}_{CEL}$ into a variational graph encoder framework (GVAE) architecture which uses constrained graph convolution to build the encoder. Our goal is to generate embedding vectors which optimally preserve graph content and topology, as well as comply to the given constraints.

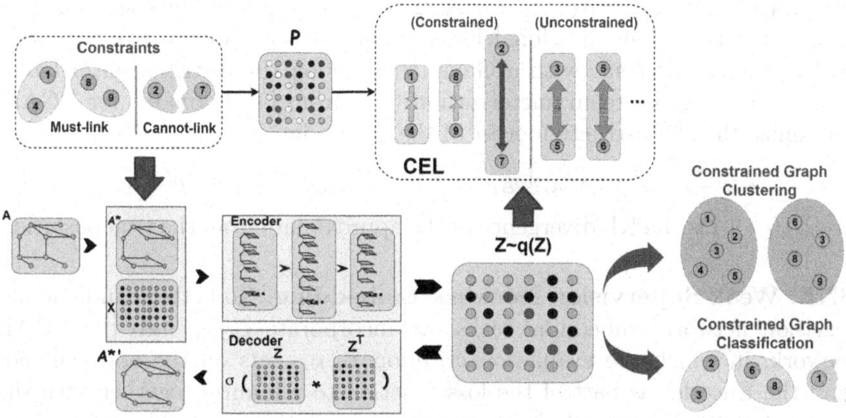

Fig. 2. The architecture of the weak supervision node embedding (WSNE) for constrained graph learning. The lower tier is the graph convolution based variational auto-encoder that reconstructs a graph $\mathcal{A}$ from $\mathcal{Z}$ which is generated by the encoder which exploits graph structure $\mathcal{A}$ and the node content matrix $\mathcal{X}$. The upper tier is the constraint embedding loss that evaluate the quality of $\mathcal{Z}$ on constrained and unconstrained graph data. The generated node embedding $\mathcal{Z}$ can be used for both constrained graph clustering and classification as right tier.

Variational Graph Auto-encoder. Graph auto-encoders (GAE) [11,19,21] are a family of models aiming at embed a graph in a low-dimensional space from which reconstructing (decoding) the graph should be possible. More precisely, the node embedding matrix $\mathcal{Z}$ is usually the output of a graph neural network (GNN) [3,6] processing $\mathcal{A}$. To reconstruct the graph, GAE stack an inner product decoder to this GNN. We have $\bar{\mathcal{A}} = \sigma(\mathcal{Z}\mathcal{Z}^\top)$, with $\sigma(\cdot)$ denoting the sigmoid function: $\sigma(x) = 1/(1+e^{-x})$. Therefore, the larger the inner product $\bar{\mathcal{A}}_{ij}$ in the embedding, the more likely nodes i and j are connected in $\mathcal{G}$ according to the GAE. Weights of the GNN are trained by gradient descent to minimize a reconstruction loss capturing the similarity of $\mathcal{A}$ and $\bar{\mathcal{A}}$, usually formulated as a weighted cross entropy loss.

VGAE extended the variational auto-encoder framework [9] to graph structure, which uses a probabilistic model involving latent variables z_i for each node $i \in \mathcal{V}$, interpreted as node representations in an embedding space. The inference model, i.e. the encoding part of the VAE, is defined as:

$$q(\mathcal{Z}|\mathcal{X},\mathcal{A}) = \prod_{i=1}^{n} q(z_i|\mathcal{X},\mathcal{A}) \tag{4}$$

where $q(z_i|\mathcal{X}, \mathcal{A}) = \mathcal{N}(z_i|\mu_i, diag(\sigma_i^2))$. Gaussian parameters are learned from two GNNs, i.e. $\mu = GNN_\mu(\mathcal{X}, \mathcal{A})$, with μ the matrix stacking up mean vectors μ_i; likewise, $log\sigma = GNN_\sigma(\mathcal{X}, \mathcal{A})$. Latent vectors z_i are samples drawn from this distribution. From these vectors, a generative model aims at reconstructing (decoding) $\mathcal{A}$, leveraging inner products: $p(\mathcal{A}|\mathcal{Z}) = \prod_{i=1}^n \prod_{j=1}^n p(\mathcal{A}_{ij}|z_i, z_j)$, where $p(\mathcal{A}_{ij} = 1|z_i, z_j) = \sigma(z_i^T z_j)$. During training, GNN weights are tuned by maximizing a tractable variational lower bound (ELBO) of the model's likelihood by gradient descent, with a Gaussian prior on the distribution of latent vectors, and using the reparameterization trick from [9]. Formally, for VGAE, we minimize the reconstruction error of the graph data by:

$$\mathcal{L}_R(\mathcal{Z}) = \mathbb{E}_{q(\mathcal{Z}|\mathcal{X},\mathcal{A})}[\log p(\mathcal{A}|\mathcal{Z})] - D_{KL}[q(\mathcal{Z}|\mathcal{X}, \mathcal{A})||p(\mathcal{Z})] \tag{5}$$

where $D_{KL}(\cdot||\cdot)$ is the KL divergence of the approximate from the true posterior.

WSNE: Weak Supervision Network Embedding. To better handle weak-supervision network embedding tasks, we incorporate $\mathcal{L}_{CEL}$ into the VGAE framework, in which the evaluation criterion $\mathcal{L}_{CEL}$ acts on the node embedding of the Encoder as part of the loss function to minimize together with the reconstruction loss. Figure 2 shows an overview of the proposed architecture. Therefore the whole network parameters are jointly trained by minimizing the following loss function as

$$\mathcal{L} = \mathcal{L}_R(\mathcal{Z}) + \lambda \mathcal{L}_{CEL}(\mathcal{Z}) \tag{6}$$

where $\mathcal{L}_R$ and $\mathcal{L}_{CEL}$ are defined in Eq. (3), respectively. Parameter $\lambda \geq 0$ is a tradeoff parameter. It is noted that, when $\lambda = 0$, the model is regressed to an original VGAE.

The proposed architecture can be used for both constrained graph clustering and classification tasks by using the learned node embeddings $\mathcal{Z}$.

Constrained Graph Clustering. We apply the linear kernel $\mathcal{K} = \mathcal{Z}\mathcal{Z}^\top$, and calculate the similarity matrix $\mathcal{S} = \frac{1}{2}(|\mathcal{K}|+|\mathcal{K}^\top|)$, where $|\cdot|$ means taking absolute value of each element of the matrix. Finally, we perform spectral clustering on $\mathcal{S}$ to obtain clustering results by computing the eigenvectors associated with the m largest eigenvalues of $\mathcal{S}$ and then applying the COP-k-means algorithm on the eigenvectors to obtain clusters.

Constrained Graph Classification. Given two nodes u and v, we define a binary operator $\otimes$ over the corresponding node embeddings $\mathcal{Z}_{u\cdot}$ and $\mathcal{Z}_{v\cdot}$ in order to generate a representation $g(u, v) \in \mathbb{R}^m$, which is the representation of the node pair $< u, v >$. We want our operator to be generally defined for any pair of nodes, even if an edge does not exist between the pair since doing so makes the representations useful for must-link/cannot-link prediction where our test set contains both true (must-link) and false (cannot-link) constraints. In this paper, we consider the Hadamard product used in [7] as the binary operator, i.e. $[g(u, v)]_i = [\mathcal{Z}_{u\cdot} \otimes \mathcal{Z}_{v\cdot}]_i = \mathcal{Z}_{ui} \times \mathcal{Z}_{vi}$. The new pairwise representation is then used to learn a binary classifier for constrained graph classification on the given constraints.

5 Constraint Assisted Topology Optimization

The given network topology induce a performance degradation if it is directly employed in classification/clustering, because it may possess high sparsity and certain noises. To make the message passing more efficient, in this section, we optimize the graph topology based upon the constraints and the GCN architecture.

Graph Convolutional Networks. Graph Convolutional Networks (GCNs) [10] achieve promising generalization in various tasks and our work is built upon the GCN module. At layer i, taking graph adjacency matrix $\mathcal{A}$ and hidden representation matrix $H^{(i)}$ as input, each GCN module outputs a hidden representation matrix $H^{(i+1)}$, which is described as:

$$H^{(i+1)} = ReLU(\hat{D}^{-\frac{1}{2}}\hat{\mathcal{A}}\hat{D}^{-\frac{1}{2}}H^{(i)}W^{(i)}) \tag{7}$$

where $H^{(0)} = \mathcal{X}$, $ReLU(a) = max(0, a)$, adjacency matrix with self-loop $\hat{\mathcal{A}} = \mathcal{A}+I$ (I is an identity matrix), $\hat{D}$ is the degree matrix of $\hat{\mathcal{A}}$, and $W^{(i)}$ is a trainable weight matrix. Then the output node embedding $\mathcal{Z} = H^{(K)}$ and $(K+1)$ is the number of layers in the network architecture.

Topology Optimization. Graph convolutional networks collectively aggregate information from graph structure, and model input and/or output consisting of elements and their dependency. The graph structure (edges) used in graph convolution architectures represents a kind of relations between nodes and guides the message passing among nodes. For example, in citation networks, a citation link between two documents is treated as an edge. While in social networks, the edges represent the interactions between users. From this point of view, the must-link/cannot-link constraints $\{\mathcal{C}^+, \mathcal{C}^-\}$ also represent a high-order relation between nodes which can be used to build the graph structure for constrained graph learning.

In order to directly utilize constraints $\{\mathcal{C}^+, \mathcal{C}^-\}$ into graph learning process, we will update the adjacency matrix $\mathcal{A}$ by using must-link and cannot-link constraints and use the updated $\mathcal{A}^*$ for graph convolution. The updated $\mathcal{A}^*$ is as follow:

$$\mathcal{A}^*_{ij} = \begin{cases} 1 & if \ <i,j> \in \mathcal{C}^+ \\ 0 & if \ <i,j> \in \mathcal{C}^- \\ \mathcal{A}_{ij} & otherwise \end{cases} \tag{8}$$

Using updated $\mathcal{A}^*$ in graph covolutions allows direct message passing between must-link nodes as they are in the same cluster and should have node embeddings consisting of elements. Using $\mathcal{A}^*$ also rejects direct message passing between cannot-link nodes as they are in different clusters and are not recommended to share similar information. By using $\mathcal{A}^*$, the GCN module should be updated accordingly as follow:

$$H^{(i+1)} = ReLU(\hat{D}^{*-\frac{1}{2}}\hat{\mathcal{A}}^*\hat{D}^{*-\frac{1}{2}}H^{(i)}W^{(i)}) \tag{9}$$

Algorithm 1: WSNE for constrained graph learning

Data: $\mathcal{A}$: The adjacency matrix; $\mathcal{X}$: The node attribute matrix; $\mathcal{C}^+$ and $\mathcal{C}^-$: The constraints;
Result: $\mathcal{Z}$: The node embedding matrix; $\mathcal{M}$: The cluster partition; and $\mathcal{Y}$: The binary classification;

```
P ← Generate constraint matrix P through Eq. (2);
A* ← Constraint assisted topology optimization through Eq. (8);
while Convergence do
    Z ← Generate latent variables Z using Eq. (4) and graph convolution framework in Eq. (9);
    L_R(Z) ← E_{q(Z|X,A)}[log p(A|Z)] − KL[q(Z|X,A)||p(Z)];
    L_CEL(Z) ← tr(Z^T L^p Z);
    L ← L_R(Z) + λ L_CEL(Z) update variational autoencoder with its gradient ;
end
if Constrained graph clustering then
    Apply the linear kernel K = Z Z^T, and calculate the similarity matrix S = 1/2(|K| + |K^T|) ;
    Obtain the cluster partition M by performing spectral clustering on S ;
end
if Constrained graph classification then
    [g(u,v)] = Z_u· × Z_v· ;
    Train a binary classifier on training C+ and C− with new representation [g(u,v)] and obtain the classification results Y ;
end
```

In this paper, we use graph convolution to process $\mathcal{A}$ and VGAE to reconstruct the graph structure for minimizing the information loss during the node embedding. Algorithm 1 lists detailed procedures of constrained graph learning.

6 Experiments

We evaluate our method on three benchmark graph datasets for both constrained graph clustering and classification tasks. **Cora**, **Citeseer** and **Pubmed** [10] are citation networks where nodes correspond to publications and are connected if one cites the other. The nodes in **Cora** and **Citeseer** are associated with binary word vectors, and nodes in **Pubmed** are associated with tf-idf weighted word vectors. Table 1 summarizes the details of the datasets. For Cora and Citeseer datasets, we randomly select 400 pairwise constraints (200 must-link pairs and 200 cannot-link pairs, respectively) as weak supervision. While for Pubmed dataset, we randomly select 3,000 pairwise constraints with 1,500 must-link pairs and 1,500 cannot-link pairs, respectively.

6.1 Constrained Graph Clustering

Baselines. As there is no existing constrained graph clustering method. We compare both embedding based approaches as well as approaches directly for graph clustering using constrained k-means for obtaining clustering results.

- **COP-k-means**: the constrained k-means algorithm [20] uses constraints as knowledge to restrict the data assignment process of the original k-means algorithm.
- **Spectral Clustering**: [18] is an effective approach for learning social embedding.
- **DeepWalk**: [16] is a network representation approach which encodes social relations into a continuous vector space.
- **GAE/VGAE**: [11] are (variational) autoencoder-based unsupervised frameworks for graph data, which naturally leverages both topological and content information.
- **ARVGA**: [15] is an adversarially regularized variational graph autoencoder for learning the node embedding.

Metrics. We employ three metrics to validate the clustering results: Accuracy (Acc), Normalized Mutual Information (NMI) and Average Rand index (ARI).

Parameter Settings. For the Cora, Citeseer and Pubmed datasets, we train all autoencoder-related models for 200 iterations and optimize them with the Adam algorithm. The learning rate is set to 0.001 and $\lambda = 0.1$. The parameters in $\mathcal{L}_{CEL}$ are set to $\alpha = \beta = 1$. We construct encoders with a 32-neuron hidden layer and a 16-neuron embedding layer for all the experiments. For the rest of the baselines, we retain the settings described in the corresponding papers.

Experimental Results. The constrained graph clustering results on the Cora, Citeseer and Pubmed data sets are given in Table 2. The results show that by incorporating the effective constraint embedding loss and constraint assisted topology optimization into our variational graph convolutional auto-encoder, WSNE achieve outstanding performance on all three metrics. Compared with the baselines, WSNE increased the Acc score from around 5.6% compared with existing node embeding methods incorporating with COP-k-means and 2.6% increased on the NMI score.

6.2 Constrained Graph Classification

Baselines. Because there is no constrained graph classification method available for comparison, we use each node embedding method, including DeepWalk, GAE, VGAE, ARVGA, and WSNE, to find node embedding. After that, we use embedding to generate vector $g(u, v)$ for node pair $< u, v >$, using constraints $\mathcal{C}^+$ and $\mathcal{C}^-$ (as we described in Sect. 4). Then we train binary classifiers using $g(u, v)$ generated from each embedding method, and report their performance in Table 3.

Table 1. Benchmark network statistics

Dataset	*# Nodes*	*# Edges*	*# Features*	*# Classes*
Cora	2,708	5,429	1,433	7
Citeseer	3,327	4,732	3,703	6
Pubmed	19,717	44,338	500	3

Table 2. Constrained graph clustering results on Cora, Citeseer and Pubmed.

Methods	CORA			CITESEER			PUBMED		
	Acc	NMI	ARI	Acc	NMI	ARI	Acc	NMI	ARI
COP-k-means	0.397	0.233	0.210	0.423	0.212	0.197	0.563	0.289	0.276
Spectral clustering	0.425	0.287	0.197	0.427	0.239	0.166	0.526	0.247	0.251
DeepWalk + COP-k-means	0.449	0.324	0.142	0.371	0.175	0.096	0.604	0.274	0.203
GAE + COP-k-means	0.557	0.406	0.290	0.451	0.277	0.213	0.627	0.269	0.175
VGAE + COP-k-means	0.570	0.424	0.332	0.471	0.259	0.124	0.615	0.193	0.095
ARVGA + COP-k-means	0.617	0.459	**0.373**	0.575	0.330	0.326	0.592	0.307	0.221
WSNE	**0.652**	**0.471**	**0.373**	**0.636**	**0.424**	**0.380**	**0.655**	**0.315**	**0.311**

Metrics. We report the results in terms of AUC score (the area under a receiver operating characteristic curve). The training set for the binary classification tasks are the provided constraints and the testing set contains 1,000 pairwise constraints for Cora and Citeseer datasets, and 5,000 pairwise constraints for Pubmed dataset to verify the performance.

Experimental Results. The constrained graph classification results on the Cora, Citeseer and Pubmed data sets are given in Table 3. The results show that WSNE achieves a significant improvement on the AUC score compared to all other baselines.

Table 3. Constrained graph classification results (AUC) on Cora, Citeseer, and Pubmed networks.

Methods	AUC values		
	Cora	Citeseer	Pubmed
DeepWalk	0.679	0.624	0.703
GAE	0.772	0.694	0.797
VGAE	0.790	0.738	0.823
ARVGA	0.793	0.754	0.836
WSNE	**0.844**	**0.802**	**0.871**

6.3 Embedding Visualization

We also visualize the Cora data in a two-dimensional space by applying the t-SNE algorithm [14] on the learned embedding. The results in Fig. 3 validate that by applying weak supervision constraints, *WSNE* is able to learn a more discriminative embedding vectors from graph data.

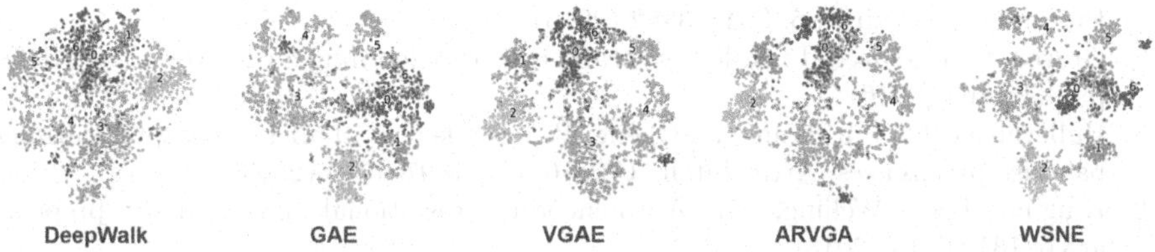

Fig. 3. Visualization comparison of embedding vectors on Cora data ($t-$SNE). From left to right: embeddings from DeepWalk, GAE, VGAE, ARVGA, and WSNE. Each point denotes a node. Nodes are color-coded based on the ground-truth class they belonging to (there are 7 classes/groups in total). The digit shows the centroid of each group, reported from $t-$SNE.

7 Conclusion

In this paper, we study a new research problem of weak supervision network embedding for constrained graph learning. We argued that existing network embedding approaches either require label information for individual nodes (strong-supervision) or do not use node labels (non-supervision). Weak supervision, such as constraints, are useful domain knowledge, but cannot be utilized in existing network embedding methods. To address the challenge, we proposed a new constraint embedding loss to quantify latent embedding vectors' loss by using both constrained and unconstrained data. Then we integrated this loss in a graph convolutional neural network and Graph Auto-Encoders combined framework to jointly model graph structures and node attributes to learn discriminative embedding vectors. Experiments and comparisons on real-world tasks show that the proposed method can effectively utilize weak supervision knowledge for constrained graph clustering and classification tasks.

Acknowledgements. This research is sponsored in part by the U. S. National Science Foundation (NSF) through Grant Nos. IIS-1763452 & CNS-1828181.

References

1. Basu, S., Davidson, I.: Clustering with constraints: theory and practice. In: KDD Tutorial (2006)
2. Basu, S., Davidson, I., Wagstaff, K.: Constrained Clustering: Advances in Algorithms, Theory, and Applications. CRC Press, Boca Raton (2008)

3. Bruna, J., Zaremba, W., Szlam, A., LeCun, Y.: Spectral networks and locally connected networks on graphs. arXiv preprint arXiv:1312.6203 (2013)
4. Cai, H., Zheng, V.W., Chang, K.C.C.: A comprehensive survey of graph embedding: problems, techniques, and applications. TKDE **30**(9), 1616–1637 (2018)
5. Davis, J.V., Kulis, B., Jain, P., Sra, S., Dhillon, I.S.: Information-theoretic metric learning. In: ICML, pp. 209–216 (2007)
6. Defferrard, M., Bresson, X., Vandergheynst, P.: Convolutional neural networks on graphs with fast localized spectral filtering. In: Advances in Neural Information Processing Systems, pp. 3844–3852 (2016)
7. Grover, A., Leskovec, J.: node2vec: scalable feature learning for networks. In: KDD, pp. 855–864 (2016)
8. Hüllermeier, E., Fürnkranz, J., Cheng, W., Brinker, K.: Label ranking by learning pairwise preferences. Artif. Intell. **172**(16–17), 1897–1916 (2008)
9. Kingma, D.P., Welling, M.: Auto-encoding variational bayes. arXiv preprint arXiv:1312.6114 (2013)
10. Kipf, T.N., Welling, M.: Semi-supervised classification with graph convolutional networks. arXiv preprint arXiv:1609.02907 (2016)
11. Kipf, T.N., Welling, M.: Variational graph auto-encoders. arXiv preprint arXiv:1611.07308 (2016)
12. Kulis, B., Sustik, M.A., Dhillon, I.S.: Low-rank kernel learning with Bregman matrix divergences. J. Mach. Learn. Res. **10**, 341–376 (2009)
13. Liu, W., Ma, S., Tao, D., Liu, J., Liu, P.: Semi-supervised sparse metric learning using alternating linearization optimization. In: KDD, pp. 1139–1148 (2010)
14. van der Maaten, L.: Accelerating t-SNE using tree-based algorithms. J. Mach. Learn. Res. **15**(1), 3221–3245 (2014)
15. Pan, S., Hu, R., Long, G., Jiang, J., Yao, L., Zhang, C.: Adversarially regularized graph autoencoder for graph embedding. IJCAI (2018)
16. Perozzi, B., Al-Rfou, R., Skiena, S.: DeepWalk: online learning of social representations. In: KDD, pp. 701–710 (2014)
17. Reagans, R., McEvily, B.: Network structure and knowledge transfer: the effects of cohesion and range. Adm. Sci. Q. **48**(2), 240–267 (2003)
18. Tang, L., Liu, H.: Leveraging social media networks for classification. Data Min. Knowl. Disc. **23**(3), 447–478 (2011)
19. Tian, F., Gao, B., Cui, Q., Chen, E., Liu, T.Y.: Learning deep representations for graph clustering. In: AAAI (2014)
20. Wagstaff, K., Cardie, C., Rogers, S., Schrödl, S., et al.: Constrained k-means clustering with background knowledge. In: ICML, pp. 577–584 (2001)
21. Wang, D., Cui, P., Zhu, W.: Structural deep network embedding. In: KDD, pp. 1225–1234 (2016)
22. Wang, F., Sun, J., Ebadollahi, S.: Integrating distance metrics learned from multiple experts and its application in patient similarity assessment. In: ICDM, pp. 59–70 (2011)
23. Wu, B., Zhang, Y., Hu, B.G., Ji, Q.: Constrained clustering and its application to face clustering in videos. In: CVPR, pp. 3507–3514 (2013)
24. Zhang, D., Yin, J., Zhu, X., Zhang, C.: Network representation learning: a survey. IEEE Trans. Big Data **6**, 3–28 (2020)
25. Zhang, H., Basu, S., Davidson, I.: A framework for deep constrained clustering - algorithms and advances. In: Proceedings of ECML/PKDD, pp. 57–72 (2019)
26. Zhou, Z.H.: A brief introduction to weakly supervised learning. Natl. Sci. Rev. **5**(1), 44–53 (2017)

RAGA: Relation-Aware Graph Attention Networks for Global Entity Alignment

Renbo Zhu[1], Meng Ma[2], and Ping Wang[1,2,3](✉)

[1] School of Software and Microelectronics, Peking University, Beijing, China
{zhurenbo,pwang}@pku.edu.cn

[2] National Engineering Research Center for Software Engineering, Peking University, Beijing, China
mameng@pku.edu.cn

[3] Key Laboratory of High Confidence Software Technologies (PKU), Ministry of Education, Beijing, China

Abstract. Entity alignment (EA) is the task to discover entities referring to the same real-world object from different knowledge graphs (KGs), which is the most crucial step in integrating multi-source KGs. The majority of the existing embedding-based entity alignment methods embed entities and relations into a vector space based on relation triples of KGs for local alignment. As these methods insufficiently consider the multiple relations between entities, the structure information of KGs has not been fully leveraged. In this paper, we propose a novel framework based on Relation-aware Graph Attention Networks to capture the interactions between entities and relations. Our framework adopts the self-attention mechanism to spread entity information to the relations and then aggregate relation information back to entities. Furthermore, we propose a global alignment algorithm to make one-to-one entity alignments with a fine-grained similarity matrix. Experiments on three real-world cross-lingual datasets show that our framework outperforms the state-of-the-art methods.

Keywords: Graph neural network · Entity alignment · Knowledge graph

1 Introduction

Knowledge graphs (KGs) have been widely applied for knowledge-driven artificial intelligence tasks, such as Question Answering [1], Recommendation [2] and Knowledge Enhancement [3]. The completeness of the KGs affects the performance of these tasks. Although lots of KGs have been constructed in recent years, none of them can reach perfect coverage due to the defects in the data sources and the inevitable manual process. A promising way to increase the completeness of KGs is integrating multi-source KGs, which includes an indispensable step entity alignment (EA). Entity alignment is the task to discover entities referring to the same real-world object from different KGs.

K. Karlapalem et al. (Eds.): PAKDD 2021, LNAI 12712, pp. 501–513, 2021.
https://doi.org/10.1007/978-3-030-75762-5_40

Recently, embedding-based methods have become the dominated approach for entity alignment. They encode entities and relations into a vector space and then find alignments between entities according to their embedding similarities. These methods can be subdivided into two categories: TransE-based methods via translating embeddings (TransE) [4] and GCNs-based methods via graph convolutional networks (GCNs) [5]. However, recent studies point out that there are still the following two critical challenges for entity alignment:

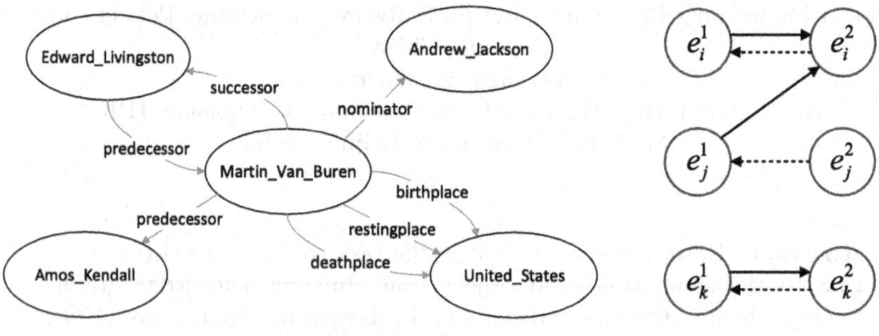

(a) An example of multiple relations. (b) An example of local alignment conflicts.

Fig. 1. Examples of two challenges for entity alignment.

Challenge 1: Sufficient Utilization of Multiple Relations. Figure 1(a) shows a mini KG, in which ellipses represent entities and directed edges represent relations. For TransE-based methods, they regard relations as the translation between two entities. However, they are limited by the uniqueness of the relation between two entities. For example, TransE-based methods fail to distinguish *brithplace*, *restingplace* and *deathplace* from *Martin_Van_Buren* to *United_States*. In real life, the above three relations are completely different, and their intersection will contain richer semantic information than any one of them. For GCNs-based methods, they model the propagation of entity information based on neighbouring entities on the graph without consideration of corresponding relation types and multiple relations. As an example in the figure, GCNs-based methods spread information of *Martin_Van_Buren* to *Edward_Livingston*, *Andrew_Jackson*, *Amos_Kendall* and *United_States* with equal weights. However, the influence of a person on a person should be different from the influence of a person on a country. Thus, the first challenge to entity alignment is how to utilize multiple relations for more reasonable entity representation sufficiently.

Challenge 2: Global Entity Alignment. Almost all entity alignment methods adopt a local alignment strategy to choose the optimal local match for each entity independently. The local alignment strategy always leads to many-to-one alignment, which means an entity may be the common best match for several entities. As an example of alignment results shown in Fig. 1(b), e_i^1, e_j^1 and e_k^1 are entities of KG_1, e_i^2, e_j^2 and e_k^2 are entities of KG_2. Solid arrows indicate the best match in KG_2 for each entity in KG_1, and dotted arrows indicate the best match in KG_1 for each entity in KG_2. Although e_k^1 and e_k^2 reach a final match with each other, the best matches of entities e_i^1, e_j^1, e_i^2 and e_j^2 lead to conflicts in bidirectional alignment. These conflicts violate the entity alignment task's essential requirement that alignments of two KGs should be interdependent. Thus, the second challenge of entity alignment is how to align entities of two KGs without conflicts from a global perspective.

Solution. To address above two challenges, we propose a framework RAGA based on Relation-aware Graph Attention Networks for Global Entity Alignment. Specially, we propose Relation-aware Graph Attention Networks to capture the interactions between entities and relations, which contributes to sufficient utilization of multiple relations between entities. We then design a global alignment algorithm based on deferred acceptance algorithm, which makes one-to-one entity alignments with a more fine-grained similarity matrix instead of the original embedding similarity matrix. Experimental results on three datasets of cross-lingual KGs demonstrate that RAGA significantly outperforms state-of-the-art baseline methods. The source code is available at https://github.com/zhurboo/RAGA.

2 Related Work

2.1 TransE-Based Entity Alignment

Most of the TransE-based entity alignment methods adopt TransE [4] to learn entity and relation embeddings. With the assumption that the relation is the translation from the head entity to the tail entity in a relation triple, TransE embeds all relations and entities into a unified vector space for a KG. MTransE [6] encodes entities and relations of each KG in separated embedding space and provides transitions to align the embedding spaces of KGs. JAPE [7] jointly embeds the structures of two KGs into a unified vector space. TransEdge [8] contextualizes relation representations in terms of specific head-tail entity pairs. BootEA [9] expands seed entity pairs in a bootstrapping way and employs an alignment editing method to reduce error accumulation during iterations. While TransE-based methods can only model fine-grained relation semantics, they cannot preserve the global structure information of KGs with multiple relations.

2.2 GCNs-Based Entity Alignment

With the insight that entities with similar neighbour structures are highly likely to be aligned, GCNs-based entity alignment approaches spread and aggregate entity information on the graph to collect neighbouring entities' representations.

GCN-Align [10] is the first attempt to generate entity embeddings by encoding information from their neighbourhoods via GCNs. NMN [11] proposes a neighbourhood matching module with a graph sampling method to effectively construct matching-oriented entity representations. MRAEA [12] directly models entity embeddings by attending over the node's incoming and outgoing neighbours and its connected relations. RREA [13] leverages relational reflection transformation to obtain relation embeddings for each entity. RDGCN [14] incorporates relation information via attentive interactions between the KGs and their dual relation counterpart. HGCN [15] applies GCNs with Highway Networks gates to embed entities and approximate relation semantics. DGMC [16] employs synchronous message passing networks to iteratively re-rank the soft correspondences to reach more accurate alignments. Since GCNs has more advantages in dealing with global structure information but ignore local semantic information, MRAEA, RREA, RDGCN, and HGCN make efforts to merge relation information into entity representations. Our framework RAGA adopts a similar idea with more effective interactions between entities and relations.

2.3 Global Entity Alignment

As each alignment decision highly correlates to the other decisions, every alignment should consider other alignments' influence. Thus, a global alignment strategy is needed for one-to-one alignments. An intuitive idea is calculating the similarity between entities and turning the global entity alignment task into a maximum weighted bipartite matching problem. The Hungarian algorithm [17] has been proven to finding the best solution for this problem with the intolerable time complexity of $O(n^4)$ for matching two KGs of n nodes.

To our best knowledge, two methods have been proposed to find an approximate solution for global entity alignment. GM-EHD-JEA [18] breaks the whole search space into many isolated sub-spaces, where each sub-space contains only a subset of source and target entities for making alignments. It requires a hyper-parameter τ of the threshold for similarity scores. CEA [19] adopts deferred acceptance algorithm (DAA) to guarantee stable matches. Although CEA achieves satisfactory performance, the similarity matrix it used lacks more fine-grained features, which are considered in our framework RAGA.

3 Problem Definition

A KG is formalized as $KG = (E, R, T)$ where E, R, T are the sets of entities, relations and relation triples, respectively. A relation triple (h, r, t) consists of a head entity $h \in E$, a relation $r \in R$ and a tail entity $t \in E$.

Given two KGs, $KG_1 = (E_1, R_1, T_1)$ and $KG_2 = (E_2, R_2, T_2)$, we define the task of entity alignment as discovering equivalent entities based on a set of seed entity pairs as $S = \{(e_1, e_2)|e_1 \in E_1, e_2 \in E_2, e_1 \leftrightarrow e_2\}$, where $\leftrightarrow$ represents equivalence.

For global entity alignment, it requires one-to-one matches, which means that each entity to be aligned has its equivalent entity, and the entity alignment results do not contain any alignment conflicts.

4 RAGA Framework

We propose our RAGA framework based on interactions between entities and relations via the self-attention mechanism. Figure 2 depicts the overall architecture of our framework. First, we adopt Basic Neighbor Aggregation Networks to obtain basic entity representations. Then, we generate enhanced entity representations via Relation-aware Graph Attention Networks, which incorporates relation information into entities. In the End-to-End Training part, the embeddings of input entities and the parameters of Relation-aware Graph Attention Networks are updated via backpropagation. Finally, the global alignment algorithm is applied to generate global alignments.

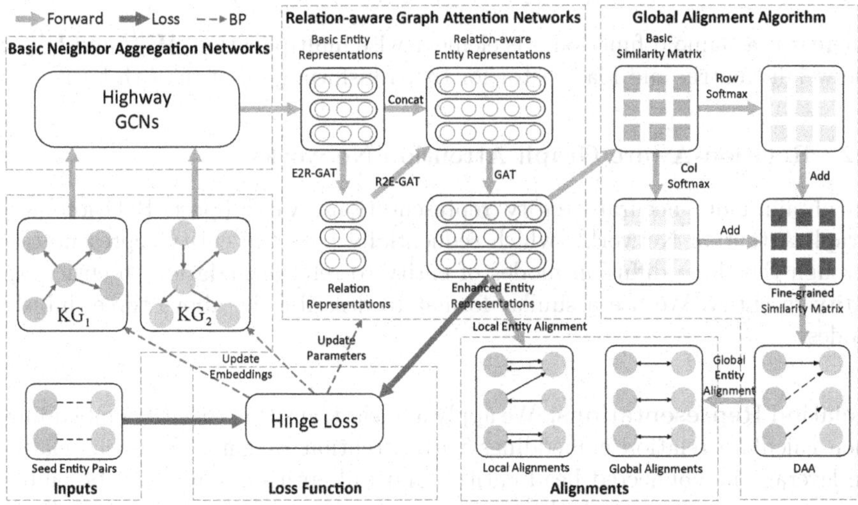

Fig. 2. Overall architecture of RAGA framework.

4.1 Basic Neighbor Aggregation Networks

To get basic entity representations, we utilize GCNs to explicitly encode entities in KGs with structure information. The input of l-th GCN layer is a set of entity embeddings $\boldsymbol{X}^{(l)} = \left\{\boldsymbol{x}_1^{(l)}, \boldsymbol{x}_2^{(l)}, \cdots, \boldsymbol{x}_n^{(l)} \mid \boldsymbol{x}_i^{(l)} \in \mathbb{R}^{d_e}\right\}$, where n is the number of entities, and d_e is the dimension of entity embeddings, the output of the l-th layer is obtained following the convolution computation:

$$\boldsymbol{X}^{(l+1)} = \text{ReLU}\left(\tilde{D}^{-\frac{1}{2}}\tilde{A}\tilde{D}^{-\frac{1}{2}}\boldsymbol{X}^{(l)}\right), \tag{1}$$

where $\tilde{A} = A + I$, A is the adjacency matrix of KG, I is an identity matrix, and $\tilde{D}$ is the diagonal node degree matrix of $\tilde{A}$. As entity embeddings are learnable, we do not apply a trainable weight matrix to change the distribution of $\boldsymbol{X}^{(l)}$, which may lead to overfitting.

Inspired by RDGCN [14], we employ layer-wise Highway Networks [20] to control the balance of the information between the entity itself and neighbour entities. The output of a Highway Network layer is the weighted sum of its input and the original output via gating weights:

$$T\left(\boldsymbol{X}^{(l)}\right) = \sigma\left(\boldsymbol{X}^{(l)}\boldsymbol{W}^{(l)} + \boldsymbol{b}^{(l)}\right), \tag{2}$$

$$\boldsymbol{X}^{(l+1)} = T\left(\boldsymbol{X}^{(l)}\right)\cdot\boldsymbol{X}^{(l+1)} + \left(1 - T\left(\boldsymbol{X}^{(l)}\right)\right)\cdot\boldsymbol{X}^{(l)}, \tag{3}$$

where σ is a sigmoid function, $\cdot$ is element-wise multiplication, $\boldsymbol{W}^{(l)}$ and $\boldsymbol{b}^{(l)}$ are the weight matrix and bias vector for the transform gate of the l-th layer.

4.2 Relation-Aware Graph Attention Networks

To obtain more accurate entity representations, we propose Relation-aware Graph Attention Networks, which sequentially pass the entity representations through the three diffusion modes of entity to relation, relation to entity, and entity to entity. We use a similar format to describe the above three diffusion modes.

Relation Representations. We apply a linear transition to entity embeddings, then calculate relation embeddings with attention weights. For each relation, we leverage its connected head entities and tail entities, which will be embedded in two vectors through their respective linear transition matrices. Different from RDGCN [14], our relation representations do not ignore the duplicate links between entities and relations, which is used to adjust the attention weights.

For relation r_k, the head entity representation $\boldsymbol{r}_k^h$ is computed as follows:

$$\alpha_{ijk} = \frac{\exp\left(\text{LeakReLU}\left(\boldsymbol{a}^T\left[\boldsymbol{x}_i\boldsymbol{W}^h \| \boldsymbol{x}_j\boldsymbol{W}^t\right]\right)\right)}{\sum_{e_{i'}\in\mathcal{H}_{r_k}}\sum_{e_{j'}\in\mathcal{T}_{e_i r_k}}\exp\left(\text{LeakReLU}\left(\boldsymbol{a}^T\left[\boldsymbol{x}_{i'}\boldsymbol{W}^h \| \boldsymbol{x}_{j'}\boldsymbol{W}^t\right]\right)\right)}, \tag{4}$$

$$\boldsymbol{r}_k^h = \text{ReLU}\left(\sum_{e_i\in\mathcal{H}_{r_k}}\sum_{e_j\in\mathcal{T}_{e_i r_k}}\alpha_{ijk}\boldsymbol{x}_i\boldsymbol{W}^h\right), \tag{5}$$

where α_{ijk} represents attention weight from head entity e_i to relation r_k based on head entity e_i and tail entity e_j, $\mathcal{H}_{r_k}$ is the set of head entities for relation r_k, $\mathcal{T}_{e_i r_k}$ is the set of tail entities for head entity e_i and relation r_k, $\boldsymbol{a}$ is a one-dimensional vector to map the $2d_r$-dimensional input into a scalar, d_r is half of the dimension of relation embeddings, and $\boldsymbol{W}^h, \boldsymbol{W}^t \in \mathbb{R}^{d_e \times d_r}$ are linear transition matrices for head and tail entity representation of relations respectively.

We can compute the tail entity representation $\boldsymbol{r}_k^t$ through a similar process, and then add them together to obtain the relation representation $\boldsymbol{r}_k$:

$$\boldsymbol{r}_k = \boldsymbol{r}_k^h + \boldsymbol{r}_k^t. \tag{6}$$

Relation-Aware Entity Representations. Based on the experience that an entity with its neighbour relations is more accurately expressing itself, we regroup the embeddings of relation adjacents into the entity representations. Specifically, for entity e_i, we adopt attention mechanism to calculate its out-relation (e_i is the head of those relations) embedding $\boldsymbol{x}_i^h$ and in-relation (e_i is the tail of those relations) embedding $\boldsymbol{x}_i^t$ separately. $\boldsymbol{x}_i^h$ is computed as follows:

$$\alpha_{ik} = \frac{\exp\left(\mathrm{LeakReLU}\left(\boldsymbol{a}^T\left[\boldsymbol{x}_i \| \boldsymbol{r}_k\right]\right)\right)}{\sum_{e_j \in \mathcal{T}_{e_i}} \sum_{r_{k'} \in \mathcal{R}_{e_i e_j}} \exp\left(\mathrm{LeakReLU}\left(\boldsymbol{a}^T\left[\boldsymbol{x}_i \| \boldsymbol{r}_{k'}\right]\right)\right)}, \tag{7}$$

$$\boldsymbol{x}_i^h = \mathrm{ReLU}\left(\sum_{e_j \in \mathcal{T}_{e_i}} \sum_{r_k \in \mathcal{R}_{e_i e_j}} \alpha_{ik} \boldsymbol{r}_k\right), \tag{8}$$

where α_{ik} represents attention weight from relation r_k to entity e_i, $\mathcal{T}_{e_i}$ is the set of tail entities for head entity e_i and $\mathcal{R}_{e_i e_j}$ is the set of relations between head entity e_i and tail entity e_j. Then the relation-aware entity representations $\boldsymbol{x}_i^{rel}$ can be expressed by concatenating $\boldsymbol{x}_i$, $\boldsymbol{x}_i^h$ and $\boldsymbol{x}_i^t$:

$$\boldsymbol{x}_i^{rel} = \left[\boldsymbol{x}_i \| \boldsymbol{x}_i^h \| \boldsymbol{x}_i^t\right]. \tag{9}$$

Enhanced Entity Representations. In relation-aware entity representations, entities only contain the information of one-hop relations. To enhance the influence of relations on two-hop entities, we adopt one layer of ordinary graph attention networks to get enhanced entity representations. This process considers bidirectional edges and does not include a linear transition matrix. For entity e_i, the final output of embedding $\boldsymbol{x}_i^{out}$ can be computed by:

$$\alpha_{ij} = \frac{\exp\left(\mathrm{LeakyReLU}\left(\boldsymbol{a}^T\left[\boldsymbol{x}_i^{rel} \| \boldsymbol{x}_j^{rel}\right]\right)\right)}{\sum_{j' \in \mathcal{N}_i} \exp\left(\mathrm{LeakyReLU}\left(\boldsymbol{a}^T\left[\boldsymbol{x}_i^{rel} \| \boldsymbol{x}_{j'}^{rel}\right]\right)\right)}, \tag{10}$$

$$\boldsymbol{x}_i^{out} = \left[\boldsymbol{x}_i^{rel} \| \mathrm{ReLU}\left(\sum_{j \in \mathcal{N}_i} \alpha_{ij} \boldsymbol{x}_i^{rel}\right)\right]. \tag{11}$$

4.3 End-to-End Training

We use Manhattan distance to calculate the similarity of entities:

$$\operatorname{dis}(e_i, e_j) = \left\|\boldsymbol{x}_i^{out} - \boldsymbol{x}_j^{out}\right\|_1. \tag{12}$$

For ent-to-end training, we regard all relation triples T in KGs as positive samples. Every p epoch, we adopt the nearest neighbour sampling to sample k negative samples from each knowledge graph for each entity. Finally, we use Hinge Loss as our loss function:

$$L = \sum_{(e_i,e_j)\in T} \sum_{(e_i',e_j')\in T'_{(e_i,e_j)}} \max\left(\operatorname{dis}(e_i, e_j) - \operatorname{dis}(e_i', e_j') + \lambda, 0\right), \tag{13}$$

where $T'_{(e_i,e_j)}$ is the set of negative sample for e_i and e_j, λ is margin.

4.4 Global Alignment Algorithm

As optimal local matches for entity alignment may lead to many-to-one alignments that reduce performance and bring ambiguity to entity alignment results, entities should be aligned globally. Thus, we design a global alignment algorithm.

Through Basic Neighbor Aggregation Networks and Relation-aware Graph Attention Networks, we obtain entity embeddings for each entity of two KGs. Then a similarity matrix $S \in \mathbb{R}^{|E_1|\times|E_2|}$ can be constructed based on the Manhattan distance between every two entity from different KGs. While CEA [19] directly applied deferred acceptance algorithm (DAA) [21] to the similarity matrix S for global entity alignment, we argue that more fine-grained features can be merged into the matrix. According to prior knowledge, entity alignment is a bidirectional match problem between two KGs. Thus, we calculate a fine-grained similarity matrix S^g by summing the weights of each entity aligned in two directions. Specifically, we adopt softmax on both rows and columns of S and add them together to get the fine-grained similarity matrix S^g:

$$S_{i,j}^g = \frac{\exp(S_{i,j})}{\sum_{j'=1}^{|E_2|}\exp(S_{i,j'})} + \frac{\exp(S_{i,j})}{\sum_{i'=1}^{|E_1|}\exp(S_{i',j})}. \tag{14}$$

Finally, we also adopt DAA to the fine-grained similarity matrix S^g to get global alignments. The detailed process of DAA can refer to [19]. The time complexity of the alignment process is $O(|E_1| \cdot |E_2| \cdot \log(|E_1| \cdot |E_2|))$, which is much smaller than that of Hungarian algorithm.

5 Experiments

5.1 Experimental Settings

Datasets. We evaluate the proposed framework on DBP15K [7]. It contains three pairs of cross-lingual KGs: ZH-EN, JA-EN, and FR-EN. Each dataset includes 15,000 alignment entity pairs. Almost all entity alignment studies based on DBP15K adopt a simplified version of DBP15K, which removes lots of unrelated entities and relations. Our experiment is also based on the simplified version of DBP15K, which is shown in the Table 1. For each dataset, we use 30% of the alignment entity pairs as seed entity pairs for training and 70% for testing.

Table 1. Statistical data of simplified DBP15K.

DBP15K		#Entities	#Relations	#Rel Triples	#Ent alignments
ZH-EN	ZH	19,388	1,700	70,414	15,000
	EN	19,572	1,322	95,142	
JA-EN	JA	19,814	1,298	77,214	15,000
	EN	19,780	1,152	93,484	
FR-EN	FR	19,661	902	105,998	15,000
	EN	19,993	1,207	115,722	

Evaluation Metrics. For local entity alignment, following [6], we use Hitratio@K (H@k) and mean reciprocal rank (MRR) to measure the performance. For global entity alignment, since one-to-one alignment results are produced, only H@1 was adopted. For all metrics, the larger, the better.

Baselines. To our best knowledge, DGMC and CEA are the state-of-the-art methods for local and global entity alignment respectively without additional information. To comprehensively evaluate our framework, we compare to both TransE-based, GCNs-based and global entity alignment methods:

- TransE-based methods: MtransE [6], JAPE [7], BootEA [9], TransEdge [8].
- GCNs-based methods: GCN-Align [10], MRAEA [12], RREA [13], RDGCN [14], HGCN [15], NMN [11], DGMC [16].
- Global methods: GM-EHD-JEA [18], CEA [19].

Implementation Details. We use the same initial entity embeddings as [14]. In Basic Neighbor Aggregation Networks, the depth of Highway-GCNs is 2. In Relation-aware Graph Attention Networks, the dimension of relation embeddings is 100. For end-to-end training, the number of epochs for updating negative samples is 5, and the negative sample number is 5. In margin-based loss function, the margin λ is 3.0.

Model Variants. In order to study the effectiveness of each component in our framework, we provide the following different variants of RAGA:

- w/o RGAT: Our framework without Relation-aware Graph Attention Networks for local entity alignment.
- w/o BNA: Our framework without Basic Neighbor Aggregation Networks for local entity alignment.
- RAGA-l: Our framework for local entity alignment.
- w/o Bi: Our framework without the fine-grained similarity matrix.

5.2 Experimental Results and Analysis

Table 2 shows the overall results of all methods. The parts of the results separated by solid line denote TransE-based methods, GCNs-based methods and global methods. The last parts below the dashed line of GCNs-based methods and global methods are the results of our models.

Table 2. Overall performance of entity alignment.

	ZH-EN			JA-EN			FR-EN		
Methods	H@1	H@10	MRR	H@1	H@10	MRR	H@1	H@10	MRR
MTransE	30.8	61.4	0.364	27.9	57.5	0.349	24.4	55.6	0.335
JAPE	41.2	74.5	0.490	36.3	68.5	0.476	32.3	66.7	0.430
BootEA	62.9	84.8	0.703	62.2	85.4	0.701	65.3	87.4	0.731
TransEdge	73.5	91.9	0.801	71.9	93.2	0.795	71.0	94.1	0.796
GCN-Align	41.3	74.4	0.549	39.9	74.5	0.546	37.3	74.5	0.532
MRAEA	63.5	88.2	0.729	63.6	88.7	0.731	66.6	91.2	0.764
RREA	71.5	92.9	0.794	71.3	93.3	0.793	73.9	94.6	0.816
RDGCN	70.8	84.6	–	76.7	89.5	–	88.6	95.7	–
HGCN	72.0	85.7	–	76.6	89.7	–	89.2	96.1	–
NMN	73.3	86.9	–	78.5	91.2	–	90.2	96.7	–
DGMC	74.8	82.5	–	80.4	86.4	–	**93.1**	95.8	–
w/o RGAT	74.7	86.4	0.790	78.5	89.6	0.826	89.9	96.0	0.922
w/o BNA	76.0	88.1	0.805	79.5	89.8	0.833	90.9	96.3	0.930
RAGA-l	**79.8**	**93.0**	**0.847**	**83.1**	**95.0**	**0.875**	91.4	**98.3**	**0.940**
GM-EHD-JEA	73.6	–	–	79.2	–	–	92.4	–	–
CEA	78.7	–	–	86.3	–	–	**97.2**	–	–
w/o Bi	84.3	–	–	86.7	–	–	94.1	–	–
RAGA	**87.3**	–	–	**90.9**	–	–	96.6	–	–

Overall EA Performance. For TransE-based methods, BootEA and TransEdge outperform MTransE and JAPE with their iterative strategies. Furthermore, by contextualizing relation representations in terms of specific head-tail entity pairs and interpreting them as translations between entity embeddings, TransEdge achieves excellent performance with random initial entity embeddings.

For GCNs-based methods, GCN-Align performs worst due to simple usage of relation triples. As MRAEA and RREA leverage more relation information, they get much better performance than GCN-Align. Based on initial entity embeddings, RDGCN, HGCN, NMN, and DGMC further improve their performance. For local alignment, our RAGA-l performs best in almost all evaluation metrics. It is noteworthy that DGMC performs 1.7% better than RAGA-l on H@1 of FR-EN. The reasons can be summarized as the following two points. First, in FR-EN dataset, due to high language similarity, the init embeddings contain rich information, which reduces the difficulty of the alignment task. Second, DGMC employs synchronous message passing networks, which contribute to the use of global information.

For global entity alignment methods, combined with the global alignment algorithm, our RAGA outperforms other methods in ZH-EN and JA-EN datasets. CEA performs slightly better than our RAGA in FR-EN. It is because CEA leverages extra entity descriptions, which are not considered in our methods.

Ablation Study. To analyze the effect of Relation-aware Graph Attention Networks, we construct two variants of RAGA-l: w/o RGAT and w/o BNA. From the results, we can see that both Basic Neighbor Aggregation Networks and Relation-aware Graph Attention Networks improve the performance significantly. Besides, Relation-aware Graph Attention Networks has a more significant effect than the former. To analyze the effect of our global alignment algorithm, we compare RAGA with w/o Bi, which adopts the same global alignment strategy as CEA. Experiments show that our global alignment algorithm with the fine-grained similarity matrix S^g further brings 2.5–4.3% improvement.

Impact of Seed Entity Pairs. To explore the impact of seed entity pairs on our framework, we compare RAGA and RAGA-l with DGMC by varying the proportion of seed entity pairs from 10% to 50% with a step size of 10%. Figure 3 depicts H@1 with respect to different proportions. It seems that when seed entity

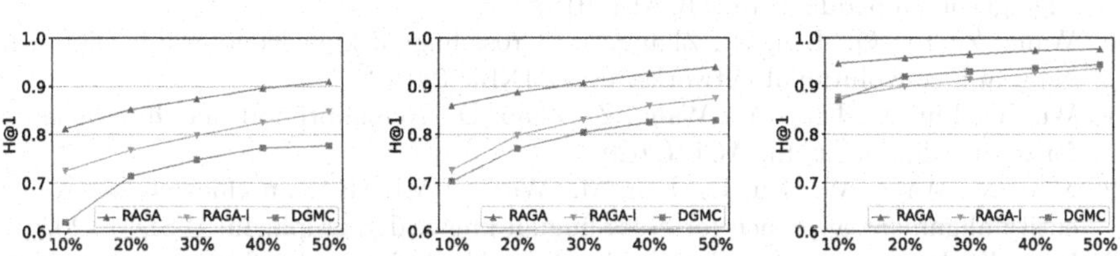

(a) ZH-EN (b) JA-EN (c) FR-EN

Fig. 3. H@1 of entity alignment results with different seed entity pairs.

pairs increase, RAGA and RAGA-l have more room for improvement while the performance of DGMC gradually reaches the bottleneck. Moreover, RAGA has a more gradual slope curve, which means the good capability of generalization.

6 Conclusion

In this paper, we have investigated the problem of entity alignment for the fusion of KGs. To address sufficient utilization of multiple relations and global entity alignment, we propose our framework RAGA to model the interactions between entities and relations for global entity alignment. Combined with Relation-aware Graph Attention Networks and global alignment algorithm, our framework outperforms the state-of-the-art entity alignment methods on three real-world cross-lingual datasets.

Acknowledgement. This work is supported by National Key Research and Development Program of China under Grant 2017YFB1200700.

References

1. Han, J., Cheng, B., Wang, X.: Open domain question answering based on text enhanced knowledge graph with hyperedge infusion. In: EMNLP (2020)
2. Xian, Y., Fu, Z., Huang, Q., Muthukrishnan, S., Zhang, Y.: Neural-symbolic reasoning over knowledge graph for multi-stage explainable recommendation. In: AAAI (2020)
3. Liu, W., et al.: K-BERT: enabling language representation with knowledge graph. In: AAAI (2020)
4. Bordes, A., Usunier, N., Garcia-Duran, A., Weston, J., Yakhnenko, O.: Translating embeddings for modeling multi-relational data. In: NeurIPS (2013)
5. Kipf, T. N., Welling, M.: Semi-supervised classification with graph convolutional networks. In: ICLR (2017)
6. Chen, M., Tian, Y., Yang, M., Zaniolo, C.: Multilingual knowledge graph embeddings for cross-lingual knowledge alignment. In: IJCAI (2017)
7. Sun, Z., Hu, W., Li, C.: Cross-lingual entity alignment via joint attribute-preserving embedding. In: ISWC (2017)
8. Sun, Z., Huang, J., Hu, W., Chen, M., Guo, L., Qu, Y.: TransEdge: translating relation-contextualized embeddings for knowledge graphs. In: ISWC (2019)
9. Sun, Z., Hu, W., Zhang, Q., Qu, Y.: Bootstrapping entity alignment with knowledge graph embedding. In: IJCAI (2018)
10. Wang, Z., Lv, Q., Lan, X., Zhang, Y.: Cross-lingual knowledge graph alignment via graph convolutional networks. In: EMNLP (2018)
11. Wu, Y., Liu, X., Feng, Y., Wang, Z., Zhao, D.: Neighborhood matching network for entity alignment. In: ACL (2020)
12. Mao, X., Wang, W., Xu, H., Lan, M., Wu, Y.: MRAEA: an efficient and robust entity alignment approach for cross-lingual knowledge graph. In: WSDM (2020)
13. Mao, X., Wang, W., Xu, H., Wu, Y., Lan, M.: Relational reflection entity alignment. In: CIKM (2020)
14. Wu, Y., Liu, X., Feng, Y., Wang, Z., Yan, R., Zhao, D.: Relation-aware entity alignment for heterogeneous knowledge graphs. In: IJCAI (2019)

15. Wu, Y., Liu, X., Feng, Y., Wang, Z., Zhao, D.: Jointly learning entity and relation representations for entity alignment. In: EMNLP (2020)
16. Fey, M., Lenssen, J. E., Morris, C., Masci, J., Kriege, N.M.: Deep graph matching consensus. In: ICLR (2020)
17. Kuhn, H.W.: The Hungarian method for the assignment problem. Naval Res. Logist. Quar. **2**(1–2), 83–97 (1955)
18. Xu, K., Song, L., Feng, Y., Song, Y., Yu, D.: Coordinated reasoning for cross-lingual knowledge graph alignment. In: AAAI (2020)
19. Zeng, W., Zhao, X., Tang, J., Lin, X.: Collective embedding-based entity alignment via adaptive features. In: ICDE (2020)
20. Srivastava, R. K., Greff, K., Schmidhuber, J.: Highway networks. arXiv preprint arXiv:1505.00387 (2015)
21. Roth, A.E.: Deferred acceptance algorithms: history, theory, practice, and open questions. Internat. J. Game Theory **36**(3–4), 537–569 (2008)

Graph Attention Networks with Positional Embeddings

Liheng Ma[1,2](✉), Reihaneh Rabbany[1,2], and Adriana Romero-Soriano[1,3]

[1] SCS McGill University, Montreal, Canada
liheng.ma@mail.mcgill.ca, rrabba@cs.mcgill.ca
[2] Mila, Montreal, Canada
[3] Facebook AI Research, Montreal, Canada
adrianars@fb.com

Abstract. Graph Neural Networks (GNNs) are deep learning methods which provide the current state of the art performance in node classification tasks. GNNs often assume homophily – neighboring nodes having similar features and labels–, and therefore may not be at their full potential when dealing with non-homophilic graphs. In this work, we focus on addressing this limitation and enable Graph Attention Networks (GAT), a commonly used variant of GNNs, to explore the structural information within each graph locality. Inspired by the positional encoding in the Transformers, we propose a framework, termed Graph Attentional Networks with Positional Embeddings (GAT-POS), to enhance GATs with positional embeddings which capture structural and positional information of the nodes in the graph. In this framework, the positional embeddings are learned by a model predictive of the graph context, plugged into an enhanced GAT architecture, which is able to leverage both the positional and content information of each node. The model is trained jointly to optimize for the task of node classification as well as the task of predicting graph context. Experimental results show that GAT-POS reaches remarkable improvement compared to strong GNN baselines and recent structural embedding enhanced GNNs on non-homophilic graphs.

Keywords: Graph Neural Networks · Attention · Positional Embedding

1 Introduction

The use of graph-structure data is ubiquitous in a wide range of applications, from social networks, to biological networks, telecommunication networks, 3D vision or physics simulations. The recent years have experienced a surge in graph representation learning methods, with Graph Neural Networks (GNNs) currently being at the forefront of many application domains. Recent advances in this direction are often categorized as spectral approaches and spatial approaches.

Spectral approaches define the convolution operator in the spectral domain and therefore capture the structural information of the graph [2]. However, such

K. Karlapalem et al. (Eds.): PAKDD 2021, LNAI 12712, pp. 514–527, 2021.
https://doi.org/10.1007/978-3-030-75762-5_41

approaches require computationally intense operations and yield filters which may not be localized in the spatial domain. A number of works have been proposed to localize spectral filters and approximate them for computation efficiency [6,10,11]. However, spectral filters based on the eigenbasis of the graph Laplacian depend on the graph structure, and thus cannot be directly applied to new graph structures, which limits their performance in the inductive setting. Spatial approaches directly define a spatially localized operator on the localities of the graph (i.e., neighborhoods), and as such are better suited to generalize to new graphs. Most spatial approaches aggregate node features within the localities, and then mix the aggregated features channel-wise through a linear transformation [3,8,14,20,29]. The spatial aggregation operator has been implemented as mean-pooling or max-pooling in [8], and sum-pooling in [14,20,29]. In order to increase the expressive power of spatial approaches, previous works have defined the aggregation operators with adaptive kernels in which the coefficients are parameterized as a function of node features. For instance, MoNets [18] define the adaptive filters based on a Gaussian mixture model formulation, and graph attention networks (GATs) [25] introduce a content-based attention mechanisms to parameterize the coefficients of the filter.

Among them, GATs [25] have become widely used and shown great performance in node classification [7,15,23]. However, the GATs' adaptive filter computation is based on node content exclusively and attention mechanisms cannot inherently capture the structural dependencies among entities at their input [12,24], considering them as a set structure. Therefore, GAT's filters cannot fully explore the structural information of the graph either. This may put the GAT framework at a disadvantage when learning on non-homophilic graph datasets, in which edges merely indicate the interaction between two nodes instead of their similarity [19]. Compared to homophilic graphs in which edges indicate similarity between the connected nodes, non-homophilic graphs are more challenging and higher-level structural patterns might be required to learn the node labels. In sequence-based [5,21,24], tree-based [26] and image-based [4,32] tasks, the lack of structural information leveraged by the attention mechanisms has been remedied by introducing handcrafted or learned positional encodings [4,5,21,24,26,32], resulting in improved performances.

Inspired by these positional encodings, and to improve learning on non-homophilic graphs, we aim to enhance GATs with structural information. Therefore, we propose a framework, called *Graph Attention Networks with Positional Embeddings* (GAT-POS), which leverages both positional information and node content information in the attention computation. More precisely, we modify the graph attention layers to incorporate a positional embedding for each node, produced by an positional embedding model predictive of the graph context. Our GAT-POS model is trained end-to-end to optimize for a node classification task while simultaneously learning the node positional embeddings. Our results on non-homophilic datasets highlight the potential of the proposed approach, notably outperforming GNN baselines as well as the recently introduced Geom-GCN [19], a method tailored to perform well on non-homophilic datasets.

Moreover, as a sanity check, we validate the proposed framework on standard homophilic datasets and demonstrate that GAT-POS can reach a comparable performance to GNN baselines. To summarize, the contributions of this paper are:

- We propose GAT-POS, a novel framework which enhances GAT with positional embeddings for learning on non-homophilic graphs. The framework enhances the graph attentional layers to leverage both node content and positional information.
- We develop a joint training scheme for the proposed GAT-POS to support end-to-end training and learn positional embeddings tuned for the supervised task.
- We show experimentally that GAT-POS significantly outperforms other baselines on non-homophilic datasets.

2 Method

In this paper, we consider the semi-supervised node classification task, and follow the problem setting of GCNs [10] and GATs [25]. Let $G = (\mathcal{V}, \mathcal{E})$ be an undirected and unweighted graph, with a set of N nodes $\mathcal{V}$ and a set of edges $\mathcal{E}$. Each node $v \in \mathcal{V}$ is represented with a D-dimensional feature vector $\mathbf{x}_v \in \mathbb{R}^D$. Similarly, each node has an associated label represented with a one-hot vector $\mathbf{y}_v \in \{0,1\}^C$, where C is the number of classes. The edges in $\mathcal{E}$ are represented as an adjacency matrix $\mathbf{A} = \{0,1\}^{N \times N}$, in which $\mathbf{A}_{vu} = 1$ iff $(v,u) \in \mathcal{E}$. The neighborhood of a node is defined as $\mathcal{N}(v) = \{u|(v,u) \in \mathcal{E}\}$.

In the rest of the section, we will present our proposed GAT-POS framework, and detail the implementation of each one of its components.

2.1 The GAT-POS Model

Multiple previous works have been proposed to incorporate positional embeddings in the attention mechanisms for sequences and grids [4,5,21,24,26,32]. However, the structure in a graph is more complicated and the positional encoding for these earlier works cannot be directly generalized to the graphs. Therefore, we propose our framework, Graph Attention Networks with Positional Embeddings (GAT-POS) to incorporate the positional embeddings in the attention mechanisms of GATs. In particular, we propose to learn the positional embeddings via an embedding model with an unsupervised objective $\mathcal{L}_u$, termed *positional embedding model*, which allows the positional embeddings to capture richer positional and structural information of the graph. We provide a detailed comparison with the previous works in Sect. 3.

In order to support end-to-end training, we propose a main-auxiliary architecture where the positional embedding model is plugged into GATs, inspired by the works of [27,30]. With this architecture, the supervised task of the GATs enhanced with positional embeddings, and the unsupervised task of the positional embedding model are trained jointly. Consequently, besides supporting end-to-end training, the positional embedding model can learn embeddings not only predictive of the graph context, but also beneficial to the supervised task.

Figure 1 provides an overview of the GAT-POS architecture, where each rectangle denotes the operation of a layer; the black arrows denote forward propagation and the green arrows denote backpropagation.

For the node classification tasks, the supervised loss function is the cross-entropy error over the set of labeled examples observed during training $\mathcal{Y}_L$, as follows,

$$\mathcal{L}_{\mathcal{S}}(\{\hat{\mathbf{y}}_v\}_{v\in\mathcal{Y}_L}, \{\mathbf{y}_v\}_{v\in\mathcal{Y}_L}) = -\sum_{v\in\mathcal{Y}_L} \mathbf{y}_v^\intercal \log \hat{\mathbf{y}}_v \tag{1}$$

where $\hat{\mathbf{y}}_v$ is node v's predicted label. The unsupervised task is designed to guide the positional embedding model to capture information beneficial to the supervised task.

In the following subsection, we will introduce a particular implementation of GAT-POS. However, it is worth noting that our framework is agnostic to the particular setup of the enhanced graph attentional layer, the architecture of the positional embedding model, the choice of the unsupervised objective as well as the way the main and auxiliary architectures are connected.

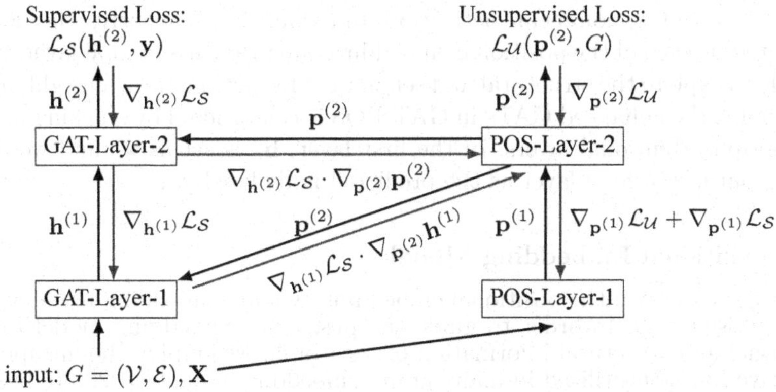

Fig. 1. A demonstration of GAT-POS architecture (the subscriptions are dropped for simplicity)

2.2 Positional Embedding Enhanced Graph Attentional Layer

We extend the graph attentional layer of [25] to leverage the node embeddings extracted from the GAT-POS positional embedding model when computing the attention coefficients. We modify the graph attentional layer to consider positional embeddings in the attention scores computation. In particular, our positional embedding enhanced graph attentional layer transforms a vector of node features $\mathbf{h}_v \in \mathbb{R}^F$ into a new vector of node features $\mathbf{h}'_v \in \mathbb{R}^{F'}$, where F and F' are the number of input and output features in each node, respectively. We start by computing the attention coefficients in the neighborhood of node v as follows,

$$\alpha^k_{vu} = \underset{u \in \mathcal{N}_v \cup v}{\text{softmax}} \left(\text{leakyrelu}(\mathbf{a}_k{}^{\intercal}[\mathbf{W}_k\mathbf{h}_v + \mathbf{U}_k\mathbf{p}_v \| \mathbf{W}_k\mathbf{h}_u + \mathbf{U}_k\mathbf{p}_u])\right) \quad (2)$$

where $\mathbf{W}_k$, $\mathbf{U}_k$ and $\mathbf{a}_k$ are the weights in the k-th attention head; $\mathbf{p}_v$ is the positional embedding for node v; $\|$ denotes the concatenation; and

$$\underset{u \in \mathcal{U}}{\text{softmax}}(e_{vu}) = \frac{\exp(e_{vu})}{\sum_{u' \in \mathcal{U}} \exp(e_{vu'})}$$

The attention coefficients, computed based on the input node features and structural information, are expected to exploit the structural and semantic information within the neighborhood. We subsequently update the features of node v by linear transforming the features of the nodes in the neighborhood with the obtained attention coefficients,

$$\mathbf{h}'_v = \begin{cases} \|_{k=1}^K \sigma(\sum_{u \in \mathcal{N}_v \cup v} \alpha^k_{vu} \cdot \mathbf{W}_k\mathbf{h}_u), & \text{if at the hidden layers,} \\ \sigma\left(\frac{1}{K}\sum_{k=1}^K \sum_{u \in \mathcal{N}_v \cup v} \alpha^k_{vu} \cdot \mathbf{W}_k\mathbf{h}_u\right), & \text{if at the output layer.} \end{cases} \quad (3)$$

where σ denotes nonlinear activation function. Consequently, the features extracted by each of the positional embedding enhanced attentional layers should be able to explore the structural and semantic information in each neighborhood.

Finally, the enhanced GATs in GAT-POS is constructed by stacking multiple such graph attentional layers. In the first layer, $\mathbf{h}_v$ is set as $\mathbf{x}_v$ and we denote the output of the final layer as the predicted node label $\hat{\mathbf{y}}_v$.

2.3 Positional Embedding Model

We propose to learn the positional embeddings via an embedding model with an unsupervised task. In order to guide the positional embedding model to learn positional and structural information of the graph, we employ the unsupervised objective function utilized in many graph embedding models [8,22,30] based on the skip-gram model [17]. This objective function is a computationally efficient approximation to the cross-entropy error of predicting first-order and second-order proximities among nodes, via a negative sampling scheme. More precisely, we define

$$\mathcal{L}_{\mathcal{U}}(\{\mathbf{p}_v\}_{v\in\mathcal{V}}, G)) = \sum_{v\in\mathcal{V}} \sum_{u\in\mathcal{N}(v)} \left(-\log\sigma(\mathbf{p}_v{}^{\intercal}\mathbf{p}_u) - Q \cdot \mathbb{E}_{u'\sim P_n(v)} \log(\sigma(-\mathbf{p}_v{}^{\intercal}\mathbf{p}_{u'}))\right), \quad (4)$$

where $P_n(v)$ denotes the distribution negative sampling for node v; and Q is the number of negative samples per edge.

With the unsupervised objective, we utilize a positional embedding model with simple architecture, which is constructed by stacking multiple fully-connected layers. The t-th layer of the positional embedding model is computed as follows,

$$\mathbf{p}^t_v = \sigma(\mathbf{W}^t_{emb}\mathbf{p}^{t-1}_v) \quad (5)$$

where $\mathbf{p}_v^t$ and $\mathbf{W}_{emb}^t$ are the learned positional representation for node v and the weight matrix at layer t, respectively; and σ denotes an arbitrary nonlinear activation. Even though more complicated embedding models may be used, embedding models with such simple architectures can still capture meaningful information of the graph structure according to previous works [17,22,30]. We only consider the transductive positional embedding model given the nature of the datasets used in our experiments. Thus, $\mathbf{p}_v^0$ is the learned initial positional embedding for node v and $\mathbf{p}_v$ is the corresponding output positional embedding from the final layer.

3 Related Works

Attention-Based Models. Attention mechanisms have been widely used in many sequence-based and vision-based tasks. Attention mechanisms were firstly proposed in the machine translation literature to overcome the information bottleneck problem of encoder-decoder RNN-based architectures [1,16]. Similar models were designed to enhance image captioning architectures [28]. Since then, self-attention and attention-based models have been extensively utilized as feature extractors, which allow for variable-sized inputs [4,5,21,24–26,32]. Among those, Transformers [24] and GATs [25] are closely related to our proposed framework. The former is a well-known language model based on attention mechanisms exclusively. Particularly, self-attention is utilized to learn word representations in a sentence, capturing their syntactic and semantic information. Note that self-attention assumes the words under the set structure, regardless their associated structural dependencies. Therefore, to explore the positional information of words in a sentence, a positional encoding was introduced in the Transformer and its variants. Following the Transformer, GATs learn node reprensentations in a graph with self-attention. More specifically, GATs mask out the interactions between unconnected nodes to somehow make use of the graph structure. This simple masking technique allows GATs to capture the co-occurrence of nodes in each neighborhood but, unfortunately, does not enable the model to fully explore the structure of the graph. Thus, we propose to learn positional embeddings to enhance the ability of GATs to fully explore the structural information of a graph. Consequently, our model, GAT-POS, can learn node representations which better capture the syntactic and semantic information in the graph.

Other Structural Embedding Enhanced GNNs. As pointed by [19], GNNs struggle to fully explore the structural information of a graph, and this shortcoming limits their effective application to non-homophilic graphs, which require increased understanding of higher-level structural dependencies. To remedy this, position-aware Graph Neural Networks (PGNNs) [31] introduced the concept of position-aware node embeddings for pairwise node classification and link prediction. In order to capture the position or location information of a given target node, in addition to the architecture of vanilla GNNs, PGNN samples sets of anchor nodes, computes the distances of the target node to each anchor-set, and learns a non-linear aggregation scheme for distances over the anchor-sets. It is worth mentioning that the learned position-aware embeddings are permutation sensitive to the order of nodes in the anchor-sets. Moreover, to learn on non-homophilic graphs, [19] proposes a framework, termed Geom-GCN, in which GNNs are enhanced with latent space embeddings, capturing structural information of the graph. As opposed to GAT-POS, Geom-GCN does not support end-to-end training, and therefore their embeddings cannot be adaptively adjusted for different supervised tasks. Similar to GAT-POS, Geom-GCN also employs adaptive filters to extract features from each locality. However, the adaptive filters in Geom-GCN are merely conditioned on the latent space embeddings, which do not leverage node content information, and might be less effective than our GAT-POS' attention mechanisms based on both node content and positional information. Moreover, Geom-GCN also introduces an extended neighborhood that goes beyond the spatial neighborhood of each node, and which corresponds to the set of nodes close to the center node in the latent space. However, according to the experimental results of our paper, the extended neighborhood of the Geom-GCN is not beneficial when dealing with undirected graphs. Finally, a very recent concurrent work [13] enhances GNNs with distance encodings for GNNs. As opposed to Geom-GCN and GAT-POS, the distance encoding is built without any training process on the graph, capturing graph properties such as shortest-path-distance and generalized PageRank scores based on an encoding process.

4 Experiments and Results

We have performed comparative evaluation experiments of GAT-POS on non-homophilic graph-structured datasets against standard GNNs (i.e., GCNs and GATs) and related embedding enhanced GNNs (i.e., Geom-GCN). Our proposed framework reaches remarkably improved performance compared to standard GNNs and outperforms Geom-GCNs. Besides, we also performed a sanity check test on homophilic datasets to validate that GAT-POS can reach comparable performance to standard GNNs.

4.1 Datasets

We utilize two Wikipedia page-page networks (i.e., Chameleon and Squirrel) and the Actor co-occurrence network (shortened as Actor) introduced in [19] as our non-homophilic datasets. We also include three widely used citation networks (i.e., Cora, Citeseer and Pubmed) [10,25,30] as representatives of homophilic datasets for our sanity check. Note that all graphs are converted to undirected graphs, following the widely utilized setup in other works [10,25].

Table 1 summarizes basic statistics of each dataset, together with their homophily levels. The homophily level of a graph dataset can be demonstrated by a measure β introduced in [19], which computes the ratio of the nodes in each neighborhood sharing the same labels as the center node.

Table 1. Summary of the datasets used in our experiments.

Dataset	Homophilic Datasets			Non-homophilic Datasets		
	Cora	Citeseer	Pubmed	Chameleon	Squirrel	Actor
Homophily(β)	0.83	0.71	0.79	0.25	0.22	0.24
# Nodes	2708	3327	19717	2277	5201	7600
# Edges*	5429	4732	44338	36101	217073	33544
# Features	1433	3703	500	2325	2089	931
# Classes	7	6	3	5	5	5

Following the evaluation setup in [19], for all graph datasets, we randomly split nodes of each class into 60%, 20%, and 20% for training, validation and testing, respectively. The final experimental results reported are the average classification accuracies of all models on the test sets over 10 random splits for all graph datasets. For each split, 10 runs of experiments are performed with different random initializations. We utilize the splits provided by [19]. Note that all datasets in our evaluation are under the transductive setting.

4.2 Methods in Comparison

In order to demonstrate that our framework can effectively enhance the performance of GATs on non-homophilic datasets, the standard GAT [25] is considered in the comparison. Furthermore, another commmonly used GNN, the GCN [10] is also included in our evaluation as a representative of spectral GNNs. We also include Geom-GCN [19] into our evaluation as a baseline of structural embedding enhanced GNNs. Specifically, three variants of Geom-GCNs coupled with different embedding methods are considered, namely Geom-GCN with Isomap (Geom-GCN-I), Geom-GCN with Poincare embedding (Geom-GCN-P), and Geom-GCN with struc2vec embedding (Geom-GCN-S). Besides, we also include the variants of Geom-GCNs without their extended neighborhood scheme, considering only spatial neighborhoods in the graph, termed Geom-GCN-I-g, Geom-GCN-P-g and Geom-GCN-S-g respectively.

4.3 Experimental Setup

We utilize the hyperparameter settings of baseline models from [19] in our experiments since they have already undergone extensive hyperparameter search on each dataset's validation set. For our proposed GAT-POS, we perform a hyperparameter search on the number of hidden units, the initial learning rate, the weight decay, and the probability of dropout. For fairness, the hyperparameter search is performed on the same searching space as the models tuned in [19]. Specifically, the number of layers of GNN architectures is fixed to 2 and the Adam optimizer [9] is used to train all models.

For GAT-POS, according to the result of the hyperparameter search, we set the initial learning rate to 5e-3, weight decay to 5e-4, a dropout of $p = 0.5$ and the number of hidden units for the positional embedding model to 64. Besides, the activation functions of the positional embedding model and the enhanced GAT architecture are set to ReLU and ELU, respectively. The number of hidden units per attention head in the main architecture is 8 (Cora, Citeseer, Pubmed and Squirrel) and 32 (Chameleon and Actor). The number of attention heads for the hidden layer is 8 (Cora, Citeseer and Pubmed) and 16 (Chameleon, Squirrel and Actor). For all datasets, the number of attention heads in the output layer is 1 and the residual connections are employed.

4.4 Results

The results of our comparative evaluation experiments on non-homophilic and homophilic datasets are summarized in Tables 2 and 3, respectively.

Table 2. The proposed positional embedding aware models outperforms their corresponding baselines in terms of accuracy on non-homophilic datasets and GAT-POS shows significantly better performance compared to Geom-GCN variations.

	Non-homophilic datasets		
Method	Chameleon	Squirrel	Actor
GCN	$65.22 \pm 2.22\%$	$45.44 \pm 1.27\%$	$28.30 \pm 0.73\%$
GAT	$63.88 \pm 2.42\%$	$41.19 \pm 3.38\%$	$28.49 \pm 1.06\%$
Geom-GCN-I	$57.35 \pm 1.85\%$	$31.92 \pm 1.04\%$	$29.14 \pm 1.15\%$
Geom-GCN-P	$60.68 \pm 1.97\%$	$35.39 \pm 1.21\%$	$31.92 \pm 0.95\%$
Geom-GCN-S	$57.89 \pm 1.65\%$	$35.74 \pm 1.47\%$	$30.12 \pm 0.92\%$
Geom-GCN-I-g	$61.97 \pm 2.01\%$	$39.91 \pm 1.77\%$	$32.98 \pm 0.78\%$
Geom-GCN-P-g	$60.31 \pm 2.07\%$	$37.14 \pm 1.19\%$	$31.60 \pm 1.06\%$
Geom-GCN-S-g	$63.86 \pm 1.78\%$	$42.73 \pm 2.16\%$	$31.96 \pm 1.97\%$
GAT-POS (ours)	$\mathbf{67.76 \pm 2.54\%}$	$\mathbf{52.90 \pm 1.55\%}$	$\mathbf{34.89 \pm 1.38\%}$

Non-homophilic Graph Datasets. The experimental results demonstrate that our framework reaches better performance than baselines on all non-homophilic graphs. Compared to GATs, GAT-POS achieves a remarkable improvement on the three non-homophilic datasets. Our proposed framework also reaches a better performance compared to GCNs and Geom-GCNs. Unexpectedly, Geom-GCNs does not show significant improvement compared to standard GNNs on undirected graphs. This is against the previous experimental results on directed graphs. One possible reason is that the Geom-GCNs' extended neighborhood scheme cannot include extra useful neighbors in the aggregation but rather introduces redundant operations and parameters to train on undirected graphs. This is also demonstrated by the better results of the instantiations of Geom-GCNs with only spatial neighborhoods.

Homophilic Graph Datasets. Recall that the evaluation of homophilic datasets is only performed as a sanity check, since GAT-POS has been tailored to exploit structural and positional information essential to non-homophilic tasks. As shown in the table, our proposed method reaches comparable results to standard GNNs. These observations are in line with our expectations on homophilic datasets since merely learning underlying group-invariances can already lead to good performance on graph datasets with high homophily.

Table 3. The proposed positional embedding aware models are on par with their corresponding baselines in terms of accuracy on homophilic datasets; in particular, GAT-POS's results are comparable to GATs.

	Homophilic datasets		
Method	Cora	Citeseer	Pubmed
GCN	85.67 ± 0.94%	73.28 ± 1.37%	88.14 ± 0.32%
GAT	**87.06 ± 0.98%**	74.79 ± 1.89%	87.51 ± 0.43%
Geom-GCN-I	84.79 ± 2.04%	78.84 ± 1.51%	89.73 ± 0.54%
Geom-GCN-P	84.68 ± 1.59%	73.77 ± 1.59%	88.15 ± 0.57%
Geom-GCN-S	85.25 ± 1.46%	74.42 ± 2.52%	84.80 ± 0.62%
Geom-GCN-I-g	85.81 ± 1.50%	**80.05 ± 1.59%**	**92.56 ± 0.33%**
Geom-GCN-P-g	86.48 ± 1.49%	75.74 ± 1.58%	88.49 ± 0.56%
Geom-GCN-S-g	86.50 ± 1.43%	75.91 ± 2.26%	88.55 ± 0.53%
GAT-POS (ours)	86.61 ± 1.13%	73.81 ± 1.27%	87.56 ± 0.48%

4.5 Ablation Study

In this section, we present an ablation study to understand the contributions of the join-training scheme. To that end, we include a variant of GAT-POS without joint-training. This variant pretrains the positional embedding model with the same unsupervised task and then freezes the learnt embeddings while training the enhanced GATs with the supervised objective. Our proposed implementation of the enhanced GAT is inspired the work applying attention mechanism on images [32]. Thus, we also introduce a variant following the architecture of Transformer from natural language processing domain, termed GAT-POS-Transformer. The difference between both architectures lies in that GAT-POS only considers the positional embeddings in the computation of attention coefficients but not in the neighborhood aggregation, while GAT-POS-Transformer directly injects the positional embeddings into the node features before feeding them to the attention module. Note that, GAT-POS-Transformer utilizes the architecture of the standard GATs since the input features have been enhanced by the positional embeddings prior to the attention computation. The results of the comparison are summarized in Table 4.

Table 4. The ablation study of GAT-POS

	GAT-POS		GAT-POS-transformer	
Joint-Training	✓	✗	✓	✗
Chameleon	**67.76 ± 2.54%**	65.75 ± 1.81%	**65.55 ± 2.38%**	65.42 ± 2.13%
Squirrel	**52.90 ± 1.55%**	50.63 ± 1.29%	**51.62 ± 1.84%**	50.79 ± 1.35%
Actor	34.89 ± 1.38%	**34.95 ± 0.95%**	**34.97 ± 1.27%**	34.66 ± 1.17%

On Chameleon and Squirrel, both instantiations of GAT-POS with joint-training reach better results than without joint-training on average. However, it is worth noting that the joint-training also leads to relatively larger standard deviations. On Actor datasets, the difference of performances between the instantiations with and without joint-training is not notable. This might be because, on Actor dataset, the positional embedding is probably less affected by the supervised signals. Overall, enabling joint training of the main model and the positional embedding model is beneficial. Moreover, it is worth noting that the variants of GAT-POS without joint-training still outperform Geom-GCNs by a notable margin, which also highlights the advantage of having an attention mechanism incorporate semantic and structural information, compared to the weighting functions in Geom-GCNs, which only considers the structural information.

In the comparison between GAT-POS and GAT-POS-Transformer, GAT-POS reaches slightly better performance than GAT-POS-Transformer on average, but the difference is not significant. A deeper research on the architecture design of the enhanced GATs will be left for the future work.

5 Conclusion

We have presented a framework to enhance the GAT models with a positional embedding to explicitly leverage both the semantic information of each node and the structural information contained in the graph. In particular, we extended the standard GAT formulation by adding a positional embedding model, predictive of the graph context, and connecting it to the enhanced GAT model through the proposed attentional layer. Although we focused on extending the original GAT formulation, our proposed framework is compatible with most current graph deep learning based models, which leverage graph attentional layers. Moreover, it is worth mentioning that this framework is agnostic to the choice of positional embedding models as well as the particular setup of the graph attentional layers.

Experiments on a number of graph-structured datasets, especially those with more complicated structures and edges joining nodes beyond simple feature similarity, suggest that this framework can effectively enhance GATs by leveraging both graph structure and node attributes for node classification tasks, resulting in increased performances when compared to baselines as well as recently introduced methods. Finally, through an ablation study, we have further emphasized the benefits of our proposed framework by showing that the performance improvements do not only come from the joint training of both parts of the model but rather from endowing GATs with node positional information.

There are several potential improvements to our proposed framework that could be addressed as future work. One is about improving the generalization ability of GAT-POS in the inductive setting. Even though GAT-POS can support the inductive setting by utilizing inductive embedding methods such as GraphSAGE, the bottleneck is still on the embedding methods, especially when the input node features are homogeneous. Another potential research direction is asynchronous training on supervised and unsupervised tasks with the overall architecture, which might be essential when learning on super large-scale graph.

References

1. Bahdanau, D., Cho, K., Bengio, Y.: Neural machine translation by jointly learning to align and translate. In: ICLR (2015)
2. Bruna, J., Zaremba, W., Szlam, A., Lecun, Y.: Spectral networks and locally connected networks on graphs. In: International Conference on Learning Representations (ICLR2014), CBLS, April 2014 (2014)
3. Chollet, F.: Xception: deep learning with depthwise separable convolutions. In: Proceedings of the IEEE Conference on Computer Vision and Pattern Recognition, pp. 1251–1258 (2017)
4. Cordonnier, J.B., Loukas, A., Jaggi, M.: On the relationship between self-attention and convolutional layers (2020)
5. Dai, Z., Yang, Z., Yang, Y., Carbonell, J.G., Le, Q., Salakhutdinov, R.: Transformer-xl: attentive language models beyond a fixed-length context. In: Proceedings of the 57th Annual Meeting of the Association for Computational Linguistics, pp. 2978–2988 (2019)

6. Defferrard, M., Bresson, X., Vandergheynst, P.: Convolutional neural networks on graphs with fast localized spectral filtering. In: Advances in Neural Information Processing Systems, pp. 3844–3852 (2016)
7. Gao, H., Ji, S.: Graph u-nets. In: International Conference on Machine Learning, pp. 2083–2092 (2019)
8. Hamilton, W., Ying, Z., Leskovec, J.: Inductive representation learning on large graphs. In: Advances in Neural Information Processing Systems, pp. 1024–1034 (2017)
9. Kingma, D.P., Ba, J.: Adam: a method for stochastic optimization. arXiv preprint arXiv:1412.6980 (2014)
10. Kipf, T.N., Welling, M.: Semi-supervised classification with graph convolutional networks. In: ICLR (Poster). OpenReview.net (2017)
11. Lanczos, C.: An iteration method for the solution of the eigenvalue problem of linear differential and integral operators. United States Governm. Press Office Los Angeles, CA (1950)
12. Lee, J., Lee, Y., Kim, J., Kosiorek, A., Choi, S., Teh, Y.W.: Set transformer: a framework for attention-based permutation-invariant neural networks. In: International Conference on Machine Learning, pp. 3744–3753 (2019)
13. Li, P., Wang, Y., Wang, H., Leskovec, J.: Distance encoding-design provably more powerful GNNs for structural representation learning. arXiv preprint arXiv:2009.00142 (2020)
14. Li, Y., Tarlow, D., Brockschmidt, M., Zemel, R.: Gated graph sequence neural networks. arXiv preprint arXiv:1511.05493 (2015)
15. Liao, R., Zhao, Z., Urtasun, R., Zemel, R.S.: LanczosNet: multi-scale deep graph convolutional networks. In: 7th International Conference on Learning Representations, ICLR 2019 (2019)
16. Luong, M.T., Pham, H., Manning, C.D.: Effective approaches to attention-based neural machine translation. In: Proceedings of the 2015 Conference on Empirical Methods in Natural Language Processing, pp. 1412–1421 (2015)
17. Mikolov, T., Sutskever, I., Chen, K., Corrado, G.S., Dean, J.: Distributed representations of words and phrases and their compositionality. In: Advances in Neural Information Processing Systems, pp. 3111–3119 (2013)
18. Monti, F., Boscaini, D., Masci, J., Rodola, E., Svoboda, J., Bronstein, M.M.: Geometric deep learning on graphs and manifolds using mixture model cnns. In: Proceedings of the IEEE Conference on Computer Vision and Pattern Recognition, pp. 5115–5124 (2017)
19. Pei, H., Wei, B., Chang, K.C.C., Lei, Y., Yang, B.: Geom-GCN: geometric graph convolutional networks. arXiv preprint arXiv:2002.05287 (2020)
20. Scarselli, F., Gori, M., Tsoi, A.C., Hagenbuchner, M., Monfardini, G.: The graph neural network model. IEEE Trans. Neural Netw. **20**(1), 61–80 (2008)
21. Shaw, P., Uszkoreit, J., Vaswani, A.: Self-attention with relative position representations. In: Proceedings of the 2018 Conference of the North American Chapter of the Association for Computational Linguistics: Human Language Technologies, Volume 2 (Short Papers), pp. 464–468 (2018)
22. Tang, J., Qu, M., Wang, M., Zhang, M., Yan, J., Mei, Q.: Line: large-scale information network embedding. In: Proceedings of the 24th International Conference on World Wide Web, pp. 1067–1077 (2015)
23. Vashishth, S., Yadav, P., Bhandari, M., Talukdar, P.P.: Confidence-based graph convolutional networks for semi-supervised learning. In: AISTATS. Proceedings of Machine Learning Research, vol. 89, pp. 1792–1801. PMLR (2019)

24. Vaswani, A., et al.: Attention is all you need. In: Advances in Neural Information Processing Systems, pp. 5998–6008 (2017)
25. Velickovic, P., Cucurull, G., Casanova, A., Romero, A., Liò, P., Bengio, Y.: Graph attention networks. In: ICLR (Poster). OpenReview.net (2018)
26. Wang, X., Tu, Z., Wang, L., Shi, S.: Self-attention with structural position representations. In: Proceedings of the 2019 Conference on Empirical Methods in Natural Language Processing and the 9th International Joint Conference on Natural Language Processing (EMNLP-IJCNLP), pp. 1403–1409 (2019)
27. Weston, J., Ratle, F., Collobert, R.: Deep learning via semi-supervised embedding. In: Proceedings of the 25th International Conference on Machine Learning, pp. 1168–1175 (2008)
28. Xu, K., et al.: Show, attend and tell: neural image caption generation with visual attention. In: International conference on machine learning, pp. 2048–2057 (2015)
29. Xu, K., Hu, W., Leskovec, J., Jegelka, S.: How powerful are graph neural networks? In: International Conference on Learning Representations (2019). https://openreview.net/forum?id=ryGs6iA5Km
30. Yang, Z., Cohen, W.W., Salakhutdinov, R.: Revisiting semi-supervised learning with graph embeddings. In: Proceedings of the 33rd International Conference on International Conference on Machine Learning-Volume 48, pp. 40–48. JMLR. org (2016)
31. You, J., Ying, R., Leskovec, J.: Position-aware graph neural networks. In: International Conference on Machine Learning, pp. 7134–7143 (2019)
32. Zhao, H., Jia, J., Koltun, V.: Exploring self-attention for image recognition. arXiv preprint arXiv:2004.13621 (2020)

Unified Robust Training for Graph Neural Networks Against Label Noise

Yayong Li[1], Jie Yin[2(✉)], and Ling Chen[1]

[1] Faculty of Engineering and Information Technology, University of Technology Sydney, Ultimo, Australia
Yayong.Li@student.uts.edu.au, Ling.Chen@uts.edu.au

[2] Discipline of Business Analytics, The University of Sydney, Sydney, Australia
jie.yin@sydney.edu.au

Abstract. Graph neural networks (GNNs) have achieved state-of-the-art performance for node classification on graphs. The vast majority of existing works assume that genuine node labels are always provided for training. However, there has been very little research effort on how to improve the robustness of GNNs in the presence of label noise. Learning with label noise has been primarily studied in the context of image classification, but these techniques cannot be directly applied to graph-structured data, due to two major challenges—*label sparsity* and *label dependency*—faced by learning on graphs. In this paper, we propose a new framework, UnionNET, for learning with noisy labels on graphs under a semi-supervised setting. Our approach provides a unified solution for robustly training GNNs and performing label correction simultaneously. The key idea is to perform label aggregation to estimate node-level class probability distributions, which are used to guide sample reweighting and label correction. Compared with existing works, UnionNET has two appealing advantages. First, it requires no extra clean supervision, or explicit estimation of the noise transition matrix. Second, a unified learning framework is proposed to robustly train GNNs in an end-to-end manner. Experimental results show that our proposed approach: (1) is effective in improving model robustness against different types and levels of label noise; (2) yields significant improvements over state-of-the-art baselines.

Keywords: Graph neural networks · Label noise · Label correction

1 Introduction

Nowadays, graph-structured data is being generated across many high-impact applications, ranging from financial fraud detection in transaction networks to gene interaction analysis, from cyber security in computer networks to social network analysis. To ingest rich information on graph data, it is of paramount importance to learn effective node representations that encode both node attributes and graph topology. To this end, graph neural networks (GNNs)

K. Karlapalem et al. (Eds.): PAKDD 2021, LNAI 12712, pp. 528–540, 2021.
https://doi.org/10.1007/978-3-030-75762-5_42

have been proposed, built upon the success of deep neural networks (DNNs) on grid-structured data (e.g., images, etc.). GNNs have abilities to integrate both node attributes and graph topology by recursively aggregating node features across the graph. GNNs have achieved state-of-the-art performance on many graph related tasks, such as node classification or link prediction.

The core of GNNs is to learn neural network primitives that generate node representations by passing, transforming, and aggregating node features from local neighborhoods [3]. As such, nearby nodes would have similar node representations [20]. By generalizing convolutional neural networks to graph data, graph convolutional networks (GCNs) [10] define the convolution operation via a neighborhood aggregation function in the Fourier domain. The convolution of GCNs is a special form of Laplacian smoothing on graphs [11], which mixes the features of a node and its nearby neighbors. However, this smoothing operation can be disrupted when the training data is corrupted with label noise. As the training proceeds, GCNs would completely fit noisy labels, resulting in degraded performance and poor generalization. Hence, one key challenge is how to improve the robustness of GNNs against label noise.

Learning with noisy labels has been extensively studied on image classification. Label noise naturally stems from inter-observer variability, human annotator's error, and errors in crowdsourced annotations [9]. Existing methods attempt to correct the loss function by directly estimating a noise transition matrix [15,19], or by adding extra layers to model the noise transition matrix [4,17]. However, it is difficult to accurately estimate the noise transition matrix particularly with a large number of classes. Alternative methods such as MentorNet [8] and Co-teaching [6] seek to separate clean samples from noisy samples, and use only the most likely clean samples to update model training. Other methods [2,16] reweight each sample in the gradient update of the loss function, according to model's predicted probabilities. However, they require a large number of labeled samples or an extra clean set for training. Otherwise, reweighting would be unreliable and result in poor performance.

The aforementioned learning techniques, however, cannot be directly applied to tackle label noise on graphs. This is attributed to two significant challenges. (1) **Label sparsity**: graphs with inter-connected nodes are arguably harder to label than individual images. Very often, graphs are sparsely labeled, with only a small set of labeled nodes provided for training. Hence, we cannot simply drop "bad nodes" with corrupted labels like previous methods using "small-loss trick" [6,8]. (2) **Label dependency**: graph nodes exhibit strong label dependency, so nodes with high structural proximity (directly or indirectly connected) tend to have a similar label. This presses a strong need to fully exploit graph topology and sparse node labels when training a robust model against label noise.

To tackle these challenges, we propose a novel approach for robustly learning GNN models against noisy labels under semi-supervised settings. Our approach provides a <u>uni</u>fied r<u>o</u>bust trai<u>n</u>ing framework for graph neural <u>net</u>works (UnionNET) that performs sample reweighting and label correction simultaneously. The core idea is twofold: (1) leverage random walks to perform label aggregation among nodes with structural proximity. (2) estimate node-level class distribution to guide sample reweighting and label correction. Intuitively, noisy labels could cause disordered predictions around context nodes, thus its derived node class distribution could in turn reflect the reliability of given labels. This provides an effective way to assess the reliability of given labels, guided by which sample reweighting and label correction are expected to weaken unreliable supervision and encourage label smoothing around context nodes. We verify the effectiveness of our proposed approach through experiments and ablation studies on real-world networks, demonstrating its superiority over competitive baselines.

2 Related Work

2.1 Learning with Noisy Labels

Learning with noisy labels has been widely studied in the context of image classification. The first line of research focuses on correcting the loss function, by directly estimating the noise transition matrix between noisy labels and ground true labels [15,19], or adding an extra softmax layer to estimate the noise transition matrix [4,17]. However, it is non-trivial to estimate the noise transition matrix accurately. [22] used the negative Box-Cox transformation to improve the robustness of standard cross entropy loss but with worse converging capacity. The second line of approaches seek to separate clean samples from noisy ones, and use only the most likely clean samples to guide network training. MentorNet [8] pre-trains an extra network on a clean set to select clean samples. Co-teaching [6] trains two peer networks to select small-loss samples to train each other. Decoupling [13] updates two networks using only samples with which they disagree. In our setting with very few labeled nodes, we cannot simply drop "bad nodes" as they are still useful to infer the labels of nearby nodes. The third category takes a reweighting approach. [2] utilized a two-component Beta Mixture Model to estimate the probability of a sample being mislabeled, which is used to reweight the sample in the gradient update. It was further improved by combining with *mixup augmentation* [21]. [16] proposed a meta-learning algorithm that allowed the network to put more weights on the samples with the closest gradient directions with the clean data. Unlike these reweighting methods that rely on the predicted probabilities, our method assigns weights to each node by leveraging topology structure, which is less prone to label noise. Several other methods are concerned with the problem of label correction. [7] chose class prototypes based on sample distance density to correct labels, incurring significant computational overhead. [18] proposed a self-training approach to correct the labels. However, this method discards the original given labels, leading to degraded performance with high noise rates. Our work integrates sample reweighting with label correction, yielding remarkable gains with high noise rates.

2.2 Graph Neural Networks

GNNs have emerged as a new class of deep learning models on graphs. Various types of GNNs, such as GCN [10], graph attention network (GAT) [20], GraphSAGE [5], are proposed in recent years. These models have shown competitive results on node classification, assuming that genuine node labels are provided for training purposes. To date, there has been little research work on robustly training a GNN against label noise. [14] studied the problem of learning GNNs with symmetric label noise. This method adopts a backward loss correction [15] for graph classification. [1] analyzed the robustness of traditional collective node classification methods on graphs (such as label propagation) towards random label noise, but it did not propose new solutions to tackle this problem. To the best of our knowledge, our work is the first to study the problem of learning robust GNNs for semi-supervised node classification on graphs with both symmetric and asymmetric label noise. Our method provides a unified learning framework and does not require explicit estimation of the noise transition matrix.

3 Problem Definition

Given an undirected graph $G = \{\mathcal{V}, \mathcal{E}, \mathbf{X}\}$, where $\mathcal{V}$ denotes a set of n nodes, and $\mathcal{E}$ denotes a set of edges connecting nodes. $\mathbf{X} = [\mathbf{x}_1, \mathbf{x}_2, \ldots, \mathbf{x}_n]^T \in \mathcal{R}^{n \times d}$ denotes the node feature matrix, where $\mathbf{x}_i \in \mathcal{R}^d$ is d-dimensional feature vector of node v_i. Let $A \in \mathcal{R}^{n \times n}$ denote the adjacent matrix.

We consider semi-supervised node classification, where only a small fraction of nodes are labeled. Let $\mathcal{L} = \{(\mathbf{x}_i, \mathbf{y}_i)\}_{i=1}^{|L|}$ denote the set of labeled nodes, where $\mathbf{x}_i$ is feature vector of node v_i, and $\mathbf{y}_i = \{y_{i1}, y_{i2}, \ldots, y_{im}\}$ is the one-hot encoding of node v_i's class label, with $y_{ij} \in \{0, 1\}$ and m being the number of classes. The rest of nodes belong to the unlabeled set $\mathcal{U}$. Under the GNN learning framework, the aim is to learn a representation $\mathbf{h}_{\mathbf{x}_i}$ for each node v_i such that its class label can be correctly predicted by $f(\mathbf{h}_{\mathbf{x}_i})$. For node classification, the standard cross entropy loss is used as the objective function:

$$\mathcal{J}(f(\mathbf{h}_{\mathbf{x}}), \mathbf{y}) = -\sum_{i \in |L|} \sum_{j \in m} \mathbf{y}_{ij} \log(f(\mathbf{h}_{\mathbf{x}_i})_j). \tag{1}$$

However, when class labels in $\mathcal{L}$ are corrupted with label noise, the standard cross entropy would cause the GNN training to overfit incorrect labels, and in turn lead to degraded classification performance. Therefore, in our work, we aim to train a robust GNN model that is less sensitive to label noise.

Formally, given a small set of noisy labeled nodes $\mathcal{L}$, we aim to: (1) learn node representations $\mathbf{h}$ for all nodes $\mathcal{V}$, and (2) learn a model $f(\mathbf{h})$ to predict the labels of unlabeled nodes in $\mathcal{U}$ with maximum classification performance.

4 The UnionNET Learning Framework

To effectively tackle label noise on graphs, one desirable solution should consider the following key aspects. First, since only a small set of labeled nodes are available for training, we cannot simply drop "bad nodes" using "small-loss trick" [6,8]. Second, graph nodes that share similar structural context exhibit label dependency. Thus, we propose a unified framework, UnionNET, for robustly training a GNN and performing label correction, as shown in Fig. 1.

Taking a given graph as input, a GNN is first applied to learn node representations and generate the predicted label for each node. Then, label information is aggregated to estimate a class probability distribution per node. This aggregation is operated on a support set constructed by collecting context nodes with high structural proximity. According to node-level class probability distributions, our algorithm generates label weights and corrected labels for each labeled node. Those corrected labels generated from the support set could potentially provide extra "correct" supervision. Taken together, both given labels reweighted by label weights and corrected labels are used to update model parameters.

4.1 Label Aggregation

On graphs, it is well studied that nodes with high structural proximity tend to have the same labels [12,23]. The supervision from noisy labels however disrupt such label smoothness around context nodes. Nevertheless, their smoothness degree could provide a reference to assess the reliability of given labels. Hence, we design a label aggregator that aggregates label information for each labeled node from its context nodes to estimate its class probability distribution. Specifically, we perform random walks to collect context nodes with higher-order proximity. For each labeled node $\hat{\mathbf{x}} \in \mathcal{L}$, called *anchor node*, we construct a *support set* of

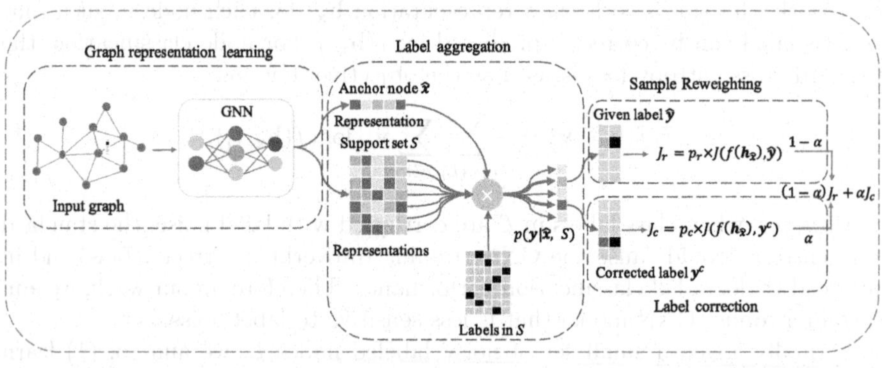

Fig. 1. Overview of the UnionNET Framework. The key idea is to infer the reliability of the given labels through estimating node-level class probability distributions via label aggregation. Based on this, the corresponding label weights and corrected labels are obtained to update model parameters during training.

size k, denoted as $S = \{(\mathbf{x}_i, \mathbf{y}_i)|\hat{\mathbf{x}}\}^k$, where $\mathbf{x}_i$ is the supportive node in S and $\mathbf{y}_i$ is one-hot encoding of $\mathbf{x}_i$'s class label. During a random walk, if node $\mathbf{x}_i \in \mathcal{L}$, the given label $\mathbf{y}_i$ is collected in S. Otherwise, the predicted label is used.

Given anchor node $\hat{\mathbf{x}}$ and its support set $\mathcal{S}$, we derive a node-level class probability distribution $P(\mathbf{y}|\hat{\mathbf{x}}, S)$ over m classes. It signifies the probabilities of the anchor node belonging to m classes in reference of its support set. Particularly, we specify a non-parametric attention mechanism given by,

$$P(\mathbf{y}|\hat{\mathbf{x}}, S) = \sum_{\mathbf{x}_i \in S} \mathcal{A}(\hat{\mathbf{x}}, \mathbf{x}_i)\mathbf{y}_i = \sum_{\mathbf{x}_i \in S} \frac{\exp(\mathbf{h}_{\mathbf{x}_i}^T \mathbf{h}_{\hat{\mathbf{x}}})}{\sum_{\mathbf{x}_j \in S} \exp(\mathbf{h}_{\mathbf{x}_j}^T \mathbf{h}_{\hat{\mathbf{x}}})} \mathbf{y}_i. \tag{2}$$

Here, the probability of the anchor node belonging to each class is calculated according to its proximity with nearby nodes in the support set. We define the proximity as the inner product in the embedding space, and apply softmax to measure the contribution made by each label in the support set to estimating the anchor node' class probability distribution. In the support set, if a node has a higher similarity with the anchor node (i.e., higher inner product), its label would contribute more to $P(\mathbf{y}|\hat{\mathbf{x}}, S)$, and vice verse. This simple yet effective mechanism estimates a class probability distribution for each node, which is used to guide sample reweighting and label correction.

4.2 Sample Reweighting

For GNNs, the standard cross entropy loss implicitly puts more emphasis on the samples for which the predicted labels disagree with the provided labels during gradient update. This mechanism enables faster convergence and better fitting to the training data. However, if there exist corrupted labels in the training set, this implicit weighting scheme would conversely push the model to overfit noisy labels, leading to degraded performance [22]. To mitigate this, we devise a reweighting scheme for each node according to the reliability of its given label, so that the loss of reliable labels could contribute more during gradient update.

Specifically, we define the reweighting score of anchor node $\hat{\mathbf{x}}$ as:

$$p_r(\hat{\mathbf{y}}|\hat{\mathbf{x}}, S) = \sum_{\mathbf{x}_i \in S, \mathbf{y}_i = \hat{\mathbf{y}}} \frac{\exp(\mathbf{h}_{\mathbf{x}_i}^T \mathbf{h}_{\hat{\mathbf{x}}})}{\sum_{\mathbf{x}_j \in S} \exp(\mathbf{h}_{\mathbf{x}_j}^T \mathbf{h}_{\hat{\mathbf{x}}})} \mathbf{y}_i. \tag{3}$$

The loss function for the labeled nodes is thus defined as:

$$\mathcal{J}_r = -\sum_{\hat{\mathbf{x}} \in \mathcal{L}} p_r(\hat{\mathbf{y}}|\hat{\mathbf{x}}, S) \times \hat{\mathbf{y}} \log(f(\mathbf{h}_{\hat{\mathbf{x}}})), \tag{4}$$

where $p_r(\hat{\mathbf{y}}|\hat{\mathbf{x}}, S)$ is the weight imposed on each labeled node $\hat{\mathbf{x}}$ according to the aggregated label information. If the given label $\hat{\mathbf{y}}$ is highly consistent with nearby labels, its gradient would be back-propagated as it is. Otherwise, it would be penalized by the weight during back-propagation.

4.3 Label Correction

The reweighting method reduces the sensitivity of the standard cross entropy to noisy labels, and boosts the robustness of the model. As labeled nodes are limited for training, we also augment the set of labeled nodes by correcting noisy labels. Accordingly, we define the label correction loss as

$$\mathcal{J}_c = -\sum_{\hat{\mathbf{x}}\in\mathcal{L}} p_c(\mathbf{y}^c|\hat{\mathbf{x}}, S) \times \mathbf{y}^c \log(f(\mathbf{h}_{\hat{\mathbf{x}}})), \tag{5}$$

$$p_c(\mathbf{y}^c|\hat{\mathbf{x}}, S) = \max_{\mathbf{y}_i} P(\mathbf{y}_i|\hat{\mathbf{x}}, S) = \max_{\mathbf{y}_i} \sum_{\mathbf{x}_i\in S} \frac{\exp(\mathbf{h}_{\mathbf{x}_i}^T \mathbf{h}_{\hat{\mathbf{x}}})}{\sum_{\mathbf{x}_j\in S}\exp(\mathbf{h}_{\mathbf{x}_j}^T \mathbf{h}_{\hat{\mathbf{x}}})}\mathbf{y}_i. \tag{6}$$

This provides additional supervision for $\hat{\mathbf{x}}$ with the corrected label $\mathbf{y}^c$, encouraging it to have the same label with the most consistent one in its support set. This approach aggregates labels from context nodes via a linear combination based on their similarity in the embedding space. It thus helps diminish the gradient update of corrupted labels, and boosts the supervision from consistent labels.

However, in the presence of extreme label noise, this approach would produce biased correction that deviates far away from its original prior distribution over the training data. This bias could exacerbate the overfitting problem caused by noisy labels. To overcome this, we employ a KL-divergence loss between the prior and predicted distributions to push them as close as possible [18]. It is given by:

$$\mathcal{J}_p = \sum_{j=1}^{m} p_j \log\frac{p_j}{\overline{f(\mathbf{h_x})}_j}, \tag{7}$$

Where p_j is the prior probability of class j in $\mathcal{L}$, and $\overline{f(\mathbf{h_x})}_j = \frac{1}{|L|}\sum_{\mathbf{x}\in\mathcal{L}} f(\mathbf{h_x})_j$ is the mean value of predicted probability distribution on the training set.

4.4 Model Training

The training of UnionNET is given in Algorithm 1, which consists of the pre-training phase (Step 1–4) and the training phase (Step 6–11). The pre-training is employed to obtain a parameterized GNN. The pre-trained GNN then generates node representations $\mathbf{h}$, which are used to compute sample weights and corrected labels. After that, model parameters are updated according to the loss function:

$$\mathcal{J}_f = (1-\alpha)\mathcal{J}_r + \alpha\mathcal{J}_c + \beta\mathcal{J}_p. \tag{8}$$

Compared with GNNs with the standard cross entropy loss, the training of UnionNET incurs an extra computational complexity of $\mathcal{O}(|L|ml)$ to estimate node-level class distributions, where $|L|$ is number of labeled nodes, m is number of classes, and l is number of nodes including context nodes in the support set.

Algorithm 1: Robust training for GNNs against label noise

Input: Graph $G = \{\mathcal{V}, \mathcal{E}, \mathbf{X}\}$, node sets $\mathcal{L}, \mathcal{U}$, α, β
Output: label predictions
Initialize network parameters;
for $t = 0; t < epoches; t = t + 1$ **do**
  if $t < start\_epoch$ **then**
    pre-train the network according to Eq. (1);
  else
    Generate node representations $\mathbf{h}_{\mathbf{x}}$;
    Construct support set S for each node $\hat{\mathbf{x}} \in \mathcal{L}$;
    Aggregate labels to produce node-level class distribution $P(\mathbf{y}|\hat{\mathbf{x}}, S)$;
    Compute weight $p_r(\hat{\mathbf{y}}|\hat{\mathbf{x}}, S)$ using Eq. (3);
    Generate corrected label $\mathbf{y}^c$ and its weight $p_c(\mathbf{y}^c|\hat{\mathbf{x}}, S)$ using Eq. (6);
    Update parameters by descending gradient of Eq. (8)
return *Label predictions*

5 Experiments

Datasets and Baselines. Three benchmark datasets are used in our experiments: Cora, Citeseer, and Pubmed[1]. We use the same data split as in [10], with 500 nodes for validation, 1000 nodes for testing, and the remaining for training. Of these training sets, only a small fraction of nodes are labeled (3.6% on Citeseer, 5.2% on Cora, 0.3% on Pubmed) and the rest of nodes are unlabeled. Details about the datasets can be found in [10].

As far as we are concerned, there has not yet been any method exclusively proposed to deal with the label noise problem on GNNs for semi-supervised node classification. We select three strong competing methods from image classification, and adapt them to work with GCN [10] under our setting as baselines.

- **Co-teaching** [6] trains two peer networks and each network selects the samples with small losses to update the other network.
- **Decoupling** [13] also trains two networks, but updates model parameters using only the samples with which two networks disagree.
- **GCE** [22] utilizes a negative Box-Cox transformation as the loss function.

As a general robust training framework, UnionNET can be applied to any semi-supervised GNNs for node classification. Hereby, we instantiate UnionNET with two state-of-the-art GNNs, GCN [10] and GAT [20], denoted as **UnionNET-GCN** and **UnionNET-GAT**, respectively.

Experimental Setup. Due to the fact that there are not yet benchmark graph datasets corrupted with noisy labels, we manually generate noisy labels on public datasets to evaluate our algorithm. We follow commonly used label noise

[1] https://linqs.soe.ucsc.edu/data.

generation methods in the domain of images [6,8]. Given a noise rate r, we generate noisy labels over all classes according to a noise transition matrix $Q^{m \times m}$, where $Q_{ij} = p(\tilde{y} = j|y = i)$ is the probability of clean label $\mathbf{y}$ being flipped to noisy label $\tilde{\mathbf{y}}$. We consider two types of noise: 1). **Symmetric noise**: label i is corrupted to other labels with a uniform random probability, s.t. $Q_{ij} = Q_{ji}$; 2). **Pairflip noise**: mislabeling only occurs between similar classes. For instance, given $r = 0.4$ and $m = 3$, the two types of noise transition matrices are given by

$$Q^{\text{symmetric}} = \begin{bmatrix} 0.6 & 0.2 & 0.2 \\ 0.2 & 0.6 & 0.2 \\ 0.2 & 0.2 & 0.6 \end{bmatrix}; \quad Q^{\text{pairflip}} = \begin{bmatrix} 0.6 & 0.4 & 0. \\ 0. & 0.6 & 0.4 \\ 0.4 & 0. & 0.6 \end{bmatrix}$$

Our experiments follow a transductive setting, where the noise transition matrix is only applied to $\mathcal{L}$, while both validation and test sets are kept clean. For UnionNET-GCN, we apply a two-layer GCN, which has 16 units of hidden layer. The hyper-parameters are set as L2 regularization of $5 * 10^{-4}$, learning rate of 0.01, dropout rate of 0.5. For UnionNET-GAT, we apply a two-layer GAT, with the first layer consisting of 8 attention heads, each computing 8 features. The learning rate is 0.005, dropout rate is 0.6, L2 regularization is $5 * 10^{-4}$.

We set the random walk length as 10 on Cora and Citeseer, and 4 on Pubmed, and the random walk is repeated for 10 times for each node to create the support set. We first pre-train the network, during which only the standard cross entropy are used, i.e. $\mathcal{J}_{pre} = \mathcal{J}(f(\mathbf{h_x}), \mathbf{y})$. After that, it proceeds to the formal training, which uses $\mathcal{J}_f$ in Eq. (8) as the loss function. And α and β are set as 0.5 and 1.0.

5.1 Comparison with State-of-the-Art Methods

Table 1 compares the node classification performance of all methods w.r.t. both the symmetric and asymmetric noise types under various noise rates. The best performer is highlighted by **bold** on each setting. For GCN-based baselines, UnionNET-GCN generally outperforms all baselines by a large margin. Compared with GCN in case of symmetric noise type, UnionNET-GCN achieves an accuracy improvement of 3.4%, 6.3%, 13.1% and 7.1% under the noise rate of 10%, 20%, 40% and 60% on Cora, respectively. Similar improvements can be seen on Citeseer and Pubmed, where the smallest improvement is 2.1% on Pubmed with a noise rate of 10%, and the largest improvement is 8.7% on Citeseer with a noise rate of 40%. In case of asymmetric noise type, UnionNET-GCN has the similar performance. Quantitatively, UnionNET-GCN outperforms GCN by an average of 3.6%, 5.0%, 5.4%, 3.8% on the four noise rates on three datasets.

In most cases, GCE is the second best performer, but its advantage comes at the cost of worse converging capability, leading to sub-optimal performance. Co-teaching and Decoupling do not exhibit robustness towards noisy labels as reported in fully supervised image classification. Their performance drops are expected, as labeled data is further reduced when they prune the training data. This exacerbates the label scarcity problem in our semi-supervised setting.

Table 1. Performance comparison (Micro-F1 score) on node classification

Dataset	Methods	Symmetric label noise				Asymmetric label noise			
		Noise rate (%)							
		10	20	40	60	10	20	30	40
Cora	GCN	0.778	0.732	0.576	0.420	0.768	0.696	0.636	0.517
	Co-teaching	0.775	0.665	0.486	0.249	0.773	0.630	0.542	0.393
	Decoupling	0.738	0.708	0.564	0.436	0.743	0.683	0.574	0.518
	GCE	0.794	0.741	0.621	0.402	0.773	0.714	0.652	0.509
	UnionNET-GCN	**0.812**	**0.795**	**0.707**	**0.491**	**0.801**	**0.771**	**0.710**	**0.584**
	GAT	0.755	0.709	0.566	0.389	0.764	0.683	0.616	0.534
	UnionNET-GAT	0.797	0.784	0.692	0.546	0.774	0.745	0.660	0.540
Citeseer	GCN	0.670	0.634	0.480	0.360	0.667	0.624	0.531	0.501
	Co-teaching	0.673	0.541	0.379	0.273	0.677	0.583	0.472	0.418
	Decoupling	0.588	0.584	0.402	0.348	0.615	0.548	0.537	0.468
	GCE	0.690	0.649	0.542	0.358	0.701	0.633	0.552	0.498
	UnionNET-GCN	**0.701**	**0.673**	**0.567**	**0.401**	**0.706**	**0.667**	**0.587**	**0.521**
	GAT	0.649	0.604	0.475	0.338	0.651	0.599	0.551	0.480
	UnionNET-GAT	0.695	0.667	0.585	0.424	0.697	0.654	0.604	0.512
Pubmed	GCN	0.748	0.672	0.508	0.367	0.739	0.686	0.618	0.528
	Co-teaching	0.769	0.660	0.478	0.345	0.761	0.634	0.576	0.472
	Decoupling	0.650	0.625	0.422	0.334	0.641	0.592	0.428	0.396
	GCE	0.750	0.699	0.561	0.393	0.753	0.696	0.609	0.567
	UnionNET-GCN	**0.769**	**0.725**	**0.588**	**0.409**	**0.776**	**0.719**	**0.649**	**0.556**
	GAT	0.736	0.670	0.525	0.381	0.737	0.657	0.594	0.536
	UnionNET-GAT	0.751	0.726	0.570	0.361	0.758	0.702	0.626	0.552

On three datasets, UnionNET-GAT also surpasses GAT w.r.t. most noise rates. Similar to UnionNET-GCN, UnionNET-GAT generally exhibits greater superiority on higher noise rates. For example, in case of symmetric noise type, UnionNET-GAT outperforms GAT by an average of 3.4%, 6.4%, 9.4% and 7.4% at the four noise rates. Such performance gains validate the generality of UnionNET on improving robustness of different GNN models against noisy labels.

5.2 Ablation Study

We conduct ablation studies to test the effectiveness of different components in UnionNET. Our ablation study is based on GCN, with two ablation versions: 1) **UnionNET-R** with only sample reweighting; 2) **UnionNET-RC** with sample reweighting and label correction. The ablation results are summarized in Table 2. When only reweighting is applied, UnionNET-R consistently exhibits advantages over GCN, though the advantageous margins vary over different noise rates and noise types. When it comes to UnionNET-RC, both sample reweighting and label correction are applied, but, surprisingly, the performance becomes worse than UnionNET-R in some extreme cases with higher noise rates. Therefore, label correction does not guarantee performance gains, whose utility is exerted only with the regularization of the prior distribution loss.

Table 2. Performance comparison of ablation experiments based on GCN

Dataset	Methods	Symmetric label noise				Asymmetric label noise			
		Noise rate (%)							
		10	20	40	60	10	20	30	40
Cora	GCN	0.778	0.732	0.576	0.420	0.768	0.696	0.636	0.517
	UnionNET-R	0.785	0.770	0.659	0.480	0.796	0.709	0.646	0.521
	UnionNET-RC	0.788	0.759	0.626	0.339	0.783	0.703	0.601	0.516
	UnionNET-GCN	**0.812**	**0.795**	**0.707**	**0.491**	**0.801**	**0.771**	**0.710**	**0.584**
Citeseer	GCN	0.670	0.634	0.480	0.360	0.667	0.624	0.531	0.501
	UnionNET-R	0.692	0.643	0.507	0.363	0.699	0.627	0.547	0.484
	UnionNET-RC	0.657	0.645	0.495	0.330	0.660	0.642	0.511	0.431
	UnionNET-GCN	**0.701**	**0.673**	**0.567**	**0.401**	**0.706**	**0.667**	**0.587**	**0.521**
Pubmed	GCN	0.748	0.672	0.508	0.367	0.739	0.686	0.618	0.528
	UnionNET-R	0.766	0.710	0.573	**0.417**	0.759	0.705	0.624	0.560
	UnionNET-RC	**0.770**	0.695	0.573	0.362	0.757	0.650	0.608	0.497
	UnionNET-GCN	0.769	**0.725**	**0.588**	0.409	**0.776**	**0.719**	**0.649**	**0.556**

5.3 Hyper-parameter Sensitivity

We further test the sensitivity of UnionNET-GCN w.r.t. the hyper-parameters (α, β) in Eq. (8) and the random walk length for the support set construction. We report the results on the three datasets at 40% symmetric noise rate in Fig. 2. α controls the trade-off between sample reweighting and label correction. When α is zero, our method is only a reweighting method. When α reaches 1, our method evolves as a self-learning based label correction method, where given labels are replaced with predicted labels after the initial epoches. On Cora and Citeseer, our method achieves the best results at a medium α value. But on Pubmed, its performance improves as α increases, and reaches its best when $\alpha = 1.0$. This is possibly because Pubmed has stronger clustering property with only three classes, enabling the predicted labels to be more reliable for correction. The performance changes w.r.t. β exhibits similar trends on the three datasets, where our method gradually improves its performance as β increases. The random walk length determines the order of proximity the support set could cover. Either too small or too large of the random walk length would impair the reliability of the supportive nodes, and thus undermine performance improvements. Empirically, our method achieves its best at a medium range of random walk lengths.

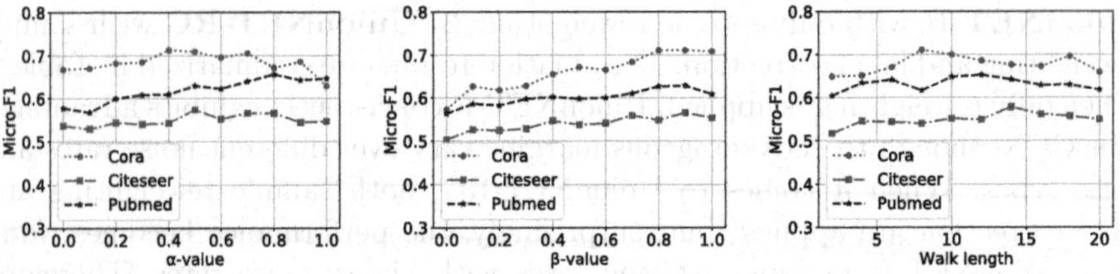

(a) α (b) β (c) Walk Length

Fig. 2. Hyper-parameter sensitivity analysis on α, β, and the random walk length

6 Conclusion

We proposed a novel semi-supervised framework, UnionNET, for learning with noisy labels on graphs. We argued that, existing methods on image classification fail to work on graphs, as they often take a fully supervised approach, and requires extra clean supervision or explicit estimation of the noise transition matrix. Our approach provides a unified solution to robustly training a GNN model and performing label correction simultaneously. UnionNET is a general framework that can be instantiated with any state-of-the-art semi-supervised GNNs to improve model robustness, and it can be trained in an end-to-end manner. Experiments on three real-world datasets demonstrated that our method is effective in improving model robustness w.r.t. different label noise types and rates, and outperform competitive baselines.

Acknowledgement. This work is supported by the USYD-Data61 Collaborative Research Project grant, the Australian Research Council under Grant DP180100966, and the China Scholarship Council under Grant 201806070131.

References

1. de Aquino Afonso, B.K., Berton, L.: Analysis of label noise in graph-based semi-supervised learning. In: SAC, pp. 1127–1134 (2020)
2. Arazo, E., Ortego, D., Albert, P., O'Connor, N.E., McGuinness, K.: Unsupervised label noise modeling and loss correction. In: ICML, pp. 312–321 (2019)
3. Gilmer, J., Schoenholz, S.S., Riley, P.F., Vinyals, O., Dahl, G.E.: Neural message passing for quantum chemistry. In: ICML, pp. 1263–1272 (2017)
4. Goldberger, J., Ben-Reuven, E.: Training deep neural-networks using a noise adaptation layer. In: ICLR (2016)
5. Hamilton, W., Ying, Z., Leskovec, J.: Inductive representation learning on large graphs. In: NeurIPS, pp. 1024–1034 (2017)
6. Han, B., et al.: Co-teaching: robust training of deep neural networks with extremely noisy labels. In: NeurIPS, pp. 8527–8537 (2018)
7. Han, J., Luo, P., Wang, X.: Deep self-learning from noisy labels. In: ICCV, pp. 5138–5147 (2019)
8. Jiang, L., Zhou, Z., Leung, T., Li, L.J., Fei-Fei, L.: MentorNet: learning data-driven curriculum for very deep neural networks on corrupted labels. In: ICML, pp. 2304–2313 (2018)
9. Karimi, D., Dou, H., Warfield, S.K., Gholipour, A.: Deep learning with noisy labels: exploring techniques and remedies in medical image analysis. Med. Image Anal. **65**, 101759 (2020)
10. Kipf, T.N., Welling, M.: Semi-supervised classification with graph convolutional networks. In: ICLR (2017)
11. Li, Q., Han, Z., Wu, X.M.: Deeper insights into graph convolutional networks for semi-supervised learning. In: AAAI, pp. 3538–3545 (2018)
12. Lu, Q., Getoor, L.: Link-based classification. In: ICML, pp. 496–503 (2003)
13. Malach, E., Shalev-Shwartz, S.: Decoupling "when to update" from "how to update". In: NeurIPS, pp. 960–970 (2017)

14. NT, H., Jin, C., Murata, T.: Learning graph neural networks with noisy labels. In: 2nd ICLR Learning from Limited Labeled Data (LLD) Workshop (2019)
15. Patrini, G., Rozza, A., Krishna Menon, A., Nock, R., Qu, L.: Making deep neural networks robust to label noise: a loss correction approach. In: CVPR, pp. 1944–1952 (2017)
16. Ren, M., Zeng, W., Yang, B., Urtasun, R.: Learning to reweight examples for robust deep learning. In: ICML, pp. 4334–4343 (2018)
17. Sukhbaatar, S., Bruna, J., Paluri, M., Bourdev, L., Fergus, R.: Training convolutional networks with noisy labels. In: ICLR (2014)
18. Tanaka, D., Ikami, D., Yamasaki, T., Aizawa, K.: Joint optimization framework for learning with noisy labels. In: CVPR, pp. 5552–5560 (2018)
19. Vahdat, A.: Toward robustness against label noise in training deep discriminative neural networks. In: NeurIPS, pp. 5596–5605 (2017)
20. Veličković, P., Cucurull, G., Casanova, A., Romero, A., Lio, P., Bengio, Y.: Graph attention networks. In: ICLR (2018)
21. Zhang, H., Cisse, M., Dauphin, Y.N., Lopez-Paz, D.: mixup: beyond empirical risk minimization. In: ICLR (2018)
22. Zhang, Z., Sabuncu, M.: Generalized cross entropy loss for training deep neural networks with noisy labels. In: NeurIPS, pp. 8778–8788 (2018)
23. Zhu, X., Ghahramani, Z., Lafferty, J.D.: Semi-supervised learning using gaussian fields and harmonic functions. In: ICML, pp. 912–919 (2003)

Graph InfoClust: Maximizing Coarse-Grain Mutual Information in Graphs

Costas Mavromatis(✉) and George Karypis

Department of Computer Science and Engineering, University of Minnesota, Minneapolis, MN 55455, USA
{mavro016,karypis}@umn.edu

Abstract. This work proposes a new unsupervised (or self-supervised) node representation learning method that aims to leverage the *coarse-grain* information that is available in most graphs. This extends previous attempts that only leverage *fine-grain* information (similarities within local neighborhoods) or *global* graph information (similarities across all nodes). Intuitively, the proposed method identifies nodes that belong to the same clusters and maximizes their mutual information. Thus, coarse-grain (cluster-level) similarities that are shared between nodes are preserved in their representations. The core components of the proposed method are (i) a jointly optimized clustering of nodes during learning and (ii) an Infomax objective term that preserves the mutual information among nodes of the same clusters. Our method is able to outperform competing state-of-art methods in various downstream tasks, such as node classification, link prediction, and node clustering. Experiments show that the average gain is between 0.2% and 6.1%, over the best competing approach, over all tasks. Our code is publicly available at: https://github.com/cmavro/Graph-InfoClust-GIC.

1 Introduction

Graph structured data naturally emerge in various real-world applications. Such examples include social networks, citation networks, and biological networks. The challenge, from a data representation perspective, is to encode the high-dimensional, non-Euclidean information about the graph structure and the attributes associated with the nodes and edges into a low dimensional embedding space. The learned embeddings (a.k.a. representations) can then be used for various tasks, e.g., node classification, link prediction, community detection, and data visualization. In this paper, we focus on *unsupervised* (or self-supervised) representation learning methods that estimate node embeddings without using any labeled data but instead employ various self-supervision approaches. These methods eliminate the need to develop task-specific graph representation models, eliminate the cost of acquiring labeled data, and can lead to better representations by using large unlabeled datasets.

K. Karlapalem et al. (Eds.): PAKDD 2021, LNAI 12712, pp. 541–553, 2021.
https://doi.org/10.1007/978-3-030-75762-5_43

Various approaches have been developed to self-supervise graph representation learning. Many of them optimize the embeddings based on a loss function that ensures that pairs of nearby nodes are closer in the embedding space compared to pairs of distant nodes, e.g., DeepWalk [16] and GraphSAGE [4]. A key difference between such methods is whether they use graph neural network (GNN) encoders to compute the embeddings. These encoders insert an additional inductive bias that nodes share similarities with their neighbors and lead to significant performance gains.

Since GNN encoders already preserve similarities between neighboring nodes, Deep Graph Infomax (DGI) [21] uses a different loss function as self-supervision. This self-supervision encourages each node to be mindful of the global graph properties, in addition to its local neighborhood properties. Specifically, DGI maximizes the mutual information (MI) between the representation of each node (*fine-grain* representations) and the *global* graph representation, which corresponds to the summary of all node representations. DGI has shown to estimate superior node representations and is considered to be among the best unsupervised node representation learning approaches. However, in many graphs, besides their global structure, there is additional structure that can be captured. For example, nodes tend to belong to (multiple) clusters that represent topologically near-by nodes; or some nodes may share similar structural roles with topologically distant nodes. In such cases, methods that simultaneously preserve these *coarse-grain* interactions allow node representations to encode richer structural information.

Motivated by this observation, we developed *Graph InfoClust* (GIC), an unsupervised representation learning method that extracts coarse-grain information by identifying nodes that belong to the same clusters. Then, GIC learns node representations by maximizing the mutual information of nodes and their cluster-derived summaries, which preserves the coarse-grain information in the embedding space. Furthermore, since the node representations can help identify better clusters (both w.r.t. communities and also their role), we do not rely on an a priori clustering solution. Instead, the cluster-level summaries are obtained and jointly optimized by a differentiable K-means clustering [23] in an end-to-end fashion. We evaluated GIC on seven standard datasets using node classification, link prediction, and clustering as the downstream tasks. Our experiments show that in eleven out of thirteen dataset-task combinations, GIC performs better than the best competing approach and its average improvement over DGI is 0.9, 2.6, and 15.5% points for node classification, link prediction, and clustering, respectively. These results demonstrate that by leveraging cluster summaries, GIC is able to improve the quality of the estimated representations.

2 Notation, Definitions, and Problem Statement

Let $G := \{\mathcal{V}, \mathcal{E}\}$ denote a graph with N nodes and $|\mathcal{E}|$ edges, where $\mathcal{V} := \{v_1, \ldots, v_N\}$ is the set of nodes and $\mathcal{E}$ is the set of edges. The connectivity is represented with the adjacency matrix $\mathbf{A} \in \mathbb{R}^{N \times N}$, with $\mathbf{A}_{i,j} = 1$ if $(v_i, v_j) \in \mathcal{E}$ and

$\mathbf{A}_{i,j} = 0$, otherwise. Let $\mathbf{x}_i \in \mathbb{R}^F$ be the feature vector associated with node v_i and $\mathbf{X} \in \mathbb{R}^{N\times F}$ be the matrix that stores these features across all nodes. Here, we denote vectors by bold lower-case letters and matrices by bold upper-case letters. We also use the terms *representation* and *embedding* interchangeably.

Let $\mathbf{H} := [\mathbf{h}_1, \ldots, \mathbf{h}_N] \in \mathbb{R}^{N\times D}$ be the *node embedding matrix* of G, where $\mathbf{h}_i \in \mathbb{R}^D$ is the *node embedding vector* for v_i The goal of *node representation learning* is to learn a function (encoder), $f : \mathbb{R}^{N\times N} \times \mathbb{R}^{N\times F} \to \mathbb{R}^{N\times D}$, such that $\mathbf{H} = f(\mathbf{A}, \mathbf{X})$. Once learned, $\mathbf{H}$ can be used as an input feature matrix to downstream tasks such as node classification, link prediction, and clustering.

3 Graph InfoClust (GIC)

3.1 Motivation and Overview

Preliminaries, DGI Framework. DGI employs a loss function that encourages node embeddings to contain information about the global graph properties. It does so by training the GNN-encoder to maximize the mutual information (MI) between the representation of each node $\mathbf{h}_i$ (fine-grain representation) and a summary representation $\mathbf{s} \in \mathbb{R}^D$ of the entire graph (global summary). Maximizing the precise value of mutual information is intractable; instead, DGI maximizes the Jensen-Shannon MI estimator that maximizes MI's lower bound [6]. This estimator acts like a binary cross-entropy (BCE) loss, whose objective maximizes the expected log-ratio of the samples from the joint distribution (positive examples) and the product of marginal distributions (negative examples). The positive examples are pairings of $\mathbf{s}$ with $\mathbf{h}_i$ of the real input graph $\mathcal{G} := (\mathbf{A}, \mathbf{X})$, but the negatives are pairings of $\mathbf{s}$ with $\tilde{\mathbf{h}}_i$, which are obtained from a fake/corrupted input graph $\tilde{\mathcal{G}} := (\tilde{\mathbf{A}}, \tilde{\mathbf{X}})$ (assuming the same number of nodes). The graph summary $\mathbf{s}$ is obtained by averaging all nodes' representations followed by the logistic sigmoid nonlinearity $\sigma(\cdot)$, as $\mathbf{s} = \sigma\left(\frac{1}{N}\sum_{k=1}^{N}\mathbf{h}_i\right)$. A discriminator $d : \mathbb{R}^D \times \mathbb{R}^D \to \mathbb{R}$ is used to assign higher scores to the positive than the negative examples. The Jensen-Shannon-based BCE objective is expressed as

$$\mathcal{L}_1 = \sum_{i=1}^{N} \mathbb{E}_{(\mathbf{X},\mathbf{A})}\Big[\log d(\mathbf{h}_i, \mathbf{s})\Big] + \sum_{i=1}^{N} \mathbb{E}_{(\tilde{\mathbf{X}},\tilde{\mathbf{A}})}\Big[\log\left(1 - d(\tilde{\mathbf{h}}_i, \mathbf{s})\right)\Big], \tag{1}$$

which corresponds to a noise-contrastive objective between positive and negative examples. This type of objective has been proven to effectively maximize mutual information between $\mathbf{h}_i$ and $\mathbf{s}$ [6,21].

GIC Framework. Optimizing the node representations based on Eq. (1) encourages the encoder to preferentially encode information that is shared across all nodes. Such approach ensures that the computed representations do not encode the noise that may exist in some neighborhoods—the noise will be very different from the global summary. However, for exactly the same reason, it will also fail to encode information that is different from the global summary but is over-represented in small parts of the graph (e.g., the neighborhoods of some

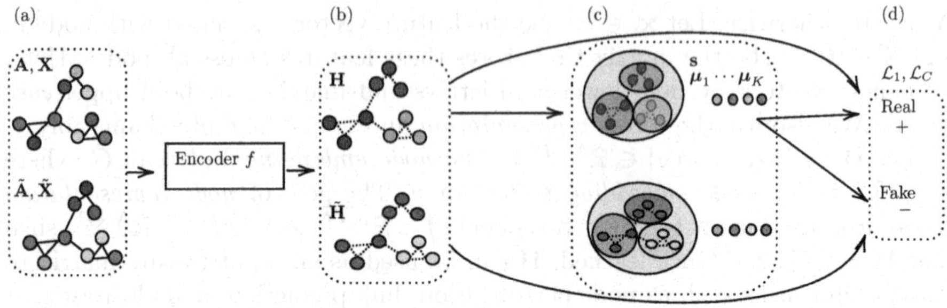

Fig. 1. GIC's framework. (a) A fake input is created based on the real one. (b) Embeddings are computed for both inputs with a GNN-encoder. (c) The global graph summary and the coarse-grain summaries are computed. (d) The goal is to discriminate between real and fake samples based on the computed summaries.

nodes). Capturing such information can be important for downstream tasks like link prediction and clustering.

Graph InfoClust (GIC) is specifically designed to address this problem. It postulates that the nodes belong to multiple clusters and learns node representations by simultaneously maximizing the MI between a node's (fine-grain) representation with that of the global graph summary and a coarse-grain summary derived from the clusters that it belongs to. Since this approach takes advantages of multiple entities within the graph, it leverages significantly more information that is present across the nodes and across different levels of the graph.

As Fig. 1 illustrates, GIC uses a GNN-based encoder to compute fine-grain node representations, which are then used to derive (i) the global summary, and (ii) the cluster-based summaries via a differentiable k-means clustering algorithm. These summaries are used in a contrastive-loss setting to define whether a node representation comes from the real or a fake graph. Specifically, GIC introduces a coarse-grain $\mathcal{L}_C$ loss to discriminate between real and fake samples based on the cluster-based summaries and uses the global-level $\mathcal{L}_1$ loss for the global summary, accordingly. GIC's overall objective is given by

$$\mathcal{L} = \alpha\mathcal{L}_1 + (1-\alpha)\mathcal{L}_C, \tag{2}$$

where $\alpha \in [0,1]$ controls the relative importance of each component. The optimization of this objective leads to the maximization of the mutual information that is present in both fine-grain and coarse-grain levels as well as the global level of the graph.

3.2 Coarse-Grain Loss

Suppose we have computed a coarse-grain/cluster-derived summary $\mathbf{z}_i \in \mathbb{R}^D$ that is associated with v_i (described later in Sect. 3.3). Then, we can simply maximize the mutual information between $\mathbf{z}_i$ and $\mathbf{h}_i$ in a similar manner with DGI. The new coarse-grain-based objective term is

$$\mathcal{L}_C = \sum_{i=1}^{N} \mathbb{E}_{(\mathbf{X},\mathbf{A})}\Big[\log g(\mathbf{h}_i, \mathbf{z}_i)\Big] + \sum_{i=1}^{N} \mathbb{E}_{(\tilde{\mathbf{X}},\tilde{\mathbf{A}})}\Big[\log\big(1 - g(\tilde{\mathbf{h}}_i, \mathbf{z}_i)\big)\Big], \tag{3}$$

where positive examples are pairings of $\mathbf{h}_i$ with $\mathbf{z}_i$ and negatives are pairings $\tilde{\mathbf{h}}_i$ with $\mathbf{z}_i$. A discriminator $g : \mathbb{R}^D \times \mathbb{R}^D \rightarrow \mathbb{R}$ is used to facilitate the optimization, as before, by assigning higher scores to the positive examples.

3.3 Coarse-Grain Summaries

Coarse-grain summaries summarize the information that is present in the corresponding (coarse-grain) clusters of nodes within the graph. Because nodes may belong to multiple clusters, and thus, may be present in multiple coarse-grain groups of the graph, it is advantageous to perform a soft-assignment to these clusters. Then, the coarse-grain summary $\mathbf{z}_i$ can be computed as a weighted average of the cluster centroids that node v_i belongs to.

Since the representations of the nodes can help identify better clusters (both w.r.t. communities and also their role), we optimize the clusters in an end-to-end fashion along with the node representations. Specifically, the cluster centroids $\boldsymbol{\mu}_k \in \mathbb{R}^D$ with $k = 1, \ldots, K$ (suppose K clusters) are obtained by a layer that implements a differentiable version of K-means clustering, as in ClusterNet [23], as follows. The clusters are updated by optimizing Eq. (3) via an iterative process by alternately setting

$$\boldsymbol{\mu}_k = \frac{\sum_i r_{ik}\mathbf{h}_i}{\sum_i r_{ik}} \quad k = 1, \ldots, K \tag{4}$$

and

$$r_{ik} = \frac{\exp(-\beta \cos(\mathbf{h}_i, \boldsymbol{\mu}_k))}{\sum_k \exp(-\beta \cos(\mathbf{h}_i, \boldsymbol{\mu}_k))} \quad k = 1, \ldots, K, \tag{5}$$

where $\cos(\cdot,\cdot)$ denotes the cosine similarity between two instances and β is an inverse-temperature hyperparameter; $\beta \rightarrow \infty$ gives a binary value for each cluster assignment. The gradients propagate only to the last iteration of the forward-pass updates in order to ensure convergence [23]. Finally, the cluster-derived summary, i.e., coarse-grain summary, associated with v_i in Eq. (3) is given by $\mathbf{z}_i = \sigma\left(\sum_{k=1}^{K} r_{ik}\boldsymbol{\mu}_k\right)$, where r_{ik} is the degree that v_i is assigned to cluster k and $\boldsymbol{\mu}_k$ is the centroid of the kth cluster, as described before, followed by a sigmoid nonlinearity.

3.4 Fake Input and Discriminators

When the input is a single graph, we opt to corrupt the graph by row-shuffling the original features $\mathbf{X}$ as $\tilde{\mathbf{X}} := \text{shuffle}([\mathbf{x}_1, \mathbf{x}_2, \ldots, \mathbf{x}_N])$ and $\tilde{\mathbf{A}} := \mathbf{A}$ (see Fig. 1a).

In the case of multiple input graphs, it may be useful to randomly sample a different graph from the training set as negative examples.

As the discriminator function d in $\mathcal{L}_1$, we use a bilinear scoring function, followed by a logistic sigmoid nonlinearity, which converts scores into probabilities, as $d(\mathbf{h}_i, \mathbf{s}) = \sigma(\mathbf{h}_i^T \mathbf{W} \mathbf{s})$, where $\mathbf{W}$ is a learnable scoring matrix. We use an inner product similarity, followed by $\sigma(\cdot)$, as the discriminator function g in $\mathcal{L}_C$, $g(\mathbf{h}_i, \mathbf{z}_i) = \sigma(\mathbf{h}_i^T \mathbf{z}_i)$. Here, we replace the bilinear scoring function used for g by an inner product, since it dramatically reduces the memory requirements and worked better in our case.

4 Related Work

Many unsupervised/self-supervised graph representation learning approaches follow a contrastive learning paradigm. Their objective is to give a higher score to positive examples and a lower to negative examples, which acts as a binary classification between positives and negatives. Based on the selection of positive/negative examples, methods are able to capture fine-grain similarities (DeepWalk [16], node2vec [3], GraphSAGE [4], GAE/VGAE [9], ARGVA [13], GMI [15]) or global similarities (DGI [21], MVGRL [5]) that are shared between nodes.

For example, in DeepWalk, node2vec and GraphSAGE, positive examples are representations of node pairs that co-occur in short random walks while negatives are representations of distant nodes. In GAE/VGAE and ARGVA, positive examples are representations of incident nodes and negatives are representations of random pairs of nodes. ARGVA uses additional positive/negative pairs as it discriminates at the same time whether a latent node representation comes from the prior (positive) or the graph encoder (negative). GMI has two types of positive/negative examples: First, by pairing a node representation with its input structure and attributes (positive) and with the input of a random node (negative), and second, by obtaining them as in GAE. MVGRL [5] works in a same manner with DGI, but it augments the real graph to get two additional versions of the graph (real and fake), which doubles the pairs of positive/negative examples. Methods that capture additional global properties (like DGI and MVGRL) are shown to estimate superior node representations.

5 Experimental Methodology and Results

5.1 Methodology and Configuration

Datasets. We evaluated the performance of GIC using seven commonly used benchmarks (Table 1). CORA, CiteSeer, and PubMed [25] are three citation networks, CoauthorCS and CoauthorPhysics are co-authorship graphs, and AmazonComputer and AmazonPhoto [18] are segments of the Amazon co-purchase graph.

Table 1. Datasets statistics of the entire graph and the largest connected component (LCC) of the graph.

	Classes	Features	Nodes	Edges	Label rate	Nodes LCC	Edges LCC	Label rate LCC
CORA	7	1,433	2,708	6,632	0.0517	2,485	5,069	0.0563
CiteSeer	6	3,703	3,327	4,614	0.0324	2,110	3,668	0.0569
PubMed	3	500	19,717	44,324	0.0030	19,717	44,324	0.0030
CoauthorCS	15	6,805	18,333	81,894	0.0164	18,333	81,894	0.0164
CoauthorPhysics	5	8,415	34,493	247,962	0.0029	34,493	247,962	0.0029
AmazonComputer	10	767	13,752	287,209	0.0145	13,381	245,778	0.0149
AmazonPhoto	8	745	7,650	143,663	0.0209	7,487	119,043	0.0214

Label rate is the fraction of nodes in the training set for node classification tasks

Hyper-parameter Tuning and Model Selection. As the encoder function f_{GNN} we use the graph convolution network (GCN) [8] with the propagation rule at layer l: $\mathbf{H}^{(l+1)} = \text{PReLU}(\hat{\mathbf{D}}^{-\frac{1}{2}}\hat{\mathbf{A}}\hat{\mathbf{D}}^{-\frac{1}{2}}\mathbf{H}^{(l)}\boldsymbol{\Theta})$, where $\hat{\mathbf{A}} = \mathbf{A} + \mathbf{I}_N$ is the adjacency matrix with self-loops, $\hat{\mathbf{D}}$ is the diagonal degree matrix of $\hat{\mathbf{A}}$, $\boldsymbol{\Theta} \in \mathbb{R}^{F \times D}$ is a learnable matrix, PReLU denotes the nonlinear parametric rectified linear unit, and $\mathbf{H}^{(0)} = \mathbf{X}$.

We use an one-layer GCN-encoder ($l = 1$) and iterate the cluster updates in Eq. (4) and (5) for 10 times. Since GIC's cluster updates are performed in the unit sphere (cosine similarity), we row-normalize the embeddings before the downstream task.

GIC's learnable parameters are initialized with Glorot [2] and the objective is optimized using the Adam [7] with a learning rate of 0.001. We train for a maximum of $2k$ epochs, but the training is early terminated if the training loss does not improve in 20 or 50 consecutive epochs. The model state is reset to the one with the best (lowest) training loss. For each dataset-task pair, we perform model selection based on the validation set of the corresponding task (except for clustering, in which we use the link prediction task, instead). We set $\alpha \in \{0.25, 0.5, 0.75\}$ (regularization of the two objective terms), $\beta = \{10, 100\}$ (softness of the cluster assignments) and $K \in \{32, 128\}$ (number of clusters) to train the model, and keep the parameters' triplet that achieved the best result on the validation set. We determine these parameters once for each dataset-task pair, so that the corresponding results are reported based on the *same* triplet.

Competing Approaches and Implementation. We compare the performance of GIC against seventeen unsupervised and six semi-supervised methods and variants. The results for all the competing methods, except DGI and sometimes GMI and MVGRL, were obtained directly from [5,13,15,18]. We name VGAE-best and ARGVA-best the best performing variant of VGAE and ARGVA methods, respectively. We implemented GIC using the Deep Graph Library (DGL) [22] and PyTorch [14], as well by modifying DGI's original implementation (https://github.com/cmavro/Graph-InfoClust-GIC). All experiments were performed on a Nvidia Geforce RTX-2070 GPU on a i5-8400 CPU and 32 GB RAM machine.

Table 2. Mean node classification accuracy (with standard deviations) in % over 20 runs and for two different train/val sets: balanced and imbalanced. The datasets are randomly split in each run.

Train/Val.	Unsupervised				Semi-supervised	
	GIC		DGI		Best	Worst
	Imbalanced	Balanced	Imbalanced	Balanced	Balanced	
CORA	**81.7**(1.5)	80.7(1.1)	80.2(1.8)	80.0(1.3)	81.8(1.3)	76.6(1.9)
CiteSeer	**71.9**(1.4)	70.8(2.0)	71.5(1.3)	70.5(1.2)	71.9(1.9)	67.5(2.3)
PubMed	77.3(1.9)	**77.4**(1.9)	76.2(2.0)	76.8(2.3)	78.7(2.3)	76.1(2.3)
CoauthorCS	**89.4**(0.4)	89.3(0.7)	89.0(0.4)	88.7(0.8)	91.3(2.3)	85.0(1.1)
Coauth.Phys	**93.1**(0.7)	92.4(0.9)	92.7(0.8)	91.8(1.0)	93.0(0.8)	90.3(1.2)
Am.Comp.	**81.5**(1.0)	79.5(1.4)	79.0(1.7)	77.9(1.8)	83.5(2.2)	78.0(19.0)
Am.Photo	**90.4**(1.0)	89.0(1.6)	88.2(1.7)	86.8(1.7)	91.4(1.3)	85.7(20.3)

Results are reported on the largest connected component (LCC) of the graph.
Semi-supervised methods: GCN [8], GAT [20], GraphSAGE [4], and MoNet [12].

5.2 Results

Node Classification. In unsupervised methods, the learned node embeddings are passed to a downstream classifier that is based on logistic regression. Following [18], we set the embedding dimensions to $D = 64$, unless otherwise stated. Also, we sample 20×#classes nodes as the train set, 30×#classes nodes as the validation set, the remaining nodes are the test set for the classifier. The sets are either uniformly drawn from each class (balanced sets) or randomly sampled (imbalanced sets). In the Planetoid split [25], 1,000 nodes of the remaining nodes are only used for testing. We use a logistic regression classifier, which is trained with a learning rate of 0.01 for 300 or 1k epochs with Adam optimizer and Glorot initialization.

Table 2 shows the performance of GIC compared to DGI and semi-supervised methods. Leveraging cluster information benefits datasets as CORA and PubMed, where GIC achieves a mean classification accuracy gain of more than 1% over DGI. In CiteSeer, CoauthorCS and CoauthorPhysics, the gain is slightly lower, but still more than 0.4%, since the abundant attributes of each node makes the cluster extraction more challenging. In AmazonComputers and AmazonPhoto, GIC performs significantly better than DGI with a gain of more than 2%, on average. Due to the large edge density of these datasets, the GNN-encoder aggregates information from multiple other nodes leading to representations very similar to the global summary. In such cases, GIC's objective term is responsible for making the representations to contain different aspects of information. In all cases, GIC performs better than the worst performing semi-supervised method with a gain of more than 1.5% and as high as 4.3%.

Table 3 further illustrates the performance of GIC compared to recently developed unsupervised methods based on MI maximization. The table shows

Table 3. Mean node classification accuracy (with standard deviations) for the Planetoid split.

	GIC		DGI		GMI		MVGRL	
	$D*$	$D = 64$	$D*$	$D = 64$	$D*$	$D = 64$	$D*$	$D = 64$
CORA	**83.2**(0.4)	81.4(0.5)	82.3(0.6)	79.2(0.7)	83.0(0.3)	77.7(0.4)	–	76.9(0.4)
CiteSeer	**73.4**(0.4)	71.6(0.6)	71.8(0.7)	69.0(0.9)	73.0(0.3)	68.4(0.9)	73.3(0.5)	70.8(0.7)
PubMed	**80.3**(0.6)	78.3(1.3)	76.8(0.6)	77.5(1.2)	80.1(0.2)	OOM	80.1(0.7)	78.5(0.8)

$D*$: Results obtained by the corresponding papers. For GIC, $D* = 300, 400, 150$.
"OOM": out of GPU memory. "–": Results were not reported for the Planetoid split.

Table 4. Link prediction scores: Mean area Under Curve (AUC) score [1] and average Precision (AP) score [19] with standard deviations over 10 runs (in %). Top: $D = 16$, Bottom: $D*$.

	CORA		CiteSeer		PubMed	
	AUC	AP	AUC	AP	AUC	AP
DeepWalk [16]	83.1	85.0	80.5	83.6	84.4	84.1
VGAE-best [9]	91.4 ± 0.01	92.6 ± 0.01	90.8 ± 0.02	92.0 ± 0.02	96.4 ± 0.00	96.5 ± 0.00
ARGVA-best [13]	92.4	93.2	92.4	93.0	**96.8**	**97.1**
DGI [21]	89.8 ± 0.8	89.7 ± 1.0	95.5 ± 1.0	95.7 ± 1.0	91.2 ± 0.6	92.2 ± 0.5
GIC	**93.5** ± 0.6	**93.3** ± 0.7	**97.0** ± 0.5	**96.8** ± 0.5	93.7 ± 0.3	93.5 ± 0.3
DGI	94.8 ± 0.7	95.2 ± 0.8	98.5 ± 0.4	98.4 ± 0.3	93.9 ± 0.4	93.9 ± 0.4
GMI [15]	95.1 ± 0.3	95.6 ± 0.2	97.8 ± 0.1	97.4 ± 0.2	OOM	OOM
GIC	**96.0** ± 0.2	**96.1** ± 0.4	**98.9** ± 0.2	**99.0** ± 0.1	95.5 ± 0.1	95.6 ± 0.2

For DGI only, we set $D = 32$ which greatly improves its results compared to $D = 16$.

that leveraging additional coarse-grain information (GIC, MVGRL) helps better than leveraging extra fine-grain information (GMI). Moreover, GIC is the only method that consistently performs better than DGI across all datasets with varying embedding size D. The performance improvement for other methods (GIC, MVGRL) is evident only for large D, e.g., $D = 512$ in the papers.

Link Prediction. In link prediction, some edges are hidden in the input graph and the goal is to predict the existence of these edges based on the computed embeddings. The probability of an edge between nodes i and j is given by $\sigma(\mathbf{h}_i^T \mathbf{h}_j)$, where σ is the logistic sigmoid function. We follow the setup described in [9]: 5% of edges and negative edges as validation set, 10% of edges and negative edges as test set.

Table 4 illustrates the benefits of GIC for link prediction tasks. GIC outperforms DGI in all three datasets, since GIC's clustering is able to preserve and reveal useful interactions between nodes which may hint the existence of links between them. GIC also outperforms VGAE and ARGVA, in CORA and CiteSeer by 1%–2% and 4.5%–5.5%, respectively, even though these methods are specifically designed for link prediction tasks. In PubMed, which has fewer attributes to exploit, the performance of GIC is slightly worse than that of VGAE

Table 5. Clustering results with respect to the true labels.

	CORA			CiteSeer			PubMed		
	Acc	NMI	ARI	Acc	NMI	ARI	Acc	NMI	ARI
K-means	49.2	31.1	23.0	54.0	30.5	27.9	39.8	0.1	0.2
DeepWalk [16]	48,4	32.7	24.3	33.7	8.8	9.2	68.4	27.9	29.9
TADW [24]	56.0	44.1	33.2	45.5	29.1	22.8	35.4	0.1	0.1
VGAE-best [9]	60.9	43.6	34.7	40.8	17.6	12.4	67.2	27.7	27.9
ARGVA-best [13]	71.1	52.6	49.5	58.1	33.8	30.1	**69.0**	30.5	**30.6**
DGI [21]	59.0	38.6	33.6	57.9	30.9	27.9	49.9	15.1	14.5
GIC	**72.5**	**53.7**	**50.8**	**69.6**	**45.3**	**46.5**	67.3	**31.9**	29.1

Acc: accuracy, NMI [11]: normalized mutual information, ARI [11]: average rand index in percents (%).

and ARGVA. It is noteworthy, that GIC's proposed objective term works better than GMI which combines a mutual information objective term with a link prediction term.

Clustering. In clustering, the goal is to cluster together related nodes (e.g., nodes that belong to the same class) without any label information. The computed embeddings are clustered into K = #classes clusters with K-means. We set $D = 32$ and the evaluation is provided by external labels, the same used for node classification.

Table 5 illustrates GIC's performance for clustering. GIC performs better than other unsupervised methods in two out of three datasets (CORA and CiteSeer), and performs almost equally with ARGVA in PubMed. The gain over DGI is significantly large in all datasets, and can be as high as 15% to 18.5% for the NMI metric. Due to its interactive clustering, GIC can be considered an efficient method when the downstream task is clustering.

Ablation and Parameter Study

Effect of the Loss Function. We provide results w.r.t. silhouette score (SIL) [17], which is a clustering evaluation metric, of the learned representation after their t-SNE 2D projection [10]. Table 6 illustrates that using the novel objective term $\mathcal{L}_C$ ($\alpha \neq 1$) outperforms the baseline DGI objective $\mathcal{L}_1$ in all experiments. Primarily, using a single global vector to optimize the node representations may lead to certain pitfalls, especially when the embedding size and thus, the global vector's capacity, is low (Table 6a). Moreover, accounting for both cluster-level and graph-level information leads to better representations, while ignoring some of this information ($\alpha = 0$ or $\alpha = 1$) worsens the quality of the representations. This is also true when hyperparameters β and K are not optimized (Avg vs. Max in Table 6b), since the soft assignment of the clusters and their joint optimization during learning alleviates this need.

Table 6. SIL scores with varying embedding size D for CORA.

(a) SIL scores with varying embedding size D for CORA.

	$D = 16$	$D = 32$	$D = 512$
$\alpha = 0.5$ (GIC)	**0.195**	**0.229**	**0.257**
$\alpha = 1$ (DGI)	−0.121	−0.012	0.222

(b) Performance gain percentage (in %) w.r.t. SIL over DGI: $\beta \in \{10,100\}$, $K \in \{\#\text{classes}, 32, 128\}$.

	CORA		**CiteSeer**		**PubMed**	
	Avg	Max	Avg	Max	Avg	Max
$\alpha = 0$	3.3	18.3	7.8	19.3	15.8	22.1
$\alpha = .25$	22.0	32.5	**13.8**	**22.0**	21.4	32.4
$\alpha = .5$	**27.8**	**38.2**	12.0	21.4	43.1	49.4
$\alpha = .75$	25.7	30.9	7.5	10.3	**47.2**	**62.3**

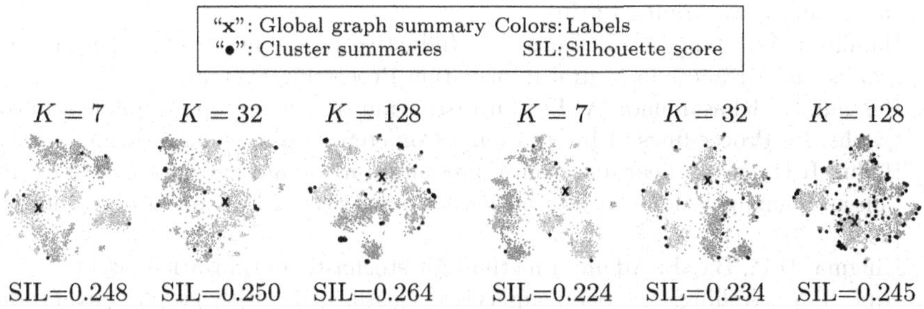

(a) $\alpha = 0.5, \beta = 10$ (b) $\alpha = 0.5, \beta = 100$

Fig. 2. t-SNE plots for CORA dataset and the corresponding silhouette scores (SIL). SIL score for $\alpha = 1$ is 0.212.

Visualization of the Clusters. In Fig. 2, we plot the t-SNE 2D projection of the learned node representations ($D = 64$) for the CORA dataset. A large β, e.g., $\beta = 100$ in Fig. 2b ($K = 128$), makes the distances between the cluster centers larger compared to a smaller one, e.g., $\beta = 10$ in Fig. 2a ($K = 128$). Increasing K generally helps, e.g., Fig. 2a ($K = 128$) compared to Fig. 2a ($K = 7$), however, that does not mean that all clusters will be distinct.

6 Conclusion

We have presented Graph InfoClust (GIC), an unsupervised graph representation learning method which relies on leveraging cluster-level content. GIC identifies nodes with similar representations, clusters them together, and maximizes their mutual information. This enables us to improve the quality of node representations with richer content and obtain better results than existing approaches for tasks like node classification, link prediction, clustering, and data visualization.

Acknowledgements. This work was supported in part by NSF (1447788, 1704074, 1757916, 1834251), Army Research Office (W911NF1810344), Intel Corp, and the Digital Technology Center at the University of Minnesota. Access to research and computing facilities was provided by the Digital Technology Center and the Minnesota Supercomputing Institute.

References

1. Bradley, A.P.: The use of the area under the roc curve in the evaluation of machine learning algorithms (1997)
2. Glorot, X., Bengio, Y.: Understanding the difficulty of training deep feedforward neural networks. In: Proceedings of the Thirteenth International Conference on Artificial Intelligence and Statistics (2010)
3. Grover, A., Leskovec, J.: node2vec: scalable feature learning for networks. In: Proceedings of the 22nd ACM SIGKDD International Conference on Knowledge Discovery and Data Mining (2016)
4. Hamilton, W., Ying, Z., Leskovec, J.: Inductive representation learning on large graphs. In: Advances in Neural Information Processing Systems (2017)
5. Hassani, K., Khasahmadi, A.H.: Contrastive multi-view representation learning on graphs. In: Proceedings of International Conference on Machine Learning (2020)
6. Hjelm, R.D., et al.: Learning deep representations by mutual information estimation and maximization. In: International Conference on Learning Representations (2019)
7. Kingma, D.P., Ba, J.: Adam: a method for stochastic optimization (2014)
8. Kipf, T.N., Welling, M.: Semi-supervised classification with graph convolutional networks. arXiv preprint arXiv:1609.02907 (2016)
9. Kipf, T.N., Welling, M.: Variational graph auto-encoders. In: NIPS Workshop on Bayesian Deep Learning (2016)
10. Maaten, L.V.D., Hinton, G.: Visualizing data using t-SNE. J. Mach. Learn. Res. (2008)
11. Manning, C.D., Raghavan, P., Schütze, H.: Introduction to Information Retrieval (2008)
12. Monti, F., Boscaini, D., Masci, J., Rodola, E., Svoboda, J., Bronstein, M.M.: Geometric deep learning on graphs and manifolds using mixture model CNNs. In: Proceedings of the IEEE Conference on Computer Vision and Pattern Recognition, pp. 5115–5124 (2017)
13. Pan, S., Hu, R., Fung, S.F., Long, G., Jiang, J., Zhang, C.: Learning graph embedding with adversarial training methods. IEEE Trans. Cybern. (2019)
14. Paszke, A., et al.: Automatic differentiation in pytorch (2017)
15. Peng, Z., et al.: Graph representation learning via graphical mutual information maximization. In: Proceedings of the Web Conference (2020)
16. Perozzi, B., Al-Rfou, R., Skiena, S.: Deepwalk: online learning of social representations. In: Proceedings of the 20th ACM SIGKDD International Conference on Knowledge Discovery and Data Mining (2014)
17. Rousseeuw, P.J.: Silhouettes: a graphical aid to the interpretation and validation of cluster analysis. J. Comput. Appl. Math. **20**, 53–65 (1987)
18. Shchur, O., Mumme, M., Bojchevski, A., Günnemann, S.: Pitfalls of graph neural network evaluation. arXiv preprint arXiv:1811.05868 (2018)
19. Su, W., Yuan, Y., Zhu, M.: A relationship between the average precision and the area under the ROC curve. In: Proceedings of the 2015 International Conference on the Theory of Information Retrieval (2015)
20. Veličković, P., Cucurull, G., Casanova, A., Romero, A., Liò, P., Bengio, Y.: Graph attention networks. In: International Conference on Learning Representations (2018)
21. Veličković, P., Fedus, W., Hamilton, W.L., Liò, P., Bengio, Y., Hjelm, R.D.: Deep graph infomax. In: International Conference on Learning Representations (2019)

22. Wang, M., et al.: Deep graph library: towards efficient and scalable deep learning on graphs. In: ICLR Workshop (2019)
23. Wilder, B., Ewing, E., Dilkina, B., Tambe, M.: End to end learning and optimization on graphs. In: Advances in Neural and Information Processing Systems (2019)
24. Yang, C., Liu, Z., Zhao, D., Sun, M., Chang, E.: Network representation learning with rich text information. In: Twenty-Fourth International Joint Conference on Artificial Intelligence (2015)
25. Yang, Z., Cohen, W.W., Salakhutdinov, R.: Revisiting semi-supervised learning with graph embeddings. In: Proceedings of the 33rd International Conference on International Conference on Machine Learning (2016)

A Deep Hybrid Pooling Architecture for Graph Classification with Hierarchical Attention

Sambaran Bandyopadhyay[1,2](✉), Manasvi Aggarwal[2], and M. Narasimha Murty[2]

[1] IBM Research AI, New Delhi, India
[2] Indian Institute of Science, Bangalore, Bengaluru, India
sambaran@alum.iisc.ac.in, {manasvia,mnm}@iisc.ac.in

Abstract. Graph classification has been a classical problem of interest in machine learning and data mining because of its role in biological and social network analysis. Due to the recent success of graph neural networks for node classification and representation, researchers started extending them for the entire graph classification purpose. The main challenge is to represent the whole graph by a single vector which can be used to classify the graph in an end-to-end fashion. Global pooling, where node representations are directly aggregated to form the graph representation and more recently hierarchical pooling, where the whole graph is converted to a smaller graph through a set of hierarchies, are proposed in the literature. Though hierarchical pooling shows promising results for graph classification, it looses a significant amount of information in the hierarchical architecture. To address this, we propose a novel hybrid graph pooling architecture, which finds the importance of different hierarchies of pooling and aggregates them accordingly. We use a series of graph isomorphism networks, along with a bi-directional LSTM with self attention to implement the proposed hybrid pooling. Experiments show the merit of the proposed architecture with respect to a diverse set of state-of-the-art algorithms on multiple datasets.

Keywords: Graph Neural Network · Hierarchical graph representation · Graph pooling · Self attention

1 Introduction

Graphs are important data types to represent different kinds of relational objects such as molecular structures, protein-protein interactions and information networks [12,16]. A graph is represented by $G = (V, E)$, where V is the set of nodes and E is the set of edges. Real-world graphs often come with a set of attributes, where each node $v_i \in V$ is also associated with an attribute (or feature) vector $x_i \in \mathbb{R}^D$. Graph classification, i.e., predicting the class label of an entire graph, is a classical problem of interest to the machine learning and data

K. Karlapalem et al. (Eds.): PAKDD 2021, LNAI 12712, pp. 554–565, 2021.
https://doi.org/10.1007/978-3-030-75762-5_44

mining community. Different practical applications such as finding anti-cancer activity, solubility, or toxicity of a molecule can be addressed by classifying the entire graph [9]. Graph classification is tackled as a supervised task. More formally, given a set of M graphs $\mathcal{G} = \{G_1, G_2, \cdots, G_M\}$, and a subset of graphs $\mathcal{G}_s \subseteq \mathcal{G}$ with each graph $G_i \in \mathcal{G}_s$ labelled with $Y_i \in \mathcal{L}_g$ (the subscript g stands for 'graphs'), the task is to predict the label of a graph $G_j \in \mathcal{G}_u = \mathcal{G} \setminus \mathcal{G}_s$ using the structure of the graphs and the node attributes, and the graph labels from $\mathcal{G}_s$. This leads to learning a function $f_g : \mathcal{G} \mapsto \mathcal{L}_g$. Here, $\mathcal{L}_g$ is the set of discrete labels for the graphs. Graph kernel algorithms [13,16] remained to be the stat-of-the-art for a long time for graph classification. Graph kernels typically rely on different types of hand crafted features such as the occurrence of some specific subgraph patterns in a graph.

Recently, graph neural networks (such as graph convolution network) and and node embedding techniques [1,4,7,17] are able to achieve promising results for node classification and representation. They map the nodes of the graph from non-Euclidean to Euclidean space by using both the link structure and the node attributes, and consequently use the vector representation for node classification in an end-to-end (or integrated) fashion. The main challenge to extend these approaches from node representation to graph representation is to design an intelligent node aggregation technique which can map a graph to a smaller graph (also known as graph pooling). There exist two types of graph pooling strategies in the literature. First, global pooling uses simple aggregation (such as averaging or concatenating) techniques on the embeddings of all the nodes to get a single vector representation of the graph [3]. Global pooling strategy works better for smaller graphs where aggregating all the nodes with a single function makes more sense. Second, hierarchical pooling recursively maps the input graph into a smaller graph (which may even contain a single node) and finally uses some aggregation technique [20].

Graphs exhibit hierarchical structures by default [2]. Nodes are the entities at the lowest of this hierarchy. Multiple nodes may form a sub-community and few sub-communities may for a community. In contrast to text documents where words, sentences and paragraphs form the hierarchy, hierarchical structure of a graph is latent in nature. Hierarchical graph neural networks, such as DIFF-POOL [20], jointly discover the hierarchical nature of the graph using graph pooling and finally convert it to a single node which is used to classify the entire graph. But different entities in the hierarchy do not play equal role to determine the label of a graph. For example, some intermediate level in the hierarchy may contain more useful information for classifying the graph rather than the final layer. This is true even for document classification, where importance of all the words and sentences are not the same to classify the entire document [19]. Besides, hierarchical pooling for graphs, though performs better, looses a significant amount of information in the multiple hierarchies of the pooling layers [10]. For example, if the graph after the final pooling layer contains a single node [20], then it is difficult for that node to encode all the structural information of the entire input graph in it.

To address the above challenges, we propose a novel graph pooling technique for graph classification, by mixing hierarchical and global pooling. Our algorithm (referred as *HybridPool*) maps the input graph to consecutive smaller graphs in multiple hierarchies (or levels) and then employs a global pooling across the multiple hierarchies (or levels). We observe that these graph hierarchies in the pooling network form a sequence. As the importance of different hierarchies are unknown, we use a bi-directional LSTM [5] with self-attention to weight them accordingly. Following are the contributions we make in this work.

- We propose a novel hybrid graph pooling algorithm *HybridPool* which employs multiple layers of graph convolution to create multiple hierarchies (or levels) of smaller graphs from an input graph, and then use bi-directional LSTM with self-attention to get the final representation of the entire graph. In contrast to existing literature, HybridPool learns different weights for different intermediate entities of a hierarchical representation of a graph and aggregates them globally for the entire graph classification. We use cross entropy loss of graph classification to jointly learn all the parameters of the entire network using back propagation.
- We conduct thorough experimentation on real-world graph datasets and achieve competitive performance with respect to state-of-the-art graph classification algorithms.

2 Proposed Solution: HybridPool

In this section, we describe the proposed algorithm HybridPool for graph classification.

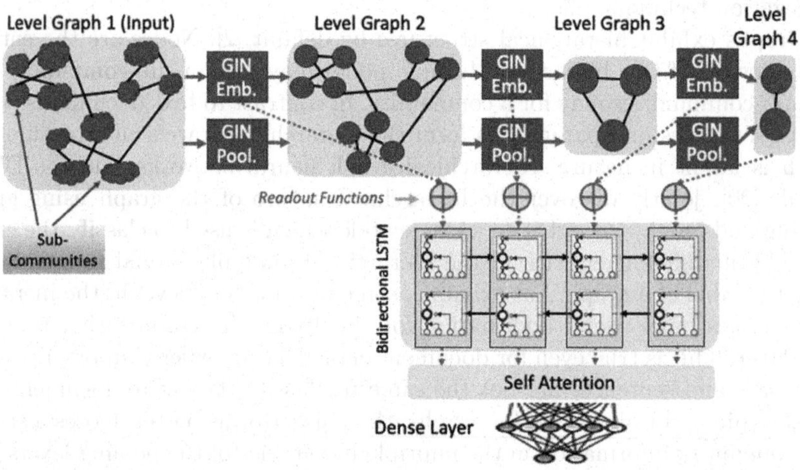

Fig. 1. Architecture of a HybridPool Network for graph classification

2.1 Overview of the Architecture

Before going to the details of the individual layers, we present a high level overview of HybridPool. Figure 1 shows the architecture of a HybridPool network with $R = 4$ level graphs. The first level is an input graph from the set of graphs $\mathcal{G}$. Next level graphs form different hierarchies in the graph, such as sub-communities, communities etc. Let us denote these level graphs (i.e., graphs at different levels) by $G^1, \cdots, G^R$. In level graph G^r, number of nodes is N_r, and the dimension of a feature vector for a node is K (except the input level graph G^1 which has D dimensional feature vector for each node). There is a GNN layer between level graph G^r (i.e., the graph at level r) and level graph G^{r+1}. This GNN layer comprises of an embedding layer which generates the embedding of the nodes of G^r and a pooling layer which maps the nodes of G^r to the nodes of G^{r+1}. A GIN [17] embedding layer and a pooling layer together convert a graph to a smaller (having lesser number of nodes) graph. We discuss the details of them in next two subsections. We refer the GNN layer between level graph G^r and G^{r+1} by rth layer of GNN, $\forall r = 1, 2, \cdots, R-1$. The last level graph G^R contains only one node, whose feature summarizes the entire input graph. Pleas note, number of nodes N_1 in the first level graph depends on the input graph, but we keep the number of nodes N_r in the consequent level graphs G^r ($\forall r = 2, \cdots, R$) fixed for all the input graphs (in a graph classification dataset), which helps us to build the next stages of the network conveniently as discussed below. Different level graphs can have different interpretation. In Fig. 1, each node in level graph 2 roughly represents a sub-community of the input graph, each node in level graph 3 can represent a community of the input graph, and finally the node in level graph 4 represents the whole input graph.

As mentioned in Sect. 1, the last level graph (for example, level graph 4 with two nodes in Fig. 1) is meant to encode all the structural and attribute information of the input graph at level 1. But as discussed before, due to the loss of some information in the formation of each level graph, the last level graph may not be able to preserve the intrinsic properties of the input graph to classify it. Besides, it is possible that the label information of the input graph depends directly on the overall sub-community structure or the community structure of the input graph, which are more prominent in some intermediate level graphs. So, we propose to learn the importance of different level graphs in the training phase. We use a bidirectional LSTM (BLSTM), which captures the ordered dependency of different level graphs. The use of bidirectional LSTM instead of a regular LSTM ensures the explicit modeling of a level graph as a function of the level graphs, both up and down in the hierarchy. First to obtain a summary vector for each level graph, we use a readout function. More precisely, a readout function takes the embeddings of all the nodes from a level graph and map them to a single vector which is invariant to the ordering of the nodes inside a level graph. The output of readout functions (one summary vector for each level graph as shown in Fig. 1) is fed to BLSTM. A self attention layer is used after the BLSTM layer to determine the importance of individual level graphs and aggregate them accordingly. Finally, this representation is fed to a dense neural network with a softmax layer at the end to classify the entire graph. This completes one forward pass of the proposed HybridPool network.

Please note that the upper layers of HybridPool network resemble a hierarchical pooling structure (DIFFPOOL [20]) where an input graph is pooled to smaller graphs recursively. But there are crucial design differences between DIFFPOOL and the upper layers of the architecture adopted by us. First, DIFFPOOL uses GCN [7] as the embedding and pooling mechanisms. We use GIN [17] as the embedding and pooling mechanisms. The representation power of GIN is theoretically more than GCN. Second, DIFFPOOL always needs to have only a single node in the last level of the hierarchy to represent the whole graph by the features of that single node. But we may have more than one node in the last level, as the final graph representation is not obtained directly from the hierarchical structure. Rather, we use readout functions to compute a summary vector from each level graph and they are fed to BLSTM and the self-attention layers to compute the graph representation in HybridPool. Further, experimental comparison with DIFFPOOL, both for graph classification and model ablation study in Sect. 3 shows the merit of such a design adopted by us.

Hybrid Nature of the Pooling Strategy in HybridPool: The pooling strategy adopted by us is a mixture of hierarchical and global pooling strategies. The upper layers of HybridPool is a hierarchical graph pooling strategy, as discussed in the last paragraph. Lower layers which consist of the readout functions, bidirectional LSTM and self attention resemble a global pooling structure, as features from all the nodes in a level graph are directly aggregated and processed further to compute the final graph representation. That is why we call the proposed architecture *HybridPool*. Individual components of HybridPool are discussed below.

2.2 GIN Embedding Layer

This subsection defines the embedding layer (referred as GIN Emb. in Fig. 1) used in HybridPool algorithm. We use the GNN discussed in Graph Isomorphism Network (GIN) [17] as that is theoretically shown to be maximally powerful GNN to represent graphs, but still simple in nature. For a graph $G = (V, E)$, with adjacency matrix $A \in \mathbb{R}^{N \times N}$ and node attribute matrix $X \in \mathbb{R}^{N \times D}$, the l-th layer of GIN can be defined as:

$$h_v^{l+1} = MLP^l\Big((1+\epsilon^l)h_v^l + \sum_{u \in \mathcal{N}(v)} h_u^k\Big) \tag{1}$$

Here, $h_v^{l+1} \in \mathbb{R}^K$ is the hidden representation of the node v in $l+1$th layer of GIN and K is the feature dimension of the hidden layers of GCN. $\mathcal{N}(v)$ is the neighbors of the node v in G. ϵ is a parameter of GIN which determine the importance of a node's own representation with respect to the aggregated representation of its neighbors. The initial representations H^0 of the nodes are initialized with their respective features, $H^0 = X$.

For most of the datasets in Sect. 3, our GIN embedding layer consists of 1 or 2 layered deep GCN. The rth GIN embedding layer (between level graph r and $r+1$) is defined as:

$$Z_r = \mathrm{GIN}_{r,embed}(A_r, X_r) \quad (2)$$

Here, $Z_r \in \mathbb{R}^{N_r \times K}$ is the final embedding matrix of the nodes of rth level graph G^r.

2.3 GIN Pooling Layer

We again use the same GIN proposed in [17] and as discussed in Sect. 2.2 as the GIN pooling layer (denoted as GIN Pool. in Fig. 1) of HybridPool. But the goal of pooling layer is to map the nodes from a previous level graph to next level graph, its output feature dimension is different from that of the embedding layer. Also we use a softmax layer after the final layer of GIN in the pooling layer as defined below.

$$P_r = \mathrm{softmax}(\mathrm{GIN}_{r,pool}(A_r, X_r)) \quad (3)$$

Here, (i,j)th element of $P_r \in \mathbb{R}^{N_r \times N_{r+1}}$ gives the probability of assigning node v_i^r in G^r to node v_j^{r+1} in G^{r+1}. The softmax in the pooling is applied row-wise.

2.4 Formulation of a Level Graph

Level graph 1 in the architecture of HybridPool is the input graph itself. This subsection discusses the formation of level graph G^{r+1} from G^r using the GIN embedding layer and the GIN pool layer. The adjacency matrix A_{r+1} of G^{r+1} is constructed as:

$$A_{r+1} = P_r^T A_r P_r \in \mathbb{R}^{N_{r+1} \times N_{r+1}} \quad (4)$$

Similarly, feature matrix X_{r+1} of G^{r+1} is constructed as:

$$X_{r+1} = P_r^T Z_r \in \mathbb{R}^{N_{r+1} \times K} \quad (5)$$

The matrix P_r contains information about how nodes in G^r are mapped to the nodes of G^{r+1}, and the adjacency matrix A_r contains information about the connection of nodes in G^r. Eq. 4 combines them to generate the link structure between the nodes (i.e., the adjacency matrix A_{r+1}) of G^{r+1}. Node feature matrix X_{r+1} of G^{r+1} is also generated similarly. Please note, G^1 is the input graph. For the intermediate level graphs, the number of nodes is determined by pooling ratio p as:

$$N_{r+1} = pN_r \, , \ \forall \, 1 \leq r \leq R-1 \quad (6)$$

Pooling ratio $p \in (0,1)$ is a hyper-parameter of HybridPool. We vary the pooling ratio from 0.05 to 0.5, depending on the size of input graph.

2.5 Attending the Important Hierarchies

In the above subsections, we discussed the generation of different level graphs from the input and their corresponding node features. As discussed in Sect. 1, graphs exhibit hierarchical structure and importance of different hierarchies are different to determine the label of the entire graph. So a simple global pool on all the nodes in the hierarchy may not be able to capture the varying importance across the levels of the hierarchy. This would become more evident on the model ablation study in Sect. 2.7.

In this subsection, we aim to determine the importance of different nodes in level graph 2 onward, for classifying the entire input graph. We employ a Bidirectional Long Short Term Memory Recurrent Neural Networks (BLSTM) and an attention layer for this purpose. BLSTM is proposed to handle the problem of vanishing gradients for recurrent neural networks [5] and they have been applied successfully to multiple NLP and sequence modeling tasks. In contrast to an LSTM, a bidirectional LSTM (or BLSTM) [21,22] processes data in both directions with two separate hidden layers, which are then fed forward to the same output layer. As shown in Fig. 1, we use a readout function to obtain the summary of a level graph. The readout function takes the GIN embeddings Z_r (from Eq. 2) of all the nodes of a level graph (except for the last level) and outputs a single vector. For the last level graph, there is no GIN embedding layer to generate the node embeddings. Also typically the number of nodes in the last level graph is really small. So instead of adding another GIN encoder, we use the readout function on the node feature matrix X_R (from Eq. 5) to generate the summary vector x^{G^R}. The readout function needs to be invariant to the ordering of the nodes in a graph. There are different types of readout functions proposed in the context of graph neural networks. They can be some simple aggregators such as sum (or mean) of the node embeddings [17], or attention-based node aggregators [10]. We use a simple readout function which just computes the sum of the node embeddings to obtain a summary vector of a level graph. This is because we further process those summary embeddings with the BLSTM and self-attention as discussed next. If we denote the summary vector of the level graph G^r as $s^r \in \mathbb{R}^K$, then

$$s^r = \begin{cases} \sum_{i=1}^{N_r} (Z_r)_{i,:} & 1 \leq r \leq R-1 \\ \sum_{i=1}^{N_R} (X_R)_{i,:} & \text{otherwise} \end{cases} \tag{7}$$

Though nodes within a level graph do not have any fixed order, but levels graphs themselves have an explicit hierarchical order. For example, a node can be a part of sub-community, the sub-community can be a part of a community, and so on. We use a BLSTM to explicitly capture this ordering of the level graphs. For each level graph G^r, $1 \leq r \leq R$, we feed its summary vectors s^r to the BLSTM as shown in Fig. 1. The BLSTM provides the forward-LSTM and backward-LSTM hidden representations h_r^f and h_r^b for each level graph G^r. We concatenate and map them to K dimensional space as $h_r = \sigma(W_L[h_r^f || h_r^b])$, $\forall$ $1 \leq r \leq R$, where $W_L \in \mathbb{R}^{K \times 2K}$ and σ is an activation function. We discuss the attention layer next.

Attention mechanism has been used in node embedding in multiple works [8,15,18]. Here we propose to use a self attention mechanism over all the hidden representations $h_{r,i} \in \mathbb{R}^K$ to determine their importance in the final graph representation. Let us use $\mathcal{H}$ to denote the matrix containing each $h_{r,i}$ as a row. The self attention mechanism is defined below:

$$e = \text{softmax}(\mathcal{H}\theta), \; h = \mathcal{H}^T e \in \mathbb{R}^K \tag{8}$$

Here, $\theta \in \mathbb{R}^K$ is a trainable attention vector. Intuitively, each element of this vector determines the importance of a feature dimension of the node representations of the level graphs. So e in turn contains the normalized attention score (importance) of the individual nodes of level graphs from 2 onward. Finally, h is the final vector representation of the input graph, which is the sum of the representations of the nodes of the level graphs from 2 onward, weighted by their normalized attention score. The final graph representation h is fed to a dense neural network, followed by a softmax layer to classify the graph. Backpropagation algorithm with ADAM optimization technique [6] by minimizing the cross entropy loss of graph classification on the training set $\mathcal{G}_s$ is used to learn all the parameters of the architecture in an end-to-end fashion.

2.6 Run Time Complexity of HybridPool

There are different components of HybridPool. The runtime of a GIN depends on the size of the trainable parameter matrix and the number of nodes. For an input graph with N nodes, the total runtime for GIN embedding and pooling layers are $O(NDKR)$ and $O(NDN_2 + N\sum_{r=2}^{R-1} N_r N_{r+1})$ respectively, where D is the input node feature dimension, K is the dimension of final graph representation, R is the number of level graphs and N_r is the number of nodes in rth level graph. Computation of the summary vectors of all the level graphs takes $O(K\sum_{r=2}^{R} N_r)$ time. Next, Bidirectional LSTM and self attention layers take $O(KR)$ time. Hence, the runtime to process an input graph in HybridPool is $O(NDKR + NDN_2 + N\sum_{r=2}^{R-1} N_r N_{r+1} + K\sum_{r=2}^{R} N_r)$. For all the experiments, values of N_r, $r > 1$ and number of levels R are small. So, HybridPool is scalable even for large graph classification datasets. Also note that some of the steps above are easy to run in parallel, which can reduce the runtime further.

2.7 Variants of HybridPool: Model Ablation Study

HybridPool has multiple components and layered architecture. So, estimating the importance of individual components of this architecture is necessary. To give more insight about it, we present the following two variants of HybridPool.

- **MaxPool**: Here we replace the BLSTM and the attention layer of HybridPool with a maximum function $\max_{r>1,i}\{X_{r,i}\} \in \mathbb{R}^K$, which selects the maximum value for each feature dimension from all the nodes. Please note, $X_{r,i}$ denotes the K dimensional embedding of the ith node in level graph $r > 1$ (Eq. 5). In contrast to HybridPool, MaxPool has a static rule which selects the most important node for each feature dimension. $\max_{r>1,i}\{X_{r,i}\}$ is fed to the dense layer followed by a softmax.
- **BLSTMPool**: Here we remove the self attention layer from HybridPool. Instead, the concatenated hidden states from the final cells of the bidirectional LSTM is directly fed to the dense layer followed by a softmax for graph classification. Please note, BLSTMPool does not give different importance to different nodes in the level graphs.

It is to be noted that, above two architectures are simpler compared to HybridPool, as they miss one or more components from it. Thus, by comparing the performance of HybridPool with both of them in Sect. 3, we are able to show the significance of different components of HybridPool.

Table 1. Different datasets used in our experiments

Dataset	#Graphs	#Max Nodes	#Labels	#Attributes
MUTAG	188	28	2	NA
PTC	344	64	2	NA
PROTEINS	1113	620	2	1
NCI1	4110	111	2	NA
NCI109	4127	111	2	NA
IMDB-BINARY	1000	136	2	NA
IMDB-MULTI	1500	89	3	NA
REDDIT-M-12K	11929	3782	11	NA

3 Experimental Evaluation

In this section, we conduct thorough experimentation to validate the merit of HybridPool for graph classification. We also experimentally show more insights about the proposed architecture.

3.1 Datasets and Baselines

We use 5 bioinformatics graph datasets and 3 social network datasets to evaluate the performance of graph classification. These datasets are MUTAG, PTC, PROTEINS, NCI1, NCI09 and IMDB-MULTI. The details of these datasets can be found at (https://bit.ly/39T079X). Table 1 contains a high-level summary of

Table 2. Classification accuracy (%) of different algorithms (23 in total) for graph classification on 8 benchmark datasets. NA denotes the case when the result of a baseline algorithm could not be found on that particular dataset from the existing literature. The last row 'Rank' is the rank (1 being the highest position) of our proposed algorithm HybridPool among all the algorithms present in the table.

Algorithms	MUTAG	PTC	PROTEINS	NCI1	NCI109	REDDIT-M-12K	IMDB-B	IMDB-M
GK	81.39 ± 1.7	55.65 ± 0.5	71.39 ± 0.3	62.49 ± 0.3	62.35 ± 0.3	31.82 ± 0.08	NA	NA
RW	79.17 ± 2.1	55.91 ± 0.3	59.57 ± 0.1	NA	NA	NA	NA	NA
PK	76 ± 2.7	59.5 ± 2.4	73.68 ± 0.7	82.54 ± 0.5	NA	NA	NA	NA
WL	84.11 ± 1.9	57.97 ± 2.5	74.68 ± 0.5	**84.46 ± 0.5**	**85.12±0.3**	39.03	NA	NA
AWE-DD	NA	NA	NA	NA	NA	39.20 ± 2.09	74.45 ± 5.8	51.54 ± 3.6
AWE-FB	87.87 ± 9.7	NA	NA	NA	NA	41.51 ± 1.98	73.13 ± 3.2	51.58 ± 4.6
node2vec	72.63 ± 10.20	58.85 ± 8.00	57.49 ± 3.57	54.89 ± 1.61	52.68 ± 1.56	NA	NA	NA
sub2vec	61.05 ± 15.79	59.99 ± 6.38	53.03 ± 5.55	52.84 ± 1.47	50.67 ± 1.50	NA	55.26 ± 1.54	36.67 ± 0.83
graph2vec	83.15 ± 9.25	60.17 ± 6.86	73.30 ± 2.05	73.22 ± 1.81	74.26 ± 1.47	NA	71.1 ± 0.54	50.44 ± 0.87
InfoGraph	89.01 ± 1.13	61.65 ± 1.43	NA	NA	NA	NA	73.03 ± 0.87	49.69 ± 0.53
DGCNN	85.83 ± 1.7	58.59 ± 2.5	75.54 ± 0.9	74.44 ± 0.5	NA	41.82	70.03 ± 0.9	47.83 ± 0.9
PSCN	88.95 ± 4.4	62.29 ± 5.7	75 ± 2.5	76.34 ± 1.7	NA	41.32 ± 0.32	71 ± 2.3	45.23 ± 2.8
DCNN	NA	NA	61.29 ± 1.6	56.61 ± 1.0	NA	NA	49.06 ± 1.4	33.49 ± 1.4
ECC	76.11	NA	NA	76.82	75.03	41.73	NA	NA
DGK	87.44 ± 2.7	60.08 ± 2.6	75.68 ± 0.5	80.31 ± 0.5	80.32 ± 0.3	32.22 ± 0.10	66.96 ± 0.6	44.55 ± 0.5
DiffPool	NA	NA	76.25	NA	NA	47.08	NA	NA
IGN	83.89 ± 12.95	58.53 ± 6.86	76.58 ± 5.49	74.33 ± 2.71	72.82 ± 1.45	NA	72.0 ± 5.54	48.73 ± 3.41
GIN	89.4 ± 5.6	64.6 ± 7.0	76.2 ± 2.8	82.7 ± 1.7	NA	NA	75.1 ± 5.1	52.3 ± 2.8
1-2-3GNN	86.1 ±	60.9 ±	75.5 ±	76.2 ±	NA	NA	74.2 ±	49.5 ±
3WL-GNN	90.55 ± 8.7	**66.17 ± 6.54**	77.2 ± 4.73	83.19 ± 1.11	81.84 ± ± 1.85	NA	72.6 ± 4.9	50 ± 3.15
MaxPool	90.28 ± 4.98	63.00 ± 7.03	74.23 ± 3.10	80.33 ± 2.08	78.56 ± 2.00	43.12 ± 4.01	74.01 ± 3.08	47.73 ± 5.53
BLSTMPool	90.44 ± 5.09	61.92 ± 9.61	75.1 ± 2.93	81.10 ± 1.99	79.81 ± 1.21	45.99 ± 3.72	74.98 ± 3.64	50.60 ± 3.43
HybridPool	**92.02 ± 4.63**	65.99 ± 5.47	**78.34 ± 2.37**	82.76 ± 1.95	80.58 ± 2.15	**48.21 ± 3.61**	**76.39 ± 5.93**	**52.60 ± 4.01**
Rank	1	2	1	3	3	1	1	1

these datasets. We compare the performance of our proposed algorithms with a diverse set of state-of-the-art baseline algorithms for graph classification. The twenty baselines algorithms can broadly be classified into **three groups**: Graph Kernel Based Algorithms, Unsupervised Graph Representation Algorithms and Graph Neural Network based Algorithms. Additionally, we also show the classification results for the two simpler variants of HybridPool - MaxPool and BLSTMPool. Comparison to these variants experimentally show the marginal contribution of different components of HybridPool. To obtain the results of existing baseline algorithms on different graph classification datasets, we check the respective papers and state-of-the-art papers from the literature [11], and report the best classification accuracy available. Thus, we avoid any degradation of the performance of baselines due to insufficient parameter tuning.

3.2 Performance Analysis for Graph Classification

Table 2 shows the graph classification performance of all the baselines, our proposed algorithm HybridPool and its two simpler variants. We have grouped the algorithms according to their types in both the tables. We marked the places to be NA where the performance of some baseline algorithm is not present on a particular dataset for graph classification in the existing literature. Otherwise, all the results are obtained from the state-of-the-art research papers [11,17] and best accuracy is reported when multiple of such results are present. As the number of algorithms presented in each of the tables are large in number, we also show the rank (1 being the best) of HybridPool among all the algorithms. For the bioinformatics graph datasets, HybridPool is able to improve the state-of-the-art for MUTAG and PROTEINS dataset. For PTC dataset, HybridPool is able to reach very close to 3WL-GNN, which is a very recently proposed graph neural network technique for graph classification. For NCI1 and NCI109, WL kernel consistently outperform other algorithms. For the social network datasets, HybridPool improves the state-of-the-art performance on all of REDDIT-M-12K, IMDB-B and IMDB-M. 3WL-GNN and GIN also perform well among the baselines.

Also, we can see that in most of the cases the HybridPool is able to outperform MaxPool and BLSTMPool. Moreover, the performance of BLSTMPool is better than MaxPool, except on PTC. Note that, MaxPool does not have the BLSTM layer and the self attention layer of HybridPool. Similarly, BLSTMPool does not have the self attention layer of HybridPool. Thus the relative performance of these three algorithms show the marginal positive contribution of both the BLSTM layer and the self attention layer to boost the performance of HybridPool. In addition, the overall standard deviation of accuracies of the HybridPool is at par or often less than many of the baseline algorithms. This shows the stability of HybridPool.

4 Discussion and Future Work

Graph neural network is an important area of research in machine learning and data mining. In this paper, we introduce a novel graph pooling strategy by combining global and hierarchical pooling techniques for a graph. Experimentally, we are able to improve the state-of-the-art performance on multiple graph datasets. Our framework can be extended in two ways. First, one can replace the GIN update rule inside the embedding and pooling layers of HybridPool with other types of node aggregation strategies [4,7]. Second, one can try different types of attention mechanisms [14] in our framework.

References

1. Bandyopadhyay, S., Lokesh, N., Murty, M.N.: Outlier aware network embedding for attributed networks. In: Proceedings of the AAAI Conference on Artificial Intelligence, vol. 33, pp. 12–19 (2019)
2. Clauset, A., Moore, C., Newman, M.E.: Hierarchical structure and the prediction of missing links in networks. Nature **453**(7191), 98 (2008)
3. Duvenaud, D.K., et al.: Convolutional networks on graphs for learning molecular fingerprints. In: Advances in Neural Information Processing Systems, pp. 2224–2232 (2015)
4. Hamilton, W., Ying, Z., Leskovec, J.: Inductive representation learning on large graphs. In: Advances in Neural Information Processing Systems, pp. 1025–1035 (2017)

5. Hochreiter, S., Schmidhuber, J.: Long short-term memory. Neural Comput. **9**(8), 1735–1780 (1997)
6. Kingma, D.P., Ba, J.: Adam: a method for stochastic optimization. arXiv preprint arXiv:1412.6980 (2014)
7. Kipf, T.N., Welling, M.: Semi-supervised classification with graph convolutional networks. In: International Conference on Learning Representations (2017)
8. Knyazev, B., Taylor, G.W., Amer, M.: Understanding attention and generalization in graph neural networks. In: Advances in Neural Information Processing Systems, pp. 4204–4214 (2019)
9. Lee, J.B., Rossi, R., Kong, X.: Graph classification using structural attention. In: Proceedings of the 24th ACM SIGKDD International Conference on Knowledge Discovery & Data Mining, pp. 1666–1674. ACM (2018)
10. Lee, J., Lee, I., Kang, J.: Self-attention graph pooling. In: International Conference on Machine Learning, pp. 3734–3743 (2019)
11. Maron, H., Ben-Hamu, H., Serviansky, H., Lipman, Y.: Provably powerful graph networks. In: Advances in Neural Information Processing Systems, pp. 2153–2164 (2019)
12. Morris, C., et al.: Weisfeiler and leman go neural: higher-order graph neural networks. In: Proceedings of the AAAI Conference on Artificial Intelligence, vol. 33, pp. 4602–4609 (2019)
13. Shervashidze, N., Schweitzer, P., Leeuwen, E.J.V., Mehlhorn, K., Borgwardt, K.M.: Weisfeiler-Lehman graph kernels. J. Mach. Learn. Res. **12**(Sep), 2539–2561 (2011)
14. Vaswani, A., et al.: Attention is all you need. In: Advances in Neural Information Processing Systems, pp. 5998–6008 (2017)
15. Veličković, P., Cucurull, G., Casanova, A., Romero, A., Lio, P., Bengio, Y.: Graph attention networks. In: International Conference on Learning Representations (2018). https://openreview.net/forum?id=rJXMpikCZ
16. Vishwanathan, S.V.N., Schraudolph, N.N., Kondor, R., Borgwardt, K.M.: Graph kernels. J. Mach. Learn. Res. **11**(Apr), 1201–1242 (2010)
17. Xu, K., Hu, W., Leskovec, J., Jegelka, S.: How powerful are graph neural networks? In: International Conference on Learning Representations (2019). https://openreview.net/forum?id=ryGs6iA5Km
18. Xu, K., Li, C., Tian, Y., Sonobe, T., Kawarabayashi, K.I., Jegelka, S.: Representation learning on graphs with jumping knowledge networks. In: International Conference on Machine Learning, pp. 5449–5458 (2018)
19. Yang, Z., Yang, D., Dyer, C., He, X., Smola, A., Hovy, E.: Hierarchical attention networks for document classification. In: Proceedings of the 2016 Conference of the North American Chapter of the Association for Computational Linguistics: Human Language Technologies, pp. 1480–1489 (2016)
20. Ying, Z., You, J., Morris, C., Ren, X., Hamilton, W., Leskovec, J.: Hierarchical graph representation learning with differentiable pooling. In: Advances in Neural Information Processing Systems, pp. 4800–4810 (2018)
21. Yu, Z., et al.: Using bidirectional LSTM recurrent neural networks to learn high-level abstractions of sequential features for automated scoring of non-native spontaneous speech. In: 2015 IEEE Workshop on Automatic Speech Recognition and Understanding (ASRU), pp. 338–345. IEEE (2015)
22. Zhou, P., et al.: Attention-based bidirectional long short-term memory networks for relation classification. In: Proceedings of the 54th Annual Meeting of the Association for Computational Linguistics (vol. 2: Short Papers), pp. 207–212 (2016)

Maximizing Explainability with SF-Lasso and Selective Inference for Video and Picture Ads

Eunkyung Park[1(✉)], Raymond K. Wong[1], Junbum Kwon[2], and Victor W. Chu[3]

[1] Computer Science and Engineering, University of New South Wales, Sydney, Australia
{eunkyung.park,ray.wong}@unsw.edu.au
[2] School of Marketing, University of New South Wales, Sydney, Australia
junbum.kwon@unsw.edu.au
[3] SPIRIT Centre, Nanyang Technological University, Singapore, Singapore
wchu@ntu.edu.sg

Abstract. There is a growing interest in explainable machine learning methods. In our investigation, we have collected heterogeneous features from two series of YouTube video ads and seven series of Instagram picture ads to form our datasets. There are two main challenges that we found in analysing such data: i) multicollinearity and ii) infrequent common features. Due to these issues, standard estimation methods, such as OLS, Lasso, and Elastic-net, are only able to find a small number of significant features. This paper proposes a method called Significant Feature Lasso (SF-Lasso) to maximize model explainability by identifying most of the significant features that affect a target outcome (such as online video and picture ad popularity). Experiments show that SF-Lasso is able to identify much more significant features while maintaining similar prediction accuracy as what Lasso and Elastic-net can obtain. The human evaluation shows that SF-Lasso is better at identifying true features that appeal to ad viewers. We also find that the number of significant features is mainly affected by the model size (i.e., the number of active variables) and the correlations among explanatory variables.

Keywords: Explainable models · Lasso · Significance test

1 Introduction

Machine learning has received massive attention due to its learning and prediction abilities. Most recently, there is a growing demand for explainable models [6], as decision-makers often need to fully understand what factors drive the outcomes in many practical applications.

For example, advertising professionals want to know as many effective content features (e.g., visual objects, spoken words, and text descriptions) to explain ad

K. Karlapalem et al. (Eds.): PAKDD 2021, LNAI 12712, pp. 566–577, 2021.
https://doi.org/10.1007/978-3-030-75762-5_45

popularity. These compelling features become key inputs to come up with better content ideas for their next ads. In this paper, we aim to identify the maximum number of statistically significant content features that affect a target outcome (i.e., the number of likes in YouTube Video and Instagram Picture ads).

Recently, there are renewed interests in feature selection to explain a target outcome in the machine learning community by making the second model after running Deep Neural Network (DNN) [2,14,16,19]. However, these recent approaches are not ideal for our data due to its small sample size. Deep learning requires enough observations. Moreover, a conservative test is required in feature selection because some features might be chosen by random chance in a small sample.

However, most existing methods only rely on the magnitude of estimated coefficients. They choose features with large coefficients [3] or non-zero coefficients from Least Absolute Shrinkage and Selection Operator (Lasso) [18]. As an alternative method, Hastie et al. [8] suggests determining the statistical strength of the selected features using p-values. The estimated coefficients need to have a big size and enough small variance to survive in this statistical test. Unfortunately, recent post-hoc feature selection methods do not provide the framework for statistical tests using p-values yet.

Therefore, we employ explainable linear models in this paper. Ordinary Least Square (OLS), Lasso, and Elastic-net are commonly used to estimate linear models. However, they have difficulty in identifying significant features from certain types of data. In particular, in video ad content analysis, we notice that the three estimation methods can only find a few or even nil significant ad content features. As such, they offer immaterial insight to creative directors and advertising managers in identifying significant ad content features.

Unfortunately, video and picture ad contents exhibit the following unfavorable properties making their analysis a challenging task.

1. **Multicollinearity**
   Small-n-large-p, which means that the number of features (p) is greater than the number of observations (n). As creating video or picture ads is expensive, the number of video or picture ads is limited. On the other hand, video and picture ads have many content features. For example, World Vision US has only 529 YouTube videos, but those videos have 7,370 unique content features. In this scenario, explanatory variables exhibit high correlations.
2. **Infrequent common features across samples**
   Since the ad contents need to be creative, the content features vary substantially across different ads. When content heterogeneity is high, there are many infrequent content features. Let us use zero to represent a missing feature value. Some features have many zeros or even all zeros across ads in the validation set. We call them "unmatched" variables because they have non-zero values in the training set but all zero values in the validation set. For example, in Apple Instagram data, 83% of variables have all zeros in their validation set.

In this paper, we propose SF-Lasso to maximize the explainability of a target outcome by maximizing the number of significant features. Considering two opposing factors: model size (i.e., the number of active variables) and the correlations among the active variables, SF-Lasso chooses λ, generating the largest number of significant features via statistical significance test using selective inference [11,17] in training data. On the other hand, Lasso chooses λ, which gives the highest prediction accuracy in validation data.

Using two YouTube video and seven Instagram picture ads as the empirical study, our proposed SF-Lasso shows significantly improved results. It identifies much more significant content features over all the nine ads datasets while OLS, Lasso, and Elastic-net obtain much less or even nil significant features. By comparing with "true" appealing features in World Vision videos from human evaluators, we find that SF-Lasso shows up to 25 times (US), 14 times (CA) better accuracy than other state-of-art models. Importantly, for predicting the number of likes with test data, the prediction error of SF-Lasso is similar to those of Lasso and Elastic-net.

2 Related Work

2.1 Explainable Models

Intrinsic explainable models such as linear models provide a straightforward relationship between input features and a target outcome based on coefficients' magnitude. Such examples are active variables with non-zero coefficients in Lasso, and significant features resulted from the statistical test. In contrast, the post-hoc explainability approach typically runs complex DNN first for high prediction accuracy and then runs the second model to explain the first model's prediction by selecting input features. For feature selection, DeepLIFT [15] decomposes the output of DNN on each input and L2X [2] maximizes the mutual information between selected features and the response variable. ACD [16] is a hierarchical algorithm for identifying clusters of features. Given that our data have small observations, which would not be enough to train DNN, we extend Lasso, which is an intrinsic explainable model. This allows us to do a statistical test for selecting features using recently developed selective inference which we introduce next.

2.2 Lasso and Its Variants

Many of Lasso's variants using regularization parameters such as Elastic-net[20] have been proposed to improve prediction accuracy. Enumerate Lasso [5] enumerates solutions with different supports. There have been attempts to explain black-box models by incorporating Lasso. Neural Lasso [13] applies $l1$ and $l2$ penalties to their input gradient. To summarize, most Lasso variants focus on improving prediction accuracy or finding more or better active variables, but not finding more significant features.

2.3 Selective Inference for Lasso

To test the significance of the estimated coefficients after Lasso, OLS post-Lasso [1] proposed to rerun OLS with the active variables resulted from Lasso. However, this naive approach results in many false significant features [8,11,17]. Covariance Test [12] pioneered for significance test after variable selection when signal variables are not too correlated with noise variables [8]. For more general explanatory variables, Lee et al. [11] drive closed-form p-values for selected active variables after fitting Lasso with a fixed value of hyperparameter λ. Taylor and Tibshirani [17] extend these results to generalized regression models such as logistic regression.

3 Intrinsic Explainable Linear Models

This paper is to identify which ad content features explain their popularity significantly. While neural network methods are often used if the goal is prediction, linear models are employed if the goal is to explain the effect of explanatory variables $X \in \mathbb{R}^{n \times p}$ (e.g., visual objects, text) on a response variable $y \in \mathbb{R}^n$ (e.g., the number of likes of YouTube video ads and Instagram ads) as follows. $y = X\beta + \epsilon$ where $\beta \in \mathbb{R}^p$ is a vector of regression coefficient and $\epsilon \in \mathbb{R}^n$ represents noises.

3.1 Predictability Maximized Methods

Among many solutions to fit the above linear model, by far, the most popular methods are OLS and Lasso. We also add Elastic-net to compare with other models. Given the values of regularization parameters, those methods estimate regression parameters (β) by purely minimizing prediction error using the following objective function of Elastic-net:

$$\min_{\beta} \parallel y - X\beta \parallel_2^2 + \lambda[\alpha \parallel \beta \parallel_1 + (1-\alpha) \parallel \beta \parallel^2], \tag{1}$$

where $\lambda > 0$ is a regularization parameter and $0 \leq \alpha \leq 1$ is a mixing parameter between Ridge ($\alpha = 0$) [9] and Lasso ($\alpha = 1$). It becomes OLS when $\lambda = 0$.

However, in this paper, we are more interested in maximizing explainability than predictability. In the next section, we illustrate why these "Predictability Maximized" methods may fail in identifying significant features with different reasons and then propose our novel "Explainability Maximized" method.

OLS ($\boldsymbol{\lambda = 0}$ in Eq. 2**).** Since we can assume that OLS's standardized coefficient follows t-distribution, one can test whether the estimated coefficient is significantly different from zero easily. Its t-statistics and p-value are

$$t = \hat{\beta}_i \;/\; \sqrt{\mathrm{Var}(\hat{\beta}_i)}, \text{and} \tag{2}$$

$$p\text{-value} = 2 \times 1 - P(T \leq t)) \text{ if } \hat{\beta}_i > 0, \;\; 2 \times P(T \leq t), \text{ otherwise.} \tag{3}$$

As t-statistics deceases, the p-value increases. If p-value is greater than a significance level (e.g., 0.05), one cannot reject the null hypothesis that $\hat{\beta} = 0$. Although the estimated coefficient0 $\hat{\beta}$ is not zero, t-statistics can be close to zero if its variance is big. In this sense, statistical tests using p-value is a stronger test than other feature selection methods such as Lasso, which selects active variables as long as their coefficients are not zero.

One could throw away some x variables (e.g., frequent low variables) to make $p < n$ as our X is not full rank, and the estimated coefficients are no longer unique [7]. However, OLS still suffers from multicollinearity, albeit a high correlation among x variables. It is because the number of x variables is still close to the number of observations. The amount of redundancy of x_i with the rest of x variables can be measured by R-squared statistics (R_i^2), which is the proportion of the variance in x_i that is predictable from the other x variables. R_i^2 never decreases as the number of x increases. Hence, when there are many x variables, the chance of multicollinearity is high.

Variance Inflation Factor (VIF) can be calculated by

$$\text{VIF}_i = 1 \;/\; (1 - R_i^2), \tag{4}$$

where $i = 1, 2, ..., p$ and R_i^2 is the square of the coefficient of the multiple correlations for the regression of x_i on the other x variables. The variance of the estimated coefficient is

$$\text{Var}(\hat{\beta}_i) = (\sigma^2 \;/\; (n-1)\text{Var}(x_i))\text{VIF}_i. \tag{5}$$

Therefore, one can see that multicollinearity (i.e., high R_i^2) causes high VIF as in Eq. (4) and so increases variance of the estimated coefficient according to Eq. (5). This finally decreases t statistics as in Eq. (2) and so increases p-value as in Eq. (3). With a high p-value (>0.05), we cannot reject the null hypothesis that the estimated coefficient is zero. When one x variable has a high VIF, many other x variables are likely to have high VIFs because redundancy is interdependently measured. Therefore, when there are many x variables and multicollinearity happens, the variance of many estimated coefficient increases together. As a result, one can not find many significant explanatory variables.

Lasso ($\alpha = 1$ in Eq. 2). When several x variables are highly correlated with each other, Lasso is likely to select only one or, at most, a few active variables and exclude others [10]. As a result, active variables resulted from Lasso tend to have low VIFs. Therefore, Lasso is free from multicollinearity among active variables and can identify more significant features than OLS. However, Lasso, which focuses on prediction accuracy from unseen validation data, may not achieve our goal to identify many significant x variables. The justification to use validation data for prediction accuracy is originated from the assumption that validation data do not possess too different patterns from training data. However, our training and validation data are quite different. Because "unmatched variables" have all zeros in the validation set, their coefficients do not contribute to the prediction value at all in the validation set. Thus, when choosing λ, "matched variables"

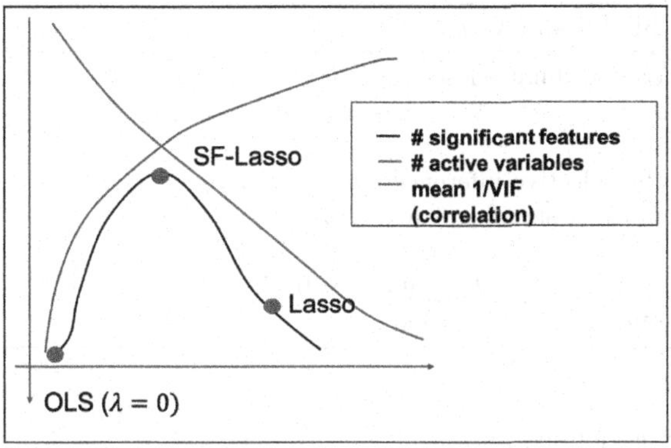

Fig. 1. # significant features is affected by # active variables and correlation among the active variables

are forced to do more significant roles. For the only subset of variables to have attention, Lasso needs to penalize parameters more (i.e., bigger λ), which, in turn, forces some of the "matched variables" inactive as well. Thus, Lasso ends up with very few active variables in the existence of many "unmatched variables". In other words, there is "inflated regularization" due to the "unmatched variables". This is problematic because it reduces the number of candidates of significant features in training data.

Elastic-net (Eq. 2). If there are many "unmatched" variables due to infrequent content features, like Lasso, Elastic-net will also suffer from "heavy" regularization, as seen in Eq. (1), and it results in only a few active variables and a small number of significant features.

4 Proposed Explainability Maximized Method

Figure 1 shows that as λ decreases, the number of active variables increases, but the correlation among the active variables also increases. This relationship suggests that too big or too small values of λ are not good at identifying significant features. When λ is very big (e.g., Lasso, Elastic-net), only a small number of active variables is generated, so multicollinearity is not likely to happen. However, only a few active variables result in even fewer significant features. On the other hand, When λ is very small (e.g., $\lambda = 0$ for OLS), there are many active variables, so the correlation among active variables is high. This multicollinearity reduces the number of significant features.

Considering those two competing forces, one can expect that the maximum number of significant features is likely to occur at some point of λ where there is a good enough number of active variables while the correlation among x variables is not too high. Based on the above behaviors, we propose a novel SF-Lasso to maximize the number of significant features. The objective function for our

Algorithm 1. SF-Lasso (X, y)

1: $\mathcal{P} \leftarrow$ Extended coordinate descent(X, y)
2: **for** $\lambda \in \mathcal{P}$ **do**
3: $A =$ Lasso (X, y)
4: Conduct the selective inference (A)
5: Count number of significant features
6: **end for**
7: $\lambda^*_{\text{SF-Lasso}} = \arg\max_\lambda \sum_{i=1}^{p} I(p\text{-value}_{\beta_i} \leq 0.05)$
8: **return** The optimal solution $\lambda^*_{\text{SF-Lasso}}$

proposed model is defined as

$$f(\lambda) = \max \sum_{i=1}^{p} I(p\text{-value}_{\beta_i} \leq 0.05),$$

where $I = 1$ if p-value$_{\beta_i} \leq 0.05$ or $I = 0$ otherwise.

For the efficient search of regularization path $\mathcal{P}$, we adopt the regularization path of coordinate descent [4]. As smaller λ generates more active variables, we search more λ values from λ_{min} to 0 with an interval, which is the difference of the last two λ values (i.e., $\lambda_{\text{min+1}}$ - λ_{min}). For a given λ, we select active variables (A) with non-zero coefficients that minimize the objective function with $\alpha = 1$ in Eq. (1) for Lasso.

For testing each selected variable's significance, we apply the selective inference [11,17]. Then, we count the number of significant features. We repeat this process within a range of λ values. Finally, we choose λ that generates the maximum number of significant features. The whole algorithm of SF-Lasso is summarized in Algorithm 1.

4.1 Selective Inference

We conduct statistical testing to find significant features using selective inference. The main idea is to account for the fact that active variables selected from adaptive methods such as Lasso have non-zero coefficients in doing a significance test [8]. The event that the Lasso selects a certain model M (i.e., a set of active variables) by assigning its non-zero coefficients and certain signs can be written as the polyhedral form with a set of linear constraints $\{Ay \leq b\}$, where the matrix A and vector b depend on X and the selected model M. The details of A and b are spelled out in the papers [11,17]. $\{y|Ay \leq b\}$ is the set of random response vectors y that would yield the same active variables and their coefficient signs with the selected model. Now suppose that $y \sim N(\mu, \sigma^2 I)$, where $\mu = X\beta$. With the reduced model M, the estimated coefficient vector is $\hat{\beta} = (X_M^T X_M)^{-1} X_M^T y$. Its population parameter is $\beta = (X_M^T X_M)^{-1} X_M^T \mu$, which is the coefficients in the projection of μ on X_M. Individual coefficient is $\beta_i = \eta^T \mu$, where η is one of the columns $X_M(X_M^T X_M)^{-1}$. Its estimated value is $\hat{\beta}_i = \eta^T y$. The main purpose of selective inference is to make inference on a linear combination of μ, which is $\beta_i = \eta^T \mu$ conditional on the event $\{Ay \leq b\}$. Conditional on the selected model, $\hat{\beta}_i|\{Ay \leq b\} = \eta^T y|\{Ay \leq b\}$ follows truncated normal distribution with support determined by A and b

$$\eta^T y|\{Ay \leq b\} \overset{\mathrm{d}}{=} TN(\eta^T \mu, \sigma^2 \|\eta\|_2^2, [\nu^-(y), \nu^+(y)]),$$

$$\{Ay \leq b\} = \{v^-(y) \leq \eta^T y \leq v^+(y), v^0(y) \geq 0\},$$

where
$$\alpha = \frac{A\eta}{\|\eta\|_2^2}, \nu^-(y) = \max_{j:\alpha<0} \frac{b_j - (Ay)_j + \alpha_j \eta^T y}{\alpha_j},$$

$$\nu^+(y) = \min_{j:\alpha>0} \frac{b_j - (Ay)_j + \alpha_j \eta^T y}{\alpha_j}, \nu^0(y) = \min_{j:\alpha=0} (b_j - (Ay)_j).$$

The cumulative distribution function of a random variable, evaluated at the value of that random variable, has uniform distribution as follows.

$$\begin{aligned} G(\eta^T y) &= F^{\nu_-,\nu_+}_{\eta^T \mu, \sigma^2 \|\eta\|_2^2}(\eta^T y)|\{Ay \leq b\} \\ &= \frac{\Phi((\eta^T y - \eta^T \mu)/\sigma\|\eta\|_2) - \Phi((\nu^- - \eta^T \mu)/\sigma\|\eta\|_2)}{\Phi((\nu^+ - \eta^T \mu)/\sigma\|\eta\|_2) - \Phi((\nu^- - \eta^T \mu)/\sigma\|\eta\|_2)} \sim U(0,1). \end{aligned}$$

Then, for testing $\beta_i = \eta^T \mu = 0$,

$$p\text{-value} = 2 \times (1 - G(\eta^T y)), \text{if } \hat{\beta}_i > 0, \;\; 2 \times G(\eta^T y)), \text{otherwise.} \tag{6}$$

Because the closed-form for p-value calculation is available in Eq. (6), the additional calculation time in each lambda value is trivial.

5 Empirical Evaluation

We conduct experiments using two YouTube video ads and seven Instagram picture ads data from diverse industries such as charity, cafe, beer, and fashion. We divide our datasets into the training (70%), validation (15%), and test (15%) sets. The validation set is used to tune hyperparameters for Lasso and Elastic-net, optimized by prediction accuracy on unseen data. The test set is used to assess the prediction abilities of all models on unseen data. Table 1 shows the size of training datasets, variables, and unmatched ratio. From YouTube video ads, we extract visual objects, speech tones, and spoken words using Microsoft Video Indexer. From Instagram picture ads, we extract color, face emotion, brand logo, landmark, text in the picture, and other objects using Google Vision. We also use text descriptions of the picture ads.

Table 1. Size of training, variables, and unmatched ratio

	# train	# vars	unmatch
US	370	7,370	68.9
CA	318	6,207	73.0
Starbucks	1,216	3,871	74.7
Corona	561	2,723	78.1
Nike	709	2,071	67.2
Apple	313	1,913	83.0
AWS	377	1,953	77.5
Google	695	3,961	76.8
Microsoft	364	3,744	74.7

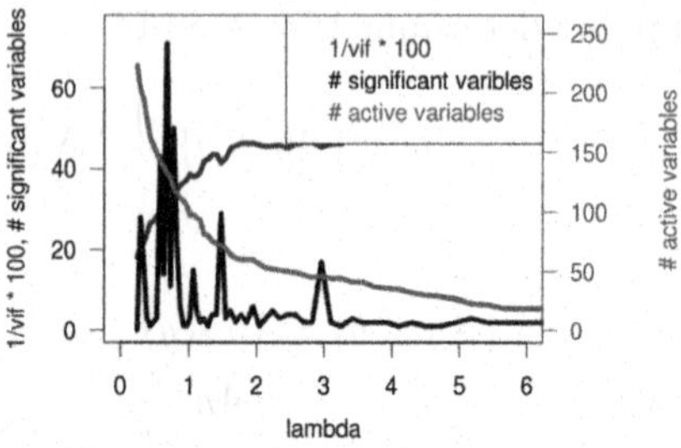

Fig. 2. Results: world vision US

5.1 Illustration of SF-Lasso

Figure 2 is the result of World Vision US YouTube Video ads. It shows that our SF-Lasso is mainly affected by two opposing factors: model size and correlations among explanatory variables. The # of active variables, and the mean of VIF, and # of significant features resemble theoretical prediction in Fig. 1. As λ decreases, because of the two opposing forces, the peaks in the # of significant features increase and then decrease after the optimum values of λ (0.68), at which SF-Lasso finds the maximum number of significant features. One exception is that # of significant features does not have a smooth function of λ. Instead, there are many low values close to zero between peaks due to the selective inference screening of some features with insufficient significance.

5.2 Experiment Results

Table 2 shows that our SF-Lasso can identify the maximum number of significant features over nine ad datasets, while others detect a small number of significant features. As seen in Fig. 1 and 2, our SF-Lasso has the optimum value of λ between OLS ($\lambda = 0$, which means zero regularization) and the other two models.

For Elastic-net, we experiment with the same ranges of λ varying α from 0.1 to 0.9 with 0.1 intervals. The optimal α values of Elastic-net in each data are 0.1, 0.1, 0.8, 0.1, 0.9, 0.6, 0.9, 0.9, 0.3, respectively.

As OLS uses the biggest number of active variables, it has the highest average VIF (multicollinearity), resulting in less significant features. On the other hand, Lasso and Elastic-net produce a relatively small number of active variables, thus less significant features despite the smallest average VIF. As discussed, Lasso and Elastic-net lose many active variables while optimizing for prediction accuracy, mainly due to heavy regularization resulted from many unmatched variables. Therefore, ad practitioners cannot get much insight from the standard models. On the contrary, our proposed SF-Lasso identifies many active variables, much

Table 2. Results from models

		λ	#act	VIF	RMSE	#sig		λ	#act	VIF	RMSE	#sig
OLS	US	0	368	1407.0	2486.5	**0**	Apple	0	311	23.5	137657.2	**37**
Lasso		34.9	4	1.3	203.0	**2**		21.3	27	1.3	2004.5	**1**
Elastic		276.4	7	1.6	204.0	**1**		32.3	32	1.4	2058.2	**10**
SFLasso		0.7	132	3.2	209.8	**71**		0.6	275	6.8	2169.0	**126**
OLS	CA	0	316	1326.6	3895.4	**0**	AWS	0	375	34.1	13459.4	**16**
Lasso		135.6	4	1.5	280.1	**0**		1.3	83	1.9	376.4	**3**
Elastic		1356.1	3	1.1	277.4	**1**		1.4	92	2.0	382.9	**1**
SFLasso		2.2	117	2.9	315.4	**40**		0.3	230	3.9	306.5	**69**
OLS	Star bucks	0	1,214	55.5	21474.5	**10**	Google	0	693	154.3	478739.8	**4**
Lasso		3.2	149	1.6	1428.3	**25**		2.8	51	1.5	422.7	**9**
Elastic		4.1	155	1.7	1432.9	**0**		3.2	52	1.5	418.7	**1**
SFLasso		0.6	275	6.8	1423.5	**72**		0.5	324	2.5	418.8	**115**
OLS	Corona	0	559	20.3	5947.6	**21**	MS	0	362	4964.3	13499.6	**0**
Lasso		33.5	4	2.4	1483.4	**1**		3.0	17	1.4	184.7	**5**
Elastic		200.9	8	4.0	1485.2	**2**		4.7	44	1.4	186.5	**0**
SFLasso		0.8	143	2.0	1450.4	**46**		0.4	140	2.0	181.5	**59**
OLS	Nike	0	707	127.3	12213.7	**4**						
Lasso		2.3	39	1.8	315.9	**5**						
Elastic		2.8	39	1.8	322.9	**11**						
SFLasso		0.3	239	2.3	369.3	**45**						

more than Lasso and Elastic-net. However, it has a much lower correlation among active variables than OLS. Because of the low correlation among relatively many active variables, SF-Lasso could identify many significant features.

Our proposed SF-Lasso has the similar prediction error as Lasso and Elastic-net. SF-Lasso has the lowest prediction error in four datasets (Starbucks, Corona Beer, AWS, Microsoft). Although SF-Lasso uses only training data in searching the maximum number of significant features, both regularization and selective inference prevent SF-Lasso from over-fitting. Lasso and Elastic-net have the lowest one in three and two test data, respectively, confirming that regularization helps avoid over-fitting. OLS has the highest error on the test set, suggesting that OLS suffers from over-fitting (Table 2).

5.3 Human Evaluations

To validate whether each model detects the right ad content features significantly appealed to ad viewers, we employ a human evaluation to validate the significant features resulted from all the four methods. One hundred video ads were randomly selected. The survey includes a URL and a list of visual objects, speech tones, and spoken words for given videos. Twenty marketing students were asked to watch video ads and circle video content features that make them like the video ad.

We use the set of content features from the survey as our ground-truth sets and compare them with significant features resulted from each model. Note that there are many video content features: 7,370 and 6,207 features across all the train set videos. The human evaluators chose 269 (4%) and 356 (6%) unique

Table 3. Significant content features - World Vision US

OLS (0/0)	
Lasso (2/2)	⟨visual objects⟩ **mammal(1)** ⟨spoken words⟩ **water(1)**
Elastic (1/1)	⟨spoken words⟩ **mammal(1)**
SF-Lasso (51/71)	⟨visual objects⟩ **outdoor(10) person(8) sky(8) women(7) plant(3) watersport(3) yellow(3) young(3) chair(2) cow(2) food(2) house(2) little(2) orange(2) standing(2) brick(1) chicken(1) clothing(1) group(1) people(1) rock(1) rollingcredits(1) suit(1) sheep(1) text(1)** car cup curtain floor glasses militaryuniform shelf sitting tattoo television ⟨spoken words⟩ **day(5) life(5) love(5) make(5) food(4) people(4) believe(3) community(3) gift(2) healthy(2) mom(2) program(2) really(2) break(1) clean(1) continue(1) end(1) happen(1) follow(1) important(1) information(1) letter(1) parent(1) stuff(1) time(3)** Anna Celisa close listen local midwife night photo Rwanda Sabina student

features from each country's videos. Therefore, accuracy between the evaluators' answers and each model's outcomes is likely to be small.

From the survey, SF-Lasso obtains 37.8 and 14.1 in F1 for the World Vision US and CA, respectively, which are almost from 14 times to 25 times bigger than Lasso and Elastic-net, respectively. As OLS (for both US and CA) and Lasso (for CA) do not detect any significant feature, precision, recall, and F1 are zero. Lasso in the US videos and Elastic-net in both US and CA show 100% in precision. However, they record very low recall and F1 as they identify only one or two significant feature(s). In other words, Lasso and Elastic-net identify the right features that are appealed to ad viewers, but a too small number of content features.

Table 3 shows statistically significant features from each model and 'true' content features by the survey participants. Our SF-Lasso hits 51 variables among 71. The numbers within parentheses mean the number of videos where the evaluators like the video feature. The most frequently selected (in more than five videos) visual objects are 'outdoor'(10), 'person'(8), 'sky'(8), and 'women'(7), and spoken words are 'day'(5), 'life'(5), 'love'(5), and 'make'(5) and only SF-Lasso identifies these features, suggesting that SF-Lasso has better ability to identify more important features to ad viewers than other methods.

6 Conclusion

Data from many applications may have small-n-large-p and contain heterogeneous features. Such data often have multicollinearity and infrequent common variables across samples. When explainability matters, traditional methods that estimate a linear model do not work well. This paper has proposed SF-Lasso to address these problems and verified its effectiveness. Our proposed model has discovered many significant content features, and our human survey has verified the identified ad features. While providing better explainability, SF-Lasso maintains a similar degree of prediction accuracy with Lasso and Elastic-net.

Acknowledgements. This research is supported by the Australian Government Research Training Program Scholarship.

References

1. Belloni, A., Chernozhukov, V.: Least squares after model selection in high-dimensional sparse models. Bernoulli **19**, 521–547 (2013)
2. Chen, J., Song, L., Wainwright, M.J., Jordan, M.I.: Learning to explain: an information-theoretic perspective on model interpretation. In: ICML, vol. 80, pp. 882–891 (2018)
3. Doshi-Velez, F., Kim, B.: Towards a rigorous science of interpretable machine learning. arXiv, Machine Learning (2017)
4. Friedman, J.H., Hastie, T.J., Tibshirani, R.: Regularization paths for generalized linear models via coordinate descent. J. Stat. Softw. **33**(1), 1–22 (2010)
5. Hara, S., Maehara, T.: Enumerate lasso solutions for feature selection. In: AAAI, pp. 1985–1991 (2017)
6. Harder, F., Bauer, M., Park, M.: Interpretable and differentially private predictions. In: AAAI, pp. 4083–4090 (2020)
7. Hastie, T., Tibshirani, R., Friedman, J.: The Elements of Statistical Learning. Springer Series in Statistics. Springer, New York (2001)
8. Hastie, T., Tibshirani, R., Wainwright, M.: Statistical Learning with Sparsity: The Lasso and Generalizations (2015)
9. Hoerl, A.E., Kennard, R.W.: Ridge regression: biased estimation for nonorthogonal problems. Technometrics **12**(1), 55–67 (1970)
10. Kim, Y., Bum Kim, S.: Collinear groupwise feature selection via discrete fusion group regression. Pattern Recognit. **83**, 1–13 (2018)
11. Lee, J.D., Sun, D.L., Sun, Y., Taylor, J.E.: Exact post-selection inference, with application to the lasso. Ann. Stat. **44**(3), 907–927 (2016)
12. Lockhart, R., Taylor, J., Tibshirani, R.J., Tibshirani, R.: A significance test for the lasso. Ann. Stat. **42**(2), 413–468 (2014)
13. Ross, A.S., Lage, I., Doshi-Velez, F.: The neural lasso: local linear sparsity for interpretable explanations. In: NIPS (2017)
14. Shrikumar, A., Greenside, P., Kundaje, A.: Learning important features through propagating activation differences. In: ICML, pp. 3145–3153 (2017)
15. Shrikumar, A., Greenside, P., Kundaje, A.: Learning important features through propagating activation differences. In: ICML, vol. 70, pp. 3145–3153 (2017)
16. Singh, C., Murdoch, W.J., Yu, B.: Hierarchical interpretations for neural network predictions. In: ICLR (2019)
17. Taylor, J., Tibshirani, R.: Post-selection inference for $l1$-penalized likelihood models. Can. J. Stat. **46**(1), 41–61 (2018)
18. Tibshirani, R.: Regression shrinkage and selection via the lasso. J. Royal Stat. Soc. Series B (Methodol.) **58**(1), 267–288 (1996)
19. Tu, M., Huang, K., Wang, G., Huang, J., He, X., Zhou, B.: Select, answer and explain: interpretable multi-hop reading comprehension over multiple documents. In: AAAI, pp. 9073–9080 (2020)
20. Zou, H., Hastie, T.: Regularization and variable selection via the elastic net. J. Roy. Stat. Soc. B **67**, 301–320 (2005)

Reliably Calibrated Isotonic Regression

Otto Nyberg(✉) and Arto Klami

University of Helsinki, Helsinki, Finland
{otto.nyberg,arto.klami}@helsinki.fi
https://www.helsinki.fi

Abstract. Using classifiers for decision making requires well-calibrated probabilities for estimation of expected utility. Furthermore, knowledge of the reliability is needed to quantify uncertainty. Outputs of most classifiers can be calibrated, typically by using isotonic regression that bins classifier outputs together to form empirical probability estimates. However, especially for highly imbalanced problems it produces bins with few samples resulting in probability estimates with very large uncertainty. We provide a formal method for quantifying the reliability of calibration and extend isotonic regression to provide reliable calibration with guarantees for width of credible intervals of the probability estimates. We demonstrate the method in calibrating purchase probabilities in e-commerce and achieve significant reduction in uncertainty without compromising accuracy.

Keywords: Isotonic regression · Calibration · E-commerce

1 Introduction

Even though classification is in academic contexts often studied in isolation, in real use the predictions are used for making decisions with associated costs and benefits. Evaluating task performance requires knowledge of the probability of each possible class, and for accurate evaluation we need *well-calibrated* probabilities [10,20,22]. A binary classifier is said to be well-calibrated if the empirical probability (relative frequency) of the positive class $p(y = 1|s(x) = x)$ converges to the output score $s(x)$ of the classifier at the limit of infinite data.

The definition extends naturally to multi-class problems, but for simplicity of notation we consider binary problems used e.g. in medical diagnosis [2], credit scoring [8] and e-commerce [9,19].

Most classifiers do not directly produce well-calibrated probabilities. Many methods like deep neural networks with logistic outputs or many decision trees formally output probabilities, but they are often poorly calibrated in practice [6,14], whereas other models like support vector machines output scores that are larger when the classifier is more certain of the result but do not even attempt to represent probabilities. The outputs of all such classifiers can be calibrated after training. Various algorithms have been proposed for this [7,12,13,16] but the

K. Karlapalem et al. (Eds.): PAKDD 2021, LNAI 12712, pp. 578–589, 2021.
https://doi.org/10.1007/978-3-030-75762-5_46

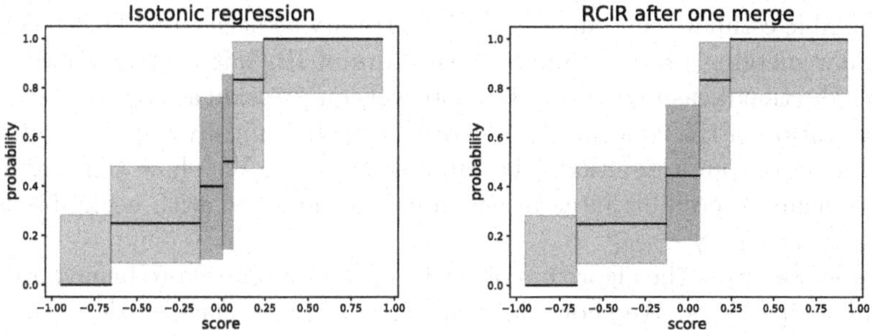

Fig. 1. Isotonic regression (left) calibrates classifier outputs by binning, but the probability estimates (horizontal lines) are unreliable as shown here using 95% credible intervals (boxes) on artificial data. Reliably calibrated isotonic regression merges bins further (right figure shows the result after one merge combining the two bins indicated by blue), retaining calibration and monotonicity while reducing uncertainty. In this example the maximum credible interval width decreased by 22% in one step, and on real data we can often achieve 70–90% reduction in uncertainty while retaining almost identical accuracy – see Fig. 2 for an example. (Color figure online)

classical method of *isotonic regression (IR)* [15,22] remains the most common approach; it works particularly well with large sample sizes and is robust over different classification problems. For an overview and empirical comparison of calibration methods in context of neural network classifiers, arguably the most relevant practical model family today, see [6].

Isotonic regression produces a function that maps the raw scores to well-calibrated probabilities, but the probabilities are calibrated only for the training data; each bin is associated with an empirical estimate for the probability but the algorithm provides no guarantees on the true value. We analyse isotonic regression from a perspective of reliability. We use Bayesian statistics to provide credible intervals for the true bin probability and observe that in many cases the credible intervals for IR outputs are extremely wide. This makes the method fragile and unreliable, even though the probability estimates are unbiased. The problem occurs in particular in imbalanced classification problems where the bins capturing the rare positive samples are the most uncertain, causing problems especially for the samples that are most interesting. As an example, in e-commerce the task is often to predict customer conversion with rates below 3%, and sometimes as small as 0.2% [3]. Quantifying the uncertainty is also important for other tasks such as medical diagnostics [20], credit scoring [8], and in modern portfolio management theory where variance is one of the fundamental building blocks [18].

To address the high uncertainty of IR, we extend isotonic regression in a manner that provides reliable estimates, coining the method *Reliably Calibrated Isotonic Regression (RCIR)*. The goal of the method is to provide well-calibrated probability estimates that additionally have low variance and probabilistic guarantees for maximum deviation, so that in downstream applications we can trust the available estimates and handle the uncertainty as necessary. This is achieved by similar merging process that is at the core of IR, but that now uses uncertainty reduction measured by credible interval compression as criterion; see Fig. 1 for illustration of the concept. The algorithm monotonically improves reliability by reducing output resolution (the number of bins). We show that significant improvements in credible interval width can be achieved with negligible loss in testing set accuracy.

We demonstrate the algorithm on real world data relating to behavioral modeling of online users. We show that for typical binary classification tasks with class-imbalance, regular isotonic regression produces bins with extremely wide credible intervals despite relatively large total sample size. Our algorithm compresses the credible intervals to a fraction of the original ones without compromising the accuracy of the classifier.

2 Background: Isotonic Regression

To provide sufficient background for the rest of the paper, we start by a recap of isotonic regression and its limitations.

Assume that we have a binary classifier $s(\mathbf{x}_i)$ that outputs scores z_i corresponding to the input features $\mathbf{x}_i$ for sample i. No assumptions on how the classifier has been trained is made, but higher scores z_i are assumed to more likely belong to the positive class. The goal of isotonic regression [22] is to calibrate the classifier so that the calibrated output, denoted by $g(z)$, matches the empirical ratio of the positive class among the training samples with the same output. That is, it seeks to find the function $g(z)$ that maps the arbitrary scores into actual probabilities of the positive class.

IR finds the function $g(\cdot)$ that minimizes the mean square error (MSE)

$$L(g) = \frac{1}{N} \sum_{i} \left(y_i - g(z_i)\right)^2 \tag{1}$$

under the constraint that g must be monotonic, i.e. $g(a) \geq g(b)$ for $a > b$. Here y_i is the dependent variable of sample i. The optimal solution is a piecewise constant function g that maps score ranges to some positive *values*. For a binary class variable $y \in \{0, 1\}$, these values are well-calibrated probabilities for the positive class [22]. The range of scores that map to a certain value is called a *bin*. At training time, a number of samples with similar scores are assigned to a bin, and the value (probability estimate) associated with the bin is the relative frequency of the positive class samples in that bin.

A global optimum of the objective (1) is obtained with the pair-adjacent violators algorithm [1] that starts by ordering all samples according to score. The algorithm then goes through all samples adjusting bin boundaries to smooth out any violations of the monotonicity, terminating when there are none left. Besides minimizing MSE, the result maximizes AUC-ROC [17]. While AUC-ROC is completely insensitive to calibration, it is an actual measure of refinement, i.e. it measures how well positive samples are separated from negative ones. As MSE can be decomposed into refinement and reliability, we think AUC-ROC together with a calibration metric is a more comprehensive measure of goodness.

Despite being frequently used, IR has limitations especially in problems with high class-imbalance. The resolution (the number of bins) of the output is determined solely by the monotonicity criterion with no user control, and for imbalanced setups some of the bins often contain very few samples [15]. For such bins there are no guarantees for the empirical estimate to accurately represent the true probability on test data. Many real-world classification problems are highly imbalanced so that the class of interest is smaller, and for such problems IR creates the smallest bins for the high probability samples. This is problematic because this happens precisely for the samples we care most about. For example, the conversion rate in online stores is typically below 3%, as is click-through rate of ads [3], and the goal is to identify users most likely to take action, yet IR provides the least reliable estimates exactly for those users.

3 Method

We first describe a formal procedure for inspecting isotonic regression in terms of reliability of the probability estimates, and then describe a practical algorithm for improving the reliability by merging bins with too high uncertainty.

3.1 Credible Intervals for Isotonic Regression

Isotonic regression assigns for each bin a single value, which in the case of binary classification corresponds to the empirical ratio of training samples in the positive class falling into this bin. This is equivalent to assuming a Bernoulli model

$$p(y = 1|s(x) = z_i) = \text{Bernoulli}(\theta_i)$$

and using maximum likelihood estimator for inferring the parameter θ_i.

We use Bayesian statistics to characterize the posterior distribution of the parameter using the conjugate prior Beta(α, β). Straightforward calculus (see, e.g. [5]) then provides the posterior

$$p(\theta_i|\text{data}) = \text{Beta}(k_i + \alpha, n_i - k_i + \beta) = \frac{\theta^{k_i+\alpha-1} \times (1-\theta)^{n_i-k_i+\beta-1}}{B(k_i + \alpha - 1, n_i - k_i + \beta - 1)},$$

where n_i indicates the number of samples falling into the ith bin, k_i is the number of samples in the positive class, and $B(a, b)$ is the beta-function. We use $\alpha = \beta = 1$ to indicate uniform prior, but other choices would be equally easy to implement. We characterize the posterior distribution using highest posterior density (HPD) credible intervals [11] $H(p(\cdot), c)$ corresponding to the smallest continuous range of parameter values capturing a given total mass $1 - c$ (where the confidence level c is often set to 0.05) of the posterior such that for all $\theta \in H(p(\cdot), c)$ we have $p(\theta|\text{data}) > p_0$ for some threshold p_0. This range is typically not symmetric around the posterior mean, but captures the intuitive idea of credible alternatives better than central intervals – every point within the region has higher posterior density than any of the points outside it.

Since the beta distribution is unimodal, we can efficiently find the HPD interval by binary search in the log domain. Defining the credible interval as $[l, h]$, we start by making an initial guess for l and find h such that $p(\theta = h|\text{data}) = p(\theta = l|\text{data})$ and $l < h$. Here the denominator can be ignored and the search is fast. We then refine l (and consequently h) using binary search until the total probability mass within the region is sufficiently close to $1 - c$.

3.2 Reliably Calibrated Isotonic Regression

The only way of obtaining narrow credible intervals is to guarantee sufficiently many samples in each bin, but the required number depends on both n and k and hence we cannot set direct thresholds upfront. We can, however, set a threshold for the width of the HPD credible interval. This provides us the formal definition of reliably calibrated isotonic regression (RCIR):

Definition 1. Reliably calibrated isotonic regression: *Given a binary classifier that outputs scores z_i for N samples with true classes y_i, minimize the objective*

$$\frac{1}{N}\sum_{i=1}^{N}(y_i - g(z_i))^2$$

under the constraints that $g(\cdot)$ is a monotonic function and for every distinct value of $g(\cdot)$ the highest posterior density credible interval at confidence level c for the estimated class probability $p(y = 1|g(z_i))$ is at most of width d.

A solution to the above problem provides well-calibrated probabilities in the same sense as regular isotonic regression, but additionally guarantees that the probability estimates provided for the bins are reliable also for future samples at a level chosen by the analyst. That is, for any given sample there are probabilistic guarantees for the estimate to be sufficiently close to the true value.

We will next present a practical approximate algorithm for solving the problem, but before that we make a general remark that motivates our algorithm. As described in Sect. 2, IR both minimizes MSE of the predictions and maximizes the AUC-ROC; it directly optimizes the former with an algorithm guaranteed to find the optimal solution under the monotonicity constraint, and the latter follows from [4]. Since RCIR incorporates additional constraints for the same objective, IR directly provides a lower bound for MSE and an upper bound for AUC-ROC for any algorithm solving the problem of Definition 1. This only holds for training data; the additional constraints may regularize the solution so that these metrics improve on test data.

3.3 Greedy Optimization Algorithm

We are not aware of an efficient algorithm guaranteed to find the global optimum of the problem in Definition 1. We can, however, derive a practical algorithm that solves it computationally efficiently and produces solutions that for practical problem instances are very close to optimal. The basic idea is to first solve the unconstrained IR problem, guaranteed to produce optimal solution without the additional constraints. We then refine the solution by merging bins with too wide confidence intervals in such a way that the monotonicity assumption is retained until all credible intervals are sufficiently narrow.

For each merge we greedily select the bin with the widest confidence interval. It can be merged either with the bin to its left or its right; both choices retain monotonicity. As we are already controlling for calibration, we wish to retain as much of the refinement as possible and hence chose the solution that maximizes AUC-ROC of the classifier after the merge. That is, we compute AUC-ROC for both alternative merges and choose the one that decreases the value less, corresponding to a better classifier.

The full algorithm is provided in Algorithm 1.[1] It is a deterministic algorithm that terminates in a finite number of steps since the number of bins is decreased by one at every step, and in practice it is computationally efficient due to merely selecting one bin at a time to be merged until all credible intervals are narrow enough. We do not have approximation bounds for the algorithm, but as illustrated in the empirical experiments the AUC-ROC of the training set of the final calibrated classifier is often extremely close (in our experiments always within 0.01%) to the AUC-ROC of the IR solution that provides an upper bound, and hence even if the solution is not optimal it cannot be far from it.

3.4 Parameter Choice

The algorithm has only two parameters, the maximum width d of credible interval for any of the bins and the associated confidence c. For c one can safely choose a standard value in the order of 0.05. The width parameter d, in turn, has a natural interpretation in terms of the decision problem since it corresponds to the maximum error one is willing to accept in probability estimates. We recommend setting this manually according to the task at hand, but would expect values between 0.1 and 0.2 to be acceptable in many problems.

In absence of domain knowledge for determining d, it can also be set automatically using cross-validation based on the intuition that good values generalize well for test samples. To find the optimal value we leave out a set of validation samples not used for training the classifier or for calibrating it, and evaluate AUC-ROC of the RCIR model on the validation set after every merge. We then choose the refinement level that corresponds to the best validation score.

4 Related Work

The value of calibration has clearly been recognized, exemplified e.g. by the observation that modern deep learning networks are typically poorly calibrated

[1] A Python implementation of the algorithm and the data used in the experiments are available at https://github.com/Trinli/calibration.

Algorithm 1: Greedy algorithm for reliably calibrated isotonic regression

Input : Real-valued scores z_i provided for N training samples with class labels $y_i \in \{0, 1\}$, a confidence level $c \in [0, 1]$, and maximum width of credible interval $d \in [0, 1]$

Output : A collection of bins B_k, each characterized by a range of input values $[a_k, b_k]$ mapped to the bin, and the associated class probabilities θ_k within highest posterior density credible interval $[l_k, h_k]$, such that $h_k - l_k \leq d$.

Initialization: Run pair-adjacent violators algorithm to produce initial bins B_k and estimate the HPD intervals $[l_k, h_k]$ for all k

```
while h_k − l_k > d for any bin k do
    Choose k = arg max h_k − l_k
    Consider merging B_k with B_{k−1} to produce new B_j:
    begin
        Set new input range to [a_{k−1}, b_k]
        Compute the new HPD interval [l_j, h_j]
        Compute AUC-ROC for a classifier that replaces b_{k−1} and b_k with b_j
    end
    Consider merging B_k with B_{k+1} to produce new B_j:
    begin
        Set new input range to [a_k, b_{k+1}]
        Compute the new HPD interval [l_j, h_j]
        Compute AUC-ROC for a classifier that replaces B_k and B_{k+1} with B_j
    end
    Choose the better of the two alternatives based on AUC-ROC
    Replace B_K and its chosen neighbor with B_J
end
```

[6], but arguably has not been receiving sufficient attention [20]. Nevertheless, in recent years a few calibration methods have been proposed to improve on the classical choices of isotonic regression, Platt scaling [16], and histogram binning. We briefly discuss these alternatives here, but note that none of these works address our main goal, the reliability of the estimates. Instead, they are motivated simply as more accurate calibration methods that explicitly optimize for calibration accuracy on the training data. Building on our work for IR, it could be possible to estimate credible intervals also for some of these other methods and develop generalizations that would optimize also for that.

Bayesian Binning by Quantiles (BBQ) [13] builds on histogram binning, forming an ensemble that combines binning models with different bin widths using Bayesian weighting for different models. Even though it achieves good calibration metrics, it has a tendency to underestimate the highest scoring samples in imbalanced cases due to using bins of constant width. The Ensemble of Near-Isotonic Regression (ENIR) [12] is also an ensemble with Bayesian weighting and builds on a variant of isotonic regression that allows small violations to the monotonicity constraint and typically achieves better empirical performance. As ensembles, both BBQ and ENIR are computationally less efficient (though this is often not a practical issue as training the classifier itself dominates the total computation) and more difficult to interpret and analyse theoretically. Importantly, neither of these optimizes for criteria that would account for reliability of the estimates and they provide no guarantees for it.

The recent scaling-binning calibration method [7] combines Platt scaling and histogram binning to achieve sample efficiency of the former with theoretical properties of the latter, achieving good calibration performance for deep neural networks and focusing in particular on sample complexity. Even though the method is motivated in part via variance reduction, they do not quantify the reliability of the probability estimates either theoretically or empirically, explicitly leaving finite sample guarantees for future work.

5 Experiments

We illustrate the method on data collected on e-commerce sites, where the classification problem concerns predicting a particular decision of the user, matching our eventual use case. For the purpose of this manuscript the specific data sets are largely irrelevant, as is the underlying classifier since we only access its outputs.

We conduct two separate experiments. First we demonstrate the method on a typical imbalanced binary classification problem, showing how we can obtain increasingly tighter bounds for the credible intervals with minimal loss in classification accuracy. The second experiment demonstrates the performance of the automatic selection of d explained in Sect. 3.4, removing the only tunable parameter in Algorithm 1.

5.1 Experiment 1

For this experiment only users that had more than two pageloads were included resulting in a generous 4.8% positive rate even though the conversion rate for the entire site is below 1.0%. We used a sample of 200,000 visitors not used to train the base classifier and randomly selected 1/3 of the data for testing.

Figure 2 illustrates how the method works, by depicting the bins and their credible intervals for both standard IR and the proposed RCIR for three choices of d. The reliability of the estimates is dramatically improved compared to regular IR, yet the refinement is sufficient as we still retain the important bins with high probability. With excessively small d (bottom right) the calibration further improves, but too much resolution is lost.

Table 1 quantifies the results using MSE and AUC-ROC evaluated on test data. The main observations are that standard IR is extremely unreliable having maximum credible interval width above $d = 0.8$, and that RCIR can dramatically improve reliability without compromising accuracy. In particular, for all $d \in [0.1, 0.3]$ AUC-ROC decreases by less than 0.004% and MSE increases by less than 0.03%. Differences this small are irrelevant in practice, yet we achieve over

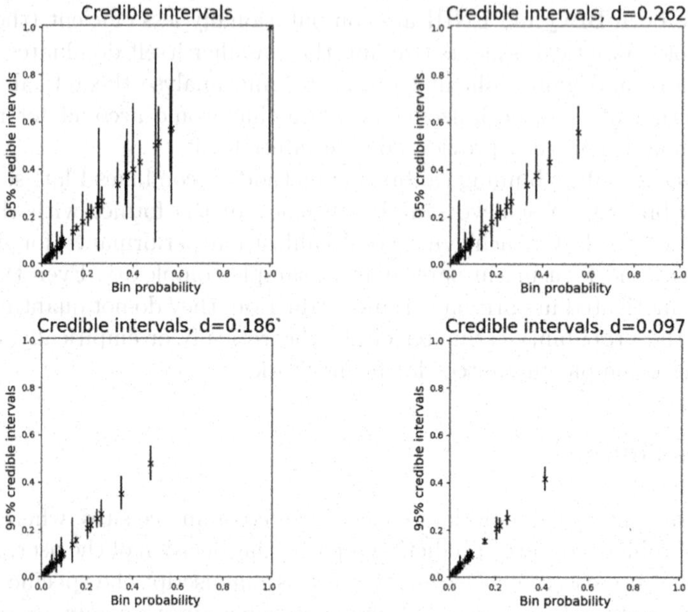

Fig. 2. The top left plot shows the result of isotonic regression with 95% credible intervals for the 45 bins produced. The remaining plots show the results of reliably calibrated isotonic regression for varying levels of reliability d. For these plots we used maximum threholds of 0.3, 0.2 and 0.1 (with $c = 0.05$), and report the actual maximum width in the plot titles.

Table 1. Goodness of fit for different algorithms demonstrating that tighter credible interval requirements (reliably calibrated isotonic regression, RCIR, with different values for d) result naturally in lower number of bins, but do not decrease the overall performance of the classifier as measured by AUC-ROC and MSE.

Model	d	Bins	AUC-ROC	MSE
Uncalibrated classifier	–	–	0.714878	–
Isotonic regression	(0.811400)	45	0.713166	0.0435589
RCIR	.30	38	0.713167	0.0435541
RCIR	.20	34	0.713138	0.0435429
RCIR	.10	28	0.713149	0.0435715
RCIR	.05	23	0.712874	0.0436705
RCIR	.01	9	0.671561	0.0447910

an 80% drop in credible interval width while retaining more than half of the bins. This indicates that the greedy algorithm is able to find a solution that is almost indistinguishable as a classifier from the original one while providing a

Table 2. Average results of 30 repeated experiments for every data set with randomization. Compared to regular isotonic regression (IR) the proposed reliably calibrated isotonic regression (RCIR) results in nearly identical AUC-ROC and MSE metrics, but reduces the maximum credible interval width more than 80%.

	Data set 1	Data set 2	Data set 3
Samples [#]	200,000	79,155	200,000
Positive rate [%]	4.799	8.040	7.482
Max credible interval (IR)	0.6586	0.7113	0.6925
AUC-ROC (IR)	0.715180	0.773431	0.760706
MSE (IR)	0.0432544	0.0634838	0.0635181
Bin merges [#] (RCIR)	7.63	11.07	12.47
Max credible interval (RCIR)	0.1266	0.08200	0.04504
AUC-ROC (RCIR)	0.715179	0.773388	0.760740
MSE (RCIR)	0.0433455	0.0644942	0.0678055

formal reliability guarantee. The accuracy starts to drop noticeably only when enforcing extremely narrow confidence intervals ($d = 0.01$).

5.2 Experiment 2

To evaluate the automatic procedure for determining d, we use three different data sets. The first is the one used in experiment 1, whereas the other two have slightly higher positive class rates. In all cases the prediction task was the same, i.e. predicting the conversion probability of a visitor.

Table 2 reports the average results over 30 random splits into training, validation, and test sets (1/3 for each). We again see that standard IR is unreliable in all cases, with the largest confidence intervals exceeding 0.65 in all cases. For binary classification intervals this wide implies the bin provides essentially no information on true probability. For RCIR the automatic procedure is able to push the maximum width down to 0.13 or below, achieving always at least an 80% reduction in credible interval width. Again this happens without sacrificing accuracy. For AUC-ROC the reduction is always below 0.01% and for the first two data set MSE increases also less than 2%. For the third one for which we have the narrowest credible intervals, AUC-ROC actually improves slightly but MSE worsens slightly more, almost 7%.

6 Conclusion

In many applications classification is just the starting point. The actual goal is to make a decision e.g. on whether to provide credit to an applicant, whether to address a potential buyer with marketing activities, what investment instruments to include in a portfolio, or which treatments to prescribe to a patient. While the need for well-calibrated probabilities for making justified decisions has been clearly identified [6,20] and several practical methods for calibrating arbitrary classifiers are available, the literature has been ignoring the reliability of the estimates. The existing methods calibrate the outputs well on training data, but do not have guarantees on performance on test data and often have very large uncertainty.

In this work we showed how the reliability of the calibration can be analysed using simple statistical analysis and demonstrated it in the context of isotonic regression [21]. We further extended isotonic regression to provide estimates with probabilistic guarantees on reliability. As isotonic regression is one of the most widely applied calibration methods, our results have clear practical value. The formulation has potential also for follow-up work on extending other calibration methods to better account for reliability, even though rigorous statistical analysis will not be trivial for e.g. the ensemble methods [12,13].

The main conclusion of our work is clear. By explicitly measuring the reliability of calibration and directly optimizing for it, we can improve reliability without notably affecting accuracy. In most cases the deterioration in accuracy metrics was less than 1% while the width of the credible intervals was reduced by 80–90%. Our method guarantees that downstream applications that use individual predictions for further processing can be sufficiently certain that the estimate will be close to the true probability. This was achieved with a deterministic algorithm that terminates after a finite number of steps, demonstrated in practice to perform near optimally even though we lack formal guarantees of optimality.

Acknowledgement. This work was supported by the Academy of Finland (Flagship programme: Finnish Center for Artificial Intelligence, FCAI).

References

1. Ayer, M., Brunk, H.D., Ewing, G.M., Reid, W.T., Silverman, E.: An empirical distribution function for sampling with incomplete information. Ann. Math. Stat. **26**(4), 641–647 (1955)
2. de Bruijne, M.: Machine learning approaches in medical image analysis: from detection to diagnosis. Med. Image Anal. **33**(5) (2016)
3. Diemert, E., Betlei, A., Renaudin, C., Massih-Reza, A.: A large scale benchmark for uplift modeling. In: Proceedings of the AdKDD and TargetAd Workshop, KDD (2018)
4. Fawcett, T., Niculescu-Mizil, A.: PAV and the ROC convex hull. Mach. Learn. **68**(1), 97–106 (2007)
5. Gelman, A., Carlin, J., Stern, H.S., Rubin, D.B.: Bayesian Data Analysis. Chapman & Hall (2004)
6. Guo, C., Pleiss, G., Sun, Y., Weinberger, K.Q.: On calibration of modern neural networks. In: Proceedings of the 34th International Conference on Machine Learning, Sydney (2017)
7. Kumar, A., Liang, P., Ma, T.: Verified uncertainty calibration. In: Advances in Neural Information Processing. No. NeurIPS (2019)

8. Louzada, F., Ara, A., Fernandes, G.B.: Classification methods applied to credit scoring: systematic review and overall comparison. Surv. Oper. Res. Manag. Sci. **21**(2), 117–134 (2016)
9. McMahan, H.B., et al.: Ad click prediction: a view from the trenches. In: Proceedings of the 19th ACM SIGKDD International Conference on Knowledge Discovery and Data Mining, pp. 1222–1230 (2013)
10. Murphy, A.H., Winkler, R.L.: Reliability of subjective probability forecasts of precipitation and temperature. Source J. Royal Stat. Soc. Series C (Appl. Stat.) **26**(1), 41–47 (1977)
11. Murphy, K.: Machine Learning: A Probabilistic Perspective. The MIT Press, Cambridge (2011)
12. Naeini, M.P., Cooper, G.F.: Binary classifier calibration using an ensemble of near isotonic regression models. In: 2016 IEEE 16th International Conference on Data Mining, pp. 360–369 (2016)
13. Naeini, M.P., Cooper, G.F., Hauskrecht, M.: Obtaining well calibrated probabilities using Bayesian binning. In: Proceedings of the 29th AAAI Conference on Artificial Intelligence, pp. 2901–2907 (2015)
14. Niculescu-Mizil, A., Caruana, R.: Obtaining calibrated probabilities from boosting. In: Proceedings of Uncertainty in Artificial Intelligence, pp. 413–420 (2005)
15. Niculescu-Mizil, A., Caruana, R.: Predicting good probabilities with supervised learning. In: Proceedings of the 22nd International Conference on Machine Learning, ICML 2005, pp. 625–632. No. 1999 (2005)
16. Platt, J.: Probabilistic outputs for support vector machines and comparisons to regularized likelihood methods. Adv. Large Margin Classif. **10**(3), 61–74 (1999)
17. Provost, F., Fawcett, T.: Robust classification for imprecise environments. Mach. Learn. **42**, 203–231 (2001)
18. Rubinstein, M.: Markowitz's "portfolio selection": a fifty-year retrospective. J. Financ. **57**(3), 1041–1045 (2002)
19. Shmueli-Scheuer, M., Roitman, H., Carmel, D., Mass, Y., Konopnicki, D.: Extracting user profiles from large scale data. In: Proceedings of the 2010 Workshop on Massive Data Analytics on the Cloud, pp. 1–6 (2010)
20. Van Calster, B., et al.: Calibration: the Achilles heel of predictive analytics. BMC Med. **17**(1), 1–7 (2019)
21. Zadrozny, B., Elkan, C.: Obtaining calibrated probability estimates from decision trees and Naive Bayesian classifiers. In: Proceedings of International Conference on Machine Learning, pp. 1–8 (2001)
22. Zadrozny, B., Elkan, C.: Transforming classifier scores into accurate multiclass probability estimates. In: Proceedings of the eighth ACM SIGKDD International Conference on Knowledge Discovery and Data Mining, pp. 694–699 (2002)

Multiple Instance Learning for Unilateral Data

Xijia Tang, Tingjin Luo(✉), Tianxiang Luan, and Chenping Hou(✉)

National University of Defense Technology, Changsha, China

Abstract. Multi-instance learning (MIL) is a popular learning paradigm rooted in real-world applications. Recent studies have achieved prominent performance with sufficient annotation data. Nevertheless, acquisition of enough labeled data is often hard and only a little or partially labeled data is available. For example, in web text mining, the concerning bags (positive) is often rare compared with the unrelated ones (negative) and unlabeled ones. This leads to a new learning scenario with little negative bags and many unlabeled bags, which we name it as unilateral data. It is a new learning problem and has received little attention. In this paper, we propose a new method called Multiple Instance Learning for Unilateral Data (MILUD) to tackle this problem. To utilize the information of bags fully, we consider statistics characters and discriminative mapping information simultaneously. The key instances of bags are determined by the distinguishability of mapped samples based on fake labels. Besides, we also employed an empirical risk minimization loss function based on the mapping results to learn the optimal classifier and analyze its generalization error bound. The experimental results show that method outperforms other existing state-of-art methods.

Keywords: Multi-instance learning · Negative and unlabeled data learning · Bag mapping · Classification

1 Introduction

Multiple instance learning (MIL) is one of the popular learning paradigms in the practical applications. In MIL, the annotations of data were only assigned to the bags. Therefore, MIL methods are able to deal with the classification problems with label ambiguity and reduce the requirements of the label information. Due

This work was supported by the NSF of China under Grants No. 61922087, 61906201 and 62006238, NSF of Hunan Province under Grant No. 2020JJ5669, and the NSF for Distinguished Young Scholars of Hunan Province under Grant No. 2019JJ20020.

Electronic supplementary material The online version of this chapter (https://doi.org/10.1007/978-3-030-75762-5_47) contains supplementary material, which is available to authorized users.

K. Karlapalem et al. (Eds.): PAKDD 2021, LNAI 12712, pp. 590–602, 2021.
https://doi.org/10.1007/978-3-030-75762-5_47

to their excellent characteristics and outstanding performance, they have been widely used in many practical applications, such as drug activity prediction[9], image classification [1] and text recognition [4] etc.

In literature, there are a lot of methods proposed to solve MIL problem. According to different characteristics, these methods can be divided into three categories: instance-level methods, bag-level methods and embedding approaches [1]. The first group is the instance-level methods [5,9,13,15,19], which all of bags split up into the instances and then learn the optimal classifier based on all instances. Ins-KI-SVM [15] is one of their representative methods and proposed to build the instance classifier by maximizing the margin between the selected key instances. The second group is bag-level methods [2,8,11,12,17], which build the classification model of the bags. Citation-kNN [17] and MI-Kernel [11] are two representatives of traditional bag-level methods. Citation-kNN [17] adapts kNN to MIL problem, which not only takes the neighbors of bag into account but also samples that count bag as neighbor during classification. Smola et al. [11] proposed MI-Kernel to obtain the high dimensional mapping of each bag via set kernels and learned a linear model by SVM. Last group is embedding methods [5–7,10,14,18], whose main idea is to extract particular kinds of information for each bag in the new latent feature space. After mapping, MIL problem is transferred as a classical supervised problem. Wu et al. [18] proposed a discriminative mapping approach named as MILDM, which maps bags into the latent feature space via discriminative instance pool. Bao et al. [3] proposed a convex classification method called PU-SKC which solves the MIL problem in positive labels only scenario based on set kernel.

Although existing methods can handle the MIL problem well, most of them require the entire accurate label information of all bag data. However, acquisition of enough labeled data is often hard and only a little partially labeled data is available in practice. For example, in web text mining, the concerning bags (positive) is often rare compared with the unrelated ones (negative) and unlabeled ones. In the spam identification, characteristic of spam (positive) changes frequently to evade our shielding, while valid emails (negative) are always stable. Traditional classifier need to recollect and mark new spam continuously to ensure its effectiveness. Another less costly approach is to adopt all the new emails (unlabeled) and fixed valid emails for classifier training. These actual cases leads to a new learning scenario with little negative bags and many unlabeled bags, which we name it as unilateral data. As far as we know, this problem has so far been little studied and cannot be suitable for existing multiple instance methods.

In this paper, we propose a new method called Multiple Instance Learning for Unilateral Data (MILUD) to tackle this problem. Specifically, to utilize the information of bags fully, we consider statistics characters and discriminative mapping information simultaneously to make the bags in different classes more separable. To preserve more data information of bags, the key instances of bags are selected by maximizing the distances between bags in different classes. After bag mapping, we propose to use the convex NU empirical risk loss to learn the optimal classifier for unilateral data problem. Besides, the generalization error bound of our method are provided. Finally, the extensive experimental results show our proposed method achieves better performance than other existing state-of-art methods. The contributions of this paper can be summarized as follows:

- To the best of our knowledge, our method MILUD is the first one proposed to solve the unilateral MIL data problem.
- We propose a novel method by incorporating the statistics characters with the discriminative mapping information to enhance its performance. Besides, we select the key instances to preserve more discriminate embedded features of bags by maximizing the distances between bags in different classes.
- The extensive experimental results on multiple public datasets verify the effectiveness of our proposed method and analyse the effect of the quantity of labeled and unlabeled data.

2 The Proposed Method

The problem of NU learning in MIL scenario comes from real life and have rarely been studied. For example, when the probability distribution of positive data changes frequently and the negative part remains constant. In this case, cost of updating the unlabeled train set is much less than the positive train set, which leads to a NU learning problem. Compared with traditional learning, difficulties in this scenario are mainly reflected in two aspects. First, to extract the information of positive and negative categories from only unlabeled and negative data. It indicates that the emphasis of NU learning is how to fully utilize unlabeled data. Second, labels of positive bags cannot express the labels of specific instances in it. This characteristic makes the unlabeled data, which is the focus point in NU learning, become ambiguous.

In this paper, we propose a novel method called MILUD, which can overcome the inexact information in NU-MIL problem. The mean ideas of our approach are corresponding to the two difficulties. We first use statistical and discriminative feature to fully extract the effective information simultaneously. After the NU-MIL problem be convert to a NU learning problem, train the classifier by minimize the empirical risk loss on the basis of dataset distribution.

Before presenting the details of our method, we describe the notations used in this paper. Denote $B_p^N = \{\boldsymbol{x}_{p_1}^N, \boldsymbol{x}_{p_2}^N, \cdots, \boldsymbol{x}_{p_i}^N\}$, $B_q^U = \{\boldsymbol{x}_{q_1}^U, \boldsymbol{x}_{q_2}^U \cdots, \boldsymbol{x}_{q_j}^U\}$ denote p_i and q_j instances in bag B_p^N and B_q^U, respectively, where $\boldsymbol{x}_{p_i}^N$, $\boldsymbol{x}_{q_j}^U \in \mathbb{R}^d$. The negative and unlabeled training sets are defined as D^N=$\{B_1^N, B_2^N, \cdots, B_{Nn}^N\}$ and $D^U = \{B_1^U, B_2^U, \cdots, B_{Nu}^U\}$, N_n and N_u represent the number of negative and unlabeled bags in D^N and D^U.

2.1 Bag Mapping

Bag mapping solves the label ambiguous problem in MIL by extracting feature mainly from two perspectives: data-based and label-based. However, in NU-MIL scenario, label-based mapping usually leads to unsatisfied performance, because labels of the unlabeled samples are invisible. Therefore, we consider the data-based mapping features first, which can be extracted without label information. Statistical feature is a great choice to be the first part of our mapping. It complements the shortcomings of label-based features and possesses low computational cost. Motivated by the idea of set kernel [11], we calculate statistic $\boldsymbol{s}_{3m}(B)$ which are consist of the maximum, minimum and mean values of each dimension $x^{(i)}$ of the instances $\boldsymbol{x}$ in bag B by the equation below:

$$\boldsymbol{s}_{3m}(B) = \left[\min_{x\in B} x^{(1)}, \cdots, \min_{x\in B} x^{(d)}, \max_{x\in B} x^{(1)}, \cdots, \max_{x\in B} x^{(d)}, \overline{x^{(1)}}, \cdots, \overline{x^{(d)}}\right]. \tag{1}$$

The maximum and minimum value describe the boundary of bags, and the average value can further give the outlier information of bags. Based on Eq. (1), the set kernel $\widetilde{\boldsymbol{k}}_{3m}$ can be calculated by

$$\widetilde{\boldsymbol{k}}_{3m}(B, C) = \widetilde{\boldsymbol{k}}\left(\boldsymbol{s}_{3\mathrm{m}}(B), \boldsymbol{s}_{3\mathrm{m}}(C)\right), \tag{2}$$

where $\widetilde{\boldsymbol{k}}$ can be any kernel function and it is a basic mapping derived from set kernel based on the limit information. Now we can map the bags to single instances by kernel centers $\{C_1, C_2, \cdots, C_M\}$ and $\widetilde{\boldsymbol{k}}_{3m}(B, C)$ mentioned in Eq. (2):

$$\boldsymbol{\Phi}_{3m}(B) := \left[\widetilde{\boldsymbol{k}}_{3\mathrm{m}}(B, C_1), \cdots, \widetilde{\boldsymbol{k}}_{3\mathrm{m}}(B, C_M)\right]^{\top} \tag{3}$$

However, the statistical features do not make use of label information, which means the extraction of dataset information is insufficient. This is why we adopte discriminative feature, which is a excellent label-based mapping strategy. The discriminative feature should meet the following two conditions:

- bags with same label are as similar as possible in mapping feature space;
- bags with different labels are as diverse as possible in mapping feature space.

These two conditions guarantees the separability of samples in the mapping space, because it makes labels directly linked to the classification performance. For PN data, training process is carried out on the existing of two categories of label. However, there is only label information of negative data for NU data, which makes it impossible to maximize the distance between different data directly. By mapping process, when keeping negative data far away positive data, it also makes negative data far away from itself. Therefore, it is not feasible to map data by only discriminative features in NU-MIL scenarios.

To solve the discriminative feature extraction problem of NU dataset thereby further obtain more complete train set information, we separate the "most likely positive samples" from unlabeled data. The specific method is to train a preliminary classifier by Eq. (3), and on this basic classifier we can give each unlabeled bag a fake label. Here we directly use the result of this strategy. The unlabeled data D^N is divided into two parts: D^{UP} and D^{UN}, which are the positive and negative bags identified from unlabeled bags. Denote $D^{N'} = \{D^N, D^{UN}\}$ and

$D^{P'} = \{D^{UP}\}$ as the new dataset based on fake labels, which is the relatively reliable information extracted from the unilateral data.

Since the label-based extraction problem is solved, next we'll explain the step details of discriminative feature. Denote $\boldsymbol{\Phi}_{DIP}(B)$ as a mapping rule based on instance similarity $\boldsymbol{s}$:

$$\boldsymbol{\Phi}_{DIP}(B) = \left[\boldsymbol{s}\left(B, \boldsymbol{x}_1^{\phi}\right), \cdots, \boldsymbol{s}\left(B, \boldsymbol{x}_m^{\phi}\right)\right]^{\top}, \tag{4}$$

where $\boldsymbol{s}\left(B, \boldsymbol{x}_k^{\phi}\right) = \max_{\boldsymbol{x}_l \in B} \exp\left(-\left\|\boldsymbol{x}_l - \boldsymbol{x}_k^{\phi}\right\|^2 / \sigma^2\right)$, which can be viewed as the similarity between bag B and $\boldsymbol{x}_k^{\phi}$. $\boldsymbol{x}_l$ is the l-th instance in the bag B, $\boldsymbol{x}_k^{\phi}$ is the instance used for mapping and m is the number of it. The m most discriminative instances used for mapping make up a collection, which called discriminative instance pool (DIP): $\mathcal{P} = \left\{\boldsymbol{x}_1^{\phi}, \cdots, \boldsymbol{x}_m^{\phi}\right\}, \boldsymbol{x}_k^{\phi} \in \left[B^{N'}, B^{U'}\right]$, where $\boldsymbol{x}_k^{\phi}$ is the k-th element in $\mathcal{P}$. The strategy of processing NU data by DIP is similar to the idea in MILDM [18]. We can map a bag into single instance by the DIP. In order to find the m instances that make the mapped samples most separable, denote $\mathcal{J}(\mathcal{P})$ as objective function:

$$\mathcal{J}(\mathcal{P}) = \frac{1}{2}\sum_{i,j} K_{\mathcal{P}}\left(B_i, B_j\right) Q_{i,j}, \; Q_{i,j} = \begin{cases} -1/|A|, y_i y_j = 1; \\ 1/|B|, y_i y_j = -1, \end{cases} \tag{5}$$

where $K_{\mathcal{P}}\left(B_i, B_j\right)$ denotes the distance between B_i and B_j after being mapped, $Q_{i,j}$ is weight factor, $|\cdot|$ is the number of elements in the set, and y_i, y_j are the labels of B_i and B_j, respectively. A and B are denoted as:

$$A = \{(i,j) \mid y_i y_j = 1\}, B = \{(i,j) \mid y_i y_j = -1\}. \tag{6}$$

By optimizing $\mathcal{J}(\mathcal{P})$, the goal of minimizing the distance between samples from same category and maximizing the distance between different ones' can be directly achieved. The function $K_{\mathcal{P}}\left(B_i, B_j\right)$ is used to describe the distance between B_i and B_j denoted as:

$$K_{\mathcal{P}}\left(B_i, B_j\right) = \left\|B_i^{\phi} - B_j^{\phi}\right\|^2 = \left\|\mathcal{I}_{\mathcal{P}} B_i^{\phi_x} - \mathcal{I}_{\mathcal{P}} B_j^{\phi_x}\right\|^2, \tag{7}$$

where $\mathcal{I}_{\mathcal{P}}$ denotes a diagonal matrix, if x_k belongs to the discriminative instance pool, the k-th diagonal element in $\mathcal{I}_{\mathcal{P}}$ is 1, otherwise it is 0. Based on $\mathcal{I}_{\mathcal{P}}$, we can choose mapped features by discriminative instance pool from all the instances.

To sum up, when maximizing $\mathcal{J}(\mathcal{P})$, we diminish the distances between instances from same class of bags and enlarge distances between different classes instances. So our next target is maximizing $\mathcal{J}(\mathcal{P})$ to get $\mathcal{P}_*$:

$$\mathcal{P}_* = \arg\max \mathcal{J}(\mathcal{P}), \quad \text{s.t.} \quad |\mathcal{P}| = m. \tag{8}$$

Let $\mathcal{X}_{\phi} = [B_1^{\phi_x}, \cdots, B_n^{\phi_x}] = [\phi_1, \cdots, \phi_p]^{\top}$ and $n = |D^{N'}| + |D^{P'}|$. $L = D - Q$ is a Laplacian matrix, where diagonal matrix D satisfy $D_{i,i} = \sum_j Q_{ij}$. $\mathcal{J}(\mathcal{P})$

can be rewritten as:

$$\begin{aligned}\mathcal{J}(\mathcal{P}) &= \frac{1}{2}\sum_{i,j}\left\|\mathcal{I}_{\mathcal{P}}B_i^{\phi_x} - \mathcal{I}_{\mathcal{P}}B_j^{\phi_x}\right\|^2 Q_{i,j} \\ &= \sum_i \left(B_i^{\phi_x}\right)^\top \mathcal{I}_{\mathcal{P}}^\top \mathcal{I}_{\mathcal{P}} B_i^{\phi_x} D_{i,i} - \sum_{i,j}\left(B_i^{\phi_x}\right)^\top \mathcal{I}_{\mathcal{P}}^\top \mathcal{I}_{\mathcal{P}} B_j^{\phi_x} Q_{i,j} \\ &= \operatorname{tr}\left(\mathcal{I}_{\mathcal{P}}^\top \mathcal{X}_\phi L \mathcal{X}_\phi^\top \mathcal{I}_{\mathcal{P}}\right) = \sum_{x_k^\phi \in \mathcal{P}} \phi_k^\top L \phi_k.\end{aligned} \tag{9}$$

Let $\phi_k^\top L\phi_k = f(\boldsymbol{x}_k^\phi, L)$, the optimal DIP (8) is equivalent to:

$$\mathcal{P}_* = \arg\max \sum_{x_k^\phi \in \mathcal{P}} f(\boldsymbol{x}_k^\phi, L), \quad \text{s.t.} \quad |\mathcal{P}| = m. \tag{10}$$

Since $\mathcal{P}_*$ is composed of the function $f(\boldsymbol{x}_k^\phi, L)$ of all instances $\boldsymbol{x}_k$ in dataset, we can optimize $\mathcal{P}$ by the searching algorithm. Under the premise of satisfying the constraints, update $f(\boldsymbol{x}_k^\phi, L)$ one by one and iterate $\mathcal{P}$. Through each iteration, instance $\boldsymbol{x}_{min}$ corresponding to the minimum value of $f(\boldsymbol{x}_{min}^\phi, L)$ will be removed. The iteration ends until all instances have been traversed.

Now we have two kinds of mapped features based on different rules: statistic rule $\boldsymbol{\Phi}_{3m}(B)$ and discriminative instance pool rule $\boldsymbol{\Phi}_{DIP}(B)$, which two are complementary to each other. The combination of two mappings is defined as

$$\boldsymbol{\Phi}_{cps}(B) := [\boldsymbol{\Phi}_{3m}(B), \boldsymbol{\Phi}_{DIP}(B)]. \tag{11}$$

Composite mapping $\boldsymbol{\Phi}_{cps}(B)$ describe the characteristics of the bag from both statistical features and discriminative features, so that the mapped results can better help the next step of NU classification.

2.2 NU Classification Based on Composite Mapping

By the mapping rule in Eq. (11), bags in datasets D_N and D_U can be mapped to single instances. The next step is to get the classifier by constructing the NU loss function. For more convenient representation, we write mapped sample $\boldsymbol{\Phi}_{cps}(B)$ as φ. In particular, $\boldsymbol{\Phi}_{cps}(B_i^N)$ and $\boldsymbol{\Phi}_{cps}(B_j^U)$ are denoted as φ_i^N and φ_j^U, respectively. Classifier $g(\varphi)$ is a linear parametric model:

$$g(\varphi) = \boldsymbol{\omega}^\top \varphi + b. \tag{12}$$

On the basis of all the samples in dataset are independent and identically distributed with probability $p(\varphi, y)$, we intend to train a classifier based on minimizing the empirical risk loss on the training dataset. The loss function is composed of the expectation over risk loss of positive and negative data. However, labels for positive class are inexistent in NU dataset, which means the positive part in loss function is not directly available. In order to train a

Algorithm 1. Multiple Instance learning for Unilateral Data (MILUD)

Input: negative dataset D^N and unlabeled dataset D^U
Output: The classifier (ω, b) for Multiple Instance Data
1: Calculate the statistic $\boldsymbol{s}_{3m}(B^{N(U)})$ by Eq. (1);
2: Compute the set kernel $\widetilde{\boldsymbol{k}}_{3m}(B^{N(U)}, C)$ by Eq. (2);
3: Learn a fake label by Eq. (3) and Eq. (15), $D^U = \{D^{UP}, D^{UN}\}$;
4: Search the optimal DIP set $\mathcal{P}_*$ by solving the problem in (10);
5: Compute the feature φ by concatenating $\boldsymbol{\Phi}_{3m}(B)$ and $\boldsymbol{\Phi}_{DIP}(B)$ in Eq. (11);
6: Learn the optimal NU classifier (ω, b) by optimizing Eq. (16).
7: **return** the optimal model (ω, b).

classifier by only negative and unlabeled data, the risk loss in positive samples can be estimated by negative and unlabeled samples:

$$\theta_P \mathbb{E}_P[l(g(\varphi)] = \mathbb{E}_U[l(g(\varphi))] - \theta_N \mathbb{E}_N[l(g(\varphi))], \tag{13}$$

where θ_P and θ_N denote the class-prior probabilities of the positive and negative class, $\theta_P + \theta_N = 1$. $g(\varphi)$ is the classifier and $l(z)$ is a loss function. $\mathbb{E}_P[\cdot]$ and $\mathbb{E}_N[\cdot]$ are the expectations over the prior distribution on positive and negative data. Inspired by the similar formulation proposed in [16], specifically, we use loss function $R(g)$ drawn from NU data by the risk:

$$\begin{aligned} R(g) &= \theta_P \mathbb{E}_P[l(g(\varphi))] + \theta_N \mathbb{E}_N[l(-g(\varphi))] \\ &= \theta_N \mathbb{E}_N[\tilde{l}(-g(\varphi))] + \mathbb{E}_U[l(g(\varphi))], \end{aligned} \tag{14}$$

where $\tilde{l}(z) = l(z) - l(-z)$ is a composite loss function.

When $\tilde{l}(z)$ satisfies the condition of $\tilde{l}(z) = l(z) - l(-z) = -z$, the minimization of NU loss function Eq. (14) is a convex optimization problem. In this paper, we choose double hinge loss $l_{DH}(z) = \max(-z, \max(0, \frac{1}{2} - \frac{1}{2}z))$ as $l(z)$. $R(g)$ in Eq. (14) is rewritten as:

$$\begin{aligned} R(g) &= \theta_N \mathbb{E}_N[g(\varphi^N)] + \mathbb{E}_U[l(g(\varphi^U))] \\ &= \frac{\theta_N}{N_n} \sum_{i=1}^{N_n} \omega^\top \varphi_i^N + \theta_N b + \frac{\lambda}{2} \omega^\top \omega + \frac{1}{N_u} \sum_{j}^{N_u} l_{DH}(\omega^\top \varphi_j^U + b) \end{aligned} \tag{15}$$

where $\omega^\top \omega$ is the regularization item, λ is the parameter used to adjust $\omega^\top \omega$. The optimization in Eq. (15) can be rewritten with the slack variable $\boldsymbol{\xi}$, which is used to bound the max operators:

$$\begin{cases} \min\limits_{\omega, b, \xi} \frac{\theta_N}{N_n} \mathbf{1}^\top \varphi_i^N \omega + \theta_N b + \frac{1}{N_u} \mathbf{1}^\top \boldsymbol{\xi} + \frac{\lambda}{2} \omega^\top \omega \\ \text{s.t.} \qquad \boldsymbol{\xi} \geq \mathbf{0}, \ \ \boldsymbol{\xi} \geq -\varphi_j^U \omega - b\mathbf{1}, \\ \qquad\qquad \boldsymbol{\xi} \geq \frac{1}{2}\mathbf{1} - \frac{1}{2}\varphi_j^U \omega - \frac{1}{2} b\mathbf{1}. \end{cases} \tag{16}$$

The problem in Eq. (16) is a quadratic program, which can be solved by interior point method. Interior point method transforms constrained optimization problem into unconstrained optimization problem by adding an obstacle function to the original target function, the original constraints will be replaced. Then the unconstrained optimization problem would be solved by Newton's method. By optimizing Eq. (16), we get the final classifier g. The main steps of MILUD are shown in Algorithm 1.

2.3 Analysis of Generalization Error Bounds

Similar to [16], we analyze the upper bound of generalization error for g. Denote $\mathcal{H}$ as the domain set, C_ω and C_φ are certain positive constants. Define $\mathcal{G} = \{g(\varphi) = \omega^\top \varphi \mid \|\omega\| \leq C_\omega, \sup_{\varphi \in \mathcal{H}} \|\varphi\| \leq C_\varphi\}$ as function class. The expected risk $\mathcal{R}(g)$ and empirical risk $\widehat{\mathcal{R}}(g)$ of classifier can be written as:

$$\begin{aligned} \mathcal{R}(g) &= \theta_N \mathbb{E}_{p(\varphi|y=-1)}[g(\varphi)] + \mathbb{E}_{p(\varphi)}[\ell(g(\varphi))], \\ \widehat{\mathcal{R}}(g) &= \frac{\theta_N}{N_n} \sum_{u=1}^{N_n} g\left(\varphi_u^N\right) + \frac{1}{N_u} \sum_{v=1}^{N_u} \ell\left(g\left(\varphi_v^U\right)\right). \end{aligned} \quad (17)$$

Theorem 1. *For any fixed g and any $\delta \in (0,1)$, the difference between $\mathcal{R}(g)$ and $\widehat{\mathcal{R}}$ satisfies:*

$$\mathcal{R}(g) - \widehat{\mathcal{R}}(g) \leq \sqrt{C_\omega^2 C_\varphi^2 \log \frac{2}{\delta} / 2} \left(\frac{2\theta_N}{\sqrt{N_n}} + \frac{1}{\sqrt{N_u}} \right). \quad (18)$$

The details of this proof is presented in the supplemental file. The results of this theorem indicates that the generalization error of our model decreases with the increase of $\sqrt{N_n}$ and $\sqrt{N_u}$. In other words, increasing the number of negative and unlabeled bags can reduce the error and improve the performance of our method. The conclusion of this proof is also verified by experimental results. both contributes to reducing the error.

3 Experiments

3.1 Experiment Settings

To verify the superiority of our approach, we compare MILUD with four classical methods, including C-kNN [17], aMILGDM [18], KISVM [15] and PU-SKC [3]. The first three are proposed to solve the problem with complete data and balanced labels. PU-SKC is one representative of PU-MIL, which train the classifier via concise statistical mapping and PU empirical risk loss. We take the experiments on eight public datasets, i.e. Corel_bm, Corel_hd, Sival_ab, Sival_bc, Atoms, Bonds, Elephant and Tiger. The specific information of these datasets is shown in Table 1.

Table 1. The main information of eight public dataset.

Name	# of Pos Bags	# of Neg Bags	Features	Avg # of Insts
Corel_bm	100	100	9	3.46
Corel_hd	100	100	9	3.85
Sival_ab	60	60	30	31.68
Sival_bc	60	60	30	31.78
Atoms	125	63	10	8.61
Bonds	125	63	16	21.25
Elephant	100	100	230	6.96
Tiger	100	100	230	6.10

Since these datasets in Table 1 are too small to evaluate NU or PU methods, similar with [3], we augment the information of datasets by increasing the number of bags. Specifically, we randomly select bags from original dataset and duplicate them with the Gaussian noise of mean zero and variance 0.01. In this way, we increase the number of negative and unlabeled bags to 20 and 180, respectively. The remaining 100 positive bags and 100 negative bags are test set. In addition, we take comparative experiments to verify the effectiveness of MILUD under different unlabeled bags composition. The ratio of negative bags in unlabeled bags are set as {0.1,0.3,0.5,0.7}. Finally, we repeat 30 times experiments and report the mean value of each method under different classification metrics.

3.2 Results

Classification Performance. We compared the performance of MILUD and other four state-of-the-art approaches on eight datasets. The classification accuracy, area under the curve (AUC) and F-measure are adopted to evaluated their performance. Under the four different class-prior probability θ_N of {0.1,0.3,0.5,0.7}, the average classification accuracy and standard deviation of 30 independent trials of different methods are presented in Table 2.

We also provide the AUC and F-measure results on Sival_ab, Corel_bm, Sival_bc and Corel_hd in Fig. 1. The results in Table 2 and Fig. 1 present that the performance of our MILUD is better than other methods in most of cases and simultaneously verify the effectiveness of our proposed method.

Effect of Mapping Combination. To verify the validity of our mapping combination, we conduct experiments on three datasets Atoms, Bonds and Sival_ab with different class-prior probability θ_N. We compare the NU empirical risk minimization classifier over three mapping rules, included the composite mapping

Table 2. Average classification accuracy (standard deviation) of different compared methods on eight public datasets. The first highest score is in bold.

Dataset	θ_N	Accuracy				
		MILUD	PU-SKC	C-kNN	aMILGDM	KI-SVM
Corel_bm	0.1	**0.694(0.048)**	0.571(0.075)	0.665(0.032)	0.647(0.031)	0.630(0.041)
	0.3	**0.662(0.040)**	0.608(0.053)	0.592(0.024)	0.600(0.032)	0.556(0.035)
	0.5	**0.602(0.067)**	0.589(0.060)	0.560(0.025)	0.567(0.029)	0.528(0.035)
	0.7	**0.562(0.059)**	0.524(0.044)	0.542(0.023)	0.546(0.016)	0.538(0.050)
Corel_hd	0.1	**0.839(0.034)**	0.74210.064)	0.776(0.041)	0.701(0.034)	0.602(0.070)
	0.3	**0.866(0.031)**	0.821(0.031)	0.662(0.036)	0.617(0.032)	0.588(0.103)
	0.5	**0.845(0.036)**	0.794(0.061)	0.611(0.025)	0.583(0.025)	0.572(0.053)
	0.7	**0.774(0.082)**	0.689(0.081)	0.560(0.023)	0.555(0.029)	0.703(0.049)
Sival_ab	0.1	**0.773(0.048)**	0.740(0.043)	0.699(0.034)	0.689(0.032)	0.763(0.046)
	0.3	**0.733(0.050)**	0.711(0.050)	0.631(0.031)	0.612(0.028)	0.523(0.059)
	0.5	**0.719(0.057)**	0.705(0.054)	0.579(0.028)	0.575(0.020)	0.524(0.076)
	0.7	**0.654(0.054)**	0.607(0.070)	0.565(0.026)	0.562(0.029)	0.515(0.072)
Sival_bc	0.1	**0.875(0.038)**	0.808(0.041)	0.747(0.035)	0.754(0.040)	0.860(0.036)
	0.3	**0.883(0.035)**	0.825(0.033)	0.643(0.030)	0.626(0.026)	0.680(0.038)
	0.5	**0.823(0.068)**	0.797(0.068)	0.599(0.035)	0.585(0.025)	0.683(0.076)
	0.7	**0.742(0.090)**	0.728(0.077)	0.569(0.037)	0.567(0.027)	0.659(0.036)
Atoms	0.1	0.6017(0.051)	0.525(0.029)	0.517(0.022)	0.605(0.036)	**0.623(0.047)**
	0.3	**0.671(0.047)**	0.576(0.043)	0.508(0.027)	0.572(0.029)	0.564(0.040)
	0.5	**0.660(0.055)**	0.556(0.052)	0.502(0.022)	0.549(0.028)	0.599(0.036)
	0.7	**0.574(0.097)**	0.524(0.054)	0.506(0.023)	0.523(0.023)	0.555(0.023)
Bonds	0.1	**0.616(0.063)**	0.520(0.058)	0.527(0.024)	0.605(0.039)	0.559(0.074)
	0.3	0.620(0.048)	0.601(0.057)	0.523(0.021)	0.551(0.023)	**0.642(0.040)**
	0.5	**0.572(0.076)**	0.572(0.063)	0.523(0.018)	0.543(0.026)	0.530(0.024)
	0.7	**0.608(0.074)**	0.564(0.063)	0.511(0.020)	0.523(0.023)	0.539(0.026)
Elephant	0.1	0.785(0.041)	**0.788(0.043)**	0.691(0.034)	0.658(0.034)	0.677(0.064)
	0.3	**0.779(0.057)**	0.752(0.069)	0.634(0.000)	0.601(0.025)	0.682(0.053)
	0.5	**0.749(0.060)**	0.730(0.049)	0.586(0.032)	0.566(0.019)	0.676(0.065)
	0.7	**0.647(0.059)**	0.643(0.060)	0.563(0.026)	0.552(0.024)	0.653(0.062)
Tiger	0.1	**0.714(0.050)**	0.712(0.056)	0.699(0.033)	0.665(0.044)	0.579(0.035)
	0.3	**0.739(0.056)**	0.728(0.058)	0.630(0.034)	0.595(0.023)	0.558(0.041)
	0.5	**0.688(0.042)**	0.676(0.046)	0.569(0.028)	0.565(0.024)	0.531(0.025)
	0.7	**0.591(0.063)**	0.587(0.064)	0.547(0.021)	0.551(0.021)	0.499(0.086)

$\boldsymbol{\Phi}_{cps}$ employed in MILUDM, statistical feature mapping $\boldsymbol{\Phi}_{3m}$ and discriminative feature mapping $\boldsymbol{\Phi}_{DIP}$. It should be noted that the $\boldsymbol{\Phi}_{DIP}$ makes negative and unlabeled samples far away directly without fake-label strategy. Experiment results in Fig. 2 are the accuracy and standard deviation on three mapping rules, which illustrates that the mapping combination improves the classifier performance by making up the shortcomings of extracting information unilaterally.

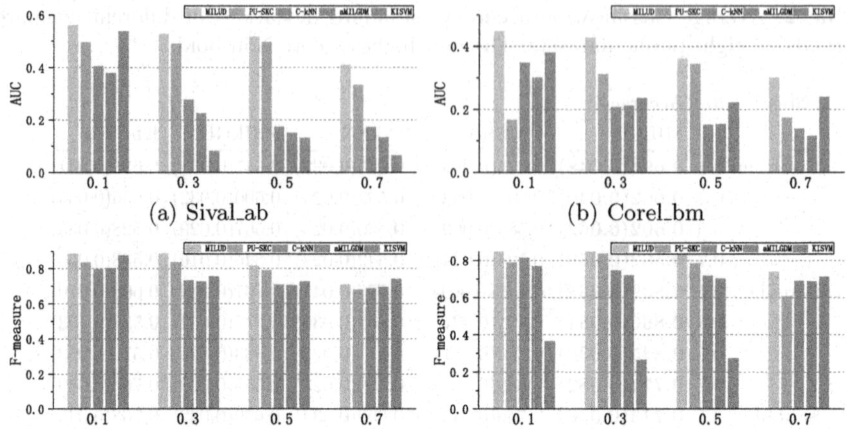

(a) Sival_ab (b) Corel_bm

(c) Sival_bc (d) Corel_hd

Fig. 1. The results of AUC and F-measure on four datasets. The first line are the AUC of Sival_ab and Corel_bm, the second line are the F-measure of Sival_bc and Corel_hd.

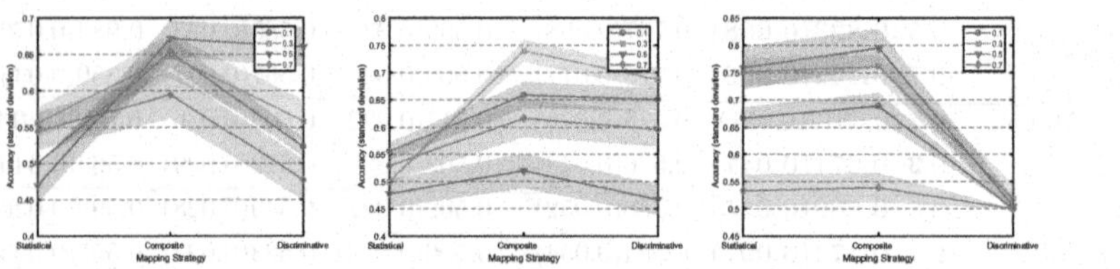

(a) Atoms (b) Bonds (c) Sival_ab

Fig. 2. The result of accuracy and standard deviation on three mapping strategy on three datasets Atoms, Bonds and Sival_ab. Each figure contains the results on four θ_N.

Impact of Negative and Unlabeled Data Increasing. To explore the impact of unlabeled and negative data size on MILUD, we take two experiments with increase of unlabeled or negative bags on Atoms, Elephant and Sival_ab. Specifically, we randomly select 20 negative and 10 unlabeled samples as initial training set. Then increase the number of unlabeled or negative samples of training set and observe the tendency of classification performance. As seen from the accuracy results in Fig. 3, performance of MILUD becomes better with the increase of two types of bags. The results verify that our method can be effective to utilize the information of unlabeled data to improve the performance of our model.

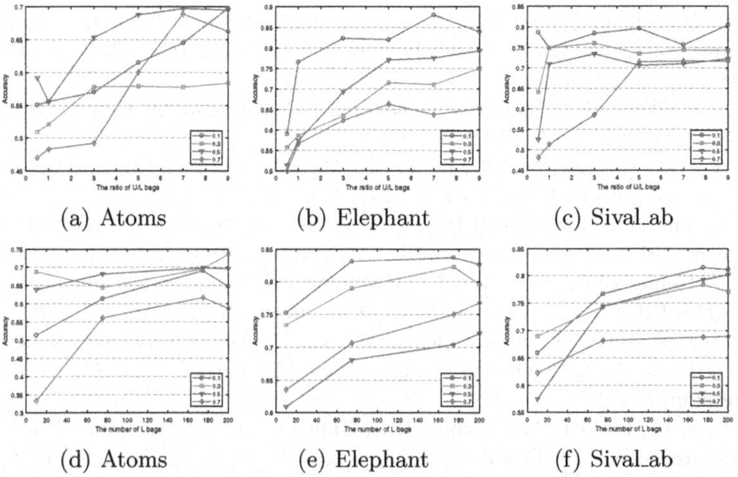

Fig. 3. The change trend of accuracy when the number of U/L bags increases at four class prior. The first line are the accuracy of increasing unlabeled bags, the second line are the accuracy of increasing L bags.

4 Conclusion

In this paper, we focus on an important but unheeded NU-MIL problem, and propose a two-stages method named MILUD to solve it. The composite mapping utilizes both data based statistical features and label based discriminative mapping information, which ensure that the information of bags can be fully preserved. Then, a convex learning model is derived by minimizing the empirical loss to solve the NU problem. Experimental results on eight public datasets indicate our method outperforms other compared methods in most of cases.

References

1. Amores, J.: Multiple instance classification: review, taxonomy and comparative study. Artif. Intell. **201**, 81–105 (2013)
2. Andrews, S., Tsochantaridis, I., Hofmann, T.: Support vector machines for multiple-instance learning. In: Advances in Neural Information Processing Systems, pp. 561–568. MIT Press (2002)
3. Bao, H., Sakai, T., Sato, I., Sugiyama, M.: Convex formulation of multiple instance learning from positive and unlabeled bags. Neural Netw. **105**, 132–141 (2018)
4. Carbonneau, M., Cheplygina, V., Granger, E., Gagnon, G.: Multiple instance learning: a survey of problem characteristics and applications. Pattern Recogn. **77**, 329–353 (2018)
5. Carbonneau, M., Granger, E., Raymond, A.J., Gagnon, G.: Robust multiple-instance learning ensembles using random subspace instance selection. Pattern Recogn. **58**, 83–99 (2016)

6. Chen, Y., Bi, J., Wang, J.Z.: MILES: multiple-instance learning via embedded instance selection. IEEE Trans. Pattern Anal. Mach. Intell. **28**(12), 1931–1947 (2006)
7. Chen, Y., Wang, J.Z.: Image categorization by learning and reasoning with regions. J. Mach. Learn. Res. **5**, 913–939 (2004)
8. Cheplygina, V., Tax, D.M.J., Loog, M.: Multiple instance learning with bag dissimilarities. Pattern Recogn. **48**(1), 264–275 (2015)
9. Dietterich, T.G., Lathrop, R.H., Lozano-Pérez, T.: Solving the multiple instance problem with axis-parallel rectangles. Artif. Intell. **89**(1–2), 31–71 (1997)
10. Fu, Z., Robles-Kelly, A., Zhou, J.: MILIS: multiple instance learning with instance selection. IEEE Trans. Pattern Anal. Mach. Intell. **33**(5), 958–977 (2011)
11. Gärtner, T., Flach, P.A., Kowalczyk, A., Smola, A.J.: Multi-instance kernels. In: Proceedings of the Nineteenth International Conference, pp. 179–186. Morgan Kaufmann (2002)
12. Leistner, C., Saffari, A., Bischof, H.: MIForests: multiple-instance learning with randomized trees. In: Daniilidis, K., Maragos, P., Paragios, N. (eds.) ECCV 2010. LNCS, vol. 6316, pp. 29–42. Springer, Heidelberg (2010). https://doi.org/10.1007/978-3-642-15567-3_3
13. Li, F., Sminchisescu, C.: Convex multiple-instance learning by estimating likelihood ratio. In: Advances in Neural Information Processing Systems, pp. 1360–1368. Curran Associates, Inc. (2010)
14. Li, W., Yeung, D.: MILD: multiple-instance learning via disambiguation. IEEE Trans. Knowl. Data Eng. **22**(1), 76–89 (2010)
15. Li, Y.-F., Kwok, J.T., Tsang, I.W., Zhou, Z.-H.: A convex method for locating regions of interest with multi-instance learning. In: Buntine, W., Grobelnik, M., Mladenić, D., Shawe-Taylor, J. (eds.) ECML PKDD 2009. LNCS (LNAI), vol. 5782, pp. 15–30. Springer, Heidelberg (2009). https://doi.org/10.1007/978-3-642-04174-7_2
16. Sakai, T., du Plessis, M.C., Niu, G., Sugiyama, M.: Semi-supervised classification based on classification from positive and unlabeled data. In: Proceedings of the 34th International Conference on Machine Learning. Proceedings of Machine Learning Research, vol. 70, pp. 2998–3006. PMLR (2017)
17. Wang, J., Zucker, J.: Solving the multiple-instance problem: a lazy learning approach. In: Proceedings of the Seventeenth International Conference on Machine Learning, pp. 1119–1126. Morgan Kaufmann (2000)
18. Wu, J., Pan, S., Zhu, X., Zhang, C., Wu, X.: Multi-instance learning with discriminative bag mapping. IEEE Trans. Knowl. Data Eng. **30**(6), 1065–1080 (2018)
19. Xiao, Y., Liu, B., Hao, Z., Cao, L.: A similarity-based classification framework for multiple-instance learning. IEEE Trans. Cybern. **44**(4), 500–515 (2014)

An Online Learning Algorithm for Non-stationary Imbalanced Data by Extra-Charging Minority Class

Sajjad Kamali Siahroudi(✉) and Daniel Kudenko

L3S Research Center, Leibniz University Hannover, Hanover, Germany
{kamali,kudenko}@l3s.de

Abstract. Online learning is one of the trending areas of machine learning in recent years. How to update the model based on new data is the core question in developing an online classifier. When new data arrives, the classifier should keep its model up-to-date by (1) learn new knowledge, (2) keep relevant learned knowledge, and (3) forget obsolete knowledge. This problem becomes more challenging in imbalanced non-stationary scenarios. Previous approaches save arriving instances, then utilize up/down sampling techniques to balance preserved samples and update their models. However, this strategy comes with two drawbacks: first, a delay in updating the models, and second, the up/down sampling causes information loss for the majority classes and introduces noise for the minority classes. To address these drawbacks, we propose the Hyper-Ellipses-Extra-Margin model (HEEM), which properly addresses the class imbalance challenge in online learning by reacting to every new instance as it arrives. HEEM keeps an ensemble of hyper-extended-ellipses for the minority class. Misclassified instances of the majority class are then used to shrink the ellipse, and correctly predicted instances of the minority class are used to enlarge the ellipse. Experimental results show that HEEM mitigates the class imbalance problem and outperforms the state-of-the-art methods.

Keywords: Online learning · Imbalanced data · Nonstationary data

1 Introduction

Today, with the spread of the Internet and the increasing power and capacity of hardware, the use of online applications is rapidly increasing. As a consequence, the analysis of online data generated by individuals, companies or devices becomes crucial in various fields such as business, medicine, management and economics [1,2]. Many of the application domains are non-stationary, with a large amount of new data arriving at a very high frequency. The source of the non-stationarity (so-called *concept drifts*) can be because of software or hardware faults, change in users' behaviours or interests, or an unknown external reason [3]. Response time, adaptive behaviour, and performance are therefore the most

K. Karlapalem et al. (Eds.): PAKDD 2021, LNAI 12712, pp. 603–615, 2021.
https://doi.org/10.1007/978-3-030-75762-5_48

crucial features of a classifier in this environment. Besides this, another challenge that often co-exists in many real-world applications is class imbalance, i.e. where there is a significant difference among the number of instances of each class. Both concept drift and class imbalance effect applications in various domains such as anomaly detection, novel class detection, social media mining, fraud detection, and software engineering [1,4–6].

Concept drift and class imbalance are a formidable challenge for online learning algorithms, because they decrease the performance of classifiers [1]. In the past decade, learning from imbalanced data and non-stationary data has been studied in the literature, see e.g. [2]. However, most of the investigations addressed these challenges separately and mainly focused on batch learning. In this paper, we consider these challenges together in the online learning context. Batch learning supposes that data points arrive in batches, and at the end of each batch the true labels of the data points are provided. In contrast, in the online learning scenario the data points arrive one by one, and the true label of a data point will be available immediately after classification [4,5,8]. A classifier in this latter scenario must address the following challenges. (1) *Updating:* since the data points are arriving over time, the classifier should be able to learn and update its models to cover new parts of the data that have not been seen. (2) *Remembering:* the classifier should be able to preserve relevant learned knowledge for the future. (3) *Forgetting:* due to concept drift, it is possible that after a while, some changes happen in the data distribution or the label distribution. (4) *Class imbalance:* the number of data points in the majority class is significantly higher than the number of data points in the minority class, so the classifier should be able to train its model with imbalanced data. (5) *Response time:* since in the online model data points are arriving with a high frequency, the classifier must respond to each data point very fast. (6) *Memory limitation:* the classifier can use only a fixed amount of memory which is not sufficient to store all the incoming data. Thus, the classifier needs to update its models incrementally [5,8].

Research on learning from online class-imbalanced data in a non-stationary environment is still in an early stage [7]. Most of the methods in this area consider either class imbalance or concept drift separately. Other methods that address both concept drift and class imbalance are working well when the imbalance is not too large. However, in more challenging situations with an extreme class imbalance (e.g. CI = 0.1%) or when several types of concept drifts occur (see Sect. 2.1), the methods show their weaknesses [2,3,5,8,9]. In this paper, we propose a new novel method that is capable of addressing all of the challenges mentioned above, even in extreme cases. To address these challenges, our method creates or enlarges a Hyper Ellipse With Extra Margin (HEEM) for each new sample of the minority class. This approach has the following main advantages: (1) Unlike existing methods that need to wait until sufficient instances of a minority class arrive, HEEM can react to a new arrived instance immediately by updating its models. (2) Instead of ignoring instances of majority class (by downsampling) and risk loosing some of the essential instances of majority class,

HEEM uses them to modify its model (margin of the ellipse). (3) HEEM does avoid adding noise to the data by not using upsampling methods. (4) HEEM adaptively learns its parameters based on the accuracy of models and the margin of recent ellipses, and thus there is no need for parameter tuning. (5) HEEM keeps an extra margin for positive samples that makes HEEM capable of predicting future samples of the minority class correctly. (6) At the end of the data stream, HEEM provides a complete track of the minority class that can be used for data augmentation.

2 Background and Related Work

Incremental learning methods that store and process data in batches (chunks) are different from online learning methods which learn from data one by one [8]. Without loss of generality in this paper, we focus on online binary classification. In online binary classification, it is assumed that at each time step t, a sample $(\mathbf{x}^t, y^t)$ is generated from an unknown non-stationary distribution $p^t(\mathbf{x}, y)$. Where $\mathbf{x}^t$ is a d-dimensional input vector that belongs to input space $\mathbf{X} \subset R^d$ and y^t is the class label of the current instance $y^t \in Y = \{0, 1\}$. At each time step t, a new sample $\mathbf{x}^t$ arrives and after the classifier makes a prediction for the sample $\hat{y}^t = h(\mathbf{x}^t), h : X \rightarrow Y$, the correct label of the sample will be available for the classifier. The classifier evaluates and updates its models by the true label [1,2,5].

2.1 Concept Drift

Concept drift refers to changes in the statistical properties of the target variable that happens over time in unforeseen ways, and occurs in many real-life problems [2]. Drifts can be categorized from their effects on learned decision rules as (1) *a change in* $p(y)$ (prior probability) (2) *a change in* $p(\mathbf{x}|y)$(likelihood) (3) *a change in* $p(y|\mathbf{x})$ (posterior probability). The latter is known as real drift because of changes in true decision boundaries. Virtual drift refers to the changes in the prior probability of incoming data without affecting decision boundaries. Drifts may be analyzed according to the speed of changes taking place through the stream. From this aspect, the four main types of drifts are *Sudden drift*, *Gradual drift*, *Incremental drift*, and *recurrent drift* [2]. *Blips*(random changes in stream characteristics) and *noise* are two other types of drifts that happen because of incorrect information in the stream. The algorithms which address concept drift are categorized as change detection-based, memory-based, and ensembling [2,5].

Change detection-based algorithms utilize an explicit method to discover concept drifts and update their model based on the drift. Most of these methods are based on monitoring and control charts, such as *cumulative sum (CUSUM)* [10], *Page-Hinkley (PH) test* [10], *adaptive windowing (ADWIN)* [11], *just-in-time (JIT) classifiers* [12]. These approaches, active detectors, are suitable for detecting abrupt drifts. However, they can not handle other types of drifts as well as abrupt drift. Memory-based algorithms typically keep a set of recent samples in a buffer. When the buffer becomes full, the algorithm updates its models based on the recent instances. *FLORA* [13] and *SAM-KNN* [14] are well-known methods for this category. However, these algorithms suffer from a critical challenge, namely sensitivity to the size of the window (memory). With a small window size, the algorithms can not detect incremental (or gradual) drifts, and a big window size causes difficulty in detecting abrupt drifts. Multiple sliding windows or adaptive sliding window are proposed to cope with this challenge. However, incorrectly keeping/forgetting old examples decreases the performance of algorithms.

Ensemble algorithms [7] can handle concept drifts by adding/removing a classifier to forget changed concepts and learn new concepts. The main challenge of ensemble methods is the size of the ensemble. Increasing the size of an ensemble increases the time, and space complexity and algorithm can not be able to satisfy the time and space limitation of online data. There are several well known methods such as *diversity for dealing drifts (DDD)* [15], *the streaming ensemble algorithm (SEA)* [16],*online bagging (OB)* [17], *Learn++.NSE* [18], and *accuracy updated ensemble (AUE)* [19].

2.2 Class Imbalance

None of the approaches mentioned above address class imbalance. The online learning algorithms which consider the joint problem of imbalanced data and concept drifts can be categorized into two groups [2].

Cost Sensitive Methods: For considering class imbalance, the algorithms in this category assign a different weight for the misclassification error of each class. The cost-sensitive online gradient descent algorithm *(CSOGD)* utilizes *SGD* (stochastic gradient descent) and perceptron classifier with cost-sensitive hinge loss function for imbalanced data [21]. Since the weights of each class are predefined, the model can not handle real concept drift (change in $p(y|x)$). *CID* (class imbalance detection) [20] and *RLSACP* [22] are proposed to address this issue by using an adaptive cost strategy. However, the perceptron method fails in the case of complex high dimensional data (when the data points are not linearly separable), or when recurrent drifts happen. *EONN* [23] is an ensemble of online neural networks that uses a cost-sensitive function to cope with imbalance and drifts. Nevertheless, training a neural network is a time-consuming procedure that needs many positive and negative samples to train a model that can not address the constraints of online imbalance data.

Resampling Methods: Approaches in this category convert an imbalanced data set to a balanced dataset by using undersampling, oversampling, or a mix of these methods. Undersampling methods decrease the number of data points in majority classes, while oversampling algorithms increase the number of samples of minority classes. The algorithms then treat the data set as being balanced. One of the most straightforward techniques is random undersampling (or oversampling). However, there are some more sophisticated algorithms. *SMOTE* [24] is one of the most popular algorithms for resampling based on the similarity of neighbours in the original data. *Generative Adversarial Networks (GAN)* [25] is another method that is proposed for oversampling by approximating the distribution of minority classes. *Uncorrelated bagging (UCB)* [26] is an ensemble method that uses all of the samples of majority class and most recent samples of minority classes that has been seen so far. *Oversampling-based online bagging (OOB)* [27] is an extension of the *online bagging* method that addresses concept drift, *Adaptive REBAlancing (AREBA)* [5] is another method for balancing data. It keeps N_B samples of majority class and last N_B samples of minority class where $2 \times N_B$ is the size of the buffer. All of the mentioned methods can not address all kinds of concept drifts. The other main drawback of these methods is that for generating new samples, they need to wait until enough samples from the majority class arrive. That causes some delay in learning new concepts (or forgetting old concepts). Another downside of these methods is information lost from important samples of the minority class that are removed during the resampling phase. Interested readers can find more details regarding learning from imbalanced datasets in [2].

Algorithm 1. HEEM

Require: initial value for $\alpha_{min}, \alpha_{max}$, Default value is 0.5
  $index \leftarrow -1$
  $N_P, N_N, k, t \leftarrow 0$
  $E \leftarrow \{\}$
  while $True$ **do**
    $\hat{y}^t \leftarrow 0$
    $\mathbf{x}^t$ = get_new_sample(t)
    for i in $range$ $len(E)$ **do**
      if $\mathbf{x}^t$ $inside$ E_i **then**
        $index \leftarrow i$
        $\hat{y}^t \leftarrow 1$
      end if
    end for
    $y^t \leftarrow get\_true\_label(t)$
    if $y^t == \hat{y}^t \& y^t == 1$ **then**
      $Enlarge(E_{index}, \mathbf{x}^t, \alpha)$ {based on Equation (5)}
    end if
    if $y^t \neq \hat{y}^t \& y^t == 0$ **then**
      $Shrink(E_{index}, \mathbf{x}^t, \beta)$ {based on Equation (4)}
    end if
    if $y^t \neq \hat{y}^t \& y^t == 1$ **then**
      $E \leftarrow add\_ellipes(E, \mathbf{x}^t, \alpha)$ {based on Equation (3)}
      $\alpha_{min} \leftarrow AVG(\sum_{k=len(E)/2}^{k=len(E)} (x^K - x_L^K))$, $\alpha_{max} \leftarrow AVG(\sum_{k=len(E)/2}^{k=len(E)} (x_U^K - x^K))$
    end if
    $N_P \leftarrow N_P + y^t$
    $N_N \leftarrow N_N - (y^t - 1)$
    $CI \leftarrow \frac{N_P}{N_N}$
    $\beta_{min}, \beta_{max} \leftarrow Update\_Beta(CI, \alpha_{min}, \alpha_{max})$ {Based on Equation (6)}
    $t \leftarrow t + 1$
  end while

3 Proposed Method

As has been mentioned in Sect. 2, in imbalanced data $N_N >> N_P$, where N_N is the number of samples from the majority class (negative samples, $y = 0$)

and N_P is the number of samples of minority class (positive samples, $y = 1$). Since in imbalanced data T_P (number of positive samples correctly predicted) is more important than T_N (number of negative samples correctly predicted). Our proposed method (HEEM) tries to find a trade-off between the accuracy on the majority class and the accuracy on the minority class to increase the geometric mean (G-Mean) metric.

$$G - Mean = \sqrt{ACC^{+} * ACC^{-}} = \sqrt{\frac{T_P}{N_P} * \frac{T_N}{N_N}} \tag{1}$$

Instead of up/down sampling, HEEM considers an extra budget for positive samples based on imbalance rate and directly tracks and predicts the positive class. For this aim, HEEM is trying to find hyper ellipse(s) that cover maximum samples of the minority class and minimum samples from the majority class.

$$\underset{\mathbf{x}^c, R}{\operatorname{argmax}} f(x) = \sum_{t=0}^{\infty} \left(I\left(\left(\sum_{j=0}^{d} I(\mathbf{x}_j^t - \mathbf{x}_j^c \leq R_j) \right) = d \right) \times (2y^t - 1) \right) \tag{2}$$

Where $\mathbf{x}^c$ and R are the centre and radius of the ellipse, respectively. d is the feature size, and y^t is true label of instance t. I(.) is an indicator function that is 1 if the condition satisfied otherwise it is 0. In online data we do not have access to the entire dataset to create/update models. This problem becomes more challenging when the class imbalance rate increases. So, instead of waiting for samples of the minority class, HEEM reserves some extra margin for future positive samples. To cover non-linear discriminate data with different unknown distributions HEEM keeps an ensemble of hyper ellipses for the minority class. If a new sample falls in an ellipse its label will be positive, otherwise the label will be negative. For each new positive sample $(\mathbf{x}^t, y^t)$ with $\hat{y}^t \neq y^t$, HEEM generates a new extended ellipse and keeps it in the ensemble. HEEM calculates a maximum and a minimum extended margin for the new point as follows:

$$\mathbf{x}_U^k = \mathbf{x}^t + |\mathbf{x}^t \times \alpha_{max}|, \quad \mathbf{x}_L^k = \mathbf{x}^t - |\mathbf{x}^t \times \alpha_{min}| \tag{3}$$

Where $0 \leq \alpha_{min}, \alpha_{max} \leq 1$ are the enlarging parameters and k is an index for the kth ellipses in the ensemble. HEEM simply keeps $\mathbf{x}_U^k$ and $\mathbf{x}_L^k$ as margins of the ellipse. Then, based on samples of the majority class, HEEM modifies the boundaries of the ellipses. When a sample from the majority class $\mathbf{x}^t \in X_{Majo}$ is predicted incorrectly ($\hat{y}^t \neq y^t$), the algorithm modifies the corresponding ellipse(s) by shrinking the boundaries as follows:

$$\begin{cases} if(\forall i \quad \mathbf{x}^{i,t} > \mathbf{x}^{i,k}) : \mathbf{x}_U^k = \mathbf{x}_U^k - |(\mathbf{x}_U^k - \mathbf{x}^t) \times \beta_{max}| \\ if(\forall i \quad \mathbf{x}^{i,t} < \mathbf{x}^{i,k}) : \mathbf{x}_L^k = \mathbf{x}_L^k + |(\mathbf{x}^t - \mathbf{x}_L^k) \times \beta_{min}| \\ otherwise : \begin{cases} \mathbf{x}_U^{i,k} = \mathbf{x}_U^{i,k} - |min((\mathbf{x}_U^{i,k} - \mathbf{x}^{i,t}), (\mathbf{x}_U^{i,k} - \mathbf{x}^{i,k}) \times \beta_{max}| \\ \mathbf{x}_L^{i,k} = \mathbf{x}_L^{i,k} + |min((\mathbf{x}^{i,t} - \mathbf{x}_L^{i,k}), (\mathbf{x}^{i,k} - \mathbf{x}_L^{i,k})) \times \beta_{min}| \end{cases} \end{cases} \tag{4}$$

Where $0 \leq \beta_{max}, \beta_{min} \leq 1$ are the shrinking parameters, i indicates i^{th} feature and $\mathbf{x}^k$ is the centre of kth ellipse. In Eq. 4 if all features of a new sample are

Table 1. The characteristics of the datasets used in the experiments.

Dataset	Fraud	Cancer	CoverType	Sine_S	Sine_G	Sea_S	Sea_G
#Instances	284807	858	214587	50K	50K	50K	50K
#Features	29	35	54	4	4	3	3
CI	0.172%	6.4%	1%	1%	1%	10%	10%
#Drift	N/A	N/A	N/A	3	3	3	3
Drift Type	N/A	N/A	N/A	Real	Real	Real	Real
Drift Speed	N/A	N/A	N/A	Sudden	Gradual	Sudden	Gradual
Noise	N/A	N/A	N/A	5%	5%	5%	5%

bigger (smaller) than the mean of the ellipse, then just the upper (lower) margin is shrunk, otherwise, both margins are shrunk. If a sample from the minority class falls inside an ellipse the predicted label for that will be positive and the boundaries of the corresponding ellipse(s) are enlarged as follows:

$$\begin{cases} \mathbf{x}_U^{new} = \mathbf{x}^t + |\mathbf{x}^t \times \alpha_{max}/2| \\ \\ \mathbf{x}_L^{new} = \mathbf{x}^t - |\mathbf{x}^t \times \alpha_{min}/2| \end{cases} \Rightarrow \begin{cases} \mathbf{x}_U^k = max(\mathbf{x}_U^{new}, \mathbf{x}_U^k) \\ \\ \mathbf{x}_L^k = min(\mathbf{x}_L^{new}, \mathbf{x}_L^k) \end{cases} \tag{5}$$

In the model $\alpha_{min/max}$ is enlarging variable. The model starts with an initial value for $\alpha_{min/max}$ and adaptivitly updates and uses it. $\beta_{min/max}$ is calculated based on $\alpha_{min/max}$ and class imbalance ratio (CI) as follows:

$$\beta_{min} = 2 \times \sqrt{CI} \times (1 + \alpha_{min}), \quad \beta_{max} = 2 \times \sqrt{CI} \times (1 + \alpha_{min}) \tag{6}$$

If $\alpha_{min/max}$ is set as a big number then $\beta_{min/max}$ also will be increased and decreasing the value of $\alpha_{min/max}$ results in a smaller value for $\beta_{min/max}$. Algorithm 1 shows our model in detail.

4 Experiments

In this section, we compare our proposed method, HEEM, with state-of-the-art algorithms. In Subsect. 4.1, these baseline algorithms and their parameters are described. We evaluate all the methods on a variety of real and synthetic data sets in Subsect. 4.2. The performance metrics for imbalanced online data is shown in Subsect. 4.3. Finally, the predictive performance of all methods is shown in Subsect. 4.4.

4.1 Competitors

Simple: A straight-forward incremental and online learning algorithm from [5] where the batch size is set to 1. This algorithm does not have any mechanism to handle class imbalance or concept drift.

Sliding Window [13]: An online memory-based method with a single sliding window. This method addresses drifts. The window size is set to W = 100. This method is not an incremental learning algorithm and needs to keep $W-1$ previously seen data samples.

Adaptive_CS [21]: A state-of-the-art adaptive cost-sensitive learning method. It utilizes the CSOGD cost function ($J = (I_{y^t=0} + I_{y^t=1}\frac{c_p}{c_n})l(y^t, \hat{y}^t)$). Where I is an indicator function. $c_n, c_p \in [0,1]$ and $c_n + c_p = 1$ are the misclassification costs for negative and positive classes respectively. These costs are adapted according to the class imbalance ratio. The time-decayed factor is set to $\theta = 0.99$.

OOB [27]: A state-of-the-art online resampling algorithm (see also Sect. 2.2). It is an online and incremental algorithm. The number of classifiers is set to 20 classifiers. At each time step, it performs multiple updates of the classifier.

AREBA [5]: An online learning algorithm with time-decayed factor $\theta = 0.99$. For a small buffer size (e.g. Buffer size = 2) the algorithm is similar to an incremental algorithm since it needs access only to a single old example. AREBA is rebalancing the number of positive and negative samples by dynamically modifying its queue length. We set the best buffer size of AREBA for each dataset as it is reported in the original paper [5].[1]

4.2 Datasets

We experimented with a variety of synthetic and real datasets summarized in terms of their dimensionality, cardinality, class ratio, and drift type in Table 1.

Real-World Datasets: The real-world datasets are used to evaluate the performance of classifiers in practice. For this aim we utilize the *Cervical Cancer*, *Fraud*, and Forest Cover Type datasets. *Cervical Cancer*(Cancer) [28] contains 36 features corresponding to the historical medical records and demographic information of 858 patients with a class imbalance rate of 6.4%. *Fraud* dataset [29] contains 492 frauds out of 284,807 (CI = 0.172%) transactions that were made by European cardholders in September 2013. This dataset is severely imbalanced and contains 30 features. *Forest Cover Type*(CoverType) is one of the well-known datasets that has been studied in many concept drift papers. In this dataset, the cartographic information of trees is represented using 55 features to predict the forest type. We considered class "1" with 200,000 records as a majority class and class "4" with 2,000 records as a minority class. The imbalanced class rate is 1% [5].

Synthetic Datasets: The main advantage of synthetic datasets is their capability for generating any desired dataset with different kinds of concept drift behaviour. We used the MOA framework [30] to generate four synthetic streams with different types of concept drift. We used the SEA generator to create two

[1] The implementations for all baselines have been taken from https://github.com/kmalialis/areba [5].

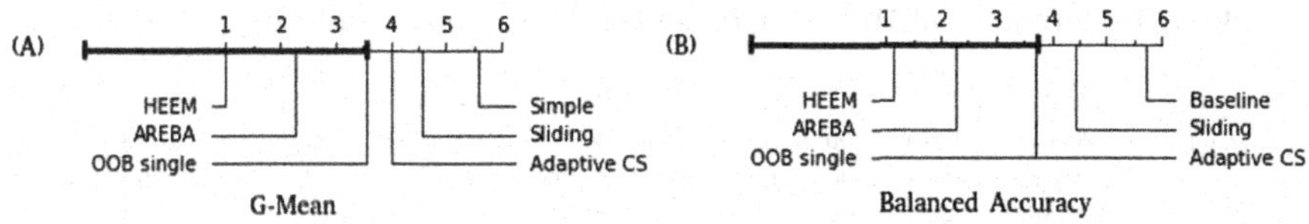

Fig. 1. Post-hoc Bonferroni-Dunn test, comparing the average ranks of HEEM on all datasets, for G-Mean (A) and balanced accuracy (B).

datasets with different type of concept drifts. Sea_S(Sea_S) contains three sudden (gradual) real concept drifts with 5% noise and CI = 10%. Similarly, SINE generator is utilized to generate Sine_G and Sine_S datasets. Sine_S (Sine_G) has three sudden (gradual) real concept drifts with 5% noise and CI = 1%.

4.3 Performance Metrics

For a balanced dataset, an appropriate performance metric considers the performance on all classes. However, such a performance metric is heavily affected by the majority class. For imbalanced datasets, sensitivity, specificity, and geometric mean (G-Mean) are often used to evaluate the performance of classifiers. Sensitivity measures the performance of a classifier only on actual positive (minority) instances, while specificity does the same thing on actual negative (majority) instances. Another evaluation metric, G-Mean (binary classification), is calculated as the square root of the product of the specificity and sensitivity. G-Mean maximizes the specificity and sensitivity while keeping both balanced. For the experiments, we used prequential evaluation which is commonly used in data stream classification. This method is specifically designed for cases where samples arrive in sequential order. The predictive model is always tested on unseen data. The fading factor is set to $\theta = 0.99$ in our empirical study. We evaluate the performance of the classifiers using the prequential evaluation and report on the average of final G-Mean and balanced accuracy over 50 runs.

4.4 Predictive Performance

Table 2 shows the predictive performance of HEEM and othe competitors on the different datasets. For *CoverType*, with $\theta = 0.99$ all the methods could reach to 1 since majority samples mostly happen at the beginning of the stream. To have a better comparison, we mentioned average G-means (T-G-Mean, and T_B_Acc, where $\theta = 1$). HEEM outperforms all the compared methods in terms of G-Mean and balanced accuracy. We used the post-hoc Bonferroni-Dunn test to further investigate the differences in the G-Mean and balanced accuracy to compute the critical difference. As it is shown in Fig. 1 the performance of HEEM is significantly better than *Simple, Sliding, and AdaptiveCS* in terms of G-Mean, and significantly better than *Simple, Sliding* in terms of balanced accuracy.

Table 2. Performance of HEEM and the baselines on real and synthetic datasets.

Dataset	Measure	HEEM	AREBA	OOBSingle	AdaptiveCS	Sliding	Simple
Fraud	G-Mean	**0.914 ± .000**	0.898 ± .005	0.897 ± .003	0.804 ± .006	0.853 ± .004	0.823 ± .012
	B_Accuracy	**0.918 ± .000**	0.901 ± .003	0.899 ± .002	0.855 ± .006	0.871 ± .005	0.856 ± .018
Cancer	G-Mean	**0.874 ± .000**	0.857 ± .005	0.663 ± .116	0.810 ± .023	0.644 ± .029	0.455 ± .022
	B_Accuracy	**0.875 ± .000**	0.860 ± .003	0.663 ± .011	0.840 ± .020	0.701 ± .025	0.541 ± .019
CoverType	T-G-Mean	**0.986 ± .000**	0.977 ± .003	0.959 ± .021	0.923 ± .003	0.979 ± .001	0.903 ± .024
	T-B_Acc	**0.982 ± .000**	0.953 ± .000	0.936 ± .013	0.903 ± .000	0.962 ± .000	0.892 ± .026
Sea_S	G-Mean	**0.845 ± .000**	0.823 ± .012	0.809 ± .010	0.767 ± .012	0.63 ± .022	0.686 ± .031
	B_Accuracy	**0.842 ± .000**	0.821 ± .009	0.802 ± .011	0.768 ± .012	0.688 ± .020	0.725 ± .027
Sea_G	G-Mean	**0.864 ± .000**	0.855 ± .018	0.824 ± .007	0.803 ± .014	0.635 ± .031	0.545 ± .014
	B_Accuracy	**0.865 ± .000**	0.860 ± .020	0.831 ± .007	0.806 ± .010	0.695 ± .025	0.650 ± .018
Sine_S	G-Mean	**0.824 ± .000**	0.803 ± .021	0.748 ± .011	0.820 ± .004	0.539 ± .021	0.434 ± .004
	B_Accuracy	**0.825 ± .000**	0.806 ± .017	0.774 ± .009	0.818 ± .006	0.645 ± .020	0.594 ± .005
Sine_G	G-Mean	**0.833 ± .000**	0.802 ± .003	0.733 ± .018	0.763 ± .009	0.485 ± .000	0.373 ± .020
	B_Accuracy	**0.836 ± .000**	0.810 ± .001	0.756 ± .015	0.767 ± .010	0.615 ± .000	0.570 ± .023

5 Model Behavior

In the previous section, we demonstrated the superior predictive performance of HEEM over state-of-the-art methods. In this section, we describe the behaviour of our model and how it can successfully handle concept drifts and noises on imbalanced data. Then we will discuss the complexity of our model in terms of parameters, time and space.

5.1 Nonstationary and Noisy Data

The problem of imbalanced online classification becomes harder when concept drift and noisy data occur in the data. HEEM can handle this challenge automatically by adding new ellipses or removing decayed ellipses. When a concept change occurs, HEEM adds a new ellipse for the new concept and modifies the ellipses based on future samples to learn the concept. On the other hand, when a concept changes, our method removes (by shrinking) the ellipse corresponding to the concept if new instances wrongly fall in the ellipses. Figure 2 shows the behaviour of our model in the face of concept drifts and noises. Both *Sea_G* and *Sine_S* contain three concept drifts and CI = 10% and CI = 1% respectively.

5.2 Parameters

The main parameter of HEEM is the enlarging rate($\alpha_{min/max}$) that indicates the extra margins for a new ellipse. However, HEEM adaptively modifies the initial value of $\alpha_{min/max}$ based on the average margin of the previous ellipses in the ensemble. Figure 3 shows the effect of the initial value of $\alpha_{min/max}$ on the performance of HEEM in the term of G-Mean, Specificity and recall on three real datasets. For our experiments we set $\alpha_{min}, \alpha_{max} = 0.5$ as initial values.

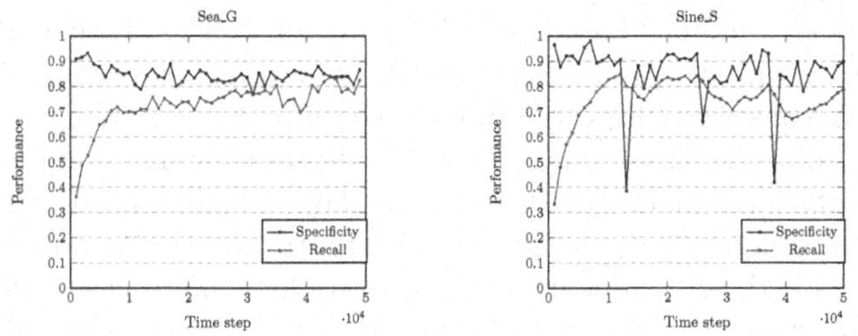

Fig. 2. Behaviour for the Sine_G (gradual drift) and Sea_S (Sudden drift) dataset.

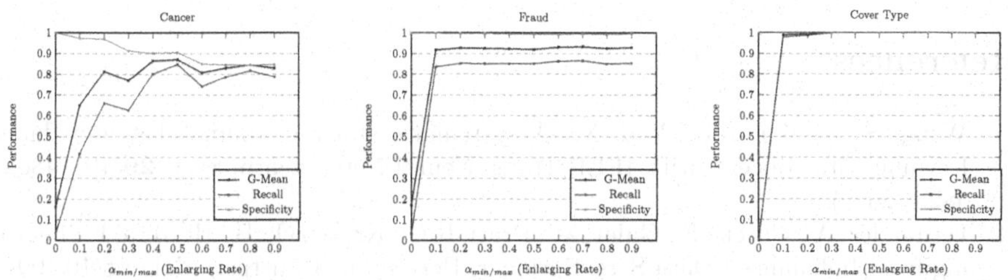

Fig. 3. The role of the initial value of enlarging rate (α) on the performance of HEEM on Cancer, Fraud and CoverType datasets.

5.3 Time and Space Complexity

Time and space complexity of an online algorithm are essential factors. The total time complexity of HEEM is $O((N_P \times (1 - ACC^+) \times d) \times N)$. Where N_P is number of samples of minority class, ACC^+ is the accuracy of minority class, d is the number of features and N is total samples of the stream. Since $N_P << N$, $d << N$ and $1 - ACC^+ \leq 1$ in imbalanced dataset, HEEM time complexity order is $O(N)$. Thus the average response time complexity per each sample is $O(1)$.

$$\begin{cases} (N_P \times (1 - ACC^+)) << N \\ d << N \end{cases} \Rightarrow \quad O((N_P \times (1 - ACC^+) \times d) \times N) = O(N) \tag{7}$$

The total space that HEEM uses is equal to $O(2 \times (1 - ACC^+) \times N_P) = O(N_P)$ However, in the case of limited memory, HEEM removes the ellipse with the smallest radius value once the memory is full.

6 Conclusion

In this paper, we proposed the Hyper- Ellipses-Extra-Margin (HEEM) algorithm that addresses the imbalanced data challenge in online data streams by utilizing extended hyper-ellipses for the minority class. Instead of waiting for instances of the majority class, HEEM predicts the future instances of the minority class and keeps extra margins for positive samples based on the performance of the model.

These ellipses are modified based on misclassified instances. Misclassified samples of the majority (minority) class are used for enlarging (shrinking) the borders. Since HEEM considers the class imbalance ratio for shrinking/enlarging the borders, it is robust on noisy data. Also, the ellipses provide a complete summary of the minority class that can be used for data augmentation. Since different datasets have different properties and distributions, HEEM adaptively learns its parameters and uses the learned parameters for enlarging and shrinking the borders. Our experimental results showed that the proposed method mitigates the class imbalance and outperforms the state-of-the-art methods in terms of G-Mean and balanced accuracy.

References

1. Wang, S., Minku, L.L., Yao, X.: A systematic study of online class imbalance learning with concept drift. IEEE Trans. Neural Netw. Learn. Syst. **29**(10), 4802–4823 (2018)
2. Fernández, A., García, S., Galar, M., Prati, R.C., Krawczyk, B., Herrera, F.: Learning from Imbalanced Data Sets. Springer, Berlin (2018). https://doi.org/10.1007/978-3-319-98074-4
3. Wang, S., Minku, L.L., Yao, X.: Dealing with multiple classes in online class imbalance learning. In: Proceedings of IJCAI, pp. 2118–2124 (2016)
4. Chrysakis, A., Moens, M.F.: Online continual learning from imbalanced data. In: Proceedings of ICML (2020)
5. Malialis, K., Panayiotou, C.G., Polycarpou, M.M.: Online learning with adaptive rebalancing in nonstationary environments. IEEE Trans. Neural Netw. Learn. Syst. (2020)
6. ZareMoodi, P., Siahroudi, S.K., Beigy, H.: A support vector based approach for classification beyond the learned label space in data streams. In: Proceedings of the 31st Annual ACM SAC, pp. 910–915 (2016)
7. Krawczyk, B., Minku, L.L., Gama, J., Stefanowski, J., Woźniak, M.: Ensemble learning for data stream analysis: a survey. Inf. Fusion **37**, 132–156 (2017)
8. Malialis, K., Panayiotou, C., Polycarpou, M.M.: Queue-based resampling for online class imbalance learning. In: Proceedings of IJCNN, pp. 498–507 (2018)
9. Ross, S., Mineiro, P., Langford, J.: Normalized online learning. arXiv preprint, pp. 1305–6646. arXiv (2013)
10. Page, E.S.: Continuous inspection schemes. Biometrika **41**(1/2), 100–115 (1954)
11. Bifet, A., Gavalda, R.: Learning from time-changing data with adaptive windowing. In: Proceedings of SIAM Conference on Data Mining, pp. 443–448 (2007)
12. Alippi, C., Roveri, M.: Just-in-time adaptive classifiers-Part I: detecting nonstationary changes. IEEE Trans. Neural Netw. **19**(7), 1145–1153 (2008)
13. Widmer, G., Kubat, M.: Learning in the presence of concept drift and hidden contexts. Mach. Learn. **23**(1), 69–101 (1996)
14. Losing, V., Hammer, B., Wersing, H.: KNN classifier with self adjusting memory for heterogeneous concept drift. In: 2016 IEEE 16th International Conference on Data Mining (ICDM), pp. 291–300 (2016)
15. Minku, L.L., Yao, X.: A new ensemble approach for dealing with concept drift. IEEE Trans. Knowl. Data Eng. **24**(4), 619–633 (2011)

16. Street, W.N., Kim, Y.: A streaming ensemble algorithm (SEA) for large-scale classification. In: Proceedings of the Seventh ACM SIGKDD International Conference on Knowledge Discovery and Data Mining, pp. 377–382 (2001)
17. Oza, N.C.: Online bagging and boosting. In: Proceedings of IEEE International Conference on Systems, Man and Cybernetics, pp. 2340–2345 (2005)
18. Ditzler, G., Polikar, R.: Incremental learning of concept drift from streaming imbalanced data. IEEE Trans. Knowl. Data Eng. **25**(10), 2283–2301 (2012)
19. Brzezinski, D., Stefanowski, J.: Combining block-based and online methods in learning ensembles from concept drifting data streams. Inf. Sci. **265**, 50–67 (2014)
20. Wang, S., Minku, L. L., Yao, X.: A learning framework for online class imbalance learning. In: IEEE Symposium Computational Intelligence and Ensemble Learning (CIEL), pp. 36–45 (2013)
21. Wang, J., Zhao, P., Hoi, S.C.H.: Cost-sensitive online classification. IEEE Trans. Knowl. Data Eng. **26**(10), 2425–2438 (2014)
22. Ghazikhani, A., Monsefi, R., Yazdi, H.S.: Recursive least square perceptron model for non-stationary and imbalanced data stream classification. Evol. Syst. **4**(2), 119–131 (2013)
23. Ghazikhani, A., Monsefi, R., Yazdi, H.S.: Ensemble of online neural networks for non-stationary and imbalanced data. Neurocomputing **122**, 535–544 (2013)
24. Chawla, N.V., Bowyer, K.W., Hall, L.O., Kegelmeyer, W.P.: SMOTE: synthetic minority over-sampling technique. J. Artif. Intell. Res. **16**, 321–357 (2002)
25. Douzas, G., Bacao, F.: Effective data generation for imbalanced learning using conditional generative adversarial networks. Expert Syst. Appl. **91**, 464–471 (2018)
26. Gao, J., Ding, B., Fan, W., Han, J., Philip, S.Y.: Classifying data streams with skewed class distributions and concept drifts. IEEE Internet Comput. **12**(6), 37–49 (2008)
27. Wang, S., Minku, L.L., Yao, X.: Resampling-based ensemble methods for online class imbalance learning. IEEE Trans. Knowl. Data Eng. **27**(5), 1356–1368 (2014)
28. Fernandes, K., Cardoso, J.S., Fernandes, J.: Transfer learning with partial observability applied to cervical cancer screening. In: Proceedings of Iberian Conference on Pattern Recognition and Image Analysis, pp. 243–250 (2017)
29. Dal Pozzolo, A., Caelen, O., Johnson, R.A., Bontempi, G.: Calibrating probability with undersampling for unbalanced classification. In: Proceedings of IEEE Symposium Series on Computational Intelligence, pp. 159–166 (2015)
30. Bifet, A., Holmes, G., Kirkby, R., Pfahringer, B.: Moa: massive online analysis. J. Mach. Learn. Res. **11**, 1601–1604 (2010)

Locally Linear Support Vector Machines for Imbalanced Data Classification

Bartosz Krawczyk(✉) and Alberto Cano

Department of Computer Science, Virginia Commonwealth University, Richmond, VA, USA
{bkrawczyk,acano}@vcu.edu

Abstract. Classification of imbalanced data is one of most challenging aspects of machine learning. Despite over two decades of progress there is still a need for developing new techniques capable to overcome numerous difficulties embedded in the nature of imbalanced datasets. In this paper, we propose Locally Linear Support Vector Machines (LL-SVMs) for effectively handling imbalanced datasets. LL-SVMs is a lazy learning approach which trains a local classifier for each new test instance using its k nearest neighbors. This way, we are able to maximize the margin in the original input features space and obtain a better adaptation to complex class boundaries. We combine LL-SVMs with local oversampling and cost-sensitive approaches to make them skew-insensitive. Working only in the local neighborhood significantly improves the generalization over the minority class and tackles instance-level difficulties, such as class overlapping, borderline and noisy instances, as well as small disjuncts. An extensive experimental study shows that our local models are able to outperform their global counterparts, especially when handling difficult, borderline, and noisy imbalanced datasets.

Keywords: Machine learning · Class imbalance · Instance-level difficulties · Support Vector Machines · Resampling · Cost-sensitive learning

1 Introduction

The problem of imbalanced data is considered among one of the biggest data-level challenges that may impair the performance of machine learning algorithms [5]. Despite more than two decades of development, we are still gaining a deeper understanding into what problems are posed by imbalanced data and how to alleviate them [10]. Initial understanding pointed to the disproportion between instances from different classes (known as imbalance ratio, IR) as the major factor that hinders the learning process. As most of classifiers are evaluated by predictive accuracy or are optimizing 0–1 loss function, each training instance has the same importance. Therefore, when the training set is strongly imbalanced the training procedure may create a strong bias towards the majority class, neglecting the minority one as a side-effect. Most methods designed for imbalanced data focus on countering the effects of IR by either generating new

K. Karlapalem et al. (Eds.): PAKDD 2021, LNAI 12712, pp. 616–628, 2021.
https://doi.org/10.1007/978-3-030-75762-5_49

minority instances, removing majority instances, modifying the cost function used during training, or employing modifications of the training procedure [5]. Recent studies point to the fact that IR is not the sole source of learning difficulties [10]. By analyzing the structure of minority class, one may see that instances may exhibit different properties [5].

In this paper, we propose a novel approach for learning from difficult imbalanced data, based on training local classifiers. Locally Linear Support Vector Machines (LL-SVM) are based on training a compact classifier for each new test instance, considering only the nearest neighborhood from the training set. This is fundamentally different than having a globally trained classifier handling test instances. Our approach allows for capturing local data properties and creating classifiers that are more flexible, achieving a better adaptation to complex class boundaries. Furthermore, training local classifiers using small instance subsets allows us to propose local data preprocessing methods that alleviate the limitations of global approaches, as data characteristics may differ across feature space. We propose to augment LL-SVM with local oversampling and cost-sensitive learning, in order to improve robustness of these methods to difficult data distributions. The main contributions of this paper are as follow:

- **Local Support Vector Machines for class imbalance.** A novel approach for handling imbalanced and difficult data by dynamically training local SVMs for each instance to be classifier. This allows to efficiently capture local data properties without a need for sophisticated sampling or cost-sensitive extensions.
- **Theoretical analysis of LL-SVM usefulness for imbalanced data.** Using Rademacher complexity we show that training local models with the proposed LL-SVM is highly beneficial for imbalanced data, leading to tighter stability bounds than when using standard SVM-based approaches.

2 Shortcomings of Global Approaches to Imbalanced Data

One of the most interesting directions in imbalanced classification is the observation that imbalance ratio is not the sole, or even main, source of difficulty for classifiers. There exist problems with high IR, yet with both classes well represented and with non-overlapping distributions - and they pose little, if any challenge for modern classifiers. Therefore, there must be an alternative source of performance degradation. Recent works show that this can be caused by the presence of difficult examples, especially within the minority class [10]. By analyzing the neighborhood of each minority class instance, we may assign it to one of four predefined groups: safe, borderline, rare and outliers [5].

Limitations of Global Oversampling. Oversampling methods are known to suffer from multiple limitations when applied uniformly over the training set [10], among which the most important are: (i) sensitiveness to multi-modal data distributions; (ii) varying properties of local data disjuncts; and (iii) lack of robustness to difficult and noisy instances. In the case of a class being composed of multiple small disjuncts (or clusters), applying oversampling will lead to injection of artificial instances in areas outside of minority class boundaries. This will not only lead to alteration of the original class distribution, but also may increase the overlapping between minority and majority classes. This is connected with the local properties of data disjuncts. Recent works suggest that the global imbalance ratio may be a misleading factor, as the disproportion between classes may vary among different data subspaces. Therefore, the oversampling ratio should be selected locally according to data characteristics, instead of being set uniformly for the whole minority class.

Limitations of Global Cost-Sensitive Learning. This family of methods suffer from similar limitations as oversampling, but from the angle of establishing the cost matrix. While there exist several efficient methods for learning effective values of cost matrix, they still consider a single fixed cost for the entire minority class. As the imbalance ratio differs among areas of the decision space and the minority class is composed of instances of varying difficulty, it seems only logical that the cost matrix should reflect these local characteristics. A single penalty factor may be insufficient. By utilizing multiple different cost matrices, each tuned to local characteristics, one should better capture the properties of minority class.

3 LL-SVM for Imbalanced Data

This section presents the procedure for training LL-SVMs and the proposed modifications for local learning, it discusses the advantages of local learning under data-level difficulties, and it shows that LL-SVM will produce a tighter bound for imbalanced classification problems than its global counterparts.

3.1 Training Procedure for LL-SVM

This paper proposes a lazy learning approach for handling imbalanced data with local SVMs. In the local SVM approach, k nearest neighbors of each new test instance are firstly selected, then a local and linear SVM (LL-SVM) is trained using only these instances, without using any kernel mapping. The class label of the new instance is assigned depending on what side of the separation hyperplane(s) the query instance lies. In choosing k nearest neighbors the number k refers to all possible classes. Hence, there can be from 0 to k members of a certain class within the neighborhood of the new instance. Please note that this method is applicable to multi-class problems. However, in this paper we restrict it only to binary imbalanced datasets.

In the testing phase k nearest neighbors of the new test instance are selected using the k smallest weighted Euclidean distances between the new instance and all the training instances. Having k-nn's chosen, a class label is predicted by analyzing the content of each individual neighborhood as:

- **Case 1: single class in the neighborhood:** if all k nearest neighbors of the new test instance belong to same class, then the new instance is labeled as belonging to this class. There is no need to train any LL-SVM model at this point, as we deal with a homogeneous decision region.
- **Case 2: two classes in the neighborhood:** if there are members of two classes among k nearest neighbors, we follow a two-step procedure:

  Step 2.1: Calculating local IR. Data characteristics may vary in different areas of feature space and thus require local analysis. LL-SVM allows for measuring the local IR in the local neighborhood, which is a much better indicator of the learning difficulty in this area than global IR. We use this information to guide balancing procedures applied during LL-SVM training.
  Step 2.2: Training local LL-SVM. a local classifier (without kernel mapping) is trained only on instances from the selected neighborhood. We use this model to predict the label of the new instance and then discard it. Additionally, in order to make it skew-insensitive, we propose to use either data oversampling or cost-sensitive learning. During these procedures, we take into account only the local IR to set oversampling ratio or cost parameter.

3.2 Advantages of LL-SVM in Imbalanced Domain

We will prove now that LL-SVM is a more stable and locally accurate classifier than its global counterpart for imbalanced data.

Theorem 1 *For a class of functions $\mathcal{F}$, let us denote the empirical loss of a decision function f_1 as $\widehat{R}(f_1)$. The decision function $f_1 \in \mathcal{F}$ is calculated using a training set with N instances. For any $\delta \in (0,1)$, the following holds with a probability at least $1-\delta$:*

$$R(f_1) \leq \widehat{R}(f_1) + \widehat{C}(\mathcal{F}) + 3\sqrt{\frac{\ln(2/\delta)}{2N}}, \tag{1}$$

where $\widehat{C}(\mathcal{F})$ is the Rademacher complexity of $\mathcal{F}$.

This allows us to derive the upper bound of the Rademacher complexity for $\mathcal{F}^{k,C}$, which stands for a class of all decision functions that can be outputted by a locally linear with respect to the used parameters k and C [8].

Proposition 1 *Let $K(\cdot,\cdot)$ be a kernel function used when training a global SVM classifier. Then, its Rademacher complexity can be calculated as:*

$$\widehat{C}(\mathcal{F}^{k,C}) \leq \frac{4\sqrt{Ck}}{N}\sqrt{\sum_{i=1}^{N} K(x_i, x_i)}. \tag{2}$$

We have bounded the value of the hyperplane coefficient norms in the feature space for each LL-SVM. By using the primal SVM formulation for a given i-th local neighborhood $\min \frac{1}{2}||w_i||^2 + C\sum_{j=1}^{k} \xi_J$, s.t. $y_j(w_i^T \Phi(x_j) + b) \geq 1 - \xi_j$ and using the fact that there is a single LL-SVM trained for each test instance, we can obtain a stability bound. A feasible solution can be reached by setting $w = 0, b = 1$, and $\xi_j = 2$ for all j. This allows to estimate the optimal objective as at most $2Ck$.

An upper bound for a global SVM for binary imbalanced data can be expressed as:

$$\|w\| \leq 2\sqrt{C \min\{N_{min}, N_{maj}\}}, \tag{3}$$

where N_{min} and N_{maj} are the numbers of instances in the minority and majority classes. As in imbalanced data the minority class is always smaller, we can rewrite the bound from Eq. 4 as:

$$\|w\| \leq 2\sqrt{CN_{min}}. \tag{4}$$

The bound presented in Proposition 1 for the Rademacher complexity will differ for global SVM and LL-SVM, as the term $\sqrt{Ck}$ for LL-SVM will be replaced in global SVM by a larger value $\sqrt{CN_{min}}$. Therefore, the stability bounds for LL-SVM over imbalanced data are tighter than the ones for a global SVM by a factor t:

$$t = \sqrt{\frac{N_{min}}{k}}. \tag{5}$$

This is very important from the imbalanced data point of view, as it allows us to train more stable and locally accurate classifiers, even for cases with very small number of minority instances in given neighborhood. The advantages of the proposed LL-SVM approach for imbalanced data classification are as follow:

Handing Local Data Irregularities. We are able to efficiently adapt to local data characteristics, such as borderline instances or small disjuncts by training LL-SVMs on a small subset of local instances. Such data characteristics pose a significant challenge for standard SVM-based classifiers as difficult instances influence the decision boundary significantly, then leading to either overfitting or increased error on the minority class.

Local Preprocessing. By conducting either oversampling or cost-sensitive learning locally, we are able to overcome limitations inherent to SMOTE-based methods that are sensitive to rare, sparse, or noisy instances. As we limit the oversampling only to the small local neighborhood, we achieve in practice a constrained introduction of artificial samples that follow underlying instance distribution. Similar argument can be made for cost-sensitive learning, where LL-SVMs allow for having multiple different cost parameters, instead of having a single cost factor for the entire dataset. This offers significantly improved flexibility and leverages cost-sensitive SVMs.

Training Diverse Classifiers. Finally, LL-SVMs can be seen as a realization of ensemble approach, where base classifiers are trained for specialized subproblems. Hence, we fully use advantages normally reserved for ensemble models, such as exploitation for diverse classifier-base of smaller, yet specialized models.

4 Experimental Study

The experimental study was designed to compare the proposed LL-SVMs in oversampling and cost-sensitive variants with their contemporary global counterparts. Our aim is to show that using locally learned classifiers is highly beneficial for imbalanced data classification, especially when they are difficulties embedded in the nature of data. We conduct two independent experiments using oversampling and cost-sensitive approaches.

4.1 Experimental Set-Up

Datasets. Datasets used in the evaluation of the proposed solutions are shown in Table 1 and sorted by their IR value. We provide information about percentage amount of safe, borderline, rare and outlier examples [5]. Datasets are divided into two groups: normal and difficult, based on their instance type distribution.

Table 1. Datasets description.

Dataset	IR	Instances	Features	Safe [%]	Borderline [%]	Rare [%]	Outlier [%]	Type
wisconsin	1.86	683	10	61.09	18.83	5.02	15.06	Normal
glass-0-1-2-3_vs_4-5-6	3.20	214	10	62.75	21.57	5.88	9.80	Normal
glass6	6.38	214	10	75.86	6.90	3.45	13.79	Normal
cleveland-0_vs_4	12.31	173	14	61.54	0.00	15.38	23.08	Normal
nursery-very_recom	38.52	12960	8	82.00	17.00	1.00	0.00	Normal
paw3$_{normal1}$	50.00	15000	10	90.00	10.00	0.00	0.00	Normal
flower5$_{normal1}$	50.00	15000	10	85.00	15.00	0.00	0.00	Normal
kddcup-buffer_vs_back	73.43	2233	42	96.67	0.00	3.33	0.00	Normal
paw3$_{normal2}$	300.00	30000	25	65.00	25.00	10.00	0.00	Normal
flower5$_{normal2}$	400.00	60000	40	80.00	10.00	10.00	0.00	Normal
glass1	1.82	214	10	2.63	11.84	6.58	78.95	Difficult
pima	1.87	768	9	32.09	41.04	13.81	13.06	Difficult
haberman	2.78	306	4	4.94	4.94	2.47	87.65	Difficult
vehicle2	2.88	846	19	34.86	9.63	55.05	0.46	Difficult
vehicle1	2.90	846	19	41.01	7.37	3.23	48.39	Difficult
vehicle3	2.99	846	19	46.70	8.02	3.30	41.98	Difficult
vehicle0	3.25	846	19	29.65	10.05	59.80	0.50	Difficult
new-thyroid1	5.14	215	6	37.14	34.29	20.00	8.57	Difficult
segment0	6.02	2308	20	55.32	31.31	8.81	4.56	Difficult
vowel0	9.98	988	14	26.67	41.11	8.89	23.33	Difficult
abalone-0-4_vs_16-29	11.54	4177	8	8.36	20.60	20.60	50.45	Difficult
seismic-bumps	14.38	2584	18	3.52	29.41	16.47	50.58	Difficult
abalone9-18	16.40	731	9	2.38	2.38	92.86	2.38	Difficult
glass5	22.78	214	10	0.00	33.33	22.22	44.44	Difficult
lymph-normal-fibrosis	23.67	148	19	0.00	50.00	16.67	33.33	Difficult
car	24.04	1728	6	47.84	39.14	8.70	4.35	Difficult
winequality-red-4	29.17	1599	12	0.00	1.89	13.21	84.91	Difficult
winequality-white-3_vs_7	44.00	900	12	0.00	10.00	20.00	70.00	Difficult
paw3$_{difficult}$	300.00	30000	25	15.00	55.00	20.00	10.00	Difficult
flower5$_{difficult}$	400.00	60000	40	10.00	40.00	40.00	10.00	Difficult

Reference Methods. For the oversampling experiment, LL-SVM uses SMOTE. As references, we use a global SVM augmented with SMOTE [3], SPIDER2 [12], kernel-ADASYN [13], LR-SMOTE [11], Optimal Transport Oversampling [15], and RBO [9]. For the cost-sensitive experiment, we use a cost-sensitive SVM [1] trained locally, where the cost is equal to the local IR. As references, we use global SVM with cost equal to the global IR [1], SVM with cost tuned using ROC analysis [2], Near-Bayesian Cost-Sensitive SVM [4], SVM with example-dependent cost [7], SVM with hinge loss [6], and SVM with self-adaptive cost [14].

Parameters. All SVM classifiers have used the following parameter set: C $\in [1.0, 1.5, \ldots, 5.0]$; training = SMO. Global versions additionally use kernel = RBF; $\gamma \in [0.001, 0.002, \ldots, 0.01]$. LL-SVM does not use a kernel mapping. LL-SVMs use $k \in [25, 50, \ldots, 200]$. Classifier parameters and neighborhood size have been established independently for each dataset using internal 3-fold cross validation.

Evaluation and Testing. We use Geometric Mean (GM), F1-measure (FM), and area under ROC curve (AUC). We have employed combined 5×2-fold CV F-test for training, testing, and pairwise statistical significance testing on a single dataset. Additionally, we use Friedman ranking test and Bayesian sign-rank test for statistical comparison over multiple datasets.

4.2 Analysis of LL-SVM Neighborhood Parameter

Firstly, we want to analyze the influence of the k neighborhood size on LL-SVM performance. In remaining experiments, we tune this parameter independently for each dataset, but here we analyze global trends. The relationships between the neighborhood size and performance metrics for LL-SVM with oversampling and cost-sensitive learning are depicted in Fig. 1.

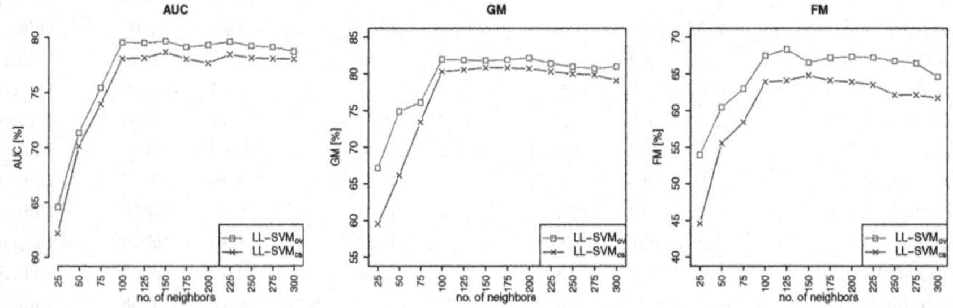

Fig. 1. Relationships between the LL-SVM neighborhood size and three skew-insensitive performance metrics.

For all three metrics increasing the neighborhood size improved the predictive power of LL-SVMs only to a certain point and reached saturation around 100–125 instances used for training each LL-SVM. Further increasing of the neighborhood side neither improved, nor decreased the performance, showing interesting stability of LL-SVM. From the experiments, we can conclude that too small neighborhood will impair LL-SVM performance. At the same time, while too large neighborhood will not have a negative impact on LL-SVM, it will significantly increase the complexity of our model.

4.3 Results for Oversampling-Based Approaches

In this experiment, we analyze the performance of LL-SVM with local oversampling. Table 2 shows detailed results according to the GM metric. Figure 2 depicts performance comparison over all data sets and metrics, while Fig. 3 illustrates visualizations of the Bayesian sign-rank test of statistical significance.

Table 2. Performance of LL-SVM + SMOTE and reference global SVMs with varying oversampling methods according to GM [%].

Dataset	LL-SVM$_{ov}$	SVM$_{ov}^{SMOTE}$	SVM$_{ov}^{SPID2}$	SVM$_{ov}^{kADA}$	SVM$_{ov}^{LRS}$	SVM$_{ov}^{OPT}$	SVM$_{ov}^{RBO}$
wisconsin	93.83	95.21	95.21	95.85	94.72	95.21	**96.88**
glass-0-1-2-3_vs_4-5-6	**89.64**	87.54	89.22	88.82	87.99	89.02	89.22
glass6	**79.47**	78.27	79.11	78.27	77.45	79.11	79.30
cleveland-0_vs_4	**66.34**	62.53	62.53	63.17	64.02	64.28	65.06
nursery-very_recom	90.00	89.26	90.10	89.90	88.19	90.27	**91.12**
paw3$_{safe1}$	87.29	86.23	87.91	86.87	87.03	86.43	**88.02**
flower5$_{safe1}$	**83.91**	80.41	81.25	80.41	82.07	80.83	80.51
kddcup-buffer_vs_back	97.03	98.33	**99.17**	98.33	97.68	97.03	97.03
paw3$_{safe2}$	**88.71**	84.98	84.98	86.26	87.19	86.29	87.92
flower5$_{safe2}$	**87.26**	83.92	84.76	83.92	81.99	85.91	85.91
glass1	**81.65**	73.91	74.83	79.83	76.82	79.57	80.24
pima	**76.88**	70.81	75.18	73.21	75.19	76.01	76.01
haberman	**73.13**	63.52	65.81	70.82	71.06	72.19	71.82
vehicle2	**97.83**	95.04	95.71	**97.83**	95.89	95.20	96.00
vehicle1	**75.19**	70.63	72.83	71.93	70.81	73.80	73.28
vehicle3	**77.06**	72.90	75.18	73.10	72.88	75.66	76.44
vehicle0	91.24	90.43	91.24	**91.41**	90.83	90.39	91.24
new-thyroid1	93.97	93.28	**94.32**	93.97	93.97	94.11	93.97
segment0	96.72	98.12	98.52	**98.77**	95.92	97.40	98.12
vowel0	**95.03**	92.87	93.92	93.12	92.49	92.88	93.92
abalone-0-4_vs_16-29	**52.98**	48.54	50.18	45.99	44.98	50.36	51.47
seismic-bumps	**40.63**	34.77	33.98	34.51	32.19	35.28	37.11
abalone9-18	**77.72**	72.87	74.08	74.62	71.99	74.62	75.31
glass5	**97.68**	94.75	92.18	95.82	94.07	95.82	95.82
lymph-normal-fibrosis	82.56	82.21	**82.99**	76.47	80.75	81.30	81.66
car	82.99	84.19	86.02	82.52	81.92	84.19	**86.58**
winequality-red-4	53.17	**53.37**	52.98	52.24	53.01	51.97	53.17
winequality-white-3_vs_7	**62.85**	61.01	53.98	55.98	56.12	60.18	60.15
paw3$_{difficult}$	**61.17**	49.85	50.92	47.31	48.04	57.29	57.27
flower5$_{difficult}$	**53.28**	38.19	37.19	35.98	40.19	46.99	47.92
Avg. values	**79.57**	76.26	76.88	76.57	76.25	77.99	78.62
Avg. rank	**2.1167**	5.3333	4.0500	4.6500	5.4167	3.7833	2.6500

Performance on Normal Datasets. When we deal with standard imbalanced datasets (i.e., predominantly composed from safe instances), LL-SVM returns statistically similar performance to its counterparts. There are some datasets on which it is able to outperform reference methods in a statistically significant way, while there is but few cases where it falls behind. This shows that LL-SVM advantages do not trigger frequently when dealing with standard datasets.

Performance on Difficult Datasets. The situation is diametrically different on difficult datasets, where LL-SVM is capable of outperforming all reference methods in a statistically significant manner. This shows that the proposed local learning is capable of proper capturing data characteristics and exploiting sub-concepts within decision space. What is interesting, we achieve this only with a standard SMOTE approach applied locally. SPIDER2, kernel-ADASYN, Optimal Transport Oversampling, and RBO are advanced methods designed specifically for handling difficult instances. SPIDER2 performs simultaneous relabeling and oversampling, while Kernel-ADASYN creates a new artificial feature space for oversampling. RBO automatically estimates the local difficulty level of each instance and adapts the strength of oversampling locally. It is interesting to note that such advanced approaches can be outperformed by a much simpler algorithm that is combined with local classifier training, proving the effectiveness of data-driven oversampling.

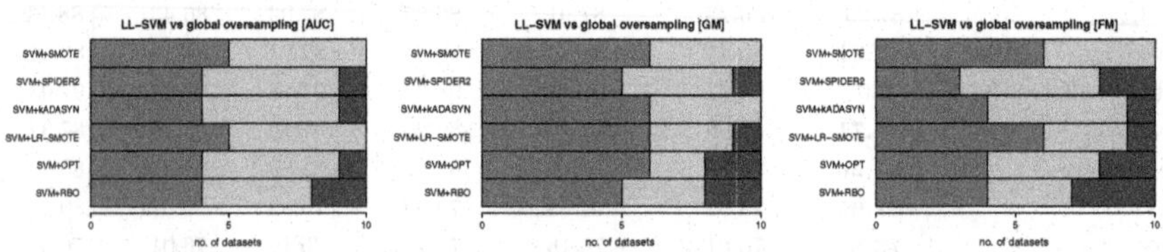

(a) Normal imbalanced datasets.

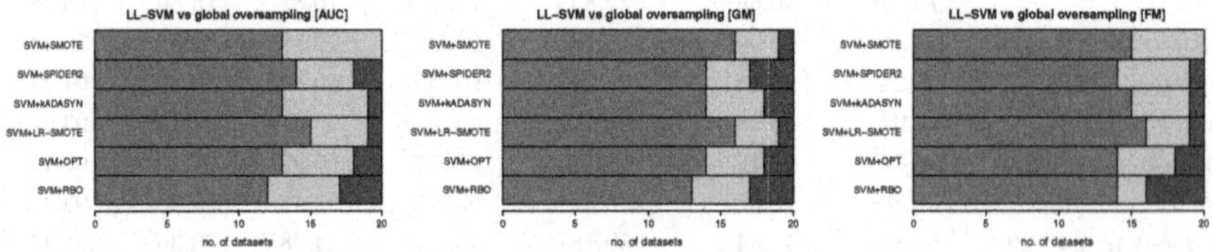

(b) Difficult imbalanced datasets.

Fig. 2. Comparison of LL-SVM + SMOTE with reference methods with respect to the number of wins (green), ties (yellow), and losses (red), according to a pairwise combined 5×2-fold CV F-test with statistical significance level $\alpha = 0.05$. (Color figure online)

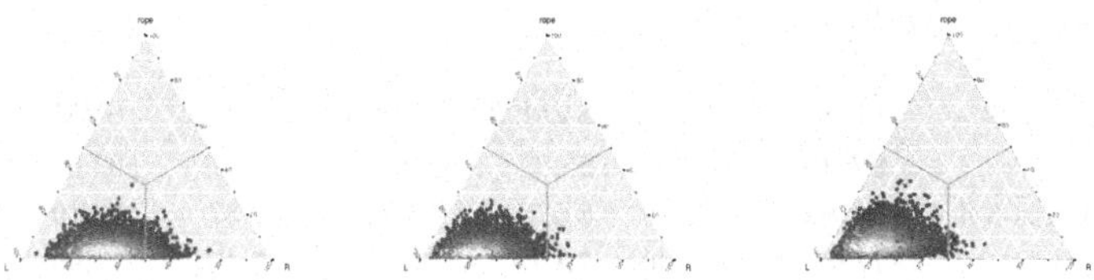

(a) AUC. (b) GM. (c) FM.

Fig. 3. Posteriors for LL-SVM vs. the best reference (SVM + RBO) for the Bayesian sign-rank test. Higher concentration of points on one of the sides of the triangle shows that a given method has a higher probability of being statistically significantly better.

4.4 Results for Cost-Sensitive Approaches

In this experiment, we analyze the performance of the cost-sensitive LL-SVM. Table 3 shows detailed results according to the GM metric. Figure 4 depicts performance comparison over all data sets and metrics, while Fig. 5 illustrates visualizations of Bayesian sign-rank test of statistical significance.

Performance on Normal Datasets. In the case of cost-sensitive methods, we observe similar behavior to the oversampling experiment. LL-SVM does not outperform in a significant manner the best performing reference methods (namely

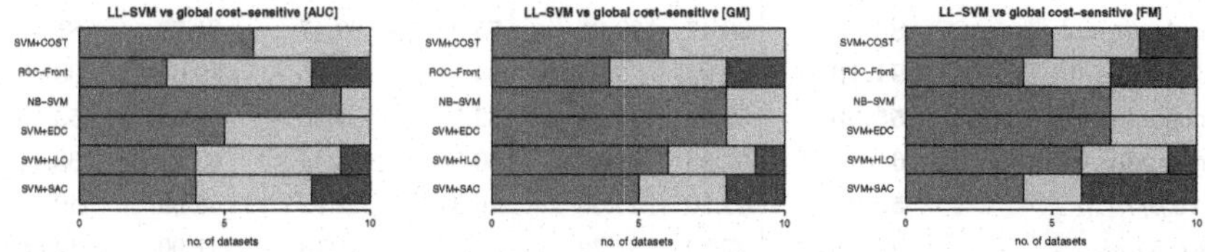

(a) Normal imbalanced datasets.

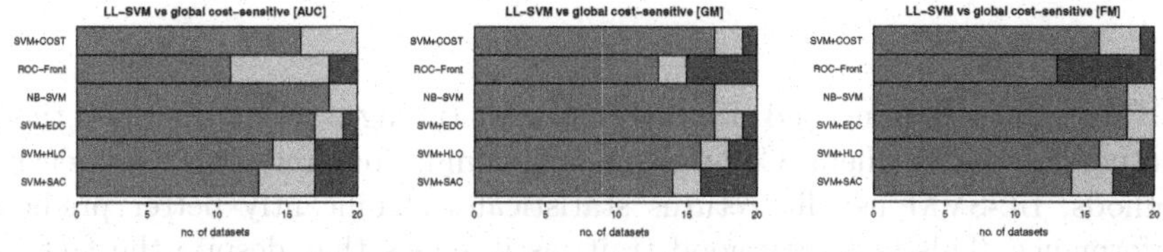

(b) Difficult imbalanced datasets.

Fig. 4. Comparison of cost-sensitive LL-SVM with reference methods with respect to the number of wins (green), ties (yellow), and losses (red), according to a pairwise combined 5×2-fold CV F-test with statistical significance level $\alpha = 0.05$. (Color figure online)

Table 3. Performance of cost-sensitive LL-SVM and reference global cost-sensitive SVMs according to GM [%].

Dataset	$LL\text{-}SVM_{cs}$	SVM_{cs}	SVM_{cs}^{ROC}	NB-SVM	SVM_{cs}^{EDC}	SVM_{cs}^{HLO}	SVM_{cs}^{SAC}
wisconsin	**97.20**	93.47	95.85	89.83	92.87	93.55	95.85
glass-0-1-2-3_vs_4-5-6	**88.82**	83.19	84.75	84.10	82.87	83.19	84.51
glass6	77.95	77.40	**79.74**	75.58	77.40	77.95	78.42
cleveland-0_vs_4	60.75	58.18	**61.30**	54.54	60.02	60.75	61.00
nursery-very_recom	**92.43**	91.00	91.64	88.27	91.00	91.00	91.64
$paw3_{safe1}$	**89.79**	87.10	87.96	86.19	88.01	87.96	87.96
$flower5_{safe1}$	83.44	81.28	**83.62**	81.28	81.01	81.28	82.44
kddcup-buffer_vs_back	96.28	94.85	**98.97**	93.03	93.98	94.85	95.02
$paw3_{safe2}$	**86.37**	81.50	84.62	76.95	81.26	82.75	83.61
$flower5_{safe2}$	**86.98**	81.31	84.43	85.62	84.43	85.11	85.52
glass1	**76.28**	71.30	72.86	71.30	70.58	74.22	74.22
pima	67.03	65.46	**67.24**	63.55	66.07	66.07	66.07
haberman	**62.86**	59.17	63.95	59.17	60.88	62.86	62.86
vehicle2	**93.72**	90.69	91.47	89.78	89.78	91.47	92.02
vehicle1	**74.18**	66.28	69.84	65.37	66.28	70.80	72.19
vehicle3	**77.91**	72.03	74.37	72.94	73.58	75.39	73.58
vehicle0	**91.12**	89.56	93.34	89.56	90.72	92.19	91.61
new-thyroid1	**93.81**	91.54	93.10	87.90	92.05	92.75	91.54
segment0	95.51	95.51	**96.29**	93.69	94.22	95.51	95.51
vowel0	92.47	90.26	**92.82**	90.26	90.26	90.98	90.26
abalone-0-4_vs_16-29	**54.33**	47.67	50.79	43.12	50.79	51.76	52.13
seismic-bumps	**40.12**	33.90	37.02	33.90	38.15	39.03	39.84
abalone9-18	**75.45**	70.26	70.26	71.17	72.04	72.49	72.99
glass5	92.07	91.27	**94.83**	89.45	90.13	90.68	91.24
lymph-normal-fibrosis	**84.85**	80.47	82.81	75.92	80.47	81.77	81.77
car	**91.19**	85.93	89.05	84.11	89.05	89.05	90.00
winequality-red-4	**55.50**	52.50	53.84	52.50	52.50	53.50	53.50
winequality-white-3_vs_7	64.28	61.88	**65.00**	60.06	63.44	63.72	63.44
$paw3_{difficult}$	**52.99**	45.50	45.50	46.41	47.92	48.69	49.14
$flower5_{difficult}$	**47.98**	37.32	37.32	38.23	41.18	41.90	42.68
avg. values	**78.06**	74.26	76.49	73.13	75.10	76.11	76.42
avg. rank	**1.6167**	5.6500	2.7667	6.2333	5.1500	3.5500	3.0333

ROC-based cost tuning and Self-adaptive cost tuning), yet at the same time is also not inferior to them. Compared to remaining four reference cost-sensitive methods, LL-SVM usually returns statistically significantly better predictive performance. This is a very good trait, as it shows that despite the fact that LL-SVM was designed mainly for difficult imbalanced data, it can effectively handle any type of imbalanced scenario. Additionally, LL-SVM does very simple and fully automatic local cost tuning, while reference methods use complex, time-consuming and parameter-depended cost tuning solutions.

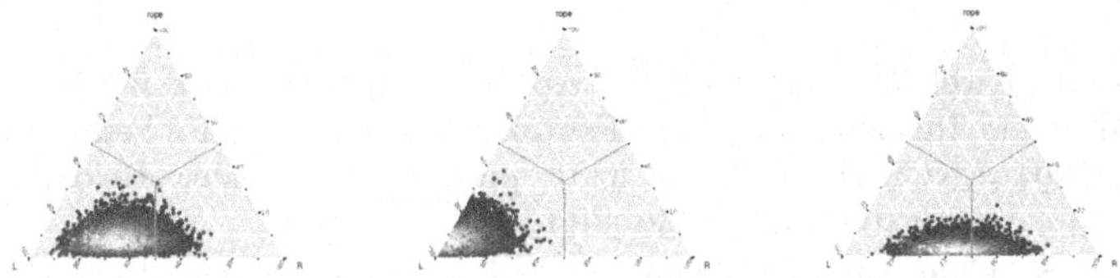

(a) AUC. (b) GM. (c) FM.

Fig. 5. Posteriors for LL-SVM vs. the best reference (SVM + ROC) for the Bayesian sign-rank test. Higher concentration of points on one of the sides of the triangle shows that a given method has a higher probability of being statistically significantly better.

Performance on Difficult Datasets. Once again, for difficult datasets LL-SVM offers excellent classification effectiveness. Regardless of the used metric, they are capable of achieving significantly better predictive power than their counterparts. This shows that using a simple local IR as a cost indicator can outperform advances ROC-based analysis, Bayesian optimization, or Hinge loss that are performed on the entire dataset (i.e., global level). In the case of difficult datasets, using a single cost parameter is too restrictive and does not capture local data variations. LL-SVM offers a possibility of tuning the cost parameter independently for each subconcept, thus achieving the desired flexibility for adapting to difficult data.

5 Conclusions and Future Works

In this paper we proposed Locally Linear Support Vector Machines, a new approach for handling imbalanced data. Instead of training a single global model, for each new test instance we trained a small local classifier using only limited nearest neighborhood of the test instance. This allowed us to relax the need for kernel usage, as we relied on simpler, linear classifiers. We showed that LL-SVM offers tighter stability bounds over imbalanced data than canonical global approaches. Furthermore, LL-SVM allowed for local data preprocessing, resulting in different oversampling or cost parameter for each local model. This allowed for excellent adaptation to data characteristics, which was provenby a theorethical analysis and followed up by a thorough experimental study. We showed that LL-SVM is an attractive choice when dealing with complex and challenging imbalanced datasets that do not require any parameter tuning or complex preprocessing algorithms. Our future works will concentrate on more detailed analysis of neighborhood during LL-SVM training, as well as LL-SVM adaptation to multi-class imbalanced data and data streams where IR may change through time.

References

1. Akbani, R., Kwek, S., Japkowicz, N.: Applying support vector machines to imbalanced datasets. In: Boulicaut, J.-F., Esposito, F., Giannotti, F., Pedreschi, D. (eds.) ECML 2004. LNCS (LNAI), vol. 3201, pp. 39–50. Springer, Heidelberg (2004). https://doi.org/10.1007/978-3-540-30115-8_7
2. Bernard, S., Chatelain, C., Adam, S., Sabourin, R.: The multiclass ROC front method for cost-sensitive classification. Pattern Recogn. **52**, 46–60 (2016)
3. Chawla, N.V., Bowyer, K.W., Hall, L.O., Kegelmeyer, W.P.: SMOTE: synthetic minority over-sampling technique. J. Artif. Intell. Res. **16**, 321–357 (2002)
4. Datta, S., Das, S.: Near-Bayesian support vector machines for imbalanced data classification with equal or unequal misclassification costs. Neural Netw. **70**, 39–52 (2015)
5. Fernández, A., García, S., Galar, M., Prati, R.C., Krawczyk, B., Herrera, F.: Learning from Imbalanced Data Sets. Springer, Cham (2018). https://doi.org/10.1007/978-3-319-98074-4
6. Gu, B., Quan, X., Gu, Y., Sheng, V.S., Zheng, G.: Chunk incremental learning for cost-sensitive hinge loss support vector machine. Pattern Recogn. **83**, 196–208 (2018)
7. Iranmehr, A., Masnadi-Shirazi, H., Vasconcelos, N.: Cost-sensitive support vector machines. Neurocomputing **343**, 50–64 (2019)
8. Kecman, V., Brooks, J.P.: Locally linear support vector machines and other local models. In: IJCNN, pp. 1–6. IEEE (2010)
9. Koziarski, M., Krawczyk, B., Wozniak, M.: Radial-based oversampling for noisy imbalanced data classification. Neurocomputing **343**, 19–33 (2019)
10. Krawczyk, B.: Learning from imbalanced data: open challenges and future directions. Prog. AI **5**(4), 221–232 (2016)
11. Liang, X.W., Jiang, A.P., Li, T., Xue, Y.Y., Wang, G.: LR-SMOTE - an improved unbalanced data set oversampling based on k-means and SVM. Knowl. Based Syst. **196** (2020)
12. Napierala, K., Stefanowski, J., Wilk, S.: Learning from imbalanced data in presence of noisy and borderline examples. In: International Conference on Rough Sets and Current Trends in Computing, pp. 158–167 (2010)
13. Tang, B., He, H.: Kerneladasyn: Kernel based adaptive synthetic data generation for imbalanced learning. In: CEC, pp. 664–671. IEEE (2015)
14. Tao, X., Li, Q., Guo, W., Ren, C., Li, C., Liu, R., Zou, J.: Self-adaptive cost weights-based support vector machine cost-sensitive ensemble for imbalanced data classification. Inf. Sci. **487**, 31–56 (2019)
15. Yan, Y., et al.: Oversampling for imbalanced data via optimal transport. In: The Thirty-Third AAAI Conference on Artificial Intelligence, AAAI 2019, Honolulu, Hawaii, USA, 27 January–1 February 1, 2019, pp. 5605–5612. AAAI Press (2019)

Low-Dimensional Representation Learning from Imbalanced Data Streams

Łukasz Korycki and Bartosz Krawczyk(✉)

Department of Computer Science, Virginia Commonwealth University, Richmond, VA, USA
{koryckil,bkrawczyk}@vcu.edu

Abstract. Learning from data streams is among the contemporary challenges in the machine learning domain, which is frequently plagued by the class imbalance problem. In non-stationary environments, ratios among classes, as well as their roles (majority and minority) may change over time. The class imbalance is usually alleviated by balancing classes with resampling. However, this suffers from limitations, such as a lack of adaptation to concept drift and the possibility of shifting the true class distributions. In this paper, we propose a novel ensemble approach, where each new base classifier is built using a low-dimensional embedding. We use class-dependent entropy linear manifold to find the most discriminative low-dimensional representation that is, at the same time, skew-insensitive. This allows us to address two challenging issues: (i) learning efficient classifiers from imbalanced and drifting streams without data resampling; and (ii) tackling simultaneously high-dimensional and imbalanced streams that pose extreme challenges to existing classifiers. Our proposed low-dimensional representation algorithm is a flexible plug-in that can work with any ensemble learning algorithm, making it a highly useful tool for difficult scenarios of learning from high-dimensional imbalanced and drifting data streams.

Keywords: Machine learning · Data stream mining · Class imbalance · Concept drift · Low-dimensional representation

1 Introduction

We define a data stream as a sequence $< S_1, S_2, ..., S_n, ... >$, in which each element S_j is a collection of instances (batch scenario) or a single instance (online scenario). Each instance is independent and randomly generated using a stationary probability distribution D_j. In this paper, we consider the supervised learning scenario that allows us to define each element as $S_j \sim p_j(x^1, \cdots, x^d, y) = p_j(\mathbf{x}, y)$, where $p_j(\mathbf{x}, y)$ is a joint distribution of j-th instance, defined by d-dimensional feature space and belonging to class y. Each instance in the stream is independent and randomly drawn from a stationary probability distribution $\Psi_j(\mathbf{x}, y)$. Data streams are also subject to a phenomenon known as concept drift

K. Karlapalem et al. (Eds.): PAKDD 2021, LNAI 12712, pp. 629–641, 2021.
https://doi.org/10.1007/978-3-030-75762-5_50

[13], where the properties of a stream evolve over time. Furthermore, two challenging problems connected with data streams are class imbalance [7] and high-dimensionality of the feature space [15]. These two problems have been analyzed disjointly, but in realistic scenarios they may appear together [7]. Therefore, there is a need for developing novel methods that can simultaneously deal with both of these issues, while being robust to concept drift.

In this paper, we propose a novel approach for learning low-dimensional representations of data streams. It is based on finding an optimal projection of data into a subspace that maximizes the Renyi's quadratic entropy per class. By weighting class representations when searching for an optimal low-dimensional manifold, we ensure that the new projection is not only highly discriminative, but also skew-insensitive. We show that the proposed technique is a universal and flexible plug-in that can be used with any ensemble algorithm created for data streams. Additionally, we highlight and address two important shortcomings of existing methods: (i) reference low-dimensional projections cannot handle increasing imbalance ratio; and (ii) existing resampling and cost-sensitive algorithms for imbalanced data streams fail with increasing dimensionality of the feature space. We show that finding a discriminative and skew-insensitive projection may actually lead to better separation between classes and outperforming resampling algorithms. The main contributions of this paper are given as follows.

- **First low-dimensional embedding for imbalanced data streams.** Novel approach to tackling extremely challenging scenario of mining imbalanced, high-dimensional and drifting data streams.
- **Theoretically and practically sound streaming low-dimensional embedding.** Efficient and theoretically grounded entropy-based low dimensional embedding that can be easily used with most streaming ensembles.
- **Extensive empirical evaluation of the proposed method.** We carry a thorough experimental study on 576 diverse generators and three difficult real-world data streams.

2 Low-Dimensional Representation

General Idea. High-dimensional data may be challenging for many machine learning algorithms, due to a phenomenon known as the curse of dimensionality. Therefore, a projection from $\mathbb{R}^d$ to $\mathbb{R}^k$, where $d >> k$, is highly attractive for both classification and visualization purposes. There are two main approaches for achieving this – linear projections or more complex embeddings.

Linear projections aim to find a matrix $\mathrm{V} \in \mathbb{R}^{d \times k}$ which for a given data set $\mathrm{X} \in \mathbb{R}^{d \times N}$ provides a projection $\mathrm{V}^T\mathrm{X}$ preserving as much of the original data characteristics as possible. This is measured by a selected information measure (which we will generally denote as IM), as well as a set of constraints φ_i. We can define a projection for obtaining a low-dimensional representation as:

$$\max_{\mathrm{V} \in \mathbb{R}^{d \times k}} \quad \mathrm{IM}(\mathrm{V}^T\mathrm{X}; \mathrm{X}, \mathrm{Y}) \tag{1}$$

subject to $\varphi_i(\mathrm{V}), i = 1, \cdots, m$, where Y is a set of additional information used during the projection, such as class labels.

Benefits of Low-dimensional Representation for Imbalanced Data Streams. The disproportion between classes is not the sole source of learning difficulties. As long as classes are well-separated skewed distributions are not going to affect the classifier. However, imbalanced data is often accompanied by a number of other learning problems [7]. We argue that by using a discriminative and low-dimensional embedding one may achieve similar or better results than when using resampling, without risking its drawbacks. In this paper, we propose a skew-insensitive embedding that is able to learn optimal subspaces that improve separation between classes and alleviate the need for any resampling or cost-sensitive learning. Additionally, as we generate new embedding for each incoming chunk of data, we are able to adapt to concept drift.

3 Low-Dimensional Projection for Imbalanced Data

Initial Assumptions. In this work, we aim at having a projection that is able to find a low-dimensional representation offering the best discrimination between minority (X_{min}) and majority classes (X_{maj}). This can be done using Cauchy-Schwarz Divergence ($\mathrm{D}_{CS}(\cdot,\cdot)$) for measuring the discriminative power and allows us to rewrite Eq. 1 as:

$$\max_{\mathrm{V}\in\mathbb{R}^{d\times k}} \quad \mathrm{D}_{CS}([[\mathrm{V}^T\mathrm{X}_{min}]].[[\mathrm{V}^T\mathrm{X}_{maj}]]) \tag{2}$$

subject to $\mathrm{V}^T\mathrm{V} = I$, where $[[\cdot]]$ stands for a density estimator.

In order to apply a low-dimensional projection to data streams, we need to choose a proper information measure. In this work, we decided to use the Renyi's quadratic entropy [5], as it has two main advantages for streaming data: (i) low computational complexity; and (ii) confirmed high usefulness for classification tasks. We can express its density f on $\mathbb{R}^k$ as:

$$\mathrm{H}_2(f) = -\log \int_{\mathbb{R}^k} f^2(x)dx. \tag{3}$$

For computing the Renyi's quadratic entropy on actual data, one needs to select a density estimator and solve an optimization problem of finding an orthonormal base $\mathbb{V}$ of k-dimensional subspace that maximizes $H_2[[\mathrm{V}^T\mathrm{X}]]$. It has been shown that the maximization of the Renyi's quadratic entropy leads to selecting a representation with a high spread, while its minimization offers condensed representation.

For the problem of imbalanced data, we can express the class-depended Cauchy-Schwarz Divergence for minority and majority classes using the mentioned Renyi's quadratic entropy (H_2) and Renyi's quadratic cross-entropy ($\mathrm{H}_2^{\times}$):

$$\begin{aligned} \mathrm{D}_{CS}(V) &= \log \int [[\mathrm{V}^T \mathrm{X}_{min}]]^2 + \log \int [[\mathrm{V}^T \mathrm{X}_{maj}]]^2 - 2log \int [[\mathrm{V}^T \mathrm{X}_{min}]][[\mathrm{V}^T \mathrm{X}_{maj}]] \\ &= -\mathrm{H}_2([[\mathrm{V}^T \mathrm{X}_{min}]]) - \mathrm{H}_2([[\mathrm{V}^T \mathrm{X}_{maj}]]) + 2\mathrm{H}_2^{\times}([[\mathrm{V}^T \mathrm{X}_{min}]], [[\mathrm{V}^T \mathrm{X}_{maj}]]). \quad (4) \end{aligned}$$

Objective Function for the Projection. Let us now formulate the objective function and its gradient for the Renyi's quadratic entropy-based projection, so it can be solved using any first-order optimization method. We want to project the entire data set X on V and compute $\mathrm{D}_{CS}(V)$ with kernel density estimators of the obtained projection:

$$\mathrm{G}^{-1}(V)[[\mathrm{V}^T \mathrm{X}_{min}]] \quad \text{and} \quad \mathrm{G}^{-1}(V)[[\mathrm{V}^T \mathrm{X}_{maj}]], \tag{5}$$

where $\mathrm{G}(V) = \mathrm{V}^T\mathrm{V}$ stands for grassmannian. We look for such a projection space V that maximizes D_{CS}. Such projections do not depend on the affine transformations of data, thus allowing us to restrict the analogous formulas for sets $\mathrm{V}^T\mathrm{X}_{min}$ and $\mathrm{V}^T\mathrm{X}_{maj}$ to such V that consist of linearly independent vectors. Additionally, we want our projection to be skew-insensitive and improve the separation between minority and majority classes. We obtain this by weighting the classes with W_{min} and W_{maj} that stand for weights assigned to the minority and majority objects, respectively. By assigning higher weights to the minority class, we obtain a better representation after low-dimensional projection that alleviates the class imbalance problem. Therefore, in order to maximize D_{CS} with respect to being skew-insensitive, we need to compute the gradient of following function:

$$\begin{aligned} \mathrm{D}_{CS}(V_{im}) &= \mathrm{D}_{CS}([[\mathrm{V}^T \mathrm{X}_{min} \mathrm{W}_{min}]], [[\mathrm{V}^T \mathrm{X}_{maj} \mathrm{W}_{maj}]]) \\ &= \log \int [[\mathrm{V}^T \mathrm{X}_{min} \mathrm{W}_{min}]]^2 + \log \int [[\mathrm{V}^T \mathrm{X}_{maj} \mathrm{W}_{maj}]]^2 \\ &\quad - 2log \int [[\mathrm{V}^T \mathrm{X}_{min} \mathrm{W}_{min}]][[\mathrm{V}^T \mathrm{X}_{maj} \mathrm{W}_{maj}]], \end{aligned} \tag{6}$$

where V_{im} stands for a skew-insensitive projection and we consider only linearly independent vectors. In order to avoid numerical instabilities, a penalty factor can be added [6]:

$$\mathrm{D}_{CS}(V_{im}) - \|\mathrm{V}^T\mathrm{V} - I\|^2, \tag{7}$$

that is used to penalize non-orthonormal V's.

This allows us to formulate the proposed *Low-Dimensional Projection for Imbalanced Data* (LDP_{IM}) objective function used to select the best possible low-dimensional projection that alleviates the effect of class imbalance:

$$\mathrm{LDP}_{IM} = \mathrm{D}_{CS}(V_{im}) - \|\mathrm{V}^T\mathrm{V} - I\|^2. \tag{8}$$

Additionally, we need to be able to compute the gradient of the objective function $\nabla \mathrm{LDP}_{IM}$. For the second term, we can compute this as:

$$\nabla\|\mathrm{V}^T\mathrm{V} - I\|^2 = 4\mathrm{V}\mathrm{V}^T\mathrm{V} - 4\mathrm{V}. \tag{9}$$

Let us now present how can we compute the gradient of the first term. For this, we will need to compute the product of kernel densities of two sets [6]. For the notation simplicity let us assume that set A stands for the minority class X_{min} and set B for the majority class X_{maj}. We can estimate the kernel density of a given set with the Gaussian kernel:

$$[[A]] = \frac{1}{|A|}\sum_{a\in A}\mathcal{N}(a, \Sigma_A), \tag{10}$$

where $\Sigma_A = (h_A^\gamma)^2\mathrm{cov}_A$, $h_A^\gamma = \gamma(\frac{4}{k+2})^{1/(k+4)}|A|^{-1/(k+4)}$, and γ is a scaling hyperparameter.

We need to obtain $\int[[A]][[B]]$ which can be calculated from the following:

$$\int \mathcal{N}(a, \Sigma_A)\mathcal{N}(b, \Sigma_B) = \mathcal{N}(a - b, \Sigma_A + \Sigma_B)(0), \tag{11}$$

leading to:

$$\int[[A]][[B]] = \tfrac{1}{|A||B|}\textstyle\sum_{w\in A-B}\mathcal{N}(w, \Sigma_A + \Sigma_B)(0) = \tfrac{1}{(2\pi)^{k/2}\mathrm{det}^{1/2}(\Sigma_{AB}|A||B|)}\textstyle\sum_{w\in A-B}\exp(-\tfrac{1}{2}\|w\|^2_{\Sigma_{AB}}), \tag{12}$$

where $A - B = \{a - b : a \in A, b \in B\}$ and $\Sigma_{AB} = (h_A^\gamma)^2\mathrm{cov}_A + (h_B^\gamma)^2\mathrm{cov}_B$.

As we deal with a sequence of linearly independent vectors $\mathrm{V} = [\mathrm{V}_1, \cdots, \mathrm{V}_k] \in \mathbb{R}^{d\times k}$, we may use the following:

$$\Sigma_{AB}(\mathrm{V}) = \mathrm{V}^T\Sigma_{AB}\mathrm{V} \quad \text{and} \quad S_{AB}(\mathrm{V}) = \Sigma_{AB}(\mathrm{V})^{-1}, \tag{13}$$

where $\Sigma_{AB}(\mathrm{V})$ and $S_{AB}(\mathrm{V})$ are square symmetric matrices storing the properties of a given projection onto space V.

Let us now calculate:

$$\phi_{AB}(\mathrm{V}) = \frac{1}{(2\pi)^{k/2}\mathrm{det}^{1/2}(\Sigma_{AB}|A||B|)}, \tag{14}$$

and compute gradient of this function as:

$$\nabla\phi_{AB}(\mathrm{V}) = -\phi_{AB}(\mathrm{V}) \cdot \Sigma_{AB} \cdot \mathrm{V} \cdot S_{AB}(\mathrm{V}). \tag{15}$$

To compute the final formula for the first term of $\nabla \mathrm{LDP}_{IM}$, we need to calculate the gradient of function $\mathrm{V} \to \mathrm{det}(\Sigma_{AB}(\mathrm{V}))$ that is given by the following formula:

$$\nabla\mathrm{det}(\Sigma_{AB}(\mathrm{V})) = 2\mathrm{det}(\mathrm{V}^T\Sigma_{AB}\mathrm{V}) \cdot \Sigma_{AB}\mathrm{V}(\mathrm{V}^T\Sigma_{AB}\mathrm{V})^{-1}. \tag{16}$$

We define the function for information potential:

$$\psi^{w}_{AB}(\mathrm{V}) = \exp(-\frac{1}{2}\|\mathrm{V}^{T}w\|^{2}_{\Sigma_{AB}(\mathrm{V})}), \tag{17}$$

where for an arbitrarily set value of w parameter we are able to compute its gradient as following:

$$\nabla\psi^{w}_{AB}(\mathrm{V}) = -\psi^{w}_{AB}(\mathrm{V}) \cdot (ww^{T}\mathrm{V}S_{AB}(\mathrm{V}) - \Sigma_{AB}(\mathrm{V})S_{AB}(\mathrm{V})\mathrm{V}^{T}ww^{T}\mathrm{V}S_{AB}(\mathrm{V})). \tag{18}$$

Finally, we need to define cross information potential function (between A and B) and its gradient:

$$\mathrm{ip}^{x}_{AB} = \phi_{AB}(\mathrm{V}) \sum_{w \in A-B} \psi^{w}_{AB}(\mathrm{V}), \tag{19}$$

$$\nabla\mathrm{ip}^{x}_{AB} = \phi_{AB}(\mathrm{V}) \sum_{w \in A-B} \psi^{w}_{AB}(\mathrm{V}) + \left(\sum_{w \in A-B} \psi^{w}_{AB}(\mathrm{V}) \right) \cdot \nabla\phi_{AB}(\mathrm{V}). \tag{20}$$

This allows us to get back to $\mathrm{D}_{CS}(V_{im})$ and rewrite it as:

$$\mathrm{D}_{CS}(V_{im}) = \log(\mathrm{ip}^{x}_{X_{min}X_{min}}(\mathrm{V})) + \log(\mathrm{ip}^{x}_{X_{maj}X_{maj}}(\mathrm{V})) - 2\log(\mathrm{ip}^{x}_{X_{min}X_{maj}}(\mathrm{V})), \tag{21}$$

and calculate its gradient as:

$$\nabla\mathrm{D}_{CS}(V_{im}) = \frac{1}{\mathrm{ip}^{x}_{X_{min}X_{min}}(\mathrm{V})}\nabla\mathrm{ip}^{x}_{X_{min}X_{min}}(\mathrm{V}) + \frac{1}{\mathrm{ip}^{x}_{X_{maj}X_{maj}}(\mathrm{V})}\nabla\mathrm{ip}^{x}_{X_{maj}X_{maj}}(\mathrm{V}) - 2\frac{1}{\mathrm{ip}^{x}_{X_{min}X_{maj}}(\mathrm{V})}\nabla\mathrm{ip}^{x}_{X_{min}X_{maj}}(\mathrm{V}). \tag{22}$$

After these steps, we may properly formulate the LDP_{IM} objective function and its gradient as:

$$\mathrm{LDP}_{IM}(V) = \mathrm{D}_{CS}(V_{im}) - \|\mathrm{V}^{T}\mathrm{V} - I\|^{2}, \tag{23}$$

$$\nabla\mathrm{LDP}_{IM}(V) = \nabla\mathrm{D}_{CS}(V_{im}) - (4\mathrm{V}\mathrm{V}^{T}\mathrm{V} - 4\mathrm{V}). \tag{24}$$

These equations can be used as an input to any first-order optimization method to find a new k-dimensional projection that can be used as a discriminative, skew-insensitive and low-dimensional projection of the original feature space.

Embedding LDP_{IM} into Ensembles for Data Streams. The proposed LDP_{IM} can be seamlessly embedded into any chunk-based ensemble learning algorithm dedicated to data streams [10]. Their general idea is based on training a new base classifier on the most recent chunk of data and using it to replace the most incompetent classifier in the pool. We propose to use LDP_{IM} as a flexible plug-in to any such ensemble, performing the low-dimensional embedding when a new chunk of data arrives and then training a new classifier in the reduced feature space. This will not only make any ensemble robust to class imbalance, but also improve the predictive power and speed of training of base classifiers.

Adaptation to Concept Drift. As LDP_{IM} will be run independently on each data chunk, this will have two interesting effects on the underlying ensemble. Firstly, it will allow for LDP_{IM} to adapt to concept drift, as each embedding will be done on the most recent data. Secondly, each base classifier will be trained on different embedding, thus positively impacting the diversity of the ensemble.

4 Experimental Study

This experimental study was designed to answer the following research questions.
RQ1: Does the proposed low-dimensional embedding offer robustness to class imbalance by providing a better representation of the minority class, and can it outperform the state-of-the-art low-dimensional projection algorithms?
RQ2: Is LDP_{IM} flexible enough to work with a variety of ensemble learning algorithms designed for drifting data streams?
RQ3: Does using the skew-insensitive low-dimensional representation for high-dimensional data streams offer better discriminative power than resampling and cost-sensitive solutions?
RQ4: Is the proposed LDP_{IM} capable of outperforming reference methods for real-world data streams with various combinations of feature space dimensionality and imbalance ratio?

4.1 Data Stream Benchmarks

For the purpose of evaluating our proposed algorithm, we generated 576 diverse and large-scale data stream benchmarks using MOA. We used four generators to generate binary imbalanced streams, each with a number of features in $[50, 100, 250, 500, 1000, 5000]$, an imbalance ratio in $[10, 30, 50, 80, 100, 150]$, and with four types of drifts [no drift, sudden, gradual, incremental]. Exhaustive combinations of these factors lead to the creation of 576 data streams with 1M–5M instances each. Additionally, we use three real-world data streams: CIFAR-100, ImageNet and SUN-397 transformed to multi-class imbalanced problems [14]. Their properties are given in Table 1.

Table 1. Properties of data generators that were used to create 576 imbalanced data stream benchmarks and three real-world data sets.

Dataset	Instances	Features	Classes	IR	Drift
Aggrawal	1 000 000	50–5000	2	10–150	n/s/g/i
Hyperplane	1 000 000	50–5000	2	10–150	n/s/g/i
RBF	5 000 000	50–5000	2	10–150	n/s/g/i
RandomTree	2 000 000	50–5000	2	10–150	n/s/g/i
CIFAR-100	60 000	1024	100	50	Unknown
ImageNet	1 200 000	4096	200	50	Unknown
SUN-397	108 753	1024	397	137	Unknown

4.2 Set-Up

Reference Algorithms for Low-dimensional Representation. We have selected three state-of-the-art reference methods for obtaining low-dimensional projections: (i) Per-Class Principal Component Analysis (pPCA); (ii) Structure-Preserving Non-Negative Matrix Factorization (NMF) [11]; and (iii) Discriminative Learning using Generalized Eigenvectors (GEM) [9].

Reference Algorithms for Handling Class Imbalance. We have selected three state-of-the art reference methods for handling imbalanced data streams: (i) Incremental Oversampling for Data Streams (IOSDS) [1]; (ii) undersampling via Selection-Based Resampling (SRE) [12]; and (iii) Online Multiple Cost-Sensitive Learning (OMCSL) [16].

Ensemble Learning Algorithms. The proposed LDP_{IM} is a flexible plug-in that can be used with any ensemble learning algorithm. Therefore, to evaluate its interplay with various ensembles, we have selected recent and popular architectures designed for streaming data: (i) Kappa Updated Ensemble (KUE) [4]; (ii) Adaptive Random Forest (ARF) [8]; (iii) Geometrically Optimum and Online-Weighted Ensemble (GOOWE) [2]; and (iv) Accuracy Updated Ensemble (AUE) [3]. Each of them worked in a block-based mode, with 10 base classifiers maintained. We used Hoeffding Trees as base learners.

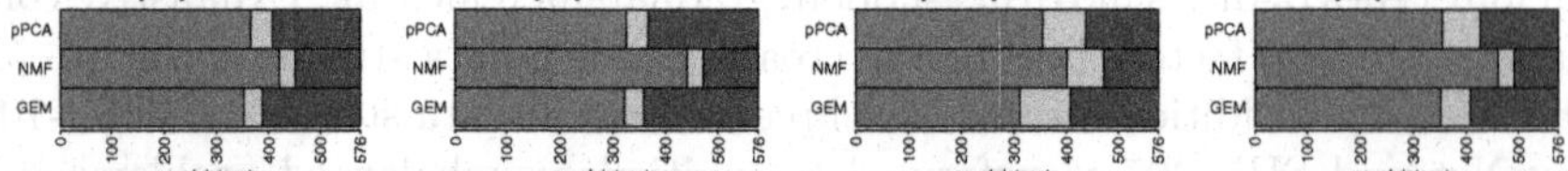

Fig. 1. Comparison of LDP_{IM} and reference algorithms for low-dimensional representation over four different ensemble architectures. Results presented with respect to the number of wins (green), ties (yellow), and losses (red) over 576 data streams. A tie was considered when the McNemar's test rejected the significance of difference between tested algorithms. (Color figure online)

Parameters. LDP_{IM} uses the inverse imbalance ratio for setting weights in Eq. 6. All algorithms for low-dimensional representation use $k = 0.05d$, so they reduce the input feature space by 95%. All of them were tuned with parameters and procedures suggested by their authors.

Evaluation Metrics. As we deal with imbalanced and drifting data streams, we evaluated the examined algorithms using prequential AUC [7].

Windows. We used a window size $\omega = 1000$ for calculating the prequential metrics and training new classifiers.

Statistical Analysis. We used the McNemar's test for the pairwise comparison and the Bonferroni-Dunn test for the multiple comparison.

4.3 Experiment 1: Low-Dimensional Representations

General Comparison. Figure 1 offers a detailed comparison of the proposed method with three reference low-dimensional embedding methods using four underlying ensemble architectures over 576 diverse data stream benchmarks. We

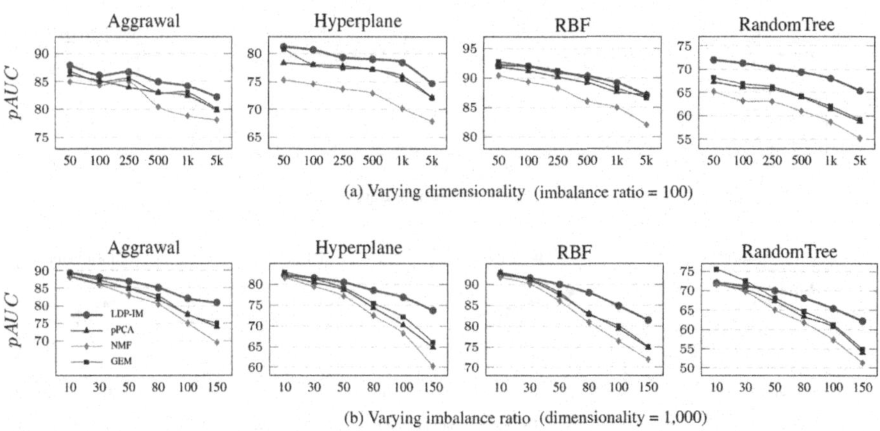

Fig. 2. Analysis of the relationship between prequential AUC and increasing imbalance ratio/dimensionality for LDP_{IM} and reference algorithms for low-dimensional representation. KUE used as the ensemble method.

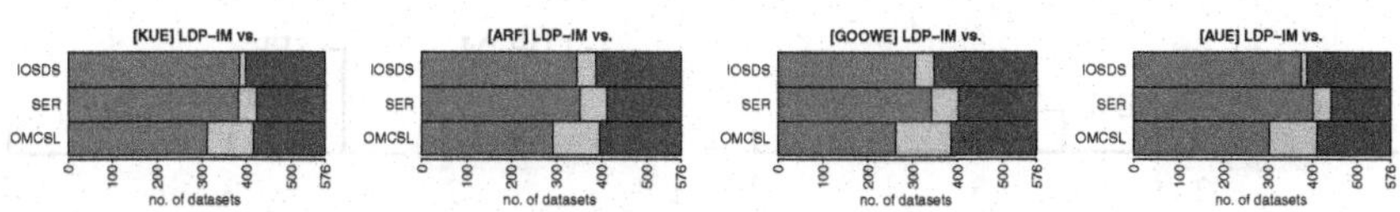

Fig. 3. Comparison of LDP_{IM} and reference algorithms for imbalanced data streams over four different ensemble architectures. Results presented with respect to the number of wins (green), ties (yellow), and losses (red) over 576 data streams. A tie was considered when the McNemar's test rejected the significance of difference between tested algorithms. (Color figure online)

can see that LDP_{IM} is capable of outperforming reference embeddings regardless of the underlying ensemble architecture. This shows that not only the proposed method is able to create more discriminative subspaces (**RQ1**), but it is also highly flexible with regard to the utilized classification scheme and can be used as a general-purpose plug-in (**RQ2**). This is further confirmed by the Bonferonii-Dunn test (see Fig. 5), proving that the differences between the proposed LDP_{IM} and reference algorithms are statistically significant.

Analysis of Robustness to High Dimensionality and Class Imbalance. Figure 2 shows a detailed analysis of the robustness of each of the examined projection methods to increasing dimensionality and imbalance ratio. What was expected, all four algorithms can handle data streams with even up to thousands of features. LDP_{IM} shows some improvements over reference methods when facing thousands of features, which can be contributed to using weighted entropies from both classes. Things are different when we analyze an increasing imbalance ratio. Here LDP_{IM} clearly outperforms all reference methods, showing that they are not capable of creating discriminative projections when dealing with highly skewed distributions.

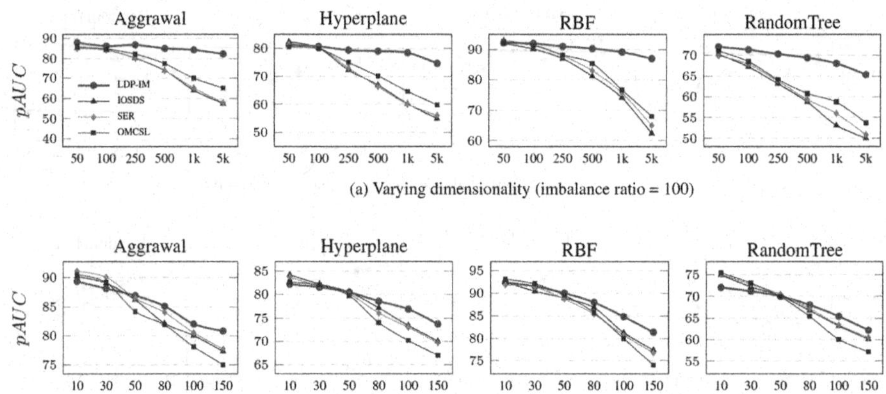

(a) Varying dimensionality (imbalance ratio = 100)

(b) Varying imbalance ratio (dimensionality = 1,000)

Fig. 4. Analysis of the relationship between prequential AUC and increasing imbalance ratio/dimensionality for LDP_{IM} and reference algorithms for handling imbalanced data streams. KUE used as the ensemble method.

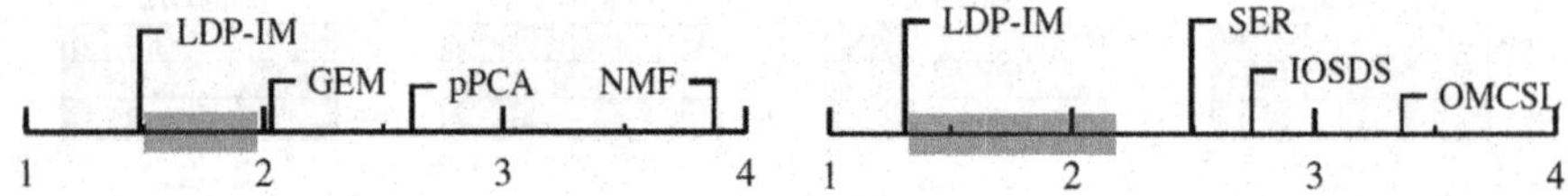

Fig. 5. The Bonferroni-Dunn tests for the comparison among methods for low-dimensional representations and handling class imbalance.

4.4 Experiment 2: Skew-Insensitive Algorithms

General Comparison. Figure 3 offers a detailed comparison of the proposed method with three reference resampling and cost-sensitive methods utilizing four underlying ensemble architectures over 576 diverse data stream benchmarks. The obtained results show that by learning a skew-insensitive representation via LDP_{IM} we may be highly competitive to traditional solutions created for handling class imbalance. Furthermore, we achieve this without artificially inflating the size of the data set or the need for meta-tuning of the cost parameter. This is further confirmed by the Bonferonii-Dunn test (see Fig. 5).

Analysis of Robustness to High Dimensionality and Class Imbalance. Figure 4 shows a detailed analysis of the robustness of each of the examined imbalance alleviation methods to increasing dimensionality and imbalance ratio. These results show a weakness of the existing solutions, as neither resampling nor cost-sensitive learning can work under high dimensionality of the feature space. LDP_{IM} offers comparable performance when the dimensionality is low, but offers excellent robustness to even thousands of features. As for the imbalance ratio, all examined methods can adapt to increasing disproportion between classes, but only LDP_{IM} can handle imbalanced and high-dimensional streams (**RQ3**). It is important to notice that LDP_{IM} also offers excellent performance when dealing with lower imbalance ratios or smaller feature dimensionality, making it a very flexible solution to learning from difficult data streams.

4.5 Experiment 3: Evaluation on Real-World Data Streams

Finally, we want to evaluate the performance of the proposed LDP_{IM} on real-world data. Table 2 presents the obtained results according to the prequential AUC metric and for four examined ensemble classifiers used as base learners. We

Table 2. Results according to prequential multi-class AUC on real-world imbalanced data sets for all reference methods and four ensemble classifiers used as base learners.

	LDP_{IM}	pPCA	NMF	GEM	IOSDS	SER	OMCSL
KUE							
CIFAR-100	**82.72** ± 2.38	56.26 ± 9.73	60.23 ± 9.18	64.34 ± 8.56	65.83 ± 7.02	66.11 ± 6.45	65.23 ± 7.98
ImageNet	**41.07** ± 5.09	16.54 ± 8.83	17.99 ± 9.02	20.04 ± 7.81	28.94 ± 6.17	31.43 ± 6.93	32.80 ± 5.98
SUN-397	**38.28** ± 6.79	9.28 ± 5.99	11.26 ± 7.03	12.36 ± 8.04	17.29 ± 8.92	18.92 ± 7.48	20.01 ± 9.01
ARF							
CIFAR-100	**80.19** ± 2.72	51.54 ± 9.29	59.57 ± 9.33	62.41 ± 8.07	62.95 ± 7.38	63.83 ± 6.18	64.01 ± 7.29
ImageNet	**40.03** ± 6.11	14.99 ± 9.02	17.58 ± 9.31	21.93 ± 7.48	27.91 ± 7.03	30.52 ± 7.28	31.06 ± 6.77
SUN-397	**36.28** ± 7.28	8.62 ± 6.58	10.24 ± 7.33	11.64 ± 7.58	16.03 ± 9.04	17.88 ± 8.04	18.84 ± 9.62
GOOWE							
CIFAR-100	**78.03** ± 2.99	50.21 ± 9.80	53.07 ± 9.94	56.61 ± 9.17	58.29 ± 7.78	60.06 ± 6.92	60.92 ± 8.64
ImageNet	**36.86** ± 5.89	11.72 ± 8.14	13.58 ± 8.22	16.22 ± 8.01	24.17 ± 8.14	26.77 ± 7.94	27.94 ± 7.80
SUN-397	**32.09** ± 4.99	5.89 ± 3.44	7.09 ± 2.99	7.82 ± 3.88	11.57 ± 5.01	13.39 ± 4.78	15.28 ± 5.02
AUE							
CIFAR-100	**79.44** ± 2.76	51.95 ± 9.41	55.11 ± 9.38	58.92 ± 8.15	60.02 ± 7.76	62.17 ± 6.35	63.88 ± 7.48
ImageNet	**37.93** ± 4.81	12.77 ± 6.62	15.02 ± 7.13	18.54 ± 6.96	25.82 ± 7.11	28.44 ± 8.09	28.98 ± 5.49
SUN-397	**34.18** ± 2.79	7.28 ± 1.99	8.99 ± 2.52	8.90 ± 3.08	13.44 ± 3.72	16.48 ± 4.08	17.24 ± 4.27

can see that LDP_{IM} offers superior performance and exceptional stability for all data sets and regardless of the used ensemble classifier. This verifies the high attractiveness of the proposed LDP_{IM} for learning from difficult data streams in various real-world scenarios (**RQ4**).

5 Conclusions and Future Works

In this paper, we have tackled a highly challenging and, to the best of our knowledge, previously unaddressed problem of learning from data streams under class imbalance and high feature space dimensionality. We have presented LDP_{IM} – a novel low-dimensional representation learning algorithm designed for imbalanced data streams. It optimizes the search for a new subspace that can effectively represent data, while maximizing its discriminative power. We achieve this by using the Renyi's quadratic entropy to evaluate the usefulness of potential subspaces. We additionally compute the individual impact of each class, taking into account the disproportions in their distributions and using this information to weight entropy computations. This allowed us to find theoretically-justified skew-insensitive projections that alleviate the need for any resampling or cost-sensitive learning. By looking for discriminative low-dimensional representations, we were able to enhance the predictive power of examined classifiers without the need of using additional pre-processing methods.

Our claims were confirmed by an extensive empirical evaluation on 576 generated streaming benchmarks and three challenging real-world data sets. We have showed that LDP_{IM} can be seamlessly integrated into any streaming ensemble.

References

1. Anupama, N., Jena, S.: A novel approach using incremental oversampling for data stream mining. Evol. Syst. **10**(3), 351–362 (2019)
2. Bonab, H.R., Can, F.: GOOWE: geometrically optimum and online-weighted ensemble classifier for evolving data streams. ACM TKDD **12**(2), 25:1–25:33 (2018)
3. Brzezinski, D., Stefanowski, J.: Reacting to different types of concept drift: the accuracy updated ensemble algorithm. IEEE Trans. Neural Netw. Learning Syst. **25**(1), 81–94 (2014)
4. Cano, A., Krawczyk, B.: Kappa updated ensemble for drifting data stream mining. Mach. Learn. **109**(1), 175–218 (2020)
5. Czarnecki, W.M., Józefowicz, R., Tabor, J.: Maximum entropy linear manifold for learning discriminative low-dimensional representation. In: Machine Learning and Knowledge Discovery in Databases - European Conference, ECML PKDD 2015, Porto, Portugal, September 7–11, 2015, Proceedings, Part I. pp. 52–67 (2015)
6. Czarnecki, W.M., Tabor, J.: Multithreshold entropy linear classifier: theory and applications. Expert Syst. Appl. **42**(13), 5591–5606 (2015)
7. Fernández, A., García, S., Galar, M., Prati, R.C., Krawczyk, B., Herrera, F.: Learning from Imbalanced Data Sets. Springer (2018). 10.1007/978-3-319-98074-4
8. Gomes, H.M., Bifet, A., Read, J., Barddal, J.P., Enembreck, F., Pfharinger, B., Holmes, G., Abdessalem, T.: Adaptive random forests for evolving data stream classification. Mach. Learn. **106**(9–10), 1469–1495 (2017)

9. Karampatziakis, N., Mineiro, P.: Discriminative features via generalized eigenvectors. In: Proceedings of the 31th International Conference on Machine Learning, ICML 2014, Beijing, China, 21–26 June 2014, pp. 494–502 (2014)
10. Krawczyk, B., Minku, L.L., Gama, J., Stefanowski, J., Woźniak, M.: Ensemble learning for data stream analysis: A survey. Inform. Fus. **37**, 132–156 (2017)
11. Li, Z., Liu, J., Lu, H.: Structure preserving non-negative matrix factorization for dimensionality reduction. Comput. Vis. Image Underst. **117**(9), 1175–1189 (2013)
12. Ren, S., Zhu, W., Liao, B., Li, Z., Wang, P., Li, K., Chen, M., Li, Z.: Selection-based resampling ensemble algorithm for nonstationary imbalanced stream data learning. Knowl.-Based Syst. **163**, 705–722 (2019)
13. Wang, S., Minku, L.L., Yao, X.: A systematic study of online class imbalance learning with concept drift. IEEE Trans. Neural Netw. Learn. Syst. **29**(10), 4802–4821 (2018)
14. Wang, Y., Ramanan, D., Hebert, M.: Learning to model the tail. In: Guyon, I., von Luxburg, U., Bengio, S., Wallach, H.M., Fergus, R., Vishwanathan, S.V.N., Garnett, R. (eds.) Advances in Neural Information Processing Systems 30: Annual Conference on Neural Information Processing Systems 2017, December 4–9, 2017, Long Beach, CA, USA. pp. 7029–7039 (2017)
15. Wang, Z., Kong, Z., Chandra, S., Tao, H., Khan, L.: Robust high dimensional stream classification with novel class detection. In: 35th IEEE International Conference on Data Engineering, ICDE 2019, Macao, China, April 8–11, 2019. pp. 1418–1429 (2019)
16. Yan, Y., Yang, T., Yang, Y., Chen, J.: A framework of online learning with imbalanced streaming data. In: Proceedings of the Thirty-First AAAI Conference on Artificial Intelligence, February 4–9, 2017, San Francisco, California, USA. pp. 2817–2823 (2017)

PhotoStylist: Altering the Style of Photos Based on the Connotations of Texts

Siamul Karim Khan(✉), Daniel (Yue) Zhang, Ziyi Kou, Yang Zhang, and Dong Wang

Department of Computer Science and Engineering, University of Notre Dame, Notre Dame, IN 46556, USA
{skhan22,yzhang40,zkou,yzhang42,dwang5}@nd.edu

Abstract. The need to modify a photo to reflect the connotations of a text can arise due to multifarious reasons (e.g., a musician might modify a photo in the album cover to better reflect the connotations in her song lyrics). An interesting observation is that different styles of photos convey different feelings. In this paper, we propose the PhotoStylist scheme to effectively modify the style of an input photo to represent the connotations in an input text. Existing methods that aim to transfer emotions into photos rely on an emotion class being provided as input and modify the overall color of photos based on the input emotion class, generating unrealistic colors for many objects in the image. To address these limitations, we design PhotoStylist, a novel deep-learning-based approach, to alter the individual style of each object in the photo in a way that the connotations of the input text are naturally and effectively embedded into the modified photos. Evaluation results on the Amazon Mechanical Turk (MTurk) show that our scheme can achieve output photos significantly closer to the connotations of the input text than the output photos from the state-of-the-art baselines.

1 Introduction

In many real world applications, an individual may need to modify a photo to match the connotations (i.e., the inherent sentiments and themes) of a given input text. For example, a writer may share an edited photo in her book to reflect the connotations of her writing. With the rise of massive data dissemination opportunities [17,18], a frequent need for such photo modifications emerge in online social media posts that pair images with user input texts [23]. In this paper, we develop an AI system that can automatically alter the style of a photo to effectively match the connotation in the text (e.g., photo caption) given by the user (as shown in Fig. 1). Previous efforts on altering photos based on emotions, known as Emotional Color Transfer, primarily focus on changing the color characteristics of the photo based on a *predefined* set of emotion classes [8,10,11,16]. In particular, these emotional color transfer techniques change the *entire* photo based on an emotion class provided by the user, which is a one-size-fits-all solution. The problem we focus on in this paper is more challenging because we are not provided a single emotion class *a priori* but rather we aim

K. Karlapalem et al. (Eds.): PAKDD 2021, LNAI 12712, pp. 642–654, 2021.
https://doi.org/10.1007/978-3-030-75762-5_51

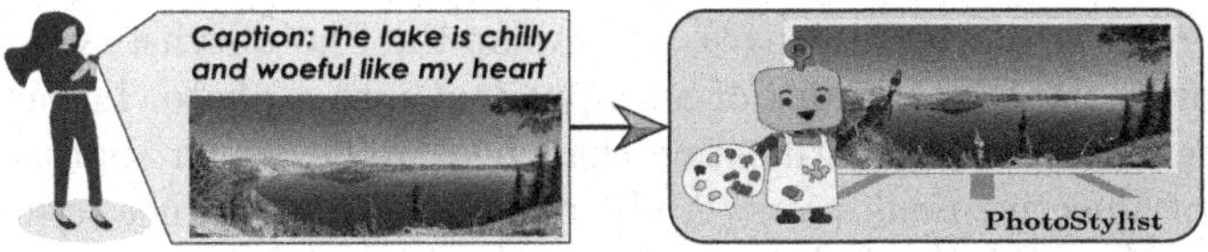

Fig. 1. An example of PhotoStylist. It alters the style of a photo to better match the connotations in the input text.

to identify the connotations in an input text and modify the look and feel of *the individual objects* in the photo accordingly which results in more visually-pleasing modifications that better incorporate the connotations of the text into the image. Furthermore, emotional color transfer solutions often utilize simple statistical values (e.g., mean and variance) in a color space to modify the colors of a photo and hence, cannot introduce new color transitions and patterns into the photo. In this paper, we develop PhotoStylist, a deep-learning-based approach to address the above limitations. However, two challenges exist in developing our solution. We elaborate them below.

Identification of Text Connotation Conforming to the Photo: When providing a caption for a photo, the user may not mention all objects in the photo explicitly in her text and, as such, we cannot identify the relevant style for the objects in a photo using only the text [22]. For example, we observe the input text the user provides in Fig. 1 has information only about the condition of the lake in the photo. However, the knowledge of how to change the other objects in the photo is also required. For example, we observe from the output photo that the mountains have a more 'snowy' feel and the sky has a more 'stormy' feel to match the photo to the connotations of the text. We can achieve such results if our PhotoStylist system could correctly identify and alter the look and feel for each object in the photo to match the connotations of the text. However, identifying the connotations of a text corresponding to the objects in a photo is more challenging than the traditional sentiment analysis which mainly aims to classify texts into a set of emotion classes and, so, is insufficient to identify styles for different objects in the photo.

Connotation Transfer from Text to Photo: The second challenge lies in transferring the connotation from the text domain to the image domain. While there has been work on text-based image generation solutions [24] that can convert information from texts to images, they do not focus on the connotations of the input text. We found that generating images to accurately capture the connotations in the text is not a trivial task. In particular, the generated images would not only need to contain the correct objects as the input image but also each object would need to have the correct style. However, the problem we address provides an input image and thus, instead of generating a new image, an image-domain representation needs to be identified that contains adequate information to guide the modification of the individual style of the objects based on the connotations in the text.

To address the above challenges, we develop the PhotoStylist system to effectively transfer the connotations of an input text into a photo. In particular, to address the connotation identification challenge, we devise an approach to identify a set of adjective-noun pairs (ANP) representing the target style, conforming to the connotations of the text, for each object in the photo. To address the connotation transfer challenge, we develop a novel method to identify the style representation for each identified ANP for the modification of the input image style by employing the large-scale Multilingual Visual Sentiment Ontology (MVSO). To the best of our knowledge, PhotoStylist is among the first *connotation-aware* text-to-image style transfer schemes that aim to bridge the gap between the visual style present in an image and the connotations present in a text. Crowdsourcing-based evaluation is carried out on the Amazon Mechanical Turk (MTurk), one of the largest crowdsourcing platforms. The results show that PhotoStylist achieves significantly better output images compared to the state-of-the-art baselines in terms of user perception of these images relative to the connotations of the input text.

2 Related Work

Image-to-Image Style Transfer: A plethora of research work has been done on image-to-image style transfer. For example, Luan et al. [9] introduced photorealism into the neural style transfer algorithm by Gatys et al. [2]. Li et al. designed an autoencoder-based approach where he modified the intermediate feature representations from the encoder to transfer style information and used unpooling layers for upsampling in the decoder [6]. Yoo et al. introduced wavelet pooling and unpooling layers to further reduce distortions in the autoencoder-based approach [21]. Different from the above image-to-image style transfer solutions, our work focuses on a connotation-aware text-to-image style transfer. In particular, we develop a method to generate gram matrices [2] as target style representations for the different objects in the input image based on the connotations of the input text.

Emotional Color Transfer: Emotional Color Transfer schemes utilize color transfer techniques to change the look and feel of images according to a predefined set of emotion classes. For example, He et al. developed a emotional color transfer framework based on color combinations [3]. Liu et al. proposed a texture-aware emotional color transfer method which can adjust an image with an emotion word or a reference image [8]. Liu et al. extracted semantic information from images to prevent unnatural changes in emotional color transfer [7]. However, these solutions suffer from two limitations: i) they cannot identify connotations from the text but rather rely on the user providing an input emotion class; ii) they cannot introduce new color transitions and patterns into the transfer process, which are critical for obtaining visually-pleasing results. Our scheme addresses these limitations by identifying the connotations in the input text and altering the style of the individual objects in the images based on these connotations.

Text-based Image Generation: There has been several efforts on generating images based on text descriptions. For example, Zhang et al. proposed a Stacked Generative Adversarial Network architecture to generate photorealistic images [24]. Xu et al. proposed an Attentional Generative Adversarial Network (AttnGAN) for fine-grained text-to-image generation [19]. However, none of the above solutions focus on the connotations in the text description. Our scheme differs in that it focuses on identifying connotations and, instead of generating an entirely new image, modifies the style of an input image by leveraging the connotations of the text.

3 Problem Statement

In this section, we define the problem of transferring the connotations of a text into a photo. Let us first define the style and content of an image I as follows.

Definition 1. ***Style of an image*** S_I: *it is defined by a feature space designed to capture the color, and its transitions and patterns, in the image* I.

Definition 2. ***Content of an image*** C_I: *it is defined by a feature space designed to capture the objects and their relative positions in the image* I.

Figure 2 illustrates the difference between content and style. The image in Fig. 2b has the similar content as Fig. 2a because it contains the same objects in the same layout, however, the images are dissimilar in style. On the other hand, the image in Fig. 2c has similar style as Fig. 2a due to the similar coloring, lighting and other texture details but are dissimilar in content as different objects are present in different layouts. Now, let us define the connotations of a text T relative to an image I.

(a) Image with the content of (b) but style of (c)

(b) Image with similar content as (a) but dissimilar style.

(c) Image with similar style as (a) but dissimilar content.

Fig. 2. Examples of styles and contents of an image.

Definition 3. ***Connotation*** $C_{T,I}$ ***of a Text*** T ***relative to an Image*** I: *it is defined as the sentiments and themes present in a piece of text beyond the literary meaning of the text conditioned upon the given image.*

In the above definition, the sentiment represents the emotion (e.g. happiness, loneliness, anger) present in the text and the theme is the overall topic of the text (e.g. love, travel, festival, farewell). The color, and its patterns and transitions, in an image can significantly impact a viewer's feelings related to these sentiments and themes. For the example in Fig. 1, the picture of the lake captured by the individual does not match the connotation of melancholy and iciness in the caption given by her. The objective is to identify the condition for each object in the image (e.g., the sky should be stormy, the lake should be icy) from the connotations of the text. Therefore, $C_{T,I}$ are the inherent sentiments and themes that must be identified from the connotations of a text T and injected to the individual style of each object in the image I. Now, let us define the Image-Style IS_T of a Text T.

Definition 4. ***Image-Style of a Text*** IS_T**:** *it is defined by a feature space to capture the style that each object in an image should possess to accurately reflect the connotations in* $C_{T,I}$.

To meet our objective of producing the output image O matching the content of the input image I and image-style of the input text T, we define two losses:

Definition 5. ***Content Loss*** $L_C^{I,O}$ **:** *it specifies the distance between the content of the input image* C_I *and the content of the output image* C_O.

Definition 6. ***Image-Text Style Loss*** $L_S^{T,O}$ **:** *it specifies the distance between the image-style representation of the input text* IS_T *and the style of the output image* S_O.

The objective of our problem is to generate an output image O that minimizes both of these losses. Formally, we can define our objective as:

$$\arg\min_{O} \; \alpha L_C^{I,O} + \beta L_S^{T,O} \tag{1}$$

where I is the input image, T is the input text, O is the output image, $L_C^{I,O}$ defines the content loss for images I and O, $L_S^{T,O}$ defines the image-text style loss for text T and image O, and the hyperparameters α and β define the priority for each type of loss, which are often application specific.

4 Solution

In this section, we present the PhotoStylist scheme to address the problem formulated in the previous section. Figure 3 shows an overview of our PhotoStylist scheme. The PhotoStylist consists of three major components: i) Image-Text ANP Crossmatcher (ITAC) module, ii) ANP-based Style Generator (ASG) module, and iii) Segmented Style Transfer (SST) module. We delve into the details of these components below.

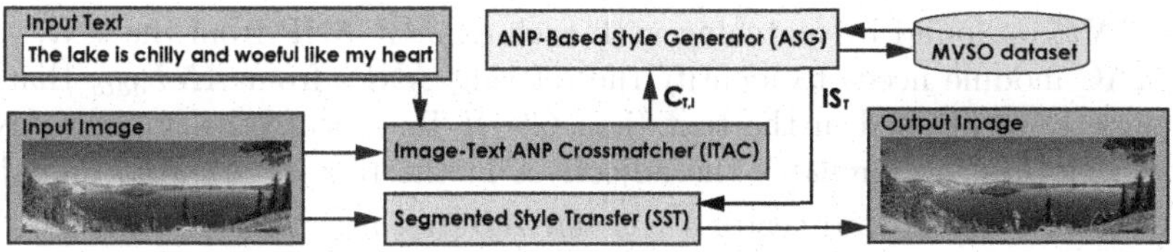

Fig. 3. An overview of the PhotoStylist scheme

4.1 Image-Text ANP Crossmatcher (ITAC) Module

To relate the connotations from the text to the different objects present in the image, it is essential to find a representation of the connotation that can be utilized to alter the style of our input image. To this end, our solution utilizes adjective-noun pairs (ANP) which has been used for sentiment analysis in texts [20] as well as for visual sentiment classification in images [4]. ANPs are capable of turning a neutral noun like "sky" into an ANP with strong sentiment like "stormy sky" by adding an adjective with a strong sentiment. However, to identify ANPs that relate the connotations present in text to the objects in the images, it is not possible to just utilize existing methods that extract ANPs from text. This is because these methods focus only on the text and, as such, many of the extracted ANPs may not relate to the objects in the image. Moreover, taking only the ANPs that relate to objects from this set would mean many of the objects in the image would not be accounted for. Therefore, the goal of our ITAC module is to crossmatch images and texts to generate a set of ANPs, reflecting the connotations of the text, where the adjectives indicate the target style for the objects in the image. To achieve this goal, we leverage the Multilingual Visual Sentiment Ontology (MVSO) [4] that links ANPs to images. We define ANP_{data} as the list of ANPs present in the leveraged dataset. Thus, we define the connotation of the text conforming to the image (i.e., $C_{T,I}$) as the set of ANPs that we find by crossmatching our input image and text. To identify the relevant ANPs, our solution utilizes both the explicit and implicit information present in the text. For our example in Fig. 1, from the input text (i.e., the lake is chilly and woeful like my heart) we have an explicit ANP (i.e., 'chilly lake') in this sentence that provides the target style for the lake. However, we also need to identify the implicit ANPs (e.g.., 'stormy sky', 'snowy mountain') corresponding to each of the other objects in the image. Therefore, $C_{T,I}$ in our scheme is divided into two parts: $C^e_{T,I}$, the relevant ANPs from ANP_{data} identified by matching with explicit ANPs from the text, and $C^i_{T,I}$, the relevant ANPs from ANP_{data} that are identified from the implicit information in the text. To identify the explicit ANPs in the input text, the module assigns the closest adjective to each noun in the text by assuming that the closest adjective to it best represents the connotation associated with it [20]. To match the explicit ANPs from the input text to those from ANP_{data}, it is essential to define a metric that represents similarity among words. Thus, the ITAC module utilizes the cosine similarity between word vectors from emotion-incorporated GloVe [14]. $C^e_{T,I}$ is the set of ANPs from ANP_{data} found by matching with each explicit ANP from the text. Now, the ITAC module needs to identify the relevant ANPs from ANP_{data} that are not directly mentioned in the text (i.e., $C^i_{T,I}$). To capture the implicit ANPs, our scheme first accumulates the adjectives in the text as ADJ_{text} and then, calculates the mean word vector $\overline{V}^{ADJ}_{text}$ for ADJ_{text} as:

$$\overline{V}^{ADJ}_{text} = \frac{1}{|ADJ_{text}|} \sum_{x \in ADJ_{text}} V(x) \tag{2}$$

By utilizing this vector, the module searches for the most similar adjective from ANP_{data} associated with each noun (object) present in the image but not in the text. The set of ANPs we identify from this search is our $C^e_{T,I}$. Figure 4 summarizes the ITAC module.

4.2 ANP-Based Style Generator (ASG) Module

The ASG module is designed to identify suitable image-style representations IS_T for the connotations of text $C_{T,I}$ that is obtained from ITAC. Our scheme represents image-style in the form of Gram matrices [2] by utilizing the VGG19 network pre-trained on ImageNet [15]. By transferring style utilizing gram matrices, our scheme introduces new color transitions into the output image.

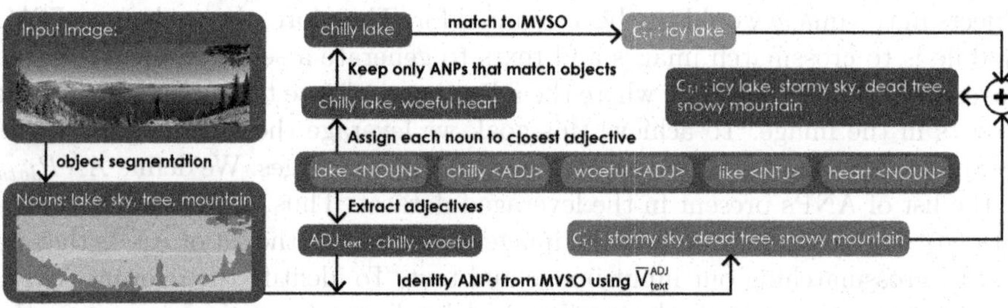

Fig. 4. Illustration of the ITAC module. The red box contains the final set of ANPs identified as the connotation $C_{T,I}$ of text T relative to image I. (Color figure online)

Thus, the objective of the ASG module is to generate relevant gram matrices for each ANP in $C_{T,I}$. Previous efforts on style transfer (in images) utilized gram matrices calculated from different segments of the same style image. Our scheme uses gram matrices from segments of different images and transfers the style from these different images onto a single output image. The ASG module identifies relevant gram matrices as follows. For each ANP in $C_{T,I}$ that is obtained from the ITAC module, the module selects an image from the MVSO dataset. We realize that the main concern when selecting these different images are the different illumination conditions, rest of the textures should represent the emotion conveyed by the adjective in the ANP. Thus, the ASG module first converts each of the images to the decorrelated $l\alpha\beta$ color space [13] where l represents luminance of the image. Then, it finds the global mean and standard deviation for the luminance channels of all the selected images. After that, it alters the individual mean and standard deviation of the luminance channel for each image to match the global mean and standard deviation. This ensures conformity between the different ANP style images. Then, the ASG module segments out the object of interest and feeds the result into the VGG19 network and calculates gram matrices at different depths. The image-style representation of the text IS_T is the collection of gram matrices for all ANPs in $C_{T,I}$.

4.3 Segmented Style Transfer (SST) Module

Finally, after obtaining the image-style representation IS_T of the input text T from ASG, the SST module transfers the obtained style into the input image. First, we have the Content Loss $L_C^{I,O}$ that minimizes the mean squared error between feature maps in selected layers that aims to ensure the contents in the output image O conforms to the input image I. Now, we define the Image-Text Style Loss $L_S^{T,O}$ to ensure the connotations obtained from the text is transferred into the style of the output image as:

$$L_S^{T,O} = \sum_{ANP \in C_{T,I}} \sum_{l \in L^S} \sum_{i,j} (G_l(ANP)_{ij} - G_l^N(O)_{ij})^2 \tag{3}$$

where $G_l(ANP)_{ij}$ is the gram matrix for layer l extracted from IS_T^{ANP} accessed via indices i and j. $G_l^N(O)_{ij}$ is the gram matrix of the output image masked according to noun N from ANP utilizing masks from the input image I. This loss results in the image-style from different ANPs to be transferred to the different segments of the image. Finally, to maintain the photorealism in the output image, we utilize a photorealism regularization loss $L_m^{I,O}$ [9]. Now, we obtain our output image by minimizing the combination of these losses. Thus, our objective is:

$$\arg\min_O \; \alpha L_C^{I,O} + \beta L_S^{T,O} + \gamma L_m^{I,O} \tag{4}$$

which is the same as the objective in our problem statement save for the photorealism regularization which we omitted in our problem statement for simplicity. The hyperparameters α, β, and γ define the priority for each type of loss, which are often application specific.

5 Evaluation

We carry out extensive evaluations of our scheme to answer the following questions: i) Can PhotoStylist provide output images that capture connotations of the text better than the output images of the state-of-the-art baselines? ii) How realistic are the output images from the PhotoStylist scheme?

5.1 Baselines

While there has been a plethora of work on sentiment classification of texts as well as altering images based on predefined emotion classes, we couldn't find any work that directly addresses the same problem as PhotoStylist which is utilizing the connotations of a given text to automatically modify the style of an image. Thus, to evaluate our scheme, we had to construct the baselines by combining state-of-the-art models for fine-grained sentiment analysis of texts with Emotional Color Transfer by He et al. [3]. For sentiment analysis part of our baseline, we use the following models: i) IBM Watson Tone Analyzer, ii) Attention-based Transformer finetuned for classifying into Plutchik's eight core emotions [5], iii) BERT-based model finetuned on GoEmotions dataset [1]. However, the classifications provided by the above sentiment analyzers cannot be directly provided as input to the emotional color transfer algorithm. Thus, to match the classification by the sentiment analyzer with the appropriate input emotion class for emotional color transfer, we utilize two different state-of-the-art similarity finding methods: i) Wpath Semantic Similarity [25] using WordNet [12], and ii) Emotion-incorporated Global vectors (GloVe) for word representation [14].

In summary, we have three different sentiment analysis methods integrated with emotional color transfer using two different similarity finding methods, giving us a total of six baselines. For maintaining simplicity in our discussion, we name these baselines according to the combination that we obtain it from: WatsonGloVe, WatsonWordnet, TransformerGloVe, TransformerWordnet, GoemotionsGloVe and GoemotionsWordnet.

5.2 Crowdsourcing-Based Evaluation

We create 40 different pairs of texts and images, where the images do *not* reflect the connotations of the text, from 30 different images and 20 different texts. For each of these 40 pairs, we test them with our PhotoStylist scheme and all compared baselines. Then, we carry out a real-world user study on Amazon Mechanical Turk (MTurk) to study the performance of all compared schemes. Each task published on MTurk is an online survey which contains a piece of text and 8 images (the original image and one from each compared method). The MTurk worker is asked to provide two scores on a scale of 1-to-10: i) *Text Faithfulness Score*: how well the image represents the connotations of the text, ii) *Realism Score*: how realistic the image is. To improve the reliability of our results, we assign each task to five different workers (not common across tasks) resulting in a total of 200 surveys.

The results of the average text faithfulness score are shown in Fig. 5a. We observe that our method achieves an average text faithfulness score of 7.77 which is significantly higher than the average scores of the baselines. This indicates that the MTurk workers found that the output images of our scheme is much closer to the connotations of the text than the output images of the baseline methods. This is because even the fine-grained sentiment analysis methods used in the baselines have relatively few output emotion classes compared to our scheme which

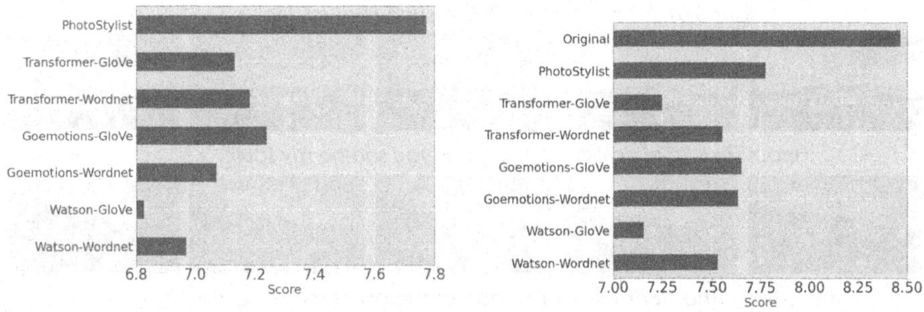

(a) Average text faithfulness score comparison. A score of '1' indicates the image does not match with the text at all, and a score of '10' indicating the image and text are a perfect match.

(b) Average realism score comparison. A score of '1' indicating the image is totally unrealistic, and a score of '10' indicating the photo is fully realistic.

Fig. 5. Evaluation results

generates more fine-grained ANPs by crossmatching the input image and text. Consequently, our scheme can better capture the connotations of the input text relative to the input image than the baselines. Moreover, the segmented style-transfer-based approach utilized in our scheme is able to effectively transfer the appropriate connotations to different objects of the input image independently which results in more pertinent styles for each of the objects in the output images compared to the output images of the baselines.

The results of the average realism score are shown in Fig. 5b. As expected, the original image receives the highest average realism score of 8.47. However, we observe that our PhotoStylist method achieves a realism score of 7.78 which is higher than all baselines. This is because of the photorealism regularization term in our total loss function. However, we also note that the difference in realism score between Goemotions-GloVe method (score: 7.65) and our method is small. This is because the modifications in color space utilized by emotional color transfer do not cause distortions in the image thus, the baselines can achieve high realism scores as well. However, emotional color transfer can cause colors to become unrealistic for certain emotion classes which is why many of the baselines have a significantly lower realism score than our method.

5.3 Visual Comparison

Figure 6 shows a comparison of PhotoStylist and the highest-rated baselines in our crowdsourcing evaluation. We observe that while the baselines modify the overall color of the picture, our method modifies textures separately for each object in the image. This is because our scheme identifies connotations from the text separately for each segment of the image and so it can introduce styles to a segment of the image independent of the other segments. For example, for the image of roses in a dark field, our scheme is able to inject colors into the sky

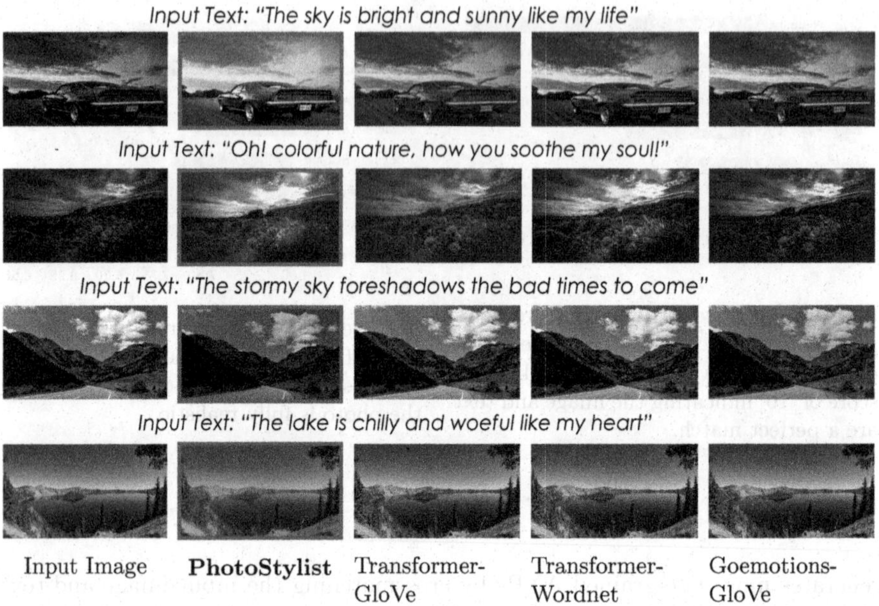

Fig. 6. A visual comparison of PhotoStylist with the highest-rated baselines.

independent of the field below it. It turns the bleak sky into a colorful one by injecting new color gradients and also, turns the grass greener.

6 Conclusion

In this paper, we develop the PhotoStylist system which is among the firsts to leverage the connotations of texts to alter the style of images. The evaluation results show that PhotoStylist can generate photorealistic results that are much closer to the connotations of the text than the state-of-the-art baselines. We envision that our PhotoStylist system will catalyse further research on connotation-aware text-based image modification and generation applications.

Acknowledgment. This research is supported in part by the National Science Foundation under Grant No. IIS-2008228, CNS-1845639, CNS-1831669, Army Research Office under Grant W911NF-17-1-0409. The views and conclusions contained in this document are those of the authors and should not be interpreted as representing the official policies, either expressed or implied, of the Army Research Office or the U.S. Government. The U.S. Government is authorized to reproduce and distribute reprints for Government purposes notwithstanding any copyright notation here on.

References

1. Demszky, D., Movshovitz-Attias, D., Ko, J., Cowen, A., Nemade, G., Ravi, S.: GoEmotions: a dataset of fine-grained emotions. In: Proceedings of the 58th Annual Meeting of the Association for Computational Linguistics, pp. 4040–4054 (2020)
2. Gatys, L.A., Ecker, A.S., Bethge, M.: Image style transfer using convolutional neural networks. In: Proceedings of the IEEE Conference on Computer Vision and Pattern Recognition, pp. 2414–2423 (2016)
3. He, L., Qi, H., Zaretzki, R.: Image color transfer to evoke different emotions based on color combinations. SIViP **9**(8), 1965–1973 (2015)
4. Jou, B., Chen, T., Pappas, N., Redi, M., Topkara, M., Chang, S.F.: Visual affect around the world: a large-scale multilingual visual sentiment ontology. In: Proceedings of the 23rd ACM International Conference on Multimedia, pp. 159–168 (2015)
5. Kant, N., Puri, R., Yakovenko, N., Catanzaro, B.: Practical text classification with large pre-trained language models. arXiv preprint arXiv:1812.01207 (2018)
6. Li, Y., Liu, M.Y., Li, X., Yang, M.H., Kautz, J.: A closed-form solution to photorealistic image stylization. In: Proceedings of the European Conference on Computer Vision (ECCV), pp. 453–468 (2018)
7. Liu, D., Jiang, Y., Pei, M., Liu, S.: Emotional image color transfer via deep learning. Pattern Recogn. Lett. **110**, 16–22 (2018)
8. Liu, S., Pei, M.: Texture-aware emotional color transfer between images. IEEE Access **6**, 31375–31386 (2018)
9. Luan, F., Paris, S., Shechtman, E., Bala, K.: Deep photo style transfer. In: Proceedings of the IEEE Conference on Computer Vision and Pattern Recognition, pp. 4990–4998 (2017)
10. Marshall, J., Wang, D.: Mood-sensitive truth discovery for reliable recommendation systems in social sensing. In: Proceedings of the 10th ACM Conference on Recommender Systems, pp. 167–174 (2016)
11. Marshall, J., Wang, D.: Towards emotional-aware truth discovery in social sensing applications. In: 2016 IEEE International Conference on Smart Computing (SMARTCOMP), pp. 1–8. IEEE (2016)
12. Miller, G.A.: Wordnet: a lexical database for English. Commun. ACM **38**(11), 39–41 (1995)
13. Reinhard, E., Adhikhmin, M., Gooch, B., Shirley, P.: Color transfer between images. IEEE Comput. Graphics Appl. **21**(5), 34–41 (2001)
14. Seyeditabari, A., Tabari, N., Gholizade, S., Zadrozny, W.: Emotional embeddings: refining word embeddings to capture emotional content of words. arXiv preprint arXiv:1906.00112 (2019)
15. Simonyan, K., Zisserman, A.: Very deep convolutional networks for large-scale image recognition. In: 3rd International Conference on Learning Representations, ICLR (2015)
16. Su, Y.Y., Sun, H.M.: Emotion-based color transfer of images using adjustable color combinations. Soft. Comput. **23**(3), 1007–1020 (2019)
17. Wang, D., Abdelzaher, T., Kaplan, L.: Social Sensing: Building Reliable Systems on Unreliable Data. Morgan Kaufmann, Massachusetts (2015)
18. Wang, D., Szymanski, B.K., Abdelzaher, T., Ji, H., Kaplan, L.: The age of social sensing. Computer **52**(1), 36–45 (2019)

19. Xu, T., Zhang, P., Huang, Q., Zhang, H., Gan, Z., Huang, X., He, X.: Attngan: fine-grained text to image generation with attentional generative adversarial networks. In: Proceedings of the IEEE Conference on Computer Vision and Pattern Recognition, pp. 1316–1324 (2018)
20. Yatani, K., Novati, M., Trusty, A., Truong, K.: Analysis of adjective-noun word pair extraction methods for online review summarization. In: Twenty-Second International Joint Conference on Artificial Intelligence, IJCAI (2011)
21. Yoo, J., Uh, Y., Chun, S., Kang, B., Ha, J.W.: Photorealistic style transfer via wavelet transforms. In: Proceedings of the IEEE International Conference on Computer Vision, pp. 9036–9045 (2019)
22. Zhang, D.Y., Ni, B., Zhi, Q., Plummer, T., Li, Q., Zheng, H., Zeng, Q., Zhang, Y., Wang, D.: Through the eyes of a poet: Classical poetry recommendation with visual input on social media. In: 2019 IEEE/ACM International Conference on Advances in Social Networks Analysis and Mining (ASONAM), pp. 333–340. IEEE (2019)
23. Zhang, D.Y., Shang, L., Geng, B., Lai, S., Li, K., Zhu, H., Amin, M.T., Wang, D.: Fauxbuster: A content-free fauxtography detector using social media comments. In: 2018 IEEE International Conference on Big Data (Big Data), pp. 891–900. IEEE (2018)
24. Zhang, H., Xu, T., Li, H., Zhang, S., Wang, X., Huang, X., Metaxas, D.N.: Stackgan: Text to photo-realistic image synthesis with stacked generative adversarial networks. In: Proceedings of the IEEE International Conference on Computer Vision, pp. 5907–5915 (2017)
25. Zhu, G., Iglesias, C.A.: Computing semantic similarity of concepts in knowledge graphs. IEEE Trans. Knowl. Data Eng. **29**(1), 72–85 (2016)

Gazetteer-Guided Keyphrase Generation from Research Papers

T. Y. S. S. Santosh[1], Debarshi Kumar Sanyal[2(✉)], Plaban Kumar Bhowmick[3], and Partha Pratim Das[1]

[1] Department of Computer Science and Engineering, IIT Kharagpur, Kharagpur, India
ppd@cse.iitkgp.ac.in

[2] Indian Association for the Cultivation of Science, Kolkata, India

[3] G. S. Sanyal School of Telecommunication, IIT Kharagpur, Kharagpur, India
plaban@cet.iitkgp.ac.in

Abstract. The task of keyphrase generation aims to generate the key phrases that capture the primary content of a document. An external domain-specific gazetteer can assist in generating keyphrases that are literally absent in the document (i.e., do not match any contiguous sub-sequence of source text) but relevant to the content of the document. In this paper, we present a technique to integrate knowledge from a gazetteer in order to improve keyphrase generation from research papers. We also present a copy mechanism that helps our model to utilize the gazetteer vocabulary to deal with the out-of-vocabulary words in keyphrases. Since constructing and maintaining relevant high-quality gazetteer by hand is very expensive, we also propose a method for automatic construction of a gazetteer given the input document, by leveraging similar documents in the training corpus. The thus constructed gazetteer helps focus on corpus-level information carried by other similar documents. Although this external information is crucial, it is never considered in previous studies. Experiments on real world datasets of research papers demonstrate that our proposed approach improves the performance of the state-of-the-art keyphrase generation models.

Keywords: Gazetteer · Keyphrase generation · Encoder-decoder · Copy mechanism · Attention

1 Introduction

Keyphrases are short informative text pieces that capture the primary content of a document. Keyphrases are common in research papers and help search engines in indexing the papers. Keyphrases are beneficial to various downstream tasks, such as document retrieval [23], document clustering [8], text summarization [20] and opinion mining [2]. This task is dealt under two settings namely (i) *extractive* [15,21,22] and (ii) *generative* [3–5,14]. Extractive methods aim to

K. Karlapalem et al. (Eds.): PAKDD 2021, LNAI 12712, pp. 655–667, 2021.
https://doi.org/10.1007/978-3-030-75762-5_52

identify the *present keyphrases*, i.e., keyphrases that fully match a part of the source text. The generative paradigm attempts to produce, rather than extract, keyphrases. Generative methods are more challenging and helpful because they can identify *absent keyphrases*, i.e., keyphrases that do not match any contiguous sub-sequence of source text but appear in the list of keyphrases mentioned in the document.

Existing methods for keyphrase generation [3–5,14] mainly adopt encoder-decoder framework in which the encoder obtains the hidden state contextual representations of the document and then the decoder generates keyphrases using the encoder representations from a fixed vocabulary regardless of whether the keyphrases are present or absent in the document. Although these methods have proven to be successful for generating both the present and absent keyphrases, they can still be enhanced by leveraging appropriate *external knowledge* that can be effectively captured through gazetteers. More recently, methods incorporating gazetteer knowledge have been explored in the area of Named Entity Recognition [12,19,25]. We believe that keyphrase generation can be improved if such an external knowledge base in the form of gazetteers is available to the keyphrase generator algorithm. Consider an example of the document provided in Table 1; one requires relevant external knowledge to identify that this document deals with *evolutionary computation* as it is not explicitly mentioned in the document. Such domain-specific external knowledge can be obtained from domain-specific gazetteers. But constructing such domain-specific gazetteers is highly expensive. On the other hand, after a closer look at the keyphrases of the documents similar to the given document, we observe that they contain phrases similar to *evolutionary computation*, and this could help us in the construction of the desired gazetteer. In other words, keyphrases from similar documents can provide useful knowledge and serve as a gazetteer, and aid in the generation of keyphrases for the given document.

Based on the above hypothesis, this paper proposes to *use a gazetteer* in addition to the input document for keyphrase generation. Since constructing and maintaining relevant high-quality gazetteers by hand is very expensive, we devise a method for *automatic construction of gazetteer* given the input document, by using keyphrases from similar documents in the training corpus. Keyphrases from similar documents – that is, the constructed gazetteer – capture useful knowledge from the whole corpus. This, in turn, can assist the model in learning topic-related salient information, similar to how humans might provide the keyphrases for a document based on their domain knowledge. We use a pointer-generator network [24] augmented with a gazetteer and trained via reinforcement learning to generate keyphrases. In addition, we also adopt a *copy mechanism* that leverages the gazetteer to generate keyphrases that might not even occur in the given document. Experiments on five real-world datasets of research papers demonstrate that our proposed approach improves the performance of the state-of-the-art keyphrase generation models.

Table 1. Example to illustrate how keyphrases from similar documents can provide useful external knowledge and serve as a gazetteer for keyphrase generation.

Title	A multi-objective genetic optimization of interpretability-oriented fuzzy rule-based classifiers
Abstract	The paper presents a multi-objective genetic approach to design interpretability-oriented fuzzy rule-based classifiers from data. The proposed approach allows us to obtain systems with various levels of compromise between their accuracy and interpretability. Original crossover and mutation operators, as well as chromosome-repairing technique to directly transform the rules are also proposed. The interpretability measure is based on the arithmetic mean of three components: the average length of rules, the number of active fuzzy sets, and the number of active inputs of the system. Effectiveness of the proposed technique in various classification problems is confirmed by experimental results
Absent Keyphrase	Evolutionary computation
Documents similar to the above document	
Title	Handling fuzzy systems' accuracy-interpretability trade-off by means of multi-objective evolutionary optimization methods– selected problems
Keyphrases	Accuracy and interpretability of fuzzy rule-based systems, multi-objective evolutionary optimization, genetic computations
Title	A multi-objective genetic optimization for fast, fuzzy rule-based credit classification with balanced accuracy and interpretability
Keyphrases	Accuracy and interpretability of credit classification systems, multi-objective evolutionary optimization, fuzzy rule-based systems, genetic computations

2 Related Work

2.1 Keyphrase Extraction

The extraction paradigm deals with the identification of keyphrases present in the document. Traditional methods use a two step approach where first they identify a set of candidate phrases using various heuristics [13] and then rank them based on supervised [7,17] or unsupervised [16] approaches. More recently, keyphrase extraction has been formulated as a sequence-labeling task [6], and various deep learning models [21,22] have been proposed to deal with the same. However, they cannot produce keyphrases that are absent in the document.

2.2 Keyphrase Generation

Unlike the extractive methods, generative techniques can produce keyphrases that are absent in the input documents. Meng et al. [14] first adopted the encoder-decoder approach and proposed CopyRNN using an attention network with copy mechanism. Chen et al. [4] proposed CorrRNN which captures the correlation among the generated keyphrases to eliminate the problem of coverage and duplicate keyphrases. Chen et al. [5] proposed TG-Net which exploits the supremacy of title information to learn a better representation for a given input document; they identify the important information in a document using a title-guided attention mechanism. But all the above methods first use beam search to generate a large number of keyphrases and then pick the top ranked ones as the final prediction, which means they can only predict a fixed number of keyphrases and lacks the ability to determine the appropriate number of keyphrases for a document. To alleviate this problem, Yuan et al. [27] proposed a new decoding setup, One2Seq paradigm, by concatenating multiple keyphrases with a delimiter and letting the model learn the length of the output sequence, which provides the ability to generate a variable number of keyphrases for the different documents. Extending on that decoding strategy, Chan et al. [3] formulated keyphrase generation using a reinforcement learning framework to encourage the model to generate the correct number of keyphrases with the aid of separate adaptive reward signals for the present and the absent keyphrases. In this work, we follow the generative paradigm and the setups proposed by [27] and [3] for decoding strategy and training setup respectively. We explore how to improve keyphrase generation by incorporating knowledge from a gazetteer. We also propose a method for automatic construction of a gazetteer given an input document, by leveraging similar documents from the training corpus.

3 Problem Definition

Given the title and the abstract of a document, the task is to generate a sequence of keyphrases, each separated from the next by a delimiter. To enable our model to generate a *variable* number of keyphrases for the documents in an end-to-end manner, we follow the training setup put forth by [3,27] that slightly rearranges the ground-truth keyphrase list for each paper as follows: the output sequence contains the *present* keyphrases followed by the *absent* ones, with all the present keyphrases rearranged according to their first appearance in the input document and all the absent keyphrases kept in their original order. The present and absent keyphrases are separated by different delimiters.

4 Baseline Architecture

Given the input title $t = \{t_1, ..., t_l\}$ consisting of l words and the abstract of the document $x = \{x_1, x_2, \ldots, x_m\}$ consisting of m words, we first convert them into their word embeddings and feed them into a bi-directional Gated Recurrent Unit (biGRU) *encoder* to obtain the hidden state representations $\{v_1, v_2, \ldots v_l\}$ and $\{u_1, u_2, \ldots u_m\}$ of the title and the abstract, respectively. We follow [5] where we use the title information to emphasize the importance of relevant information in the abstract based on the fact that the title-related information in the abstract reflects core information; in particular, we obtain enhanced abstract representations using *title-guided attention* as shown in Eq. (1).

$$s_{i,j} = (u_i)^\top W v_j \qquad \eta_{i,j} = \frac{\exp(s_{i,j})}{\sum_{k=1}^{l} \exp(s_{i,k})} \qquad \tilde{u}_i = \sum_{j=1}^{l} \eta_{i,j} v_j \tag{1}$$

where $\tilde{u}_i$ is the enhanced vector for u_i, and $\eta_{i,j}$ and $s_{i,j}$ denote the normalized and unnormalized attention scores between u_i and v_j, respectively. We pass the enhanced vector through a biGRU layer to incorporate the local contextual information and finally, we merge the original contextual vector u_i of the abstract and the enhanced vector p_i using a residual connection as shown in Eq. (2).

$$p_i = \text{biGRU}(\tilde{u}_i) \qquad a_i = \lambda u_i + (1-\lambda)p_i \tag{2}$$

where λ is a hyperparameter. We use a single-layered GRU as the *decoder* which produces the decoder hidden state s_t using previous hidden state s_{t-1}, predicted word from previous time stamp y_{t-1} and the decoder context vector h_t^{*D}.

$$s_t = \text{GRU}(s_{t-1}, y_{t-1}, h_t^{*D}) \tag{3}$$

s_0 is initialized with the last hidden state of the encoder representation a_m. Note y_0 and h_1^{*D} are initialized with zeros. We compute h_t^{*D} by applying attention mechanism [1] over the decoder hidden states before time step t i.e. $(s_1, s_2, \ldots, s_{t-1})$ with respect to decoder hidden state at previous time stamp (s_{t-1}) as shown in Eq. (4).

$$f_j^t = v_1 \tanh(W_1 s_j + W_2 s_{t-1} + b_1) \quad \beta_j^t = \frac{\exp(f_j^t)}{\sum_{k=1}^{t-1}\exp(f_k^t)} \qquad h_t^{*D} = \sum_{j=1}^{t-1}\beta_j^t s_j \tag{4}$$

where v_1, W_1, W_2, b_1 are trainable parameters, β_j^t and f_j^t represents normalized and unnormalized attention scores between s_j and s_{t-1}, respectively. This decoder context vector is expected to help reduce the generation of duplicate keyphrases. We incorporate the well-known *coverage mechanism* [4] to generate more diverse phrases and reduce the information redundancy among the generated keyphrases. Concretely, we maintain a coverage vector c^t which is the sum of the encoder attentive distributions $\{\alpha^i\}$ over all previous decoder time steps. Intuitively, c^t (with $c^0 = 0$) represents the degree of coverage that the words in the source document have received from the attention mechanism so far. It is used in computing the attention over the encoder hidden representations and the encoder context vector h_t^{*E} as follows.

$$c^t = \sum_{i=0}^{t-1}\alpha^i \qquad e_j^t = v_2 \tanh(W_3 h_j + W_4 s_t + W_5 c_j^t + b_2) \tag{5}$$

$$\alpha^t = \text{softmax}(e^t) \qquad h_t^{*E} = \sum_{j=1}^{n}\alpha_j^t h_j \tag{6}$$

where v_2, W_3, W_4, W_5, b_2 are trainable parameters. Then, the probability distribution over all words in the vocabulary is P_{vocab}, generated by another two layer network:

$$P_{vocab} = \text{softmax}(W_7 \tanh(W_6[s_t, y_{t-1}, h_t^{*D}, h_t^{*E}] + b_3) + b_4) \tag{7}$$

where W_6, W_7, b_3, b_4 are trainable parameters. The fixed vocabulary only contains the most frequently occurring words and it does not change with different inputs. This causes the out-of-vocabulary (OOV) problem as as some of the keyphrases come from the input document but are absent in the fixed vocabulary *vocab*. Inspired by the *copy mechanism* in [5,24], for each input document, we build an extended vocabulary by merging *vocab* and all words that appear in the input document. The probability distribution of words w over this extended vocabulary is calculated as

$$P(w) = p_{gen} P_{vocab}(w) + (1 - p_{gen}) \sum_{i:w_i=w} \alpha_i^t \tag{8}$$

where p_{gen} is a soft switch to choose between generating a word from the fixed vocabulary by sampling from P_{vocab} and copying a word directly from the corresponding source document by sampling from attention distribution α_t. The latter addresses the OOV problem as $P_{vocab}(w) = 0$ for OOV words w. The generation probability p_{gen} is calculated as:

$$p_{gen} = \sigma(W_8 h_t^{*E} + W_9 h_t^{*D} + W_{10} s_t + W_{11} y_{t-1} + b_5) \tag{9}$$

where $W_8, W_9, W_{10}, W_{11}, b_5$ are the trainable parameters and σ denotes the sigmoid function. We train the model using the reinforcement learning approach proposed by [3], in which an agent interacts with an environment in discrete time steps. At each time step $t = 1, \ldots, T$, the agent produces an action y_t sampled from the policy $\pi(y_t|y_{1:t-1}, x; \theta)$ which is the generation model where $y_{1:t-1}$ denotes the sequence generated by the agent from time step 1 to $t-1$. The environment also produces a reward r_t to the agent and transits to the next step $t+1$ with a new state s_{t+1}. They give separate reward signals to present keyphrase predictions and absent keyphrase predictions. Once the agent generates the delimiter token separating the present and the absent keyphrases, then the environment computes a reward using an adaptive reward function $RF1$ by comparing the generated keyphrases with the ground-truth present keyphrases. Again once the model generates the EOS token, the environment compares the generated absent keyphrases with the ground-truth absent keyphrases. The reward to the agent is 0 for all other time steps. We calculate the sum of the future rewards starting from time step t as $R_t = \sum_{t=t}^{T} r_i$ where T denotes the length of sequence. The goal of the agent is to maximize the expected initial return $\mathbb{E}_{y \sim \pi(.|x;\theta)}[R_1]$. An adaptive reward function RF_1 is defined as follows.

$$RF_1 = \begin{cases} recall & N < G \\ F_1 score & otherwise \end{cases} \tag{10}$$

where N is the number of predicted keyphrases and G is the number of ground-truth keyphrases. The loss function is defined to maximize the expected initial return as $L(\theta) = -\mathbb{E}_{y \sim \pi(.|x,\theta)}[R_1(y)]$ and the gradient is computed using Reinforce algorithm [26]

5 Gazetteer-Enhanced Architecture

An external domain-specific gazetteer can assist in generating keyphrases that are absent in a document but are still relevant to the content of the document. We design a novel module that integrates the knowledge acquired from a gazetteer into the basic model through a *gazetteer-guided attention* mechanism. Given an input document, the gazetteer is automatically constructed using the keyphrases from other *semantically similar* documents in the corpus. This helps to enrich the keyphrase generation process with corpus-level information carried by other similar documents.

5.1 Automatic Gazetteer Construction

To construct the gazetteer for an input document, we compute the embedding-based similarity called the *Word Mover's Distance* [11] between the input document and all other documents in training corpus. Given a preset threshold, this identifies the documents that are semantically similar to the input document. The gazetteer is the collection of all the keyphrases from the above obtained similar documents.

5.2 Integrating Gazetteer Knowledge

Let the gazetteer obtained be $G_x = \{g_1, g_2, \ldots, g_l\}$ where l is the number of keyphrases present in the gazetteer constructed for the source document x. Denote the ith phrase in the gazetteer as $g_i = \{g_{i_1}, g_{i_2}, \ldots, g_{i_K}\}$ where K is the number of words in g_i. The words of a phrase g_i are fed one-by-one into a Bidirectional GRU-based encoder and then, we concatenate the last forward hidden state and the first backward hidden state to construct the hidden representation h_{g_i} for phrase g_i. Thus, we obtain the hidden representations of all the keyphrases of the gazetteer as $H_{G_x} = \{h_{g_1}, h_{g_2}, \ldots, h_{g_l}\}$. Then we need to identify context-appropriate keyphrases from the gazetteer. We achieve this by an attention distribution γ^l over the above hidden representations of the keyphrases, given the context of the input document.

$$\gamma_j^t = \text{softmax}(W_{12}h_t^{*D} + W_{13}h_t^{*E} + W_{14}h_{g_j} + b_6) \qquad g_t^* = \sum_{j=1}^{l} \gamma_j^t h_{g_j} \tag{11}$$

where W_9, W_{10}, b_5 are trainable parameters. We posit that this corpus-level information in the gazetteer will assist the model to generate absent keyphrases. Therefore, we exploit the gazetteer information when calculating the vocabulary probability distribution for keyphrase generation, and change Eq. (7) to

$$P_{vocab} = \text{softmax}(W_7 \tanh(W_6[s_t, y_{t-1}, h_t^{*D}, h^{*E}, g_t^*] + b_3) + b_4) \tag{12}$$

In order to actually generate words that are part of the gazetteer but not the fixed vocabulary of the decoder, the vocabulary of the keyphrase generator must include all the words in the gazetteer. But the fixed vocabulary is much smaller. Recollect that we solved a similar problem in Sect. 4 where we used the copy mechanism to enable the decoder to directly copy words from the input document. Now we extend the effective vocabulary further to include the words in the gazetteer (that depends on the input document) so that the decoder can copy from the input document or the gazetteer. Therefore, revising Eq. (8), the probability distribution over the final extended vocabulary is calculated as:

$$P(w) = p_{gen} P_{vocab}(w) + (1 - p_{gen})\left(\sum_{i:w_i=w} \alpha_i^t + \sum_{i:w_i=w} \gamma_i^t\right) \tag{13}$$

We also use the gazetteer contextual representation in the computation of the generation probability p_{gen}, which is used to choose between generating a word from *vocab* and copying a word; therefore, changing Eq. (9) to

$$p_{gen} = \sigma(W_6 h_t^* + W_7 s_t + W_8 e_{t-1} + W_{11} g_t^* + b_4) \tag{14}$$

6 Experiments and Results

6.1 Datasets

We conduct experiments on five research article datasets namely KP20k [14], Inspec [7], Krapivin [10], NUS [17] and SemEval [9]. Each instance from these datasets consists of the title, abstract and a set of keyphrases. Detailed splits for training, validation, testing and other statistics are provided in Table 2.

Table 2. Statistics of datasets. Avg. KP indicates the average of keyphrases per each instance, %absent indicates percentage of absent keyphrases

Dataset	KP20k	Inspec	Krapivin	NUS	SemEval
Training	513,918	–	–	–	–
Valid	19,992	1500	1844	169	144
Testing	19,992	500	460	42	100
Avg. KP	5.3	9.6	5.2	11.5	15.7
%absent	36.7	21.5	43.8	48.7	55.5

6.2 Baselines and Evaluation Metrics

We compare our model with the state-of-the-art keyphrase generation models under One2Seq decoding strategy [27] where all keyphrases are concatenated into a single target sequence: (i) catSeq [27] which uses encoder-decoder framework with copy mechanism, (ii) catSeqD [27] which is built on catSeq by incorporating orthogonal regularization, (iii) catSeqCorr which is derived from [4] to fit this One2Seq setting, (iv) catSeqTG which is derived from [5] and adopted to this setting and their reinforcement learning (RL) implementations namely (v) catSeq-RL, (vi) catSeqD-RL, (vii) catSeqCorr-RL, (viii) catSeqTG-RL [3]. We implement our neural model, **GazGuid**, with maximum likelihood loss. We also build another version of our model, **GazGuid-RL** that is trained with RL. To demonstrate the effectiveness of the proposed gazetteer guidance, we also implement variants of our model without gazetteer guidance, indicated by **w/oGazGuid** and **w/oGazGuid-RL**, respectively. Following [3,27], we evaluate the performance of our model by comparing the predicted keyphrases with the ground-truth keyphrases using macro-averaged F1-measure with a variable cutoff F1@M (M = number of predicted keyphrases) and with a fixed cutoff F1@5 .

6.3 Implementation

We use pre-trained word embeddings from GloVe [18] and set the hidden size dimension to be 100. For reinforced models, we first pre-train each model using maximum likelihood loss and then apply RL approach to train each of them. We initialize all the model parameters using a uniform distribution within the interval [−0.1,0.1]. We use a dropout rate of 0.1 and gradient clipping of 1.0. We use the Adam optimizer with a batch size of 32 and an initial learning rate of 0.0001.

Table 3. Performance of present keyphrase prediction on five datasets.

	KP20K		Inspec		Krapivin		NUS		SemEval	
Model	F1@5	F1@M	F1@5	F1@M	F1@5	F1@M	F1@5	F1@M	F1@5	F1@M
CatSeq	0.291	0.367	0.225	0.262	0.269	0.354	0.323	0.397	0.242	0.283
CatSeqD	0.285	0.363	0.219	0.263	0.264	0.349	0.321	0.394	0.233	0.274
CatSeqCorr	0.289	0.365	0.227	0.269	0.265	0.349	0.319	0.390	0.246	0.290
CatSeqTG	0.292	0.366	0.229	0.270	0.282	0.366	0.325	0.393	0.246	0.290
CatSeq-RL	0.310	0.383	0.250	0.300	0.287	0.362	0.364	0.426	0.285	0.327
CatSeqD-RL	0.305	0.379	0.242	0.292	0.282	0.360	0.353	0.419	0.272	0.316
CatSeqCorr-RL	0.308	0.382	0.240	0.291	0.286	0.369	0.349	0.414	0.278	0.322
CatSeqTG-RL	0.321	0.386	0.253	0.301	0.300	0.369	0.375	0.433	0.287	**0.329**
GazGuid	0.312	0.374	0.255	0.283	0.303	0.369	0.372	0.419	0.268	0.312
GazGuid-RL	**0.344**	**0.388**	**0.278**	**0.303**	**0.326**	**0.372**	**0.397**	**0.435**	**0.308**	**0.329**
w/oGazGuid	0.289	0.367	0.231	0.265	0.289	0.366	0.332	0.395	0.244	0.288
w/oGazGuid-RL	0.324	0.383	0.255	0.300	0.313	0.369	0.378	0.424	0.282	**0.329**

6.4 Performance Comparison

We report the results on *present* keyphrases and *absent* keyphrases in Tables 3 and 4, respectively. We observe that our model with RL approach, **GazGuid-RL** performs better than all of the baselines in both variable and fixed cut-off settings on all datasets for both present and absent keyphrase prediction. It performs better than its counterpart without gazetteer guidance in both present and absent keyphrase prediction. This demonstrates the ability of gazetteer to distil the corpus-level knowledge into the model. It helps the model to understand the salient information of the document more effectively, thereby increasing its extractive power, showing an improvement in present keyphrases prediction. Gazetteer guidance also helps the decoder select appropriate words to form absent keyphrases. Similarly, the copy mechanism with the extended vocabulary from the gazetteer assists in generating OOV words. Our result is also in line with previous studies showing the capability of the RL approach to generate both sufficient and accurate keyphrases using adaptive reward signals. The performance improvement is, however, modest because the gazetteer while enhancing the vocabulary in an adaptive way, is unable to fully compensate for the OOV words and absent keyphrases.

Table 4. Performance of absent keyphrase prediction on five datasets.

	KP20K		Inspec		Krapivin		NUS		SemEval	
Model	F1@5	F1@M	F1@5	F1@M	F1@5	F1@M	F1@5	F1@M	F1@5	F1@M
CatSeq	0.015	0.032	0.004	0.008	0.018	0.036	0.016	0.028	0.020	0.028
CatSeqD	0.015	0.031	0.007	0.011	0.018	0.037	0.014	0.024	0.016	0.024
CatSeqCorr	0.015	0.032	0.005	0.009	0.020	0.038	0.014	0.024	0.018	0.026
CatSeqTG	0.015	0.032	0.005	0.011	0.018	0.034	0.011	0.018	0.019	0.027
CatSeq-RL	0.024	0.047	0.009	0.017	0.026	0.046	0.019	0.031	0.018	0.027
CatSeqD-RL	0.023	0.046	0.010	**0.021**	0.026	0.048	**0.022**	**0.037**	0.021	0.030
CatSeqCorr-RL	0.022	0.045	0.010	0.020	0.022	0.040	**0.022**	**0.037**	0.021	**0.031**
CatSeqTG-RL	0.027	0.050	0.012	**0.021**	0.030	0.053	0.019	0.031	0.021	0.030
GazGuid	0.018	0.038	0.008	0.011	0.022	0.040	0.019	0.031	0.021	0.026
GazGuid-RL	**0.030**	**0.052**	**0.014**	**0.021**	**0.033**	**0.055**	**0.022**	**0.037**	**0.022**	**0.031**
w/oGazGuid	0.015	0.032	0.007	0.011	0.018	0.038	0.016	0.028	0.019	0.024
w/oGazGuid-RL	0.025	0.046	0.011	0.020	0.030	0.051	0.019	0.031	0.021	**0.031**

6.5 Analysis of the Number of Generated Keyphrases

We now analyze the ability of models to predict the correct number of keyphrases per document. We report the mean absolute error (MAE) between the number of generated keyphrases and the number of ground-truth keyphrases for all documents in the KP20k validation dataset, where a lower MAE indicates a better generation performance in Table 5. We also report the average number of generated keyphrases per document denoted as Avg.#; a value closer to the oracle indicates a better performance. From Table 5, we observe that our model, GuidGen-RL has the lowest MAE and the closest Avg.# to the oracle on both the present and the absent keyphrases compared to all other baselines.

Table 5. Prediction of correct number of keyphrases on the KP20k valid dataset.

	Present		Absent	
Model	MAE	Avg.#	MAE	Avg.#
Oracle	0.000	2.837	0.000	2.432
CatSeq	2.271	3.781	1.943	0.659
CatSeqD	2.225	3.694	1.961	0.629
CatSeqCorr	2.292	3.790	1.914	0.703
CatSeqTG	2.276	3.780	1.956	0.638
CatSeq-RL	2.118	3.733	1.494	1.574
CatSeqD-RL	2.087	3.666	1.541	1.455
CatSeqCorr-RL	2.107	3.696	1.557	1.409
CatSeqTG-RL	2.204	3.865	1.439	1.749
GazGuid	2.126	3.746	1.766	0.986
GazGuid-RL	**2.024**	**3.659**	**1.427**	**1.760**
w/oGazGuid	2.223	3.746	1.883	0.845
w/oGazGuid-RL	2.096	3.664	1.524	1.578

Table 6. Example to show the effectiveness of our constructed gazetteer.

Title	A geometric proof of the upper bound on the size of partial spreads in $H(4n+1, q^2)$
Abstract	We give a geometric proof of the upper bound of $q(2n+1)+1$ on the size of partial spreads in the polar space $H(4n+1, q(2))$. This bound is tight and has already been proved in an algebraic way. Our alternative proof also yields a characterization of the partial spreads of maximum size in $H(4n+1, q(2))$
Golden Keyphrases	Partial spreads, Hermitian varieties
Gazetteer	Spread, polar space, Hermitian variety, complete partial spread

6.6 Case Study

We perform a case study to better understand our model performance. In Table 6, we show an example document with gold-standard keyphrases: 'partial spreads' and 'Hermitian varieties'. For a fair evaluation, we only compare RL-based models. We observe that the keyphrase 'partial spread', which is present in title, is correctly predicted by CatSeqTG-RL and GazGuid-RL while the other models such as CatSeq-RL, CatSeqD-RL and CatSeqCorr-RL fail to do so. In case of the keyphrase 'Hermitian varieties', only GuizGuid-RL generates it while all the other models fail to produce it. The former can be attributed to the title-guided attention mechanism incorporated in our model inspired from TG-Net. The latter shows the effectiveness of how external knowledge can be leveraged from the gazetteer built with keyphrases from similar documents.

7 Conclusion

We presented a new technique for keyphrase generation that integrates knowledge from an external gazetteer. We built a gazetteer automatically using keyphrases from similar documents in the corpus. We also expanded the dynamic vocabulary in the copy mechanism with vocabulary from the gazetteer. Our method achieves better performance than competitive baselines in predicting both the present and the absent keyphrases. Thus, our work validates the idea of using gazetteers to enrich the vocabulary and improve keyphrase generation. In future we aim to enhance the performance with external knowledge from other sources such as cited documents and citation contexts.

Acknowledgements. This work is supported by *National Digital Library of India* Project sponsored by Ministry of Human Resource Development, Government of India at IIT Kharagpur and Faculty Research Grant, IACS.

References

1. Bahdanau, D., Cho, K., Bengio, Y.: Neural machine translation by jointly learning to align and translate. In: Proceedings of 3rd International Conference on Learning Representations, ICLR 2015 (2015)
2. Berend, G.: Opinion expression mining by exploiting keyphrase extraction. In: Proceedings of 5th International Joint Conference on Natural Language Processing, pp. 1162–1170 (2011)
3. Chan, H.P., Chen, W., Wang, L., King, I.: Neural keyphrase generation via reinforcement learning with adaptive rewards. In: Proceedings of the 57th Annual Meeting of the Association for Computational Linguistics, pp. 2163–2174 (2019)
4. Chen, J., Zhang, X., Wu, Y., Yan, Z., Li, Z.: Keyphrase generation with correlation constraints. In: Proceedings of the 2018 Conference on Empirical Methods in Natural Language Processing, pp. 4057–4066 (2018)
5. Chen, W., Gao, Y., Zhang, J., King, I., Lyu, M.R.: Title-guided encoding for keyphrase generation. Proc. AAAI Conf.Artif. Intell. **33**, 6268–6275 (2019)
6. Gollapalli, S.D., Li, X.L., Yang, P.: Incorporating expert knowledge into keyphrase extraction. In: Thirty-first AAAI Conference on Artificial Intelligence (2017)
7. Hulth, A.: Improved automatic keyword extraction given more linguistic knowledge. In: Proceedings of EMNLP (2003)
8. Hulth, A., Megyesi, B.B.: A study on automatically extracted keywords in text categorization. In: Proceedings of the 21st International Conference on Computational Linguistics and the 44th Annual Meeting of the Association for Computational Linguistics, pp. 537–544 (2006)
9. Kim, S.N., Medelyan, O., Kan, M.Y., Baldwin, T.: Semeval-2010 task 5: automatic keyphrase extraction from scientific articles. In: Proceedings of the 5th International Workshop on Semantic Evaluation, pp. 21–26 (2010)
10. Krapivin, M., Autaeu, A., Marchese, M.: Large dataset for keyphrases extraction. University of Trento, Technical report (2009)
11. Kusner, M., Sun, Y., Kolkin, N., Weinberger, K.: From word embeddings to document distances. In: International Conference on Machine Learning (2015)

12. Lin, H., Lu, Y., Han, X., Sun, L., Dong, B., Jiang, S.: Gazetteer-enhanced attentive neural networks for named entity recognition. In: Proceedings of the 2019 Conference on Empirical Methods in Natural Language Processing and the 9th International Joint Conference on Natural Language Processing (EMNLP-IJCNLP), pp. 6233–6238 (2019)
13. Medelyan, O., Frank, E., Witten, I.H.: Human-competitive tagging using automatic keyphrase extraction. In: Proceedings of the 2009 Conference on Empirical Methods in Natural Language Processing, pp. 1318–1327 (2009)
14. Meng, R., Zhao, S., Han, S., He, D., Brusilovsky, P., Chi, Y.: Deep keyphrase generation. In: Proceedings of the 55th Annual Meeting of the Association for Computational Linguistics: Long Papers, vol. 1, pp. 582–592 (2017)
15. Merrouni, Z.A., Frikh, B., Ouhbi, B.: Automatic keyphrase extraction: a survey and trends. J. Intell. Inf. Syst. **54**, 1–34 (2019)
16. Mihalcea, R., Tarau, P.: Textrank: bringing order into text. In: Proceedings of the 2004 Conference on Empirical Methods in Natural Language Processing (2004)
17. Nguyen, T.D., Kan, M.Y.: Keyphrase extraction in scientific publications. In: Goh, D.H.L., Cao, T.H., Sølvberg, I.T., Rasmussen, E. (eds.) Asian Digital Libraries: Looking Back 10 Years and Forging New Frontiers. Lecture Notes in Computer Science, vol. 4822, pp. 317–326. Springer, Berlin, Heidelberg (2007). https://doi.org/10.1007/978-3-540-77094-7_41
18. Pennington, J., Socher, R., Manning, C.D.: GloVe: global vectors for word representation. In: Proceedings of EMNLP, pp. 1532–1543 (2014)
19. Peshterliev, S., Dupuy, C., Kiss, I.: Self-attention gazetteer embeddings for named-entity recognition. arXiv preprint arXiv:2004.04060 (2020)
20. Qazvinian, V., Radev, D., Özgür, A.: Citation summarization through keyphrase extraction. In: Proceedings of the 23rd International Conference on Computational Linguistics (COLING 2010), pp. 895–903 (2010)
21. Santosh, T.Y.S.S., Sanyal, D.K., Bhowmick, P.K., Das, P.P.: DAKE: document-level attention for keyphrase extraction. In: Jose, J., et al. (eds.) Advances in Information Retrieval (ECIR 2020). Lecture Notes in Computer Science, vol. 12036, pp. 392–401. Springer, Cham (2020). https://doi.org/10.1007/978-3-030-45442-5_49
22. Santosh, T., Sanyal, D.K., Bhowmick, P.K., Das, P.P.: Sasake: syntax and semantics aware keyphrase extraction from research papers. In: Proceedings of the 28th International Conference on Computational Linguistics, pp. 5372–5383 (2020)
23. Sanyal, D.K., Bhowmick, P.K., Das, P.P., Chattopadhyay, S., Santosh, T.Y.S.S.: Enhancing access to scholarly publications with surrogate resources. Scientometrics **121**(2), 1129–1164 (2019)
24. See, A., Liu, P.J., Manning, C.D.: Get to the point: Summarization with pointer-generator networks. In: Proceedings of the 55th Annual Meeting of the Association for Computational Linguistics: Long Papers, vol. 1, pp. 1073–1083 (2017)
25. Song, C.H., Lawrie, D., Finin, T., Mayfield, J.: Improving neural named entity recognition with gazetteers. UMBC Faculty Collection (2020)
26. Williams, R.J.: Simple statistical gradient-following algorithms for connectionist reinforcement learning. Mach. Learn. **8**(3–4), 229–256 (1992)
27. Yuan, X., Wang, T., Meng, R., Thaker, K., Brusilovsky, P., He, D., Trischler, A.: One size does not fit all: generating and evaluating variable number of keyphrases. In: Proceedings of the 58th Annual Meeting of the Association for Computational Linguistics: Long Papers, vol. 1 (2020)

Minits-AllOcc: An Efficient Algorithm for Mining Timed Sequential Patterns

Somayah Karsoum[1(✉)], Clark Barrus[1], Le Gruenwald[1], and Eleazar Leal[2]

[1] University of Oklahoma, Norman, OK 73019, USA
{somayah.karsoum,clark.barrus,ggruenwald}@ou.edu
[2] University of Minnesota Duluth, Duluth, MN 55812, USA
eleal@d.umn.edu

Abstract. Sequential pattern mining aims to find the subsequences in a sequence database that appear together in the order of timestamps. Although there exist sequential pattern mining techniques, they ignore the temporal relationship information between the itemsets in the subsequences. This information is important in many real-world applications. For example, even if healthcare providers know that symptom Y frequently occurs after symptom X, it is also valuable for them to be able to estimate when Y will occur after X so that they can provide treatment at the right time. Considering temporal relationship information for sequential pattern mining raises new issues to be solved, such as designing a new data structure to save this information and traversing this structure efficiently to discover patterns without re-scanning the database. In this paper, we propose an algorithm called Minits-AllOcc (MINIng Timed Sequential Pattern for All-time Occurrences) to find sequential patterns and the transition time between itemsets based on all possible occurrences of a pattern in the database. We also propose a parallel multicore CPU version of this algorithm, called MMinits-AllOcc (Multicore Minits-AllOcc), to deal with Big Data. Extensive experiments on real and synthetic datasets show the advantages of this approach over the brute-force method. Also, the multicore CPU version of the algorithm is shown to outperform the single-core version on Big Data by 2.5X.

Keywords: Sequential pattern mining · Timed sequential pattern · Multicore · Parallel sequential pattern mining

1 Introduction

Sequential pattern mining [1] is a data mining task that discovers frequent subsequences in a sequence database of time-ordered transactional data. Finding interesting, useful, and unexpected patterns is beneficial for a wide range of real-world applications, such as illness symptom pattern prediction [12], network intrusion detection [18], and customer shopping behaviors [1]. Existing sequential pattern mining algorithms such as [9, 17, 22] use implicit timestamps to order the itemsets within a sequential pattern but the transition time between these itemsets is not kept. In many applications, it is important to know the time to move from one itemset to another in the pattern. For example, in

K. Karlapalem et al. (Eds.): PAKDD 2021, LNAI 12712, pp. 668–685, 2021.
https://doi.org/10.1007/978-3-030-75762-5_53

healthcare applications, knowing when the next symptom of heart attack will occur helps healthcare providers in forming diagnoses, providing treatments at the right time, and intervening earlier in critical cases. For monitoring weather forecasts in Oklahoma during the tornado season, we want to be able to track the transition time range between cities, when a tornado hits multiple cities in the timestamp order. With sequential patterns that also contain the temporal relationship about the transition time, which indicates when the next symptoms will appear, we are answering not only a question like in which order the symptoms for heart attack frequently occur, but also questions like when the symptoms for heart attack frequently occur.

Let us suppose that we have the historical health information of temperature (T) and blood pressure (BP) of patients who have had a heart attack. The time is recorded when each of the measurements is taken for each patient. Since sequential pattern mining algorithms do not deal with continuous data, we need to apply a discretization technique in order to segment the data into classes that have similar features or fall within the same group. For instance, the blood pressure (BP) has five levels [4]: 1.) Normal (BP $<$ 120), 2.) Elevated (120 $\leqslant$ BP $\leqslant$ 129), 3.) High Stage 1 (131 $\leqslant$ BP$\leqslant$ 139), 4.) High Stage 2 (140 $\leqslant$ BP $\leqslant$ 180), and 5.) Crisis (BP $>$ 181). Therefore, we refer to the blood pressure with the abbreviation BP followed by the class number in which the blood pressure falls. Since the temporal information is available in time-ordered transactional data, we can discover more informative sequential patterns that not only show the symptoms that frequently occur among patients but also include the typical transition times between symptoms (in terms of number of weeks in our example). We call this special type of sequential patterns Timed Sequential Patterns (TSP). For example, <{T1, BP3} [2, 7] {T2}> is a TSP that has two itemsets: itemset 1 consisting of two items T1 and BP3, and itemset 2 consisting of item T2. Itemset 2 occurs within 2 to 7 weeks after itemset 1. In our notations, all items enclosed within braces { } occur at the same time and constitute an itemset, and the square brackets [] indicates the time duration to move from one itemset to the next. Thus, the previous example of TSP shows that when patients have a temperature in class 1 (T1) and a blood pressure in class 3 (BP3), then within 2 to 7 weeks, the patients will have a temperature in class 2 (T2). If we apply traditional sequential mining, then this pattern will only be <{T1, BP3} {T2}>, which does not include the transition time [2, 7].

Incorporating the temporal information in a sequential pattern raises additional challenges when compared to the regular sequential pattern mining. First, while both sequential pattern mining and timed sequential pattern mining need to find out whether a pattern occurs in a sequence database tuple, timed sequential pattern mining also needs to find out how many times the pattern occurs in that tuple to compute the temporal relationship between the itemsets in the pattern. So, to find all possible occurrences of the pattern, the naïve mechanism is required to scan each tuple in the database from the beginning until the end. Unlike sequential pattern mining, an algorithm will stop checking the rest of the tuple in the database as soon as the pattern is found. In other words, in the best case, sequential pattern mining techniques do not need to check until the end of each sequence in the database. However, timed sequential pattern mining requires checking all the sequences in the database. Suppose we have a tuple of a patient P1 in the database that has all measurements within six months and with the following symptoms occurring many times: high temperature followed by low blood pressure after some time. Since the timed sequential patterns mining problem wants to know when the low blood pressure occurs, it is not sufficient to find only the first position of this pattern and report the temporal relation. First, it is necessary to consider all possible occurrences of that pattern and also all the different timestamps of each occurrence and find the temporal relation. Second, the temporal information must be updated for each pattern that is discovered based on the timestamps of the tuples that contain the pattern. For example, after we discover the pattern from the patient P1 and calculate the temporal relations, we also find the same pattern--high temperature followed by low blood pressure--in another tuple for another patient Pi in the database. That means the temporal relations need to be updated to represent the actual time duration. So, we need to capture all the occurrences of that pattern for Pi and re-calculate the temporal relations. On top of that, we need to keep track of timestamps of all occurrences of the patterns for both patients P1 and Pi to finalize the temporal relationships. The brute force technique needs to scan the database again to retrieve or store the required information for P1. Thus, for every pattern, we need to scan the whole database many times to make sure that we have the correct temporal relations. Of course, this requires more space and time, impacting the performance of any algorithm.

The existing techniques [3,8,21], which will be discussed in detail in Sect. 3, do not address these challenges. To fill this gap, in this paper we propose a timed sequential pattern mining algorithm, called Minits-AllOcc (MINIng Timed Sequential Pattern for All-time Occurrences), that addresses all these challenges. We also validate the proposed algorithm through a set of empirical experiments. The contributions of this paper are the following:

1. The idea of incorporating transition time between itemsets in a sequential pattern, which indicates all possible time occurrences of the pattern within whole timed sequence database. The temporal relations require time to move from one itemset to the next in the timestamp order. The time can be any descriptive statistic based on the user's preference, such as range, average, etc.
2. The parallel implementation of the Minits-AllOcc algorithm and the extensive experiments comparing the single-core against multi-core algorithms on real and synthetic datasets.

The remainder of this paper is organized as follows: Sect. 2 defines the timed sequential pattern mining problem. Section 3 reviews the related works. Section 4 explains the proposed techniques for the Minits-AllOcc algorithm. Section 5 presents the results of the experiments on the dataset. Finally, the conclusion and future work are presented in Sect. 6.

2 Problem Definitions

In this section, we review the definitions of the sequential pattern mining problem and introduce new definitions for the timed sequential pattern mining problem. Recalling the traditional sequential pattern mining problem [1], we define an **itemset** I as a set of **items,** such that $I \subseteq X$, where $X = \{x_1, x_2, \ldots x_l\}$ is a set of items. A **sequence** (tuple) *s* is an ordered list (based on timestamps) of itemsets. A sequence A = <{a_1}, {a_2}, …{a_n}> is **contained in** another sequence B = <{b_1}, {b_2}, …{b_m}> and B is **super-sequence** of A, if there exists a set of integers, $1 \leq j_1 < j_2 < \ldots < j_n \leq m$, such that $a_1 \subseteq b_{j_1}, a_2 \subseteq b_{j_2}, \ldots, a_n \subseteq b_{jn}$.

A **sequence database** S is a set of sequences (tuples) <sid, s_i>, where sid is a sequence identifier and s_i is a sequence. A tuple <sid, s_i> is said to contain a sequence α, if α is a subsequence of s_i. Since our problem considers the temporal data too, we incorporate timestamps explicitly in the database and introduce new definitions.

Definition 1. A timed ***event*** is a pair *e* = (*I, t*), where *I* is an itemset that occurs at the timestamp *t*. We use e.I and e.t to indicate, respectively, the itemset I and the timestamp t associated with the event e. The list of events that is sorted in the timestamp order is called a ***timed sequenceTS*** = <{e_1}, {e_2},…, {e_n}>, such that $e_i.x \subseteq I\ (1 \leq I \leq n)$. A ***timed sequence database TSDB*** is a set of sequences <*TS_id, TS*>, where *TS_id* is a timed-sequence identifier and *TS* is a timed sequence.

Example 1 (Running example). The timed sequence database in Fig. 1 is used as an illustrative example in this paper. For simplicity, we will use letters to refer to items which represent different properties of objects in the database (e.g., temperature and blood pressure for patients), and integer numbers to refer to timestamps, which represent the times when those properties are collected. In this example, there are four timed sequences with IDs from TS1 to TS4. Each timed sequence consists of a set of events ordered in the events' timestamps. For example, TS1 consists of two events: the first event {a, b, 5}, which occurred at timestamp 5, followed by the second event {d, g, 12}, which occurred at timestamp 12.

Definition 2. Given a sequence A = <{I_1}, {I_2}, …{I_n}> and a timed sequence TS = <{e_1}, {e_2}, …, {e_m}>, the ***All-time Occurrences*** of A in TS in the timed sequence database TSDB is defined as an ordered list of indices $1 \leq j_1 < j_2 < \ldots < j_n \leq m$, such that: $I_1 \subseteq e_{j_1}.I, I_2 \subseteq e_{j_2}.I, \ldots I_n \subseteq e_{j_n}.I$. The ***deltas*** Δ are defined as $= \Delta e_{p.j_{i-1}}.t - e_{p.j_i}.t$.

Example 2. Let sequence A = <{a}{b}> and timed sequence TS4 = <{a, 10}, {b, f, 19}, {d, 20}, {b, 30}>, as shown in Fig. 1. The indices of the events for the first occurrence of sequence A in TS4 are {e_1, e_2}, as shown by the solid arrow in Fig. 2. The delta Δ is the difference between the timestamps of these two consecutive events, which is $e_{1}.t_1 = 10$ and $e_{2}.t_2 = 19$. Thus, the Δ = 19 – 10 = 9. Then, the second occurrence of sequence A in TS4, as shown by the dotted arrow in Fig. 2, has the following events' indices {e_1, e_4}. The delta Δ is the difference between the timestamps of these two consecutive events, which is $e_{1}.t_1 = 10$ and $e_{4}.t_2 = 30$. Thus, the Δ = 30 – 10 = 20. Similarly, we can find the rest of the All-time Occurrence. The **support** of a sequence A in a sequence database, or a timed sequence database, is the percentage of the number of sequences in the database that contains A, such that sup(A) = (#sequences that contains A/#sequences in DB) * 100. If the support of sequence A is greater than or equal to a user-defined threshold called minimum support (min_sup), then it is called a **sequential pattern** [1].

Definition 3. A sequence A is called a **timed sequential pattern TSP**, if and only if it is a sequential pattern and accompanied by **temporal relationships** τ_i between itemsets where it represents any descriptive statistic, such as average of transition time or range, calculated based on values of delta Δ. TSP is denoted as: TSP = $<\{I_0\}\ [\tau_1]\ \{I_1\}\ [\tau_2]\ \{I_2\} \ldots\ldots [\tau_n]\ \{I_n\}>$. For brevity, when we mention a pattern, we refer to a timed sequential pattern.

Example 3. Let us assume the min-sup = 50%; since the support of sequence A = $<\{a\}\{b\}>$ is 50%, the sequence is a sequential pattern. In this paper, we assume that a user chooses the temporal relation to be presented as a range of time [min, max]. Thus, the timed sequential pattern version is $<\{a\}\ [9, 20]\ \{b\}>$. The timed sequential patterns are originally sequential patterns that satisfy the min_sup condition and clearly state the transition time between itemsets.

Timed Sequence ID	Timed Sequences TS
TS1	< {a,b,5} , {d,g,12} >
TS2	< {e,g,21} >
TS3	< {a, 2} , {a,b,19} , {d,25} >
TS4	< {a, 10} , {b,f,19} , {d,20} , {b,30} >

Fig. 1. An example of timed sequence database

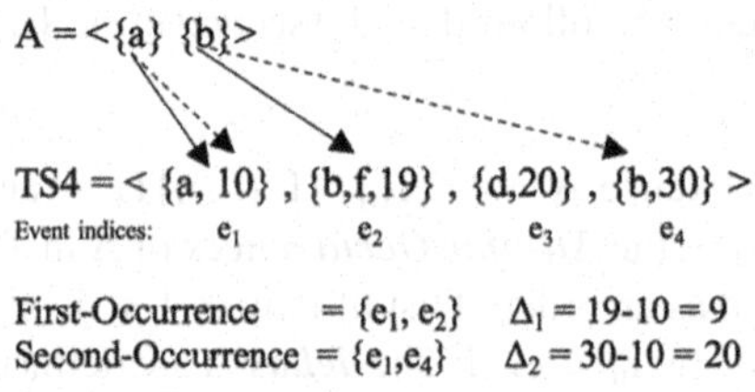

A = <{a} {b}>
TS4 = < {a, 10} , {b,f,19} , {d,20} , {b,30} >
Event indices: e1 e2 e3 e4
First-Occurrence = {e1, e2} Δ1 = 19-10 = 9
Second-Occurrence = {e1,e4} Δ2 = 30-10 = 20

Fig. 2. All-time occurrence of A in TS4

3 Related Works

The concept of sequential pattern mining was first introduced in [1], where three algorithms were proposed: AprioriSome, DynamicSome, and AprioriAll algorithms for discovering sequential patterns. AprioriAll was the basis of many other efficient algorithms that had been proposed to improve its performance. Those algorithms inspired [19] to propose a technique to generate fewer candidates called GSP. Since all algorithms were based on the Apriori algorithm, they were classified as Apriori-based algorithms. Other algorithms such as SPADE [22] adopted a vertical id-list database format that reduced the number of database scans. In contrast, pattern-growth based algorithms, such as FreeSpan [9] and PrefixSpan [17], used the concept of database projection, which made them more efficient than other Apriori-based algorithms, especially when they dealt with a large database. These algorithms generated a smaller database for their next pass because the sequence database was projected into a set of smaller databases, and then sequential patterns in each of them were explored. Thus, they were more efficient. More literature reviews about the state-of-the-art sequential pattern mining algorithms can be found in [5].

Recently, with the existence of a large volume of data in many applications, several sequential pattern mining algorithms have been proposed to handle large databases consisting of huge amounts of sequences efficiently using different platforms. For example, [11] used the multi-core processor architecture for implementing pDBV-SPM to improve processing speed for mining sequential patterns. Ha-GSP [16] adopted the principles of GSP and implemented them on the Hadoop platform for solving the limited computing capacity and insufficient performance with massive data of the traditional GSP. MR-PrefixSpan [20] used the MapReduce platform to implement the parallel version of PrefixSpan to mine sequential patterns on a large database. More literature reviews about the state-of-the-art parallel sequence mining algorithms are in [6].

In a sequential pattern, objects have an ordinal correlation based on the timestamp precedence. We can obtain a sequence by sorting all these objects based on the order of their timestamps. However, the time between itemsets is discarded. This kind of sequential pattern that incorporates temporal relations is more informative for some applications. Some techniques were proposed to specify some timing constraints, such as the time gaps between adjacent itemsets in sequential patterns. For example, [3] modified the Apriori [1] and PrefixSpan [17] algorithms to discover the time-interval sequential patterns that satisfied the interval duration boundaries. The I-PrefixSpan algorithm in [3] has another input called a set of time-intervals TI, where each time-interval has a range. [10] extended the work of [3] and proposed two algorithms: MI-Apriori and MI-Prefix. The time-intervals incorporated in the patterns revealed the time between all pairs of items in a pattern, which is called multi-time-interval sequential patterns. A list of intervals (ti_3, ti_2, ti_1) before item d in a pattern like <a, ti_1, b, (ti_2, ti_1), c, (ti_3, ti_2, ti_1), d> means the intervals between items a, b, and c and item d are ti_3, ti_2 and ti_1, respectively. [2] also extracted the sequential patterns of diseases from a medical dataset within user-specified time intervals. The drawback of these methods is that their results will miss some frequent patterns that do not fulfill the time constraint. To decide if a pattern is frequent, two conditions must be satisfied: the support of the pattern must be greater than or equal to min_sup, and the time range between itemsets must lie within the defined time intervals. Therefore, if a pattern fulfills the first condition, which means it is frequent but does not fulfill the second condition, the algorithm will not report it.

[7] incorporated the temporal dimension in the sequential pattern by defining temporally annotated sequences (TAS), and [8] proposed the Trajectory Pattern algorithm (T-pattern) to extract a set of TAS to produce trajectory patterns with a fixed amount of time to travel between places. The algorithm only worked with one-dimensional data that did not represent cases in real life, such as healthcare applications. Also, the time between events in trajectory patterns was strict, which did not consider different cases of traveling between locations, such as transportation types. [21] relaxed the travel time to be a realistic range for traveling time. Nevertheless, the algorithm still could not deal

with multidimensional data because it was dealing only with locations in trajectory data. Also, all the previous techniques did not consider all possible occurrences of a pattern in an individual sequence in the database, which meant the temporal relations were calculated based on the first occurrence of a pattern. The issue of calculating the time-intervals of the first occurrence of a pattern and ignoring other occurrences was addressed in [15]. However, this approach is beneficial for a limited number of applications. For example, if a developer wants to evaluate the ease of use of a navigation system, the time of moving from A to B is tested when the users visit those locations for the first time. In contrast, in other applications such as healthcare, mentioned above, we must consider all possible occurrences to provide accurate time-intervals. To the best of our knowledge, there is no existing algorithm that can find the complete set of timed sequential patterns, in which each pattern represents the transition time between successive itemsets in a pattern and considers all possible occurrences.

4 The Proposed Algorithm: Minits-AllOcc

To discover timed sequential patterns, we propose a timed sequential pattern mining algorithm called Minits-AllOcc. We first describe the occurrence tree, which is the core data structure of the algorithm, in Sect. 4.1. Then, in Sect. 4.2 and 4.3, we introduce an overview of Minits-AllOcc and how it works in detail. In Sect. 4.4, we propose some enhancements to improve the efficiency of the algorithm's performance.

4.1 Occurrence Tree (O-Tree)

The occurrence tree *O-tree* is a data structure used to represent the different possible occurrences of a pattern in a timed sequence in TSDB. This tree is the seed of the algorithm because it helps to generate timed sequence patterns without scanning the timed sequence database many times. The O-tree has two types of nodes: root node, which contains a timed sequence ID, and regular nodes. A regular node has the following information 1.) the event ID e_{ID} and 2.) its timestamp $e_{ID}.t$. The edge between two regular nodes represents the difference Δ between the timestamps associated with the two nodes. For example, sequence $<\{a\}>$ appears twice in TS3; thus, its O-tree in Fig. 3 has two nodes connected to the root. However, sequence $<\{a\}\ \{a\}>$ appears once in TS3 that has two nodes too, but one is connected to the root and the other is connected to a regular node via an edge Δ, which is $19 - 2 = 17$. Since each sequence has an O-tree for each timed sequence in TSDB that contains it, the sequence will have a collection of O-trees that identify its occurrence in the whole TSDB. Thus, we give the following definition:

Definition 4. Given a sequence A and timed sequence database TSDB, ***A- forest*** is a collection of all O-trees that identify all possible occurrences of the sequence A in TSDB. Figure 5 demonstrates the forests of four sequences <{a}>, <{b}>, <{a} [9, 20] {b}>, and <{a, b}>. Each forest is surrounded by a dotted rectangle, which has a group of O-trees that indicates all time-occurrences of a sequence in TSDB.

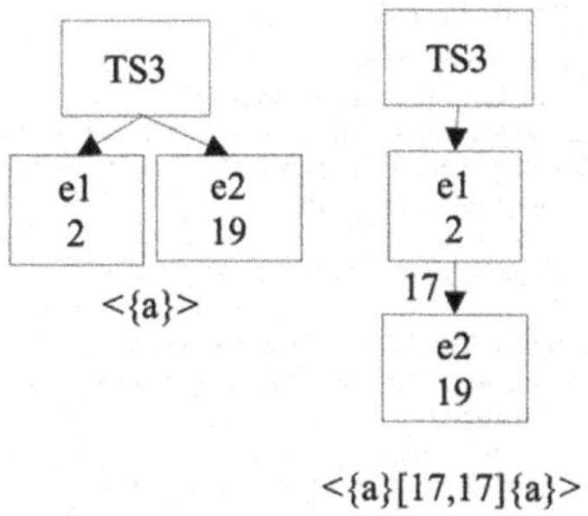

Fig. 3. An O-tree for sequences <{a}> and <{a}{a}> in TS3

4.2 Overview

The main goal of Minits-AllOcc is to find the complete set of the timed sequential patterns that satisfy the min_sup threshold condition from a given TSDB. To achieve this goal, Minits-AllOcc utilizes the forests to store all required information from timed sequences in TSDB. The following steps are performed: 1.) Scan TSDB to build a I_j-forest for each distinct item I_j; 2.) Find frequent items by counting the number of O-trees in each forest, compare it against the min_sup threshold, and remove the infrequent items; 3.) Merge all O-trees that have the same root from different forests to build a new forest for a candidate sequence. It should be noted that there are two different relations between itemsets considered while merging the steps *event-relation* and *sequence-relation,* which are defined as:

Definition 5. Given two items X and Y, it is said that X and Y have an ***Event-relation*** e-relation between them, denoted as <{X, Y}> if X and Y occur in the same event.

Definition 6. Given two items X and Y, it is said that X and Y have a ***Sequence-relation*** s-relation between them, denoted as <{X} {Y}> if X and Y occur in two different events and the event of X occurs before the event of Y; 4.) Find the timed sequential patterns among candidate sequences by counting the number of O-trees in each forest, compare it against the min_sup threshold, and ignore the infrequent sequences. By doing step 4, Minits-AllOcc avoids scanning TSDB for each candidate to calculate the support; 5.) Compute the temporal relation of the suffix, the new appending part of the pattern, and update the temporal relation of the prefix, the previous part of the pattern; 6.) Repeat steps 3, 4, and 5 until the algorithm cannot identify any new timed sequential pattern. Minits-All Occ's pseudocode is presented in Fig. 4.

4.3 The Proposed Algorithm: Minits-AllOcc

In this section, we describe the abovementioned steps in detail using the running example shown in Fig. 1. At first, the algorithm starts reading the TSDB row by row and builds the associated O-tree for each distinct item until all forests are completed (line 1). As shown in Fig. 5 for instance, after the algorithm finishes scanning TSDB, <{a}> -forest has three O-trees because sequence <{a}> appears in three timed sequences TS1, TS3,

```
Algorithm: (Minits-AllOcc)
Input: Timed Sequence Database TSDB,
       minimum support threshold min-sup
Output: The complete set of Timed sequential patterns TSPs
1  Scan TSDB and build 1-sequence forest for each distinct item
2  Add frequent 1-seq into TSPs
3   for each frequent 1-seq
4     Call find-TSPs (1-seq forest, 1-seq forests)
5   end for
6  function find-TSPs (k-seq forest, 1-seq forests)
7     for each possible combination between k-seq and 1-seq
8       if (s-relation)
9         for each common root (TSi)
10          if (O-tree(k-seq). node.EventId < O-tree(1-seq).node.EventId)
11             Append node(1-seq) to O. tree(k-seq)
12             Delta = O-tree(k-seq).node.timestamp -
                     O- tree(1seq).node.timestamp
13          end if
14        end for
15        if (# O-trees (k+1-seq) in forest ≥ min-sup)
16             Traverse k+1-seq forest to update temporal stamp
17             Add k+1-seq into TSPs
18             Call find-TSPs (k+1-seq forest, 1-seq forests)
19        else
20          return
21        end if
22      else if (e-relation)
23        for each common root (TSi)
24          if (O-tree(k-seq).node.EventId == O-tree(1-seq).node.EventId)
25             Append node(1-seq) to O.tree(k-seq)
26             Delta = 0
27          end if
28        end for
29        if (# O-trees (k+1-seq) in forest ≥ min-sup)
30             Add k+1-seq into TSPs
31             Call find-TSPs (k+1-seq forest, 1-seq forests)
32        else
33          return
34        end if
35      end if
36    end for
37 end function
```

Fig. 4. Pseudo-code of the Minits-AllOcc algorithm

and TS4. Then, the algorithm excludes the infrequent sequences by calculating their supports using the number of O-trees in each forest. The two sequences <{e}> and <{f}> are not frequent because their forests have only one tree, which means they appear in one TS; therefore, their support is 25%. The following is the set of 1-timed sequential patterns = <{a},{b},{d},{g}> (line 2). The third step is generating candidates by merging the O-trees of all 1-timed sequential patterns, so the algorithm calls function *find-TSPs* (line 3). The mechanism of merging is as follows: if the relation is s-relation, the appended node must have an event ID e_i that is greater than the parent(line 11–14). Then, the edge holds the difference between the timestamps of the parent and its child(line 15). In contrast, if the relation is e-relation, the appended node must have the same event ID e_i of its parent(line 23–26). For instance, the forest of the two candidates <{a}[]{b}>, which represents the s-relation, and <{a, b}>, which represents the e-relation, is shown in Fig. 5. The first <{a}[]{b}> -forest has two O-trees that are generated by combining the <{a}>-forest and <{b}>-forest. Even though both forests have an O-tree that has a root TS1, the O-tree of <{b}> does not contain a node that has an event ID greater than e_1; thus, it was removed from the <{a}[]{b}>-forest. In contrast, the node that has e_2

from <{b}>-forest is attached to the node that has e_1 from <{a}>-forest, and the Δ is calculated between those nodes, which is $19-2=17$. However, the node that has e_2 from <{a}> -forest does not connect to any node. Since the algorithm is looking for all possible occurrences, the node in TS4 that has e_1 is connected to the two nodes, which has event ID e_2 and e_4, from <{b}> O-tree, and each link carries the difference between the timestamps of the two connected nodes. Because in this example we consider the temporal relations as a range of [min, max], the algorithm chooses the minimum and maximum values among all O-trees in <{a}[]{b}>-forest, which is [9, 20]. The second <{a, b}>-forest has two O-trees that are generated by combining the <{a}> -forest and <{b}>-forest. The difference between the technique of merging the trees from the previous case and this one is the condition of appending nodes. Since this is an e-relation, all added nodes must have the same event ID e_i as their parents. Also, the Δ is always 0 because the nodes have the same timestamps. Both patterns <{a} [9, 20] {b}> and <{a, b}> are considered to be timed sequential patterns and they are added to TSP set because their supports are 50% (line18–20, 30–31). The supports are calculated as follows: # O-trees in the forest/#timed sequences in TSDB * 100,(2/4) * 100 = 50%. The algorithm now repeats the same steps, by calling function *find-TSPs* recursively in line 21 and 32, to extend the pattern by merging O-trees, extracting TSPs, and computing temporal relations until no more TSPs can be found. As shown in Fig. 6, pattern <{a} [9, 17] {b} [1, 6] {d}> is a result of merging between <{a} [9, 20] {b}> -forest and <{d}> -forest. The forest displays only the O-trees that represent the pattern, then the time between the prefix <{a} [] {b}> and suffix <{d}> is calculated as defined before (the range). Also, it should emphasize that the time of prefix <{a} [] {b}> is updated based on the current forest. Before, it was <{a} [9, 20] {b}> but now it is <{a} [9, 17] {b}...>. Again, the <{a} [9, 17] {b} [1, 6] {d}> is TSPs because its support is 50%. Minits-AllOcc continues repeating the steps until the complete set of TSPs is discovered. The reader can verify that the TSPs in this example is = {<{a} [9, 20] {b}>, <{a} [6, 23] {d}>, <{b} [1, 7] {d}>, <{a, b} [6, 7] {d}>, <{a} [9, 17] {b} [1, 6] {d}>}.

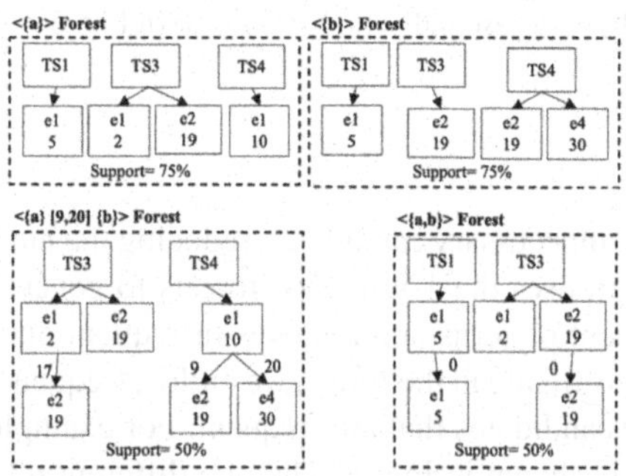

Fig. 5. Merging O-trees of <{a}> and <{b}> to generate <{a, b}> -forest and <{a} [1, 6] {b}> -forest

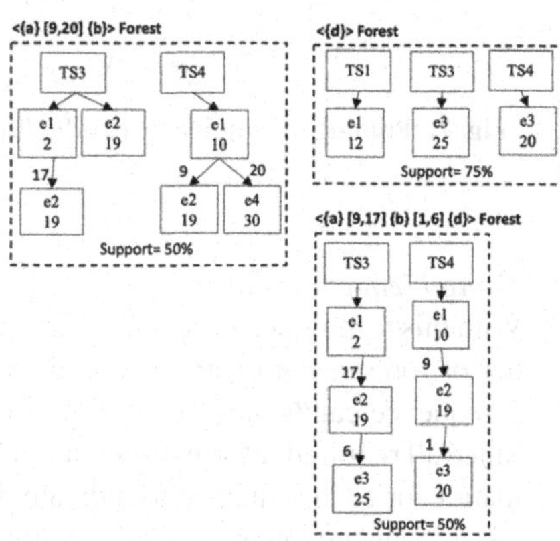

Fig. 6. Merging <{a} [9, 20] {b}> -forest and <{d}> to generate <{a} [9, 17] {b} [1, 6] {d}> -forest

4.4 The Proposed Enhancement

In this section, we describe some effective mechanisms to improve the efficiency of Minits-AllOcc.

1. *Pruning the Forests*
   This technique is meant to refine a sequence's forest after merging the O-trees. So, when those O-trees are used in the next step for generating candidates, they carry only the necessary information and therefore save space by removing some nodes and save time by avoiding traversing needless branches in trees. Any branch in an O-tree that does not have a new appended node will be removed after the merging step is executed. Figure 7 represents the idea by marking the deleted branch of O-trees with a cross symbol. For example, the O-tree that has TS3 root is a result of merging TS3 O-tree from <{a}> and <{b}>-forests. Since there is no appended node to the right branch of <{a}>-forest, this node is removed from <{a} [9, 20] {b}>-forest. Those branches do not exist anymore in the O-trees.

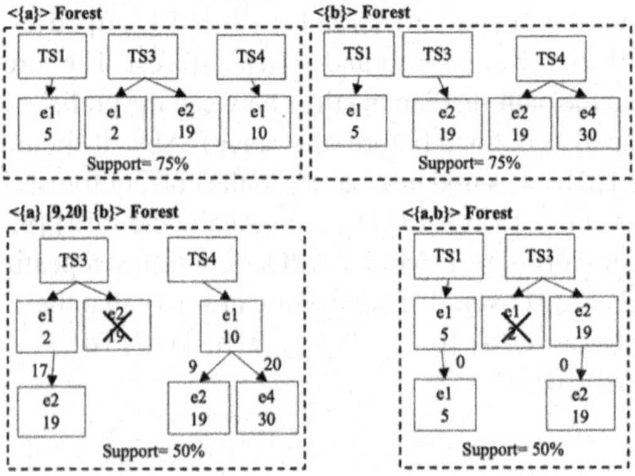

Fig. 7. Pruning the original <{a} [9, 20] {b}> *-forest* and <{a, b}> *-forest* in Fig. 5

2. *Using Frequency Matrix*
   With this technique, we avoid generating unnecessary candidates, reducing the number of forests. For example, the algorithm uses the 1-sequence-forests to generate 2-sequence candidates, then keeps frequent patterns and removes infrequent ones. Since all required information is already available in the forest, we build a frequency matrix for each sequence to indicate the candidates that are frequent. For example, the frequency matrix of <{a}> pattern is shown in Fig. 8. The two different relations, event and sequence (the rows) and all 1-timed sequential patterns that can be combined with {a} (the columns), are considered. The cells under <{b}> column represent the frequency of the two relations between <{a}> and <{b}>. This frequency is calculated from the forests of those patterns, as shown in Fig. 7. For s-relation, there are two O-trees (TS3 and TS4) in which the <{a}> and <{b}>

occur at the different timestamps within the same timed sequence. For e-relation, there are two O-trees (TS1 and TS3) in which the <{a}> and <{b}> occur at the same timestamps within the same timed sequence. From the matrix, we can infer that <{g}> is not frequent either with s-relation or e-relation; thus, we do not need to build the forest of sequence <{a}[]{g}> or <{a, g}>.

<{a}>	a	b	d	g
s-relation	25%	50%	75%	25%
e-relation	0%	50%	0%	0%

Fig. 8. Frequency matrix for <{a}>

3. *Using Multicore CPUs*
   Another enhancement is using multicore CPUs for implementing Minits-AllOcc, which we call MMinits-AllOcc. A queue is created to hold all jobs of the algorithm and as soon as one thread becomes idle, the next job in the queue is assigned to it. The first mechanism of parallelism is all threads work in parallel when the algorithm recursively generates the patterns. In the serial version, the algorithm starts with the pattern <{a}> and keeps extending it until no more patterns can be found that have prefix <{a}>, for example. Then, it starts with the pattern <{b}> and so on. With the multi-core version, the algorithm works on all patterns <{a}>, <{b}., …etc. at the same time.

5 Performance Analysis

In this section, we describe the environment of experiments and report our evaluation results based on the performance of Minits-AllOcc and MMinits-AllOcc, considering the impact of different parameters.

5.1 Experimental Setup

All experiments were performed on a computer with a 2.10 GHz Intel Xeon(R) processor with 64 GB of RAM, running Ubuntu 18.04.1 LTS CPU with 12 cores. The Minits-AllOcc and MMinits-AllOcc algorithms are implemented in Java 1.8.

5.2 Datasets and Experimental Parameters

We use real-life and synthetic datasets. The real dataset is T-Drive [13, 14] and the synthetic dataset was generated by using a tool provided by the SPMF Library [4]. Also, we set several parameters to conduct the experiments on the dataset. There are two types of parameters: static and dynamic parameters. The values of the static parameters are not changed in experiments. In contrast, the values of the dynamic parameters are changed from one experiment to another. In this experiment, we have four dynamic parameters. The first one is the minimum support threshold (min_sup). It is a user-defined threshold that is applied to find all timed sequential patterns in a timed sequence database TSDB. The second parameter is the number of timed sequences TS in TSDB (#Seq). The third parameter is the length of TS in TSDB, which can also be represented as the number of events per TS (# Events). The last parameter is the number of items in each event (#items). It should be noted that the timestamp is a fixed attribute in all events. When it is said that the number of items per event is 3, for instance, it means three items plus the timestamp. We study the impacts of all four parameters shown in Table 1 on the synthetic dataset. However, for the T-Drive dataset, the only valid dynamic parameter is the min-sup. Thus, all other three parameters are considered static. Now, we explain the range of the parameters and the default values of this analysis, as summarized in Tables 1. When the experiment was conducted, we chose various values of one parameter within its range and assigned the default value to the other parameters. The min-sup parameter has a range from 20% to 80% with the default value = 50%, which is the median of the interval. The range of the number of timed sequences parameter is from 1 to 100,000, and its median value of 50,000, the default value. For the number of events per sequence, the default value is 25 because the range is from 5 to 50. The number of items in the last parameter range has been set at 1 to 10 items per event; thus, the default value is 5, which is the median.

5.3 Competing Algorithms

Since no existing algorithm can discover the timed sequential patterns and consider All-time Occurrences, we cannot compare Minits-AllOcc against any technique. We will compare it against MMinits-AllOcc.

5.4 Evaluation Metrics

The evaluation metrics include two measurements: 1.) Execution Time (ET) of algorithms (Minits-AllOcc, and MMintis-AllOcc) and 2.) Number of Patterns (#patterns) that are generated by these algorithms.

5.5 Experimental Results

In this section, we present the performance of the two algorithms, Minits-AllOcc and MMinits-AllOcc, in terms of execution time (ET) and the number of discovered patterns (#patterns) for the real and synthetic datasets.

Table 1. Parameter list for the synthetic dataset

Parameter name	Range of values	Default value
min_sup	20%–80%	50%
# sequences	1–100,000	50,000
# events per sequence	1–50	25
# items per event	1–10	5

1. *Accuracy*
   In order to validate that Minits-AllOcc always gives the same sequential patterns in terms of the numbers and contents excluding the temporal relation, PrefixSpan was used [17]. PrefixSpan was chosen because it is one of the well-known algorithms for discovering sequential patterns. It has been proven to produce complete and correct sequential patterns. First, all temporal relations were removed from the patterns that were generated by Minits-AllOcc. Then these patterns were compared to the patterns that were generated by PrefixSpan to make sure that each sequential pattern generated by PrefixSpan has a matching one generated by Minits-AllOcc and MMinits-AllOcc. For example, a sequential pattern X = <{a} {b} {a, b}> was generated by PrefixSpan, and a timed sequential pattern Y = <{a} [2, 5] {b} [3, 7] {a, b}> was generated by Minits-AllOcc and MMinits-AllOcc. We took away the temporal relations from Y and compared it with the pattern X. In case the order of at least one itemset was different, the pattern X was not matching the pattern Y. For instance, Z = <{b} [2, 5] {a} [3, 7] {b, a}> was not matching pattern X because the item <{b}> occurred before <{a}>. However, within the last itemset {a, b} the order did not matter because all the items appeared at the same timestamp. At the end of this experiment, we found that the two algorithms--Minits-AllOcc and MMinits-AllOc--discovered the exact patterns that were produced by PrefixSpan. In other words, all algorithms produced the complete and correct set of sequential patterns.

Table 2. Average execution time ET and #patterns

Datasets	Minits-AllOcc		MMintis-AllOcc	
	ET	# patt	ET	# patt
T-Drive	12.05 (h)	126	5.97 (h)	126
Synthetic data	27.319 (min)	3780	10.825 (min)	3780

2. *Execution Time*
   The execution time was recorded from the moment that a dataset had been read to the moment that an algorithm produced the timed sequential patterns. Table 2

shows the average performance of the two algorithms: Minits-AllOcc and MMinits-AllOcc. The execution time (ET) of MMinits-AllOcc decreases by about 50% and 60% for T-Drive and synthetic datasets, respectively, compared to the execution time of Minits-AllOcc.

3. *Impact of Minimum Support*
   In this set of experiments, we compared execution time (ET) and the number of patterns (#patterns) for different values of minimum support threshold (min_sup) for datasets T-Drive and synthetic. From Fig. 9(a) and Fig. 10(a), we can see that when the minimum support increased, the execution time of all algorithms decreased. This is because the algorithms generate fewer timed sequential patterns when the min-sup is high, due to fewer candidate sequences that satisfy the min-sup condition. With a large amount of data and a huge number of discovered timed sequential patterns, MMinits-AllOcc outperformed Minits-AllOcc, as shown in Fig. 9(a) and Fig. 10(a). Therefore, multicore CPUs ought to be used when the size of the timed sequence database is huge.
   The multicore CPU version was also efficient when we had low min-sup. As shown in Fig. 10(b), the ETs of both Minits-AllOcc and MMinits-AllOcc were very close when the min-sup was greater than 60%. This is because the number of candidate sequences, and thus the number of timed sequential patterns, was getting smaller, so most of the threads were idle. Therefore, MMinits-AllOcc did not need to use all the available threads and behaved almost like a single-core version Minits-AllOcc. Another observation was made based on the number of timed sequential patterns that were generated by these algorithms. All algorithms discovered the same number of patterns; thus, their curves were overlapping in Fig. 9(b), 10(b), 10(d), 11(b) and 11(d). When the min-sup increased, the number of timed sequential patterns decreased because the patterns that satisfied the min-sup condition became fewer. By increasing the threshold min_sup, the percentage of timed sequences in the timed sequence database that was supposed to contain a candidate sequence decreased, as shown in Fig. 9(b) and Fig. 10(b).
4. *Impact of the Number of Sequences in the Database*
   In this set of experiments, we compared the execution time (ET) and the number of discovered timed sequential patterns (#patterns) according to the number of the timed sequences (#Seq). From Fig. 10(c), we can see that when the number of timed sequences increased, the execution times of all algorithms increased. This is because the algorithms needed more time to check the extra timed sequences that were added in the timed sequence database to decide if they contained a timed sequential pattern or not. We observed that the number of timed sequential patterns, which were generated by these algorithms, increased when the number of timed sequences increased, as shown in Fig. 10(d). The number of timed sequential patterns that were discovered by the algorithms also increased because the possibility of finding more patterns in the new timed sequences that satisfy the min-sup (50% as the default value) condition also increased. With an increased number of timed sequences in the database, the algorithms needed to check if some new patterns could occur and did not exist in the old timed sequences. Next, the algorithm checked their support against the threshold (min-sup). It is possible that the support of some old

patterns in the database before new sequences were added did not satisfy the min-sup condition because they were not supported by a sufficient number of timed sequences; but with a new timed sequence database, these patterns became timed sequential pattern. Thus, the number of newly discovered timed sequential patterns would increase. For example, if a database had 1000 sequences in the synthetic dataset, the number of timed sequential patterns was 3720, while the number of timed sequential patterns was 3780 when the timed sequence database had 10,000 timed sequences.

5. *Impact of the Number of Events Per Sequence*
Figure 11(a) and (b) show the impact of the number of events (#Events) per timed sequence on the execution time (ET) and the number of discovered sequential patterns (#patterns). There was a strong relationship between the length of a timed sequence and the number of discovered patterns. Increasing the length of timed sequences (#Events) drove discovering more patterns because the algorithm could extend a pattern up to the length of the timed sequence. In other words, if we have a timed sequence that contains *n* events, we can discover a set of timed sequential patterns such that their length varies from 1 to *n*. Subsequently, the required time of discovering those patterns will increase.

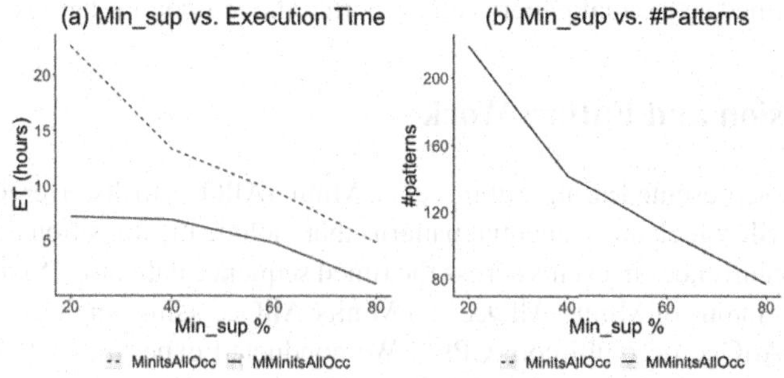

Fig. 9. Parameter study for t-drive

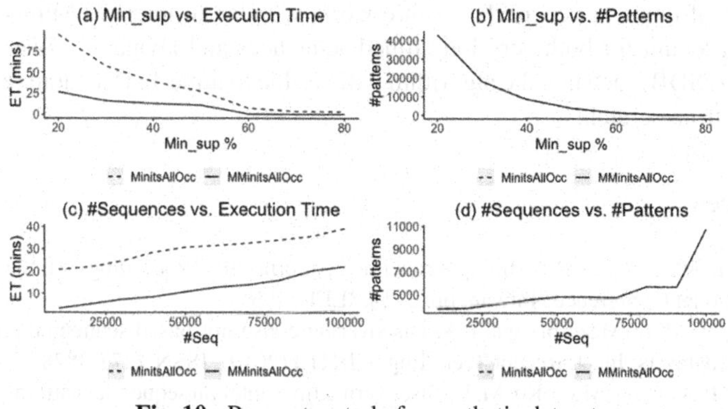

Fig. 10. Paramater study for synthetic dataset

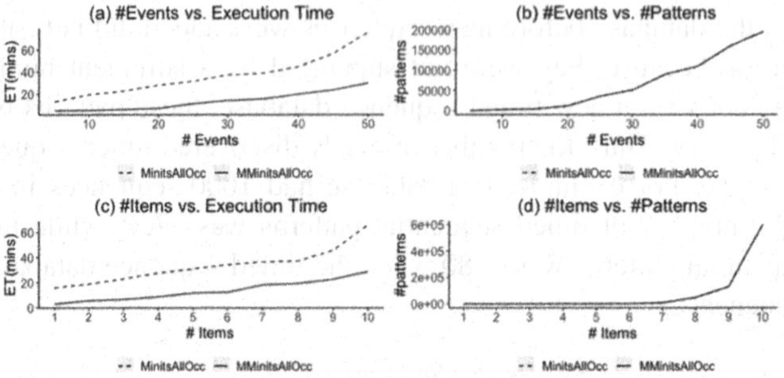

Fig. 11. Paramater study for synthetic dataset

6. *Impact of the Number of Items Per Event*
In the last experiment, we increased the number of unique items in each event. That means many new items appear in timed sequence database TSDB that lead to detecting more new timed sequential patterns. When the number of items increases, the number of possible combinations between those items to generate candidates also increases. Thus, the number of patterns increased, as shown in Fig. 11(d). Growing the length of events led to the growth of the number of candidates, which means the algorithms needed more time, as shown in Fig. 11(c), to check those events, generate candidates, and determine if they were timed sequential patterns and reported the temporal relations.

6 Conclusion and Future Work

In this paper, we presented an algorithm, called Minits-AllOcc, to discover timed sequential patterns TSP, which are sequential patterns that include the transition times between all possible occurrences in events across the timed sequence database TSDB. We implemented two versions of Minits-AllOcc: 1.) Minits-AllOcc using single-core CPUs, and 2.) MMinits-AllOcc on multi-core CPUs. We conducted experiments to compare the accuracy and execution time of the algorithms. The experiments showed that the algorithms produced accurate patterns but MMinits-AllOcc outperformed Minits-AllOcc when the dataset was large in terms of the size of TSDB, length of timed sequences, or the number of items per event. For future work, we plan to improve Minits-AllOcc to be able to account for both very long timed sequences and Dynamic Timed Sequence Database DTSDB, such that the algorithm will be able to mine TSP without re-executing everything from scratch.

References

1. Agrawal, R., Srikant, R.: Mining sequential patterns. In: Proceedings of the 11th IEEE International Conference, Taiwan, pp. 3–14. IEEE (1995)
2. AlZahrani, M.Y., Mazarbhuiya, F.A. Discovering constraint-based sequential patterns from medical datasets. Int. J. Recent Tech. Eng. (IJRTE) (2019). ISSN 2277-3878
3. Chen, Y.L., Chiang, M.C., Ko, M.T.: Discovering time-interval sequential patterns in sequence databases. Expert Syst. Appl. **25**(3), 343–354 (2003)

4. Fournier-Viger, P., et al.: The SPMF open-source data mining library version 2. In: Berendt, B., et al. (eds.) ECML PKDD 2016. LNCS (LNAI), vol. 9853, pp. 36–40. Springer, Cham (2016). https://doi.org/10.1007/978-3-319-46131-1_8
5. Fournier-Viger, P., Lin, J.C.W., Kiran, R.U., Koh, Y.S., Thomas, R.: A survey of sequential pattern mining. Data Sci. Pattern Recognit. **1**(1), 54–77 (2017)
6. Gan, W., Lin, J.C.W., Fournier-Viger, P., Chao, H.C., Yu, P.S.: A survey of parallel sequential pattern mining. ACM Trans. Knowl. Discov. Data (TKDD) **13**(3), 1–34 (2019)
7. Giannotti, F., Nanni, M., Pedreschi, D.: Efficient mining of temporally annotated sequences. In: Proceedings of the 2006 SIAM International Conference on Data Mining, pp. 348–359 (2006)
8. Giannotti, F., Nanni, M., Pinelli, F., Pedreschi, D.: Trajectory pattern mining. In: Proceedings of the 13th ACM SIGKDD International Conference on Knowledge Discovery and Data Mining, pp. 330–339 (2007)
9. Han, J., Pei, J., Mortazavi-Asl, B., Chen, Q., Dayal, U., Hsu, M.-C.: FreeSpan: frequent pattern-projected sequential pattern mining. In: Proceedings of the Sixth ACM SIGKDD International Conference on Knowledge Discovery and Data Mining, pp. 355–359 (2000)
10. Hu, Y.H., Huang, T.C.K., Yang, H.R., Chen, Y.L.: On mining multi-time-interval sequential patterns. Data Knowl. Eng. **68**(10), 1112–1127 (2009)
11. Huynh, B., Vo, B., Snasel, V.: An efficient method for mining frequent sequential patterns using multi-core processors. Appl. Intell. **46**(3), 703–716 (2017)
12. Jay, N., Herengt, G., Albuisson, E., Kohler, F., Napoli, A.: Sequential pattern mining and classification of patient path. Medinfo 1667 (2004)
13. Yuan, J., et al.: T-drive: driving directions based on taxi trajectories. In: Proceedings of the 18th SIGSPATIAL International Conference on Advances in Geographic Information Systems, GIS 2010, pp. 99–108. ACM, New York (2010)
14. Yuan, J., Zheng, Y., Xie, X., Sun, G.: Driving with knowledge from the physical world. In: The 17th ACM SIGKDD International Conference on Knowledge Discovery and Data Mining, KDD 2011. ACM, New York (2011)
15. Karsoum, S., Gruenwald, L., Barrus, C., Leal, E.: Using timed sequential patterns in the transportation industry. In: 2019 IEEE International Conference on Big Data (Big Data), pp. 3573–3582. IEEE, December 2019
16. Li, H., Zhou, X., Pan, C.: Study on GSP algorithm based on Hadoop. In: 5th IEEE International Conference Electronics Information and Emergency Communication (ICEIEC), pp. 321–324 (2015)
17. Pei, J., et al.: PrefixSpan: mining sequential patterns efficiently by prefix-projected pattern growth. In: ICCCN, p. 0215 (2001)
18. Pramono, Y.W.T.: Anomaly-based intrusion detection and prevention system on website usage using rule-growth sequential pattern analysis: case study: statistics of Indonesia (BPS) website. In: 2014 International Conference of Advanced Informatics: Concept, Theory, and Application (ICAICTA), pp. 203–208. IEEE, August 2014
19. Srikant, R., Agrawal, R.: Mining sequential patterns: generalizations and performance improvements. In: Apers, P., Bouzeghoub, M., Gardarin, G. (eds.) EDBT 1996. LNCS, vol 1057, pp. 1–17. Springer, Heidelberg (1996). https://doi.org/10.1007/BFb0014140
20. Wei, Y.-Q., Liu, D., Duan, L.-S.: Distributed prefixspan algorithm based on mapreduce. In: International Symposium on IEEE Information Technology in Medicine and Education (ITME), vol. 2, pp. 901–904 (2012)
21. Yang, H., Gruenwald, L., Boulanger, M.: A novel real-time framework for extracting patterns from trajectory data streams. In: Proceedings of the 4th ACM SIGSPATIAL International Workshop on GeoStreaming, pp. 26–32 (2013)
22. Zaki, M.: SPADE: an efficient algorithm for mining frequent sequences. Mach. Learn. **42**(1–2), 31–60 (2001)

T^3N: Harnessing Text and Temporal Tree Network for Rumor Detection on Twitter

Nikhil Pinnaparaju(✉), Manish Gupta, and Vasudeva Varma

IIIT Hyderabad, Hyderabad, India
nikhil.pinnaparaju@research.iiit.ac.in, {manish.gupta,vv}@iiit.ac.in

Abstract. Social media platforms have democratized the publication process resulting into easy and viral propagation of information. However, spread of rumors via such media often results into undesired and extremely impactful political, economic, social, psychological and criminal consequences. Several manual as well as automated efforts have been undertaken in the past to solve this critical problem. Existing automated methods are text based, user credibility based or use signals from the tweet propagation tree. We aim at using the text, user, propagation tree and temporal information jointly for rumor detection on Twitter. This involves several challenges like how to handle text variations on Twitter, what signals from user profile could be useful, how to best encode the propagation tree information, and how to incorporate the temporal signal. Our novel architecture, T^3N (Text and Temporal Tree Network), leverages deep learning based architectures to encode text, user and tree information in a temporal-aware manner. Our extensive comparisons show that our proposed methods outperform the state-of-the-art techniques by ∼7 and ∼6% points respectively on two popular benchmark datasets, and also lead to better early detection results.

1 Introduction

Social media portals provide a rich platform to share, forward, vote and review to encourage users to discuss online news. While this allows for faster and democratized publication of news, it also in-turn allows for malicious users to spread misinformation. Misinformation events have had immense economic impact in the past as evidenced by $130B stock market fluctuation due to "Barack Obama injured in explosion" rumor in 2017 [24], $4B drop in Apple market capitalization due to the "iPhone and Leopard delay email" rumor in 2008 [14], etc. Misinformation can even have criminal consequences. E.g., in 2016, a man was arrested after he walked into a popular pizza restaurant in Northwest Washington carrying an assault rifle and fired one or more shots. The man told police he had come to the restaurant to "self-investigate" an election-related rumor ("Mrs. Clinton is kidnapping, molesting and trafficking children in the back rooms of a D.C.

M. Gupta—The author is also a Principal Applied Scientist at Microsoft.

K. Karlapalem et al. (Eds.): PAKDD 2021, LNAI 12712, pp. 686–700, 2021.
https://doi.org/10.1007/978-3-030-75762-5_54

pizzeria") that spread online during her presidential campaign. Lastly, spreading rumor and sharing misinformation undermines your credibility; in the past, news reporters have been suspended or fired for sharing misinformation [28].

Manual credibility verification on websites like Snopes, Politifact, etc. is challenging and time consuming. Automatic detection of misinformation from social media posts (especially tweets) is also challenging because (1) They are very short and do not typically follows English grammar rules. (2) Even real news appears on Twitter much faster than other news media like TV new channels, which means that it is difficult to cross-check such news with other sources. (3) Since these posts pertain to very fresh news, it is difficult to fact check against relatively stale information in knowledge bases. (4) Difficult to handle text variations on Twitter, identify best way to encode the tweet propagation tree information, and incorporate the temporal signal. Lastly, it is very important to detect such misinformation early, which makes the task arduous.

Previous work on rumor detection has mainly leveraged the post content or the network structure [38]. Content based methods have focused on (1) matching facts extracted from content with knowledge base facts, or (2) textual feature engineering [2,31], or (3) image information [13]. However, textual representations prove to be insufficient for platforms where the amount of characters and therefore words is limited. Additionally, they rely on a very quickly updated knowledge base which is difficult. Hence, recently, network based approaches have been proposed. Such approaches are broadly of two types: propagation tree based [21,33,36] and user credibility based [9,12,27]. Overall, previous work has studied the importance of individual factors for predicting the credibility of social media posts.

We aim to study such factors jointly. We extract semantics from the post text content using deep recurrent models as well as Transformer models. We extract multiple features from the user profile. We combine such features along with the propagation tree structure to learn a semantic representation of the tree. Further, we learn representations of multiple temporal snapshots of the propagation tree. Representations of such snapshots are combined using another recurrent network to obtain a representation for the temporal tree. The text representation and the temporal tree representation are combined to finally predict credibility of the social media post. Figure 1 shows the architecture of our proposed system, T^3N (Text and Temporal Tree Network).

Overall, in this paper, we make the following main contributions. (1) We propose to use the text content, propagation tree, user profile as well as temporal signals for early misinformation detection on Twitter. (2) We propose a novel deep learning architecture, T^3N (Text and Temporal Tree Network), which consists of a text encoder and a temporal tree encoder to jointly model the above signals. (3) Our experiments with Twitter-15 and Twitter-16 benchmark datasets show that T^3N outperforms state-of-the-art methods by ~7 and ~6% points respectively.

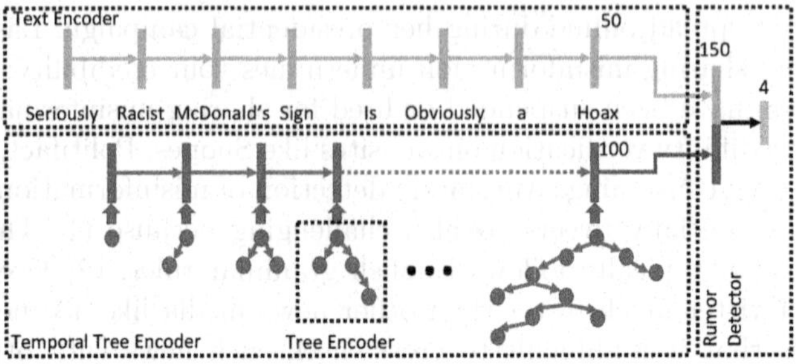

Fig. 1. T^3N System Architecture. T^3N could use a variety of text and tree encoders as discussed in Sect. 3. Here we show an instance with LSTM text encoder and temporal tree encoder.

2 Related Work

Content Based Rumor Detection: Content based methods have focused on (1) matching facts extracted from content with knowledge base facts [10], (2) textual feature engineering for style-based misinformation detection [2,3,22,23]. Fact checking can be manual or automated. Popular manual fact checking websites include Snopes, Politifact, FactCheck, HoaxSlayer, TruthOrFiction, etc. Accuracy of automated fact checking [10,26] is rather limited because they rely on a very quickly updated knowledge base for factual checking. While early studies used linguistic features [2,23], more recently, deep neural networks (DNNs) [3,21] have been explored for rumor detection. Content based representations prove to be insufficient for platforms where the amount of characters and therefore words is limited. Hence, our approach combines text with tree based signals.

Rumor Detection Using Propagation Trees: To address drawbacks of pure content based methods (following the observations made by Vosoughi et al. [32]), recently, propagation tree based methods [17,20,21,33] have been proposed. Earlier efforts used graph kernels [20,33] to capture high-order patterns differentiating different types of rumors by evaluating the similarities between their propagation tree structures. Recent DNNs like recursive neural networks (RNNs) [21], long short term memory (LSTM) networks [34], combination of recurrent and convolutional networks [17] have been explored. While it is beneficial to use propagation tree signals, it is important to carefully choose the right tree representation. We explore temporal trees.

Network Based Rumor Detection: Apart from the simple propagation trees with only user nodes, rumor detection community has also studied other richer networks – both homogeneous [12,25] as well as heterogeneous [9,12,27,36]. Heterogeneous network based rumor detection is orthogonal to the direction we follow in this work. This is mainly due to lack of benchmark heterogeneous network datasets. We plan to explore whether this line of work complements our findings, as part of future work.

Fake Profile Detection on Twitter: As another relevant but orthogonal line of work, previous studies have attempted automated fake profile detection on Twitter. There are broadly two types of work in this area: (1) methods that analyze user behavior patterns [4] or content posted [6] or both [1], and (2) methods which jointly learn the user credibility along with credibility of other node types like tweets, events, etc.

3 T^3N: A System for Rumor Detection on Twitter

3.1 Problem Definition and T^3N System Overview

Both the benchmark Twitter-15 [16] and Twitter-16 [18] datasets have instances consisting of an origin tweet, its propagation tree and a class label. Each propagation tree is represented using tree edges which link a tweet to its reply or retweet. Each node (tweet) in the tree is represented by its userid, tweet id, and post time delay (in minutes). The class label is one of the following four classes: true news, rumor, debunking of rumor and unverified news[1]. Given this dataset, we model the problem of rumor detection as a 4-class classification problem. The classifier should take a new origin tweet with its propagation tree as input and output the appropriate class label with high accuracy.

We propose a novel deep text and temporal tree encoder network (T^3N) classifier. The basic idea behind T^3N is to jointly learn a unified representation of both the tweet content as well as the tweet's propagation tree as shown in Fig. 1. It has the following four main components: (1) Text Encoder: It encodes the information from the tweet text into a latent vector. (2) Tree Encoder: It encodes the information from a propagation tree snapshot into a latent vector. (3) Temporal Tree Encoder: It encodes the information from a series of historical snapshots of the propagation tree into a latent vector. (4) Rumor Detector: It uses the learned combined text + temporal tree representation (latent vector) to predict one of the four classes. We discuss these components in detail in the remainder of this section.

3.2 Text Encoder

We use two kinds of text encoder architectures: recurrent architecture and Transformers.

RNNs like Gated Recurrent Units (GRUs) [5], LSTMs [11], bi-directional LSTMs [8] and their attention based variants have been found to be very effective in modeling text sequences across a large variety of natural language processing tasks mainly due to their representational power and effectiveness at capturing long-term dependencies. Attention allows the model to give more importance to certain set of words in the tweet while ignoring the others, effectively learning the focus points to better predict the correct rumor class for the tweet. The resultant

[1] The label "debunking of rumor" denotes a news story that tells people that a certain news story is rumorous.

tweet embedding, v_i, learned using any of these models, which is jointly learned during the training process captures the essential information from the tweet text. Figure 1 shows an attention-based recurrent text encoder.

Another way of encoding text which has become popular in the past two years is using Transformer [30] based architectures like Bidirectional Encoder Representations from Transformers (BERT) [7]. The post text sequence is prepended with a "CLS" token. The representation C for the "CLS" token from the last encoder layer is used as the tweet text embedding. We also finetune the pre-trained model using labeled training data for the rumor prediction task.

3.3 Tree Encoder

Every propagation tree consists of the origin tweet as the root node. Replies and retweets of a tweet are represented using its children. Each node in the tree takes a user representation as input corresponding to the user who replied to or retweeted the parent tweet node, like in [17]. Tai et al. [29] extended regular linear LSTMs to child sum Tree-LSTMs. These allow for tree structured data (be it parse trees or propagation trees) to be trained without the loss of the data's inherent structure. While the standard LSTM composes its hidden state from the input at the current time step and the hidden state of the LSTM unit in the previous time step, the tree-structured LSTM, or Tree-LSTM, composes its state from an input vector and hidden states of arbitrarily many child units.

Given a tree, if we denote children of a node j by $C(j)$, the Child-Sum tree-LSTM transition equations can be written as follows.

$$\tilde{h}_j = \sum_{k \in C(j)} h_k; \quad i_j = \sigma(W_i x_j + U_i \tilde{h}_j + b_i); o_j = \sigma(W_o x_j + U_o \tilde{h}_j + b_o) \tag{1}$$

$$f_{jk} = \sigma(W_f x_j + U_f h_k + b_f); \quad c_j = i_j \circ tanh(W_c x_j + U_c \tilde{h}_j + b_c) + \sum_{k \in C(j)} f_{jk} \circ c_k \tag{2}$$

$$h_j = o_j \circ tanh(c_j) \tag{3}$$

Here, h_k and c_k denote the hidden layer and the cell state outputs of the kth child respectively. x_j denotes the user profile vector for the node j. However, this method inspired by Tai et al. [29] does not handle temporal aspects of the propagation tree. Hence, we propose novel temporal-aware tree encoder methods in Sects. 3.4 and 3.5.

3.4 Decayed Tree Encoder

Typically, the first few tweets in a rumor event are the most malicious in nature. Later tweets are often copies of the initial tweets. Also, intuitively, in a propagation tree for a rumor, the first few users (closer to the root of the tree) are malicious while users at lower levels (who reply to or retweet the original tweet) are naïve (who unintentionally engage in rumor propagation). Hence, we propose a modified Tree-LSTM variant that incorporates a temporal decay factor into our model. The variant provides less weight to propagation tree nodes that are sharing this misinformation at a later point of time. In some ways our time decay idea is also inspired by a similar notion as described in [20].

Each node k in the tree is associated with a timestamp $ts(k)$. We compute the decay factor corresponding to the node by comparing its timestamp with that of the original tweet (i.e., the root node). The modified Tree-LSTM transition equations can then be written as follows. Equations for $\tilde{h}_j$, i_j, o_j, c_j, h_j remain the same as in Eqs. 1 to 3.

$$td_k = ts(k) - ts(root); \quad f_{jk} = \sigma(g(td_k) * (W_f x_j + U_f h_k + b_f)) \tag{4}$$

We experimented with two different forms of decay function g: (1) Linear: $g(td_k) = td_k$ (2) Exponential: $g(td_k) = e^{-td_k}$. Exponential decay function performed better compared to the linear function and hence we report results using exponential one.

3.5 Temporal Tree Encoder

We hypothesize that the order in which the nodes get added to a propagation tree depends on whether it contains true information or a rumor. Hence, we wish to learn a vector representation across multiple snapshots of the tree using a temporal tree encoder discussed as follows. Consider a tree with N nodes where each node has an associated timestamp. Such a tree can be represented as a time series of N snapshots $S_1, S_2, ..., S_N$ where S_{i+1} has exactly one node ($(i+1)^{th}$) more than S_i and $(i+1)^{th}$ node has a timestamp larger than all of the nodes from 1 to i and smaller than all of the nodes from $i+2$ to N.

For each such tree snapshot, we use a tree encoder or a decayed tree encoder to get a latent representation. Parameters are shared across all the trees in the sequence. The temporal sequence of such latent representations across all snapshots S_1 to S_N are then combined using an LSTM. The last hidden layer output from the LSTM can then be considered as a latent representation for the temporal tree. Note that we use an LSTM rather than a BiLSTM because at test time, we will not have future tree snapshots.

If N is large, memory needed to learn the temporal tree representation could be large. Also, the number of computations needed are high. Hence, to generate the temporal tree representation we take $M+1$ equidistant snapshots from within the overall sequence, and then design an LSTM with just $M+1$ inputs. The gap between each unit in the LSTM input sequence is N/M. Note that $M=1$ is equivalent to learning from all snapshots of the temporal tree sequence. We experiment with different values of M and present results in Sect. 4.

3.6 Putting It All Together

Overall, our proposed T^3N system first encodes the tweet text to obtain a text representation. Next, it creates a $M+1$-sized temporal tree sequence. Each tree is encoded using a tree encoder or a decayed tree encoder. Further, these individual $M+1$ tree representations are combined using a BiLSTM to obtain a temporal tree representation. Finally, the text and the temporal tree representations are concatenated, and connected to an output softmax layer with four neurons corresponding to the four class labels.

We use the cross entropy loss. The parameters of the text encoder, the temporal tree encoder and the misinformation detector are all initialized randomly and trained jointly using back propagation. We also experiment with separate pre-training and fine-tuning phases. In this way of training, we first train the text encoder using the labeled training data by directly connecting the output of the text encoder with an output layer in a fully connected manner. Next, we separately pre-train the tree encoder using the labeled training data by directly connecting the output of the temporal tree encoder with an output layer in a fully connected manner. Once the separate pre-training of text encoder and the temporal tree encoder is done, we use the end-to-end architecture shown in Fig. 1 and fine-tune all the weights using the same labeled training data. We observed that such pre-training followed by fine-tuning provided better results compared to training end-to-end from scratch.

4 Experiments

4.1 Datasets, Experimental Settings and Baselines

Following the work in [2,15,17,19–21,33,35,37], we experiment with two popular benchmark datasets: Twitter-15 [16] and Twitter-16 [18] which are also the only two datasets that provide temporal network information along with the tweets. We eliminated (1) repeated edges from tree files (2) nodes that appear as children but appear before parent from the original datasets, and (3) nodes with negative post time delay). Table 1 shows statistics of the two datasets. Further Figs. 2, 3, 4, 5 show distributions of important parameters of the two datasets. Note that the URLs in the tweets have been replaced by the keyword URL in the original dataset. The original dataset contains user IDs but not their profile information. Hence, we crawled user profiles using the Twitter API[2]. From these profiles we extract user profile attributes which could be correlated to their credibility. These features include length of the user profile description in words, length of username in characters, followers count, friends count, statuses count, registration age, whether the user is verified and whether the user is geo enabled. These features were also proposed in [17]. This feature vector of eight features is used as the data input vector for the tree-LSTMs in the tree encoder. The datasets can be downloaded from here[3].

[2] https://developer.twitter.com/en/docs/accounts-and-users/follow-search-get-users/overview.

[3] https://www.dropbox.com/s/7ewzdrbelpmrnxu/rumdetect2017.zip?dl=0.

Table 1. Statistics of the Twitter-15 and Twitter-16 Datasets

Statistic	#stories	#true news	#rumors	#debunking of rumor	#unverified news	#users	#posts	Avg Tweet length	Avg Tree Depth	Avg #tree nodes
Twitter-15	1490	374	370	372	374	276663	331612	15.48	4.74	405.1
Twitter-16	818	205	205	207	201	173487	204820	15.19	4.89	428.9

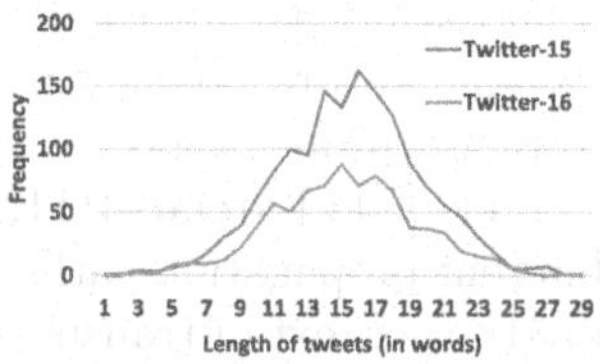

Fig. 2. Length distribution of tweets (in words)

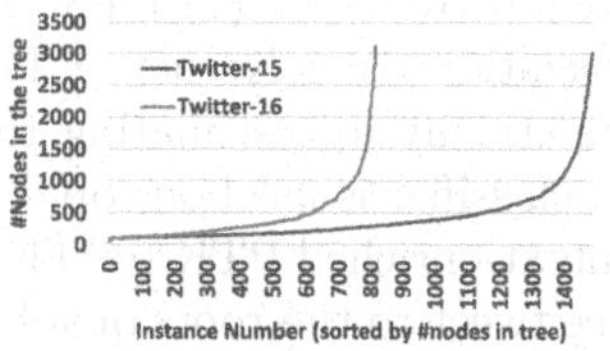

Fig. 3. Distribution of #nodes in the propagation trees

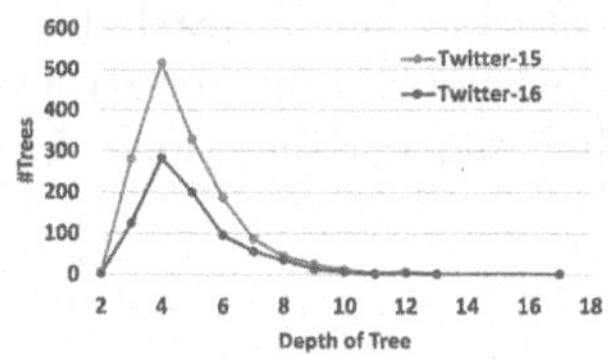

Fig. 4. Depth distribution of propagation trees

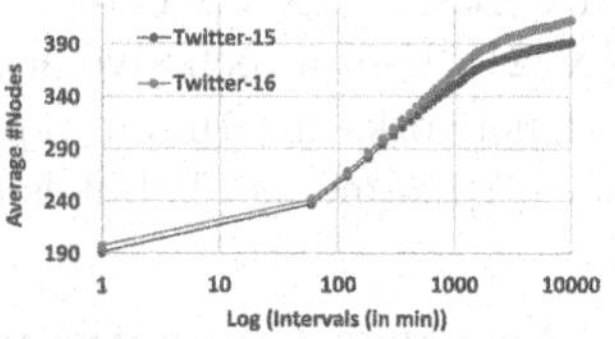

Fig. 5. Distribution of arrival time of nodes in the propagation trees (up to 1 week)

For our recurrent text encoders (GRUs and LSTMs), we used GloVe 100D embeddings and Adagrad optimizer with learning rate of 0.01. For BERT, we initialized using bert-base-uncased with standard 768D hidden size, and used AdamW optimizer with learning rate of 2e−5. No other specific pre-processing steps were performed. We train all models for 10 epochs. For other settings for reproducibility, please refer to our code[4]. We perform 4-fold experiments and report the average.

Baselines: We compare with the following 9 previously proposed traditional machine learning methods: (1) SVM-BOW: Linear Support Vector Machines (SVM) with bag-of-words features extracted from the text in each tree. (2) DTC [2]: The information credibility model using a Decision-Tree Classifier based

[4] https://www.dropbox.com/sh/nw14d4qd3zhm3mb/AAB843fUKQIXxVqsWnrRW5mfa?dl=0.

on various hand-crafted statistical features of the tweets. (3) SVM-RBF [35]: Same as (2) but uses SVM classifier with Radial Basis Function (RBF) kernel. (4) SVM-TS [19]: A linear SVM classification model that uses time-series to model the variation of a set of hand-crafted features. (5) DTR [37]: A Decision-Tree-based Ranking method to identify trending rumors. (6) RFC [15]: A Random Forest Classifier using three parameters to fit the temporal properties and an extensive set of hand-crafted features related to the user, linguistic and structure characteristics. (7) PTK [20]: Used kernel-based data-driven method called Propagation Tree Kernel (PTK) to generate relevant features (i.e., subtrees) automatically for estimating the similarity between two propagation trees. An SVM classifier is used on top of such features. (8) cPTK [20]: Extends PTK into a context-enriched PTK (cPTK) by considering different propagation paths from source tweet to the roots of subtrees, which capture the context of transmission. (9) SVM-HK [33]: An SVM classifier using features derived from propagation structures with Hybrid kernel.

Further, we compare with the following 5 previously proposed deep learning methods. (1) PPC_RNN+CNN [17]: Uses propagation path construction and transformation, RNN and CNN-based propagation path representation and propagation path classification. (2) PPC_RNN [17]: Same as PPC_RNN+CNN where CNN based representation is not used. (3) PPC_CNN [17]: Same as PPC_RNN+CNN where RNN based representation is not used. (4) BU-RvNN [21]: Uses a recursive neural model based on a bottom-up tree-structured neural networks for rumor representation learning and classification. (5) TD-RvNN [21]: Same as (4) but is top-down.

4.2 Accuracy Comparison

Results in Table 2 show that cPTK is the best traditional ML method across both the datasets. Multiple DL methods are better than the ML methods. Our proposed method, T^3N, outperforms all the other methods by a large margin. This particular variant of our system used BERT as the text encoder and Decay Temporal Tree encoder for encoding the propagation tree. Also, our method performs well across all the classes for both the datasets, except for the "rumor" class for Twitter-16 where the PPC_RNN+CNN is better. We believe this is because for Twitter-16, all of our text encoders perform relatively weakly for the "rumor" class (Table 3). On the other hand, this is not a concern for PPC_RNN+CNN since it uses a CNN also.

4.3 Ablation Studies

To understand the degree of contribution of various components towards rumor detection, we perform a series of ablation tests: using only text encoder, only tree encoder and multiple combinations of various text and tree encoders in Tables 3, 4 and 5 resp. In Table 3, we compare the accuracy obtained using multiple text encoders: GRU and LSTM, and their bi-directional variants. We also tried adding attention to these models, but it did not lead to significantly

Table 2. Accuracy and class-wise F1 comparison across various methods for Twitter-15 and Twitter-16 datasets. Last row is our proposed method. (T = true, R = Rumor, D = debunk, U = unverified)

	Model	Twitter-15					Twitter-16				
		Acc.	T F1	R F1	D F1	U F1	Acc.	T F1	R F1	D F1	U F1
Traditional ML	SVM-BOW	0.548	0.564	0.524	0.582	0.512	0.585	0.553	0.556	0.655	0.578
	DTC	0.454	0.733	0.355	0.317	0.415	0.465	0.643	0.393	0.419	0.403
	SVM-RBF	0.318	0.455	0.037	0.218	0.225	0.321	0.423	0.085	0.419	0.037
	SVM-TS	0.544	0.796	0.472	0.404	0.483	0.574	0.755	0.420	0.571	0.526
	DTR	0.409	0.501	0.311	0.364	0.473	0.414	0.394	0.273	0.630	0.344
	RFC	0.565	0.810	0.422	0.401	0.543	0.585	0.752	0.415	0.547	0.563
	cPTK	0.750	0.804	0.698	0.765	0.733	0.732	0.740	0.709	0.836	0.686
	PTK	0.710	0.825	0.685	0.688	0.647	0.722	0.784	0.690	0.786	0.644
	SVM-HK	0.493	0.650	0.439	0.342	0.336	0.511	0.648	0.434	0.473	0.451
Deep learning	PPC_RNN	0.811	0.759	0.842	0.765	0.787	0.842	0.809	0.865	0.836	0.839
	PPC_CNN	0.803	0.737	0.835	0.751	0.775	0.847	0.812	0.871	0.833	0.841
	PPC_RNN+CNN	0.842	0.811	0.875	0.790	0.818	0.863	0.820	**0.898**	0.837	0.843
	BU-RvNN	0.708	0.695	0.728	0.759	0.653	0.718	0.723	0.712	0.779	0.659
	TD-RvNN	0.723	0.682	0.758	0.821	0.654	0.737	0.662	0.743	0.835	0.708
	Best T^3N	**0.912**	**0.905**	**0.915**	**0.912**	**0.914**	**0.927**	**0.957**	0.875	**0.970**	**0.909**

Table 3. Accuracy and class-wise F1 values for methods which use only tweet text information. (T = true, R = Rumor, D = debunk, U = unverified)

Model	Twitter-15					Twitter-16				
	Acc.	T F1	R F1	D F1	U F1	Acc.	T F1	R F1	D F1	U F1
GRU	0.6040	0.6795	0.6438	0.5756	0.4894	0.6332	0.7086	0.5119	0.6870	0.4723
LSTM	0.5637	0.6881	0.5488	0.5445	0.4631	0.6528	0.7351	0.5169	0.6804	0.5127
Bi-GRU	0.5839	0.6833	0.6098	0.5659	0.4525	0.6235	0.7325	0.5146	0.6330	0.4508
Bi-LSTM	0.5496	0.6914	0.5430	0.5276	0.4223	0.6039	0.7448	0.4815	0.6018	0.4479
BERT	**0.7795**	**0.8342**	**0.7530**	**0.7554**	**0.7721**	**0.7744**	**0.8992**	**0.5553**	**0.7383**	**0.7278**

better results. We observe that bidirectional models do not lead to better results either. We also present the accuracy using BERT. Across both datasets, BERT clearly performs much better than any of the recurrent models.

In Table 4, we compare the accuracy obtained using multiple tree encoders: Standard, Decay, Temporal, Decay Temporal. We also vary the number of trees as 1, 5, 10, 15 and 20. Overall, we observe that temporal tree representation gives better results compared to the standard one (where we just use the last snapshot of the tree). We also observe that as we increase the number of trees, typically the accuracy increases slightly. It is also important to note that the accuracy values obtained using tree only encoders are typically smaller than those using text only encoders. This could be because the text content has relatively much stronger signals compared to the tree.

In Table 5, we compare the accuracy obtained using multiple text and tree encoder combinations. For temporal and decay temporal tree encoders, we fixed

Table 4. Accuracy and class-wise F1 values for methods which use only propagation tree information. (T = true, R = Rumor, D = debunk, U = unverified)

Tree Encoder	# Trees	Twitter-15					Twitter-16				
		Acc.	T F1	R F1	D F1	U F1	Acc.	T F1	R F1	D F1	U F1
Standard	1	0.386	0.086	0.618	0.000	0.444	0.507	0.120	0.576	0.520	0.592
Decay	1	0.400	0.057	**0.664**	0.021	0.443	0.527	0.000	0.520	**0.587**	0.661
Temporal	5	0.437	0.475	0.610	0.378	0.203	0.522	0.077	0.537	0.544	0.649
Decay Temporal	5	0.458	0.462	0.621	0.376	0.358	0.571	0.290	0.644	0.579	0.640
Temporal	10	0.424	0.424	0.578	0.335	0.258	0.551	0.290	0.487	0.585	**0.671**
Decay Temporal	10	0.445	0.495	0.593	0.391	0.288	0.576	0.310	**0.673**	0.581	0.612
Temporal	15	0.464	**0.568**	0.605	0.352	0.113	0.517	0.246	0.546	0.496	0.639
Decay Temporal	15	**0.485**	0.514	0.372	0.541	0.540	0.551	0.478	0.500	0.520	0.644
Temporal	20	0.472	0.398	0.318	**0.587**	**0.578**	0.551	0.369	0.584	0.509	0.656
Decay Temporal	20	0.477	0.483	0.335	0.559	0.542	**0.576**	**0.500**	0.592	0.538	0.619

Table 5. Accuracy and class-wise F1 values for various combinations of the text and tree encoders, i.e., variants of the T^3N system. (T = true, R = Rumor, D = debunk, U = unverified)

Tree Encoder	Text Encoder	Twitter-15					Twitter-16				
		Acc.	T F1	R F1	D F1	U F1	Acc.	T F1	R F1	D F1	U F1
Standard	LSTM	0.625	0.699	0.549	0.559	0.710	0.629	0.745	0.547	0.633	0.596
Standard	GRU	0.625	0.701	0.568	0.597	0.652	0.659	0.764	0.598	0.674	0.621
Standard	BiLSTM	0.585	0.675	0.495	0.515	0.674	0.659	0.674	0.672	0.639	0.646
Standard	BiGRU	0.652	0.756	0.577	0.631	0.653	0.590	0.681	0.609	0.568	0.500
Standard	BERT	0.509	0.000	0.212	0.866	0.494	0.683	0.774	0.598	0.687	0.685
Temporal	LSTM	0.582	0.683	0.448	0.554	0.648	0.629	0.605	0.696	0.679	0.679
Temporal	GRU	0.619	0.696	0.561	0.570	0.647	0.683	0.725	0.631	0.685	0.703
Temporal	BiLSTM	0.595	0.573	0.467	0.688	0.679	0.600	0.651	0.569	0.489	0.684
Temporal	BiGRU	0.649	0.696	0.555	0.647	0.706	0.659	0.651	0.696	0.631	0.654
Temporal	BERT	0.895	0.913	0.899	0.876	0.902	0.751	0.860	0.905	0.571	0.667
Decay Temporal	LSTM	0.614	0.675	0.520	0.541	0.714	0.634	0.777	0.511	0.547	0.698
Decay Temporal	GRU	0.646	0.734	0.558	0.597	0.696	0.683	0.714	0.638	0.661	0.721
Decay Temporal	BiLSTM	0.619	0.674	0.570	0.591	0.639	0.663	0.667	0.633	0.626	0.726
Decay Temporal	BiGRU	0.614	0.651	0.532	0.590	0.688	0.644	0.607	0.696	0.661	0.600
Decay Temporal	BERT	**0.912**	**0.905**	**0.915**	**0.912**	**0.914**	**0.927**	**0.957**	**0.875**	**0.970**	**0.909**

the number of trees to 20. For standard tree encoder, of course, the number of trees is 1, i.e., the latest tree snapshot. As expected, results for text+tree methods (Table 5) are far better than results with just text (Table 3) or just tree (Table 4). The best combination is Decay Temporal tree encoder and BERT based text encoder. It is interesting to note that decay temporal usually performs better than temporal tree encoder in most cases. Further, we observed that the best model is confused the most between true and debunking of rumor classes. Intuitively, these classes are very similar and hence this is expected (Table 6).

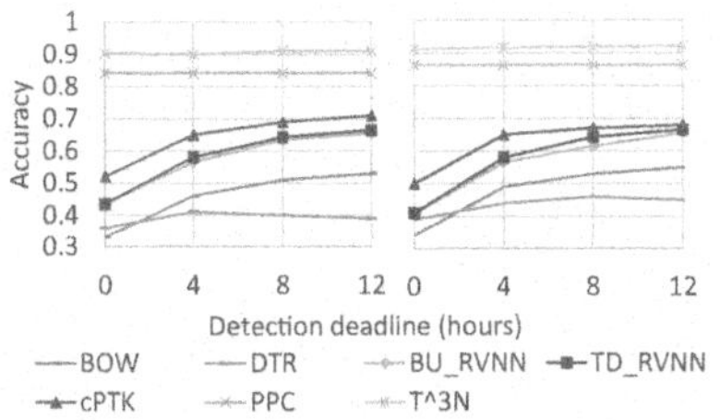

Fig. 6. Accuracy versus detection deadline (hours). Left: Twitter-15 and right: Twitter-16.

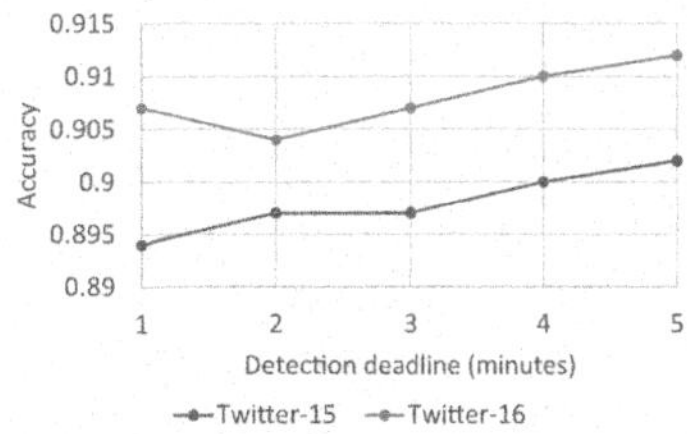

Fig. 7. Accuracy versus detection deadline for our best T^3N method (minutes).

Table 6. Examples where the joint text+tree model provided correct prediction, and one of text-only or tree-only encoders provided correct prediction.

Dataset	Tweet Text	Tree only classifier prediction	Text only classifier prediction	Actual Label	Text contribution in Best Model	Tree contribution in Best Model
Twitter-15	ca kkk grand wizard endorses @hillaryclinton #neverhillary #trump2016 URL	Unverified	Rumor	Unverified	0.820	0.180
Twitter-15	Florida woman pays $20,000 for third breast — URL URL	Debunk	Rumor	Rumor	0.934	0.066
Twitter-15	One person dead, many taken to hospital after shootings, stabbing at denver coliseum, police say. URL	Rumor	Debunk	Debunk	0.922	0.078
Twitter-15	Reporter charlo green quit on air to start a marijuana business URL will quitting on tv pay off? URL	Debunk	True	True	0.925	0.075
Twitter-16	Chick-fil-a to open on sundays URL	Rumor	Unverified	Rumor	0.844	0.156
Twitter-16	Dna confirms hakeem from 'empire' is jay-z's biological son [read details URL URL]	Rumor	Debunk	Rumor	0.633	0.367
Twitter-16	72 dhs employees on terrorist watch list URL	Unverified	Rumor	Rumor	0.907	0.093
Twitter-16	#backtothefuture fans, it's october 21, 2015 – the future is finally here! URL URL	Debunk	Rumor	Rumor	0.901	0.099
Twitter-16	#ripnathancirillo rt @ABC: soldier killed at war memorial identified as cpl. nathan cirillo #ottawashooting URL	Rumor	True	True	0.968	0.032

4.4 Early Detection Results

Further, we wanted to check how early can we predict rumorous tweets. Figure 6 shows the variation in accuracy wrt. time from original tweets in hours for various methods for the two datasets. We have ensured that we take the data points only until the time to detection. Note that our method shows an almost flat curve in

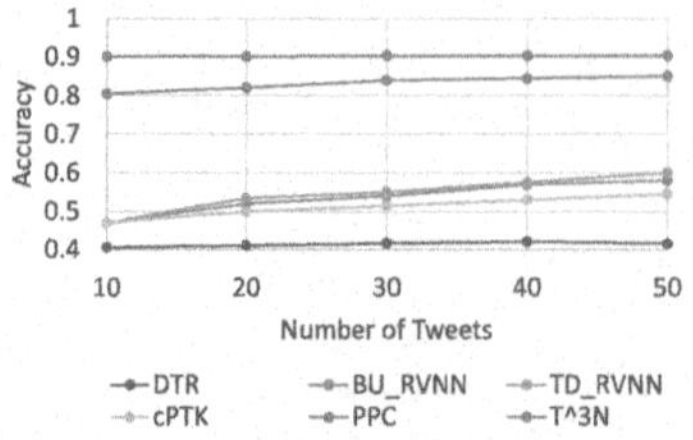

Fig. 8. Accuracy versus #tweets (Twitter-15).

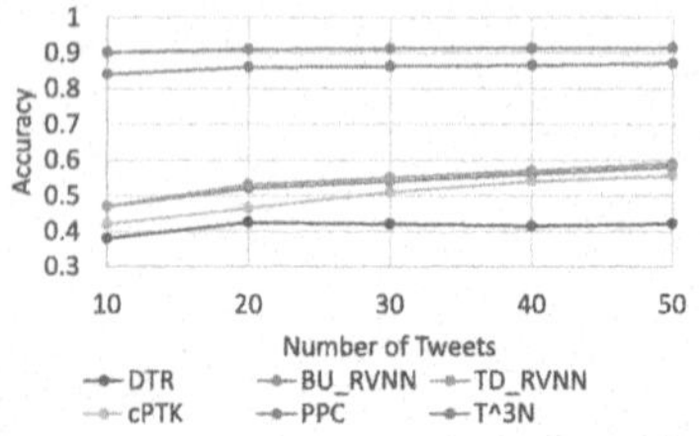

Fig. 9. Accuracy versus #tweets (Twitter-16).

Fig. 6 since it reaches a high accuracy within the first 5 min of the original tweet itself (as shown in Fig. 7). We believe this is because of the best use of text, network and user profile data in our method. Although the accuracy using just the text encoder (at time = 0) is relatively lower, the tree generated within 1–2 min is good enough for accuracy to be high. The best baseline is PPC. We also plot the variation in accuracy wrt. #tweets after the original tweet in Figs. 8 and 9 for the two datasets respectively. Again, we observe that our proposed method reaches high accuracy within a very few number of tweets.

5 Conclusion

In this paper, we discussed the critical problem of rumor detection on Twitter. To the best of our knowledge this is the first work to explore the application of Transformer based models for this task. Besides using a BERT based text encoder, our system T^3N couples it with a temporal tree based encoder. Using two datasets and extensive comparisons with numerous previously proposed methods, we show the efficacy of our method.

References

1. Cai, C., Li, L., Zeng, D.: Detecting social bots by jointly modeling deep behavior and content information. In: CIKM, pp. 1995–1998 (2017)
2. Castillo, C., Mendoza, M., Poblete, B.: Information credibility on twitter. In: WWW, pp. 675–684 (2011)
3. Chen, T., Li, X., Yin, H., Zhang, J.: Call attention to rumors: deep attention based recurrent neural networks for early rumor detection. In: Ganji, M., Rashidi, L., Fung, B., Wang, C. (eds.) Trends and Applications in Knowledge Discovery and Data Mining (PAKDD 2018). Lecture Notes in Computer Science, vol. 11154, pp. 40–52. Springer, Cham (2018). https://doi.org/10.1007/978-3-030-04503-6_4
4. Cheng, J., Bernstein, M., Danescu-Niculescu-Mizil, C., Leskovec, J.: Anyone can become a troll: causes of trolling behavior in online discussions. In: CSCW, pp. 1217–1230 (2017)
5. Cho, K., et al.: Learning phrase representations using rnn encoder-decoder for statistical machine translation. arXiv:1406.1078 (2014)

6. Chu, Z., Gianvecchio, S., Wang, H., Jajodia, S.: Detecting automation of twitter accounts: are you a human, bot, or cyborg? Trans. Dependable Secure Comput. **9**(6), 811–824 (2012). https://doi.org/10.1109/TDSC.2012.75
7. Devlin, J., Chang, M.W., Lee, K., Toutanova, K.: Bert: pre-training of deep bidirectional transformers for language understanding. arXiv:1810.04805 (2018)
8. Graves, A., Jaitly, N., Mohamed, A.: Hybrid speech recognition with deep bidirectional lstm. In: Workshop on Automatic Speech Recognition and Understanding, pp. 273–278 (2013)
9. Gupta, M., Zhao, P., Han, J.: Evaluating event credibility on twitter. In: ICDM, pp. 153–164 (2012)
10. Hassan, N., Arslan, F., Li, C., Tremayne, M.: Toward automated fact-checking: detecting check-worthy factual claims by claimbuster. In: KDD, pp. 1803–1812 (2017)
11. Hochreiter, S., Schmidhuber, J.: Long short-term memory. Neural Comput. **9**(8), 1735–1780 (1997)
12. Jin, Z., Cao, J., Zhang, Y., Luo, J.: News verification by exploiting conflicting social viewpoints in microblogs. In: AAAI (2016)
13. Khattar, D., Goud, J.S., Gupta, M., Varma, V.: MVAE: multimodal variational autoencoder for fake news detection. In: WWW, pp. 2915–2921 (2019)
14. Krazit, T.: Engadget sends apple stock plunging on iphone rumor. CNET.com, May 16 (2007)
15. Kwon, S., Cha, M., Jung, K.: Rumor detection over varying time windows. PLoS ONE **12**(1), e0168344 (2017)
16. Liu, X., Nourbakhsh, A., Li, Q., Fang, R., Shah, S.: Real-time rumor debunking on twitter. In: CIKM, pp. 1867–1870 (2015)
17. Liu, Y., Wu, Y.F.B.: Early detection of fake news on social media through propagation path classification with recurrent and convolutional networks. In: AAAI, pp. 354–361 (2018)
18. Ma, J., et al.: Detecting rumors from microblogs with recurrent neural networks. In: AAAI (2016)
19. Ma, J., Gao, W., Wei, Z., Lu, Y., Wong, K.F.: Detect rumors using time series of social context information on microblogging websites. In: CIKM, pp. 1751–1754 (2015)
20. Ma, J., Gao, W., Wong, K.F.: Detect rumors in microblog posts using propagation structure via kernel learning. In: ACL, pp. 708–717 (2017)
21. Ma, J., Gao, W., Wong, K.F.: Rumor detection on twitter with tree-structured recursive neural networks. In: ACL, pp. 1980–1989 (2018)
22. Ma, J., Gao, W., Wong, K.F.: Detect rumors on twitter by promoting information campaigns with generative adversarial learning. In: WWW, pp. 3049–3055. ACM (2019)
23. Popat, K., Mukherjee, S., Strötgen, J., Weikum, G.: Credibility assessment of textual claims on the web. In: CIKM, pp. 2173–2178 (2016)
24. Rapoza, K.: Can 'fake news' impact the stock market? by Forbes (2017)
25. Rath, B., Gao, W., Ma, J., Srivastava, J.: From retweet to believability: utilizing trust to identify rumor spreaders on twitter. In: ASONAM, pp. 179–186. ACM (2017)
26. Shi, B., Weninger, T.: Discriminative predicate path mining for fact checking in knowledge graphs. Knowl. Based Syst. **104**, 123–133 (2016)
27. Shu, K., Wang, S., Liu, H.: Exploiting tri-relationship for fake news detection (2017). arXiv:1712.07709

28. Steel, E., Somaiya, R.: Brian williams suspended from NBC for 6 months without pay. The New York Times (2015)
29. Tai, K.S., Socher, R., Manning, C.D.: Improved semantic representations from tree-structured long short-term memory networks. In: ACL, pp. 1556–1566 (2015)
30. Vaswani, A., et al.: Attention is all you need. In: NIPS, pp. 5998–6008 (2017)
31. Volkova, S., Shaffer, K., Jang, J.Y., Hodas, N.: Separating facts from fiction: linguistic models to classify suspicious and trusted news posts on twitter. In: ACL, pp. 647–653 (2017)
32. Vosoughi, S., Roy, D., Aral, S.: The spread of true and false news online. Science **359**(6380), 1146–1151 (2018)
33. Wu, K., Yang, S., Zhu, K.Q.: False rumors detection on sina weibo by propagation structures. In: ICDE, pp. 651–662 (2015)
34. Wu, L., Liu, H.: Tracing fake-news footprints: characterizing social media messages by how they propagate. In: WSDM, pp. 637–645 (2018)
35. Yang, F., Liu, Y., Yu, X., Yang, M.: Automatic detection of rumor on sina weibo. In: KDD Workshop on Mining Data Semantics, pp. 1–7 (2012)
36. Zhang, J., Cui, L., Fu, Y., Gouza, F.B.: Fake news detection with deep diffusive network model (2018). arXiv:1805.08751
37. Zhao, Z., Resnick, P., Mei, Q.: Enquiring minds: early detection of rumors in social media from enquiry posts. In: WWW, pp. 1395–1405 (2015)
38. Zhou, X., Zafarani, R.: Fake news: a survey of research, detection methods, and opportunities (2018). arXiv:1812.00315

AngryBERT: Joint Learning Target and Emotion for Hate Speech Detection

Md Rabiul Awal[1(✉)], Rui Cao[2(✉)], Roy Ka-Wei Lee[3(✉)], and Sandra Mitrović[4(✉)]

[1] University of Saskatchewan, Saskatoon, SK, Canada
mda219@usask.ca
[2] Singapore Management University, Singapore 188065, Singapore
ruicao.2020@phdcs.smu.edu.sg
[3] Singapore University of Technology and Design, Singapore 487372, Singapore
roy_lee@sutd.edu.sg
[4] Dalle Molle Institute for Artificial Intelligence, Lugano, Switzerland
sandra.mitrovic@idsia.ch

Abstract. Automated hate speech detection in social media is a challenging task that has recently gained significant traction in the data mining and Natural Language Processing community. However, most of the existing methods adopt a supervised approach that depended heavily on the annotated hate speech datasets, which are imbalanced and often lack training samples for hateful content. This paper addresses the research gaps by proposing a novel multitask learning-based model, AngryBERT, which jointly learns hate speech detection with sentiment classification and target identification as secondary relevant tasks. We conduct extensive experiments to augment three commonly-used hate speech detection datasets. Our experiment results show that AngryBERT outperforms state-of-the-art single-task-learning and multitask learning baselines. We conduct ablation studies and case studies to empirically examine the strengths and characteristics of our AngryBERT model and show that the secondary tasks are able to improve hate speech detection.

Keywords: Hate speech detection · Social media · Multitask learning

1 Introduction

Motivation. The sharp increase in online hate speeches has raised concerns globally as the spread of such toxic content and misbehavior have not only sowed discord among individuals or communities online but also resulted in violent hate crimes. Therefore, it is a pressing issue to detect and curb hate speech in online social media. Researchers have proposed many traditional and deep learning hate speech classification methods to detect hate speeches in online social media automatically [9]. Specifically, the existing deep learning methods have achieved promising performance in the hate speech detection task [5]. However, most of these supervised methods depended heavily on the annotated

K. Karlapalem et al. (Eds.): PAKDD 2021, LNAI 12712, pp. 701–713, 2021.
https://doi.org/10.1007/978-3-030-75762-5_55

hate speech datasets, which are imbalanced and often lack training samples for hateful content [1]. A potential solution to address the challenges of imbalanced datasets is to perform data augmentation for the class with fewer training samples [4]. Nevertheless, the existing data augmentation methods have shown limited improvement in hate speech detection.

Research Objectives. In this paper, we adopt a different approach to address the research gaps. We propose a novel multitask learning-based model, AngryBERT[1], which jointly learns hate speech detection with secondary relevant tasks. Multitask learning (MTL) [20] is a machine learning paradigm that aims to leverage useful information in multiple related tasks to help improve the generalization performance of all the tasks. Earlier studies have shown that MTL improved the performance of text classification tasks even when training with inadequate samples [20]. Similarly, the intuition of our AngryBERT model is that the auxiliary datasets from the secondary relevant tasks supplement the limited hateful samples of the datasets used for the main hate speech detection task. Specifically, we utilize emotion classification [13] and hateful target identification [8,19] as the secondary tasks in our proposed model. Emotion classification is a relevant task as previous studies have demonstrated that sentiments are useful features in hate speech classification [5,9]. Hateful target identification is an extension to the hate speech detection task where it aims to identify the target group or individual victim of the hateful content. Another key component in our AngryBERT model is the BERT transformer model [7], which is fine-tuned and used as the layer to share knowledge across various tasks. To the best of our knowledge, AngryBERT is the first model that uses a pre-trained and fine-tuned language model in a MTL framework for hate speech detection.

Contributions. We summarize our paper contribution as follows: (i) We propose a novel MTL and BERT-based model call AngryBERT, which jointly learns hate speech detection with secondary relevant tasks. (ii) We conduct extensive experiments on three commonly-used hate speech detection datasets. Our experiment results show that AngryBERT outperforms the state-of-the-art single-task and multitask baselines in hate speech detection. (iii) We identify case studies to demonstrate that AngryBERT is able to detect hate speeches accurately and identify the target of the hate speech and the emotion expressed. This showcases AngryBERT's potential to provide some form of explainability to the hate speech detection task.

2 Related Work

In this section, we reviewed two groups of literature relevant to our study, namely, (i) existing studies on automated hate speech detection and (ii) multitask learning (MTL) for natural language processing (NLP) tasks.

Automatic detection of hate speech has received considerable attention from data mining, information retrieval, and NLP research communities. Earlier works

[1] Code implementation: https://gitlab.com/bottle_shop/safe/angrybert.

have explored hand-crafted and canonical NLP features for automatic hate detection [6,9,16,17]. In recent years, researchers have proposed deep learning methods to extract latent features more effectively for hate speech detection [3,9,14,21]. Most of these methods adopt a supervised approach that heavily depends on labeled datasets for training, which is a challenge as existing hate speech datasets are highly imbalanced and lack training examples for hateful content.

MTL is a popular machine learning paradigm that has been explored and applied in various NLP problems, such as text classification [11,12], etc. MTL has also been applied to abusive speech detection [15,18]. Waseem et al. [18] proposed a fully-shared MTL model, which all tasks utilize the same fully shared features, to performed hate speech detection on three hate speech datasets. Unlike [18], our proposed AngryBERT model adopts the shared-private scheme, which model distinguishes between task-dependent and task-invariant (shared) features to perform the primary and secondary tasks. Furthermore, unlike [18] that only considered hate speech detection task and datasets, our proposed model used other relevant auxiliary tasks and dataset to improve the primary hate speech detection task. Closer to our study, Rajamanickam et al. [15] proposed a shared-private MTL framework that utilized a stacked BiLSTM encoder as the shared layer and attention mechanism for intra-task learning. The framework is trained on a hate speech detection dataset for the primary task and emotion detection as the secondary relevant task. Different from [15], our AngryBERT model adopted BERT [7] as the shared layer, and is trained on both emotion classification and hateful target identification as secondary tasks to aid hate speech detection.

3 Datasets and Tasks

Previous studies have shown that the relevance of tasks in an MTL framework affects the model's stability of training and performance [20]. According to the definition of hate speech, there are two main characteristics of hate speech: (i) offensive language that (ii) targets individuals or groups. Considering the two aspects, we select two secondary tasks relevant to hate speech detection: emotion classification and target identification. Offensive language usually involves negative sentiments. Therefore, emotions in tweets can serve as complementary information for hate speech detection [5]. Our goal is to train a network that can extract emotions hidden in tweets using the emotion classification task. For the target identification task, we aim to train the model to identify targets in a text. Co-trained with these two secondary tasks, MTL models will be capable of extracting emotions and target groups or individuals in tweets, which facilitates hate speech detection indirectly. In the remaining parts of this section, we discuss the datasets involved to train the AngryBERT model and MTL baselines. Table 1 shows the statistical summary of the datasets.

3.1 Primary Task and Datasets

The primary task of AngryBERT is hate speech detection. Therefore, we train and evaluate our proposed model on three publicly available hateful and abusive speech datasets, namely, WZ-LS [14], DT [6], and FOUNTA [10].

Table 1. Statistic information about datasets in experiments

Dataset	#tweets	Classes (#tweets)
DT	24,783	hate(1,430), offensive(19,190), neither(4,163)
WZ-LS	16,035	racism(1923), sexism(3,079), neither(11,033)
FOUNTA	89,990	normal (53,011), abusive (19,232), spam (13,840), hate (3,907)
HateLingo	5,680	disability(257), ethnicity(351), gender(2841), religion(1590), sexual orientation(641)
SemEval_A	10,983	anger(2544), anticipation(978), disgust(2602), fear(1242), joy(2477), love(700), optimism(1984), pessimism(795), sadness(2008), surprise(361), trust(357)
OffensEval_C	4,089	individual(2,507), group(1,152), other(430)

WZ-LS [14]: Park et al. [14] combined two Twitter datasets [16,17] to form the WZ-LS dataset. We retrieve the tweets' text using Twitter's APIs and the tweet ids release in [14]. However, some of the tweets have been deleted by Twitter due to their inappropriate content. Thus, our dataset is slightly smaller than the original dataset reported in [14].

DT [6]: Davidson et al. [6] The researchers constructed the DT Twitter dataset, which manually labeled and categorized tweets into three categories: offensive, hate, and neither.

FOUNTA [10]: The FOUNTA dataset is a human-annotated dataset that went through two rounds of annotations. Awal et al. [2] found that there were duplicated tweets in FOUNTA dataset as the dataset annotators have included retweets in their dataset. For our experiments, we remove the retweets resulting in the distribution in Table 1.

3.2 Secondary Tasks and Datasets.

Three publicly available Twitter datasets are selected for the secondary tasks: SemEval_A [13], HateLingo [8], and OffensEval_C [19].

SemEval_A [13]: Mohammad et al. collected and annotated a Twitter dataset that supported array of subtasks on inferring the affectual state of a person from their tweet. We perform emotion classification task using this Twitter dataset.

HateLingo [8]: ElSherief et al. collected the HateLingo dataset that identifies the target of hate speeches. We perform hate speech target group identification task using the HateLingo dataset. Specifically, the task aims to identify the target group in a given hateful tweet.

OffensEval_C [19]: Zampieri et al. proposed the OffenEval_C dataset, which categorize the targets of abusive tweets into *individual, group*, or *other*. Similarly, our proposed model is trained on this dataset for target identification task.

4 Proposed Model

4.1 Problem Formulation

Essentially, hate speech detection (i.e., primary task) and the relevant secondary tasks can be generalized as text classification tasks. Therefore, we define a general problem formulation of text classification tasks under the MTL setting. Assume we have K tasks and the input for the i-th task is: $S_i = \{s_1^i, s_2^i \ldots, s_n^i\}, i \in \{1, 2, \ldots, K\}$, where n is the length of the sentence. For the i-th task, the goal is to correctly classify the input text into: $C = \{c_1^i, c_2^i \ldots, c_m^i\}$, where m is the number of classes of task i.

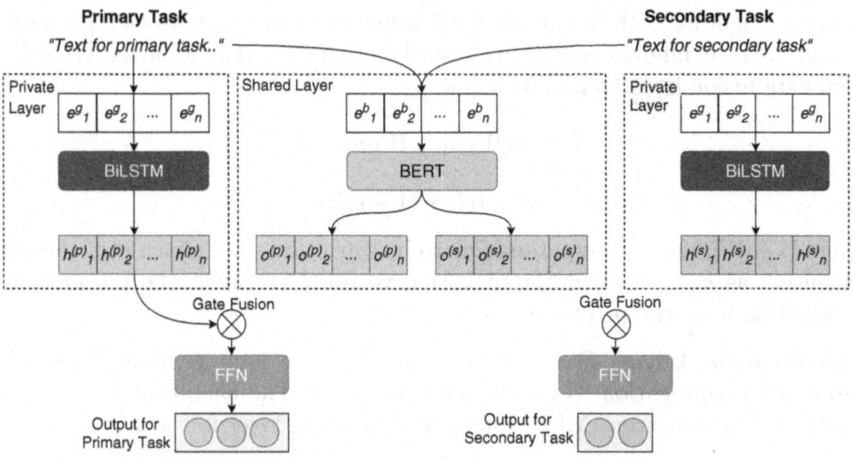

Fig. 1. The overall architecture of AngryBERT model

4.2 Architecture of AngryBERT

Figure 1 illustrates the overall architecture of AngryBERT model. We adopt a shared-private MTL setting where a shared layer is encouraged to learn the task-invariant features while the private layers aim to learn the task-specific representations. The gate fusion mechanism aggregates the shared and private information. Finally, the joint representation of each task is fed into their classification layer, respectively. To simplify our discussion, we ignore the superscript for each task in the rest of this section.

Shared Layer. Here we exploit the pre-trained BERT model as the shared layer. Given a sentence, it is first tokenized using the default tokenizer of BERT then transformed into pre-trained BERT embeddings: $E_B = \{e_1^b, e_2^b, \ldots, e_n^b\}$, $e_i^b \in R^{768}$. These embeddings are sent to a pre-trained BERT model. We use the output from the [CLS] token as the representation from the shared layer, denoted as $o_1 \in R^d$:

$$o_1 = BERT(E_B) \tag{1}$$

Private Layer. For each task, a private layer is used to learn the task-specific representation. In order to fully exploit contexts of each word, a Bi-directional Long-Short Term Memory Network (Bi-LSTM) is applied. Each word of the sentence is first embedded using GloVe Embedding: $E_G = \{e_1^g, e_2^g, \ldots, e_n^g\}$, $e_i^g \in R^{300}$. The embeddings are sent to the Bi-LSTM to learn the sequential information. The concatenation of final hidden states from forward and backward path is used as the latent representation learnt from the private layer, denoted as $h_n \in R^d$:

$$h_n = Bi - LSTM(E_G) \tag{2}$$

Gate Fusion. After learning the respective representations from shared and private layers, we exploit the gate mechanism for feature fusion. Instead of directly assign a weight for each vector, the gate fusion mechanism allows each position of vectors to have different contribution to the prediction. The joint representation from gate fusion is computed as below:

$$\alpha = \sigma(W_L o_1 + W_B h_n + b_g) \tag{3}$$

$$J = \alpha L + (1 - \alpha)B \tag{4}$$

where W_L, W_B and b_g are parameters to be learnt. $\alpha \in R^d$, which is of the same dimension as h_n and o_1, is the attention vector. It controls the proportion of information from the private and shared flow.

Classification Layer. For each task, we feed the joint representation after information aggregation to its classification layer. The classification layer is a Multi-Layer Perceptron (MLP) follow by a Softmax layer for normalization.

$$M = ReLU(W_f J + b_f) \tag{5}$$

$$O = softmax(W_e M + b_e) \tag{6}$$

where W_f, W_e and b_f, b_e are weights and biases to be learnt. The final prediction is $O \in R^m$ and each position of O denotes the confidence score for each class. Non linear activation function and weight normalization are used between two linear projection layers. Dropout is applied in order to avoid overfitting in the classification layers.

4.3 Training of AngryBERT

In this part, we describe the loss function of individual tasks and the training for AngryBERT under the MTL setting.

Single Task Loss. For each task, cross entropy is used as the loss function. The loss of the i-th task is:

$$M_i = \sum_{t=1}^{N_i} Cross - Entropy(O_t^i, \hat{O}_t^i) \tag{7}$$

where $\hat{O_t^i}$ is the ground-truth class for the t-th instance of task i and N_i is the number of training instances for the i-th task. For all tasks, we obtain: $M = \{M_1, M_2, \ldots, M_K\}$.

Multi-task Loss. There are several objectives involved: the primary task and secondary tasks. Rather than averaging all losses, we consider different speeds of divergence of tasks. The objective function is a weighted average of losses from different tasks:

$$\Phi = \sum_{i=1}^{K} \beta_i M_i \tag{8}$$

where weights β_i are learnt end-to-end, which represents the contribution from task i to the multitask loss. By exploiting multitask loss, tasks have different importance for parameter updating, which mitigates the issue of different speeds of convergence. All tasks are trained with the same number of epochs.

5 Experiments

In this section, we will first describe the settings of experiments conducted to evaluate our AngryBERT model. Next, we discuss the experiment results and assess how AngryBERT fares against other state-of-the-art baselines. We conduct more in-depth ablation studies on the various tasks co-trained with the primary hate speech detection task in the AngryBERT. We demonstrate interesting case studies where the tweets' predicted labels for various tasks co-trained in AngryBERT presented.

5.1 Baselines

We compare AngryBERT with the state-of-the-art hate speech classification baselines and multitask learning text classification models:

- **CNN:** Previous studies have utilized CNN to achieve good performance in hate speech detection [3]. We train a CNN model with word embeddings as input.
- **LSTM:** The LSTM model, is another model that was commonly explored in previous hate speech detection studies [3]. Similarly, we train a LSTM model with word embeddings as input.
- **HybridCNN:** We replicate the HybridCNN model proposed by Park and Fung [14] for comparison. The HybridCNN model trains CNN over both word and character embeddings for hate speech detection.
- **CNN-GRU:** The CNN-GRU model was proposed in a recent study by Zhang et al. [21] is also replicated in our study as a baseline. The CNN-GRU model takes word embeddings as input.
- **DeepHate:** The DeepHate model was proposed in a recent study by Cao et al. [5]. The DeepHate model trains on semantics, sentiment, and topical features for hate speech detection.

- **BERT:** BERT [7] is a contextualized word representation model that is based on a masked language model and pre-trained using bidirectional transformers. For our study, we fine-tune the pre-trained BERT model using the train set and subsequently perform classification on tweets in the test set.
- **SP-MTL:** Liu et al. [11] proposed the SP-MTL model, which is a Recurrent Neural Network (RNN) based multitask learning model for text classification tasks. We trained the SP-MTL model with the same tasks as our AngryBERT model.
- **MT-DNN:** Liu et al. [12] proposed the Multi-Task Deep Neural Network (MT-DNN), which combined multitask learning and language model pre-training for language representation learning. We replicated the MT-DNN as a baseline in our study. Similarly, we trained the MT-DNN with the same tasks as our AngryBERT model.
- **MTL-GatedDEncoder:** Rajamanickam et al. [15] proposed a shared-private MTL framework that utilized a stacked BiLSTM encoder as the shared layer and attention mechanism for hate speech detection. This is the state-of-the-art MTL baseline for hate speech detection.

Table 2. Experiment results of AngryBERT and baselines on DT, WZ-LS, and FOUNTA datasets. "#" denotes MTL models that co-trained with other secondary tasks.

	DT			WZ-LS			FOUNTA		
Model	**Prec**	**Rec**	**F1**	**Prec**	**Rec**	**F1**	**Prec**	**Rec**	**F1**
CNN	89.32	90.07	89.35	80.63	78.35	78.21	79.97	80.35	79.84
LSTM	89.58	90.26	89.56	80.43	77.54	77.27	80.24	81.18	80.22
HybridCNN	88.65	89.91	88.85	80.71	78.91	78.3	79.86	80.52	79.86
CNN-GRU	88.89	89.80	88.91	80.85	77.05	77.12	79.96	80.73	79.99
DeepHate	89.97	90.39	89.92	77.95	79.48	78.19	78.95	80.43	79.09
BERT	90.35	90.53	90.34	**83.25**	80.05	79.95	79.69	80.03	79.79
SP-MTL#	89.44	90.22	89.44	81.11	**81.59**	80.68	80.46	81.65	80.66
MT-DNN#	90.29	90.69	90.31	83.05	80.25	80.18	80.66	81.64	80.72
MTL-GatedDEncoder#	89.20	89.55	89.22	81.33	78.62	78.18	80.00	81.33	80.08
AngryBERT#	**90.71**	**91.14**	**90.71**	83.19	81.45	**81.25**	**81.00**	**81.82**	**81.08**

5.2 Evaluation Metrics

Similar to most existing hate speech detection studies, we use micro averaging precision, recall, and F1 score as the evaluation metrics. Five-fold cross-validation is used in our experiments, and the average results are reported.

5.3 Experiment Results

Table 2 shows the experiment results on DT, WZ-LS, and FOUNTA datasets. In the table, the highest figures are highlighted in **bold**. We observe that AngryBERT outperformed the state-of-the-art single and multitask baselines in Micro-F1 scores. We observed that the single task BERT model is able to achieve

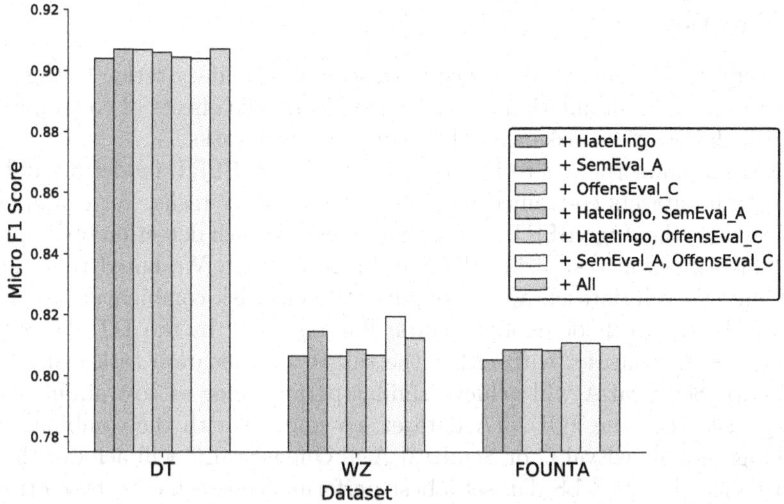

Fig. 2. Micro-F1 scores of AngryBERT model for various hate speech datasets co-trained with various combinations of secondary task datasets

good performance in hate speech detection, outperforming the other single task baselines for DT and WZ-LS datasets. Nevertheless, AngryBERT outperformed the BERT baseline by leveraging the BERT language model to learn shared knowledge across tasks.

Comparing the single-task baselines with the MTL-based models, we noted that the MTL-based models are able to outperform most single-task baselines across the three hate speech datasets. The observation shows the advantage to co-train the hate speech detection task with other secondary tasks in a multi-task setting. AngryBERT is observed to outperform the state-of-the-art MTL hate speech detection model, MTL-GatedDEncoder, and other MTL text classification models. The good performance demonstrates BERT's strength as the shared layer in the multitask learning architecture.

It is worth noting that there are differences between HybridCNN and CNN-GRU models in our experiments and the results reported in previous studies [14,21]. For instance, earlier studies for HybridCNN [14] and CNN-GRU [21] had conducted experiments on the WZ-LS dataset. However, we did not cite the previous scores directly as some of the tweets in WZ-LS have been deleted. Similarly, CNN-GRU was also previously applied to the DT dataset. However, in the previous work [21], the researchers have cast the problem into binary classification by re-labeling the offensive tweets as non-hate. In our experiment, we perform the classification based on the original DT dataset [6]. Therefore, we replicated the HybridCNN and CNN-GRU models and applied them to the updated WZ-LS dataset and original DT dataset.

5.4 Ablation Study

The AngryBERT model is co-trained with several secondary tasks. In this evaluation, we perform an ablation study to investigate the effects of co-training the hate speech detection tasks with different secondary tasks.

Figure 2 shows the Micro-F1 scores of the AngryBERT model for different hate speech datasets co-trained with various secondary tasks. For example, the red bars show the AngryBERT co-training on hate speech detection tasks and the target identification task using the HateLingo dataset. We noted that the different hate speech datasets would require different task combination to achieve the best hate speech detection results. For instance, in the DT dataset, co-training the DT dataset with either the target identification task using OffensEval_C or SemEval_A will achieve similar performance as co-training all secondary tasks. For the FOUNTA dataset, co-training with the combinations of HateLingo + OffensEval_C or SemEval_A + OffensEval_C will achieve the best performance. The WZ-LS dataset's best performance is achieved by co-training with the SemEval_A + OffensEval_C, and co-training with only SemEval_A outperforms co-training with all secondary task datasets. Nevertheless, co-training with any combinations of the secondary tasks in AngryBERT outperforms the single-task methods in hate speech detection. These observations highlighted

Table 3. Samples of AngryBERT predictions on DT dataset

Tweet	DT (Actual)	DT (Predict)	SemEval_A (Predict)	HateLingo (Predict)	OffenEval_C (Predict)
[*USER*] f*ck outta here and go put some more trash a*s ink on your faggot a*s self p*ssy	hateful	hateful	anger, disgust	sexual orientation	individual
RT [*USER*] We Muslims have no military honour whatsoever we are sub human savages that slaughter unarmed men women and children	hateful	hateful	anger, disgust	religion	group
RT [*USER*] I hate these Mone Davis commercials B*tch is gonna end up either a dyke or a loser like every other female	offensive	hateful	anger, disgust	gender	individual

Table 4. Samples of AngryBERT predictions on WZ-LS dataset

Tweet	WZ-LS (Actual)	WZ-LS (Predict)	SemEval_A (Predict)	HateLingo (Predict)	OffenEval_C (Predict)
[*USER*] Of course Muslim religious bigots like you think that is okay	racism	racism	disgust, fear	religion	group
[*USER*] And if you are going to follow a prophet that approved of collateral damage then do not complain about collateral damage	racism	racism	anger, disgust, fear	religion	individual
[*USER*] a lying taquiyya b*tch with zero followers opened an account to feed me bullshit	sexism	neither	disgust, anger	gender	individual

the intricacy of task selections for performing hate speech detection in a MTL setting. For future works, we will explore developing better approaches to automatically select the optimal combination of co-training tasks for hate speech detection.

5.5 Case Studies

To better understand how the secondary tasks could help in the hate speech detection task, we qualitatively examine some sample predictions of the **AngryBERT** model. Table 3, 4, and 5 shows example posts from DT, WZ-LS, and FOUNTA datasets respectively. In each example post, we display the actual label and predicted label from **AngryBERT** model. The correct predictions are marked in green font, while the incorrect predictions are represented in red font. Besides the hate speech predictions, we also display the predicted labels of various secondary tasks. Specifically, we highlighted the keywords in the given post that might have influenced the predicted target in HateLingo dataset.

Table 5. Samples of AngryBERT predictions on FOUNTA datasets.

Tweet	FOUNTA (Actual)	FOUNTA (Predict)	SemEval_A (Predict)	HateLingo (Predict)	OffenEval_C (Predict)
RT [*USER*] I hope a tornado destroys your house you f*cking Jew	hateful	hateful	anger, disgust	ethnicity	individual
[*USER*] and added to his discomfort he is mightily pissed off at having to pay tampon tax the cheek	normal	normal	anger, disgust	gender	individual
RT [*USER*] They are F*CKING EVIL I DESPISE liberals They KILL RAPE ASSAULT	abusive	hateful	anger, disgust	gender	group

From the example posts, we observed that the secondary tasks profoundly impact AngryBERT's hate speech detection performance. For instance, we noted that most of the predicted hateful posts are also predicted to contain "anger" and "disgust" emotions using the secondary task emotion classifier co-trained using the SemEval_A dataset. We postulate that the emotions captured by the AngryBERT model have helped the model in identifying hateful content as the two emotions are commonly exhibited in online hate speeches and abusive tweets. Another interesting observation is the identification of targets in hate speeches. We observe that the secondary task of target identification classifier co-trained using HateLingo dataset is able to predict the target in a hate speech reasonably. For example, the second tweet in Table 3 is a hateful tweet against Muslims, and the target identification classifier predicted "religion" as the target in this tweet. Although the AngryBERT has outperformed the state-of-the-art baselines in hate speech predictions, the model also made some incorrect predictions. However, we noted that as the ground truth labels of the incorrect predictions look contestable. For example, the last tweet in Table 5 seems hateful, but it was instead annotated as abusive.

The interesting predictions from secondary relevant tasks seems to provide a form of explanation that could help us understand the context when a tweet is predicted to be hateful. For future work, we will explore building explainable models that utilize the prediction of secondary tasks as supplementary information to aid explaining hate speech detection.

6 Conclusion

This paper proposed a novel multitask learning-based model, AngryBERT, which jointly learns hate speech detection with emotion classification and target identification as secondary relevant tasks. We evaluated AngryBERT on three publicly available real-world datasets, and our extensive experiments have shown that AngryBERT outperforms the state-of-the-art single-task and multitask baselines in the hate speech detection tasks. We identify case studies to demonstrate that AngryBERT is able to detect hate speeches accurately and identify the target of the hate speech and the emotion expressed. For future works, we will explore developing better approaches to automatically select the optimal combination of co-training tasks for hate speech detection. We will also explore developing explainable hate speech detection methods that utilized the predictions of secondary tasks as supplementary information.

References

1. Arango, A., Pérez, J., Poblete, B.: Hate speech detection is not as easy as you may think: a closer look at model validation. In: ACM SIGIR (2019)
2. Awal, M.R., Cao, R., Lee, R.K.W., Mitrović, S.: On analyzing annotation consistency in online abusive behavior datasets. arXiv preprint arXiv:2006.13507 (2020)
3. Badjatiya, P., Gupta, S., Gupta, M., Varma, V.: Deep learning for hate speech detection in tweets. In: WWW (2017)
4. Cao, R., Lee, R.K.W.: HateGAN: adversarial generative-based data augmentation for hate speech detection. In: COLING (2020)
5. Cao, R., Lee, R.K.W., Hoang, T.A.: DeepHate: hate speech detection via multi-faceted text representations. In: ACM WebSci (2020)
6. Davidson, T., Warmsley, D., Macy, M., Weber, I.: Automated hate speech detection and the problem of offensive language. In: ICWSM (2017)
7. Devlin, J., Chang, M.W., Lee, K., Toutanova, K.: BERT: pre-training of deep bidirectional transformers for language understanding. In: NAACL (2019)
8. ElSherief, M., Kulkarni, V., Nguyen, D., Wang, W.Y., Belding-Royer, E.M.: Hate Lingo: a target-based linguistic analysis of hate speech in social media. In: ICWSM (2018)
9. Fortuna, P., Nunes, S.: A survey on automatic detection of hate speech in text. ACM Comput. Surv. (CSUR) **51**(4), 1–30 (2018)
10. Founta, A.M., et al.: Large scale crowdsourcing and characterization of twitter abusive behavior. In: ICWSM (2018)

11. Liu, P., Qiu, X., Huang, X.: Recurrent neural network for text classification with multi-task learning. In: IJCAI (2016)
12. Liu, X., He, P., Chen, W., Gao, J.: Multi-task deep neural networks for natural language understanding. In: ACL (2019)
13. Mohammad, S., Bravo-Marquez, F., Salameh, M., Kiritchenko, S.: Semeval-2018 task 1: affect in tweets. In: SemEval (2018)
14. Park, J.H., Fung, P.: One-step and two-step classification for abusive language detection on Twitter. In: Workshop on Abusive Language Online (2017)
15. Rajamanickam, S., Mishra, P., Yannakoudakis, H., Shutova, E.: Joint modelling of emotion and abusive language detection. In: ACL (2020)
16. Waseem, Z.: Are you a racist or am I seeing things? Annotator influence on hate speech detection on Twitter. In: Workshop on NLPCSS (2016)
17. Waseem, Z., Hovy, D.: Hateful symbols or hateful people? Predictive features for hate speech detection on Twitter. In: NAACL (2016)
18. Waseem, Z., Thorne, J., Bingel, J.: Bridging the gaps: multi task learning for domain transfer of hate speech detection. In: Online Harassment (2018)
19. Zampieri, M., Malmasi, S., Nakov, P., Rosenthal, S., Farra, N., Kumar, R.: Semeval-2019 task 6: identifying and categorizing offensive language in social media (offenseval). In: SemEval (2019)
20. Zhang, Y., Yang, Q.: A survey on multi-task learning. arXiv preprint arXiv:1707.08114 (2017)
21. Zhang, Z., Robinson, D., Tepper, J.: Detecting hate speech on Twitter using a convolution-GRU based deep neural network. In: ESWC (2018)

SCARLET: Explainable Attention Based Graph Neural Network for Fake News Spreader Prediction

Bhavtosh Rath[1(✉)], Xavier Morales[2(✉)], and Jaideep Srivastava[1(✉)]

[1] University of Minnesota, Minneapolis, USA
{rathx082,srivasta}@umn.edu
[2] Harvard College, Cambridge, USA
xavier_morales@college.harvard.edu

Abstract. False information and true information fact checking it, often co-exist in social networks, each competing to influence people in their spread paths. An efficient strategy here to contain false information is to proactively identify if nodes in the spread path are likely to endorse false information (i.e. further spread it) or refutation information (thereby help contain false information spreading). In this paper, we propose SCARLET (truSt andCredibility bAsed gRaph neuraLnEtwork model using aTtention) to predict likely action of nodes in the spread path. We aggregate trust and credibility features from a node's neighborhood using historical behavioral data and network structure and explain how features of a spreader's neighborhood vary. Using real world Twitter datasets, we show that the model is able to predict false information spreaders with an accuracy of over 87%.

1 Introduction

Social network platforms like Twitter, Facebook and Whatsapp are used by millions around the world to share information and opinions. Often, the veracity of content shared on these platforms is not confirmed. This gives rise to scenarios where information having conflicting veracity, i.e. false information and its refutation, co-exist. Refutation can be defined as true information which fact checks claims made by a false information. A typical scenario is that false information originates at time t_1, and starts propagating. Once it is identified, its refutation information is created at time t_2 ($t_1 < t_2$). Both pieces of information propagate simultaneously, with many nodes lying in their common spreading paths.

While detecting false information is an important and widely researched problem, an equally important problem is that of preventing the impact of false information spreading. Techniques involve containment/suppression of false information, as well as accelerating the spread of its refutation. *Being able to predict the likely action of such users before they are exposed to false information is an important aspect of such a strategy.* Nodes identified as vulnerable to believing false information can thus 1) be cautioned about the presence of the

K. Karlapalem et al. (Eds.): PAKDD 2021, LNAI 12712, pp. 714–727, 2021.
https://doi.org/10.1007/978-3-030-75762-5_56

false information so that they do not propagate it, and 2) be urged to propagate its refutation. While optimization models based on information diffusion theories have been proposed in the past for misinformation containment, recent advancements in deep learning on graphs serve as the motivation to explore false information control models which use components that exist even before false information starts spreading, namely the underlying network structure and people's historical behavioral data.

Trust and *Credibility* are important psychological and sociological concepts respectively, that have subtle differences in their meanings. While trust represents the confidence one person has in another person, credibility represents generalized confidence in a person based on their perceived performance record [14]. Thus, in a graph representation of a social network, trust is a property of a (directed) edge, while credibility is a property of an individual node. Metzger et al. [7] showed that *the interpretation of a neighbor's credibility by a node relies on its perception of the neighbor based on their trust dynamics.* Motivated with this idea, we propose a graph neural network model that integrates people's credibility and interpersonal trust features in a social network to predict whether a node is likely to spread false information or not. We make the following contributions in this paper:

1) We propose *SCARLET*, a novel user-centric model using graph neural network with attention mechanism to predict whether a node will most likely spread false information, its refutation or be a non-spreader.
2) We demonstrate that a person's decision to spread a false information is sensitive to its perception of neighbor's credibility, and this perception is a function of trust dynamics with the neighbor.
3) To the best of our knowledge, this is the first model being evaluated on real world Twitter datasets of co-existing false and refutation information.

Related Work: Social science research in the past has explored the aspects of people's behavior that cause false information spreading. Jaeger et al. [5] was one of the first to study what makes rumors believable when told by peers instead of authority figures. While it focused on modelling people's anxiety, it served as motivation to explore other sociological features that are relevant to information spreading. Petty and Cacioppo [10] found credibility perception to be an important factor for believing false information. Rosnow et al. [15] proposed that *interpersonal trust* also played an important role in rumor transmission. The idea was further enforced by Morris et al. [8] where they claimed that people assess credibility based on trust relationships with their neighbors in a social network. Motivated by these ideas, there has been much interest in computational models for false information spreader detection using trust, which has shown promising results [12,13]. Many computational techniques to combat false information spreading have been explored over the past decade, as summarized by Sharma et al. [17]. Most models rely on generating relevant features from the information that help distinguish false information from true. Our proposed model is based on recent advances in graph neural networks [22]. In addition, our work proposes an explainable attention based model, inspired from recent work [23,24].

Qui et al. [11] focuses on influence in general, while our model integrates people's psychological and sociological features to identify false information spreaders.

Models inspired by information diffusion models for false information mitigation have also been proposed. Budak et al. [1] proposed an optimization strategy to identify false information spreaders in a network who, when convinced by its refutation, would minimize the number of people receiving the false information. Nguyen et al. [9] proposed greedy approaches to a similar problem of limiting the spread of false information in social networks. More recently, Tong et al. [19] studied the problem as a multiple cascade diffusion problem.

2 Interpersonal Trust and User Credibility Features

2.1 Trust-Based Features

1. **Global Trust (Tr^G):** Global trust are trust scores that are computed on the directed follower-followee network around information spreaders. It is called global because an individual's trust score is sensitive to changes in the network structure. Using the Trust in Social Media (TSM) algorithm [16], we quantify the likelihood of *trusting others* and being *trusted by others.* The TSM algorithm uses a directed graph $\mathcal{G}(\mathcal{V}, \mathcal{E})$ as input, together with a specified convergence criteria, and computes trustingness and trustworthiness scores using the equations: $ti(v) = \sum_{\forall x \in out(v)} \left(\frac{w(v,x)}{1+(tw(x))^s} \right)$ and $tw(u) = \sum_{\forall x \in in(u)} \left(\frac{w(x,u)}{1+(ti(x))^s} \right)$ where $u, v, x \in \mathcal{V}$ are nodes, $ti(v)$ and $tw(u)$ are the trustingness and trustworthiness scores of v and u, respectively, $w(v, x)$ is the weight of edge from v to x, $out(v)$ is the set of out-edges of v, $in(u)$ is the set of in-edges of u, and s is the involvement score of the network. The involvement score is basically the potential risk an actor takes when creating a link in the network. Details of the algorithm are excluded due to space constraints and can be found in [16].
2. **Local Trust (Tr^L):** Local trust is computed based on the retweeting behavior of an individual. It is termed local because the trust score depends on node's behavior, and not on the network structure. We consider the proxy for *trusting others* as the fraction of tweets of x that are retweets (RT_x) denoted by $\sum_{\forall i \in t}\{1 \text{ if } i = RT_x \text{ else } 0\}/n(t)$. Meanwhile, we consider the proxy for *trusted by others* as the average number of times x's tweets are retweeted ($n(RT)$) denoted by $\sum_{\forall i \in t} i_{n(RT_x)}/n(t)$. ($t$ represents the most recent tweets posted in x's timeline).

2.2 Credibility-Based Features

Credibility of users is generalized based on features extracted from information posted on their timeline and are obtained from [2]. We generate relevant credibility features for nodes in the network, which can be categorized into two types: user-based and content-based.

1. **User-based Credibility (Cr^U):** User credibility features are extracted from user metadata of nodes in the network. Features used in our model are summarized below:
   A. Registration age (U1): Registration age denotes the time that has transpired since a user created their account. Older accounts tend to be associated with more credible users.
   B. Overall activity count (U2): Activity or statuses count is the number of tweets issued by a user. Low credibility is associated with users who have less activity on their timeline.
   C. Is verified (U3): This label suggests whether a user account is marked as authentic or not by Twitter. Verified accounts are more likely to be credible.
2. **Content-based Credibility (Cr^C):** These features are obtained by aggregating a user's timeline activity. It is important to note that, unlike Castillo's assumption, we do not make a distinction between information that is specifically related to news or not, as that process would require manually assessing newsworthiness of the tweets. The following relevant features are extracted:
   A. Emotions conveyed by user (M1): Emotions represent positive or negative sentiments associated with a tweet. Content with negative sentiments is usually associated with non-credible users [2].
   B. Level of uncertainty (M2): Level of uncertainty is quantified as the fraction of user's tweets that are questioning in nature. Tweets with a high level of uncertainty tend to be less credible.
   C. External source citation (M3): External source citation is quantified as the fraction of user's tweets that cite an external URL. tweets which do not include URLs tend to be related to non-credible news [2].

3 Proposed Approach

This section explains how we integrate both credibility and trust features in an attention based graph neural network model to predict whether a person would likely be a spreader of false information or its refutation. The problem formulation is as follows:

Problem formulation: Let $\mathcal{G}(\mathcal{V}, \mathcal{E})$ be a directed social network containing false information spreaders ($\mathcal{V}_F$), refutation information spreaders ($\mathcal{V}_T$) and non-spreaders ($\mathcal{V}_{\hat{S}p}$) at a time instance t ($\{\mathcal{V}_F \cup \mathcal{V}_T \cup \mathcal{V}_{\hat{S}p}\} \subset \mathcal{V}$). By assigning importance score using global (Tr^G) and local (Tr^L) trust features ($Tr = Tr^G || Tr^L$), and aggregating user-based (Cr^U) and content-based (Cr^C) credibility features ($Cr = Cr^U || Cr^C$) of node i and its neighborhood nodes ($\mathcal{N}_i{}^K$) sampled till depth K, we predict whether i is more likely to spread false information, refutation information or be non-spreader at future time $t + \Delta t$.

The proposed graph neural network framework can be broadly divided into two steps:

1. We assign an importance score to neighborhood nodes ($\mathcal{N}_i{}^K$) sampled till depth K based on trust (Tr) features. This is done using an attention mechanism.

2. We learn representations using Graph Convolutional Networks by aggregating credibility (Cr) features proportional to the importance scores assigned for the neighborhood nodes based on step 1.

An overview of the proposed model architecture is shown in Fig. 1. The following subsections explain the framework in detail.

3.1 Importance Score Using Attention:

We apply a graph attention mechanism [21] which attends over the neighborhood of i and, based on their trust features, assigns an importance score to every j ($j \in \mathcal{N}_i$). First, every node is assigned a parameterized weight matrix (**W**) to perform linear transformation. Then, self-attention is performed using a shared attention mechanism a (a single layer feed-forward neural network) which computes trust-based importance scores. The unnormalized trust score between i,j is represented as:

$$e_{ij} = a(\mathbf{W}_{Tr_i}, \mathbf{W}_{Tr_j}) \tag{1}$$

where e_{ij} quantifies j's importance to i in the context of interpersonal trust. We perform masked attention by only considering nodes in $\mathcal{N}_i$. This way we aggregate features based only on the neighborhood's structure. To make the importance scores comparable across all neighbors we normalize them using the softmax function:

$$\alpha_{ij} = softmax(e_{ij}) = \frac{exp(e_{ij})}{\sum_{k \in \mathcal{N}_i} exp(e_{ik})} \tag{2}$$

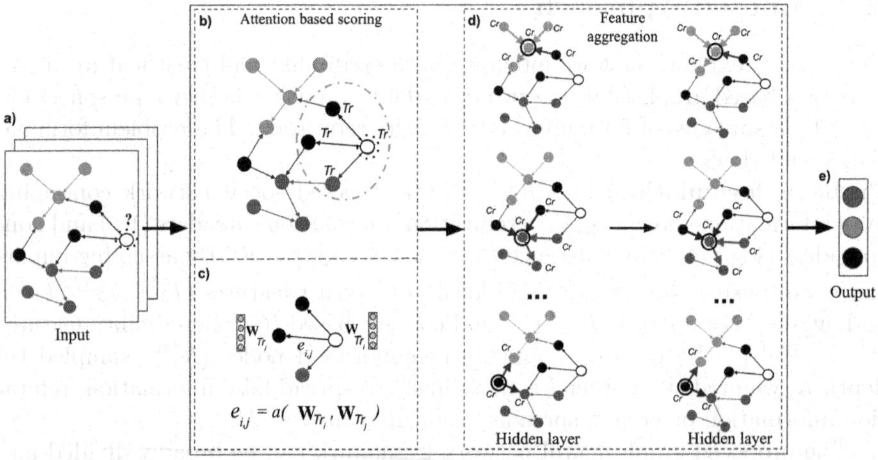

Fig. 1. Architecture overview. Importance score e is assigned to neighbors based on trust features (Tr). Credibility (Cr) features are aggregated proportional to neighbors' importance scores using graph convolution networks for node classification.

The attention layer a is parameterized by weight vector $\mathbf{a}$ and applied using LeakyReLU nonlinearity. Normalized neighborhood edge weights can be represented as:

$$\alpha_{ij} = \frac{exp(LeakyReLU(\mathbf{a}^T[\mathbf{W}_{Tr_i}||\mathbf{W}_{Tr_j}]))}{\sum_{k\in\mathcal{N}_i} exp(LeakyReLU(\mathbf{a}^T[\mathbf{W}_{Tr_i}||\mathbf{W}_{Tr_k}]))} \tag{3}$$

α_{ij} thus represents trust between i and j with respect to all nodes in $\mathcal{N}_i$. Each α_{ij} obtained for the edges is used to create an attention-based adjacency matrix $\hat{A}_{atn} = [\alpha_{ij}]_{|\mathcal{V}|\times|\mathcal{V}|}$ which is later used to aggregate credibility features.

3.2 Feature Aggregation

The Graph Convolution Network [6] is a graph neural network model that efficiently aggregates features from a node's neighborhood. It consists of multiple neural network layers where the information propagation between layers can be generalized by Eq. 4. Here, H represents the hidden layer and A represents the adjacency matrix representation of the subgraph ($A = \hat{A}_{atn}$). $H^{(0)} = Cr$ and $H^{(L)} = Z$, where Z denotes node-level output during transformation.

$$H^{(l+1)} = f(H^{(l)}, A) \tag{4}$$

We implement a Graph Convolution Network with two hidden layers using a propagation rule as explained in [6].

$$H^{(l+1)} = \sigma(\hat{D}^{-1/2}\hat{A}\hat{D}^{-1/2}H^{(l)}W^{(l)}) \tag{5}$$

Here, $\hat{A} = A + I$, where I is the identity matrix of the neighborhood subgraph. This operation ensures that we include self-features during aggregation of neighbor's credibility features. $\hat{D}$ is the diagonal matrix of node degrees for $\hat{A}$, where $\hat{D}_{ii} = \sum_j \hat{A}_{ij}$. $W^{(l)}$ is the layer weight matrix, and σ denotes the activation function. Symmetric normalization of $\hat{D}$ ensures our model is not sensitive to varying scale of the features being aggregated.

3.3 Node Classification

Using credibility features and network structure for nodes in i's neighborhood, node representations are learned from the graph using a symmetric adjacency matrix with attention-based edge weights ($\hat{A} = \hat{D}^{-1/2}\hat{A}_{atn}\hat{D}^{-1/2}$). Following forward propagation model is applied:

$$Z = f(X, \hat{A}_{atn}) = softmax(\hat{A}ReLU(\hat{A}XW^{(0)})W^{(1)}) \tag{6}$$

X represents the credibility features. $W^{(0)}$ and $W^{(1)}$ are input-to-hidden and hidden-to-output weight matrices respectively, and are learnt using gradient descent learning. Classification is performed using the following cross entropy loss function:

$$\mathcal{L} = \sum_{l\in\mathcal{Y}_L}\sum_{f\in Cr} Y_{lf} ln Z_{lf} \tag{7}$$

where $\mathcal{Y}_L$ represents indices of labeled vertices, f represents each of the credibilty features being used in the model, and $Y \in R^{|\mathcal{Y}_L| \times |Cr|}$ is the label indicator matrix.

Table 1. Network dataset statistics for news events N1-N10.

	N1			N2			N3			N4			N5		
	$\|\mathcal{V}\|$	$\|\mathcal{E}\|$	$\|Sp\|$	$\|\mathcal{V}\|$	$\|\mathcal{E}\|$	$\|Sp\|$	$\|\mathcal{V}\|$	$\|\mathcal{E}\|$	$\|Sp\|$	$\|\mathcal{V}\|$	$\|\mathcal{E}\|$	$\|Sp\|$	$\|\mathcal{V}\|$	$\|\mathcal{E}\|$	$\|Sp\|$
F	1,797,059	5,316,114	2,584	885,598	1,824,585	943	1,228,479	2,477,986	1,313	2,607,629	7,146,454	4,552	2,150,820	5,215,120	3,344
T	1,164,162	2,283,160	437	453,537	879,854	403	1,169,681	1,988,576	425	433,616	773,778	467	1,168,820	1,543,513	305
F ∪ T	2,677,924	7,562,503	3,017	1,230,559	2,641,513	1,337	2,198,524	4,458,228	1,738	2,900,925	7,882,019	5,015	3,019,066	6,631,032	3,627
F ∩ T	283,297	8,956	4	108,576	59,912	9	199,636	376	0	140,320	3,273	5	300,574	112,098	22
	N6			**N7**			**N8**			**N9**			**N10**		
	$\|\mathcal{V}\|$	$\|\mathcal{E}\|$	$\|Sp\|$	$\|\mathcal{V}\|$	$\|\mathcal{E}\|$	$\|Sp\|$	$\|\mathcal{V}\|$	$\|\mathcal{E}\|$	$\|Sp\|$	$\|\mathcal{V}\|$	$\|\mathcal{E}\|$	$\|Sp\|$	$\|\mathcal{V}\|$	$\|\mathcal{E}\|$	$\|Sp\|$
F	2,387,610	5,356,288	3,498	627,147	1,071,120	696	2,036,162	2,876,783	894	1,197,935	2,139,912	2,317	2,174,023	4,280,962	2,323
T	1,297,371	1,727,503	481	1,166,528	2,524,907	847	1,058,482	1,513,404	489	2,999,865	6,317,032	1,833	704,006	1,314,996	741
F ∪ T	2,449,434	5,691,728	3,769	1,606,924	3,577,449	1,534	2,663,392	4,082,373	1,365	4,064,545	8,443,888	4,151	2,729,312	5,584,915	3,063
F ∩ T	1,235,547	1,379,510	212	186,751	11,131	9	431,252	305,358	20	133,255	722	1	148,717	699	1

4 Experimental Analysis

4.1 Data Collection

We evaluate our proposed model using real world Twitter datasets. The ground truth of false information and the refuting true information was obtained from *www.altnews.in*, a popular fact checking website based in India and are based around politics in India. The source tweet related to the information was obtained directly as a tweet embedded in the website. From that source tweet, we used the Twitter API to determine the source tweeter and retweeters (proxy for spreaders), the follower-following network of the spreaders (proxy for social network), and user activity data (100 most recent tweets) for all nodes in the network. Trust and credibility scores extracted from the activity data are summarized in Fig. 2 are directly used as feature vectors. Besides evaluating our model on the false information (F) and true information (T) spreading networks separately, we also evaluated our model on the combined information spreading networks (F ∪ T). Details regarding the number of nodes ($|\mathcal{V}|$), edges ($|\mathcal{E}|$), and spreaders ($|Sp|$) for the networks of 10 different news events (N1-N10) is detailed in Table 1.

4.2 Analysis of F ∩ T

F ∩ T in Table 1 denotes the section of the network that was exposed to both the false and its refutation information. An interesting observation is the spreaders who decided to spread both types of information. Figure 3 (a) denotes the distribution of spreaders in F ∩ T who spread false information followed by its refutation (FT) and those whose spread refutation followed by the false information (TF). N1 and N9 is excluded from the analysis as our dataset as we did not have the spreaders' timestamp information. An interesting observation is that the majority of spreaders belong to FT. Intuitively, these are spreaders

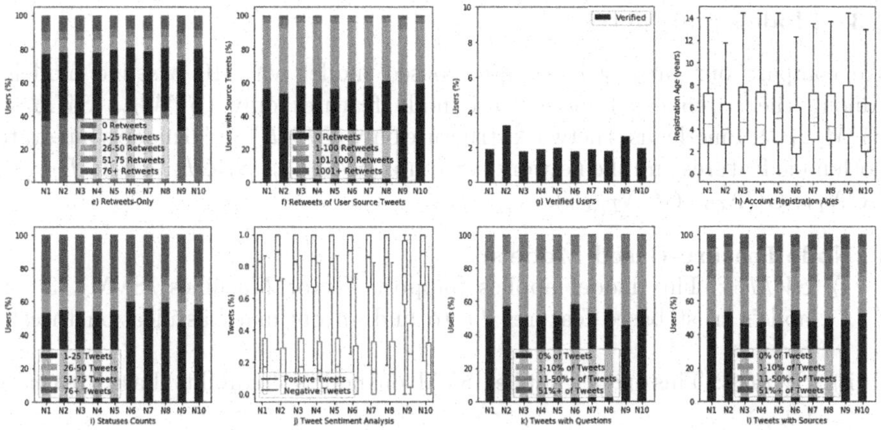

Fig. 2. Trust and credibility feature analysis from networks N1-N10.

who trusted the endorser without verifying the information and later corrected their position, thereby implying that they did not intentionally want to spread false information. Consequently, the proposed model can help identify such people proactively in order to take measures to prevent them from endorsing false information in the first place. While spreaders belonging to TF are comparatively fewer (whose intentions are not certain) the proposed model can help identify them and effective containment strategies can be adopted. Figure 3 (b) shows the time that transpired between spreading refutation and false information for FT spreaders. Once the false information is endorsed, large portions of the network must have already been exposed to false information before the endorser corrected themselves after a significant amount of time (~1 day). This serves as a strong motivation to have a spreader prediction model which proactively identifies likely future spreaders.

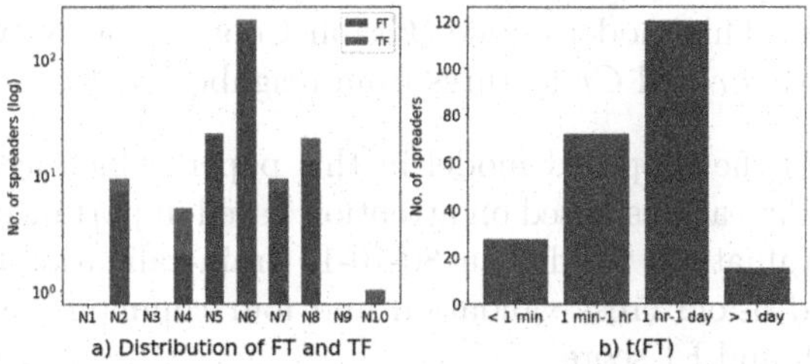

Fig. 3. Analysis of spreaders in F ∩ T.

4.3 Models and Metrics

We compare our proposed attention based model with 10 baseline models. Among the baselines, 3 models use node features only (SVM_{Tr}, SVM_{Cr}, $SVM_{Tr,Cr}$), 1 model uses network structure only ($LINE$) and 6 models integrate both node features and the network structure ($SAGE_{Tr}$, $SAGE_{Cr}$, $SAGE_{Tr,Cr}$, GCN_{Tr}, GCN_{Cr}, $GCN_{Tr,Cr}$).

1. **Node Feature-Based Models:**
   - i). SVM_{Tr}: This model applies Support Vector Machines (SVM) [3] on node's trust based features Tr to find an optimal classification threshold.
   - ii). SVM_{Cr}: This model applies SVM on node's credibility based features Cr.
   - iii). $SVM_{Tr,Cr}$: This model applies SVM by combining node's trust based and credibility based features.
2. **Network Structure-Based Models:**
   - iv). $LINE$: Applies the Large-scale Information Network Embedding [18] as a transductive representation learning baseline, where node embeddings are generated after optimization is performed on the entire graph structure.
3. **Network Structure + Node Feature-Based Models:**
   - v). $SAGE_{Tr}$: GraphSAGE [4] serves as the inductive learning baseline where node embeddings are generated by aggregating Tr features from neighborhoods.
   - vi). $SAGE_{Cr}$: This inductive representation learning baseline generates node embeddings by aggregating Cr features from neighborhoods.
   - vii). $SAGE_{Tr,Cr}$: This inductive representation learning baseline generates node embeddings by aggregating both Tr and Cr features from neighborhoods.
   - viii). GCN_{Tr}: This model applies Graph Convolution Networks [6] to learn node embeddings by aggregating Tr features from neighborhoods.
   - ix). GCN_{Cr}: This model applies Graph Convolution Networks by aggregating Cr features from neighborhoods.
   - x). $GCN_{Tr,Cr}$: This model applies Graph Convolution Networks by aggregating both Tr and Cr features from neighborhoods.

$SCARLET$ is the proposed model in this paper, which aggregates a node neighborhood's Cr features based on attention based importance scores assigned using Tr. For evaluation, we did an 80-10-10 train-validation-test split of the dataset. We used 5-fold cross validation and four common metrics: Accuracy, Precision, Recall, and F1 score.

4.4 Implementation Details

We obtained Global Trust features by running the TSM algorithm on the follower-following network of the spreaders. We used the generic settings for

TSM parameters (number of iterations = 100, involvement score = 0.391) based on [16]. The size of sampled neighborhood was set to 50 and depth was set to 1. We considered neighbors with higher degrees in order to generate denser adjacency matrices. The number of epochs, batch size, learning rate and dropout rate were set to 200, 64, 0.001 and 0.2, respectively. The code implementation is also available[1].

Table 2. Model performance evaluation ($\mathcal{V}_F$): False information spreader, ($\mathcal{V}_T$): Refutation spreader.

	F ($\mathcal{V}_F$)				T ($\mathcal{V}_T$)				F ∪ **T** ($\mathcal{V}_F$)			
	Accu.	Prec.	Rec.	F1	Accu.	Prec.	Rec.	F1	Accu.	Prec.	Rec.	F1
SVM_{Tr}	0.497	0.512	0.468	0.478	0.473	0.472	0.452	0.445	0.398	0.19	0.465	0.229
SVM_{Cr}	0.508	0.517	0.517	0.509	0.501	0.477	0.565	0.509	0.408	0.196	0.542	0.272
$SVM_{Tr,Cr}$	0.516	0.514	0.579	0.53	0.52	0.513	0.598	0.545	0.444	0.193	0.489	0.267
$LINE$	0.686	0.626	0.896	0.733	0.635	0.608	0.881	0.717	0.688	0.71	0.896	0.786
$SAGE_{Tr}$	0.734	0.762	0.691	0.722	0.680	0.698	0.719	0.705	0.752	0.743	0.859	0.793
$SAGE_{Cr}$	0.747	0.772	0.710	0.736	0.714	0.692	0.764	0.725	0.764	0.747	0.881	0.805
$SAGE_{Tr,Cr}$	0.779	0.831	0.720	0.763	**0.755**	**0.787**	0.732	0.755	0.785	0.764	0.878	0.814
GCN_{Tr}	0.784	0.726	0.947	0.821	0.718	0.675	0.916	0.767	0.753	0.783	0.930	0.845
GCN_{Cr}	0.800	0.742	0.953	0.834	0.731	0.697	0.906	0.773	0.762	0.786	0.940	0.851
$GCN_{Tr,Cr}$	0.824	0.774	0.942	0.848	0.743	0.702	0.916	0.783	0.776	**0.788**	0.954	0.861
$SCARLET$	**0.876**	**0.834**	**0.966**	**0.893**	0.734	0.674	**0.981**	**0.794**	**0.789**	0.785	**0.972**	**0.866**

4.5 Performance Evaluation

Classification results of the baselines and proposed model are summarized in Table 2. The results are averaged over the 10 news events. We report the precision, recall, and F1 scores of the false information spreaders class ($\mathcal{V}_F$) in F and F ∪ T networks, and of the refutation spreaders class ($\mathcal{V}_T$) in T network. Due to class imbalance, we undersample the majority class to obtain balanced class distribution. We observe that structure only baseline performs better than feature only baselines, and models that combine both node features and network structure show further improvement in performance. Additionally, we observe that Cr features perform better than Tr features (because there are more number of Cr features than Tr features) and the model performance increases when we use Tr and Cr features together. $LINE$, the structure only baseline, performs better than feature only baselines by a substantial margin, which suggests that network structure plays an important role in identifying false information spreaders. In terms of accuracy, the $LINE$ model shows an increase of 32.9%, 22.1% and 54.9% for F, T and F ∪ T networks, respectively, over $SVM_{Tr,Cr}$. Graph neural network baselines that combine both network structure and node features show a significant improvement in performance. GCN models perform better than *GraphSAGE* models on all metrics for F networks, while that is not the case for T and F ∪ T networks. This is because Tr and Cr features

[1] https://github.com/BhavtoshRath/GAT-GCN-SpreaderPrediction.

for neighborhood of refutation information spreaders and non-spreaders do not differ much from each other. Our proposed model *SCARLET* shows an increase in performance for all three networks. However, $SAGE_{Tr,Cr}$ shows better accuracy and precision on T networks because the specific news events on which it performed better involved religious tones, and so decision to refute them is more sensitive to neighborhood's Cr than Tr. Precision on F ∪ T networks is highest for $GCN_{Tr,Cr}$, though it is still comparable to the proposed model's performance. More importantly, in the F ∪ T network we observe highest accuracy and F1 scores of 78.9% and 86.6% , thus supporting our hypothesis that false information spreading is very sensitive to trust and credibility.

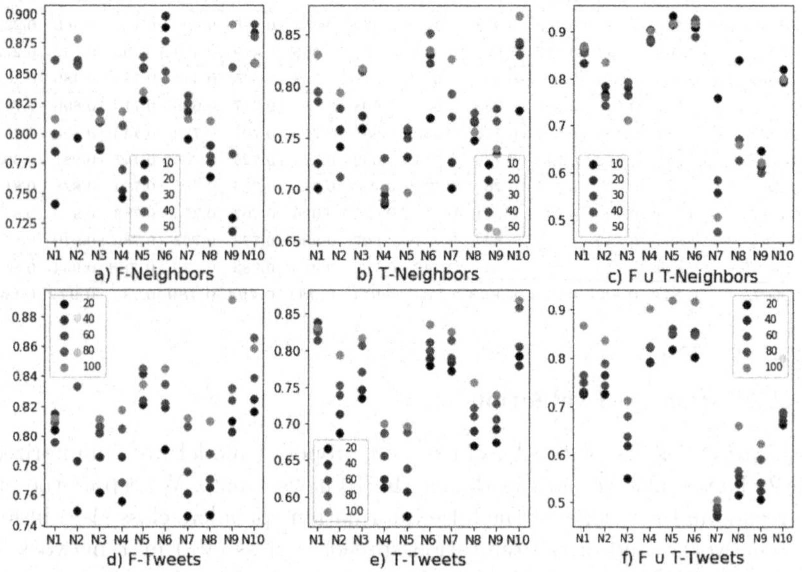

Fig. 4. Sensitivity analysis: Neighborhood size (Neighbors) and features (Tweets). (x-axis: News events N1-N10, y-axis: F1 scores for spreader prediction).

4.6 Sensitivity Analysis

Figure 4 shows the sensitivity analysis of F1 scores of the proposed model on two important parameters: the size of neighborhoods (Neighbors), and the number of recent tweets from user timeline (Tweets).

Neighbors: We evaluated our model on n-neighbors, where n = 10, 20, 30, 40, 50. Figure 4(a), (b), and (c) show results on F, T and F ∪ T networks, respectively. We observe that model performance is not very sensitive to varying neighborhood size, which could be attributed to the fact that since we have only the immediate follower-following network (sampling depth=1) we are not able to entirely capture meaningful dynamics (i.e. the decision to retweet might depend less on the immediate neighbors, and more on the source tweeter).

Tweets: We also evaluated our model on the n-most recent timeline tweets, where n = 20, 40, 60, 80, 100. Figure 4(d), (e), and (f) shows results on F, T and F ∪ T networks, respectively. We observe that for all three networks, prediction performance tends to increase as the number of timeline tweets used to aggregate features increases. This is probably because using more behavioral data helps us estimate trust and credibility features better.

4.7 Explainability Analysis of Trust and Credibility

Figure 5 shows importance scores that false ($\mathcal{V}_F$) and refutation ($\mathcal{V}_T$) spreader's neighbors (size = 10) assign each other based on trust dynamics (softmax attention score) and credibility score (euclidean norm of normalized feature vector) for neighbors with both high and low modularity. Node 0 is the neighbor that the spreader endorses. We observe that $\mathcal{V}_T$'s neighbors have higher credibility than $\mathcal{V}_F$'s neighbors because of network homophily. Also low magnitude of importance scores for neighbors of node 0 of $\mathcal{V}_F$ suggest that it's neighbors trust each other less compared to $\mathcal{V}_T$'s neighbors. We observe in Fig. 5(a) and (b) that node 0 in $\mathcal{V}_F$'s neighbor has strong trust dynamics with its followers (i.e. incoming edges) because it has more incoming edges than outgoing edges and also retweets and gets retweeted substantially more by the neighbors, unlike who $\mathcal{V}_T$ endorses in Fig. 5 c) and d), because $\mathcal{V}_T$'s decision to endorse depends more on information source, which is usually a fact checker.

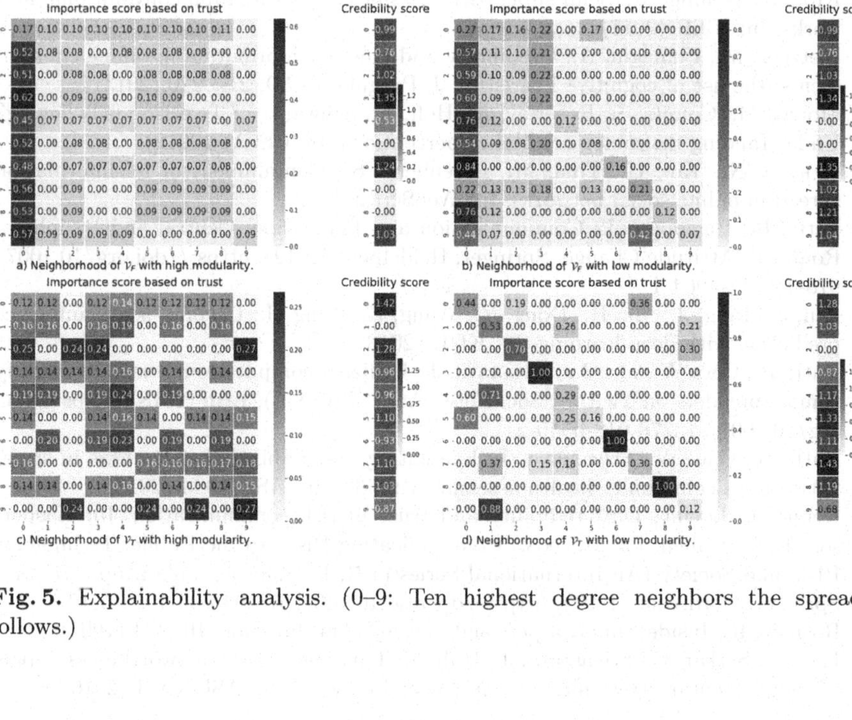

Fig. 5. Explainability analysis. (0–9: Ten highest degree neighbors the spreader follows.)

5 Conclusions and Future Work

We propose *SCARLET*, an attention-based explainable graph neural network model to predict whether a node is likely to spread false information or not. The model learns node embeddings by first assigning trust-based importance scores and then aggregating its neighborhood's credibility features proportionally. What makes this model different from most existing research is that it does not rely on features extracted from the information itself. Thus it can be used to predict spreaders even before information spreading begins. As part of future work, we would like to analyze our model on more news events comprising larger networks in order to sample and aggregate features at greater sampling depths.

References

1. Budak, C., Agrawal, D., Abbadi, A.: Limiting the spread of misinformation in social networks. In: WWW (2011)
2. Castillo, C., Mendoza, M., Poblete, B.: Information credibility on Twitter. In: WWW (2011)
3. Cortes, C., Vapnik, V.: Support-vector networks. Mach. Learn. **20**(3), 273–297 (1995)
4. Hamilton, W., Ying, Z., Leskovec, J.: Inductive representation learning on large graphs. In: NeurIPS (2017)
5. Jaeger, M., Anthony, S., Rosnow, R.: Who hears what from whom and with what effect: a study of rumor. Pers. Soc. Psychol. Bull. **6**, 473–478 (1980)
6. Kipf, T., Welling, M.: Semi-supervised classification with graph convolutional networks. In: ICLR (2017)
7. Metzger, M., Flanagin, A.: Credibility and trust of information in online environments: the use of cognitive heuristics. J. Pragmatics **59**, 210–220 (2013)
8. Morris, M., Counts, S., Roseway, A., Hoff, A., Schwarz, J.: Tweeting is believing? Understanding microblog credibility perceptions. In: CSCW (2012)
9. Nguyen, N., Yan, G., Thai, M., Eidenbenz, S.: Containment of misinformation spread in online social networks. In: WebSci (2012)
10. Petty, R., Cacioppo, J.: Communication and Persuasion: Central and Peripheral Routes to Attitude Change. Springer, Heidelberg (2012). https://doi.org/10.1007/978-1-4612-4964-1
11. Qiu, J., Tang, J., Ma, H., Dong, Y., Wang, K., Tang, J.: Deepinf: social influence prediction with deep learning. In: KDD (2018)
12. Rath, B., Gao, W., Ma, J., Srivastava, J.: Utilizing computational trust to identify rumor spreaders on Twitter. Soc. Netw. Anal. Min. **8**(1), 1–16 (2018). https://doi.org/10.1007/s13278-018-0540-z
13. Rath, B., Gao, W., Srivastava, J.: Evaluating vulnerability to fake news in social networks: a community health assessment model. In: ASONAM (2019)
14. Renn, O., Levine, D.: Credibility and trust in risk communication. In: Kasperson, R.E., Stallen, P.J.M. (eds.) Communicating Risks to the Public. Technology, Risk, and Society (An International Series in Risk Analysis), vol. 4, pp. 175–217. Springer, Dordrecht (1991). https://doi.org/10.1007/978-94-009-1952-5_10
15. Rosnow, R.: Inside rumor: a personal journey. Am. Psychol. **46**, 484 (1991)
16. Roy, A., Sarkar, C., Srivastava, J., Huh, J.: Trustingness & trustworthiness: a pair of complementary trust measures in a social network. In: ASONAM (2016)

17. Sharma, K., Qian, F., Jiang, H., Ruchansky, N., Zhang, M., Liu, Y.: Combating fake news: a survey on identification and mitigation techniques. In: TIST (2019)
18. Tang, J., Qu, M., Wang, M., Zhang, M., Yan, J., Mei, Q.: Line: large-scale information network embedding. In: WWW (2015)
19. Tong, A., Du, D., Wu, W.: On misinformation containment in online social networks. In: NeurIPS (2018)
20. Vaswani, A., et al.: Attention is all you need. In: NeurIPS (2017)
21. Veličković, P., Cucurull, G., Casanova, A., Romero, A., Lio, P., Bengio, Y.: Graph attention networks. In: ICLR (2018)
22. Wu, Z., Pan, S., Chen, F., Long, G., Zhang, C., Philip, S.: A comprehensive survey on graph neural networks. Trans. Neural Netw. Learn. Syst. (2020)
23. Lu, Y., Li, C.: GCAN: graph-aware co-attention networks for explainable fake news detection on social media. In: ACL (2020)
24. Shu, K., Cui, L., Wang, S., Lee, D., Liu, H.: defend: explainable fake news detection. In: KDD (2019)

Content Matters: A GNN-Based Model Combined with Text Semantics for Social Network Cascade Prediction

Yujia Liu, Kang Zeng, Haiyang Wang, Xin Song, and Bin Zhou(✉)

National University of Defense Technology, ChangSha, China
{liuyujia,zengkang,wanghaiyang19,songxin,binzhou}@nudt.edu.cn

Abstract. Effectively modeling and predicting the size of information cascades is essential for downstream tasks such as rumor detection and epidemic prevention. Traditional methods normally rely on tedious handcrafted feature engineering efforts, which is inefficient in complex diffusion processes such as social network (SN) cascades. In recent years, graph neural network methods have been successfully used in cascade prediction tasks. Most of these methods make use of the structural features of SNs, while the effect of textual user-generated content (UGC) is far from clear or fully utilized. In this paper, we focus on the questions of how the textual UGC affect user activation state and trigger their retweet behaviors. We propose a novel GNN-based model named TSGNN, which jointly model the textual content and structure features. It uses recurrent neural networks with attentions to learn content feature representations that potentially affect information propagation. We find that tweets of fewer high coherent topics are more likely to trigger user retweet behaviors, and we also design a gate mechanism to model the activation state of users under the combined influence of content, structure, and other self-activation. Experimental results demonstrate that TSGNN significantly outperforms all the state-of-the-art methods in multiple metrics.

Keywords: Cascade prediction · Graph neural network · Text semantics

1 Introduction

With the popularity of online social platforms (e.g. Twitter, Sina Weibo, etc.), network users communicate on topics of interest by posting and sharing various user-generated content (UGC) with each others, which makes information propagation much easier. Meanwhile, the huge amount of information brings a series of challenges to platform managers, such as viral marketing identification [9], influence prediction [12], rumor detection [1] and epidemic prevention [18], etc. It becomes challenging and vital to know the factors behind information diffusion and to predict cascading size. Most of the previous works mainly rely on dissemination structure and content and most of these methods relies on feature selection within the machine learning pipelines, which is inefficient

K. Karlapalem et al. (Eds.): PAKDD 2021, LNAI 12712, pp. 728–740, 2021.
https://doi.org/10.1007/978-3-030-75762-5_57

in complex diffusion processes such as social network (SN) cascades. In order to measure the role of content in information diffusion, researchers have made some useful attempts. Existing works use quite a few vector or language models such as TF-IDF and Latent Dirichlet Allocation (LDA), combined with typical machine learning models (e.g., SVM) to analyze the textual UGC [5]. In addition, textual content and linguistic features, including comment length, number of verbs/nouns, content entropy, readability [8], emoticons, etc., are the most well studied [14]. However, opinions on the effectiveness of content features still differ [21]. The question of how the textual UGC affect user activation states and trigger their retweet behaviors remains unclear [4]. With the success of graph representation learning technology, various graph neural networks are used to capture the structure of information diffusion in a natural and effective manner. Related methods represent the observed cascade graph as a series of subgraphs [3], or model the cascade graph as a dynamic graph [6]. However, most of these GNN-based methods only concentrate on the representation of graph structure, and ignore the hidden role of text semantics in cascading diffusion. Will the content benefit GNN-related methods? To what extend will it be helpful?

To solve the above problems, we propose a novel GNN-based model called TSGNN, which shows that the accuracy of cascade prediction improves by taking text semantic into account. TSGNN is devided into two parts, social network structure modeling and text semantic modeling. For social network structure modeling, we use a GNN to characterize the underlying social structure, where the neighborhood aggregation strategy can effectively capture the influence from social networks. For a target user, we feed the topic embedding, his/her influence from social networks and self-activation (reflect the influence of offline communication, etc.) into a gated activation unit to simulate his/her activation state.

For text semantic modeling, [20] showed that topic models with a lower content complexity lead to a better cascade prediction accuracy. Inspired by it, we bias the topic representation to some fewer key ones and neglecting the rest. In details, we designed a cascade text Topic Embedding (TE) module to fulfill the task. It contains the follow steps. (1) A short-text topic model, BTM [17], is used for topic discovery on the entire cascade corpus in order to obtain a *topic encoding path* for each cascade text; (2) All of the topic encoding paths are fed into a RNN to get a topic embedding; (3) An attention mechanism is used to bias the focus on those key parts of the topic embeddings found in (2).

As far as we know, TSGNN is the first GNN-based model that combines cascade text and social structure for social information cascading prediction. To sum up, the contributions of this work are:

- We introduce a Topic Embedding (TE) module, which uses RNN and attention mechanism to effectively characterize the textual semantics affecting social information diffusion.
- We propose a novel model named TSGNN based on graph neural network to jointly model cascaded content features and network structure. Our result also shows that textual content with coherent semantics contribute to users' retweet behaviors.

– We conducted experiments on different social network datasets, demonstrating that TSGNN significantly outperforms the state-of-the-art baselines.

2 Related Work

In this section, we review related work on information cascade prediction and related research on graph neural networks.

2.1 Information Cascade Prediction

Existing research has proved that the information cascade is predictable to some extent [5,8,20]. Feature-based methods usually analyze and extract effective features related to the popularity of social network content, including hashtags [13], the number of forwarded messages and certain features related to users [7]. Generally, the performance of feature-based methods is highly dependent on the extracted features and is usually only suitable to particular data. Recently, DFTC model [11] uses Hierarchical Attention Network (HAN) to model the content characteristics of online articles. However, when applied to short texts like tweets, the above method may not be effective. The difference between our proposed model and above is that we utilize the Recurrent Neural Network to directly model fewer key topics representation of text and could be extended to other datasets in theory.

2.2 GNNs-Based Approaches

In the aspect of information cascade structure modeling, researchers have made extensive attempts [21]. Inspired by the huge success of graph neural network in graph representation learning, CasCN [3] samples the cascaded graphs according to time slices for subgraphs and merges GCN and LSTM to learn the cascade structure and the time attenuation effect in cascade propagation. Cascade2vec [6] designed the graph residual block to learn more effective graph-level representation of cascades. However, most of these models tend to model the network structure, and seldom do they focus on the semantic information of cascade text. In our work, we suggest jointly model cascade text semantics and cascade structure to better adapt to the scenario of social information diffusion.

3 Methods

In this section, we introduce the proposed model TSGNN. It takes a topic embedding of the cascading text and the corresponding information cascading network graph as input (shown as Fig. 1(a)) and outputs the cascade size prediction of the social information. After an overview of TSGNN, we will focus on the details in each section. The main part of the TSGNN model includes three components: (1) **Topic embedding module**, embedding text topic into a low-dimensional space to represent the semantics of the text (2) **The TSGNN layer**, it uses the neighborhood aggregation strategy to get the influence of the entire social network on the node. (3) **Gated activation unit**, simulating the activation status of nodes under the influence of text, social networks, and self-activation.

3.1 Problem Definition

We cast information cascade size prediction as a regression problem, and the goal is to predict the final size of the social information cascade. For a piece of information m in the social network, we record the list of active users $C_T^m = \{u_1, u_2, ..., u_n\}$ who took actions on the information m within the observation time window T, the cascade size prediction can be formally defined as:

Definition 1. *Cascade size prediction. Given N information items* $\{m_1, m_2, ..., m_n\}$ *and the list of active users* $C_T^{m_i} = \{(u_j, t_j) \mid j \in [1, N], t_j \leq t_p\}$ *who took action on each information item* m_i *in the observation window T. Given a known social network* $\mathcal{G} = (\mathcal{V}, \mathcal{E})$ *where* $\mathcal{V}$ *is the set of all users, and* $\mathcal{E} \subseteq \mathcal{V} \times \mathcal{V}$ *is the set of relationships among all users. The purpose of cascade size prediction is to give the final cascade size* $\tilde{S}_i$ *of the information item* m_i *in the future.*

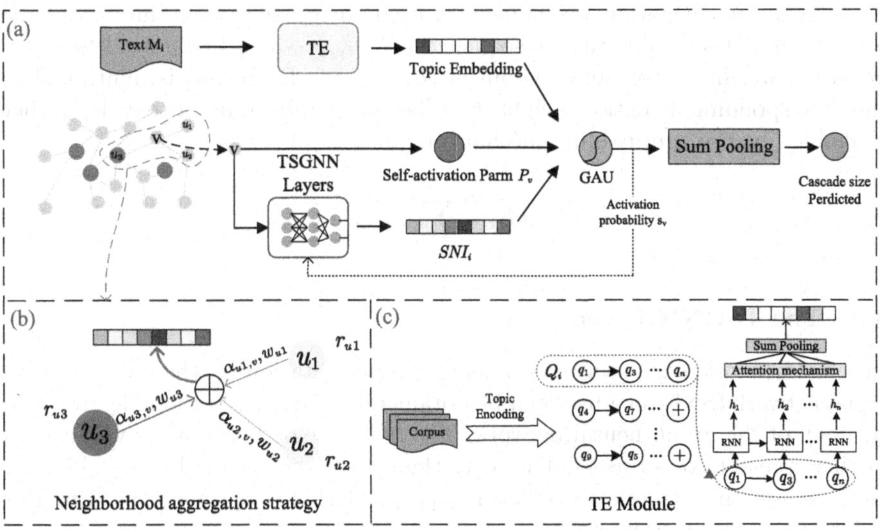

Fig. 1. (a) The overall model of TSGNN: it first takes the social network graph and cascading text as input, and then outputs the social network influence (SNI), topic embedding and self-activation influence of each node (user v and text M_i in this case). Finally, the sum pooling mechanism summarizes the activation probability of each node as the predicted cascade size. (b) The neighborhood aggregation strategy at TSGNN Layer. (c) TE module: calculating the topic embedding of each text.

3.2 Topic Embedding Module

For a retweet cascade, semantic components of the tweet may trigger user retweet behaviors, e.g., controversial topics, disinformation and misinformation attract significantly more attention than normal content [21]. Specifically, we input N pieces of texts as a corpus into the BTM [17] to generate the Document-Topic matrix, and each row $L_i = \{l_{i1}, l_{i2}, ..., l_{ij}, ...l_{in}\}$ represents a topic distribution corresponding to the text M_i, n represents the total number of topics. Each element $l_{ij} = (k_{ij}, d_{ij})$ in the list L_i, where k_{ij} is the topic index, and d_{ij} is the probability of topic k_{ij}. Assuming that the proportion of these n topics in the text M_i satisfies $d'_{i1} > d'_{i2} > d'_{i3} \cdots > d'_{in}$, the topic path of the message M_i is expressed as a set of sequences $P_i = \{k'_{i1}, k'_{i2} \cdots k'_{in}\}$. In order to obtain the representation of each topic k'_{ij} in the topic path P_i, we first compute the word embeddings of the keywords of topic k'_{ij} and then do the concatenate operation to obtain the topic encoding q_i of topic k'_{ij}. In order to focus on fewer topics, we select the top x (a hyper-parameter which will be discussed in Sect. 5.3) valid topics in P_i, and the remaining topics are expressed as all-zero vectors. Finally, each topic path P_i is encoded as $Q_i = \{q_1, q_2, \cdots q_n\}$. Then we feed Q_i into the RNN units to capture the importance of topics and the dependence between topics. The hidden state of the output by the t-th RNN unit is $h_{it} \in \mathbb{R}^H$, where H is the dimension of the hidden state.

In addition, attention mechanism is used to measure topic importance in our module. Specifically, the subsequence $\{d'_{i1}, d'_{i2}, \cdots d'_{in}\}$ in L_i is taken as the attention weight vector, and the output h_{it} of each RNN unit is multiplied by the corresponding attention weight d'_{it}. The topic embedding of text M_i is then assembled by the sum pooling mechanism:

$$TE_i = \sum_{t=1}^{n} d'_{it} h_{it} \quad (1)$$

3.3 The TSGNN Layer

In social network, each user v has a feature representation f_v, the influence from social network (SNI) on a user can be obtained by the neighborhood aggregation strategy of the graph neural network. The influence of active users will usually further spread to other users along with the network structure. Inspired by GAT [16], we use the attention coefficient α_{uv} in GAT to represent the attention weight from user u to user v.

Then the influence of user v from the social network can be expressed as:

$$SNI_v^{(k)} = \sum_{u \in \mathcal{N}(v)} s_u^{(k)} \alpha_{uv}^{(k)} W^{(k)} f_u^{(k)} \quad (2)$$

Where $s_u^{(k)} \in \mathbb{R}$ is the activation probability of user u at the k-th GNN layer output by the gated activation unit, which will be introduced in Sect. 3.4. $W^{(k)} \in \mathbb{R}^{h^{(k)} \times h^{(k)}}$ is a weight matrix, $f_u^{(k)} \in \mathbb{R}^{h^{(k)}}$ is the user u's feature representation at the k-th layer. Then update the user u's feature representation at the $(k+1)$-th layer:

$$f_v^{(k+1)} = \sigma\left(\zeta_r^{(k)} W^{(k)} f_v^{(k)} + W^{(k)}\left[SNI_v^{(k)} \| TE_i\right]\right) \tag{3}$$

Where $W^{(k)} \in \mathbb{R}^{h^{(k)} \times (h^{(k)}+H)}$ is a weight matrix, $\zeta_r^{(k)} \in \mathbb{R}$ is a weight parameter, σ is the nonlinear activation function, $TE_i \in \mathbb{R}^H$ is obtained by the Eq. (1) and $\|$ represents the concatenate operation. The user's initial feature representation mainly consists of two parts: (1) node embedding (2) node attributes (e.g., the PageRank score, the out-degree of the node, the authority score and the hub score, etc.)

3.4 Gated Activation Unit

Gated activation unit (GAU) is used to model the activation states of users in social networks. Similarly to [2], each user v in TSGNN is associated with an activation probability s_v. The GAU takes the combined influence of text, structure, and other self-activation as input to get the user's activation probability. The combined of the expected influence that the target user v receives is:

$$a_v^{(k)} = \sum_{u \in \mathcal{N}(v)} \beta^{(k)}\left[W^{(k)} f_u^{(k)} \| W^{(k)} f_v^{(k)}\right] s_u^{(k)} + p_v \tag{4}$$

Where $W^{(k)} \in \mathbb{R}^{h^{(k)} \times h^{(k)}}$ represents the weight matrix, $\beta^{(k)} \in \mathbb{R}^{2h^{(k)}}$ is a weight vector, $s_u^{(k)}$ represents the activation probability of user u and $p_v \in \mathbb{R}$ represents the self-activation parameters.

The activation probability of user v at the $(k+1)$-th layer can be updated to:

$$s_v^{(k+1)} = \sigma\left(\eta_s^{(k)} s_v^{(k)} + \eta_a^{(k)} a_v^{(k)}\right) \tag{5}$$

Where $\eta_s^{(k)}, \eta_a^{(k)} \in \mathbb{R}$ is the weight parameter and σ is the nonlinear activation function. The initial activation probability of the user in the observation cascade is set to 1 (black nodes in Fig. 1(a)), otherwise 0 (gray nodes in Fig. 1(a)).

3.5 Cascade Size Prediction

After the last GNN layer, the activation probability of each user output by the GAU is $s^{(K)} \in [0, 1]$, and the size of the information cascade can be expressed as the sum of all users' activation probability:

$$\tilde{S}_i = \sum_{u \in \mathcal{V}} s_u^{(K)} \tag{6}$$

Our final task is to predict the size of the information cascade, which we choose mean relative square error (MRSE) as the optimized loss function:

$$\ell\left(S_i, \tilde{S}_i\right) = \frac{1}{N}\sum_{i=1}^{N}\left(\frac{\tilde{S}_i - S_i}{S_i}\right)^2 \tag{7}$$

Where N is the total number of cascade texts, S_i is the ground truth and $\tilde{S}_i$ is the final predicted cascade size of TSGNN.

In order to accelerate model convergence and avoid overfitting during model training, we use the DropEdge mechanism [15] which has been proven to play a good role in preventing GNN-based model training from overfitting. Similarly, We randomly drop certain edges with probability p when entering the adjacency matrix.

4 Experiments

In this section, we compare the prediction performance of our proposed TSGNN model with several baselines and variants of the TSGNN model and performed it on the real-world Sina Weibo and Twitter datasets experiment.

4.1 Datasets

To comprehensively evaluate the effectiveness of our model, we selected Sina Weibo and Twitter data sets to conduct experiments. The topic distributions of tweets in the datasets are shown in Table 1.

Sina Weibo Dataset. Sina Weibo is the most popular microblogging platform in China. The Sina Weibo dataset we use can be downloaded from here.[1] This is a user-following network, which records 300,000 popular microblog retweet information cascade in a month. In order to construct a forwarding path based on the information cascade, we extracted the corresponding cascade path in the network. Specifically, we randomly select some users in the network and then traverse the forwarding path of each information cascade, record the selected users that appear in the cascaded forwarding path, and then obtain the following relationships from the global network. We set the observation time window to T = 1.5 h and T = 3 h.

Twitter Dataset. In addition, We quoted the twitter dataset used in [1] which can be downloaded here[2]. Note that there is no complete following network among users in this dataset. In order to extract the following relationship between users, we first counted the number of times that the user directly reposted another user's tweets, and then assumed that if user v directly reposted the tweets of user u up to 4 times, we would assume that user v has followed user u. Based on this assumption, we extracted a total of 3579 users and 13154 edges. We set the observation time window to T = 2 h.

1 https://www.aminer.cn/Influencelocality.

2 https://www.dropbox.com/s/7ewzdrbelpmrnxu/rumdetect2017.zip.

Table 1. Topic distribution of datasets

Topic category	Sina Weibo	Twitter
	Tweets number	
Social	5314	1029
Economic	9715	295
Sports	4615	191
Entertainment	6365	421
Political	3115	599
Daily life	12304	611
Education	8975	284

4.2 Baselines

The baselines and their implementation details are as follows:

Feature-Based: In our implementations, We consider the degree distribution of the network and the number of leaf nodes as the *structural features*, and for *temporal features*, we select the time difference between the forwarding time of the node in the cascade and the publishing time of the original tweet, finally use the logistic regression classifier to make the cascade prediction.

SEISMIC [19]**:** SEISMIC is a typical method for cascading prediction using a self-excited process generation model, which takes the historical interval forwarded by the node and the degree of the node as input.

DeepCas [10]**:** The model first performs a random walk on the cascade graph, and expresses the cascade graph as the path obtained by the walk, then input node embeddings into the RNN to obtain a sequence representation.

CoulpedGNN [2]**:** CoupledGNN designed a state graph neural network and an influence graph neural network that are coupled to each other, which effectively captures the cascade effect in information diffusion.

TSGNN-LSTM/GRU: The LSTM/GRU unit is used to replace the RNN unit in the TSGNN.

TSGCN: TSGCN uses basic GCN layer to replace the TSGNN layer.

4.3 Evaluation Metrics

According to the existing work, we have selected several evaluation indicators to judge the effect of the model.

- *Wrong Percentage Error* (*WroPerc*). WroPerc is defined as the percentage of cascade size that are incorrectly predicted for a given error tolerance ε.

- *Mean Absolute Percentage Error (MAPE)*. MAPE is used to measure the average error between the predicted size of the cascade and the actual size of the cascade.
- *Median Relative Square Error (mRSE)*. mRSE is defined as the 50th percentile of RSE in the data distribution of the test set.

5 Results

In this section, We first show the performance of our model and comparison methods, then we compare different variants of TSGNN, and finally, we analyze the parameter settings in TSGNN.

Table 2. Cascade size prediction.

Dataset	Sina Weibo								Twitter			
Observation time	1.5 h				3 h				2 h			
Evaluation metric	MRSE	MAPE	mRSE	WroPerc	MRSE	MAPE	mRSE	WroPerc	MRSE	MAPE	mRSE	WroPerc
Features-Based	0.2207	0.3713	0.1254	34.27%	0.1641	0.3147	0.0912	20.11%	0.1588	0.3177	0.1159	19.52%
SEISMIC	–	–	0.1911	44.38%	–	–	0.0882	26.88%	–	–	0.1235	24.77%
DeepCas	0.2077	0.3633	0.1716	31.45%	0.1456	0.2844	0.0489	18.23%	0.1267	0.2989	0.0914	17.34%
CoupledGNN	0.1916	0.3523	0.1437	28.66%	0.1239	0.2711	0.0455	14.33%	0.0926	**0.2368**	0.0511	13.58%
TSGNN	**0.1773**	**0.3502**	**0.1286**	**27.84%**	**0.1143**	**0.2587**	**0.0366**	**13.44%**	**0.0830**	0.2442	**0.0496**	**12.64%**
TSGNN-LSTM	0.1834	0.3520	0.1337	27.93%	0.1189	0.2703	0.0432	14.32%	0.0950	0.2596	0.0764	14.03%
TSGNN-GRU	0.1893	0.3527	0.1335	29.01%	0.1225	0.2712	0.0456	14.61%	0.1160	0.2738	0.0697	13.51%
TSGCN	0.1901	0.3517	0.1592	28.11%	0.1251	0.2748	0.0471	14.78%	0.1188	0.2749	0.0637	14.08%

5.1 Performance Comparison

Table 2 shows the experiment results on the Sina Weibo and Twitter datasets. First of all, for the method based on feature engineering, due to the model effect depends on the selected features, and the features are difficult to summarize the key factors in cascading information diffusion, the effect is not ideal from the experimental results. For SEISMIC, it does not model the complex network structure but only takes the time and number of reposts of the user's post and the average number of fans of the user as input, it is difficult to capture the network information of the underlying propagation structure and compare it with the true value error Large, limited forecasting performance. DeepCas uses a deep learning method to improve the prediction effect, and its performance is better than the other effect of the feature-based method and the generation process method. CoulpedGNN captures the information cascade effect, and achieves good results, which shows the importance of cascade effects under the influence information diffusion. As for TSGNN, it outperforms all baselines on Sina Weibo and Twitter datasets. Compared with CoupledGNN, the MRSE metric on the Weibo dataset reduces by about 7% and 10% on Twitter dataset. These results show that the cascade text semantics is meaningful for popularity prediction.

5.2 Variants of TSGNN

In order to study and prove the effectiveness of each component of our model, we have designed several variants of TSGNN. It can be seen from Fig. 2(a) that compared with the variant with the theme path embedding module replaced, the effect of TSGNN is better than that of LSTM and GRU, note that both of them are better than TSGNN-noRNN which does not use the RNN structure. This shows that our text semantics is effective, and basic RNN is better than LSTM and GRU in capturing topic semantics. Basic RNN has fewer parameters than LSTM and GRU, and the model is not easy to overfit. Compared with TSGCN, TSGNN essentially adopts a different domain aggregation strategy of updating node feature representation. The results show that the effect of a basic GCN is not as good as TSGNN layer. In other words, considering the activation state of node neighbors in the neighborhood aggregation strategy is useful to adapt to the scenario of information diffusion.

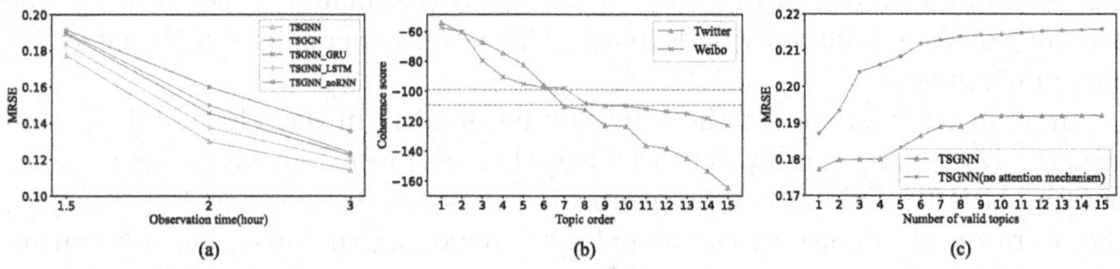

Fig. 2. Result analysis. (a) Comparison of TSGNN variants. (b) The coherence score of topics. (The dotted line is the mean value) (c) Ablative analysis

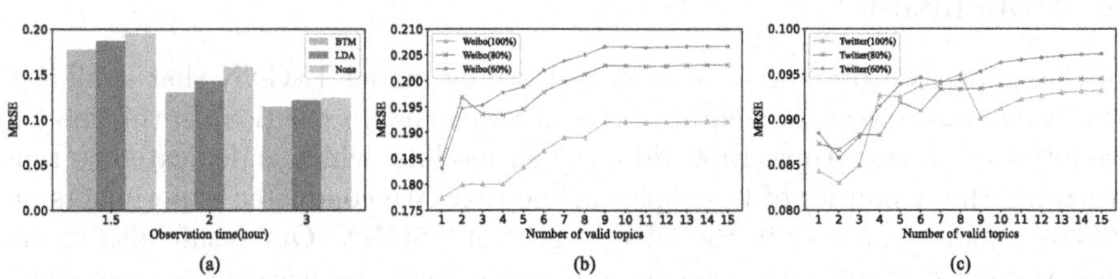

Fig. 3. The results of parameter analysis.

5.3 Parameter Analysis

The Topic Model. Considering that our model contains semantic modeling of text topics, we select the classic LDA model in TE module for comparative experiments. Compared with LDA, BTM can find high coherent topics for modeling topics in short texts [17]. The results (shown as Fig. 3(b)) also shows that cascade texts with high coherent topics can improve prediction accuracy.

The Number of Valid Topics. To analyze the impact of the semantics of key topics in the text, we adjusted the number(x) of valid topics in *topic encoding path*. The value of x we select is from 1 to the total number of topics n. We also conducted experiments on datasets that randomly discarded 20% and 40% to verify the stability of our algorithm. We can see from the results in Fig. 3(b) & Fig. 3(c) that the prediction effect is best when the number of valid topic is one or two, increasing the number of valid topics will not further improve the model performance. Also, we calculated the average coherence scores on different datasets. The result is shown in the Fig. 2(b). The coherence score on the Sina Weibo dataset is higher than that on the Twitter dataset. As a result, the high coherent topic embeddings can improve prediction accuracy as well as further supports the effectiveness of our proposed model.

5.4 Ablation Experiment

We removed the TE module in the TSGNN model and performed experiments. The results are shown in Fig. 3(a). When the text semantic is not utilized, the accuracy of the prediction will decrease. This also further supports the effectiveness of our model.

In addition, we removed the attention mechanism in the TE module (set all the attention weights in Eq. (1) to 1), and then performed an experimental comparison with TSGNN. It can be seen from the results in shown as Fig. 2(c) that the overall performance of the model after removing the attention mechanism becomes worse. This shows that different topics in a tweet play different roles due to different distributions, and the attention mechanism can well promote the model to learn the most important topic impact that causes the promotion.

6 Conclusion

In this paper, we propose a novel deep learning model TSGNN that combines text semantics into graph neural networks for cascade prediction. In this model, we introduce a text Topic Embedding (TE) module, which is dedicated to representing the semantics of key topics in the text. We conducted experiments on different datasets to verify the effectiveness of TSGNN. Our result also shows that textual content with coherent semantics has a positive impact on users' retweet behaviors. Moreover, it is beneficial to integrate text into GNN-based methods for cascade prediction tasks. As for future work, we will devote to better predict the scale of the cascade by combining the micro and macro scales.

Acknowledgement. This work is supported by the National Key R&D Program of China under Grants No. 2018YFC0831703.

References

1. Bian, T., et al.: Rumor detection on social media with bi-directional graph convolutional networks. In: Proceedings of the AAAI Conference on Artificial Intelligence, vol. 34, pp. 549–556 (2020)

2. Cao, Q., Shen, H., Gao, J., Wei, B., Cheng, X.: Popularity prediction on social platforms with coupled graph neural networks. In: Proceedings of the 13th International Conference on Web Search and Data Mining, pp. 70–78 (2020)
3. Chen, X., Zhou, F., Zhang, K., Trajcevski, G., Zhong, T., Zhang, F.: Information diffusion prediction via recurrent cascades convolution. In: 2019 IEEE 35th International Conference on Data Engineering (ICDE), pp. 770–781. IEEE (2019)
4. Cheng, J., Adamic, L., Dow, P.A., Kleinberg, J.M., Leskovec, J.: Can cascades be predicted? In: Proceedings of the 23rd International Conference on World Wide Web, pp. 925–936 (2014)
5. Hong, L., Dan, O., Davison, B.D.: Predicting popular messages in Twitter. In: Proceedings of the 20th International Conference Companion on World Wide Web, pp. 57–58 (2011)
6. Huang, Z., Wang, Z., Zhang, R.: Cascade2vec: learning dynamic cascade representation by recurrent graph neural networks. IEEE Access **7**, 144800–144812 (2019)
7. Jenders, M., Kasneci, G., Naumann, F.: Analyzing and predicting viral tweets. In: Proceedings of the 22nd International Conference on World Wide Web, pp. 657–664 (2013)
8. Khabiri, E., Hsu, C.F., Caverlee, J.: Analyzing and predicting community preference of socially generated metadata: a case study on comments in the digg community. In: ICWSM (2009)
9. Leskovec, J., Adamic, L.A., Huberman, B.A.: The dynamics of viral marketing. ACM Trans. Web (TWEB) **1**(1), 5-es (2007)
10. Li, C., Ma, J., Guo, X., Mei, Q.: DeepCas: an end-to-end predictor of information cascades. In: Proceedings of the 26th International Conference on World Wide Web, pp. 577–586 (2017)
11. Liao, D., Xu, J., Li, G., Huang, W., Liu, W., Li, J.: Popularity prediction on online articles with deep fusion of temporal process and content features. In: Proceedings of the AAAI Conference on Artificial Intelligence, vol. 33, pp. 200–207 (2019)
12. Liu, Y., Wu, Y.F.B.: Early detection of fake news on social media through propagation path classification with recurrent and convolutional networks. In: Thirty-Second AAAI Conference on Artificial Intelligence (2018)
13. Ma, Z., Sun, A., Cong, G.: On predicting the popularity of newly emerging hashtags in Twitter. J. Am. Soc. Inform. Sci. Technol. **64**(7), 1399–1410 (2013)
14. Naveed, N., Gottron, T., Kunegis, J., Alhadi, A.C.: Bad news travel fast: a content-based analysis of interestingness on Twitter. In: Proceedings of the 3rd International Web Science Conference, pp. 1–7 (2011)
15. Rong, Y., Huang, W., Xu, T., Huang, J.: Dropedge: towards deep graph convolutional networks on node classification. In: International Conference on Learning Representations (2019)
16. Veličković, P., Cucurull, G., Casanova, A., Romero, A., Lio, P., Bengio, Y.: Graph attention networks. arXiv preprint arXiv:1710.10903 (2017)
17. Yan, X., Guo, J., Lan, Y., Cheng, X.: A biterm topic model for short texts. In: Proceedings of the 22nd International Conference on World Wide Web, pp. 1445–1456 (2013)
18. Zhao, L., et al.: Online flu epidemiological deep modeling on disease contact network. GeoInformatica **24**(2), 443–475 (2019). https://doi.org/10.1007/s10707-019-00376-9
19. Zhao, Q., Erdogdu, M.A., He, H.Y., Rajaraman, A., Leskovec, J.: Seismic: a self-exciting point process model for predicting tweet popularity. In: Proceedings of the 21th ACM SIGKDD International Conference on Knowledge Discovery and Data Mining, pp. 1513–1522 (2015)

20. Zhao, Y., Wang, C., Chi, C.H., Lam, K.Y., Wang, S.: A comparative study of transactional and semantic approaches for predicting cascades on twitter. In: IJCAI, pp. 1212–1218 (2018)
21. Zhou, F., Xu, X., Trajcevski, G., Zhang, K.: A survey of information cascade analysis: models, predictions and recent advances. arXiv preprint arXiv:2005.11041 (2020)

TERMCast: Temporal Relation Modeling for Effective Urban Flow Forecasting

Hao Xue(✉) and Flora D. Salim

RMIT University, Melbourne, Australia
{hao.xue,flora.salim}@rmit.edu.au

Abstract. Urban flow forecasting is a challenging task, given the inherent periodic characteristics of urban flow patterns. To capture the periodicity, existing urban flow prediction approaches are often designed with *closeness*, *period*, and *trend* components extracted from the urban flow sequence. However, these three components are often considered separately in the prediction model. These components have not been fully explored together and simultaneously incorporated in urban flow forecasting models. We introduce a novel urban flow forecasting architecture, TERMCast. A Transformer based long-term relation prediction module is explicitly designed to discover the periodicity and enable the three components to be jointly modeled This module predicts the periodic relation which is then used to yield the predicted urban flow tensor. To measure the consistency of the predicted periodic relation vector and the relation vector inferred from the predicted urban flow tensor, we propose a consistency module. A consistency loss is introduced in the training process to further improve the prediction performance. Through extensive experiments on three real-world datasets, we demonstrate that TERMCast outperforms multiple state-of-the-art methods. The effectiveness of each module in TERMCast has also been investigated.

Keywords: Urban flow prediction · Relation modeling · Transformer

1 Introduction

Forecasting urban flows is essential for a wide range of applications from public safety control, urban planning to intelligent transportation systems. For example, taxi and ride sharing companies are capable of providing better services with more accurate flow prediction. From the user's perspective, based on the predicted urban flows, commuters and travelers are able to avoid traffic congestion and arrange driving routes. Unlike other temporal sequence formats like language sentences and pedestrian trajectories, urban flow data has a unique inherent feature, that is, the periodicity, e.g., morning rush hours are more likely to occur during weekdays instead of weekends. To explore this periodicity feature, *closeness*, *period*, and *trend* components are widely used as the input of a urban flow prediction model. Two different settings of the three components

K. Karlapalem et al. (Eds.): PAKDD 2021, LNAI 12712, pp. 741–753, 2021.
https://doi.org/10.1007/978-3-030-75762-5_58

are illustrated in Fig. 1 (a) and (b). Figure 1 (a) is widely used in existing methods [6,9,18,20,21] and Fig. 1 (b) is recently proposed by Jiang et al. [4]. In these settings, the *closeness* component corresponds to the most recent observations, while the *period* and *trend* components reflect the daily and weekly periodicity, respectively.

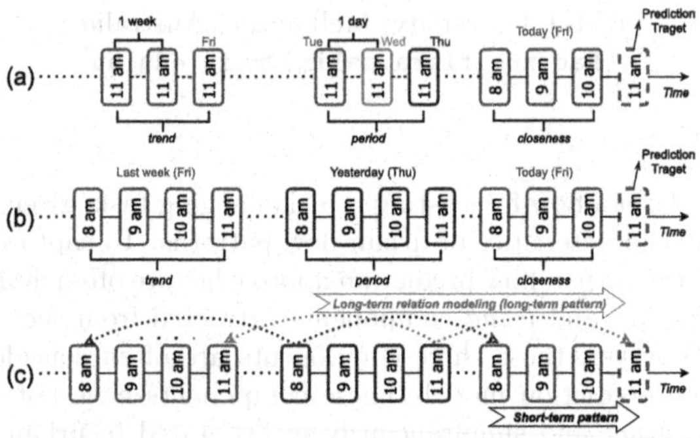

Fig. 1. Illustration of the *closeness*, *period*, and *trend* components. Figure (a) and (b) are two widely used construction methods of these components in the literature. In the proposed TERMCast, to predict urban flows at the prediction target, we focus on exploring the long-term relation among these components as shown in (c). (Color figure online)

Nevertheless, we notice that the relation of the three components is not well researched in the literature. Existing methods always process these input components separately (Fig. 1 (a) and (b)) to extract features of *closeness*, *period*, and *trend*. These features are then combined for urban flow prediction through a simple weighted fusion step such as ST-ResNet [18] and PCRN [21] or network structures like the *ResPlus* in DeepSTN+ [9] and the attention mechanism in VLUC-Net [4]. Although these methods have shown good abilities to predict urban flows, there are still unsolved questions such as what is the relation of each time interval among the *closeness*, *period*, and *trend* component and how can these periodic relations facilitate the prediction? For example, in Fig. 1 (c), if the relations at *8 am*, *9 am*, and *10 am* (blue arrows) are modeled, can we predict the periodic relation at *11 am* (pink arrows) and use this predicted periodic relation to help the prediction of urban flow at 11 am?

Our research is motivated by the aforementioned questions. We propose a novel urban flow prediction network TERMCast in this paper. Specifically, in the proposed TERMCast, a short-term prediction module is developed to explore the short-term pattern and generate an initial predicted flow. This module handles the most up-to-date *closeness* component which is a strong cue for the prediction target. For better modeling the long-term pattern, rather than processing *closeness*, *period*, and *trend* individually, we design a Transformer [14] based long-term relation prediction module to model the periodic relation per time step among urban flows from the three components at the same time interval (e.g., 8 am, 9 am, and 10 am shown in Fig. 1 (c)) and predict the relation of our target time interval (e.g., 11 am). This predicted periodic relation and the initial flow from the short-term prediction module are then combined together to yield the predicted urban flow tensor. So far, for the prediction target, there are two types of relations available: the *predicted periodic relation* generated by the long-term relation prediction module; and the *inferred relation* that can be calculated based on the predicted urban flow tensor. Intuitively, these two relations should be consistent with each other as they both belong to the same time interval. To this end, we also design a prediction consistency module and propose a consistency loss to reflect such consistency in the training process of our network. In summary, our contributions are: (i) We propose a novel Transformer based long-term relation module to explore the relation among the *closeness*, *period*, and *trend* components. It captures the long-term periodicity in the urban flow sequence and works together with the short-term prediction module to yield urban flow predictions. (ii) We design a prediction consistency module to build a connection between the predicted urban flow tensor and the predicted periodic relation generated by the long-term relation module. Furthermore, we propose a consistency loss upon the prediction consistency module to improve the accuracy of the prediction. (iii) We conduct extensive experiments on three publicly available real-world urban flow datasets. The results show that TERMCast achieves state-of-the-art performance.

2 Related Work

The urban flow prediction problem and similar problems such as crowd flow prediction and taxi demand prediction have been the focus of researchers for a quite long time. Based on the famous time series prediction model Auto Regressive Integrated Moving Average (ARIMA) and its variants such as Seasonal ARIMA, numerous traditional approaches including [5,8,10,13,15] have been designed and proposed. In the last few years, deep learning based neural networks such as ResNet [3] and Transformer [14] have been widely used in the areas of computer vision and neural language processing. In the area of temporal sequence modeling, deep learning based methods have also been proposed and successfully applied to many time series prediction problems such as human mobility prediction [2] and high-dimensional time series forecasting [12].

More specifically, for urban flow prediction task, to model the periodic pattern of the urban flow sequence, Deep-ST [19] firstly proposes to use *closeness*, *period*, and *trend* three components (e.g., Fig. 2 (a)) to form the input instance of the prediction network. Based on Deep-ST, ST-ResNet [18] employs convolution-based residual networks to further model the spatial dependencies in the city scale. In ST-ResNet, three residual networks with the same structure are used to extract spatio-temporal features from three components respectively. To yield predictions, the weighted fusion of these three features are then combined with the extracted extra features from an external component, such as weather conditions and events. To capture spatial and temporal correlations, a pyramidal convolutional recurrent network is introduced in Periodic-CRN (PCRN) [21].

Yao et al. [17] proposed a unified multi-view model that jointly considers the spatial, temporal, and semantic relations for flow prediction. STDN [16] further improves the DMVST by proposing a periodically shifted attention mechanism to incorporate the long-term periodic information that is ignored in the DMVST. Following the trend of fusing periodic representations (i.e., the *period* and *trend* components), more approaches are designed and proposed for the prediction of urban flow. DeepSTN+ [9] proposes an architecture of *ResPlus* unit, whereas VLUC-Net [4] utilizes the attention mechanism in the fusion process. However, these methods only consider and fuse periodic components after processing the three components separately. This late fusion manner cannot fully explore the intrinsic relation in these components. For example, the relation at 8 am (the blue dash arrow in Fig. 1 (c)) should be able to provide cues for the prediction at 11 am but it is overlooked by existing methods.

In summary, the proposed TERMCast differs from other methods in the aspect of modeling the periodicity, i.e. the long-term relation. We also introduce a novel consistency loss to connect the inferred relation and the predicted periodic relation, which is different from the training loss functions of other urban flow prediction networks. Our work is also different from [7] that uses CNN and RNN to extract short- and long-term patterns for multivariate time series forecasting.

3 Preliminaries

Following the widely-used grid-based urban flow definition from [4,18], we divide a city area into $H \times W$ disjoint regions based on the longitude and latitude.

Inflow/Outflow [18]**:** At the ith time interval, the inflow and outflow of the region (h, w) are defined as follows: $x_i^{in,h,w} = \sum_{T_r \in \mathbb{P}} |\{j > 1 | g_{j-1} \notin (h,w) \ \& \ g_j \in (h,w)\}|$, $x_i^{out,h,w} = \sum_{T_r \in \mathbb{P}} |\{j > 1 | g_{j-1} \in (h,w) \ \& \ g_j \notin (h,w)\}|$, where $\mathbb{P}$ is a set of trajectories at the ith time interval. For each trajectory $T_r : g_1 \to g_2 \to \cdots \to g_{T_r}$ in the trajectory set $\mathbb{P}$, $g_j \in (h,w)$ indicates that location g_j lies insider the region (h,w) , and vice versa. Tensor $\mathbf{X}_i \in \mathbb{R}^{2 \times H \times W}$ ($(\mathbf{X}_i)_{0,h,w} = x_i^{in,h,w}$ and $(\mathbf{X}_i)_{1,h,w} = x_i^{out,h,w}$) can be used to represent the urban flow of the whole city are. For simplification, in this paper, we use $\mathbf{X}_t^c$, $\mathbf{X}_t^p$, and $\mathbf{X}_t^q$ to represent the urban flow tensor in the *closeness*, *period*, and *trend* that correspond to the recent time intervals, daily periodicity, and weekly trend. Specifically, for the same t value, the time in between $\mathbf{X}_t^c$ and $\mathbf{X}_t^p$ is a day apart, whereas the time in between $\mathbf{X}_t^c$ and $\mathbf{X}_t^q$ is a week apart (see Fig. 1 (c)).

Flow Prediction: Given *trend* $\{\mathbf{X}_1^q, \mathbf{X}_2^q, \cdots, \mathbf{X}_T^q\}$, *period* $\{\mathbf{X}_1^p, \mathbf{X}_2^p, \cdots, \mathbf{X}_T^p\}$, and *closeness* urban flow tensors $\{\mathbf{X}_1^c, \mathbf{X}_2^c, \cdots, \mathbf{X}_{T-1}^c\}$, the goal is to predict the flow in the next time interval, i.e., $\hat{\mathbf{X}}_T^c$. Here, $T - 1$ is the observation length.

4 Methodology

The architecture of our method for urban flow prediction is illustrated in Fig. 2. It consists of three major modules: (i) *Short-Term Prediction* module (the light

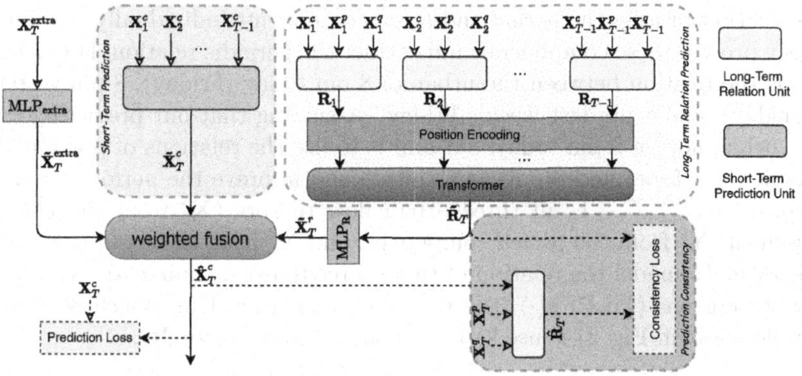

Fig. 2. The architecture of the proposed TERMCast. Dash arrows indicate operations used in the training phase only. (Color figure online)

green part in Fig. 2) to yield a preliminary prediction based on the recent observations, i.e., the *closeness* component. (ii) A *Long-Term Relation Prediction* module (the light yellow part in Fig. 2) to model the long-term relation per time step based on the *period* component (daily periodicity) and the *trend* component (weekly periodicity). Such long-term periodic influence is then incorporated into the urban flow prediction. (iii) A *Prediction Consistency* module (the light blue part in Fig. 2) to measure the consistency between the inferred relation calculated from the predicted $\hat{\mathbf{X}}_T^c$ and the predicted periodic relation from the above long term relation prediction module.

4.1 Short-Term Prediction

Intuitively, the most up-to-date data plays an important role in the prediction of the next time interval urban flow. This module focuses on the short-term pattern with the recent time intervals flow tensors, i.e., the *closeness* component. With tensors $\{\mathbf{X}_1^c, \mathbf{X}_2^c, \cdots, \mathbf{X}_{T-1}^c\}$ as input, we can get an initial predicted urban flow tensor $\tilde{\mathbf{X}}_T^c$ which serves as the foundation of the final prediction $\hat{\mathbf{X}}_T^c$. This process is modeled as:

$$\tilde{\mathbf{X}}_T^c = \mathcal{F}(\mathbf{X}_1^c, \mathbf{X}_2^c, \cdots, \mathbf{X}_{T-1}^c), \tag{1}$$

where $\mathcal{F}(\cdot)$ represents the prediction function (the Short-Term Prediction Unit that is shown as the green box in Fig. 2). In the proposed TERMCast, the widely-used *Residual Unit* proposed by ST-ResNet [18] is selected as the base model for the Short-Term Prediction Unit. It has shown a good ability to model the spatial dependencies in the city scale. Note that using the *Residual Unit* is the only similarity between our TERMCast and ST-ResNet.

4.2 Long-Term Relation Prediction

In this module, we focus on the daily periodicity and weekly trend and exploring how to model the long-term pattern for urban flow prediction. Unlike existing methods that process the *period* and *trend* components individually, we simultaneously process these components and extract the periodic relation. For example, what is the relation between the urban of 8 pm today (Friday), 8 pm yesterday (Thursday), and 8 pm last week's Friday? Assuming that our predicting target is the urban flow of 9 pm today, our aim is to use the relations of previous time intervals (e.g., 5 pm and 8 pm) to facilitate and improve the performance.

For $1 \leq t \leq T-1$, all three urban flow tensors ($\mathbf{X}_t^c$ from the *closeness* component, $\mathbf{X}_t^p$ from the *period* component, and $\mathbf{X}_t^q$ from the *trend* component) are known. To model the relation of these three tensors, similar to [11], a multi-layer perceptrons (MLP) $g(\cdot)$ (the Long-Term Relation Unit which is shown as the yellow box in Fig. 2) is used to learn the relation vector $\mathbf{R}_t$:

$$\mathbf{R}_t = g(\mathbf{X}_t^c \oplus \mathbf{X}_t^p \oplus \mathbf{X}_t^q), \tag{2}$$

where $\oplus$ is the concatenation operation. Given the sequence of relation vectors from $t=1$ to $t=T-1$, the relation for $t=T$ can be predicted through:

$$\tilde{\mathbf{R}}_T = \text{Transformer}(\mathbf{R}_1, \mathbf{R}_2, \cdots, \mathbf{R}_{T-1}). \tag{3}$$

Here, we take the merit of the *Transformer* [14] a popular architecture for sequence prediction. The self-attention mechanisms of the transformer is suitable for modeling pairwise interactions between two relation vectors of two time intervals in the input sequence. Based on the predicted periodic relation vector $\tilde{\mathbf{R}}_T$, a tensor $\tilde{\mathbf{X}}_T^r \in \mathbb{R}^{2\times H\times W}$ is generated via:

$$\tilde{\mathbf{X}}_T^r = \text{MLP}_{\text{R}}(\tilde{\mathbf{R}}_T), \tag{4}$$

where $\text{MLP}_{\text{R}}(\cdot)$ stands for a MLP architecture.

The tensor $\tilde{\mathbf{X}}_T^r$ can be seen as a supplement to the initial prediction $\tilde{\mathbf{X}}_T^c$ from the short-term prediction module. The rationale of this long-term relation prediction process is that given the predicted periodic relation, we should be able to "decode" a predicted tensor. This decoded tensor helps the prediction of $\hat{\mathbf{X}}_T^c$ as the relation vector $\tilde{\mathbf{R}}_T$ models the long-term periodicity at T.

Position Encoding. Given that there is no recurrence operation in the Transformer architecture, position encodings are often added to each input vector of the Transformer to compensate the missing sequential order information [14]. These position encodings are modeled as sine and cosine functions:

$$PE_{(pos,2k)} = \sin(pos/10000^{2k/d_{\mathbf{R}}}), \tag{5}$$

$$PE_{(pos,2k+1)} = \cos(pos/10000^{2k/d_{\mathbf{R}}}), \tag{6}$$

where pos is the position index and k is the dimension index. $d_{\mathbf{R}}$ is the dimension of the relation vector $\mathbf{R}_t$ (the input of the Transformer TERMCast).

Unlike other Transformer based applications such as translation [14] and object detection [1], in the proposed TERMCast, relation vector at each time interval (the input sequence of Eq. (3)) has its own unique periodic position to indicate the absolute temporal position in one day. Thus, we explicitly design a periodic position encoding strategy to model the periodic position information. Since it represents the absolute position in one day of each time interval, the maximum *pos* value in Eq. (5) and Eq. (6) depends on the total number of time intervals in one day (e.g., 24 for 1 h interval and 48 for 0.5 h interval).

4.3 Prediction with Consistency

In the urban flow prediction task, extra information such as the date information and time in the day (e.g., the prediction target is *11 am* on *Friday* in the example shown in Fig. 1) also have influences on the prediction. In TERMCast, such influence is modeled as a tensor $\tilde{\mathbf{X}}_T^{\text{extra}} \in \mathbb{R}^{2\times H\times W}$ through a MLP ($\text{MLP}_{\text{extra}}$ in Fig. 2) with the input of the extra information $\mathbf{X}_T^{\text{extra}}$ at the prediction target T.

$$\hat{\mathbf{X}}_T^c = \mathbf{W}_1 \circ \tilde{\mathbf{X}}_T^c + \mathbf{W}_2 \circ \tilde{\mathbf{X}}_T^r + \mathbf{W}_3 \circ \tilde{\mathbf{X}}_T^{\text{extra}} \tag{7}$$

As given in Eq. (7) where $\circ$ represents the element-wise multiplication, the final predicted urban flow tensor $\hat{\mathbf{X}}_T^c$ is a weighted fusion of three parts: (1) the initial predicted $\tilde{\mathbf{X}}_T^c$ from the short-term prediction; (2) the $\tilde{\mathbf{X}}_T^r$ based on the long-term relation modeling; and (3) the influence $\tilde{\mathbf{X}}_T^{\text{extra}}$ from the extra information. More detailed discussion of this weighted combination step is given in Sect. 5.5.

Given two known history observations $\mathbf{X}_T^p$ and $\mathbf{X}_T^q$, we can infer a relation based on the predicted $\hat{\mathbf{X}}_T^c$ by:

$$\hat{\mathbf{R}}_T = g(\hat{\mathbf{X}}_T^c \oplus \mathbf{X}_T^p \oplus \mathbf{X}_T^q) \tag{8}$$

Intuitively, this inferred relation $\hat{\mathbf{R}}_T$ should be consistent with the predicted periodic relation $\tilde{\mathbf{R}}_T$ given by Eq. (3). To further build a connection between these two vectors, we propose to use the cosine similarly as a scoring function to model the consistency between the inferred relation vector $\hat{\mathbf{R}}_T$ and the predicted periodic relation vector $\tilde{\mathbf{R}}_T$ at T.

Loss Function. Unlike existing urban flow prediction methods that only use the mean squared error between the predicted flow and the ground truth flow as the loss function, the loss function in our TERMCast contains two terms:

$$\mathcal{L} = \alpha \|\mathbf{X}_T^c - \hat{\mathbf{X}}_T^c\|_2^2 + \beta\left(1 - \frac{\hat{\mathbf{R}}_T \cdot \tilde{\mathbf{R}}_T}{\|\hat{\mathbf{R}}_T\| \|\tilde{\mathbf{R}}_T\|}\right), \tag{9}$$

where α and β are the weights of two terms. The first term is the widely used mean squared error and the second term is the above cosine similarly based consistency loss. It gives more penalty if the inferred relation and the predicted periodic relation are not consistent.

5 Experiments

5.1 Datasets and Metrics

In this paper, we focus on spatio-temporal raster urban flow data and conduct experiments on the following three real-world urban flow datasets: (1) BikeNYC [9]: This dataset is extracted from the New York City Bike system from the time period: 2014-04-01 to 2014-09-30. (2) TaxiNYC [16]: This datasets consists of taxi trip records of New York City from the period: 2015-01-01 to 2015-03-01. (3) TaxiBJ [18]: This dataset is the taxicab GPS data collected in Beijing from time periods: 2013-07-01 to 2013-10-30, 2014-05-01 to 2014-06-30, 2015-03-01 to 2015-06-30, and 2015-11-01 to 2016-04-10. Since there is no connectivity information such as road maps, graph based spatio-temporal traffic prediction models are irrelevant. For each dataset, the first 80% data is used for training and the rest 20% data is used for testing. The Min-Max normalization is adopted to transform the urban flow values into the range $[0, 1]$. During the evaluation process, we rescale predicted values back to the normal values to compare with the ground truth data. We evaluate our method with two commonly used metrics: the Rooted Mean Squared Error (RMSE) and the Mean Absolute Error (MAE).

5.2 Implementation Details

The hyperparameters are set based on the performance on the validation set (10% of the training data). The Adam optimizer is used to train our models. The base unit used in our short-term prediction module consists of three 3×3 convolutional layers with 32 filters and one 3×3 convolutional layer with 2 filters (corresponding to the in and out flow in each urban flow tensor). The dimension of each relation vector is 256, i.e., $\mathbf{R}_t \in \mathbb{R}^{256}$. As for the loss function (9), both weights α and β are set to 1. Following VLUC-Net [4], we set the observation length to 6. That is, the length of the *closeness* component is 6, whereas the length of the *period* and *trend* components are 7. For the extra information, considering that information such as weather is not available for all three datasets, only the temporal metadata (i.e., the time interval of a day and the day of a week) is used. The reported results of our TERMCast and its variants in the experiments are the average of 5 runs.

5.3 Comparison Against Other Methods

In our experiments, we compare TERMCast against the following urban flow prediction methods: HistoricalAverage (HA); Convolutional LSTM (ConvLSTM); ST-ResNet [18]; DMVST [17]; DeepSTN+ [9]; STDN [16]; VLUC-Net [4].

The upper half of Table 1 shows the results of the proposed TERMCast as compared to other methods on three datasets. For each column, the best result is given in bold. Specifically, compared to other deep learning based prediction methods, the HA has the worst performance as it can only make predictions by simply averaging historical observations. In general, TERMCast achieves the best performance on all three public datasets for both RMSE and MAE. STDN and VLUC-Net share the second best performance depending on the dataset and the metric. Compared to our TERMCast, in other methods, the three components are connected merely through attention based fusion, which overlooks the periodic relation between the *closeness*, *period*, and *trend* components. These comparison results demonstrate the superior of the proposed method.

Table 1. The RMSE/MAE results on three datasets. The upper half and the latter half list the performance of existing methods and variants of our TERMCast, respectively.

	BikeNYC		TaxiNYC		TaxiBI	
	RMSE	MAE	RMSE	MAE	RMSE	MAE
HA	15.676	4.882	21.535	7.121	45.004	24.475
ConvLSTM	6.616	2.412	12.143	4.811	19.247	10.816
ST-ResNet	6.106	2.360	11.553	4.535	18.702	10.493
DMVST	7.990	2.833	13.605	4.928	20.389	11.832
DeepSTN+	6.205	2.489	11.420	4.441	18.141	10.126
STDN	5.783	2.410	11.252	4.474	17.826	9.901
VLUC-Net	5.831	2.175	10.654	4.157	18.378	10.325
V1	10.743	3.462	19.292	6.298	23.529	13.273
V2	6.331	2.462	11.726	4.816	18.998	11.197
V3	5.971	2.215	10.695	4.077	17.442	10.068
TERMCast	**5.729**	**2.139**	**10.395**	**3.955**	**17.017**	**9.778**

5.4 Ablation Studies

To explore the effectiveness of each module in our proposed method, we consider the following three variants: (i) V1: We discard the short-term prediction module for this variant. That is, the *closeness* component is not used in the prediction process. (ii) V2: The long-term relation module is removed. Since the consistency loss is based on the long-term relation module, it is disabled in this variant as well. (iii) V3: For this variant, we only remove the consistency loss in TERMCast, which means the second term (used to measure the cosine similarity between the inferred relation and the predicted periodic relation in Eq. (9)) is dropped.

The results of these variants on three datasets are presented in the lower half of Table 1. We also include the results of our TERMCast in the table for comparison. In general, V1 has the worst results in these variants. Actually, it is only better than HA in Table 1. This is as expected because the *closeness* component contains the most up-to-date urban flow information for prediction. We found that the usage of only historical data (i.e., the *period* component of the previous day and the *trend* component of the previous week) is not enough for accurate urban flow prediction.

The prediction performance of V2 leads V1 by a relatively large margin but is worse than V3 in which the long-term prediction module is enabled. Furthermore, if we compare V2 against other methods reported in Table 1, it can be noticed that the performance of V2 is close to ST-ResNet. Since the Residual Unit from ST-ResNet is used as the forecasting function in the short-term prediction module to predict urban flow based on the *closeness* component, it can be seen as a simplified version of ST-ResNet where the *period* and *trend* components are disabled. Results of V3 outperform the other two variants but are worse than TERMCast where the consistency loss is applied. The proposed consistency loss can add more constraints during the training and stabilize the training process. In a nutshell, it harmonizes the relation inferred from the predicted urban flow tensor and the predicted periodic relation at the same time interval so that the predicted urban flow is more realistic. Comparing the results from the three variants against TERMCast, we observe that TERMCast outperforms all the variants on all three datasets. It demonstrates the effectiveness of each module and justifies the need for all modules to be included in our TERMCast.

Table 2. Six different configurations of the weighted fusion. C0 is our default configuration used in other experiments.

Configurations	$\mathbf{W}_1$	$\mathbf{W}_2$	$\mathbf{W}_3$
C0	✓	✓	✓
C1	✓	✓	×
C2	✓	×	✓
C3	×	✓	✓
C4	×	×	×
C5	✓ (softmax)	✓ (softmax)	✓ (softmax)

5.5 Different Weighted Fusion Configurations

The weighted fusion (Eq. (7)) is an important step in TERMCast as it combines the initial predicted $\tilde{\mathbf{X}}_T^c$ with the influence from the long-term relation and extra information. To fully research this fusion step, six different fusion configurations are explored. Table 2 summarizes the setting of each configuration. A ✓ indicates that the corresponding weight matrix is enabled whereas a × means that the corresponding weight matrix is removed from Eq. (7). Taken the configuration C4 as an example, without all three weighting terms, the weighted fusion prediction step given in Eq. (7) will be degraded to a plain summation operation, i.e., $\hat{\mathbf{X}}_T^c = \tilde{\mathbf{X}}_T^c + \tilde{\mathbf{X}}_T^r + \tilde{\mathbf{X}}_T^{\text{extra}}$. Note that the last row (C5) in the table is the default configuration of TERMCast. The softmax function is applied to make sure that the sum of three weights with the same index in the three weight matrices equals to 1. C0 can be seen as a simplified configuration of C5. The prediction results of these different fusion configurations are demonstrated in Fig. 3.

For configurations (C1-C4) where one or more weight matrices are disabled, their prediction performance are worse than C0. It indicates that each weight matrix is necessary for the fusion step. To be more specific, C4 has the worst results on all three datasets as it simply adds the periodic relation influence $\tilde{\mathbf{X}}_T^r$ and the extra influence $\tilde{\mathbf{X}}_T^{\text{extra}}$ to the initial predicted $\tilde{\mathbf{X}}_T^c$. If only one weight matrix is removed (C1, C2, and C3), C3 cannot yield predictions as accurate as of the other two configurations. C3 discards $\mathbf{W}_1$ which is the weight matrix of the initial predicted $\tilde{\mathbf{X}}_T^c$ generated through the short-term prediction module. This result is consistent with the above ablation results that V1 does not perform well. It further confirms that the short-term prediction module is a significant part of the traffic flow prediction. For the configuration C5, its performance stands out in the figure and outperforms all the other configurations. It reveals that applying an extra softmax operation on these weight matrices helps to further improve the traffic flow prediction performance.

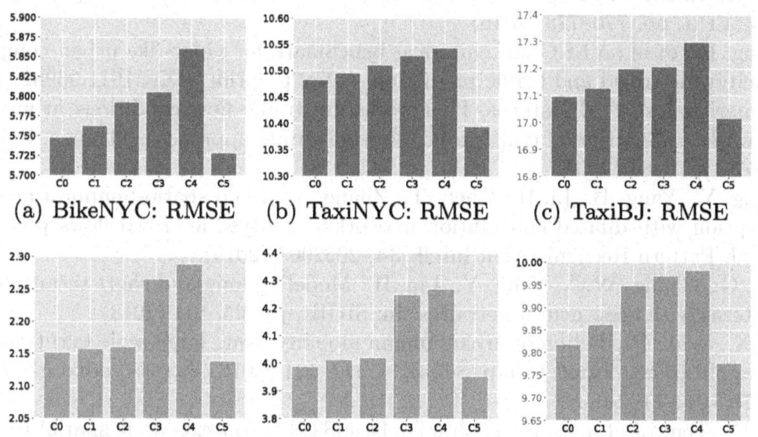

(a) BikeNYC: RMSE (b) TaxiNYC: RMSE (c) TaxiBJ: RMSE

(d) BikeNYC: MAE (e) TaxiNYC: MAE (f) TaxiBJ: MAE

Fig. 3. The RMSE and MAE results of different configurations listed in Table 2.

6 Conclusion

In this paper, we present a novel method called TERMCast for the urban flow prediction task. We explicitly design a long-term relation prediction module to better capture the periodicity in the urban flow sequence. Through the prediction consistency module and the consistency loss, TERMCast is able to make the predicted urban flow tensor and the predicted periodic relation be consistent with each other so that the prediction performance can be improved further. We conduct extensive experiments on three real-world datasets. The experimental results demonstrate that the proposed TERMCast outperforms other prediction methods and show the effectiveness of each module in TERMCast. In addition, we explore and compare different weighted fusion methods in our experiments. These studies and findings would be of interest to other researchers in related areas.

Acknowledgments. We acknowledge the support of Australian Research Council Discovery Project *DP190101485*.

References

1. Carion, N., Massa, F., Synnaeve, G., Usunier, N., Kirillov, A., Zagoruyko, S.: End-to-end object detection with transformers. In: Vedaldi, A., Bischof, H., Brox, T., Frahm, J.-M. (eds.) ECCV 2020. LNCS, vol. 12346, pp. 213–229. Springer, Cham (2020). https://doi.org/10.1007/978-3-030-58452-8_13
2. Feng, J., et al.: Deepmove: predicting human mobility with attentional recurrent networks. In: WWW (2018)
3. He, K., Zhang, X., Ren, S., Sun, J.: Deep residual learning for image recognition. In: CVPR, pp. 770–778 (2016)
4. Jiang, R., et al.: VLUC: an empirical benchmark for video-like urban computing on citywide crowd and traffic prediction. arXiv preprint arXiv:1911.06982 (2019)
5. Kamarianakis, Y., Prastacos, P.: Forecasting traffic flow conditions in an urban network: comparison of multivariate and univariate approaches. Transp. Res. Rec. **1857**(1), 74–84 (2003)
6. Kang, Y., Yang, B., Li, H., Chen, T., Zhang, Y.: Deep spatio-temporal modified-inception with dilated convolution networks for citywide crowd flows prediction. Int. J. Pattern Recognit Artif Intell. **34**, 2052003 (2019)
7. Lai, G., Chang, W.C., Yang, Y., Liu, H.: Modeling long-and short-term temporal patterns with deep neural networks. In: SIGIR, pp. 95–104 (2018)
8. Li, X., et al.: Prediction of urban human mobility using large-scale taxi traces and its applications. Front. Comp. Sci. **6**(1), 111–121 (2012). https://doi.org/10.1007/s11704-011-1192-6
9. Lin, Z., Feng, J., Lu, Z., Li, Y., Jin, D.: DeepSTN+: context-aware spatial-temporal neural network for crowd flow prediction in metropolis. In: AAAI, vol. 33, pp. 1020–1027 (2019)
10. Lippi, M., Bertini, M., Frasconi, P.: Short-term traffic flow forecasting: an experimental comparison of time-series analysis and supervised learning. IEEE T-ITS **14**(2), 871–882 (2013)
11. Santoro, A., et al.: A simple neural network module for relational reasoning. In: NeurIPS, pp. 4967–4976 (2017)
12. Sen, R., Yu, H.F., Dhillon, I.S.: Think globally, act locally: a deep neural network approach to high-dimensional time series forecasting. In: NeurIPS (2019)
13. Shekhar, S., Williams, B.M.: Adaptive seasonal time series models for forecasting short-term traffic flow. Transp. Res. Rec. **2024**(1), 116–125 (2007)
14. Vaswani, A., et al.: Attention is all you need. In: NeurIPS, pp. 5998–6008 (2017)
15. Williams, B.M., Hoel, L.A.: Modeling and forecasting vehicular traffic flow as a seasonal ARIMA process: theoretical basis and empirical results. J. Transp. Eng. **129**(6), 664–672 (2003)
16. Yao, H., Tang, X., Wei, H., Zheng, G., Li, Z.: Revisiting spatial-temporal similarity: a deep learning framework for traffic prediction. In: AAAI (2019)

17. Yao, H., et al.: Deep multi-view spatial-temporal network for taxi demand prediction. In: AAAI (2018)
18. Zhang, J., Zheng, Y., Qi, D.: Deep spatio-temporal residual networks for citywide crowd flows prediction. In: AAAI (2017)
19. Zhang, J., Zheng, Y., Qi, D., Li, R., Yi, X.: DNN-based prediction model for spatio-temporal data. In: ACM SIGSPATIAL, pp. 1–4 (2016)
20. Zhang, J., Zheng, Y., Qi, D., Li, R., Yi, X., Li, T.: Predicting citywide crowd flows using deep spatio-temporal residual networks. Artif. Intell. **259**, pp. 147–166 (2018)
21. Zonoozi, A., Kim, J.j., Li, X.L., Cong, G.: Periodic-CRN: a convolutional recurrent model for crowd density prediction with recurring periodic patterns. In: IJCAI, pp. 3732–3738 (2018)

Traffic Flow Driven Spatio-Temporal Graph Convolutional Network for Ride-Hailing Demand Forecasting

Hao Fu, Zhong Wang, Yang Yu, Xianwei Meng, and Guiquan Liu(✉)

School of Computer Science and Technology, University of Science and Technology of China, Hefei, China
{hfu,gracewz,koinu}@mail.ustc.edu.cn, gqliu@ustc.edu.cn

Abstract. Accurately predicting the demand for ride-hailing in the region is important for transportation and the economy. Prior works are devoted to mining the spatio-temporal correlations between regions limited to historical demand data, weather data, and event data, ignoring rich traffic flow information related to citizens' travel. However, due to the dynamic characteristics of traffic flow and the irregularity of the road network structure, it is difficult to utilize traffic flow information directly. In this paper, we propose a framework called traffic flow driven spatio-temporal graph convolutional network (TST-GCN) to forecast ride-hailing demand. Specifically, we construct a novel region graph based on point of interest (POI) information to model the association between different regions. Besides, we design a stacked traffic-region demand graph convolutional network (TRGCN) module, which is composed of two kinds of nested graph convolutional network structures, effectively modeling the spatial dynamic dependency between regions. Then, the convolution long short-term memory (ConvLSTM) layer is further adopted to obtain spatio-temporal features. We evaluate the proposed model on two real datasets, and the experimental results show that our model outperforms many state-of-art methods.

Keywords: Ride-hailing demand prediction · Graph neural networks · Spatial-temporal feature extraction

1 Introduction

Predicting passenger pickup demands based on historical observations is one of the high-leverage points in intelligent transportation systems (ITS). Together with accurate ride-hailing demand prediction and excellent scheduling algorithms, companies like Uber and Didi are committed to providing a high level of service and satisfaction, winning the loyalty of its passengers. According to the scope of the target area, the demand forecast of ride-hailing can be divided into three categories, station level, region level, and city level respectively [13,26,28].

K. Karlapalem et al. (Eds.): PAKDD 2021, LNAI 12712, pp. 754–765, 2021.
https://doi.org/10.1007/978-3-030-75762-5_59

In this paper, we focus on the region level demand prediction. In detail, the underlying network structure of the city is aligned component in a grid format, with requiring the grids be the same size.

Plenty of works have been devoted to addressing the problem of ride-hailing demand forecasting and achieved good results [13,22,24,26]. These methods adopted convolutional neural network (CNN) to capture the spatial dependency for their ability of extracting the spatial features on grid-based data and applied long short-term memory (LSTM) to capture regular temporal patterns. Later, a general graph convolution neural network (GCN) framework on the basis of Graph Laplacian proposed by [4] is widespread concern among scholars, emerging a number of methods apply GCN to capture non-Euclidean structural information and achieve state-of-the-art results [7,21]. Current methods [6,17,18,20] have achieved great success in the problem of demand prediction by applying social temporal factors from other rich resource data, such as weather, holidays, taxi app version, and price preference information. Although these advanced methods model the spatio-temporal relationship of demand information and consider many event influence factors, they ignore the impact of traffic flow information on demand, leading to their unsatisfactory results.

Traffic flow prediction is also a hot research direction of ITS. The traffic flow information of the whole city can effectively reflect the travel rules of citizens. For example, more people travel from residential regions to workplaces in the morning peak, and vice versa in the evening peak. Ride-hailing is one of the most important ways for people to travel, so it is feasible to extract features from traffic flow information for rife-hailing demand forecasting. However, the use of traffic flow information for demand forecasting has the following challenges: 1) Traffic flow information is usually reflected by roads, and it is difficult to directly extract features that are useful for demand forecasting in the region. 2) Traffic flow and ride-hailing demand information are both spatio-temporal data, but it is a great challenge to model two different types of data at the same time.

In this work, we propose a new deep learning model called traffic flow driven spatio-temporal graph convolutional network to fuse the traffic flow information for ride-hailing demand forecasting. Specifically, we use traffic topological network and region graph to represent traffic flow and demand information respectively, and then design a GCN-based stackable TRGCN network structure to integrate traffic flow and demand information for multiple time periods. Finally, The ConvLSTM network is used to extract regional representations with temporal and spatial neighborhood features for prediction. We verify the performance of the model on two datasets, and compared with a number of current advanced models to prove the effectiveness of the proposed model.

2 Related Work

2.1 Graph Convolution Network

Graph convolution neural network (GCN) is a feature extractor just like CNN, but its process object is graph data. GCN ingeniously designs a method to extract features from non-Euclidean space, so that we can use these features to carry out node classification, graph classification, link prediction on the graph.

In the past few years, there has been a lot of work devoted to the research of GCN. There are two mainstreams graph convolution methods, spatial domain and spectral domain respectively. [3] proposes a spectral graph theoretical formulation through creatively applies Chebyshev polynomials to approximate eigenvalue decomposition. The inductive presentation learning approach GraphSAGE [10] learns a function that generates embeddings by sampling and aggregating features from a node's local neighborhood, which breaks through the limitation that the existing GCN can not express the new nodes quickly. [19] proposes the GAT model, which introduces an attention mechanism based on GCN. Each node in the graph can be assigned different weights according to the characteristics of its neighbors.

2.2 Deep Learning on Traffic Prediction

Recently, deep learning has been greatly developed in traffic prediction for its ability to acquire non-linear spatio-temporal relationship [16,26]. Early work uses a stacked autoencoder model to learn generic traffic flow features[15]. Later, many works tend to adopt LSTM and CNN to capture the dependency of traffic on spatial and temporal [14,30]. [27] uses LSTM based model for extreme traffic prediction. [9] proposes a novel end-to-end deep learning model based on 3D CNN for traffic raster data prediction. [25] uses CNN and LSTM simultaneously to learn the dynamic similarity between positions. However, these methods can only capture spatial and temporal features based on Euclidean distance.

The information of traffic network is graph structure, so in recent years, the research of graph convolution neural network in the field of traffic is also emerging in endlessly [1,5,7,11]. [29] proposes a temporal graph convolutional network model for traffic forecasting, which combines with the graph convolutional network and the gated recurrent unit. [8] proposes a novel attention-based spatio-temporal graph convolutional network model to model the recent, daily-periodic and weekly-periodic dependencies of the historical data. [7] constructs a variety of graph from different perspectives, and then uses GCN to capture non-Euclidean information to predict ride-hailing demand. Nevertheless, the existing methods do not fully integrate the traffic flow information into the demand forecasting.

3 Preliminaries

Definition 1 (Traffic Topological Network). *As shown on the left side of Fig. 1(B), we define the traffic topological network as undirected graph* $G_T = (V_T, A_T)$*, where* V_T *is a finite set of* $|V_T| = N$ *nodes, each node represents a road segment,* $A_T \in R^{N \times N}$ *is the adjacency matrix of* G_T*. Suppose that there are* S_f *types of traffic flow data (e.g., speed and volume), then the historical traffic flow data at each time interval constitutes a tensor* $F \in R^{|V_T| \times S_f}$*.*

Definition 2 (Region Division). *As shown on the right side of Fig. 1(B), we divide the city region into disjoint $l_1 * l_2$ grids of the same size according to longitude and latitude, then we use $M = \{m_i | i = 0, 1, ..., l_1 * l_2 - 1\}$ to represent the set of regions. In addition, we characterize each region m_i with the POI types represented as vector $P_i \in R^{S_p}$, where S_p is the number of POI categories.*

Definition 3 (Region Graph). *We construct a novel region graph $G_R = (V_R, E_R)$ to associate regions with different distances, where V_R and E_R represent node set and directed edge set respectively. Each grid divided in Definition 2 is a region node, the attribute $X \in R^{|V_R| \times 1}$ of region nodes denotes the order demand. The concatenate vector $[P_i, P_j]$ represented by POI of any two regions $m_i \in M$ and $m_j \in M$ is the attribute of $E_R \in R^{|V_R| \times 2 * S_p}$ between the corresponding region nodes.*

Definition 4 (Problem Definition). *We define $I_t = [F_t, X_t]$, where F_t and X_t are the traffic flow information and order demand of all regions at the t-th time interval, respectively. Then, the problem can be defined as to learn a function g, whose parameter is θ, the given L continuous historical time intervals information are mapped to the demands in the next time step:*

$$X_{t+1} = g(\theta, (I_{t-L+1}, I_{t-L+2}, \cdots, I_t)). \quad (1)$$

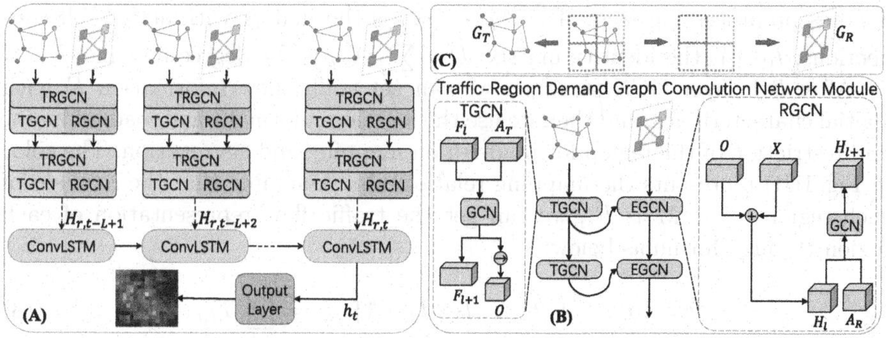

Fig. 1. Architecture of TST-GCN, where → represents the mapping and aggregation operation.

4 Methodology

4.1 Overview

Figure 1(A) provides an overview of our TST-GCN network for the task. We input the historical traffic flow information and historical demand information with graphical data G_T and G_R respectively, and use the TRGCN module to integrate road traffic observations and demand information. Further, the spatio-temporal feature is captured by ConvLSTM to generate predictive tensors.

4.2 Traffic-Region Demand Graph Convolution Network Module for Spatial Modeling

As shown in Fig. 1(C), TRGCN is a stackable module, which consists of traffic flow graph convolution network (TGCN) and region graph convolution network (RGCN) to model spatial correlation of traffic flow and capture spatial dependence of demand respectively.

Traffic Flow Graph Convolution Network Block. The traffic state of the road network structure closely correlated to crowds can reflect the demand of ride-hailing to a certain extent. [7] considers the transportation connectivity between regions in the task of ride-hailing demand prediction, which neglects the potential impact of the dynamic change of traffic flow information. Here, we use the Chebyshev polynomial GCN in the TGCN block to model the spatial dependence of road traffic flow information [3]. Given the traffic flow network graph G_T, we can integrate the spatial characteristics of traffic state by Formula 2.

$$F_{l+1} = \sigma(f(\widehat{A_T}, \theta)F_l W_{T,l}). \tag{2}$$

Where $F_l \in R^{|V_T| \times P_l}$ and $F_{l+1} \in R^{|V_T| \times P_{l+1}}$ are the feature tensors of the roads at l and l+1 layers respectively. $W_{T,l} \in R^{P_l \times P_{l+1}}$ is the learnable parameter of the l-th layer, $\sigma(\cdot)$ and $f(\cdot)$ are the activation function. $\widehat{A_T} = \widehat{D}^{-1/2}\widetilde{A}\widehat{D}^{-1/2}$ denotes preprocessing step, $\widetilde{A} = A_T + I_{|V_T|}$ is the matrix with added self connections, $I_{|V_T|}$ is the identity matrix, $\widehat{D} = \sum_j \widetilde{A}_{i,j}$ is the degree matrix.

After l-th layer GCN stacking, we can get traffic flow information F_l with spatial characteristics, and then we use the location information of road midpoint and restriction of the region for road-region mapping and aggregation. The color in Fig. 1(B) represents the mapping relationship. Assuming that the road set of each region $m_i \in M$ is R_i, we can get the traffic flow representation of each region through formulas below:

$$O = \{o_i | i = 1, 2, \cdots, l_1 * l_2 - 1\}, o_i = \bigsqcup_{r \in R_i} F_{l,r}, \tag{3}$$

where $\bigsqcup$ is the aggregate function, $F_{l,r}$ represents the traffic flow information of road r.

Regin Graph Convolution Network Block. The region nodes attribute X represent demand information of each region. To further capture the dynamic traffic conditions correlated to the prediction result, we integrate the traffic flow information into the region nodes:

$$H_{region} = \text{concatenate}(X, O). \tag{4}$$

The edge attribute between region nodes in G_R is represented by the POI types of two regions. Noted that the edge we construct is a directed edge, therefore, there are two vectors $[P_{m_i}, P_{m_j}]$ and $[P_{m_j}, P_{m_i}]$ between regions m_i and m_j

in different sides. Here we use the full connection layer to calculate the region connective degree.

$$A_R = \sigma(E_R W_R + b_R), \tag{5}$$

where $\sigma(\cdot)$ is the activation function and $W_R \in R^{2|S_p|}$, b_R are the learnable parameters. Then we can learn the representation of each region through the following formulas:

$$H_{l+1} = \sigma(f(\text{softmax}(A_R), \theta) H_l W_{R,l}). \tag{6}$$

Where $W_{R,l}$ is the learnable parameter of the l-th layer, $H_0 = H_{region}$. The basic TRGCN structural elements are obtained by the above operations. Subsequently, we can get deeper spatial dependence by stacking multiple TRGCN elements.

4.3 ConvLSTM for Temporal Correlation Modeling

Suppose the output through RGCN in the last layer of TRGCN is H_r, then the region features of L time intervals can be obtained by transferring the data of multiple consecutive time intervals into the stacked multi-layer TRGCN structure:

$$H = [H_{r,t-L+1}, H_{r,t-L+2}, \cdots, H_{r,t}], \tag{7}$$

where $H_{r,t}$ is the representation of regions at the t-th time interval.

ConvLSTM is a variant of LSTM, which extends the weight calculation in each unit cell to convolution operation [23]. Like LSTM, ConvLSTM consists of an internal memory unit and three gates: input gate i, forget gate f and output gate o. Here, in order to capture the regional spatio-temporal characteristics of L time intervals, we input H into ConvLSTM. For the t-th neuron of ConvLSTM, the input is: input value $H_{r,t}$ at time t, output value h_{t-1} at time $t-1$ and state c_{t-1} of gate control unit at time $t-1$. The outputs of ConvLSTM is: the output value h_t and the state c_t of gate unit of ConvLSTM at time t. The formulas of forgetting gate, input gate and output gate are shown as bellow.

$$\begin{aligned}
i_t &= \sigma(W_{xi} * H_{r,t} + W_{hi} * h_{t-1} + W_{ci} \circ C_{t-1} + b_i),\\
f_t &= \sigma(W_{xf} * H_{r,t} + W_{hf} * h_{t-1} + W_{cf} \circ C_{t-1} + b_f),\\
C_t &= f_t \circ C_{t-1} + i_t \circ tanh(W_{xc} * H_{r,t} + W_{hc} * h_{t-1} + b_c),\\
o_t &= \sigma(W_{xo} * H_{r,t} + W_{ho} * h_{t-1} + W_{co} \circ C_t + b_o),\\
h_t &= o_t \circ tanh(C_t),
\end{aligned} \tag{8}$$

where $*$ denotes the convolutional operator, and $\circ$ is the Hadamard product. Different kernel matrices W and biases b are trainable parameters.

4.4 Demand Forecasting

The final output h_t of ConvLSTM can get the spatial and temporal representation of regions. For the representation $h_{t,i}$ of the i-th region, we get the demand prediction value y_i' by feeding into a full connection layer containing one neuron.

$$y_i' = \sigma(h_{t,i}w + b), \tag{9}$$

where w and bias b are trainable parameters. For the training phase, we use the Mean Squard Error (MSE) as the loss function.

5 Experiments

5.1 Datasets

Datasets. We verify the effect of the model on two real-world ride-hailing datasets in **Hefei** and **Chengdu**, China. Among them, the longitude range of Hefei is [117.1994, 117.3795] and the latitude range is [31.7759, 31.9137]; the longitude range of Chengdu is [104.020824, 104.115581], and the latitude range is [30.6191, 30.6950]. The field information of experimental data includes order data and location data. Among them, the positioning data includes the longitude and latitude of each vehicle positioning point, the positioning time and the driving speed; the order data includes the order number, the departure time of the order, the longitude and latitude of departure position in the order. Mapping the location data and road network can get the traffic flow information of the roads. The dataset of Hefei is collected from November 1, 2019 to December 25, 2019. We choose data on the first 40 days as the training set, and the remains as the test set. The dataset of Chengdu is a competition open dataset from August 3, 2014 to August 30, 2014. For data split, the first 20 days are used as the training set and the rest is the test set. As described in Definition 2, the cities Hefei and Chengdu are divided into grids of 15×17 and 9×9 respectively according to longitude and latitude. Each grid represents a region with a size of about $1\,\text{km} \times 1\,\text{km}$.

POI Data. We sample school, shopping mall, theatre, hospital, cinema and subway station as POI data, and extract 14291 and 3225 POI data for Hefei and Chengdu respectively. For these POI data, we encode them with One-Hot encoding.

5.2 Experiment Settings

Implementation Details. We implement the TST-GCN model with the Keras framework based on Python language, and use Tensorflow as the backend of Keras [2]. All of our experiments are conducted on a server with 1 T K80.

In the experiment, we set the sampling frequency to 30 min and the number of observation intervals to 6. For TRGCN module, the output dimension of region is set to 5. Subsequently, we choose sum as the aggregation function. The convolution kernel size of ConvLSTM is $5 * 5$, and the number of filters is set to 32. In our TST-GCN model, we stack 2 layers in TRGCN module and choose rectified linear unit (ReLU) as the activation function. All of the other compared deep neural networks are implemented with Keras, we choose Adam as their optimizer [12], and the learning rate is $\alpha = 1 \times e^{-3}$. In the evaluation phase, we only count the regions with more than 5 orders, and use Root Mean Squard Error (RMSE) and Mean Absolute Error (MAE) as the metrics.

Compared Methods. We compare our proposed method with both basic and advanced methods as follows. We tune the parameters of all methods and report their best performance. The optimal parameters of these methods are found by grid search in scikit-learn on the test set.

- History Average model (HA): This method uses the mean value of historical data in the same time interval as the prediction value.
- LSTM: LSTM network is a special RNN model, which can effectively extract the temporal characteristics of historical data in each region.
- DMVST-Net [26]: DMVST-Net is a model which uses multi view to model the spatial and temporal characteristics of traffic data to predict the demand of taxi. Specifically, the model uses LSTM, CNN and graph embedding to build temporal view, spatial view, and semantic view respectively.
- CSTN [13]: CSTN is a ConvLSTM based taxi demand prediction model. First, it uses convolutional neural network and ConvLSTM to capture and maintain the local context of taxi demand, and then uses convolutional filter to capture global features.
- ST-MGCN [7]: ST-MGCN is a graph convolution based model for online car hailing demand forecasting, which constructs adjacency matrix from adjacency relationship, road connection relationship and functional similarity, and obtains non Euclidean correlation between regions through GCN.

5.3 Results

Performance Comparison. Table 1 reports the comparison of prediction results of all models under the optimal parameters. We can draw the following conclusions: (1) the prediction performance of LSTM is significantly improved compared with HA by extracting the temporal characteristics; (2) DMVST-Net, CSTN and ST-MGCN use CNN, ConvLSTM and GCN respectively to capture the spatio-temporal characteristics, which make a promising performance compared with LSTM; (3) TST-GCN model has achieved the best results under each evaluation standard, with 6.47% RMSE decrease and 7.06% MAE decrease in Hefei dataset when comparing with state-of-the-art model ST-MGCN, showing the effectiveness of our proposed model.

Table 1. Performance comparison with different baselines.

Method	Hefei		Chengdu	
	RMSE	MAE	RMSE	MAE
HA	10.51	7.28	44.30	28.97
LSTM	7.86	5.25	26.47	17.11
DMVST-Net	6.84	4.83	21.16	14.94
CSTN	6.47	4.69	21.30	15.03
ST-MGCN	6.03	4.25	20.61	14.73
TST-GCN	**5.64**	**3.95**	**19.98**	**14.56**

Table 2. The influence of traffic flow to TST-GCN. (Hefei)

Method	RMSE	MAE
TST-GCN-1	5.79	3.98
TST-GCN-2	5.80	4.07
TST-GCN	**5.64**	**3.95**

Table 3. The influence of ConvLSTM to TST-GCN. (Hefei)

Method	RMSE	MAE
TRGCN + FC	7.86	5.25
TRGCN + LSTM	6.54	4.63
TST-GCN	**5.64**	**3.95**

Effectiveness of Traffic GCN. In order to explore the influence of traffic flow information in the model, we compare experiment results on several variants of TST-GCN, including: TST-GCN-1, which removes the traffic GCN in TRGCN module and directly feeds the traffic flow data into the aggregation function, and TST-GCN-2, which takes no account of traffic flow information. As is shown in Table 2, TST-GCN-2 makes relatively poor results under the absence of TGCN. TST-GCN uses the GCN to capture the spatial dependency based on traffic topological network, it achieves the value 5.64 RMSE and 3.95 MAE, which are much better than above mentioned models.

Effectiveness of ConvLSTM. ConvLSTM can capture the spatial domain features and temporal features. In order to verify its impact on the prediction, we compare it with the following two models: (1) TRGCN + FC, which is directly predicted by using the full connection neural network through the region node representation obtained by TRGCN; (2) TRGCN + LSTM that replaces ConvLSTM in TST-GCN model with LSTM.

The experimental results are shown in Table 3. The prediction performance of TRGCN + FC model without considering the temporal characteristics is the worst, which shows that in order demand prediction, the prediction results rely heavily on the data in a long period of history. The prediction performance of TRGCN + LSTM model is 16.79% and 11.81% lower than that of TRGCN + FC model under the metric of RMSE and MAE, respectively, and TST-GCN has achieved the best effect, which shows that it is effective to improve the prediction performance of the model after using ConvLSTM to aggregate the representation of the geographical region.

Effectiveness of Model Parameters. In this part, we further evaluate the performance of TRGCN module with different stacking layer numbers and convolution kernel sizes. In Fig. 2(a), We can observe the changes caused by the varying stacking numbers: when the number of stacking layers is less than 2, the effects present a positive relationship with the number of layers. While the number of stacking layers is greater than 3, the performance reverses. In Fig. 2(b), the convolution kernels selected by 5 × 5 with 32 filters in ConvLSTM can achieve the best performance. In Fig. 3, the thermodynamic diagram of the model for the demand prediction of each region is shown.

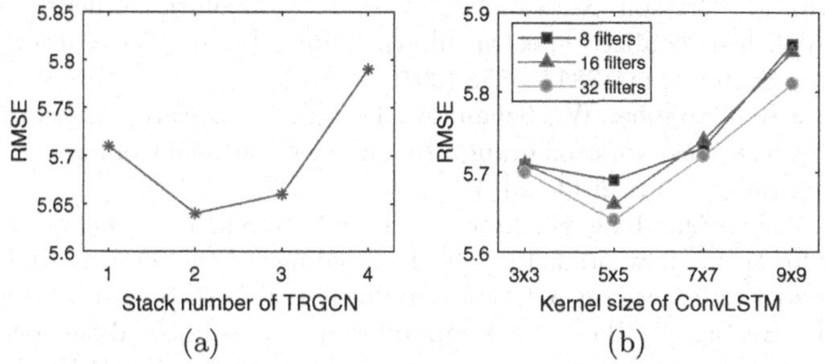

Fig. 2. (a) The influence of different stacked numbers of TRGCN on TST-GCN. (b) The influence of kernel size with different filters in ConvLSTM on TST-GCN. (Hefei)

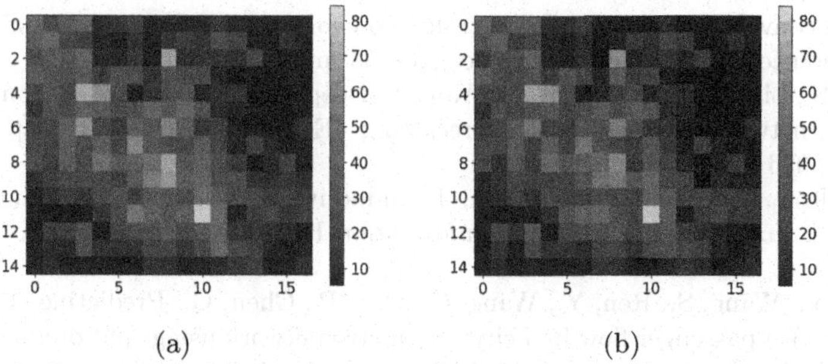

Fig. 3. Display the regional thermal diagram corresponding to the order demand, where (a) is the real value and (b) is the predicted value of TST-GCN. (Hefei)

6 Conclusion

In this paper, we introduce a more worth-exploring task, traffic flow information driven demand forecast of ride-hailing. We solve this problem by constructing a new spatio-temporal graph convolution network, which consists of a stackable graph convolution network module and a ConvLSTM network. Extensive experiments demonstrate that our proposed method can be effectively used for regional-level riding demand forecasting.

References

1. Chai, D., Wang, L., Yang, Q.: Bike flow prediction with multi-graph convolutional networks. In: Proceedings of the 26th ACM SIGSPATIAL International Conference on Advances in Geographic Information Systems, pp. 397–400. ACM (2018)
2. Chollet, F., et al.: Keras (2015). https://github.com/fchollet/keras
3. Defferrard, M., Bresson, X., Vandergheynst, P.: Convolutional neural networks on graphs with fast localized spectral filtering. In: Advances in Neural Information Processing Systems, pp. 3844–3852 (2016)
4. Estrach, J.B., Zaremba, W., Szlam, A., LeCun, Y.: Spectral networks and deep locally connected networks on graphs. In: 2nd International Conference on Learning Representations, ICLR 2014 (2014)
5. Fang, S., Zhang, Q., Meng, G., Xiang, S., Pan, C.: GSTNet: global spatial-temporal network for traffic flow prediction. In: Proceedings of the Twenty-Eighth International Joint Conference on Artificial Intelligence, IJCAI, pp. 10–16 (2019)
6. Ge, L., Li, H., Liu, J., Zhou, A.: Temporal graph convolutional networks for traffic speed prediction considering external factors. In: 2019 20th IEEE International Conference on Mobile Data Management (MDM), pp. 234–242. IEEE (2019)
7. Geng, X., Li, Y., Wang, L., Zhang, L., Yang, Q., Ye, J., Liu, Y.: Spatiotemporal multi-graph convolution network for ride-hailing demand forecasting. In: 2019 AAAI Conference on Artificial Intelligence (AAAI 2019) (2019)
8. Guo, S., Lin, Y., Feng, N., Song, C., Wan, H.: Attention based spatial-temporal graph convolutional networks for traffic flow forecasting. Proceedings of the AAAI Conference on Artificial Intelligence, vol. 33, pp. 922–929 (2019)
9. Guo, S., Lin, Y., Li, S., Chen, Z., Wan, H.: Deep spatial-temporal 3D convolutional neural networks for traffic data forecasting. IEEE Trans. Intell. Transp. Syst. **20**, 3913–3926 (2019)
10. Hamilton, W., Ying, Z., Leskovec, J.: Inductive representation learning on large graphs. In: Advances in Neural Information Processing Systems, pp. 1024–1034 (2017)
11. Han, Y., Wang, S., Ren, Y., Wang, C., Gao, P., Chen, G.: Predicting station-level short-term passenger flow in a citywide metro network using spatiotemporal graph convolutional neural networks. ISPRS Int. J. Geo-Inf. **8**(6), 243 (2019)
12. Kingma, D.P., Ba, J.: Adam: a method for stochastic optimization. arXiv preprint arXiv:1412.6980 (2014)
13. Liu, L., Qiu, Z., Li, G., Wang, Q., Ouyang, W., Lin, L.: Contextualized spatial-temporal network for taxi origin-destination demand prediction. IEEE Trans. Intell. Transp. Syst. **20**, 3875–3887 (2019)
14. Liu, Y., Zheng, H., Feng, X., Chen, Z.: Short-term traffic flow prediction with Conv-LSTM. In: 2017 9th International Conference on Wireless Communications and Signal Processing (WCSP), pp. 1–6. IEEE (2017)

15. Lv, Y., Duan, Y., Kang, W., Li, Z., Wang, F.Y.: Traffic flow prediction with big data: a deep learning approach. IEEE Trans. Intell. Transp. Syst. **16**(2), 865–873 (2014)
16. Ma, X., Dai, Z., He, Z., Ma, J., Wang, Y., Wang, Y.: Learning traffic as images: a deep convolutional neural network for large-scale transportation network speed prediction. Sensors **17**(4), 818 (2017)
17. Rodrigues, F., Markou, I., Pereira, F.C.: Combining time-series and textual data for taxi demand prediction in event areas: a deep learning approach. Inf. Fusion **49**, 120–129 (2019)
18. Tong, Y., Chen, Y., Zhou, Z., Lei, C., Lv, W.: The simpler the better: a unified approach to predicting original taxi demands based on large-scale online platforms. In: the 23rd ACM SIGKDD International Conference (2017)
19. Veličković, P., Cucurull, G., Casanova, A., Romero, A., Liò, P., Bengio, Y.: Graph attention networks. In: International Conference on Learning Representations (2018). https://openreview.net/forum?id=rJXMpikCZ. Accepted as poster
20. Wang, D., Cao, W., Li, J., Ye, J.: DeepSD: supply-demand prediction for online car-hailing services using deep neural networks. In: 2017 IEEE 33rd International Conference on Data Engineering (ICDE), pp. 243–254. IEEE (2017)
21. Wang, Y., Yin, H., Chen, H., Wo, T., Xu, J., Zheng, K.: Origin-destination matrix prediction via graph convolution: a new perspective of passenger demand modeling. In: Proceedings of the 25th ACM SIGKDD International Conference on Knowledge Discovery & Data Mining, pp. 1227–1235 (2019)
22. Wu, W., Liu, T., Yang, J.: CACRNN: a context-aware attention-based convolutional recurrent neural network for fine-grained taxi demand prediction. In: Lauw, H.W., Wong, R.C.-W., Ntoulas, A., Lim, E.-P., Ng, S.-K., Pan, S.J. (eds.) PAKDD 2020. LNCS (LNAI), vol. 12084, pp. 636–648. Springer, Cham (2020). https://doi.org/10.1007/978-3-030-47426-3_49
23. Shi, X, Chen, Z., Wang, H., Yeung, D.Y., Wong, W.K., Woo, W.: Convolutional LSTM network: a machine learning approach for precipitation nowcasting. In: Advances in Neural Information Processing Systems, pp. 802–810 (2015)
24. Xu, J., Rahmatizadeh, R., Bölöni, L., Turgut, D.: Real-time prediction of taxi demand using recurrent neural networks. IEEE Trans. Intell. Transp. Syst. **19**(8), 2572–2581 (2017)
25. Yao, H., Tang, X., Wei, H., Zheng, G., Li, Z.: Revisiting spatial-temporal similarity: a deep learning framework for traffic prediction. In: AAAI Conference on Artificial Intelligence (2019)
26. Yao, H., et al.: Deep multi-view spatial-temporal network for taxi demand prediction. In: Thirty-Second AAAI Conference on Artificial Intelligence (2018)
27. Yu, R., Li, Y., Shahabi, C., Demiryurek, U., Liu, Y.: Deep learning: a generic approach for extreme condition traffic forecasting. In: Proceedings of the 2017 SIAM International Conference on Data Mining, pp. 777–785. SIAM (2017)
28. Zhang, J., Zheng, Y., Qi, D.: Deep spatio-temporal residual networks for citywide crowd flows prediction. In: Thirty-First AAAI Conference on Artificial Intelligence (2017)
29. Zhao, L., et al.: T-GCN: a temporal graph convolutional network for traffic prediction. IEEE Tran. Intell. Transp. Syst. **21**, 3848–3858 (2019)
30. Zhao, Z., Chen, W., Wu, X., Chen, P.C., Liu, J.: LSTM network: a deep learning approach for short-term traffic forecast. IET Intell. Transp. Syst. **11**(2), 68–75 (2017)

A Proximity Forest for Multivariate Time Series Classification

Yue Zhang, Zhihai Wang, and Jidong Yuan(✉)

School of Computer and Information Technology,
Beijing Jiaotong University, Beijing 100044, China
yuanjd@bjtu.edu.cn

Abstract. Multivariate time series (MTS) classification has gained attention in recent years with the increase of multiple temporal datasets from various domains, such as human activity recognition, medical diagnosis, etc. The research on MTS is still insufficient and poses two challenges. First, discriminative features may exist on the interactions among dimensions rather than individual series. Second, the high dimensionality exponentially increases computational complexity. For that, we propose a novel proximity forest for MTS classification (MTSPF). MTSPF builds an ensemble of proximity trees that are split through the proximity between unclassified time series and its exemplar one. The proximity of trees is measured by locally slope-based dynamic time warping (DTW), which enhances traditional DTW by considering regional slope information. To extract the interaction among dimensions, several dimensions of an MTS instance are randomly selected and converted into interrelated sequences as the input of trees. After constructing each tree independently, the weight of each tree is calculated for weighted classifying. Experimental results on the UEA MTS datasets demonstrate the efficiency and accuracy of the proposed approach.

Keywords: Multivariate time series classification · Interrelated sequence · Proximity forest · Dynamic time warping · Local slope feature

1 Introduction

Time series classification is a specific field of machine learning whose input is sequences ordered in time. Time series data are widely distributed in various fields of human life, such as climate, finance, medical care, etc. Univariate time series classifications have had enormous progress over the last decades. One of the most accurate algorithms, such as, Hierarchical Vote Collective of Transformation-based Ensembles (HIVE-COTE) [1] combines multiple classifiers with different feature extraction methods, making it hard to beat. However, it is hardly to suit for a task of multivariate time series (MTS) classification. A time series in MTS contains multiple interrelated sequence, which increases the difficulty of feature extraction. Discriminative features in MTS could not only appear within a single sequence, but also exist among multiple sequences. As a concrete example, in the MTS ERing dataset [2], the researchers collected the electric signal from four sensors on a finger ring, which can be used to detect six kinds of finger gestures

K. Karlapalem et al. (Eds.): PAKDD 2021, LNAI 12712, pp. 766–778, 2021.
https://doi.org/10.1007/978-3-030-75762-5_60

by electric field sensing. The four electric sensors are distinguished by the distance from the hand. When classifying this dataset, although the electrical series generated by individual sensor is significant, the interactions between the sensors of different positions cannot be ignored. However, there is not any existing algorithm that can handle the issues. Figure 1 demonstrates some interactions among multi-dimensions. The differences between sequences from two classes are more obvious after conversion than before.

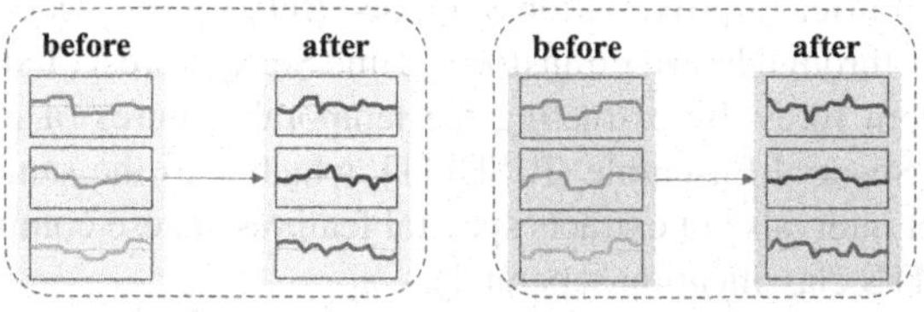

Fig. 1. Series before and after conversion in the ERing dataset

In this paper, we provide a novel MTS learning model based on proximity forest which ensembles proximity trees [3], named by MTSPF. The difference between proximity tree and conventional decision tree is related to splitting functions. Proximity tree splits through the proximity between unclassified time series and its exemplar series, whereas a decision tree splits on the threshold of the attribute value. When building a tree, several dimensions are randomly selected to increase the discrimination among the trees and reduce time complexity. In order to extract the interaction between dimensions, we propose a novel method to convert two original time series into an interrelated sequence, so that the interaction can be treated a general time series (As shown in Fig. 1). Interrelated series or original series is randomly selected as the input of the tree. After determining the input of proximity tree, the similarity is measured by local slope dynamic time warping (LSDTW) [4], which enhances traditional DTW by taking regional slope information into consideration. Each tree is built independently and produces separate feature representation, so we calculate the weight of each tree for weighted voting.

The rest of the paper is organized as follows: Sect. 2 summarizes related work. Then we introduce our approach in Sect. 3. Section 4 evaluates the efficiency and accuracy of our proposed approach. Finally, Sect. 5 provides conclusions.

2 Related Work

Basically, 1-Nearest neighborhood (1-NN) model with DTW or its variants is a baseline of MTS classification [5]. DTW is a distance measure that uses dynamic programming theory to find the optimal matching path between two sequences. When extending the DTW method to multiple dimensions, there are three strategies. The first category, like DTW independent (DTWI) [6], independently calculates the distance of each dimension in the MTS, then adds them up as the final result. The second method, such as DTW dependent (DTWD), regards dimensions as dependent and considers the attribute values of all dimensions in a dynamic programming. The third one, like DTW Adaptive (DTWA) [7], combines the above two methods, and determines which method to use through the threshold calculated by cross validation.

The most advanced algorithm HIVE-COTE combined with multiple algorithms can be used for MTS classification after building classifier for each dimension of each component. The combined algorithms in COTE include Elastic Ensemble (EE) [8], which combines various distance measures with 1-NN classifiers; Shapelet Transform (ST) [9], which converts the time series into the distance between the shapelet and the series; Bag of Symbolic Fourier Approximation Symbols (BOSS) [10], which extracts discrete dictionary features through Fourier transform; Time Series Forest (TSF) [11, 12], which constructs a random forest by extracting the temporal features of each interval; and Random Interval Spectral Ensemble (RISE) [1], which uses the same method as TSF to randomly select intervals but extracts spectral features. These component algorithms can be used for MTS classification separately.

Except the above methods, some algorithms specifically focus on MTS classification task. Symbolic representation for MTS (SMTS) [13] constructs random forest with bag-of-words as features. Generalized Random Shapelet Forests (gRSF) [14] generates shapelet-based forest, where instances and shapelets for building trees are randomly selected. Learned Pattern Similarity (LPS) [15] learns representations from segments of time series by training regression trees. word extraction for time series classification plus multivariate unsupervised symbols and derivatives (WEASEL+MUSE) [16] classifies by logistic regression after adding dimension identifier to each extracted discrete feature.

In recent years, deep learning algorithms for MTS classification are popular. Multi-Channel Deep Convolutional Neural Network (MCDCNN) [17] constructs convolutional layers for individual dimension in each channel, combining all channels into a multilayer perceptron for classification. Multivariate Long Short Term Memory Fully Convolutional Network (MLSTM-FCN) [18] combines LSTM for temporal feature learning and FCN with a squeeze-and-excitation block, followed by a shared dense layer for prediction. Time Series Attentional Prototype Network (TapNet) [19] design a random dimension permutation method combined with CNN to learn the low-dimensional features from MTS data. The performance of deep learning methods surpasses traditional methods in many field, but the experiment in [5] indicates that the existing deep learning methods are not better than DTW on MTS problem.

3 Proximity Forest for Multivariate Time Series

In this section, we introduce our novel algorithm MTSPF based on proximity forest for MTS classification. We start with the training phase of MTSPF, which is the construction of the forest, and then explain the classification phase, especially the weighted voting strategy. Finally, the distance measure LSDTW used for proximity tree branches is explained in detail.

3.1 Construction of Forest

During the training phase, MTSPF builds a forest composed of randomly generated proximity trees. As shown in Fig. 2, The input of the tree is determined by random dimension selection and interrelated feature extraction. Original sequences and interrelated sequences after interrelated feature extraction are randomly selected to build the tree. We explain these two steps separately, and finally introduce the construction of each tree.

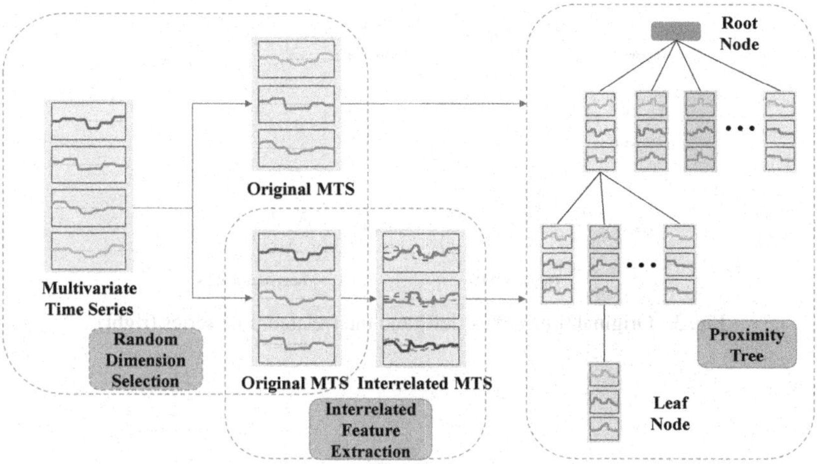

Fig. 2. MTSPF architecture

Random Dimension Selection. One of the challenges of MTS problem is that the increase in dimensions exponentially increases the time complexity. By using the random forest method, d dimensions in the D-dimensional time series are randomly selected, and a d-dimensional time series is used to construct each tree. This method reduces the computational complexity while increasing the difference between each tree, thereby improve the performance of the forest.

Interrelated Feature Extraction. In order to extracting the interaction between dimensions in the d-dimensional time series, we calculate the difference between two sequences point by point to generate an interrelated sequence with the same length as the original time series. The calculation of the weight is as follows:

$$S_{InterrelatedAB}[i] = S_{OriginalA}[i] - S_{OriginalB}[i], \quad 0 \leq i < S.length \tag{1}$$

After the extraction of d-dimensional time series, the dimension of interrelated time series is $d' = d(d-1)/2$. Assuming that $d = 3$, the conversion from the original time series to the interrelated time series is shown in Fig. 3. We believe that the original sequence and the interrelated sequence are equally important, so we randomly select whether to employ this strategy.

Construction of Tree. Unlike traditional decision trees that splits on the threshold of attribute value, proximity trees split through the proximity between time series and exemplar series. As shown in Fig. 3, a set of exemplars are randomly selected one per class at each node. By calculating the LSDTW distances between each time series and each exemplar of the current node, the time series split into the branch of the closest exemplar. The initial training set is continuously split until all instances in a node belong to the same class, and then this node is regarded as a leaf node.

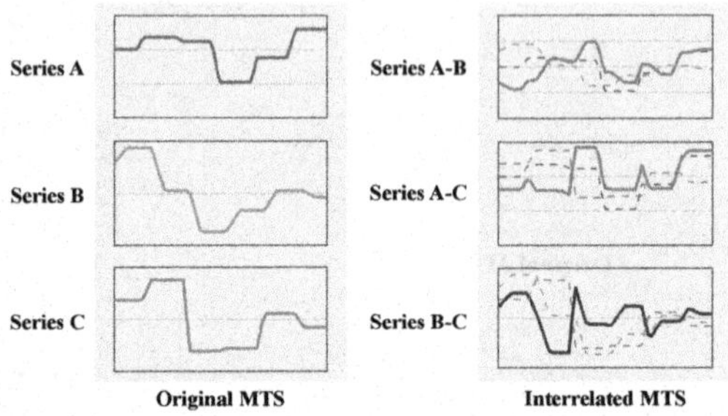

Fig. 3. Original time series (left) and interrelated time series (right)

3.2 Weighted Classifying

Different multivariate time series datasets have different discriminatory dimensions and local slope features. If each tree has the same weight, the difference between the trees may be ignored. Weight calculated based on the structure of each tree can reflect whether the dimensions and local slope features randomly selected by each tree are suitable for the current dataset, thereby improving the classification accuracy.

The weight of decision tree is measured by measuring the purity of the dataset before and after the split of each node in the tree. We use Gini impurity to measure the purity of the dataset. Gini impurity represents the probability of any instance in the dataset being classified into the wrong class. The smaller the value of Gini impurity, the lower the probability of incorrect classification of the instance.

In this method, each node of the tree contains multiple branches, so when calculating the weight of node, it is necessary to calculate the Gini impurity of each branch separately. Then multiply the Gini impurity of each branch by the proportion of each branch's subset to the total data of the node (Eq. (4)), and add up to get the weight of the node (Eq. (3)). The weight of the entire tree is obtained by summing the weights of each node. Finally, these weights are standardized by Eq. (2). The calculation of the weight is as follows:

$$w_t = \text{softmax}(W_t) = \frac{e^{-W_t}}{\sum_{k=0}^{K} e^{-W_k}}, \tag{2}$$

$$W_t = \sum\nolimits_{i=0}^{p} W_{gi}, \tag{3}$$

$$W_{gj} = \sum\nolimits_{j=0}^{n} Gini(S_j)\frac{Size(S_j)}{Size(S_{all})}, \tag{4}$$

where w_t is the final proximity tree weight, p is the number of nodes of the tree, and n is the number of branches of the current node. After constructing all trees and calculating the weight of each tree, the training phase is completed.

In the classifying phase, for each tree, a time series starts from the root node, selects the branch of the closest exemplar depends on LSDTW distance, and repeats this process until reaching the leaf node. This series is labelled with the class of the leaf node. Finally, the class of the sequence is obtained by weighted voting.

3.3 Locally Slope-Based DTW

For the similarity measure of the proximity tree, we adopted local slope DTW which considering the local slope characteristics based on DTW. The slope of the time series is calculated and discretized for feature encoding. The encoded features are used to improve the traditional DTW.

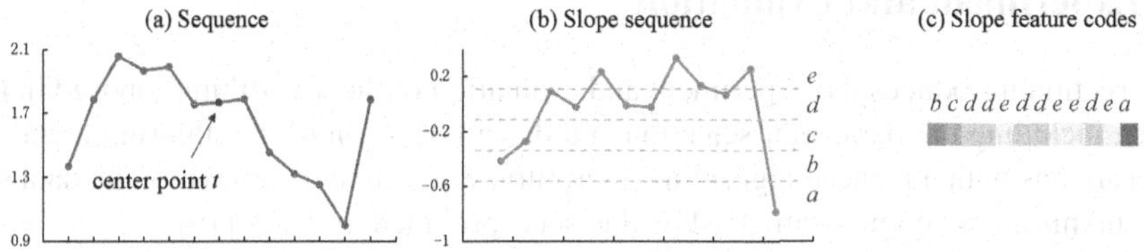

Fig. 4. Local slope feature coding

Given a time series of length n $T = \{t_1, t_2, \ldots, t_n\}$. Firstly, subsequence S_i of length l is extracted from each temporal point t_i, which is the center of S_i as shown in the Fig. 4(a). The length l defines the number of neighborhoods around t_i. Secondly, the subsequence S_i is converted into a slope sequence of length $l - 1$ as shown in the Fig. 4(b). The slope of two points is calculated as:

$$s_i = \frac{t_{i+1} - t_i}{\Delta t}, 1 \leq i < n, \tag{5}$$

where Δt represents the time interval of these two points. Here simply make $\Delta t = 1$. Finally, the obtained continuous slopes are discretized to symbolic slopes through equal depth division strategy, and represented by a grey scale map visually as shown in the Fig. 4(c). Since each point t_i is transformed to discrete feature of length $l - 1$, the original time series T will be represented by a local slope feature coding matrix of $n \times (l - 1)$ symbols.

Given two time series $T^A, T^B \in \ominus\mathcal{R}^n$. The generated local slope matrices are $\boldsymbol{M}^A = \{M_1^A, M_2^A, \ldots, M_n^A\}$ and $\boldsymbol{M}^B = \{M_1^B, M_2^B, \ldots, M_n^B\}$, where $M_i^A, M_j^B \in \mathcal{R}^{l-1}$, $\boldsymbol{M}^A, \boldsymbol{M}^B \in \mathcal{R}^{n \times (l-1)}$. Then the LSDTW is calculated as:

$$d_{ij} = \alpha Sim(M_i^A, M_j^B) + (1 - \alpha)\left|t_i^A - t_j^B\right|^2, \tag{6}$$

$$c(i,j) = d_{ij} + \min\{c(i-1,j-1), c(i-1,j), c(i,j-1)\}, \tag{7}$$

$$LSDTW(T^A, T^B) = c(n, n), \quad (8)$$

where α in Eq. (6) is a balance factor, which represents the importance of local slope features during sequence alignment. $Sim(M_i^A, M_i^B)$ defines the local shape similarity measure of t_i^A and t_i^B as:

$$Sim(M_i^A, M_j^B) = \sum_{g=1}^{l-1} \left| m_{ig}^A - m_{jg}^B \right| \quad (9)$$

In the original LSDTW algorithm, the values of the parameters, the balance factor α, the alphabet size k and the length of subsequence l are fixed. However, since the best parameters on multiple datasets have different values, we have determined the optimal range through experiments, and randomly selected parameter values within this range. The experimental results are explained in Sect. 4.1.

4 Experiment and Evaluation

This section introduces the experiment and evaluation of the algorithm proposed in this paper, including the parameter selection and design decision of our algorithm, and the comparisons with advanced algorithms in accuracy and time complexity. The data sets used in this article come from 20 MTS datasets[1] provided by UEA [19].

4.1 Parameter Analysis

The parameter setting has an important influence on the accuracy of the algorithm. There are 5 parameters in MTSPF that need to be considered. Figure 5 shows the effect of different parameter values on the classification accuracy.

Tree Size. In a forest algorithm, the number of trees is obviously an important factor. It is generally believed that the accuracy of the algorithm will increase as the number of trees increases, but at the same time the computational expense will also increase. To maintain the balance between accuracy and efficiency, this value needs to be determined through experiments. The experimental range of tree size T is {20, 40, 60, 80, 100}. As shown in Fig. 5, accuracy rate is relatively stable when $T = 60$, and then slowly rises until $T = 100$. In order to reduce computational expense, in the experiments in Sects. 4.1 and 4.2, we set $T = 60$. In the comparative experiment with other algorithms in Sect. 4.3, we set $T = 100$ with the highest accuracy.

Dimension Size. In the random dimension selection stage, the number of dimensions that can be selected is a fixed value, which needs to be determined experimentally. The number of dimensions after interrelated feature extraction increases exponentially with the number of original sequences, so this value should be small enough. The experimental range of dimension size d is $\{3, 4, 5\}$. It can be seen from Fig. 5 that the value of dimension size has little effect on accuracy, and $d = 4$ is relatively the best choice.

[1] UEA provides 30 datasets, while for all the comparable algorithms, we can only get the results on 20 of them.

Parameters of LSDTW. There are 3 parameters in LSDTW. The experimental range of these parameters is as follows. The balance factor $\alpha \in [0 : 0.02 : 1]$, the alphabet size $k \in [3, 8]$, the length of subsequence $l \in [3, 21]$. When one parameter changes, other two takes the largest random range. As illustrated in Fig. 5, in the experiment of α and l, the accuracy rate gradually decreases as the parameter increases. The situation of k is complicated, and the best value is different on different datasets. We try to make multiple datasets gain better performance. The range of the final decision is the balance factor $\alpha \in [0 : 0.02 : 1]$, the alphabet size $k \in [6, 8]$, the length of subsequence $l \in [3, 11]$.

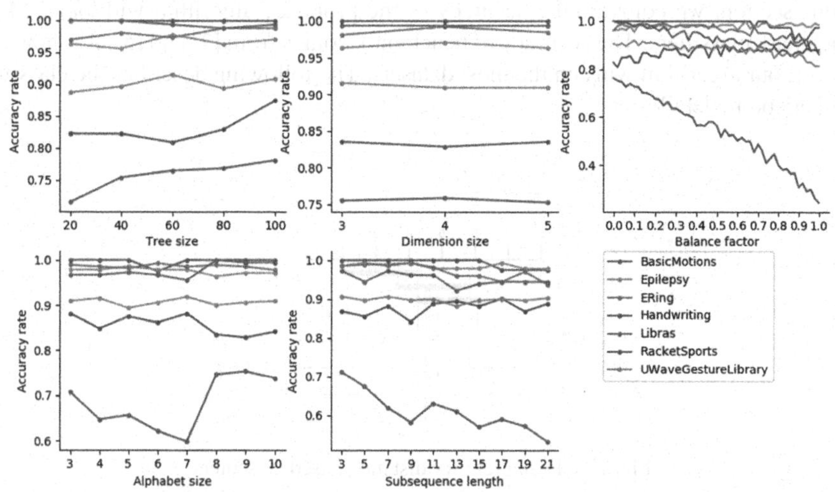

Fig. 5. Accuracy rates under different parameters

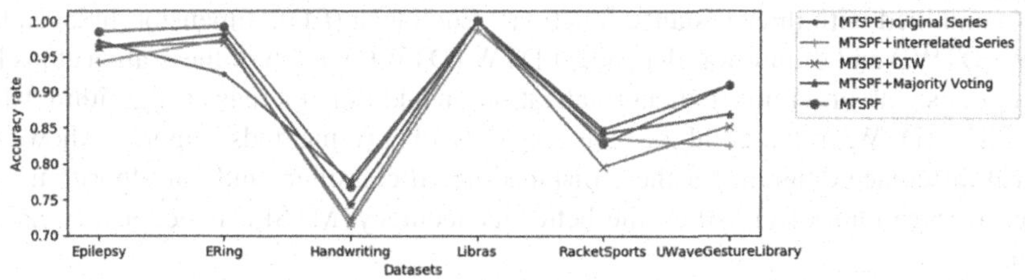

Fig. 6. Accuracy comparison of various decisions

4.2 Impact of Design Decisions

In the MTSPF algorithm, the interrelated feature extraction, distance measure of tree, and weight calculation of proximity tree are all part of the design decision. Therefore, we conducted four comparative experiments to prove the effectiveness of the algorithm. In interrelated feature extraction stage, we compared two algorithms, one considers only the original sequence and nothing about interrelated features at all, and the other is just the opposite. When determining the distance measure, we compare LSDTW with traditional DTW. In the voting phase, we experimented with the common majority voting strategies. Due to the limitations of the experimental conditions, we only conducted experiments on 6 small-scale datasets used in parameter selection, but the experimental results are sufficient to prove that our algorithm has more advantages in accuracy as shown in Fig. 6.

4.3 Classification Accuracy

In this section, we compare the accuracy of the proposed algorithm with other MTS benchmark classifiers. The accuracy of benchmark comes from [2] and [5]. As shown in Table 1, our algorithm wins on the most datasets. The following describes the classifier comparison in detail.

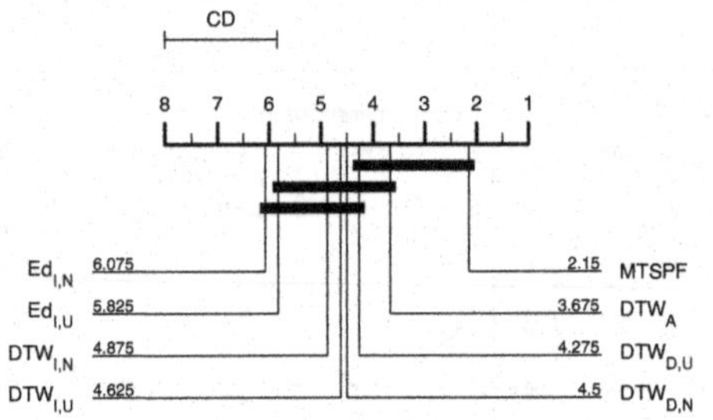

Fig. 7. CD diagram for distance-based classifiers

Compared with Distance-Based Classifiers. The basic distance-based classifiers are 1NN combined with three distance functions: Euclidean (ED), dimension-independent DTW (DTWI) and dimension-dependent DTW (DTWD). Each of these algorithms has two options with or without data normalization. In addition, there is an algorithm Adaptive DTW (DTWA) that combines the above two DTW methods. Figure 7 shows the critical difference diagram for the 7 distance-based classifiers and our algorithm. The lower average rank of classifier, the better its accuracy. MTSPF is better than any of them.

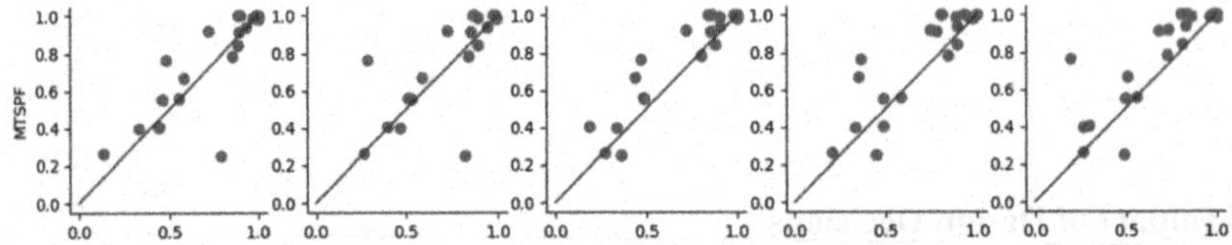

Fig. 8. Accuracy comparison between MTSPF and expanded univariate time series classifier

Compared with HIVE-COTE and Its Components. HIVE-COTE and its components are univariate time classifier, which are used for MTS classification after simple expansion. Since the MTS classification is still in the early stage of research, these algorithms are still competitive. Figure 8 shows the accuracy comparison between MTSPF and HIVE-COTE, ST, BOSS, TSF and RISE. The point above the diagonal line indicates that the MTSPF algorithm has a higher accuracy on this dataset. MTSPF is more accurate than HIVE-COTE on 14 of the 20 datasets (with 1 tie). Compared with other classifiers, MTSPF has more obvious advantages.

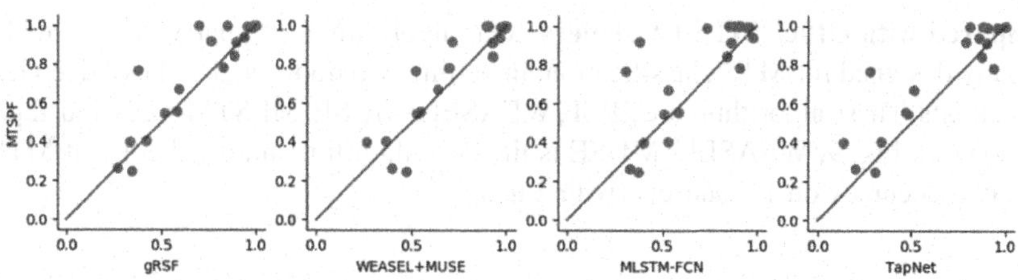

Fig. 9. Accuracy comparison between MTSPF and other MTS classifiers

Table 1. Accuracy for each classifier. The best approaches are highlighted using a bold font.

Dataset	MTSPF	COTE	ST	BOSS	TSF	RISE	gRSF	MUSE	MLCN	TapNet
AWR	**0.993**	0.990	0.990	0.990	0.953	0.963	0.983	**0.993**	0.957	0.957
AF	0.267	0.133	0.267	0.267	0.200	0.267	0.267	**0.400**	0.333	0.200
BM	**1.000**	**1.000**	0.975	**1.000**	**1.000**	**1.000**	**1.000**	**1.000**	0.875	**1.000**
CR	**1.000**	0.986	0.986	0.986	0.931	0.986	0.986	0.986	0.917	**1.000**
EP	0.986	**1.000**	0.993	**1.000**	0.978	**1.000**	0.978	0.993	0.732	0.957
EC	0.251	0.791	**0.821**	0.361	0.445	0.487	0.346	0.475	0.373	0.308
ER	**0.993**	0.970	0.889	0.907	0.881	0.859	0.952	0.974	0.941	0.904
FM	0.560	0.550	0.510	0.480	**0.580**	0.560	**0.580**	0.550	**0.580**	0.470
HMD	0.405	0.446	0.392	0.189	0.486	0.297	0.419	0.365	**0.527**	0.338
HW	**0.768**	0.482	0.288	0.472	0.366	0.194	0.375	0.522	0.309	0.281
HB	**0.917**	0.722	0.722	0.722	0.741	0.732	0.761	0.712	0.380	0.790
LIB	**1.000**	0.900	0.861	0.844	0.806	0.806	0.694	0.894	0.850	0.878
LSST	**0.672**	0.575	0.587	0.435	0.350	0.509	0.588	0.640	0.528	0.513
NATO	**1.000**	0.889	0.872	0.861	0.800	0.839	0.844	0.906	0.900	0.811
PD	0.939	0.934	0.941	0.908	0.892	0.832	0.935	0.967	**0.979**	0.856
RS	0.842	0.888	0.888	0.882	0.888	0.809	0.882	**0.928**	0.842	0.875
SRS1	0.782	0.853	0.840	0.805	0.840	0.724	0.823	0.696	0.908	**0.935**
SRS2	**0.550**	0.461	0.533	0.489	0.483	0.494	0.517	0.528	0.506	0.483

(continued)

Table 1. (*continued*)

Dataset	MTSPF	COTE	ST	BOSS	TSF	RISE	gRSF	MUSE	MLCN	TapNet
SWJ	0.400	0.333	**0.467**	0.333	0.333	0.267	0.333	0.267	0.400	0.133
UW	0.913	0.891	0.850	0.856	0.775	0.684	0.897	**0.931**	0.859	0.900
Wins	10	2	2	2	2	2	2	5	3	3

Compared with Other MTS Classifiers. Our algorithm is compared with 4 methods specially designed for MTS classification, including two traditional and two deep learning methods. These algorithms are gRSF, WEASEL+MUSE, MLSTM-FCN and TapNet. As shown in Fig. 9, WEASEL+MUSE is the best algorithm among them, but MTSPF beats it in accuracy on 12 datasets (with 2 ties).

4.4 Complexity Analysis

In comparative complexity analysis, we first consider the splitting function LSDTW. Note that the time complexity of DTW is $O(n^2)$, since LSDTW is a general case of DTW, its complexity can be deduced as $O(n^2)$[4].

During the training phase, when generating the nodes of the tree, we calculate the distance between each instance of the current node and c exemplars from c class. If there are m time series in training set, the time complexity of each node is $O(cn^2m)$. In the worst case, the depth of the tree will be $O(m)$, but in general, the average depth of the tree is $O(\log(m))$. Thus, the time complexity of generating a tree is $O(log(m)cn^2m)$, and the comparative complexity of generating a proximity forest composed of T trees is $O(Tlog(m)cn^2m)$. Table 2 shows the comparison of training time complexity between MTSPF and several algorithms. It is easy to find that MTSPF is definitely the most efficient one. During the classifying phase, each instance is compared with $\log(m)$ nodes in T trees on average, and the distance between it and c exemplars is calculated on each node, so the time complexity of the classification stage is $O(T\log(m)cn^2)$.

Table 2. The training time complexities of each classifier

Classifier	Comparative complexity	Parameters
MTSPF	$O(Tlog(m)cn^2m)$	
HIVE-COTE	$O(m^2n^4)$	
WEASEL+MUSE	$O(\min[mn^2, s^{2l}n] \times d)$	*s*: number of symbols, *l*: word length, *d*: number of dimensions
gRSF	$O(n^2m^2\log(mn^2))$	

5 Conclusion

We introduced a novel MTS classifier MTSPF based on proximity forest ensemble of proximity trees, which is building by MTS data after random dimension selection and interrelated feature extraction. Experimental results on the UEA MTS archive demonstrate the efficiency and accuracy of the proposed approach.

Acknowledgement. This work is supported by National Natural Science Foundation of China (No. 61771058), Beijing Natural Science Foundation (No. 4214067).

References

1. Lines, J., Taylor, S., Bagnall, A.: Time series classification with HIVE-COTE: the hierarchical vote collective of transformation-based ensembles. ACM Trans. Knowl. Disc. Data **12**(5), 1–35 (2018)
2. Bagnall, A., Dau, H.A., Lines, J., et al.: The UEA multivariate time series classification archive. arXiv preprint arXiv:1811.00075 (2018)
3. Lucas, B., Shifaz, A., Pelletier, C., et al.: Proximity forest: an effective and scalable distance-based classifier for time series. Data Min. Knowl. Disc. **33**(3), 607–635 (2019). https://doi.org/10.1007/s10618-019-00617-3
4. Yuan, J., Lin, Q., Zhang, W., Wang, Z.: Locally slope-based dynamic time warping for time series classification. In: Proceedings of the 28th ACM International Conference on Information and Knowledge Management, pp. 1713–1722 (2019)
5. Ruiz, A.P., Flynn, M., Bagnall, A.: Benchmarking multivariate time series classification algorithms. arXiv preprint arXiv:2007.13156 (2020)
6. Shokoohi-Yekta, M., Wang, J., Keogh, E.: On the non-trivial generalization of dynamic time warping to the multi-dimensional case. In: Proceedings of the 2015 SIAM International Conference on Data Mining, pp. 289–297. Society for Industrial and Applied Mathematics (2015)
7. Shokoohi-Yekta, M., Hu, B., Jin, H., Wang, J., Keogh, E.: Generalizing DTW to the multi-dimensional case requires an adaptive approach. Data Min. Knowl. Disc. **31**(1), 1–31 (2016). https://doi.org/10.1007/s10618-016-0455-0
8. Lines, J., Bagnall, A.: Time series classification with ensembles of elastic distance measures. Data Min. Knowl. Disc. **29**(3), 565–592 (2014). https://doi.org/10.1007/s10618-014-0361-2
9. Hills, J., Lines, J., Baranauskas, E., Mapp, J., Bagnall, A.: Classification of time series by shapelet transformation. Data Min. Knowl. Disc. **28**(4), 851–881 (2013). https://doi.org/10.1007/s10618-013-0322-1
10. Schäfer, P.: The BOSS is concerned with time series classification in the presence of noise. Data Min. Knowl. Disc. **29**(6), 1505–1530 (2014). https://doi.org/10.1007/s10618-014-0377-7
11. Deng, H., Runger, G., Tuv, E., et al.: A time series forest for classification and feature extraction. Inf. Sci. **239**, 142–153 (2013)
12. Shi, M., Wang, Z., Yuan, J., Liu, H.: Random pairwise shapelets forest. In: Phung, D., Tseng, V.S., Webb, G.I., Ho, B., Ganji, M., Rashidi, L. (eds.) PAKDD 2018. LNCS (LNAI), vol. 10937, pp. 68–80. Springer, Cham (2018). https://doi.org/10.1007/978-3-319-93034-3_6
13. Baydogan, M., Runger, G.: Learning a symbolic representation for multivariate time series classification. Data Min. Knowl. Disc. **29**(2), 400–422 (2014). https://doi.org/10.1007/s10618-014-0349-y

14. Karlsson, I., Papapetrou, P., Boström, H.: Generalized random shapelet forests. Data Mining and Knowledge Discovery **30**(5), 1053–1085 (2016). https://doi.org/10.1007/s10618-016-0473-y
15. Baydogan, M., Runger, G.: Time series representation and similarity based on local autopatterns. Data Min. Knowl. Disc. **30**(2), 476–509 (2015). https://doi.org/10.1007/s10618-015-0425-y
16. Schäfer, P., Leser, U.: Multivariate time series classification with WEASEL+ MUSE. arXiv preprint arXiv:1711.11343 (2017)
17. Zheng, Y., Liu, Q., Chen, E., Ge, Y., Zhao, J.: Exploiting multi-channels deep convolutional neural networks for multivariate time series classification. Front. Comput. Sci. **10**(1), 96–112 (2016). https://doi.org/10.1007/s11704-015-4478-2
18. Karim, F., Majumdar, S., Darabi, H., et al.: Multivariate LSTM-FCNs for time series classification. Neural Netw. **116**, 237–245 (2019)
19. Zhang, X., Gao, Y., Lin, J., et al.: TapNet: multivariate time series classification with attentional prototypical network. In: AAAI, pp. 6845–6852 (2020)

C²-Guard: A Cross-Correlation Gaining Framework for Urban Air Quality Prediction

Yu Chu[1], Lin Li[1(✉)], Qing Xie[1], and Guandong Xu[2]

[1] Wuhan University of Technology, Wuhan, China
{chory7437,cathylilin,felixxq}@whut.edu.cn
[2] University of Technology Sydney, Ultimo, Australia
guandong.xu@uts.edu.au

Abstract. Predicting air quality is increasingly important for protecting people's daily health and helping government decision-making. The multistep air quality prediction largely depends on the correlations of air quality-related factors. How to model the correlations among factors is a big challenge. In this paper, we propose a cross-correlation gaining framework (C²-Guard) consisting of a temporal correlation module, factor correlation module, and cross gaining module for air quality (mainly PM2.5) prediction. Specifically, the temporal correlation module is used to extract the temporal dependence of air pollutant time series to gain their distributed representation. In the factor correlation module, a novel convolution and recalibration block is designed for air quality factor correlations extraction to gain their distributed representation in the factor dimension. In the cross gaining module, a joint-representation block is proposed to learn the cross-correlations between time and factor dimensions. Finally, extensive experiments are conducted on two real-world air quality datasets. The results demonstrate that our C²-Guard outperforms the state-of-the-art methods of air pollutants prediction in terms of RMSE and MAE.

Keywords: Air quality prediction · Temporal correlation · Factor correlation · Cross correlation learning

1 Introduction

With the development of industrialization and urbanization, the air pollution problem has become increasingly serious. According to the Health Effects Institute (HEI), air pollution (PM2.5, ozone, and household air pollution) is the fifth leading risk factor for mortality worldwide. In 2017, air pollution is estimated to have contributed to close to 5 million deaths globally – nearly 1 in every 10 deaths[1]. Therefore, predicting changes in air pollutants that seriously affect

[1] https://www.healtheffects.org/announcements/state-global-air-2019-air-pollution-significant-risk-factor-worldwide.

K. Karlapalem et al. (Eds.): PAKDD 2021, LNAI 12712, pp. 779–790, 2021.
https://doi.org/10.1007/978-3-030-75762-5_61

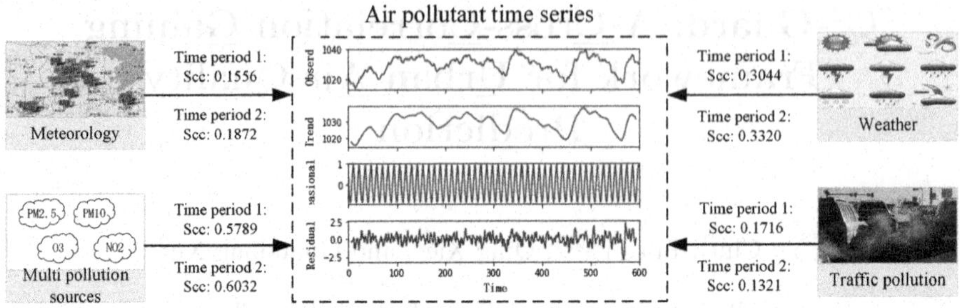

Fig. 1. The correlation analysis of air pollution time series. The center part is STL decomposition of air pollution time series. The two sides are Spearman correlation coefficients(Scc) of multiple air quality influential factors in different time periods.

urban air quality is of great significance for protecting people's daily health and helping government decision-making.

In recent years, great efforts have been made on air quality prediction. First, air quality prediction can be regarded as a standard time series prediction problem. Traditional methods like autoregressive integrated moving average(ARIMA) [2], long short-term memory neural network (LSTM) [8,9,16], Gated Recurrent Unit [17] and temporal convolutional network [20] are used to predict air quality. However, as shown in Fig. 1, air pollutants are not only trend and seasonal in nature, but also related to other influencing factors such as weather and multiple pollutants. It is not comprehensive enough to predict air quality only by the temporal correlation of time series.

Second, several attempts have been made to predict air quality from multiple dimensions [5,12,18,19] including time and factor. These methods usually take the hidden information output from one dimension module as input to another dimension module, or weight and fuse the outputs of several unique dimension modules. However, it is necessary to take into account the fact that air quality factors have different influence degrees in different time periods. As shown in Fig. 1, the Spearman correlation coefficients of meteorology and traffic pollution in the first time period are 0.1556 and 0.1716, respectively, and they are changed to 0.1872 and 0.1321 in the second time period. Therefore, if multidimensional hidden information is not considered at each prediction time step, the multistep prediction results will be greatly compromised.

To address these challenges, in this paper, we propose a cross-correlation gaining framework for predicting urban air quality such as PM2.5, entitled C^2-Guard. First, a temporal correlation module based on Encoder-Decoder unit is utilized to learn the long temporal dependency of air pollutant time series by encoding the air pollutant values on historical time slots. Second, a novel factor correlation module is designed to extract and recalibrate the correlations among air quality-related factors. Multivariate air quality time series are inputted through different channels, and interdependence among different factors is learned through this module. Finally, a cross gaining module is employed to learn the cross-correlations between time and factor dimensions. Joint representation learning is applied to obtain the cross representation at each prediction time step, aiming at reducing the error accumulation of multistep prediction. The main contributions of this work are summarized as follows:

- We propose a novel air quality prediction framework named C^2-Guard that models the complex correlations of temporal features and factor features of air pollutants to predict multistep air quality.
- To learn the correlations of time and factor, a factor correlation module is developed to extract and recalibrate the correlations of related factors affecting air quality. What's more, factor correlations are jointly learned with temporal correlations to gain cross-correlations in a novel and effective cross gaining module.
- Comprehensive experiments are conducted on two real-world air quality datasets. These results indicate that our C^2-Guard performs better than state-of-the-art methods in terms of RMSE and MAE.

The remaining part of the paper is organized as follows. Section 2 discusses the related works. Section 3 formulates the problem of air quality prediction. Section 4 describes our model C^2-Guard. Section 5 presents the evaluation results. In Sect. 6, we conclude this paper and talk about future work.

2 Related Work

Air quality prediction has always been a hot topic in society. First, air quality prediction can be regarded as a standard time series prediction problem. Conventionally, some classic time series forecasting methods like ARIMA [2] and variants of the recurrent neural network including LSTM [8] and GRU [17] are developed for air quality prediction and other tasks [21,22]. As a variant of a convolutional neural network, TCN [3] has a flexible receptive field and a stable gradient with good performance in time series modeling. Jorge *et al.* attempt to use TCN to evaluate air quality levels [20]. Ong *et al.* propose a deep recurrent neural network (DRNN) for air pollution prediction by using the auto-encoder model as a novel pre-training method [14]. However, these methods take only historical time data as input, and it has been widely recognized that air pollutants are related to other influencing factors.

To solve this problem, in recent years, several efforts have been made to introduce other air quality factors to enhance prediction performances. Li *et al.* propose a spatiotemporal deep learning (STDL) based air quality prediction method that inherently considers spatial and temporal correlations [11]. Qi *et al.* develop a general and effective approach to improve the performance of the interpolation and the prediction [15]. Zheng *et al.* consider meteorological data, weather forecasts, and air quality data of the station and that of other stations within a few hundred kilometers [18,19]. Du *et al.* propose a novel deep learning framework named DAQFF [5] that is the state-of-the-art air quality (mainly PM2.5) forecasting method and outperforms the above methods in real datasets. We observe that these methods mainly study the spatiotemporal modeling of air quality and all of them have local receptive fields of factors, so the learned relationships of factors are not comprehensive enough. In addition, temporal correlation information and factor correlation information are in different dimensions. If the information of two dimensions is not simultaneously considered in each prediction time step, the accuracy of multistep prediction will be affected. To the best of our knowledge, our C^2-Guard is the first to learn the correlations of time and factors at each prediction time step in multi-step prediction, thus reducing prediction errors.

3 Problem Formulation

Before formulating the problem of air quality prediction, some necessary mathematical notations are given first. Suppose an urban air quality monitoring station S can detect I types of relevant factors that affect air quality, defined as $\boldsymbol{S} = \{s_1, ..., s_j, ..., s_I\}$. Given an urban air pollutant s_i and a time window with a length of L, a vector $\boldsymbol{s_i} = [s_i^1, ..., s_i^j, ..., s_i^L]$ is defined as the historical values of air pollutant s_i.

Air Quality Prediction. Based on the above notations, the prediction process is formally defined as follows. Given an air quality factor matrix $X \in N^{I \times T_1}$, where N represents the number of data samples, I represents channels composed of factors that affect air quality and T_1 represents the length of a historical time window. Our task is to learn a predictive model $M : X \rightarrow Y$ from historical air quality factor matrix $\boldsymbol{X}$ to future air pollutant time series $\boldsymbol{Y} = [y_1, ..., y_i, ..., y_{T_2}]$ where T_2 means the prediction range of future air pollutant values.

4 The Proposed Framework

In this section, the proposed framework for the task of air quality prediction is described. The architecture is presented in Fig. 2 where three modules work together. The inputs of temporal correlation module and factor correlation module are historical air pollutant time series and air quality impact factor matrix $\boldsymbol{X^0}$ (shown in the upper part of Fig. 2), respectively from real data. Through the above two modules, the temporal correlations of air pollutants and air quality factor correlations can be modeled. The generated temporal hidden correlation matrix $\boldsymbol{T_{out}}$ and factor hidden correlation vector $\boldsymbol{F_{out}}$ are then fed into the cross gaining module for integration at each prediction time step (shown in the lower right part of Fig. 2). The final output of our framework is future air pollutant values (shown in the lower right corner of Fig. 2). The details of different modules are described in the following sections.

4.1 Temporal Correlation Module

As shown in the upper left corner of Fig. 2, the original input of the temporal correlation module is historical air pollutant time series that are extracted from

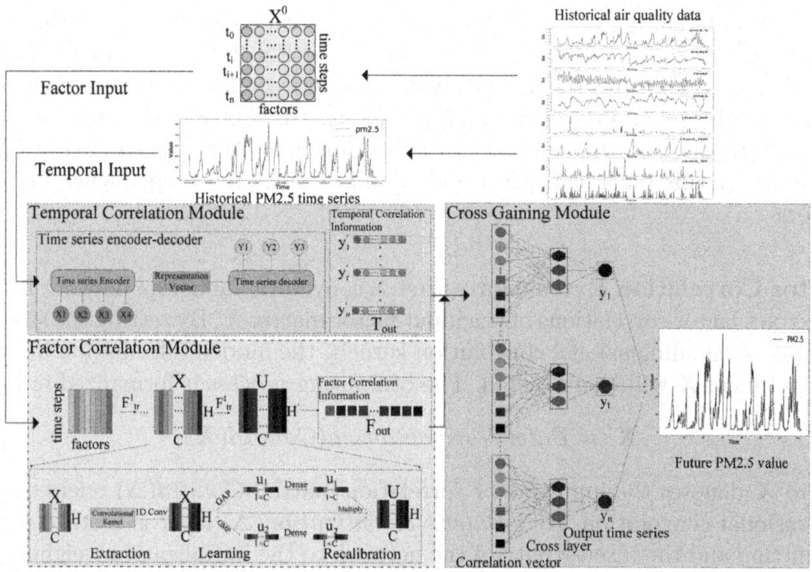

Fig. 2. The overall architecture of the proposed air quality prediction framework C^2-Guard. The prediction of PM2.5 is taken as an example. The upper part is the factor input and the temporal input that come from historal air quality data. The left parts are temporal correlation module and factor correlation module respectively. The right part is cross gaining module that outputs the future PM2.5 data.

the time series of various factors affecting air quality detected at an air monitoring station. To predict multistep time series, we adopt Enc-Dec units (shown in the temporal correlation module of Fig. 2) to learn the long-term temporal correlations of historical air pollutant time series. Since the main work of this paper is to learn the interrelation between air quality factors (discussed in Sect. 4.2) and how to effectively integrate the temporal correlations and factor correlations (discussed in Sect. 4.3), the encoding and decoding unit in this module is LSTM that can be replaced by GRU, Transformer and so on.

The final output of this module is a 2-dimension temporal correlation matrix $\boldsymbol{T_{out}}$.

$$\boldsymbol{T_{out}} = [\boldsymbol{y'_1}, ..., \boldsymbol{y'_t}, .., \boldsymbol{y'_n}] \tag{1}$$

where $\boldsymbol{y'_t}$ represents the hidden temporal correlation vector at time step t. n represents the prediction range of air pollutant values. The temporal hiding information at each time step generated in this module will be fused with the factor hiding information in subsequent modules.

4.2 Factor Correlation Module

Essentially, the factors affecting air quality are interdependent, hence the air pollutant to be predicted is related to other air quality factors (shown in Fig. 1). In this part, different factors are input through different channels to form the factor matrix $\boldsymbol{X^0}$ as shown in the upper left corner of Fig. 2. This module explicitly models and recalibrates the interdependencies between factors by superimposing novel convolution and recalibration block F_{tr}^t that is shown in the lower-left part of Fig. 2. In the following content of this section, we describe each part of F_{tr}^t to explain why this block can learn and recalibrate the interdependence between factors.

Factor Correlation Extraction. First, a convolutional neural network is used to extract factor correlations of the input factor matrix X. By setting the size and number of one-dimensional convolution kernels, the hidden information between the factors of X will be extracted. The computing process is formalized in Eq. 2.

$$\boldsymbol{X'} = BatchNormalization(Conv1d(\boldsymbol{X})) \tag{2}$$

where $\boldsymbol{X}$ denotes the input factor correlation matrix. $Conv1d(\Delta)$ refers to one-dimensional convolution. And $BatchNormalization(\Delta)$ layer is used to reduce overfitting and the insensitivity of the network to the initialization weights. The relationship of different factors will be extracted by this operation. Next, the generated new factor correlation matrix $\boldsymbol{X'} \in \mathbb{R}^{H \times C}$ is inputted into the following block.

Factor Correlation Learning. Second, in order to learn the factor correlations effectively, we need to integrate the temporal information in each factor dimension of $\boldsymbol{X'}$ to expand the local receptive field. For aggregating temporal information, Hu *et al.* proposed the use of GAP (GlobalAveragePooling) methods to shrink dimensions [6]. As another pooling method, GMP(GlobalMaxPooling) can also gather temporal information by selecting the maximum value of historical time series. It is reliable in terms of shrinking temporal dimensions. Thus, we use GAP and GMP to shrinking temporal information in each factor channel. Formally, two vectors ($\boldsymbol{u_1}$, $\boldsymbol{u_2}$) are generated by shrinking $\boldsymbol{X'}$ through its temporal dimension H. $\boldsymbol{u_1} \in \mathbb{R}^{1 \times C}$ and $\boldsymbol{u_2} \in \mathbb{R}^{1 \times C}$ are calculated by:

$$\begin{aligned} \boldsymbol{u_1} &= F_{GAP}(\boldsymbol{X'}) \\ &= [\frac{1}{H}\sum_{i=1}^{H} \boldsymbol{X'}_{C_1}(i), ..., \frac{1}{H}\sum_{i=1}^{H} \boldsymbol{X'}_{C_j}(i), ..., \frac{1}{H}\sum_{i=1}^{H} \boldsymbol{X'}_{C_n}(i)] \end{aligned} \tag{3}$$

$$\begin{aligned} \boldsymbol{u_2} &= F_{GMP}(\boldsymbol{X'}) \\ &= [MAX(\boldsymbol{X'}_{C_1}(i)), ..., MAX(\boldsymbol{X'}_{C_j}(i)), ..., MAX(\boldsymbol{X'}_{C_n}(i))] \end{aligned} \tag{4}$$

where C_j refers to the jth factor channel of factor matrix $\boldsymbol{X'}$.

In order to make use of the information summarized in the shrinking operation, we follow $\boldsymbol{u_1}$ and $\boldsymbol{u_2}$ with an excitation block (two nonlinear fully connected layers) to fully capture factor correlations.

$$\boldsymbol{u_1'} = F_{ex}(\boldsymbol{u_1}, \boldsymbol{W}) = \delta(\boldsymbol{W_2}\delta(\boldsymbol{W_1}\boldsymbol{u_1})) \tag{5}$$

$$\boldsymbol{u'_2} = F_{ex}(\boldsymbol{u_2}, \boldsymbol{W}) = \delta(\boldsymbol{W_2}\delta(\boldsymbol{W_1}\boldsymbol{u_2})) \tag{6}$$

where $\boldsymbol{W_1} \in \mathbb{R}^{\frac{C}{r} \times C}$ and $\boldsymbol{W_2} \in \mathbb{R}^{C \times \frac{C}{r}}$ are weight matrixs of the two layers. To reduce the complexity of the block, the dimensional reduction rate r is set in the first fully connected layer. δ refers to the ReLU [13] function.

The outputs $\boldsymbol{u'_1} \in \mathbb{R}^{1 \times C}$ and $\boldsymbol{u'_2} \in \mathbb{R}^{1 \times C}$ of this block learn and amplify the relationship between different factors.

Factor Correlation Recalibration. Third, we multiply the learned relationship vectors $\boldsymbol{u'_1}$ and $\boldsymbol{u'_2}$ with the intermediate factor matrix $\boldsymbol{X'}$ to obtain a new factor matrix $\boldsymbol{U}$. The calculation process is as follows:

$$\boldsymbol{U} = F_{mul}(\boldsymbol{X'}, \boldsymbol{u'_1}, \boldsymbol{u'_2}) = \boldsymbol{X'}\boldsymbol{u'_1}\boldsymbol{u'_2} \tag{7}$$

The new factor matrix $\boldsymbol{U}$ has the same size as the intermediate factor matrix $\boldsymbol{X'}$, and the correlations between factors have been learned and enlarged.

In order to recalibrate the learned factor hiding information with the temporal hiding information at each prediction time step, we need to squeeze the factor matrix U:

$$\begin{aligned} F_{out} &= [\frac{1}{H'}\sum_{i=1}^{H'} U_{C'_1}(i), ..., \frac{1}{H'}\sum_{i=1}^{H'} U_{C'_j}(i), ..., \frac{1}{H'}\sum_{i=1}^{H'} U_{C'_n}(i)] \\ &= [c_1, ..., c_n] \end{aligned} \tag{8}$$

By squeezing the temporal information of each factor channel C_j of $\boldsymbol{U}$, the final factor correlation information vector $\boldsymbol{F_{out}} = [c_1, ..., c_n]$ represents the hidden correlation informations between the factors of the original input matrix $\boldsymbol{X^0}$.

4.3 Cross Gaining Module

In order to gain the cross-correlation at each prediction time step, cross gaining module is proposed to combine the hidden information of the two above modules.

Temporal correlations and factor correlations are fed into the joint layers that combine the correlations of temporal information and factor information into the common space to obtain more accurate prediction results.

$$C_{out} = Joint(T_{out}, F_{out}) = Joint([y'_1, ..., y'_t, .., y'_n], F_{out}) = \begin{bmatrix} y'_1 \ F_{out} \\ .. \quad .. \\ y'_t \ F_{out} \\ .. \quad .. \\ y'_n \ F_{out} \end{bmatrix} \tag{9}$$

$$Y = W'_{out}\delta(W_{out}C_{out}) = [y_1, ..., y_t, ..., y_n] \tag{10}$$

The learned factor hidden information vector T_{out} and the temporal hidden informatin vector F_{out} are concatenated at each prediction time step shown as Eq. 9. To gain the final prediction vector, two fully connected layers are used to process the mixed hidden vector at each time step defined as Eq. 10. δ refers to the ReLU function. As shown in the lower right part of Fig. 2, the final output $Y = [y_1, ..., y_t, ..., y_n]$ represents the predicted value of PM2.5 at n time steps in the future.

5 Experiment

5.1 Datasets

Our experiments are based on two real public datasets from UCI[2]. The details of the two experimental datasets are given as follows:

Beijing PM2.5 Dataset. [3]This hourly dataset contains the PM2.5 data of US Embassy in Beijing and other related data including meteorological data, wind speed, and so on. The time period of this dataset is between 1/1/2010 to 12/31/2014, and it has 43824 records.

The Temple of Heaven Air Quality Dataset. [4]This dataset includes hourly air pollutants data from the temple of heaven air-quality monitoring site, where the data items include PM2.5, PM10, SO2, NO2, CO, O3, and meteorological data. The time period is from March 1st, 2013 to February 28th, 2017.

5.2 Experimental Setup

Baselines. The following three categories of baselines are used to compare with our proposed C^2-Guard: (*i*) the classical machine learning time series prediction methods (i.e., ARIMA); (*ii*) the traditional sequence modeling neural network (i.e., LSTM, GRU, TCN) based on encoder-decoder structure; (*iii*) the state-of-the-art PM2.5 prediction methods(i.e., DAQFF) which learn the interdependence of multivariate air quality-related time series data. More details are listed as follows:

- **Auto-Regressive Integrated Moving Average (ARIMA)**: It is one of the most common statistical models used for time series forecasting and also used in air quality prediction [2,4]. Scikit-learn is used to build this model in experiment.
- **Long Short-Term Memory Network (LSTM)**: LSTM is widely used in time series prediction tasks and has proposed to predict air quality [8,16]. A one-layer LSTM network with 100 hidden units is built in our experiments.
- **Gated Recurrent Unit (GRU)**: As an effective variant of LSTM, GRU [17] is implemented in Keras and set the parameters as the same as LSTM.

---

2 https://archive.ics.uci.edu/ml/index.php.
3 https://archive.ics.uci.edu/ml/datasets/Beijing+PM2.5+Data.
4 https://archive.ics.uci.edu/ml/datasets/Beijing+Multi-Site+Air-Quality+Data.

- **Temporal Convolutional Network (TCN)** [3]: TCN is a popular convolutional neural network capable of processing time series data and widely used to perform fine-grained action segmentation or detection [10] and predict air quality [20]. We utilize the implementation version that is released in GitHub[5] and set the filters and kernel_size are set to 100 and 2.
- **Deep Air Quality Forecasting Framework (DAQFF)** [5]: DAQFF is a state-of-the-art air quality prediction model that consists of Bi-LSTM layers and convolution layers. We implement DAQFF in Keras and set the parameters as the same as mentioned in [5].

Evaluation Methodology. To evaluate the performance of C^2-Guard, one year's data are used from each of the two datasets. For Beijing PM2.5 dataset, we select ten-month data for training and validation (01/01/2014-10/31/2014) and two-month data for testing (11/01/2014-12/31/2014). For the temple of heaven air quality dataset, the data from January to October 2016 are used for training and validation, the last two months' data are used for testing. In each experiment, we predict the value of PM2.5 in the next 12 time steps. The Root Mean Squared Error (RMSE) and Mean Absolute Error (MAE) are adopted as the evaluation metrics. [5]. The adaptived moment estimation algorithm (Adam) [7] is imployed to optimize the parameters.

Implementation Details. Experiments are conducted on an NVIDIA GTX 1060 GPU with 8 GB memory. We implement our C^2-Guard through Keras based on Tensorflow [1]. The initial learning rate is 0.01. In addition, the number of neurons in the LSTM and the kerner_size of 1D-CNN are 128 and 64, respectively. Finally, the number of neurons in the dense layer is 16.

5.3 Performance Comparison

To analyze the comprehensive performance of our C^2-Guard, multistep PM2.5 prediction experiments are performed on two datasets, and the experimental results are shown in Table 1.

By comparison, it is not difficult to find that our framework C^2-Guard performs better than all the baselines on both evaluation indicators. It should be noted that the error value in the table is the average of all results on the test dataset, and each value represents the sum of the prediction error value in the next 1–6 h or 7–12 h. Compared to the state-of-the-art air quality prediction model DAQFF, for the first six hours, our framework C^2-Guard improves prediction quality on two data sets by 12.4% and 10.3%, respectively. And for the next six hours, C^2-Guard improves by 7.3% and 7.4%, respectively.

The prediction error at the farther time step is larger because of the forward propagation of the error. To verify that the prediction effect of our framework on various time spectrums in the future is the best, more specific experiments are

[5] https://github.com/philipperemy/keras-tcn.

Table 1. In the experiments on two datasets, the errors among different models for the multistep prediction of PM2.5 values in the next 12 h. The smaller error, the better performance.

Model	Beijing PM2.5 dataset				The temple of heaven air quality dataset			
	RMSE		MAE		RMSE		MAE	
	1 h–6 h	7 h–12 h	1 h–6 h	7 h–12 h	1 h–6 h	7 h–12 h	1 h–6 h	7 h–12 h
ARIMA [2]	60.3432	84.3432	55.3816	59.3816	58.3432	70.3432	54.3816	66.3816
LSTM [8]	43.0802	63.6328	39.2571	60.5136	44.0545	66.4585	40.0375	63.1094
GRU [17]	40.7278	62.5122	36.7438	59.2638	50.3635	71.1009	46.6650	67.9050
TCN [3]	40.6655	64.7542	34.9552	59.6885	45.9718	71.3350	39.7683	65.8612
DAQFF [5]	40.4987	62.1307	36.5149	58.8346	44.6469	65.9207	40.3688	62.4490
C²-Guard	**35.4614**	**57.5892**	**31.3690**	**54.1637**	**40.0465**	**61.0232**	**35.6896**	**57.0689**

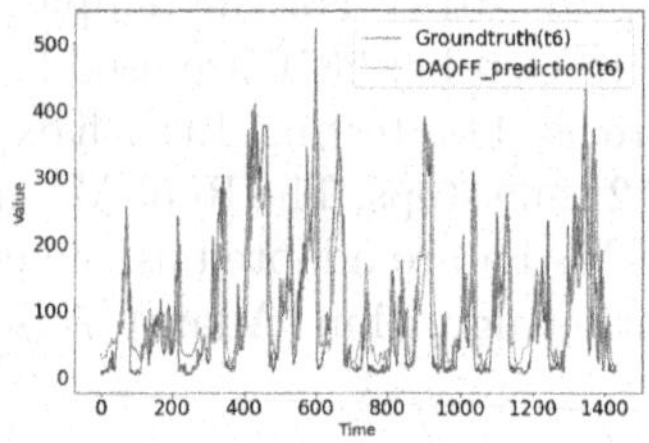

(a) DAQFF(t6)

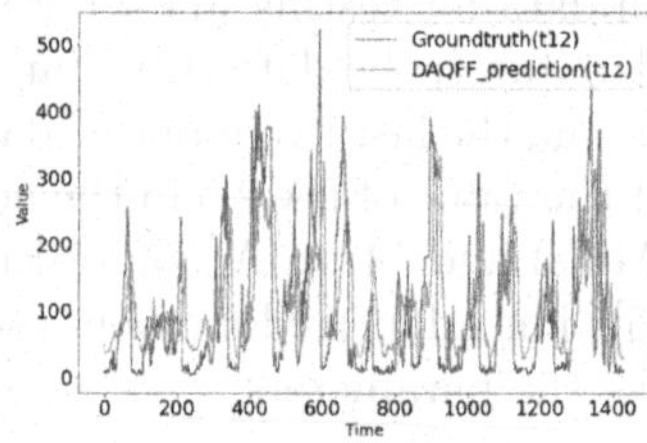

(b) DAQFF(t12)

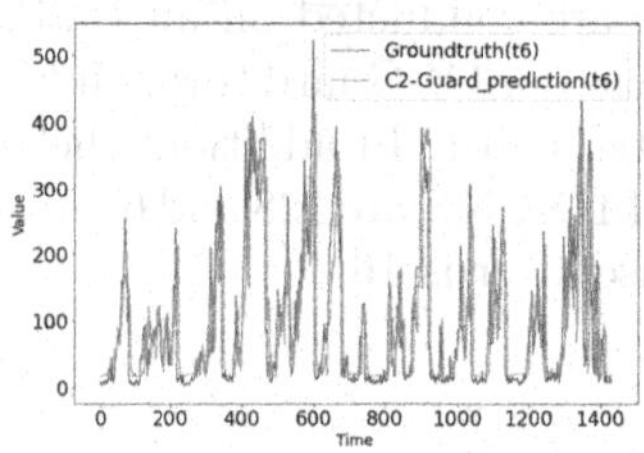

(c) C^2-Guard(t6)

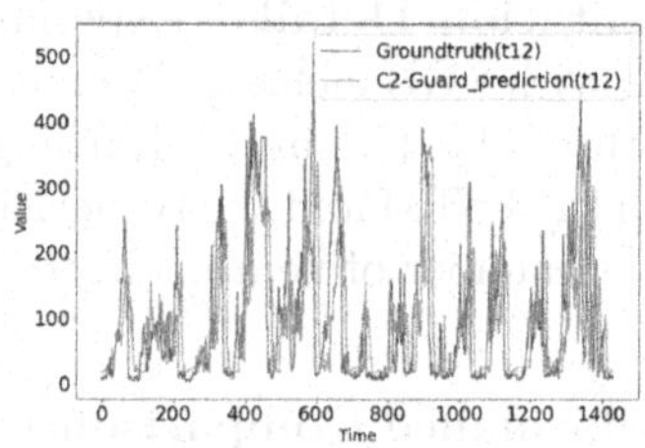

(d) C^2-Guard(t12)

Fig. 3. In the experiments on Beijing PM2.5 dataset, a comparison of the sixth hour and the twelfth hour ground truth and predicted PM2.5 value of DAQFF and C^2-Guard. (a) DAQFF for the next 6th h(t6) prediction; (b) DAQFF for the next 12th h(t12) prediction; (c) C^2-Guard for the next 6th h(t6) prediction; (d) C^2-Guard for the next 12th h(t12) prediction.

conducted in the predicted 12 h. The prediction results at the sixth time point and the twelfth time point in the future are shown in Fig. 3.

Figure 3 (a)–(d) shows the comparison results of the predicted PM2.5 data and ground truth values of DAQFF and C^2-Guard on the Beijing PM2.5 dataset at different time points (the sixth hour and the twelfth hour). As shown in these figures, the predicted value of our C^2-Guard is closer to the true value at each step. Especially where the extreme value is taken in figures, our predicted value curve is obviously closer to the true value curve, which further reflects the superiority of our model in multi-step prediction. Due to C^2-Guard relies on more time to effectively learn factor correlations, C^2-Guard takes 500 s while DAQFF requires 454 s in training experiments. It is important to note that the time we spend is within reasonable limits.

5.4 Ablation Study

In this section, an ablation experiment based on two datasets is conducted to gain a better understanding of the effect of factor correlation module. We compare the complete C^2-Guard with the framework without the factor correlation module. The experimental results are shown in Table 2.

Table 2. Comparison of prediction effects of C^2-Guard and C^2-Guard that without factor correlation module

Model	Beijing PM2.5 dataset				The temple of heaven air quality dataset			
	RMSE		MAE		RMSE		MAE	
	1 h–6 h	7 h–12 h	1 h–6 h	7 h–12 h	1 h–6 h	7 h–12 h	1 h–6 h	7 h–12 h
C^2-Guard(w/o factor)	38.5199	60.1601	34.5194	56.8523	44.5105	63.0524	39.9345	59.5993
C^2-Guard	**35.4614**	**57.5892**	**31.3690**	**54.1637**	**40.0465**	**61.0232**	**35.6896**	**57.0689**

It can be observed that the errors of the complete C^2-Guard are smaller than C^2-Guard (without factor) on two datasets. Therefore, it is reasonable to learn the correlations of air quality factors.

6 Conclusion and Future Work

In this paper, a novel cross-correlation gaining framework C^2-Guard for urban air quality prediction is explored. This framework, consisting of three modules, performs well on two real datasets. Taking the PM2.5 prediction as an example, experimental results demonstrate that C^2-Guard outperforms the state-of-the-art methods. Some future works are laid in our work. The spatial correlations of air pollution monitoring sites can be added to the learning of our model. Our C^2-Guard is a general framework that can be applied to more prediction tasks.

References

1. Abadi, M., et al.: Tensorflow: a system for large-scale machine learning. In: USENIX Symposium on Operating Systems Design and Implementation, pp. 265–283 (2016)
2. Abhilash, M.S.K., Thakur, A., Gupta, D., Sreevidya, B.: Time series analysis of air pollution in Bengaluru using ARIMA model. In: Perez, G.M., Tiwari, S., Trivedi, M.C., Mishra, K.K. (eds.) Ambient Communications and Computer Systems. AISC, vol. 696, pp. 413–426. Springer, Singapore (2018). https://doi.org/10.1007/978-981-10-7386-1_36

3. Bai, S., Kolter, J.Z., Koltun, V.: An empirical evaluation of generic convolutional and recurrent networks for sequence modeling. CoRR abs/1803.01271 (2018)
4. Díaz-Robles, L.A., Ortega, J.C.: A hybrid ARIMA and artificial neural networks model to forecast particulate matter in urban areas: The case of Temuco, Chile. Atmos. Environ. **42**(35), 8331–8340 (2008)
5. Du, S., Li, T., Yang, Y., Horng, S.J.: Deep air quality forecasting using hybrid deep learning framework. IEEE Trans. Knowl. Data Eng. 1 (2019)
6. Hu, J., Shen, L., Sun, G.: Squeeze-and-excitation networks. In: CVPR, pp. 7132–7141 (2018)
7. Kingma, D.P., Ba, J.: Adam: a method for stochastic optimization. In: ICLR (Poster) (2015)
8. Kok, I., Simsek, M.U., Özdemir, S.: A deep learning model for air quality prediction in smart cities. In: BigData, pp. 1983–1990 (2017)
9. Krishan, M., Jha, S., Das, J., Singh, A., Goyal, M.K., Sekar, C.: Air quality modelling using long short-term memory (lSTM) over NCT-Delhi, India. Air Qual. Atmos. Health **12**(8), 899–908 (2019)
10. Lea, C., Flynn, M.D., Vidal, R., Reiter, A., Hager, G.D.: Temporal convolutional networks for action segmentation and detection. In: CVPR, pp. 1003–1012 (2017)
11. Li, X., Peng, L., Hu, Y., Shao, J., Chi, T.: Deep learning architecture for air quality predictions. Environ. Sci. Pollut. Res. **23**(22), 22408–22417 (2016). https://doi.org/10.1007/s11356-016-7812-9
12. Liu, D., Lee, S., Huang, Y., Chiu, C.: Air pollution forecasting based on attention-based LSTM neural network and ensemble learning. Expert Syst. J. Knowl. Eng. **37**(3), e12511 (2020)
13. Nair, V., Hinton, G.E.: Rectified linear units improve restricted Boltzmann machines. In: ICML, pp. 807–814 (2010)
14. Ong, B.T., Sugiura, K., Zettsu, K.: Dynamically pre-trained deep recurrent neural networks using environmental monitoring data for predicting PM2.5. Neural Comput. Appl. **27**(6), 1553–1566 (2016)
15. Qi, Z., Wang, T., Song, G., Hu, W., Li, X., Zhang, Z.: Deep air learning: interpolation, prediction, and feature analysis of fine-grained air quality. IEEE Trans. Knowl. Data Eng. **30**(12), 2285–2297 (2018)
16. Tong, W., Li, L., Zhou, X., Hamilton, A., Zhang, K.: Deep learning PM 2.5 concentrations with bidirectional LSTM RNN. Air Qual. Atmos. Health **12**(4), 411–423 (2019). https://doi.org/10.1007/s11869-018-0647-4
17. Wang, B., Yan, Z., Lu, J., Zhang, G., Li, T.: Deep multi-task learning for air quality prediction. In: Cheng, L., Leung, A.C.S., Ozawa, S. (eds.) ICONIP 2018. LNCS, vol. 11305, pp. 93–103. Springer, Cham (2018). https://doi.org/10.1007/978-3-030-04221-9_9
18. Yi, X., Zhang, J., Wang, Z., Li, T., Zheng, Y.: Deep distributed fusion network for air quality prediction. In: KDD, pp. 965–973 (2018)
19. Zheng, Y., et al.: Forecasting fine-grained air quality based on big data. In: KDD, pp. 2267–2276 (2015)
20. Loy-Benitez, J., Heo, S., Yoo, C.: Imputing missing indoor air quality data via variational convolutional autoencoders: implications for ventilation management of subway metro systems. Build. Environ. **182**, 107135 (2020)
21. Vo, N.N., He, X., Liu, S., Xu, G.: Deep learning for decision making and the optimization of socially responsible investments and portfolio. Decis. Support Syst. **124**, 113097 (2019)
22. Vo, N.N., Liu, S., Li, X., Xu, G.: Leveraging unstructured call log data for customer churn prediction. Knowl.-Based Syst. **212**, 106586 (2021)

Simultaneous Multiple POI Population Pattern Analysis System with HDP Mixture Regression

Yuta Hayakawa[1(✉)], Kota Tsubouchi[2], and Masamichi Shimosaka[1]

[1] Tokyo Institute of Technology, Tokyo, Japan
{hayakawa,simosaka}@miubiq.cs.titech.ac.jp
[2] Yahoo Japan Corporation, Tokyo, Japan
ktsubouc@yahoo-corp.jp

Abstract. In recent years, the use of smartphone Global Positioning System (GPS) logs has accelerated the analysis of urban dynamics. Predicting the population of a city is important for understanding the land use patterns of specific areas of interest. The current state-of-the-art predictive model is a variant of bilinear Poisson regression models. It is independently optimized for each point of interest (POI) using the GPS logs captured at that single POI. Thus, it is prone to instability during fine-scale POI analysis. Inspired by the success of topic modeling, in this study, we propose a novel approach based on the hierarchical Dirichlet process mixture regression to capture the relationship between POIs and upgrade the prediction performance. Specifically, the proposed model enables mixture regression for each POI, while the parameters of each regression are shared across the POIs owing to the hierarchical Bayesian property. The empirical study using 32 M GPS logs from mobile phones in Tokyo shows that our model for large-scale finer-mesh analysis outperforms the state-of-the-art models. We also show that our proposed model realizes important applications, such as visualizing the relationship between cities or abnormal population increase during an event.

1 Introduction

The analysis of urban dynamics is of significant importance in urban planning, especially in the placement of restaurants, shopping centers, or open spaces, and for local services such as transportation. Owing to the increase in smartphone usage, extensive Global Positioning System (GPS) logs, which are sufficient to reflect real-world population flow, are stored. Hence, several researchers have recently begun to analyze urban dynamics using these GPS logs.

For example, Fan et al. [4] assumed that population flow patterns represent the features of the cities and extracted patterns from GPS logs. This pattern extraction is an important approach in urban dynamics analyses, and has been studied widely [8,13]. The extracted patterns help in understanding the urban

K. Karlapalem et al. (Eds.): PAKDD 2021, LNAI 12712, pp. 791–803, 2021.
https://doi.org/10.1007/978-3-030-75762-5_62

characteristics of the city, and improve planning efficiency for store openings or local commercial distributions.

Population prediction is another important topic in urban dynamics research. Shimosaka et al. [12] predicted active populations in the areas including big stations and amusement parks using external variables such as weather or day of the week. These predictions enable practical applications such as traffic prediction [9], sales prediction based on external variables, and the detection of anomalies by evaluating the difference between a predicted and an actual population.

Moreover, predicting or interpreting urban dynamics in small individual meshes (e.g., 200 m $\times$ 200 m) enables effective responses to social needs, such as promotion of location-specific products and services . In the state-of-the-art research [12], they independently constructed a predictive model for each point of interest (POI). However, it is difficult for the model to predict urban dynamics accurately in the smaller meshes. This is because the observed number of logs can be affected by its noise and increasing the sample size can suppress the effects of the noise but it is not always possible to achieve such an increase.

However, similarity of functions in various cities can be used to overcome this limitation. Cities may have certain functions in common. For example, if a considerable part of one city is a business district and some parts of another city are also business districts, it means that these two cities have the same urban function. It can be assumed that such cities with certain similar functions share similar urban dynamics. Therefore, to effectively learn the predictive model for a city, data from other cities having partially similar functions to one of the functions of the target city can be utilized.

Inspired by the success of hierarchical Bayesian models that extract latent urban dynamics patterns across cities [8,13], in this study, we propose a hierarchical Dirichlet process (HDP) regression mixture model that utilizes data of other areas having functions similar to those of the target city to achieve stable prediction in small areas This utilization of data from other areas virtually increases the sample size for learning a predictive model of the target area and stabilizes the learning even for small areas. The HDP regression mixture model also achieves parameter reduction. This is because the model for each city consists of a mixture of latent models; only the mixture coefficients and the parameters of latent models are learned.

The contributions of this work are as follows.

- We proposed the HDP regression model to make stable predictions of urban dynamics even for smaller areas by utilizing the data in the other areas and reduce the total parameter size by sharing the parameters among the cities.
- We conducted experiments using a real-world dataset, 32 M GPS logs from smartphones, to show that the proposed model predicts the urban dynamics more accurately than previous predictive models.
- We show two important industrial applications that can be realized using the model; detection of the abnormal congestion of an event and a visualization of the mixing coefficient of our proposed model to better understand the relationship.

2 Related Work

In this section, we describe work related to urban dynamics modeling. Research on extracting urban dynamics patterns have been conducted actively in recent years. In the present work, tensor factorization [4,14,18,19,21] or mixture modeling [8,13] has been used to extract patterns. In the tensor factorization approaches, many researchers have modeled the urban dynamics data as a tensor with cities, time of the day, and date axes, and they extract latent patterns by factorization. For example, claims of city noise [21], check-in activities [14], and GPS logs [4] were modeled as a tensor to analyze their latent patterns. Nishi et al. [8] modeled the active population transition in one day using the Dirichlet mixture model and extracted the latent patterns shared by cities. Shimosaka et al. [13] also extracted patterns by mixture modeling. Moreover, they simultaneously clustered the cities using their proposed hierarchical Bayesian framework. This clustering provides an explicit understanding of the similarities between cities. However, these to extract latent patterns in past datasets do not predict future urban dynamics.

Several researchers have attempted to predict the activities in cities using external variables [2,12,17]. Wang et al. [17] proposed the negative binomial regression model by using external variables such as population or weather to predict traffic volume. Bogomolov et al. [2] constructed a model to predict the number of crimes using a random forest with demographic information. The research by Shimosaka et al. [12] is state-of-the-art research in urban dynamics prediction.

However, they constructed predictive models for each city using data specific to that city. Therefore, learning may be unstable, particularly for small areas. To solve these problems, some studies [11,20] that utilized data from other cities to learn the model of the target area and improve accuracy. Zheng et al. [20] modeled the flow of people in the cities by a convolutional neural network (CNN). They trained the model while sharing the dataset between neighboring areas by convolution. However, the CNN only shared the data were among neighboring areas. Shimosaka et al. [11] also proposed a predictive model called SPF, which shared parameters between meshes while retaining spatial preservation and reduced the number of parameters using a factorization approach. In terms of model expression, however, the method proposed herein is a mixture regression and can model complex multimodal data more effectively than SPF, which is a single regression model.

3 Urban Population Pattern Analysis System with HDP Regression

This section presents an outline of the proposed urban population pattern analysis system with HDP regression using GPS logs from smartphones. Further, the applications realized by the proposed system are discussed. As shown on the left side of the system diagram in Fig. 1, the number of GPS logs from smartphone use is sufficient to represent the populations of a city, thus we regard the counting logs as the population in each area. These population counts, as well as external factors such as weather and holiday information, were used as the datasets for the urban dynamics prediction system. In particular, for the population count, one day was segmented into S parts; then, it was assumed that the number of GPS logs in area l within the τth time segment was an active population. We modeled the transition of the active population $y^{(l)}_{\boldsymbol{c},\tau,n}$ of the n-th day with conditions $\boldsymbol{c}$ (e.g., weather, day of the week, national holiday) through the time segments $\tau = \{1, ..., S\}$. The transition of the population is commonly used for the application of urban dynamics analyses. The proposed model realizes three main areas of urban dynamics: population prediction, anomaly detection, and inter-city relationship analysis that are widely studied by researchers. In the following subsections, we describe the detailed setting of each application.

Active population prediction is the problem of predicting the active population $y^{(l)}_{\boldsymbol{c},\tau,n}$ in the area l, in the τ-th time segment on the n-th day using external factors $\boldsymbol{c}$. Active population prediction helps to provide real-world applications such as predictions of sales or traffic volume. As can be expected, prediction accuracy is quite important for the quality of the application. However, prediction accuracy can be unstable owing to the lack of a sufficient sample size for training, particularly in fine-grained meshes. Thus, it is necessary to share a dataset over the meshes to enable improved prediction.

Anomaly detection is one of the hot topics in urban dynamics research [6,10,12]. It is also useful in several practical applications, including the detection of city events and traffic obstacles. One method for detecting anomalies in the population is evaluating a difference between a predicted and an actual population. Some studies [6,12] evaluated the anomalous population counts in a city based on the irregularity index, $\frac{1}{\hat{y}^{(l)}_{\boldsymbol{c},\tau}}(y^{(l)}_{\boldsymbol{c},\tau,n} - \hat{y}^{(l)}_{\boldsymbol{c},\tau})$,, where $y^{(l)}_{\boldsymbol{c},\tau,n}$ is an actual population and $\hat{y}^{(l)}_{\boldsymbol{c},\tau}$ is the prediction. Owing to the accurate population estimate by the bilinear Poisson regression, they successfully detected or predict a large event and unexpected heavy rain. Note that the anomaly detection significantly depends on the prediction accuracy; unstable prediction in the fine-grained mesh leads to impractical anomaly detection. Thus, stable prediction is essential for anomaly detection in fine-grained meshes.

Understanding the relationships between the different areas of a city is another important aspect of urban dynamics analysis. Finding areas that have similar features within a region is helpful in deciding the placement of new commercial entities such as restaurants or stores. As mentioned above, the relationships between areas were investigated based on pattern extraction using generative models in previous research [8,13], and these aspects of generative models that is extracting latent pattern is helpful for updating predictive performance. We proposed the combination model of a generative and discriminative model described in the following section.

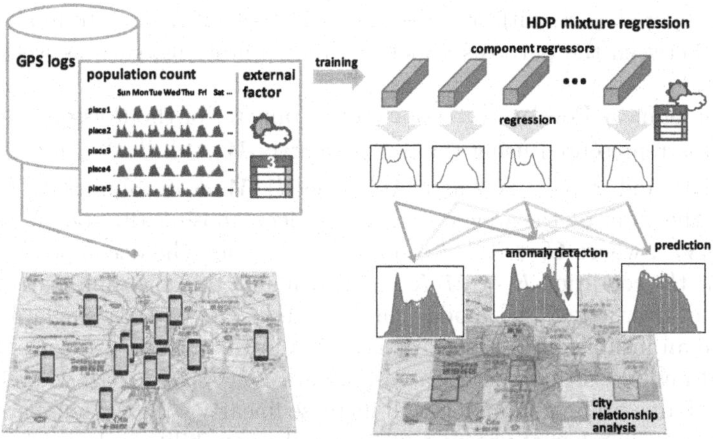

Fig. 1. Urban population pattern analysis system proposed in this paper.

4 Definition of Proposed Method: HDP Mixture Regression

In this research, we proposed the regression mixture model for accurate prediction in fine-grained meshes such as 200 m × 200 m mesh while the existing state-of-the-art model [12] predicts in 900 m × 900 m mesh. A model for an area consists of a mixture of latent models shared by all cities; each area has a specific mixing coefficient. LDA [1] and HDP [15] are widely known methods for latent allocation. We proposed a predictive model using HDP mixture regression, which uses the HDP as the prior of the mixture model.

In HDP mixture regression, the model is learned by Bayesian inference. However, it is difficult to analytically infer the Poisson regression under Bayesian inference. To infer the Poisson regression in the Bayesian framework, the Poisson distribution can be approximated by a Gaussian distribution [3]. Approximation around zero is inadequate and can result in poor prediction. Here, Gaussian regression, which can be analytically applied to Bayesian inference, is used. We confirm the prediction accuracy of the Gaussian regression, as well as the Poisson regression in the experiment described in Sect. 5. This section first explains prediction by the state-of-the-art model (bilinear Poisson regression) and then describes prediction by the proposed HDP mixture regression.

4.1 Urban Dynamics Prediction by Bilinear Poisson Regression

As mentioned earlier, the predictive model using bilinear Poisson regression [12] is the state-of-the-art model in urban dynamics prediction research. They assumed that the active population $y^{(l)}_{c,\tau,n}$ follows the Poisson distribution $\mathcal{P}(y^{(l)}_{c,\tau,n}|\lambda^{(l)}_{c,\tau})$, where $\lambda^{(l)}_{c,\tau}$ is the mean parameter of the Poisson distribution. They considered a combination of the time feature, which drastically influences the active population, with other features. This enabled them to model the differences between the patterns under different conditions, such as weekday and weekend.

In the bilinear Poisson regression, the parameter of the Poisson distribution $\lambda_{c,\tau}^{(l)} > 0$ is represented using the weight matrix $\boldsymbol{W}_l \in \mathbb{R}^{M \times S}$, the time feature $\boldsymbol{\phi}(\tau) \in \mathbb{R}^S$, and $\boldsymbol{\varphi}(\boldsymbol{d}) \in \mathbb{R}^M$ as $\ln \lambda_{c,\tau}^{(l)} = \boldsymbol{\varphi}(\boldsymbol{d})^\top \boldsymbol{W}_l \boldsymbol{\phi}(\tau)$. Shimosaka et al. [12] reduced the rank of the weight matrix by decomposing the weight matrix as $\boldsymbol{W}_l = \boldsymbol{U}_l \boldsymbol{V}_l^\top$, where $\boldsymbol{U}_l \in \mathbb{R}^{M \times K}$ and $\boldsymbol{V}_l \in \mathbb{R}^{S \times K}$ are the decomposed matrices satisfying the condition $K \ll M, K \ll S$. Although this low-rank matrix reduces the risk of overfitting, the model only uses the dataset in a single area. This results in an insufficient sample size as shown in the next subsection. Moreover, the total number of parameters across all areas is $K(M+S)L$ where L is the number of areas. The parameters will increase linearly with the number of areas, and this large number of parameters causes the instability in learning.

4.2 Definition of HDP Mixture Regression

We formulate herein the HDP mixture regression to utilize the data from other areas to learn the model of the target area while suppressing the increase in the number of parameters. Following the work by Wang et al. [16], we used Sethuraman's construction was used to represent the stick-breaking process that realizes the HDP. The latent variable $z_{l,n,m}$ represents an active population $\boldsymbol{y_c}^{(n,l)}$ as defined in the previous chapter, and is assigned to the cluster m. It follows the condition $z_{l,n,m} \in \{1, 0\}, \sum_m z_{l,n,m} = 1$ and m is an area-level cluster. This area-level cluster is defined to analytically apply the variational inference as the document-level cluster in the previous work [16]. $z_{l,n,m} = 1$ indicates that the active population in the area l on the day n is assigned to cluster m. The latent variable $r_{l,m,k}$ represents the correspondence between an area-level cluster m of the area l and a global cluster k. The global clusters are shared between areas; $r_{l,m,k}$ also follows the requirement $r_{l,m,k} \in \{1, 0\}, \sum_k r_{l,m,k} = 1$. $r_{l,m,k} = 1$ indicates the area-level cluster m of the area l, and corresponds to the global cluster k.

To utilize the data from other areas focusing on the peak of the urban dynamics pattern rather than the volume of the active population, we normalize the active population in the learning process as $\tilde{y}_{c,\tau}^{(n,l)} = \frac{1}{\eta_l} y_{c,\tau,n}^{(l)}$, where η_l is calculated by $\eta_l = \frac{1}{N} \sum_{n=1}^{N} \sum_{\tau=1}^{T} y_{c,\tau,n}^{(l)}$ using the training dataset. The joint distribution $\tilde{y}_{c,\tau}^{(n,l)}$, $z_{l,n,m}$, $r_{l,m,k}$ is represented as follows,

$$p(\tilde{y}_{c,\tau}^{(n,l)}, z_{l,n,m}, r_{l,m,k}) = \prod_m \pi_{l,m}^{z_{l,n,m}} \prod_k \rho_k^{r_{l,m,k}} \mathcal{N}(\tilde{y}_{c,\tau}^{(n,l)} | \lambda_{c,\tau,k}, \sigma_k^2)^{z_{l,n,m} r_{l,m,k}} \quad (1)$$

where $\pi_{l,m}$ is the mixing coefficient for the area l, and ρ_k is the global mixing coefficient. $\lambda_{c,\tau,k} = \boldsymbol{\varphi}(\boldsymbol{c})^\top \boldsymbol{W}_k \boldsymbol{\phi}(\tau)$, where $\boldsymbol{\phi}(\tau)$ is the time feature and $\boldsymbol{\varphi}(\boldsymbol{c})^\top$ is the feature with other conditions. $\pi_{l,m}$ and ρ_k are generated by the stick-breaking process as follows,

$$p(\pi'_{l,m}) = \mathcal{B}(1, \beta_0), p(\rho'_k) = \mathcal{B}(1, \gamma_0)$$
$$\pi_{l,m} = \pi'_{l,m} \prod_{s=1}^{t-1}(1 - \pi'_{l,s}), \rho_k = \rho'_k \prod_{j=1}^{k-1}(1 - \rho'_j) \tag{2}$$

The priors of $z_{l,n,m}$, $r_{l,m,k}$, and the weight parameter $\boldsymbol{W}_k$ are represented as follows,

$$p(\boldsymbol{z}_l|\boldsymbol{\pi}_l) = \prod_{n,m} \pi_{l,m}^{z_{l,n,m}}, p(\boldsymbol{r}) = \prod_{l,m,k} \rho_k^{r_{l,m,k}}$$
$$p(\boldsymbol{W}_k) = \mathcal{N}(\mathrm{vec}(\boldsymbol{W}_k)|\boldsymbol{\mu}_k, \boldsymbol{\Sigma}_k), p(\sigma^2) = \mathrm{Gamma}(\sigma^2|a_0, b_0), \tag{3}$$

where $\mathrm{vec}(\cdot)$ is the vectorization of the matrix. For example, for the $A = [\boldsymbol{a}_1, \boldsymbol{a}_2, ...\boldsymbol{a}_K]$, $\mathrm{vec}(A) = [\boldsymbol{a}_1^\top, \boldsymbol{a}_2^\top, ...\boldsymbol{a}_K^\top]^\top$. Generally, the posterior of the parameters or latent variables cannot be analytically estimated in the model using HDP. A sampling method such as Gibbs sampling [5,7] or variational inference [16] is used to estimate the posterior approximately. In this research, we use variational inference to estimate the approximated posterior of the parameters and latent variables.

Our proposed mixture model using HDP as a prior reduces the number of parameters compared to the previous method. As mentioned earlier, in the previous predictive method [12], the number of parameters is $K(M+S)L$ where K is the dimension of the low-rank matrix, M is the dimension of the time feature, and S is the dimension of the other feature. If we set $L = 100 \times 100 = 10000$, $M = 48$, $S = 28$, and $K = 5$, the number of all the parameters is 3.8M. In our proposed method, the number of parameters is dependent on the number of global clusters B. Considering the number of mixing coefficients, the total number of parameters is $BMS + LT + B$ where B is the maximum number of global clusters and T is the maximum number of area-specific clusters. In the aforementioned setting, the number of parameters is suppressed to 567K with the setting $B = T = 50$.

4.3 Prediction by HDP Mixture Regression

It is assumed that the posterior estimated with the dataset $\boldsymbol{Y} = \{\tilde{y}_{c,\tau}^{(n,l)}\}_{n,l,\tau}$ in all areas is approximated as follows,

$$p(\boldsymbol{\rho}, \boldsymbol{\pi}, \boldsymbol{r}, \boldsymbol{z}, \boldsymbol{W}, \boldsymbol{\sigma}|\boldsymbol{Y}) = q(\boldsymbol{\rho})q(\boldsymbol{\pi})q(\boldsymbol{r})q(\boldsymbol{z})q(\boldsymbol{W}, \boldsymbol{\sigma}). \tag{4}$$

The predictive distribution of the HDP mixture regression is a mixture and is multimodal. The predictive distribution $p^*(\tilde{y}_{\boldsymbol{c},\tau}^{(l)*}|\boldsymbol{Y})$ and the prediction value of the model $\hat{y}_{\boldsymbol{c},\tau}^{(l)*}$ are defined as follows small

$$p^*(\tilde{y}_{\boldsymbol{c},\tau}^{(l)*}|\boldsymbol{Y}) \simeq \sum_m \mathbb{E}_{q(\boldsymbol{\pi})}[\pi_{l,m}] \sum_k \mathbb{E}_{q(\boldsymbol{r})}[r_{l,m,k}] \mathbb{E}_{q(\boldsymbol{W},\sigma)}[p(\tilde{y}_{\boldsymbol{c},\tau}^{(l)*}|\lambda_{\boldsymbol{c},\tau,k}, \sigma_k^2)], \tag{5}$$
$$\hat{y}_{\boldsymbol{c},\tau}^{(l)*} = \eta_l \arg\max_{\tilde{y}_{\boldsymbol{c},\tau}^{(l)*}} (p^*(\tilde{y}_{\boldsymbol{c},\tau}^{(l)*}|\boldsymbol{Y})). \tag{6}$$

4.4 Urban Dynamics Prediction Systems Using HDP Mixture Regression

We describe the outline of the urban dynamics prediction system using the proposed model, the HDP mixture regression shown in Fig. 1. The proposed method shares the datasets of all areas of interest during the training (see the upper part of the figure). Because only the component regressors (except for the mixing coefficients) have parameters, the proposed method can suppress the number of parameters (see the upper right part of the figure). As shown in the right lower part of Fig. 1, the HDP mixture regression model realizes important applications. The model provides a prediction for each area because each area has its own mixing coefficient (see the right part of the figure). Consequently, owing to the prediction, the model can also realize anomaly detection for each area. The mixture coefficients represent the manner in which the model for each area depends on each component regressor; thus, users can find the similarities between cities in terms of active population transition by visualizing the value of one of the mixture coefficients.

5 Experimental Results

To evaluate the performance of the proposed urban dynamics prediction method, we conducted experiments comparing the proposed method with existing methods used to model urban dynamics.

5.1 Dataset

This experiment utilized the GPS logs obtained by the smartphone app *Bosai Sokuho*[1] released by Yahoo Japan Corporation. The GPS logs were collected from users across Japan who consented to providing their location information, and they have all been anonymized. In the Kanto region alone, 15 M logs were collected per day. Each GPS log includes a time stamp, longitude, and latitude and is collected when the user moves. Thus, the logs represent human activities. Logs collected from July 1st, 2013 to June 30th, 2014 were used in this experiment. The number of logs in each mesh in the 3 km × 3 km square area shown in Fig. 2 were counted at 30 min intervals. We used the dataset for this experiment. In this experiment, we divided the target area into two different sizes of meshes: one was 600 m × 600 m mesh and the other was 200 m × 200 m mesh as shown in Fig. 2.

5.2 Evaluation Metric

We used Mean Negative Log Likelihood (MNLL), Mean Absolute Error (MAE) as evaluation metrics. They were also used in the existing state-of-the-art research [12]. MNLL is defined as $\mathrm{MNLL} = \frac{1}{NT}\sum_{n=1}^{N}\sum_{\tau=1}^{T}(-\ln p(y_{c,\tau,n}^{(l)}|\lambda_{c,\tau}^{(l)}))$, MAE is defined as $\mathrm{MAE} = \frac{1}{NT}\sum_{n=1}^{N}\sum_{\tau=1}^{T}|y_{c,\tau,n}^{(l)} - \hat{y}_{c,\tau}^{(l)}|$.

[1] https://emg.yahoo.co.jp/.

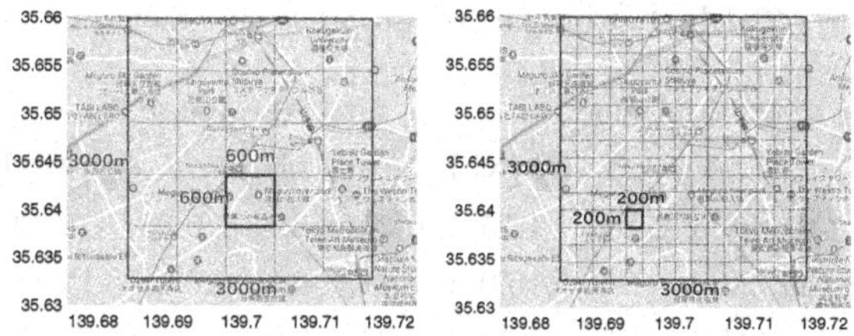

Fig. 2. Left: First mixing coefficient in each mesh l, right: Second mixing coefficient in each mesh l

5.3 Comparison Methods

We compared the proposed HDP regression mixture model (HDP-reg) with the bilinear Poisson regression model (BP) [12], and the bilinear Gaussian regression model (BG), and SPF [11] in terms of MAE and MNLL. The bilinear Gaussian regression model used in this experiment, models urban dynamics with Gaussian distribution as the HDP-reg does, and it was used to compare with MNLL fairly. We also verified that the bilinear Gaussian regression model has accuracy equivalent to that of the bilinear Poisson regression model to confirm the validity of the comparison between HDP-reg and the bilinear Gaussian regression model.

In the bilinear Gaussian regression setting, we assumed population count $y^{(l)}_{c,\tau,n}$ was sampled from a Gaussian distribution, $\mathcal{N}(y^{(l)}_{c,\tau,n}|\mu^{(l)}_{c,\tau},\sigma^2)$ and the mean parameter $\mu^{(l)}_{c,\tau}$ was estimated by the bilinear form, $\hat{\mu}^{(l)}_{c,\tau} = \boldsymbol{\varphi}(\boldsymbol{d})^\top \boldsymbol{W}^{(l)}\boldsymbol{\phi}(\tau)$. We conducted the experiments using two types of bilinear regression (Poisson or Gaussian) as shown below,

1. **BP/BG 1 for All**: One regressor for all meshes. These models can be learned using all the datasets in all meshes, and this can stabilize the learning. However, these models provide the same predictive results for All meshes, which implies that they cannot represent the differences between cities.
2. **BP/BG 1 for 1**: One regressor for each mesh. Each model can model the characteristics of the urban dynamics for each mesh; however, the dataset used during the learning is specific to the respective mesh, and this small dataset can cause overfitting.

We use the expectation of the model as the prediction value for the bilinear Poisson/Gaussian regression model. In this experiment, we used two types of features; one is the time feature and the other is the weekday feature.

5.4 Comparison with Previous Predictive Methods

We compared our proposed model with previous predictive models using datasets in two different meshes, as shown in Fig. 2. MAE and MNLL were used as metrics on the five-fold cross-validation in this experiment. We did not compare the

Table 1. Comparison on predictive metrics

Mesh size	600 m × 600 m		200 m × 200 m	
	MAE	MNLL	MAE	MNLL
BG 1 for All	26.0 ± 8.6	1.55 ± 0.25	3.73± 0.83	2.14 ± 0.19
BP 1 for All	26.0 ± 8.6	10.0 ± 3.4[a]	3.73 ±0.83	2.88 ± 0.45[a]
BG 1 for 1	24.2 ± 8.7	1.44 ± 0.27	3.45 ± 0.83	1.94 ± 0.17
BP 1 for 1	23.8 ± 8.9	8.73 ± 3.3[a]	3.42 ± 0.86	2.67 ± 0.42[a]
SPF	24.4 ± 8.7	9.68 ± 3.1[a]	3.44 ± 0.83	2.80 ± 0.39[a]
Proposed	**23.1 ± 7.2**	**1.37 ± 0.25**	**3.38 ± 0.68**	**1.92 ± 0.24**

[a]The performance of bilinear Poisson regression in MNLL should be compared only between BP 1 for '1' and BP 1 for All because the measurement of a Poisson distribution and that of a Gaussian distribution are different.

MNLL between the bilinear Gaussian regression and the bilinear Poisson regression because each measurement in each model was different. We used the 30-day dataset for training and the 180-day dataset for testing.

The experimental results are shown in Table 1. With this experiment, we confirmed that the performance of BP and BG on MAE was almost equivalent. This result ensured the validity of comparison of HDP-reg and BG. The performance of the proposed model was better than any of the bilinear regression models in MAE The smaller mesh size made the number of datasets in each mesh small, and made the MAE smaller. Proportionally, the performance difference in the 200 m meshes was bigger than for the 600 m meshes. The performance of HDP-reg in the MNLL was better than those of BG 1 for '1' and BG 1 for All.

5.5 Application Using HDP Mixture Regression

In this section, we show that the proposed model has the potential to realize the applications mentioned in Sect. 3, especially, Anomaly detection and City relationship analysis.

We evaluated the congestion caused by cherry blossom viewing using prediction by HDP mixture regression. The actual populations and predictions every Saturday from March 8 to April 12 around Nakameguro Station (Tokyo, Japan) are shown in Fig. 3. It is famous for cherry blossom viewing along the Meguro River. Cherry blossoms come into full bloom from the end of March to the beginning of April and the figure also shows that the congestion on April 5 is the heaviest. This abnormal congestion can be automatically detected by setting the threshold of the anomaly metric shown in Eq. (3). This anomaly detection also can be used for evaluating the effect of the events in terms of the increase or decrease of visits, compared to normal.

We also visualized the mixing coefficients for understanding the relationship between meshes. Mixing coefficients represent how each mesh area relies on each component regressor, and the cities that have similar mixing coefficients have

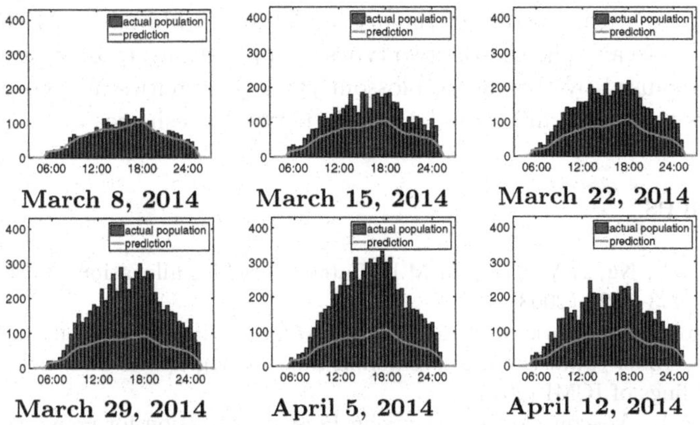

Fig. 3. The actual populations and predictions in cherry blossom season in Tokyo.

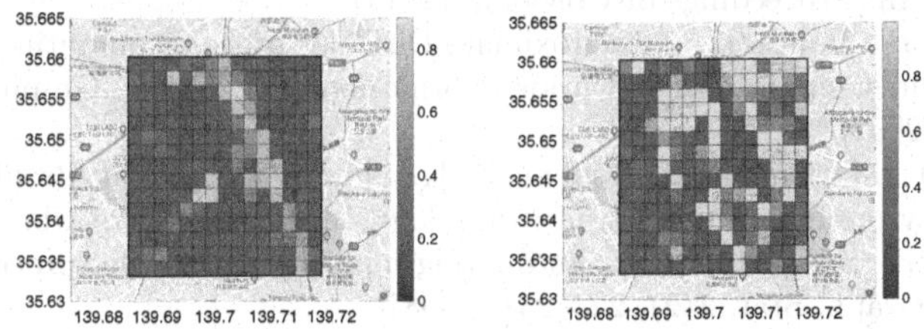

Fig. 4. Left: 1-st mixing coeffieint in each mesh l, right: 2-nd mixing coefficient in each mesh l

similar active population transition. The mixing coefficient for the visualization was calculated as $\zeta_{l,k} = \sum_m \mathbb{E}_{q(\pi)}[\pi_{l,m}]\mathbb{E}_{q(r)}[r_{l,m,k}]$. $\zeta_{l,k}$ represents how a mesh area l relies on k-th component regressor. The left figure of Fig. 4 represents the k = 1-st and k = 2-nd patterns from $K = 12$ patterns. The right figure of Fig. 4 shows a large number of models because the areas on the railway or station area rely on the k = 1-st component regressor. The k = 2-nd components are relied on by the downtown areas around stations. From these visualizations, it can be said that the proposed model can represent the relationship between mesh areas.

6 Conclusion

In this research, we modeled urban dynamics on large-scale high-resolution meshes using GPS logs of mobile phones for urban dynamics analysis systems. To predict future population stably even in smaller areas, we proposed using an HDP mixture regression model that uses the datasets in all the meshes for the training phase and predicted the urban dynamics for each mesh stably. We conducted the experiments using smartphone GPS logs in 3 km × 3 km squares in the Tokyo region, and the number of datasets was 32 M. The proposed model was compared with an state-of-the-art model and achieved an MAE improvement of 1.3. We also showed the two types of applications: evaluating anomalous congestion caused by the cherry blossom viewing activities and visualizing the coefficients to understand the relationship between mesh areas.

References

1. Blei, D.M., Ng, A.Y., Jordan, M.I.: Latent Dirichlet allocation. J. Mach. Learn. Res. **3**, 993–1022 (2003)
2. Bogomolov, A., Lepri, B., Staiano, J., Oliver, N., Pianesi, F., Pentland, A.: Once upon a crime: towards crime prediction from demographics and mobile data. In: Proceedings of ICMI (2014)
3. Chan, A.B., Vasconcelos, N.: Bayesian Poisson regression for crowd counting. In: Proceedgins of ICCV (2009)
4. Fan, Z., Song, X., Shibasaki, R.: CitySpectrum: a non-negative tensor factorization approach. In: Proceedings of UbiComp (2014)
5. Ishwaran, H., James, L.F.: Approximate Dirichlet process computing in finite normal mixtures: smoothing and prior information. J. Comput. Graph. Stat. **11**(3), 508–532 (2002)
6. Konishi, T., Maruyama, M., Tsubouchi, K., Shimosaka, M.: CityProphet: city-scale irregularity prediction using transit app logs. In: Proceedigs of UbiComp (2016)
7. MacEachern, S.N., Müller, P.: Estimating mixture of Dirichlet process models. J. Comput. Graph. Stat. **7**(2), 223–238 (1998)
8. Nishi, K., Tsubouchi, K., Shimosaka, M.: Extracting land-use patterns using location data from smartphones. In: Proceedings of the First International Conference on IoT in Urban Space (2014)
9. Okawa, M., Kim, H., Toda, H.: Online traffic flow prediction using convolved bilinear Poisson regression. In: Proceedings of MDM (2017)
10. Pan, B., Zheng, Y., Wilkie, D., Shahabi, C.: Crowd sensing of traffic anomalies based on human mobility and social media. In: Proceedings of SIGSPATIAL (2013)
11. Shimosaka, M., Hayakawa, Y., Tsubouchi, K.: Spatiality preservable factored Poisson regression for large scale fine grained GPS-based population analysis. In: Proceedings of AAAI (2019)
12. Shimosaka, M., Maeda, K., Tsukiji, T., Tsubouchi, K.: Forecasting urban dynamics with mobility logs by bilinear Poisson regression. In: Proceedings of UbiComp (2015)
13. Shimosaka, M., Tsukiji, T., Tominaga, S., Tsubouchi, K.: Coupled hierarchical Dirichlet process mixtures for simultaneous clustering and topic modeling. In: Proceedings of ECML-PKDD (2016)
14. Takeuchi, K., Tomioka, R., Ishiguro, K., Kimura, A., Sawada, H.: Non-negative multiple tensor factorization. In: Proceedings of ICDM (2013)
15. Teh, Y.W., Jordan, M.I., Beal, M.J., Blei, D.M.: Hierarchical Dirichlet processes. J. Am. Stat. Assoc. **101**, 1566–1581 (2006)
16. Wang, C., Paisley, J., Blei, D.: Online variational inference for the hierarchical Dirichlet process. In: Proceedings of AISTATS (2011)
17. Wang, X., Lindsey, G., Hankey, S., Hoff, K.: Estimating mixed-mode urban trail traffic using negative binomial regression models. J. Urban Plan. Devel. **140**(1), 04013006 (2013)

18. Yuan, J., Zheng, Y., Xie, X.: Discovering regions of different functions in a city using human mobility and POIs. In: Proceedings of KDD (2012)
19. Zhang, F., Yuan, N.J., Wilkie, D., Zheng, Y., Xie, X.: Sensing the pulse of urban refueling behavior: a perspective from taxi mobility. ACM Trans. Intell. Syst. Technol. **6**, 1–23 (2015)
20. Zhang, J., Zheng, Y., Qi, D.: Deep spatio-temporal residual networks for citywide crowd flows prediction. In: Proceedings of AAAI (2017)
21. Zheng, Y., Liu, T., Wang, Y., Zhu, Y., Liu, Y., Chang, E.: Diagnosing New York city's noises with ubiquitous data. In: Proceedings of UbiComp (2014)

Interpretable Feature Construction for Time Series Extrinsic Regression

Dominique Gay[1(✉)], Alexis Bondu[2], Vincent Lemaire[2], and Marc Boullé[2]

[1] LIM-EA2525, Université de La Réunion, Saint-Denis, France
dominique.gay@univ-reunion.fr
[2] Orange Labs, Lannion, France
{alexis.bondu,vincent.lemaire,marc.boulle}@orange.com

Abstract. Supervised learning of time series data has been extensively studied for the case of a categorical target variable. In some application domains, e.g., energy, environment and health monitoring, it occurs that the target variable is numerical and the problem is known as *time series extrinsic regression* (TSER). In the literature, some well-known time series classifiers have been extended for TSER problems. As first benchmarking studies have focused on predictive performance, very little attention has been given to interpretability. To fill this gap, in this paper, we suggest an extension of a Bayesian method for robust and interpretable feature construction and selection in the context of TSER. Our approach exploits a relational way to tackle with TSER: *(i)*, we build various and simple representations of the time series which are stored in a relational data scheme, then, *(ii)*, a propositionalisation technique (based on classical aggregation/selection functions from the relational data field) is applied to build interpretable features from secondary tables to "flatten" the data; and *(iii)*, the constructed features are filtered out through a Bayesian Maximum A Posteriori approach. The resulting transformed data can be processed with various existing regressors. Experimental validation on various benchmark data sets demonstrates the benefits of the suggested approach.

1 Introduction

Time series analysis has attracted much effort of research in the past decade, driven largely by the wide spread of sensors and their emerging applications in various domains ranging from medicine to IoT industry. The literature about supervised time series classification is abundant [2] and dozens of algorithms have been designed to predict a discrete class label for time series data. However, in some application domains, like sentiment analysis, forecasting, and energy monitoring [20], the target variable is numeric: e.g., the task of predicting the total energy usage in kWh of a house given historical records of temperature and humidity measurements in rooms and weather measurements. This problem is known as *time series extrinsic regression* (TSER [21]). For an incoming time series $\tau = \langle (t_1, X_1), (t_2, X_2), \dots, (t_m, X_m) \rangle$, which is a time-ordered collection

K. Karlapalem et al. (Eds.): PAKDD 2021, LNAI 12712, pp. 804–816, 2021.
https://doi.org/10.1007/978-3-030-75762-5_63

of m pairs of time stamps t_i and measurements $X_i \in \mathbb{R}^d$, the goal is to predict the value of a numeric target variable, given a training set of n series, $\mathcal{D} = \{(\tau_1, y_1), (\tau_2, y_2), \ldots, (\tau_n, y_n)\}$, where $y_i \in \mathbb{R}$ are the known target values for series τ_i.

Classical regression algorithms like, e.g., linear regression, regression tree, random forest or support vector regression can deal with TSER, provided that potential multiple dimensions of the input series are concatenated into a single feature vector. Beside k nearest neighbors models using popular distance metrics, like Euclidean distance (ED) and Dynamic Time Warping (DTW), Tan et al. [21] suggest a TSER benchmarking study involving also three recent deep learning approaches [12] (FCN, ResNet, InceptionTime [13]) and an adaptation of Random Convolutional Kernel Transform (Rocket [9]) for regression tasks. The first benchmarking study in [21] evaluates 13 TSER algorithms with a focus on predictive performance comparison with root mean squared error (RMSE) as performance measure. As a result, Rocket scores the best mean rank although no significant difference of performance is observed compared with classical regression ensembles like XGBoost [8] or random forest [7].

In this paper, we exploit a relational machine learning approach [6] for interpretable feature construction and selection and suggest an extension for TSER problems. As a motivating example, we consider the *AppliancesEnergy* data [20]. The goal is to predict the total energy usage in kWh in a house given 24-dimensional time series recording historical temperature and humidity measurements in 9 rooms in addition to 6 other weather and climate data series. In this context, let us consider *(i)*, the constructed variable $v = StdDev(Derivative(Dim5))$, i.e., the standard deviation of the derivative transform of dimension 5, and its discretisation into three informative intervals and *(ii)*, the discretisation of the target variable y into three intervals (see Fig. 1(a)). Plotting frequency histograms in this 2D-grid discretisation (i.e., contingency table of *intervals of* v $\times$ *intervals of* y) directly highlights that variations of measurements related by dimension 5 are characteristic of total power usage and the interpretation is straightforward. Indeed, low v values (below 0.0503) mainly means low power usage (below 11.98), higher v values (above 0.0846) means higher power usage (above 16.13) and in between values of v are characteristic of target interval $]11.98; 16.13]$.

Our approach, called `iFx`, brings a methodological contribution to TSER problems as it aims at generalizing the underlying concepts of the above intuitive example to efficiently extract simple and interpretable features as follows: *(i)*, firstly, we transform the original series into multiple representations which are stored in secondary tables as in relational data scheme; *(ii)*, then, informative and robust descriptors are extracted from relational data through propositionalisation and selected using a regularized Bayesian method. Classical regression algorithms can be trained on the obtained flattened data.

The rest of the paper successively presents the main concepts of our approach in Sect. 2, the experimental validation in Sect. 3 and opens future perspectives after concluding in Sect. 4.

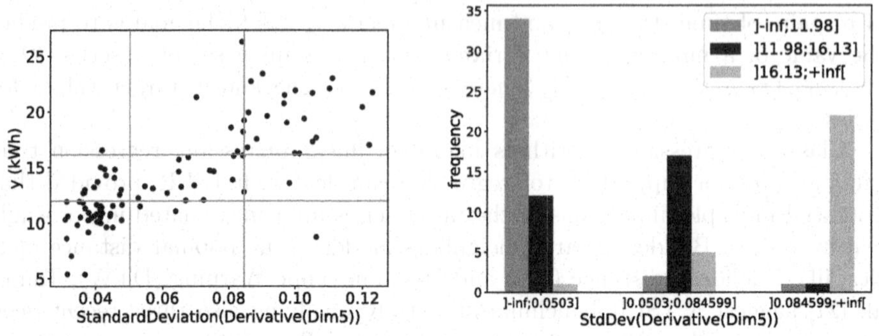

(a) (b)

Fig. 1. (a) Scatter plot of $v = StdDev(Derivative(Dim5))$ versus target y. One point represents a training time series τ_i. (b) Histograms of frequencies from the contingency table of *intervals of* $v \times$ *intervals of* y

2 TSER via a Relational Way

To extract features such as in the illustrative example, our approach is based on *(i)* the computation of multiple yet simple representations of time series, and their storage in a relational data scheme, *(ii)* a recent approach for feature engineering through propositionalisation [6] and its extension for regression problems. In the following, we describe these two steps in order to make the paper self-contained.

Multiple Representations of Series in a Relational Scheme - Enriching time series data with multiple transformations at the first stages of the learning process has demonstrated a significant enhancement of the predictive performance for the case of supervised classification [1,3,14,17]. As transforming time series from the time domain to an alternative data space is a good way for accuracy improvement, we also build six simple transformations commonly used in the literature in addition to the original representation: local derivatives (D) and second-order derivatives (DD), cumulative and double cumulative sums (S and SS), the auto-correlation transform (ACF) and the power spectrum (PS).

To embed all the computed representations in the same data, we use a relational data scheme. The root table is made of two attributes (columns), the series ID and the Class value. Each representation of a dimension of a series is stored in a secondary table in which there are three attributes, the series ID, linked with the the series ID of the root table by a foreign key, the Time attribute (or frequency for PS transform) and the Value attribute (or power for PS transform); thus, each tuple (line) of a secondary table is a single data point. For the introductory Appliances Energy 24-dimensional time series data, the resulting relational scheme is made of 7×24 secondary tables.

Feature Construction Through Propositionalisation - Propositionalisation [11,16] is the natural way to build features from secondary tables. It simply adds columns (variables) containing information extracted from secondary tables to the root table. For TSER data, propositionalisation may generate different aggregate features from various representations of the multiple dimensions. The introductory variable $v = StdDev(Derivative(Dim5))$, i.e., *"the standard deviation of the derivative transform of dimension 5"* is an example of the type of variables we want to build. It involves an aggregate function (or operator), *StdDev*, the standard deviation and the Value attribute of the table containing the derivative transform of the original fifth dimension that we build in the previous step.

To generalize from this example and build various types of interpretable features while avoiding intractable search space, we suggest propositionalisation through a restricted language, i.e., we will use a finite set of functions. As in a programming language, a function is defined by its name, the list of its operand and its return value and is expressed n the form $fname([operand, \ldots]) \rightarrow value$ – the operands and return value being typed. The operands can be a column of a table or the output of another function, i.e. another feature. Since time series data are inherently numeric the language of functions is made of:

- well-known and interpretable aggregate functions coming from relation data base domain
  - $Count(Table) \rightarrow Num$; count of records in a table
  - $Mean(Table, NumFeat) \rightarrow Num$; mean value of a numerical feature in a table,
  - $Median(Table, NumFeat) \rightarrow Num$; median value,
  - $Min(Table, NumFeat) \rightarrow Num$; min value,
  - $Max(Table, NumFeat) \rightarrow Num$; max value,
  - $StdDev(Table, NumFeat) \rightarrow Num$; standard deviation,
  - $Sum(Table, NumFeat) \rightarrow Num$; sum of values.
- $Selection(Table, selectioncriterion) \rightarrow Table$; for the selection of records from the table according to a conjunction of selection terms (membership in a numerical interval, on a column of the operand table or on a feature built from tables related to the operand table). For TSER data, the selection function allows restriction to intervals of timestamp or value in secondary tables.

Let us consider the variable $w = Min(Selection(derivative, 0 < timestamp < 10), ValueDim3)$, i.e., *the minimum value of the derivative transform of dimension 3 in the time interval* $[0; 10]$ as an example of the use of the selection function. Here, the Min function is applied on the output of the *selection* function used to select a specific time period.

Given the aforementioned language, to construct a given number K of variables, we use simultaneous random draws to efficiently sample the search space [6].

Feature Selection Through Bayesian Maximum A Posteriori - The randomized facet of the propositionalisation step does not guarantee that the K aggregate features of the main table are relevant for target variable prediction. All generated features are numerical due to the nature of aggregate functions. In order to select informative ones, we proceed a supervised pre-processing step which is 2D-discretisation, i.e., similarly to the 2D-grid in Fig. 1, we jointly partition each pair (v, y), where v is an aggregate feature and y the target variable.

In the Bayesian framework [15], 2D-discretisation is turned into a model selection problem and solved in Bayesian way through optimization algorithms. According the Maximum A Posteriori (MAP) approach, the best discretisation model $M^*_{v,y}$ is the one that maximizes *the probability of a discretisation model given the input data D*, which is:

$$P(M_{v,y} \mid D) \propto P(M_{v,y}) \times P(D \mid M_{v,y}) \tag{1}$$

Switching to negative logarithm refers to information theory and coding lengths. We define a *cost* criterion, noted c:

$$c(M_{v,y}) = -\log(P(M_{v,y})) - \log(P(D \mid M_{v,y})) = L(M_{v,y}) + L(D|M_{v,y}) \tag{2}$$

In terms of information theory, this criterion is interpreted as coding lengths [19]: the term $L(M_{v,y})$ represents the number of bits used to describe the model and $L(D|M_{v,y})$ represents the number of bits used to encode the target variable with the model, given the model $M_{v,y}$.

The prior $P(M_{v,y})$ and the likelihood $P(D \mid M_{v,y})$ are both computed with the parameters of a specific discretisation which is entirely identified by:

- a number of intervals I and J for v and y,
- a partition of v in intervals, specified on the ranks of the values of v,
- for each interval i of v, the distribution of instances over the intervals j of y, specified by N_{ij}, the instance counts locally to each interval of v.

Therefore, according to [15], using a prior that exploits the hierarchy of parameters that is uniform at each stage of the hierarchy, allows us to obtain an exact analytical expression of the cost criterion:

$$c(M_{v,y}) = 2\log(N) + \log\binom{N+I-1}{I-1} + \sum_{i=1}^{i=I}\log\binom{N_{i.}+J-1}{J-1} \tag{3}$$

$$+\sum_{i=1}^{i=I}\log\frac{N_{i.}!}{N_{i1}!N_{i2}!\ldots N_{iJ}!} + \sum_{j=1}^{j=J}\log N_{.j}! \tag{4}$$

where $N_{i.}$ is the number of instances in interval i of v and $N_{.j}$ the number of instances in interval j of y. The prior part (Eq. 3) of the cost criterion favors simple models with few intervals, and the likelihood part (Eq. 4) favors models that fit the data regardless of their complexity. Since the magnitude of the cost criterion depends on the size of the data N, we define a normalized version, which can be interpreted as a compression rate and is called *level*:

$$level(M_{v,y}) = 1 - \frac{c(M_{v,y})}{c(M^{\emptyset}_{v,y})} \tag{5}$$

where $c(M^{\emptyset}_{v,y})$ is the cost of the null model (i.e. when v and y are partitioned into only one interval). The cost of the null model can be deduced from previous formula and is formally $c(M^{\emptyset}_{v,y})) = 2\log(N) + \log(N!)$.

For example, again for the Appliances Energy data of Fig. 1, for $N = 95$, $c(M^{\emptyset}_{v,y})) \simeq 505$; $c(M_{v,y}) \simeq 13 + 12 + 27 + 72 + 349 = 473$; thus the level of v is positive as $level(M_{v,y}) \simeq 0.0614$.

A variable v whose discretisation model $M_{v,y}$ obtains a positive level value will be considered as informative whereas negative level value indicates spurious variables. Indeed, with negative level, a discretisation model $M_{v,y}$ is less probable than the null model, thus irrelevant for the regression task. When $0 < level(M_{v,y}) < 1$, we reach the most probable models that highlight a correlation between v and y. To reach those models, we use classical greedy bottom-up algorithms [4] that allows to efficiently find the most probable model given the input data in $O(N \log N)$ time complexity, where N is the number of time series.

At the end of our iFx process, we obtain a classical tabular data format, i.e., the series are now described by informative and interpretable features selected among the K variables extracted from multiple representations of original series. And on-the-shelf regressions algorithms can be trained.

As an example, for Appliances Energy data, one can find below the 17 selected variables (with positive *level* values) among the $K = 1000$ extracted, and their corresponding optimal number of intervals J and I for target variable y and v.

Feature v	$Level(v)$	#TargetIntervals	#vIntervals
$StdDev(TS5D.Value5D)$	0.0485	3	3
$StdDev(TS6DD.Value6DD)$	0.0395	3	2
$Max(TS6DD.Value6DD)$	0.0360	2	2
$StdDev(TS6D.Value6D)$	0.0345	3	2
$Max(TS5D.Value5D)$	0.0302	2	2
$StdDev(TS5DD.Value5DD)$	0.0299	2	2
$Min(TS6DD.Value6DD)$	0.0275	3	2
$StdDev(TS5.Value5)$	0.0253	2	2
$Mean(TS5PS.Value5PS)$	0.0242	2	2
$Sum(TS5PS.Value5PS)$	0.0242	2	2
$Max(TS6D.Value6D)$	0.0213	2	2
$Min(TS6D.Value6D)$	0.0211	2	2
$Max(TS5DD.Value5DD)$	0.0207	2	2
$Max(TS5PS.Value5PS)$	0.0171	2	2
$StdDev(TS5PS.Value5PS)$	0.0171	2	2
$Min(TS5D.Value5D)$	0.0162	2	2
$Min(TS5DD.Value5DD)$	0.0077	3	2

3 Experimental Validation

The experimental evaluation of our approach iFx are performed to discuss the following questions:

Q_1 Concerning iFx, how does the predictive performance evolve w.r.t. the number generated features? How many relevant features are selected? Are there preferred dimensions/representations for feature selection? And what about the time efficiency of the whole process?

Q_2 Are the performance of iFx comparable with current TSER methods?

Experimental Protocol and Data Sets - Tan et al. [20] has recently released 19 TSER data sets that are publicly available. The repository exhibits a large variety of TSER application domains with various numbers of dimensions and series' lengths. Predefined train/test sets are provided and we used it per se. As end classifiers, we use Python implementations of regression tree (RT) and random forest (RF) from scikit-learn library [18], Gradient Boosting Trees[1] (XGB) and the C++ implementation of the Selective Naive Bayes (SNB) [5] since iFx and SNB is part of the same tool[2]. All implementations are used with default parameters except for RF and XGB for which the number of trees is set to 100. Notice that SNB is parameter-free. The Root Mean Squared Error (RMSE) serves as the predictive performance measure. All experiments are run under a laptop with Ubuntu 20.04 using an Intel Core i7-5500U CPU@ 2.40 GHz x4 and 12Go RAM.

Table 1. RMSE results of our iFx method with $K = 10, 100, 1000$ and with a regression tree (RT), random forest (RF), gradient boosting trees (XGB) and Selective Naive Bayes (SNB) as end regressors.

Data	RT	10-RT	100-RT	1000-RT	RF	1000-RF	XGB	1000-XGB	1000-SNB
AE	5.903	3.455	4.158	2.7	3.397	1.999	4.024	2.202	2.463
HPC1	429.067	473.538	143.29	118.692	256.14	58.729	278.086	107.817	56.431
HPC2	58.754	59.043	58.99	46.874	46.941	36.51	48.571	36.813	38.529
BC	0.808	9.978	5.545	4.39	0.838	3.187	0.607	4.429	4.119
BPM10	140.825	142.955	136.61	127.906	94.759	94.596	95.542	96.486	109.163
BPM25	85.091	95.461	85.551	87.066	62.787	63.812	62.325	62.891	75.799
LFMC	62.169	58.499	59.418	61.764	44.589	45.477	47.549	48.109	44.551
FM1	0.022	0.017	0.009	0.01	0.016	0.007	0.017	0.008	0.009
FM2	0.019	0.008	0.007	0.007	0.015	0.006	0.018	0.008	0.015
FM3	0.021	0.017	0.01	0.009	0.021	0.008	0.02	0.007	0.009
AR	14.62	11.46	51.578	43.011	8.526	14.873	8.918	12.64	8.266
PPGDalia	24.964	24.165	24.899	22.008	17.487	16.083	16.144	15.985	16.819
IEEEPPG	39.75	49.169	41.041	43.031	32.11	38.51	31.716	37.931	41.875
BIDMC32HR	19.233	19.994	18.013	15.116	15.069	17.085	13.524	15.112	13.729
BIDMC32RR	4.709	5.888	4.728	5.377	4.362	4.511	4.313	4.473	3.910
BIDMC32SpO2	4.924	5.03	5.273	4.825	4.555	4.382	4.538	4.604	4.961
NHS	0.192	0.142	0.142	0.142	0.148	0.142	0.144	0.142	0.142
NTS	0.186	0.14	0.177	0.183	0.143	0.143	0.14	0.14	0.138
C3M	0.05	0.053	0.061	0.058	0.043	0.045	0.051	0.053	0.045
Win vs Orig	–	9	11	13	–	11	–	8	–

[1] https://xgboost.readthedocs.io/en/latest/python/index.html.
[2] http://www.khiops.com *(available as a shareware for research purpose)*.

Performance Evolution w.r.t. the Number of Features - We study the predictive performance evolution w.r.t. K, the number of extracted features on the 19 data sets [20]. For this experiment, we use a simple DecisionTreeRegressor from scikit-learn library [18] as the end regressor. In Table 1, we report RMSE results of our approach for increasing $K = 10, 100, 1000$. As expected, adding more informative features generally brings better predictive performance. Similar results are observed for RF but due to text width, we only reports results for $K = 1000$.

With $K = 1000$, the regression tree iFx1000-RT achieves better RMSE results than regression tree on original data for 13/19 data sets. Thus, using the new representation induced by iFx improves the predictive performance of the regression tree. The same observations stand for RF (11/19), and surprisingly, it is not the case for XGB (only 8/19). Notice also that iFx1000-RF seems to be the best combination as, in terms of Win-Tie-Loss, it scores 17-1-1 vs iFx1000-RT, 10-1-8 vs iFx1000-XGB and 11-2-6 vs iFx1000-SNB.

While more features give better predictive performance, it also means higher computational time. As the time complexity of the greedy bottom-up optimisation algorithms, that filter informative features, is supra-linear, then for a fixed K, the overall time time complexity is $\mathcal{O}(K.Nlog(N))$, where N is the number of training series. In practice, with a small computational time overhead to compute the 7 representations, iFx is time-efficient as shown in Fig. 2. For data sets with less than 1000 series, 1000 s are enough and for data with up to 100000 series it runs in 10000 s; except for the PPGDalia data which demands 46000 s ($\simeq 7\,h$). We conjecture that this particular high computational time is due to the high precision of the values of the series' data and target variable, that leads to a huge search space for 2D-discretisation model optimisation. Finally, for $K = 1000$, iFx process the whole benchmark in 20 h.

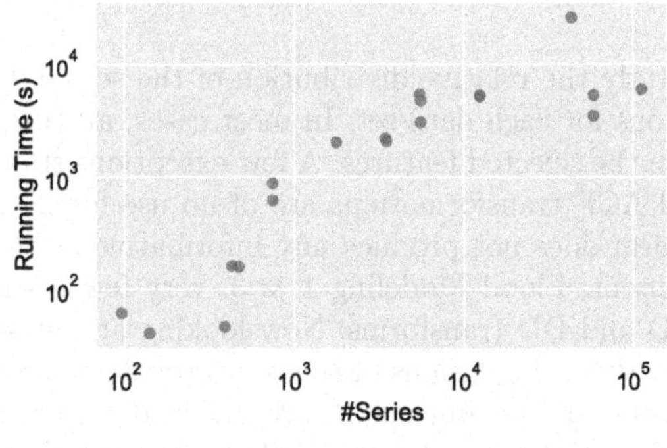

Fig. 2. iFx1000 - Running time in seconds w.r.t. the number of training series for $K = 1000$.

Distribution of Selected Features and Representations - As the iFx's feature extraction step by propositionalisation is randomized, we study the proportion of selected informative features (with positive *level*) among the K extracted for each data set in Fig. 3. For most of the data sets, hundreds of the aggregate features are considered as informative, except for Appliances Energy, Flood Modeling 2, News Title Sentiment and Covid3Month. Notice the particular case of News Headline Sentiment (NHS) data, for which none of the 1000 features is informative (i.e., negative *level* value). We conjecture that neither the present language of aggregate functions nor the transforms are adequate for predictive modeling of NHS data. In this case of no relevant feature, the default way to predict the target value is to predict the mean of the training target values. For NHS data, the default prediction leads to a RMSE value of 0.142 (see Table 1) on test set and we remark that this is the best score over all contenders in [21].

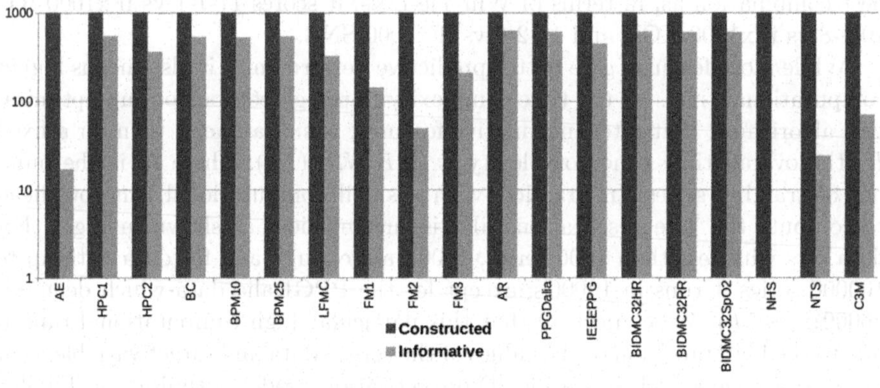

Fig. 3. Number of informative features (with positive level) among the $K = 1000$ generated features, for each data set.

In Fig. 4, we study the relative distribution of the selected features into the seven representations for each data set. In most cases, all the seven representations are present in the selected features. A few exceptions stand: for Appliances Energy, S, SS and ACF transformations are of no use for iFx; for Flood Modeling 2, SS transform does not produce any informative attribute and for Live Fuel Moisture Content, Flood Modeling 1 & 3, very few interesting attributes are coming from D and DD transforms. Now looking at the distribution of the selected features over the dimensions of the series (in the multivariate cases), for most of the data sets, all the dimensions are involved in the selected variables, except for Appliances Energy and News Title Sentiment data where for 22/24 (resp. 2/3) dimensions, no informative attribute has been found.

As for some cases, depending on the data at hand, the transforms of dimensions end up with no informative attributes, there is a potential for further investigations to identify the dimensions and their transforms that are relevant for our language of functions, and then focus the search on them. We postpone this idea for future work.

Predictive Performance Comparison with State-of-the-Art - In order to compare the predictive performance of iFx with state-of-the-art methods, we use the RMSE results of the contenders from Tan et al.'s benchmarking study [21] and integrate ours for comparison. In Fig. 5, with the critical difference diagrams [10] stemming from Friedman test with post-hoc Nemenyi test, we compare the predictive performance of ifX1000 ended with previous regressors with other contenders.

Whereas no method is significantly singled out, we observe that Rocket still scores the best mean rank. iFx1000-SNB reaches the second place (Fig. 5 (up right)) while iFx1000-RF takes the 4th place (Fig. 5 (up left)). These two approaches are then comparable to the best state-of-the-art TSER methods while XGBoost seems to not benefit from our feature engineering method (Fig. 5 (down left)).

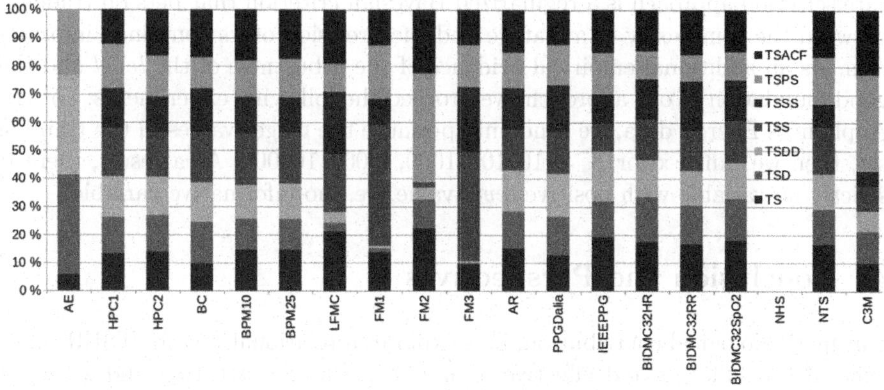

Fig. 4. Relative distribution of the seven representations among features selected through iFx (with positive *level*) for each data set.

On the other hand (Fig. 5 (down right)), using a single regression tree, the combination iFx1000-RT is 8th with a mean rank of about 7.39 and a significant loss of performance is observed compared with the best ranked approach Rocket. Thus, iFx1000-RT is the most interpretable model of our study but, here, the interpretability is at the cost of performance loss.

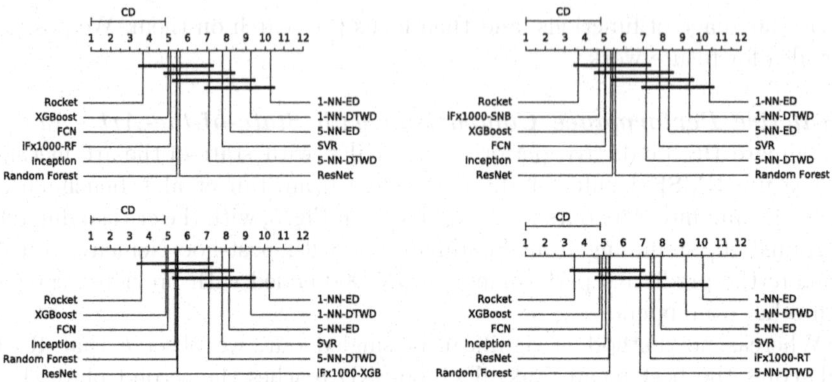

Fig. 5. Critical difference diagrams for iFx1000-RF, iFx1000-SNB, iFx1000-XGB and iFx1000-RT versus state-of-the-art regression methods [21].

About the Robustness of iFx - The *level* criterion at the core of our feature selection approach is a regularized Bayesian criterion that bets on trade-off between the complexity of a feature and the precision of its contained information. As an additional empirical evidence of the robustness of the *level* and the good foundation of our approach, we proceed the following experiments. For the Appliances Energy data, we randomly permute the target values in the training set, then we run iFx for $K = 10, 100, 1000, 10000, 100000$. As a result, there is strictly no variable with positive *level* value, i.e., no informative variable.

4 Conclusion and Perspectives

Our methodological contribution, iFx, explores a relational way for TSER problems. iFx is efficient and effective as it is capable of extracting and selecting and interpretable features from simple representations of original time series. Learning classical regression techniques on the new feature set generally results in better predictive performance than with the raw original time series. While Rocket still scores the best mean rank on the TSER benchmark, iFx is comparable with state-of-the-art TSER methods, depending on the end regressor at use.

As future work, we envision two ways of improvement for iFx: *(i)*, enriching the language of functions used in the propositionalisation step; *(ii)*, exploring the space of series dimensions and representations with a feed-forward/feed-backward strategy to reduce the field of extraction of aggregate features to the relevant representations and dimensions.

References

1. Bagnall, A.J., Davis, L.M., Hills, J., Lines, J.: Transformation based ensembles for time series classification. In: Proceedings of the Twelfth SIAM International Conference on Data Mining, (SDM 2012), Anaheim, California, USA, 26-28 April 2012, pp. 307–318 (2012)
2. Bagnall, A.J., Lines, J., Bostrom, A., Large, J., Keogh, E.J.: The great time series classification bake off: a review and experimental evaluation of recent algorithmic advances. Data Min. Knowl. Disc. **31**(3), 606–660 (2017). https://doi.org/10.1007/s10618-016-0483-9
3. Bondu, A., Gay, D., Lemaire, V., Boullé, M., Cervenka, E.: FEARS: a feature and representation selection approach for time series classification. In: Proceedings of The 11th Asian Conference on Machine Learning, ACML 2019, Nagoya, Japan, 17–19 November 2019, pp. 379–394 (2019)
4. Boullé, M.: MODL: a Bayes optimal discretization method for continuous attributes. Mach. Learn. **65**(1), 131–165 (2006). https://doi.org/10.1007/s10994-006-8364-x
5. Boullé, M.: Compression-based averaging of selective Naive Bayes classifiers. J. Mach. Learn. Res. **8**, 1659–1685 (2007)
6. Boullé, M., Charnay, C., Lachiche, N.: A scalable robust and automatic propositionalization approach for Bayesian classification of large mixed numerical and categorical data. Mach. Learn. **108**(2), 229–266 (2019). https://doi.org/10.1007/s10994-018-5746-9
7. Breiman, L.: Random forests. Mach. Learn. **45**(1), 5–32 (2001). https://doi.org/10.1023/A:1010933404324
8. Chen, T., Guestrin, C.: XGBoost: a scalable tree boosting system. In: Proceedings of the 22nd ACM SIGKDD International Conference on Knowledge Discovery and Data Mining, San Francisco, CA, USA, 13–17 August 2016, pp. 785–794. ACM (2016)
9. Dempster, A., Petitjean, F., Webb, G.I.: ROCKET: exceptionally fast and accurate time series classification using random convolutional kernels. Data Min. Knowl. Disc. **34**(5), 1454–1495 (2020)
10. Demšar, J.: Statistical comparisons of classifiers over multiple data sets. JMLR **7**, 1–30 (2006)
11. Dzeroski, S., Lavrac, N.: Relational Data Mining. Springer, Heidelberg (2001)
12. Fawaz, H.I., Forestier, G., Weber, J., Idoumghar, L., Muller, P.: Deep learning for time series classification: a review. Data Min. Knowl. Disc. **33**(4), 917–963 (2019). https://doi.org/10.1007/s10618-019-00619-1
13. Fawaz, H.I., et al.: InceptionTime: finding AlexNet for time series classification. Data Min. Knowl. Disc. **34**(6), 1936–1962 (2020). https://doi.org/10.1007/s10618-020-00710-y
14. Gay, D., Bondu, A., Lemaire, V., Boullé, M., Clérot, F.: Multivariate time series classification: a relational way. In: Song, M., Song, I.-Y., Kotsis, G., Tjoa, A.M., Khalil, I. (eds.) DaWaK 2020. LNCS, vol. 12393, pp. 316–330. Springer, Cham (2020). https://doi.org/10.1007/978-3-030-59065-9_25
15. Hue, C., Boullé, M.: A new probabilistic approach in rank regression with optimal Bayesian partitioning. J. Mach. Learn. Res. **8**, 2727–2754 (2007)
16. Lachiche, N.: Propositionalization. In: Encyclopedia of Machine Learning and Data Mining, pp. 1025–1031. Springer (2017)

17. Lines, J., Taylor, S., Bagnall, A.J.: Time series classification with HIVE-COTE: the hierarchical vote collective of transformation-based ensembles. ACM Trans. Knowl. Disc. Data **12**(5), 52:1-52:35 (2018)
18. Pedregosa, F., et al.: Scikit-learn: machine learning in Python. J. Mach. Learn. Res. **12**, 2825–2830 (2011)
19. Shannon, C.E.: A Mathematical theory of communication. ACM SIGMOBILE Mob. Comput. Commun. Rev. **5**(1), 3–55 (2001)
20. Tan, C.W., Bergmeir, C., Petitjean, F., Webb, G.I.: Monash University, UEA, UCR time series regression archive. CoRR abs/2006.10996 (2020). https://arxiv.org/abs/2006.10996
21. Tan, C.W., Bergmeir, C., Petitjean, F., Webb, G.I.: Time series regression. CoRR abs/2006.12672 (2020). https://arxiv.org/abs/2006.12672

SEPC: Improving Joint Extraction of Entities and Relations by Strengthening Entity Pairs Connection

Jiapeng Zhao[1,2], Panpan Zhang[1,2], Tingwen Liu[1,2(✉)], and Jinqiao Shi[3]

[1] Institute of Information Engineering Chinese Academy of Sciences, Beijing, China
{zhaojiapeng,zhangpanpan,liutingwen}@iie.ac.cn
[2] School of Cyber Security, University of Chinese Academy of Sciences, Beijing, China
[3] National Engineering Laboratory for Information Security Technologies, Beijing, China
shijinqiao@bupt.edu.cn

Abstract. Joint extraction of entities and relations aims at recognizing relational triples (subject s, relation r, object o) from unstructured text. For any entity pair (s, o) in correct relational triples, they do not appear independently, but depending on each other. While existing approaches usually model entity pairs only by sharing the encoder layer, which is insufficient to exploit entity pair intrinsic connection. To solve this problem, we propose to strengthen entity pairs connection (SEPC) by utilizing the duality property of entity pairs, which can further improve the joint extraction. The entity pairs recognization is transformed to finding subject conditioned on the object and finding object conditioned on the subject, and the dual supervised learning is introduced to model their connection. We finally demonstrate the effectiveness of our proposed method on two widely used datasets NYT and WebNLG (Code and data available: https://github.com/zjp9574/SEPC).

Keywords: Joint extraction · Entity pair recognization · Dual supervised learning · Cycle-consistent

1 Introduction

Joint extraction of entities and relations aims at recognizing entity mentions and their semantic relations simultaneously. Usually, they are in the form of (subject-s, relation-r, object-o), referred as relational triples. As it plays an important role or even an essential step for many applications of natural language

Supported by the National Key Research and Development Program of China (grant 2016YFB0801003), the Strategic Priority Research Program of Chinese Academy of Sciences (grant XDC02040400), the Key Research and Development Program for Guangdong Province (grant No.2019B010137003) and the National Natural Science Foundation of China (grant No.61902394).

K. Karlapalem et al. (Eds.): PAKDD 2021, LNAI 12712, pp. 817–828, 2021.
https://doi.org/10.1007/978-3-030-75762-5_64

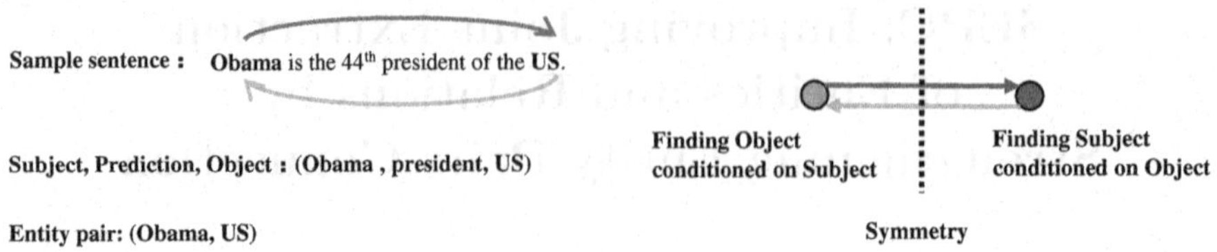

Fig. 1. The symmetry of an entity pair.

processing, such as automatic knowledge base construction [1], knowledge base population [2] and knowledge discovery [3], it has attracted considerable attention from researchers over the past decades.

Traditionally, the **extract-then-classify** approach is adopted to the joint extraction. These works [4–6] first identify all entities in a sentence with an entity recognization module and then perform relation classification for each entity pair with a relation classification module. The major problem of this approach is error propagation. It means errors that occurred in the entity recognization module can't be corrected by the relation classification module. Although subsequent works [7–9] are proposed to alleviate the error propagation problem by representing both entities and relations with shared parameters in a single model, the error propagation problem has not really been solved. And most extracted entity pairs can't form valid relations, generating too many negative examples leads the class distribution highly imbalanced caused by the "no relation". But if the entity pairs could be extracted accurately, not only the problem of error propagation would be alleviated, but also the highly imbalanced class distribution will be reduced. The extract-then-classify approach will also achieve a promising performance.

The **sequence tagging** approach [10] is designed to encode the information of entities and relations into a sequence to handle the error propagation problem. But this approach will suffer from the overlapping triples since the tagging scheme often assumes each token bears only one tag. For example, the sentence "The new president of the New York City, De Blasio is a native-born." contains (New York City, mayor, De Blasio) and (De Blasio, born in, New York City) can't be tagged simultaneously. Subsequent works [9,11] pay much attention on solving the overlapping problems. The most effective framework [12] is modeling relations as functions that map subjects to objects. They make the triple extraction become a two-step process: first identify all possible subjects s instead of entities in a sentence with a subject tagger; then for each subject, the relation-specific taggers $f_r(s) \rightarrow o$ will identify all possible relations and the corresponding objects. It not only performs well in dealing with overlapping triples but also alleviating data imbalance of the relation classification module.

Although this framework achieves the state-of-the-art result, both the training and testing of the subject tagger miss connections with relation and object. The mis-tagged subjects can't be corrected by relation-specific taggers. Since the relation-specific tagger is only trained on the sampled correct subjects, it can't judge the correctness of recognized entities.

To address this problem, we propose an improved jointly extraction method which improves the joint extraction of entities and relations by strengthening entity pairs connection, named **SEPC**. Intuitively, for any entity pairs (s, o) in correct triples, as we all know, they are not appear independently but depending on each other. As shown in Fig. 1, finding subject conditioned on the object and finding object conditioned on the subject is a pair of dual problem [13]. Hence, a dual probabilistic constraint could be introduced to model entity pairs, which can improve the subject tagger by strengthening the intrinsic connections between subject and object. The core idea is optimizing $P(s, o)$ in two equivalent ways, ideally the conditional distributions should satisfy the following equality:

$$P(s, o) = P(s)P(o|s; \theta_{so}) = P(o)P(s|o; \theta_{os}) \tag{1}$$

If $P(o|s; \theta_{so})$ and $P(s|o; \theta_{os})$ are learned separately by minimizing their own loss functions, there is no guarantee that the equation will hold. **SEPC** aims to jointly learn the two models θ_{so} and θ_{os} by minimizing their loss functions subject to the constraint of Eq. 1. By doing so, the intrinsic probabilistic connection between θ_{so} and θ_{os} are explicitly strengthened, which is supposed to push the learning process towards the right direction.

Taking advantage of SEPC, when an entity can't make pair with other entities in the sentence, it will be filtered out effectively. In summary, this paper improves the joint extraction of entities and relations by improving the recognization of entity pairs. Our main contributions are as follows:

- We propose a novel joint extraction framework named SEPC, which mainly strengthen entity pairs connection by utilizing the duality property of entity pairs to improve the joint extraction of entities and relations.
- SEPC allows only sampling correct triples, which makes it can save two key advantages of previous works: (1) alleviating data imbalance of the relation classification module by only sampling correct triples; (2) performs well on processing overlapping relational triples by building multiple mapping functions $f_r(s) \rightarrow o$ from subject to object.
- We first deploy SEPC on the latest sequence tagging approach. After that, we make SEPC as independent entity pair recognization module, which makes it become an extract-then-classify approach. On two widely used datasets NYT and WebNLG, both of them present improved performance over the state-of-the-art approach.

2 Related Work

2.1 Joint Extraction of Entities and Relations

The extraction of entities and relations could be divided into two types, namely the extract-then-classify one and the sequence tagging one. The extract-then-classify approach [6,14] first extract entity pairs and then perform relation classification on the extracted entity pairs. But they face the problem of error propagation. The sequence tagging approach [10] is proposed to solve the problem of error propagation. But this approach faces serious difficulties when processing overlapping triples since it assumes each token pairs bears only one tag. Zeng [9] proposed to solve this problem by a sequence-to-sequence model with a copy mechanism. MultiHead [15] solve the overlapping problem by multi-head attention. It first identifies all candidate entities, then the multi-head attention is used to extract relational triples. GraphRel [16] first use graph convolutional networks to extract overlapping triples and then considering all entity pairs for prediction. OrderCopyRE [11] solve the overlapping problem by generating all triples of a sentence, the encoder-decoder architecture is introduced to the extraction.

ETL-Span [17] and CasRel [12] are two most effective models in processing overlapping triples. They all first identify all possible subjects s in a sentence with an entity recognizer, then for each subject, the relation-specific taggers $f_r(s) \rightarrow o$ will identify all possible relations and the corresponding objects. But this new framework is maximizing $p(s, r, o)$ by maxmizing $p(s)$ and $p((r, o)|s)$ separately, both the training and testing of the entity recognizer module misses connections with (s, o). In hence, strengthening entity pairs connection could improve this approach.

2.2 Dual Supervised Learning

The idea of using the symmetry of dual tasks has been well studied [13], such as machine translation (MT), which formulates MT as a sequence-to-sequence learning problem, with the sentences in the source language as inputs and those in the target language as outputs. The input space and output space are symmetric, and there is almost no information loss while mapping from x to y or from y to x. In image processing, the prime task is image classification, the dual task is image generation conditioned on category labels. In sentiment analysis, the primal task is sentiment classification and the dual one is sentence generation with the given sentiment. In SEPC, the primal task is recognizing objects conditioned on the subject, the dual task is recognizing subjects conditioned on the object. And using transitivity as a way to regularize structured data has a long history. In visual tracking, enforcing simple forward-backward consistency has been a standard trick for decades [18]. In the language domain, verifying and improving translations via "back translation and reconciliation" is a technique used by human translators [19], as well as by machines [13]. Our work is the first attempt in joint extraction of entities and relations.

3 Methods and Technical Solutions

The goal of relational triple extraction is to identify all possible (subject-s, relation-r, object-o) triples in a sentence. Given an annotated sentence x_i from the training set D and the relational triples $T_i = \{(s, r, o)\}$ in x_i. It contains three key components, namely entity tagger, entity pair recognizer and relation tagger. In the training stage, they share the same BERT [20] encoder layers and were trained together. In the testing stage, the entity tagger first predicts candidate subjects and objects and the entity pair recognizer recognizes entity pairs based on predicted subjects and objects. Two types of relation taggers are used for predicting relations of entity pairs.

3.1 Entity Tagger

We first set four binary classifiers to tag the start and end positions of the subject or object in the sentence x_i respectively. These binary classifiers are built by directly decoding the encoded vector $\boldsymbol{h}_N$ produced by the N-layer BERT encoder. They share the same network structure but are trained to have different parameters:

$$p_i^{(\cdot)} = \sigma(\mathbf{W}_{(\cdot)}\boldsymbol{h}_i + \boldsymbol{b}_{(\cdot)}) \tag{2}$$

$(\cdot)$ could be one of subject start position, subject end position, object start position and object end position. $p_i^{(\cdot)}$ is the probability of identifying the i-th token in the input sequence as the start or end token position of subject or object. $\boldsymbol{h}_i = \boldsymbol{h}_N[i]$, where $\mathbf{W}_{(\cdot)}$ represents the trainable weight, $\boldsymbol{b}_{(\cdot)}$ is the bias and σ is the sigmoid activation function. The corresponding token will be assigned with a tag 1 if the probability exceeds a certain threshold or with a tag 0 otherwise. For multiple entities detection, we adopt the nearest start-end pair match principle to decide the span of any entity based on the results of the start and end position taggers. This entity recognition method has been widely used in existing works [12,17].

3.2 Entity Pair Recognizer: Strengthening Entity Pairs Connection

The entity pair recognizer aims at strengthening intrinsic connections between subject and object. More formally, it aims at maximizing the data likelihood of all entity pairs in the training set D: The Eq. (3) is maximized by maximizing Eq. (4) and Eq. (5) simultaneously.

$$\prod_{i=1}^{|D|}\left[\prod_{(s,o)\in T_i} p((s,o)|x_i)\right] \tag{3}$$

$$= \prod_{i=1}^{|D|}\left[\prod_{s\in T_i} p(s|x_i) \prod_{o\in T_i|s} p(o|s,x_i)\right] \tag{4}$$

$$= \prod_{i=1}^{|D|}\left[\prod_{o\in T_i} p(o|x_i) \prod_{s\in T_i|o} p(s|o,x_i)\right] \tag{5}$$

Inspired by the cycle-consistent unit in image-to-image translation [21], we designed a cycle-consistent unit to calculate $p(s,o)$, as shown in Fig. 2. Instead of enumerating all possible token pairs, entity pair recognizer only sample correct

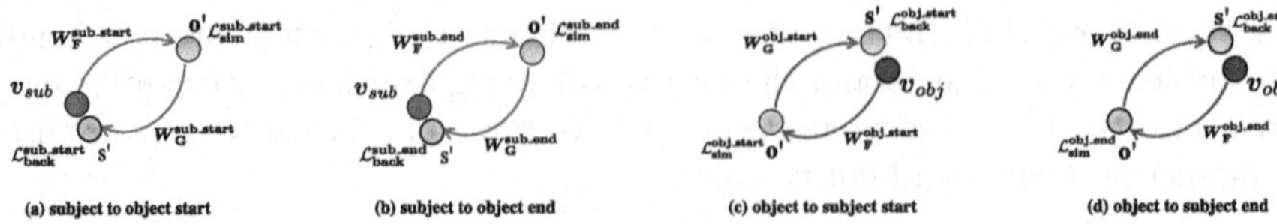

Fig. 2. The illustration of four groups cycle-consistent unit. $\boldsymbol{v}_{sub}$ and $\boldsymbol{v}_{obj}$ are the averages of the start and end position representations of subject and object. $\boldsymbol{o}'$ and s are generated vectors and they will be used to calculate the similarity with all $\boldsymbol{h}_i$. When the similarity exceeds a certain threshold, the corresponding position will be regarded as the start or end position.

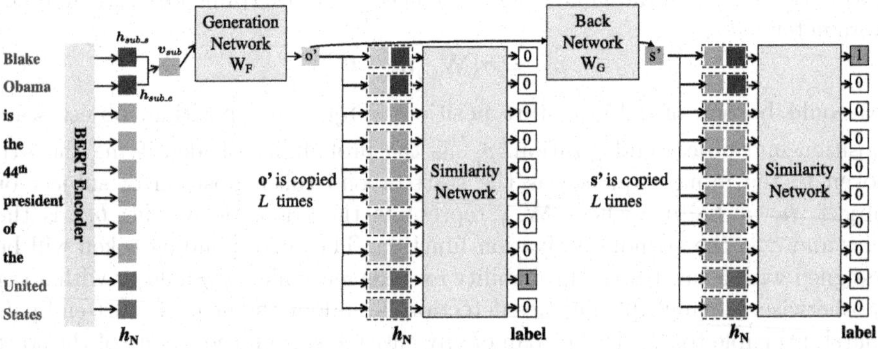

Fig. 3. The network structure of the subject to the object start position.

(subject, object) pairs when training. This allows SEPC could be trained in an end-to-end model with the entity tagger and the relation tagger. In testing, the entity pair recognizer uses the subject and object predicted by the entity tagger. The entity pair recognizer contains four mappings in total, included subject to the object start position, subject to the object end position, object to the subject start position and object to the subject end position.

As these four mappings share the same network structure, we only describe the subject to the object start position in detail. As shown in Fig. 3, the key components contain three networks, namely generation network, similarity network and backward network.

The representation of subject $\boldsymbol{v}_{sub}$ is the average of the start and end position representations $\boldsymbol{h}_{sub\_s}$ and $\boldsymbol{h}_{sub\_e}$. The generation network generates an embedding representation $\boldsymbol{o}'$ conditioned on $\boldsymbol{v}_{sub}$, $\mathbf{W}_F^{sub\_start}$ is corresponding network.

$$\boldsymbol{o}' = \mathbf{W}_F^{sub\_start} \cdot \boldsymbol{v}_{sub} \tag{6}$$

The similarity network $sim(*, *)$ is to calculate the similarity of $\boldsymbol{o}'$ and $\boldsymbol{h}_i$. The loss function is shown in Eq. 7, y_i equals 1 if i is the object start index, else 0. L is the length of tokens.

$$\mathcal{L}_{sim} = \frac{1}{L}\sum_{i=1}^{L} - [y_i \cdot \log sim(\boldsymbol{o}', \boldsymbol{h}_i) + (1 - y_i) \cdot (1 - \log sim(\boldsymbol{o}', \boldsymbol{h}_i))] \tag{7}$$

The backward network $G(*)$ generate $\boldsymbol{s}'$ conditioned on $\boldsymbol{o}'$, the detail operation is as follows:

$$\boldsymbol{s}' = \mathbf{W}_G^{sub\_start} \cdot \boldsymbol{o}' \tag{8}$$

The loss function is Eq. 9, y_i equals 1 if i is the subject start index, else 0.

$$\mathcal{L}_{back} = \frac{1}{L}\sum_{i=1}^{L} - [y_i \cdot \log sim(\boldsymbol{s}', \boldsymbol{h}_{sub\_s}) + (1 - y_i) \cdot (1 - \log sim(\boldsymbol{s}', \boldsymbol{h}_{sub\_s}))] \tag{9}$$

Let's define $\mathcal{L}_{sim}^{sub} = \mathcal{L}_{sim}^{sub\_start} + \mathcal{L}_{sim}^{sub\_end}$ and $\mathcal{L}_{back}^{sub} = \mathcal{L}_{back}^{sub\_start} + \mathcal{L}_{back}^{sub\_end}$. Then the loss function of the whole cycle-consistent unit is:

$$\mathcal{L}_{entity\_pair} = \mathcal{L}_{sim}^{sub} + \mathcal{L}_{sim}^{obj} + \mathcal{L}_{back}^{sub} + \mathcal{L}_{back}^{obj} \tag{10}$$

In summary, the core idea of the cycle-consistent unit is using transitivity as a way to regularize structured data. Only the embeddings of two entities are reachable to each other (the probability exceeds a certain threshold), the two entities are judged as a valid entity pair. In visual tracking, enforcing simple forward-backward consistency has been a standard trick for decades [18].

3.3 Relation Tagger

We have tried two types of relation taggers to impose relational labels on extracted entity pairs. One is relation-specific tagger [12], which could be formulated as $f_r(s) \rightarrow o$. It performs better when some relation types have few instances. The other is the relation classifier, which could be formulated as $f(s, o) \rightarrow r$. It's more suitable for each relation type has enough instances.

Relation-Specific Tagger. We named this relation tagging method as $SEPC_{SL}$. The detail operations of the relation-specific tagger is as follows:

$$p_i^{(\cdot)} = \sigma(\mathbf{W}_{(\cdot)}^r(\boldsymbol{h}_i + \boldsymbol{v}_{sub}^k) + \boldsymbol{b}_{(\cdot)}^r) \tag{11}$$

each relation have a independent relation-specific tagger, r represents a relation type, $\mathbf{W}_{(\cdot)}^r$ represents the trainable weight, $\boldsymbol{b}_{(\cdot)}^r$ is the bias and σ is the sigmoid activation function. $\boldsymbol{v}_{sub}^k$ represents the encoded representation vector of the k-th recognized subject. For each subject, they iteratively apply the same decoding process on it. In the testing stage, we first label subject and object with the entity tagger, and then label entity pairs with the cycle-consistent network, after that, label relation with the relation-specific tagger.

Relation Classifier. Since adding a relation classifier makes this method become an extract-then-classify one, we named this relation tagging method as $SEPC_{ETC}$. The relation classification module is a quite simple but effective network, especially for each relation type that has enough instances.

$$p(r|\boldsymbol{x}_i, \boldsymbol{v}_{sub}, \boldsymbol{v}_{obj}) = \sigma(\mathbf{W}_{rel}[\frac{1}{L}\sum_{i=1}^{L}(\boldsymbol{h}_i + \boldsymbol{v}_{sub} + \boldsymbol{v}_{obj})]) \tag{12}$$

The final learning objective is shown in Eq. 13. $\mathcal{L}_{RC}$ is K_{rel} (number of relation types) binary cross entropy loss of ground truth relation labels r and the prediction $p(r|\boldsymbol{x}_i, \boldsymbol{v}_{sub}, \boldsymbol{v}_{obj})$. Notice that we use multiple binary cross-entropy instead of a softmax cross-entropy is that no mutually exclusive in relation types.

$$\mathcal{L}_{SEPC_{ETC}} = \mathcal{L}_{entity} + \mathcal{L}_{entity\_pair} + \mathcal{L}_{RC} \tag{13}$$

4 Empirical Evaluation

4.1 Dataset and Evaluation Metrics

We evaluate SEPC on two public datasets NYT [22] and WebNLG [23]. They are widely used for evaluating the performance of joint extraction methods. NYT contains 24 predefined relation types, 56195 sentences for training, 5000 sentences for validation, and 5000 sentences for the test. WebNLG contains 211 predefined relation types, 5019 sentences for training, 500 sentences for validation and 703 sentences for the test. We follow the previous work [12] in the evaluation method. An extracted relational triple (subject, relation, object) is regarded as correct only if the relation types and the heads of both subject and object are all correct. For a fair comparison, we utilize the standard micro Precision(P.), Recall(R.), and F1-score(F.) as in line with baselines.

4.2 Experimental Settings

For comparison, we employ the following models as baselines. (1) NovelTagging [10] is the first sequence-tagging work to extract relational triples. They design a novel tagging strategy that labels entity and relation together, thus the joint extraction task is converted to a sequence tagging problem. But this approach faces difficulties in processing the overlapping triples; (2) MultiHead [15] first identifies all candidate entities, then adopt multi-head attention to extract relational triples; (3) GraphRel [16] first use graph convolutional network to extract overlapping triples and then consider all entity pairs for prediction. (4) CopyRE [9] adopts the encoder-decoder architecture for this task, they solve the overlapping problems by generating all possible relational triples of a sentence; (5) OrderCopyRE [11] applies the reinforcement learning to the encoder-decoder architecture for generating high-quality relational triples, it can be seen as an improvement of CopyRE; (6) ETL-Span [17] applies a span-based tagging strategy, they aims at modeling the internal dependencies by decoding triplets hierarchically; (7) CasRel [12] is the state-of-art method on NYT and WebNLG datasets based on the BERT, which first identifies all possible subjects in a sentence then identifies all possible relations and corresponding objects for each subject.

4.3 Implementation Detail

To make a fair comparison, we follow the setting of CasRel [12]. We adopt the mini-batch mechanism to train our model with the batch size as 6 and the learning rate is set to 1e−5 for both NYT and WebNLG. The hyper-parameters are determined on the validation set. The number of stacked bidirectional Transformer blocks N is 12 and the size of hidden state $\boldsymbol{h}_N$ is 768. The pre-trained BERT model is [BERT-Base, Cased][1], which contains 110M parameters. The max length of the input sentence is set to 100/300 words. The network weights are optimized with Adam [24]. We use Tesla V100 to train the model for at most 100/500 epochs on NYT/WebNLG. All the similarity thresholds are set to 0.5. The best model is selected with the best performance on the validation set and then is used for predicting the test set.

4.4 Experimental Result Analysis

Main Results. The overall performance of SEPC is shown in Table 1. SEPC outperforms all the baselines in F1-value. It achieves 2.5% and 0.5% improvements over the state-of-the-art method CasRel_{BERT} [12] on NYT and WebNLG datasets respectively. CasRel_{LSTM} is the method instantiated on a LSTM-based structure with pre-trained Glove embedding [25]. It has shown improved performance in comparison with other baseline methods. The F1-value of CasRel_{BERT} indicates adding pre-trained BERT weights bring considerable improvements. The performance improvements from CasRel_{LSTM} to CasRel_{BERT} highlight the importance of BERT. In hence, we directly deploy our SEPC on BERT-based structure, it shows stable improvements over CasRel_{BERT}, which demonstrates that strengthening entity pairs connection does help to the joint extraction of entities and relations.

Table 1 shows SEPC_{ETC} performs a little worse than CasRel_{BERT} on the WebNLG dataset. Through our analysis of mis-extracted triples, we find that errors mainly occur on the relation classification. The WebNLG contains many relation types and most relation types only contain few instances, these bring difficulties to the relation classification module. While the CasRel framework performs better in processing this problem. By comparing the F1-value of SEPC_{SL} and CasRel_{BERT}, SEPC_{SL} can save this advantage of CasRel_{BERT}. In summary, we could select SEPC_{SL} or SEPC_{ETC} according to their performance on the validation set.

Detail Results on Triples. To further study the capability of the proposed SEPC in recognizing entity pairs, we conduct a detailed evaluation on the subject, object, entity pairs and relations types, and compare the performance with CasRel_{BERT}. Detailed comparison of CasRel_{BERT}, SEPC_{SL} and SEPC_{ETC} on NYT and WebNLG are presented in Fig. 4. It can be seen that the performance of recognizing subject, object and entity pairs presents an increasing trend in comparison with CasRel_{BERT}. These reveal the effectiveness of strengthening

[1] https://storage.googleapis.com/bertmodels/20181018/casedL-12H-768A-12.zip.

Table 1. Results of different methods on NYT and WebNLG datasets.

Methods	NYT			WebNLG		
	P.(%)	R.(%)	F1(%)	P.(%)	R.(%)	F1(%)
MultiHead [15]	60.7	58.6	59.6	57.5	54.1	55.7
GraphRel [16]	63.9	60.0	61.9	44.7	41.1	42.9
NovelTagging [10]	62.4	31.7	42.0	52.5	19.3	28.3
CopyRE [9]	61.0	56.6	58.7	37.7	36.4	37.1
OrderCopyRE [11]	77.9	67.2	72.1	63.3	59.9	61.6
ETL-Span [17]	84.9	72.3	78.1	84.0	91.5	87.6
CasRel$_{LSTM}$ [12]	84.2	83.0	83.6	86.9	80.6	83.7
CasRel$_{BERT}$ [12]	89.7	89.5	89.6	93.4	90.1	91.8
SEPC$_{SL}$	92.6	89.6	**91.1**	95.0	89.7	**92.3**
SEPC$_{ETC}$	92.5	91.7	**92.1**	93.4	90.1	91.7
SEPC	92.5	91.7	**92.1**	95.0	89.7	**92.3**

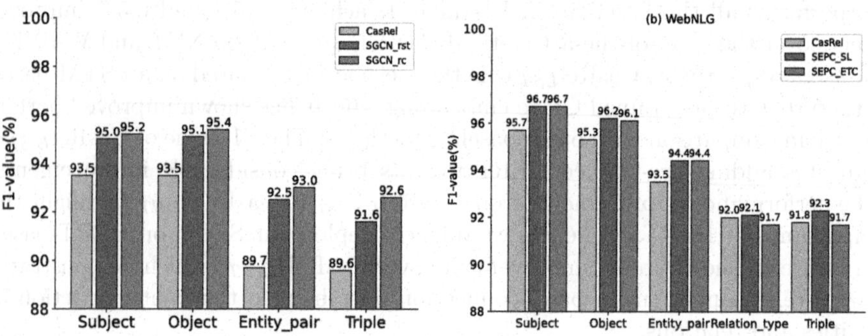

Fig. 4. Comparison of CasRel, SEPC$_{SL}$ and SEPC$_{ETC}$ on NYT and WebNLG.

entity pairs connection. By comparing the F1-value of relation_type recognization, we can see that SEPC$_{ETC}$ performs worse than CasRel$_{BERT}$ mainly lead by the relation classification module. In summary, through our comprehensive analysis of the results on two datasets, strengthening entity pair connection does help to improve the recognization of entities and entity pairs.

5 Conclusion and Inspiration

In this paper, we propose to improve the joint extraction of entities and relations by strengthening entity pairs connection. We realize this by utilizing the symmetry property of correct entity pairs. A dual probabilistic constraint, named the cycle-consistent unit, is introduced to model the intrinsic probability connections between subject and object. We conduct extensive experiments on two widely used datasets to validate the effectiveness of SEPC. Experimental results show that our SEPC outperforms state-of-art baselines on relational triples, and performs better on the extraction of entity pairs. Besides, $SEPC_{SL}$ saves the advantage of CasRel in alleviating data imbalance by only sampling correct triples, and well processing overlapping relational triples by building multiple mapping functions $f_r(s) \rightarrow o$ from subject to object.

There are multiple directions to explore in the further. First, the start and end position of an entity is also symmetrical, which may help the identification of entity boundaries. Second, the object function of CasRel and SEPC are both segmented, given (s, r) to find o and given (o, r) to find s is also a symmetrical problem, making the representation of relation labels r could be tuned in the triple-level objective function.

References

1. De Sa, C., et al.: DeepDive: declarative knowledge base construction. ACM SIGMOD Rec. **45**(1), 60–67 (2016)
2. Getman, J., Ellis, J., Strassel, S., Song, Z., Tracey, J.: Laying the groundwork for knowledge base population: nine years of linguistic resources for TAC KBP. In: Proceedings of the Eleventh International Conference on Language Resources and Evaluation (LREC 2018) (2018)
3. Singh, B., Dubey, V., Sheetlani, J.: A review and analysis on knowledge discovery and data mining techniques. Int. J. Adv. Technol. Eng. Explor. **5**(41), 70–77 (2018)
4. Zelenko, D., Aone, C., Richardella, A.: Kernel methods for relation extraction. J. Mach. Learn. Res. **3**(Feb), 1083–1106 (2003)
5. Zhou, G., Su, J., Zhang, J., Zhang, M.: Exploring various knowledge in relation extraction. In: Proceedings of the 43rd Annual Meeting of the Association for Computational Linguistics (ACL 2005), pp. 427–434 (2005)
6. Chan, Y.S., Roth, D.: Exploiting syntactico-semantic structures for relation extraction. In: Proceedings of the 49th Annual Meeting of the Association for Computational Linguistics: Human Language Technologies, pp. 551–560 (2011)
7. Gupta, P., Schütze, H., Andrassy, B.: Table filling multi-task recurrent neural network for joint entity and relation extraction. In: Proceedings of COLING 2016, the 26th International Conference on Computational Linguistics: Technical Papers, pp. 2537–2547 (2016)
8. Katiyar, A., Cardie, C.: Going out on a limb: joint extraction of entity mentions and relations without dependency trees. In: Proceedings of the 55th Annual Meeting of the Association for Computational Linguistics (Volume 1: Long Papers), pp. 917–928 (2017)
9. Zeng, X., Zeng, D., He, S., Liu, K., Zhao, J.: Extracting relational facts by an end-to-end neural model with copy mechanism. In: Proceedings of the 56th Annual Meeting of the Association for Computational Linguistics, Melbourne, Australia, (Volume 1: Long Papers), pp. 506–514. Association for Computational Linguistics, July 2018
10. Zheng, S., Wang, F., Bao, H., Hao, Y., Zhou, P., Xu, B.: Joint extraction of entities and relations based on a novel tagging scheme. In: Proceedings of the 55th Annual Meeting of the Association for Computational Linguistics (Volume 1: Long Papers), pp. 1227–1236 (2017)

11. Zeng, X., He, S., Zeng, D., Liu, K., Liu, S., Zhao, J.: Learning the extraction order of multiple relational facts in a sentence with reinforcement learning. In: Proceedings of the 2019 Conference on Empirical Methods in Natural Language Processing and the 9th International Joint Conference on Natural Language Processing (EMNLP-IJCNLP), Hong Kong, China, pp. 367–377. Association for Computational Linguistics, November 2019
12. Wei, Z., Su, J., Wang, Y., Tian, Y., Chang, Y.: A novel cascade binary tagging framework for relational triple extraction. In: Proceedings of the 58th Annual Meeting of the Association for Computational Linguistics, pp. 1476–1488 (2020)
13. Xia, Y., Qin, T., Chen, W., Bian, J., Yu, N., Liu, T.Y.: Dual supervised learning. In: Proceedings of the 34th International Conference on Machine Learning, vol. 70, pp. 3789–3798 (2017)
14. Gormley, M.R., Yu, M., Dredze, M.: Improved relation extraction with feature-rich compositional embedding models. In: Proceedings of the 2015 Conference on Empirical Methods in Natural Language Processing, pp. 1774–1784 (2015)
15. Bekoulis, G., Deleu, J., Demeester, T., Develder, C.: Joint entity recognition and relation extraction as a multi-head selection problem (2018)
16. Fu, T.J., Li, P.H., Ma, W.Y.: GraphRel: modeling text as relational graphs for joint entity and relation extraction. In: Proceedings of the 57th Annual Meeting of the Association for Computational Linguistics, pp. 1409–1418 (2019)
17. Yu, B., et al.: Joint extraction of entities and relations based on a novel decomposition strategy. In: Proceedings of ECAI (2020)
18. Sundaram, N., Brox, T., Keutzer, K.: Dense point trajectories by GPU-accelerated large displacement optical flow. In: Daniilidis, K., Maragos, P., Paragios, N. (eds.) ECCV 2010. LNCS, vol. 6311, pp. 438–451. Springer, Heidelberg (2010). https://doi.org/10.1007/978-3-642-15549-9_32
19. Brislin, R.W.: Back-translation for cross-cultural research. J. Cult. Psychol. **1**(3), 185–216 (1970)
20. Devlin, J., Chang, M.W., Lee, K., Toutanova, K.: BERT: pre-training of deep bidirectional transformers for language understanding. In: NAACL-HLT (1) (2019)
21. Zhu, J. Y., Park, T., Isola, P., Efros, A.A.: Unpaired image-to-image translation using cycle-consistent adversarial networks. In: Proceedings of the IEEE International Conference on Computer Vision, pp. 2223–2232 (2017)
22. Riedel, S., Yao, L., McCallum, A.: Modeling relations and their mentions without labeled text. In: Balcázar, J.L., Bonchi, F., Gionis, A., Sebag, M. (eds.) ECML PKDD 2010. LNCS (LNAI), vol. 6323, pp. 148–163. Springer, Heidelberg (2010). https://doi.org/10.1007/978-3-642-15939-8_10
23. Gardent, C., Shimorina, A., Narayan, S., Perez-Beltrachini, L.: Creating training corpora for NLG micro-planning. In: 55th Annual Meeting of the Association for Computational Linguistics (ACL) (2017)
24. Kingma, D.P., Ba, J.: Adam: a method for stochastic optimization. arXiv preprint arXiv:1412.6980 (2014)
25. Pennington, J., Socher, R., Manning, C.D.: Glove: Global vectors for word representation. In: Proceedings of the 2014 Conference on Empirical Methods in Natural Language Processing (EMNLP), pp. 1532–1543 (2014)

Author Index